a LANGE medical book

Review of
Medical
Physiology

sixteenth edition

William F. Ganong, MD
Jack and DeLoris Lange Professor of Physiology Emeritus
University of California
San Francisco

APPLETON & LANGE
Norwalk, Connecticut

0-8385-8426-8

Prentice Hall International (UK) Limited, *London*
Prentice Hall of Australia Pty. Limited, *Sydney*
Prentice Hall Canada, Inc., *Toronto*
Prentice Hall Hispanoamericana, S.A., *Mexico*
Prentice Hall of India Private Limited, *New Delhi*
Prentice Hall of Japan, Inc., *Tokyo*
Simon & Schuster Asia Pte. Ltd., *Singapore*
Editora Prentice Hall do Brasil Ltda., *Rio de Janeiro*
Prentice Hall, *Englewood Cliffs, New Jersey*

ISBN: 0-8385-8426-8
ISSN: 0892-1253

Acquisitions Editor: John Dolan
Production Editor: Christine Langan
Designer: Janice Bielawa

PRINTED IN THE UNITED STATES OF AMERICA

Table of Contents

Preface xi

SECTION I. INTRODUCTION — 1

1. The General & Cellular Basis of Medical Physiology 1

Introduction 1
Body Fluid Compartments 1
Units for Measuring Concentration of Solutes 3
Functional Morphology of the Cell 4
Lysosomes 8
Transport Across Cell Membranes
    & Membrane Potentials 20
Forces Producing Movement of Substances Across
    Cell Membranes 25

The Capillary Wall 29
Sodium & Potassium Distribution
    & Total Body Osmolality 29
Intercellular Communication 30
Homeostasis 40
Aging 40

Section I References: 40

SECTION II. PHYSIOLOGY OF NERVE & MUSCLE CELLS — 43

2. Excitable Tissue: Nerve 43

Introduction 43
Nerve Cells 43
Excitation & Conduction 45
Measurement of Electrical Events 45
Ionic Basis of Excitation & Conduction 50

Properties of Mixed Nerves 52
Nerve Fiber Types & Function 53
Nerve Growth Factor 54
Glia 55

3. Excitable Tissue: Muscle 56

Introduction 56
Skeletal Muscle 56
  Morphology 56
  Electrical Phenomena & Ionic Fluxes 59
  Contractile Responses 59
  Energy Sources & Metabolism 63
  Properties of Muscles in the Intact Organism 65
Cardiac Muscle 67
  Morphology 67

Electrical Properties 67
Mechanical Properties 69
Metabolism 70
Pacemaker Tissue 70
Smooth Muscle 71
  Morphology 71
  Visceral Smooth Muscle 71
  Multi-Unit Smooth Muscle 73

4. Synaptic & Junctional Transmission 74

Introduction 74
Synaptic Transmission 74
  Functional Anatomy 74
  Electrical Events at Synapses 77

Inhibition & Facilitation at Synapses 80
Chemical Transmission of Synaptic Activity 83
Synaptic Plasticity & Learning 98
Neuromuscular Transmission 99

Neuromuscular Junction 99
Nerve Endings in Smooth & Cardiac Muscle 101

Denervation Hypersensitivity 101

5. **Initiation of Impulses in Sense Organs** **103**

Introduction 103
Sense Organs & Receptors 103
The Senses 103

Electrical & Ionic Events in Receptors 105
Ionic Basis of Excitation 106
"Coding" of Sensory Information 107

Section II References: 109

SECTION III. FUNCTIONS OF THE NERVOUS SYSTEM —————————— **111**

6. **Reflexes** **111**

Introduction 111
Monosynaptic Reflexes: The Stretch Reflex 111

Polysynaptic Reflexes: The Withdrawal Reflex 117
General Properties of Reflexes 118

7. **Cutaneous, Deep & Visceral Sensation** **120**

Introduction 120
Pathways 120
Touch 123
Proprioception 123
Temperature 124
Pain 124

Differences Between Somatic
    & Visceral Sensory Mechanisms 125
Visceral Pain 126
Referral of Pain 127
Inhibition & Facilitation of Pain 128
Other Sensations 130

8. **Vision** **132**

Introduction 132
Anatomic Considerations 132
The Image-Forming Mechanism 136
The Photoreceptor Mechanism: Genesis
    of Action Potentials 139

Electrical Responses of Retinal Cells 140
Responses in the Visual Pathways & Cortex 144
Color Vision 146
Other Aspects of Visual Function 148
Eye Movements 151

9. **Hearing & Equilibrium** **153**

Introduction 153
Anatomic Considerations 153
Hair Cells 156

Hearing 157
Vestibular Function 164

10. **Smell & Taste** **166**

Introduction 166
Smell 166
    Receptors & Pathways 166
    Physiology of Olfaction 167

Taste 170
    Receptor Organs & Pathways 170
    Physiology of Taste 171

11. **Arousal Mechanisms, Sleep, & the Electrical Activity of the Brain** **173**

Introduction 173
The Reticular Formation & the Reticular
    Activating System 173
The Thalamus & the Cerebral Cortex 173

Evoked Cortical Potentials 174
The Electroencephalogram 175
Physiologic Basis of the EEG, Consciousness,
    & Sleep 177

12. Control of Posture & Movement 183

Introduction 183
General Principles 183
Corticospinal & Corticobulbar System 184
  Anatomy 184
  Function 186
Posture-Regulating Systems 187

Spinal Integration 189
Medullary Components 190
Midbrain Components 192
Cortical Components 192
Basal Ganglia 193
Cerebellum 196

13. The Autonomic Nervous System 201

Introduction 201
Anatomic Organization of Autonomic Outflow 201
Chemical Transmission at Autonomic Junctions 201

Responses of Effector Organs to Autonomic
  Nerve Impulses 203
Cholinergic Discharge 204

14. Central Regulation of Visceral Function 208

Introduction 208
Medulla Oblongata 208
Hypothalamus 209
  Anatomic Considerations 209
  Hypothalamic Function 210
  Relation of Hypothalamus
    to Autonomic Function 210

Relation to Sleep 211
Relation to Cyclic Phenomena 211
Hunger 212
Thirst 215
Control of Posterior Pituitary Secretion 216
Control of Anterior Pituitary Secretion 222
Temperature Regulation 225

15. Neural Basis of Instinctual Behavior & Emotions 231

Introduction 231
Anatomic Considerations 231
Limbic Functions 231
Sexual Behavior 232

Fear & Rage 235
Motivation 236
Brain Chemistry, Behavior, & Synaptic Transmission
  in the Central Nervous System 237

16. "Higher Functions of the Nervous System": Conditioned Reflexes, Learning, & Related Phenomena 241

Introduction 241
Methods 241

Learning & Memory 241
Functions of the Neocortex 247

Section III References: 251

SECTION IV. ENDOCRINOLOGY, METABOLISM, & REPRODUCTIVE FUNCTION _____ 253

17. Energy Balance, Metabolism, & Nutrition 253

Introduction 253
Energy Metabolism 253
Intermediary Metabolism 256
Carbohydrate Metabolism 259

Protein Metabolism 266
Fat Metabolism 271
Nutrition 282

18. The Thyroid Gland 287

Introduction 287
Anatomic Considerations 287
Formation & Secretion of Thyroid Hormones 288
Transport & Metabolism of Thyroid Hormones 290

Effects of Thyroid Hormones 293
Regulation of Thyroid Secretion 296
Clinical Correlates 297

19. Endocrine Functions of the Pancreas & the Regulation of Carbohydrate Metabolism 302

Introduction 302
Islet Cell Structure 302
Structure, Biosynthesis, & Secretion of Insulin 303
Fate of Secreted Insulin 303
Effects of Insulin 305
Mechanism of Action 307
Consequences of Insulin Deficiency 308

Insulin Excess 313
Regulation of Insulin Secretion 313
Glucagon 316
Other Islet Cell Hormones 318
Endocrine Regulation of Carbohydrate Metabolism 319
Hypoglycemia & Diabetes Mellitus in Humans 320

20. The Adrenal Medulla & Adrenal Cortex 323

Introduction 323
Adrenal Morphology 323
Adrenal Medulla 325
  Structure & Function of Medullary Hormones 325
  Regulation of Adrenal Medullary Secretion 327
Adrenal Cortex 328
  Structure & Biosynthesis of Adrenocortical Hormones 328
  Transport, Metabolism, & Excretion of Adrenocortical Hormones 332
  Effects of Adrenal Androgens & Estrogens 334

Physiologic Effects of Glucocorticoids 334
Pharmacologic & Pathologic Effects of Glucocorticoids 336
Regulation of Glucocorticoid Secretion 338
Effects of Mineralocorticoids 341
Regulation of Aldosterone Secretion 343
Role of Mineralocorticoids in the Regulation of Salt Balance 346
Summary of the Effects of Adrenocortical Hyper- & Hypofunction in Humans 346

21. Hormonal Control of Calcium Metabolism & the Physiology of Bone 348

Introduction 348
Calcium & Phosphorus Metabolism 348
Bone Physiology 349
Vitamin D & the Hydroxycholecalciferols 353

The Parathyroid Glands 355
Calcitonin 357
Effects of Other Hormones & Humoral Agents on Calcium Metabolism 359

22. The Pituitary Gland 361

Introduction 361
Morphology 361
Intermediate Lobe Hormones 363
Growth Hormone 364

Physiology of Growth 370
Pituitary Insufficiency 372
Pituitary Hyperfunction in Humans 374

23. The Gonads: Development & Function of the Reproductive System 375

Introduction 375
Sex Differentiation & Development 375
  Chromosomal Sex 375
  Embryology of the Human Reproductive System 377
  Aberrant Sexual Differentiation 381
  Puberty 382
  Precocious & Delayed Puberty 383
  Menopause 384
Pituitary Gonadotropins & Prolactin 385
The Male Reproductive System 387
  Structure 387

Gametogenesis & Ejaculation 388
Endocrine Function of the Testes 391
Control of Testicular Function 393
Abnormalities of Testicular Function 395
The Female Reproductive System 395
  The Menstrual Cycle 395
  Ovarian Hormones 400
  Control of Ovarian Function 405
  Abnormalities of Ovarian Function 407
Pregnancy 408
Lactation 411

24. Other Endocrine Organs 414

Introduction 414
The Renin-Angiotensin System 414
Erythropoietin 418

The Endocrine Function of the Heart: Atrial Natriuretic Peptide 419
Pineal 422

Section IV References 424

SECTION V. GASTROINTESTINAL FUNCTION _____ 427

25. Digestion & Absorption 427

Introduction 427
Carbohydrates 427
Proteins & Nucleic Acids 431

Lipids 432
Absorption of Water & Electrolytes 433
Absorption of Vitamins & Minerals 435

26. Regulation of Gastrointestinal Function 438

Introduction 438
General Considerations 438
Gastrointestinal Hormones 439
Mouth & Esophagus 444
Stomach 446

Exocrine Portion of the Pancreas 452
Liver & Biliary System 454
Small Intestine 459
Colon 462

Section V References 466

SECTION VI. CIRCULATION _____ 469

27. Circulating Body Fluids 469

Introduction 469
Blood 469
 Bone Marrow 469
 White Blood Cells 470
 Immune Mechanisms 473
 Platelets 481

Red Blood Cells 481
Blood Types 485
Plasma 488
Hemostasis 489
Lymph 493

28. Origin of the Heartbeat & the Electrical Activity of the Heart 494

Introduction 494
Origin & Spread of Cardiac Excitation 494
The Electrocardiogram 497

Cardiac Arrhythmias 501
Electrocardiographic Findings in Other Cardiac
 & Systemic Diseases 506

29. The Heart as a Pump 510

Introduction 510
Mechanical Events of the Cardiac Cycle 510

Cardiac Output 515

30. Dynamics of Blood & Lymph Flow 521

Introduction 521
Anatomic Considerations 521
Biophysical Considerations 523
Arterial & Arteriolar Circulation 529

Capillary Circulation 532
Lymphatic Circulation & Interstitial
 Fluid Volume 534
Venous Circulation 535

31. Cardiovascular Regulatory Mechanisms 538

Introduction 538
Local Regulatory Mechanisms 538
Substances Secreted by the Endothelium 539

Systemic Regulation by Hormones 541
Systemic Regulation by the Nervous System 542

32. Circulation Through Special Regions **551**

Introduction 551
Cerebral Circulation 551
  Anatomic Considerations 551
  Cerebrospinal Fluid 552
  The Blood-Brain Barrier 553
  Cerebral Blood Flow 556

Regulation of Cerebral Circulation 558
  Brain Metabolism & Oxygen Requirements 560
Coronary Circulation 561
Splanchnic Circulation 564
Circulation of the Skin 566
Placental & Fetal Circulation 567

33. Cardiovascular Homeostasis in Health & Disease **571**

Introduction 571
Compensations for Gravitational Effects 571
Exercise 573
Inflammation & Wound Healing 576

Shock 576
Hypertension 581
Heart Failure 583

Section VI References 585

SECTION VII. RESPIRATION _____ **587**

34. Pulmonary Function **587**

Introduction 587
Properties of Gases 587
Anatomy of the Lungs 588
Mechanics of Respiration 590

Gas Exchange in the Lungs 598
Pulmonary Circulation 599
Other Functions of the Respiratory System 602

35. Gas Transport Between the Lungs & the Tissues **604**

Introduction 604
Oxygen Transport 604

Buffers 607
Carbon Dioxide Transport 609

36. Regulation of Respiration **611**

Introduction 611
Neural Control of Breathing 611
Regulation of Respiratory Activity 612

Chemical Control of Breathing 613
Nonchemical Influences on Respiration 617

37. Respiratory Adjustments in Health & Disease **620**

Introduction 620
Effects of Exercise 620
Hypoxia 622
Hypoxic Hypoxia 623
Other Forms of Hypoxia 628

Oxygen Treatment 629
Hypercapnia & Hypocapnia 630
Other Respiratory Abnormalities 630
Effects of Increased Barometric Pressure 631
Artificial Respiration 632

Section VII References 634

SECTION VIII. FORMATION AND EXCRETION OF URINE _____ **635**

38. Renal Function & Micturition **635**

Introduction 635
Functional Anatomy 635

Renal Circulation 639
Glomerular Filtration 641

Tubular Function 644
Water Excretion 648
Acidification of the Urine & Bicarbonate
    Excretion 654
Regulation of Na^+ & Cl^- Excretion 657

Regulation of K^+ Excretion 658
Diuretics 659
Effects of Disordered Renal Function 660
Filling of the Bladder 661
Emptying of the Bladder 661

39. Regulation of Extracellular Fluid Composition & Volume 664

Introduction 664
Defense of Tonicity 664
Defense of Volume 664

Defense of Specific Ionic Composition 665
Defense of H^+ Concentration 665

Section VIII References 671

Appendix 672

General References 672
Normal Values & the Statistical Evaluation
    of Data 672
Abbreviations & Symbols Commonly Used
    in Physiology 675

Some Standard Respiratory Symbols 679
Equivalents of Metric, United States,
    & English Measures 679
Greek Alphabet 680

Self-Study: Objective, Essay, & Multiple Choice Questions 681

Answers to Quantitative & Multiple Choice Questions 730

Index 735

Tables

Atomic Weights .... Inside Front Cover
Ranges of Normal Values in Human Whole Blood, Plasma, or Serum .... Inside Back Cover

Preface

This book is designed to provide a concise summary of mammalian and, particularly, of human physiology, which medical students and others can use as a text by itself or supplement with readings in current texts, monographs, and reviews. Pertinent aspects of general and comparative physiology are also included. Summaries of relevant anatomic considerations will be found in each section, but this book is written primarily for those who have some knowledge of anatomy, chemistry, and biochemistry. Examples from clinical medicine are given where pertinent to illustrate physiologic points. In many of the chapters, physicians desiring to use this book as a review will find short discussions of important symptoms produced by disordered function.

A new feature in this edition is a Self-Study Section, which provides for each chapter a list of objectives, some general essay-type questions, and a selection of multiple-choice questions. Most of the multiple-choice questions are similar to the questions used in the National Board Examinations in the USA, although some K and other types are also included to provide breadth and variety. Answers are provided at the end of the section. The material in the section is a truncated, revised version of the material that was formerly published separately in my Study Guide. Students found *Physiology: A Study Guide,* the third edition of which was published by Appleton & Lange in 1989, to be helpful in organizing their thinking and preparing for board examinations. Publication of this material as part of *Review of Medical Physiology* should increase its value and availability.

Review of Medical Physiology also includes an appendix that contains, among other things, general references, a discussion of statistical methods, a glossary of abbreviations and symbols commonly used in physiology, and several useful tables. The index is comprehensive and specifically designed for ease in locating important terms, topics, and concepts.

In writing this book, it has not been possible to be complete and concise without also being dogmatic. I believe, however, that the conclusions presented without a detailed discussion of the experimental data on which they are based are supported by the bulk of the currently available evidence. Much of this evidence can be found in the papers cited in the credit lines of the illustrations. Further discussions of particular subjects and information on subjects not considered in detail in this book can be found in the references listed at the end of each section. Information about serial review publications that provide up-to-date discussions of various physiologic subjects is included in the note on general references in the appendix.

In the interest of brevity and clarity, I have in most instances omitted the names of the many investigators whose work made possible the view of physiology presented here. This is in no way intended to slight their contributions, but including their names and specific references to original papers would greatly increase the length of the book.

In this sixteenth edition, as in previous editions, the entire book has been thoroughly revised, eliminating errors, incorporating suggestions of readers, updating concepts, and discarding material that is no longer relevant. In this way, the book has been kept as up-to-date and accurate as possible. Since the last edition, there have been continued advances in knowledge about intracellular signaling. To accommodate this new information and present it in the most effective way, Chapter 1 has been reorganized and now has expanded coverage of heterotrimeric and other types of G proteins, guanylate cyclases, the various families of protein kinases, and the interplay between phosphorylation and dephosphorylation of proteins in the regulation of cell function. The exploding field of growth factors, which now permeates all aspects of physiology, is covered in general terms in Chapter 1 and in terms of effects on specific systems in Chapters 19, 21, 22, and 27. The molecular motors responsible for cell movement and muscle contraction are now considered in more detail in Chapters 1 and 3. The section on the basal ganglia has been revised and updated, with emphasis on the role of the subthalamic nucleus, striasomes, and the abnormalities in Huntington's disease as well as Parkinson's disease. The section on learning and memory has been rewritten to emphasize both the growing knowledge about the molecular bases of the processes in simple neural systems and new ideas about the role of the hippocampus, limbic cortex, and neocortical association areas in short-term memory in humans. Chapter 31 has been expanded to give greater consideration to the cardiovascular and other functions of substances produced in the vascular endothelium, especially nitric oxide and the endothelins. The section on inflammation in Chapter 33 has been expanded to include wound healing and the role of integrins and other cell adhesion molecules in this process. The sixteenth edition also includes new information on many other topics, including immediate early genes and transcription factors, sex determination, olfactory discrimination, new multiple receptors for angiotensin II and the atrial natiuretic polypeptides, reperfusion-induced injury, autoimmune disease, and tubuloglomerular feedback and glomerulotubular balance in the kidneys.

I am greatly indebted to may individuals who helped with the preparation of this book. Those to whom I extend special thanks for their help with the sixteenth edition include Drs David Gardner, Otto Hansen, John Karam, and Ralph Kellogg. Katherine Cure provided invaluable secretarial assistance, and as always, my wife labored long hours typing and inserting corrections. Yvonne Strong did an excellent job of copyediting. Many associates and friends provided unpublished illustrative materials, and numerous authors and publishers generously granted permission to reproduce illustrations from other books and journals. I also thank all the students and others who took the time to write me offering helpful criticisms and suggestions. Such comments are always welcome, and I solicit additional corrections and criticisms, which may be addressed to me at

Department of Physiology
University of California
San Francisco, CA 94143-0444, USA

Since this book was first published in 1963, the following translations have been published and are currently available: Portuguese (fourth edition), German (fourth edition), Italian (sixth edition), Spanish (twelfth edition), Japanese (twelfth edition), Indonesian (second edition), Greek (second edition), French, Chinese, and Hungarian. Translations have also been published in Czech, Polish, Serbo-Croatian, and Turkish but are no longer in print. New translations into Serbo-Croatian and Polish are under way, and the book is being translated into Malay. It has also appeared in various foreign English-language editions and has been recorded in English on tape for use by the blind. The tape recording is available from Recording for the Blind, Inc., 20 Roszel Road, Princeton, NJ 08540, USA. Finally, for computer users, the book is now available, along with several other Lange Medical Books, in STAT!-Ref, a searchable CD-ROM format, from Teton Data Systems.

San Francisco
March, 1993

William F. Ganong

The General & Cellular Basis of Medical Physiology

1

INTRODUCTION

In unicellular organisms, all vital processes occur in a single cell. As the evolution of multicellular organisms has progressed, various cell groups have taken over particular functions. In humans and other vertebrate animals, the specialized cell groups include a gastrointestinal system to digest and absorb food; a respiratory system to take up O_2 and eliminate CO_2; a urinary system to remove wastes; a cardiovascular system to distribute food, O_2, and the products of metabolism; a reproductive system to perpetuate the species; and nervous and endocrine systems to coordinate and integrate the functions of the other systems. This book is concerned with the way these systems function and the way each contributes to the functions of the body as a whole.

This chapter presents general concepts and principles that are basic to the function of all the systems. It also reviews fundamental aspects of cell physiology. Additional aspects of cellular and molecular biology are considered in the relevant chapters on the various organs.

BODY FLUID COMPARTMENTS

Organization of the Body

The cells that make up the bodies of all but the simplest multicellular animals, both aquatic and terrestrial, exist in an "internal sea" of **extracellular fluid (ECF)** enclosed within the integument of the animal. From this fluid, the cells take up O_2 and nutrients; into it, they discharge metabolic waste products. The ECF is more dilute than present-day sea water, but its composition closely resembles that of the primordial oceans in which, presumably, all life originated.

In animals with a closed vascular system, the ECF is divided into 2 components: the **interstitial fluid** and the circulating **blood plasma.** The plasma and the cellular elements of the blood, principally red blood cells, fill the vascular system, and together they constitute the **total blood volume.** The interstitial fluid is that part of the ECF that is outside the vascular system, bathing the cells. The special fluids lumped together as transcellular fluids are discussed below. About a third of the **total body water (TBW)** is extracellular; the remaining two-thirds are intracellular **(intracellular fluid).**

Size of the Fluid Compartments

In the average young adult male, 18% of the body weight is protein and related substances, 7% is mineral, and 15% is fat. The remaining 60% is water. The distribution of this water is shown in Fig 1–1.

The intracellular component of the body water accounts for about 40% of body weight and the extracellular component for about 20%. Approximately 25% of the extracellular component is in the vascular system (plasma = 5% of body weight) and 75% outside the blood vessels (interstitial fluid = 15% of body weight). The total blood volume is about 8% of body weight.

Measurement of Body Fluid Volumes

It is theoretically possible to measure the size of each of the body fluid compartments by injecting substances that will stay in only one compartment and then calculating the volume of fluid in which the test substance is distributed (the **volume of distribution** of the injected material). The volume of distribution is equal to the amount injected (minus any that has been removed from the body by metabolism or excretion during the time allowed for mixing) divided by the concentration of the substance in the sample. *Example:* 150 mg of sucrose is injected into a 70-kg man. The plasma sucrose level after mixing is 0.01 mg/mL, and 10 mg has been excreted or metabolized during the mixing period. The volume of distribution of the sucrose is

$$\frac{150\ mg - 10\ mg}{0.01\ mg/mL} = 14,000\ mL$$

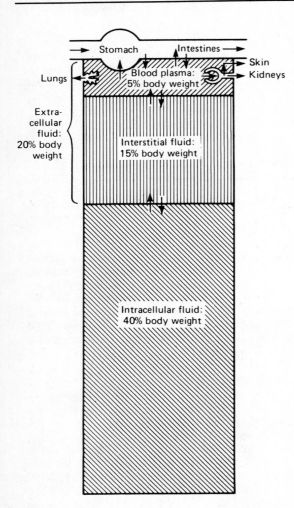

Figure 1–1. Body fluid compartments. Arrows represent fluid movement. Transcellular fluids, which constitute a very small percentage of total body fluids, are not shown. (Modified and reproduced, with permission, from Gamble JL: *Chemical Anatomy, Physiology, and Pathology of Extracellular Fluid,* 6th ed. Harvard Univ Press, 1954.)

Since 14,000 mL is the space in which the sucrose was distributed, it is also called the **sucrose space.**

Volumes of distribution can be calculated for any substance that can be injected into the body provided the concentration in the body fluids and the amount removed by excretion and metabolism can be accurately measured.

Although the principle involved in such measurements is simple, a number of complicating factors must be considered. The material injected must be nontoxic, must mix evenly throughout the compartment being measured, and must have no effect of its own on the distribution of water or other substances in the body. In addition, either it must be unchanged by the body during the mixing period, or the amount

changed must be known. The material also should be relatively easy to measure.

Plasma Volume, Total Blood Volume & Red Cell Volume

Plasma volume has been measured by using dyes that become bound to plasma protein—particularly Evans blue (T-1824). Plasma volume can also be measured by injecting serum albumin labeled with radioactive iodine. Suitable aliquots of the injected solution and plasma samples obtained after injection are counted in a scintillation counter. An average value is 3500 mL (5% of the body weight of a 70-kg man, assuming unit density).

If one knows the plasma volume and the hematocrit (ie, the percentage of the blood volume that is made up of cells), the **total blood volume** can be calculated by multiplying the plasma volume by

$$\frac{100}{100 - \text{hematocrit}}$$

Example: The hematocrit is 38 and the plasma volume 3500 mL. The total blood volume is

$$3500 \times \frac{100}{100 - 38} = 5645 \text{ mL}$$

The **red cell volume** (volume occupied by all the circulating red cells in the body) can be determined by subtracting the plasma volume from the total blood volume. It may also be measured independently by injecting tagged red blood cells and, after mixing has occurred, measuring the fraction of the red cells that is tagged. A commonly used tag is ^{51}Cr, a radioactive isotope of chromium that is attached to the cells by incubating them in a suitable chromium solution. Isotopes of iron and phosphorus (^{59}Fe and ^{32}P) and antigenic tagging have also been employed.

Extracellular Fluid Volume

The ECF volume is difficult to measure because the limits of this space are ill defined and because few substances mix rapidly in all parts of the space while remaining exclusively extracellular. The lymph cannot be separated from the ECF and is measured with it. Many substances enter the cerebrospinal fluid (CSF) slowly because of the blood-brain barrier (see Chapter 32). Equilibration is slow with joint fluid and aqueous humor and with the ECF in relatively avascular tissues such as dense connective tissue, cartilage, and some parts of bone. Substances that distribute in ECF appear in glandular secretions and in the contents of the gastrointestinal tract. Because they are separated from the rest of the ECF, these fluids—as well as CSF, the fluids in the eye, and a few other special fluids—are called **transcellular fluids.** Their volume is relatively small.

Perhaps the most accurate measurement of ECF

volume is that obtained by using inulin, a polysaccharide with a molecular weight of 5200. Radioactive inulin has been prepared by substituting ^{14}C for one of the carbon atoms of the molecule; and when this material is used, inulin levels are easily determined by counting the samples with suitable radiation detectors. Mannitol and sucrose have also been used to measure ECF volume. A generally accepted value for ECF volume is 20% of the body weight, or about 14 L in a 70-kg man (3.5 L = plasma; 10.5 L = interstitial fluid).

Interstitial Fluid Volume

The interstitial fluid space cannot be measured directly, since it is difficult to sample interstitial fluid and since substances that equilibrate in interstitial fluid also equilibrate in plasma. The volume of the interstitial fluid can be calculated by subtracting the plasma volume from the ECF volume. The ECF volume/intracellular fluid volume ratio is larger in infants and children than it is in adults, but the absolute volume of ECF in children is, of course, smaller than in adults. Therefore, dehydration develops more rapidly and is frequently more severe in children than in adults.

Intracellular Fluid Volume

The intracellular fluid volume cannot be measured directly, but it can be calculated by subtracting the ECF volume from the total body water (TBW). TBW can be measured by the same dilution principle used to measure the other body spaces. Deuterium oxide (D_2O, heavy water) is most frequently used. D_2O has slightly different properties from those of H_2O, but in equilibration experiments for measuring body water it gives accurate results. Tritium oxide and aminopyrine have also been used for this purpose.

The water content of lean body tissue is constant at 71–72 mL/100 g of tissue, but since fat is relatively free of water, the ratio of TBW to body weight varies with the amount of fat present. In young men, water constitutes about 60% of body weight. The values for women are somewhat lower. In both sexes, the values tend to decrease with age (Table 1–1).

UNITS FOR MEASURING CONCENTRATION OF SOLUTES

In considering the effects of various physiologically important substances and the interactions be-

Table 1–1. TBW (as percentage of body weight) in relation to age and sex.

Age	Male	Female
10–18	59%	57%
18–40	61%	51%
40–60	55%	47%
Over 60	52%	46%

tween them, the number of molecules, electrical charges, or particles of a substance per unit volume of a particular body fluid are often more meaningful than simply the weight of the substance per unit volume. For this reason, concentrations are frequently expressed in moles, equivalents, or osmoles.

Moles

A mole is the gram-molecular weight of a substance, ie, the molecular weight of the substance in grams. Each mole (mol) consists of approximately 6×10^{23} molecules. The millimole (mmol) is 1/1000 of a mole, and the micromole (μmol) is 1/1,000,000 of a mole. Thus, 1 mol of NaCl = 23 + 35.5 g = 58.5 g, and 1 mmol = 58.5 mg. The mole is the standard unit for expressing the amount of substances in the SI unit system (see Appendix).

It is worth noting that the molecular weight of a substance is the ratio of the mass of one molecule of the substance to the mass of $1/12$ the mass of an atom of carbon-12. Since molecular weight is a ratio, it is dimensionless. The dalton (Da) is a unit of mass equal to $1/12$ the mass of an atom of carbon-12, and 1000 Da = 1 kilodalton (kDa). The kilodalton, which is often expressed simply as K, is a useful unit for expressing the molecular mass of proteins. Thus, for example, one can speak of a 64K protein or state that the molecular mass of the protein is 64,000 Da. However, since molecular weight is a dimensionless ratio, it is incorrect to say that the molecular weight of the protein is 64kDa.

Equivalents

The concept of electrical equivalence is important in physiology because many of the important solutes in the body are in the form of charged particles. One equivalent (eq) is 1 mol of an ionized substance divided by its valence. One mole of NaCl dissociates into 1 eq of Na^+ and 1 eq of Cl^-. One equivalent of Na^+ = 23 g/l = 23 g; but 1 eq of Ca^{2+} = 40 g/2 = 20 g. The milliequivalent (meq) is 1/1000 of 1 eq.

Electrical equivalence is not necessarily the same as chemical equivalence. A gram equivalent is the weight of a substance that is chemically equivalent to 8.000 g of oxygen. The normality (N) of a solution is the number of gram equivalents in 1 liter. A 1 N solution of hydrochloric acid contains 1 + 35.5 g/L = 36.5 g/L.

Osmoles

When one is dealing with concentrations of osmotically active particles, the amounts of these particles are usually expressed in osmoles. Osmosis is discussed in detail in a later section of this chapter. One osmole (osm) equals the molecular weight of the substance in grams divided by the number of freely moving particles each molecule liberates in solution. The milliosmole (mosm) is 1/1000 of 1 osm.

The **osmolal concentration** of a substance in a

fluid is measured by the degree to which it depresses the freezing point, 1 mol/L of ideal solute depressing the freezing point 1.86 Celsius degrees. The number of milliosmoles per liter in a solution equals the freezing point depression divided by 0.00186. The **osmolarity** is the number of osmoles per liter of solution—eg, plasma—whereas the **osmolality** is the number of osmoles per kilogram of solvent. Therefore, osmolarity is affected by the volume of the various solutes in the solution and the temperature, while the osmolality is not. Osmotically active substances in the body are dissolved in water, and the density of water is 1, so osmolal concentrations can be expressed as osmoles per liter (osm/L) of water. In this book, osmolal (rather than osmolar) concentrations are considered, and osmolality is expressed in milliosmoles per liter (of water).

pH

The maintenance of a stable hydrogen ion concentration in the body fluids is essential to life. The pH of a solution is the logarithm to the base 10 of the reciprocal of the H$^+$ concentration ([H$^+$]), ie, the negative logarithm of the [H$^+$]. The pH of water at 25 °C, in which H$^+$ and OH$^-$ ions are present in equal numbers, is 7.0 (Fig 1–2). For each pH unit less than 7.0, the [H$^+$] is increased tenfold; for each pH unit above 7.0, it is decreased tenfold.

Buffers

Intracellular and extracellular pH are generally maintained at very constant levels. For example, the pH of the ECF is 7.40, and in health, this value usually varies less than ± 0.05 pH unit. Body pH is stabilized by the **buffering capacity** of the body fluids. A buffer is a substance that has the ability to bind or release H$^+$ in solution, thus keeping the pH of the solution relatively constant despite the addition of considerable quantities of acid or base. One buffer in the body is carbonic acid. This acid is only partly disso-

ciated into H$^+$ and bicarbonate: $H_2CO_3 \rightleftarrows H^+ + HCO_3^-$. If H$^+$ is added to a solution of carbonic acid, the equilibrium shifts to the left and most of the added H$^+$ is removed from solution. If OH$^-$ is added, H$^+$ and OH$^-$ combine, taking H$^+$ out of solution. However, the decrease is countered by more dissociation of H_2CO_3, and the decline in H$^+$ concentration is minimized. Other buffers include the blood proteins and the proteins in cells. The quantitative aspects of buffering and the respiratory and renal adjustments that operate with buffers to maintain a stable ECF pH of 7.40 are discussed in Chapters 35 and 39.

FUNCTIONAL MORPHOLOGY OF THE CELL

Revolutionary advances in the understanding of cell structure and function have been made through use of the techniques of modern cellular and molecular biology. There have also been major advances in the study of embryology and development at the cellular level. Developmental biology and the details of cell biology are beyond the scope of this book. However, a basic knowledge of cell biology is essential to an understanding of the organ systems in the body and the way they function.

The specialization of the cells in the various organs is very great, and no cell can be called "typical" of all cells in the body. However, a number of structures (**organelles**) are common to most cells. These structures are shown in Fig 1–3.

Cell Membrane

The membrane that surrounds the cell is a remarkable structure. Not only is it semipermeable, allowing some substances to pass through it and excluding others, but also its permeability can be varied. It is generally referred to as the **plasma membrane.** The nucleus is also surrounded by a membrane, and the organelles are surrounded by or made up of a membrane.

Although the chemical structure of membranes and their properties vary considerably from one location to another, they have certain common features. They are generally about 7.5 nm (75 Angstrom units) thick. They are made up primarily of protein and lipids. The chemistry of proteins and lipids is discussed in Chapter 17. The major lipids are phospholipids such as phosphatidylcholine and phosphatidylethanolamine. The shape of the phospholipid molecule is roughly that of a clothespin (Fig 1–4). The head end of the molecule contains the phosphate portion and is relatively soluble in water (polar, **hydrophilic**). The tails are relatively insoluble (nonpolar, **hydrophobic**). In the membrane, the hydrophilic ends of the molecules are exposed to the aqueous environment that bathes the exterior of the cells and the aqueous cytoplasm; the hydrophobic ends meet in the

H$^+$ concentration (mol/L)

ACIDIC		pH
10^{-1}		1
10^{-2}		2
10^{-3}		3
10^{-4}		4
10^{-5}		5
10^{-6}		6
For pure water, [H$^+$] = 10^{-7} mol/L → 10^{-7}		7
10^{-8}		8
ALKALINE 10^{-9}		9
10^{-10}		10
10^{-11}		11
10^{-12}		12
10^{-13}		13
10^{-14}		14

Figure 1–2. pH. (Reproduced, with permission, from Alberts B et al: *Molecular Biology of the Cell.* Garland, 1983.)

Figure 1–3. Diagram showing a hypothetical cell in the center as seen with the light microscope. It is surrounded by various organelles. (After Bloom and Fawcett. Reproduced, with permission, from Junqueira LC, Carneiro J, Kelley, RO: *Basic Histology,* 7th ed. Appleton & Lange, 1992.)

water-poor interior of the membrane. In **prokaryotes** (cells such as bacteria in which there is no nucleus), phospholipids are generally the only membrane lipids, but in **eukaryotes** (cells containing nuclei), cell membranes also contain cholesterol (in animals) or other steroids (in plants).

There are many different proteins embedded in the membrane. They exist as separate globular units and many pass through the membrane (**integral proteins**), whereas others (**peripheral proteins**) stud the inside and outside of the membrane (Fig 1–4). The amount of protein varies with the function of the membrane but makes up on average 50% of the mass of the membrane; ie, there is about one protein molecule per 50 of the much smaller phospholipid molecules. The uncharged, hydrophobic portions of the proteins are usually located in the interior of the membrane, whereas the charged, hydrophilic portions are located on the surfaces. Peripheral proteins are attached to the surfaces of the membrane in various ways. One common way is attachment to glycosylated forms of phosphatidylinositol. Proteins held by these **glycosyl-phosphatidylinositol anchors** include enzymes such as alkaline phosphatase,

various antigens, and a number of cell adhesion molecules.

The proteins in the membranes carry out at least 7 functions. In addition to **structural proteins,** there are proteins that function as **pumps,** actively transporting ions across the membrane. Other proteins function as **carriers,** transporting substances down electrochemical gradients by facilitated diffusion. Still others are **ion channels,** which, when activated, permit the passage of ions into or out of the cell. The role of the pumps, carriers, and ion channels in transport across the cell membrane is discussed below. Proteins in a fifth group function as **receptors** that bind neurotransmitters and hormones, initiating physiologic changes inside the cell. Proteins in a sixth group function as **enzymes,** catalyzing reactions at the surfaces of the membrane. A seventh group is made up of glycoproteins that function in antibody processing and distinguishing self from nonself (see Chapter 27).

The protein structure—and particularly the enzyme content—of biologic membranes varies not only from cell to cell but also within the same cell. For example, there are different enzymes embedded

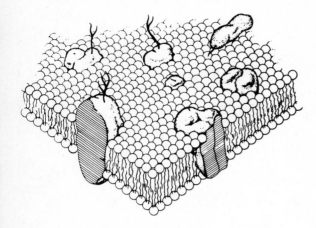

Figure 1–4. Biologic membrane. The phospholipid molecules each have 2 fatty acid chains (wavy lines) attached to a phosphate head (open circle). Proteins are shown as irregular shaded globules. Many are integral proteins, which extend through the membrane, but peripheral proteins are attached to the inside (not shown) and outside of the membrane, sometimes by glycosylphosphatidylinositol anchors (top). (Modified from Singer SJ, Nicolson GL: The fluid mosaic model of the structure of cell membranes. *Science* 1972;**175:**720.)

Figure 1–5. Intercellular junctions in the mucosa of the small intestine. The various types of desmosomes are not shown in detail. (Modified from Weinstein RS, McNutt SN: Cell junctions. *N Engl J Med* 1972;**286:**521.)

in cell membranes than in mitochondrial membranes. In epithelial cells, the enzymes in the cell membrane on the mucosal surface differ from those in the cell membrane on the lateral margins of the cells; ie, the cells are **polarized.** The membranes are dynamic structures, and their constituents are being constantly renewed at different rates. Some proteins move laterally in the membrane. For example, receptors move in the membrane and aggregate at sites of endocytosis (see below).

Underlying most cells is a thin, fuzzy layer plus some fibrils that collectively make up the **basement membrane** or, more properly, the **basal lamina.** The material that makes up the basal lamina is made up of a collagen derivative plus glycoproteins.

Intercellular Connections

Two types of junctions form between the cells that make up tissues: junctions that fasten the cells to one another and to surrounding tissues, and junctions that permit transfer of ions and other molecules from one cell to another. The junctions that tie cells together and endow tissues with strength and stability include **tight junctions,** which are also known as the **zonulae occludens. Desmosomes, hemidesmosomes,** and probably **zonulae adherens** (Fig 1–5) also hold tissues together. The junctions by which molecules are transferred are **gap junctions.**

Tight junctions characteristically surround the apical margins of the cells in epithelia such as the intestinal mucosa, the walls of the renal tubules, and the choroid plexus. They are made up of ridges—half from one cell and half from the other—that adhere so strongly at cell junctions that they almost obliterate the space between the cells. The ridges are made up of proteins, and 2 of these, **cingulin** and **ZO-1,** have been isolated and studied. In addition to tying the cells together, tight junctions form a barrier to the movement of ions and other solutes from one side of the epithelium to the other. The magnitude of this barrier varies, and some epithelia are more "leaky" to solutes than others. Another function of tight junctions is maintenance of cell polarity. The ridges prevent the lateral movement of proteins in the cell membrane and consequently act as a fence, keeping proteins inserted into the apical membrane in the apical region.

In epithelial cells, the zonula adherens is usually a continuous structure on the basal side of the zonula occludens, and it is a major site of attachment for intracellular **actin fibers** (see below and Chapter 3).

Desmosomes are spotlike patches characterized by apposed thickenings of the membranes of 2 adjacent cells. Attached to the thickened area in each cell are cytoplasmic fibrils made up of actin, some running parallel to the membrane and others radiating away from it. Between the 2 membrane thickenings, there is filamentous material in the intercellular space. Hemidesmosomes are half desmosomes that attach cells to an underlying basal lamina without apparent specialization of the basal lamina.

Cell Adhesion Molecules

Cells are attached to the basal lamina and to each other by **cell adhesion molecules (CAMs).** These adhesion proteins gave attracted great attention in recent years because they are important in embryonic development and formation of the nervous system and other tissues, in holding tissues together in adults, in inflammation and wound healing, and in the metastasis of tumors. Many pass through the cell membrane and are anchored to the cytoskeleton (see below). Many CAMs have been characterized biochemically, and their functions are being investigated. Some bind to like molecules on other cells (homophilic binding), whereas others bind to other molecules (heterophilic binding). Many bind to **laminin,** a large cross-shaped molecule with multiple receptor domains in the basal lamina.

Nomenclature in the CAM field is somewhat chaotic, partly because the field is growing so rapidly and partly because of the extensive use of acronyms, as in other areas of modern biology. However, the CAMs can be divided into 5 broad families: (1) the **integrins,** heterodimers that bind to various receptors; (2) adhesion molecules of the **IgG superfamily** of immunoglobulins, some of which bind to other molecules and some of which bind homophilically; (3) the **cadherins,** Ca^{2+}-dependent molecules that mediate cell-to-cell adhesion by homophilic reactions; (4) the **LEC-CAMs,** which have lectin-like domains and are involved in the interactions between white blood cells and endothelial cells; and (5) **homing molecules,** which target lymphocytes to specific lymphoid tissues. The relation of CAMs to platelets is discussed in Chapter 27 and their relation to wound healing in Chapter 33.

Gap Junctions

At gap junctions, the intercellular space narrows from 25 nm to 3 nm, and the hexagonal arrays of protein units called **connexons** in the membrane of each cell are lined up with one another (Fig 1–6). Each connexon is made up of 6 subunits surrounding a channel that, when lined up with the channel in the corresponding connexon in the adjacent cell, permits substances to pass between the cells without entering the ECF. The diameter of the channel is normally about 2 nm, which permits the passage of ions, sugars, amino acids, and other solutes with molecular weights up to about 1000. Gap junctions thus permit the rapid propagation of electrical activity from cell to cell (see Chapter 4) and the exchange of various chemical messengers. The diameter of each channel is regulated by intracellular Ca^{2+}, and an increase in Ca^{2+} causes the subunits to slide together, reducing the diameter of the channel. The diameter may also be regulated by pH and voltage.

Mitochondria

Although their morphology varies somewhat from

Figure 1–6. Gap junction. Note that each connexon is made up of 6 subunits and that each connexon in the membrane of one cell lines up with a connexon in the membrane of the neighboring cell, forming a channel through which substances can pass from one cell to another without entering the ECF. (Reproduced, with permission, from Kandel ER, Schwartz JH, [editors]: *Principles of Neural Science,* 2nd ed. Elsevier, 1985.)

cell to cell, each mitochondrion (Fig 1–3) is in essence a sausage-shaped structure. It is made up of an outer membrane and an inner membrane that is folded to form shelves **(cristae).** The space between the 2 membranes is called the intracristal space, and the space inside the inner membrane is called the matrix space. The mitochondria are the power-generating units of the cell and are most plentiful and best developed in parts of cells where energy-requiring processes take place. The chemical reactions occurring in them are discussed in detail in Chapter 17. The outer membrane of each mitochondrion is studded with the enzymes concerned with biologic oxidations, providing raw materials for the reactions occurring inside the mitochondrion. In the interior, the matrix space contains the enzymes that convert the products of carbohydrate, protein, and fat metabolism to CO_2 and water via the citric acid cycle. In this process, electrons are transferred along the respiratory-enzyme chain and the high-energy phosphate compound **adenosine triphosphate (ATP)** is synthesized by the process of **oxidative phosphorylation.** ATP is the principal energy source for energy-requiring actions in plants and animals. The enzymes responsible for electron transfer and oxidative phosphorylation are embedded in the inner membrane, and they form characteristic repeating units visible under the electron microscope. Each unit is made up of a basepiece, a stalk, and a spherical headpiece (Fig 1–3). The basepieces contain the enzymes of the electron transfer chain, and the stalks and headpieces contain adenosine triphosphatase and other enzymes concerned with the synthesis and metabolism of ATP.

Most of the DNA in cells is in the nucleus (see below), but a small amount is also found in mitochondria. Many investigators believe that mitochondria were once autonomous microorganisms that developed a symbiotic relation with and became incorporated into ancestral eukaryotic cells. Human mitochondrial DNA contains 16,569 nucleotides, which code for 13 of the proteins in the mitochondrial respiratory chain and ATP-generating system, but most of the mitochondrial proteins are encoded by nuclear DNA, produced in the cytoplasm, and transported into the mitochondria. Mitochondrial DNA, which also codes for 2 ribosomal RNAs (rRNAs) and 22 transfer RNAs (tRNAs) (see below), is unique in that it has a slightly different genetic code than that in the nucleus. In addition, the sperm contributes little if any cytoplasm and no mitochondria to the zygote. Mitochondria come from the ovum; thus, mitochondrial DNA is transmitted solely from the mother to the offspring.

Lysosomes

In the cytoplasm of the cell, there are large, somewhat irregular structures surrounded by membrane. The interior of these structures, which are called **lysosomes,** is more acidic than the rest of the cytoplasm, and they may contain fragments of other cell structures. Some of the granules of the granulocytic white blood cells are lysosomes. Each lysosome contains a variety of enzymes (Table 1–2) that would cause the destruction of most cellular components if the enzymes were not separated from the rest of the cell by the membrane of the lysosome.

The lysosomes function as a form of digestive system for the cell. Exogenous substances such as bacteria that become engulfed by the cell end up in membrane-lined vacuoles. A vacuole of this type **(phagocytic vacuole)** may merge with a lysosome, permitting the contents of the vacuole and the lysosome to mix within a common membrane. Some of the products of the "digestion" of the engulfed material are absorbed through the walls of the vacuole, and the remnants are dumped from the cell (exocytosis; see below). The lysosomes also engulf worn-out components of the cell in which they are located, forming **autophagic vacuoles.** When a cell dies, lysosomal enzymes cause autolysis of the remnants.

Table 1–2. Some of the enzymes found in lysosomes and the cell components that are their substrates.

Enzyme	Substrate
Ribonuclease	RNA
Deoxyribonuclease	DNA
Phosphatase	Phosphate esters
Glycosidases	Complex carbohydrates: glycosides and polysaccharides
Arylsulfatases	Sulfate esters
Collagenase	Proteins
Cathepsins	Proteins

There is evidence that in gout, phagocytes ingest uric acid crystals and such ingestion triggers the extracellular release of lysosomal enzymes which contribute to the inflammatory response in the joints. When one of the lysosomal enzymes is congenitally absent, the lysosomes become engorged with one of the materials they normally degrade. This eventually disrupts the cells that contain the defective lysosomes and leads to one of the **lysosomal storage diseases.** More than 25 such diseases have been described. They are generally rare, but they include such widely known disorders as Tay-Sachs disease.

Microfilaments, Intermediate Filaments, & Microtubules

Most, if not all, eukaryotic cells contain **microtubules,** long hollow structures (Fig 1–7) with 5-nm walls surrounding a cavity 15 nm in diameter. They also contain **microfilaments,** long solid fibers 4–6 nm in diameter.

Microtubules are made up of 2 globular protein subunits, α and β tubulin. A third subunit, γ tubulin, is apparently associated with the production of microtubules by the centrosomes (see below). The α and β subunits form heterodimers (Fig 1–8), and the dimers aggregate to form long tubes made up of stacked rings, with each ring usually containing 13 subunits. The tubules also contain other proteins that facilitate their formation. The assembly of microtubules is facilitated by warmth and various other factors, and disassembly is facilitated by cold and other factors. The end where assembly predominates is called the + end, and the end where disassembly predominates is the – end. Both processes occur simultaneously in vitro. Assembly is prevented by colchicine and vinblastine.

Microtubules have been called the skeleton of the cell, but because of their constant assembly and disassembly, they are a very dynamic skeleton. They provide the tracks for transport of vesicles, organelles such as secretory granules, and mitochondria from one part of the cell to another. They also form the spindle, which moves the chromosomes in mitosis (see below). Microtubules can transport in both directions, and, indeed, the same microtubule has been seen transporting 2 particles in opposite directions.

Cells also contain **intermediate filaments** that are 8–14 nm in diameter and are made up of various subunits. Some of these filaments connect the nuclear membrane to the cell membrane. They appear to be part of the cytoskeleton, but their exact function is not known.

Microfilaments are made up of **actin,** the protein that by its interaction with **myosin** brings about contraction of muscle (see Chapter 3). Actin and its mRNA (see below) are present in all types of cells. All cells also contain a variety of actin-binding proteins, including myosin. Myosin molecules consist of a globular head, which has ATPase activity, and a

Figure 1–7. **Left:** Electron micrograph of the cytoplasm of a fibroblast, showing microfilaments (MF) and microtubules (MT). (Reproduced, with permission, from Junqueira LC, Carneiro J, Kelley RO: *Basic Histology,* 7th ed. Appleton & Lange, 1992.) **Right:** Distribution of microtubules in fibroblasts. The cells are treated with a fluorescently labeled antibody to tubulin, making microtubules visible as the light-colored structures. (Reproduced, with permission, from Connolly J et al: Immunofluorescent staining of cytoplasmic and spindle microtubules in mouse fibroblasts with antibody to τ protein. *Proc Natl Acad Sci* USA 1977;**74:**2437.)

tail. There are 2 general types (Fig 1–9). **Myosin-I** molecules have a single head and a tail that varies in length. Along with actin, myosin-I is often associated with cell membranes. **Myosin-II** molecules have 2 heads and are characteristically found in muscle, although they also occur elsewhere (Table 1–3).

Actin microfilaments reach to the tips of the microvilli on the epithelial cells of the intestinal mucosa and bring about their contraction, probably via interaction with myosin filaments located at their bases. Microfilaments are also found in association with desmosomes and zonulae adherens (see above), in bundles under the plasma membrane, and scattered in a seemingly random fashion in the cytoplasm. Actin molecules polymerize and depolymerize in vivo, with polymerization often occurring at one end of a microfilament (the + end, as in microtubules) and depolymerization at the other (the – end). They are abundant in lamellipodia, the processes that cells put out when they crawl along surfaces. Polymerization within these processes causes the cell to move.

Considerable attention is currently being focused on the **molecular motors** responsible for producing movement of organelles within cells or contraction of muscle (Table 1–3). They include, in addition to myosin-I and myosin-II, dynein, kinesin, and dynamin. They are 100–500-kDa ATPases that form cross bridges and convert ATP into energy that produces bending of these cross-bridges. The formation and movement of the cross-bridges in the case of myosin-II filaments binding to actin to produce muscle contrac-

tion are considered in detail in Chapter 3. Axonemal dynein is responsible for the beating of flagella and cilia. Cytoplasmic dynein and kinesin are involved in the movement of chromosomes along the microtubules of the spindle in mitosis and are also responsible for movement of vesicles and other organelles along microtubules in neurons. In general, kinesin mediates anterograde transport to the nerve endings and cytoplasmic dynein mediates transport in the other direction (see Chapter 2). Dynamin produces sliding of one microtubule on another.

Centrioles

In the cytoplasm of most cells is a **centrosome** made up of 2 **centrioles.** The centrioles are short cylinders located near the nucleus, and they are arranged so that they are at right angles to each other. Microtubules in groups of 3 run longitudinally in the walls of each centriole (Fig 1–3). There are 9 of these triplets spaced at regular intervals around the circumference.

The centrosomes are organizing centers for microtubules, establishing their polarity. They duplicate themselves, and the pairs move apart to form the poles of the mitotic spindle, which is made up of microtubules. In multinucleate cells, there is a centrosome near each nucleus.

Cilia

There are various types of projections from cells. True **cilia** are motile processes that are used by unicellular organisms to propel themselves through the

Figure 1–8. Assembly and disassembly of a microtubule by aggregation and disaggregation of dimers made up of α and β tubulin. (From Sloboda RD: The role of microtubules in cell structure and cell division. *Am Sci* 1980;**68**:290. Reprinted by permission of *American Scientist,* journal of Sigma Xi, the Scientific Research Society.)

water by and by multicellular organisms to propel mucus and other substances over the surface of various epithelia. They resemble centrioles in having an array of 9 tubular structures in their walls, but they have in addition a pair of microtubules in the center, and there are 2 rather than 3 microtubules in each of

Figure 1–9. Diagrammatic representation of myosin-I and myosin-II. The tail of myosin-I molecules are short and variable, and there is some uncertainty about their structure. (From Pollard TD, Doberstein SK, Zot HG: Myosin-I. *Annu Rev Physiol* 1991;**53**:653. Reproduced, with permission, from the *Annual Review of Physiology,* vol. 53, ©1991 by Annual Reviews Inc.)

Table 1–3. Molecular motors from eukaryotic cells.[1]

Motor	Effect
Actin-based Myosin-I	Moves membranes on actin filaments; moves one actin filament on another
Myosin-II	Causes muscle contraction, cortical tension, capping surface molecules, cell polarity, cytokinesis
Microtubule-based Axonemal dynein	Causes sliding of one flagellar microtubule on another, moves particles toward "minus" end of microtubule
Cytoplasmic dynein	Moves particles and membranes toward "minus" end of microtubule
Kinesin	Moves particles and membranes toward "plus" end of microtubule
Dynamin	Drives active sliding of one microtubule on another

[1]Modified and reproduced, with permission, from Scholey JM: Multiple microtubule motors. Reprinted by permission from *Nature* **343**:118,1990. Copyright © 1990. Macmillan Magazines Ltd.

the 9 circumferential structures. The **basal granule,** on the other hand, is the structure to which each cilium is anchored. It has 9 circumferential triplets, like a centriole.

Peroxisomes

If cells are homogenized and the resulting suspension is centrifuged, various cellular components can be isolated. The nuclei sediment first, followed by the mitochondria. High-speed centrifugation that generates forces of 100,000 times gravity or more causes a fraction made up of granules called the **microsomes** to sediment. This fraction includes the ribosomes, but it also contains other granular material that can be isolated by further ultracentrifugation and other techniques. The **peroxisomes** are found in the granular material. These organelles are about 0.5 μm in diameter and are surrounded by a membrane. They contain various oxidases which catalyze reactions that generate hydrogen peroxide, and **catalase,** an enzyme that catalyzes the conversion of hydrogen peroxide to water. Various peroxisomes have various substrates. In mammals, peroxisomes are most common in the liver and kidneys. Their numbers are increased by **peroxisome proliferators,** drugs that include hypolipidemic agents such as clofibrate, herbicides, and industrial plasticizers. The drugs appear to activate receptors in the steroid hormone superfamily (see below), and their use may be related to hepatic cancer.

Nucleus & Related Structures

A nucleus is present in all eukaryotic cells that divide. If a cell is cut in half, the anucleate portion eventually dies without dividing. The nucleus is made up in large part of the **chromosomes,** the structures in the nucleus that carry a complete blue-print for all the heritable species and individual characteristics of the animal. Except in germ cells, the chromosomes occur in pairs, one originally from each parent (see Fig 23–2). Each chromosome is made up of a giant molecule of **deoxyribonucleic acid (DNA)** that in multicellular organisms is virtually covered with proteins. The molecule has been likened to a string of beads; the beads are balls of histone proteins with DNA wrapped around them, and the string is more DNA with other proteins attached to it. This complex of DNA and proteins is called **chromatin.** During cell division, the pairs of chromosomes become visible, but between cell divisions, only clumps of chromatin can be discerned in the nucleus. The ultimate units of heredity are the **genes** on the chromosomes (see below), and each gene is a portion of the DNA molecule.

During normal cell division by **mitosis,** the chromosomes duplicate themselves and then divide in such a way that each daughter cell receives a full complement (**diploid number**) of chromosomes. During their final maturation, germ cells undergo a division in which half the chromosomes go to each daughter cell (see Chapter 23). This reduction division (**meiosis**) is actually a 2-stage process; but the important consideration is that, as a result of it, mature sperms and ova contain half the normal number (the **haploid number**) of chromosomes. When a sperm and ovum unite, the resultant cell (**zygote**) has a full diploid complement of chromosomes, one-half from the female parent and one-half from the male. The chromosomes undergo recombination, which mixes maternal and paternal genes.

The nucleus of most cells contains a **nucleolus** (Fig 1–3), a patchwork of granules rich in **ribonucleic acid (RNA).** In some cells, the nucleus contains several of these structures. Nucleoli are most prominent and numerous in growing cells. They are the site of synthesis of ribosomes, the structures in the cytoplasm in which proteins are synthesized (see below).

The nucleus is surrounded by a **nuclear membrane (envelope)** (Fig 1–3). This membrane is double, and spaces between the 2 folds are called **perinuclear cisterns.** The nuclear membrane contains pores, some of which clearly open into the endoplasmic reticulum. There is considerable interchange between the nucleus and the cytoplasm, with molecules as large as the various forms of RNA, for example, moving out of the nucleus. The nuclear pores contain 2 rings, each made up of 8 subunits, and it has been suggested that each ring opens and closes like an iris diaphragm, permitting the passage of fairly large structures.

Endoplasmic Reticulum

The **endoplasmic reticulum** is a complex series of tubules in the cytoplasm of the cell (Fig 1–3). The tubule walls are made up of membrane. In **rough** or **granular endoplasmic reticulum,** granules called **ribosomes** are attached to the cytoplasmic side of the membrane, whereas in **smooth** or **agranular endoplasmic reticulum,** the granules are absent. Free ribosomes are also found in the cytoplasm. The granular endoplasmic reticulum is concerned with protein synthesis and the initial folding of polypeptide chains with the formation of disulfide bonds. The agranular endoplasmic reticulum is the site of steroid synthesis in steroid-secreting cells and the site of detoxification processes in other cells. As the sarcoplasmic reticulum (see Chapter 3), it plays an important role in skeletal and cardiac muscle.

Ribosomes

The ribosomes in eukaryotes measure approximately 22 by 32 nm. Each is made up of a large and a small subunit called, on the basis of their rates of sedimentation in the ultracentrifuge, the 60S and 40S subunits. The ribosomes contain about 65% RNA and 35% protein. They are the sites of protein synthesis. The ribosomes that become attached to the endoplasmic reticulum synthesize proteins such as hormones that are secreted by the cell, proteins that are segregated in lysosomes, and proteins that are inserted in cell membranes. The polypeptide chains that form these proteins are extruded into the endoplasmic reticulum. The free ribosomes synthesize cytoplasmic proteins such as hemoglobin (see Chapter 27) and the proteins found in peroxisomes and mitochondria.

Structure & Function of DNA & RNA

DNA is found in bacteria, in the nuclei of eukaryotic cells, and in mitochondria. It is made up of 2 extremely long nucleotide chains containing adenine (A), guanine (G), thymine (T), and cytosine (C) (Fig 1–10). The chemistry of these purine and pyrimidine bases and of nucleotides is discussed in Chapter 17. The chains are bound together by hydrogen bonding between the bases, with adenine bonding to thymine and guanine to cytosine. The resultant double-helical structure of the molecule is shown in Fig 1–11. An indication of the complexity of the molecule is the fact that the DNA in the human haploid genome (the total genetic message) is made up of 3×10^9 base pairs.

DNA is the component of the chromosomes that carries the "genetic message," the blueprint for all the heritable characteristics of the cell and its descendants. Each chromosome contains a segment of the DNA double helix. The genetic message is encoded by the sequence of purine and pyrimidine bases in the nucleotide chains. The text of the message is the order in which the amino acids are lined up in the proteins manufactured by the cell. The message is trans-

Figure 1–10. Segment of the structure of the DNA molecule in which the purine and pyrimidine bases adenine (A), thymine (T), cytosine (C), and guanine (G) are held together by a phosphodiester backbone between 2′-deoxyribosyl moieties attached to the nucleobases by an N-glycosidic bond. Note that the backbone has a polarity (ie, a 5′ and a 3′ direction). (Reproduced, with permission, from Murray RK et al: *Harpers Biochemistry,* 22nd ed. Appleton & Lange, 1990.)

Figure 1–11. Double-helical structure of DNA, with adenine (A) bonding to thymine (T) and cytosine (C) to guanine (G). (Reproduced, with permission, from Murray RK et al: *Harpers Biochemistry,* 22nd ed. Appleton & Lange, 1990.)

ferred to the sites of protein synthesis in the cytoplasm by RNA. The proteins formed include all the enzymes, and these in turn control the metabolism of the cell. A gene has been defined as the amount of information necessary to specify a single peptide molecule, although this is an oversimplification, since it is now known that single genes can code for more than one protein. Genes also contain promoters and enhancers, DNA sequences that facilitate the formation of RNA. Mutations occur when the base sequence in the DNA is altered by x-rays, cosmic rays, or other mutagenic agents. There are estimated to be 50,000–100,000 genes in the 3 billion base pairs that make up the human genome. As of 1991, approximately 2150 of these had been mapped, so there is obviously much more to do to map the whole genome. Abnormalities of about 450 of the genes have been linked to human diseases.

Over 99% of the human genome is preserved from individual to individual, but there are also **hypervariable regions,** which contain so many variations in base pairs that the DNA of each person becomes almost unique for that individual. These variations cause variations in the locations where DNA is cut by various restriction enzymes. Therefore, when the pieces of DNA left after exposure to the enzymes are spread out and identified, they vary markedly in length; ie, they come from areas of **restriction fragment length polymorphism (RFLP).** Consequently, analysis of RFLPs gives a pattern that is in effect a **DNA fingerprint.** The value of DNA fingerprinting has been improved by additional specialized techniques. The chance of obtaining identical DNA patterns by using these techniques in individuals who are not identical twins varies somewhat with the number of enzymes used and other factors, and currently there is debate about the appropriate statistics to use for analysis. However, some estimate the chance of identity to be as low as one in 40 billion. Furthermore, RFLP analysis can be carried out on small specimens of semen, blood, or other tissue, and multiple copies of pieces of DNA can be made by using the **polymerase chain reaction, (PCR),** an ingenious technique for making DNA copy itself. Therefore, DNA fingerprinting is of obvious value in investigating crimes and determining paternity, although reliable techniques must be used and the results interpreted with care. RFLP analysis is also of value in studying animal and human evolution and in identifying the chromosomal location of genes causing inherited diseases.

At the time of each somatic cell division (**mitosis),** the 2 DNA chains separate, each serving as a template for the synthesis of a new complementary chain. **DNA polymerase** catalyzes this reaction. One of the double helices thus formed goes to one daughter cell and one to the other, so the amount of DNA in each daughter cell is the same as that in the parent cell. In germ cells, reduction division (**meiosis)** takes place during maturation. The net result is that one of each pair of chromosomes ends up in each mature germ cell; consequently, each mature germ cell contains half the amount of chromosomal material found in somatic cells. Therefore, when a sperm unites with an ovum, the resulting zygote has the full complement of DNA, half of which came from the father and half from the mother. The chromosomal events that occur at the time of fertilization are discussed in detail in Chapter 23. The term "ploidy" is sometimes used to refer to the DNA content of cells. Normal resting diploid cells are **euploid,** and become **tetraploid** just before division. **Aneuploidy** is the condition in which a cell contains other than the haploid amount of DNA or an exact multiple of it, and this condition is common in cancerous cells.

Obviously, normal cell division depends on the orderly occurrence of events during what has come to be called the **cell cycle.** A current view of these events is shown in Fig 1–12. There is intense interest in the biochemical machinery that produces mitosis, in part because of the obvious possibility of its relation to cancer, Entry into mitosis is initiated when one of a group of proteins called **cyclins** combines with another protein, p34^{cdc2}, to form **cdc2 kinase.** This in turn phosphorylates other proteins. p34^{cdc2} combines with other cyclins to regulate other phases

Figure 1–12. Sequence of events during the cell cycle.

of the cycle. However, regulation of the cycle is complex, and a detailed consideration is beyond the scope of this book.

The strands of the DNA double helix not only replicate themselves, but also serve as templates by lining up complementary bases for the formation in the nucleus of **messenger RNA (mRNA)** and **soluble (transfer) RNA (tRNA).** This process is called **transcription** (Fig 1–13) and is catalyzed by **RNA polymerase.** In prokaryotes, the RNAs move without modification out of the nucleus to the ribosomes. However, in eukaryotes, the mRNA that is transcribed is really a pre-mRNA, and it generally undergoes considerable **posttranscriptional modification** before moving out of the nucleus and becoming associated with the ribosomes. In the ribosomes, mRNA dictates the formation of the polypeptide chain of a protein (**translation).** tRNA attaches the amino acids to mRNA. The mRNA molecules are smaller than the DNA molecules, and each represents a transcription of a small segment of the DNA chain. There are additional small RNAs in the cell. The molecules of tRNA contain only 70–80 nitrogenous bases, compared with hundreds in mRNA and 3 billion in DNA. tRNA and mRNA are single-stranded, and they contain the base uracil (U) in place of thymine. A third distinct type of RNA is the RNA in the ribosomes (**ribosomal RNA, RNA).**

It is worth noting that DNA is responsible for the maintenance of the species; it passes from one generation to the next in germ cells. RNA, on the other hand, is responsible for the production of the individual; it transcribes the information coded in the DNA and forms a mortal individual, a process that has been called "budding off from the germ line."

Genes

Information is accumulating at an accelerating rate about the structure of genes and their regulation. A typical gene is shown in Fig 1–14. It is made up of a strand of DNA that includes a transcription unit and flanking regions at either end of this unit. In eukary-

Figure 1–13. Diagrammatic outline of protein synthesis. The nucleic acids are represented as lines with multiple short projections representing the individual bases.

otes, unlike prokaryotes, the portions of the genes that dictate the formation of proteins are usually broken into several segments (**exons**) separated by segments that are not translated (**introns**). A pre-mRNA is formed from the DNA, and then the introns and sometimes some of the exons are eliminated in the nucleus by posttranscriptional processing, so that the final mRNA which enters the cytoplasm is made up of exons (Fig 1–15). Introns are eliminated and exons are joined by unique units called **spliceosomes**, which are made up of RNA and proteins. More than

one mRNA can be formed from the same gene, the difference in mRNAs being the inclusion of different exons. The physiologic function of the introns is still unsettled, although they may foster changes in the genetic message and thus aid evolution.

The **transcription unit** of the gene starts with the **cap site** and extends about 20 bases beyond an AATAAA sequence to a poly(A) addition site. The **promoter region** on the 5' side of the cap site contains various DNA segments that regulate gene expression. These include the sequence TATAA called

Figure 1–14. Diagram showing organization of a gene. The segment of DNA, from the 5' end to the 3' end, is divided into a transcription unit and a promoter region. Key base sequences are indicated along the segment. CAT, sequences related to amplitude of transcription; E, enhancer sequences. (Reproduced, with permission, from Habener JF: Regulation of polypeptide-hormone biosynthesis at the level of the genome. *Am J Physiol* 1985;**249**:C191.)

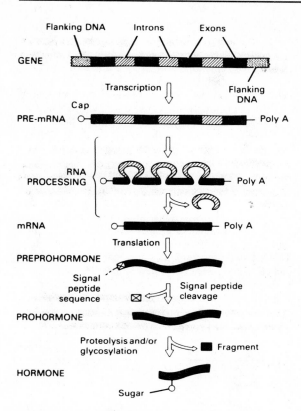

Figure 1–15. Transcription, posttranscriptional modification of mRNA, translation in the ribosomes, and posttranslational processing in the formation of hormones and other proteins. (Modified and reproduced, with permission, from Baxter JD: Principles of endocrinology. In: *Cecil Textbook of Medicine,* 16th ed. Wyngaarden JB, Smith LH Jr [editors]. Saunders, 1982.)

the **TATA box,** which ensures that transcription starts at the proper point. There are sequences that regulate the amplitude of transcription (quantifiers) and enhancer and silencer (inhibitor) sequences, as well as regulator sequences that respond to specific regulatory signals such as steroid hormones. Finally, there is a region, called z-DNA, that may determine tissue-specific expression.

There is currently a great deal of interest in **oncogenes,** genes which are carried in the genomes of cancer cells and seem to be responsible for producing many of their malignant properties. These genes are derived by somatic mutation from closely related **proto-oncogenes,** which are normal genes that control growth. Many different proto-oncogenes have been described. They are involved in all the steps in growth regulation from activation of growth factor receptors on the cell surface to the formation of second messengers (see below) and from activation of kinases in the cytoplasm to regulation of events that occur in the nucleus.

Regulation of Gene Expression

Each nucleated somatic cell in the body contains the full genetic message, yet there is great differentiation and specialization in the functions of the various types of adult cells, and only small parts of the message are normally transcribed. Thus, the genetic message is normally maintained in a repressed state. What turns on genes in one cell and not in other cells? What turns on genes in a cell at one stage of development and not at other, inappropriate stages? What maintains orderly growth in cells and prevents the uncontrolled growth that we call cancer? Obviously, the various parts of the promoter sequences in genes play an important role. Promoter regions in a given gene that affect transcription of that gene (*cis* regulation) are turned on by steroid hormones and other signals. In addition, **transcription factors,** proteins encoded by other genes in the cell, can bind to DNA and affect transcription (*trans* regulation). Genes are expressed to a lesser degree when they are condensed into chromatin ("closed") in any given cell. Another factor is the degree to which the DNA is methylated. Methylation of DNA can be inhibited by 5-azacytidine, and this drug has been shown to bring about transcription of previously inactive genes.

It is common for stimuli such as neurotransmitters that bind to the cell membrane to initiate chemical events which activate **immediate-early genes.** These in turn produce transcription factors that act on other genes. The best characterized immediate-early genes are c-*fos,* and c-*jun.* The proteins produced by these genes, c-Fos, c-Jun, and several related proteins, form homodimers or heterodimers which bind to a specific DNA regulatory sequence called an **AP-1 site.** Some of the dimers enhance transcription and other inhibit it. The appearance of c-Fos and c-Jun and related proteins is such a common sign of cell activation that immunocytochemistry for them or measurement of their mRNAs is used to determine which cells in the nervous system and elsewhere are activated by a particular stimulus.

The subunits that form the dimers have α-helical regions with regularly spaced leucine residues which interact with leucines on another subunit to form a coiled coil. This dimerizing arrangement has been called a **leucine zipper.** Extensions of the subunits beyond the zippered region are rich in arginine and lysine, which bind to DNA.

Other motifs common to proteins that bind to DNA have been identified. Hundreds of transcription factors and related proteins contain characteristically shaped complexes in which Zn^{2+} is coordinately bound between 2 cysteines and 2 histidines, or some instances between 4 cysteines, forming a **zinc finger** (Fig 1–16). It is these zinc fingers that bind to DNA.

Protein Synthesis

The process of protein synthesis is a complex but fascinating one that, as noted above, involves 4 steps:

Figure 1–16. ***Top:*** Activation of genes by second messengers. Increased protein kinase C causes production of c-Fos and c-Jun by immediate-early genes. The c-Fos–c-Jun heterodimer binds to an AP-1 site, in this case activating RNA polymerase II (Pol II) and increasing transcription of other genes. (Courtesy of DG Gardner.) ***Bottom:*** Zinc Fingers. The curved lines represent polypeptide chains of proteins that bind to DNA, and the straight lines indicate coordinate binding of zinc to cysteines (C) or cysteines and histidines (H). (Reproduced, with permission, from Murray RK et al: *Harper's Biochemistry,* 22nd ed. Appleton & Lange, 1990.)

transcription, posttranscriptional modification, translation, and posttranslational modification. The various steps are summarized in simplified form in Figs 1–13 and 1–15.

When suitably activated, transcription of the gene starts at the **cap site** and ends about 20 bases beyond the signal sequence AATAAA. The RNA transcript is capped in the nucleus by addition of 7-methylguanosine triphosphate to the 5′ end; this cap is necessary for proper binding to the ribosome (see below). A **poly(A) tail** of about 100 bases is added to the untranslated segment at the 3′ end. The function of the poly(A) tail is unsettled, but it may help maintain the stability of the mRNA. The pre-mRNA formed by capping and addition of the poly(A) tail is

then processed by elimination of the introns (Fig 1–15), and once this posttranscriptional modification is complete, the mature mRNA moves to the cytoplasm. Posttranscriptional modification of the pre-mRNA is a regulated process, and, as noted above, differential splicing can occur, with the formation of more than one mRNA from a single pre-mRNA.

When a definitive mRNA reaches a ribosome in the cytoplasm, it dictates the formation of a polypeptide chain. Amino acids in the cytoplasm are activated by combination with an enzyme and adenosine monophosphate (adenylate), and each **activated amino acid** then combines with a specific molecule of tRNA. There is at least one tRNA for each of the 20 unmodified amino acids found in large quantities in the body proteins of animals (see Chapter 17), but there is more than one tRNA for some amino acids. The tRNA-amino acid-adenylate complex is next attached to the mRNA template, a process that occurs in the ribosomes. This process is shown diagrammatically in Fig 1–13. The tRNA "recognizes" the proper spot to attach on the mRNA template because it has on its active end a set of 3 bases that are complemen-

Figure 1–17. Translation of protein into endoplasmic reticulum according to the signal hypothesis. The ribosomes synthesizing a protein move along the mRNA from the 5′ to the 3′ end. When the signal peptide of a protein destined for secretion, the cell membrane, or lysosomes emerges from the large unit of the ribosome, it binds to a signal recognition particle (SRP), and this arrests further translation until it binds to the signal receptor on the endoplasmic reticulum. This frees the SRP, which is recycled in the cytoplasm. Binding also occurs to the ribosome receptor, and a tunnel opens, permitting the growing protein chain to enter the endoplasmic reticulum. The signal peptide is removed by signal peptidase. At the termination of protein synthesis, the 2 subunits of the ribosome dissociate, the carboxy terminus enters the endoplasmic reticulum, and the tunnel units disassemble. N, amino end of protein; C, carboxy end of protein. (Reproduced, with permission, from Perara E, Linappa VR: Transport of proteins into and across the endoplasmic reticulum membrane. In: *Protein Transfer and Organelle Biogenesis.* Das RC, Robbins PW [editors]. Academic Press, 1988.)

tary to a set of 3 bases in a particular spot on the mRNA chain. The genetic code is made up of such **triplets,** sequences of 3 purine or pyramidine bases or both; each triplet stands for a particular amino acid.

Translation starts in the ribosomes with an AUG (transcribed from ATG in the gene) which codes for methionine. The N-terminal amino acid is then added, and the chain is lengthened one amino acid at a time. The mRNA attaches to the 40S subunit of the ribosome during protein synthesis; the polypeptide chain being formed attaches to the 60S subunit; and the tRNA attaches to both. As the amino acids are added in the order dictated by the triplet code, the ribosome moves along the mRNA molecule like a bead on a string. Translation stops at one of 3 stop, or nonsense, codons—UGA, UAA, or UAG—and the polypeptide chain is released. The tRNA molecules are used again. The mRNA molecules are also reused approximately 10 times before being replaced.

Typically, there is more than one ribosome on a given mRNA chain at one time. The mRNA chain plus its collection of ribosomes is visible under the electron microscope as an aggregation of ribosomes called a **polyribosome (polysome).**

Posttranslational Modification

After the polypeptide chain is formed, it is modified to the final protein by one or more of a combination of reactions that include hydroxylation, carboxylation, glycosylation, or phosphorylation of amino acid residues; cleavage of peptide bonds that converts a larger polypeptide to a smaller form; and the folding and packaging of the protein into its ultimate, often complex configuration.

It has been claimed that a typical eukaryotic cell synthesizes about 10,000 different proteins during its lifetime. How do these proteins get to the right locations in the cell? Synthesis starts in the free ribosomes. Peptides enter the endoplasmic reticulum if the developing peptide chain has at its N terminal a **leader sequence (signal peptide)** made up of 15–30 predominantly hydrophobic amino acid residues. The signal peptide, once synthesized, binds to a **signal recognition particle (SRP),** a complex molecule made up of 6 polypeptides and 7S RNA, one of the small RNAs. The SRP stops translation until it binds to an SRP receptor in the endoplasmic reticulum. The ribosome also binds, and the signal peptide leads the growing peptide chain into the cavity of the endoplasmic reticulum (Fig 1–17). The signal peptide is next cleaved from the rest of the peptide by a signal peptidase while the rest of the peptide chain is still being synthesized.

The signals that direct nascent proteins to some of the other parts of the cell are fashioned in the Golgi apparatus and, as discussed above, involve specific modifications of the carbohydrate residues on glycoproteins.

Many and perhaps all proteins that are secreted by

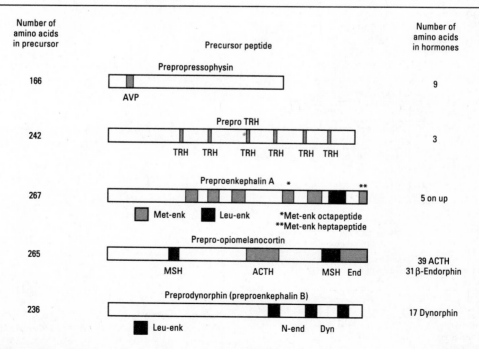

Number of amino acids in precursor	Precursor peptide	Number of amino acids in hormones
166	Prepropressophysin · AVP	9
242	Prepro TRH · TRH TRH TRH TRH TRH TRH	3
267	Preproenkephalin A * ** · ■ Met-enk ■ Leu-enk *Met-enk octapeptide **Met-enk heptapeptide	5 on up
265	Prepro-opiomelanocortin · MSH ACTH MSH End	39 ACTH 31 β-Endorphin
236	Preprodynorphin (preproenkephalin B) · ■ Leu-enk N-end Dyn	17 Dynorphin

Figure 1–18. Examples of large precursors (preprohormones) for small peptide hormones. TRH, thyrotropin-releasing hormone. AVP, arginine vasopressin; Met-enk, met-enkephalin; Leu-enk, leu-enkephalin; MSH, melanocyte-stimulating hormone; ACTH, adrenocorticotropic hormone; End, β-endorphin; Dyn, dynorphin; N-end neoendorphin.

cells are synthesized as larger proteins, and polypeptide sequences are cleaved off from them during maturation. In the case of the hormones, these larger forms are called **preprohormones** and **prohormones** (Figs 1–15 and 1–18). Parathyroid hormone is an example. It is synthesized as a molecule containing 115 amino acid residues (preproparathyroid hormone). The leader sequence, 25 amino acid residues at the N terminal, is rapidly removed to form proparathyroid hormone. Before secretion, an additional 6 amino acids are removed from the N terminal to form the secreted molecule. The function of the 6-amino-acid fragment is unknown.

Protein folding is a complex process that is dictated partly by the sequence of the amino acids in the polypeptide chain. In some instances, however, nascent proteins associate with other proteins called chaperones, which prevent inappropriate contacts with other proteins and ensure that the final "proper" conformation of protein is reached.

Effects of Antibiotics on Protein Synthesis

Antibiotics act in a variety of ways, but many act by inhibiting protein synthesis at one or another of the steps described above (Table 1–4). Some of them have this effect primarily in bacteria, but others inhibit protein synthesis in the cells of other animals, including mammals. This fact makes antibiotics of great value for research as well as for treatment of infections.

Golgi Apparatus

The Golgi apparatus is a collection of membrane-enclosed sacs (cisterns) that are stacked like dinner plates (Fig 1–3). There are usually about 6 sacs in each apparatus, but there may be more. One or more Golgi apparatuses are present in all eukaryotic cells, usually near the nucleus. The Golgi apparatus is a polarized structure, with *cis* and *trans* sides. Membranous vesicles containing newly synthesized proteins bud off from the granular endoplasmic reticulum and fuse with the cistern on the *cis* side of the apparatus. The proteins are then passed, probably via other vesicles, to the middle cisterns and finally to the cistern on the *trans* side, from which vesicles branch off into the cytoplasm (Fig 1–19). These vesicles are coated, like those involved in endocytosis, but they do not contain clathrin (see below). The initial glycosylation of proteins occurs with the attachment of preformed oligosaccharides in the endoplasmic reticulum, but these oligosaccharides are altered to a variety of different carbohydrate moieties in the Golgi apparatus. The alterations provide the code for the ultimate destination of the protein. Proteins destined for lysosomes have phosphate groups added and then bind to one of the 2 types of **mannose-6-phosphate receptors (MPRs)** in the Golgi. They shuttle in clathrin-coated vesicles from the Golgi to a vesicular acidified compartment (prelysosome), where the proteins are

Table 1–4. Mechanisms by which antibiotics and related drugs inhibit protein synthesis.

Agent	Effect
Chloramphenicol	Prevents normal association of mRNA with ribosomes.
Streptomycin, neomycin, kanamycin	Cause misreading of genetic code.
Cycloheximide, tetracycline	Inhibit transfer of tRNA-amino acid complex to polypeptide.
Puromycin	Puromycin-amino acid complex substitutes for tRNA-amino acid complex and prevents addition of further amino acids to polypeptide.
Mithramycin, mitomycin C, dactinomycin (actinomycin D)	Bind to DNA, preventing polymerization of RNA on DNA.
Chloroquine, colchicine, novobiocin	Inhibit DNA polymerase.
Nitrogen mustards, eg, mechlorethamine (Mustargen)	Bind to guanine in base pairs.
Diphtheria toxin	Prevents ribosome from moving on mRNA.

Figure 1–19. Pathways involved in protein processing in the Golgi apparatus, secretion by exocytosis, and membrane recovery by endocytosis. ER, endoplasmic reticulum with attached ribosomes. Note the different pathways for regulated (nonconstitutive) and constitutive secretion by exocytosis.

released and transferred to lysosomes. The MPRs return to the Golgi to be used again. Proteins destined to be secreted by the cell have their sugar residues modified in a more extensive fashion, and glycoproteins destined for the cell membrane have their sugar residues modified in a different, equally extensive way.

Evidence is accumulating that fusion of vesicles inside cells is catalyzed by small GTP-binding proteins (sometimes called smgs). The proteins bind GTP, hydrolysis of which provides the energy for fusion. The GDP that is formed is then reconverted to GTP, and the smg recycles.

Exocytosis

Proteins that are secreted by cells move from the endoplasmic reticulum to the Golgi apparatus, and from the *trans* Golgi, they are extruded into secretory granules or vesicles (Fig 1–19). The granules and vesicles move to the cell membrane. Their membrane then fuses to the cell membrane (Fig 1–20), and the area of fusion breaks down. This leaves the contents of the granules or vesicles outside the cell and the cell membrane intact. The extrusion process is called exocytosis. It requires Ca^{2+} and energy, but the details of the mechanism responsible for the breakdown of the membrane are unknown.

Note that there are 2 pathways by which secretion from the cell occurs (Fig 1–19). In the **nonconstitutive pathway,** proteins initially enter secretory granules coated with clathrin (see below). Processing of prohormones to the mature hormones occurs in these granules. The clathrin coat is then lost, and the secretory granules move to the cell membrane. The other pathway, the **constitutive pathway,** involves the prompt transport of proteins to the cell membrane in vesicles that are not coated with clathrin, with little or no processing or storage. The nonconstitutive path-

way is sometimes called the **regulated pathway,** but this term is misleading because constitutive secretion is also regulated.

Endocytosis

Endocytosis is the reverse of exocytosis. One form of endocytosis, called **phagocytosis** ("cell eating"), is the process by which bacteria, dead tissue, or other bits of material visible under the microscope are engulfed by cells such as the polymorphonuclear leukocytes of the blood. The material makes contact with the cell membrane, which then invaginates. The invagination is pinched off, leaving the engulfed material in the membrane-enclosed vacuole and the cell membrane intact. **Pinocytosis** ("cell drinking") is essentially the same process, the only difference being that the substances ingested are in solution and hence not visible under the microscope.

There are 2 kinds of endocytosis: **constitutive endocytosis,** a continuous process that is not induced; and **receptor-mediated endocytosis,** which is produced for the most part via specialized **coated pits** on the cell membrane and triggered by various ligands binding to their receptors on the cell surface. Some receptors are localized to the coated pits. Others move laterally in the cell membrane once a ligand has bound to them and aggregate in the coated pits. The pit then pinches off, forming a coated vesicle. The vesicle then becomes an **early endosome.** Many of the ingested receptors recycle back to the membrane from the early endosome (Fig 1–19). The pH in this organelle is lowered by protein pumps, and this frees the receptors. In other situations, the early endosome moves deeper into the cell and becomes a **late endosome.** The late endosomes fuse with lysosomes, and their contents are digested by the lysosomal enzymes.

The coated appearance of pits and vesicles concerned with receptor-mediated endocytosis transport from the *trans* Golgi to lysomes, and the formation of secretory granules is due to the presence in their wall of **clathrin.** This polypeptide has a molecular weight of 18,000. It helps select and concentrate the proteins that are found in the vesicles it lines, although it is not essential for transport to occur. Receptor-mediated endocytosis can occur more rapidly than constitutive endocytosis, and the former is more specific in the sense that the particular molecules which triggered it are concentrated on their receptors in the coated pits and hence the coated vesicles. Receptor-mediated endocytosis is responsible for the internalization of low-density lipoproteins—an important part of cellular metabolism of cholesterol (see Chapter 17)—and for the uptake of many substances, including insulin, epidermal growth factor, nerve growth factor, diphtheria toxin, and a number of different viruses.

It is apparent that exocytosis adds to the total amount of membrane surrounding the cell, and if membrane were not removed elsewhere at an equivalent rate, the cell would enlarge. However, removal

Lumen

Secretion granule

Fusion

Elimination of layers

Fission

Figure 1–20. Sequence of events in exocytosis.

of cell membrane occurs by endocytosis, and such ex-ocytosis-endocytosis coupling (Fig 1–19) maintains the surface area of the cell at its normal size.

A good deal is now known about the alterations in proteins that occur in each of the cellular organelles discussed above. The sites of many of the processes that occur seriatim in the formation and discharge of proteins via the nonconstitutive pathway are summarized in Fig 1–21.

TRANSPORT ACROSS CELL MEMBRANES & MEMBRANE POTENTIALS

Distribution of Ions and Other Substances Across Cell Membranes

The unique properties of the cell membranes are responsible for the differences in the composition of intracellular and interstitial fluid. Average values for the composition of intracellular fluid in humans are shown in Fig 1–22, and specific values for ions in mammalian neurons are shown in Table 1–5. Figure 1–22 also shows the differences in composition of plasma and interstitial fluid.

Membrane Potentials

There is a potential difference across the membranes of most if not all cells, with the inside of the

Table 1–5. Concentration of some ions inside and outside mammalian spinal motor neurons.[1]

Ion	Concentration (mmol/L of H₂O)		Equilibrium Potential (mV)
	Inside Cell	Outside Cell	
Na⁺	15.0	150.0	+60
K⁺	150.0	5.5	−90
Cl⁻	9.0	125.0	−70

Resting membrane potential = −70 mV

[1] Data from Mommaerts WFHM, in: *Essentials of Human Physiology.* Ross G (editor). Year Book, 1978.

cells negative relative to the exterior. By convention, this **resting membrane potential (steady potential)** is written with a minus sign, signifying that the inside is negative relative to the exterior. Its magnitude varies considerably from tissue to tissue, ranging from –9 to –100 mV.

Membrane Permeability & Membrane Transport Proteins

Cell membranes are practically impermeable to intracellular protein and other organic anions, which make up most of the intracellular anions and are usually represented by the symbol A⁻. When considering membrane permeability to smaller molecules, it is

Figure 1–21. Steps in the maturation of preproproteins into secreted proteins such as hormones via the nonconstitutive pathway, and the locations in which they occur. (Reproduced, with permission, from Mains RE et al: Cellular and molecular aspects of peptide hormone biosynthesis. *Front Neuroendocrinol* 1990;**11**:52.)

Figure 1–22. Electrolyte composition of human body fluids. Note that the values are in meq/L of water, not of body fluid. (Reproduced, with permission, from Johnson LR [editor]: *Essential Medical Physiology.* Raven, 1992.

important to distinguish between the lipid bilayer itself and the numerous and various transport proteins embedded in it. The lipid bilayer is highly permeable to water. Its permeability to other substances depends on their size (Table 1–6), lipid solubility, and charge. Molecules such as O_2 and N_2 that are nonpolar, ie, hydrophobic, dissolve in the bilayer and cross with ease. Small uncharged polar (hydrophilic) molecules such as CO_2 also diffuse rapidly across lipid bilayers, whereas the diffusion of large uncharged polar molecules such as glucose and of charged particles, ie, ions, is extremely slow. However, ions, glucose, urea, and many other small molecules also utilize ion channels and transport proteins to cross cell mem-

branes. These transmembrane proteins are of various types (Fig 1–23).

An important technique that has permitted major advances in knowledge about channels and carriers is **patch clamping.** A micropipette is placed on the membrane of a cell and forms a tight seal to the membrane. The patch of membrane under the pipette tip usually contains only a few channels, and they can be studied in detail. The cell can be left intact (**cell-attached patch clamp**). Alternatively, the patch can be pulled loose from the cell, forming an **inside-out patch.** A third alternative is to suck out the patch with the micropipette still attached to the rest of the cell membrane, providing direct access to the interior of the cell.

Some channels are continuously open, whereas others are **gated;** ie, they have gates that open or close. Some are gated by alterations in membrane potential (**voltage-gated**), whereas others are opened or closed when they bind a ligand (**ligand-gated**). The ligand is often external, eg, a neurotransmitter or a hormone. However, it can also be internal; intracellular Ca^{2+}, cyclic AMP, or one of the G proteins produced in cells (see below) can bind directly to channels and activate them. Some cation channels are also opened by mechanical stretch. A typical voltage-gated channel is the Na^+ channel (see below), and a typical ligand-gated channel is the acetylcholine receptor (see Chapter 4). Other transport proteins are **carriers** that bind ions and other molecules and then change their configuration, moving the bound molecule from one side of the cell membrane to the other. Molecules move from areas of high concentration to areas of low concentration (down their **chemical gra-**

Table 1–6. Size of hydrated ions and other substances of biologic interest.[1]

Substance	Atomic or Molecular Weight	Radius (nm)
Cl^-	35	0.12
K^+	39	0.12
H_2O	18	0.12
Ca^{2+}	40	0.15
Na^+	23	0.18
Urea	60	0.23
Li^+	7	0.24
Glucose	180	0.38
Sucrose	342	0.48
Inulin	5000	0.75
Albumin	69,000	7.50

[1]Data from Moore EW: *Physiology of Intestinal Water and Electrolyte Absorption.* American Gastroenterological Association, 1976.

Figure 1–23. Transport of ions and other small molecules across cell membranes. ***Top:*** Diffusion pathways, showing slow diffusion through membrane or rapid diffusion through voltage-gated or ligand-gated channel proteins that span the membrane. ***Bottom:*** Transport proteins that carry substances down their electrochemical gradients (facilitated diffusion) or against these gradients (active transport). ISF, interstitial fluid.

dient), and cations move to negatively charged areas whereas anions move to positively charged areas (down their **electrical gradient**). When carrier proteins move substances in the direction of their chemical or electrical gradients, no energy input is required and the process is called **facilitated diffusion.** A typical example is glucose transport by the glucose transporter, which moves glucose down its concentration gradient from the ECF to the cytoplasm of the cell (see Chapter 19). Other carriers transport substances against their electrical and chemical gradients. This form of transport requires energy and is called **active transport.** In animal cells, the energy is provided almost exclusively by hydrolysis of ATP (see above and Chapter 17). Not surprisingly, therefore, the carrier molecules are ATPases, enzymes that catalyze the hydrolysis of ATP. One of these ATPases is **sodium-potassium-activated adenosine triphosphatase** (Na^+-K^+ ATPase), which is also known as the Na^+-K^+ pump. There are also H^+-K^+ ATPases in the gastric mucosa (see Chapter 26) and the renal tubules (see Chapter 38). V-ATPases are proton-ATPases that acidify many intracellular organelles, including parts of the Golgi complex and lysomes. F-ATPases are present in mitochondria and

synthesize ATP from ADP by using the free energy of NADH and FADH (see Chapter 17). Some cell membranes contain ATPases that transport Ca^{2+}.

Some of the carrier proteins are called **uniports** because they transport only one substance. Others are called **symports,** because transport requires the binding of the more than one substance to the transport protein (Fig 1–24) and the substances are transported across the membrane together. An example is the symport in the intestinal mucosa that is responsible for the cotransport by facilitated diffusion of Na^+ and glucose from the intestinal lumen into mucosal cells (see Chapter 25). Other transporters are called **antiports** because they exchange one substance for another. The Na^+-K^+ ATPase mentioned above is a typical antiport; it moves 3 Na^+ out of the cell in exchange for each 2 K^+ that it moves into the cell.

Sodium & Other Ion Channels

The chemistry of the voltage-gated Na^+ channel has been studied in considerable detail. In the electric eel, the Na^+ channel is a single protein made up of 1820 amino acid residues, but in rat and rabbit skeletal muscle and in rat brain, this protein, which has a molecular weight of 200,000, is associated with one or 2 smaller peptides that have molecular weights of 33,000–45,000. The large protein has 4 subunits, each made up of 6 membrane-spanning segments (Fig 1–25). The subunits surround an aqueous pore about 0.5 nm in diameter. The toxins tetrodotoxin (TTX) and saxitoxin (STX) bind to Na^+ channels and block them. The number and distribution of Na^+ channels in tissue can be determined by tagging TTX or STX with a suitable label and analyzing the distribution of the label.

There are many different K^+ channels in the membranes of various types of cells (see Table 2–1). Some of these can be blocked by tetraethylammonium (TEA) or 4-aminopyridine (4AP). There are also many types of Ca^{2+} and Cl^- channels, as well as other

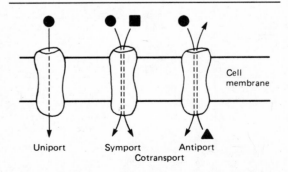

Figure 1–24. Transport proteins. The substance transported (●) may cross the membrane by itself, or its transport may be coupled to that of another substance transported in the same direction (■; symport) or the opposite direction (▲; antiport).

Figure 1–25. Diagrammatic representation of the structure of the principal subunits of 3 ion channels. The Arabic numbers identify the α-helical domains that cross the cell membrane. **Top:** Rat brain Na⁺ channel. **Middle:** Rabbit skeletal muscle Ca²⁺ channel. **Bottom:** K⁺ channel from *Drosophila* Shaker mutant. Note the hairpin transmembrane structure between helical loops 5 and 6. This is probably the ion pore, and similar structures probably occur in the Na⁺ and Ca²⁺ channels. (Na⁺ and Ca²⁺ channels after WK Catterall. Modified and reproduced from Hall ZW: *An Introduction to Molecular Neurobiology.* Sinauer, 1992.)

types of channels. One of the Cl⁻ channels that is voltage-sensitive crosses the cell membrane 12 or possibly 13 times. Another Cl⁻ channel is abnormal in cystic fibrosis; its unique characteristics are discussed in Chapter 37.

Na⁺-K⁺ ATPase

As noted above, Na⁺-K⁺ ATPase catalyzes the hydrolysis of ATP to ADP and uses the energy to extrude 3 Na⁺ from the cell and take 2 K⁺ into the cell for each mole of ATP hydrolyzed. Consequently, the pump is said to have a **coupling ratio** of 3/2. Its activity is inhibited by ouabain and related digitalis glycosides used in the treatment of heart failure. It is a heterodimer made up of an α subunit with a molecular weight of approximately 100,000 and a β subunit with a molecular weight of approximately 55,000. Both extend through the cell membrane (Fig 1–26). Separation of the subunits eliminates activity. However, the β subunit is a glycoprotein, whereas Na⁺ and K⁺ transport occur though the α subunit. The β subunit has a single membrane-spanning domain and 3 extracellular glycosylation sites, all of which appear to have attached carbohydrate residues. These residues account for one third of its molecular weight. The α subunit probably spans the cell membrane 8 times, with the NH_2 and COOH ends both located intracellularly. This subunit has intracellular Na⁺- and ATP-binding sites and a phosphorylation site; it also has extracellular binding sites for K⁺ and ouabain. When Na⁺ binds to the α subunit, ATP also binds and is converted to ADP, with a phosphate being transferred to Asp 376, the phosphorylation site. This causes a change in the configuration of the protein,

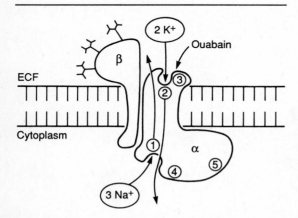

Figure 1–26. Na⁺-K⁺ ATPase. The intracellular portion of the α subunit has a Na⁺-binding site (1), a phosphorylation site (4), and an ATP-binding site (5). The extracellular portion has a K⁺-binding site (2) and an ouabain-binding site (3). (From Horisberger J-D et al: Structure-function relationship of Na-K-ATPase. *Annu Rev Physiol* 1991;**53**:565. Reproduced, with permission, from the *Annual Review of Physiology*, vol. 53, ©1991 by Annual Reviews Inc.)

extruding Na⁺ in to the ECF. K⁺ then binds extracellularly, dephosphorylating the α subunit, which returns to its previous conformation, releasing K⁺ into the cytoplasm.

It is now known that the α and β subunits are heterogeneous, with α_1, α_2, and α_3 subunits and β_1 and β_2 subunits described so far. The α_1 isoform is found in the kidneys, and the α_2 and α_3 subunits in the brain. The β_1 subunit is ubiquitous, whereas the β_2 subunit is found in the brain and the pineal gland. The physiologic significance of the occurrence of the α and β isoforms is unsettled.

Effects of Na⁺ & K⁺ Transport

It should be noted that since Na⁺-K⁺ ATPase transports 3 Na⁺ out of the cell for each 2 K⁺ it transports in, it is an **electrogenic pump;** ie, it produces net movement of positive charge out of the cell. The amount of Na⁺ provided to the pump can be the rate-limiting factor in its operation; consequently, the amount of Na⁺ extruded by the cell is partly regulated in a feedback fashion by the amount of Na⁺ in the cell.

Na⁺-K⁺ ATPase is found in all parts of the body. In some tissues, the active transport of Na⁺ is coupled to the transport of other substances (**secondary active transport**). For example, the luminal membranes of mucosal cells in the small intestine contain a symport that transports glucose into the cell only if Na⁺ binds to the protein and is transported down its electrochemical gradient at the same time. The Na⁺-glucose symport has been identified by cloning and sequencing of its cDNA. It is a 662-amino-acid protein that spans the cell membrane 12 times. From the cells, the glucose enters the blood. The electrochemical gradient for Na⁺ is maintained by the active transport of Na⁺ out of the mucosal cell into ECF (see Chapter 25). Other nutrients are reabsorbed in a similar fashion. In the brain, transport by Na⁺-K⁺ ATPase is coupled to the reuptake of neurotransmitters. In the heart, Na⁺-K⁺ ATPase indirectly affects Ca^{2+} transport because an antiport in the membranes of cardiac muscle cells exchanges intracellular Ca^{2+} for extracellular Na⁺ on an electrically neutral 1 for 2 basis. The rate of this exchange is proportionate to the concentration gradient for Na⁺ across the cell membrane. If the operation of the Na⁺-K⁺ ATPase is inhibited (eg, by ouabain), less intracellular Ca^{2+} is extruded and the intracellular Ca^{2+} concentration is increased. This facilitates the contraction of cardiac muscle and is the probable explanation of the positively inotropic effect of digitalis glycosides (see Chapter 3).

Active transport of Na⁺ and K⁺ is one of the major energy-using processes in the body. On the average, it accounts for 33% of the energy utilized by cells, and in neurons it accounts for 70%. Thus, it accounts for a large part of the basal metabolism. Furthermore, there is a direct link between Na⁺ and K⁺ transport and metabolism; the greater the rate of pumping, the more ADP is formed, and the available supply of ADP determines the rate at which ATP is formed by oxidative phosphorylation (see Chapter 17).

In animals, the maintenance of normal cell volume and pressure depends on Na^+ and K^+ pumping (see below).

FORCES PRODUCING MOVEMENT OF SUBSTANCES ACROSS CELL MEMBRANES

The primary forces producing movement of water and other molecules across these cell membranes are diffusion, facilitated diffusion, solvent drag, osmosis, active transport, and exocytosis and endocytosis.

It should be noted that next to biologic membranes in living animals, there is a layer of relatively unstirred fluid. Solutes cross this **unstirred layer** by diffusion, and diffusion across it must be taken into account in considering transport across membranes. Its role is currently a matter of some debate, but in some instances it represents a significant portion of the resistance to movement of a given solute across a membrane.

Diffusion

Diffusion is the process by which a gas or a substance in solution expands, because of the motion of its particles, to fill all of the available volume. The particles (molecules or atoms) of a substance dissolved in a solvent are in continuous random movement. A given particle is equally likely to move into or out of an area in which it is in high concentration. However, since there are more particles in the area of high concentration, the total number of particles moving to areas of lower concentration is greater; ie, there is a **net flux** of solute particles from areas of high to areas of low concentration. The time required for equilibrium by diffusion is proportionate to the square of the diffusion distance. The magnitude of the diffusing tendency from one region to another is directly proportionate to the cross-sectional area across which diffusion is taking place and the **concentration gradient,** or **chemical gradient,** which is the difference in concentration of the diffusing substance divided by the thickness of the boundary **(Ficks law of diffusion).** Thus

$$J = -DA \frac{\Delta c}{\Delta x}$$

where J is the net rate of diffusion, D is the diffusion coefficient, A is the area, and $\Delta c/\Delta x$ is the concentration gradient. The minus sign indicates the direction of diffusion. When considering movement of molecules from a higher to a lower concentration, $\Delta c/\Delta x$ is negative, so multiplying by $-DA$ gives a positive value. The permeabilities of the boundaries across which dif-

fusion occurs in the body vary, but diffusion is still a major force affecting the distribution of water and solutes. As noted above, diffusion of ions is also affected by their **electrical gradient.**

Solvent Drag

When there is net solvent flow in one direction **(bulk flow),** the solvent tends to drag along some molecules of solute. This force is called **solvent drag.** In most situations in the body, its effects are very small.

Osmosis

When a substance is dissolved in water, the concentration of water molecules in the solution is less than that in pure water, since the addition of solute to water results in a solution that occupies a greater volume than does the water alone. If the solution is placed on one side of a membrane that is permeable to water but not to the solute and an equal volume of water is placed on the other, water molecules diffuse down their concentration gradient into the solution (Fig 1–27). This process—the diffusion of **solvent** molecules into a region in which there is a higher concentration of a **solute** to which the membrane is impermeable—is called **osmosis.** It is an immensely important factor in physiologic processes. The tendency for movement of solvent molecules to a region of greater solute concentration can be prevented by applying pressure to the more concentrated solution. The pressure necessary to prevent solvent migration is the **osmotic pressure** of the solution.

Osmotic pressure, like vapor pressure lowering, freezing-point depression, and boiling-point elevation, depends upon the number rather than the type of particles in a solution; ie, it is a fundamental colliga-

Figure 1–27. Diagrammatic representation of osmosis. Water molecules are represented by small open circles, solute molecules by large solid circles. In the diagram on the left, water is placed on one side of a membrane permeable to water but not to solute, and an equal volume of a solution of the solute is placed on the other. Water molecules move down their concentration gradient into the solution, and, as shown in the diagram on the right, the volume of the solution increases. As indicated by the arrow on the right, the osmotic pressure is the pressure that would have to be applied to prevent the movement of the water molecules.

tive property of solutions. In an **ideal solution,** osmotic pressure (P) is related to temperature and volume in the same way as the pressure of a gas:

$$P = \frac{nRT}{V}$$

when n is the number of particles, R the gas constant, T the absolute temperature, and V the volume. If T is held constant, it is clear that the osmotic pressure is proportionate to the number of particles in solution per unit volume of solution. If the solute is a nonionizing compound such as glucose, the osmotic pressure is a function of the number of glucose molecules present. If the solute ionizes and forms an ideal solution, each ion is an osmotically active particle. For example, NaCl would dissociate into Na^+ and Cl^- ions, so that each mole in solution would supply 2 osm. One mole of Na_2SO_4 would dissociate into Na^+, Na^+, and SO_4^{2-}, supplying 3 osm. However, the body fluids are not ideal solutions, and although the dissociation of strong electrolytes is complete, the number of particles free to exert an osmotic effect is reduced owing to interactions between the ions. Thus, it is actually the effective concentration **(activity)** in the body fluids rather than the number of equivalents of an electrolyte in solution that determines its osmotic effect. This is why, for example, 1 mmol of NaCl per liter in the body fluids contributes somewhat less than 2 mosm of osmotically active particles per liter. The more concentrated the solution, the greater the deviation from an ideal solution.

It should be noted that although a homogeneous solution contains osmotically active particles and can be said to have an osmotic pressure, it can exert an osmotic pressure only when it is in contact with another solution across a membrane permeable to the solvent but not to the solute.

Osmolal Concentration of Plasma: Tonicity

The freezing point of normal human plasma averages –0.54 °C, which corresponds to an osmolal concentration in plasma of 290 mosm/L. This is equivalent to an osmotic pressure against pure water of 7.3 atmospheres. The osmolality might be expected to be higher than this, because the sum of all the cation and anion equivalents in plasma is over 300. It is not this high because plasma is not an ideal solution, and ionic interactions reduce the number of particles free to exert an osmotic effect. Except when there has been insufficient time after a sudden change in composition for equilibrium to occur, all fluid compartments of the body are apparently in or nearly in osmotic equilibrium. The term **tonicity** is used to describe the osmolality of a solution relative to plasma. Solutions that have the same osmolality as plasma are said to be **isotonic;** those with greater osmolality are **hypertonic;** and those with lesser osmolality are **hypotonic.** All solutions that are initially isosmotic with plasma—ie, that have the same actual osmotic pressure or freezing point depression as plasma—would remain isotonic if it were not for the fact that some solutes diffuse into cells and others are metabolized. Thus, a 0.9% saline solution remains isotonic because there is no net movement of the osmotically active particles in the solution into cells and the particles are not metabolized. On the other hand, a 5% glucose solution is isotonic when initially infused intravenously, but glucose is metabolized, so the net effect is that of infusing a hypotonic solution.

It is important to note the relative contributions of the various plasma components to the total osmolal concentration of plasma. All but about 20 of the 290 mosm in each liter of normal plasma are contributed by Na^+ and its accompanying anions, principally Cl^- and HCO_3^-. Other cations and anions make a relatively small contribution. Although the concentration of the plasma proteins is large when expressed in grams per liter, they normally contribute less than 2 mosm/L because of their very high molecular weights. The major nonelectrolytes of plasma are glucose and urea, which in the steady state are in equilibrium with cells. Their contributions to osmolality are normally about 5 mosm/L each but can become quite large in hyperglycemia or uremia. The total plasma osmolality is important in assessing dehydration, overhydration, and other fluid and electrolyte abnormalities. Hyperosmolality can cause coma (hyperosmolar coma; see Chapter 19). Because of the predominant role of the major solutes and the deviation of plasma from an ideal solution, one can ordinarily approximate the plasma osmolality within a few milliosmoles per liter by using the following formula, in which the constants convert the clinical units to millimoles of solute per liter:

$$\text{Osmolality} = 2[Na^+] + 0.055[\text{Glucose}] + 0.36\,[\text{BUN}]$$
$$\text{(mosm/L)} \quad \text{(meq/L)} \quad \text{(mg/dL)} \quad \text{(mg/dL)}$$

BUN is the blood urea nitrogen. The formula is also useful in calling attention to abnormally high concentrations of other solutes. An observed plasma osmolality (measured by freezing-point depression) that greatly exceeds the value predicted by this formula probably indicates the presence of a foreign substance such as ethanol, mannitol (sometimes injected to shrink swollen cells osmotically), or poisons such as ethylene glycol or methanol (components of antifreeze).

Nonionic Diffusion

Some weak acids and bases are quite soluble in cell membranes in the undissociated form, whereas in the ionic form they cross membranes with difficulty. Consequently, if molecules of the undissociated substance diffuse from one side of the membrane to the other and then dissociate, there is appreciable net movement of the undissociated substance from one side of the membrane to another. This phenomenon, which occurs in the gastrointestinal tract (see Chapter 25) and kidneys (see Chapter 38), is called **nonionic diffusion.**

Donnan Effect

When there is an ion on one side of a membrane that cannot diffuse through the membrane, the distribution of other ions to which the membrane is permeable is affected in a predictable way. For example, the negative charge of a nondiffusible anion hinders diffusion of the diffusible cations and favors diffusion of the diffusible anions. Consider the following situation,

$$
\begin{array}{c|c}
\underline{X} & \underline{Y} \\
& m \\
K^+ & K^+ \\
Cl^- & Cl^- \\
Prot^- & \\
\end{array}
$$

in which the membrane (m) between compartments X and Y is impermeable to $Prot^-$ but freely permeable to K^+ and Cl^-. Assume that the concentrations of the anions and of the cations on the 2 sides are initially equal. Cl^- diffuses down its concentration gradient from Y to X, and K^+ moves with the negatively charged Cl^-, maintaining electroneutrality on side Y. Therefore, at equilibrium,

$$[K^+{}_X] > [K^+{}_Y]$$

Furthermore,

$$[K^+{}_X] + [Cl^-{}_X] + [Prot^-{}_X] > [K^+{}_Y] + [Cl^-{}_Y]$$

ie, there are more osmotically active particles on side X than on side Y.

Donnan and Gibbs showed that in the presence of a nondiffusible ion, the diffusible ions distribute themselves so that at equilibrium, their concentration ratios are equal:

$$\frac{[K^+{}_X]}{[K^+{}_Y]} = \frac{[Cl^-{}_Y]}{[Cl^-{}_X]}$$

Cross-multiplying,

$$[K^+{}_X][Cl^-{}_X] = [K^+{}_Y][Cl^-{}_Y]$$

This is the **Gibbs-Donnan equation.** It holds for any pair of cations and anions of the same valence.

The Donnan effect on the distribution of ions has 3 effects in the body. First, because of proteins ($Prot^-$) in cells, there are more osmotically active particles in cells than in interstitial fluid, and since animal cells have flexible walls, osmosis would make them swell and eventually rupture if it were not for Na^+-K^+ ATP-

ase pumping ions back out of cells. Thus, normal cell volume and pressure depend on Na^+-K^+ ATPase. Second, because at equilibrium there is an asymmetric distribution of permeant ions across the membrane (m in the example used here), there will be an electrical difference across the membrane whose magnitude can be determined by the Nernst equation (see below). In the example used here, side X will be negative relative to side Y. It should be emphasized that any macroscopic region of a solution has an equal number of positive and negative charges. However, in this situation, charges line up along the membrane with the concentration gradient for Cl^- exactly balanced by the oppositely directed electrical gradient, and the same holds true for K^+. Third, since there are more proteins in plasma than in interstitial fluid, there is a Donnan effect on ion movement across the capillary wall (see below).

Forces Acting on Ions

With the background provided in the preceding paragraphs, the forces acting across the cell membrane on each ion can be analyzed. Chloride ions are present in higher concentration in the ECF than in the cell interior, and they tend to diffuse along this **concentration gradient** into the cell. The interior of the cell is negative relative to the exterior, and chloride ions are pushed out of the cell along this **electrical gradient.** An equilibrium is reached at which Cl^- influx and Cl^- efflux are equal. The membrane potential at which this equilibrium exists is the **equilibrium potential.** Its magnitude can be calculated from the **Nernst equation,** as follows:

$$E_{Cl} = \frac{RT}{FZ_{Cl}} \ln \frac{[Cl_o{}^-]}{[Cl_i{}^-]}$$

where

E_{Cl} = **equilibrium potential for Cl^-**
R = **gas constant**
T = **absolute temperature**
F = **the faraday (number of coulombs per mole of charge)**
Z_{Cl} = **valence of Cl^- (–1)**
$[Cl_o{}^-]$ = **Cl^- concentration outside the cell**
$[Cl_i{}^-]$ = **Cl^- concentration inside the cell**

Converting from the natural log to the base 10 log and replacing some of the constants with numerical values, the equation becomes

$$E_{Cl} = 61.5 \log \frac{[Cl_i{}^-]}{[Cl_o{}^-]} \text{ at 37 °C}$$

Note that in converting to the simplified expression

the concentration ratio is reversed because the −1 valence of Cl⁻ has been removed from the expression.

E_{Cl}, calculated from the values in Table 1–5, is −70 mV, a value identical to the measured resting membrane potential of 70 mV. Therefore, no forces other than those represented by the chemical and electrical gradients need be invoked to explain the distribution of Cl⁻ across the membrane.

A similar equilibrium potential can be calculated for K⁺.

$$E_K = \frac{RT}{FZ_K} \ln \frac{[K_o^+]}{[K_i^+]} = 61.5 \log \frac{[K_o^+]}{[K_i^+]} \text{ at } 37\,°C$$

where

E_K = equilibrium potential for K⁺

Z_K = valence of K⁺(+1)

$[K_o^+]$ = K⁺ concentration outside the cell

$[K_i^+]$ = K⁺ concentration inside the cell

R, T, and F as above

In this case, the concentration gradient is outward and the electrical gradient inward. In mammalian spinal motor neurons. E_k is 90 mV (Table 1–5). Since the resting membrane potential is −70 mV, there is somewhat more K⁺ in the neurons than can be accounted for by the electrical and chemical gradients.

The situation for Na⁺ is quite different from that for K⁺ and Cl⁻. The direction of the chemical gradient for Na⁺ is inward, to the area where it is in lesser concentration, and the electrical gradient is in the same direction. E_{Na} is +60 mV (Table 1–5). Since neither E_K nor E_{Na} is at the membrane potential, one would expect the cell to gradually gain Na⁺ and lose K⁺ if only passive electrical and chemical forces were acting across the membrane. However, the intracellular concentration of Na⁺ and K⁺ remain constant because there is active transport of Na⁺ out of the cell against its electrical and concentration gradients, and this transport is coupled to active transport of K⁺ into the cell. The passive and active fluxes across the membrane of a motor neuron at rest are summarized in Fig 1–28.

The magnitude of the membrane potential at any given time depends, of course, upon the distribution of Na⁺, K⁺, and Cl⁻ and the permeability of the membrane to each of these ions. An equation that describes this relationship with considerable accuracy is the **Goldman constant-field equation:**

$$V = \frac{RT}{F} \ln \frac{P_K \cdot [K_o^+] + P_{Na} \cdot [Na_o^+] + P_{Cl}^- [Cl_i^-]}{P_K \cdot [K_i^+] + P_{Na} \cdot [Na_i^+] + P_{Cl}^- [Cl_o^-]}$$

where V = membrane potential, R = gas constant, T = absolute temperature, F = faraday, and P_{K+}, P_{Na+}, and

Figure 1–28. Summary of the net passive fluxes of ions across the membrane of mammalian motor neurons at various membrane potentials. RMP, resting membrane potential. (Courtesy of AM Thompson.)

P_{Cl^-} = permeability of the membrane to K⁺, Na⁺, and Cl⁻. The brackets signify concentration, and i and o refer to the inside and outside of the cell, respectively. Since P_{Na^+} is low relative to P_{K^+} in the resting cells, Na⁺ contributes little to the value of V. As would be predicted from the Goldman equation, changes in external Na⁺ produce little change in the resting membrane potential whereas increases in external K⁺ decrease it.

The situation in muscle and various other types of cells is similar. In muscle, for example, the resting membrane potential is about −90 mV. E_{Cl} is −86 mV, E_K is −100 mV, and E_{Na} is +55 mV. In red blood cells, on the other hand, the membrane potential is low (−9 mV) and near the Cl⁻ equilibrium potential, whereas both Na⁺ and K⁺ are actively transported across the cell membrane.

Ca²⁺ should also be mentioned in the present context. In mammals, the ionized calcium concentration in ECF is about 1.2 mmol/L, whereas the intracellular free Ca²⁺ concentration is normally much lower, about 100 nmol/L. Thus, both the electrical and chemical gradients are directed inward.

Genesis of the Membrane Potential

The distribution of ions across the cell membrane and the nature of this membrane provide the explanation for the membrane potential. Two transport proteins are primarily responsible for the resting membrane potential: a K⁺ leak channel that permits K⁺ to diffuse out of the cell, and Na⁺-K⁺ ATPase. The concentration gradient for K^+ facilitates its diffusion out of the cell, but its electrical gradient is in the opposite (inward) direction. Consequently, an equilibrium is reached in which the tendency of K^+ to move out of the cell is balanced by its tendency to move into the cell, and at that equilibrium there is a slight excess of cations on the outside and anions on the inside. This condition is maintained by Na^+-K^+ ATPase, which pumps K^+ back into the cell and keeps the intracellular concentration of Na^+ low. The Na^+-K^+ pump is also electrogenic, because it pumps 3 Na^+ out of the cell for every 2 K^+ it pumps in; thus, it also contributes a small amount to the membrane potential by itself. It should be emphasized that the number of ions responsible for the membrane potential is a minute fraction of the total number present and that the total concentrations of positive and negative ions are equal everywhere except along the membrane. Na^+ influx does not compensate for the K^+ efflux because, owing to the K^+ leak channel, the membrane is more permeable to K^+ than to Na^+.

Variations in Membrane Potential

If the resting membrane potential is decreased by the passage of a current through the membrane, the electrical gradient that keeps K^+ inside the cell is decreased and there is an increase in K^+ diffusion out of the cell. This K^+ efflux and the simultaneous movement of Cl^- into the cell result in a net movement of positive charge out of the cell, with consequent restoration of the resting membrane potential. When the membrane potential increases, these ions move in opposite directions. These processes occur in all polarized cells and tend to keep the resting membrane potential of the cells constant within narrow limits. However, in nerve and muscle cells, reduction of the membrane potential triggers a voltage-dependent increase in Na^+ permeability. This unique feature permits these cells to generate self-propagating impulses that are transmitted along their membranes for great distances. These impulses are considered in detail in Chapter 2.

THE CAPILLARY WALL

Filtration

The capillary wall separating plasma from interstitial fluid is different from the cell membranes separating interstitial fluid from intracellular fluid because the pressure difference across it makes **filtration** a significant factor in producing water and solute.

By definition, filtration is the process by which fluid is forced through a membrane or other barrier because of a difference in pressure on the 2 sides. The amount of fluid filtered in a given interval is proportionate to the difference in pressure, the surface area

of the membrane, and the permeability of the membrane.

Oncotic Pressure

The structure of the capillary wall varies from one vascular bed to another (see Chapter 30). However, in skeletal muscle and many other organs, water and relatively small solutes are the only substances that cross the wall with ease. The apertures in the wall (the junctions between the endothelial cells) are too small to permit plasma proteins and other colloids to pass through in significant quantities. The colloids have a high molecular weight but are present in large amounts. Small amounts cross the capillary wall by vesicular transport (see below), but their effect is slight. Therefore, the capillary wall behaves like a membrane impermeable to colloids, which exert an osmotic pressure of about 25 mm Hg. The colloid osmotic pressure due to the plasma colloids is called the **oncotic pressure.** Filtration across the capillary membrane as a result of the hydrostatic pressure head in the vascular system is opposed by the oncotic pressure. The way the balance between the hydrostatic and oncotic pressures controls exchanges across the capillary wall is considered in detail in Chapter 30.

Transcytosis

There are vesicles in the cytoplasm of endothelial cells, and tagged protein molecules injected into the bloodstream have been found in the vesicles and in the interstitium. This indicates that small amounts of protein are transported out of capillaries across endothelial cells by endocytosis followed by exocytosis on the interstitial side of the cells. The transport mechanism makes use of coated vesicles and is called **transcytosis, vesicular transport,** or **cytopempsis.**

SODIUM & POTASSIUM DISTRIBUTION & TOTAL BODY OSMOLALITY

The foregoing discussion of the various compartments of the body fluids and the barriers between them facilitates consideration of the total body stores of the principal cations, Na^+ and K^+.

Total Body Sodium

The total amount of exchangeable Na^+ in the body (Na_E)—as opposed to its concentration in any particular body fluid—can be determined by applying the same dilution principle used to measure the body fluid compartments. A radioactive isotope of sodium (usually ^{24}Na) is injected, and after equilibration, the fraction of the sodium in the body that is radioactive is determined. The fraction of a substance that is radioactive—ie, the concentration of radioactive molecules divided by the concentration of radioactive plus nonradioactive molecules—is the **specific activity**

(SA) of the substance. The SA of sodium in plasma after the injection of ^{24}Na, for example, is

$$\frac{^{24}\text{Na (counts/min/L)}}{^{24}\text{Na + nonradioactive Na (meq/L)}}$$

The total exchangeable body sodium (Na_E) =

$$\frac{^{24}\text{Na injected} - {}^{24}\text{Na excreted}}{\text{SA of plasma}}$$

The average normal value for Na_E in healthy adults is 41 meq/kg, whereas the total amount of Na$^+$ in the body is about 58 meq/kg. Therefore, approximately 17 meq/kg is not available for exchange. The vast majority of this nonexchangeable Na$^+$ is the hydroxyapatite crystal lattice of bone. The distribution of body Na$^+$, which is mostly extracellular, is summarized in Table 1–7.

Total Body Potassium

The total exchangeable body K$^+$ can be determined in the same way as total exchangeable body Na$^+$, using radioactive potassium (^{42}K). The average value in young adult men is about 45 meq/kg body weight. It is somewhat less in women, and it declines slightly with advancing age. The distribution of body K$^+$, which is mostly intracellular, is shown in Table 1–7. About 10% of the total body K$^+$ is bound, and the remaining 90% is exchangeable.

Plasma K$^+$ levels are not a good indicator of the total body K$^+$, since most of the K$^+$ is in the cells. There is a correlation between the K$^+$ and H$^+$ and content of plasma, the 2 rising and falling together.

Interrelationships of Sodium & Potassium

Since the salts of sodium and potassium dissociate in the body to such a great extent, and since they are so plentiful, they determine in large part the osmolality of the body fluids. A change in the amount of electrolyte in one compartment is followed by predictable changes in the volume and electrolyte concentration in the others because the various compartments are in osmotic equilibrium. Fig 1–29 shows, for example, the changes in osmolal concentration and volume of the intracellular and extracellular fluid spaces that follow removal of 500 mosm of extracellular electrolyte. Loss of electrolyte in excess of water from the ECF leads to hypotonicity of the ECF relative to the intracellular fluid. Consequently, water moves into the cells by osmosis until osmotic equilibrium between the ECF and the intracellular fluid is again attained. The net result is a decrease in ECF volume and an increase in intracellular fluid volume.

Total Body Osmolality

Since Na$^+$ is the principal cation of the plasma, the plasma osmotic pressure correlates well with the plasma Na$^+$ level. However, plasma Na$^+$ does not necessarily correlate well with Na_E. Total body osmolality reflects the electrolyte concentration in the body, ie, the total exchangeable body sodium plus the total exchangeable body potassium divided by the total body water:

$$\frac{Na_E + K_E}{TBW}$$

Because this latter figure has been found to correlate well with plasma Na$^+$, plasma Na$^+$ is a relatively accurate measure of total body osmolality. It is worth noting that plasma Na$^+$ can be affected by changes in any of 3 body components: Na_E itself, K_E, or TBW. A detailed discussion of the effects of disease on electrolyte and water balance is beyond the scope of this book, but the facts discussed above are basic to any consideration of pathologic water and electrolyte changes.

INTERCELLULAR COMMUNICATION

Cells communicate with each other via chemical messengers. Within a given tissue, some messengers move from cell to cell via gap injunctions (see above) without entering the ECF. In addition, cells are affected by chemical messengers secreted into the ECF. These chemical messengers bind to protein receptors on the surface of the cell or, in some instances, in the cytoplasm or the nucleus, triggering sequences of intracellular changes that produce their physiologic effects. There are 3 general types of intercellular communication mediated by messengers in the ECF: (1) **neural communication,** in which neurotransmitters are released at synaptic junctions from nerve cells and act across a narrow synaptic cleft on a postsynaptic cell (see Chapter 4); (2) **endocrine communication,** in which hormones reach cells via the circulating blood (see Chapters 18–24); and (3) **paracrine communication,** in which the products of cells diffuse in the ECF to affect neighboring cells

Table 1–7. Distribution of sodium and potassium in the body.

Component	Amount (% of Total) of:	
	Na$^+$	K$^+$
Total intracellular	9.0	89.6
Total extracellular	91.0	10.4
Plasma	11.2	0.4
Interstitial fluid	29.0	1.0
Dense connective tissue and cartilage	11.7	0.4
Bone	36.5[1]	7.6
Transcellular locations	2.6	1.0

[1]Of this, 11.0 is exchangeable and 25.5 is nonexchangeable.

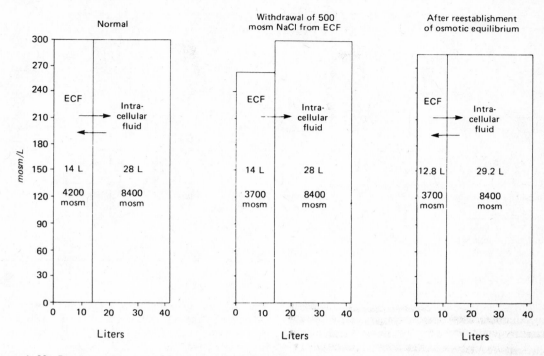

Figure 1–29. Response of body fluid to removal of 500 mosm from ECF without change in total body water in a 70-kg man. The hypotonicity of the ECF causes water to enter cells until osmotic equilibrium is reestablished. Total mosm in body after removal of 500 mosm = $(14 \times 300) + (28 \times 300) - 500 = 12,100$ mosm. New osmotic equilibrium = $12,100/(28 + 14) = 288$ mosm/L. This means that the new volume of the ECF will be $(4200 - 500)/288 = 12.8$ L and the new volume of the intracellular fluid will be $8400/288 = 29.2$ L. (After Darrow DC and Yannet H. Modified and reproduced, with permission, from Gamble JL: *Chemical Anatomy, Physiology, and Pathology of Extracellular Fluid,* 4th ed. Harvard Univ Press, 1942.)

that may be some distance away (Fig 1–30). In addition, cells secrete chemical messengers that in some situations bind to receptors on the same cell, ie, the cell that secreted the messenger (**autocrine communication**). The chemical messengers include amines, amino acids, steroids, polypeptides, and in a few instances, other substances. It is worth noting that in various parts of the body, the same chemical messenger can function as a neurotransmitter, a paracrine mediator, a hormone secreted by neurons into the blood (neural hormone), and a hormone secreted by gland cells into the blood.

Recently, an additional form of intercellular communication called **juxtacrine communication** has been described. Some cells express multiple repeats of growth factors such as **transforming growth factor alpha (TGFα)** extracellularly on transmembrane proteins that provide an anchor to the cell. The units of TGFα are hydrolyzed from the proteins and eventually enter the circulation, but the rate of hydrolysis is low. Other cells have TGFα receptors. Consequently, TGFα anchored to a cell can bind to a TGFα receptor on another cell, linking the two. This could be important in producing local foci of growth in tissues.

Radioimmunoassay

Many of the chemical messengers are polypeptides or proteins, whereas others are amino acids, amines, or steroids. Antibodies to the polypeptides and proteins are readily produced, and, by using special techniques, it is possible to make antibodies to the other messengers as well. The antibodies can then be used to measure the messengers in body fluids and in tissue extracts by **radioimmunoassay.** This technique depends on the fact that the naturally occurring, unlabeled ligand and added radioactive ligand compete to bind to an antibody to the ligand. The greater the amount of unlabeled ligand in the specimen being analyzed, the more it competes and the smaller the amount of radioactive ligand that binds to the antibody. Radioimmunoassays are extensively used in research and in clinical medicine.

Receptors for Hormones, Neurotransmitters, & Other Ligands

Many of the receptors for chemical messengers have now been isolated and characterized. These proteins are not static components of the cell, but their numbers increase and decrease in response to various stimuli, and their properties change with changes in

	GAP JUNCTIONS	SYNAPTIC	PARACRINE	ENDOCRINE
Message transmission	Directly from cell to cell	Across synaptic cleft	By diffusion in inter-stitial fluid	By circulating body fluids
Local or general	Local	Local	Locally diffuse	General
Specificity depends on	Anatomic location	Anatomic location and receptors	Receptors	Receptors

Figure 1–30. Intercellular communication by chemical mediators. (Reproduced, with permission, from Greenspan FS, [editor]. *Basic & Clinical Endocrinology,* 3rd ed. Appleton & Lange, 1991.)

physiologic conditions. When a hormone or neuro-transmitter is present in excess, the number of active receptors generally decreases **(down regulation)**, whereas in the presence of a deficiency of the chemical messenger, there is an increase in the number of active receptors **(up regulation)**. Angiotensin II in its actions on the adrenal cortex is an exception; it increases rather than decreases the number of its receptors in the adrenal. These effects on receptors are important in explaining such phenomena as denervation hypersensitivity (see Chapter 4) and tolerance to morphine (see Chapter 7). In the case of receptors in the membrane, receptor-mediated endocytosis is responsible for down regulation in some instances; ligands bind to their receptors, and the ligand-receptor complexes move laterally in the membrane to coated pits, where they are taken into the cell by endocytosis **(internalization)**. This decreases the number of receptors in the membrane. Some receptors are recycled after internalization, whereas others are re-placed by de novo synthesis in the cell. Another type of down regulation is desensitization, in which receptors are chemically modified in ways that make them less responsive (see Chapter 4).

Mechanisms by Which Chemical Messengers Act

The principal mechanisms by which chemical messengers exert their intracellular effects are summarized in Table 1–8. Some ligands bind directly to ion channels in the cell membrane, changing their conductance. Thyroid and steroid hormones and some other lipid-soluble ligands enter cells and act on one or another member of a family of structurally related cytoplasmic or nuclear receptors. The activated receptor binds to DNA and increases transcription of selected mRNAs. Almost all other ligands in the ECF bind to receptors on the surface of cells, and many of them trigger the release of intracellular mediators such as cyclic AMP, IP_3, and DAG (see below) that

Table 1–8. Principal mechanisms by which chemical messengers in the ECF bring about changes in cell function.

Mechanism	Examples
Open or close ion channels in cell membrane	Acetylcholine on nicotinic cholinergic receptor; norepinephrine on K^+ channel in the heart.
Act via cytoplasmic or nuclear receptors to increase transcription of selected mRNAs	Thyroid hormones, retinoic acid, steroid hormones.
Activate phospholipase C with intracellular production of DAG, IP_3, and other inositol phosphates	Angiotensin II, norepinephrine via α_1-adrenergic receptor, vasopressin via V_1 receptor.
Activate or inhibit adenylate cyclase, causing increased or decreased intracellular production of cyclic AMP	Norepinephrine via β_1-adrenergic receptor (increased cyclic AMP); norepinephrine via α_2-adrenergic receptor (decreased cyclic AMP).
Increase cyclic GMP in cell	ANP; NO (EDRF).
Increased tyrosine kinase activity of cytoplasmic portions of transmembrane receptors	Insulin, EGF, PDGF, M-CSF.

initiate changes in cell function. Consequently, the extracellular ligands are called "first messengers" and the intracellular mediators are called "second messengers." When activated, many of the membrane receptors initiate release of second messengers or other intracellular events via GTP-binding proteins (G proteins; see below). The second messengers generally activate **protein kinases,** enzymes that catalyze the phosphorylation of amino acids on proteins. Approximately 100 protein kinases have been described, but the principal ones that are important in mammals are summarized in Table 1–9. Addition of phosphate groups changes the configuration of the proteins, altering their functions and consequently the functions of the cell. In some instances, the intracellular portions of the receptors themselves are protein kinases.

Stimulation of Transcription

When thyroid and steroid hormones bind to their receptors inside cells, the conformation of the receptor protein is changed and a DNA-binding domain is exposed (Fig 1–31). The receptor-hormone complex moves to DNA, where it binds to enhancer elements in the untranslated 5'-flanking portions of certain genes. The estrogen and the triiodothyronine (T_3) receptors bind hormones in the nucleus. The T_3 receptor also binds thyroxine (T_4), but with less affinity. The glucocorticoid receptor is located mainly in the cytoplasm but migrates promptly to the nucleus as soon as it binds its ligand. The initial location of the other receptors that act in this fashion is not yet known. In any case, binding of the receptor-hormone complex to DNA increases the transcription of mRNAs encoded by the gene to which it binds. The mRNAs are translated in the ribosomes, with the production of increased quantities of proteins that alter cell function.

At least for the glucocorticoid, estrogen, and progesterone receptors, the receptor is bound to a **heat shock protein** in the absence of the steroid, and it appears that the heat shock protein covers the DNA-binding domain. When the steroid binds to the recep-

Figure 1–31. Mechanism of action of steroid and thyroid hormones. H, hormone; R, receptor.

tor, the conformation change is release of the heat shock protein, exposing the DNA-binding domain.

Structure of Receptors

The structures of the human glucocorticoid, mineralocorticoid, and estrogen receptors are shown in Fig 1–32. One of the 2 known T_3 receptors is also shown in the figure; for unknown reasons, there are 2 similar T_3 receptors, encoded by genes on separate chromosomes. All these receptors are part of a large family of receptors that have in common a highly conserved cysteine-rich DNA-binding domain; a ligand-binding domain at or near the C terminal of

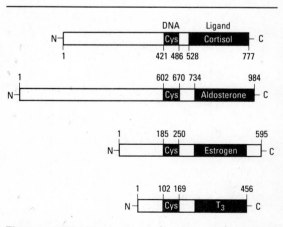

Figure 1–32. Structure of human glucocorticoid, mineralocorticoid, and estrogen receptors, and one of 2 T_3 receptors, hT_3R$_\beta$. Note that each receptor has a cysteine-rich DNA-binding domain and a ligand-binding domain at or near the C terminus, with considerable variability in the N-terminal part of the protein. The numbers identify amino acid residues.

Table 1–9. Principal protein kinases.

Phosphorylate serine and/or threonine residues
 Calmodulin-dependent
  Myosin light-chain kinase
  Phosphorylase kinase
  Ca^{2+}/calmodulin kinase I
  Ca^{2+}/calmodulin kinase II
  Ca^{2+}/calmodulin kinase III
 Calcium-phospholipid-dependent
  Protein kinase C (7 subspecies)
 Cyclic nucleotide-dependent
  cAMP-dependent kinase (protein kinase A; 2 subspecies)
  cGMP-dependent kinase
Phosphorylate tyrosine residues
 Insulin receptor, EGF receptor, PDGF receptor, and
  M-CSF receptor have tyrosine kinase activity

the receptor; and a relatively variable, poorly conserved N-terminal region. Binding to DNA occurs via zinc fingers (see above). Other receptors in the family include the other T_3 receptor (hT_3R_α) and the receptors for progesterone, androgen, and 1,25-dihydroxycholecalciferol. Retinoic acid, which is a derivative of retinol (vitamin A) produced in some cells, dictates the formation of limbs during embryonic life, and several receptors for retinoic acid and related compounds are members of this family. Molecules such as retinoic acid that determine the form of the body during development are called **morphogens.** The protein encoded by the *erb*A oncogene is closely related to the T_3 receptors. Other factors that regulate genes have also been shown to be receptors of this type, including one receptor that is activated by dopamine. Thus, this family of receptors occupies a prominent place in regulatory physiology.

Rapid Action of Steroids

Some of the actions of steroids are much more rapid than those known to be mediated via binding to DNA. Examples include the rapid increase in the Ca^{2+} concentration in sperm heads that is produced by progesterone and prompt steroid-induced alteration in the functions of various neurons. This has led to the hypothesis that steroids can also bind to membrane receptors, and there is some evidence for the existence of such receptors. Steroids also bind to $GABA_A$ receptors, facilitating their action (see Chapter 4).

Intracellular Ca^{2+}

The free Ca^{2+} concentration in the cytoplasm is maintained at about 100 nmol/L. The Ca^{2+} concentration in the interstitial fluid is about 1,200,000 nmol/L, so there is a marked inwardly directed concentration gradient as well as an inwardly directed electrical gradient. Much of the intracellular Ca^{2+} is bound by the endoplasmic reticulum and other organelles (Fig 1–33), and these organelles provide a store from which Ca^{2+} can be mobilized to increase the concentration of free Ca^{2+} in the cytoplasm. Increased cytoplasmic Ca^{2+} binds to and activates calcium-binding proteins, and these in turn activate a number of protein kinases.

Ca^{2+} enters cells through 2 kinds of channels: voltage-gated Ca^{2+} channels and ligand-gated Ca^{2+} channels. The voltage-gated Ca^{2+} channels, of which there are at least 4 types, are activated by depolarization, whereas the ligand-gated channels are activated by many different neurotransmitters and hormones. In addition, there appear to be Ca^{2+} channels that are activated by stretch.

Ca^{2+} is also pumped out of cells in exchange for $2H^+$ by a Ca^{2+}-H^+ ATPase, and it is transported out of cells by an antiport driven by the Na^+ gradient that exchanges $3Na^+$ for each Ca^{2+}.

Some second messengers act by increasing the

Figure 1–33. Ca^{2+} metabolism in mammalian cells. Cytoplasmic Ca^{2+} is in equilibrium with Ca^{2+} bound to mitochondria and the endoplasmic reticulum. Calcium-binding proteins (CaBP) bind cytoplasmic Ca^{2+} and, when activated in this fashion, bring about a variety of physiologic effects. Ca^{2+} enters the cell via voltage-gated (volt) and ligand-gated (lig) Ca^{2+} channels. It is transported out of the cell by a Ca^{2+} channels. It is transported out of the cell by a Ca^{2+}-H^+ ATPase and a Na^+-Ca^{2+} antiport.

cytoplasmic Ca^{2+} concentration. The increase is produced by releasing Ca^{2+} from intracellular stores—primarily the endoplasmic reticulum—or by increasing the entry of Ca^{2+} into cells, or by both mechanisms. In many instances, the increase starts in one part of the cell and spreads to other parts of the cytoplasm. In addition, it is now clear that in many instances, hormones or other ligands initiate cyclic oscillations in cytoplasmic Ca^{2+} rather than a steady increase. However, the physiologic role of these oscillations is unsettled.

Calcium-Binding Proteins

Many different Ca^{2+}-binding proteins have been described, including **troponin, calmodulin,** and **calbindin.** Troponin is the Ca^{2+}-binding protein involved in contraction of skeletal muscle (see Chapter 3). Calmodulin contains 148 amino acid residues (Fig 1–34) and has 4 Ca^{2+}-binding domains. It is unique in that residue 115 is trimethylated, and it is extensively conserved, being found in plants as well as animals. When calmodulin binds Ca^{2+}, it is capable of activating 5 different calmodulin-dependent kinases (Table 1–9). One of these is **myosin light-chain kinase,** which phosphorylates myosin. This brings about contraction in smooth muscle. Another is **phosphorylase kinase,** which activates phosphorylase (see Chapter 17). Ca^{2+}/calmodulin kinases I and II are concerned with synaptic function, and Ca^{2+}/calmodulin kinase III is concerned with protein synthesis.

G Proteins

A common way to translate a signal to a biologic

Figure 1–34. Structure of calmodulin from bovine brain. The abbreviations for the amino acid residues are those identified in Table 17–2 and in the appendix. Note the 4 calcium domains (dark residues) flanked on either side by stretches of α helix. (Reproduced, with permission, from Cheung WY: Calmodulin: An overview. *Fed Proc* 1982;**41**:2253.)

Figure 1–35. Signal transduction by heterotrimeric G proteins. ***Top:*** Summary of overall reaction. ***Bottom:*** When the ligand (square) binds to the serpentine receptor in the cell membrane, GTP replaces GDP on the α subunit. GTP-α separates from the βγ subunit and activates various effectors, producing physiologic effects. The intrinsic GTPase activity of GTP-α then connects GTP to GDP, and the α, β, and γ subunits reassociate. The βγ subunit also activates certain effectors.

effect inside cells is by way of nucleotide regulatory proteins **(G proteins)** that bind GTP. GTP is the guanosine analogue of ATP (see Chapter 17). When the signal reaches a G protein, the protein exchanges GDP for GTP. The GTP-protein complex brings about the effect. The inherent GTPase activity of the protein then converts GTP to GDP, restoring the resting state.

One family of **small G proteins** is related to the *ras* oncogene. They are involved in the regulation of cell growth and multiplication, but the details of their functional roles are still unsettled. Other small G proteins carry out related functions.

A third family of G proteins, the larger **heterotrimeric G proteins,** couple cell surface receptors to catalytic units that catalyze the intracellular formation of second messengers or couple the receptors directly to ion channels (Fig 1–35). These G proteins are made up of 3 subunits designated α, β, and γ. The α subunit is bound to GDP. When a ligand binds to a G-coupled receptor, this GDP is exchanged for GTP and the α subunit separates from the combined β and γ subunits. In most instances, it is the separated α subunit that brings about biologic effects. The β and γ units do not separate, but there is evidence that in some instances, they can also activate effectors. The intrinsic GTPase activity of the α subunit then converts GTP and GDP, and this leads to

reassociation of the α with the βγ subunit and termination of effector activation.

The large number of heterotrimeric G proteins characterized to date fall into 4 families (Fig 1–36). They are encoded by at least 16 different genes for the α subunit, 4 different genes for the β subunit, and many different genes for the γ subunit.

All the heterotrimeric G protein-coupled receptors that have been characterized to date are proteins that span the cell membrane 7 times **(serpentine receptors).** A very large number have been cloned (Table 1–10), and their functions are multiple and diverse. The structures of 2 of them are shown in Fig 1–37. In general, small ligands bind to the amino acid residues in the membrane, whereas large polypeptide and protein ligands bind to the extracellular domains, which are bigger and better developed in the receptors for polypeptides and proteins. It is generally amino acid residues in the third cytoplasmic loop, the loop nearest the C terminal, that interact with the G proteins.

Inositol Triphosphate & Diacylglycerol as Second Messengers

The link between membrane binding of a ligand that acts via Ca^{2+} and the prompt increase in the

Figure 1–36. The 4 families of α subunits of heterotrimeric G proteins plotted in terms of their percent amino acid identity to emphasize their relations to one another. (Reproduced, with permission, from Simon MI, Strathman MP, Gautam N: Diversity of G proteins in signal transduction. *Science* 1991;**252:**802.)

Table 1–10. Some of the known ligands for cloned G protein-coupled receptors.[1]

Biogenic amines	Epinephrine
	Norepinephrine
	Dopamine
	Acetylcholine
	Histamine
	5-Hydroxytryptamine
	Adenosine
Tachykinins	Substance P
	Substance K
	Neuromedin K
Glycoprotein hormones	Thyrotropin
	Follicle-stimulating hormone
	Lutropin/choriogonadotropin
Polypeptide hormones	Parathyroid hormone
Brain/gut peptide hormones	Angiotensin
	Arginine-vasopressin
	Vasoactive intestinal polypeptide
	Bombesin/gastrin-releasing hormone
	Thyrotropin-releasing hormone
Arachidonic acid derivatives	Thromboxane A_2
Sensory stimuli	Light (ie retinal)
	Odorants
Miscellaneous	Thrombin
	Endothelins
	Platelet-activating factor
	N-formyl peptide
	Mating factors (yeast)
	cAMP (*Dictyostelium*)
	Cannabinoids
	IL-8

[1]Reproduced, with permission, from Lefkowitz RJ: Variations on a theme. *Nature* 1991;**351:**353. Copyright © 1991 MacMillan Magazines Limited.

cytoplasmic Ca^{2+} concentration is often **inositol triphosphate** (inositol 1,4,5-triphosphate, IP_3). When one of these ligands binds to its receptor, activation of the receptor produces activation of phospholipase C on the inner surface of the membrane via a G_q or G_{11}.

Phospholipase C catalyzes the hydrolysis of phosphatidylinositol 4,5-diphosphate (PIP_2) to form IP_3 and diacylglycerol (Fig 1–38). The IP_3 diffuses to the endoplasmic reticulum, where it triggers the release of Ca^{2+} into the cytoplasm (Fig 1–39). The IP_3 receptor is very similar to the ryanodine receptor, which is involved in Ca^{2+} metabolism (see Chapter 3). Other inositol phosphates are also formed, and they may increase the permeability of the cell membrane to Ca^{2+}. The diacylglycerol also plays a role as a second messenger; it stays in the cell membrane, where it activates one of the 7 subspecies of **protein kinase C** (Table 1–9). Examples of ligands that act via these second messengers are listed in Table 1–8.

The precursor of PIP_2 is phosphatidylinositol (Fig 1–38). This phospholipid is found in relatively small amounts in the inner lamella of the cell membrane. It is first converted to phosphatidyl 4-phosphate (PIP) and then to PIP_2, the derivative that is hydrolyzed to form IP_3 and the diacylglycerol. The IP_3 is metabolized by stepwise dephosphorylation to inositol. Diacylglycerol is converted to phosphatidic acid and

then to cytosine diphosphate (CDP) diacylglycerol, which combines with inositol to form phosphatidylinositol, completing the cycle.

Cyclic AMP

Another important second messenger is cyclic AMP (Fig 1–40). Some of the many ligands that act via this compound are listed in Table 1–8. Cyclic AMP, which is also known as **cAMP,** is cyclic adenosine 3′,5′-monophosphate. It is formed from ATP by the action of the enzyme **adenylate cyclase** and converted to physiologically inactive 5′-AMP by the action of the enzyme phosphodiesterase. Cyclic AMP activates a cyclic nucleotide-dependent protein kinase (**protein kinase A**) that, like protein kinase C, catalyzes the phosphorylation of proteins, changing their conformation and altering their activity. A typical example is the activation of phosphorylase kinase in the liver by epinephrine via cyclic AMP and protein kinase A (see Fig 17–13). Phosphorylase kinase is also a substrate for calmodulin.

Cyclic AMP is metabolized by a phosphodi-

Figure 1–37. Structure of the β_2-adrenergic receptor and rhodopsin. The individual amino acid residues are identified by their single-letter codes, and the black residues are sites of phosphorylation. The y-shaped symbols on N residues identify glycosylation sites. Note the extracellular N terminal, the intracellular C terminal, and the 7 membrane-spanning portions of each protein. (Reproduced, with permission, from Benovic JL et al: Light-dependent phosphorylation of rhodopsin by β-adrenergic receptor kinase. Reprinted by permission from *Nature* 1986;**321**:869. Copyright ©1986, Macmillan Journals Limited.)

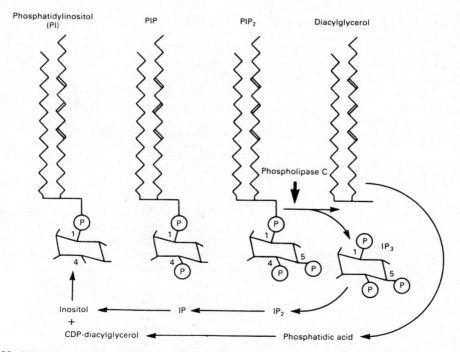

Figure 1–38. Metabolism of phosphatidylinositol in cell membranes. Phosphatidylinositol is successively phosphorylated to form phosphatidylinositol 4-phosphate (PIP), then phosphatidylinositol 4,5-diphosphate (PIP$_2$). Phospholipase C catalyzes the breakdown of PIP$_2$ to inositol 1,4,5-trisphosphate (IP$_3$) and diacylglycerol. IP$_3$ can be converted to inositol tetrakisphosphate (inositol 1,3,4,5-P$_4$), and this can be converted in turn to 1,3,4-P$_3$ (not shown). IP$_3$ is also dephosphorylated to inositol, and diacylglycerol is metabolized to cytosine diphosphate (CDP) diacylglycerol. CDP-diacylglycerol and inositol then combine to form phosphatidylinositol, completing the cycle. (Modified from Berridge MJ: Inositol triphosphate and diacylglycerol as second messengers. *Biochem J* 1984;**220**:345.)

Figure 1–39. Diagrammatic representation of release of inositol triphosphate (IP$_3$) and diacylglycerol (DAG) as second messengers. Binding of ligand to receptor (R) activates phospholipase C (PLC) via G$_q$ or G$_{11}$. The resulting hydrolysis of PIP$_2$ produces IP$_3$, which releases Ca^{2+} from the endoplasmic reticulum (ER), and DAG, which activates protein kinase C (PKC). CaBP, Ca^{2+}-binding proteins.

Figure 1–41. The cyclic AMP system. Activation of the catalytic unit (C), adenylate cyclase, catalyzes the conversion of ATP to cyclic AMP. Cyclic AMP activates protein kinase A, which phosphorylates proteins, producing physiologic effects. Stimulatory ligands bind to stimulatory receptors (R$_s$) and activate C via G$_s$. Inhibitory ligands inhibit C via inhibitory receptors (R$_i$) and G$_i$. (Modified and reproduced, with permission, from Spiegel AM et al: Clinical implications of guanine nucleotide-binding proteins as receptor-effector couplers. Reprinted by permission of the *New England Journal of Medicine* 1985;**312:**26.)

Adenine — Adenosine triphosphate (ATP)

↓ Adenylate cyclase

Adenine — Cyclic AMP

↓ Phosphodiesterase

Adenine — 5'-Adenosine monophosphate (5'-AMP)

Figure 1–40. Formation and metabolism of cyclic AMP.

esterase. This phosphodiesterase is inhibited by methylxanthines such as caffeine and theophylline; consequently, these compounds augment hormonal and transmitter effects mediated via cyclic AMP.

Activation of Adenylate Cyclase

Five components are involved in the mechanism by which ligands bring about changes in the intracellular concentration of cyclic AMP: a catalytic unit, **adenylate cyclase,** which catalyzes the conversion of ATP to cyclic AMP; stimulatory and inhibitory receptors; and stimulatory and inhibitory G proteins that link the receptor to the catalytic unit (Fig 1–41). Like the receptors, adenylate cyclase is a transmembrane protein, and it crosses the membrane 12 times. When the appropriate ligand binds to a stimulatory receptor, a G$_s$ α subunit activates adenylate cyclase. Conversely, when the appropriate ligand binds to the inhibitory receptor, a G$_i$ α subunit inhibits adenylate cyclase. Heterotrimeric G proteins mediate the stimulatory and inhibitory effects on adenylate cyclase that are produced by many different ligands, yet the responses are specific. The specificity depends on the specificity of the receptor, which responds at low threshold to only one or a select group of related ligands.

Some cyclic AMP escapes from cells upon stimulation by certain hormones, but the amounts are small compared with the intracellular concentration, and only small amounts of extracellular cyclic AMP enter cells.

Ca^{2+} can interact with cyclic AMP, and the diacylglycerol-protein kinase C system interacts with the cyclic AMP-protein kinase A system. In some instances, the cyclic AMP system inhibits the diacylglycerol system, whereas in others, the diacylglycerol system facilitates the cyclic AMP system.

Guanylate Cyclase

Another cyclic nucleotide of physiologic importance is **cyclic guanosine monophosphate** (cyclic GMP, cGMP). Cyclic GMP is important in vision. Light acts on rhodopsin in rods, rhodopsin is linked to phosphodiesterase by G_{t1}, and activation of phosphodiestrase accelerates conversion of cyclic GMP to 5'-GMP (see Chapter 8). A similar process occurs in the cones.

Increases in intracellular cyclic GMP activate cyclic GMP-dependent kinase (Table 1–9), producing physiologic effects. Guanylate cyclases are a family of enzymes that catalyze the formation of cyclic GMP. They exist in 2 forms (Fig 1–42). In one form, there is an extracellular N-terminal domain that is a receptor, a single transmembrane domain, and a cytoplasmic C terminal that has a tyrosine kinase-like and a guanylate cyclase catalytic domain. Three such guanylate cyclases have been characterized. Two are receptors for ANP (ANPR-A and ANPR-B; see Chapter 24), and a third binds an *Escherichia coli* enterotoxin. The other form of guanylate cyclases is soluble, contains heme, and is totally intracellular. There appear to be 2 isoforms of the intracellular enzyme. They are activated by nitric oxide (NO) and NO-containing compounds. NO is the endothelium-derived relaxing factor (EDRF), an intercellular messenger formed from arginine that plays an important role in regulating blood vessel diameter (see Chapter 31). It is also involved in brain function (see Chapter 4).

Growth Factors

Growth factors are polypeptides and proteins that have become increasingly important in many different aspects of physiology. They are conveniently divided into 3 groups. One group is made up of agents that foster the multiplication and/or development of various types of cells; nerve growth factor (see Chapter 2), insulinlike growth factor I (IGF-I; see Chapter 22), and epidermal growth factor (EGF) are examples, and more than 20 have been described. The lymphokines and cytokines are a second group. These factors are produced by macrophages and lymphocytes and are important in regulation of the immune system (see Chapter 27). Again, more than 20 have been described. The third group is made up of the colony-stimulating factors that regulate prolifera-

Figure 1–42. Diagrammatic representation of guanylate cyclases, tyrosine kinases, and tyrosine phosphatases **(top)** compared with the receptor-heterotrimeric G protein-adenylate cyclase system **(bottom).** ANP, atrial natiuretic peptide; C, cytoplasm; cyc, cyclase domain; EGF, epidermal growth factor; ICF, interstitial fluid; M, cell membrane; PDGF, platelet-derived growth factor; PTK, tyrosine kinase domain; PTP, tyrosine phosphatase domain; ST, *E coli* enterotoxin. (Modified and reproduced, with permission, from Koesling D, Böhme E, Schultz G: Guanylyl cyclases, a growing family of signal transducing enzymes. *FASEB J* 1991;**5:**2785.)

tion and maturation of red and white blood cells. They are also discussed in Chapter 27.

Many of the growth factors bind to receptors on the cell surface which have intracellular domains that are tyrosine kinases (Fig 1–42). When the ligand binds, the activated receptors trigger phosphorylation of tyrosine residues in the receptor itself and in neighboring proteins. This leads to changes in configuration of the proteins that alter cell function. Phospholipase C can be activated by phosphorylation, as well as, more commonly, by interaction with a G protein. The tyrosine kinase activity of the insulin receptor is discussed in Chapter 19.

A number of tyrosine phosphatases have also been characterized. Some of these are intracellular and presumably dephosphorylate proteins that are phosphorylated by tyrosine kinases. However, others cross the membrane and have extensive extracellular domains that resemble those in adhesion molecules (see above). They may be receptors, but their function is unsettled. In addition, various serine and threonine kinases and phosphatases have been described. Finally, there are receptors for cytokines and hematopoietic factors (see Chapter 27) that cross the membrane one time and have cytoplasmic domains that produce changes in cell function by mechanisms that are as yet unknown.

Receptor Diseases

Disordered function of receptors and the membrane proteins with which they interact causes a variety of diseases. For example, in pseudohypoparathyroidism and nephrogenic diabetes insipidus, parathyroid hormone and vasopressin fail to produce the increase in cyclic AMP in their target organs that they produce in normal individuals. There are several forms of pseudohypoparathyroidism, and one of them appears to be due to a single-base substitution in the gene encoding the α subunit of G_s. This reduces the ability of parathyroid hormone to increase cyclic AMP and produce its physiologic effects. A different activating mutation in G_s α is associated with the rough-bordered areas of skin pigmentation in the condition called the McCune-Albright syndrome. A third mutation in G_s α reduces its intrinsic GTPase activity. As a result, it is much more active in somatotropes of the pituitary, which develop constantly elevated levels of cyclic AMP. This causes them to become neoplastic, producing pituitary adenomas that cause acromegaly (see Chapter 22).

Certain other diseases are caused by production of antibodies against receptors. Thus, antibodies against TSH receptors cause Grave's disease (see Chapter 18), and antibodies against nicotinic acetylcholine receptors cause myasthenia gravis (see Chapter 18).

HOMEOSTASIS

The actual environment of the cells of the body is the interstitial component of the ECF. Since normal cell function depends upon the constancy of this fluid, it is not surprising that in multicellular animals, an immense number of regulatory mechanisms have evolved to maintain it. To describe "the various physiologic arrangements which serve to restore the normal state, once it has been disturbed," W.B. Cannon coined the term **homeostasis.** The buffering properties of the body fluids and the renal and respiratory adjustments to the presence of excess acid or alkali are examples of homeostatic mechanisms. There are countless other examples, and a large part of physiology is concerned with regulatory mechanisms that act to maintain the constancy of the internal environment. Many of these regulatory mechanisms operate on the principle of negative feedback; deviations from a given normal set point are detected by a sensor, and signals from the sensor trigger compensatory changes that continue until the set point is again reached.

AGING

Aging is a general physiologic process that is as yet poorly understood. It affects cells and the systems made up of them, as well as tissue components such as collagen, and numerous theories have been advanced to explain it. Normal cells stop dividing and then eventually die, probably under genetic control. Cancer cells continue to divide and apparently escape senescence. One theory holds that tissues age as a result of random mutations in the DNA of somatic cells, with consequent introduction of cumulative abnormalities. Others hold that cumulative abnormalities are produced by increased cross-linkage of collagen and other proteins, possibly as the end result of the nonenzymatic combination of glucose with amino groups on these molecules. A third theory envisions aging as the cumulative result of damage to tissues by free radicals formed in them. Some investigators have speculated that in mammals, there is a biologic clock, possibly located in the hypothalamus, that is responsible for aging via hormonal or other pathways. However, it seems fair to say that at present, despite intensive research, the mechanisms of aging are still unknown.

REFERENCES

Section I: Introduction

Alberts B et al: *Molecular Biology of the Cell,* 2nd ed. Garland, 1989.

Antonarakis SE: Diagnosis of genetic disorders at the DNA level. *N Engl J Med* 1989;**320:**153.

Beck K, Hunter I, Engel J: Structure and function of laminin: Anatomy of a multidomain glycoprotein. *FASEB J* 1990;**4:**148.

Bennett MVL, Spray DC (editors): *Gap Junctions.* Cold Spring Harbor, 1985.

Bennett V: Spectin-based membrane skeleton: A multi-potential adaptor between plasma membrane and cytoplasm. *Physiol Rev* 1990;**70:**1029.

Berridge MJ, Irvine RF: Inositol phosphates and cell signaling. *Nature* 1989;**341:**197.

Brodsky FM: Living with clathrin: Its role in intracellular membrane traffic. *Science* 1988;**242:**1396.

Cannon WB: *The Wisdom of the Body.* Norton, 1932.

Cooper GM: *Oncogenes.* Jones and Bartlett, 1990.

Cosman D et al: A new cytokine receptor superfamily. *Trends Biochem Sci* 1990;**15:**265.

Cross GAM: Glycolipid anchoring of plasma membrane proteins. *Annu Rev Cell Biol* 1990;**6:**1.

Darnell J, Lodish H, Baltimore D: *Molecular Cell Biology,* 2nd ed. Freeman, 1990.

Drlica K: *Understanding DNA and Gene Cloning: A Guide for the Curious.* Wiley, 1984.

Evans RM: The steroid and thyroid hormone receptor superfamily. *Science* 1988;**240:**889.

Finch C: *Longevity, Senescence, and the Genome.* Univ Chicago Press, 1991.

Gumbiner B: Structure, biochemistry, and assembly of epithelial tight junctions. *Am J Physiol* 1987;**253:** C749.

Hill WE et al (editors): *The Ribosome: Structure, Function, and Evolution.* Am Soc Microbiol, 1990.

Holtzman E: *Lysosomes.* Plenum, 1989.

Horisberger J-D et al: Structure-function relationship of Na,K-ATPase. *Annu Rev Physiol* 1991;**12:**71.

Howard J, Hudspeth AJ, Vale RD: Movement of microtubules by single kinesin molecules. *Nature* 1989; **342:**154.

Jan LY, Jan YN: Structural elements involved in specific K$^+$ channel functions. *Annu Rev Physiol* 1992;**54:**537.

Kirby LT: *DNA Fingerprints: An Introduction.* Freeman, 1990.

Koesling D, Böhme E, Schultz G: Guanylyl cyclases, a growing family of signal-transducing enzymes. *FASEB J* 1991;**5:**2785.

Levitan M: *Textbook of Human Genetics,* 3rd ed. Oxford Univ Press, 1988.

Maniatis T, Reed R: The role of small nuclear ribonucleoprotein particles in pre-mRNA splicing. *Nature* 1987;**325:**673.

Massague J: Transforming growth factor-α: A model for membrane-anchored growth factors. *J Biol Chem* 1990;**265:**21393.

McIntosh JR, Koonca MP: Mitosis. *Science* 1989;**246:** 622.

Miller AD: Human gene therapy comes of age. *Nature* 1992;**357:**455.

Mullis KB: The unusual origin of the polymerase chain reaction. *Sci Am* (Apr) 1990;**262:**56.

Nelson N: Structure and pharmacology of the proton-ATPases. *Trends Pharmacol Sci* 1991;**12:**71.

Petersen OH, Maruyama Y: Calcium-activated potassium channels and their role in secretion. *Nature* 1984;**307:**693.

Pollard TD, Cooper JA: Actin and actin-binding proteins: A critical evaluation of mechanisms and function. *Annu Rev Biochem* 1986;**55:**987.

Rodriquez-Boulan E, Nelson WJ: Morphogenesis of the polarized epithelial cell phenotype. *Science* 1989; **245:**718.

Rothman JE, Orci L: Molecular dissection of the secretory pathway. *Nature* 1992;**355:**409.

Samuels HH et al: Regulation of gene expression by thyroid hormone. *Annu Rev Physiol* 1989;**51:**623.

Schumacher M: Rapid membrane effects of steroid hormones: An emerging concept in neuroendocrinology. *Trends Neurosci* 1990;**13:**359.

Stossel TP: How cells crawl. *Am Sci* 1990;**78:**408.

Unwin N, Henderson R: The structure of proteins in biological membranes. *Sci Am* (Feb) 1984;**250:**78.

Verkman A: Water channels in cell membranes. *Annu Rev Physiol* 1992;**54:**97.

Walter P, Lingappa VR: Mechanisms of protein translocation across the endoplasmic reticulum. *Annu Rev Cell Biol* 1986;**2:**499.

Watson JD, Tooze J, Kuntz DT: *Recombinant DNA: A Short Course.* Freeman, 1984.

Watson JD, et al: *Molecular Biology of the Gene,* 4th ed. 2 vols. Benjamin/Cummings, 1987.

Watson PA: Function follows form: Generation of intracellular signals by cell deformation. *FASEB J* 1991;**5:**2013.

Weinberg RA: Tumor suppressor genes. *Science* 1991;**254:**1138.

Westphal H: Transgenic mammals and biotechnology. *FASEB J* 1989;**3:**117.

Yuen PST, Garbers DL: Guanylyl cyclase-linked receptors. *Annu Rev Neurosci* 1992;**15:**193.

Section II.
Physiology of Nerve & Muscle Cells

Excitable Tissue: Nerve

2

INTRODUCTION

The human nervous system contains about 10^{12} (1 trillion) neurons. It also contains 10–50 times this number of glial cells. The neurons, the basic building blocks of the nervous system, have evolved from primitive neuroeffector cells that respond to various stimuli by contracting. In more complex animals, contraction has become the specialized function of muscle cells, whereas integration and transmission of nerve impulses have become the specialized functions of neurons. This chapter is concerned with the ways these neurons are excited and the way they integrate and transport impulses.

NERVE CELLS

Morphology

Neurons in the mammalian central nervous system come in many different shapes and sizes (Fig 2–1). However, most have the same parts as the typical spinal motor neuron illustrated in Fig 2–2. This cell has 5–7 processes called **dendrites** that extend out from the cell body and arborize extensively. It also has a long fibrous **axon** that originates from a somewhat thickened area of the cell body, the **axon hillock.** The first portion of the axon is called the **initial segment.** The axon divides into terminal branches, each ending in a number of **synaptic knobs.** The knobs are also called **terminal buttons** or **axon telodendria.** They contain granules or vesicles in which the synaptic transmitters secreted by the nerves are stored (see Chapter 4). The neuron is **myelinated;** ie, outside the CNS, the axon acquires a sheath of **myelin,** a protein-lipid complex made up of many layers of the cell membrane of Schwann cells (Fig 2–3). Schwann cells are glialike cells found along peripheral nerves. The myelin sheath envelopes the axon except at its ending and at the **nodes of Ranvier,** periodic 1-μm constrictions that are about 1 mm apart. The insulating function of myelin is discussed below. Not all mammalian neurons are myelinated; some are **unmyelinated,** ie, are simply surrounded by Schwann cells without the wrapping of the Schwann cell membrane around the axon that produces myelin. Most neurons in invertebrates are unmyelinated. In the central nervous system of mammals, most neurons are myelinated, but the cells that form the myelin are oligodendrogliocytes rather than Schwann cells (Fig 2–3). Unlike the Schwann cell, which forms the myelin between 2 nodes of Ranvier on a single neuron, oligodendrogliocytes send off multiple processes that form myelin on many neighboring axons.

The dimensions of some neurons are truly remarkable. For spinal neurons supplying the muscles of the foot, for example, it has been calculated that if the cell body were the size of a tennis ball, the dendrites of the cell would fill an average-sized living room and the axon would be up to 1.6 km (almost a mile) long although only 13 mm (½ in) in diameter.

In multiple sclerosis, a crippling disease of unknown cause, there is patchy destruction of myelin in the central nervous system. The loss of myelin is associated with delayed or blocked conduction in the demyelinated axons.

The conventional terminology used above for the parts of a neuron works well enough for spinal motor neurons and interneurons, but there are problems in terms of "dendrites" and "axons" when it is applied to other types of neurons found in the nervous system. From a functional point of view (see below and Chapters 4 and 5), neurons generally have 4 important zones. There is a receptor, or dendritic zone, where multiple local potential changes generated by synaptic connections are integrated (Fig 2–4); a site where propagated action potentials are generated (the initial segment in spinal motor neurons, the initial node of Ranvier in cutaneous sensory neurons); an axonal process that transmits propagated impulses to the nerve endings; and the nerve endings, where action potentials cause the release of synaptic transmitters. The cell body is often located at the dendritic zone end of the axon, but it can be within the axon (eg, auditory neurons) or attached to the side of the axon (eg, cutaneous neurons; see Fig 2–1). Its location makes no difference as far as the receptor function of the dendritic zone and the transmission func-

Figure 2–1. Some of the types of neurons in the mammalian nervous system. (Reproduced, with permission, from Bodian D: Introductory survey of neurons. *Cold Spring Harbor Symp Quant Biol* 1952;**17:**1.)

tion of the axon are concerned. It should be noted that integration of activity is not the only function of dendrites. Some neurons in the central nervous system have no axons, and local potentials pass from one dendrite to another (see below).

Protein Synthesis & Axoplasmic Transport

Nerve cells are secretory cells, but they differ from other secretory cells in that the secretory zone is often at the end of the axon, far removed from the cell body. There are no ribosomes in axons and nerve terminals, and all necessary proteins are synthesized in the endoplasmic reticulum and Golgi apparatus of the

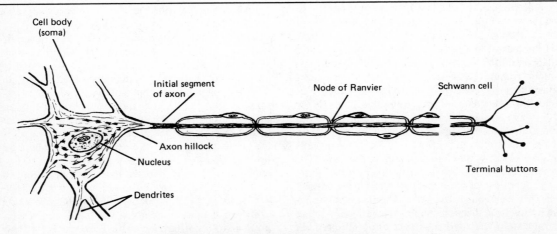

Figure 2–2. Motor neuron with myelinated axon.

Figure 2–3. *Top:* Relation of Schwann cells to axons in peripheral nerves. On the left is an unmyelinated axon, and on the right is a myelinated axon. Note that the cell membrane of the Schwann cell has wrapped itself around and around the axon. ***Bottom:*** Myelination of axons in the central nervous system by oligodendrogliocytes. One oligodendrogliocyte sends processes to up to 40 axons.

Figure 2–4. Functional organization of neurons. Nonconducted local potentials are integrated in the receptor zone, and action potentials are initiated at a site close to the receptor zone (arrow). The action potentials are conducted along the axon to the nerve endings, where they cause release of synaptic transmitters.

cell body and then transported along the axon to the synaptic knobs by the process of **axoplasmic flow.** Thus, the cell body maintains the functional and anatomic integrity of the axon; if the axon is cut, the part distal to the cut degenerates (**wallerian degeneration**). **Anterograde transport** from cell bodies to axons can be divided into 2 components: fast transport, at about 400 mm/d, which moves organelles along microtubules and is brought about by kinesin (see Table 1–3); and slow transport, at 0.5–10 mm/d, which probably depends on polymerization of cytoskeletal units at one end and their depolymerization at the other. There is also **retrograde transport** from the endings to the cell body at about 200 mm/d. This form of transport along microtubules is mediated by cytoplasmic dynein. Synaptic vesicles recycle in the membrane, but some used vesicles are carried back to the cell body and deposited in lysosomes. Some of the material taken up at the ending by endocytosis, including nerve growth factor (see below) and various viruses, is also transported back to the cell body.

EXCITATION & CONDUCTION

Nerve cells have a low threshold for excitation. The stimulus may be electrical, chemical, or mechanical. Two types of physiochemical disturbances are produced: local, nonpropagated potentials called, depending on their location, **synaptic, generator,** or **electrotonic potentials;** and propagated disturbances, the **action potentials** (or **nerve impulses**). These are the only electrical responses of neurons and other excitable tissues, and they are the main language of the nervous system. They are due to changes in the conduction of ions across the cell membrane that are produced by alterations in ion channels.

The impulse is normally transmitted (**conducted**) along the axon to its termination. Nerves are not "telephone wires" that transmit impulses passively; conduction of nerve impulses, although rapid, is much slower than that of electricity. Nerve tissue is in fact a relatively poor passive conductor, and it would take a potential of many volts to produce a signal of a fraction of 1 V at the other end of a 1-m axon in the absence of active processes in the nerve. Conduction is an active, self-propagating process, and the impulse moves along the nerve at a constant amplitude and velocity. The process is often compared to what happens when a match is applied to one end of a train of gunpowder; by igniting the powder particles immediately in front of it, the flame moves steadily down the train to its end.

MEASUREMENT OF ELECTRICAL EVENTS

The electrical events in neurons are rapid, being measured in **milliseconds (ms);** and the potential changes are small, being measured in **millivolts (mV).** In addition to development of microelectrodes with a tip diameter of less than 1 μm, the principal advances that made detailed study of the electrical activity in nerves possible were the development of electronic amplifiers and the cathode-ray oscilloscope. Modern amplifiers magnify potential changes 1000 times or more, and the cathode-ray oscilloscope provides an almost inertialess and almost instantaneously responding "lever" for recording electrical events.

The Cathode-Ray Oscilloscope

The **cathode-ray oscilloscope (CRO)** is used to measure the electrical events in living tissue. A cathode emits electrons when a high voltage is applied across it and a suitable anode in a vacuum. In the CRO, the electrons are directed into a focused beam that strikes the face of the glass tube in which the cathode is located (Fig 2–5). The face is coated with one of a number of substances (phosphors) that emit light when struck by electrons. A vertical metal plate is placed on either side of the electron beam. When a voltage is applied across these plates, the negatively charged electrons are drawn toward the positively charged plate and repelled by the negatively charged plate. If the voltage applied to the vertical plates (X plates) is increased slowly and then reduced suddenly and increased again, the beam moves steadily toward the positive plate, snaps back to its former position, and moves toward the positive plate again. Application of a "saw-tooth voltage" of this type thus causes the beam to sweep across the face of the tube, and the speed of the sweep is proportionate to the rate of rise of the applied voltage.

Figure 2–5. Cathode-ray oscilloscope. Simplified diagram of the principal connections when arranged to record potential changes in a nerve.

Another set of plates (Y plates) is arranged horizontally, with one plate above and one below the beam. Voltages applied to these plates deflect the beam up and down as it sweeps across the face of the tube, and the magnitude of the vertical deflection is proportionate to the potential difference between the horizontal plates. When these plates are connected to electrodes on a nerve, any changes in potential occurring in the nerve are recorded as vertical deflections of the beam as it moves across the tube.

Recording From Single Axons

With the CRO, the electrical events occurring in a piece of peripheral nerve dissected from a laboratory animal can be demonstrated. However, such preparations contain many axons, and it is important to be able to study the properties of a single axon. Mammalian axons are relatively small (20 μm or less in diameter) and are difficult to separate from other axons, but giant unmyelinated nerve cells exist in a number of invertebrate species. Such giant cells are found, for example, in crabs *(Carcinus)* and cuttlefish *(Sepia),* but the largest known axons are found in the squid *(Loligo).* The neck region of the muscular mantle of the squid contains single axons up to 1 mm in diameter. The fundamental properties of these long axons are similar to those of mammalian axons.

Resting Membrane Potential

When 2 electrodes are connected through a suitable amplifier to a CRO and placed on the surface of a single axon, no potential difference is observed. However, if one electrode is inserted into the interior of the cell, a constant potential difference is observed, with the inside negative relative to the outside of the cell at rest. This **resting membrane potential** is found in almost all cells. Its genesis is discussed in Chapter 1. In neurons, it is usually about –70 mV.

Latent Period

If the axon is stimulated and a conducted impulse occurs, a characteristic series of potential changes known as the **action potential** is observed as the impulse passes the exterior electrode.

When the stimulus is applied, there is a brief irregular deflection of the baseline, the **stimulus artifact.** This artifact is due to current leakage from the stimulating electrodes to the recording electrodes. It usually occurs despite careful shielding, but it is of value because it marks on the cathode-ray screen the point at which the stimulus was applied.

The stimulus artifact is followed by an isopotential interval (**latent period**) that ends with the start of the action potential and corresponds to the time it takes the impulse to travel along the axon from the site of stimulation to the recording electrodes. Its duration is proportionate to the distance between the stimulating and recording electrodes and the speed of conduction of the axon. If the duration of the latent period and the distance between the electrodes are known, the speed of conduction in the axon can be calculated. For example, assume that the distance between the cathodal stimulating electrode and the exterior electrode in Fig 2–6 is 4 cm. The cathode is normally the stimulating electrode (see below). If the latent period is 2 ms, the speed of conduction is 4 cm/2 ms, or 20 m/s.

Action Potential

The first manifestation of the approaching action potential is a beginning depolarization of the membrane. After an initial 15 mV of depolarization, the rate of depolarization increases. The point at which this change in rate occurs is called the **firing level** or sometimes the **threshold**. Thereafter, the tracing on the oscilloscope rapidly reaches and **overshoots** the isopotential (zero potential) line to approximately + 35 mV. It then reverses and falls rapidly toward the resting level. When repolarization is about 70% completed, the rate of repolarization decreases and the tracing approaches the resting level more slowly. The sharp rise and rapid fall are the **spike potential** of the axon, and the slower fall at the end of the process is the **after-depolarization.** After reaching the previous resting level, the tracing overshoots slightly in the hyperpolarizing direction to form the small but prolonged **after-hyperpolarization.** When recorded with one electrode in the cell, the action potential is called **monophasic,** because it is primarily in one direction.

The proportions of the tracing in Fig 2–6 are intentionally distorted to illustrate the various components of the action potential. A tracing with the components plotted on exact temporal and magnitude scales for a mammalian neuron is shown in Fig 2–7. Note that the rise of the action is so rapid that it fails to show clearly the change in depolarization rate at the firing level, and also that the after-hyperpolarization is only about 1–2 mV in amplitude although it lasts about 40 ms. The duration of the after-depolarization is about 4 ms in this instance. It is shorter and less prominent in many other neurons. Changes may occur in the after-polarizations without changes in the rest of the action potential. For example, if the nerve has been conducting repetitively for a long time, the after-hyperpolarization is usually quite large.

"All or None" Law

If an axon is arranged for recording as shown in Fig 2–6, with the recording electrodes at an appreciable distance from the stimulating electrodes, it is possible to determine the minimal intensity of stimulating current (**threshold intensity**) that, acting for a given duration, will just produce an action potential. The threshold intensity varies with the duration; with weak stimuli it is long, and with strong stimuli it is short. The relation between the strength and the duration of a threshold stimulus is called the **strength-duration curve.** Slowly rising currents fail to fire the nerve because the nerve adapts to the applied stimulus, a process called **accommodation.**

Once threshold intensity is reached, a full-fledged action potential is produced. Further increases in the intensity of a stimulus produce no increment or other change in the action potential as long as the other experimental conditions remain constant. The action potential fails to occur if the stimulus is subthreshold in magnitude, and it occurs with a constant amplitude and form regardless of the strength of the stimulus if the stimulus is at or above threshold intensity. The

Figure 2–6. Action potential in a neuron recorded with one electrode inside the cell.

Figure 2–7. Diagram of the complete action potential of a large mammalian myelinated fiber, drawn without time or voltage distortion to the show proportions of the components.

action potential is therefore "all or none" in character and is said to obey the **all or none law.**

Electrotonic Potentials, Local Response, & Firing Level

Although subthreshold stimuli do not produce an action potential, they do have an effect on the membrane potential. This can be demonstrated by placing recording electrodes within a few millimeters of a stimulating electrode and applying subthreshold stimuli of fixed duration. Application of such currents with a cathode leads to a localized depolarizing potential change that rises sharply and decays exponentially with time. The magnitude of this response drops off rapidly as the distance between the stimulating and recording electrodes is increased. Conversely, an anodal current produces a hyperpolarizing potential change of similar duration. These potential changes are called **electrotonic potentials,** those produced at a cathode being **catelectrotonic** and those at an anode **anelectrotonic.** They are passive changes in membrane polarization caused by addition or subtraction of charge by the particular electrode. At low current intensities producing up to about 7 mV of depolarization or hyperpolarization, their size is proportionate to the magnitude of the stimulus. With stronger stimuli, this relationship remains constant for anelectrotonic responses but not for responses at the cathode. The cathodal responses are greater than would be expected from the magnitude of the applied current. Finally, when the cathodal stimulation is great enough to produce about 15 mV of depolarization, ie, at a membrane potential of –55 mV, the membrane potential suddenly begins to fall rapidly, and a propagated action potential occurs. The disproportionately greater response at the cathode to stimuli of sufficient strength to produce 7–15 mV of depolarization is produced when voltage-gated Na^+ channels begin to open (see below) and is called the **local response** (Fig 2–8). The point at which a runaway spike potential is initiated is the **firing level.** Thus, cathodal currents that produce up to 7 mV of depolarization have a purely passive effect on the membrane as a result of addition of negative charges. Those producing 7–15 mV of depolarization also initiate a slight active contribution to the depolarizing process. However, the repolarizing forces are still stronger than the depolarizing forces, and the potential decays. At 15 mV of depolarization, the depolarizing forces are strong enough to overwhelm the repolarizing processes, and an action potential results.

Stimulation normally occurs at the cathode, because cathodal stimuli are depolarizing. Anodal currents, by taking the membrane potential farther away from the firing level, inhibit impulse formation. However, cessation of an anodal current may lead to an overshoot of the membrane potential in the depolarizing direction. This rebound is sometimes large

Figure 2–8. Electrotonic potentials and local response. The changes in the membrane potential of a neuron following application of stimuli of 0.2, 0.4, 0.6, 0.8, and 1.0 times threshold intensity are shown superimposed on the same time scale. The responses below the horizontal line are those recorded near the anode, and the responses above the line are those recorded near the cathode. The stimulus of threshold intensity was repeated twice. Once it caused a propagated action potential (top line), and once it did not.

enough to cause the nerve to fire at the end of an anodal stimulus.

Changes in Excitability During Electrotonic Potentials & the Action Potential

During the action potential as well as during catelectrotonic and anelectrotonic potentials and the local response, there are changes in the threshold of the neuron to stimulation. Hyperpolarizing anelectrotonic responses elevate the threshold and catelectrotonic potentials lower it as they move the membrane potential closer to the firing level. During the local response, the threshold is lowered, but during the rising and much of the falling phases of the spike potential, the neuron is refractory to stimulation. This **refractory period** is divided into an **absolute refractory period,** corresponding to the period from the time the firing level is reached until repolarization is about one-third complete, and a **relative refractory period,** lasting from this point to the start of after-depolarization. During the absolute refractory period, no stimulus, no matter how strong, will excite the nerve, but during the relative refractory period, stronger than normal stimuli can cause excitation. During after-depolarization, the threshold is again decreased, and during after-hyperpolarization, it is increased. These changes in threshold are correlated with the phases of the action potential in Fig 2–9.

Electrogenesis of the Action Potential

The nerve cell membrane is polarized at rest, with positive charges lined up along the outside of the

Figure 2–9. Relative changes in excitability of a nerve cell membrane during the passage of an impulse. Note that excitability is the reciprocal of threshold. (Modified and reproduced, with permission, from Morgan CT: *Physiological Psychology,* McGraw-Hill, 1943.)

Figure 2–10. Local current flow (movement of positive charges) around an impulse in axon. ***Top:*** Unmyelinated axon. ***Bottom:*** Myelinated axon.

membrane and negative charges along the inside. During the action potential, this polarity is abolished and for a brief period is actually reversed (Fig 2–10). Positive charges from the membrane ahead of and behind the action potential flow into the area of negativity represented by the action potential ("current sink"). By drawing off positive charges, this flow decreases the polarity of the membrane ahead of the action potential. Such electrotonic depolarization initiates a local response, and when the firing level is reached, a propagated response occurs that in turn electrotonically depolarizes the membrane in front of it. This sequence of events moves regularly along an unmyelinated axon to its end. Thus, the self-propagating nature of the nerve impulse is due to circular current flow and successive electrotonic depolarizations to the firing level of the membrane ahead of the action potential. Once initiated, a moving impulse does not depolarize the area behind it to the firing level, because this area is refractory.

The action potentials produced at synaptic junctions and sensory receptors also depend upon electrotonic depolarization of the nerve cell membrane to the firing level. The details of the process by which the receptor zone of a neuron serves as a large current sink that draws off positive charges from the axon until the firing level is reached are described in Chapter 4.

Saltatory Conduction

Conduction in myelinated axons depends upon a similar pattern of circular current flow. However,

myelin is an effective insulator, and current flow through it is negligible. Instead, depolarization in myelinated axons jumps from one node of Ranvier to the next, with the current sink at the active node serving to electrotonically depolarize to the firing level the node ahead of the action potential (Fig 2–10). This jumping of depolarization from node to node is called **saltatory conduction.** It is a rapid process, and myelinated axons conduct up to 50 times faster than the fastest unmyelinated fibers.

Orthodromic & Antidromic Conduction

An axon can conduct in either direction. When an action potential is initiated in the middle of it, 2 impulses traveling in opposite directions are set up by electrotonic depolarization on either side of the initial current sink.

In a living animal, impulses normally pass in one direction only, ie, from synaptic junctions or receptors along axons to their termination. Such conduction is called **orthodromic.** Conduction in the opposite direction is called **antidromic.** Since synapses, unlike axons, permit conduction in one direction only, any antidromic impulses that are set up fail to pass the first synapse they encounter (see Chapter 4) and die out at that point.

Biphasic Action Potentials

The descriptions of the resting membrane potential and action potential outlined above are based on recording with 2 electrodes, one on the surface of the axon and the other inside the axon. If both recording electrodes are placed on the surface of the axon, there is no potential difference between them at rest. When the nerve is stimulated and an impulse is conducted past the 2 electrodes, a characteristic sequence of potential changes results. As the wave of depolarization reaches the electrode nearest the stimulator, this electrode becomes negative relative to the other electrode (Fig 2–11). When the impulse passes to the portion of the nerve between the 2 electrodes, the potential returns to zero, and then, as it passes the second electrode, the first electrode becomes positive relative to the second. It is conventional to connect the leads in such a way that when the first electrode becomes negative relative to the second, an upward deflection is recorded. Therefore, the record shows an upward deflection followed by an isoelectric interval and then a downward deflection. This sequence is called a **biphasic action potential** (Fig 2–11). The duration of the isoelectric interval is proportionate to the speed of conduction of the nerve and the distance between the 2 recording electrodes.

Conduction in a Volume Conductor

Because the body fluids contain large quantities of electrolytes, the nerves in the body function in a conducting medium that is often called a **volume conductor.** The monophasic and biphasic action potentials described above are those seen when an axon is stimulated in a nonconducting medium outside the body. The potential changes observed during extracellular recording in a volume conductor are basically similar to these action potentials, but they are

complicated by the effects of current flow in the volume conductor. These effects are complex and are influenced by such factors as the orientation of the electrodes relative to the direction the action potential is moving and the distance between the recording electrode over active tissue and the indifferent electrode. In general, when an action potential is recorded in a volume conductor, there are positive deflections before and after the negative spike.

IONIC BASIS OF EXCITATION & CONDUCTION

The cell membranes of nerves, like those of other cells, contain many different types of ion channels (Table 2–1). Some of these are passive, ie, continually open, whereas others are voltage-gated and still others are ligand-gated. It is the behavior of these channels, and particularly Na$^+$ and K$^+$ channels, that explains the electrical events in nerves.

Ionic Basis of Resting Membrane Potential

The ionic basis of the resting membrane potential is discussed in Chapter 1. In nerves, as in other tissues, Na$^+$ is actively transported out of the cell and K$^+$ is actively transported into the cell. K$^+$ diffuses out of the cell down its concentration gradient through K$^+$ leak channels, and Na$^+$ diffuses back in, but because of the leak channels, the permeability of the membrane to K$^+$ is much greater than it is to Na$^+$ at rest. Therefore, passive K$^+$ efflux is much greater than passive Na$^+$ influx. Since the membrane is impermeable to most of the anions in the cell, the K$^+$ efflux is not accompanied by an equal flux of anions and the membrane is maintained in a polarized state, with the outside positive relative to the inside.

Ionic Fluxes During the Action Potential

The changes in membrane conductance of Na$^+$ and K$^+$ that occur during the action potentials are shown in Fig 2–12. The conductance of an ion is the reciprocal of its electrical resistance in the membrane and is a measure of the membrane permeability to that ion.

A slight decrease in resting membrane potential leads to increased K$^+$ efflux and Cl$^-$ influx, restoring the resting membrane potential. However, when depolarization exceeds 7 mV, the voltage-gated Na$^+$ channels start to open at an increased rate (**Na$^+$ channel activation**), and when the firing level is reached, the influx of Na$^+$ along its inwardly directed concentration and electrical gradients is so great that it temporarily swamps the repolarizing forces.

The equilibrium potential for Na$^+$ in mammalian neurons, calculated by using the Nernst equation, is about +60 mV. The membrane potential moves toward this value but does not reach it during the action potential, primarily because the increase in Na$^+$ conductance is short-lived. The Na$^+$ channels rapidly

Figure 2–11. Biphasic action potential. Both recording electrodes are on the outside of the nerve membrane.

Table 2–1. Some of the major types of ion channels found in neurons.[1]

Type	Characteristics[2]
Na⁺	Depolarization opens, then inactivates; blocked by TTX, STX.
Ca²⁺	Depolarization opens slowly.
K⁺ (delayed rectifier)	Depolarization opens slowly, then inactivates; blocked by TEA, 4AP.
K⁺ (inward rectifier)	Hyperpolarization opens; blocked by Ca²⁺, Ba²⁺.
K⁺ ("A" current)	Depolarization opens, inactivates rapidly; blocked by 4AP.
K⁺ ("M")	Depolarization opens slowly; blocked by muscarinic agents.
K⁺ (Ca²⁺-activated)	Depolarization, cytoplasmic Ca²⁺ opens.
K⁺ (s)	Closed when phosphorylated via serotonin and cyclic AMP in *Aplysia*.
Cation (Ca²⁺-activated)	Cytoplasmic Ca²⁺ opens.
Cl⁻ (Ca²⁺-activated)	Cytoplasmic Ca²⁺ opens.
Cl⁻	Hyperpolarization opens.
H⁺	Depolarization opens; pH-sensitive.
Transmitter-activated	Numerous types.

[1] Courtesy of MD Cahalan.
[2] TTX, tetrodotoxin; STX, saxitoxin; TEA, tetraethylammonium; 4AP, 4-aminopyridine.

enter a closed state called the **inactivated state** and remain in this state for a few milliseconds before returning to the resting state. In addition, the direction of the electrical gradient for Na⁺ is reversed during the overshoot because the membrane potential is reversed, and this limits Na⁺ influx. A third factor producing repolarization is the opening of voltage-gated K⁺ channels. This opening is slower and more prolonged than the opening of the Na⁺ channels, and consequently, much of the increase in K⁺ conductance comes after the increase in Na⁺ conductance. The net movement of positive charge out of the cell due to K⁺ efflux at this time helps complete the process of repolarization. The slow return of the K⁺ channels to the closed state also explains the after-hyperpolarization.

Decreasing the external Na⁺ concentration decreases the size of the action potential but has little effect on the resting membrane potential. The lack of much effect on the resting membrane potential would be predicted from the Goldman equation (see Chapter 1), since the permeability of the membrane to Na⁺ at rest is relatively low. Conversely, increasing the external K⁺ concentration decreases the resting membrane potential.

Although Na⁺ enters the nerve cell and K⁺ leaves it during the action potential, the number of ions involved is minute relative to the total numbers present. The fact that the nerve gains Na⁺ and loses K⁺ during activity has been demonstrated experimentally, but significant differences in ion concentrations can be measured only after prolonged, repeated stimulation.

The slower opening and delayed closing of the voltage-gated K⁺ channels also explains accommodation. If depolarization occurs rapidly, the opening of the Na⁺ channels overwhelms the repolarizing forces, but if the induced depolarization is produced slowly, the opening of K⁺ channels balances the gradual opening of Na⁺ channels, and an action potential does not occur.

A decrease in extracellular Ca²⁺ increases the excitability of nerve and muscle cells by decreasing the amount of depolarization necessary to initiate the changes in the Na⁺ and K⁺ conductance that produce the action potential. Conversely, an increase in extracellular Ca²⁺ "stabilizes the membrane" by decreasing excitability.

Distribution of Ions Channels in Myelinated Neurons

As noted in Chapter 1, various substances that bind to Na⁺ and K⁺ channels can be labeled and used to identify the locations of the channels in the cell membrane. Na⁺ channels are highly concentrated in the nodes of Ranvier and the initial segment in myelinated neurons. The initial segment and, in sensory neurons, the first node of Ranvier are the sites where impulses are normally generated, and the other nodes

Figure 2–12. Changes in Na⁺ and K⁺ conductance during the action potential in giant squid axon. The dotted line represents the action potential superimposed on the same time coordinate. Note that the initial electrotonic depolarization initiates the change in Na⁺ conductance, which in turn adds to the depolarization. (Redrawn and reproduced, with permission, from Hodgkin AL: Ionic movements and electrical activity in giant nerve fibers. *Proc R Soc Lond Ser B* 1958;**143**:1.)

of Ranvier are the sites to which the impulses jump during saltatory conduction. The number of Na⁺ channels per square micrometer of membrane in myelinated mammalian neurons has been estimated to be 50–75 in the cell body, 350–500 in the initial segment, less than 25 on the surface of the myelin, 2000–12,000 at the nodes of Ranvier, and 20–75 at the axon terminals. Along the axons of unmyelinated neurons, the number is about 110. In many myelinated neurons, the Na⁺ channels are flanked by K⁺ channels that are involved in repolarization.

Energy Sources & Metabolism of Nerve

The major part of the energy requirement of nerve—about 70%—is the portion used to maintain polarization of the membrane by the action of Na⁺-K⁺ ATPase. During maximal activity, the metabolic rate of the nerve doubles; by comparison, the metabolic rate of skeletal muscle increases as much as 100-fold. Inhibition of lactic acid production does not influence nerve function.

Like muscle, nerve has a resting heat while inactive, an initial heat during the action potential, and a recovery heat that follows activity. However, in nerve, the recovery heat after a single impulse is about 30 times the initial heat. There is some evidence that the initial heat is produced during the after-depolarization rather than the spike. The metabolism of muscle is discussed in detail in Chapter 3.

PROPERTIES OF MIXED NERVES

Peripheral nerves in mammals are made up of many axons bound together in a fibrous envelope called the **epineurium.** Potential changes recorded from such nerves therefore represent an algebraic summation of the all or none action potentials of many axons. The thresholds of the individual axons in the nerve and their distance from the stimulating electrodes vary. With subthreshold stimuli, none of the axons are stimulated and no response occurs. When the stimuli are of threshold intensity, axons with low thresholds fire and a small potential change is observed. As the intensity of the stimulating current is increased, the axons with higher thresholds are also discharged. The electrical response increases proportionately until the stimulus is strong enough to excite all of the axons in the nerve. The stimulus that produces excitation of all the axons is the **maximal stimulus,** and further application of greater, **supramaximal** stimuli produces no further increase in the size of the observed potential.

Compound Action Potentials

Another property of mixed nerves, as opposed to single axons, is the appearance of multiple peaks in the action potential. The multipeaked action potential is called a **compound action potential.** It has a unique shape because a mixed nerve is made up of families of fibers with varying speeds of conduction. Therefore, when all the fibers are stimulated, the activity in fast-conducting fibers arrives at the recording electrodes sooner than the activity in slower fibers; and the farther away from the stimulating electrodes the action potential is recorded, the greater is the separation between the fast and slow fiber peaks. The number and size of the peaks vary with the types of fibers in the particular nerve being studied. If less than maximal stimuli are used, the shape of the compound action potential also depends upon the number and type of fibers stimulated.

Erlanger and Gasser divided mammalian nerve fibers into A, B, and C groups, further subdividing the A group into α, β, γ, and δ fibers. The relative latencies of the electrical activity caused by each of these components are shown in Fig 2–13. It should be emphasized that the drawing is not the compound action

Figure 2–13. Compound action potential. **Left:** Record obtained with recording electrodes at various distances from the stimulating electrodes along a mixed nerve. **Right:** Reconstruction of a compound action potential to show relative sizes and time relationships of the components. (Redrawn and reproduced, with permission, from Erlanger J, Gasser HS: *Electrical Signs of Nervous Activity.* Univ of Pennsylvania Press, 1937.)

potential of any particular peripheral nerve; none of the peripheral nerves show all the components illustrated in this composite diagram because none contain all of the fiber types.

NERVE FIBER TYPES & FUNCTION

By comparing the neurologic deficits produced by careful dorsal root section and other nerve-cutting experiments with the histologic changes in the nerves, the functions and histologic characteristics of each of the families of axons responsible for the various peaks of the compound action potential have been established. In general, the greater the diameter of a given nerve fiber, the greater its speed of conduction. The larger axons are concerned primarily with proprioceptive sensation and somatic motor function, while the smaller axons subserve pain and temperature sensations and autonomic function. In Table 2–2, the various fiber types are listed with their diameters, electrical characteristics, and functions. The dorsal root C fibers conduct impulses generated by touch and other cutaneous receptors in addition to pain and temperature receptors, but only pain and temperature are relayed to consciousness. The other fibers presumably are concerned with reflex responses integrated in the spinal cord and brain stem.

Further research has shown that not all the classically described lettered components are homogeneous, and a numerical system (Ia, Ib, II, III, IV) has been used by some physiologists to classify sensory fibers. Unfortunately, this has led to confusion. A comparison of the number system and the letter system is shown in Table 2–3.

In addition to variations in speed of conduction and fiber diameter, the various classes of fibers in peripheral nerves differ in their sensitivity to hypoxia and anesthetics (Table 2–4). This fact has clinical as well as physiologic significance. Local anesthetics

Table 2-3. Numerical classification sometimes used for sensory neurons.

Number	Origin	Fiber Type
Ia	Muscle spindle, annulo-spinal ending.	A α
Ib	Golgi tendon organ.	A α
II	Muscle spindle, flower-spray ending; touch, pressure.	A β
III	Pain and cold receptors; some touch receptors.	A δ
IV	Pain, temperature, and other receptors.	Dorsal root C

depress transmission in the group C fibers before they affect the touch fibers in the A group. Conversely, pressure on a nerve can cause loss of conduction in motor, touch, and pressure fibers while pain sensation remains relatively intact. Patterns of this type are sometimes seen in individuals who sleep with their arms under their heads for long periods, causing compression of the nerves in the arms. Because of the association of deep sleep with alcoholic intoxication, the syndrome is commonest on weekends and has acquired the interesting name Saturday night or Sunday morning paralysis.

Human peripheral nerve fibers are also classified on a physio-anatomic basis. This classification (Table 2–5) divides nerves into afferent and efferent categories and further subdivides them according to whether they have somatic or visceral and general or special functions. The term "special" is applied to nerves that supply the organs of the special senses and the musculature of branchiomeric origin, ie, musculature arising from the branchial arches during embryonic development. Thus, for example, special somatic efferent (SSE) fibers are the efferents to the eye and ear.

Table 2-2. Nerve fiber types in mammalian nerve.[1]

Fiber Type	Function	Fiber Diameter (μm)	Conduction Velocity (m/s)	Spike Duration (ms)	Absolute Refractory Period (ms)
A					
α	Proprioception; somatic motor	12–20	70–120		
β	Touch, pressure	5–12	30–70	0.4–0.5	0.4–1
γ	Motor to muscle spindles	3–6	15–30		
δ	Pain, cold, touch	2–5	12–30		
B	Preganglionic autonomic	<3	3–15	1.2	1.2
C					
Dorsal root	Pain, temperature, some mechanoreception, reflex responses	0.4–1.2	0.5–2	2	2
Sympathetic	Postganglionic sympathetics	0.3–1.3	0.7–2.3	2	2

[1]A and B fibers are myelinated; C fibers are unmyelinated.

Table 2–4. Relative susceptibility of mammalian A, B, and C nerve fibers to conduction block produced by various agents.

Susceptibility to:	Most Susceptible	Interme-diate	Least Susceptible
Hypoxia	B	A	C
Pressure	A	B	C
Local anes-thetics	C	B	A

Table 2–5. Types of fibers in peripheral and cranial nerves.

Afferent
  Somatic
    General: sensory from cutaneous and muscle receptors
    Special: vision, hearing
  Visceral
    General: sensation from viscera
    Special: smell, taste

Efferent
  Somatic
    General: motor to skeletal muscle
    Special: efferent to eye, ear
  Visceral
    General: autonomic motor
    Special: motor to branchiomeric musculature

NERVE GROWTH FACTOR

Nerve growth factor (NGF) is a protein growth factor that is necessary for the growth and maintenance of sympathetic neurons and some sensory neurons. It is present in a broad spectrum of animal species, including humans, and is found in many different tissues. In male mice, there is a particularly high concentration in the submaxillary salivary glands, and the level is reduced by castration to that seen in females. The factor is a dimer made up of 2 α, 2 β, and 2 γ subunits. The β subunits, each of which has a molecular weight of 13,200, have all the nerve growth-promoting activity, the α subunits have trypsinlike activity, and the γ subunits are serine proteases. The function of the proteases is unknown. The structure of the β unit of NGF resembles that of insulin. The NGF receptor is a tyrosine kinase, and its high affinity form is probably a heterodimer made up of a 75K protein and a 140K protein that is encoded by the *trk* proto-oncogene.

NGF is picked up by neurons in the extracerebral organs they innervate and is transported in retrograde fashion from the endings of the neurons to their cell bodies. It is also present in the brain and appears to be

Protoplasmic astrocyte

Fibrous astrocyte

Microglia

Oligodendrocytes

Figure 2–14. Glial cells in the brain. (Reproduced, with permission, from Junqueira LC, Carneiro J, Kelley RO: *Basic Histology,* 7th ed. Appleton & Lange, 1992.)

responsible for the growth and maintenance of cholinergic neurons in the basal forebrain and striatum. Injection of antiserum against NGF in newborn animals leads to near total destruction of the sympathetic ganglia; it thus produces an **immunosympathectomy.**

Several other substances that maintain certain kinds of nerve cells have also been isolated, including fibroblast growth factor, purpurin, brain-derived neurotrophic factor (BDNF), neurotropin 3 (NT-3), ciliary neurotrophic factor, and, interestingly, activins and inhibins (see Fig 23–21).

GLIA

In addition to neurons, the nervous system contains glial cells (neuroglia). Glial cells are very numerous; as noted above, there are 10–50 times as many glial cells as neurons. The Schwann cells that invest axons in peripheral nerves are classified as glia. In the central nervous system, there are 3 main types of glia. **Microglia** (Fig 2–14) are scavenger cells that enter the nervous system from the blood vessels. **Oligodendrogliocytes** are involved in myelin formation (Fig 2–3). **Astrocytes,** which are found throughout the brain, are of 2 subtypes. **Fibrous astrocytes,** which contain many intermediate filaments, are found primarily in white matter. **Protoplasmic astrocytes** are found in gray matter and have granular cytoplasm. Both types send processes to blood vessels, where they induce capillaries to form the tight junctions that form the blood-brain barrier (see Chapter 32). They also send processes that envelop synapses and the surface of nerve cells. They have a membrane potential that varies with the external K^+ concentration but do not generate propagated potentials. They produce substances that are trophic to neurons, and they probably help maintain the appropriate concentration of substances in the interstitial fluid by taking up K^+ and the neurotransmitters glutamate and γ-aminobutyrate (GABA; see Chapter 4).

3

Excitable Tissue: Muscle

INTRODUCTION

Muscle cells, like neurons, can be excited chemically, electrically, and mechanically to produce an action potential that is transmitted along their cell membrane. Unlike neurons, they have a contractile mechanism that is activated by the action potential. The contractile proteins actin and myosin are abundant in muscle, where they bring about contraction. However, they are not unique. They are found in all cells, and in different forms they are responsible for cell motility, mitosis, and movement of components within cells (see Chapter 1).

Muscle is generally divided into 3 types, **skeletal, cardiac,** and **smooth,** although smooth muscle is not a homogeneous single category. Skeletal muscle makes up the great mass of the somatic musculature. It has well-developed cross-striations, does not normally contract in the absence of nervous stimulation, lacks anatomic and functional connections between individual muscle fibers, and is generally under voluntary control. Cardiac muscle also has cross-striations, but it is functionally syncytial and contracts rhythmically in the absence of external innervation owing to the presence in the myocardium of pacemaker cells that discharge spontaneously. Smooth muscle lacks cross-striations. The type found in most hollow viscera is functionally syncytial and contains pacemakers that discharge irregularly. The type found in the eye and in some other locations is not spontaneously active and resembles skeletal muscle.

SKELETAL MUSCLE

MORPHOLOGY

Organization

Skeletal muscle is made up of individual muscle fibers that are the "building blocks" of the muscular

system in the same sense that the neurons are the building blocks of the nervous system. Most skeletal muscles begin and end in tendons, and the muscle fibers are arranged in parallel between the tendinous ends, so that the force of contraction of the units is additive. Each muscle fiber is a single cell, multinucleated, long, cylindric, and surrounded by a cell membrane, the **sarcolemma** (Fig 3–1). There are no syncytial bridges between cells. The muscle fibers are made up of myofibrils, which are divisible into individual filaments. The filaments are made up of the contractile proteins.

The contractile mechanism in skeletal muscle depends on the proteins **myosin** (molecular weight 460,000), **actin** (molecular weight 43,000), **tropomyosin** (molecular weight 70,000), and **troponin.** Troponin is made up of 3 subunits, **troponin I, troponin T,** and **troponin C.** The 3 subunits have molecular weights ranging from 18,000 to 35,000. Another protein, α-actinin, which has a molecular weight of 190,000, binds actin to the Z lines (see below), and other proteins play important roles in linking excitation to contraction.

Striations

Differences in the refractive indexes of the various parts of the muscle fiber are responsible for the characteristic cross-striations seen in skeletal muscle. The parts of the cross-striations are identified by letters (Fig 3–2). The light I band is divided by the dark Z line, and the dark A band has the lighter H band in its center. A transverse M line is seen in the middle of the H band, and this line plus the narrow light areas on either side of it are sometimes called the pseudo-H zone. The area between 2 adjacent Z lines is called a **sarcomere.** The arrangement of thick and thin filaments that is responsible for the striations is diagrammed in Fig 3–3. The thick filaments, which are about twice the diameter of the thin filaments, are made up of myosin; the thin filaments are made up of actin, tropomyosin, and troponin. The thick filaments are lined up to form the A bands, whereas the array of thin filaments forms the less dense I bands. The lighter H bands in the center of the A bands are the regions where, when the muscle is relaxed, the thin filaments do not overlap the thick filaments. The Z lines transect the fibrils and connect to the thin filaments.

Figure 3–1. Mammalian skeletal muscle. A single muscle fiber surrounded by its sarcolemma has been cut away to show individual myofibrils (1). The cut surface of the myofibrils shows the arrays of thick and thin filaments. The sarcoplasmic reticulum (2) with its terminal cisterns (3) surrounds each myofibril. The T system of tubules (4), which invaginates from the sarcolemma, contacts the myofibrils between the A and I bands twice in every sarcomere. The T system and the adjacent cisterns of the sarcoplasmic reticulum constitute a triad. A basal lamina (5) surrounds the sarcolemma. (6) Mitochondria. (Modified and reproduced, with permission, from Krstic RV: *Ultrastructure of the Mammalian Cell.* Springer-Verlag, 1979.)

If a transverse section through the A band is examined under the electron microscope, each thick filament is found to be surrounded by 6 thin filaments in a regular hexagonal pattern.

Myosin is a complex actin-binding protein. The form in muscle is myosin-II, with 2 globular heads and a long tail (see Fig 1–9). As noted in Chapter 1, myosin-II is also found in other cells along with single-headed myosin-I. Myosin-II is made up of 2

heavy chains and 4 light chains. The light chains and the N-terminal portions of the heavy chains combine to form the globular heads (Fig 3–3). These heads contain an actin-binding site and a catalytic site that hydrolyzes ATP (see below). The molecules are arranged as shown in Fig 3–3, with the heads of the myosin molecules forming cross-bridges to the actin molecules. The myosin molecules are arranged symmetrically on either side of the center of the

Figure 3–2. Electron micrograph of human gastrocnemius muscle. The various bands and lines are identified at the top. (× 13,500.) (Courtesy of SM Walker and GR Schrodt.)

Figure 3–3. Top left: Arrangement of thin (actin) and thick (myosin) filaments in skeletal muscle. **Top right:** Sliding of actin on myosin during contraction so that Z lines move closer together. **Middle right:** Detail of relation of myosin to actin. Note that myosin thick filaments reverse polarity at the M line in the middle of the sarcomere. (Modified from Alberts B et al: *Molecular Biology of the Cell,* 2nd ed. Garland, 1989.) **Bottom left:** Myosin-II molecule made up of 2 intertwined heavy chains and 4 light chains. (Modified from Sheetz MP: Muscle-bound bacteria and weak worms. *Nature* 1988;**331**:212.) **Bottom right:** Diagrammatic representation of the arrangement of actin, tropomyosin, and the 3 subunits of troponin (I, C, and T).

sarcomere, and it is this arrangement that creates the light areas in the pseudo-H zone. The M line is due to a central bulge in each of the thick filaments. At these points, there are slender cross-connections that hold the thick filaments in proper array. There are several hundred myosin molecules in each thick segment.

The thin filaments are made up of 2 chains of globular units that form a long double helix. Tropomyosin molecules are long filaments located in the groove between the 2 chains in the actin (Fig 3–3). Each thin filament contains 300–400 actin molecules and 40–60 tropomyosin molecules. Troponin molecules are small globular units located at intervals along the tropomyosin molecules. Troponin T binds the other troponin components to tropomyosin, troponin I inhibits the interaction of myosin with actin (see below), and troponin C contains the binding sites for the Ca^{2+} that initiates contraction.

Sarcotubular System

The muscle fibrils are surrounded by structures made up of membrane that appear in electron photomicrographs as vesicles and tubules. These structures form the **sarcotubular system,** which is made up of a **T system** and a **sarcoplasmic reticulum.** The T system of transverse tubules, which is continuous with the membrane of the muscle fiber, forms a grid perforated by the individual muscle fibrils (Fig 3–1). The space between the 2 layers of the T system is an extension of the extracellular space. The sarcoplasmic reticulum, which forms an irregular curtain around each of the fibrils, has enlarged **terminal cisterns** in close contact with the T system at the junctions between the A and I bands. At these points of contact, the arrangement of the central T system with a cistern of the sarcoplasmic reticulum on either side has led to the use of the term **triads** to describe the system. The function of the T system, which is continuous with the sarcolemma, is the rapid transmission of the action potential from the cell membrane to all the fibrils in the muscle. The sarcoplasmic reticulum is concerned with Ca^{2+} movement and muscle metabolism (see below).

ELECTRICAL PHENOMENA & IONIC FLUXES

Electrical Characteristics of Skeletal Muscle

The electrical events in skeletal muscle and the ionic fluxes underlying them are similar to those in nerve, although there are quantitative differences in timing and magnitude. The resting membrane potential of skeletal muscle is about –90 mV. The action potential lasts 2–4 ms and is conducted along the muscle fiber at about 5 m/s. The absolute refractory period is 1–3 ms long, and the after-polarizations, with their related changes in threshold to electrical stimulation, are relatively prolonged. The initiation of impulses at the myoneural junction is discussed in Chapter 4.

Although the electrical properties of the individual fibers in a muscle do not differ sufficiently to produce anything resembling a compound action potential, there are slight differences in the thresholds of the various fibers. Furthermore, in any stimulation experiment, some fibers are farther from the stimulating electrodes than others. Therefore, the size of the action potential recorded from a whole-muscle preparation is proportionate to the intensity of the stimulating current between threshold and maximal current intensities.

Ion Distribution & Fluxes

The distribution of ions across the muscle fiber membrane is similar to that across the nerve cell membrane. The values for the various ions and their equilibrium potentials are shown in Table 3–1. As in nerves, depolarization is a manifestation of Na^+ influx, and repolarization is a manifestation of K^+ efflux (as described in Chapter 2 for nerves).

CONTRACTILE RESPONSES

It is important to distinguish between the electrical and mechanical events in muscle. Although one response does not normally occur without the other, their physiologic basis and characteristics are different. Muscle fiber membrane depolarization normally starts at the motor end-plate, the specialized structure under the motor nerve ending (see Chapter 4); the action potential is transmitted along the muscle fiber and initiates the contractile response.

The Muscle Twitch

A single action potential causes a brief contraction followed by relaxation. This response is called a **muscle twitch.** In Fig 3–4, the action potential and the twitch are plotted on the same time scale. The twitch starts about 2 ms after the start of depolarization of the membrane, before repolarization is complete. The duration of the twitch varies with the type of muscle being tested (see below). "Fast" muscle fibers, primarily those concerned with fine, rapid, precise movement, have twitch durations as short as 7.5 ms. "Slow" muscle fibers, principally those involved in strong, gross, sustained movements, have twitch durations up to 100 ms.

Molecular Basis of Contraction

The process by which the shortening of the contractile elements in muscle is brought about is a sliding of the thin filaments over the thick filaments. The width of the A bands is constant, whereas the Z lines move closer together when the muscle contracts and farther apart when it is stretched (Fig 3–3). As the muscle shortens, the thin filaments from the opposite

Table 3–1. Steady-state distribution of ions in the intracellular and extracellular compartments of mammalian skeletal muscle, and the equilibrium potentials for these ions.[1]

| Ion[2] | Concentration, mmol/L | | Equilibrium Potential (mV) |
	Intracellular Fluid	Extracellular Fluid	
Na^+	12	145	+65
K^+	155	4	–95
H^+	13×10^{-5}	3.8×10^{-5}	–32
Cl^-	3.8	120	–90
HCO_3^-	8	27	–32
A^-	155	0	...

Membrane potential = –90 mV

[1]Data from Ruch TC, Patton HD (editors): *Physiology and Biophysics,* 19th ed. Saunders, 1965.
[2]A^- represents organic anions. The value for intracellular Cl^- is calculated from the membrane potential, using the Nernst equation.

Figure 3–4. The electrical and mechanical responses of a mammalian skeletal muscle fiber to a single maximal stimulus. The electrical response (mV potential change) and the mechanical response (T, tension in arbitrary units) are plotted on the same abscissa (time).

ends of the sarcomere approach each other; when the shortening is marked, these filaments overlap.

The sliding during muscle contraction is produced by breaking and re-forming of the cross-linkages between actin and myosin. The heads of the myosin molecules link to actin at a 90-degree angle, produce movement of myosin on actin by swiveling, and then disconnect and reconnect at the next linking site, repeating the process in serial fashion (Fig 3–5). Each single cycle of attaching, swiveling, and detaching shortens the muscle 1%. Each thick filament has about 500 myosin heads, and each of these cycles about 5 times per second during a rapid contraction.

The immediate source of energy for muscle contraction is ATP. Hydrolysis of the high-energy phosphate in this molecule bonds is catalyzed by the **adenosine triphosphatase (ATPase)** activity in the heads of the myosin molecules, where they are in contact with actin.

The process by which depolarization of the muscle fiber initiates contraction is called **excitation-contraction coupling.** The action potential is transmitted to all the fibrils in the fiber via the T system. It triggers the release of Ca^{2+} from the terminal cisterns, the lateral sacs of the sarcoplasmic reticulum next to the T system. The Ca^{2+} initiates contraction.

Ca^{2+} initiates contraction by binding to troponin C. In resting muscle, troponin I is tightly bound to actin and tropomyosin covers the sites where myosin heads bind to actin. Thus, the troponin-tropomyosin complex constitutes a "relaxing protein" that inhibits the interaction between actin and myosin. When the Ca^{2+} released by the action potential binds to troponin C, the binding of troponin I to actin is presumably weakened, and this permits the tropomyosin to move laterally (Fig 3–5). This movement uncovers binding sites for the myosin heads. ATP is then split and contraction occurs. Seven myosin-binding sites are uncovered for each molecule of troponin that binds a calcium ion.

Shortly after releasing Ca^{2+}, the sarcoplasmic reticulum begins to reaccumulate it by actively transporting it into the longitudinal portions of the reticulum. From there, it diffuses into the terminal cisterns, where it is stored until released by the next action potential. Once the Ca^{2+} concentration outside the reticulum has been lowered sufficiently, chemical interaction between myosin and actin ceases and the muscle relaxes. If transport of Ca^{2+} into the reticulum is inhibited, relaxation does not occur even though there are no more action potentials; the resulting sustained contraction is called a **contracture.** ATP provides the energy for the active transport of Ca^{2+} into the sarcoplasmic reticulum. Thus, both contraction and relaxation of muscle require ATP.

Depolarization of the T tubule membrane activates the sarcoplasmic reticulum via **dihydropyridine receptors,** which are voltage-gated Ca^{2+} channels in the T tubule membrane. In cardiac muscle, influx of Ca^{2+} via these channels triggers the release of Ca^{2+} stored in the sarcoplasmic reticulum, but in skeletal muscle, Ca^{2+} entry from the ECF by this route is not required for Ca^{2+} release. Instead, the dihydropyridine receptor serves as the voltage sensor and trigger that unlocks release of Ca^{2+} from the nearby sarcoplasmic reticulum. The dihydropyridine receptor gets its name from the fact that it is blocked by the drug dihydropyridine. It has 4 homologous regions, each of which spans the T tubule membrane 6 times. The Ca^{2+} channel in the sarcoplasmic reticulum that opens to permit the outpouring of Ca^{2+} is called the **ryanodine receptor** because it is locked in the open position by the plant alkaloid ryanodine. It is closely related to the IP_3 receptor, a ligand-gated Ca^{2+} channel that, when it binds IP_3, permits Ca^{2+} to enter the cytoplasm from the endoplasmic reticulum (see

Figure 3–5. Initiation of muscle contraction by Ca^{2+}. The cross-bridges (heads of myosin molecules) attach to binding sites on actin (striped areas) and swivel when tropomyosin is displaced laterally by binding of Ca^{2+} to troponin C.

Chapter 1). The pump that actively transports Ca^{2+} back into the sarcoplasmic reticulum, ending muscle contraction, is Ca^{2+}-Mg^{2+} ATPase.

The events involved in muscle contraction and relaxation are summarized in Table 3–2.

Types of Contraction

Muscular contraction involves shortening of the contractile elements, but because muscles have elastic and viscous elements in series with the contractile mechanism, it is possible for contraction to occur without an appreciable decrease in the length of the whole muscle. Such a contraction is called **isometric** ("same measure" or length). Contraction against a constant load, with approximation of the ends of the muscle, is **isotonic** ("same tension").

A whole-muscle preparation arranged for recording isotonic contractions is shown in Fig 3–6A. The muscle lifts the lever, and the distance the lever moves indicates the degree of shortening. In this situation, the muscle does external work, since the lever is being moved a certain distance. The muscle preparation in Fig 3–6B is arranged for recording isometric contractions. It is attached to an electronic force transducer, and the force that is generated is recorded. Since the product of force times distance in this situation is essentially zero, no external work is done by the muscle. In other situations, it is possible for muscles to do negative work while contracting (Fig 3–7). This happens, for example, when a heavy weight is lowered onto a table. In this case, the biceps muscle actively resists the descent of the object, but the net effect of the effort is to lengthen the biceps muscle while it is contracting.

Figure 3–6. **A:** Muscle preparation arranged for recording isotonic contractions. **B:** Preparation arranged for recording isometric contractions. In **A,** the muscle is fastened to a writing lever that swings on a pivot. In **B,** it is attached to an electronic transducer that measures the force generated without permitting the muscle to shorten.

Table 3–2. Sequence of events in contraction and relaxation of skeletal muscle.

Steps in contraction[1]
- (1) Discharge of motor neuron.
- (2) Release of transmitter (acetylcholine) at motor end-plate.
- (3) Binding of acetylcholine to nicotinic acetylcholine receptors.
- (4) Increased Na^+ and K^+ conductance in end-plate membrane.
- (5) Generation of end-plate potential.
- (6) Generation of action potential in muscle fibers.
- (7) Inward spread of depolarization along T tubules.
- (8) Release of Ca^{2+} from terminal cisterns of sarcoplasmic reticulum and diffusion to thick and thin filaments.
- (9) Binding of Ca^{2+} to troponin C, uncovering myosin binding sites on actin.
- (10) Formation of cross-linkages between actin and myosin and sliding of thin on thick filaments, producing shortening.

Steps in relaxation
- (1) Ca^{2+} pumped back into sarcoplasmic reticulum.
- (2) Release of Ca^{2+} from troponin.
- (3) Cessation of interaction between actin and myosin.

[1]Steps 1–6 in contraction are discussed in Chapter 4.

Summation of Contractions

The electrical response of a muscle fiber to repeated stimulation is like that of nerve. The fiber is electrically refractory only during the rising and part of the falling phase of the spike potential. At this time, the contraction initiated by the first stimulus is just beginning. However, because the contractile mechanism does not have a refractory period, repeated stimulation before relaxation has occurred produces additional activation of the contractile elements and a response that is added to the contraction already present. This phenomenon is known as **summation of contractions.** The tension developed during summation is considerably greater than that during the single muscle twitch. With rapidly repeated stimulation, activation of the contractile mechanism occurs repeatedly before any relaxation has occurred, and the individual responses fuse into one continuous contraction. Such a response is called a **tetanus (tetanic contraction).** It is a **complete tetanus** when there is no relaxation between stimuli and an **incom-**

Figure 3–7. Top: Isotonic (free) contraction. The biceps shortens freely, and the weight is lifted. **Bottom:** Isometric contraction. The biceps generates force but cannot shorten and raise weight.

plete tetanus when there are periods of incomplete relaxation between the summated stimuli. During a complete tetanus, the tension developed is about 4 times that developed by the individual twitch contractions. The development of an incomplete and a complete tetanus in response to stimuli of increasing frequency is shown in Fig 3–8.

The stimulation frequency at which summation of contractions occurs is determined by the twitch duration of the particular muscle being studied. For example, if the twitch duration is 10 ms, frequencies less than 1/10 ms (100/s) cause discrete responses interrupted by complete relaxation, and frequencies greater than 100/s cause summation.

Treppe

When a series of maximal stimuli is delivered to skeletal muscle at a frequency just below the tetanizing frequency, there is an increase in the tension developed during each twitch until, after several contractions, a uniform tension per contraction is reached. This phenomenon is known as **treppe,** or the "staircase" phenomenon (German *Treppe,* staircase).

It also occurs in cardiac muscle. Treppe is believed to be due to increased availability of Ca^{2+} for binding to troponin C. It should not be confused with summation of contractions and tetanus.

Relation Between Muscle Length, Tension, & Velocity of Contraction

Both the tension that a muscle develops when stimulated to contract isometrically (the **total tension**) and the **passive tension** exerted by the unstimulated muscle vary with the length of the muscle fiber. This relationship can be studied in a whole skeletal muscle preparation such as that shown in Fig 3–6B. The length of the muscle can be varied by changing the distance between its 2 attachments. At each length, the passive tension is measured, the muscle is then stimulated electrically, and the total tension is measured. The difference between the 2 values at any length is the amount of tension actually generated by the contractile process, the **active tension.** The records obtained by plotting passive tension and total tension against muscle length are shown in Fig 3–9. Similar curves are obtained when single muscle fibers are studied. Passive tension rises slowly at first and then rapidly as the muscle is stretched. The muscle ruptures when it is stretched to about 3 times its **equilibrium length,** ie, the length of the relaxed muscle cut free from its bony attachments.

The total tension curve rises to a maximum and then declines until it reaches the passive tension curve, ie, until no additional tension is developed upon further stimulation. The length of the muscle at which the active tension is maximal is usually called its **resting length.** The term comes originally from experiments demonstrating that the length of many of the muscles in the body at rest is the length at which they develop maximal tension. Other definitions of resting length are sometimes used, but that given here is the most widely accepted and has the greatest physiologic validity.

The observed length-tension relation in skeletal muscle is explained by the sliding filament mechanism of muscle contraction. When the muscle fiber contracts isometrically, the tension developed is proportionate to the number of cross-linkages between the actin and the myosin molecules. When muscle is stretched, the overlap between actin and myosin is re-

Figure 3–8. Tetanus. Isometric tension of a single muscle fiber during continuously increasing and decreasing stimulation frequency. Dots at the top are at intervals of 0.2 seconds. (Reproduced, with permission, from Buchthal. F: *Dan Biol Med* 1942;**17:**1.)

Figure 3–9. Length-tension relationship for the human triceps muscle. The passive tension curve measures the tension exerted by this skeletal muscle at each length when it is not stimulated. The total tension curve represents the tension developed when the muscle contracts isometrically in response to a maximal stimulus. The active tension is the difference between the two.

duced and the number of cross-linkages is reduced. Conversely, when the muscle is appreciably shorter than resting length, the thin filaments overlap, and this also reduces the cross-linkages.

The velocity of muscle contraction varies inversely with the load on the muscle. At a given load, the velocity is maximal at the resting length and declines if the muscle is shorter or longer than this length.

ENERGY SOURCES & METABOLISM

Muscle contraction requires energy, and muscle has been called "a machine for converting chemical energy into mechanical work." The immediate source of this energy is the energy-rich organic phosphate derivatives in muscle; the ultimate source is the intermediary metabolism of carbohydrate and lipids. The hydrolysis of ATP to provide the energy for contraction has been discussed above.

Phosphorylcreatine

ATP is resynthesized from ADP by the addition of a phosphate group. Some of the energy for this endothermic reaction is supplied by the breakdown of glucose to CO_2 and H_2O, but there also exists in muscle another energy-rich phosphate compound that can supply this energy for short periods. This compound is **phosphorylcreatine** (see Fig 17–21), which is hydrolyzed to creatine and phosphate groups with the release of considerable energy. At rest, some ATP in the mitochondria transfers its phosphate to creatine,

so that a phosphorylcreatine store is built up. During exercise, the phosphorylcreatine is hydrolyzed at the junction between the myosin heads and the actin, forming ATP from ADP and thus permitting contraction to continue.

Carbohydrate & Lipid Breakdown

At rest and during light exercise, muscles utilize lipids in the form of free fatty acids (FFA; see Chapter 17) as their energy source. As the intensity of exercise increases, lipids alone cannot supply energy fast enough and so utilization of carbohydrate becomes the predominant component in the muscle fuel mixture. Thus, during exercise, much of the energy for phosphorylcreatine and ATP resynthesis comes from the breakdown of glucose to CO_2 and H_2O. An outline of the major metabolic pathways involved is presented in Chapter 17. For the purposes of the present discussion, it is sufficient to point out that glucose in the bloodstream enters cells, where it is degraded through a series of chemical reactions to pyruvate. Another source of intracellular glucose, and consequently of pyruvate, is glycogen, the carbohydrate polymer that is especially abundant in liver and skeletal muscle. When adequate O_2 is present, pyruvate enters the citric acid cycle and is metabolized—through this cycle and the so-called respiratory enzyme pathway—to CO_2 and H_2O. This process is called **aerobic glycolysis.** The metabolism of glucose or glycogen to CO_2 and H_2O liberates sufficient energy to form large quantities of ATP from ADP. If O_2 supplies are insufficient, the pyruvate formed from glucose does not enter the tricarboxylic acid cycle but is reduced to lactate. This process of **anaerobic glycolysis** is associated with the net production of much smaller quantities of energy-rich phosphate bonds, but it does not require the presence of O_2. The various reactions involved in supplying energy to skeletal muscle are summarized in Fig 3–10.

The Oxygen Debt Mechanism

During muscular exercise, the muscle blood vessels dilate and blood flow is increased so that the available O_2 supply is increased. Up to a point, the increase in O_2 consumption is proportionate to the energy expended, and all the energy needs are met by aerobic processes. However, when muscular exertion is very great, aerobic resynthesis of energy stores cannot keep pace with their utilization. Under these conditions, phosphorylcreatine is still used to resynthesize ATP. Some ATP synthesis is accomplished by using the energy released by the anaerobic breakdown of glucose to lactate. However, use of the anaerobic pathway is self-limiting because in spite of rapid diffusion of lactate into the bloodstream, enough accumulates in the muscles to eventually exceed the capacity of the tissue buffers and produce an enzyme-inhibiting decline in pH. However, for short periods, the presence of an anaerobic pathway for

$$ATP + H_2O \rightarrow ADP + H_3PO_4 + 7.3 \, kcal$$

$$Phosphorylcreatine + ADP \rightleftharpoons Creatine + ATP$$

Glucose + 2 ATP (or glycogen + 1 ATP)

$$\xrightarrow[\text{Anaerobic}]{} 2 \, Lactic \, acid + 4 \, ATP$$

Glucose + 2 ATP (or glycogen + 1 ATP)

$$\xrightarrow[\text{Oxygen}]{} 6 \, CO_2 + 6 \, H_2O + 40 \, ATP$$

$$FFA \xrightarrow[\text{Oxygen}]{} CO_2 + H_2O + ATP$$

Figure 3–10. Energy released by hydrolysis of 1 mol of ATP and reactions responsible for resynthesis of ATP. The amount of ATP formed per mole of free fatty acid (FFA) oxidized is large but varies with the size of the FFA. For example, complete oxidation of 1 mol of palmitic acid generates 140 mol of ATP.

glucose breakdown permits muscular exertion of a far greater magnitude than would be possible without it. For example, in a 100-m dash that takes 10 seconds, 85% of the energy consumed is derived anaerobically; in a 2-mile race that takes 10 minutes, 20% of the energy is derived anaerobically; and in a long-distance race that takes 60 minutes, only 5% of the energy comes from anaerobic metabolism.

After a period of exertion is over, extra O_2 is consumed to remove the excess lactate, replenish the ATP and phosphorylcreatine stores, and replace the small amounts of O_2 that have come from myoglobin. The amount of extra O_2 consumed is proportionate to the extent to which the energy demands during exertion exceeded the capacity for the aerobic synthesis of energy stores, ie, the extent to which an **oxygen debt** was incurred. The O_2 debt is measured experimentally by determining O_2 consumption after exercise until a constant, basal consumption is reached and subtracting the basal consumption from the total. The amount of this debt may be 6 times the basal O_2 consumption, which indicates that the subject is capable of 6 times the exertion that would have been possible without it. Obviously, the maximal debt can be incurred rapidly or slowly; violent exertion is possible for only short periods, whereas less strenuous exercise can be carried on for longer periods.

Trained athletes are able to increase the O_2 consumption of their muscles to a greater degree than untrained individuals and are able to utilize FFA more effectively. Consequently, they are capable of greater exertion without depleting their glycogen stores and increasing their lactate production. Because of this, they contract smaller oxygen debts for a given amount of exertion. They have also learned to gorge on carbohydrates for several days before a competitive event, increasing their muscle stores of glycogen. This alone can greatly increase their endurance.

Rigor

When muscle fibers are completely depleted of ATP and phosphorylcreatine, they develop a state of extreme rigidity called **rigor.** When this occurs after death, the condition is called rigor mortis. In rigor, almost all of the myosin heads attach to actin but in an abnormal, fixed, and resistant way.

Heat Production in Muscle

Thermodynamically, the energy supplied to a muscle must equal its energy output. The energy output appears in work done by the muscle, in energy-rich phosphate bonds formed for later use, and in heat. The overall mechanical efficiency of skeletal muscle (work done/total energy expenditure) ranges up to 50% while lifting a weight during isotonic contraction and is essentially 0% during isometric contraction. Energy storage in phosphate bonds is a small factor. Consequently, heat production is considerable. The heat produced in muscle can be measured accurately with suitable thermocouples.

Resting heat, the heat given off at rest, is the external manifestation of basal metabolic processes. The heat produced in excess of resting heat during contraction is called the **initial heat.** This is made up of **activation heat,** the heat that muscle produces whenever it is contracting, and **shortening heat,** which is proportionate in amount to the distance the muscle shortens. Shortening heat is apparently due to some change in the structure of the muscle during shortening.

Following contraction, heat production in excess of resting heat continues for as long as 30 minutes. This **recovery heat** is the heat liberated by the metabolic processes that restore the muscle to its precontraction state. The recovery heat of muscle is approximately equal to the initial heat; ie, the heat produced during recovery is equal to the heat produced during contraction.

If a muscle that has contracted isotonically is restored to its previous length, extra heat in addition to recovery heat is produced (**relaxation heat**). External work must be done on the muscle to return it to its previous length, and relaxation heat is mainly a manifestation of this work.

Fiber Types in Muscle

There are variations in the myosin ATPase activity, metabolism, and contractile properties of the different fibers that make up muscles. The properties of the 3 main types of fibers found in skeletal muscle in mammals are summarized in Table 3–3. In humans, type IIA fibers are rare, and in general, slow muscles are type I and fast muscles are type IIB.

Muscles containing many type I fibers are called **red muscles** because they are darker than other muscles. The red muscles, which respond slowly and have a long latency, are adapted for long, slow, posture-maintaining contractions. The long muscles of

Table 3–3. Classification of fiber types in skeletal muscles.[1]

	Type I	Type IIB	Type IIA
Other names	Slow oxidative; red.	Fast glycolytic; white.	Fast oxidative; red.
Myosin isoenzyme ATPase activity	Slow.	Fast.	Fast.
Ca^{2+} pumping capacity of sarcoplasmic reticulum	Moderate.	High.	High.
Diameter	Moderate.	Large.	Small.
Glycolytic capacity	Moderate.	High.	High.
Oxidative capacity (correlates with content of mitochondria, capillary density, myoglobin content)	High.	Low.	High.

[1] Modified from Murphy RA: Muscle. In: *Physiology*, 2nd ed. Berne RM, Levy MN (editors). Mosby, 1988.

the back are red muscles. **White muscles,** which contain mostly type IIB fibers, have short twitch durations and are specialized for fine, skilled movement. The extraocular muscles and some of the hand muscles contain many type IIB fibers and are generally classified as white muscles.

Protein Isoforms in Muscle & Their Genetic Control

The differences in the fibers that make up muscles stem from differences in the proteins in them. Myosin, actin, tropomyosin, and troponin T all exist in multiple **isoforms,** ie, forms with similar biologic activity but different amino acid composition. There is also heterogeneity in cardiac and smooth muscle. Actin and tropomyosin each have α and β isoforms, and there are at least 7 different isoforms of the myosin heavy chains (MHC). The MHC isoforms, which differ somewhat in myosin ATPase activity, are encoded by a family of genes. In addition, an α MHC and a β MHC are found in the heart in humans. The expression of the MHCs is precisely regulated during development, and alterations in expression play a major role in the assembly of muscles. In adults, the proportions of the various isoforms determine the functional characteristics of each muscle. In addition, changes in muscle function can be produced by alterations in activity, innervation, and hormonal milieu, and these changes are generally produced by alterations in the transcription of MHC genes. Examples include conversion of fast to slow skeletal muscle when muscle innervation is changed (see below) and marked alterations in the isoforms in skeletal and cardiac muscle in hypothyroidism. Finally, since the isoforms have similar functions, the existence of mul-

tiple isoforms encoded by different genes protects against a deleterious mutation in one of the genes.

PROPERTIES OF MUSCLES IN THE INTACT ORGANISM

Effects of Denervation

In the intact animal or human, healthy skeletal muscle does not contract except in response to stimulation of its motor nerve supply. Destruction of this nerve supply causes muscle atrophy. It also leads to abnormal excitability of the muscle and increases its sensitivity to circulating acetylcholine (denervation hypersensitivity; see Chapter 4). Fine, irregular contractions of individual fibers (**fibrillations**) appear. If the motor nerve regenerates, these disappear. Such contractions usually are not visible grossly and should not be confused with **fasciculations,** which are jerky, visible contractions of groups of muscle fibers that occur as a result of pathologic discharge of spinal motor neurons.

The Motor Unit

Since the axons of the spinal motor neurons supplying skeletal muscle each branch to innervate several muscle fibers, the smallest possible amount of muscle that can contract in response to the excitation of a single motor neuron is not one muscle fiber but all the fibers supplied by the neuron. Each single motor neuron and the muscle fibers it innervates constitute a **motor unit.** The number of muscle fibers in a motor unit varies. In muscles such as those of the hand and those concerned with motion of the eye— ie, muscles concerned with fine, graded, precise movement—there are 3–6 muscle fibers per motor unit. On the other hand, values of 120–165 fibers per unit have been reported in cat leg muscles, and some of the large muscles of the back in humans probably contain even more.

Each spinal motor neuron innervates only one kind of muscle fiber, so that all of the muscle fibers in a motor unit are of the same type. On the basis of the type of muscle fiber they innervate (Table 3–3), and thus on the basis of the duration of their twitch contraction, motor units are divided into fast and slow units. In general, slow muscle units are innervated by small, slowly conducting motor neurons and fast units by large, rapidly conducting motor neurons (**size principle**). In large limb muscles, the small, slow units are first recruited in most movements, are resistant to fatigue, and are the most frequently used units. The fast units, which are more easily fatigued, are generally recruited with more forceful movements.

The differences between types of muscle units are not inherent but are determined by, among other things, their activity (see below). When the nerve to a slow muscle is cut and replaced with the nerve to a fast muscle, the nerve regenerates and innervates the slow muscle. However, the muscle becomes fast and there are corresponding changes in its MHC isoforms and myosin ATPase activity. There has been speculation that this change is due to trophic substances flowing from the nerve to the muscle, but it is apparently due instead to changes in the pattern of activity of the muscle. In stimulation experiments, it has been demonstrated that changes in the expression of MHC genes and consequently of MHC isoforms can be produced by changes in the pattern of electrical activity used to stimulate the muscle.

Denervation of skeletal muscle leads to atrophy and flaccid paralysis, with the appearance of fibrillations. These effects are the classic consequences of a **lower motor neuron lesion.** The muscle also becomes hypersensitive to acetylcholine (denervation hypersensitivity; see Chapter 4).

Electromyography

Activation of motor units can be studied by **electromyography,** the process of recording the electrical activity of muscle on a cathode-ray oscilloscope. This may be done in unanesthetized humans by using small metal disks on the skin overlying the muscle as the pick-up electrodes or by using hypodermic needle electrodes. The record obtained with such electrodes is the **electromyogram (EMG).** With needle electrodes, it is usually possible to pick up the activity of single muscle fibers. A typical EMG is shown in Fig 3–11.

Factors Responsible for Grading of Muscular Activity

It has been shown by electromyography that there is little if any spontaneous activity in the skeletal muscles of normal individuals at rest. With minimal voluntary activity a few motor units discharge, and with increasing voluntary effort more and more are brought into play. This process is sometimes called **recruitment of motor units.** Gradation of muscle response is therefore in part a function of the number of motor units activated. In addition, the frequency of discharge in the individual nerve fibers plays a role, the tension developed during a tetanic contraction being greater than that during individual twitches. The length of the muscle is also a factor. Finally, the motor units fire asynchronously, ie, out of phase with each other. This asynchronous firing causes the individual muscle fiber responses to merge into a smooth contraction of the whole muscle.

The Strength of Skeletal Muscles

Human skeletal muscle can exert 3–4 kg of tension per cm^2 of cross-sectional area. This figure is about the same as that obtained in a variety of experimental animals and seems to be constant for all mammalian species. Since many of the muscles in humans have a relatively large cross-sectional area, the tension they can develop is quite large. The gastrocnemius, for example, not only supports the weight of the whole body during climbing but resists a force several times this great when the foot hits the ground during running or jumping. An even more striking example is the gluteus maximus, which can exert a tension of 1200 kg. The total tension that could be developed by all muscles in the body of an adult man is approximately 22,000 kg (nearly 25 tons).

Body Mechanics

Body movements are generally organized in such a way that they take maximal advantage of the physiologic principles outlined above. For example, the attachments of the muscles in the body are such that many of them are normally at or near their resting length when they start to contract. In muscles that extend over more than one joint, movement at one joint may compensate for movement at another in such a way that relatively little shortening of the muscle occurs during contraction. Nearly isometric contractions of this type permit development of maximal tension per contraction. The hamstring muscles extend from the pelvis over the hip joint and the knee joint to the tibia and fibula. Hamstring contraction produces flexion of the leg on the thigh. If the thigh is flexed on the pelvis at the same time, the lengthening of the hamstrings across the hip joint tends to compensate for the shortening across the knee joint. In the course of various activities, the body moves in a way that takes advantage of this. Such factors as momentum and balance are integrated into body movement in ways that make possible maximal motion with minimal muscular exertion. One net effect is that the stress put on tendons and bones is rarely over 50% of their failure strength, protecting them from damage.

In walking, each limb passes rhythmically through a support or stance phase when the foot is on the

Biceps

500 μV

Triceps

0.5 s

Figure 3–11. Electromyographic tracings from human biceps and triceps muscles during alternate flexion and extension of the elbow. (Courtesy of BC Garoutte.)

ground and a swing phase when the foot is off the ground. The support phases of the 2 legs overlap, so that there are 2 periods of double support during each cycle. There is a brief burst of activity in the leg flexors at the start of each step, and then the leg is swung forward with little more active muscular contraction. Therefore, the muscles are active for only a fraction of each step, and walking for long periods causes relatively little fatigue.

A young adult walking at a comfortable pace moves at a velocity of about 80 m/min and generates a power output of 150–175 watts per step. A group of young adults asked to walk at their most comfortable rate selected a velocity close to 80 m/min, and it was found that they had selected the velocity at which their energy output was minimal. Walking more rapidly or more slowly took more energy.

Even though walking is a complex activity, it is common knowledge that it is carried out more or less automatically. Experiments in animals indicate that it is organized in preprogrammed nerve pathways within the spinal cord and that it is activated by some sort of command signal in a fashion that is probably analogous to the initiation of patterns of activity by the discharge of command neurons in invertebrates.

Muscular Dystrophy

Duchenne's muscular dystrophy and a similar but milder disease, **Becker's muscular dystrophy,** are X-linked congenital diseases characterized by progressive muscle weakness. Duchenne's muscular dystrophy occurs in about 1 of every 3000 male infants and is fatal. A major recent advance was the isolation of the gene that is defective in these diseases and the characterization of its product, the protein **dystrophin.** Dystrophin is a large protein (molecular weight 427,000) located on the inner surface of the sarcolemma near the triads. Its function is unknown. However, it is absent in Duchenne's muscular dystrophy and present but at low levels or abnormal in size in Becker's muscular dystrophy. In Duchenne's muscular dystrophy, muscle Ca^{2+} channels are open more of the time than they are in normal muscle, and Ca^{2+} gradually accumulates in the muscles, denaturing their protein and causing necrosis. Thus, dystrophin may be involved in the regulation of Ca^{2+} channels. The dystrophin gene is also expressed in cardiac and smooth muscle and, with an alternative promoter and first exon, in the brain.

CARDIAC MUSCLE

MORPHOLOGY

The striations in cardiac muscle are similar to those in skeletal muscle, and Z lines are present. The muscle fibers branch and interdigitate, but each is a complete unit surrounded by a cell membrane. Where the end of one muscle fiber abuts on another, the membranes of both fibers parallel each other through an extensive series of folds. These areas, which always occur at Z lines, are called **intercalated disks** (Fig 3–12). They provide a strong union between fibers, maintaining cell-to-cell cohesion, so that the pull of one contractile unit can be transmitted along its axis to the next. Along the sides of the muscle fibers next to the disks, the cell membranes of adjacent fibers fuse for considerable distances, forming gap junctions. These junctions provide low-resistance bridges for the spread of excitation from one fiber to another (see Chapter 1). They permit cardiac muscle to function as if it were a syncytium, even though there are no protoplasmic bridges between cells. The T system in cardiac muscle is located at the Z lines rather than at the A-I junction, where it is located in mammalian skeletal muscle.

Like skeletal muscle, cardiac muscle contains myosin, actin, tropomyosin, and troponin in various forms. It also contains large numbers of elongated mitochondria in close contact with the fibrils.

ELECTRICAL PROPERTIES

Resting Membrane & Action Potentials

The resting membrane potential of individual mammalian cardiac muscle cells is about −90 mV (interior negative to exterior). Stimulation produces a propagated action potential that is responsible for initiating contraction. Depolarization proceeds rapidly, and an overshoot is present, as in skeletal muscle and nerve, but this is followed by a plateau before the membrane potential returns to the baseline (Fig 3–13). In mammalian hearts, depolarization lasts about 2 ms, but the plateau phase and repolarization last 200 ms or more. Repolarization is therefore not complete until the contraction is half over. With extracellular recording, the electrical events include a spike and a later wave that resemble the QRS complex and T wave of the ECG.

As in other excitable tissues, changes in the external K^+ concentration affect the resting membrane potential of cardiac muscle, whereas changes in the external Na^+ concentration affect the magnitude of the action potential. The initial rapid depolarization and

Figure 3–12. ***Top:*** Electron photomicrograph of cardiac muscle. The fuzzy thick lines are intercalated disks (× 12,000). (Reproduced, with permission, from Bloom W. Fawcett DW: *A Textbook of Histology,* 10th ed. Saunders, 1975). ***Bottom:*** Diagram of cardiac muscle as seen under the light microscope ***(A)*** and the electron microscope ***(B).*** N, nucleus. (Reproduced, with permission, from Braunwald E. Ross J. Sonnenblick EH: Mechanisms of contraction of the normal and failing heart. *N Engl J Med* 1967;**277:**794. Courtesy of Little, Brown, Inc.)

Figure 3–13. Action potentials and contractile response of mammalian cardiac muscle fiber plotted on the same time axis. ARP, absolute refractory period; RRP, relative refractory period.

the overshoot (phase 0) are due to a rapid increase in Na$^+$ conductance similar to that occurring in nerve and skeletal muscle (Fig 3–14). The initial rapid repolarization (phase 1) is due to closure of Na$^+$ channels and Cl$^-$ influx. The subsequent prolonged plateau (phase 2) is due to a slower but prolonged opening of voltage-gated Ca^{2+} channels. Final repolarization (phase 3) is due to closure of the Ca^{2+} chan-

Figure 3–14. Phases of the action potential of a cardiac muscle fiber and the corresponding changes in ionic conductances across the muscle membrane. 0, depolarization; 1, initial rapid repolarization; 2, plateau phase; 3, late rapid repolarization; 4, baseline. (Modified from Shepherd JT, Vanhoutte PM: *The Human Cardiovascular System: Facts and Concepts.* Raven Press, 1979.)

nels and K$^+$ efflux through 2 kinds of K$^+$ channels. This restores the resting potential (phase 4). The fast Na$^+$ channel in cardiac muscle probably has 2 gates, an outer gate that opens at the start of depolarization, at a membrane potential of –70 to –80 mV, and a second inner gate that then closes and precludes further influx until the action potential is over (Na$^+$ channel inactivation). The slow Ca^{2+} channel is activated at a membrane potential of –30 to –40 mV. The types of ion channels in cardiac muscle are summarized in Table 3–4.

In cardiac muscle, the repolarization time decreases as the cardiac rate increases. At a cardiac rate of 75 beats per minute, the duration of the action potential (0.25 second) is almost 70% longer than it is at a cardiac rate of 200 beats per minute (0.15 second).

MECHANICAL PROPERTIES

Contractile Response

The contractile response of cardiac muscle begins just after the start of depolarization and lasts about 1½ times as long as the action potential (Fig 3–13). The role of Ca^{2+} in excitation-contraction coupling is similar to its role in skeletal muscle (see above). However, as noted above, it is the influx of Ca^{2+} that is triggered by activation of the dihydropyridine channels in the T system and not simply the depolarization that triggers release of stored Ca^{2+} from the sarcoplasmic reticulum. As in skeletal muscle, this release via the ryanodine channels initiates contraction.

During phases 0–2 and about half of phase 3 (until the membrane potential reaches approximately –50 mV during repolarization), cardiac muscle cannot be excited again; ie, it is in its **absolute refractory period** (Fig 3–13). It remains relatively refractory until phase 4. Therefore, tetanus of the type seen in skeletal muscle cannot occur. Of course, tetanization of

Table 3–4. Types of gating of cardiac ion channels.[1]

Voltage-gated[2]
Na$^+$ channel
T Ca^{2+} channel
L Ca^{2+} channel
K$^+$ channels
Ligand-gated, G protein-dependent
Acetylcholine-sensitive K$^+$ channel
Adenosine-sensitive K$^+$ channel
Ligand-gated, G protein-independent
ATP-sensitive K$^+$ channel
Stretch-activated channels

[1]Modified from Opie LH: *The Heart: Physiology and Metabolism,* 2nd ed. Raven, 1991.
[2]T (transient) Ca^{2+} channels open when the muscle is more depolarized, whereas the L (long-lasting) channels require less depolarization.

cardiac muscle for any length of time would have lethal consequences, and in this sense, the fact that cardiac muscle cannot be tetanized is a safety feature. Ventricular muscle is said to be in the "vulnerable period" just at the end of the action potential, because stimulation at this time will sometimes initiate ventricular fibrillation.

Isoforms

Cardiac muscle is generally slow and has relatively low ATPase activity. Its fibers are dependent on oxidative metabolism and hence on a continuous supply of O_2. In rats, there are α and β MHCs in the cardiac muscle fibers. In the ventricle, α MHC predominates in adulthood, but in hypothyroid animals it is largely replaced by β MHC, which has a lower myosin ATPase activity. Thyroid deficiency also alters the MHC patterns in skeletal muscle, but the pattern of change varies from muscle to muscle. The human heart also contains α MHC and β MHC; α MHC predominates in the atria and β MHC in the ventricles.

Correlation Between Muscle Fiber Length & Tension

The relation between initial fiber length and total tension in cardiac muscle is similar to that in skeletal muscle; there is a resting length at which the tension developed upon stimulation is maximal. In the body, the initial length of the fibers is determined by the degree of diastolic filling of the heart, and the pressure developed in the ventricle is proportionate to the total tension developed **(Starling's law of the heart;** see Chapter 29). As the diastolic filling increases, the force of contraction of the ventricles is increased (Fig 3–15). The homeo-

static value of this response is discussed in Chapter 29.

The force of contraction of cardiac muscle is also increased by catecholamines (see Chapters 13 and 20), and this increase occurs without a change in muscle length. The increase, which is called the positively inotropic effect of catecholamines, is mediated via innervated β_1-adrenergic receptors and cyclic AMP (see Chapter 1). The heart also contains noninnervated β_2-adrenergic receptors, which also act via cyclic AMP, but their inotropic effect is smaller and is maximum in the atria. Cyclic AMP activates protein kinase A, and this leads to phosphorylation of the voltage-dependent Ca^{2+} channels, causing them to spend more time in the open state. Cyclic AMP also increases the active transport of Ca^{2+} to the sarcoplasmic reticulum, thus accelerating relaxation and consequently shortening systole. This is important when the cardiac rate is increased, because it permits adequate diastolic filling (see Chapter 29). Digitalis glycosides increase cardiac contractions by inhibiting the Na^+-K^+ ATPase in cell membranes of the muscle fibers. The resultant increase in intracellular Na^+ increases the efflux of Na^+ in exchange for Ca^{2+} via the Ca^{2+}-Na^+ antiport in the cell membrane (see Chapter 1).

METABOLISM

Mammalian hearts have an abundant blood supply, numerous mitochondria, and a high content of myoglobin, a muscle pigment that may function as an O_2 storage mechanism (see Chapter 35). Normally, less than 1% of the total energy liberated is provided by anaerobic metabolism. During hypoxia, this figure may increase to nearly 10%; but under totally anaerobic conditions, the energy liberated is inadequate to sustain ventricular contractions. Under basal conditions, 35% of the caloric needs of the human heart are provided by carbohydrate, 5% by ketones and amino acids, and 60% by fat. However, the proportions of substrates utilized vary greatly with the nutritional state. After ingestion of large amounts of glucose, more lactate and pyruvate are used; during prolonged starvation, more fat is used. Circulating free fatty acids normally account for almost 50% of the lipid utilized. In untreated diabetics, the carbohydrate utilization of cardiac muscle is reduced and that of fat increased. The factors affecting the O_2 consumption of the human heart are discussed in Chapter 29.

PACEMAKER TISSUE

The heart continues to beat after all nerves to it are sectioned; indeed, if the heart is cut into pieces, the pieces continue to beat. This is because of the presence in the heart of specialized pacemaker tissue that

Figure 3–15. Length-tension relationship for cardiac muscle. The values are for dog heart.

can initiate repetitive action potentials. The pacemaker tissue makes up the conduction system that normally spreads impulses throughout the heart. It is characterized by an unstable membrane potential that slowly decreases after each impulse until the firing level is reached and another impulse is generated (see Chapter 28).

SMOOTH MUSCLE

MORPHOLOGY

Smooth muscle is distinguished anatomically from skeletal and cardiac muscle because it lacks visible cross-striations. Actin and myosin-II are present, and they slide on each other to produce contraction. However, they are not arranged in regular arrays, as in skeletal and cardiac muscle, and so the striations are absent. Smooth muscle also contains tropomysin, but troponin appears to be absent. The isoforms of actin and myosin differ from those in skeletal muscle. There is a sarcoplasmic reticulum, but it is poorly developed. In general, smooth muscles contain few mitochondria and depend to a large extent on glycolysis for their metabolic needs.

Types

There are various types of smooth muscle in the body. In general, smooth muscle can be divided into **visceral smooth muscle** and **multi-unit smooth muscle.** Visceral smooth muscle occurs in large sheets, has low-resistance bridges between individual muscle cells, and functions in a syncytial fashion. The bridges, like those in cardiac muscle, are junctions where the membranes of the 2 adjacent cells fuse to form gap junctions. Visceral smooth muscle is found primarily in the walls of hollow viscera. The musculature of the intestine, the uterus, and the ureters are examples. Multi-unit smooth muscle is made up of individual units without interconnecting bridges. It is found in structures such as the iris of the eye, in which fine, graded contractions occur. It is usually not under voluntary control, but it has many functional similarities to skeletal muscle.

VISCERAL SMOOTH MUSCLE

Electrical & Mechanical Activity

Visceral smooth muscle is characterized by the instability of its membrane potential and by the fact that it shows continuous, irregular contractions that are independent of its nerve supply. This maintained state of partial contraction is called **tonus** or **tone.**

The membrane potential has no true "resting" value, being relatively low when the tissue is active and higher when it is inhibited, but in periods of relative quiescence it averages about –50 mV. Superimposed on the membrane potential are waves of various types (Fig 3–16). There are slow sine wave-like fluctuations a few millivolts in magnitude and spikes that sometimes overshoot the zero potential line and sometimes do not. In many tissues, the spikes have a duration of about 50 ms. However, in some tissues the action potentials have a prolonged plateau during repolarization, like the action potentials in cardiac muscle. The spikes may occur on the rising or falling phases of the sine wave oscillations. There are, in addition, pacemaker potentials similar to those found in the cardiac pacemakers. However, in visceral smooth muscle, these potentials are generated in multiple foci that shift from place to place. Spikes generated in the pacemaker foci are conducted for some distance in the muscle. Because of the continuous activity, it is difficult to study the relation between the electrical and mechanical events in visceral smooth muscle, but in some relatively inactive preparations, a single spike can be generated. The muscle starts to contract about 200 ms after the start of the spike and 150 ms after the spike is over. The peak contraction is reached as long as 500 ms after the spike. Thus, the excitation-contraction coupling in visceral smooth muscle is a very slow process compared with that in skeletal and cardiac muscle, in which the time from initial depolarization to initiation of contraction is less than 10 ms.

Molecular Basis for Contraction

Ca^{2+} is involved in the initiation of contraction of

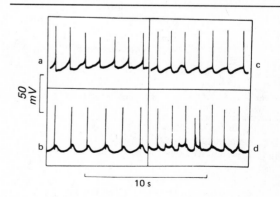

Figure 3–16. Spontaneous electrical activity in individual smooth muscle cells of teniae coli of guinea pig colon. **a:** Pacemaker type; **b:** sinusoidal waves with action potentials on the rising phases; **c:** sinusoidal waves with action potentials on the falling phases; **d:** mixture of pacemaker, sinusoidal, and action potentials. (Reproduced, with permission, from Bulbring E: Physiology and pharmacology of intestinal smooth muscle. *Lectures on the Scientific Basis of Medicine* 1957;**7**:374.)

smooth muscle, as it is in skeletal muscle. However since visceral smooth muscle generally has a poorly developed sarcoplasmic reticulum, the increase in intracellular Ca^{2+} concentration that initiates contraction is due primarily to Ca^{2+} influx from the ECF via voltage-gated Ca^{2+} channels. In addition, the myosin in smooth muscle must be phosphorylated for activation of the myosin ATPase. Phosphorylation and dephosphorylation of myosin also occur in skeletal muscle, but phosphorylation is not necessary for activation of the ATPase. In smooth muscle, Ca^{2+} binds to calmodulin (see Chapter 1), and the resulting complex activates **calmodulin-dependent myosin light chain kinase,** the enzyme that catalyzes the phosphorylation of myosin. Actin then slides on myosin, producing contraction. This is in contrast to skeletal and cardiac muscle, where contraction is triggered by the binding of Ca^{2+} to troponin C.

Myosin is dephosphorylated by phosphatases in the cell. However, this does not necessarily lead to relaxation of the smooth muscle. Instead, it appears that smooth muscle has a **latch bridge** mechanism by which dephosphorylated myosin cross bridges remain attached to actin for some time after the cytoplasmic Ca^{2+} concentration falls. This produces sustained contraction, which is especially important in vascular smooth muscle. Relaxation of the muscle presumably occurs when there is final dissociation of the Ca^{2+}-calmodulin complex or when some other mechanism comes into play. The events leading to contraction and relaxation of visceral smooth muscle are summarized in Table 3–5. The events in multiunit smooth muscle are generally similar.

It is worth highlighting the differences between cardiac and vascular smooth muscle, since both are involved in cardiovascular control. In the heart, the responses are phasic, with contraction alternating with relaxation, whereas in smooth muscle, contraction is often tonic because of the latch bridge mechanism. Furthermore, increases in intracellular cyclic AMP increase the force of contraction of cardiac muscle, whereas cyclic AMP relaxes vascular smooth muscle because it inhibits the phosphorylation of myosin light chain kinase.

Stimulation

Visceral smooth muscle is unique in that, unlike other types of muscle, it contracts when stretched in the absence of any extrinsic innervation. Stretch is followed by a decline in membrane potential, an increase in the frequency of spikes, and a general increase in tone.

If epinephrine or norepinephrine is added to a preparation of intestinal smooth muscle arranged for recording of intracellular potentials in vitro, the membrane potential usually becomes larger, the spikes decrease in frequency, and the muscle relaxes (Fig 3–17). Norepinephrine is the chemical mediator

Table 3–5. Sequence of events in contraction and relaxation of visceral smooth muscle.[1]

(1)	Binding of acetylcholine to muscarinic receptors
(2)	Increased influx of Ca^{2+} into the cell
(3)	Activation of calmodulin-dependent myosin light chain kinase
(4)	Phosphorylation of myosin
(5)	Binding of myosin to actin and increased myosin ATPase activity
(6)	Contraction
(7)	Dephosphorylation of myosin by various phosphatases
(8)	Relaxation, or sustained contraction due to the latch bridge mechanism

[1]Compare with Table 3–2. Smooth muscle can also remain contracted for longer periods because of the formation of latch bridges.

INTESTINAL SMOOTH MUSCLE

UTERINE SMOOTH MUSCLE

Figure 3–17. ***Top:*** Effects of various agents on the membrane potential of intestinal smooth muscle. ***Bottom:*** Effects of estrogens and of progesterone on the membrane potential of the uterus of the ovariectomized rabbit or rat.

released at noradrenergic nerve endings (see Chapter 4), and stimulation of the noradrenergic nerves to the preparation produces inhibitory potentials (see Chapter 4). Stimulation of the noradrenergic nerves to the intestine inhibits contractions in vivo. Norepinephrine exerts both α and β actions (see Chapter 4) on the muscle. The β action, reduced muscle tension in response to excitation, is mediated via cyclic AMP (see Chapter 1) and is probably due to increased intracellular binding of Ca^{2+}. The α action, which is also inhibition of contraction, is associated with increased Ca^{2+} efflux from the muscle cells.

Acetylcholine has an effect opposite to that of norepinephrine on the membrane potential and contractile activity of intestinal smooth muscle. If acetylcholine is added to the fluid bathing a smooth muscle preparation in vitro, the membrane potential decreases and the spikes become more frequent (Fig 3–17). The muscle becomes more active, with an increase in tonic tension and the number of rhythmic contractions. The effect is mediated by phospholipidase C and IP_3, which increases the intracellular Ca^{2+} concentration. In the intact animal, stimulation of cholinergic nerves causes release of acetylcholine, excitatory potentials (see Chapter 4), and increased intestinal contractions. In vitro, similar effects are produced by cold and stretch.

Function of the Nerve Supply to Smooth Muscle

The effects of acetylcholine and norepinephrine on visceral smooth muscle serve to emphasize 2 of its important properties: (1) its spontaneous activity in the absence of nervous stimulation, and (2) its sensitivity to chemical agents released from nerves locally or brought to it in the circulation. In mammals, visceral muscle usually has a dual nerve supply from the 2 divisions of the autonomic nervous system. The structure and function of the contacts between these nerves and smooth muscle are discussed in Chapter 4. The function of the nerve supply is not to initiate activity in the muscle but rather to modify it. Stimulation of one division of the autonomic nervous system usually increases smooth muscle activity, whereas stimulation of the other decreases it. However, in some organs, noradrenergic stimulation increases and cholinergic stimulation decreases smooth muscle activity; in others, the reverse is true.

Other chemical agents also affect smooth muscle. An interesting example is the uterus. Uterine smooth muscle is relatively inexcitable during diestrus and in the ovariectomized animal. During estrus or in the estrogen-treated ovariectomized animal, excitability is enhanced and tonus and spontaneous contractions occur. However, estrogen increases rather than decreases the membrane potential (Fig 3–17). Progesterone increases the membrane potential even further and inhibits the electrical and contractile activity of uterine muscle (see Chapter 23).

Relation of Length to Tension; Plasticity

Another special characteristic of smooth muscle is the variability of the tension it exerts at any given length. If a piece of visceral smooth muscle is stretched, it first exerts increased tension (see above). However, if the muscle is held at the greater length after stretching, the tension gradually decreases. Sometimes the tension falls to or below the level exerted before the muscle was stretched. It is consequently impossible to correlate length and developed tension accurately, and no resting length can be assigned. In some ways, therefore, smooth muscle behaves more like a viscous mass than a rigidly structured tissue, and it is this property that is referred to as the **plasticity** of smooth muscle.

The consequences of plasticity can be demonstrated in intact humans. For example, the tension exerted by the smooth muscle walls of the bladder can be measured at different degrees of distention as fluid is infused into the bladder via a catheter, as shown in Fig 38–27. Initially there is relatively little increase in tension as volume is increased, because of the plasticity of the bladder wall. However, a point is eventually reached at which the bladder contracts forcefully.

MULTI-UNIT SMOOTH MUSCLE

Unlike visceral smooth muscle, multi-unit smooth muscle is nonsyncytial and contractions do not spread widely through it. Because of this, the contractions of multi-unit smooth muscle are more discrete, fine, and localized than those of visceral smooth muscle. Like visceral smooth muscle, multi-unit smooth muscle is very sensitive to circulating chemical substances and is normally activated by chemical mediators (acetylcholine and norepinephrine) released at the endings of its motor nerves. Norepinephrine in particular tends to persist and to cause repeated firing of the muscle after a single stimulus rather than a single action potential. Therefore, the contractile response produced is usually an irregular tetanus rather than a single twitch. When a single twitch response is obtained, it resembles the twitch contraction of skeletal muscle except that its duration is 10 times as long.

4

Synaptic & Junctional Transmission

INTRODUCTION

The all or none type of conduction seen in axons and skeletal muscle has been discussed in Chapters 2 and 3. Impulses are transmitted from one nerve cell to another cell at **synapses** (Fig 4–1). These are the junctions where the axon or some other portion of one cell (the **presynaptic cell)** terminates on the dendrites, soma, or axon of another neuron (Fig 4–2) or in some cases a muscle or gland cell (the **postsynaptic cell).** It is worth nothing noting that in some situations dendrites as well as axons can be presynaptic as well as postsynaptic. Transmission at most synaptic junctions is chemical; the impulse in the presynaptic axon causes secretion of **neurotransmitter** such as acetylcholine or serotonin. This chemical mediator binds to receptors on the surface of the postsynaptic cell, and this triggers events that open or close channels in the membrane of the postsynaptic cell. At some of the junctions, however, transmission is electrical, and at a few conjoint synapses it is both electrical and chemical. In any case, transmission is not a simple jumping of one action potential from the presynaptic to the postsynaptic cell. The effects of discharge at individual synaptic endings can be excitatory or inhibitory, and when the postsynaptic cell is a neuron, the summation of all the excitatory and inhibitory effects determines whether an action potential is generated. Thus, synaptic transmission is a complex process that permits the grading and adjustment of neural activity necessary for normal function.

In electrical synapses, the membranes of the presynaptic and postsynaptic neurons come close together, and gap junctions form between the cells (see Chapter 1). Like the intercellular junctions in other tissues, these junctions form low-resistance bridges through which ions pass with relative ease. Electrical and conjoint synapses occur in mammals, and there is electrical coupling, for example, between some of the neurons in the lateral vestibular nucleus. However, most synaptic transmission is chemical. Unless other-wise specified, consideration in this chapter is limited to chemical transmission.

Transmission from nerve to muscle resembles chemical synaptic transmission from one neuron to another. The **neuromuscular junction,** the specialized area where a motor nerve terminates on a skeletal muscle fiber, is the site of a stereotyped transmission process. The contacts between autonomic neurons and smooth and cardiac muscle are less specialized, and transmission in these locations is a more diffuse process.

SYNAPTIC TRANSMISSION

FUNCTIONAL ANATOMY

There is considerable variation in the anatomic structure of synapses in various parts of the mammalian nervous system. The ends of the presynaptic fibers are generally enlarged to form **terminal buttons (synaptic knobs)** (Fig 4–1). Endings are commonly located on dendrites (Fig 4–2) and sometimes on **dendritic spines**, which are small knobs projecting from dendrites (Fig 4–3). In some instances, the terminal branches of the axon of the presynaptic neuron form a basket or net around the soma of the postsynaptic cell ("basket cells" of the cerebellum and autonomic ganglia). In other locations, they intertwine with the dendrites of the postsynaptic cell (climbing fibers of the cerebellum) or end on the dendrites directly (apical dendrites of cortical pyramids) or on the axons (axo-axonal endings). On average, each neuron divides to form 1000 synaptic endings, and since there are 10^{12} neurons in the human brain, it follows that there are about 10^{15} synapses. Obviously, the number of pathways that an impulse can take in such a complex network is extremely large. In the spinal cord, the presynaptic endings are closely applied to the soma and the proximal portions of the dendrites

Figure 4–1. Synapses on a typical motor neuron. The neuron has dendrites (1), an axon (2), and a prominent nucleus (3). Note that rough endoplasmic reticulum extends into the dendrites but not into the axon. Many different axons converge on the neuron, and their terminal buttons form axodendritic (4) and axosomatic (5) synapses. (6) Myelin sheath. (Reproduced, with permission, from Krstic RV: *Ultrastructure of the Mammalian Cell.* Springer-Verlag, 1979.)

Figure 4–2. Electron photomicrograph of synaptic knob (S) ending on a dendrite (D) in the central nervous system, P, postsynaptic thickening; M, mitochondrion. (× 56,000; courtesy of DM McDonald.)

of the postsynaptic neuron. The number of synaptic knobs applied to a single spinal motor neuron has been calculated to be about 10,000, with 2000 on the cell body and 8000 on the dendrites. Indeed, there are so many knobs that the neuron appears to be encrusted with them. The portion of the membrane covered by any single synaptic knob is small, but the synaptic knobs are so numerous that, in aggregate, the area covered by them all is often 40% of the soma membrane area (Fig 4–1) and 75% of the dendritic membrane area. It has been calculated that in the cerebral cortex, 98% of the synapses are on dendrites and only 2% are on cell bodies. The ratio of synapses to neurons in the human forebrain has been calculated to be 40,000:1.

Synaptic Knobs

Under the electron microscope, the synaptic knobs at synapses where transmission is chemical are found to be separated from the soma of the postsynaptic cell by a definite **synaptic cleft** 30–50 nm wide. The synaptic knob and the soma each have an intact membrane. Inside the knob, there are many mitochondria and small clear and granulated vesicles (Figs 4–2 and 4–3), the clear vesicles being especially numerous in the part of the knob closest to the synaptic cleft. The vesicles contain small "packets" of the chemical transmitter responsible for synaptic transmission (see below). They vary in morphology depending on the particular transmitter they contain. At most synapses, there is also some degree of postsynaptic specialization in the form of a postsynaptic density or web under the membrane of the postsynaptic cell.

The transmitter is released from the synaptic knobs when action potentials pass along the axon to the endings. The membranes of the vesicles or granules fuse to the nerve cell membrane, and the area of fusion breaks down, releasing the contents by the process of exocytosis (see Chapter 1). Ca^{2+} triggers this process, and the action potential opens voltage-gated Ca^{2+} channels in the synaptic knob. Ca^{2+} enters the ending, and exocytosis is increased. The action of Ca^{2+} is terminated by its rapid sequestration within the ending. The amount of transmitter released is proportionate to the Ca^{2+} influx.

At the neuromuscular junction (see below) and other cholinergic synapses, there appear to be 2 types of transmitter release. Granulated vesicles that con-

Figure 4–3. Axodendritic, axo-axonal, and axosomatic synapses. Many presynaptic neurons terminate on dendritic spines, as shown at the top, but some also end directly on the shafts of dendrites. Note the presence of clear and granulated synaptic vesicles in endings and clustering of clear vesicles at active zones.

tain polypeptides are located in all parts of the nerve terminals and discharge their contents by exocytosis from many different parts of the ending. Clear vesicles that contain acetylcholine are located close to the synaptic cleft. Discharge is by exocytosis, but granule contents are released very rapidly into the cliff along dense bars called **active zones.**

The walls of clear synaptic vesicles contain **synapsins,** a group of at least 4 closely related proteins with highly homologous N-terminal domains and divergent C-terminal domains. There is reason to believe that at least one of these tethers synaptic vesicles to the cytoskeleton near active zones and that Ca^{2+} acts via Ca^{2+}/calmodulin kinase II (see Table 1–9) to phosphorylate this synapsin. The phosphorylation probably loosens the association to the cytoskeleton, permitting the synaptic vesicles to move to the membrane, where they fuse and release their contents. Rapid release of other small neurotransmitter molecules may occur in a similar fashion. However, granulated vesicles and secretory granules in endocrine glands do not contain synapsins.

Convergence & Divergence

Only a few of the synaptic knobs on a postsynaptic neuron are endings of any single presynaptic neuron. The inputs to the cell are multiple. In spinal motor neurons, for example, some inputs come directly from the dorsal root, some from the long descending spinal tracts, and many from **interneurons,** the short interconnecting neurons of the spinal cord. Thus, many presynaptic neurons **converge** on any single postsynaptic neuron. Conversely, the axons of most presynaptic neurons divide into many branches that **diverge** to end on many postsynaptic neurons. Convergence and divergence are the anatomic substrates for facilitation, occlusion, and reverberation (see below).

One-Way Conduction

Synapses generally permit conduction of impulses in one direction only, from the presynaptic to the postsynaptic neurons. An impulse conducted antidromically up the axons of the ventral root dies out after depolarizing the cell bodies of the spinal motor neurons. Since axons will conduct in either direction with equal facility, the one-way gate at the synapses is necessary for orderly neural function. Chemical mediation at synaptic junctions explains one-way conduction. The mediator is located in the synaptic knobs of the presynaptic fibers but not in the postsynaptic membrane. Therefore, an impulse arriving at the postsynaptic membrane cannot release synaptic mediator. Progression of impulse traffic occurs only when the action potential arrives in the presynaptic terminals and causes secretion of stored chemical transmitter.

Synaptic Development

A fascinating question that has attracted a great deal of attention is how, during development, neurons find the "right" targets and make the "right" synaptic connections. It now seems clear that a number of different factors are involved. It is well established that chemical signals permit neurons to recognize their targets and ignore other neurons in the same region, and there are trophic interactions between pre- and postsynaptic cells once the connections have been made. However, it has been demonstrated that in many situations, many neurons make synaptic connections with other neurons or skeletal muscles and then the "inappropriate" connections disappear, leaving only the connections found in the adult. Many neurons die during development, and it may be the most active neurons and synaptic junctions that persist. It is now established that genetically programmed cell death (**apoptosis**) is part of normal development in many different species. In addition, there is competition between neurons for synaptic sites, and nerve endings from adjacent neurons grow into a brain area that has been denervated.

ELECTRICAL EVENTS AT SYNAPSES

Synaptic activity in the spinal cord has been studied in detail by inserting a microelectrode into the soma of a motor neuron and recording the electrical events that follow stimulation of the excitatory and inhibitory inputs to these cells. Activity at other synapses has also been studied, and many of the events occurring at these synapses are similar to those occurring at spinal synapses.

Penetration of an anterior horn cell is achieved by advancing a microelectrode through the ventral portion of the spinal cord. Puncture of a cell membrane is signaled by the appearance of a steady 70-mV potential difference between the microelectrode and an electrode outside the cell. The cell can be identified as a spinal motor neuron by stimulating the appropriate ventral root and observing the electrical activity of the cell. Such stimulation initiates an antidromic impulse (see Chapter 2) that is conducted to the soma and stops at this point. Therefore, the presence of an action potential in the cell after antidromic stimulation indicates that the cell that has been penetrated is a motor neuron rather than an interneuron. Activity in some of the presynaptic terminals impinging on the impaled spinal motor neuron (Fig 4–4) can be initiated by stimulating the dorsal roots.

Excitatory Postsynaptic Potentials

Single stimuli applied to the sensory nerves in the experimental situation described above characteristically do not lead to the formation of a propagated action potential in the postsynaptic neuron. Instead, the

Figure 4–4. Arrangement of recording electrodes and stimulators for studying synaptic activity in spinal motor neurons in mammals. One stimulator (S_2) is used to produce antidromic impulses for identifying the cell; the other (S_1) is used to produce orthodromic stimulation via reflex pathways.

stimulation produces either a transient partial depolarization or a transient hyperpolarization.

The initial depolarizing response produced by a single stimulus to the proper input begins about 0.5 ms after the afferent impulse enters the spinal cord. It reaches its peak 1–1.5 ms later and then declines exponentially, with a **time constant** (time required for the response to decay to 1/e, or 1/2.718, of its maximum) that varies depending on the transmitter and the properties of the postsynaptic membrane. During this potential, the excitability of the neuron to other stimuli is increased, and consequently the potential is called an **excitatory postsynaptic potential (EPSP).**

The EPSP is due to depolarization of the postsynaptic cell membrane immediately under the active synaptic knob. The area of inward current flow thus created is so small that it will not drain off enough positive charges to depolarize the whole membrane. Instead, an EPSP is inscribed. The EPSP due to activity in one synaptic knob is small, but the depolarizations produced by each of the active knobs summate.

Summation may be **spatial** or **temporal.** When activity is present in more than one synaptic knob at the same time, spatial summation occurs and activity in one synaptic knob is said to **facilitate** activity in another to approach the firing level. Temporal summation occurs if repeated afferent stimuli cause new EPSPs before previous EPSPs have decayed. Obviously, the longer the time constant for the EPSP, the greater the opportunity for summation. Spatial and temporal facilitation are illustrated in Fig 4–5. The EPSP is therefore not an all or none response but is proportionate in size to the strength of the afferent stimulus. If the EPSP is large enough to reach the firing level of the cell, a full-fledged action potential is produced.

Synaptic Delay

When an impulse reaches the presynaptic terminals, there is an interval of at least 0.5 ms, the **synaptic delay,** before a response is obtained in the postsynaptic neuron. The delay following maximal stimulation of the presynaptic neuron corresponds to the latency of the EPSP and is due to the time it takes for the synaptic mediator to be released and to act on the membrane of the postsynaptic cell. Because of it, conduction along a chain of neurons is slower if there are many synapses in the chain than if there are only a few. Since the minimum time for transmission across one synapse is 0.5 ms, it is also possible to determine whether a given reflex pathway is monosynaptic or polysynaptic (contains more than one synapse) by measuring the delay in transmission from the dorsal to the ventral root across the spinal cord.

Ionic Basis of EPSPs

When transmitters that exert an excitatory effect cross the synaptic cleft and bind to appropriate postsynaptic receptors, they usually trigger the opening of chemically gated ion channels clustered in large numbers in the membrane of the postsynaptic cell under the synaptic knob. It is important to remember that whereas axons typically have only voltage-gated Na^+ channels and K^+ channels, the cell bodies, dendrites, and axonal endings have many different kinds of chemically gated channels. The type of postsynaptic response produced by a transmitter depends on the type of channel that is associated with and activated by that particular transmitter.

The production of EPSPs by acetylcholine at nicotinic synapses where acetylcholine is an excitatory transmitter provides a good example of the operation of these mechanisms. Acetylcholine binds to nicotinic receptors, and this triggers the opening of channels that permit Na^+ and other small cations to pass with relative ease. Na^+ moves along its concentration and electrical gradients into the cell (see Chapter 2), and an EPSP is produced (Fig 4–6). However, the area in which this influx occurs is so small that the repolarizing forces are able to overcome its influence, and runaway depolarization of the whole membrane does not result. If more excitatory synaptic knobs are active, more Na^+ enters and the depolarizing potential is greater. If Na^+ influx is great enough, the firing level is reached and a propagated action potential results.

EPSPs can also be produced by agents that close K^+ channels.

Inhibitory Postsynaptic Potentials

An EPSP is usually produced by afferent stimulation, but stimulation of certain presynaptic fibers regularly initiates a hyperpolarizing response in spinal motor neurons. This response often begins 1–1.25 ms after the afferent stimulus enters the cord, reaches its peak in 1.5–2 ms, and declines exponentially with a time constant of about 3 ms (Fig 4–7). Longer hyperpolarizing potentials are also observed (see below). During this potential, the excitability of the

Figure 4–5. Spatial summation **(A–C)** and temporal summation **(D–F)** of EPSPs. Records are potential changes recorded with one electrode inside the postsynaptic cell. In A–C, afferent volleys of increasing strength were delivered. More and more synaptic knobs were activated, and in C, the firing level was reached and an action potential generated. In D–F, 2 different volleys of the same strength were delivered, but the time interval between them was shortened. In F, the firing level was reached and an action potential generated.

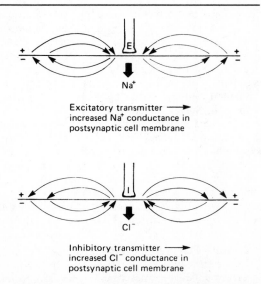

Figure 4–6. Summary of some of the events occurring at synapses in mammals.

Figure 4–7. IPSPs produced by stimulating inhibitory input to a neuron when its membrane potential is set at various values with a voltage clamp. RMP, resting membrane potential of this neuron. Note that when the voltage is set at E_{Cl}, the IPSP disappears and that at higher membrane voltages it becomes positive.

neuron to other stimuli is decreased; consequently, it is called an **inhibitory postsynaptic potential (IPSP).** Spatial summation of IPSPs occurs, as shown by the increasing size of the response as the strength of an inhibitory afferent volley is increased. Temporal summation also occurs. This type of inhibition is called **postsynaptic** or **direct inhibition.**

Ionic Basis of IPSPs

The IPSP is often due to a localized increase in membrane permeability to Cl^- but not to Na^+. When an inhibitory synaptic knob becomes active, the released transmitter triggers the opening of Cl^- channels in the area of the postsynaptic cell membrane under the knob. Cl^- moves down its concentration gradient. The net effect is the transfer of negative charge into the cell, so that the membrane potential increases. However, the permeability change is short-lived, and resting conditions are rapidly restored.

The decreased excitability of the nerve cell during the IPSP is due in part to moving the membrane potential away from the firing level. Consequently, more excitatory (depolarizing) activity is necessary to reach the firing level. The fact that an IPSP is mediated by Cl^- can be demonstrated by repeating the stimulus while varying the resting membrane potential of the postsynaptic cell and holding it with a voltage clamp. When the membrane potential is set at E_{Cl}, the potential disappears (Fig 4–7), and at more negative membrane potentials, it becomes positive. However, it is still inhibitory, because the total Cl^- conductance is increased, and it moves only toward E_{Cl},

which is still greater than the resting membrane potential.

IPSPs can also be produced by opening of K^+ channels, with movement of K^+ out of the postsynaptic cell. In addition, they can be produced by closure of Na^+ or Ca^{2+} channels.

Neurons Responsible for Postsynaptic Inhibition

Stimulation of certain sensory nerve fibers known to pass directly to motor neurons in the spinal cord produces EPSPs in these neurons and IPSPs in other neurons. In the inhibitory pathways, a single interneuron is inserted between the afferent dorsal root fiber and the motor neuron. This special interneuron, called a Golgi bottle neuron, is short and plump and has a thick axon. Its synaptic transmitter is glycine, and when this amino acid is secreted from its synaptic knobs to the proximal dendrites or cell body of the postsynaptic neuron, an IPSP is produced owing to an increase in the conductance of the postsynaptic cell membrane to Cl^-. In this way, excitatory input is "converted" into inhibitory input.

Slow Postsynaptic Potentials

In addition to the EPSPs and IPSPs described above, slow EPSPs and IPSPs have been described in autonomic ganglia, cardiac and smooth muscle, and cortical neurons. These postsynaptic potentials have a latency of 100–500 ms and last several seconds. The slow EPSPs are generally due to decreases in K^+ conductance, and the slow IPSPs are due to increases in K^+ conductance. In sympathetic ganglia, there is also a late slow EPSP that has a latency of 1–5 seconds and lasts 10–30 minutes. This potential is also due, at least in part, to decreased K^+ conductance, and the transmitter responsible for the potential is a peptide very closely related to GnRH, the hormone secreted by neurons in the hypothalamus that stimulates LH secretion (see Chapter 14).

Generation of the Action Potential in the Postsynaptic Neuron

The constant interplay of excitatory and inhibitory activity on the postsynaptic neuron produces a fluctuating membrane potential that is the algebraic sum of the hyperpolarizing and depolarizing activity. The soma of the neuron thus acts as a sort of integrator. When the 10–15 mV of depolarization sufficient to reach the firing level is attained, a propagated spike results. However, the discharge of the neuron is slightly more complicated than this. In motor neurons, the portion of the cell with the lowest threshold for the production of the cell of a full-fledged action potential is the **initial segment,** the portion of the axon at and just beyond the axon hillock. This unmyelinated segment is depolarized or hyperpolarized electrotonically by the current sinks and sources under the excitatory and inhibitory synaptic knobs. It is the first part of the neuron to fire, and its discharge is propagated in 2 directions: down the axon and back into the soma. Retrograde firing of the soma in this fashion probably has value in "wiping the slate clean" for subsequent renewal of the interplay of excitatory and inhibitory activity on the cell.

Function of the Dendrites

Dendrites usually do not conduct like the axons. Action potentials are generated in some of the dendrites, but they are usually part of the "receptor membrane" of the neuron (see Chapter 2), the site of current sources or sinks that electrotonically change the membrane potential of the initial segment or other locus from which the action potentials are generated. In the central nervous system and the retina, there are in addition neurons with dendrites but no axons. These cells spread EPSPs and IPSPs from one neuron to another without a propagated action potential. In the olfactory bulb, for example, action potentials in mitral cells activate granule cell dendrites via dendrodendritic synapses. The dendrite then discharges an inhibitory transmitter on the same mitral cell, and as the electrical activity spreads in the granule cell, transmitter is released on other nearby mitral cells. Such microcircuits appear to be common in various parts of the brain.

When the dendritic tree of a neuron is extensive and has multiple presynaptic knobs ending on it, there is room for a great interplay of inhibitory and excitatory activity. Current flow to and from the dendrites waxes and wanes. The role of the dendrites in the genesis of the EEG is discussed in Chapter 11.

Electrical Transmission

At synaptic junctions where transmission is electrical, the impulse reaching the presynaptic terminal generates an EPSP in the postsynaptic cell that, because of the low-resistance bridge between the two, has a much shorter latency than the EPSP at a synapse where transmission is chemical. In conjoint synapses, there is both a short-latency response and a longer-latency, chemically mediated, postsynaptic response.

INHIBITION & FACILITATION AT SYNAPSES

Direct & Indirect Inhibition

Postsynaptic inhibition during the course of an IPSP is also called **direct inhibition** because it is not a consequence of previous discharges of the postsynaptic neuron. Various forms of **indirect inhibition,** inhibition due to the effects of previous postsynaptic neuron discharge, also occur. For example, the postsynaptic cell can be refractory to excitation because it has just fired and is in its refractory period. During after-hyperpolarization it is also less excitable. In spi-

nal neurons, especially after repeated firing, this after-hyperpolarization may be large and prolonged.

Postsynaptic Inhibition in the Spinal Cord

The various pathways in the nervous system that are known to mediate postsynaptic inhibition are discussed in Chapter 6, but one illustrative example is presented here. Afferent fibers form the muscle spindles (stretch receptors) in skeletal muscle are known to pass directly to the spinal motor neurons of the motor units supplying the same muscle. Impulses in this afferent supply cause EPSPs and, with summation, propagated responses in the postsynaptic motor neurons. At the same time, IPSPs are produced in motor neurons supplying the antagonistic muscles. This latter response is mediated by branches of the afferent fibers that end on Golgi bottle neurons. These interneurons, in turn, secrete the inhibitory transmitter glycine at synapses on the proximal dendrites or cell bodies of the motor neurons that supply the antagonist (Fig 4–8). Therefore, activity in the afferent fibers from the muscle spindles excites the motor neurons supplying the muscle from which the impulses come and inhibits those supplying its antagonists (**reciprocal innervation**).

Presynaptic Inhibition & Facilitation

Another type of inhibition occurring in the central

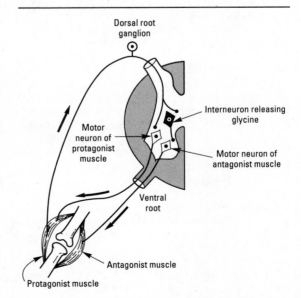

Figure 4–8. Diagram illustrating the probable anatomic connections responsible for inhibiting the antagonists to a muscle contracting in response to stretch. Activity is initiated in the spindle in the protagonist muscle. Impulses pass directly to the motor neurons supplying the same muscle and, via branches, to inhibitory interneurons that end on the motor neurons of the antagonist muscle.

nervous system is **presynaptic inhibition,** a process mediated by neurons that end on excitatory endings, forming **axo-axonal synapses** (Fig 4–3). Activation of the receptors mediating presynaptic inhibition reduces the amount of neurotransmitter secreted when action potentials arrive at excitatory synaptic knobs. The neurons responsible for postsynaptic and presynaptic inhibition are compared in Fig 4–9. An action potential that reaches the ending of the excitatory neuron opens voltage-gated Ca^{2+} channels in the ending, and the resulting Ca^{2+} influx triggers exocytosis of the neurotransmitter-containing vesicles or granules. In some instances, activation of presynaptic receptors reduces the opening of Ca^{2+} channels directly. In others, presynaptic receptor stimulation increases Cl^- and/or K^+ conductance. This makes the membrane more permeable, so that the size of the action potentials reaching the ending is reduced (Fig 4–10). As a consequence, there is less opening of voltage-gated Ca^{2+} channels. In either case, Ca^{2+} influx is reduced and less transmitter is released.

The first transmitter to be shown to produce synaptic inhibition was GABA. Acting by $GABA_A$ receptors (see below), GABA increases Cl^- conductance. $GABA_B$ receptors are also present in the spinal cord and may also mediate presynaptic inhibition via a G protein that produces an increase in K^+ conductance. In a similar fashion, other transmitters mediate presynaptic inhibition by G protein-mediated effects on Ca^{2+} channels and K^+ channels.

Conversely, **presynaptic facilitation** is produced when the action potential is prolonged (Fig 4–10) and the Ca^{2+} channels are open for a longer period. The molecular events responsible for the production of

Figure 4–10. Effects of presynaptic inhibition and facilitation on presynaptic action potential, Ca^{2+} current in the presynaptic ending, and EPSP in the postganglionic cell. In each case, the solid lines are the controls and the dashed lines the changes seen during inhibition and facilitation.

presynaptic facilitation mediated by serotonin in the sea snail *Aplysia* have been worked out in detail (Fig 4–11). Serotonin released at an axo-axonal ending increases intraneuronal cyclic AMP, and the resulting phosphorylation of one group of K^+ channels closes the channels, slowing repolarization and prolonging the action potential.

Organization of Inhibitory Systems

Presynaptic and postsynaptic inhibition are usually produced by stimulation of certain systems converging on a given postsynaptic neuron ("afferent inhibition"). Neurons may also inhibit themselves in a negative feedback fashion ("negative feedback inhibition"). For instance, spinal motor neurons regularly give off a recurrent collateral, which synapses with an inhibitory interneuron that terminates on the cell body of the spinal neuron and other spinal motor neurons (Fig 4–12). This particular inhibitory neuron is sometimes called the Renshaw cell after its discoverer. Impulses generated in the motor neuron activate the inhibitory interneuron to secrete inhibitory mediator, and this slows or stops the discharge of the motor neuron. Similar inhibition via recurrent collaterals is seen in the cerebral cortex and limbic system.

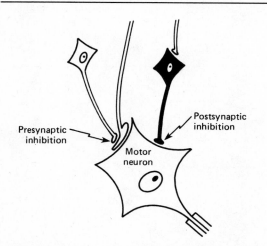

Figure 4–9. Arrangement of neurons producing presynaptic and postsynaptic inhibition. The neuron producing presynaptic inhibition is shown ending on an excitatory synaptic knob. Many of these neurons actually end higher up along the axon of the excitatory cell.

Serotonin from axoaxonic
ending binds to receptor
on presynaptic terminal

↓

Increased intracellular cyclic AMP
via G_s and adenylate cyclase

↓

Activation of protein kinase A

↓

Phosphorylation and consequent
inactivation of $K^+(s)$ channels

↓

Prolonged action potentials

↓

Increased Ca^{2+} influx via
voltage-gated Ca^{2+} channels

↓

More transmitter released
into synaptic cleft

Figure 4–11. Sequence of events producing presynaptic facilitation in *Aplysia*.

Presynaptic inhibition due to descending pathways that terminate on afferent pathways in the dorsal horn may be involved in the "gating" of pain transmission (see Chapter 7).

Another type of inhibition is seen in the cerebellum. In this part of the brain, stimulation of basket cells produces IPSPs in the Purkinje cells. However, the basket cells and the Purkinje cells are excited by the same excitatory input. This arrangement, which has been called "feed-forward inhibition," presumably limits the duration of the excitation produced by any given afferent volley.

Summation & Occlusion

The interplay between excitatory and inhibitory influences at synaptic junctions in a nerve net illustrates the integrating and modulating activity of the nervous system.

In the hypothetic nerve net shown in Fig 4–13, neurons A and B converge on X, and neuron B diverges on X and Y. A stimulus applied to A or to B will set up an EPSP in X. If A and B are stimulated at the same time, 2 areas of depolarization will be produced in X and their actions will sum. The resultant EPSP in X will be twice as large as that produced by

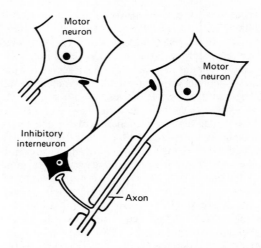

Figure 4–12. Negative feedback inhibition of a spinal motor neuron via an inhibitory interneuron (Renshaw cell).

stimulation of A or B alone, and the membrane potential may well reach the firing level of X. The effect of the depolarization caused by the impulse in A is facilitated by that due to activity in B, and vice versa; spatial facilitation has taken place. In this case, Y has not fired, but its excitability has been increased, and it is easier for activity in neuron C to fire Y during the EPSP. Y is therefore said to be in the **subliminal fringe** of X. More generally stated, neurons are in the subliminal fringe if they are not discharged by an afferent volley (not in the **discharge zone**) but do have their excitability increased. The neurons that have few active knobs ending on them are in the subliminal fringe, and those with many are in the discharge zone. Inhibitory impulses show similar temporal and spatial facilitation and subliminal fringe effects.

If neuron B in Fig 4–13 is stimulated repetitively,

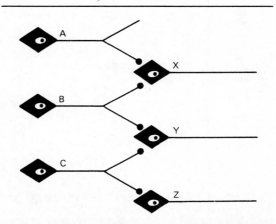

Figure 4–13. Simple nerve net. Neurons A, B, and C have excitatory endings on neurons X, Y, and Z.

X and Y will discharge as a result of temporal summation of the EPSPs produced. If C is stimulated repetitively, Y and Z will discharge. If B and C are fired repetitively at the same time, X, Y, and Z will discharge. Thus, the response to stimulation of B and C together is not as great as the sum of responses to stimulation of B and C separately, because B and C both end on neuron Y. This decrease in expected response, due to presynaptic fibers sharing postsynaptic neurons, is called **occlusion.**

Excitatory and inhibitory subliminal effects and occlusive phenomena can have pronounced effects on transmission in any given pathway. Because of these effects, temporal patterns in peripheral nerves are usually altered as they pass through synapses on the way to the brain. These effects may also explain such important phenomena as referred pain (see Chapter 7).

CHEMICAL TRANSMISSION OF SYNAPTIC ACTIVITY

Implications

The fact that transmission at most if not all synapses is chemical is of great physiologic and pharmacologic importance. Nerve endings have been called biologic transducers that convert electrical energy into chemical energy. In broad terms, this conversion process involves the synthesis of the transmitter agents, their storage in the synaptic knobs, and their release by the nerve impulses into the synaptic cleft. The secreted transmitters then act on appropriate receptors on the membrane of the postsynaptic cell and are rapidly removed from the synaptic cleft by diffusion, metabolism, and in many instances, at least for amine transmitters, reuptake into the presynaptic neuron. All these processes, plus the postreceptor events in the postsynaptic neuron, are regulated by many physiologic factors and at least in theory can be altered by drugs. Therefore pharmacologists should be able to develop drugs that regulate not only somatic and visceral motor activity but also emotions, behavior, and all the other complex functions of the brain.

Chemistry of Transmitters

One suspects that a substance is a transmitter if it is unevenly distributed in the nervous system, its distribution parallels that of its receptors, and it is synthesizing and catabolizing enzymes. Additional evidence includes demonstration that it is released from slices of appropriate brain regions in vitro and that it produces effects on single target neurons when applied to their membranes by means of a micropipette (microiontophoresis). Many transmitters have now been localized in nerve endings by special staining techniques. One of the most useful and important techniques has been **immunocytochemistry,** in which antibodies to a given substance are labeled and applied to brain and other tissues. The antibodies bind to the substance, and the location of the substance is then determined by locating the label with the light microscope or electron microscope. Histochemical and immunocytochemical techniques have been used to localize serotonin, dopamine, norepinephrine, epinephrine, the enzymes that synthesize and metabolize these substances, and many different peptides.

Identified neurotransmitters can be divided into broad categories or families based on their chemical structure; some are amines, some are amino acids, and some are purines. Many are polypeptides. It is worth noting that most of these substances are not only released into synaptic clefts, where they produce highly localized effects. In other situations, they diffuse in the ECF and exert effects at some distance from their site of release (paracrine communication; see Chapter 1). In some cases, they are also released by neurons into the bloodstream as hormones. A somewhat arbitrary compilation of most of the substances currently known or suspected to be synaptic mediators is presented in Table 4–1.

It is now known that many and possibly all neu-

Table 4-1. Neurotransmitters in the nervous system of mammals.[1]

Substance	Locations
Acetylcholine	Myoneural junction; preganglionic autonomic endings, postganglionic parasympathetic endings, postganglionic sympathetic sweat gland and muscle vasodilator endings; many parts of brain; endings of some amacrine cells in retina.
Amines Dopamine	SIF cells in sympathetic ganglia; striatum, median eminence, and other parts of hypothalamus; limbic system; parts of neocortex; endings of some interneurons in retina.
Norepinephrine	Most postganglionic sympathetic endings; cerebral cortex, hypothalamus, brain stem, cerebellum, spinal cord.
Epinephrine	Hypothalamus, thalamus, periaqueductal gray, spinal cord.
Serotonin	Hypothalamus, limbic system, cerebellum, spinal cord; retina.
Histamine	Hypothalamus.

(continued)

Table 4–1 (cont'd). Neurotransmitters in the nervous system of mammals.[1]

Substance	Locations
Excitatory amino acids	
Glutamate	Cerebral cortex, brain stem.
Aspartate	Spinal cord, other parts of CNS?
Inhibitory amino acids	
Glycine	Neurons mediating direct inhibition; retina.
γ-Aminobutyrate (GABA)	Cerebellum; cerebral cortex; neurons mediating presynaptic inhibition; retina.
Purines	
Adenosine	Neocortex, olfactory cortex, hippocampus, cerebellum.
ATP	Primary afferents?
Polypeptides	
Substance P	Endings of primary afferent neurons mediating nociception; many parts of brain; retina.
Vasopressin	Posterior pituitary; medulla; spinal cord.
Oxytocin	Posterior pituitary; medulla; spinal cord.
CRH	Median eminence of hypothalamus; other parts of brain.
TRH	Median eminence of hypothalamus; other parts of brain; retina.
GRH	Median eminence of hypothalamus.
Somatostatin	Median eminence of hypothalamus; other parts of brain; substantia gelatinosa; retina.
GnRH	Median eminence of hypothalamus; circumventricular organs; preganglionic autonomic endings; retina.
Endothelins	Posterior pituitary, brain stem.
Enkephalins	Substantia gelatinosa, many other parts of CNS; retina.
β-Endorphin, other derivatives of pro-opiomelanocortin	Hypothalamus, thalamus, brain stem; retina.
Cholecystokinin (CCK) octapeptide	Cerebral cortex; hypothalamus; retina.
Vasoactive intestinal polypeptide (VIP)	Postganglionic cholinergic neurons; some sensory neurons; hypothalamus; cerebral cortex; retina.
Neurotensin	Hypothalamus; retina.
Gastrin-releasing peptide (GRP)	Hypothalamus.
Gastrin	Hypothalamus; medulla oblongata.
Glucagon	Hypothalamus; retina.
Motilin	Neurohypophysis; cerebral cortex, cerebellum.
Secretin	Hypothalamus, thalamus, olfactory bulb, brain stem, cerebral cortex, septum, hippocampus, striatum.
Calcitonin-gene-related peptide alpha (CGRPα)	Endings of primary afferent neurons; taste pathways; sensory nerves; medial forebrain bundle.
Neuropeptide Y	Noradrenergic, adrenergic, and other neurons in medulla, periaqueductal gray, hypothalamus; autonomic nervous system.
Activins	Brain stem.
Inhibins	Brain stem.
Angiotensin II	Hypothalamus, amygdala, brain stem, spinal cord.
FMRF amide	Hypothalamus, brain stem.
Galanin	Hypothalamus.
Atrial natriuretic peptide (ANP)	Hypothalamus, brain stem.
Brain natriuretic peptide (BNP)	Hypothalamus, brain stem.

[1]Transmitter functions have not been proved for some of the polypeptides.

rons contain more than one transmitter; ie, they contain **cotransmitters.** Often, but not always, an amine is colocalized with one or more polypeptides. This phenomenon is discussed in more detail below.

Receptors

Cloning and related molecular biology techniques have permitted spectacular recent advances in knowledge about the structure and function of receptors for neurotransmitters and other chemical messengers. The individual receptors, along with their ligands, are discussed in the following parts of this chapter. However, 4 themes have emerged that should be mentioned in this introductory discussion.

First, in every instance studied in detail to date, it has become clear that for each ligand there are many subtypes of receptors. Thus, for example, norepinephrine acts on α_1 and α_2 receptors, and 3 of each subtype have been cloned, although the identities of the second messengers by which they act are unsettled. In addition, there are β_1, β_2, and β_3 receptors. Obviously, this multiplies and makes more selective in any given cell the possible effects of a given ligand.

Second, there are receptors on the pre- as well as the postsynaptic elements at many synaptic junctions. These **presynaptic receptors,** or **autoreceptors,** often inhibit further secretion of the ligand, providing feedback control. For example, norepinephrine acts on α_2 presynaptic receptors to inhibit norepinephrine secretion (Fig 4–14). However, autoreceptors can also facilitate the release of neurotransmitters.

Third, although there are many ligands and many subtypes of receptors for each ligand, the receptors tend to group in large families as far as structure and function are concerned. Many are serpentine receptors that act via G proteins and protein kinases (see Chapter 1) to produce their effects. Others are ion channels. The receptors, their principal second messengers, and, where established, the net effect on channels for a group of selected, established neurotransmitters are listed in Table 4–2. It should be noted

that this table is an oversimplification. For example, activation of α_2 receptors decreases intracellular cyclic AMP, but there is evidence that the G protein activated by α_2 presynaptic receptors acts directly on Ca^{2+} channels to inhibit norepinephrine release by decreasing Ca^{2+} increase.

Fourth, prolonged exposure to their ligands causes most receptors to become unresponsive, ie, to undergo **desensitization.** This can be of 2 types: **homologous desensitization,** with loss of responsiveness only to the particular ligand and maintained responsiveness of the cell to other ligands; and **heterologous desensitization,** in which the cell becomes unresponsive to other ligands as well. Desensitization in β-adrenergic receptors has been studied in great detail. Heterologous desensitization appears to involve both phosphorylation of the β_2-adrenergic receptor and the catalytic unit of adenylate cyclase. In homologous desensitization, an important event is phosphorylation of the β_2-adrenergic receptor by a specific β-adrenergic receptor kinase (β-ARK). However, other factors are also involved, in both forms, and the details remain obscure.

It is obvious that synaptic physiology is a rapidly expanding, complex field that cannot be covered in detail in this book. However, it is appropriate to summarize information about some of the principal neurotransmitters and their receptors.

Acetylcholine

The relatively simple structure of acetylcholine, which is the acetyl ester of choline, is shown in Fig 4–15. It exists, largely enclosed in small, clear synaptic vesicles, in high concentration in the terminal buttons of cholinergic neurons. Neurons that release acetylcholine are known as **cholinergic** neurons.

Acetylcholine Receptors

Historically, acetylcholine receptors have been divided into 2 main types on the basis of their pharmacologic properties. Muscarine, the alkaloid responsible for the toxicity of toadstools, has little effect on the receptors in autonomic ganglia but mimics the stimulatory action of acetylcholine on smooth muscle and glands. These actions of acetylcholine are therefore called **muscarinic actions,** and the receptors involved are **muscarinic receptors.** They are blocked by the drug atropine. In sympathetic ganglia, small amounts of acetylcholine stimulate postganglionic neurons and large amounts block transmission of impulses from pre- to postganglionic neurons. These actions are unaffected by atropine but mimicked by nicotine. Consequently, these actions of acetylcholine are **nicotinic actions** and the receptors are **nicotinic receptors.** Nicotinic receptors are further subdivided into those found at neuromuscular junctions and those found at junctions between neurons. Both muscarinic and nicotinic acetylcholine receptors are found in large numbers in the brain.

Figure 4–14. Presynaptic and postsynaptic receptors at the ending of a noradrenergic neuron. The presynaptic receptor that is shown is α_2, the postsynaptic receptors can be α_1, α_2, β_1, or β_2.

Table 4–2. Mechanism of action of selected neurotransmitters.

Transmitter	Receptor	Second Messenger	Net Channel Effects
Acetylcholine	N M_1 M_2 (cardiac) M_3 M_4 (glandular M_5	- - - ↑IP_3, DAG ↓Cyclic AMP ↓Cyclic AMP ↑IP_3, DAG ↑IP_3, DAG	↑Na^+, other small ions ↑Ca^{2+} ↓Ca^{2+}
Dopamine[1]	D_1 D_2	↑Cyclic AMP ↓Cyclic AMP	↑K^+, ↓Ca^{2+}
Norepinephrine[2]	α_1 α_2 β_1 β_2 β_3	↑IP_3, DAG ↓Cyclic AMP ↑Cyclic AMP ↑Cyclic AMP ↑Cyclic AMP	↓K^+ ↑K^+, ↓Ca^{2+}
5HT	$5HT_{1A}$ $5HT_{1B}$ $5HT_{1C}$ $5HT_{1D}$ $5HT_2$ $5HT_3$ $5HT_4$	↓Cyclic AMP ↓Cyclic AMP ↑IP_3, DAG ↓Cyclic AMP ↑IP_3, DAG - - - ↑Cyclic AMP	↑K^+ ↓K^+ ↓K^+ ↑Na^+
Adenosine	A_1 A_2	↓Cyclic AMP ↑Cyclic AMP	
Glutamate, aspartate	NMDA AMPA Quisqualate Kainate	- - - - - - ↑IP_3, DAG - -	↑Na^+, Ca^{2+} ↑Na^+ - - - ↑Na^+
GABA	$GABA_A$ $GABA_B$	- - - ↑IP_3, DAG	↑Cl^- ↑K^+, ↓Ca^{2+}

[1]D_2 receptors exist in 2 forms produced by differential splicing of mRNA, but their drug specificities are the same. D_3, D_4, and D_5, receptors have also been cloned.
[2]Three subtypes of α_1 and 3 subtypes of α_2 receptors have been cloned.

Muscle-type nicotinic acetylcholine receptors in mammalian skeletal muscle and in the electric organ of the electric eel, an organ made up a large collection of motor end-plates, have been studied in detail. In the electric eel and in fetal muscle and extrajunctional regions of muscle in mammals, the nicotinic acetylcholine receptor is a protein with a molecular weight of slightly more than 250,000. It is made up of

Figure 4–15. Biosynthesis and catabolism of acetylcholine.

5 subunits: 2 identical α subunits and one β, one γ, and one δ subunit. These 5 subunits extend through the cell membrane (Fig 4–16) and are arranged in a nearly symmetric fashion around a channel that is wide outside the cell and narrows as it goes through the cell membrane. There is a binding site for acetylcholine on each α subunit, and when an acetylcholine molecule binds to each of them, they induce a configurational change in the protein so that the channel opens. This increases the conductance of Na^+ and other cations, and the resulting influx of Na^+ produces a depolarizing potential.

The subunits of the nicotinic receptor described above are encoded by 4 different genes. In adult mammals, the γ subunit in receptors at neuromuscular junctions is replaced by the ε subunit, which is apparently encoded by yet another gene. This change does not alter its pentagonal structure but decreases its open time and increases its conductance. The nature of the signal that brings about the switch from the γ to the ε is unsettled. The half-life of nicotinic receptors in the cell membrane is about 1 week.

Neuronal-type nicotinic acetylcholine receptors also have 5 subunits arranged in a pentagonal array, but these receptors are made up of 2 α and 3 β subunits. There are at least 5 different subtypes of α subunits and at least 4 different subtypes of β subunits,

Figure 4–16. Diagram of fetal nicotinic acetylcholine receptor as viewed from the side *(above)* and from the top *(below)*. α, β, γ, δ: receptor subunits. (From McCarthy MP et al: Molecular biology of the acetylcholine receptor. *Annu Rev Neurosci* 1986;**9**:383. Reproduced with permission, from the *Annual Review of Neuroscience*, vol 9. © 1986 by Annual Reviews Inc.)

and different subtypes combine to form the pentagonal nicotinic acetylcholine receptors that are located presynaptically as well as postsynaptically in different parts of the brain. Another difference between muscle-type and neuronal-type nicotinic receptors is that the former bind the snake toxin α-bungarotoxin whereas the latter do not. Instead, the brain contains a separate population of α-bungarotoxin-binding proteins in the deep layers of the cerebral cortex, the hippocampus, the hypothalamus, and the colliculi. The functions of these proteins are unknown.

The muscle-type and neuronal-type acetylcholine receptors are members of a superfamily of ligand-gated ion channels that also includes the GABA$_A$ and glycine receptors and some of the glutamate receptors (see below).

Muscarinic cholinergic receptors are very different from nicotinic cholinergic receptors. Five types, encoded by 5 separate genes, have been cloned, and all are serpentine receptors coupled via G proteins to adenylate cyclase or phospholipase C (Table 4–2). The nomenclature of these receptors has not been standardized, but the receptor designated M$_1$ in Table 4–2 is abundant in the brain. The M$_2$ receptor is found in the heart (see Chapter 28). The M$_4$ receptor is found in pancreatic acinar and islet tissue, where it mediates increased secretion of pancreatic enzymes and insulin. The M$_2$ and M$_4$ receptors are both found in smooth muscle. The M$_3$ and M$_5$ receptors have not yet been characterized in terms of their pharmacologic properties and physiologic effects.

Cholinesterases

Acetylcholine must be rapidly removed from the synapse if repolarization is to occur. The removal occurs by way of hydrolysis of acetylcholine to choline and acetate, a reaction catalyzed by the enzyme **acetylcholinesterase.** This enzyme is also called **true** or **specific cholinesterase.** Its greatest affinity is for acetylcholine, but it also hydrolyzes other choline esters. There are a variety of esterases in the body. One found in plasma is capable of hydrolyzing acetylcholine but has different properties from acetylcholinesterase. It is therefore called **pseudocholinesterase** or **nonspecific cholinesterase.** The plasma moiety is partly under endocrine control and is affected by variations in liver function. On the other hand, the specific cholinesterase at nerve endings is highly localized. It contains 537 amino acid residues, has its active site in a deep cleft, and is anchored to the cell membrane by a glycolipid. Hydrolysis of acetylcholine by this enzyme is rapid enough to explain the observed changes in Na$^+$ conductance and electrical activity during synaptic transmission.

Acetylcholine Synthesis

Synthesis of acetylcholine involves the reaction of choline with acetate. There is an active uptake of choline into cholinergic neurons (Fig 4–17). Choline is also synthesized in neurons. The acetate is activated by the combination of acetate groups with reduced coenzyme A. The reaction between active acetate (acetyl-coenzyme A, acetyl-CoA) and choline is catalyzed by the enzyme **choline acetyltransferase.** This enzyme is found in high concentration in the cytoplasm of cholinergic nerve endings; indeed, its localization is so specific that the presence of a high concentration in any given neural area has been taken as evidence that the synapses in that area are cholinergic.

Norepinephrine & Epinephrine

The chemical transmitter present at most sympathetic postganglionic endings is norepinephrine (levarterenol). It is stored in the synaptic knobs of the neurons that secrete it in characteristic vesicles which have a dense core (granulated vesicles). There are 2 populations of granulated vesicles: small vesicles about 40 nm in diameter and large vesicles about 75

Figure 4–17. Comparison of the biochemical events at cholinergic endings **(left)** with those at noradrenergic endings **(right).** ACh, acetylcholine; ACE, acetylcholinesterase; NE, norepinephrine; X, receptor. Note that monoamine oxidase (MAO) is intracellular, so that some norepinephrine is being constantly deaminated in noradrenergic endings. Catechol-O-methyltransferase (COMT) acts on norepinephrine after it is secreted.

nm in diameter. Norepinephrine and its methyl derivative, epinephrine, are secreted by the adrenal medulla (see Chapter 20), but epinephrine is not a mediator at postganglionic sympathetic endings. The endings of sympathetic postganglionic neurons in smooth muscle are discussed below; each neuron has multiple varicosities along its course, and each of these varicosities appears to be a site at which norepinephrine is liberated. There are also norepinephrine-secreting, dopamine-secreting, and epinephrine-secreting neurons in the brain (see Chapter 15). Norepinephrine-secreting neurons are properly called **noradrenergic neurons,** although the term **adrenergic neurons** is also applied. However, it seems appropriate to reserve the latter term for epinephrine-secreting neurons. Dopamine-secreting neurons are called **dopaminergic neurons.**

Biosynthesis & Release of Catecholamines

The principal **catecholamines** found in the body—norepinephrine, epinephrine, and dopamine—are formed by hydroxylation and decarboxylation of the amino acids phenylalanine and tyrosine (Fig 4–18). **Phenylalanine hydroxylase** is found primarily in the liver. Tyrosine is transported into catecholamine-secreting neurons and adrenal medullary cells by a concentrating mechanism. It is converted to dopa and then to dopamine in the cytoplasm of the cells by **tyrosine hydroxylase** and **dopa decarboxylase.** The decarboxylase, which is also called aromatic L-amino acid decarboxylase, is very similar but probably not identical to 5-hydroxytryptophan decarboxylase. The dopamine then enters the granulated vesicles, within which it is converted to norepinephrine by **dopamine β-hydroxylase.** L-Dopa is the isomer involved, but the norepinephrine that is formed is in the D configuration. This is true even though it is levorotatory (–). Dextrorotatory (+) norepinephrine is much less active. The rate-limiting step in synthesis is the conversion of tyrosine to dopa. Tyrosine hydroxylase, which catalyzes this step, is subject to feedback inhibition by dopamine and norepinephrine, thus providing internal control of the synthetic process. The cofactor for tyrosine hydroxylase is **tetrahydrobiopterin,** which is converted to dihydrobiopterin when tyrosine is converted to dopa.

Some neurons and adrenal medullary cells also contain the cytoplasmic enzyme **phenylethanolamine-N-methyltransferase (PNMT),** which catalyzes the conversion of norepinephrine to epinephrine. In these cells, norepinephrine apparently leaves the vesicles, is converted to epinephrine, and then enters other storage vesicles.

In granulated vesicles, norepinephrine and epinephrine are bound to ATP and associated with a protein called **chromogranin A,** whose function is unknown. In some but not all noradrenergic neurons, the large granulated vesicles also contain neuropeptide Y (see below). Chromogranin A is a 49-kDa protein that is also found in many other endocrine and

Figure 4–18. Biosynthesis of catecholamines. The dashed lines indicate inhibition of tyrosine hydroxylase by norepinephrine and dopamine. Tetrahydrobiopterin is a cofactor for the action of phenylalanine hydroxylase and tyrosine hydroxylase.

neuroendocrine cells, and it may play some sort of general role in hormone storage or secretion. Plasma levels of chromogranin A are increased in patients with a variety of different endocrine tumors, in some instances even in the absence of an increase in the associated hormone. Measurement of plasma chromogranin A is of value in the diagnosis of these tumors.

The catecholamines are held in the granulated vesicles by an active transport system, and the action of this transport system is inhibited by the drug reserpine.

Catecholamines are released from autonomic neurons and adrenal medullary cells by exocytosis (see Chapter 1). Since they are present in the granulated vesicles, ATP, chromogranin A, and the dopamine β-hydroxylase that is not membrane-bound are released with norepinephrine and epinephrine. The half-life of circulating dopamine β-hydroxylase is much longer than that of the catecholamines, and circulating levels of this substance are affected by genetic and other factors in addition to the rate of sympathetic activity. Circulating levels of chromogranin A appear to be a better index of sympathetic activity.

Reuptake

Some of the norepinephrine in nerve endings is manufactured there, but some has been secreted and then taken up again into the noradrenergic neurons. An active **reuptake mechanism** is characteristic of

noradrenergic neurons. Circulating norepinephrine and epinephrine are also picked up in small amounts by noradrenergic neurons in the autonomic nervous system. In this regard, noradrenergic neurons differ from cholinergic neurons. Acetylcholine is not taken up to any appreciable degree, but the choline formed by the action of acetylcholinesterase is actively taken up and recycled (Fig 4–17).

Reuptake is also prominent in neurons that secrete dopamine, serotonin, and GABA. Transport into noradrenergic neurons is mediated by a **norepinephrine transporter,** a 69K protein that crosses the cell membrane 12 times and has its N- and C-terminals located intracellularly. The **serotonin, dopamine,** and **GABA transporters** have also been cloned, and their structures are similar to that of the norepinephrine transporter. It now appears that reuptake, presumably by a family of related transporters, is a main mechanism of inactivation for all transmitters except acetylcholine and neuropeptides. The biogenic amine transporters are of pharmacologic interest because they are inactivated by tricyclic antidepressants and by cocaine, which is believed to exert is psychogenic effects by increasing the amount of dopamine at synaptic junctions in the nucleus accumbens (see Chapter 15).

Phenylpyruvic Oligophrenia

The pathways involved in the metabolism of phenylalanine are of considerable clinical importance because they are the site of several **inborn errors of metabolism,** diseases caused by congenital absence of various specific enzymes. **Phenylpyruvic oligophrenia,** or phenylketonuria, is a disorder characterized by severe mental deficiency and the accumulation in the blood, tissues, and urine of large amounts of phenylalanine and its keto acid derivatives. It is usually due to congenital deficiency or absence of phenylalanine hydroxylase (Fig 4–18). The gene for this enzyme is located on the long arm of chromosome 12. Numerous different mutations in the gene have been shown to cause phenylketonuria. The condition can also be produced by deficiency or absence or pteridine reductase, the enzyme that regenerates the biopterin cofactor needed for the hydroxylation of tyrosine, or by deficient biopterin synthesis. The mental retardation is largely due to accumulation of phenylalanine and its derivatives in the blood, and it can be treated with considerable success by markedly reducing the amount of phenylalanine in the diet.

Catabolism of Catecholamines

Epinephrine and norepinephrine are metabolized to biologically inactive products by oxidation and methylation. The former reaction is catalyzed by **monoamine oxidase (MAO)** and the latter by **catechol-O-methyltransferase (COMT)** (Figs 4–17 and 4–19). MAO is located on the outer surface of the mitochondria. It is widely distributed, being particularly plentiful in the nerve endings at which catecholamines are secreted. COMT is also widely distributed, particularly in the liver, kidneys, and smooth muscles, but it is not found in nerve endings. Consequently, there are 2 different patterns of catecholamine metabolism.

Circulating epinephrine and norepinephrine are for the most part O-methylated, and measurement of the concentrations of the O-methylated derivatives normetanephrine and metanephrine in the urine is a good index of the rate of secretion of norepinephrine and epinephrine. The O-methylated derivatives that are not excreted are largely oxidized, and VMA (Fig 4–19) is the most plentiful catecholamine metabolite in the urine. Small amounts of the O–methylated derivatives are also conjugated to sulfates and glucuronides.

In the noradrenergic nerve endings, on the other hand, some of the norepinephrine is being constantly converted by MAO to the physiologically inactive deaminated derivatives, 3,4-dihydroxymandelic acid (DOMA) and its corresponding glycol (DHPG). These compounds enter the circulation and may subsequently be converted to their corresponding O-methyl derivatives, VMA and MHPG (Fig 4–19).

The reuptake of released norepinephrine mentioned above is a major mechanism by which norepinephrine is removed from the vicinity of autonomic endings. The hypersensitivity of the sympathetically denervated structures (see below) is probably explained in part on this basis. After the noradrenergic neurons are cut, their endings degenerate; consequently, there is no reuptake, and more of a given dose of norepinephrine is available to stimulate the receptors of the autonomic effectors.

Alpha & Beta Receptors

Epinephrine and norepinephrine both act on α and β receptors, with norepinephrine having a greater affinity for α adrenergic receptors and epinephrine for β adrenergic receptors. As noted above, the α and β receptors are typical serpentine receptors linked to G proteins, and there are multiple forms of each. In addition, they are closely related to the cloned receptors for many other cationic amines, eg, dopamine, serotonin, and muscarinic acetylcholine receptors.

Dopamine

In the small intensely fluorescent (SIF) cells in autonomic ganglia (see Chapter 13) and in certain parts of the brain (see Chapter 15), catecholamine synthesis stops at dopamine (Fig 4–18), and this catecholamine is secreted as a synaptic transmitter. Released dopamine acts on various types of dopamine receptors (Table 4–2). The D_1 receptor, activates dopamine-sensitive adenylate cyclase via G_S; a second type, the D_2 receptor, inhibits adenylate cyclase via G_i. However, the D_2 receptor also inhibits phosphatidylinositol turnover. It has recently been demonstrated that there are 2 forms of the D_2 receptor that differ in their G protein-binding domains but have identical drug specificities. D_3, D_4, and D_5 receptors have also been cloned. All the dopamine receptors are coupled to G proteins, but D_3 is more concentrated in limbic areas associated with emotion and D_4 and D_5 have greater affinity for certain antipsychotic drugs.

Released dopamine is recaptured via an active reuptake mechanism and inactivated by MAO and COMT (Fig 4–20) in a manner analogous to the inactivation of norepinephrine. DOPAC and HVA are also conjugated, primarily to sulfates.

Serotonin

Serotonin (5-hydroxytryptamine, 5-HT) is present in highest concentration in blood platelets and in the gastrointestinal tract, where it is found in the enterochromaffin cells and the myenteric plexus (see Chapter 26). Lesser amounts are found in the brain and in the retina.

Serotonin is formed in the body by hydroxylation and decarboxylation of the essential amino acid tryptophan (Fig 4–21). Normally, the hydroxylase is not saturated; consequently, increased intake of tryptophan in the diet can increase brain serotonin content. After release from serotonergic neurons, much of the released serotonin is recaptured by an active reuptake mechanism and inactivated by monoamine oxidase (Figs 4–21 and 4–22) to form 5-hydroxyindoleacetic

Figure 4–19. Top: Catabolism of circulating epinephrine and norepinephrine. The main site of catabolism is the liver. The conjugates are mostly glucuronides and sulfates. MHPG is also conjugated. **Bottom:** Catabolism of norepinephrine in noradrenergic nerve endings. The acid and the glycol enter the circulation and may be subsequently O-methylated to VMA and MHPG. Epinephrine in nerve endings is presumably catabolized in the same way.

Figure 4–20. Catabolism of dopamine. As in other oxidative deaminations catalyzed by monoamine oxidase, aldehydes are formed first and then oxidized in the presence of aldehyde dehydrogenase to the corresponding acids (DOPAC and HVA). The aldehydes are also reduced to 3,4-dihydroxyphenylethanol (DOPET) and 3-methoxy-4-hydroxyphenylethanol. Some of the DOPAC and HVA form sulfate conjugates.

acid (5-HIAA). This substance is the principal urinary metabolite of serotonin, and urinary output of 5-HIAA is used as an index of the rate of serotonin metabolism in the body. In the pineal gland, serotonin is converted to melatonin (see Chapter 24).

Seven types of serotonin receptors have now been described: $5HT_{1A}$, $5HT_{1B}$, $5HT_{1C}$, and $5HT_{1D}$ receptors, at least some of which are presynaptic; $5HT_2$ receptors, which mediate platelet aggregation, smooth muscle contraction, and various effects in the brain; $5HT_3$ receptors, which are also present in peripheral tissues and the brain and presumably mediate other brain functions; and $5HT_4$ receptors, which have yet to be assigned a detailed physiologic role. There are many $5HT_{1C}$ receptors in the choroid plexus. $5HT_1$, $5HT_2$, and $5HT_4$ receptors are serpentine receptors that act via G proteins to affect adenylate cyclase or phospholipase C (Table 4–2). However, the $5HT_3$ receptors, like the nicotinic cholinergic receptors, are ion channels.

Histamine

Histaminergic neurons have their cell bodies in the tuberomammillary nucleus of the posterior hypothalamus (see Fig 15–6), and their axons project to all parts of the brain, including the cerebral cortex and the spinal cord. Thus the histaminergic system resem-

Figure 4–21. Biosynthesis and metabolism of serotonin (5-hydroxytryptamine). The enzyme that catalyzes the decarboxylation of 5-hydroxytryptophan is very similar but probably not identical to the enzyme that catalyzes the decarboxylation of dopa. Tetrahydrobiopterin is a cofactor for the action of tryptophan hydroxylase. The details of the formation of melatonin are shown in Fig 24–11.

Figure 4–22. Biochemical events at serotonergic synapses. Compare with Fig 4–17. 5-HTP, 5-hydroxytryptophan; 5-HT, 5-hydroxytryptamine (serotonin); 5-HIAA, 5-hydroxyindoleacetic acid; X, serotonin receptor.

bles the noradrenergic, adrenergic, dopaminergic, and serotonergic systems, with projections from relatively few cells to all parts of the central nervous system.

Histamine is also found in cells in the gastric mucosa (see Chapter 6) and in heparin-containing cells called **mast cells** that are plentiful in the anterior and posterior lobes of the pituitary gland.

Histamine is formed by decarboxylation of the amino acid histidine (Fig 4–23). The enzyme that catalyzes this step differs from the L-aromatic amino acid decarboxylases that decarboxylate 5-hydroxytryptophan and L-dopa. Histamine is converted to methylhistamine or, alternatively, to imidazoleacetic acid. The latter reaction is quantitatively less important in humans. It requires the enzyme **diamine oxidase (histaminase)** rather than monoamine oxidase, even though monoamine oxidase catalyzes the oxidation of methylhistamine to methylimidazoleacetic acid.

There are 3 known types of histamine receptors, H_1, H_2, and H_3, and all 3 are found in peripheral tissues and the brain. Most, if not all, of the H_3 receptors are presynaptic, and they mediate inhibition of the release of histamine and other transmitters via a G protein. H_1 receptors activate phospholipase C, and H_2 receptors increase intracellular cyclic AMP concentration. The function of the histaminergic pathway systems in the brain is uncertain, but histamine has been related to arousal, sexual behavior, regulation of the secretion of some anterior pituitary hormones, blood pressure, drinking, and pain thresholds.

Excitatory Amino Acids: Glutamate & Aspartate

Glutamate and **aspartate** depolarize many different mammalian neurons when delivered directly on their cell membranes by iontophoresis. Glutamate is very widely used as an excitatory transmitter in the

Figure 4–23. Synthesis and catabolism of histamine.

brain, and it has been calculated that it is the transmitter responsible for 75% of the excitatory transmission in the brain.

Glutamate receptors are of 2 types: **metabotropic receptors** and **ligand-gated ion channels.** A metabotropic receptor is one that acts by G proteins and second messengers. One member of this family has been cloned, and it is a serpentine receptor coupled by a G protein to phospholipase C, so that it increases intracellular IP_3 and protein kinase C concentrations. Three ion channels that are activated by glutamate have been cloned and are named for the pharmacologic congeners of glutamate to which they respond in maximum fashion. They are the **kainate receptor,** the **AMPA receptor** (for α-amino-3-hydroxy-5-methyl-4-isoxazole propionic acid), and the **NMDA receptor** (for N-methyl-D-aspartate). All probably have multiple subtypes. There is also evidence for an additional ion channel L-AP4 receptor (for L-2-amino-4-phosphonobutyrate). All the glutamate receptors that are ion channels appear to have 4 transmembrane domains. The AMPA and metabotropic receptors are found in the cell membranes of glia as well as neurons.

The AMPA and kainate receptors are cation channels, permitting Na^+ influx and K^+ efflux. The NMDA receptor is also a cation channel, but it is unique in several ways. First, glycine binds to it to facilitate its function, and recent research indicates that glycine is essential for its normal response to glutamate. Second, at normal membrane potential, the channel is blocked by an Mg^{2+} ion and becomes unblocked only when the membrane becomes partially depolarized. Third, phencyclidine and ketamine, which produce amnesia and a feeling of dissociation from the environment, bind to another site inside the channel and are also released only upon depolarization.

There is a high concentration of NMDA receptors in the hippocampus, and blockade of these receptors prevents **long-term potentiation,** a long-lasting facilitation of transmission in neural pathways following a brief period of high-frequency stimulation (see below). Thus, these receptors may well be involved in memory and learning. Their unique voltage dependence also raises the possibility that they are involved in epilepsy.

Glutamate and some of its synthetic cogeners are unique in that when they act on neuronal cell bodies, they can produce such marked stimulation that neurons die. This is the reason why microinjections of these "excitotoxins" are used to produce discrete lesions that destroy neuronal cell bodies without affecting neighboring axons. An interesting hypothesis holds that when a part of the brain is ischemic, eg, after thrombosis of a cerebral artery, there is marked release of glutamate. Much of the brain damage appears to be due to the action of this transmitter as an excitotoxin, rather than to a lack of O_2 per se. Evidence in favor of this hypothesis is the sparing of brain tissue that occurs when vascular obstruction is produced after administration of agents that block NMDA receptors.

Inhibitory Amino Acids: γ-Aminobutyrate

Gamma-aminobutyric acid (GABA) is an inhibitory mediator in the brain and the retina and is the mediator responsible for presynaptic inhibition (see above).

GABA, which exists as γ-aminobutyrate in the body fluids, is formed by decarboxylation of glutamate (Fig 4–24). The enzyme that catalyzes this reaction is **glutamate decarboxylase** (GAD), which has been demonstrated by immunocytochemical techniques to be present in nerve endings in many parts of the brain. This enzyme is also present in B cells in pancreatic islets, and circulating antibodies to it are common in insulin-dependent diabetes mellitus, an autoimmune disease (see Chapter 19) GABA is metabolized primarily by transamination to succinic semialdehyde and thence to succinate in the citric acid cycle (see Chapter 17). **GABA transaminase(GABA-T)** is the enzyme that catalyzes the transamination. Pyridoxal phosphate, a derivative of the B complex vitamin pyridoxine, is a cofactor for GAD and GABA-T. However, the decarboxylation, unlike the transamination, is essentially irreversible. Consequently, the GABA content of the brain is reduced in pyridoxine deficiency. Pyridoxine deficiency is associated with signs of neural hyperexcitability and convulsions, although pyridoxine treatment is unfortunately of no value in most clinical cases of idiopathic epilepsy.

There are 2 very different classes of GABA receptors. The $GABA_B$ receptors act via a G protein to increase conductance in a K^+ channel. At least in some situations, they are presynaptic receptors. On the other hand, $GABA_A$ receptors are made up of subunits that form a Cl^- channel. They are somewhat homologous to the nicotinic acetylcholine receptors and the glycine receptors (see above), which are also Cl^- channels. The $GABA_A$ receptor is made up of 4 or 5 subunits, each of which has 4 membrane-spanning domains (Fig 4–25). To date, α, β, γ, and δ subunits have been identified, and there are several different kinds of α and γ subunits. Subunit composition varies in different parts of the brain.

In the $GABA_A$ receptors, GABA binds to the β subunits, increasing Cl^- conductance. This action is antagonized by picrotoxin. On the other hand the effects of GABA on Cl^- conductance are facilitated by the benzodiazepines, drugs that have marked antianxiety activity and are also effective muscle relaxants, anticonvulsants, and sedatives. Among the most widely used benzodiazepines in the USA are chlordiazepoxide (Librium) and flurazepam (Dalmane), but in these and other forms, they are used throughout the world. The benzodiazepines exert their facilitatory

Figure 4–24. Formation and metabolism of GABA.

Figure 4–25. Probable structure of the α and β subunits of the GABA$_A$ receptor in a cell membrane. The triangles identify potential sites for N-glycosylation. The receptor, which is made up of 2 α and 2 β subunits or, alternatively, of 4 or 5 other subunits, forms a Cl⁻ channel. (Reproduced, with permission, from Schofield PR et al: Sequence and functional expression of the GABA$_A$ receptor shows a ligand-gated receptor superfamily. *Nature* 1987;**328**:221.)

effect by binding to the α subunits. At least in part, barbiturates and alcohol also act by facilitating Cl⁻ conductance through the Cl⁻ channel. In addition, metabolites of the steroid hormones progesterone and deoxycorticosterone bind to GABA$_A$ receptors and increase Cl⁻ conductance. It has been known for may years that progesterone and deoxycorticosterone are sleep-inducing and anesthetic in large doses, and these effects are due to their action on GABA$_A$ receptors.

Glycine

By its action on the NMDA receptors, glycine has an excitatory effect in the brain. However, glycine is the mediator secreted by interneurons that is responsible for direct inhibition in the brain stem and spinal cord. Like GABA, it acts by increasing Cl⁻ conductance. Its action is antagonized by strychnine and tetanus toxin. The clinical picture of convulsions and muscular hyperactivity produced by tetanus toxin and strychnine emphasizes the importance of postsynaptic inhibition in normal neural function. The glycine receptor responsible for inhibition is a Cl⁻ channel formed by a large glycoprotein that contains 3 polypeptide subunits. One of these binds strychnine and has an appreciable amino acid sequence homology to proteins in the nicotinic acetylcholine receptors.

Substance P & Other Tachykinins

Substance P is a polypeptide containing 11 amino acid residues that is found in the intestine, various peripheral nerves, and many parts of the central nervous system. Its structure is shown in Table 26–2. It is one of a family of 6 mammalian polypeptides called tachykinins that differ at the N-terminal end but have in common the C-terminal sequence of Phe-X-Gly-Leu-Met-NH$_2$, where X is Val, His, Lys, or Phe. The members of the family are listed in Table 4–3. There are many related tachykinins in other vertebrates and in invertebrates.

The mammalian tachykinins are encoded by 2 genes. The **neurokinin B gene** encodes only one known polypeptide, neurokinin B. The **substance P/neurokinin A gene** encodes the remaining 5 polypeptides by differential splicing, which forms 3 different mRNAs, and further posttranslational processing.

There are 3 neurokinin receptors. Two of these, the substance P and the neuropeptide K receptors, have been cloned and found to be serpentine receptors related to G proteins. Activation of the substance P receptor causes activation of phospholipase C and increased formation of IP$_3$ and DAG.

Substance P is found in high concentration in the endings of primary afferent neurons in the spinal cord, and it is probably the mediator at the first synapse in the pathways for slow pain (see Chapter 7). It is also found in high concentration in the nigrostriatal system, where its concentration is proportionate to that of dopamine, and in the hypothalamus, where it may play a role in neuroendocrine regulation. Upon injection into the skin, it causes redness and swelling, and it is probably the mediator released by nerve fibers that is responsible for the axon reflex (see Chapter 32). In the intestine, it appears to be involved in the myenteric reflex (see Chapter 26). The functions of the other tachykinins are unsettled, although given their common C-terminal structure, they probably have overlapping actions.

Opioid Peptides

The brain and the gastrointestinal tract contain receptors that bind morphine. The search for endogenous ligands for these receptors led to the discovery of 2 closely related pentapeptides, called **enkephalins** (Table 4–4), that bind to these opiate receptors. One contains methionine (**met-enkephalin**), and one contains isoleucine (**leu-enkephalin**). These and other peptides that bind to opiate receptors are called **opioid peptides.** The enkephalins are found in nerve endings in the gastrointestinal tract and many different parts of the brain, and they appear to function as synaptic transmitters. They are found in the substantia gelatinosa and have analgesic activity when injected into the brain stem. They also decrease intestinal motility (see Chapter 26).

Like other small peptides, the opioid peptides are synthesized as part of larger precursor molecules (see Chapter 1). At least 18 active opioid peptides have been identified. Unlike other peptides, however, the opioid peptides have a number of different precursors. Each has a prepro form and a pro form from which the signal peptide has been cleaved. The 3 precursors that have been characterized, and the opioid peptides they produce are shown in Table 4–4. **Proenkephalin** was first identified in the adrenal medulla (see Chapter 20), but it is also the precursor for met-enkephalin and leu-enkephalin in the brain. Each proenkephalin molecule contains 4 met-enkephalins, one leu-eukephalin, one octapeptide, and one heptapeptide. **Pro-opiomelanocortin,** a large precursor molecule found in the anterior and intermediate lobes of the pituitary gland and the brain, contains β-endorphin, a polypeptide of 31 amino acid residues that has met-enkephalin at its N terminus (see Chapter 22). Other shorter endorphins may also be produced, and the precursor molecule also produces ACTH and MSHs. There are separate enkephalin-secreting and β-endorphin-secreting systems of neurons in the brain (see Chapter 15). β-Endorphin is also secreted into the bloodstream by the pituitary gland. A third precursor molecule is **prodynorphin,** a protein that contains 3 leu-enkephalin residues associated with dynorphin and neoendorphin. Dynorphin 1-17 is found in the duodenum and dynorphin 1-8 in the posterior pituitary and hypothalamus. α- and β-neoendorphins are also found in the hypothalamus. The reasons for the existence of multiple opioid peptide precursors and for the presence of the peptides in the circulation as well as in the brain and the gastrointestinal tract are presently unknown.

Opiate receptors have been studied in detail, and 3 are now established: μ, κ, and δ. They differ in pharmacologic properties, distribution in the brain and elsewhere, and affinity for various opioid peptides. All 3 appear to be coupled to G proteins. Activation of μ receptors increases K$^+$ conductance, hyperpolarizing central neurons and primary afferents. Activation of κ receptors and δ receptors closes Ca^{2+} channels. However, as yet none of the receptors has been closed. The main physiologic effects of stimulation of each of these receptors are summarized in Table 4–5. The natural ligand for μ receptors is probably β-endorphin, whereas the natural ligands for δ receptors are enkephalins and the natural ligands for κ receptors are dynorphins.

It is interesting that in addition to opioid peptides, small amounts of endogenously produced morphine

Table 4–3. Mammalian tachykinins.

Gene	Polypeptide Products	Receptors
SP/NKA	Substance P Neurokinin A Neuropeptide K Neuropeptide α Neurokinin A (3–10)	Substance P (NK-1) Neuropeptide K (NK-2)
NKB	Neurokinin B	Neurokinin B (NK-3)

Table 4–4. Opioid peptides and their precursors.[1]

Precursor	Opioid Peptides	Structures
Proenkephalin (see Chapter 20)	Met-enkephalin Leu-enkephalin Octapeptide Heptapeptide	$Tyr-Gly-Gly-Phe-Met_5$ $Tyr-Gly-Gly-Phe-Leu_5$ $Tyr-Gly-Gly-Phe-Met-Arg-Gly-Leu_8$ $Tyr-Gly-Gly-Phe-Met-Arg-Phe_7$
Pro-opiomelanocortin (see Chapter 22)	β-Endorphin Other endorphins	See Chapter 22 See Chapter 22
Prodynorphin	Dynorphin 1–8 Dynorphin 1–17 α-Neoendorphin β-Neoendorphin	$Tyr-Gly-Gly-Phe-Leu-Arg-Arg-Ile_8$ $Try-Gly-Gly-Phe-Leu-Arg-Arg-Ile-Arg-Pro-Lys-Leu-Lys-Trp-Asp-Asn-Gln_{17}$ $Tyr-Gly-Gly-Phe-Leu-Arg-Lys-Tyr-Pro-Lys_{10}$ $Tyr-Gly-Gly-Phe-Leu-Arg-Lys-Tyr-Pro_9$

[1]Modified from Rossier J: Opioid peptides have found their roots. *Nature* 1981;**298**:221.

and codeine are found in the mammalian brain; however, the significance of this finding is unknown.

Enkephalins are metabolized primarily by 2 peptidases: enkephalinase A, which splits the Gly-Phe bond, and enkephalinase B, which splits the Gly-Gly bond. Aminopepidase, which splits the Tyr-Gly bond, also contributes to their metabolism.

Other Polypeptides

Numerous other polypeptides are found in the brain. The hypophyseotropic hormones (see Chapter 14) are found in different parts of the nervous system, and it seems clear that many (perhaps all) of them function as neurotransmitters as well as hormones. Vasopressin and oxytocin are not only secreted as hormones but are also present in neurons that project to the brain stem and spinal cord. The brain also contains bradykinin, angiotensin II, and endothelin (see Chapters 24 and 31). Insulinlike molecules are found in the brain, but the structures of the mRNAs that produce them indicate that they do not have the same structure as insulin. They may be growth factors. The gastrointestinal hormones VIP and the C-terminal octapeptide fragment of CCK are also found in the brain, the former in highest concentration in the hypothalamus and the latter in highest concentration in

Table 4–5. Physiologic effects produced by stimulation of opiate receptors.[1]

Receptor	Effect
μ	Analgesia Respiratory depression Constipation Euphoria Dependence Miosis
κ	Analgesia Diuresis Sedation Miosis Dysphoria
δ	Analgesia

[1]Modified from McKnight AT, Rees DC: *Neurotransmission* 1991; (Oct) **7**:1.

the cerebral cortex. Gastrin, neurotensin, and gastrin-releasing peptide are also found in the gastrointestinal tract and brain. The neurotensin and the VIP receptors have been cloned and shown to be serpentine receptors. The hypothalamus contains both gastrin 17 and gastrin 34 (see Chapter 26). VIP produces vasodilation and is found in vasomotor nerve fibers. The functions of these peptides in the nervous system are unknown.

Calcitonin gene-related peptide (CGRP) is a polypeptide that in rats and humans exists in 2 forms, CGRPα and CGRPβ. These 2 forms differ by only one amino acid residue, yet they are encoded by different genes. In rats, and presumably in humans, CGRPβ is present in the gastrointestinal tract, whereas CGRPα is found in primary afferent neurons, neurons by which taste impulses project to the thalamus, and neurons in the medial forebrain bundle. It is also present along with substance P in the branches of primary afferent neurons that end near blood vessels. CGRPα and the calcium-lowering hormone calcitonin (see Chapter 21) are both products of the calcitonin gene. However, in the thyroid gland, splicing produces the mRNA that codes for calcitonin, whereas in the brain, alternative splicing produces the mRNA that codes for CGRPα.

Neuropeptide Y is a polypeptide containing 36 amino acid residues that is closely related to pancreatic polypeptide (see Chapter 19). It is present in many parts of the brain and the autonomic nervous system. In the autonomic nervous system, although not in the brain, much of it is located in noradrenergic neurons, from which it is released by high-frequency stimulation. It augments the vasoconstrictor effects of norepinephrine. Circulating neuropeptide Y from sympathetic nerves increases with severe exercise in humans.

Purinergic Transmitters

In ECF, the purine **adenosine** (see Chapter 17) binds to at least 2 different receptors, the A_1 receptor, which inhibits adenylate cyclase, and the A_2 receptor, which activates it. Adenosine has widespread effects in the body, including regulation of coronary blood flow and effects on the kidney and the immune system. In the brain there are A_2 receptors in the striatum, olfactory tuberule, and nucleus accumbens. Adenosine is a CNS depressant, and its actions are antagonized by caffeine and theophylline.

Much of the adenosine in the ECF comes from ATP, which is released with norepinephrine, dopamine, GABA, glutamate, acetylcholine, and histamine when they are secreted by neurons. ATP itself also functions as a transmitter, and 2 ATP receptors, P_1 and P_2, have been characterized pharmacologically.

Other Possible Transmitters

Nitric oxide (NO), a compound released by the endothelium of blood vessels as endothelium-derived relaxing factor (EDRF), is also produced in the brain. It activates guanylate cyclase (see Chapter 1) and appears to be a "transmitter" with unique properties. It is not contained in granules or vesicles but is released in neurons and may be the signal by which postsynaptic neurons communicate with presynaptic endings in LTP and LTD (see below). Its role in the control of vascular smooth muscle is discussed in Chapter 31. Carbon monoxide (CO) is also produced in the body and may be a similar "transmitter" in the nervous system.

Receptors that have a high affinity for Δ^9-tetrahydrocannabinol, the psychoactive ingredient in **marijuana,** have been cloned. These receptors trigger a decrease in the intracellular cyclic AMP concentration and are common in parts of the cerebellum, hippocampus, and cerebral cortex. However, their endogenous ligand and their physiologic role remain to be determined.

Prostaglandins, fatty acid derivatives found in many cells (see Chapter 17), are also found in the nervous system. They are present in nerve-ending fractions of brain homogenates and are released from neural tissue in vitro. However, they appear to exert their effects by modulating reactions mediated by cyclic AMP rather than by functioning as synaptic transmitters.

Finally, steroids affect brain function, although they are not neurotransmitters in the usual sense. Circulating steroids enter the brain with ease, and there are numerous sex steroid and glucocorticoid receptors in neurons. In addition to acting in the established fashion by binding to DNA, some steroids seem to act rapidly by a direct effect on cell membranes. The role of steroids in neuroendocrine control is discussed in Chapter 14. The effects of steroids on GABA receptors is discussed above. Evidence has now accumulated that the brain can produce some hormonally active steroids from simpler steroid precursors, and the term **neurosteroids** has been coined to refer to these products. However, the exact role of these steroids in the regulation of brain function remains to be determined.

Cotransmitters

Numerous examples have now been described in which neurons contain and secrete 2 and even 3 transmitters. The **cotransmitters** in these situations are often a catecholamine or serotonin plus a polypeptide, and examples of coexistence of a polypeptide with GABA or acetylcholine have been described. Coexistence of 2 polypeptides and coexistence of GABA with various catecholamines or acetylcholine also occur. Some neurons in the brain stem contain serotonin, substance P, and TRH. Many cholinergic neurons contain VIP, and many noradrenergic and adrenergic neurons contain ATP and neuropeptide Y. Neurons containing multiple transmitters often exist side by side with neurons containing a single transmitter. It has been suggested that low-frequency discharge of the neurons releases the small transmitter molecules, whereas high-frequency stimulation releases the polypeptide cotransmitters as well; however, this has not been proved. The physiologic significance of cotransmitters is still obscure. However, the VIP secreted with acetylcholine potentiates the postsynaptic actions of acetylcholine, and neuropeptide Y potentiates some of the actions of norepinephrine.

SYNAPTIC PLASTICITY & LEARNING

Long-term changes in synaptic function can occur as a result of the history of discharge at a synapse; ie, synaptic conduction can be strengthened or weakened on the basis of past experience. These changes are of great interest because they obviously represent forms of learning and memory. They can be presynaptic or postsynaptic in location (Fig 4–26)

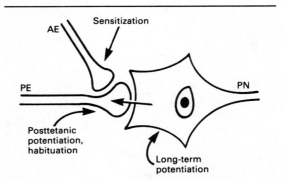

Figure 4–26. Sites of processes producing changes in the strength of synaptic transmission that resemble learning. AE, axo-axonal ending mediating presynaptic facilitation; PE, ending of presynaptic neuron. PN, postsynaptic neuron. In sensitization, discharge of AE brings about long-term changes in PE. There is evidence that long-term potentiation has a presynaptic as well as a postsynaptic component, which is mediated by a chemical messenger from the post- to the presynaptic neuron (arrow).

Posttetanic Potentiation

One form of plastic change is **posttetanic potentiation,** the production of enhanced postsynaptic potentials in response to stimulation. This enhancement lasts for minutes to hours and occurs after a brief (tetanizing) train of stimuli in the presynaptic neuron. The tetanizing stimulation causes Ca^{2+} to accumulate in the presynaptic neuron to such a degree that the intracellular binding sites that keep cytoplasmic Ca^{2+} low are overwhelmed.

Habituation

When a stimulus is benign and is repeated over and over, the response to the stimulus gradually disappears **(habituation).** This is associated with decreased release of neurotransmitter from the presynaptic terminal because of decreased intracellular Ca^{2+}. The decrease in intracellular Ca^{2+} is due to a gradual inactivation of Ca^{2+} channels. It can be short-term, or it can be prolonged if exposure to the benign stimulus is repeated many times.

Sensitization

Sensitization is the prolonged occurrence of augmented postsynaptic responses after a stimulus to which an animal has become habituated is paired once or several times with a noxious stimulus. At least in the sea snail *Aplysia,* the noxious stimulus causes discharge of serotonergic neurons that end on the presynaptic endings of sensory neurons. This initiates the sequence of events summarized in Fig 4–11. Thus, sensitization is due to presynaptic facilitation.

Sensitization may occur as a transient response, or if it is reinforced by additional pairings of the noxious stimulus and the initial stimulus, it can exhibit features of short-term or long-term memory. The short-term prolongation of sensitization is due to a Ca^{2+}-mediated change in adenylate cyclase that leads to a greater production of cyclic AMP. The long-term potentiation also involves protein synthesis and growth of the presynaptic and postsynaptic neurons and their connections.

Long-Term Potentiation

Long-term potentiation (LTP) is a rapidly developing persistent enhancement of the response to synaptic transmission produced by brief, repetitive activation of presynaptic neurons. It resembles posttetanic potentiation, but is much more prolonged and can last for days. Unlike posttetanic potentiation, it is due, at least in part, to an increase in intracellular Ca^{2+} in the postsynaptic rather than the presynaptic neuron. It has been the focus of much attention because it appears to represent a form of learning and memory. It occurs in various parts of the nervous system, but has been studied in greatest detail in the hippocampus. In the hippocampus, glutamate released from presynaptic terminals causes depolarization of

the postsynaptic neuron via AMPA and kainate receptors. This relieves the Mg^{2+} block of NMDA receptors, and Ca^{2+} enters the neurons with Na^+. The subsequent events are not completely understood, but it appears that the Ca^{2+} acts by both Ca^{2+}/calmodulin kinase II and protein kinase C (Table 1–9) to alter phosphorylation of intracellular proteins in a way that leads to greater EPSPs in response to stimulation of presynaptic terminals. However, there appears to be a presynaptic as well as a postsynaptic component in the production of LTP. It has been suggested that the postsynaptic neuron sends a signal to the presynaptic neuron, possibly by releasing arachidonic acid (arachidonate) or NO (see Chapter 1). The signal leads to increased quantal release of glutamate upon afferent stimulation.

Long-term Depression

Long-term depression (LTD), a condition that is the opposite of LTP, has also been observed to occur in the central nervous system. It occurs when neurons are partially depolarized by afferent input, but not by as much as 20 mV, the approximate threshold at which NMDA receptors open and permit Ca^{2+} influx. Thus, neurons have a narrow window for this phenomenon: synaptic use that maintains the neurons near the resting potential or below has no effect on synaptic strength, LTD develops with depolarization up to about 20 mV, and LTP develops with greater depolarization. However, the details of LTD and its exact physiologic role remain to be determined. In the cerebellum, simultaneous firing of climbing and parallel fibers causes a long-term depression of conduction of the parallel fiber-Purkinje cell synapses that is apparently mediated by NO produced by stimulation of the climbing fibers. Note that in this situation, the putative effect of NO is inhibitory, whereas in LTP in the hippocampus, it is stimulatory. In both cases, the mechanism by which the effect of NO is produced is unknown.

NEUROMUSCULAR TRANSMISSION

NEUROMUSCULAR JUNCTION

Anatomy

As the axon supplying a skeletal muscle fiber approaches its termination, it loses its myelin sheath and divides into a number of terminal buttons or end-feet (Fig 4–27). The end-feet contain many small, clear vesicles that contain acetylcholine, the transmitter at these junctions. The endings fit into depressions in the **motor end-plate,** the thickened portion of the muscle membrane of the junction. Underneath the

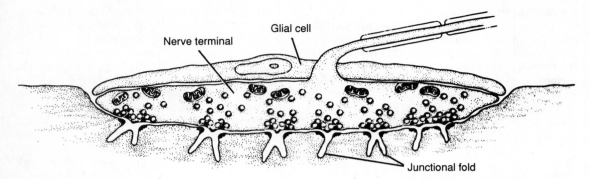

Figure 4–27. Neuromuscular junction. Note that the clear vesicles containing acetylcholine are most numerous at active zones in the nerve terminal. The zones are located over junctional folds in the motor end-plate. (Reprinted by permission of the publishers from Dowling JE: *Neurons and Networks: An Introduction to Neuroscience.* The Belknap Press of Harvard University Press. Copyright ©1992 by the President and Fellows of Harvard College.)

nerve ending, the muscle membrane of the end-plate is thrown into **junctional folds.** The space between the nerve and the thickened muscle membrane is comparable to the synaptic cleft at synapses. The whole structure is known as the **neuromuscular** or **myoneural junction.** Only one nerve fiber ends on each end-plate, with no convergence of multiple inputs.

Sequence of Events During Transmission

The events occurring during transmission of impulses from the motor nerve to the muscle (see Table 3–2) are somewhat similar to those occurring at other synapses. The impulse arriving in the end of the motor neuron increases the permeability of the endings to Ca^{2+}. Ca^{2+} enters the endings and triggers a marked increase in exocytosis of the acetylcholine-containing vesicles. The acetylcholine diffuses to the nicotinic acetylcholine receptors (Fig 4–16), which are concentrated on the junctional folds of the membrane of the motor end-plate. Binding of acetylcholine to these receptors increases the Na^+ and K^+ conductance of the membrane, and the resultant influx of Na^+ produces a depolarizing potential, the **end-plate potential.** The current sink created by this local potential depolarizes the adjacent muscle membrane to its firing level. Action potentials are generated on either side of the end-plate and are conducted away from the end-plate in both directions along the muscle fiber. The muscle action potential, in turn, initiates muscle contraction, as described in Chapter 3.

End-Plate Potential

An average end-plate contains about 50 million acetylcholine receptors. Each nerve impulse releases about 60 acetylcholine vesicles, and each vesicle contains about 10,000 molecules of the neurotransmitter. This amount is enough to activate about 10 times the number of acetylcholine receptors needed to produce a full end-plate potential. Therefore, a propagated response in the muscle is regularly produced, and this large response obscures the end-plate potential. However, the end-plate potential can be seen if the 10-fold safety factor is overcome and the potential is reduced to a size that is insufficient to fire the adjacent muscle membrane. This can be accomplished by administration of small doses of curare, a drug that competes with acetylcholine for binding to nicotinic acetylcholine receptors. The response is then recorded only at the end-plate region and decreases exponentially away from it. Under these conditions, end-plate potentials can be shown to undergo temporal summation.

Quantal Release of Transmitter

Small quanta ("packets") of acetylcholine are released randomly from the nerve cell membrane at rest, each producing a minute depolarizing spike called a **miniature end-plate potential,** which is about 0.5 mV in amplitude. The size of the quanta of acetylcholine released in this way varies directly with the Ca^{2+} concentration and inversely with the Mg^{2+} concentration at the end-plate. When a nerve impulse reaches the ending, the number of quanta released increases by several orders of magnitude, and the result is the large end-plate potential that exceeds the firing level of the muscle fiber.

Quantal release of acetylcholine similar to that seen at the myoneural junction has been observed at other cholinergic synapses, and similar processes are presumably operating at noradrenergic and other synaptic junctions.

Myasthenia Gravis and Lambert-Eaton Syndrome

Myasthenia gravis is a serious and sometimes fatal disease in which skeletal muscles are weak and tire easily. It is now known that the disease is caused by the formation of circulating antibodies to the nicotinic acetylcholine receptors. These antibodies destroy some of the receptors and bind others to neighboring receptors, triggering their removal by endocytosis (see Chapter 1). The reason for the development of autoimmunity to acetylcholine receptors in this disease is still unknown.

Another condition that resembles myasthenia gravis is **Lambert-Eaton Syndrome.** In this condition, muscle weakness is caused by antibodies against one of the Ca^{2+} channels in the nerve endings at the neuromuscular junction. This decreases the normal Ca^{2+} influx that causes acetylcholine release.

NERVE ENDINGS IN SMOOTH & CARDIAC MUSCLE

Anatomy

The postganglionic neurons in the various smooth muscles that have been studied in detail branch extensively and come in close contact with the muscle cells (Fig 4–28). Some of these nerve fibers contain clear vesicles and are cholinergic, whereas others contain the characteristic dense-core granules within vesicles of 2 sizes that are known to contain norepinephrine. There are no recognizable end-plates or other post-

synaptic specializations. The nerve fibers run along the membranes of the muscle cells and sometimes groove their surfaces. The multiple branches of the noradrenergic and, presumably, the cholinergic neurons are beaded with enlargements (**varicosities**) that are not covered by Schwann cells and contain synaptic vesicles (Fig 4–28). In noradrenergic neurons, the varicosities are about 5 μm apart, with up to 20,000 varicosities per neuron. Transmitter is apparently liberated at each varicosity, ie, at many locations along each axon. This arrangement permits one neuron to innervate many effector cells. The type of contact in which a neuron forms a synapse on the surface of another neuron or a smooth muscle cell and then passes on to make similar contacts with other cells is called a **synapse en passant.**

In the heart, cholinergic and noradrenergic nerve fibers end on the sinoatrial node, the atrioventricular node, and the bundle of His. Noradrenergic fibers also innervate the ventricular muscle. The exact nature of the endings on nodal tissue is not known. In the ventricle, the contacts between the noradrenergic fibers and the cardiac muscle fibers resemble those found in smooth muscle.

Electrical Responses

In smooth muscles in which noradrenergic discharge is excitatory, stimulation of the noradrenergic nerves produces discrete partial depolarizations that look like small end-plate potentials and are called **excitatory junction potentials (EJPs).** These potentials summate with repeated stimuli. Similar EJPs are seen in tissues excited by cholinergic discharges. In tissues inhibited by noradrenergic stimuli, hyperpolarizing **inhibitory junction potentials (IJPs)** have been produced by stimulation of the noradrenergic nerves.

These electrical responses are observed in many smooth muscle cells when a single nerve is stimulated, but their latency varies. This finding is consistent with the synapse en passant arrangement described above, but it could also be explained by transmission of the junction responses from cell to cell across low-resistance junctions or by diffusion of transmitter from its site of release to many smooth muscle cells. Miniature excitatory junction potentials similar to the miniature end-plate potentials in the skeletal muscle have been observed in some smooth muscle preparations, but they show considerable variation in size and duration. They may represent responses to single packets of transmitter, with the variation due to diffusion of the transmitter for variable distances.

Figure 4–28. Endings of postganglionic autonomic neurons on smooth muscle. (Reproduced, with permission, from Kandel ER, Schwartz JH [editors]: *Principles of Neural Science,* 2nd ed. Elsevier, 1985.)

DENERVATION HYPERSENSITIVITY

When the motor nerve to skeletal muscle is cut and allowed to degenerate, the muscle gradually becomes extremely sensitive to acetylcholine. This **denervation hypersensitivity** or **supersensitivity** is also seen in smooth muscle. Smooth muscle, unlike skeletal muscle, does not atrophy when denervated, but it becomes hyperresponsive to the chemical mediator that normally activates it. Denervated exocrine glands, except for sweat glands, also become hypersensitive. A good example of denervation hypersensitivity is the response of the denervated iris. If the postganglionic sympathetic nerves to one pupil are cut in an experimental animal and, after several weeks, norepinephrine is injected intravenously, the denervated pupil dilates widely. A much smaller, less prolonged response is observed on the intact side.

The reactions triggered by section of an axon are summarized in Fig 4–29. Hypersensitivity of the postsynaptic structure to the transmitter previously secreted by the axon endings is a general phenomenon, largely due to the synthesis or activation of more receptors. There is in addition orthograde degeneration (**wallerian degeneration;** see Chapter 2) and retrograde degeneration of the axon stump to the nearest collateral (**sustaining collateral**). A series of changes occur in the cell body that include a decrease in Nissl substance (chromatolysis). The nerve then starts to regrow, with multiple small branches projecting along the path the axon previously followed (regenerative sprouting).

When higher centers in the nervous system are destroyed, the activity of the lower centers they control is generally increased ("release phenomenon"). The increased activity may be due in part to denervation hypersensitivity of the lower centers. Indeed, one theory holds that many of the signs and symptoms of neurologic disease are due to denervation hypersensitivity of various groups of neurons in the brain.

Hypersensitivity is limited to the structures immediately innervated by the destroyed neurons and fails to develop in neurons and muscle farther "downstream." Suprasegmental spinal cord lesions do not lead to hypersensitivity of the paralyzed skeletal muscles to acetylcholine, and destruction of the preganglionic autonomic nerves to visceral structures does not cause hypersensitivity of the denervated viscera. This fact has practical implications in the treatment of diseases due to spasm of the blood vessels in the extremities. For example, if the upper extremity is sympathectomized by removing the upper part of the ganglion chain and the stellate ganglion, the hypersensitive smooth muscle in the vessel walls is stimu-

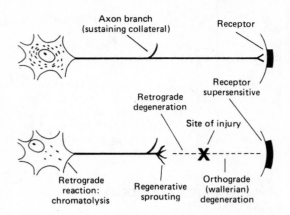

Figure 4–29. Summary of changes occurring in a neuron and the structure it innervates when its axon is crushed or cut at the point marked X. (Modified from D Ries.)

lated by circulating norepinephrine, and episodic vasospasm continues to occur. However, if preganglionic sympathectomy of the arm is performed by cutting the ganglion chain below the third ganglion (to interrupt ascending preganglionic fibers) and the white rami of the first 3 thoracic nerves, no hypersensitivity results.

Denervation hypersensitivity has multiple causes. As noted in Chapter 1, a deficiency of a given chemical messenger generally produces an increase in the number of its receptors. In denervated skeletal muscle, there is an increase in the area of the muscle membrane sensitive to acetylcholine. Normally, only the end-plate region contains nicotinic acetylcholine receptors, and they are of the adult ε subunit-containing type (see above). After denervation, the sensitivity of the end-plate is no greater, but acetylcholine receptors of the fetal δ subunit-containing type appear over large portions of the muscle membrane. These disappear and the sensitivity returns to normal if the nerve regrows. A similar spread of acetylcholine sensitivity has been demonstrated in denervated postganglionic cholinergic neurons. At endings where norepinephrine is normally secreted, another factor is lack of reuptake of secreted catecholamines; the nerve endings in normal tissue take up large amounts of norepinephrine, and after they have degenerated, norepinephrine reaching the receptors from other sites has a greater effect than it otherwise would.

Initiation of Impulses in Sense Organs

5

INTRODUCTION

Information about the internal and external environment reaches the central nervous system via a variety of **sensory receptors.** These receptors are transducers that convert various forms of energy in the environment into action potentials in neurons. The characteristics of these receptors, the way they generate impulses in afferent neurons, and the general principles of "laws" that apply to sensation are considered in this chapter.

SENSE ORGANS & RECEPTORS

It is worth noting that the term receptor is used in physiology to refer not only to sensory receptors but also, in a very different sense, to proteins that bind neurotransmitters, hormones, and other substances with great affinity and specificity as a first step in initiating specific physiologic responses (see Chapter 1). The sensory receptor may be part of a neuron or a specialized cell that generates action potentials in neurons. The receptor is often associated with nonneural cells that surround it, forming a **sense organ.** The forms of energy converted by the receptors include, for example, mechanical (touch-pressure), thermal (degrees of warmth), electromagnetic (light), and chemical energy (odor, taste, and O_2 content of blood). The receptors in each of the sense organs are adapted to respond to one particular form of energy at a much lower threshold than other receptors respond to this form of energy. The particular form of energy to which a receptor is most sensitive is called its **adequate stimulus.** The adequate stimulus for the rods and cones in the eye, for example, is light. Receptors do respond to forms of energy other than their adequate stimulus, but the threshold for these nonspecific responses is much higher. Pressure on the eyeball will stimulate the rods and cones, for example, but the threshold of these receptors to pressure is much higher than the threshold of the pressure receptors in the skin.

THE SENSES

Sensory Modalities

Because the sensory receptors are specialized to respond to one particular form of energy and because many variables in the environment are perceived, it follows that there must be many different types of receptors. We learn in elementary school that there are "5 senses," but the inadequacy of this dictum is apparent if we list the major sensory modalities and their receptors in humans. The list (Table 5–1) includes at least 11 conscious senses. There are, in addition, a large number of sensory receptors which relay information that does not reach consciousness. For example, the muscle spindles provide information about muscle length, and other receptors provide information about such variables as the arterial blood pressure, the temperature of the blood in the head, and the pH of the cerebrospinal fluid. The existence of other receptors of this type is suspected, and future research will undoubtedly add to the list of "unconscious senses." Furthermore, any listing of the senses is bound to be arbitrary. The rods and cones, for example, respond maximally to light of different wavelengths, and there are different cones for each of the 3 primary colors. There are 4 different modalities of taste—sweet, salt, sour, and bitter—and each is subserved by a more or less distinct type of taste bud. Sounds of different pitches are heard primarily because different groups of hair cells in the organ of Corti are activated maximally by sound waves of different frequencies. Whether these various responses to light, taste, and sound should be considered separate senses is a semantic question that in the present context is largely academic.

Classifications of Sense Organs

Numerous attempts have been made to classify the senses into groups, but none has been entirely successful. Traditionally, the special senses are smell, vision, hearing, rotational and linear acceleration, and taste; the cutaneous senses are those with receptors in the skin; and the visceral senses are those concerned with perception of the internal environment. Pain from visceral structures is usually classified as a vis-

Table 5–1. Principal sensory modalities. (The first 11 are conscious sensations.)

Sensory Modality	Receptor	Sense Organ
Vision	Rods and cones	Eye
Hearing	Hair cells	Ear (organ of Corti)
Smell	Olfactory neurons	Olfactory mucous membrane
Taste	Taste receptor cells	Taste bud
Rotational acceleration	Hair cells	Ear (semicircular canals)
Linear acceleration	Hair cells	Ear (utricle and saccule)
Touch-pressure	Nerve endings	Various[1]
Warmth	Nerve endings	Various[1]
Cold	Nerve endings	Various[1]
Pain	Naked nerve endings	
Joint position and movement	Nerve endings	Various[1]
Muscle length	Nerve endings	Muscle spindle
Muscle tension	Nerve endings	Golgi tendon organ
Arterial blood pressure	Nerve endings	Stretch receptors in carotid sinus and aortic arch
Central venous pressure	Nerve endings	Stretch receptors in walls of great viens, atria
Inflation of lung	Nerve endings	Stretch receptors in lung parenchyma
Temperature of blood in head	Neurons in hypothalamus	
Arterial PO_2	Glomus cells	Carotid and aortic bodies
pH of CSF	Receptors on ventral surface of medulla oblongata	
Osmotic pressure of plasma	Cells in OVLT and possibly other circumventricular organs in anterior hypothalamus	
Arteriovenous blood glucose difference	Cells in hypothalamus (glucostats)	

[1]See text.

ceral sensation. Another classification of the various receptors divides them into (1) teleceptors ("distance receivers"), which are concerned with events at a distance; (2) exteroceptors, which are concerned with the external environment near at hand; (3) interoceptors, which are concerned with the internal environment; and (4) proprioceptors, which provide information about the position of the body in space at any given instant. However, the conscious component of proprioception ("body image") is actually synthesized from information coming not only from receptors in and around joints but also from cutaneous touch and pressure receptors. Certain other special terms are sometimes used. Because pain fibers have connections that mediate strong and prepotent withdrawal reflexes (see Chapter 6) and because pain is initiated by potentially noxious or damaging stimuli, pain receptors are sometimes called **nociceptors.** The term **chemoreceptor** is used to refer to those receptors that are stimulated by a change in the chemical composition of the environment in which they are located. These include receptors for taste and smell as well as visceral receptors such as those sensitive to changes in the plasma level of O_2, pH, and osmolality.

Cutaneous Sense Organs

There are 4 cutaneous senses: touch-pressure (pressure is sustained touch), cold, warmth, and pain. The skin contains various types of sensory endings. These include naked nerve endings, expanded tips on sensory nerve terminals, and encapsulated endings.

The expanded endings include Merkel's disks and Ruffini endings (Fig 5–1), whereas the encapsulated endings include pacinian corpuscles, Meissner's corpuscles, and Krause's end-bulbs. Ruffini endings and pacinian corpuscles are also found in deep fibrous tissues. In addition, sensory nerves end around hair follicles. However, none of the expanded or encapsulated endings appear to be necessary for cutaneous

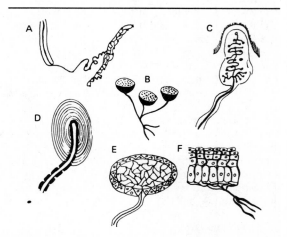

Figure 5–1. Sensory receptors in the skin. Ruffini endings **(A)** and Merkel's disks **(B)** are expanded ends of sensory nerve fibers. Meissner's corpuscles **(C),** pacinian corpuscles **(D),** and Krause's end-bulbs **(E)** are encapsulated endings. **(F)** Naked curve endings between cells in tissue.

sensation. Their distribution varies in different regions of the body, and it has been repeatedly demonstrated that all 4 cutaneous sensory modalities can be elicited from areas that on histologic examination contain only naked nerve endings. Where they are present, the expanded or encapsulated endings appear to function as mechanoreceptors that respond to tactile stimuli; Meissner's and pacinian corpuscles are rapidly adapting touch receptors, and Merkel's disks and Ruffini endings are slowly adapting touch receptors. The nerve endings around hair follicles mediate touch, and movements of hairs initiate tactile sensations. It should be emphasized that although cutaneous sensory receptors lack histologic specificity, they are physiologically specific. Thus, any given ending signals one and only one kind of cutaneous sensation.

ELECTRICAL & IONIC EVENTS IN RECEPTORS

Anatomic Relationships

The problem of how receptors convert energy into action potentials in the sensory nerves has been the subject of intensive study. In the complex sense organs such as those concerned with vision, hearing, equilibrium, and taste, there are separate receptor cells and synaptic junctions between receptors and afferent nerves. However, in most of the cutaneous sense organs, the receptors are specialized, histologically modified ends of sensory nerve fibers.

Pacinian corpuscles, which are touch receptors, have been studied in detail. Because of their relatively large size and accessibility in the mesentery of experimental animals, they can be isolated, studied with microelectrodes, and subjected to microdissection. Each capsule consists of the straight, unmyelinated ending of a sensory nerve fiber, 2 μm in diameter, surrounded by concentric lamellas of connective tissue that give the organ the appearance of a minute cocktail onion. The myelin sheath of the sensory nerve begins inside the corpuscle. The first node of Ranvier is also located inside, whereas the second is usually near the point at which the nerve fiber leaves the corpuscle (Fig 5–2).

Generator Potentials

Recording electrodes can be placed on the sensory nerve as it leaves a pacinian corpuscle and graded pressure applied to the corpuscle. When a small amount of pressure is applied, a nonpropagated depolarizing potential resembling an EPSP is recorded. This is called the **generator potential** or **receptor potential**. As the pressure is increased, the magnitude of the receptor potential increases. When the magnitude of the generator potential is about 10 mV, an action potential is generated in the sensory nerve.

Figure 5–2. Demonstration that the generator potential in a pacinian corpuscle originates in the nonmyelinated nerve terminal. **1:** The electrical responses to a pressure of x (record a), 2x (b), 3x (c), and 4x (d) were recorded. The strongest stimulus produced an action potential in the sensory nerve (e). **2:** Similar responses persisted after removal of the connective tissue capsule, except that the responses were more prolonged because of partial loss of adaptation. **3:** The generator responses persisted but the action potential was absent when the first node of Ranvier was blocked with pressure or narcotics (arrow). **4:** All responses disappeared when the sensory nerve was cut and allowed to degenerate before the experiment. (Modified from Loewenstein WR: Biological transducers. *Sci Am* [Aug] 1960;**203**:98. Copyright ©1960 by Scientific American, Inc. All rights reserved.)

As the pressure is further increased, the generator potential becomes even larger and the sensory nerve fires repetitively.

Source of the Generator Potential

By microdissection techniques, it has been shown that removal of the connective tissue lamellas from the unmyelinated nerve ending in a pacinian corpuscle does not abolish the generator potential. When the first node of Ranvier is blocked by pressure or narcotics, the generator potential is unaffected but conducted impulses are abolished (Fig 5–2). When the sensory nerve is sectioned and the nonmyelinated terminal is allowed to degenerate, no generator potential is formed. These and other experiments have established that the generator potential is produced in the unmyelinated nerve terminal. The nerve terminals also appear to be the site of origin of the generator potential in Merkel's disks. The receptor therefore converts mechanical energy into an electrical response, the magnitude of which is proportionate to the intensity of the stimulus. The generator potential in turn depolarizes the sensory nerve at the first node of Ranvier. Once the firing level is reached, an action potential is produced and the membrane repolarizes. If the generator potential is great enough, the neuron fires again as soon as it repolarizes, and it continues to fire as long as the generator potential is large enough to bring the membrane potential of the node to the firing level. Thus, the node converts the graded response of the receptor into action potentials, the frequency of which is proportionate to the magnitude of the applied stimuli.

Similar generator potentials in the muscle spindle have been studied. The relation between muscle length, which determines stimulus intensity in the spindle, and the size of the generator potential is shown in Fig 5–3, along with the relation between muscle length and frequency of action potentials in the afferent nerve fiber from the spindle. The frequency of the action potentials is generally related to the intensity of the stimulus by a power function (see below). Generator potentials occur in other sense organs, and in all, they appear to initiate depolarization in the sensory nerve fibers from the organ.

IONIC BASIS OF EXCITATION

In some receptors, mechanical distortion opens channels in the receptor membrane. The resultant influx of Na^+ produces the generator potential, and presumably the number of channels opened is proportionate to the intensity of the stimulus. This is probably the case in the pacinian corpuscles. In other receptors, such as the rods and cones, different mechanisms are responsible, and in some instances, the mechanisms are still unknown.

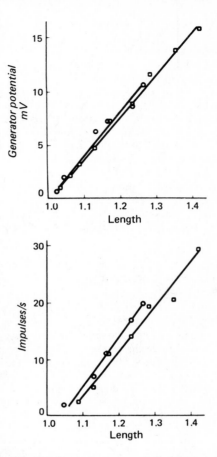

Figure 5–3. Relation between muscle length and size of generator potential **(top)** and impulse frequency **(bottom)** in crayfish stretch receptor. Squares and circles indicate values in 2 different preparations. (Reproduced, with permission, from Terzuolo CA, Washizu Y: Relation between stimulus strength, generator potential, and impulse frequency in stretch receptor of crustacea. *J Neurophysiol* 1962:**25**:56.)

Adaptation

When a maintained stimulus of constant strength is applied to a receptor, the frequency of the action potentials in its sensory nerve declines over time. This phenomenon is known as **adaptation.** The degree to which adaptation occurs varies with the type of sense organ (Fig 5–4). Touch adapts rapidly, and its receptors are called **phasic receptors.** On the other hand, the carotid sinus, the muscle spindles, and the organs for cold, pain, and lung inflation adapt very slowly and incompletely; the receptors involved are termed **tonic receptors.**

Pressure on a pacinian corpuscle produces a generator potential that decays rapidly despite continued application of the pressure. If the connective tissue capsule is removed and pressure is applied to the naked nerve ending, a generator potential is still pro-

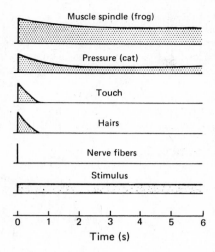

Figure 5–4. Adaptation. The height of the curve in each case indicates the frequency of the discharge in afferent nerve fibers at various times after beginning sustained stimulation. (Reproduced, with permission, from Adrian ED: *Basis of Sensation.* Christophers, 1928.)

duced, but it decays more slowly (Fig 5–2). The reason for the rapid decay when pressure is applied to the intact corpuscle is presumably lateral displacement of its connective tissue lamellas, lessening the pressure on the nerve ending. However, this is not the only factor producing adaptation; there is still a slow decline in the number of action potentials generated over time by a steady stimulus after removal of the outer lamellas from the pacinian corpuscle. This decline is due to accommodation of the sensory nerve fiber to the generator potential.

The slow, incomplete adaptation of the muscle spindle correlates with the fact that its generator potential is prolonged and decays very slowly when a steady stimulus is applied. Slow adaptation of the muscle spindles, carotid sinuses, and pain and cold receptors has some value to the individual. Muscle stretch plays a role in prolonged postural adjustments. The sensations of pain and cold are initiated by potentially noxious stimuli, and they would lose some of their warning value if their receptors showed marked adaptation. Carotid and aortic receptors operate continuously in the regulation of blood pressure, and adaptation of these receptors would limit the precision with which the regulatory system operates.

"CODING" OF SENSORY INFORMATION

There are variations in the speed of conduction and other characteristics of sensory nerve fibers (see Chapter 2), but action potentials are similar in all nerves. The action potentials in the nerve from a touch receptor, for example, are essentially identical to those in the nerve from a warmth receptor. This raises the question of why stimulation of a touch receptor causes a sensation of touch and not of warmth. It also raises the question of how it is possible to tell whether the touch is light or heavy.

Doctrine of Specific Nerve Energies

The sensation evoked by impulses generated in a receptor depends upon the specific part of the brain they ultimately activate. The specific sensory pathways are discrete from sense organ to cortex. Therefore, when the nerve pathways from a particular sense organ are stimulated, the sensation evoked is that for which the receptor is specialized no matter how or where along the pathway the activity is initiated. This principle, first enunciated by Müller, has been given the rather cumbersome name of the **doctrine of specific nerve energies.** For example, if the sensory nerve from a pacinian corpuscle in the hand is stimulated by pressure at the elbow or by irritation from a tumor in the brachial plexus, the sensation evoked is touch. Similarly, if a fine enough electrode could be inserted into the appropriate fibers of the dorsal columns of the spinal cord, the thalamus, or the postcentral gyrus of the cerebral cortex, the sensation produced by stimulation would be touch. This doctrine has been questioned from time to time, especially by those who claim that pain is produced by overstimulation of a variety of receptors. However, the overstimulation hypothesis has been largely discredited, and the principle of specific nerve energies remains one of the cornerstones of sensory physiology.

Projection

No matter where a particular sensory pathway is stimulated along its course to the cortex, the conscious sensation produced is referred to the location of the receptor. This principle is called the **law of projection.** Cortical stimulation experiments during neurosurgical procedures on conscious patients illustrate this phenomenon. For example, when the cortical receiving area for impulses from the left hand is stimulated, the patient reports sensation in the left hand, not in the head. Another dramatic example is seen in amputees. These patients may complain, often bitterly, of pain and proprioceptive sensations in the absent limb ("phantom limb"). The sensations are due in part to pressure on the stump of the amputated limb. This pressure initiates impulses in nerve fibers that previously came from sense organs in the amputated limb, and the sensations evoked are projected to where the receptors used to be.

Intensity Discrimination

There are 2 ways in which information about intensity of stimuli is transmitted to the brain: by variation in the frequency of the action potentials generated by the activity in a given receptor, and by variation in the number of receptors activated. It has long been taught that the magnitude of the sensation felt is proportionate to the log of the intensity of the stimulus **(Weber-Fechner law)**. It now appears, however, that a power function more accurately describes this relation. In other words,

$$R = KS^A$$

where R is the sensation felt, S is the intensity of the stimulus, and, for any specific sensory modality, K and A are constants. The frequency of the action potentials generated by a stimulus to a sensory nerve fiber is also related to the intensity of the initiating stimulus by a power function. An example of this relation is shown in Fig 5–3, in which the exponent is approximately 1.0. Another example is shown in Fig 5–5, in which the calculated exponent is 0.52. Current evidence indicates that in the CNS the relation between stimulus and sensation is linear; consequently, it appears that for any given sensory modality, the relation between sensation and stimulus intensity is determined primarily by the properties of the peripheral receptors themselves.

Sensory Units

The term sensory unit is applied to a single sen-

sory axon and all its peripheral branches. These branches vary in number, but may be numerous, especially in the cutaneous senses. The **receptive field** of a sensory unit is the area from which a stimulus produces a response in that unit. In the cornea and adjacent sclera of the eye, the surface area supplied by a single sensory unit is 50–200 mm². Generally, the areas supplied by one unit overlap and interdigitate with the areas supplied by others.

Recruitment of Sensory Units

As the strength of a stimulus is increased, it tends to spread over a large area and generally not only activates the sense organs immediately in contact with it but "recruits" those in the surrounding area. Furthermore, weak stimuli activate the receptors with the lowest thresholds, whereas stronger stimuli also activate those with higher thresholds. Some of the receptors activated are part of the same sensory unit, and impulse frequency in the unit therefore increases. Because of overlap and interdigitation of one unit with another, however, receptors of other units are also stimulated, and consequently more units fire. In this way, more afferent pathways are activated, and this is interpreted in the brain as an increase in intensity of the sensation.

Figure 5–5. Relation between magnitude of touch stimulus (S) and frequency of action potentials in sensory nerve fibers (R). Dots are individual values from cats and are plotted on linear coordinates *(left)* and log-log coordinates *(right).* The equation shows the calculated power function relationship between R and S. (Reproduced, with permission, from Werner G, Mountcastle VB: Neural activity in mechanoreceptive cutaneous afferents. Stimulus-response relations, Weber functions, and information transmission. *J Neurophysiol* 1965:**28:**359.)

REFERENCES

Section II: Physiology of Nerve & Muscle Cells

Akil H et al: Endogenous opioids: Biology and function. *Annu Rev Neurosci* 1984;**7**:223.

Ashcroft PM: Adenosine-5'-triphosphate sensitive K-channels. *Annu Rev Neurosci* 1988;**11**:97.

Benowitz NL: Drug Therapy: Pharmacologic aspects of cigarette smoking and nicotine addiction. *N Engl J Med* 1988;**319**:1318.

Bessman SP, Geiger PJ: Transport of energy in muscle: The phosphorylcreatine shuttle. *Science* 1981;**211**:448.

Biewener AA: Biomechanics of mammalian terrestrial locomotion. *Science* 1990;**250**:1097.

Bülbring E, Tomita T: Catecholamine action on smooth muscle. *Pharm Rev* 1987;**39**:49.

Catterall WA: The molecular basis of neural excitability. *Science* 1984;**223**:653.

Celotti F, Melcangi RC, Martini L: The 5α-reductase in the brain: Molecular aspects and relation to brain function. *Front in Neuroendocrinol* 1992;**13**:163.

Chesselet M-F: Presynaptic regulation of neurotransmitter release in the brain: Facts and hypothesis. *Neuroscience* 1984;**12**:347.

Cooper JR, Bloom FE, Roth RH: *The Biochemical Basis of Neuropharmacology*, 6th ed. Oxford Univ Press, 1991.

Cowan WM, Cuénod M (editors): *Use of Axonal Transport for Studies of Neuronal Connectivity*. Elsevier, 1975.

De Camilli P et al: The synapsins. *Annu Rev Cell Biol* 1990;**6**:433.

Freed WJ, de Medinaceli L, Wyatt RJ: Promoting functional plasticity in the damaged nervous system. *Science* 1985;**227**:1544.

Goyal RK: Muscarinic receptor subtypes: Physiology and clinical implications. *N Engl J Med* 1989;**321**:1022.

Hall ZW: *An Introduction to Molecular Neurobiology*. Sinauer, 1992.

Hille B: *Ionic Channels of Excitable Membranes*. 2nd ed Sinauer Associates, 1992.

Iggo A, Andres KH: Morphology of cutaneous receptors. *Annu Rev Neurosci* 1982;**5**:1.

Jessell TM: Adhesion molecules and the hierarchy of neural development. *Neuron* 1988;**1**:3.

Jones DA, Round JM: *Skeletal Muscle in Health and Disease: A Textbook of Muscle Physiology*. Manchester Univ Press, 1990.

Julius D: Molecular biology of serotonin receptors. *Annu Rev Neurosci* 1991;**14**:335.

Katz B: Quantal mechanism of neural transmitter release. *Science* 1971;**173**:123.

Luetje CM, Patrick O, Séguela P: Nicotinic receptors in the mammalian brain. *FASEB J* 1990;**4**:2753.

Levi-Montalcini R: The nerve growth factor 35 years later. *Science* 1987;**237**:1154.

Loewenstein WR (editor): Principles of receptor physiology. In: *Handbook of Sensory Physiology*. Vol. 1. Springer-Verlag, 1971.

Madison DV, Malenka RC, Nicoll RA: Mechanisms underlying long-term potentiation of synaptic transmission. *Annu Rev Neurosci* 1991;**14**:379.

Matthews PB: Where does Sherrington's "muscular sense" originate? Muscles, joints, corollary discharges? *Annu Rev Neurosci* 1982;**5**:189.

McMahon TA: *Muscles, Reflexes, and Locomotion*. Princeton Univ Press, 1984.

Milhorn DE, Hökfelt T: Chemical messengers and their coexistence in individual neurons. *News Physiol Sci* 1988;**3**:1.

Morell P, Norton WT: Myelin. *Sci Am* (May) 1980;**242**:88.

Morgan JI, Curran T: Stimulus-transcription coupling in the nervous system: Involvement of the inducible proto-oncogenes *fos* and *jun*. *Annu Rev Neurosci* 1991;**14**:421.

Nakanishi S: Substance P precursor and kininogen: their structure, gene organizations, and regulation. *Physiol Rev* 1987;**67**:1117.

Nicoll RA, Malenka RC, Kauer JA: Functional comparison of neurotransmitter receptor subtypes in mammalian central nervous system. *Physiol Rev* 1990;**70**:513.

Noble D, Powell T: *Electrophysiology of Single Cardiac Cells*. Academic Press, 1987.

Olsen RW, Tobin AJ: Molecular biology of $GABA_A$ receptors. *FASEB J* 1990;**4**:1469.

Palka J (editor): Developmental neurobiology. *Bioscience* 1984;**34**:294.

Raymond JR et al: Adrenergic receptors: Model for regulation of signal transduction processes. *Hypertension* 1990;**15**:119.

Rehm H, Tempel BL: Voltage-gated K^+ channels of the mammalian brain. *FASEB J* 1991;**5**:164.

Salpeter MM (editor): *The Vertebrate Neuromuscular Junction*. Plenum, 1987.

Schwartz J-C et al: Histaminergic transmission in the mammalian brain. *Physiol Rev* 1991;**71**:1.

Smith TW: Digitalis mechanism of action and clinical use. *N Engl J Med* 1988;**318**:358.

Stevens CF: The neuron. *Sci Am* (Sept) 1979;**241**:54.

Stracher A (editor): *Muscle and Nonmuscle Motility*. Academic Press, 1983.

Valee RB, Bloom G: Mechanisms of fast and slow axonal transport. *Annu Rev Neurosci* 1991;**14**:59.

Watanabe T, Wada H: *Histaminergic Neurons*. CRC Press, 1991.

Watson WE: Physiology of neuroglia. *Physiol Rev* 1974;**54**:245.

Williams M (editor): *Adenosine and Adenosine Receptors*. Humana, 1990.

Section III.
Functions of the Nervous System

Reflexes 6

INTRODUCTION

The basic unit of integrated neural activity is the **reflex arc.** This arc consists of a sense organ, an afferent neuron, one or more synapses in a central integrating station or sympathetic ganglion, an efferent neuron, and an effector. In mammals, the connection between afferent and efferent somatic neurons is generally in the brain or spinal cord. The afferent neurons enter via the dorsal roots or cranial nerves and have their cell bodies in the dorsal root ganglia or in the homologous ganglia on the cranial nerves. The efferent fibers leave via the ventral roots or corresponding motor cranial nerves. The principle that in the spinal cord the dorsal roots are sensory and the ventral roots are motor is known as the **Bell-Magendie law.**

Activity in the reflex arc starts in a sensory receptor with a receptor potential whose magnitude is proportionate to the strength of the stimulus (Fig 6–1). This generates all or none action potentials in the afferent nerve, the number of action potentials being proportionate to the size of the generator potential. In the CNS, the responses are again graded in terms of EPSPs and IPSPs at the synaptic junctions (see Chapter 4). All or none responses are generated in the efferent nerve. When these reach the effector, they again set up a graded response. When the effector is smooth muscle, responses summate to produce action potentials in the smooth muscle, but when the effector is skeletal muscle, the graded response is always adequate to produce action potentials that bring about muscle contraction. It should be emphasized that the connection between the afferent and efferent neurons is usually in the central nervous system, and activity in the reflex arc is modified by the multiple inputs converging on the efferent neurons.

The simplest reflex arc is one with a single synapse between the afferent and efferent neurons. Such arcs are **monosynaptic,** and reflexes occurring in them are **monosynaptic reflexes.** Reflex arcs in which one or more interneurons are interposed between the afferent and efferent neurons are **polysynaptic,** the number of synapses in the arcs varying from 2 to many hundreds. In both types, but especially in polysynaptic reflex arcs, activity is modified by spatial and temporal facilitation, occlusion, subliminal fringe effects, and other effects.

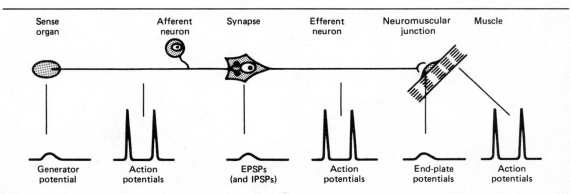

Figure 6–1. The reflex arc. Note that at the receptor and in the CNS there is a nonpropagated graded response that is proportionate to the magnitude of the stimulus, whereas in the portions of the arc specialized for transmission (afferent and efferent axons, muscle membrane), the responses are all or none action potentials.

MONOSYNAPTIC REFLEXES: THE STRETCH REFLEX

When a skeletal muscle with an intact nerve supply is stretched, it contracts. This response is called the **stretch reflex.** The stimulus that initiates the reflex is stretch of the muscle, and the response is contraction of the muscle being stretched. The sense organ is the muscle spindle. The impulses originating in the spindle are conducted in the central nervous system by fast sensory fibers that pass directly to the motor neurons which supply the same muscle. The neurotransmitter at the central synapse is glutamate. Stretch reflexes are the best known and studied monosynaptic reflexes in the body.

Clinical Examples

Tapping the patellar tendon elicits the **knee jerk,** a stretch reflex of the quadriceps femoris muscle, because the tap on the tendon stretches the muscle. A similar contraction is observed if the quadriceps is stretched manually. Stretch reflexes can also be elicited from most of the large muscles of the body. Tapping on the tendon of the triceps brachii, for example, causes an extensor response at the elbow as a result of reflex contraction of the triceps; tapping on the Achilles tendon causes an ankle jerk due to reflex contraction of the gastrocnemius; and tapping on the side of the face causes a stretch reflex in the masseter. Other examples of stretch reflexes are listed in neurology textbooks.

Structure of Muscle Spindles

Each muscle spindle consists of up to about 10 muscle fibers enclosed in a connective tissue capsule. These fibers are more embryonal in character and have less distinct striations than the rest of the fibers in the muscle. They are called **intrafusal fibers** to distinguish them from the **extrafusal fibers,** the regular contractile units of the muscle. The intrafusal fibers are in parallel with the rest of the muscle fibers because the ends of the capsule of the spindle are attached to the tendons at either end of the muscle or to the sides of the extrafusal fibers.

There are 2 types of intrafusal fibers in mammalian muscle spindles. The first type contains many nuclei in a dilated central area and is therefore called a **nuclear bag fiber** (Fig 6–2). Typically there are 2 nuclear bag fibers per spindle, nuclear bag fiber 1 with a low level of myosin ATPase activity and nuclear bag fiber 2 with a high level of myosin ATPase activity. The second fiber type, the **nuclear chain fiber,** is thinner and shorter and lacks a definite bag. There are 4 or more of these fibers per spindle. Their ends connect to the sides of the nuclear bag fibers. The ends of the intrafusal fibers are contractile, whereas the central portions probably are not.

There are 2 kinds of sensory endings in each spindle. The **primary,** or **annulospiral, endings** are the terminations of rapidly conducting group Ia afferent fibers. One branch of the Ia fiber innervates nuclear bag fiber 1, whereas another branch innervates nuclear bag fiber 2 and nuclear chain fibers. These sensory fibers wrap around the center of the nuclear bag and nuclear chain fibers. The **secondary,** or **flower-spray, endings** are terminations of group II sensory fibers and are located nearer the ends of the intrafusal fibers, but only on nuclear chain fibers.

The spindles have a motor nerve supply of their own. These nerves are 3–6 μm in diameter, constitute about 30% of the fibers in the ventral roots, and belong in Erlanger and Gasser's A γ group. Because of their characteristic size, they are called the γ efferents of Leksell or the **small motor nerve system.** They go exclusively to the spindles. In addition, larger β motorneurons innervate both intrafusal and ex-

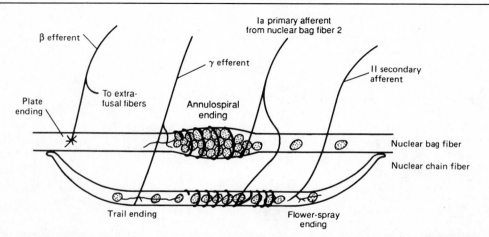

Figure 6–2. Diagrammatic representation of the main components of mammalian muscle spindle. Each spindle has a capsule and usually contains 2 nuclear bag fibers and 4 or more nuclear chain fibers.

trafusal fibers. The endings of the γ efferent fibers are of 2 histologic types. There are motor end-plates (**plate endings**) on the nuclear bag fibers, and there are endings that form extensive networks (**trail endings**) primarily on the nuclear chain fibers.

The spindles produce 2 kinds of sensory nerve patterns, dynamic and static (see below), and both γ and β motor axons produce 2 functional types of responses. Stimulation of one type increases dynamic responses (**dynamic fusimotor axons**), and stimulation of the other increases static discharge at constant length (**static fusiform axons**).

Central Connections of Afferent Fibers

It can be proved experimentally that the Ia fibers from the primary endings end directly on motor neurons supplying the extrafusal fibers of the same muscle. The time between the application of the stimulus and the response is the **reaction time.** In humans, the reaction time for a stretch reflex such as the knee jerk is 19–24 ms. Weak stimulation of the sensory nerve from the muscle, known to stimulate only Ia fibers, causes a contractile response with a similar latency. Since the conduction velocities of the afferent and efferent fiber types are known and the distance from the muscle to the spinal cord can be measured, it is possible to calculate how much of the reaction time was taken up by conduction to and from the spinal cord. When this value is subtracted from the reaction time, the remainder, called the **central delay,** is the time taken for the reflex activity to traverse the spinal cord. In humans, the central delay for the knee jerk is 0.6–0.9 ms, and figures of similar magnitude have been found in experimental animals. Since the minimal synaptic delay is 0.5 ms (see Chapter 4), only one synapse could have been traversed.

Muscle spindles also make connections that cause muscle contraction via polysynaptic pathways, and the afferents involved are probably those from the secondary endings. However, group II fibers also make monosynaptic connections to the motor neurons and make a small contribution to the stretch reflex.

Function of Muscle Spindles

When the muscle spindle is stretched, its sensory endings are distorted and receptor potentials are generated. These in turn set up action potentials in the sensory fibers at a frequency proportionate to the degree of stretching. The spindle is in parallel with the extrafusal fibers, and when the muscle is passively stretched, the spindles are also stretched. This initiates reflex contraction of the extrafusal fibers in the muscle. On the other hand, the spindle afferents characteristically stop firing when the muscle is made to contract by electrical stimulation of the nerve fibers to the extrafusal fibers because the muscle shortens while the spindle does not (Fig 6–3).

Thus, the spindle and its reflex connections constitute a feedback device that operates to maintain muscle length; if the muscle is stretched, spindle discharge increases and reflex shortening is produced, whereas if the muscle is shortened without a change in γ efferent discharge, spindle discharge decreases and the muscle relaxes.

Primary endings on the nuclear bag fibers and nuclear chain fibers are both stimulated when the spindle is stretched, but the pattern of response differs. The nerves from the endings in the nuclear bag region

Figure 6–3. Effect of various conditions on muscle spindle discharge.

show a **dynamic response;** ie, they discharge most rapidly while the muscle is being stretched and less rapidly during sustained stretch (Fig 6–4). The nerves from the primary endings on the nuclear chain fibers show a **static response;** ie, they discharge at an increased rate throughout the period when a muscle is stretched. Thus, the primary endings respond to both changes in length and changes in the rate of stretch. The response of the primary ending to the phasic as well as the static events in the muscle is important because the prompt, marked phasic response helps to dampen oscillations caused by conduction delays in the feedback loop regulating muscle length. There is normally a small oscillation in this feedback loop. This **physiologic tremor** has a frequency of approximately 10 Hz. However, the tremor would be worse if it were not for the sensitivity of the spindle to velocity of stretch.

Effects of Gamma Efferent Discharge

Stimulation of the γ efferent system produces a very different picture from that produced by stimulation of the extrafusal fibers. Such stimulation does not lead directly to detectable contraction of the muscles because the intrafusal fibers are not strong enough or plentiful enough to cause shortening. However, stimulation does cause the contractile ends of the intrafusal fibers to shorten and therefore stretches the nuclear bag portion of the spindles, deforming the annulospiral endings and initiating impulses in the Ia fibers. This in turn can lead to reflex contraction of the muscle. Thus, muscle can be made to contract via stimulation of the α motor neurons that innervate the extrafusal fibers or the γ efferent neurons that initiate contraction indirectly via the stretch reflex.

When the rate of γ efferent discharge is increased, the intrafusal fibers are shorter than the extrafusal ones. If the whole muscle is stretched during stimulation of the γ efferents, additional action potentials are generated by the additional stretch of the nuclear bag region, and the rate of discharge in the Ia fibers is further increased (Fig 6–3). Increased γ efferent discharge thus increases spindle sensitivity, and the sensitivity of the spindles to stretch varies with the rate of γ efferent discharge.

There is considerable evidence of increased γ efferent discharge along with the increased discharge of the α motor neurons that initiates movements. Because of this "α–γ linkage," the spindle shortens with the muscle, and spindle discharge may continue throughout the contraction. In this way, the spindle remains capable of responding to stretch and reflexly adjusting motor neuron discharge throughout the contraction.

The existence of dynamic and static γ and β efferents is mentioned above. Stimulation of the dynamic efferents increases spindle sensitivity to the rate of change of stretch. Stimulation of the static efferents increases spindle sensitivity to steady, maintained stretch. It is thus possible to adjust separately the spindle responses to phasic and static events.

Control of Gamma Efferent Discharge

The motor neurons of the γ efferent system are regulated to a large degree by descending tracts from a number of areas in the brain. Via these pathways, the sensitivity of the muscle spindles and hence the threshold of the stretch reflexes in various parts of the body can be adjusted and shifted to meet the needs of postural control (see Chapter 12).

Other factors also influence γ efferent discharge. Anxiety causes an increased discharge, a fact that probably explains the hyperactive tendon reflexes sometimes seen in anxious patients. In addition, unexpected movement is associated with a greater efferent discharge. Stimulation of the skin, especially by noxious agents, increases γ efferent discharge to ipsilateral flexor muscle spindles while decreasing that to extensors and produces the opposite pattern in the opposite limb. It is well known that trying to pull the hands apart when the flexed fingers are hooked together facilitates the knee jerk reflex (Jendrassik's maneuver), and this may many also be due to increased γ efferent discharge initiated by afferent impulses from the hands.

Reciprocal Innervation

When a stretch reflex occurs, the muscles that antagonize the action of the muscle involved (antagonists) relax. This phenomenon is said to be due to **reciprocal innervation.** Impulses in the Ia fibers from the muscle spindles of the protagonist muscle cause postsynaptic inhibition of the motor neurons to the antagonists. The pathway mediating this effect appears to be bisynaptic. A collateral from each Ia fiber passes in the spinal cord to an inhibitory interneuron (Golgi bottle neuron) that synapses directly on one of the motor neurons supplying the antagonist muscles. This example of postsynaptic inhibition is discussed

Figure 6–4. Response of spindle afferents to muscle stretch. The bottom 2 lines represent the number of discharges in afferent nerves from the primary endings on nuclear bag and nuclear chain fibers as the muscle is stretched and then permitted to return to its original length.

in Chapter 4, and the pathway is illustrated in Fig 4–8.

Inverse Stretch Reflex

Up to a point, the harder a muscle is stretched, the stronger is the reflex contraction. However, when the tension becomes great enough, contraction suddenly ceases and the muscle relaxes. This relaxation in response to strong stretch is called the **inverse stretch reflex** or **autogenic inhibition.**

The receptor for the inverse stretch reflex is in the **Golgi tendon organ** (Fig 6–5). This organ consists of a netlike collection of knobby nerve endings among the fascicles of a tendon. There are 3–25 muscle fibers per tendon organ. The fibers from the Golgi tendon organs make up the Ib group of myelinated, rapidly conducting sensory nerve fibers (Table 2–3). Stimulation of these Ib fibers leads to the production of IPSPs on the motor neurons that supply the muscle from which the fibers arise. The Ib fibers end in the spinal cord on inhibitory interneurons that, in turn, terminate directly on the motor neurons (Fig 6–6). They also make excitatory connections with motor neurons supplying antagonists to the muscle.

Since the Golgi tendon organs, unlike the spindles, are in series with the muscle fibers, they are stimulated by both passive stretch and active contraction of the muscle. The threshold of the Golgi tendon organs is low. The degree of stimulation by passive stretch is not great, because the more elastic muscle fibers take up much of the stretch, and this is why it takes a strong stretch to produce relaxation. However, discharge is regularly produced by contraction of the muscle, and the Golgi tendon organ thus functions as a transducer in a feedback circuit that regulates muscle force in a fashion analogous to the spindle feedback circuit that regulates muscle length.

The importance of the primary endings in the spindles and the Golgi tendon organs in regulating the velocity of the muscle contraction, muscle length, and muscle force is illustrated by the fact that section of the afferent nerves to a limb causes the limb to hang

Figure 6–6. Diagram illustrating the pathways responsible for the stretch reflex and the inverse stretch reflex. Stretch stimulates the spindle, and impulses pass up the Ia fiber to excite the motor neuron. It also stimulates the Golgi tendon organ, and impulses passing up the Ib fiber activate the interneuron to release the inhibitory mediator glycine. With strong stretch, the resulting hyperpolarization of the motor neuron is so great that it stops discharging.

loosely at the side in a semiparalyzed state. The organization of the system is shown in Fig 6–7, and the interaction of spindle discharge, tendon organ discharge, and reciprocal innervation in determining the rate of discharge of a motor neuron is shown in Fig 6–8.

Muscle Tone

The resistance of a muscle to stretch is often referred to as its **tone** or **tonus.** If the motor nerve to a

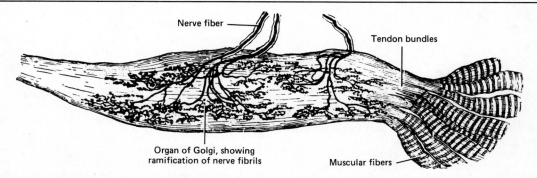

Figure 6–5. Golgi tendon organ. (Reproduced, with permission, from Goss CM [editor]: *Gray's Anatomy of the Human Body,* 29th ed. Lea & Febiger, 1973.)

Figure 6–7. Block diagram of peripheral motor control system. The dashed line indicates the nonneural feedback from muscle that limits length and velocity via the inherent mechanical properties of muscle. γ_d, dynamic γ motor neurons; γ_s, static γ motor neurons. (Reproduced, with permission, from Houk J in: *Medical Physiology*, 13th ed. Mountcastle VB [editor]. Mosby, 1974.)

muscle is cut, the muscle offers very little resistance and is said to be **flaccid.** A **hypertonic (spastic)** muscle is one in which the resistance to stretch is high because of hyperactive stretch reflexes. Somewhere between the states of flaccidity and spasticity is the ill-defined area of normal tone. The muscles are generally **hypotonic** when the rate of γ efferent discharge is low and hypertonic when it is high.

Lengthening Reaction

When the muscles are hypertonic, the sequence of moderate stretch→muscle contraction, strong stretch→muscle relaxation is clearly seen. Passive flexion of the elbow, for example, meets immediate resistance as a result of the stretch reflex in the triceps muscle. Further stretch activates the inverse stretch reflex. The resistance to flexion suddenly collapses, and the arm flexes. Continued passive flexion stretches the muscle again, and the sequence may be repeated. This sequence of resistance followed by give when a limb is moved passively is known clinically as the **clasp-knife effect** because of its resemblance to the closing of a pocket knife. The physiologic name for it is the **lengthening reaction** because it is the response of a spastic muscle (in the example cited, the triceps) to lengthening.

Clonus

Another finding characteristic of states in which increased γ efferent discharge is present is **clonus.** This neurologic sign is the occurrence of regular, rhythmic contractions of a muscle subjected to sudden, maintained stretch. Ankle clonus is a typical example. This is initiated by brisk, maintained dorsiflexion of the foot, and the response is rhythmic plantar flexion at the ankle. The stretch reflex–inverse stretch reflex sequence described above may contribute to this response. However, it can occur on the basis of synchronized motor neuron discharge without Golgi tendon organ discharge. The spindles of the tested muscle are hyperactive, and the burst of impulses from them discharges all the motor neurons supplying the muscle at once. The consequent muscle contraction stops spindle discharge. However, the stretch has been maintained, and as soon as the mus-

Figure 6–8. Discharge frequency of a motor neuron supplying a leg muscle in a cat. Discharge frequency is plotted against muscle length at various magnitudes of inhibition produced by stimulation of the nerve from the antagonist to the muscle. (Reproduced, with permission, from Henneman E et al: Excitability and inhibitability of motor neurons of different sizes. *J Neurophysiol* 1965;**28**:599.)

cle relaxes it is again stretched and the spindles stimulated.

POLYSYNAPTIC REFLEXES: THE WITHDRAWAL REFLEX

Polysynaptic reflex paths branch in a complex fashion (Fig 6–9). The number of synapses in each of their branches is variable. Because of the synaptic delay incurred at each synapse, activity in the branches with fewer synapses reaches the motor neurons first, followed by activity in the longer pathways. This causes prolonged bombardment of the motor neurons from a single stimulus and consequently prolonged responses. Furthermore, as shown in Fig 6–9, at least some of the branch pathways turn back on themselves, permitting activity to reverberate until it becomes unable to cause a propagated transsynaptic response and dies out. Such **reverberating circuits** are common in the brain and spinal cord.

Withdrawal Reflex

The withdrawal reflex is a typical polysynaptic reflex that occurs in response to a noxious and usually painful stimulation of the skin or subcutaneous tissues and muscle. The response is flexor muscle contraction and inhibition of extensor muscles, so that the part stimulated is flexed and withdrawn from the stimulus. When a strong stimulus is applied to a limb, the response includes not only flexion and withdrawal of that limb but also extension of the opposite limb. This **crossed extensor response** is properly part of the withdrawal reflex. It can also be shown that strong stimuli in experimental animals generate activity in the interneuron pool which spreads to all 4

extremities. This is difficult to demonstrate in normal animals but is easily demonstrated in an animal in which the modulating effects of impulses from the brain have been abolished by prior section of the spinal cord (**spinal animal**). For example, when the hind limb of a spinal cat is pinched, the stimulated limb is withdrawn, the opposite hind limb extended, the ipsilateral forelimb extended, and the contralateral forelimb flexed. This spread of excitatory impulses up and down the spinal cord to more and more motor neurons is called **irradiation of the stimulus,** and the increase in the number of active motor units is called **recruitment of motor units.**

Importance of the Withdrawal Reflex

Flexor responses can be produced by innocuous stimulation of the skin or by stretch of the muscle, but strong flexor responses with withdrawal are initiated only by stimuli that are noxious or at least potentially harmful to the animal. These stimuli are therefore called **nociceptive stimuli.** Sherrington pointed out the survival value of the withdrawal response. Flexion of the stimulated limb gets it away from the source of irritation, and extension of the other limb supports the body. The pattern assumed by all 4 extremities puts the animal in position to run away from the offending stimulus. Withdrawal reflexes are **prepotent;** ie, they preempt the spinal pathways from any other reflex activity taking place at the moment.

Many of the characteristics of polysynaptic reflexes can be demonstrated by studying the withdrawal reflex in the laboratory. A weak noxious stimulus to one foot evokes a minimal flexion response; stronger stimuli produce greater and greater flexion as the stimulus irradiates to more and more of the motor neuron pool supplying the muscles of the limb. Stronger stimuli also cause a more prolonged response. A weak stimulus causes one quick flexion movement; a strong stimulus causes prolonged flexion and sometimes a series of flexion movements. This prolonged response is due to prolonged, repeated firing of the motor neurons. The repeated firing is called **after-discharge** and is due to continued bombardment of motor neurons by impulses arriving by complicated and circuitous polysynaptic paths.

As the strength of a noxious stimulus is increased, the reaction time is shortened. Spatial and temporal facilitation occurs at synapses in the polysynaptic pathway. Stronger stimuli produce more action potentials per second in the active branches and cause more branches to become active; summation of the EPSPs to the firing level therefore occurs more rapidly.

Local Sign

The exact flexor pattern of the withdrawal reflex in a limb varies with the part of the limb that is stimulated. If the medial surface of the limb is stimulated, for example, the response will include some abduc-

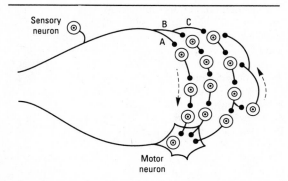

Figure 6–9. Diagram of polysynaptic connections between afferent and efferent neurons in the spinal cord. The dorsal root fiber activates pathway A with 3 interneurons, pathway B with 4 interneurons, and pathway C with 4 interneurons. Note that one of the interneurons in pathway C connects to a neuron that doubles back to other interneurons, forming reverberating circuits.

tion, whereas stimulation of the lateral surface will produce some adduction with flexion. The reflex response in each case generally serves to effectively remove the limb from the irritating stimulus. This dependence of the exact response on the location of the stimulus is called **local sign.** The degree to which local sign determines the particular pattern is illustrated in Fig 6–10.

Fractionation & Occlusion

Another characteristic of the withdrawal response is the fact that supramaximal stimulation of any of the sensory nerves from a limb never produces as strong a contraction of the flexor muscles as that elicited by direct electrical stimulation of the muscles themselves. This indicates that the afferent inputs **fractionate** the motor neuron pool, ie, each input goes to only part of the motor neuron pool for the flexors of that particular extremity. On the other hand, if all the sensory inputs are dissected out and stimulated one after the other, the sum of the tension developed by stimulation of each is greater than that produced by direct electrical stimulation of the muscle or stimulation of all inputs at once. This indicates that the various afferent inputs share some of the motor neurons and that occlusion (see Chapter 4) occurs when all inputs are stimulated at once.

Other Polysynaptic Reflexes

There are many polysynaptic reflexes in addition to the withdrawal reflex, all with similar properties. Such reflexes as the abdominal and cremasteric reflexes are forms of the withdrawal reflex. Others include visceral components. Numerous polysynaptic reflexes that relate to specific regulatory functions are described in other sections of this book, and comprehensive lists can be found in neurology textbooks.

GENERAL PROPERTIES OF REFLEXES

It is apparent from the preceding description of the properties of monosynaptic and polysynaptic reflexes that reflex activity is stereotyped and specific in terms of both the stimulus and the response; a particular stimulus elicits a particular response.

Adequate Stimulus

The stimulus that triggers a reflex is generally very precise. This stimulus is called the **adequate stimulus** for the particular reflex. A dramatic example is the scratch reflex in the dog. This spinal reflex is adequately stimulated by multiple linear touch stimuli such as those produced by an insect crawling across the skin. The response is vigorous scratching of the area stimulated. (Incidentally, the precision with which the scratching foot goes to the site of the irritant is a good example of local sign.) If the multiple touch stimuli are widely separated or not in a line, the adequate stimulus is not produced and no scratching occurs. Fleas crawl, but they also jump from place to place. This jumping separates the touch stimuli so that an adequate stimulus for the scratch reflex is not produced. It is doubtful if the flea population would long survive without the ability to jump.

Final Common Path

The motor neurons that supply the extrafusal fibers in skeletal muscles are the efferent side of many reflex arcs. All neural influences affecting muscular contraction ultimately funnel through them to the muscles, and they are therefore called the **final common paths.** Numerous inputs converge on them. Indeed, the surface of the average motor neuron accommodates about 10,000 synaptic knobs. There are at least 5 inputs from the same spinal segment to a typical spinal motor neuron. In addition to these, there are excitatory and inhibitory inputs, generally relayed via interneurons, from other levels of the spinal cord and multiple long descending tracts from the brain. All of these pathways converge on and determine the activity in the final common paths.

Central Excitatory & Inhibitory States

The spread up and down the spinal cord of subliminal fringe effects from excitatory stimulation has already been mentioned. Direct and presynaptic inhibitory effects can also be widespread. These effects are generally transient. However, the spinal cord also shows prolonged changes in excitability, possibly because of activity in reverberating circuits or prolonged effects of synaptic mediators. The terms **cen-**

Figure 6–10. The importance of local sign in determining the character of the withdrawal response in a leg. When afferent fibers in each of the 3 nerves on the left were stimulated, hip, knee, and ankle flexors contracted but the relative tension that developed in each case (shaded bars) varied.

tral excitatory state and central inhibitory state have been used to describe prolonged states in which excitatory influences overbalance inhibitory influences and vice versa. When the central excitatory state is marked, excitatory impulses irradiate not only to many somatic areas of the spinal cord but also to autonomic areas. In chronically paraplegic humans, for example, a mild noxious stimulus may cause, in addition to prolonged withdrawal-extension patterns in all 4 limbs, urination, defecation, sweating, and blood pressure fluctuations (**mass reflex**).

Habituation & Sensitization of Reflex Responses

The fact that reflex responses are stereotyped does not exclude the possibility of their being modified by experience. Examples include habituation and sensitization, which are discussed in terms of synaptic function in Chapter 4 and in terms of their relation to learning and memory in Chapter 16.

7

Cutaneous, Deep, & Visceral Sensation

INTRODUCTION

The sense organs for mechanical stimulation (touch and pressure), warmth, cold, and pain have been discussed in Chapter 5, and the types of neurons that carry impulses generated in them to the central nervous system are listed in Table 2–2. These primary afferent neurons have their cell bodies in the dorsal root ganglia or equivalent ganglia in cranial nerves. They enter the spinal cord or brain stem and make polysynaptic reflex connections to motor neurons at many levels as well as connections that relay impulses to the cerebral cortex. Each of the sensations they mediate is considered in this chapter.

PATHWAYS

The dorsal horns are divided on the basis of histologic characteristics into laminas I–VI, with I being the most superficial and VI the deepest. Lamina II and part of lamina III make up the **substantia gelatinosa,** a lightly stained area near the top of each dorsal horn. There are 3 types of primary afferent fibers that mediate cutaneous sensation: large myelinated A β fibers that transmit impulses generated by mechanical stimuli; small myelinated A δ fibers, some of which transmit impulses from cold receptors and nociceptors that mediate fast pain (see below) and some of which transmit impulses from mechanoreceptors; and small unmyelinated C fibers that are concerned primarily with pain and temperature. However, there are also a few C fibers that transmit impulses from mechanoreceptors. The orderly distribution of these fibers in the dorsal columns and the various layers of the dorsal horn is shown in Fig 7–1.

The principal direct pathways to the cerebral cortex for the cutaneous senses are shown in Fig 7–2. Fibers mediating fine touch and proprioception ascend in the dorsal columns to the medulla, where they synapse in the gracile and cuneate nuclei. The second-order neurons from the gracile and cuneate nuclei cross the midline and ascend in the medial lemniscus to end in the ventral posterior nucleus and related specific sensory relay nuclei of the thalamus

(see Chapter 11). This ascending system is frequently called the **dorsal column** or **lemniscal system.**

Other touch fibers, along with those mediating temperature and pain, synapse on neurons in the dorsal horn. The axons from these neurons cross the midline and ascend in the anterolateral quadrant of the spinal cord, where they form the **anterolateral system** of ascending fibers. Others ascend more dorsally. In general, touch is associated with the ventral spinothalamic tract whereas pain and temperature are associated with the lateral spinothalamic tract, but there is no rigid localization of function. Some of the fibers of the anterolateral system end in the specific relay nuclei of the thalamus; others project to the midline and intralaminar nonspecific projection nuclei. There is a major input from the anterolateral systems into the mesencephalic reticular formation. Thus, sensory input activates the reticular activating system, which in turn maintains the cortex in the alert state (see Chapter 11).

Collaterals from the fibers that enter the dorsal columns pass to the dorsal horn. These collaterals may modify the input into other cutaneous sensory systems, including the pain system. The dorsal horn represents a "gate" in which impulses in the sensory nerve fibers are translated into impulses in ascending tracts, and it now appears that passage through this gate is dependent on the nature and pattern of impulses reaching the substantia gelatinosa and its environs. This gate is also affected by impulses in descending tracts from the brain. The relation of the gate to pain is discussed below.

Axons of the spinothalamic tracts from sacral and lumbar segments of the body are pushed laterally by axons crossing the midline at successively higher levels. On the other hand, sacral and lumbar dorsal column fibers are pushed medially by fibers from higher segments (Fig 7–3). Consequently, both of these ascending systems are laminated, with cervical, thoracic, lumbar, and sacral segments represented from medial to lateral in the anterolateral pathways and sacral to cervical segments from medial to lateral in the dorsal columns. Because of this lamination, tumors arising outside the spinal cord first compress the spinothalamic fibers from sacral and lumbar areas, causing the early symptom of loss of pain and tem-

Figure 7–1. Schematic representation of the terminations of the 3 types of primary afferent neurons in the various layers of the dorsal horn of the spinal cord. (Modified from Salt TH, Hill RE: Neurotransmitter candidates of somatosensory primary afferent fibers. *Neuroscience* 1983;**10**:1033.)

perature sensation in the sacral region. Intraspinal tumors cause anesthesia first in higher segments.

The fibers within the lemniscal and anterolateral systems are joined in the brain stem by fibers mediating sensation from the head. Pain and temperature impulses are relayed via the spinal nucleus of the trigeminal nerve, and touch and proprioception mostly via the main sensory and mesencephalic nuclei of this nerve.

Cortical Representation

From the specific sensory nuclei of the thalamus, neurons project in a highly specific way to the 2 somatic sensory areas of the cortex: somatic sensory area I (SI) in the postcentral gyrus and somatic sensory area II (SII) in the wall of the sylvian fissure (lateral cerebral sulcus). In addition, SI projects to SII. SI corresponds to Brodmann's areas 1, 2, and 3. Brodmann was a histologist who painstakingly divided the cerebral cortex into numbered areas based on their histologic characteristics.

The arrangement of the thalamic fibers in SI is such that the parts of the body are represented in order along the postcentral gyrus, with the legs on top and the head at the foot of the gyrus (Fig 7–4). Not only is there detailed localization of the fibers from the various parts of the body in the postcentral gyrus, but also the size of the cortical receiving area for impulses from a particular part of the body is proportionate to the number of receptors in the part. The relative sizes of the cortical receiving areas are shown dramatically in Fig 7–5, in which the proportions of the homunculus have been distorted to correspond to the size of the cortical receiving areas for each. Note that the cortical areas for sensation from the trunk and

Figure 7–2. Touch, pain, and temperature pathways from the trunk and limbs. The anterolateral system (ventral and lateral spinothalamic and related ascending tracts) also projects to the mesencephalic reticular formation and the nonspecific thalamic nuclei.

back are small, whereas very large areas are concerned with impulses from the hand and the parts of the mouth concerned with speech.

Studies of the sensory receiving area emphasize the very discrete nature of the point-for-point localization of peripheral areas in the cortex and provide further evidence for the validity of the doctrine of specific nerve energies (see Chapter 5). Stimulation of the various parts of the postcentral gyrus gives rise to sensations projected to appropriate parts of the body. The sensations produced are usually numbness, tingling, or a sense of movement, but with fine enough electrodes it has been possible to produce relatively pure sensations of touch, warmth, cold, and pain. The cells in the postcentral gyrus are organized in vertical columns, like cells in the visual cortex (see Chapter 8). The cells in a given column are all acti-

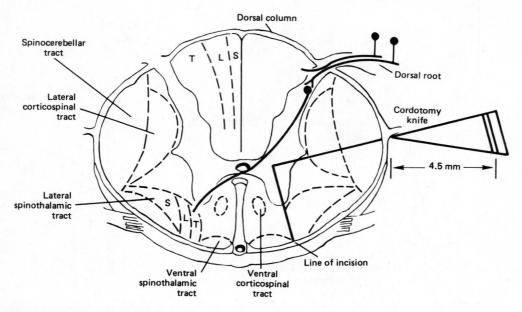

Figure 7–3. Major spinal pathways. The solid line on the right represents the line of incision in performing an anterolateral cordotomy. Note the lamination of the tracts. S, sacral; L, lumbar; T, thoracic.

vated by afferents from a given part of the body, and all respond to the same sensory modality.

It is now clear that projections of afferents on the postcentral gyrus and other cortical areas are not innate and immutable but can be changed by experience and reflect use of the represented area. For example, if a digit is amputated in a monkey, the cortical representation of the neighboring digits spreads into the cortical area that was formerly occupied by the representation of the amputated digit. Conversely, if the cortical area representing a digit is removed, the somatosensory map of the digit moves

to the surrounding cortex. Extensive, long-term deafferentation of limbs leads to even more dramatic shifts in somatosensory representation in the cortex. These findings illustrate the plasticity of the brain and its ability to adapt.

SII is located in the superior wall of the sylvian

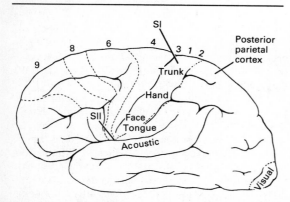

Figure 7–4. Brain areas concerned with somatic sensation, and some of the cortical receiving areas for other sensory modalities in the human brain. The numbers are those of Brodmann's cortical areas. The auditory (acoustic) area is actually located in the sylvian fissure on the top of the superior temporal gyrus and is not normally visible.

Figure 7–5. Sensory homunculus, drawn overlying a coronal section through the postcentral gyrus. Gen., genitalia. (Reproduced, with permission, from Penfield W, Rasmussen G: *The Cerebral Cortex of Man,* Macmillan, 1950.)

fissure. The head is represented at the inferior end of the postcentral gyrus, and the feet at the bottom of the sylvian fissure. The representation of the body parts is not as complete or detailed as it is in the postcentral gyrus.

Effects of Cortical Lesions

Ablation of SI in animals causes deficits in position sense and ability to discriminate size and shape. Ablation of SII causes deficits in learning based on tactile discrimination. Ablation of SI causes deficits in sensory processing in SII, whereas ablation of SII has no gross effect on processing in SI. Thus, it seems clear that SI and SII process sensory information in series rather than in parallel and that SII is concerned with further elaboration of sensory data. SI also projects to the posterior parietal cortex (Fig 7–4), and lesions of this association area produce complex abnormalities of spatial orientation on the contralateral side of the body (see Chapter 16).

It is worth emphasizing that in experimental animals and humans, cortical lesions do not abolish somatic sensation. Proprioception and fine touch are most affected by cortical lesions. Temperature sensibility is less affected, and pain sensibility is only slightly affected. Thus, perception is possible in the absence of the cortex. Upon recovery, pain sensibility returns first, followed by temperature sense and, finally, proprioception and fine touch.

Principles of Sensory Physiology

The important general principles that relate to the physiology of sensory systems have been discussed in detail in Chapter 5. Each sense organ is specialized to convert one particular form of energy into action potentials in the sensory nerves. Each modality has a discrete pathway to the brain, and the sensation perceived as well as the part of the body to which it is localized is determined by the particular part of the brain activated. Differences in intensity of a given sensation are signaled in 2 ways: by changes in the frequency of action potentials in the sensory nerves, and by changes in the number of receptors activated. An increase in the intensity of stimulation of a sense organ has very little (if any) effect on the quality of the sensation produced.

Another principle that applies to cutaneous sensation is that of punctate representation. If the skin is carefully mapped, millimeter by millimeter, with a fine hair, a sensation of touch is evoked from spots overlying touch receptors. None is evoked from the intervening areas. Similarly, pain and temperature sensations are produced by stimulation of the skin only over the spots where the sense organs for these modalities are located.

TOUCH

As noted in Chapter 5, touch is present in areas that have no specialized receptors. However, in areas where they exist, Meissner's and pacinian corpuscles are rapidly adapting and Merkel's disks and Ruffini endings are slowly adapting touch receptors. Touch receptors are most numerous in the skin of the fingers and lips and relatively scarce in the skin of the trunk. There are many receptors around hair follicles in addition to those in the subcutaneous tissues of hairless areas. When a hair is moved, it acts as a lever with its fulcrum at the edge of the follicle, so that slight movements of the hairs are magnified into relatively potent stimuli to the nerve endings around the follicles. The stiff vibrissae on the snouts of some animals are highly developed examples of hairs that act as levers to magnify tactile stimuli. The pacinian corpuscles are found in the subcutaneous tissues, muscles, and joints as well as the skin. They also abound in the mesentery, where their function is not known.

The A β sensory fibers that transmit impulses from touch receptors to the central nervous system are 5–12 μm in diameter and have conduction velocities of 30–70 m/s. Some touch impulses are also conducted via C fibers.

Touch information is transmitted in both the lemniscal and anterolateral pathways, so that only very extensive lesions completely interrupt touch sensation. However, there are differences in the type of touch information transmitted in the 2 systems. When the dorsal columns are destroyed, vibratory sensation and proprioception are lost, the touch threshold is elevated, and the number of touch-sensitive areas in the skin is decreased. In addition, localization of touch sensation is impaired. An increase in touch threshold and a decrease in the number of touch spots in the skin are also observed after interrupting the spinothalamic tracts, but the touch deficit is slight and touch localization remains normal. The information carried in the lemniscal system is concerned with the detailed localization, spatial form, and temporal pattern of tactile stimuli. The information carried in the spinothalamic tracts, on the other hand, is concerned with poorly localized, gross tactile sensations.

PROPRIOCEPTION

Proprioceptive information is transmitted up the spinal cord in the dorsal columns. A good deal of the proprioceptive input goes to the cerebellum, but some passes via the medial lemnisci and thalamic radiations to the cortex. Diseases of the dorsal columns produce ataxia because of the interruption of proprioceptive input to the cerebellum.

There is some evidence that proprioceptive information passes to consciousness in the anterolateral columns of the spinal cord. Conscious awareness of the positions of the various parts of the body in space depends in part upon impulses from sense organs in and around the joints. The organs involved are slowly adapting "spray" endings, structures that resemble Golgi tendon organs, and probably pacinian corpuscles in the synovia and ligaments. Impulses from these organs, touch receptors in the skin and other tissues, and muscle spindles are synthesized in the cortex into a conscious picture of the position of the body in space. Microelectrode studies indicate that many of the neurons in the sensory cortex respond to particular movements, not just to touch or static position. In this regard, the sensory cortex is organized like the visual cortex (see Chapter 8).

TEMPERATURE

There are 2 types of temperature sense organs: those responding maximally to temperatures slightly above body temperature, and those responding maximally to temperatures slightly below body temperature. The former are the sense organs for what we call warmth, and the latter for what we call cold. However, the adequate stimuli are actually 2 different degrees of warmth, since cold is not a form of energy.

Mapping experiments show that there are discrete cold-sensitive and warmth-sensitive spots in the skin. There are 4–10 times as many cold spots as warm. The temperature sense organs are naked nerve endings that respond to absolute temperature, not the temperature gradient across the skin. Cold receptors respond from 10 to 38 °C and warm receptors from 30 to 45 °C. The afferents for cold are A δ and C fibers, whereas the afferents for warmth are C fibers (see Table 2–2). These afferents relay information to the postcentral gyrus via the lateral spinothalamic tract and the thalamic radiation.

Because the sense organs are located subepithelially, it is the temperature of the subcutaneous tissues that determines the responses. Cool metal objects feel colder than wooden objects of the same temperature because the metal conducts heat away from the skin more rapidly, cooling the subcutaneous tissues to a greater degree. Below a skin temperature of 20 °C and above 40 °C, there is no adaptation, but between 20 and 40 °C there is adaptation, so that the sensation produced by a temperature change gradually fades to one of thermal neutrality. Above 45 °C, tissue damage begins to occur, and the sensation becomes one of pain.

PAIN

The sense organs for pain are the naked nerve endings found in almost every tissue of the body. Pain impulses are transmitted to the central nervous system by 2 fiber systems. One nociceptor system is made up of small myelinated A δ fibers 2–5 μm in diameter, which conduct at rates of 12–30 m/s. The other consists of unmyelinated C fibers 0.4–1.2 μm in diameter. These latter fibers are found in the lateral division of the dorsal roots and are often called dorsal root C fibers. They conduct at the low rate of 0.5–2 m/s. Both fiber groups end in the dorsal horn; A δ fibers terminate primarily on neurons in laminas I and V, whereas the dorsal root C fibers terminate on neurons in laminas I and II. There is abundant evidence that the synaptic transmitter secreted by primary afferent fibers subserving pain is substance P.

Some of the axons of the dorsal horn neurons end in the spinal cord and brain stem. Others enter the anterolateral system, including the lateral spinothalamic tract. A few ascend in the posterolateral portion of the cord. Some of the ascending fibers project to the specific sensory relay nuclei of the thalamus and from there to the cerebral cortex. PET and MRI studies (see Chapter 32) in normal humans indicate that pain activates 3 cortical areas: SI, SII. and the cingulate gyrus on the side opposite the stimulus. The cingulate gyrus is involved in emotion, and cingulate gyrectomy has been reported to lessen the distress associated with chronic pain.

Many fibers activated by pain end in the reticular system, which projects to the midline and intralaminar nonspecific projection nuclei of the thalamus and from there to many different parts of the cortex. Others project to the hypothalamus, and some end in the periaqueductal gray, an area known to be concerned with pain (see below).

Fast & Slow Pain

The presence of 2 pain pathways, one slow and one fast, explains the physiologic observation that there are 2 kinds of pain. A painful stimulus causes a "bright," sharp, localized sensation followed by a dull, intense, diffuse, and unpleasant feeling. These 2 sensations are variously called fast and slow pain or first and second pain. The farther from the brain the stimulus is applied, the greater the temporal separation of the 2 components. This and other evidence make it clear that fast pain is due to activity in the A δ pain fibers, while slow pain is due to activity in the C pain fibers.

Subcortical Perception & Affect

There is considerable evidence that sensory stimuli are perceived in the absence of the cerebral cortex, and this is especially true of pain. The cortical receiving areas are apparently concerned with the discriminative, exact, and meaningful interpretation of pain and some of its emotional components, but perception alone does not require the cortex.

Pain was called by Sherrington the "physical adjunct of an imperative protective reflex." Stimuli that are painful generally initiate potent withdrawal and avoidance responses. Furthermore, pain is unique among the senses in that it is associated with a strong emotional component. Information transmitted via the special senses may secondarily evoke pleasant or unpleasant emotions, depending largely upon experience, but pain alone has a "built-in" unpleasant affect. Present evidence indicates that this affective response depends in part on the cingulate and frontal portions of the cerebral cortex, but it appears that the thalamus is also involved. Damage to the thalamus may be associated with a peculiar overreaction to painful stimuli known as the **thalamic syndrome.** In this condition, usually due to blockage of the thalamogeniculate branch of the posterior cerebral artery with consequent damage to the posterior thalamic nuclei, minor stimuli lead to prolonged, severe, and very unpleasant pain. Such bouts of pain may occur spontaneously or at least without evident external stimuli.

Deep Pain

The main difference between superficial and deep sensibility is the different nature of the pain evoked by noxious stimuli. Unlike superficial pain, deep pain is poorly localized, nauseating, and frequently associated with sweating and changes in blood pressure. Pain can be elicited experimentally from the periosteum and ligaments by injecting hypertonic saline into them. The pain produced in this fashion initiates reflex contraction of nearby skeletal muscles. This reflex contraction is similar to the muscle spasm associated with injuries to bones, tendons, and joints. The steadily contracting muscles become ischemic, and ischemia stimulates the pain receptors in the muscles (see below). The pain in turn initiates more spasm, setting up a vicious circle.

Adequate Stimulus

Pain receptors are specific, and pain is not produced by overstimulation of other receptors. On the other hand, the adequate stimulus for pain receptors is not as specific as that for others, because they can be stimulated by a variety of strong stimuli. For example, pain receptors respond to warmth, but it has been calculated that their threshold for thermal energy is over 100 times that of the warmth receptors. Pain receptors also respond to electrical, mechanical, and, especially, chemical energy.

It has been suggested that pain is chemically mediated and that stimuli which provoke it have in common the ability to liberate a chemical agent that stimulates the nerve endings. The chemical agent might be a kinin (see Chapter 31) or histamine, both of which cause pain on local injection.

Muscle Pain

If a muscle contracts rhythmically in the presence of an adequate blood supply, pain does not usually result. However, if the blood supply to a muscle is occluded, contraction soon causes pain. The pain persists after the contraction until blood flow is reestablished. If a muscle with a normal blood supply is made to contract continuously without periods of relaxation, it also begins to ache because the maintained contraction compresses the blood vessels supplying the muscle.

These observations are difficult to interpret except in terms of the release during contraction of a chemical agent (Lewis's **"P factor"**) that causes pain when its local concentration is high enough. When the blood supply is restored, the material is washed out or metabolized. The identity of the P factor is not settled, but it could be K^+.

Clinically, the substernal pain that develops when the myocardium becomes ischemic during exertion (angina pectoris) is a classic example of the accumulation of P factor in a muscle. Angina is relieved by rest because this decreases the myocardial O_2 requirement and permits the blood supply to remove the factor. Intermittent claudication, the pain produced in the leg muscles of persons with occlusive vascular disease, is another example. It characteristically comes on while the patient is walking and disappears upon resting.

Hyperalgesia

In pathologic conditions, the sensitivity of the pain receptors is altered. There are 2 important types of alteration, primary and secondary hyperalgesia. In the area surrounding an inflamed or injured area, the threshold for pain is lowered so that trivial stimuli cause pain. This phenomenon, **primary hyperalgesia,** is seen around the area of the **flare,** the region of vasodilation around the injury. In the area of actual tissue damage, the vasodilation and, presumably, the pain are due to substances liberated from injured cells, but the flare in surrounding undamaged tissue is due to substance P liberated by antidromic impulses in primary afferent fibers (see Chapter 32).

Another aberration of sensation following injury is **secondary hyperalgesia.** In the area affected, the threshold for pain is actually elevated, but the pain produced is unpleasant, prolonged, and severe. The area from which this response is obtained extends well beyond the site of injury, and the condition does not last as long as primary hyperalgesia. It is probably due to some sort of central facilitation by impulses from the injured area of the pathways responsible for the unpleasant affect component of pain.

DIFFERENCES BETWEEN SOMATIC & VISCERAL SENSORY MECHANISMS

The autonomic nervous system, like the somatic, has afferent components, central integrating stations, and effector pathways. The visceral afferent mechanisms play a major role in homeostatic adjustments. In the viscera, there are a number of special receptors—osmoreceptors, baroreceptors, chemoreceptors, etc—that respond to changes in the internal environment. The afferent nerves from these receptors make reflex connections that are intimately concerned with regulating the function of the various systems with which they are associated, and their physiology is discussed in the chapters on these systems.

The receptors for pain and the other sensory modalities present in the viscera are similar to those in skin, but there are marked differences in their distribution. There are no proprioceptors in the viscera, and few temperature and touch sense organs. If the abdominal wall is infiltrated with a local anesthetic, the abdomen can be opened and the intestines can be handled, cut, and even burned without eliciting any discomfort. Pain receptors are present in the viscera, however, and although they are more sparsely distributed than in somatic structures, certain types of stimuli cause severe pain.

Afferent fibers from visceral structures reach the central nervous system via sympathetic and parasympathetic pathways (Fig 7–6). Their cell bodies are located in the dorsal roots and the homologous cranial nerve ganglia. Specifically, there are visceral afferents in the facial, glossopharyngeal, and vagus nerves; in the thoracic and upper lumbar dorsal roots;

Figure 7–6. Schematic representation of sensory substance P-containing fiber from colon. Note the collateral that synapses with a postganglionic neuron in the inferior mesenteric ganglion (IMG) and the cell body in the dorsal root ganglion (DRG). SG, substantia gelatinosa; DH, dorsal horn; VH, ventral horn; LSG, lumbar paravertebral sympathetic ganglion. (Reproduced, with permission, from Matthews MR, Cuello AC: Substance P-immunoreactive peripheral branches of sensory neurons innervate guinea pig sympathetic neurons. *Proc Natl Acad Sci USA* 1982;**79:**1668.)

and in the sacral roots (Fig 7–7). There may also be visceral afferent fibers from the eye in the trigeminal nerve. It is worth noting that at least some substance P-containing afferents make connections via collaterals to postganglionic sympathetic neurons in collateral sympathetic ganglia such as the inferior mesenteric ganglion, as shown in Fig 7–6. These connections may play a part in reflex control of the viscera independent of the central nervous system.

In the central nervous system, visceral sensation travels along the same pathways as somatic sensation in the spinothalamic tracts and thalamic radiations, and the cortical receiving areas for visceral sensation are intermixed with the somatic receiving areas.

VISCERAL PAIN

Pain from visceral structures is poorly localized, unpleasant, and associated with nausea and autonomic symptoms. It often radiates or is referred to other areas.

Stimulation of Pain Fibers

Because there are relatively few pain receptors in the viscera, visceral pain is poorly localized. However, as almost everyone knows from personal experience, visceral pain can be very severe. The receptors in the walls of the hollow viscera are especially sensitive to distention of these organs. Such distention can be produced experimentally in the gastrointestinal tract by inflation of a swallowed balloon attached to a tube. This produces pain that waxes and wanes (intestinal colic) as the intestine contracts and relaxes on the balloon. Similar colic is produced in intestinal obstruction by the contractions of the dilated intestine above the obstruction. When a viscus is inflamed or hyperemic, relatively minor stimuli cause severe pain. This is probably a form of primary hyperalgesia similar to that which occurs in somatic structures. Traction on the mesentery is also claimed to be painful, but the significance of this observation in the production of visceral pain is not clear. Visceral pain is particularly unpleasant not only because of the affective component it has in common with all pain but also because so many visceral afferents excited by the same process that causes the pain have reflex connections that initiate nausea, vomiting, and other autonomic effects.

Muscle Spasm & Rigidity

Visceral pain, like deep somatic pain, initiates reflex contraction of nearby skeletal muscle. This reflex spasm is usually in the abdominal wall and makes the abdominal wall rigid. It is most marked when visceral inflammatory processes involve the peritoneum. However, it can occur without such involvement. The anatomic details of the reflex pathways by which impulses from diseased viscera initi-

Figure 7–7. Pain innervation of the viscera. Pain afferents from structures above the thoracic pain line and below the pelvic pain line traverse parasympathetic pathways. (After White JC. Reproduced, with permission, from Ruch TC in: *Physiology and Biophysics,* 19th ed. Ruch TC, Patton HD [editors]. Saunders, 1965.)

ate skeletal muscle spasm are still obscure. The spasm protects the underlying inflamed structures from inadvertent trauma. Indeed, this reflex spasm is sometimes called "guarding."

The classic signs of inflammation in an abdominal viscus are pain, tenderness, autonomic changes such as hypotension and sweating, and spasm of the abdominal wall. From the preceding discussion, the genesis of each of these signs is apparent. The tenderness is due to the heightened sensitivity of the pain receptors in the viscus, the autonomic changes due to activation of visceral reflexes, and the spasm due to reflex contraction of skeletal muscle in the abdominal wall.

REFERRAL OF PAIN

Referred Pain

Irritation of a viscus frequently produces pain which is felt not in the viscus but in some somatic structure that may be a considerable distance away. Such pain is said to be **referred** to the somatic struc-

ture. Deep somatic pain may also be referred, but superficial pain is not. When visceral pain is both local and referred, it sometimes seems to spread (**radiate**) from the local to the distant site.

Obviously, a knowledge of referred pain and the common sites of pain referral from each of the viscera is of great importance to the physician. Perhaps the best known example is referral of cardiac pain to the inner aspect of the left arm. Other dramatic examples include pain in the tip of the shoulder caused by irritation of the central portion of the diaphragm and pain in the testicle due to distention of the ureter. Additional instances abound in the practice of medicine, surgery, and dentistry. However, sites of reference are not stereotyped, and unusual reference sites occur with considerable frequency. Heart pain, for instance, may be purely abdominal, may be referred to the right arm, and may even be referred to the neck. Referred pain can be produced experimentally by stimulation of the cut end of a splanchnic nerve.

Dermatomal Rule

When pain is referred, it is usually to a structure

that developed from the same embryonic segment or dermatome as the structure in which the pain originates. This principle is called the **dermatomal rule.** For example, during embryonic development, the diaphragm migrates from the neck region to its adult location in the abdomen and takes its nerve supply, the phrenic nerve, with it. One-third of the fibers in the phrenic nerve are afferent, and they enter the spinal cord at the level of the second to fourth cervical segments, the same location at which afferents from the tip of the shoulder enter. Similarly, the heart and the arm have the same segmental origin, and the testicle has migrated with its nerve supply from the primitive urogenital ridge from which the kidney and ureter also developed.

Role of Experience

Experience also plays an important role in referred pain. Although pain originating in an inflamed abdominal viscus is usually referred to the midline, in patients who have had previous abdominal surgery the pain of abdominal disease is frequently referred to their surgical scars. Pain originating in the maxillary sinus is usually referred to nearby teeth, but in patients with a history of traumatic dental work such pain is regularly referred to the previously traumatized teeth. This is true even when the teeth are a considerable distance away from the sinus.

Role of Convergence

Not only do the nerves from the visceral structures and the somatic structures to which pain is referred enter the nervous system at the same level, but also there are more sensory fibers in the peripheral nerves than there are axons in the spinothalamic tracts. Therefore, there is presumably a considerable degree of convergence of peripheral sensory fibers on the spinothalamic neurons. One theory of the mechanism underlying referred pain is based on this premise. It holds that somatic and visceral afferents converge on the same spinothalamic neurons (Fig 7–8). Since somatic pain is much more common than visceral pain, the brain has "learned" that activity arriving in a given pathway is caused by a pain stimulus in a particular somatic area. When the same pathway is stimulated by activity in visceral afferents, the signal reaching the brain is no different and the pain is projected to the somatic area.

Facilitation Effects

Another theory of the origin of referred pain holds that because of subliminal fringe effects, incoming impulses from visceral structures lower the threshold of spinothalamic neurons receiving afferents from somatic areas, so that minor activity in the pain pathways from the somatic areas—activity which would normally die out in the spinal cord—passes on to the brain.

If convergence alone were the explanation for re-

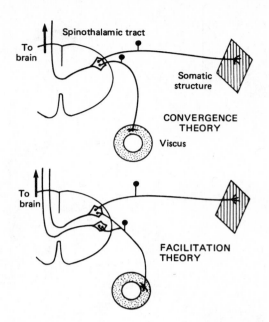

Figure 7–8. Diagram of convergence and facilitation theories of referred pain.

ferred pain, anesthetizing the somatic area of reference with procaine should have no effect on the pain, whereas if subliminal fringe effects were responsible, the pain should disappear. The effects of local anesthesia in the area of reference vary. Severe pain is usually unaffected, but mild pain may be completely abolished. Therefore, it appears that both convergence and facilitation play a role in the pathogenesis of referred pain.

INHIBITION & FACILITATION OF PAIN

Central Inhibition & Counterirritants

It is well known that soldiers wounded in the heat of battle may feel no pain until the battle is over (**stress analgesia**). Many people have learned from practical experience that touching or shaking an injured area decreases the pain of the injury. Stimulation with an electric vibrator at the site of pain also gives some relief. Acupuncture has been used for 4000 years to prevent or relieve pain, and it is possible in some instances to perform major surgery with acupuncture as the only type of anesthesia. These and other observations make it clear that pain transmission and perception are subject to inhibition or modification.

Inhibition in central sensory pathways may explain the efficacy of counterirritants. Stimulation of the skin over an area of visceral inflammation produces some relief of the pain due to the visceral disease. The old-fashioned mustard plaster works on this principle.

Gating in the Dorsal Horn

One site of inhibition of pain transmission is the dorsal horn, the "gate" through which pain impulses reach the lateral spinothalamic system. Stimulation of large fiber afferents from an area from which pain is being initiated reduces the pain. Collateral fibers from the dorsal column touch fibers enter the substantia gelatinosa, and it has been postulated that impulses in these collaterals or interneurons on which they end inhibit transmission from the dorsal root pain fibers to the spinothalamic neurons. The mechanism involved appears to be presynaptic inhibition (see Chapter 4) at the endings of the primary afferents that transmit pain impulses.

Action of Morphine & Enkephalins

Morphine relieves pain and produces euphoria. It is particularly effective when given intrathecally. The receptors that bind morphine and the "body's own morphines," the opioid peptides, are discussed in Chapter 4. There are opioid receptors in the substantia gelatinosa that may be on the substance P-containing terminals of nociceptive afferents and inhibit substance P release. It has been argued that enkephalin-containing neurons terminate presynaptically at the site of the receptors on these afferents, but no presynaptic terminals have been identified. Therefore, the mechanism by which opioids act remains unsettled. In addition, injections of morphine into the periaqueductal gray of the midbrain relieve pain by activating descending pathways that produce inhibition of primary afferent transmission in the dorsal horn. There is evidence that this activation occurs via projections from the periaqueductal gray to the nearby raphe magnus nucleus and related serotonergic nuclei (see Chapter 15) and that descending serotonergic fibers from the raphe magnus nucleus mediate the inhibition. However, the mechanism by which serotonin inhibits transmission in the dorsal horn is still unknown.

Placebos appear to be capable of producing the release of endogenous opioids, and this helps to relieve pain. Their effects are inhibited in part by morphine antagonists such as naloxone. Acupuncture at a location distant from the site of a pain also acts by releasing endorphins. Acupuncture at the site of the pain appears to act primarily in the same way as touching or shaking (see above). There appears to be a component of stress analgesia that is mediated by endogenous opioids, because in experimental animals, some forms of stress analgesia are prevented by naloxone. However, other forms are unaffected, and so other components are also involved.

Causes & Treatment of Chronic Pain

Cancer, skeletal disorders such as herniated inter-

vertebral disks, and other conditions can cause long-lasting severe hyperalgesic-like pain that is difficult to manage. In addition, 4 kinds of neuropathy are painful: diabetic neuropathy, nerve irritation due to herpes zoster, causalgia, and reflex sympathetic dystrophy. Chronic irritation of nerves apparently increases the number of substance P receptors in the spinal cord on the side of the injury and causes A β fibers from mechanoreceptors to sprout and grow into the areas where the nociceptors terminate. Both these changes could increase responses to painful stimuli. In addition, irritation of nerves can increase sympathetic discharge, and there is evidence that the resulting activation of α_1-adrenergic receptors makes pain more intense and unpleasant.

Pain can often be handled by administration of analgesic drugs in adequate doses, but this is not always the case. Chronic stimulation of the dorsal columns with implanted electrodes has been used clinically as a method of relieving intractable pain. The electrodes can be attached to a portable stimulator, and patients can stimulate themselves as needed to control their pain. The pain relief is presumably due to antidromic rather than orthodromic conduction, with impulses passing via the collaterals to the "gate" in the dorsal horn, since section of the dorsal columns enhances rather than reduces responses to noxious stimuli. Self stimulation of the periaqueductal gray also helps to relieve intractable pain, probably by increasing the release of β-endorphin.

In some instances, severe pain has been dissociated from its unpleasant affect by cutting the deep connections between the frontal lobes and the rest of the brain (prefrontal lobotomy). After this operation patients report that they feel pain but that it "doesn't bother" them. However, the operation causes extensive personality changes (see Chapter 16). The relief of distress produced by cingulate gyrectomy has been mentioned above. Other operations that have been used for the relief of pain interrupt pain pathways, as summarized in Fig 7–9. One of these is **anterolateral cordotomy.** In this procedure, a knife is inserted into the lateral aspect of the spinal cord and swept anteriorly and laterally as shown in Fig 7–3. When properly performed, this procedure cuts the lateral spinothalamic and other anterolateral system pain fibers while leaving most of the ventral spinothalamic touch fibers intact. Even if there is damage to the touch fibers in the anterolateral quadrants, touch is impaired very little, because the dorsal column touch system is intact. However, the pain can recur months to years later, indicating that other pathways can take over pain transmission.

Figure 7–9. Diagram of various surgical procedures designed to alleviate pain. 1, nerve section; 2, sympathectomy (for visceral pain); 3, myelotomy to section spinothalamic fibers in anterior white commissure; 4, posterior rhizotomy; 5, anterolateral cordotomy; 6, medullary tractotomy; 7, mesencephalic tractotomy; 8, thalamotomy; 9, cingulate gyrectomy; 10, prefrontal lobotomy.

OTHER SENSATIONS

Itch & Tickle

Relatively mild stimulation, especially if produced by something that moves across the skin, produces itch and tickle. Itch spots can be identified on the skin by careful mapping; like the pain spots, they are in regions where there are many naked endings of unmyelinated fibers. Itch persists along with burning pain in nerve block experiments when only C fibers are conducting, and itch, like pain, is abolished by section of the spinothalamic tracts. However, the distributions of itch and pain are different; itching occurs only in the skin, eyes, and certain mucous membranes and not in deep tissues or viscera. Furthermore, low-frequency stimulation of pain fibers produces pain, not itch, and high-frequency stimulation of itch spots on the skin may merely increase the intensity of the itching without producing pain. These observations indicate that the C fiber system responsible for itching is not the same as that responsible for pain. It is interesting that a tickling sensation is usually regarded as pleasurable, whereas itching is annoying and pain is unpleasant.

Itching can be produced not only by repeated local mechanical stimulation of the skin but also by a variety of chemical agents. Histamine produces intense itching, and injuries cause its liberation in the skin. However, in most instances of itching, histamine does not appear to be the responsible agent; doses of histamine that are too small to produce itching still produce redness and swelling on injection into the skin, and severe itching frequently occurs without any visible change in the skin. The kinins cause severe itching. It is interesting in this regard that itch powder, which is made up of the spicules from the pods of the tropical plant cowhage, contains a proteolytic enzyme, and the powder presumably acts by liberating itch-producing peptides.

"Synthetic Senses"

The cutaneous senses for which separate receptors exist are touch, warmth, cold, pain, and possibly itching. Combinations of these sensations, patterns of stimulation, and, in some cases, cortical components are synthesized into the sensations of vibratory sensation, 2-point discrimination, and stereognosis.

Vibratory Sensibility

When a vibrating tuning fork is applied to the skin, a buzzing or thrill is felt. The sensation is most marked over bones, but it can be felt when the tuning fork is placed in other locations. The receptors involved are the receptors for touch, especially pacinian corpuscles, but a time factor is also necessary. A pattern of rhythmic pressure stimuli is interpreted as vibration. The impulses responsible for the vibrating sensation are carried in the dorsal columns. Degeneration of this part of the spinal cord occurs in poorly controlled diabetes, pernicious anemia, some vitamin deficiencies, and occasionally other conditions; depression of the threshold for vibratory stimuli is an early symptom of this degeneration. Vibratory sensation and proprioception are closely related; when one is depressed, so is the other.

Two-Point Discrimination

The minimal distance by which 2 touch stimuli must be separated to be perceived as separate is called the **2-point threshold.** It depends upon touch plus the cortical component of identifying one or 2 stimuli. Its magnitude varies from place to place on the body and is smallest where the touch receptors are most abundant. Points on the back, for instance, must be separated by 65 mm or more before they can be distinguished as separate points, whereas on the fingers 2 stimuli can be resolved if they are separated by as little as 3 mm. On the hands, the magnitude of the 2-point threshold is approximately the diameter of the area of skin supplied by a single sensory unit. However, the peripheral neural basis of discriminating 2 points is not completely understood, and in view of the extensive interdigitation and overlapping of the sensory units, it is probably complex.

Stereognosis

The ability to identify objects by handling them without looking at them is called **stereognosis.** Normal persons can readily identify objects such as keys and coins of various denominations. This ability obviously depends upon relatively intact touch and pressure sensation, but it also has a large cortical component. Impaired stereognosis is an early sign of damage to the cerebral cortex and sometimes occurs in the absence of any defect in touch and pressure sensation when there is a lesion in the parietal lobe posterior to the postcentral gyrus.

8

Vision

INTRODUCTION

The eyes are complex sense organs that have evolved from primitive light-sensitive spots on the surface of invertebrates. Within its protective casing, each eye has a layer of receptors, a lens system that focuses light on these receptors, and a system of nerves that conducts impulses from the receptors to the brain. The way these components operate to set up conscious visual images is the subject of this chapter.

ANATOMIC CONSIDERATIONS

The principal structures of the eye are shown in Fig 8–1. The outer protective layer of the eyeball, the **sclera,** is modified anteriorly to form the transparent **cornea,** through which light rays enter the eye. Inside the sclera is the **choroid,** a pigmented layer that contains many of the blood vessels which nourish the structures in the eyeball. Lining the posterior two-thirds of the choroid is the **retina,** the neural tissue containing the receptor cells.

The **crystalline lens** is a transparent structure held in place by a circular **lens ligament (zonule).** The zonule is attached to the thickened anterior part of the choroid, the **ciliary body.** The ciliary body contains circular muscle fibers and longitudinal fibers that attach near the corneoscleral junction. In front of the lens is the pigmented and opaque **iris,** the colored portion of the eye. The iris contains circular muscle fibers that constrict and radial fibers that dilate the **pupil.** Variations in the diameter of the pupil can produce up to 5-fold changes in the amount of light reaching the retina.

The space between the lens and the retina is filled primarily with a clear gelatinous material called the **vitreous (vitreous humor). Aqueous humor,** a clear liquid, is produced in the ciliary body by diffusion and active transport and flows through the pupil to fill the anterior chamber of the eye. It is normally reabsorbed through a network of trabeculae into the **canal of Schlemm,** a venous channel at the junction between the iris and the cornea (anterior chamber angle). Obstruction of this outlet leads to increased intraocular pressure and the serious eye disease **glau-**

coma. One cause is decreased permeability through the trabeculae **(open-angle glaucoma),** and another is forward movement of the iris, obliterating the angle **(angle-closure glaucoma).**

Retina

The retina extends anteriorly almost to the ciliary body. It is organized in 10 layers and contains the **rods** and **cones,** which are the visual receptors, plus 4 types of neurons: **bipolar cells, ganglion cells, horizontal cells,** and **amacrine cells** (Fig 8–2). The rods and cones, which are next to the choroid, synapse with bipolar cells, and the bipolar cells synapse with ganglion cells. The axons of the ganglion cells converge and leave the eye as the optic nerve. Horizontal cells connect receptor cells to the other receptor cells in the outer plexiform layer. Amacrine cells connect ganglion cells to one another in the inner plexiform layer and in some instances may be inserted between

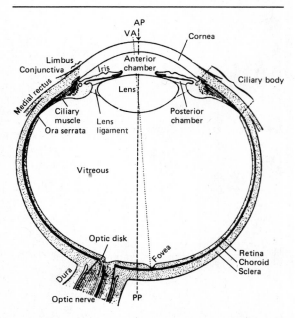

Figure 8–1. Horizontal section of the right eye. AP, anterior pole; PP, posterior pole; VA, visual axis. (Reproduced, with permission, from Warwick R: *Eugene Wolff's Anatomy of the Eye and Orbit,* 7th ed. Saunders, 1977.)

Pigment epithelium

Rod and cone

Outer segments

Inner segments

Outer nuclear layer

Outer plexiform layer

Inner nuclear layer

Inner plexiform layer

Ganglion cell layer

Optic nerve fibers

Figure 8–2. Neural components of the retina. C, cone; R, rod; MB, RB, and FB, midget, rod, and flat bipolar cells; DG and MG, diffuse and midget ganglion cells; H, horizontal cells; A, amacrine cells. (Reproduced, with permission, from Dowling JE, Boycott, BB: Organization of the primate retina: Electron microscopy. *Proc R Soc Lond Ser B [Biol]* 1966;**166:**80.)

bipolar cells and ganglion cells. They have no axons, and their processes make both pre- and postsynaptic connections with neighboring neural elements. There is considerable overall convergence of receptors on bipolar cells and of bipolar cells on ganglion cells (see below).

Since the receptor layer of the retina is apposed to the choroid, light rays must pass through the ganglion cell and bipolar cell layers to reach the rods and cones. The pigmented layer of choroid next to the retina absorbs light rays, preventing the reflection of rays back through the retina. Such reflection would produce blurring of the visual images.

The neural elements of the retina are bound together by glial cells called Müller cells. The processes of these cells form an internal limiting membrane on the inner surface of the retina and an external limiting membrane in the receptor layer.

The optic nerve leaves the eye and the retinal blood vessels enter it at a point 3 mm medial to and slightly above the posterior pole of the globe. This region is visible through the ophthalmoscope as the

optic disk (Fig 8–3). There are no visual receptors overlying the disk, and consequently this spot is blind (the **blind spot**). At the posterior pole of the eye, there is a yellowish pigmented spot, the **macula lutea.** This marks the location of the **fovea centralis,** a thinned-out, rod-free portion of the retina where the cones are densely packed and there are very few cells and no blood vessels overlying the receptors. The fovea is highly developed in humans. It is the point where visual acuity is greatest. When attention is attracted to or fixed on an object, the eyes are normally moved so that light rays coming from the object fall on the fovea.

The arteries, arterioles, and veins in the superficial layers of the retina near its vitreous surface can be seen through the ophthalmoscope. Since this is the one place in the body where arterioles are readily visible, ophthalmoscopic examination is of great value in the diagnosis and evaluation of diabetes mellitus, hypertension, and other diseases that affect blood vessels. The retinal vessels supply the bipolar and ganglion cells, but the receptors are nourished for the

Figure 8–3. Retina seen through the ophthalmoscope in a normal human. The diagram on the left identifies the landmarks in the photograph on the right. (Reproduced, with permission, from Vaughan D, Asbury T, Tabbara KF: *General Ophthalmology,* 12th ed. Appleton & Lange, 1989.)

most part by the capillary plexus in the choroid. This is why retinal detachment is so damaging to the receptor cells.

Neural Pathways

The axons of the ganglion cells pass caudally in the **optic nerve** and **optic tract** to end in the **lateral geniculate body,** a part of the thalamus (Fig 8–4). The fibers from each nasal hemiretina decussate in the **optic chiasm.** In the geniculate body, the fibers from the nasal half of one retina and the temporal half of the other synapse on the cells whose axons form the **geniculocalcarine tract.** This tract passes to the occipital lobe of the cerebral cortex.

The primary visual receiving area (**visual cortex,** Brodmann's area 17), is located principally on the sides of the calcarine fissure (Fig 8–5). The organization of the primary visual cortex is discussed below.

Branches of the ganglion cell axons pass from the optic tract to the pretectal region of the midbrain and the superior colliculus, where they form connections that mediate visual reflexes. Other axons pass directly from the optic chiasm to the suprachiasmatic nuclei in the hypothalamus, where they form connections that synchronize a variety of endocrine and other circadian rhythms with the light-dark cycle (see Chapter 14).

The brain areas activated by visual stimuli have been investigated in monkeys by means of radioactive 2-deoxyglucose (see Chapter 32). Activation occurs not only in the occipital lobe but also in parts of the inferior temporal cortex, the posteroinferior parietal cortex, and portions of the frontal lobe. The subcortical structures activated in addition to the lateral geniculate body include the superior colliculus, pulvinar, caudate nucleus, putamen, claustrum, and amygdala.

Receptors

Each rod and cone is divided into an outer segment, an inner segment that includes a nuclear region, and a synaptic zone (Fig 8–6). The outer segments are modified cilia and are made up of regular stacks of flattened saccules or disks composed of membrane. These saccules and disks contain the photosensitive compounds that react to light, initiating action potentials in the visual pathways (see below). The inner segments are rich in mitochondria. The rods are named for the thin, rodlike appearance of their outer segments. Cones generally have thick inner segments and conical outer segments, although their morphology varies from place to place in the retina. In cones, the saccules are formed in the outer segments by infoldings of the cell membrane, but in rods, the disks are separated from the cell membrane.

Rod outer segments are being constantly renewed by formation of new disks at the inner edge of the segment and phagocytosis of old disks and from the outer tip by cells of the pigment epithelium. Cone renewal is a more diffuse process and appears to occur at multiple sites in the outer segments.

The fovea contains no rods, and each foveal cone has a single midget bipolar cell connecting it to a single ganglion cell, so that each foveal cone is connected to a single fiber in the optic nerve. In other portions of the retina, rods predominate (Fig 8–7), and there is a good deal of convergence. Flat bipolar cells (Fig 8–2) make synaptic contact with several cones, and rod bipolar cells make synaptic contact with several rods. Since there are approximately 6

Upper peripheral quadrant of retina

Upper quadrant of macula

Lower peripheral quadrant of retina

Lower quadrant of macula

Figure 8–4. Visual pathways. Transection of the pathways at the locations indicated by the letters causes the visual field defects shown in the diagrams on the right (see text). Occipital lesions may spare the fibers from the macula (as in D) because of the separation in the brain of these fibers from the others subserving vision (see Fig 8–5).

Figure 8–5. Medial view of the human cerebral hemisphere showing projection of the retina on the calcarine fissure.

Figure 8–6. Schematic diagram of a rod and a cone. (Reproduced, with permission, from Lamb TD: Electrical responses of photoreceptors. In: *Recent Advances in Physiology*. No. 10. Baker PF [editor]. Churchill Livingstone, 1984.)

Figure 8–7. Rod and cone density along the horizontal meridian through the human retina. A plot of the relative acuity of vision in the various parts of the light-adapted eye would parallel the cone density curve; a similar plot of relative acuity of the dark-adapted eye would parallel the rod density curve. (Modified and reproduced, with permission, from Osterberg G: Topography of the layer of rods and cones in the human retina. *Acta Ophthalmol [Kbh]* 1935;**13**:[Suppl 6].)

million cones and 120 million rods in each human eye but only 1.2 million nerve fibers in each optic nerve, the overall convergence of receptors through bipolar cells on ganglion cells is about 105:1.

The rods are extremely sensitive to light and are the receptors for night vision **(scotopic vision).** The scotopic visual apparatus is not capable of resolving the details and boundaries of objects or determining their color. The cones have a much higher threshold, but the cone system has a much greater acuity and is the system responsible for vision in bright light **(photopic vision)** and for color vision. There are thus 2 kinds of inputs to the central nervous system from the eye: input from the rods and input from the cones. The existence of these 2 kinds of input, each working maximally under different conditions of illumination, is called the **duplicity theory.**

Eye Muscles

The eye is moved within the orbit by 6 ocular muscles (Fig 8–8). These are innervated by the oculomotor, trochlear, and abducens nerves. The muscles and the directions in which they move the eyeball are discussed at the end of this chapter.

Protection

The eye is well protected from injury by the body walls of the orbit. The cornea is moistened and kept clear by tears that course from the **lacrimal gland** in the upper portion of each orbit across the surface of the eye to empty via the **lacrimal duct** into the nose. Blinking helps keep the cornea moist.

THE IMAGE-FORMING MECHANISM

The eyes convert energy in the visible spectrum into action potentials in the optic nerve. The wavelengths of visible light range from approximately 397 nm to 723 nm. The images of objects in the environment are focused on the retina. The light rays striking the retina generate potentials in the rods and cones. Impulses initiated in the retina are conducted to the cerebral cortex, where they produce the sensation of vision.

Figure 8–8. Muscles of the right orbit. The 6 muscles that move the eyeball and the levator palpebrae superioris, which raises the upper lid, are shown. (Reproduced, with permission, from Goss CM [editor]: *Gray's Anatomy of the Human Body,* 29th ed. Lea & Febiger, 1973.)

Principles of Optics

Light rays are bent (refracted) when they pass from one medium into a medium of a different density, except when they strike perpendicular to the interface. Parallel light rays striking a biconvex lens (Fig 8–9) are refracted to a point (**principal focus**) behind the lens. The principal focus is on a line passing through the centers of curvature of the lens, the **principal axis.** The distance between the lens and the principal focus is the **principal focal distance.** For practical purposes, light rays from an object that strike a lens more than 20 ft (6 m) away are considered to be parallel. The rays from an object closer than 20 ft are diverging and are therefore brought to a focus farther back on the principal axis than the principal focus (Fig 8–9). Biconcave lenses cause light rays to diverge.

The greater the curvature of a lens, the greater its refractive power. The refractive power of a lens is conveniently measured in **diopters,** the number of diopters being the reciprocal of the principal focal distance in meters. For example, a lens with a principal focal distance of 0.25 m has a refractive power of 1/0.25, or 4 diopters. The human eye has a refractive power of approximately 66.7 diopters at rest.

Accommodation

When the ciliary muscle is relaxed, parallel light rays striking the optically normal (**emmetropic**) eye are brought to a focus on the retina. As long as this relaxation is maintained, rays from objects closer than 6 m from the observer are brought to a focus behind the retina, and consequently the objects appear blurred. The problem of bringing diverging rays from close objects to a focus on the retina can be solved by increasing the distance between the lens and the retina or by increasing the curvature or refractive power of the lens. In bony fish, the problem is solved by increasing the length of the eyeball, a solution analogous to the manner in which the images of objects closer than 6 m are focused on the film of a camera by moving the lens away from the film. In mammals, the problem is solved by increasing the curvature of the lens.

The process by which the curvature of the lens is increased is called **accommodation.** At rest, the lens is held under tension by the lens ligaments. Because the lens substance is malleable and the lens capsule has considerable elasticity, the lens is pulled into a flattened shape. When the gaze is directed at a near object, the ciliary muscle contracts. This decreases the distance between the edges of the ciliary body and relaxes the lens ligaments, so that the lens springs into a more convex shape. In young individuals, the change in shape may add as many as 12 diopters to the refractive power of the eye. The relaxation of the lens ligaments produced by contraction of the ciliary muscle is due partly to the sphincterlike action of the circular muscle fibers in the ciliary body and partly to the contraction of longitudinal muscle fibers that attach anteriorly, near the corneo-scleral junction. When these fibers contract, they pull the whole ciliary body forward and inward. This motion brings the edges of the ciliary body closer together.

The change in lens curvature during accommodation affects principally the anterior surface of the lens (Fig 8–10). This can be demonstrated by a simple experiment first described many years ago. If an observer holds an object in front of the eyes of an individual who is looking into the distance, 3 reflections

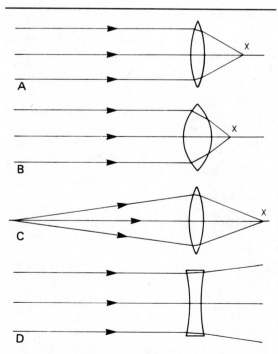

Figure 8–9. Refraction of light rays by lenses: **A:** Biconvex lens. **B:** Biconvex lens of greater strength than **A. C:** Same lens as **A,** showing effect on light rays from a near point. **D:** Biconcave lens. The center line in each case is the principal axis. X is the principal focus.

Figure 8–10. Accommodation. The solid lines represent the shape of the lens, iris, and ciliary body at rest, and the dotted lines represent the shape during accommodation.

of the object are visible in the subject's eye. A clear, small upright image is reflected from the cornea; a larger, fainter upright image is reflected from the anterior surface of the lens; and a small inverted image is reflected from the posterior surface of the lens. If the subject then focuses on an object nearby, the large, faint upright image becomes smaller and moves toward the other upright image, whereas the other 2 images change very little. The change in size of the image is due to the increase in curvature of the reflecting surface, the anterior surface of the lens (Fig 8–10). The fact that the small upright image does not change and the inverted image changes very little shows that the corneal curvature is unchanged and that the curvature of the posterior lens surface is changed very little by accommodation.

Near Point

Accommodation is an active process, requiring muscular effort, and can therefore be tiring. Indeed, the ciliary muscle is one of the most used muscles in the body. The degree to which the lens curvature can be increased is, of course, limited, and light rays from an object very near the individual cannot be brought to a focus on the retina even with the greatest of effort. The nearest point to the eye at which an object can be brought into clear focus by accommodation is called the **near point of vision.** The near point recedes throughout life, slowly at first and then rapidly with advancing age, from approximately 9 cm at age 10 to approximately 83 cm at age 60. This recession is due principally to increasing hardness of the lens, with a resulting loss of accommodation (Fig 8–11) due to the steady decrease in the degree to which the curvature of the lens can be increased. By the time a normal individual reaches age 40–45, the loss of accommodation is usually sufficient to make reading and close work difficult. This condition, which is known as **presbyopia,** can be corrected by wearing glasses with convex lenses.

The Near Response

In addition to accommodation, the visual axes converge and the pupil constricts when an individual looks at a near object. This 3-part response—accommodation, convergence of the visual axes, and pupillary constriction—is called the **near response.**

Other Pupillary Reflexes

When light is directed into one eye, the pupil constricts (**pupillary light reflex**). The pupil of the other eye also constricts (**consensual light reflex**). The optic nerve fibers that carry the impulses initiating pupillary responses end in the pretectal region and the superior colliculi. The pathway for the light reflex, which presumably passes from the pretectal region to the oculomotor nuclei (Edinger-Westphal nuclei) bilaterally, is different from that for accommodation. In some pathologic conditions, the pupil-

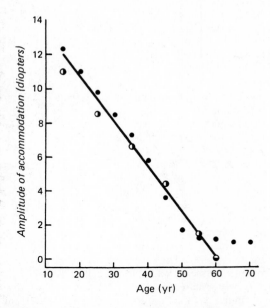

Figure 8–11. Decline in the amplitude of accommodation in humans with advancing age. The different symbols identify data from different studies. (Reproduced, with permission, from Fisher RF: Presbyopia and the changes with age in the human crystalline lens. *J Physiol* 1973;**228:**765.)

lary response to light may be absent while the response to accommodation remains intact. This phenomenon, the so-called **Argyll Robertson pupil,** is said to be due to a destructive lesion in the tectal region.

Retinal Image

In the eye, light is actually refracted at the anterior surface of the cornea and at the anterior and posterior surfaces of the lens. The process of refraction can be represented diagrammatically, however, without introducing any appreciable error, by drawing the rays of light as if all refraction occurs at the anterior surface of the cornea. Fig 8–12 is a diagram of such a "reduced" or "schematic" eye. In this diagram, the **nodal point** (optical center of the eye) coincides with the junction of the middle and posterior third of the lens, 15 mm from the retina. This is the point through which the light rays from an object pass without refraction. All other rays entering the pupil from each point on the object are refracted and brought to a focus on the retina. If the height of the object (AB) and its distance from the observer (Bn) are known, the size of its retinal image can be calculated, because AnB and anb in Fig 8–12 are similar triangles. The angle AnB is the **visual angle** subtended by object AB. It should be noted that the retinal image is inverted. The connections of the retinal receptors are such that from birth any inverted image on the retina is viewed right side up and projected to the visual

Figure 8–12. Reduced eye. n, nodal point. AnB and anb are similar triangles. In this reduced eye, the nodal point is 15 mm from the retina. All refraction is assumed to take place at the surface of the cornea, 5 mm from the nodal point, between a medium of density 1.000 (air) and a medium of density 1.333 (water). The dotted lines represent rays of light diverging from A and refracted at the cornea so that they are focused on the retina at a.

field on the side opposite to the retinal area stimulated. This perception is present in infants and is innate. If retinal images are turned right side up by means of special lenses, the objects viewed look as if they were upside down.

Common Defects of the Image-Forming Mechanism

In some individuals, the eyeball is shorter than normal and the parallel rays of light are brought to a focus behind the retina. This abnormality is called **hyperopia** or farsightedness (Fig 8–13). Sustained accommodation, even when viewing distant objects, can partially compensate for the defect, but the prolonged muscular effort is tiring and may cause headaches and blurring of vision. The prolonged convergence of the visual axes associated with the accommodation may lead eventually to squint (**strabismus**) (see below). The defect can be corrected by using glasses with convex lenses, which aid the refractive power of the eye in shortening the focal distance.

In **myopia** (nearsightedness), the anteroposterior diameter of the eyeball is too long. Myopia is said to be genetic in origin. However, in experimental animals it can be produced by changing refraction during development. Thus, the shape of the eye appears to be determined in part by the refraction presented to it. In young adult humans the extensive close work involved in activities such as studying accelerates the development of myopia. This defect can be corrected by glasses with biconcave lenses, which make parallel light rays diverge slightly before they strike the eye.

Astigmatism is a common condition in which the curvature of the cornea is not uniform. When the curvature in one meridian is different from that in others, light rays in that meridian are refracted to a different focus, so that part of the retinal image is blurred. A similar defect may be produced if the lens is pushed out of alignment or the curvature of the lens is not uniform, but these conditions are rare. Astigmatism can usually be corrected with cylindric lenses placed in such a way that they equalize the refraction in all meridians. **Presbyopia** has been mentioned above.

THE PHOTORECEPTOR MECHANISM: GENESIS OF ACTION POTENTIALS

The potential changes that initiate action potentials in the retina are generated by the action of light on photosensitive compounds in the rods and cones. When light is absorbed by these substances, their structure changes, and this change triggers a sequence of events that initiates neural activity.

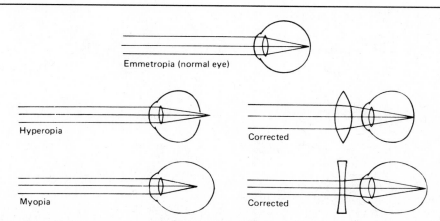

Figure 8–13. Common defects of the optical system of the eye. In hyperopia, the eyeball is too short and light rays come to a focus behind the retina. A biconvex lens corrects this by adding to the refractive power of the lens of the eye. In myopia, the eyeball is too long and light rays focus in front of the retina. Placing a biconcave lens in front of the eye causes the light rays to diverge slightly before striking the eye, so that they are brought to a focus on the retina.

ELECTRICAL RESPONSES OF RETINAL CELLS

The eye is unique in that the receptor potentials of the photoreceptors and the electrical responses of most of the other neural elements in the retina are local, graded potentials, and it is only in the ganglion cells that all or none action potentials transmitted over appreciable distances are generated. The responses of the rods, cones, and horizontal cells are hyperpolarizing (Fig 8–14), and the responses of the bipolar cells are either hyperpolarizing or depolarizing, whereas amacrine cells produce depolarizing potentials and spikes that may act as generator potentials for the propagated spikes produced in the ganglion cells.

The cone receptor potential has a sharp onset and offset, whereas the rod receptor potential has a sharp onset and slow offset. The curves relating the ampli-tude of receptor potentials to stimulus intensity have similar shapes in rods and cones, but the rods are much more sensitive. Therefore, rod responses are proportionate to stimulus intensity at levels of illumination that are below the threshold for cones. On the other hand, cone responses are proportionate to stimulus intensity at high levels of illumination when the rod responses are maximal and cannot change. This is why cones generate good responses to changes in light intensity above background but do not represent absolute illumination well, whereas rods detect absolute illumination.

Ionic Basis of Photoreceptor Potentials

Na^+ channels in the outer segments of the rods and cones are open in the dark, so current flows from the inner to the outer segment (Fig 8–15). Current also flows to the synaptic ending of the photoreceptor. The Na^+-K^+ exchange pump in the inner segment maintains ionic equilibrium. Release of synaptic transmitter is steady in the dark. When light strikes the outer segment, some of the Na^+ channels are closed, and the result is a hyperpolarizing receptor potential. The hyperpolarization reduces the release of synaptic transmitter, and this generates a signal that ultimately leads to action potentials in ganglion cells. The action potentials are transmitted to the brain.

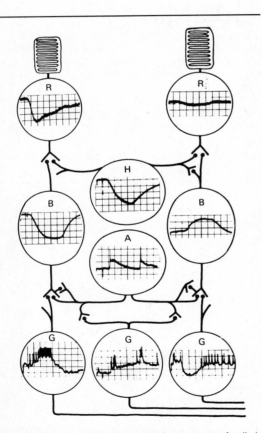

Figure 8–14. Intracellularly recorded responses of cells in the retina to light. The synaptic connections of the cells are also indicated. The rod (R) on the left is receiving a light flash, whereas the rod on the right is receiving steady, low-intensity illumination. H, horizontal cell; B, bipolar cell; A, amacrine cell; G, ganglion cell. (Reproduced, with permission, from Dowling JE: Organization of vertebrate retinas. *Invest Ophthalmol* 1970;**9**:655.)

Dark   Light

Figure 8–15. Effect of light on current flow in visual receptors. In the dark, Na^+ channels in the outer segment are held open by cyclic GMP. Light leads to increased conversion of cyclic GMP to 5'-GMP, and some of the channels close. This produces hyperpolarization of the synaptic terminal of the photoreceptor. (Modified from O'Brien DF: The chemistry of vision. *Science* 1982;**218**:961.)

Photosensitive Compounds

The photosensitive compounds in the eyes of humans and most other mammals are made up of a protein called an **opsin,** and **retinene₁,** the aldehyde of vitamin A_1 (Fig 8–16). The term retinene₁ is used to distinguish this compound from retinene₂, which is found in the eyes of some animal species. Since the retinenes are aldehydes, they are also called **retinals.** The A vitamins themselves are alcohols and are therefore called **retinols.**

Rhodopsin

The photosensitive pigment in the rods is called **rhodopsin** or **visual purple.** Its opsin is called **scotopsin.** Rhodopsin has a peak sensitivity to light at a wavelength of 505 nm.

Human rhodopsin has a molecular weight of 41,000. It is found in the membranes of the rod disks and makes up 90% of the total protein in these membranes. It is one of the many serpentine receptors coupled to G proteins (see Fig 1–37). Retinene₁ is attached to 3 of the transmembrane domains in a position parallel to the surface of the membrane (Fig 8–16).

In the dark, the retinene₁ in rhodopsin is in the 11-*cis* configuration. The only action of light is to change the shape of the retinene, converting it to the all-*trans* isomer (Fig 8–17). This activation of rhodopsin triggers formation of a series of intermediates, one of which, **metarhodopsin II,** appears to be the key compound in initiating the closure of the Na^+ channels. The final step is the separation of retinene₁ from the opsin (bleaching). Some of the rhodopsin is

Figure 8–17. Top: Structure of retinene₁, showing the 11-*cis* configuration (unbroken lines) and the all-*trans* configuration produced by light (dashed lines). **Bottom:** Effects of light on rhodopsin.

regenerated directly, while some of the retinene₁ is reduced by the enzyme alcohol dehydrogenase in the presence of NADH to vitamin A_1, and this in turn reacts with scotopsin to form rhodopsin (Fig 8–17). All of these reactions except the formation of the all-*trans* isomer of retinene₁ are independent of light, proceeding equally well in light or darkness. The amount of rhodopsin in the receptors therefore varies inversely with the incident light.

Linkage Between Rhodopsin & Na⁺ Channels

Activation of rhodopsin activates **transducin,** a G protein also known as G_{t1} (see Chapter 1). Transducin binds GTP (Fig 8–18), and this in turn activates a phosphodiesterase that catalyzes the conversion of cyclic GMP (cGMP) to 5′-GMP. cGMP in the cytoplasm of photoreceptors acts directly on Na^+ channels to maintain them in the open position, and the reduction in cGMP leads to channel closure and hyperpolarization. This cascade of reactions amplifies the light signal and helps explain the remarkable sensitivity of rod photoreceptors: these receptors are capable of producing a detectable response to as little as one photon of light.

The details of the interaction between rhodopsin, transducin, and phosphodiesterase have now been

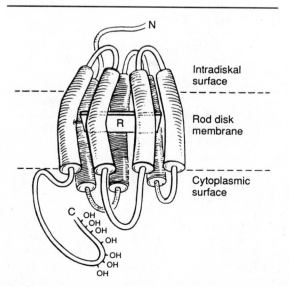

Figure 8–16. Diagrammatic representation of the structure of rhodopsin, showing the position of retinene₁ (R) in the rod disk membrane. (Modified from Lolley RN, Lee RH: Cyclic GMP and photoreceptor function. *FASEB J* 1990;**4**:3001.)

Figure 8–18. Initial steps in phototransduction in rods. Light activates rhodopsin (RHO), which activates transducin (G_{t1} to bind GTP. This activates phosphodiesterase (PDE), which catalyzes conversion of cyclic GMP to 5′-GMP (Reproduced, with permission, from Spiegel AM et al: Clinical implications of guanine nucleotide-binding proteins as receptor-effector couplers. Reprinted by permission of the *New England Journal of Medicine* 1985;**312:**26.)

Figure 8–19. Sequence of events involved in phototransduction in rods and cones.

elucidated. Activation of rhodopsin triggers the replacement of GDP with GTP on the α subunit of transducin. The GTP-containing α subunit separates from the βγ subunit and activates phosphodiesterase by removing an inhibitory constraint on the molecule. The GTP on the α subunit is then cleaved to GDP, and the units reassociate, terminating the reaction. Amplification occurs at the first and third steps of the cascade; each activated rhodopsin activates 500 molecules of transducin, and each activated phosphodiesterase hydrolyzes several thousand molecules of cGMP.

Cone Pigments

There are 3 different kinds of cones in primates. These receptors subserve color vision and respond maximally to light at wavelengths of 440, 535, and 565 nm (see below). Each contains retinene$_1$ and an opsin. The opsin resembles rhodopsin and spans the cone membrane 7 times but has a characteristic structure in each type of cone. As noted above, the cell membrane of cones is invaginated to form the saccules, but there are no separate intracellular disks like those in rods. The details of the responses of cones to light are probably similar to those in rods. Light activates retinene$_1$, and this activates G_{t2}, a G protein that differs somewhat from rod transducin. G_{t2} in turn activates phosphodiesterase, catalyzing the conversion of cGMP to 5′-GMP. This results in closure of Na$^+$ channels between the extracellular fluid and the cone cytoplasm, a rise in intracellular Na$^+$ concentration, and hyperpolarization of the cone outer segments.

The sequence of events in photoreceptors by which incident light leads to production of a signal in the next succeeding neural unit in the retina is summarized in Fig 8–19.

Resynthesis of Cyclic GMP

Light reduces the concentration of Ca^{2+} as well as that of Na$^+$ in photoreceptors. The resulting decrease in Ca^{2+} concentration activates guanylate cyclase, which generates more cyclic GMP. This in turn restores the Na$^+$ channels to their open position. The link between Ca^{2+} concentration and guanylate cyclase is a Ca^{2+}-binding protein that has been named **recoverin** because it promotes recovery of the dark state. It is a subunit of guanylate cyclase, and when it binds Ca^{2+}, it dissociates from the rest of the protein, inhibiting guanylate cyclase activity.

Retinitis Pigmentosa

Congenital defects in rod and cone opsins lead to **retinitis pigmentosa,** a condition in which pigment is not removed by phagocytosis in the normal manner. Therefore it accumulates in the retina and there is eventual blindness. This condition is now known to be the end result of a variety of different inherited defects. One documented cause in a single group of related individuals is a point mutation in which a single-base change causes a histidine to replace a proline in the N-terminal domain of rhodopsin.

Synaptic Mediators in the Retina

A great variety of different synaptic transmitters are found in the retina. These include acetylcholine, dopamine, serotonin, GABA, glycine, substance P, somatostatin, TRH, GnRH, enkephalins, β-endorphin, CCK, VIP, neurotensin, and glucagon (see Chapter 4). Amacrine cells are the only cells that secrete acetylcholine in the retina. Peptides are found in different populations of amacrine cells, and there are in addition cholinergic, dopaminergic, and serotonergic amacrine cells, each with a different shape and, presumably, function.

Image Formation

In a sense, the processing of visual information in the retina involves the formation of 3 images. The first image, formed by the action of light on the photoreceptors, is changed to a second image in the bipolar cells, and this in turn is converted to a third image in the ganglion cells. In the formation of the second image, the signal is altered by the horizontal cells, and in the formation of the third, it is altered by the amacrine cells. There is little change in the impulse pattern in the lateral geniculate bodies, so the third image reaches the occipital cortex.

A characteristic of the bipolar and ganglion cells (as well as the lateral geniculate cells and the cells in layer 4 of the visual cortex) is that they respond best to a small, circular stimulus and that, within their receptive field, an annulus of light around the center (surround illumination) inhibits the response to the central spot (Fig 8–20). The center can be excitatory with an inhibitory surround (an "on-center" cell) or inhibitory with an excitory surround (an "off-center" cell). The inhibition of the center response by the surround is probably due to inhibitory feedback from one photoreceptor to another mediated via horizontal cells. Thus, activation of nearby photoreceptors by addition of the annulus triggers horizontal cell hyperpolarization, which in turn inhibits the response of the centrally activated photoreceptors. The inhibition of the response to central illumination by an increase in surrounding illumination is an example of **lateral** or **afferent inhibition**—that form of inhibition in which activation of a particular neural unit is associated with inhibition of the activity of nearby units. It is a general phenomenon in mammalian sensory systems and helps to sharpen the edges of a stimulus and improve discrimination.

Electroretinogram

The electrical activity of the eye has been studied by recording fluctuations in the potential difference between an electrode in the eye and another on the back of the eye. At rest, there is a 6-mV potential difference between the front and the back of the eye, with the front positive. When light strikes the eye, a characteristic sequence of potential changes follows. The record of this sequence is the **electroretinogram (ERG)**. Turning on the light stimulus elicits **a** and **b waves** and also a **c wave,** which is so slow that with short stimuli its peak occurs after the end of the stimulus. When the stimulus is turned off, there is also a small negative off-deflection. The c wave is generated in the pigment epithelium. The neural components summate to give the other deflections of the ERG.

Electroretinograms can be recorded in humans with one electrode on the cornea and the other on the skin of the head. Electroretinographic records are helpful in the diagnosis of diseases in which visualization of the retina is difficult because the ocular fluids are cloudy. The ERG is also useful in congenital retinal dystrophies in which the retina appears normal by ophthalmoscopy.

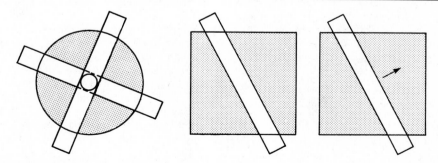

Figure 8–20. Receptive fields of cells in visual pathways. **Left:** Ganglion cells, lateral geniculate cells, and cells in layer 4 of cortical area 17 have circular fields with an excitatory center and an inhibitory surround, or an inhibitory center and an excitatory surround. There is no preferred orientation of a linear stimulus. **Center:** Simple cells respond best to a linear stimulus with a particular orientation in a particular part of the cell's receptive field. **Right:** Complex cells respond to linear stimuli with a particular orientation, but they are less selective in terms of location in the receptive field and often respond maximally when the stimulus is moved laterally, as indicated by the arrow. (Modified from Hubel DH: The visual cortex of normal and deprived monkeys. *Am Sci* 1979;**67**:532.)

RESPONSES IN THE VISUAL PATHWAYS & CORTEX

Pathways to the Cortex

The ganglion cell axons project a detailed spatial representation of the retina on the lateral geniculate body. Each geniculate body contains 6 well-defined layers (Fig 8–21). Layers 3–6 have small cells and are called parvocellular, whereas layers 1 and 2 have large cells and are called magnocellular. On each side, layers 1, 4, and 6 receive input from the contralateral eye, whereas layers 2, 3, and 5 receive input from the ipsilateral eye. In each layer, there is a precise point-for-point representation of the retina, and all 6 layers are in register so that along a line perpendicular to the layers, the receptive fields of the cells in each layer are almost identical.

Two kinds of ganglion cells can be distinguished in the retina: large ganglion cells (Y cells), which add responses from different kinds of cones and are concerned with movement and stereopsis; and small ganglion cells (X cells), which subtract input from one type of cone from input from another and are concerned with color vision. The large ganglion cells project to the magnocellular portion of the lateral geniculate, whereas the small ganglion cells project to the parvocellular portion (Fig 8–22). In the lateral geniculate, the magnocellular layers carry signals for

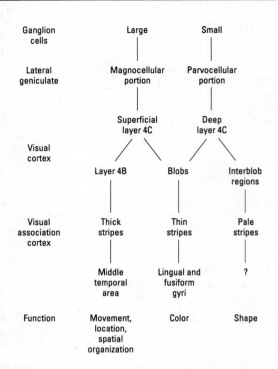

Figure 8–22. Organization of visual pathways. Two parallel pathways from the retina to the visual cortex become 3, which further process information in parallel. Information from the system that detects movement, the system that detects color, and the system that detects shape is then integrated into a single visual perception.

Figure 8–21. Cross-section of the lateral geniculate nucleus. Layers 1 and 2 are the inferomedial lateral magnocellular portions of the nucleus, and layers 3–6 the pariocellular portions. (Reproduced, with permission, from Mollon JD: The club-sandwich mystery. *Nature* 1990;**343:** 16. Copyright ©1990 Macmillan Magazines Ltd.)

detection of movement and flicker. The parvocellular layers carry the signals for color vision, texture, shape, and fine depth vision.

Visual Cortex

Just as the ganglion cell axons project a detailed spatial representation of the retina on the lateral geniculate body, the body projects a similar point-for-point representation on the visual cortex (Fig 8–5). In the visual cortex, there are many nerve cells associated with each fiber. Like the rest of the neocortex, the visual cortex has 6 layers (see Fig 11–2). The axons from the lateral geniculate neurons end on pyramidal cells in layer 4 and specifically in its deepest part, layer 4 C. The axons from the magnocellular portion of the geniculate body end more superficially in layer 4 C than the axons from the parvocellular portion.

The cells in layer 4 C in turn project primarily to more superficial layers of the cortex, particularly layer 4 B and layers 2 and 3. Layers 2 and 3 contain clusters of cells about 0.2 mm in diameter that, unlike the neighboring cells, contain a high concentration of the mitochondrial enzyme cytochrome oxidase. The clusters have been named **blobs.** They are arranged in a mosaic in the visual cortex and are concerned with color vision (see below). The areas between the blobs are called interblob regions.

Like the ganglion cells, the lateral geniculate neurons and the neurons in layer 4 of the visual cortex respond to stimuli in their receptive fields with on centers and inhibitory surrounds or off centers and excitatory surrounds. A bar of light covering the center is an effective stimulus for them because it stimulates all the center and relatively little of the surround (Fig 8–20). However, the bar has no preferred orientation and, as a stimulus, is equally effective at any angle.

The responses of the neurons in other layers of the visual cortex are strikingly different. So-called **simple cells** in these locations respond to bars of light, lines, or edges, but only when they have a particular orientation. When, for example, a bar of light is rotated as little as 10 degrees from the preferred orientation, the firing rate of the simple cell is usually decreased, and if the stimulus is rotated much more, the response disappears. There are also **complex cells,** which resemble simple cells in requiring a preferred orientation of a linear stimulus but are less dependent upon the location of a stimulus in the visual field than the simple cells and the cells in layer 4. They often respond maximally when a linear stimulus is moved laterally without a change in its orientation (Fig 8–20).

If a microelectrode is inserted perpendicularly into the visual cortex and passed through the various layers, the orientation preference of the neurons is the same. Thus, the visual cortex, like the somatosensory cortex (see Chapter 7), is arranged in vertical columns that are concerned with orientation (**orientation columns**). Each is about 1 mm in diameter. However, the orientation preferences of neighboring columns differ in a systematic way; as one moves from column to column across the cortex, there are sequential changes in orientation preference of 5–10 degrees (Fig 8–23). Thus, it is possible to speculate that for each ganglion cell receptive field in the visual field, there is a collection of columns in a small area of visual cortex representing the possible preferred orientations at small intervals throughout the full 360 degrees. The simple and complex cells have been called **feature detectors** because they respond to and analyze certain features of the stimulus. Feature detectors are also found in the cortical areas for other sensory modalities.

The orientation columns can be mapped with the aid of radioactive 2-deoxyglucose. The uptake of this glucose derivative is proportionate to the activity of neurons (see Chapter 32). When this technique is employed in animals exposed to uniformly oriented visual stimuli such as vertical lines, the brain shows a remarkable array of intricately curved but evenly spaced orientation columns over a large area of the visual cortex.

Figure 8–23. Top: Orientation preferences of 15 neurons encountered as a microelectrode penetrates the visual cortex obliquely. The preferred orientation changes steadily in a counterclockwise direction. **Bottom:** Results of a similar experiment plotted against distance traveled by the electrode. In this case, there are a number of reversals in the direction of rotation. (Modified and reproduced, with permission, from Hubel DH, Wiesel TN: Sequence regularity of orientation columns in the monkey striate cortex. *J Comp Neurol* 1974;**158:**267.)

Another feature of the visual cortex is the presence of **ocular dominance columns.** The geniculate cells, the cells of layer 4, and the simple cells receive input from only one eye. However, about half the complex cells receive an input from both eyes. The inputs are identical or nearly so in terms of the portion of the visual field involved and the preferred orientation. However, they differ in strength, so that between the cells to which the input is totally from the ipsilateral or the contralateral eye, there is a spectrum of cells influenced to varying degrees by both eyes. The cells influenced by one eye are in vertical ocular dominance columns that alternate with columns of cells influenced by the other eye. The ocular dominance columns can be mapped by injecting a large amount of a radioactive amino acid into one eye. The amino acid is incorporated into protein and transported by axoplasmic flow to the ganglion cell terminals, across the geniculate synapses, and along the geniculocalcarine fibers to the visual cortex. Layer 4 becomes evenly labeled, but above and below this layer in the cortex, labeled columns 0.8 mm in diameter alternate with unlabeled columns receiving input from the uninjected eye. The result is a vivid pattern of stripes that covers much of the visual cortex (Fig 8–24) and is separate from and independent of the grid of orientation columns. Ocular dominance columns can also be demonstrated by radioautography following injection of 2-deoxyglucose with one eye of the subject closed. The reason for the existence of the rigorously ordered, complex binocular input to some complex cells is unknown but may have something to do with binocular stereoscopic vision.

COLOR VISION

Characteristics of Color

Colors have 3 attributes: **hue, intensity,** and **saturation** (degree of freedom from dilution with white). For any color there is a **complementary color** that, when properly mixed with it, produces a sensation of white. Black is the sensation produced by the absence of light, but it is probably a positive sensation, because the blind eye does not "see black"; it "sees nothing." Such phenomena as successive and simultaneous contrasts, optical tricks that produce a sensa-

tion of color in the absence of color, negative and positive after-images, and various psychologic aspects of color vision are also pertinent. Detailed discussion of these phenomena is beyond the scope of this book.

Another observation of basic importance is the demonstration that the sensation of white, any spectral color, and even the extraspectral color, purple, can be produced by mixing various proportions of red light (wavelength 723–647 nm), green light (575–492 nm), and blue light (492–450 nm). Red, green, and blue are therefore called the **primary colors.** A third important point is that, as shown by Land, the color perceived depends in part on the color of other objects in the visual field. Thus, for example, a red object is seen as red if the field is illuminated with green or blue light but as pale pink or white if the field is illuminated with red light.

Retinal Mechanisms

The **Young-Helmholtz theory** of color vision in humans postulates the existence of 3 kinds of cones, each containing a different photopigment and maximally sensitive to one of the 3 primary colors, with the sensation of any given color being determined by the relative frequency of the impulses from each of these cone systems. The correctness of this theory has now been demonstrated by the identification and chemical characterization of each of the 3 pigments by recombinant DNA techniques. One pigment (the blue-sensitive or short-wave pigment) absorbs light maximally in the blue-violet portion of the spectrum (Fig 8–25). Another (the green-sensitive or middle-wave pigment) absorbs maximally in the green portion. The third (the red-sensitive or long-wave pigment) absorbs maximally in the yellow portion. Blue, green, and red are the primary colors, but the cones with their maximal sensitivity in the yellow portion

Figure 8–24. Reconstruction of ocular dominance columns in a subdivision of layer 4 of a portion of the right visual cortex of a rhesus monkey. Dark stripes represent one eye, light stripes the other. (Reproduced, with permission, from LeVay S, Hubel DH, Wiesel TN: The pattern of ocular dominance columns in macaque visual cortex revealed by a reduced silver stain. *J Comp Neurol* 1975;**159:**559.)

Figure 8–25. Absorption spectra of the 3 cone pigments in the human retina. The pigment that peaks at 440 nm senses blue, and the pigment that peaks at 535 nm senses green. The remaining pigment peaks in the yellow portion of the spectrum, at 565 nm, but its spectrum extends far enough into the long wavelengths to sense red. (Reproduced, with permission, from Michael CR: Color vision. *N Engl J Med* 1973;**288:**724.)

of the spectrum are sensitive enough in the red portion to respond to red light at a lower threshold than green. This is all the Young-Helmholtz theory requires.

The gene for human rhodopsin is on chromosome 3, and the gene for the blue-sensitive cone pigment is on chromosome 7. The other 2 cone pigments are encoded by genes arranged in tandem on the q arm of the X chromosome. The green-sensitive and red-sensitive pigments are very similar in structure; their opsins show 96% homology of amino acid sequences, whereas each of these pigments has only about 43% homology with the opsin of blue-sensitive pigment, and all 3 have about 41% homology with rhodopsin. Many mammals are **dichromats;** ie, they have only 2 cone pigments, a short-wave and a long-wave pigment. Old World monkeys, apes, and humans are trichromats, with separate middle- and long-wave pigments—in all probability because there was duplication of the ancestral long-wave gene followed by divergence.

It now appears that, in addition, there is variation in the human population in the red, long-wave pigment. It has been known for some time that responses to the **Rayleigh match,** the amounts of red and green light that a subject mixes to match a monochromatic orange, are bimodal. This correlates with new evidence that 62% of otherwise color-normal individuals have serine at site 180 of their long-wave cone opsin, whereas 38% have alanine. The absorption curve of the subjects with serine at position 180 peaks at 556.7 nm, and they are more sensitive to red light, whereas the absorption curve of the subjects with alanine at position 180 peaks at 552.4 nm.

Neural Mechanisms

As noted above, color sensation is mediated by the small ganglion cells, which subtract input from one type of cone from input from another type of cone. Information carried by the small cells is relayed to the parvocellular portion of the internal geniculate body, and from there to the deep portion of layer 4 C of the visual cortex. From there, the information passes to the blobs in layers 2 and 3. The neurons in the blobs lack orientation specificity but respond to colors. Like the ganglion and geniculate cells, they are center-surround cells. Many are **double opponent cells,** which, for example, are stimulated by a green center and inhibited by a green surround and are inhibited by a red center and stimulated by a red surround. From the blobs, color information is relayed to thin stripes in the visual association area and from there to a separate area concerned with color, which in humans is in the lingual and fusiform gyri of the occipital lobe (Fig 8–22). The homologous area in monkeys contains color-specific columns. The cells respond weakly when the visual field is illuminated by light of only one color but respond vigorously to a center stimulus when the surround is illuminated by light of a different wavelength.

Color Blindness

There are numerous tests for detecting color blindness. The most commonly used routine tests are the yarn-matching test and the Ishihara charts. In the former test, the subject is presented with a skein of yarn and asked to pick out the ones that match it from a pile of variously colored skeins. The Ishihara charts and similar polychromatic plates are plates on which are printed figures made up of colored spots on a background of similarly shaped colored spots. The figures are intentionally made up of colors that are liable to look the same as the background to an individual who is color-blind.

Some color-blind individuals are unable to distinguish certain colors, whereas others have only a color weakness. The suffix -anomaly denotes color weakness and the suffix -anopia color blindness. The prefixes prot-, deuter-, and tri- refer to defects of the red, green, and blue cone systems, respectively. Individuals with normal color vision and those with protanomaly, deuteranomaly, and tritanomaly are called **trichromats;** they have all 3 cone systems, but one may be weak. **Dichromats** are individuals with only 2 cone systems; they may have protanopia, deuteranopia, or tritanopia. **Monochromats** have only one cone system or, in other instances, congenital absence of all cones. Dichromats can match their color spectrum by mixing only 2 primary colors, and monochromats match theirs by varying the intensity of only one.

Inheritance of Color Blindness

Abnormal color vision is present as an inherited abnormality in Caucasian populations in about 8% of the males and 0.4% of the females. Tritanomaly and tritanopia are rare and show no sexual selectivity. However, about 2% of the color-blind males are dichromats who have protanopia or deuteranopia, and about 6% are anomalous trichromats in whom the red-sensitive or the green-sensitive pigment is shifted in its spectral sensitivity. These abnormalities are inherited as recessive and X-linked characteristics; ie, they are due to an abnormal gene on the X chromosome. Since all of the male's cells except germ cells contain one X and one Y chromosome in addition to the 44 somatic chromosomes (see Chapter 23), color blindness is present in males if the X chromosome has the abnormal gene. On the other hand, the normal female's cells have 2 X chromosomes, one from each parent, and since these abnormalities are recessive, females show a defect only when both X chromosomes contain the abnormal gene. However, female children of a man with X-linked color blindness are carriers of the color blindness and pass the defect on to half of their sons. Therefore, X-linked color blindness skips generations and appears in males of every second generation. Hemophilia, Duchenne's muscular dystrophy, and many other inherited disorders are caused by mutant genes on the X chromosome.

The common occurrence of deuteranomaly and protanomaly is probably due to the arrangement of the genes for the green-sensitive and red-sensitive cone pigments. They are located near each other in a head-to-tail tandem array on the q arm of the X chromosome and are prone to recombination (unequal crossing over) during development of the germ cells. Different combinations of introns may also occur, with both processes rendering one pigment inactive or producing opsins that have shifted spectral sensitivities.

OTHER ASPECTS OF VISUAL FUNCTION

Parallel Processing of Visual Information

The visual association areas next to the primary visual cortex (primarily Brodmann's area 18) can be divided into thick stripes, thin stripes, and pale stripes on the basis of characteristic staining reactions. Layer 4 B of the visual cortex projects to thick stripes, the blobs to thin stripes, and the interblob regions to pale stripes. Many other brain areas are also involved in vision.

Although much is still unknown about visual perception, information from a number of different disciplines permits consolidation of the data outlined in the preceding sections of this chapter into a hypothesis that a 3-part system is responsible for visual sensations. One system is concerned with the perception of shape; a second is concerned with the perception of color; and a third is concerned with the perception of movement, location, and spatial organization. The pathways involved are summarized in Fig 8–22. It is interesting that from the visual association cortex, the pathways for each of the systems appear to project to different parts of the brain: the movement system to the middle portion of the temporal lobe (MT area), the color system to a special color region (see above), and the shape system to an as yet unknown but presumably separate area. Somewhere else in the brain, the information from these 3 areas is combined into a single integrated visual perception.

Dark Adaptation

The truly remarkable range of luminance to which the human eye responds is summarized in Fig 8–26. If a person spends a considerable length of time in brightly lighted surroundings and then moves to a dimly lighted environment, the retinas slowly become more sensitive to light as the individual becomes "accustomed to the dark." This decline in visual threshold is known as **dark adaptation.** It is nearly maximal in about 20 minutes, although there is some further decline over longer periods. On the other hand, when one passes suddenly from a dim to

Figure 8–26. Range of luminance to which the human eye responds, with the receptive mechanisms involved. (Reproduced, with permission, by courtesy of Campbell FW, from Bell GH, Emslie-Smith D, Paterson CR: *Textbook of Physiology and Biochemistry,* 9th ed. Churchill Livingstone, 1976.)

a brightly lighted environment, the light seems intensely and even uncomfortably bright until the eyes adapt to the increased illumination and the visual threshold rises. This adaptation occurs over a period of about 5 minutes and is called **light adaptation,** although, strictly speaking, it is merely the disappearance of dark adaptation.

There are actually 2 components to the dark adaptation response (Fig 8–27). The first drop in visual threshold, rapid but small in magnitude, is known to be due to dark adaptation of the cones because when only the foveal, rod-free portion of the retina is tested, the decline proceeds no further. In the peripheral portions of the retina, a further drop occurs as a result of adaptation of the rods. The total change in threshold between the light-adapted and the fully dark-adapted eye is very great.

Radiologists, aircraft pilots, and others who need maximal visual sensitivity in dim light can avoid having to wait 20 minutes in the dark to become dark-adapted if they wear red goggles when in bright light. Light wavelengths in the red end of the spectrum stimulate the rods to only a slight degree while permitting the cones to function reasonably well. Therefore, a person wearing red glasses can see in bright

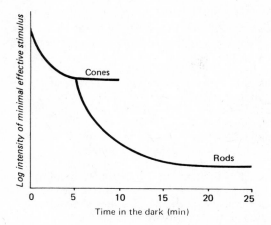

Figure 8–27. Dark adaptation. The curve shows the change in the intensity of a stimulus necessary to just excite the retina in dim light as a function of the time the observer has been in the dark.

light during the time it takes for the rods to become dark-adapted.

The time required for dark adaptation is determined in part by the time required to build up the rhodopsin stores. In bright light, much of the pigment is continuously being broken down, and some time is required in dim light for accumulation of the amounts necessary for optimal rod function. However, dark adaptation also occurs in the cones, and additional factors are undoubtedly involved.

Effect of Vitamin Deficiencies on the Eye

In view of the importance of vitamin A in the synthesis of retinene$_1$, it is not surprising that avitaminosis A produces visual abnormalities. Among these, one of the earliest to appear is night blindness (**nyctalopia**). This fact first called attention to the role of vitamin A in rod function, but it is now clear that concomitant cone degeneration occurs as vitamin A deficiency develops. Prolonged deficiency is associated with anatomic changes in the rods and cones followed by degeneration of the neural layers of the retina. Treatment with vitamin A can restore retinal function if given before the receptors are destroyed.

Other vitamins, especially those of the B complex, are necessary for the normal functioning of the retina and other neural tissues. Nicotinamide is part of the nicotinamide adenine dinucleotide (NAD$^+$) molecule, and this coenzyme plays a role in the interconversion of retinene and vitamin A in the rhodopsin cycle.

Physiologic Nystagmus

Even when a subject stares fixedly at a stationary object, the eyeballs are not still; there are continuous jerky motions and other movements. This **physiologic nystagmus** appears to have an important function. Although individual visual receptors do not adapt rapidly to constant illumination, their neural connections do. Indeed, it has been shown that if, by means of an optical lever system, the image of an object is fixed so that it falls steadily on the same spot in the retina, the object disappears from view. Continuous visualization of objects apparently requires that the retinal images be continuously and rapidly shifted from one receptor to another.

Visual Acuity

Physiologic nystagmus is one of the many factors that determine **visual acuity.** This parameter of vision should not be confused with **visual threshold.** Visual threshold is the minimal amount of light that elicits a sensation of light; visual acuity is the degree to which the details and contours of objects are perceived. Although there is evidence that other measures are more accurate, visual acuity is usually defined in terms of the **minimum separable**—ie, the shortest distance by which 2 lines can be separated and still be perceived as 2 lines. Clinically, visual acuity is often determined by use of the familiar Snellen letter charts viewed at a distance of 20 ft (6 m). The individual being tested reads aloud the smallest line distinguishable. The results are expressed as a fraction. The numerator of the fraction is 20, the distance at which the subject reads the chart. The denominator is the greatest distance from the chart at which a normal individual can read the smallest line the subject can read. Normal visual acuity is 20/20; a subject with 20/15 visual acuity has better than normal vision (not farsightedness); and one with 20/100 visual acuity has subnormal vision. The Snellen charts are designed so that the height of the letters in the smallest line a normal individual can read at 20 feet subtends a visual angle of 5 minutes. Each line in the letters subtends 1 minute of arc, and the lines in the letters are separated by 1 minute of arc. Thus, the minimum separable in a normal individual corresponds to a visual angle of about 1 minute.

Visual acuity is a complex phenomenon and is influenced by a large variety of factors. These include optical factors such as the state of the image-forming mechanisms of the eye, retinal factors such as the state of the cones, and stimulus factors including illumination, brightness of the stimulus, contrast between the stimulus and the background, and length of time the subject is exposed to the stimulus.

Critical Fusion Frequency

The time-resolving ability of the eye is determined by measuring the **critical fusion frequency (CFF),** the rate at which stimuli can be presented and still be perceived as separate stimuli. Stimuli presented at a higher rate than the CFF are perceived as a continuous stimulus. Motion pictures move because the frames are presented at a rate above the CFF. Consequently, movies begin to flicker when the projector slows down.

Visual Fields & Binocular Vision

The visual field of each eye is the portion of the external world visible out of that eye. Theoretically, it should be circular, but actually it is cut off medially by the nose and superiorly by the roof of the orbit (Fig 8–28). Mapping the visual fields is important in neurologic diagnosis. The peripheral portions of the visual fields are mapped with an instrument called a **perimeter,** and the process is referred to as **perimetry.** One eye is covered while the other is fixed on a central point. A small target is moved toward this central point along selected meridians, and, along each, the location where the target first becomes visible is plotted in degrees of arc away from the central point (Fig 8–28). The central visual fields are mapped with a **tangent screen,** a black felt screen across which a white target is moved. By noting the locations where the target disappears and reappears, the blind spot and any **objective scotomas** (blind spots due to disease) can be outlined.

The central parts of the visual fields of the 2 eyes coincide; therefore, anything in this portion of the field is viewed with **binocular vision.** The impulses set up in the 2 retinas by light rays from an object are fused at the cortical level into a single image (**fusion**). The points on the retina on which the image of an object must fall if it is to be seen binocularly as a single object are called **corresponding points.** If one eye is gently pushed out of the line while staring fix-

edly at an object in the center of the visual field, double vision (**diplopia**) results; the image on the retina of the eye that is displaced no longer falls on the corresponding point.

Binocular vision is often assigned an important role in the perception of depth. Actually, depth perception also has monocular components based on the relative sizes of objects, their shadows, and, for moving objects, their movement relative to one another (movement parallax). However, binocular vision adds appreciation of depth and proportion.

Effect of Lesions in the Optic Pathways

The anatomy of the pathways from the eyes to the brain is shown in Fig 8–4. Lesions along these pathways can be localized with a high degree of accuracy by the effects they produce in the visual fields.

The fibers from the nasal half of each retina decussate in the optic chiasm, so that the fibers in the optic tracts are those from the temporal half of one retina and the nasal half of the other. In other words, each optic tract subserves half of the field of vision. Therefore, a lesion that interrupts one optic nerve causes blindness in that eye, but a lesion in one optic tract causes blindness in half of the visual field (Fig 8–4). This defect is classified as a **homonymous** (same side of both visual fields) **hemianopia** (half-blindness). Lesions affecting the optic chiasm, such as pituitary tumors expanding out of the sella turcica, cause destruction of the fibers from both nasal hemiretinas and produce a **heteronymous** (opposite sides of the visual fields) **hemianopia.** Since the fibers from the maculas are located posteriorly in the optic chiasm, hemianopic scotomas develop before there is complete loss of vision in the 2 hemiretinas. Selective visual field defects are further classified as bitemporal, binasal, and right or left.

The optic nerve fibers from the upper retinal quadrants subserving vision in the lower half of the visual field terminate in the medial half of the lateral geniculate body, while the fibers from the lower retinal quadrants terminate in the lateral half. The geniculocalcarine fibers from the medial half of the lateral geniculate terminate on the superior lip of the calcarine fissure, while those from the lateral half terminate on the inferior lip. Furthermore, the fibers from the lateral geniculate body that subserve macular vision separate from those that subserve peripheral vision and end more posteriorly on the lips of the calcarine fissure (Fig 8–5). Because of this anatomic arrangement, occipital lobe lesions may produce discrete quadrantic visual field defects (upper and lower quadrants of each half visual field). **Macular sparing,** ie, loss of peripheral vision with intact macular vision, is also common with occipital lesions (Fig 8–4), because the macular representation is separate from that of the peripheral fields and very large relative to that of the peripheral fields. Therefore, occipital lesions must extend considerable distances to de-

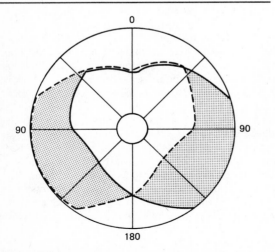

Figure 8–28. Monocular and binocular visual fields. The dashed line encloses the visual field of the left eye; the solid line, that of the right eye. The common area (heart-shaped clear zone in the center) is viewed with binocular vision. The shaded areas are viewed with monocular vision.

stroy macular as well as peripheral vision. Bilateral destruction of the occipital cortex in humans causes subjective blindness. However, there is appreciable **blindsight,** ie, residual responses to visual stimuli even though they do not reach consciousness. For example, individuals asked to guess where a stimulus is located during perimetry respond with much more accuracy than can be explained by chance. There is considerable discrimination of movement, flicker, orientation, and even color. Similar biasing of responses can be produced by stimuli in the blind areas in patients with hemianopia due to lesions in the visual cortex.

The fibers to the pretectal region that subserve the reflex pupillary constriction produced by shining a light into the eye leave the optic tracts near the geniculate bodies. Therefore, blindness with preservation of the pupillary light reflex is usually due to a lesion behind the optic tracts.

EYE MOVEMENTS

The direction in which each of the eye muscles moves the eye and the definitions of the terms used in describing eye movements are summarized in Table 8–1. Since the oblique muscles pull medially (Fig 8–8), their actions vary with the position of the eye. When the eye is turned nasally, the obliques elevate and depress it, whereas the superior and inferior recti rotate it; when the eye is turned temporally, the superior and inferior recti elevate and depress it and the obliques rotate it.

Since much of the visual field is binocular, it is clear that a very high order of coordination of the movements of the 2 eyes is necessary if visual images are to fall at all times on corresponding points in the 2 retinas and diplopia is to be avoided.

There are 4 types of eye movements, each con-

trolled by a different neural system but sharing the same final common path, the motor neurons that supply the external ocular muscles (Fig 8–29). **Saccades,** sudden jerky movements, occur as the gaze shifts from one object to another. **Smooth pursuit movements** are tracking movements of the eyes as they follow moving objects. **Vestibular movements,** adjustments that occur in response to stimuli initiated in the semicircular canals, maintain visual fixation as the head moves. **Convergence movements** bring the visual axes toward each other as attention is focused on objects near the observer. The similarity to a man-made tracking system on an unstable platform such as a ship is apparent: saccadic movements seek out visual targets; pursuit movements follow them as they move about; and vestibular movements stabilize the tracking device as the platform on which the device is mounted (ie, the head) moves about. In primates, these eye movements depend on an intact visual cortex. Saccades are programmed in the superior colliculi and pursuit movements in the cerebellum.

Strabismus

Abnormalities of the coordinating mechanisms can be due to a variety of causes. When visual images no longer fall on corresponding retinal points, **strabismus** (squint) is said to be present. Successful treatment of some types of strabismus is possible by careful surgical shortening of some of the eye mus-

Table 8–1. Actions of the external ocular muscles.[1]

Muscle	Primary Action	Secondary Action
Lateral rectus	Abduction	None
Medial rectus	Adduction	None
Superior rectus	Elevation	Adduction, intorsion
Inferior rectus	Depression	Adduction, extorsion
Superior oblique	Depression	Intorsion, abduction
Inferior oblique	Elevation	Extorsion, abduction

[1] Abduction and adduction refer to rotation of the eyeball around the vertical axis, with the pupil moving away from or toward the midline, respectively; elevation and depression refer to rotation around the transverse horizontal axis, with the pupil moving up or down; and torsion refers to rotation around the anteroposterior horizontal axis, with the top of the pupil moving toward the nose (intorsion) or away from the nose (extorsion). (Reproduced, with permission, from Vaughan D, Asbury T, Tabbara KF: *General Ophthalmology,* 12th ed. Appleton & Lange, 1989.)

Figure 8–29. Types of eye movements. (Modified and reproduced, with permission, from Robinson DA: Eye movement control in primates. *Science* 1968;**161:**1219. Copyright ©1968 by the American Association for the Advancement of Science.)

cles, by eye muscle training exercises, and by the use of glasses with prisms that bend the light rays sufficiently to compensate for the abnormal position of the eyeball. However, subtle defects in depth perception persist, and it has been suggested that congenital abnormalities of the visual tracking mechanisms may cause both the strabismus and the defective depth perception.

When visual images chronically fall on noncorresponding points in the 2 retinas in children under age 6, one is eventually suppressed (**suppression scotoma**) and diplopia disappears. This suppression is a cortical phenomenon, and it usually does not develop in adults. It is important to institute treatment before age 6 in children with one visual image suppressed, because if the suppression persists, there is permanent loss of visual acuity in the eye generating the suppressed image. A similar suppression with subsequent permanent loss of visual acuity can occur in children in whom vision in one eye is blurred or distorted owing to a refractive error. The loss of vision in these cases is called **amblyopia ex anopsia,** a term that refers to uncorrectable loss of visual acuity which is not directly due to organic disease of the eye. In infant monkeys, covering one eye with a patch for 3 months causes a loss of ocular dominance columns; input from the remaining eye spreads to take over all the cortical cells, and the patched eye becomes functionally blind. Comparable changes presumably occur in children with strabismus.

Hearing & Equilibrium

9

INTRODUCTION

Receptors for 2 sensory modalities, hearing and equilibrium, are housed in the ear. The external ear, the middle ear, and the cochlea of the inner ear are concerned with hearing. The semicircular canals, the utricle, and the saccule of the inner ear are concerned with equilibrium. Receptors in the semicircular canals detect rotational acceleration, receptors in the utricle detect linear acceleration in the horizontal direction, and receptors in the saccule detect linear acceleration in the vertical direction. The receptors for hearing and equilibrium are hair cells, and there are 6 groups of hair cells in each inner ear: one in each of the 3 semicircular canals, one in the utricle, one in the saccule, and one in the cochlea.

ANATOMIC CONSIDERATIONS

External & Middle Ear

The external ear funnels sound waves to the **exter-nal auditory meatus.** In some animals, the ears can be moved like radar antennas to seek out sound. From the meatus, the **external auditory canal** passes inward to the **tympanic membrane** (eardrum) (Fig 9–1).

The middle ear is an air-filled cavity in the temporal bone that opens via the **auditory (eustachian) tube** into the nasopharynx and through the nasopharynx to the exterior. The tube is usually closed, but during swallowing, chewing, and yawning it opens, keeping the air pressure on the 2 sides of the eardrum equalized. The 3 **auditory ossicles,** the **malleus, incus,** and **stapes,** are located in the middle ear. The **manubrium** (handle of the malleus) is attached to the back of the tympanic membrane. Its head is attached to the wall of the middle ear, and its short process is attached to the incus, which in turn articulates with the head of the stapes. The stapes is named for its resemblance to a stirrup. Its **foot plate** is attached by an annular ligament to the walls of the **oval window** (Fig 9–2). Two small skeletal muscles, the **tensor tympani** and the **stapedius,** are also located in the

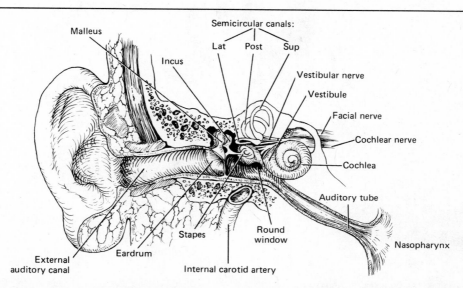

Figure 9–1. The human ear. To make the relationships clear, the cochlea has been turned slightly and the middle ear muscles have been omitted. Sup, superior; Post, posterior; Lat, lateral.

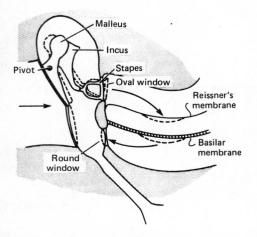

Figure 9–2. Schematic representation of the auditory ossicles and the way their movement translates movements of the tympanic membrane into a wave in the fluid of the inner ear. The wave is dissipated at the round window. The movements of the ossicles, the membranous labyrinth, and the round window are indicated by dashed lines. (Redrawn and reproduced, with permission, from an original by Netter FH in *Ciba Clinical Symposia*. Copyright © Ciba Pharmaceutical Co., 1962.)

middle ear. Contraction of the former pulls the manubrium of the malleus medially and decreases the vibrations of the tympanic membrane; contraction of the latter pulls the foot plate of the stapes out of the oval window.

Inner Ear

The inner ear (**labyrinth**) is made up of 2 parts, one within the other. The **bony labyrinth** is a series of channels in the petrous portion of the temporal bone. Inside these channels, surrounded by a fluid called **perilymph,** is the **membranous labyrinth** (Fig 9–3). This membranous structure more or less duplicates the shape of the bony channels. It is filled with a fluid called **endolymph,** and there is no communication between the spaces filled with endolymph and those filled with perilymph.

Cochlea

The cochlear portion of the labyrinth is a coiled tube which in humans is 35 mm long and makes 2¾ turns. Throughout its length, the basilar membrane and Reissner's membrane divide it into 3 chambers (**scalae**) (Fig 9–4). The upper **scala vestibuli** and the lower **scala tympani** contain perilymph and commu-

Figure 9–3. Human membranous labyrinth, with enlargements of the structures in which hair cells are embedded. (Reproduced, with permission, from Hudspeth AJ: How the ear's works work. *Nature* 1989;**341**:397. Copyright © 1989 Macmillan Magazines Ltd.)

Figure 9–4. Top: Cross section of the cochlea, showing the organ of Corti and the 3 scala of the cochlea. **Bottom:** Structure of the organ of Corti, as it appears in the basal turn of the cochlea. DC, outer phalangeal cells (Deiters' cells) supporting outer hair cells; IPC, inner phalangeal cell supporting inner hair cell. (Reproduced, with permission, from Pickels JO: *An Introduction to the Physiology of Hearing,* 2nd ed. Academic Press, 1988.)

Organ of Corti

Located on the basilar membrane is the **organ of Corti,** the structure that contains the hair cells which are the auditory receptors. This organ extends from the apex to the base of the cochlea and consequently has a spiral shape. The processes of the hair cells pierce the tough, membranelike **reticular lamina** that is supported by the **rods of Corti** (Fig 9–4). The hair cells are arranged in 4 rows: 3 rows of **outer hair cells** lateral to the tunnel formed by the rods of Corti, and one row of **inner hair cells** medial to the tunnel. There are 20,000 outer hair cells and 3500 inner hair cells in each human cochlea. Covering the rows of hair cells is a thin, viscous, but elastic **tectorial membrane** in which the tips of the hairs of the outer but not the inner hair cells are embedded. The cell bodies of the afferent neurons that arborize around the bases of the hair cells are located in the **spiral ganglion** within the **modiolus,** the bony core around which the cochlea is wound. Ninety to 95% of these afferent neurons innervate the inner hair cells; only 5–10% innervate the more numerous outer hair cells, and each neuron innervates several of these outer cells. In addition, most of the efferent fibers in the auditory nerve (see below) terminate on the outer hair cells rather than on the inner hair cells. The axons of the neurons that innervate the hair cells form the auditory (cochlear) division of the vestibulocochlear acoustic nerve and terminate in the **dorsal** and **ventral cochlear nuclei** of the medulla oblongata. The total number of afferent and efferent fibers in each auditory nerve is approximately 28,000.

In the cochlea, there are tight junctions between the hair cells and the adjacent phalangeal cells; these prevent endolymph from reaching the bases of the cells. However, the basilar membrane is relatively permeable to perilymph in the scala tympani, and consequently, the tunnel of the organ of Corti and the bases of the hair cells are bathed in perilymph. Because of similar tight junctions, the arrangement is similar for the hair cells in other parts of the middle ear; ie, the processes of the hair cells are bathed in endolymph, whereas their bases are bathed in perilymph.

Central Auditory Pathways

From the cochlear nuclei, axons carrying auditory impulses pass via a variety of pathways to the **inferior colliculi,** the centers for auditory reflexes, and via the **medial geniculate body** in the thalamus to the **auditory cortex.** Others enter the reticular formation (Fig 9–5). Information from both ears converges on each superior olive, and at all higher levels most of the neurons respond to inputs from both sides. The primary auditory cortex, Brodmann's area 41, is in the superior portion of the temporal lobe. In humans,

nicate with each other at the apex of the cochlea through a small opening called the **helicotrema.** At the base of the cochlea, the scala vestibuli ends at the oval window, which is closed by the foot plate of the stapes. The scala tympani ends at the **round window,** a foramen on the medial wall of the middle ear that is closed by the flexible **secondary tympanic membrane.** The **scala media,** the middle cochlear chamber, is continuous with the membranous labyrinth and does not communicate with the other 2 scalae. It contains endolymph (Figs 9–3 and 9–4).

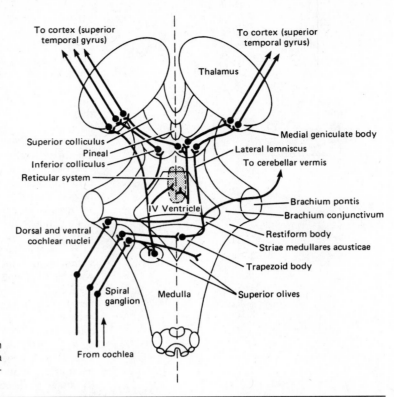

Figure 9–5. Simplified diagram of main auditory pathways superimposed on a dorsal view of the brain stem. Cerebellum and cerebral cortex removed.

Figure 9–6. Diagrammatic representation of ampullar crista. The cupula on the top of the crista closes off the ampulla and is flexible. Because of its inertia, the endolymph is displaced in a direction opposite to the direction of rotation during rotational acceleration. This bends the hair cell processes, altering their permeability and changing the membrane potential of the hair cells.

it is located in the floor of the lateral cerebral fissure (see Fig 7–4) and is not normally visible on the surface of the brain. There are several additional auditory receiving areas, just as there are several receiving areas for cutaneous sensation (see Chapter 7). The auditory association areas adjacent to the primary auditory receiving area are widespread, extending onto the insula. The **olivocochlear bundle** is a prominent bundle of efferent fibers in each auditory nerve that arises from both the ipsilateral and the contralateral superior olivary complex and ends primarily around the bases of the outer hair cells of the organ of Corti.

Semicircular Canals

On each side of the head, the semicircular canals are perpendicular to each other, so that they are oriented in the 3 planes of space. Inside the bony canals, the membranous canals are suspended in perilymph. A receptor structure, the **crista ampullaris,** is located in the expanded end (**ampulla**) of each of the membranous canals. Each crista consists of hair cells and sustentacular cells surmounted by a gelatinous partition (**cupula**) that closes off the ampulla (Fig 9–6). The processes of the hair cells are embedded in the cupula, and the bases of the hair cells are in close contact with the afferent fibers of the vestibular division of the vestibulocochlear nerve.

Utricle & Saccule

Within each membranous labyrinth, on the floor of the utricle, there is an **otolithic organ (macula).** Another macula is located on the wall of the saccule in a semivertical position. The maculas contain sustentacular cells and hair cells, surmounted by an otolithic membrane in which are embedded crystals of calcium carbonate, the **otoliths** (Fig 9–3). The otoliths, which are also called **otoconia** or **ear dust,** range from 3 to 19 μm in length in humans and are more dense than the endolymph. The processes of the hair cells are embedded in the membrane. The nerve fibers from the hair cells join those from the cristae in the vestibulocochlear nerve.

Neural Pathways

The cell bodies of the 19,000 neurons supplying the cristae and maculas on each side are located in the vestibular ganglion. Each vestibular nerve terminates in the ipsilateral 4-part vestibular nucleus and in the flocculonodular lobe of the cerebellum. Second-order neurons pass down the spinal cord from the vestibular nuclei in the vestibulospinal tracts and ascend through the **medial longitudinal fasciculi** to the motor nuclei of the cranial nerves concerned with the control of eye movement. There are also anatomically poorly defined pathways by which impulses from the vestibular receptors are relayed via the thalamus to the cerebral cortex (Fig 9–7).

HAIR CELLS

Structure

The hair cells in the inner ear have a common structure (Fig 9–8). Each is embedded in an epithelium made up of supporting or sustentacular cells. The basal end is in close contact with afferent neurons. Projecting from the apical end are 30–150 rod-shaped processes, or hairs. Except in the cochlea, one of these, the **kinocilium,** is a true but nonmotile cilium with 9 pairs of microtubules around its circumference and a central pair of microtubules (see Chapter 1). It is one of the largest processes and has a clubbed end. The kinocilium is lost in the hair cells of the cochlea in adult mammals. However, the other processes, which are called **stereocilia,** are present in all hair cells. They have cores composed of parallel filaments of actin. Within the circular clump of processes on each cell, there is an orderly structure. Along an axis toward the kinocilium, the stereocilia increase progressively in height; along the perpendicular axis, all the stereocilia are the same height.

Electrical Responses

The membrane potential of the hair cells is about –60 mV. When the stereocilia are pushed toward the kinocilium, the membrane potential is decreased to

Figure 9–7. Principal vestibular pathways superimposed on a dorsal view of the brain stem. Cerebellum and cerebral cortex removed.

about –50 mV. When the bundle of processes is pushed in the opposite direction, the cell is hyperpolarized. Displacing the processes in a direction perpendicular to this axis provides no change in membrane potential, and displacing the processes in directions that are intermediate between these 2 directions produces depolarization or hyperpolarization that is proportionate to the degree to which the direction is toward or away from the kinocilium. Thus, the hair processes provide a mechanism for generating changes in membrane potential proportionate to the direction of displacement.

Genesis of Action Potentials in Afferent Nerve Fibers

As noted above, the processes of the hair cells project into the endolymph whereas the bases are bathed in perilymph. This arrangement is necessary for the normal production of generator potentials, although it is not known exactly why. The perilymph is formed mainly from plasma. Entry of mannitol and sucrose from plasma into perilymph in the scala tympani is slower than entry into the perilymph in the

Figure 9–8. *Left:* Structure of a hair cell in the saccule of a frog, showing its relation to the otolithic membrane (OM). K, kinocilium; S, stereocilia; RC, hair cell with afferent (A) and efferent (E) nerve fibers; OL, otolith; SC, supporting cell. (Reproduced, with permission, from Hillman DE: Morphology of peripheral and central vestibular systems. In: Llinas R, Precht W [editors]: *Frog Neurobiology.* Springer-Verlag, 1976.) *Right:* Scanning electron photomicrograph of processes on a hair cell in the saccule of a frog. The otolithic membrane has been removed. The small projections around the hair cell are microvilli on supporting cells. (Courtesy of AJ Hudspeth.)

scala vestibuli, and there are small differences in composition between these 2 fluids, but both resemble extracellular fluid (Fig 9–9). On the other hand, endolymph is formed by the stria vascularis and has a high concentration of K^+ and a low concentration of Na^+. Cells in the stria vascularis have a high concentration of Na^+-K^+ ATPase. In addition, it appears that there is a unique electrogenic K^+ pump in the stria vascularis, which accounts for the fact that the scala media is electrically positive relative to the scala vestibuli and scala tympani.

Current evidence indicates that the stereocilia have mechanosensitive channels at their apices, with about one channel per stereocilium. Displacement of these processes toward the kinocilium increases the open time of these channels, and displacement away from the kinocilium decreases their open time. The channels, which have a diameter of at least 0.7 nm, are relatively nonspecific cation channels, but since they are bathed in endolymph, which has a high K^+

concentration, K^+ enters the hair cell when they are open, producing depolarization. Ca^{2+} also enters, and a synaptic transmitter is released that depolarizes the afferent neuron or neurons in contact with the hair cell. The identity of the transmitter has not been established, but there is solid evidence for its existence. Conversely displacement of the stereocilia away from the kinocilium decreases the resting permeability. K^+ influx is decreased, the cell becomes hyperpolarized, and less transmitter is released.

HEARING

Sound Waves

Sound is the sensation produced when longitudinal vibrations of the molecules in the external environment, ie, alternate phases of condensation and rarefaction of the molecules, strike the tympanic

Figure 9–9. Composition of perilymph in the scala vestibuli, perilymph in the scala tympani, and endolymph. Values for Na⁺, K⁺, and Cl⁻ are in mmol/L. (Modified and reproduced, with permission, from Sterkers O, Ferrary E, Amiel C: How are inner ear fluids formed? *News Physiol Sci* 1987;**2**:176.)

membrane. A plot of these movements as changes in pressure on the tympanic membrane per unit of time is a series of waves (Fig 9–10), and such movements in the environment are generally called sound waves. The waves travel through air at a speed of approximately 344 m/s (770 miles per hour) at 20 °C at sea level. The speed of sound increases with temperature and with altitude. Other media in which humans occasionally find themselves also conduct sound waves, but at different speeds. For example, the speed of sound is 1450 m/s at 20 °C in fresh water and is even greater in salt water. It is said that the whistle of the blue whale is as loud as 188 decibels (see below) and is audible for 500 miles.

Generally speaking, the **loudness** of a sound is correlated with the **amplitude** of a sound wave and its **pitch** with the **frequency** (number of waves per unit of time). The greater the amplitude, the louder the sound; and the greater the frequency, the higher the pitch. However, pitch is determined by other poorly understood factors in addition to frequency, and frequency affects loudness, since the auditory threshold is lower at some frequencies than others (see below). Sound waves that have repeating patterns, even though the individual waves are complex, are perceived as musical sounds; aperiodic nonrepeating vibrations cause a sensation of noise. Most musical sounds are made up of a wave with a primary frequency that determines the pitch of the sound plus a number of harmonic vibrations (**overtones**) that give the sound its characteristic **timbre** (quality). Variations in timbre permit us to identify the sounds of the various musical instruments even though they are playing notes of the same pitch.

The amplitude of a sound wave can be expressed in terms of the maximum pressure change at the eardrum, but a relative scale is more convenient. The **decibel scale** is such a scale. The power of a sound in **bels** is the logarithm of the ratio of the power of that sound and a standard sound:

$$\text{bel} = \log \frac{\text{power of sound}}{\text{power of standard sound}}$$

Acoustic power is proportionate to the square of sound pressure. Therefore,

$$\text{bel} = 2 \log \frac{\text{power of sound}}{\text{power of standard sound}}$$

A decibel is 0.1 bel. The standard sound reference level adopted by the Acoustical Society of America corresponds to 0 decibels at a pressure level of 0.000204 dyne/cm², a value that is just at the auditory threshold for the average human. In Fig 9–11, the decibel levels of various common sounds are compared. It is important to remember that the decibel scale is a log scale. Therefore, a value of 0 decibels does not mean the absence of sound but a sound level of an intensity equal to that of the standard. Furthermore, the 0–140 decibel range from threshold pressure to a pressure that is potentially damaging to the organ of Corti actually represents a 10⁷ (10 million)-fold variation in sound pressure. Put another way, at-

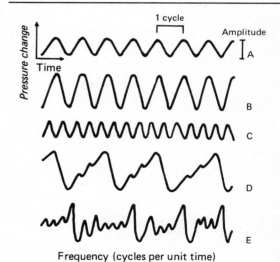

Figure 9–10. Characteristics of sound waves. **A** is the record of a pure tone. **B** has a greater amplitude and is louder than A. **C** has the same amplitude as **A** but a greater frequency, and its pitch is higher. **D** is a complex wave form that is regularly repeated. Such patterns are perceived as musical sounds, whereas waves like that shown in **E**, which have no regular pattern, are perceived as noise.

Decibels

160 — Jet plane with afterburner

Pain

120 — Discomfort

Subway, live rock music

80 — Heavy traffic

Normal conversation

40 —

Whisper

0 — Threshold of hearing (0.0002 dyne/cm²)

Figure 9–11. Decibel scale for common sounds.

mospheric pressure at sea level is 15 lb/in² or 1 bar, and the range from the threshold of hearing to potential damage to the cochlea is 0.00002–2000 µbars.

The sound frequencies audible to humans range from about 20 to a maximum of 20,000 cycles per second (cps, Hz). In other animals, notably bats and dogs, much higher frequencies are audible. The threshold of the human ear varies with the pitch of the sound (Fig 9–12), the greatest sensitivity being in the 1000–4000 Hz range. The pitch of the average male voice in conversation is about 120 Hz and that of the average female voice about 250 Hz. The number of pitches that can be distinguished by an average individual is about 2000, but trained musicians can improve on this figure considerably. Pitch discrimina-

Figure 9–12. Human audibility curve. The middle curve is that obtained by audiometry under the usual conditions. The lower curve is that obtained under ideal conditions. At about 140 decibels (top curve), sounds are felt as well as heard.

tion is best in the 1000–3000 Hz range and is poor at high and low pitches.

Masking

It is common knowledge that the presence of one sound decreases an individual's ability to hear other sounds. This phenomenon is known as **masking.** It is believed to be due to the relative or absolute refractoriness of previously stimulated auditory receptors and nerve fibers to other stimuli. The degree to which a given tone masks other tones is related to its pitch. The masking effect of the background noise in all but the most carefully soundproofed environments raises the auditory threshold by a definite and measurable amount.

Sound Transmission

The ear converts sound waves in the external environment into action potentials in the auditory nerves. The waves are transformed by the eardrum and auditory ossicles into movements of the foot plate of the stapes. These movements set up waves in the fluid of the inner ear. The action of the waves on the organ of Corti generates action potentials in the nerve fibers.

Functions of the Tympanic Membrane & Ossicles

In response to the pressure changes produced by sound waves on its external surface, the tympanic membrane moves in and out. The membrane therefore functions as a **resonator** that reproduces the vibrations of the sound source. It stops vibrating almost immediately when the sound wave stops; ie, it is very nearly **critically damped.** The motions of the tympanic membrane are imparted to the manubrium of the malleus. The malleus rocks on an axis through the junction of its long and short processes, so that the short process transmits the vibrations of the manubrium to the incus. The incus moves in such a way that the vibrations are transmitted to the head of the stapes. Movements of the head of the stapes swing its foot plate to and fro like a door hinged at the posterior edge of the oval window. The auditory ossicles thus function as a lever system that converts the resonant vibrations of the tympanic membrane into movements of the stapes against the perilymph-filled scala vestibuli of the cochlea (Figs 9–2 and 9–13). This system increases the sound pressure that arrives at the oval window, because the lever action of the malleus and incus multiplies the force 1.3 times and the area of the tympanic membrane is much greater than the area of the foot plate of the stapes. There are losses of sound energy as a result of resistance, but it has been calculated that, at frequencies below 3000 Hz, 60% of the sound energy incident on the tympanic membrane is transmitted to the fluid in the cochlea.

Tympanic Reflex

When the middle ear muscles—the tensor tympani

Figure 9–13. Diagrammatic representation of the transmission of vibrations from the outer to the inner ear. (Reproduced, with permission, from Lippold OCJ, Winton FR: *Human Physiology,* 6th ed. Churchill, 1972.)

Figure 9–14. Traveling waves. ***Top:*** The solid and the short-dashed lines represent the wave at 2 instants of time. The long-dashed line shows the "envelope" of the wave formed by connecting the wave peaks at successive instants. ***Bottom:*** Displacement of the basilar membrane by the waves generated by stapes vibration of the frequencies shown at the top of each curve.

and the stapedius—contract, they pull the manubrium of the malleus inward and the foot plate of the stapes outward. This decreases sound transmission. Loud sounds initiate a reflex contraction of these muscles generally called the **tympanic reflex.** Its function is protective, preventing strong sound waves from causing excessive stimulation of the auditory receptors. However, the reaction time for the reflex is 40–160 ms, so it does not protect against brief intense stimulation such as that produced by gunshots.

Bone & Air Conduction

Conduction of sound waves to the fluid of the inner ear via the tympanic membrane and the auditory ossicles, the main pathway for normal hearing, is called **ossicular conduction.** Sound waves also initiate vibrations of the secondary tympanic membrane that closes the round window. This process, unimportant in normal hearing, is **air conduction.** A third type of conduction, **bone conduction,** is the transmission of vibrations of the bones of the skull to the fluid of the inner ear. Considerable bone conduction occurs when tuning forks or other vibrating bodies are applied directly to the skull. This route also plays a role in transmission of extremely loud sounds.

Traveling Waves

The movements of the foot plate of the stapes set up a series of traveling waves in the perilymph of the scala vestibuli. A diagram of such a wave is shown in Fig 9–14. As the wave moves up the cochlea, its height increases to a maximum and then drops off rapidly. The distance from the stapes to this point of maximum height varies with the frequency of the vibrations initiating the wave. High-pitched sounds generate waves that reach maximum height near the base of the cochlea; low-pitched sounds generate waves that peak near the apex. The bony walls of the scala vestibuli are rigid, but Reissner's membrane is

flexible. The basilar membrane is not under tension, and it also is readily depressed into the scala tympani by the peaks of waves in the scala vestibuli. Displacements of the fluid in the scala tympani are dissipated into air at the round window. Therefore, sound produces distortion of the basilar membrane, and the site at which this distortion is maximal is determined by the frequency of the sound wave. The tops of the hair cells in the organ of Corti are held rigid by the reticular lamina, and the hairs of the outer hair cells are embedded in the tectorial membrane (Fig 9–4). When the stapes moves, both membranes move in the same direction, but they are hinged on different axes, so there is a shearing motion that bends the hairs. The hairs of the inner hair cells are probably not attached to the tectorial membrane, but they are apparently bent by fluid moving between the tectorial membrane and the underlying hair cells.

Functions of the Inner & Outer Hair Cells

The inner hair cells are the primary sensory cells that generate action potentials in the auditory nerves, and presumably they are stimulated by the fluid movements noted above.

The outer hair cells, on the other hand, are innervated by cholinergic efferent fibers from the superior olivary complexes. These cells are motile, shortening when depolarized and lengthening when hyperpolarized. It appears that these hair cells have little direct receptive function, but they improve hearing by influencing the vibration patterns of the basilar membrane. However, the exact way they alter the patterns is unknown.

Action Potentials in Auditory Nerve Fibers

The frequency of the action potentials in single auditory nerve fibers is proportionate to the loudness of the sound stimuli. At low sound intensities, each axon discharges to sounds of only one frequency, and this frequency varies from axon to axon depending upon the part of the cochlea from which the fiber originates. At higher sound intensities, the individual axons discharge to a wider spectrum of sound frequencies (Fig 9–15)—particularly to frequencies lower than that at which threshold simulation occurs.

The major determinant of the pitch perceived when a sound wave strikes the ear is the place in the organ of Corti that is maximally stimulated. The traveling wave set up by a tone produces peak depression of the basilar membrane, and consequently maximal receptor stimulation, at one point. As noted above, the distance between this point and the stapes is inversely related to the pitch of the sound, low tones producing maximal stimulation at the apex of the cochlea and high tones producing maximal stimulation at the base. The pathways from the various parts of the cochlea to the brain are distinct. An additional factor involved in pitch perception at sound frequencies of less than 2000 Hz may be the pattern of the action potentials in the auditory nerve. When the frequency is low enough, the nerve fibers begin to respond with an impulse to each cycle of a sound wave. The importance of this **volley effect,** however, is limited; the frequency of the action potentials in a given auditory nerve fiber determines principally the loudness, rather than the pitch, of a sound.

Although the pitch of a sound depends primarily on the frequency of the sound wave, loudness also plays a part; low tones (below 500 Hz) seem lower and high tones (above 4000 Hz) seem higher as their loudness increases. Duration also affects pitch to a minor degree. The pitch of a tone cannot be perceived unless it lasts for more than 0.01 second, and with durations between 0.01 and 0.1 second, pitch rises as duration increases. Finally, the pitch of complex sounds that include harmonics of a given frequency is still perceived even when the primary frequency (missing fundamental) is absent.

Auditory Responses of Neurons in the Medulla Oblongata

The responses of individual second-order neurons in the cochlear nuclei to sound stimuli are like those of the individual auditory nerve fibers. The frequency at which sounds of the lowest intensity evoke a response varies from unit to unit; with increased sound intensities, the band of frequencies to which a response occurs becomes wider. The major difference between the responses of the first- and second-order neurons is the presence of a sharper "cutoff" on the low-frequency side in the medullary neurons. This greater specificity of the second-order neurons is probably due to some sort of inhibitory process in the brain stem, but how it is achieved is not known.

Auditory Cortex

The pathways from the cochlea to the auditory cortex are described in the first section of this chapter. Impulses ascend from the dorsal and ventral cochlear nuclei through complex paths that are both crossed and uncrossed. In animals, there is an organized pattern of tonal localization in the primary auditory cortex, as if the cochlea had been unrolled upon it. In humans, low tones are represented anterolaterally and high tones posteromedially in the auditory cortex. However, it is pitch and not frequency per se that is coded in the auditory cortex, because when a complex sound with a missing fundamental is presented (see above), the part of the cortex that is stimulated is the part corresponding to the perceived pitch. Thus, processing of pure frequencies into pitch must occur at a subcortical level.

Individual neurons in the auditory cortex respond to such parameters as the onset, duration, and repetition rate of an auditory stimulus and particularly to the direction from which it came. In this way, they are analogous to some of the neurons in the visual cortex (see Chapter 8). In laboratory mammals, destruction of the auditory cortex not only fails to cause deafness but also fails to obliterate established conditioned responses to sound of a given frequency. However, the auditory cortex is concerned with recogni-

Figure 9–15. Relation of discharge rate (spike rate) in auditory nerve fiber to frequency and intensity of sound. Because the diagram represents the type of response seen in many different individual fibers, no numbers are given on the scales. (Modified and reproduced, with permission, from Kiang NYS: Peripheral neural processing of auditory information. In: *Handbook of Physiology.* Section 1, *The Nervous System,* vol 3, part 2. Brookhart JM, Mountcastle VB [editors]. American Physiological Society, 1984.)

tion of tonal patterns, with analysis of properties of sounds, and with sound localization. Lesions of the auditory association cortex in the superior temporal gyrus in monkeys impairs auditory short-term memory without impairing visual memory. Thus, this region is analogous to the interior temporal cortex, lesions of which cause loss of visual short-term memory without affecting auditory short-term memory.

Sound Localization

Determination of the direction from which a sound emanates in the horizontal plane depends upon detecting the difference in time between the arrival of the stimulus in the 2 ears and the consequent difference in phase of the sound waves on the 2 sides; it also depends upon the fact that the sound is louder on the side closest to the source. The detectable time difference, which can be as little as 20 μs, is said to be the most important factor at frequencies below 3000 Hz and the loudness difference the most important at frequencies above 3000 Hz. Many neurons in the auditory cortex receive input from both ears, and they respond maximally or minimally when the time of arrival of a stimulus at one ear is delayed by a fixed period relative to the time of arrival at the other ear. This fixed period varies from neuron to neuron. Sounds coming from directly in front of the individual differ in quality from those coming from behind, because each pinna (the visible portion of the exterior ear) is turned slightly forward. In addition, reflections of the sound waves from the pinnal surface change as sounds move up or down. Thus, the pinna plays an important role in locating sounds in the vertical plane. Sound localization is markedly disrupted by lesions of the auditory cortex in laboratory mammals and humans.

Deafness

Clinical deafness may be due to impaired sound transmission in the external or middle ear (**conduction deafness**) or to damage to the hair cells or neural pathways (**nerve deafness**). Among the causes of conduction deafness are plugging of the external auditory canals with wax or foreign bodies, destruction of the auditory ossicles, thickening of the eardrum following repeated middle ear infections, and abnormal rigidity of the attachments of the stapes to the oval window. Aminoglycoside antibiotics such as streptomycin and gentamicin obstruct the mechanosensitive channels in the stereocilia of hair cells and can cause the cells to degenerate, producing nerve deafness and abnormal vestibular function. Damage to the outer hair cells by prolonged exposure to noise is associated with hearing loss. Other causes include tumors of the vestibulocochlear nerve and cerebellopontine angle, and vascular damage in the medulla.

Conduction and nerve deafness can be distinguished by a number of simple tests with a tuning fork. Three of these tests, named for the men who developed them, are outlined in Table 9–1. The Weber and Schwabach tests demonstrate the important masking effect of environmental noise on the auditory threshold.

Audiometry

Auditory acuity is commonly measured with an **audiometer.** This device presents the subject with pure tones of various frequencies through earphones. At each frequency, the threshold intensity is determined and plotted on a graph as a percentage of normal hearing. This provides an objective measurement of the degree of deafness and a picture of the tonal range most affected.

Fenestration Procedures

A common form of conduction deafness is that due to **otosclerosis,** a disease in which the attachments of the foot plate of the stapes to the oval window become abnormally rigid. In patients with this disease, air conduction (see above) can be utilized to bring about some restoration of hearing. A membrane-covered outlet from the bony labyrinth is cre-

Table 9–1. Common tests with a tuning fork to dintinguish between nerve and conduction deafness.

	Weber	Rinne	Schwabach
Method	Base of vibrating tuning fork placed on vertex of skull.	Base of vibrating tuning fork placed on mastoid process until subject no longer hears it, then held in air next to ear.	Bone conduction of patient compared with that of normal subject.
Normal	Hears equally on both sides.	Hears vibration in air after bone conduction is over.	
Conduction deafness (one ear)	Sound louder in diseased ear because masking effect of environmental noise is absent on diseased side.	Vibrations in air not heard after bone conduction is over.	Bone conduction better than normal (conduction defect excludes masking noise).
Nerve deafness (one ear)	Sound louder in normal ear.	Vibration heard in air after bone conduction is over, as long as nerve deafness is partial.	Bone conduction worse than normal.

ated so that waves set up by vibrations of the secondary tympanic membrane can be dissipated. In the "fenestration operation," an outlet is created by drilling a hole in the horizontal semicircular canal and covering it with skin.

VESTIBULAR FUNCTION

Responses to Rotational Acceleration

Rotational acceleration in the plane of a given semicircular canal stimulates its crista. The endolymph, because of its inertia, is displaced in a direction opposite to the direction of rotation. The fluid pushes on the cupula, deforming it. This bends the processes of the hair cells (Fig 9–6). When a constant speed of rotation is reached, the fluid spins at the same rate as the body and the cupula swings back into the upright position. When rotation is stopped, deceleration produces displacement of the endolymph in the direction of the rotation, and the cupula is deformed in a direction opposite to that during acceleration. It returns to midposition in 25–30 seconds. Movement of the cupula in one direction commonly causes increased impulse traffic in single nerve fibers from its crista, whereas movement in the opposite direction commonly inhibits neural activity (Fig 9–16).

Rotation causes maximal stimulation of the semicircular canals most nearly in the plane of rotation. Since the canals on one side of the head are a mirror image of those on the other side, the endolymph is displaced toward the ampulla on one side and away from it on the other. The pattern of stimulation reaching the brain therefore varies with the direction as well as the plane of rotation. Linear acceleration probably fails to displace the cupula and therefore does not stimulate the cristae. However, there is con-

siderable evidence that when one part of the labyrinth is destroyed, other parts take over its functions. Experimental localization of labyrinthine functions is therefore difficult.

The tracts that descend from the vestibular nuclei into the spinal cord are concerned primarily with postural adjustments (see Chapter 12); the ascending connections to cranial nerve nuclei are largely concerned with eye movements.

Nystagmus

The characteristic jerky movement of the eye observed at the start and end of a period of rotation is called **nystagmus**. It is actually a reflex that maintains visual fixation on stationary points while the body rotates, although it is not initiated by visual impulses and is present in blind individuals. When rotation starts, the eyes move slowly in a direction opposite to the direction of rotation, maintaining visual fixation (**vestibulo-ocular reflex, VOR**). When the limit of this movement is reached, the eyes quickly snap back to a new fixation point and then again move slowly in the other direction. The slow component is initiated by impulses from the labyrinths; the quick component is triggered by a center in the brain stem. Nystagmus is frequently horizontal (ie, the eyes move in the horizontal plane), but it can also be vertical, when the head is tipped sidewise during rotation, or rotatory, when the head is tipped forward. By convention, the direction of eye movement in nystagmus is identified by the direction of the quick component. The direction of the quick component during rotation is the same as that of the rotation, but the **postrotatory nystagmus** that occurs owing to displacement of the cupula when rotation is stopped is in the opposite direction. Clinically, nystagmus is seen at rest in patients with lesions of the brain stem.

Responses to Linear Acceleration

In mammals, the utricular and saccular maculas respond to linear acceleration. In general, the utricle responds to horizontal acceleration and the saccule to vertical acceleration. The otoliths are more dense than the endolymph, and acceleration in any direction causes them to be displaced in the opposite direction, distorting the hair cell processes and generating activity in the nerve fibers. The maculas also discharge tonically in the absence of head movement, because of the pull of gravity on the otoliths. The impulses generated from these receptors are partly responsible for reflex righting of the head and other important postural adjustments discussed in Chapter 12.

Although most of the responses to stimulation of the maculas are reflex in nature, vestibular impulses also reach the cerebral cortex. These impulses are presumably responsible for conscious perception of motion and supply part of the information necessary for orientation in space. The nausea, blood pressure changes, sweating, pallor, and vomiting that are the

Figure 9–16. Ampullary responses to rotation. Average time course of impulse discharge from ampulla of 2 semicircular canals during rotational acceleration, steady rotation, and deceleration. (Reproduced, with permission, from Adrian ED: Discharges from vestibular receptors in the cat. *J Physiol [Lond]* 1943;**101**:389.)

well-known accompaniments of excessive vestibular stimulation are probably due to reflexes mediated via vestibular connections in the brain stem. **Vertigo** is the sensation of rotation in the absence of actual rotation.

Caloric Stimulation

The semicircular canals can be stimulated by instilling water that is hotter or colder than body temperature into the external auditory meatus. The temperature difference sets up convection currents in the endolymph, with consequent motion of the cupula. This technique of **caloric stimulation,** which is sometimes used diagnostically, causes some nystagmus, vertigo, and nausea. To avoid these symptoms when irrigating the ear canals in the treatment of ear infections, it is important to be sure that the fluid used is at body temperature.

Orientation in Space

Orientation in space depends in large part upon input from the vestibular receptors, but visual cues are also important. Pertinent information is also supplied by impulses from proprioceptors in joint capsules, which supply data about the relative position of the various parts of the body, and impulses from cutaneous exteroceptors, especially touch and pressure receptors. These 4 inputs are synthesized at a cortical level into a continuous picture of the individual's orientation in space.

Effects of Labyrinthectomy

The effects of unilateral labyrinthectomy are presumably due to unbalanced discharge from the remaining normal side. They vary from species to species and with the speed at which the labyrinth is destroyed. Rats develop abnormal body postures and roll over and over as they continuously attempt to right themselves. Postural changes are less marked in humans, but the symptoms are distressing. Motion aggravates the symptoms, but efforts to minimize stimulation by lying absolutely still are constantly thwarted by waves of nausea, vomiting, and even diarrhea. Fortunately, compensation for the loss occurs, and after 1–2 months the symptoms disappear completely.

Symptoms of this type are absent after bilateral destruction of the labyrinths, but defects in orientation are present. For this reason, diving is hazardous for persons who lack vestibular function. The only means these individuals have of finding the surface is vision, because in the water the pressure sensed by the cutaneous exteroceptors is equal all over the body. If for any reason vision is obscured, a person whose vestibular apparatus is impaired has no mechanism for self-orientation. It is therefore possible to swim away from rather than toward the surface and, consequently, to drown.

10

Smell & Taste

INTRODUCTION

Smell and taste are generally classified as visceral senses because of their close association with gastrointestinal function. Physiologically, they are related to each other. The flavors of various foods are in large part a combination of their taste and smell. Consequently, food may taste "different" if one has a cold that depresses the sense of smell. Both taste and smell receptors are chemoreceptors that are stimulated by molecules in solution in mucus in the nose and saliva in the mouth. However, these 2 senses are anatomically quite different. The smell receptors are distance receptors (telereceptors); the smell pathways have no relay in the thalamus; and there is no neocortical projection area for olfaction. The taste pathways pass up the brain stem to the thalamus and project to the postcentral gyrus along with those for touch and pressure sensibility from the mouth.

SMELL

RECEPTORS & PATHWAYS

Olfactory Mucous Membrane

The olfactory receptors are located in a specialized portion of the nasal mucosa, the yellowish-pigmented **olfactory mucous membrane.** In dogs and other animals in which the sense of smell is highly developed (macrosmatic animals), the area covered by this membrane is large; in microsmatic animals such as humans, it is small, covering an area of 5 cm² in the roof of the nasal cavity near the septum (Fig 10–1). It contains supporting cells and neuroblastlike progenitor cells that form the olfactory receptor neurons. Interspersed between these cells are 10–20 million receptor cells. Each olfactory receptor is a neuron, and the olfactory mucous membrane is said to be the

place in the body where the nervous system is closest to the external world. The neurons have short, thick dendrites with expanded ends called olfactory rods (Fig 10–2). From these rods, cilia project to the surface of the mucus. The cilia are unmyelinated processes about 2 μm long and 0.1 μm in diameter. There are 10–20 cilia per receptor neuron. The axons of the olfactory receptor neurons pierce the cribriform plate of the ethmoid bone and enter the olfactory bulbs.

The olfactory mucous membrane is constantly covered by mucus. This mucus is produced by Bowman's glands, which are just under the basal lamina of the membrane.

Olfactory Bulbs

In the olfactory bulbs, the axons of the receptors terminate among the dendrites of the **mitral cells** (Fig 10–3) to form the complex globular synapses called **olfactory glomeruli.** An average of 26,000 receptor cell axons converge on each glomerulus. The tufted cells and the periglomerular short axon cells participate in the formation of glomeruli. In the next layer, the dendrites of the mitral and granule cells form extensive reciprocal synapses. The axons of the mitral cells pass posteriorly through the **intermediate olfactory stria** and **lateral olfactory stria** to the **olfactory cortex.** The mitral cell axons terminate on the apical dendrites of pyramidal cells in the olfactory cortex. The olfactory cortex includes the anterior olfactory nucleus, piriform cortex, olfactory tubercle, corticomedial amygdala, and transitional entorhinal cortex. These structures are part of the limbic system (see Chapter 15). The anterior olfactory nucleus is concerned with coordination of input from the contralateral olfactory cortex and, presumably, transfer of olfactory memories from one side to the other. The piriform cortex, which projects via the mediobasal thalamus to the frontal lobe, is concerned with olfactory discrimination and, presumably, conscious perception. The amygdala is probably involved with the emotional responses to olfactory stimuli and the entorhinal cortex with olfactory memories.

There are 3 inputs from other parts of the brain into the olfactory bulb in addition to the extrinsic input from the olfactory mucous membrane via the olfactory nerves. The terminations of these afferent

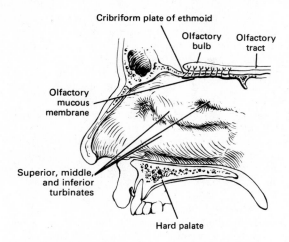

Figure 10–1. Olfactory mucous membrane. (Reproduced, with permission, from deGroot J: *Correlative Neuroanatomy*, 21st ed. Appleton & Lange, 1991.)

fibers are shown in Fig 10–3. One of the central inputs arises from the nucleus of the horizontal limb of the diagonal band (centrifugal fibers). Another arises from the ipsilateral anterior olfactory nucleus just posterior to the bulb. Stimulation of efferent fibers in the olfactory striae decreases the electrical activity of the olfactory bulbs. Presumably, therefore, inhibitory mechanisms analogous to those in the eye, ear, and other sensory systems exist in the olfactory system. The third afferent input arises from the contralateral

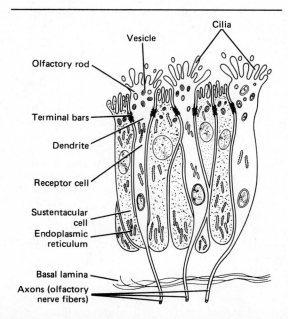

Figure 10–2. Structure of the olfactory mucous membrane.

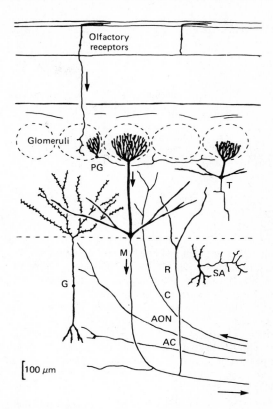

Figure 10–3. Neural elements in the olfactory bulb. In addition to olfactory nerve fibers, which terminate in the olfactory glomeruli, inputs include centrifugal fibers (C) from the nucleus of the diagonal band; ipsilateral fibers from the anterior olfactory nucleus (AON); and contralateral fibers from the anterior commissure (AC). Neurons include mitral cells (M) with recurrent collaterals (R), tufted cells (T), granule cells (G), periglomerular short-axon cells (PG), and deep short-axon cells (SA). (Reproduced, with permission, from Shepherd GM: *The Synaptic Organization of the Brain*. Oxford Univ Press, 1974.)

anterior olfactory nucleus, reaching the olfactory bulb via the anterior commissure.

PHYSIOLOGY OF OLFACTION

Olfactory Thresholds & Discrimination

Olfactory receptors respond only to substances that are in contact with the olfactory epithelium and are dissolved in the thin layer of mucus that covers it. The olfactory thresholds for the representative substances shown in Table 10–1 illustrate the remarkable sensitivity of the olfactory receptors to some substances. For example, methyl mercaptan, the substance that gives garlic its characteristic odor, can be smelled at a concentration of less than 500 pg/L of air. In addition, humans can distinguish between

Table 10-1. Some olfactory thresholds.[1]

Substance	mg/L of Air
Ethyl ether	5.83
Chloroform	3.30
Pyridine	0.03
Oil of peppermint	0.02
Iodoform	0.02
Butyric acid	0.009
Propyl mercaptan	0.006
Artificial musk	0.00004
Methyl mercaptan	0.0000004

[1] Data from Allison VC, Katz SH; *J Ind Chem* 1919;**11**:336.

2000 and 4000 different odors. On the other hand, discrimination of differences in the intensity of any given odor is poor. The concentration of an odor-producing substance must be changed by about 30% before a difference can be detected. The comparable visual discrimination threshold is a 1% change in light intensity. The direction from which a smell comes appears to be indicated by the slight difference in the time of arrival of odoriferous molecules in the 2 nostrils.

Odor-producing molecules are generally small, containing from 3–4 to 18–20 carbon atoms, and molecules with the same number of carbon atoms but different structural configurations have different odors. Relatively high water and lipid solubility are characteristic of substances with strong odors.

Signal Transduction

The olfactory system has received considerable attention in recent years because of the intriguing biologic question of how a simple sense organ such as the olfactory mucosa and its brain representation that apparently lacks a high degree of complexity can mediate discrimination of more than 2000 different odors. Indeed, the olfactory system has been likened to the immune system, in which a morphologically simple system produces specific responses to millions of different antigens (see Chapter 27). Obviously in the olfactory system, one possible explanation of the ability to discriminate so many odors is a multiplicity of different odorant receptors; another is analysis of incoming signals in the brain; and a third is a combination of the two.

Odorant receptors have recently been cloned and found to be members of a large family of similar serpentine receptors encoded by multiple genes. The receptors are coupled to heterotrimeric G proteins. Patch clamp experiments have shown that they are located on the cilia of the olfactory receptor cells. G_{olf}, a G protein that is closely related to G_s (see Chapter 1) but is apparently unique to the olfactory system, has been shown to link odorant receptors to adenylate cyclase. The cyclic AMP that is produced upon activation binds to and opens cation channels (Fig 10–4), permitting Na^+ to enter the receptor cells

and produce depolarization. Some of the odorant receptors may be coupled to phospholipase C rather than adenylate cyclase, with the products of phosphoinositol hydrolysis opening cation channels, but this is unsettled. It is clear that there is a large family of closely related serpentine odorant receptors that trigger cation channel opening by cyclic AMP.

On the other hand, it seems likely that the neural pathways subserving olfaction also play a role in olfactory discrimination. There is a pronounced degree of inhibitory control within the olfactory pathways. The reciprocal synaptic connections between mitral and granule cell dendrites mediate inhibitory control of mitral cell output. In the olfactory cortex, the response to an odor is excitation of pyramidal cells followed by inhibition. The pyramidal cells are then subject to self-reexcitation via long axon collaterals, and this may explain the propensity for rhythmic activity and seizures in the olfactory cortex.

Odorant-Binding Proteins

In contrast to the low threshold for olfactory stimulation when the olfactory mucous membrane is intact, single olfactory receptors that have been patch clamped have a relatively high threshold and a long latency. In addition, lipophilic odor-producing molecules must traverse the hydrophilic mucus in the nose to reach the receptors. These facts led to the suggestion that the olfactory mucus might contain one or more odorant-binding proteins (OBP) that concentrate the odorants and transfer them to the receptors. An 18K OBP that is unique to the nasal cavity has now been isolated, and other related proteins probably exist. The protein has considerable homology to other proteins in the body that are known to be carriers for small lipophilic molecules. A similar binding protein appears to be associated with taste (see below).

Relation to Sex & Memory

There is a close relationship between smell and sexual function in many species of animals, and the perfume ads are ample evidence that a similar relationship exists in humans. The sense of smell is said to be more acute in women than in men, and in women it is most acute at the time of ovulation. Smell and, to a lesser extent, taste have a unique ability to trigger long-term memories, a fact noted by novelists and documented by experimental psychologists.

Sniffing

The portion of the nasal cavity containing the olfactory receptors is poorly ventilated. Most of the air normally moves smoothly over the turbinates with each respiratory cycle, although eddy currents pass some air over the olfactory mucous membrane. These eddy currents are probably set up by convection as cool air strikes the warm mucosal surfaces. The amount of air reaching this region is greatly increased

Figure 10–4. Signal transduction in olfactory mucosa. Odoriferous molecules bind to receptors on cilia of olfactory receptors *(a)*, producing an influx of Na^+ *(b)*. The molecular events that trigger the influx are shown in *(c)*. An odor-producing molecule binds to the receptor (R). It in turn activates G_{olf} (G), and this activates adenylate cyclase (A_c). The cyclic AMP (cA) that is formed binds to and opens the Na^+ channel. (Reproduced, with permission, from Anonymous: Another cyclic-nucleotide-gated conductance. *Nature* 1987;**325**:389.)

by sniffing, an action that includes contraction of the lower part of the nares on the septum to help deflect the airstream upward. Sniffing is a semireflex response that usually occurs when a new odor attracts attention.

Role of Pain Fibers in the Nose

Naked endings of many trigeminal pain fibers are found in the olfactory mucous membrane. They are stimulated by irritating substances, and an irritative, trigeminally mediated component is part of the characteristic "odor" of such substances as peppermint, menthol, and chlorine. These endings are also responsible for initiating sneezing, lacrimation, respiratory inhibition, and other reflex responses to nasal irritants.

Adaptation

It is common knowledge that when one is continuously exposed to even the most disagreeable odor, perception of the odor decreases and eventually ceases. This sometimes beneficent phenomenon is due to the fairly rapid adaptation that occurs in the olfactory system and is primarily central in origin. It

is specific for the particular odor being smelled, and the threshold for other odors is unchanged.

Abnormalities

Abnormalities of olfaction include **anosmia** (absence of the sense of smell), **hyposmia** (diminished olfactory sensitivity), and **dysosmia** (distorted sense of smell). Several dozen different anosmias have been detected in humans; they are presumably due in each case to absence or disrupted function of one of the many members of the odorant receptor family. Olfactory thresholds increase with advancing age, and more than 75% of humans over the age of 80 have an impaired ability to identify smells.

TASTE

RECEPTOR ORGANS & PATHWAYS

Taste Buds

The taste buds, the sense organs for taste, are ovoid bodies measuring 50–70 μm. Each taste bud is made up of 4 types of cells (Fig 10–5): basal cells; types 1 and 2 cells, which are sustentacular cells; and type 3 cells, which are the gustatory receptor cells that make synaptic connections to sensory nerve fibers. The type 1, 2, and 3 cells have microvilli, which project into the taste pore, an opening in the lingual epithelium. The necks of all these cells are connected to each other and to the surrounding epithelial cells by tight junctions, so that the only part of the gustatory receptor exposed to the fluids in the oral cavity is its apical crown of microvilli. Each taste bud is innervated by about 50 nerve fibers, and conversely, each nerve fiber receives input from an average of 5 taste buds. The basal cells arise from the epithelial cells surrounding the taste bud. They differentiate into new receptor cells, and the old receptor cells are continuously replaced with a half time of about 10 days. If the sensory nerve is cut, the taste buds it innervates degenerate and eventually disappear. However, if the nerve regenerates, the cells in the neighborhood become organized into new taste buds, presumably as a result of some sort of chemical inductive effect from the regenerating fiber.

In humans, the taste buds are located in the mucosa of the epiglottis, palate, and pharynx and in the walls of the **fungiform** and **vallate papillae** of the tongue. The fungiform papillae are rounded structures most numerous near the tip of the tongue; the vallate papillae are prominent structures arranged in a V on the back of the tongue. There are up to 5 taste buds per fungiform papilla, and they are usually located at the top of the papilla (Fig 10–5). The larger vallate papillae each contain up to 100 taste buds, usually located along the sides of the papillae. The small conical **filiform papillae** that cover the dorsum of the tongue do not usually contain taste buds. There are a total of about 10,000 taste buds.

Taste Pathways

The sensory nerve fibers from the taste buds on the anterior two-thirds of the tongue travel in the chorda tympani branch of the facial nerve, and those from the posterior third of the tongue reach the brain stem via the glossopharyngeal nerve (Fig 10–6). The fibers from areas other than the tongue reach the brain stem via the vagus nerve. On each side, the myelinated but relatively slow-conducting taste fibers in these 3 nerves unite in the medulla oblongata to enter the **nucleus of the tractus solitarius** (Fig 10–7). There they synapse on second-order neurons, the axons of which cross the midline and join the medial lemniscus, ending with the fibers for touch, pain, and temperature sensibility in the specific sensory relay nuclei of the thalamus. Impulses are relayed from there to the taste projection area in the cerebral cortex at the foot of the postcentral gyrus. Taste does not have a separate cortical projection area but is represented in the portion of the postcentral gyrus that subserves cutaneous sensation from the face.

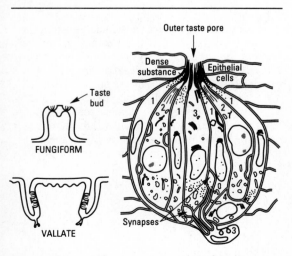

Figure 10–5. Taste bud, showing type 1, 2, and 3 cells. (Modified and reproduced, with permission, from Shepherd GM: *Neurobiology,* 2nd ed. Oxford Univ Press, 1988.)

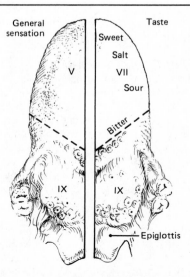

Figure 10–6. Sensory innervation of the tongue. The numbers refer to cranial nerves.

Foot of post-
central gyrus

Thalamus

Thalamic
radiation

MIDLINE

Medial lemniscus

Nucleus of the
tractus solitarius

Taste
pore

Tractus solitarius

Taste bud

Taste fiber to brain via facial,
glossopharyngeal, and vagus
nerves

Figure 10–7. Diagram of taste pathways.

PHYSIOLOGY OF TASTE

Basic Taste Modalities

In humans there are 4 basic tastes: sweet, sour, bitter, and salt. Bitter substances are tasted on the back of the tongue, sour along the edges, sweet at the tip, and salt on the dorsum anteriorly (Fig 10–6). Sour and bitter substances are also tasted on the palate along with some sensitivity to sweet and salt. All 4 modalities can be sensed on the pharynx and epiglottis. The taste buds are not histologically different in the different areas, but the existence of physiologic differences has been demonstrated by recording the electrical activity of nerve fibers from single taste buds in animals. These studies show that some taste buds respond best to bitter stimuli whereas others respond best to salt, sweet, or sour stimuli.

Receptor Stimulation

The gustatory receptor cells are chemoreceptors that respond to substances dissolved in the oral fluids bathing them. These substances act on the exposed microvilli in the taste pore to evoke generator potentials in the receptor cells, which generate action po-

tentials in the sensory neurons. The way the molecules in solution produce generator potentials varies from one gustatory modality to another. Salt stimuli probably depolarize salt receptor cells by influx of Na^+ through passive, ungated apical channels, because application of the Na^+ channel-blocking diuretic amiloride (see Chapter 38) directly to the tongue in humans abolishes the ability to taste salt. Acids, which taste sour, probably depolarize receptor cells by H^+ blocking of apical K^+ channels. Substances that taste sweet appear to bind to membrane receptors and, via G_s, activate adenylate cyclase, with a resulting increase in intracellular cyclic AMP. The cyclic AMP acts via protein kinase A to reduce K^+ conductance by phosphorylating K^+ channels on the basolateral membranes of the taste cells. Substances that taste bitter appear to act in a different fashion. Unlike the others, they do not increase current flow in patch clamp preparations, and they probably act via G protein-coupled receptors and phospholipase C to trigger release of Ca^{2+} from the endoplasmic reticulum. A novel G protein α-subunit named **α-gusducin** has recently been isolated and shown to occur in taste buds. It resembles the transducins (see Chapter 8).

A protein that binds taste-producing molecules has recently been cloned. It is produced by Ebner's glands, glands that secrete mucus into the cleft around vallate papillae (Fig 10–5), and probably has a concentrating and transport function similar to that of the OBP described in the section on olfaction.

Taste Thresholds & Intensity Discriminations

The ability of humans to discriminate differences in the intensity of tastes, like intensity discrimination in olfaction, is relatively crude. A 30% change in the concentration of the substance being tasted is necessary before an intensity difference can be detected. The threshold concentrations of substances to which the taste buds respond vary with the particular substance (Table 10–2).

Substances Evoking Primary Taste Sensations

Acids taste **sour**. The H^+, rather than the associated anion, stimulates the receptors. For any given acid, sourness is generally proportionate to the H^+

Table 10–2. Some taste thresholds.

Substance	Taste	Threshold Concentration (μmol/L)
Hydrochloric acid	Sour	100
Sodium chloride	Salt	2000
Strychnine hydrochloride	Bitter	1.6
Glucose	Sweet	80,000
Sucrose	Sweet	10,000
Saccharin	Sweet	23

concentration, but organic acids are often more sour for a given H+ concentration than are mineral acids. This is probably because they penetrate cells more rapidly than the mineral acids do.

A **salty** taste is produced by Na+. However, NaCl tastes saltier than Na acetate or Na gluconate, possibly because the smaller Cl- ions penetrate the tight junctions around the receptor cells better and limit receptor cell hyperpolarization when Na+ is pumped out of the bases of the cells by Na+-K+ ATPase. Some organic compounds also taste salty; for example, the dipeptides lysyltaurine and ornithyltaurine taste salty, and on a weight basis, lysyltaurine is more potent than NaCl.

The substance usually used to test the **bitter** taste is quinine sulfate. This compound can be detected in a concentration of 8 μmol/L, although the threshold for strychnine hydrochloride is even lower (Table 10–2). Other organic compounds, especially morphine, nicotine, caffeine, and urea, taste bitter. Inorganic salts of magnesium, ammonium, and calcium also taste bitter. The taste is due to the cation. Thus, there is no apparent common feature of the molecular structure of substances that taste bitter.

Most **sweet** substances are organic. Sucrose, maltose, lactose, and glucose are the most familiar examples, but polysaccharides, glycerol, some of the alcohols and ketones, and a number of compounds with no apparent relation to any of these, such as chloroform, beryllium salts, and various amides of aspartic acid, also taste sweet. Artificial sweeteners such as saccharin and aspartame are in demand as sweetening agents in reducing diets because they produce satisfactory sweetening in amounts that are a minute fraction of the amount of calorie-rich sucrose required for the same purpose. Lead salts also taste sweet.

Two proteins isolated from African berries, thaumatin and morellin, are 100,000 times as sweet as sucrose on a molar basis. The structures of these 2 proteins are very different, yet antibodies to one cross-react with antibodies to the other. Therefore, they must have some sort of common 3-dimensional structure, and it may be this common feature that binds to the sweetness receptor.

Flavor

The almost infinite variety of tastes so dear to the gourmet are mostly synthesized from the 4 basic taste components. In some cases, a desirable taste includes an element of pain stimulation (eg, "hot" sauces). In addition, smell plays an important role in the overall sensation produced by food, and the consistency (or texture) and temperature of foods also contribute to their "flavor."

Variation & After-Effects

There is considerable variation in the distribution of the 4 basic taste buds in various species and, within a given species, from individual to individual. In humans, there is an interesting variation in ability to taste **phenylthiocarbamide (PTC)**. In dilute solution, PTC tastes sour to about 70% of the Caucasian population but is tasteless to the other 30%. Inability to taste PTC is inherited as an autosomal recessive trait. Testing for this trait is of considerable value in studies of human genetics.

Taste exhibits after-reactions and contrast phenomena that are similar in some ways to visual afterimages and contrasts. Some of these are chemical "tricks," but others may be true central phenomena. A taste modifier protein, **miraculin,** has been discovered in a plant. When applied to the tongue, this protein makes acids taste sweet.

Animals, including humans, form particularly strong aversions to novel foods if eating the food is followed by illness. The survival value of such aversions is apparent in terms of avoiding poisons. The facility with which this type of learning occurs is so great that it has led some investigators to argue that its occurrence is reason to challenge classic learning theory.

Abnormalities

Abnormalities of taste include **ageusia** (absence of the sense of taste), **hypogeusia** (diminished taste sensitivity), and **dysgeusia** (disturbed sense of taste). Many different diseases can produce hypogeusia. In addition, drugs such as captopril and penicillamine, which contain sulfhydryl groups, cause temporary loss of taste sensation. The reason for this effect of sulfhydryl compounds is not known.

Arousal Mechanisms, Sleep, & the Electrical Activity of the Brain

11

INTRODUCTION

Most of the various sensory pathways described in Chapters 7–10 relay impulses from sense organs via 3- and 4-neuron chains to particular loci in the cerebral cortex. The impulses are responsible for perception and localization of individual sensations. Impulses in these systems also relay via collaterals to the **reticular activating system (RAS)** in the brain stem reticular formation. Activity in this system produces the conscious, alert state that makes perception possible. Other systems are responsible for drowsiness and sleep. This chapter is concerned with the way these systems operate and the correlations between behavioral states and the electroencephalogram (EEG).

THE RETICULAR FORMATION & THE RETICULAR ACTIVATING SYSTEM

The **reticular formation,** the phylogenetically old reticular core of the brain, occupies the midventral portion of the medulla and midbrain. It is made up of myriads of small neurons, many of which are arranged in complex, intertwining nets. It is primarily an anatomic area made up of various neural clusters and fibers with discrete functions. For example, it contains the cell bodies and fibers of many of the serotonergic, noradrenergic, and adrenergic systems that are discussed in Chapter 15. It also contains many of the areas concerned with regulation of heart rate, blood pressure, and respiration that are discussed in Chapters 31 and 36. Some of the descending fibers in it inhibit transmission in sensory pathways in the spinal cord (see Chapter 7). Various reticular areas and the pathways from them are concerned with spasticity and adjustment of stretch reflexes (see Chapter 12). The RAS and related reticular components concerned with consciousness and sleep are considered in this chapter.

The RAS is a complex polysynaptic pathway. Collaterals funnel into it not only from the long ascending sensory tracts but also from the trigeminal, audi-

tory, and visual systems and the olfactory system. The complexity of the neuron net and the degree of convergence in it abolish modality specificity, and most reticular neurons are activated with equal facility by different sensory stimuli. The system is therefore **nonspecific,** whereas the classic sensory pathways are **specific** in that the fibers in them are activated only by one type of sensory stimulation. Part of the RAS bypasses the thalamus to project diffusely to the cortex. Another part ends in the intralaminar and related thalamic nuclei and, from them, is projected diffusely and nonspecifically to the whole neocortex (Fig 11–1). The RAS is intimately concerned with the electrical activity of the cortex.

THE THALAMUS & THE CEREBRAL CORTEX

Thalamic Nuclei

On developmental and topographic grounds, the thalamus can be divided into 3 parts: the epithalamus, the dorsal thalamus, and the ventral thalamus. The

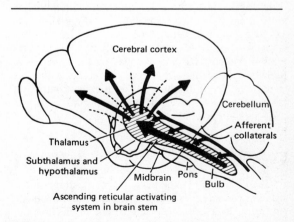

Figure 11–1. Diagram of the ascending reticular system projected on a sagittal section of the cat brain. (Reproduced, with permission, from Starzl TE, Taylor CW, Magoun HW: Collateral afferent excitation of reticular formation of brain stem. *J Neurophysiol* 1951;**14:**479.)

epithalamus has connections to the olfactory system, and the projections and functions of the **ventral thalamus** are undetermined. The **dorsal thalamus** can be divided into nuclei that project diffusely to the whole neocortex and the nuclei that project to specific discrete portions of the neocortex and limbic system. The nuclei that project to all parts of the neocortex are the midline and intralaminar nuclei. They are called collectively the **nonspecific projection nuclei.** They receive input from the RAS. Impulses responsible for the diffuse secondary response and the alerting effect of reticular activation (see below) are relayed through them. The nuclei of the dorsal thalamus that project to specific areas can be divided into 3 groups: the specific sensory relay nuclei, the nuclei concerned with efferent control mechanisms, and the nuclei concerned with complex integrative functions. The **specific sensory relay nuclei** include the medial and lateral geniculate bodies, which relay auditory and visual impulses to the auditory and visual cortices, and the ventrobasal group of nuclei, which relay somatesthetic information to the postcentral gyrus. The **nuclei concerned with efferent control mechanisms** include several nuclei that are concerned with motor function. They receive input from the basal ganglia and the cerebellum and project to the motor cortex. Also included in this group are the anterior nuclei, which receive afferents from the mamillary bodies and project to the limbic cortex. This is part of the limbic circuit, which appears to be concerned with recent memory and emotion (see Chapters 15 and 16). The **nuclei concerned with complex integrative functions** are the dorsolateral nuclei that project to the cortical association areas and are concerned primarily with functions such as language.

Cortical Organization

The neocortex is generally arranged in 6 layers (Fig 11–2). The neurons are mostly pyramidal cells with extensive vertical dendritic trees (Figs 11–2 and 11–3) that may reach to the cortical surface. The axons of these cells usually give off recurrent collaterals that turn back and synapse on the superficial portions of the dendritic trees. Afferents from the specific nuclei of the thalamus terminate primarily in cortical layer 4 whereas the nonspecific afferents are distributed to layers 1–4.

EVOKED CORTICAL POTENTIALS

The electrical events that occur in the cortex after stimulation of a sense organ can be monitored with an exploring electrode connected to another electrode at an indifferent point some distance away. A characteristic response is seen in animals under barbiturate anesthesia. If the exploring electrode is over the primary receiving area for the particular sense, a surface-positive wave appears with a latency of 5–12

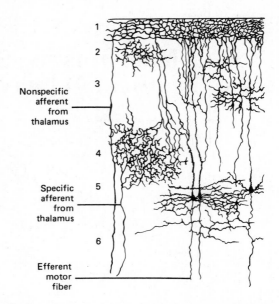

Figure 11–2. Neuronal connections in the neocortex. On the left are afferent fibers from the thalamus. The numbers identify the cortical layers. Note the extensive dendritic processes of the cells, especially those in the deep layers.

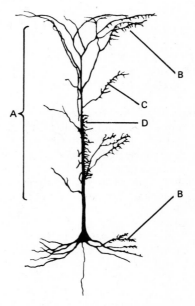

Figure 11–3. Neocortical pyramidal cell, showing the distribution of neurons that terminate on it. ***A:*** nonspecific afferents from the reticular formation and the thalamus; ***B:*** recurrent collaterals of pyramidal cell axons; ***C:*** commissural fibers from mirror image sites in the contralateral hemisphere; ***D:*** specific afferents from thalamic sensory relay nuclei. (Reproduced, with permission, from Chow KL, Leiman AL: The structural and functional organization of the neocortex. *Neurosci Res Program Bull* 1970;**8**:157.)

ms. This is followed by a small negative wave and then by a larger, more prolonged positive deflection with a latency of 20–80 ms. This sequence of potential changes is illustrated in Fig 11–4. The first positive-negative wave sequence is the **primary evoked potential;** the second is the **diffuse secondary response.**

Primary Evoked Potential

The primary evoked potential is highly specific in its location and can be observed only where the pathways from a particular sense organ end. Indeed, it is so discrete that it has been used to map the specific cortical sensory areas. An electrode on the pial surface of the cortex samples activity to a depth of only 0.3–0.6 mm. The primary response is negative rather than positive when it is recorded with a microelectrode inserted in layers 2–6 of the underlying cortex, and the negative wave within the cortex is followed by a positive wave. This indicates depolarization on the dendrites and somas of the cells in the cortex, followed by hyperpolarization. The positive-negative wave sequence recorded from the surface of the cortex occurs because the superficial cortical layers are positive relative to the initial negativity, then negative relative to the deep hyperpolarization. In unanesthetized animals, the primary evoked potential is largely obscured by the spontaneous activity of the brain, but it can be demonstrated with special techniques. It is somewhat more diffuse in unanesthetized animals but still well localized compared with the diffuse secondary response.

Diffuse Secondary Response

The surface-positive diffuse secondary response is sometimes followed by a negative wave or series of waves. Unlike the primary response, the secondary response is not highly localized. It appears at the same time over most of the cortex and in many other parts of the brain. The uniform latency in different parts of the cortex and the fact that it is not affected by a circular cut through the cortical gray matter that isolates an area from all lateral connections indicate that the secondary response cannot be due to lateral spread of the primary evoked response. Instead, it is due to activity ascending from below the cortex. The pathway involved is the nonspecific thalamic projection system from the midline and related thalamic nuclei.

THE ELECTROENCEPHALOGRAM

The background electrical activity of the brain in unanesthetized animals was first described in the 19th century. Subsequently it was analyzed in a systematic fashion by the German psychiatrist Hans Berger, who introduced the term **electroencephalogram (EEG)** to denote the record of the variations in potential recorded from the brain. The EEG can be recorded with scalp electrodes through the unopened skull or with electrodes on or in the brain. The term **electrocorticogram (ECoG)** is sometimes used to refer to the record obtained with electrodes on the pial surface of the cortex.

EEG records may be **bipolar** or **unipolar.** Bipolar records show fluctuations in potential between 2 cortical electrodes; unipolar records show potential differences between a cortical electrode and a theoretically indifferent electrode on some part of the body distant from the cortex.

Alpha Rhythm

In an adult human at rest with mind wandering and eyes closed, the most prominent component of the EEG is a fairly regular pattern of waves at a frequency of 8–12/s and an amplitude of about 50 μV when recorded from the scalp. This pattern is the **alpha rhythm.** It is most marked in the parieto-occipital area, although it is sometimes observed in other locations. A similar rhythm has been observed in a wide variety of mammalian species (Fig 11–5). In the cat it is slightly more rapid than in the human, and there are other minor variations from species to species, but in all mammals the pattern is remarkably similar.

Other Rhythms

In addition to the dominant rhythm, 18–30/s patterns of lower voltage are sometimes seen over the frontal regions. This rhythm, the **beta rhythm,** may be a harmonic of the alpha. A pattern of large, regular 4–7/s waves called the **theta rhythm** occurs in children and is generated in the hippocampus in experimental animals (see below). Large, slow waves with a frequency of less than 4/s are sometimes called **delta waves.**

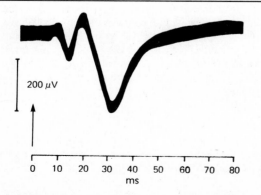

200 μV

| 0 | 10 | 20 | 30 | 40 | 50 | 60 | 70 | 80 |

ms

Figure 11–4. Response evoked in the contralateral sensory cortex by stimulation (at the arrow) of the sciatic nerve in a cat under barbiturate anesthesia. Upward deflection is surface-negative.

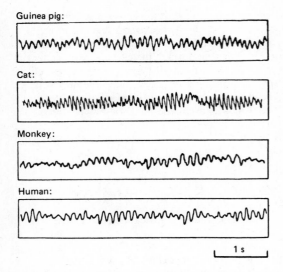

Figure 11–5. EEG records showing alpha rhythm from 4 different species. (Reproduced, with permission, from Brazier MAB: *Electrical Activity of the Nervous System,* 2nd ed. Pitman, 1960.)

Variations in the EEG

In humans, the frequency of the dominant EEG rhythm at rest varies with age. In infants, there is fast, betalike activity, but the occipital rhythm is a slow 0.5–2/s pattern. During childhood this latter rhythm speeds up, and the adult alpha pattern gradually appears during adolescence. The frequency of the alpha rhythm is decreased by a low blood glucose level, a low body temperature, a low level of adrenal glucocorticoid hormones, and a high arterial partial pressure of CO_2 (P_{aCO_2}). It is increased by the reverse conditions. Forced overbreathing to lower the P_{aCO_2} is sometimes used clinically to bring out latent EEG abnormalities.

Alpha Block

When the eyes are opened, the alpha rhythm is replaced by fast, irregular low-voltage activity with no dominant frequency. This phenomenon is called **alpha block.** A breakup of the alpha pattern is also produced by any form of sensory stimulation (Fig 11–6) or mental concentration such as solving arithmetic problems. A common term for this replacement of the regular alpha rhythm with irregular low-voltage activity is **desynchronization,** because it represents a breaking up of the synchronized activity of neural elements responsible for the wave pattern. Because desynchronization is produced by sensory stimulation and is correlated with the aroused, alert state, it is also called the **arousal** or **alerting response.**

Sleep Patterns

There are 2 different kinds of sleep: **rapid eye movement (REM) sleep** and **non-REM (NREM)** or **slow-wave sleep.** NREM sleep is divided into 4 stages. A person falling asleep first enters stage 1, which is characterized by low-amplitude, high-frequency EEG activity (Fig 11–7). Stage 2 is marked by the appearance of **sleep spindles.** These are bursts of alphalike, 10–14/s, 50–μV waves. In stage 3, the pattern is one of lower frequency and increased amplitude of the EEG waves. Maximum slowing with large waves is seen in stage 4. Thus, the characteristic of deep sleep is a pattern of rhythmic slow waves, indicating **synchronization.**

REM Sleep

The high-amplitude slow waves seen in the EEG during sleep are sometimes replaced by rapid, low-voltage, irregular EEG activity, which resembles that seen in alert animals and humans. However, sleep is not interrupted; indeed, the threshold for arousal by sensory stimuli and by stimulation of the reticular formation is sometimes elevated. The condition is **REM sleep,** or **paradoxical sleep.** There are rapid, roving movements of the eyes during paradoxical sleep, and it is for this reason that it is also called REM sleep. There are no such movements in slow-wave sleep, and consequently it is often called NREM sleep. Another characteristic of REM sleep is the occurrence of large phasic potentials, in groups of 3–5, that originate in the pons and pass rapidly to the lateral geniculate body and thence to the occipital cortex. For this reason, they are called **ponto-geniculo-occipital (PGO) spikes.** There is a marked reduction in skeletal muscle tone during REM sleep (Fig 11–7), despite the rapid eye movements, because the PGO spikes activate the reticular inhibiting area in the medulla (see Chapter 12). This brings about decreases in stretch and polysynaptic reflexes by way of both pre- and post-synaptic inhibition, producing the hypotonia.

Olf ⟶

Figure 11–6. Desynchronization of the cortical EEG of a rabbit by an olfactory stimulus (indicated by the line after Olf). (Reproduced, with permission, from Green JD, Arduini A: Hippocampal electrical activity during arousal. *J Neurophysiol* 1954;**17**:533.)

Figure 11–7. EEG and muscle activity during various stages of sleep. Sleep stages, EOG, electro-oculogram registering eye movements; EMG, electromyogram registering skeletal muscle activity; central, frontal, occip, 3 EEG leads. Note the low muscle tone with extensive eye movements in REM sleep. (Reproduced, with permission, from Kales A et al: Sleep and dreams: Recent research on clinical aspects. *Ann Intern Med* 1968;**68**:1078.)

PHYSIOLOGIC BASIS OF THE EEG, CONSCIOUSNESS, & SLEEP

The presence of rhythmic waves in the alpha and slow-wave sleep patterns indicates a priori that neural components of some sort are discharging rhythmically, since random discharges of individual units would cancel out and no waves would be produced. The main questions to be answered are which units are discharging synchronously and what mechanisms are responsible for desynchronization and synchronization.

Source of the EEG

The activity recorded in the EEG is mostly that of the most superficial layers of the cortical gray substance, and there are relatively few cell bodies in these layers. The potential changes in the cortical EEG are due to current flow in the fluctuating dipoles formed on the dendrites of the cortical cells and the cell bodies. The dendrites of the cortical cells are a forest of similarly oriented, densely packed units in the superficial layers of the cerebral cortex (Fig 11–2). As pointed out in Chapter 4, dendrites may be conducting processes, but they generally do not produce propagated all or none spikes; rather, they are the site of nonpropagated hypopolarizing and hyperpolarizing local potential changes. As excitatory and inhibitory endings on the dendrites of each cell become active, current flows into and out of these current sinks and sources from the rest of the dendritic processes and the cell body. The cell-dendrite relationship is therefore that of a constantly shifting dipole. Current flow in this dipole would be expected to produce wavelike potential fluctuations in a volume conductor (Fig 11–8). When the sum of the dendritic activity is negative relative to the cell, the cell is hypopolarized and hyperexcitable; when it is positive, the cell is hyperpolarized and less excitable. Information is also transmitted electrotonically from one dendrite to another (see Chapter 4). The cerebellar cortex and the hippocampus are 2 other parts of the central nervous system where many complex, parallel dendritic processes are located subpially over

Wave activity

I 100 μV

Dendritic tree

I 200 mV

Axon

Action potentials

Figure 11–8. Diagrammatic comparison of the electrical responses of the axon and the dendrites of a large cortical neuron. Current flow to and from active synaptic knobs on the dendrites produces wave activity, while all or none action potentials are transmitted down the axon.

a layer of cells. In both areas, there is a characteristic rhythmic fluctuation in surface potential similar to that observed in the cortical EEG.

Desynchronizing Mechanisms

Desynchronization, the replacement of a rhythmic EEG pattern with irregular low-voltage activity, is produced by stimulation of the specific sensory systems up to the level of the midbrain, but stimulation of these systems above the midbrain, stimulation of the specific sensory relay nuclei of the thalamus, or stimulation of the cortical receiving areas themselves does not produce desynchronization. On the other hand, high-frequency stimulation of the reticular formation in the midbrain tegmentum and of the nonspecific projection nuclei of the thalamus desynchronizes the EEG and arouses a sleeping animal. Large lesions of the lateral and superior portions of the midbrain that interrupt the medial lemnisci and other ascending specific sensory systems fail to prevent the desynchronization produced by sensory stimulation, but lesions in the midbrain tegmentum that disrupt the RAS without damaging the specific systems are associated with a synchronized pattern that is unaffected by sensory stimulation. Animals with the former type of lesion are awake; those with the latter type are comatose for long periods. Thus, the ascending activity responsible for desynchronization following sensory stimulation passes up the specific sensory systems to the midbrain, enters the RAS via collaterals, and continues through the thalamus

and the nonspecific thalamic projection system to the cortex.

It should be pointed out that although arousal and EEG desynchronization generally coexist, they do not always do so. The presence of desynchronization in paradoxical sleep is discussed in detail below. In addition, strong nociceptive peripheral stimulation can produce arousal without desynchronization in animals with lesions of the midbrain tegmentum and desynchronization without arousal in animals with lesions in the posterior hypothalamus.

Effect of RAS Stimulation on Cortical Neurons

It is not known how the impulses ascending in the RAS that break up synchronized cortical activity produce arousal. Stimulation of the RAS inhibits burst activity and the recruiting response (see below). Since RAS stimulation arouses and alerts the animal, one might expect it to produce a heightened rather than a depressed excitability of cortical neurons. However, some of the neurons in the cortical sensory projection areas fire randomly during sleep, whereas they discharge only in response to specific stimuli when the animal is awake. It is possible that RAS activity abolishes this random firing by producing some degree of cortical inhibition, leaving these neurons free to respond only to specific sensory signals. The net effect of RAS activity, therefore, might be an increase in the "signal/noise ratio" at the expense of a slight decrease in absolute excitability.

Arousal Following Cortical Stimulation

Electrical stimulation of certain portions of the cerebral cortex causes increased reticular activity and EEG arousal. The most effective cortical loci in the monkey are the superior temporal gyrus and the orbital surface of the frontal lobe. Stimulation of these regions wakes a sleeping animal but causes no movements and has few visible effects in the conscious animal. Stimulation of other cortical areas, even with strong currents, does not produce electrical or behavioral arousal. These observations indicate that a system of **corticofugal fibers** passes from the cortex to the reticular formation, providing a pathway by which intracortical events can initiate arousal. The system may be responsible for the alerting responses to emotions and related psychic phenomena that occur in the absence of any apparent external stimulus.

Synchronizing Mechanisms

The wave pattern of the alpha rhythm indicates that the activity of many of the dendritic units is synchronized. Two factors contribute to this synchronization: the synchronizing effect on each unit of activity in its neighbors, and rhythmic discharge of impulses from the thalamus.

Slabs of cortical tissue maintained in vitro exhibit

rhythmic synaptic potentials. This may be due in part to the fact that the neurons are connected to one another by inhibitory pathways. Each time a given unit discharges, it produces IPSPs on its neighbors (**recurrent collateral inhibition**). The neurons are also postulated to be hyperexcitable after the inhibition. Therefore, the synaptic activity they generate tends to become synchronized. A mechanism of this type has been demonstrated in the nucleus reticularis of the thalamus.

Another mechanism that may promote synchrony is the effect of parallel neural processes on each other in a volume conductor. If 2 nerve fibers are placed side by side in a volume conductor, depolarization of one of them rarely if ever causes a propagated discharge in the other but does influence its excitability. This is because external current flow into the area of depolarization in the active fiber hyperpolarizes the membrane of the neighboring fiber at 2 points and partially depolarizes it at another. Current flow in the opposite direction creates areas of positivity in fibers adjacent to sites where inhibitory endings are active (Fig 11–9). In a large population of similarly oriented neural processes, this formation of current sinks and sources in neighboring fibers tends to synchronize current flow.

The dendritic potentials of the cerebral cortex are also influenced by projections from the thalamus. A circular cut around a piece of cortex does not alter its synchronized activity if the blood supply is intact, but the rhythmicity is decreased if the deep connections of the cortical "button" are severed. Large lesions of the thalamus disrupt the EEG synchrony on the side of the lesion. Stimulation of the thalamic nuclei at a frequency of about 8/s produces a characteristic 8/s response throughout most of the ipsilateral cortex. Because the amplitude of this response waxes and wanes, it is called the **recruiting response** (Fig 11–10). The recruiting response has many similarities to the alpha rhythm and to **burst activity,** the occurrence of short trains of alphalike waves superimposed on slower rhythms. The records of burst activity such as sleep spindles are essentially identi-

Figure 11–10. Recruiting response in the cortex, produced by 8/s stimulation of the intralaminar thalamic region.

cal when recorded simultaneously from the cortex and the thalamus, which suggests that they are due to activity projected from the thalamus or reverberating activity between the thalamus and the cortex.

EEG synchrony and slow-wave sleep (see below) can be produced by stimulation of at least 3 subcortical regions. The region discussed in the preceding paragraph is the **diencephalic sleep zone** in the posterior hypothalamus and the nearby intralaminar and anterior thalamic nuclei. The stimulus frequency must be about 8/s; faster stimuli produce arousal. This finding need not be confusing; the important point is that low-frequency stimulation produces one response, whereas high-frequency stimulation produces another. The second zone is the **medullary synchronizing zone** in the reticular formation of the medulla oblongata at the level of the nucleus of the tractus solitarius. Stimulation of this zone, like stimulation of the diencephalic sleep zone, produces synchrony and sleep if the frequency is low but arousal if the frequency is high. The mechanism by which synchrony is produced is not known, but it presumably involves pathways that ascend to the thalamus. The third synchronizing region is the **basal forebrain sleep zone.** This zone includes the preoptic area and the diagonal band of Broca. It differs from the other 2 zones in that stimulation of the basal forebrain zone produces synchrony and sleep whether the stimulating frequency is high or low. There is some evidence that its effects are mediated by descending pathways which inhibit neurons in the RAS.

The RAS & General Anesthesia

One mechanism by which sedative and anesthetic drugs produce their effects is activation of $GABA_A$ receptors, with consequent increased Cl^- influx and hyperpolarization of neurons. Conduction at synaptic junctions is depressed. This raises the possibility that general anesthetics can produce unconsciousness by depressing conduction in the RAS. The effect of ether on the response of the reticular formation and the medial lemnisci to a sensory stimulus is shown in Fig 11–11. Similar blocking effects are produced by anesthetic doses of barbiturates. There are hundreds of synapses in the reticular pathways, whereas the specific systems have only 2–4. At each synapse, the EPSPs must build up until the firing level of the postsynaptic neurons is reached (see Chapter 4). Any agent that has a moderate depressant effect on this process will block conduction in multisynaptic paths

Figure 11–9. Effect of activity in one nerve fiber on the activity of a neighboring fiber in a volume conductor. Current flows into the area of depolarization (S) on fiber 2 from the surrounding membrane. At X and Z on the membrane of fiber 1, positive charges pile up, hyperpolarizing the membrane; at Y, positive charges are pulled off, partly depolarizing it.

Figure 11–11. Effect of ether anesthesia on lemniscal (ML) and reticular (RAS) response to stimulation of the sciatic nerve.

before blocking conduction in those with only a few synapses.

Concomitants of REM Sleep

Humans aroused at a time when they show the EEG characteristics of REM sleep generally report that they were dreaming, whereas individuals wakened from slow-wave sleep do not. This observation and other evidence indicate that REM sleep and dreaming are closely associated. The teeth-grinding (**bruxism**) that occurs in some individuals is also associated with dreaming. REM sleep is found in all species of mammals and birds that have been studied, but it probably does not occur in other classes of animals.

If humans are awakened every time they show REM sleep, then permitted to sleep without interruption, they show a great deal more than the normal amount of REM sleep for a few nights. Relatively prolonged REM deprivation does not seem to have adverse psychological effects. However, experimental animals completely deprived of REM sleep for long periods lose weight in spite of increased caloric intake and eventually die, indicating that REM sleep has some not yet understood but important homeostatic role.

Distribution of Sleep Stages

In a typical night of sleep, a young adult first enters NREM sleep, passes through stages 1 and 2, and spends 70–100 minutes in stages 3 and 4. Sleep then lightens, and a REM period follows. This cycle is repeated at intervals of about 90 minutes throughout the night (Fig 11–12). The cycles are similar, although there is less stage 3 and 4 sleep and more REM sleep toward morning. Thus, there are 4–6 REM periods per night. At all ages, REM sleep constitutes about 25% of total sleep time. Children have more stage 3 and 4 sleep than young adults, and old people have much less.

Genesis of Slow-Wave Sleep

Slow-wave sleep is produced in part by the absence of desynchronizing activity transmitted via the ascending reticular system. However, it is actively produced as well by the synchronizing zones. The synchronizing activity of the diencephalon is apparently influenced by ascending activity from the pons and hindbrain, since stimulation of the medullary sleep zone produces synchrony and sleep (see above). The role of the basal forebrain zone in the production

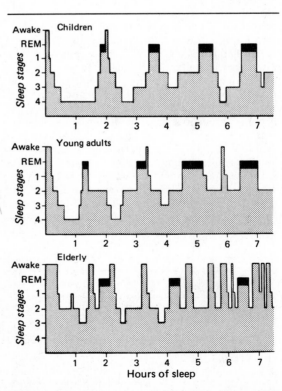

Figure 11–12. Normal sleep cycles at various ages. REM sleep is indicated by the dark areas. (Reproduced, with permission, from Kales AM, Kales JD: Sleep disorders. *N Engl J Med* 1974;**290**:487.)

of normal sleep is unknown. Stimulation of afferents from mechanoreceptors in the skin at rates of 10/s or less also produces sleep in animals, apparently via the brain stem, and it is of course common knowledge that regularly repeated monotonous stimuli put humans to sleep.

There has been considerable debate about the relation of serotonergic neurons in the brain (see Chapter 15) to sleep, but it now appears that serotonin agonists suppress sleep and that the serotonin antagonist ritanserin increases slow-wave sleep in humans. Another hypothesis holds that release of prostaglandin D_2 (PGD$_2$; see Chapter 17) in the medial preoptic area of the hypothalamus causes increased slow-wave sleep and REM sleep, whereas release of PGE$_2$ causes wakefulness. Several investigators have argued that a peptide produced in the brain is responsible for sleep. However, there is disagreement about the chemistry of this putative **sleep peptide,** and its physiologic role, if any, is uncertain.

Genesis of REM Sleep

The mechanism that triggers REM sleep is located in the pontine reticular formation. PGO spikes originate in the lateral pontine tegmentum. The spikes are due to discharge of cholinergic neurons. It now appears that discharge of noradrenergic neurons in the locus ceruleus and serotonergic neurons in the midbrain raphe contributes to wakefulness, and that these neurons are silent when cholinergic PGO spike discharge initiates REM sleep. Reserpine, which depletes serotonin and catecholamines, blocks slow-wave sleep and the hypotonia and EEG desynchronization characteristic of REM sleep but increases PGO spike activity. Barbiturates decrease the amount of REM sleep.

Sleep Disorders

Insomnia, which may be defined as the subjective problem of insufficient or nonrestorative sleep despite an adequate opportunity for sleep, occurs at one time or another in almost all adults. Persistent insomnia can be due to many different mental and medical conditions. It can be relived temporarily by "sleeping pills," especially benzodiazepines, but prolonged use of any of these pills is unwise because they compromise daytime performance and can be habit-forming.

Sleepwalking **(somnambulism),** bed-wetting **(nocturnal enuresis),** and night terrors occur during slow-wave sleep or, more specifically, during arousal from slow-wave sleep. They are not associated with REM sleep. Episodes of sleepwalking are more common in children than in adults and occur predominantly in males. They may last several minutes. Somnambulists walk with their eyes open and avoid obstacles, but when awakened they cannot recall the episodes.

Narcolepsy is a disease of unknown cause in which there is an eventually irresistible urge to sleep during daytime activities. In some cases it has been shown to start with the sudden onset of REM sleep. REM sleep almost never occurs without previous slow-wave sleep in normal individuals.

Sleep apnea is caused by obstruction of the airway during inspiration by the tongue or other relaxed muscles. Efforts to overcome the obstruction finally wake or partially wake the sleeper. When this condition occurs repeatedly, as it frequently does in elderly patients, the loss of sleep causes tiredness and poor performance during the daytime. Sleep apnea can be relieved by teaching subjects not to sleep on their backs, by avoidance of respiratory depressants such as hypnotics and alcohol, and in more severe cases, by positive airway pressure during sleep.

REM behavior disorder is a newly recognized condition in which hypotonia fails to occur during REM sleep. Consequently, patients with this condition "act out their dreams." They thrash about and may even jump out of bed, ready to do battle with imagined aggressors. This disorder usually responds to treatment with benzodiazepines.

Clinical Uses of the EEG

Patients with tumors or other lesions that interrupt the RAS are generally comatose. This system is especially vulnerable at the top of the midbrain and in the posterior hypothalamus, where it is pushed medially by other fiber systems. Relatively small lesions in this location may cause prolonged coma while producing very few other symptoms.

The EEG is sometimes of value in localizing pathologic processes. When a collection of fluid overlies a portion of the cortex, activity over this area may be damped. This fact may aid in diagnosing and localizing conditions such as subdural hematomas. Lesions in the cortex cause local formation of irregular or slow waves that can be picked up in the EEG leads. Epileptogenic foci sometimes generate high-voltage waves that can be localized.

In some forms of epilepsy, there are characteristic EEG patterns during seizures; between attacks, however, abnormalities are often difficult to demonstrate. Seizures are now divided into those that arise from one cerebral hemisphere (**partial** or **local seizures**) and those that involve both hemispheres simultaneously (**general-onset seizures**). Each category is further subdivided. Generalized seizures of the tonic-clonic (**grand mal**) type are characterized by loss of consciousness, which usually occurs without warning. This is followed by a tonic phase, with sustained contraction of limb muscles, followed by a clonic phase with symmetric jerking of the limbs as a result of alternating contraction and relaxation. There is fast EEG activity during the tonic phase. Slow waves, each preceded by a spike, occur at the time of each clonic jerk. For a while after the attack, slow waves are present. Similar changes are seen in experimental animals during convulsions produced by electric

1 sec

Figure 11–13. Petit mal epilepsy. Record of 4 cortical EEG leads from a 6-year-old boy who, during the recording, had one of his "blank spells" in which he was transiently unaware of his surroundings and blinked his eyelids. Time is indicated by the horizontal calibration line. (Reproduced, with permission, from deGroot J: *Correlative Neuroanatomy,* 21st ed. Appleton & Lange, 1991.)

shocks. Absence **(petit mal)** seizures are one of the forms of generalized seizures characterized by a momentary loss of responsiveness. They are associated with 3/s doublets, each consisting of a typical spike and rounded wave (Fig 11–13).

Control of Posture & Movement 12

INTRODUCTION

Somatic motor activity depends ultimately upon the pattern and rate of discharge of the spinal motor neurons and homologous neurons in the motor nuclei of the cranial nerves. These neurons, the final common paths to skeletal muscle, are bombarded by impulses from an immense array of pathways. There are many inputs to each spinal motor neuron from the same spinal segment (see Chapter 6). Numerous suprasegmental inputs also converge on these cells from other spinal segments, the brain stem, and the cerebral cortex. Some of these inputs end directly on the motor neurons, but many exert their effects via interneurons or via the γ efferent system to the muscle spindles and back through the Ia afferent fibers to the spinal cord. It is the integrated activity of these multiple inputs from spinal, medullary, midbrain, and cortical levels that regulates the posture of the body and makes coordinated movement possible.

The inputs converging on the motor neurons subserve 3 semidistinct functions: they bring about voluntary activity; they adjust body posture to provide a stable background for movement; and they coordinate the action of the various muscles to make movements smooth and precise. The patterns of voluntary activity are planned within the brain, and the commands are sent to the muscles primarily via the **corticospinal** and **corticobulbar system.** Posture is continually adjusted not only before but also during movement by **posture-regulating systems.** Movement is smoothed and coordinated by the medial and intermediate portions of the cerebellum (**spinocerebellum**) and its connections. The **basal ganglia** and the lateral portions of the cerebellum (**neocerebellum**) are part of a feedback circuit to the premotor and motor cortex that is concerned with planning and organizing voluntary movement.

GENERAL PRINCIPLES

Organization

Although much is still unknown about the control of voluntary movement, there is considerable evidence in monkeys and humans for the general control scheme shown in Fig 12–1. Commands for voluntary movement originate in cortical association areas. The movements are planned in the cortex as well as in the basal ganglia and the lateral portions of the cerebellar hemispheres, and the basal ganglia and cerebellum both funnel information to the premotor and motor cortex by way of the thalamus. Motor commands from the motor cortex are relayed in large part via the corticospinal tracts and the corresponding corticobulbar tracts to motor neurons in the brain stem. However, collaterals from these pathways and a few direct connections from the motor cortex end on brain stem nuclei, which also project to motor neurons in the brain stem and spinal cord, and these pathways can also mediate voluntary movement. Movement sets up alterations in sensory input from muscles, tendons, joints, and the skin. This feedback information, which adjusts and smoothes movement, is relayed directly to the motor cortex and to the spinocerebellum. The spinocerebellum projects in turn to the brain stem. The main brain stem pathways that are concerned with posture and coordination are the rubrospinal, reticulospinal, tectospinal, and vestibulospinal tracts and corresponding projections to motor neurons in the brain stem.

Encephalization

In species in which the cerebral cortex is highly developed, the process of **encephalization,** ie, the relatively greater role of the cerebral cortex in various functions, is an important phenomenon. For this reason, the effects of cortical ablation in primates are generally more severe and sometimes quite different from those in cats, dogs, and other laboratory animals. Because this chapter is concerned primarily with motor function in humans, pathologic and clinical observations in humans and experiments in laboratory primates are emphasized.

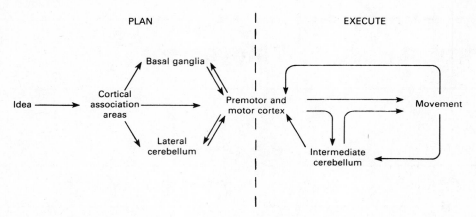

PLAN   EXECUTE

Figure 12–1. Control of voluntary movement. (Modified from McGeer PL, McGeer EG: The control of movement by the brain. *Trends Neurosci* [Nov] 1980; pages 3–4.)

Control of Axial & Distal Muscles

Another theme that is important in motor control is that in the brain stem and spinal cord, medial or ventral pathways and neurons are concerned with the control of muscles of the trunk and proximal portions of the limbs, whereas lateral pathways are concerned with the control of muscles in the distal portions of the limbs. The axial muscles are concerned with postural adjustments and gross movements, whereas the distal limb muscles are those that mediate fine, skilled movements. Thus, for example, the neurons in the medial portion of the ventral horn innervate the proximal limb muscles, particularly the flexors, whereas the lateral ventral horn neurons innervate the distal limb muscles. Similarly, the ventral corticospinal tract and the medial descending paths from the brain stem (the testospinal, reticulospinal, and vestibulospinal tracts) are concerned with adjustments of proximal muscles and posture, whereas the lateral corticospinal tract and the rubrospinal tract are concerned with distal limb muscles and, particularly in the case of the lateral corticospinal tract, with skilled voluntary movements. Phylogenetically, the medial pathways are old, whereas the lateral pathways are new.

Other Terms

Because the fibers of the lateral corticospinal tract form the pyramids in the medulla, the corticospinal pathways have often been referred to as the **pyramidal system.** The rest of the descending brain stem and spinal pathways that do not pass through the pyramids and are concerned with postural control have been called the **extrapyramidal system.** However, the ventral corticospinal pathway does not go through the pyramids, many pyramidal fibers are concerned with other functions, and the system that used to be called extrapyramidal is made up of many different pathways with multiple functions. Consequently, the terms pyramidal and extrapyramidal are misleading, and it seems wise to drop them.

In addition, the motor system has often been divided into **upper** and **lower motor neurons.** Lesions of the lower motor neurons—the spinal and cranial motor neurons that directly innervate the muscles— are associated with flaccid paralysis, muscular atrophy, and absence of reflex responses. The syndrome of spastic paralysis and hyperactive stretch reflexes in the absence of muscle atrophy is said to be due to destruction of the "upper motor neurons," the neurons in the brain and spinal cord that activate the motor neurons. However, there are 3 types of "upper motor neurons" to consider. Lesions in many of the posture-regulating pathways cause spastic paralysis, but lesions limited to the corticospinal and corticobulbar tracts produce weakness **(paresis)** rather than paralysis, and the affected musculature is generally hypotonic. Cerebellar lesions produce incoordination. The unmodified term "upper motor neuron" is therefore confusing.

CORTICOSPINAL & CORTICOBULBAR SYSTEM

ANATOMY

Tracts

The nerve fibers that cross the midline in the medullary pyramids and form the **lateral corticospinal tract** make up about 80% of the fibers in the corticospinal pathway. The remaining 20% make up the **anterior** or **ventral corticospinal tract** (Fig 12–2), which does not cross the midline until the

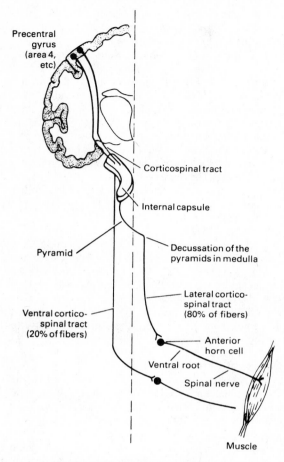

Figure 12–2. The corticospinal tracts.

Labels in figure: Precentral gyrus (area 4, etc); Corticospinal tract; Internal capsule; Pyramid; Decussation of the pyramids in medulla; Lateral cortico-spinal tract (80% of fibers); Ventral cortico-spinal tract (20% of fibers); Anterior horn cell; Ventral root; Spinal nerve; Muscle

Figure 12–3. Medial *(above)* and lateral *(below)* views of the human cerebral cortex showing the motor cortex (Brodmann's area 4) and other areas concerned with control of voluntary movement, along with the numbers assigned to the regions by Brodmann. (Reproduced, with permission, from Kandel ER, Schwartz JH [editors]: *Principles of Neural Science,* 2nd ed. Elsevier, 1985.)

Labels in figure: Supplementary motor area; Motor cortex; Premotor cortex; Primary somatic sensory cortex; Posterior parietal cortex; Prefrontal cortex

level at which it synapses with motor neurons. In addition, this tract contains corticospinal neurons that end on the same side of the body. The ventral pathway, which is the oldest phylogenetically, ends primarily on interneurons. These interneurons synapse on neurons in the medial portion of the ventral horn that control axial and proximal limb muscles. Conversely, the lateral corticospinal pathway innervates lateral neurons in the ventral horn that are concerned with distal limb muscles and hence with skilled movements. In humans, the neurons of this phylogenetically new system end directly on the lateral motor neurons.

Cortical Motor Areas

The cortical areas from which the corticospinal and corticobulbar system originates are generally held to be those where stimulation produces prompt discrete movement. The best known is the **motor cortex (MI)** in the precentral gyrus (Fig 12–3). However, there is a **supplementary motor area** on and above the superior bank of the cingulate sulcus on the

medial side of the hemisphere that reaches to the **premotor cortex** on the lateral surface of the brain (Fig 12–3). Motor responses are also produced by stimulation of somatic sensory area I in the postcentral gyrus and by stimulation of somatic sensory area II in the wall of the sylvian fissure (see Chapter 7). These observations fit with the fact that 30% of the fibers making up the corticospinal and corticobulbar tracts come from the motor cortex but 30% come from the premotor cortex and 40% from the parietal lobe, especially the somatic sensory area.

By means of stimulation experiments in patients undergoing craniotomy under local anesthesia, it has been possible to outline most of the motor projection from the motor cortex. The various parts of the body are represented in the precentral gyrus, with the feet at the top of the gyrus and the face at the bottom (Fig 12–4). The facial area is represented bilaterally, but the rest of the representation is unilateral, the cortical motor area controlling the musculature on the opposite side of the body. The cortical representation of each body part is proportionate in size to the skill with which the part is used in fine, voluntary movement. The areas involved in speech and hand movements are especially large in the cortex; use of the pharynx, lips, and tongue to form words and of the fingers and apposable thumbs to manipulate the environment are activities in which humans are especially skilled.

The conditions under which the human stimulation studies were performed precluded stimulation of the banks of the sulci and other inaccessible areas. Meticulous study has shown that in monkeys, there is

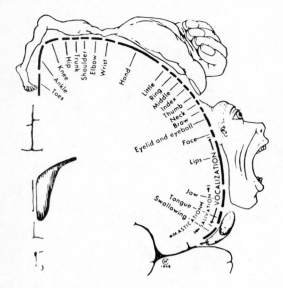

Figure 12–4. Motor homunculus. The figure represents, on a coronal section of the percentral gyrus, the location of the cortical representation of the various parts. The size of the various parts in proportionate to the cortical area devoted to them. (Reproduced, with permission, from Penfield W, Rasmussen G: *The Cerebral Cortex of Man.* Macmillan, 1950.)

a regular representation of the body, with the axial musculature and the proximal portions of the limbs represented along the anterior edge of the precentral gyrus and the distal part of the limbs along the posterior edge. Another feature of MI is the presence of considerable overlap in the muscles that are innervated and, at least in some instances, innervation of synergic muscles separated by considerable distances. These and other observations indicate that MI is organized in terms of movements rather than muscles. The cells in the cortical motor areas are arranged in columns. The cells in each column receive fairly extensive sensory input from the peripheral area in which they produce movement, providing the basis for feedback control of movement. Some of this input may be direct, and some is relayed from somatic sensory area I in the postcentral gyrus.

Supplementary Motor Area

For the most part, the supplementary motor area projects to the motor cortex. It appears to be involved primarily in programming motor sequences. Lesions of this area in monkeys produce awkwardness in performing complex activities and difficulty with bimanual coordination.

Studies of local blood flow in the brains of unanesthetized humans during motor activity, using the xenon washout or 2-deoxyglucose technique (see Chapter 32), provide additional information about this part of the brain. Increased blood flow correlates with increased neuronal activity. When one foot is

moved, for example, the foot region of the contralateral motor cortex shows increased blood flow, and when the fingers are moved, the hand area shows increased flow. When subjects count to themselves without speaking, the motor cortex is quiescent, but when they speak the numbers aloud as they count, blood flow increases in the motor cortex and the supplementary motor area. Thus, the supplementary motor area as well as the motor cortex is involved in voluntary movement when the movements being performed are complex and involve planning. Blood flow increases whether or not the planned movement is carried out. The increase occurs whether the movement is performed by the contralateral or the ipsilateral hand.

Premotor Cortex

The premotor cortex projects to the brain stem areas concerned with postural control and to the motor cortex as well as providing part of the corticospinal and corticobulbar output. Its function is still incompletely understood, but it may be concerned with setting posture at the start of a planned movement and with getting the individual ready to perform.

Posterior Parietal Cortex

In addition to providing fibers that run in the corticospinal and corticobulbar tracts, the somatic sensory area and related portions of the posterior parietal lobe project to the premotor area. Lesions of the somatic sensory area cause defects in motor performance that are characterized by inability to execute learned sequences of movements such as eating with a knife and fork. Some of the neurons in area 5 (Fig 12–3) are concerned with aiming the hands toward an object and manipulating it, whereas some of the neurons in area 7 are concerned with hand-eye coordination.

FUNCTION

Role in Movement

The corticospinal and corticobulbar system is the primary pathway for the initiation of skilled voluntary movement. This does not mean that movement—even skilled movement—is impossible without it. Nonmammalian vertebrates have essentially no corticospinal and corticobulbar system, but they move with great agility. Cats and dogs stand, walk, run, and even eat if food is presented to them after complete destruction of this system. Only in primates are relatively marked deficits produced.

Careful section of the pyramids producing highly selective destruction of the lateral corticospinal tract in laboratory primates produces prompt and sustained loss of the ability to grasp small objects between 2 fingers and to make isolated movements of the wrists.

However, the animal can still use the hand in a gross fashion and can stand and walk. These deficits are consistent with loss of control of the distal musculature of the limbs, which is concerned with fine skilled movements. On the other hand, lesions of the ventral corticospinal tract produce axial muscle deficits that cause difficulty with balance, walking, and climbing. The corticospinal fibers from the parietal lobe, which are phylogenetically the oldest, end in the dorsal rather than the ventral horn and are presumably concerned with direct sensory-motor coordination.

Effects on Stretch Reflexes

Section of the pyramids in monkeys produces prolonged hypotonia and flaccidity rather than spasticity. The anatomic arrangements in humans are such that disease processes rarely, if ever, damage the corticospinal and corticobulbar tracts without also destroying posture-regulating pathways. When spasticity is present, it is probably caused by damage to these latter pathways rather than to the corticospinal and corticobulbar tracts.

Damage to the lateral corticospinal tract in humans produces the **Babinski sign:** dorsiflexion of the great toe and fanning of the other toes when the lateral aspect of the sole of the foot is scratched. Except in infancy, the normal response to this stimulation is plantar flexion of all the toes. The Babinski sign is believed to be a flexor withdrawal reflex that is normally held in check by the lateral corticospinal system. It is of value in the localization of disease processes, but its physiologic significance is unknown.

Transcortical Load-Compensating Reflex

When a human is performing a motor task such as pushing a lever and the resistance to pushing is suddenly increased, there is a prompt increase in the force of the movement. This increase is due in part to monosynaptic stretch reflexes, but most of it has a latency of more than 50 ms, considerably longer than the latency of stretch reflexes. On the other hand, its latency is shorter than the time required for a voluntary response, which is about 100 ms. In conscious monkeys, the increase is associated with increased firing of motor neurons in the motor cortex, and it appears to be a transcortical load-compensating reflex initiated by feedback information to the cortex from the affected limb. This long-loop reflex is probably important in adjusting motor responses and making them appropriate.

POSTURE-REGULATING SYSTEMS

The posture-regulating mechanisms are multiple. They involve a whole series of nuclei and many structures, including the spinal cord, the brain stem, and the cerebral cortex. They are concerned not only with static posture but also, in concert with the corticospinal and corticobulbar systems, with the initiation and control of movement.

Integration

At the spinal cord level, afferent impulses produce simple reflex responses. At higher levels in the nervous system, neural connections of increasing complexity mediate increasingly complicated motor responses. This principle of levels of motor integration is illustrated in Table 12–1. In the intact animal, the individual motor responses are fitted into or "submerged" in the total pattern of motor activity. When the neural axis is transected, the activities integrated below the section are cut off or **released** from the "control of higher brain centers" and often appear to be accentuated. Release of this type, long a cardinal principle in neurology, may in some situations be due to removal of an inhibitory control by higher neural centers. A more important cause of the apparent hyperactivity is loss of differentiation of the reaction, so that it no longer fits into the broader pattern of motor activity. An additional factor may be denervation hypersensitivity of the centers below the transection, but the role of this component remains to be determined.

Postural Control

It is impossible to separate postural adjustments from voluntary movement in any rigid way, but it is possible to differentiate a series of postural reflexes (Table 12–2) that not only maintain the body in an upright, balanced position but also provide the constant adjustments necessary to maintain a stable postural background for voluntary activity. These adjustments include maintained **static** reflexes and dynamic, short-term **phasic** reflexes. The former involve sustained contraction of the musculature, whereas the latter involve transient movements. Both are integrated at various levels in the central nervous system from the spinal cord to the cerebral cortex and are effected largely through various motor pathways. A major factor in postural control is variation in the threshold of the spinal stretch reflexes, which is caused in turn by changes in the excitability of motor neurons and, indirectly, by changes in the rate of discharge in the γ efferent neurons to muscle spindles.

Table 12–1. Summary of levels involved in various neural functions.

Functions	Preparation[1]						Level of Integration
	Normal	Decorticate[2]	Midbrain	Hindbrain (Decerebrate)[3]	Spinal	Decerebellate	
Initiative, memory, etc.	+	0	0	0	0	+	Cerebral cortex required
Conditioned reflexes	+	+[4]	0	0	0	+	Cerebral cortex facilitates
Emotional responses	+	+	0	0	0	+	Hypothalamus, limbic system
Locomotor reflexes	+	+	+	0	0	Incoordinate	Midbrain, thalamus
Righting reflexes	+	+	+	0	0	Incoordinate	Midbrain
Antigravity reflexes	+	+	+	+	0	Incoordinate	Medulla
Respiration	+	+	+	+	0	+	Lower medulla
Spinal reflexes[5]	+	+	+	+	+	+	Spinal cord

[1]0 = absent; + = present; ++ = accentuated.
[2]Cerebral cortex removed.
[3]Brain stem transected at the top of the pons.
[4]Conditioned reflexes are more difficult to establish in decorticate than in normal animals.
[5]Other than stretch reflexes.

Table 12-2. Principal postural reflexes.

Reflex	Stimulus	Response	Receptor	Integrated In
Stretch reflexes	Stretch	Contraction of muscle	Muscle spindles	Spinal cord, medulla
Positive supporting (magnet) reaction	Contact with sole or palm	Foot extended to support body	Proprioceptors in distal flexors	Spinal cord
Negative supporting reaction	Stretch	Release of positive supporting reaction	Proprioceptors in extensors	Spinal cord
Tonic labyrinthine reflexes	Gravity	Contraction of limb extensor muscles	Otolithic organs	Medulla
Tonic neck reflexes	Head turned: (1) To side (2) Up (3) Down	Change in pattern of extensor contraction (1) Extension of limbs on side to which head is turned (2) Hind legs flex (3) Forelegs flex	Neck proprioceptors	Medulla
Labyrinthine righting reflexes	Gravity	Head kept level	Otolithic organs	Midbrain
Neck righting reflexes	Stretch of neck muscles	Righting of thorax and shoulders, then pelvis	Muscle spindles	Midbrain
Body on head righting reflexes	Pressure on side of body	Righting of head	Exteroceptors	Midbrain
Body on body righting reflexes	Pressure on side of body	Righting of body even when head held sideways	Exteroceptors	Midbrain
Optical righting reflexes	Visual cues	Righting of head	Eyes	Cerebral cortex
Placing reactions	Various visual, exteroceptive, and proprioceptive cues	Foot placed on supporting surface in position to support body	Various	Cerebral cortex
Hopping reactions	Lateral displacement while standing	Hops, maintaining limbs in position to support body	Muscle spindles	Cerebral cortex

SPINAL INTEGRATION

The responses of animals and humans after spinal cord transection of the cervical region illustrate the integration of reflexes at the spinal level. The individual spinal reflexes are discussed in detail in Chapter 6.

Spinal Shock

In all vertebrates, transection of the spinal cord is followed by a period of **spinal shock** during which all spinal reflex responses are profoundly depressed. During this period the resting membrane potential of the spinal motor neurons is 2–6 mV greater than normal. Subsequently, reflex responses return and become relatively hyperactive. The duration of spinal shock is proportionate to the degree of encephalization of motor function in the various species. In frogs and rats it lasts for minutes; in dogs and cats it lasts for 1–2 hours; in monkeys it lasts for days; and in humans it usually lasts for a minimum of 2 weeks.

The cause of spinal shock is uncertain. Cessation of tonic bombardment of spinal neurons by excitatory impulses in descending pathways undoubtedly plays a role, but the subsequent return of reflexes and their eventual hyperactivity also have to be explained. The recovery of reflex excitability may be due to the development of denervation hypersensitivity to the mediators released by the remaining spinal excitatory endings. Another possibility for which there is some evidence is the sprouting of collaterals from existing neurons, with the formation of additional excitatory endings on interneurons and motor neurons.

The first reflex response to appear as spinal shock wears off in humans is frequently a slight contraction of the leg flexors and adductors in response to a noxious stimulus. In some patients, the knee jerks come back first. The interval between cord transection and the beginning return of reflex activity is about 2 weeks in the absence of any complications, but if complications are present it is much longer. It is not known why infection, malnutrition, and other complications of cord transection inhibit spinal reflex activity.

Complications of Cord Transection

Management of paraplegic and quadriplegic humans presents complex problems. Like all immobilized patients, they develop a negative nitrogen balance and catabolize large amounts of body protein. The weight of the body compresses the circulation to the skin over bony prominences, so that unless the patient is moved frequently the skin breaks down at these points and **decubitus ulcers** form. The ulcers heal poorly and are prone to infection because of body protein depletion. The tissues that are broken down include the protein matrix of bone, and calcium is released in large amounts. This leads to hypercalcemia and hypercalciuria, and calcium

stones often form in the urinary tract. The stones and the paralysis of bladder function both cause urinary stasis, which predisposes to urinary tract infection. Therefore, the prognosis in patients with transected spinal cords used to be very poor, and death from septicemia, uremia, or inanition occurred in up to 80% of cases. Since World War II, however, the use of antibiotics and meticulous attention to nutrition, fluid balance, skin care, bladder function, and general nursing care have made it possible for many of these patients to survive and lead meaningful lives. Two recent treatments that foster recovery and minimize loss of function after spinal cord injury are administration of large doses of glucocorticoids and administration of ganglioside GM-1. Modern treatment has reduced mortality in major centers to about 6%.

Responses in Chronic Spinal Animals & Humans

Once the spinal reflexes begin to reappear after spinal shock, their threshold steadily drops. In chronically quadriplegic humans, the threshold of the withdrawal reflex is especially low. Even minor noxious stimuli may cause not only prolonged withdrawal of one extremity but marked flexion-extension patterns in the other 3 limbs. Repeated flexion movements may occur for prolonged periods, and contractures of the flexor muscles develop. Stretch reflexes are also hyperactive, as are more complex reactions based on this reflex as well. For example, if a finger is placed on the sole of the foot of an animal after the spinal cord has been transected (**spinal animal**), the limb usually extends, following the examining finger. This **magnet reaction (positive supporting reaction)** involves proprioceptive as well as tactile afferents and transforms the limb into a rigid pillar to resist gravity and support the animal. Its disappearance is also in part an active phenomenon (**negative supporting reaction**) initiated by stretch of the extensor muscles. On the basis of the positive supporting reaction, spinal cats and dogs can be made to stand, albeit awkwardly, for as long as 2–3 minutes.

The flexor spasms initiated by noxious stimuli can be associated with bursts of pain that are particularly bothersome. They can be treated with considerable success with baclofen, a $GABA_B$ receptor agonist that crosses the blood-brain barrier and facilitates inhibition (see Chapter 4). This treatment is also of benefit in patients with spasticity due to lesions of the brain stem or internal capsule (see below).

Locomotion Generator

Not only can spinal cats and dogs be made to stand, but a circuit intrinsic to the spinal cord produces walking movements when stimulated in a suitable fashion. Thus, the **pattern generator** for walking is located in the spinal cord. However, this does not mean that spinal animals can walk without stimulation; the pattern generator has to be turned on by tonic discharge of a discrete area in the midbrain, the mesencephalic locomotor region. In addition, of course, walking in the intact animal is altered, adjusted, and made appropriate by other motor pathways that descend from the brain.

Autonomic Reflexes

Reflex contractions of the full bladder and rectum occur in spinal animals and humans, although the bladder is rarely emptied completely. Hyperactive bladder reflexes can keep the bladder in a shrunken state long enough for hypertrophy and fibrosis of its wall to occur. Blood pressure is generally normal at rest, but the precise feedback regulation normally supplied by the baroreceptor reflexes is absent and wide swings in pressure are common. Bouts of sweating and blanching of the skin also occur.

Sexual Reflexes

Other reflex responses are present in the spinal animal, but in general they are only fragments of patterns that are integrated in the normal animal into purposeful sequences. The sexual reflexes are an example. Coordinated sexual activity depends upon a series of reflexes integrated at many neural levels and is absent after cord transection. However, genital manipulation in male spinal animals and humans produces erection and even ejaculation. In female spinal dogs, vaginal stimulation causes tail deviation and movement of the pelvis into the copulatory position.

Mass Reflex

In chronic spinal animals, afferent stimuli irradiate from one reflex center to another. When even a relatively minor noxious stimulus is applied to the skin, it may irradiate to autonomic centers and produce evacuation of the bladder and rectum, sweating, pallor, and blood pressure swings in addition to the withdrawal response. This distressing **mass reflex** can sometimes be used to give paraplegic patients a degree of bladder and bowel control. They can be trained to initiate urination and defecation by stroking or pinching their thighs, thus producing an intentional mass reflex.

MEDULLARY COMPONENTS

In experimental animals in which the hindbrain and spinal cord are isolated from the rest of the brain by transection of the brain stem at the superior border of the pons, the most prominent finding is marked spasticity of the body musculature. The operative procedure is called **decerebration,** and the resulting pattern of spasticity is called **decerebrate rigidity.** Decerebration produces no phenomenon akin to spinal shock, and the rigidity develops as soon as the brain stem is transected.

Mechanism of Decerebrate Rigidity

On analysis, decerebrate rigidity is found to be spasticity due to diffuse facilitation of stretch reflexes (see Chapter 6). The facilitation is due to 2 factors: increased general excitability of the motor neuron pool and increase in the rate of discharge in the γ efferent neurons.

Supraspinal Regulation of Stretch Reflexes

The brain areas that facilitate and inhibit stretch reflexes are shown in Fig 12–5. These areas generally act by increasing or decreasing spindle sensitivity (Fig 12–6). The large facilitatory area in the brain stem reticular formation discharges spontaneously, possibly in response to afferent input like the RAS. However, the smaller brain stem area that inhibits γ efferent discharge does not discharge spontaneously but is driven instead by fibers from the cerebral cortex and the cerebellum. The inhibitory area in the basal ganglia may act through descending connections, as shown in Fig 12–5, or by stimulating the cortical inhibitory center. From the reticular inhibitory and facilitatory areas, impulses descend in the lateral funiculus of the spinal cord. When the brain stem is transected at the level of the top of the pons, 2 of the 3 inhibitory areas that drive the reticular inhibitory center are removed. Discharge of the facilitatory area continues, but that of the inhibitory area is decreased. Consequently, the balance of facilitatory and inhibitory impulses converging on the γ efferent neurons shifts toward facilitation. Gamma efferent discharge is increased, and stretch reflexes become hyperactive. The cerebellar inhibitory area is still present, and in decerebrate animals, removal of the cerebellum increases the rigidity. The influence of the cerebellum

Figure 12–5. Areas in the cat brain where stimulation produces facilitation (plus signs) or inhibition (minus signs) of stretch reflexes. **1,** motor cortex; **2,** basal ganglia; **3,** cerebellum; **4,** reticular inhibitory area; **5,** reticular facilitatory area; **6,** vestibular nuclei. (Reproduced, with permission, from Lindsley DB, Schreiner LH, Magoun HW: An electromyographic study of spasticity. *J Neurophysiol* 1949; **12:**197.)

Figure 12–6. Response of a single afferent fiber from a muscle spindle to various degrees of muscle stretch. Numbers at left indicate the amount of stretch. The upward deflections are action potentials; the downward deflections are stimulus artifacts. Records were obtained before brain stimulation (first column), during stimulation of brain areas facilitating (second column) and inhibiting (third column) stretch reflexes, and after section of the motor nerve (fourth column). (Reproduced, with permission, from Eldred E, Granit R, Merton PA: Supraspinal control of muscle spindles. *J Physiol [Lond]* 1953;**122:**498.)

is complex, however, and destruction of the cerebellum in humans produces hypotonia rather than spasticity.

The vestibulospinal and some related descending pathways are also facilitatory to stretch reflexes and promote rigidity. Unlike the reticular pathways, they pass primarily in the anterior funiculus of the spinal cord, and the rigidity due to increased discharge in them is not abolished by deafferentation of the muscles. This indicates that this rigidity is due to a direct action on the α motor neurons to increase their excitability, rather than an effect mediated through the small motor nerve system, which would, of course, be blocked by deafferentation.

Significance of Decerebrate Rigidity

In dogs and cats, the spasticity produced by decerebration is most marked in the extensor muscles. Sherrington pointed out that these are the muscles with which the cat and dog resist gravity; the decerebrate posture in these animals is, as he put it, "a caricature of the normal standing position." What has been uncovered by decerebration, then, are the tonic, static postural reflex mechanisms that support the animal against gravity. Additional evidence that this is the correct interpretation of the phenomenon comes from the observation that decerebration in the sloth, an arboreal animal that hangs upside down from branches most of the time, causes rigidity in flexion. In humans, the pattern in true decerebrate rigidity is extensor in all 4 limbs, like that in cats and dogs. Apparently, human beings are not far enough removed from their quadruped ancestors to have changed the pattern in their upper extremities even though the main antigravity muscles of the arms in the upright position are flexors. However, decerebrate rigidity is rare in humans, and the defects that produce it are usually incompatible with life. The more common pattern of extensor rigidity in the legs and moderate flexion in the arms is actually **decorticate rigidity** due to lesions of the cerebral cortex, with most of the brain stem intact (Fig 12–7).

Tonic Labyrinthine Reflexes

In the decerebrate animal, the pattern of rigidity in the limbs varies with the position. No righting responses are present, and the animal stays in the position in which it is put. If the animal is placed on its back, the extension of all 4 limbs is maximal. As the animal is turned to either side, the rigidity decreases, and when it is prone, the rigidity is minimal though still present. These changes in rigidity, the **tonic labyrinthine reflexes,** are initiated by the action of gravity on the otolithic organs and are effected via the vestibulospinal tracts. They are rather surprising in view of the role of rigidity in standing, and their exact physiologic significance remains obscure.

Tonic Neck Reflexes

If the head of a decerebrate animal is moved relative to the body, changes in the pattern of rigidity occur. If the head is turned to one side, the limbs on that side ("jaw limbs") become more rigidly extended while the contralateral limbs become less so. This is the position often assumed by a normal animal looking to one side. Flexion of the head causes flexion of the forelimbs and continued extension of the hindlimbs, the posture of an animal looking into a

Figure 12–7. Human decorticate rigidity **(A–C)** and true decerebrate rigidity **(D).** In A the patient is lying supine with the head unturned. In B and C, the tonic neck reflex patterns produced by turning of the head to the right or left are shown. (Reproduced, with permission, from Fulton JF [editor]: *Textbook of Physiology,* 17th ed. Saunders, 1955.)

hole in the ground. Extension of the head causes flexion of the hindlimbs and extension of the forelimbs, the posture of an animal looking over an obstacle. These responses are the **tonic neck reflexes.** They are initiated by stretch of the proprioceptors in the upper part of the neck, and they can be sustained for long periods.

MIDBRAIN COMPONENTS

After section of the neural axis at the superior border of the midbrain (**midbrain animal**), extensor rigidity like that seen in the decerebrate animal is present only when the animal lies quietly on its back. In the decerebrate animal, the rigidity, which is a static postural reflex, is prominent because there are no modifying phasic postural reflexes. Chronic midbrain animals can rise to the standing position, walk, and right themselves. While the animals are engaged in these phasic activities, the static phenomenon of rigidity is not seen.

Righting Reflexes

Righting reflexes operate to maintain the normal standing position and keep an animal's head upright. These reflexes are a series of responses integrated for the most part in the nuclei of the midbrain.

When the midbrain animal is held by its body and tipped from side to side, the head stays level in response to the **labyrinthine righting reflexes.** The stimulus is tilting of the head, which stimulates the otolithic organs; the response is compensatory contraction of the neck muscles to keep the head level. If the animal is laid on its side, the pressure on that side of the body initiates reflex righting of the head even if the labyrinths have been destroyed. This is the **body on head righting reflex.** If the head is righted by either of these mechanisms and the body remains tilted, the neck muscles are stretched. Their contraction rights the thorax and initiates a wave of similar stretch reflexes that pass down the body, righting the abdomen and the hindquarters (**neck righting reflexes**). Pressure on the side of the body may cause body righting even if the head is prevented from righting (**body on body righting reflex**).

In cats, dogs, and primates, visual cues can initiate **optical righting reflexes** that right the animal in the absence of labyrinthine or body stimulation. Unlike the other righting reflexes, these responses depend upon an intact cerebral cortex.

In intact humans, the operation of these reflexes maintains the head in a stable position and the eyes fixed on visual targets despite movements of the body and the jerks and jolts of everyday life. The responses are initiated by vestibular stimulation, stretching of neck muscles, and movement of visual images on the retina, and the responses are the vestibulo-ocular reflex (VOR; see Chapter 8) and other remarkably precise reflex contractions of the neck and extraocular muscles.

Grasp Reflex

When a primate in which the brain tissue above the thalamus has been removed lies on its side, the limbs next to the supporting surface are extended. The upper limbs are flexed, and the hand on the upper side grasps firmly any object brought in contact with it (grasp reflex). This whole response is probably a supporting reaction that steadies the animal and aids in pulling it upright.

Other Midbrain Responses

Animals with intact midbrains show papillary light reflexes if the optic nerves are also intact. Nystagmus, the reflex response to rotational acceleration described in Chapter 9, is also present. If a blindfolded animal is lowered rapidly, its forelegs extend and its toes spread. This response to linear acceleration is a **vestibular placing reaction** that prepares the animal to land on the floor.

CORTICAL COMPONENTS

Effects of Decortication

Removal of the cerebral cortex (**decortication**) produces little motor deficit in many species of mammals. In primates, the deficit is more severe but movement is still possible. Decorticate animals have all the reflex patterns of midbrain animals. In addition, decorticate animals are easier to maintain than midbrain animals because temperature regulation and other visceral homeostatic mechanisms integrated in the hypothalamus (see Chapter 14) are present. The most striking defect is inability to react in terms of past experience. With certain special types of training, conditioned reflexes can be established in the absence of the cerebral cortex, but under normal laboratory conditions, there is no evidence that learning or conditioning occurs.

Decorticate Rigidity

Moderate rigidity is present in the decorticate animal as a result of the loss of the cortical area that inhibits γ efferent discharge via the reticular formation. Like the rigidity present after transection of the neural axis anywhere above the top of the midbrain, this **decorticate rigidity** is obscured by phasic postural reflexes and is seen only when the animal is at rest. Decorticate rigidity is seen on the hemiplegic side in humans after hemorrhages or thromboses in the internal capsule. Probably because of their anatomy, the small arteries in the internal capsule are especially prone to rupture or thrombotic obstruction, so this type of decorticate rigidity is common. Sixty percent of intracerebral hemorrhages occur in the internal capsule, as opposed to 10% in the cerebral cortex,

10% in the pons, 10% in the thalamus, and 10% in the cerebellum.

The exact site of origin in the cerebral cortex of the fibers that inhibit stretch reflexes is a subject of debate. Under certain experimental conditions, stimulation of the anterior edge of the precentral gyrus is said to cause inhibition of stretch reflexes and cortically evoked movements. This region, which also projects to the basal ganglia, has been named area 4s or the **suppressor strip.**

Hopping & Placing Reactions

Two types of postural reactions, the **hopping** and **placing reactions,** are seriously disrupted by decortication. The former are the hopping movements that keep the limbs in position to support the body when a standing animal is pushed laterally. The latter are the reactions that place the foot firmly on a supporting surface. They can be initiated in a blindfolded animal held suspended in the air by touching the supporting surface with any part of the foot. Similarly, when the snout or vibrissae of a suspended animal touch a table, the animal immediately places both forepaws on the table; and if one limb of a standing animal is pulled out from under it, the limb is promptly replaced on the supporting surface. The vestibular placing reaction has already been mentioned. In cats, dogs, and primates, the limbs are extended to support the body when the animal is lowered toward a surface it can see.

BASAL GANGLIA

Anatomic Considerations

The term **basal ganglia** is generally applied to 5 structures on each side of the brain: the **caudate nucleus, putamen,** and **globus pallidus,** 3 large nuclear masses underlying the cortical mantle (Fig 12–8), and the functionally related **subthalamic nucleus** (body of Luys) and **substantia nigra.** The globus pallidus is divided into an external and an internal segment. The substantia nigra is divided into a **pars compacta** and a **pars reticularis.** Parts of the thalamus are intimately related to the basal ganglia. The caudate nucleus and the putamen are frequently called the **striatum;** the putamen and the globus pallidus are sometimes called the **lenticular nucleus** (Table 12–3).

The main afferent connections to the basal ganglia terminate in the striatum (Fig 12–9). They include the **corticostriate projection** from all parts of the cerebral cortex. There is also a projection from the centromedian nucleus of the thalamus to the striatum (not shown in Fig 12–9).

The connections between the parts of the basal ganglia include a dopaminergic nigrostriatal projection from the pars compacta of the substantia nigra to the striatum and a corresponding GABA-ergic pro-

Figure 12–8. The basal ganglia.

Table 12-3. The basal ganglia.

Lenticular nucleus	Caudate nucleus	Striatum
	Putamen	
	Globus pallidus (pallidum)	
	Subthalamic nucleus (body of Luys)	
	Substantia nigra	

jection from the striatum to the pars reticularis of the substantia nigra. The caudate nucleus and the putamen project to both segments of the globus pallidus. The external segment of the globus pallidus projects to the subthalamic nucleus, which in turn projects to both segments of the globus pallidus and the substantia nigra.

The principal output from the basal ganglia is from the internal segment of the globus pallidus via the **thalamic fasciculus** to the ventral lateral, ventral anterior, and centromedian nuclei of the thalamus. From the thalamic nuclei, fibers project to the prefrontal and premotor cortex. The substantia nigra also projects to the thalamus. There are a few additional

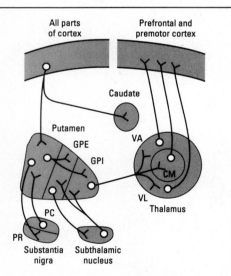

Figure 12–9. Principal connections of the basal ganglia. All parts of the cortex project to the striatum. There is a dopaminergic pathway from the pars compacta (PC) of the substantia nigra to the striatum and a return GABA-ergic pathway to the pars reticularis (PR) of the substantia nigra. In addition, the external segment of the globus pallidus (GPE) projects to the subthalamic nucleus, and this nucleus projects to GPE and the internal segment of the globus pallidus (GPI). GPI projects via the thalamic fasciculus to the ventral anterior (VA), ventral lateral (VL), and centromedian (CM) nuclei of the thalamus. These nuclei in turn project to the prefrontal and premotor cortex, completing the corticocortical loop.

projections to the habenula and the superior colliculus. However, the main feature of the connections of the basal ganglia is that the cerebral cortex projects to the striatum, the striatum to the internal segment of the globus pallidus, the internal segment of the globus pallidus to the thalamus, and the thalamus back to the cortex, completing a loop.

The striatum is made up of a unique mosaic of **patches** or **striosomes** composed of nerve endings in a **matrix** that receives other endings. The neurons of the corticostriate projection that originate in the deep portion of layer 5 of the cortex terminate in the patches, whereas the neurons that originate in layers 2 and 3 and the superficial part of layer 5 end primarily in the matrix. Neurons with their cell bodies in patches project in large part to dopaminergic neurons in the pars compacta of the substantia nigra, whereas many of the neurons with their cell bodies in the matrix project to GABA-ergic neurons in the pars reticularis of the substantia nigra. However, the physiologic significance of these connections is uncertain.

Metabolic Considerations

The metabolism of the basal ganglia is unique in a number of ways. These structures have a high O_2 consumption. The copper content of the substantia nigra and the nearby locus ceruleus is particularly high. In Wilson's disease, a genetic autosomal recessive disorder of copper metabolism in which the plasma copper-binding protein **ceruloplasmin** is usually low, there is chronic copper intoxication and severe degeneration of the lenticular nucleus.

Function

Our knowledge of the precise functions of the basal ganglia is still rudimentary. Lesions in the basal ganglia of animals have relatively little effect. However, recording studies have made it clear that neurons in the basal ganglia, like those in the lateral portions of the cerebellar hemispheres, discharge before movements begin. These observations, plus careful analysis of the effects of diseases of the basal ganglion in humans and the effects of drugs that destroy dopaminergic neurons in animals (see below), have led to the concept that the basal ganglia are involved in the planning and programming of movement or, more broadly, in the processes by which an abstract thought is converted into voluntary action (Fig 12–1). They discharge via the thalamus to areas related to the motor cortex, and the corticospinal pathways provide the final common pathway to the motor neurons.

The basal ganglia also play a role in some cognitive processes, and these are particularly the province of the caudate nucleus. Possibly because of the interconnections of this nucleus with the orbitofrontal and dorsolateral prefrontal portions of the neocortex, lesions of the caudate disrupt performance on tests involving object reversal and delayed alternation. In addition, lesions of the head of the left but not the right caudate nucleus and nearby white matter in humans are associated with a dysarthric form of aphasia that resembles but is different from Wernicke's aphasia (see Chapter 16).

Diseases of the Basal Ganglia in Humans

It is interesting that even though lesions in the basal ganglia in experimental animals have little apparent effect, disease processes affecting these ganglia in humans produce marked and characteristic abnormalities of motor function. Disorders of movement associated with diseases of the basal ganglia in humans are of 2 general types: **hyperkinetic** and **hypokinetic.** The hyperkinetic conditions, those in which there is excessive and abnormal movement, include chorea, athetosis, and ballism. Hypokinetic abnormalities include akinesia and bradykinesia.

Chorea is characterized by rapid, involuntary "dancing" movements. **Athetosis** is characterized by continuous, slow writhing movements. Choreiform and athetotic movements have been likened to the start of voluntary movements occurring in an involuntary, disorganized way. In **ballism,** there are involuntary movements that are flailing, intense, and violent. They appear when the subthalamic nuclei are damaged, and a sudden onset of the movements on one side of the body (**hemiballism**) as a result of hemorrhage in the contralateral subthalamic nucleus is one of the most dramatic syndromes in clinical medicine.

Akinesia is difficulty in initiating movement and decreased spontaneous movement. **Bradykinesia** is slowness of movement.

Huntington's Disease

Huntington's disease is characterized by hyperkinetic choreiform movements that gradually increase until they incapacitate the patient. Speech is slurred and then incomprehensible, and there is progressive dementia followed by death, usually within 10–15 years after the onset of symptoms. The initial event in Huntington's disease is selective degeneration of the striatal GABA-ergic and cholinergic neurons that project to the putamen. This pathway is inhibitory, and its loss permits the hyperkinetic features of the disease to develop.

Huntington's disease is inherited as an autosomal dominant disorder, and its onset is usually between the ages of 30 and 50. There is no effective treatment, and the disease is uniformly fatal. However, the abnormal gene that is responsible for the disease is now known to be located on the short arm of chromosome 4, and there is reason to hope that it will be cloned in the near future. Its product can then be identified, and understanding of the biochemical abnormality may lead to a cure.

Parkinson's Disease (Paralysis Agitans)

Parkinson's disease has both hypokinetic and hyperkinetic features. In this condition, which was originally described by James Parkinson and is named for him, the nigrostriatal system of dopaminergic neurons degenerates. There is also some loss of noradrenergic neurons, but this is not as great as the dopaminergic loss. Parkinsonism was a common late complication of the type of influenza that was epidemic during World War I, and it occurs today in idiopathic form in elderly individuals. There is a steady loss of dopamine and dopamine receptors with age in the basal ganglia in normal individuals, and it is apparently an acceleration of these losses that precipitates parkinsonism. Parkinsonism is also seen as a complication of treatment with the phenothiazine group of tranquilizer drugs, and other drugs that block D_2 dopamine receptors. It can be produced in rapid and dramatic form by injection of MPTP (Fig 12–10). This effect was discovered by chance when a drug dealer in northern California supplied some of his clients with a homemade preparation of "synthetic heroin" that contained MPTP. MPTP is metabolized by the enzyme monoamine oxidase B in various brain cells to produce a potent oxidant, MPP^+. In rodents, MPP^+ is rapidly removed from the brain, but in primates, it is removed more slowly and is taken up by dopaminergic neurons in the substantia nigra, which it destroys without affecting other dopaminergic neurons to any appreciable degree. Consequently, MPTP can be used to produce parkinsonism in monkeys, and its availability has accelerated research on the function of the basal ganglia.

The hypokinetic features of Parkinson's disease are akinesia and bradykinesia, and the hyperkinetic features are **rigidity** and **tremor.** The absence of motor activity and the difficulty in initiating voluntary movements are striking. There is a decrease in **associated movements,** the normal, unconscious movements such as swinging of the arms during walking, the panorama of facial expressions related to the emotional content of thought and speech, and the multiple "fidgety" actions and gestures that occur in all of us. The rigidity is different from spasticity because there is increased motor neuron discharge to both the agonist and antagonist muscles. Passive mo-

Figure 12–10. Conversion of 1-methyl-4-phenyl-1,2,5,6-tetrahydropyridine (MPTP) to 1-methyl-4-phenylpyridinium (MPP^+) by monoamine oxidase B (MAO-B).

tion of an extremity meets with a plastic, dead-feeling resistance that has been likened to bending a lead pipe and is therefore called **lead-pipe rigidity.** Sometimes there is a series of "catches" during passive motion (**cog-wheel rigidity**), but the sudden loss of resistance seen in a spastic extremity is absent. The tremor, which is present at rest and disappears with activity, is due to regular, alternating, 8/s contractions of antagonistic muscles.

An important consideration in Parkinson's disease is the balance between the output of cholinergic neurons and the output of dopaminergic neurons in the striatum. Some improvement is produced by decreasing the cholinergic influence with anticholinergic drugs. More dramatic improvement is produced by administration of L-dopa. Unlike dopamine, this dopamine precursor crosses the blood-brain barrier (see Chapter 15) and helps repair the dopamine deficiency. Bromocriptine and several other synthetic dopamine agonists that penetrate brain tissue have also proved to be of value.

Some patients have shown improvement when the medulla of one of their adrenal glands is transplanted to the ventricular surface of the caudate nucleus. The transplanted gland cells may provide an ongoing source of dopamine or they may provide a growth factor that induces neural repair. Implantation of tissue from the basal ganglia of fetuses is also being explored. However, all these transplantation procedures are still in the experimental stage, and their therapeutic value is not yet established.

Finally, deprenyl, a drug that inhibits monoamine oxidase B, appears to slow the progress of the disease in patients with no known exposure to MPTP. This raises the interesting possibility that ongoing production of an MPP$^+$-like metabolite may be contributing to the destruction of the dopaminergic neurons that causes the disease.

Role of the Subthalamic Nucleus

The flailing involuntary movements produced by lesions of the subthalamic nuclei have been mentioned above. The subthalamic nuclei provide excitatory input to the globus pallidus, the origin of the main efferent pathways from the basal ganglia. Activation of the efferent pathways in turn smoothes and moderates motor responses. It now appears that in all forms of hyperkinetic abnormalities originating in the basal ganglia, the activity of the circuit through the subthalamic nuclei (Fig 12–9) is subnormal. Conversely, it is greater than normal in Parkinson's disease, and subtotal lesions of the subthalamic nuclei in monkeys with MPTP-induced parkinsonism lead to improvement not only in the akinesia but also, interestingly, in the tremor and rigidity. Thus, the circuit through the subthalamic nuclei appears to be an important regulator of the basal ganglion output that normally maintains movement in a smooth and appropriate state.

CEREBELLUM

Anatomic Divisions

The cerebellum sits astride the main sensory and motor systems in the brain stem (Fig 12–11). It is connected to the brain stem on each side by a **superior peduncle** (brachium conjunctivum), **middle peduncle** (brachium pontis), and **inferior peduncle** (restiform body). The medial **vermis** and lateral **cerebellar hemispheres** are more extensively folded and fissured than the cerebral cortex; the cerebellum weighs only 10% as much as the cerebral cortex, but its surface area is about 75% of that of the cerebral cortex. Anatomically, the cerebellum is divided into 3 parts by 2 transverse fissures. The posterolateral fissure separates the medial nodulus and the lateral flocculus on either side from the rest of the cerebellum, and the primary fissure divides the remainder into an anterior and a posterior lobe. Lesser fissures divide the vermis into smaller sections, so that it contains 10 primary lobules numbered I–X from superior to inferior. These lobules are identified by name and number in Fig 12–12.

Functional Divisions

From a functional point of view, the cerebellum is also divided into 3 parts, but in a different way (Fig 12–13). The nodulus in the vermis and the flanking flocculus in the hemisphere on each side form the

Figure 12–11. Diagrammatic representation of the principal parts of the brain. The parts are distorted to show the cerebellar peduncles and the way the cerebellum, pons, and middle peduncle form a "napkin ring" around the brain stem. (Reproduced, with permission, from Goss CM [editor]: *Gray's Anatomy of the Human Body,* 27th ed. Lea & Febiger, 1959.)

Figure 12–12. Superior and inferior views and sagittal section of the human cerebellum. The 10 principal lobules are identified by name and by number (I–X).

Figure 12–13. Functional division of the cerebellum. (Reproduced, with permission, from Kandel ER, Schwartz JH [editors]: *Principles of Neural Science,* 2nd ed. Elsevier, 1985.)

flocculonodular lobe or **vestibulocerebellum.** This lobe, which is phylogenetically the oldest part of the cerebellum, has vestibular connections and is concerned with equilibrium and learning-induced changes in the VOR (see Chapter 8). The rest of the vermis and the adjacent medial portions of the hemispheres form the **spinocerebellum,** the region that receives proprioceptive input from the body as well as a copy of the "motor plan" from the motor cortex. By comparing plan with performance, it smoothes and coordinates movements that are ongoing. The vermis projects to the brain stem area concerned with control of axial and proximal limb muscles, whereas the hemispheres project the brain stem areas concerned with control of distal limb muscles. The lateral portions of the cerebellar hemispheres are called the **neocerebellum.** They are the newest from a phylogenetic point of view, reaching their greatest development in humans. They interact with the motor cortex in planning and programming movements.

Connections

The cerebellum has an external **cerebellar cortex** separated by white matter from the **deep cerebellar nuclei.** Its afferent input goes to the cortex, and via collaterals, to the deep nuclei. There are 4 deep nuclei: the **dentate,** the **globose,** the **emboliform,** and the **fastigial** nuclei. The globose and the emboliform nuclei are sometimes lumped together as the **interpositus nucleus.** The cerebellar cortex projects to these nuclei. Some of the vestibulocerebellar output passes directly to the brain stem, but the deep nuclei provide the only output from the spinocerebellum and the neocerebellum. The medial portion of the spinocerebellum projects to the fastigial nuclei and from there to the brain stem. The adjacent hemispheric portions of the spinocerebellum project to the emboliform and globose nuclei and from them to the brain stem. The neocerebellum projects to the dentate nucleus and from there either directly or indirectly to the ventrolateral nucleus of the thalamus.

The afferent pathways to the cerebellum are summarized in Table 12–4. They transmit proprioceptive and sensory information from all parts of the body. One portion of the proprioceptive input is relayed via the inferior olive, and the olivo-cerebellar fibers form the excitatory climbing fiber input (see below). In addition, information is relayed to the cerebellum from all the motor areas in the cerebral cortex via the pontine nuclei. There is considerable localization of the sensory input, with at least 2 point-for-point representations in the spinocerebellum.

Neural Circuits

The cerebellar cortex contains only 5 types of neu-

Table 12–4. Function of principal afferent systems to the cerebellum.[1]

Afferent Tracts	Transmits
Vestibulocerebellar	Vestibular impulses from labyrinths direct and via vestibular nuclei
Dorsal spinocerebellar	Proprioceptive and exteroceptive impulses from body
Ventral spinocerebellar	Proprioceptive and exteroceptive impulses from body
Cuneocerebellar	Proprioceptive impulses, especially from head and neck
Tectocerebellar	Auditory and visual impulses via inferior and superior colliculi
Olivocerebellar	Proprioceptive input from whole body via relay in inferior olive
Pontocerebellar	Impulses from motor and other parts of cerebral cortex via pontine nuclei

[1]Several other pathways transmit impulses from nuclei in the brain stem to the cerebellar cortex and to the deep nuclei, including a serotonergic input from the raphe nuclei and a noradrenergic input from the locus ceruleus.

rons: Purkinje, granule, basket, stellate, and Golgi cells. It has 3 layers (Fig 12–14): an external molecular layer, a Purkinje cell layer that is only one cell thick, and an internal granular layer. The **Purkinje cells** are among the biggest neurons in the body. They have very extensive dendritic arbors that extend throughout the molecular layer. Their axons, which are the only output from the cerebellar cortex, generally pass to the deep nuclei. The cerebellar cortex also contains **granule cells,** which receive input from the mossy fibers (see below) and innervate the Purkinje cells. The granule cells have their cell bodies in the granular layer. Each sends an axon to the molecular layer, where the axon bifurcates to form a T. The branches of the T are straight and run long distances. Consequently, they are called **parallel fibers.** The dendritic trees of the Purkinje cells are markedly flattened (Fig 12–14) and oriented at right angles to the parallel fibers. The parallel fibers thus make synaptic contact with the dendrites of many Purkinje cells, and the parallel fibers and Purkinje dendritic trees form a grid of remarkably regular proportions.

The other 3 types of neurons in the cerebellar cortex are in effect inhibitory interneurons. The **basket cells** (Fig 12–14) are located in the molecular layer. They receive input from the parallel fibers, and each projects to many Purkinje cells. Their axons form a basket around the cell body and axon hillock of each Purkinje cell they innervate. The **stellate cells** are similar to the basket cells but more superficial in location. The **Golgi cells** are located in the granular layer. Their dendrites, which project into the molecular layer, receive input from the parallel fibers. Their cell bodies receive input via collaterals from the incoming mossy fibers (see below) and the Purkinje cells. Their axons project to the dendrites of the granule cells.

There are 2 main sources of input to the cerebellar cortex: **climbing fibers** and **mossy fibers.** The climbing fibers come from the inferior olivary nuclei, and each projects to the primary dendrites of a Purkinje cell, around which it entwines like a climbing plant. The mossy fibers constitute most of the other incoming fibers. They end on the dendrites of granule cells in complex synaptic junctions called **glomeruli.** The glomeruli also contain the inhibitory endings of the Golgi cells mentioned above.

The fundamental circuits of the cerebellar cortex are thus relatively simple (Fig 12–15). Climbing-fiber inputs exert a strong excitatory effect on single Purkinje cells, whereas mossy-fiber inputs exert a weak excitatory effect on many Purkinje cells via the granule cells. The basket and stellate cells are also excited by granule cells via the parallel fibers, and their output inhibits Purkinje cell discharge. Golgi cells are excited by the mossy-fiber collaterals, Purkinje cell collaterals, and parallel fibers, and they inhibit transmission from mossy fibers to granule cells. The transmitter secreted by the stellate, basket, Golgi, and

Figure 12–14. Location and structure of neurons in the cerebellar cortex. (Reproduced, with permission, from Kuffler SW, Nicholls JG, Martin AR: *From Neuron to Brain,* 2nd ed. Sinauer Associates, 1984.)

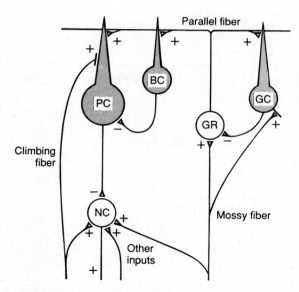

Figure 12–15. Diagram of neural connections in the cerebellum. Shaded neurons are inhibitory, and + and − signs indicate whether endings are excitatory or inhibitory. BC, basket cell; GC, Golgi cell; GR, granule cell; NC, nuclear cell; PC, Purkinje cell. The connections of the stellate cells are similar to those of the basket cells, except that they end for the most part on Purkinje cell dendrites. (Modified from Eccles JC, Itoh M, Szentágothai J: *The Cerebellum as a Neuronal Machine.* Springer, 1967.)

Purkinje cells appears to be GABA, whereas the granule cells probably secrete glutamate.

The output of the Purkinje cells is in turn inhibitory to the deep cerebellar nuclei. These nuclei also receive excitatory inputs via collaterals from the mossy and climbing fibers, and they may also receive other excitatory inputs. It is interesting, in view of their inhibitory Purkinje cell input, that the output of the deep cerebellar nuclei to the brain stem and thalamus is always excitatory. Thus, the entire cerebellar circuitry seems to be concerned solely with modulating or timing the excitatory output of the deep cerebellar nuclei to the brain stem and thalamus.

Flocculonodular Lobe

Animals in which the **flocculonodular lobe** has been destroyed walk in a staggering fashion on a broad base. They tend to fall and are reluctant to move without support. Similar defects are seen in children as the earliest signs of a midline cerebellar tumor arises from cell rests in the nodule and early in its course produces damage that is generally localized to the flocculonodular lobe.

Motion Sickness

Selective ablation of the flocculonodular lobe in dogs abolishes the syndrome of **motion sickness,** whereas extensive lesions in other parts of the cerebellum and the rest of the brain fail to affect it. This condition, a ubiquitous if minor complication of modern travel, is a consequence of excessive and repetitive labyrinthine stimulation due to the motion of the vehicle. It is also a considerable problem in space flight. In humans, it can generally be controlled by a number of antiemetic drugs.

Effects on Stretch Reflexes

Stimulation of the cerebellar areas that receive proprioceptive input sometimes inhibits and sometimes facilitates movements evoked by stimulation of the cerebral cortex. Lesions in folia I–VI and the paramedian areas in experimental animals cause spasticity localized to the part of the body that is represented in the part of the cerebellum destroyed. However, hypotonia is characteristic of cerebellar destruction in humans.

Effects on Movement

Except for the changes in stretch reflexes, experimental animals and humans with lesions of the cerebellar hemispheres show no abnormalities as long as they are at rest. However, pronounced abnormalities are apparent when they move. There is no paralysis and no sensory deficit, but all movements are characterized by a marked **ataxia,** a defect defined as incoordination due to errors in the rate, range, force, and direction of movement. With circumscribed lesions, the ataxia may be localized to one part of the body. If only the cortex of the cerebellum is involved, the movement abnormalities gradually disappear as **compensation** occurs. Lesions of the cerebellar nuclei produce more generalized defects, and the abnormalities are permanent. For this reason, care should be taken to avoid damaging the nuclei when surgical removal of the parts of the cerebellum is necessary.

Other signs of cerebellar deficit in humans provide additional illustrations of the importance of the cerebellum in the control of movement. Ataxia is manifest not only in the wide-based, unsteady, "drunken" gait of patients but also in defects of the skilled movements involved in the production of speech, so that slurred or **scanning speech** results. Other voluntary movements are also highly abnormal. For example, attempting to touch an object with a finger results in overshooting to one side or the other. This **dysmetria,** which is also called **past-pointing,** promptly initiates a gross corrective action, but the correction overshoots to the other side. Consequently, the finger oscillates back and forth. This oscillation is the **intention tremor** of cerebellar disease. Unlike the resting tremor of parkinsonism, it is absent at rest; however, it appears whenever the patient attempts to perform some voluntary action. Another characteristic of cerebellar disease is inability to "put on the brakes," to stop movement promptly. Normally, for example, flexion of the forearm against resistance is quickly

checked when the resistance force is suddenly broken off. The patient with cerebellar disease cannot brake the movement of the limb, and the forearm flies backward in a wide arc. This abnormal response is known as the **rebound phenomenon,** and similar impairment is detectable in other motor activities. This is one of the important reasons these patients show **adiadochokinesia,** the inability to perform rapidly alternating opposite movements such as repeated pronation and supination of the hands. Finally, patients with cerebellar disease have difficulty performing actions that involve simultaneous motion at more than one joint. They dissect such movements and carry them out one joint at a time, a phenomenon known as **decomposition of movement.**

The Cerebellum and Learning

The cerebellum is concerned with learned adjustments that make coordination easier when a given task is performed over and over. The basis of this learning is probably the input via the olivary nuclei. It is worth noting in this regard that each Purkinje cell receives inputs from 250,000 to 1 million mossy fibers, but each has only a single climbing fiber from the inferior olive, and this fiber makes 2000–3000 synapses on the Purkinje cell. Climbing-fiber activation produces a large, complex spike in the Purkinje cell; and this spike in some way produces long-term modification of the pattern of mossy-fiber input to that particular Purkinje cell. Climbing-fiber activity is increased when a new movement is being learned, and selective lesions of the olivary complex abolish the ability to produce long-term adjustments in certain motor responses. The role of the cerebellum in adjusting the VOR and other forms of reflexive memory is discussed in Chapter 16.

Mechanisms

Although the functions of the flocculonodular lobe, spinocerebellum, and neocerebellum are relatively clear and the cerebellar circuits are simple, the exact ways their different parts carry out their functions are still unknown. The relation of the electrical events in the cerebellum to its function in motor control is another interesting problem. The cerebellar cortex has a basic, 150–300/s, 200-μV electrical rhythm and, superimposed on this, a 1000–2000/s component of smaller amplitude. The frequency of the basic rhythm is thus more than 10 times as great as that of the similarly recorded cerebral cortical EEG, and the amplitude is considerably less. Incoming stimuli generally alter the amplitude of the cerebellar rhythm, like a broadcast signal modulating a carrier frequency in radio transmission. However, the significance of these electrical phenomena in terms of cerebellar function is not known.

The Autonomic Nervous System

13

INTRODUCTION

The autonomic nervous system, like the somatic nervous system, is organized on the basis of the reflex arc. Impulses initiated in visceral receptors are relayed via afferent autonomic pathways to the central nervous system, integrated within it at various levels, and transmitted via efferent pathways to visceral effectors. This organization deserves emphasis because the functionally important afferent components have often been ignored. The visceral receptors and afferent pathways have been considered in Chapters 5 and 7 and the major autonomic effector, smooth muscle, in Chapter 3. The efferent pathways to the viscera are the subject of this chapter. Autonomic integration in the central nervous system is considered in Chapter 14.

ANATOMIC ORGANIZATION OF AUTONOMIC OUTFLOW

The peripheral motor portions of the autonomic nervous system are made up of **preganglionic** and **postganglionic neurons** (Figs 13–1). The cell bodies of the preganglionic neurons are located in the intermediolateral gray (visceral efferent) column of the spinal cord or the homologous motor nuclei of the cranial nerves. Their axons are mostly myelinated, relatively slow-conducting B fibers. The axons synapse on the cell bodies of postganglionic neurons that are located in all cases outside the central nervous system. Each preganglionic axon diverges to an average of 8–9 postganglionic neurons. In this way, autonomic output is diffused. The axons of the postganglionic neurons, mostly unmyelinated C fibers, end on the visceral effectors.

Anatomically, the autonomic outflow is divided into 2 components: the **sympathetic** and **parasympathetic divisions** of the autonomic nervous system.

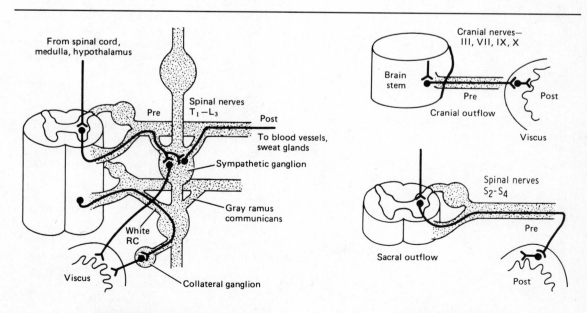

SYMPATHETIC DIVISION

PARASYMPATHETIC DIVISION

Figure 13–1. Autonomic nervous system. Pre, preganglionic neuron; Post, postganglionic neuron; RC, ramus communicans.

Many physiologists add the **enteric nervous system** as a third division (see Chapter 26).

Sympathetic Division

The axons of the sympathetic preganglionic neurons leave the spinal cord with the ventral roots of the first thoracic to the third or fourth lumbar spinal nerves. They pass via the **white rami communicantes** to the **paravertebral sympathetic ganglion chain,** where most of them end on the cell bodies of the postganglionic neurons. The axons of some of the postganglionic neurons pass to the viscera in the various sympathetic nerves. Others reenter the spinal nerves via the **gray rami communicantes** from the chain ganglia and are distributed to autonomic effectors in the areas supplied by these spinal nerves. The postganglionic sympathetic nerves to the head originate in the **superior, middle,** and **stellate ganglia** in the cranial extension of the sympathetic ganglion chain and travel to the effectors with the blood vessels. Some preganglionic neurons pass through the paravertebral ganglion chain and end on postganglionic neurons located in **collateral ganglia** close to the viscera. Parts of the uterus and the male genital tract are innervated by a special system of **short noradrenergic neurons** with cell bodies in ganglia in or near these organs, and the preganglionic fibers to these postganglionic neurons presumably go all the way to the organs (Fig 13–2).

Parasympathetic Division

The **cranial outflow** of the parasympathetic division supplies the visceral structures in the head via the oculomotor, facial, and glossopharyngeal nerves, and those in the thorax and upper abdomen via the vagus nerves. The **sacral outflow** supplies the pelvic viscera via the pelvic branches of the second to fourth sacral spinal nerves. The preganglionic fibers in both outflows end on short postganglionic neurons located on or near the visceral structures (Fig 13–2).

CHEMICAL TRANSMISSION AT AUTONOMIC JUNCTIONS

Transmission at the synaptic junctions between pre- and postganglionic neurons and between the postganglionic neurons and the autonomic effectors is chemically mediated. The principal transmitter agents involved are **acetylcholine** and **norepinephrine,** although **dopamine** is also secreted by interneurons in the sympathetic ganglia and **GnRH** is secreted by some of the preganglionic neurons. GnRH mediates a slow excitatory response (see below). In addition, there are cotransmitters in autonomic neurons, and VIP is released with acetylcholine and ATP and neuropeptide Y with norepinephrine. The chemistry of all these transmitters and the receptors on which they act are discussed in Chapters 4 and 14.

VIP causes bronchodilation, and there may be a separate VIP-secreting **nonadrenergic noncholinergic nervous system** innervating bronchial smooth muscle (see Chapter 34).

Chemical Divisions of the Autonomic Nervous System

On the basis of the chemical mediator released, the autonomic nervous system can be divided into **cholinergic** and **noradrenergic divisions.** The neurons that are cholinergic are (1) all preganglionic neurons, (2) the anatomically parasympathetic postganglionic neurons, (3) the anatomically sympathetic postganglionic neurons which innervate sweat glands, and (4) the anatomically sympathetic neurons which end on blood vessels in skeletal muscles and produce vasodilation when stimulated (sympathetic vasodilator nerves; see Chapter 31). The remaining postganglionic sympathetic neurons are noradrenergic. The adrenal medulla is essentially a sympathetic ganglion in which the postganglionic cells have lost their axons and secrete norepinephrine, epinephrine, and some dopamine directly into the bloodstream. The cholinergic preganglionic neurons to these cells have consequently become the secretomotor nerve supply of this gland.

Transmission in Sympathetic Ganglia

At least in experimental animals, the responses produced in postganglionic neurons by stimulation of their preganglionic innervation include not only a rapid depolarization (**fast EPSP**) that generates action potentials (Table 13–1) but also a prolonged inhibitory postsynaptic potential (**slow IPSP**), a prolonged excitatory postsynaptic potential (**slow EPSP**), and a **late slow EPSP** (see Chapter 4). The late slow EPSP is very prolonged, lasting minutes rather than milliseconds. These 3 responses apparently modulate and regulate transmission through the sympathetic ganglia. The initial depolarization is produced by acetylcholine via a nicotinic receptor. The slow IPSP is probably produced by dopamine, which is secreted by an interneuron within the ganglion. The interneuron is excited by activation of an M_1 muscarinic receptor. The interneurons that secrete dopamine are the small, intensely fluorescent cells (**SIF**

Table 13-1. Fast and slow responses of postganglionic neurons in sympathetic ganglia.

Potential	Duration	Mediator	Receptor
Fast EPSP	30 ms	Acetylcholine	Nicotinic cholinergic
Slow IPSP	2 s	Dopamine	D_2
Slow EPSP	30 s	Acetylcholine	M_1 cholinergic
Late slow EPSP	4 min	GnRH	GnRH

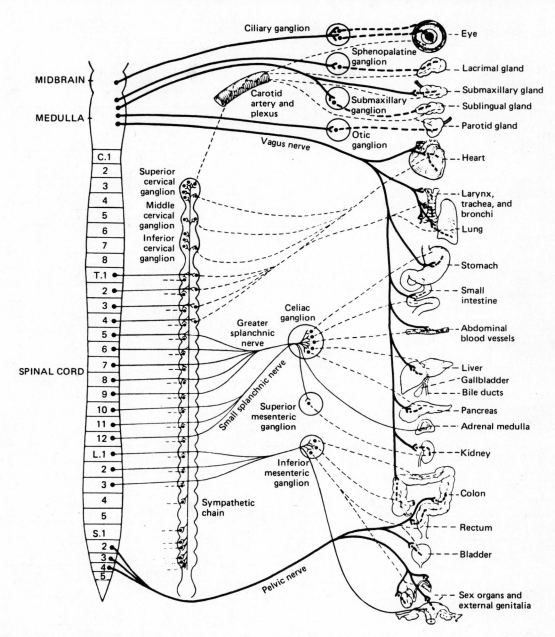

Figure 13–2. Diagram of the efferent autonomic pathways. Preganglionic neurons are shown as solid lines, and postganglionic neurons as dashed lines. The heavy lines are parasympathetic fibers; the light lines are sympathetic. (Modified and reproduced, with permission, from Youmans W: *Fundamentals of Human Physiology,* 2nd ed. Year Book, 1962.)

cells) in the ganglia. The production of the slow IPSP does not appear to be mediated via cyclic AMP, suggesting that a D_2 receptor is involved (see Chapter 4). The slow EPSP is produced by acetylcholine acting on a muscarinic receptor on the membrane of the postganglionic neuron. The late slow EPSP is produced by GnRH or a peptide closely resembling it.

RESPONSES OF EFFECTOR ORGANS TO AUTONOMIC NERVE IMPULSES

General Principles

The effects of stimulation of the noradrenergic and cholinergic postganglionic nerve fibers to the viscera are listed in Table 13–2. The smooth muscle in the

Table 13–2. Responses of effector organs to autonomic nerve impulses and circulating catecholamines.[1]

Effector Organs	Cholinergic Impulses Response	Noradrenergic Impulses Receptor Type[2]	Noradrenergic Impulses Response
Eye			
Radial muscle of iris	. . .	α_1	Contraction (mydriasis)
Sphincter muscle of iris	Contraction (miosis)		. . .
Ciliary muscle	Contraction for near vision	β_2	Relaxation for far vision
Heart			
S-A node	Decrease in heart rate, vagal arrest	β_1	Increase in heart rate
Atria	Decrease in contractility and (usually) increase in conduction velocity	β_1	Increase in contractility and conduction velocity
A-V node	Decrease in conduction velocity	β_1	Increase in conduction velocity
His-Purkinje system	Decrease in conduction velocity	β_1	Increase in conduction velocity
Ventricles	Decrease in conduction velocity	β_1	Increase in contractility
Arterioles			
Coronary	Constriction	α_1, α_2	Constriction
		β_2	Dilation
Skin and mucosa	Dilation	α_1, α_2	Constriction
Skeletal muscle	Dilation	α_1	Constriction
		β_2	Dilation
Cerebral	Dilation	α_1	Constriction
Pulmonary	Dilation	α_1	Constriction
		β_2	Dilation
Abdominal viscera	. . .	α_1	Constriction
		β_2	Dilation
Salivary glands	Dilation	α_1, α_2	Constriction
Renal	. . .	α_1, α_2	Constriction
		β_1, β_2	Dilation
Systemic veins	. . .	α_1	Constriction
		β_2	Dilation
Lung			
Bronchial muscle	Contraction	β_2	Relaxation
Bronchial glands	Stimulation	α_1	Inhibition
		β_2	Stimulation
Stomach			
Motility and tone	Increase	$\alpha_1, \alpha_2, \beta_2$	Decrease (usually)
Sphincters	Relaxation (usually)	α_1	Contraction (usually)
Secretion	Stimulation		Inhibition
Intestine			
Motility and tone	Increase	$\alpha_1, \alpha_2, \beta_2$	Decrease (usually)
Sphincters	Relaxation (usually)	α_1	Contraction (usually)
Secretion	Stimulation	α_2	Inhibition
Gallbladder and ducts	Contraction	β_2	Relaxation
Urinary bladder			
Detrusor	Contraction	β_2	Relaxation (usually)
Trigone and sphincter	Relaxation	α_1	Contraction
Ureter			
Motility and tone	Increase (?)	α_1	Increase (usually)
Uterus	Variable[3]	α_1, β_2	Variable[3]

Table 13–2. Responses of effector organs to autonomic nerve impulses and circulating catecholamines.[1] (continued)

Effector Organs	Cholinergic Impulses Response	Noradrenergic Impulses	
		Receptor Type[2]	Response
Male sex organs	Erection	α_1	Ejaculation
Skin			
Pilomotor muscles	. . .	α_1	Contraction
Sweat glands	Generalized secretion	α_1	Slight, localized secretion[4]
Spleen capsule	. . .	α_1	Contraction
		β_2	Relaxation
Adrenal medulla	Secretion of epinephrine and norepinephrine		. . .
Liver	. . .	$\alpha_1 \beta_2$	Glycogenolysis
Pancreas			
Acini	Increased secretion	α	Decreased secretion
Islets	Increased insulin and glucagon secretion	α_2	Decreased insulin and glucagon secretion
		β_2	Increased insulin and glucagon secretion
Salivary glands	Profuse, watery secretion	α_1	Thick, viscous secretion
		β_2	Amylase secretion
Lacrimal glands	Secretion	α	Secretion
Nasopharyngeal glands	Secretion		. . .
Adipose tissue	. . .	β_1, β_3	Lipolysis
Juxtaglomerular cells	. . .	β_1	Increased renin secretion
Pineal gland	. . .	β	Increased melatonin synthesis and secretion

[1]Modified from Gilman AG et al (editors): *Goodman and Gilman's The Pharmacological Basis of Therapeutics,* 8th ed. Pergamon Press, 1990.
[2]Where a receptor subtype is not specified, data are as yet inadequate for characterization.
[3]Depends on stage of menstrual cycle, amount of circulating estrogen and progesterone, pregnancy, and other factors.
[4]On palms of hands and in some other locations ("adrenergic sweating").

walls of the hollow viscera is generally innervated by both noradrenergic and cholinergic fibers, and activity in one of these systems increases the intrinsic activity of the smooth muscle whereas activity in the other decreases it. However, there is no uniform rule about which system stimulates and which inhibits. In the case of sphincter muscles, both noradrenergic and cholinergic innervations are excitatory, but one supplies the constrictor component of the sphincter and the other the dilator.

There is usually no acetylcholine in the circulating blood, and the effects of localized cholinergic discharge are generally discrete and of short duration because of the high concentration of acetylcholinesterase at cholinergic nerve endings. Norepinephrine spreads farther and has a more prolonged action than acetylcholine. Norepinephrine, epinephrine, and dopamine are all found in plasma (see Chapter 20). The epinephrine and some of the dopamine come from the adrenal medulla, but most of the norepinephrine diffuses into the bloodstream from noradrenergic nerve endings. Metabolites of norepinephrine and dopamine also enter the circulation, some from the sympathetic nerve endings and some from smooth muscle cells (Fig 13–3). The metabolism of catecholamines is discussed in Chapter 4. It is worth noting that even when MAO and COMT are both inhibited, the metabolism of norepinephrine is still rapid. However, inhibition of reuptake prolongs its half-life.

CHOLINERGIC DISCHARGE

In a general way, the functions promoted by activity in the cholinergic division of the autonomic nervous system are those concerned with the vegetative aspects of day-to-day living. For example, cholinergic action favors digestion and absorption of food by increasing the activity of the intestinal musculature, increasing gastric secretion, and relaxing the pyloric sphincter. For this reason, and to contrast it with the "catabolic" noradrenergic division, the cholinergic division is sometimes called the **anabolic nervous system.**

The function of the VIP released from postganglionic cholinergic neurons is unsettled, but there is ev-

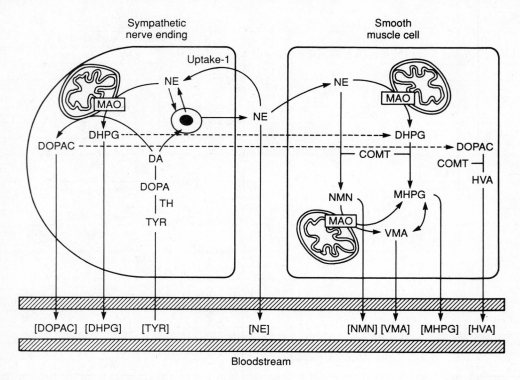

Figure 13–3. Catecholamine metabolism in the sympathetic nervous system. TYR, tyrosine; TH, tyrosine hydroxylase; DA, dopamine; NE, norepinephrine; NMN, normetanephrine. For other abbreviations, see Figs 4–19 and 4–20. (Courtesy of DS Goldstein.)

idence that it facilitates the postsynaptic actions of acetylcholine. Since VIP is a vasodilator, it may also increase blood flow in target organs.

Noradrenergic Discharge

The noradrenergic division discharges as a unit in emergency situations. The effects of this discharge are of considerable value in preparing the individual to cope with the emergency, although it is important to avoid the teleologic fallacy involved in the statement that the system discharges in order to do this. For example, noradrenergic discharge relaxes accommodation and dilates the pupils (letting more light into the eyes), accelerates the heartbeat and raises the blood pressure (providing better perfusion of the vital organs and muscles), and constricts the blood vessels of the skin (which limits bleeding from wounds). Noradrenergic discharge also leads to lower thresholds in the reticular formation (reinforcing the alert, aroused state) and to elevated blood glucose and free fatty acid levels (supplying more energy). On the basis of effects like these, Cannon called the emergency-induced discharge of the noradrenergic nervous system the "preparation for flight or fight."

The emphasis on mass discharge in stressful situations should not obscure the fact that the nor-

adrenergic autonomic fibers also subserve other functions. For example, tonic noradrenergic discharge to the arterioles maintains arterial pressure, and variations in this tonic discharge are the mechanism by which carotid sinus feedback regulation of blood pressure is effected. In addition, sympathetic discharge is decreased in fasting animals and increased when fasted animals are refed. These changes may explain the decrease in blood pressure and metabolic rate produced by fasting and the opposite changes produced by feeding.

The small granulated vesicles in postganglionic noradrenergic neurons contain ATP and norepinephrine, and the large granulated vesicles contain neuropeptide Y. There is evidence that low-frequency stimulation promotes release of ATP whereas high-frequency stimulation causes release of neuropeptide Y. However, the functions of the released ATP and neuropeptide Y are unsettled.

Autonomic Pharmacology

The junctions in the peripheral autonomic motor pathways are a logical site for pharmacologic manipulation of visceral function because transmission across them is chemical. The transmitter agents are synthesized, stored in the nerve endings, and released

Table 13–3. Some drugs and toxins that affect autonomic activity. Only the principal actions are listed. Note that guanethidine is believed to have 2 principal actions.

Site of Action	Compounds That Augment Autonomic Activity	Compounds That Depress Autonomic Activity
Sympathetic and parasympathetic ganglia	**Stimulate postganglionic neurons** Nicotine Dimethyphenylpiperazinium **Inhibit acetylcholinesterase** DFP (diisopropyl fluorophosphate) Physostigmine (Eserine) Neostigmine (Prostigmin) Parathion	**Block conduction** Hexamethonium (C-6) Mecamylamine (Inversine) Pentolinium Trimethaphan (Arfonad) High concentrations of acetylcholine Anticholinesterase drugs
Endings of postganglionic noradrenergic neurons	**Release norepinephrine** Tyramine Ephedrine Amphetamine	**Block norepinephrine synthesis** Metyrosine (Demser) **Interfere with norepinephrine storage** Reserpine Guanethidine (Ismelin) **Prevent norepinephrine release** Bretylium (Bretylol) Guanethidine (Ismelin) **Form false transmitters** Methyldopa (Aldomet)
α receptors	**Stimulate α_1 receptors** Methoxamine (Vasoxyl) Phenylephrine (Neo-Synephrine) **Stimulate α_2 receptors** Clonidine[1] (Catapres)	**Block α receptors** Phenoxybenzamine (Dibenzyline) Phentolamine (Regitine) Prazosin (Minipress) (blocks α_1) Yohimbine (blocks α_2)
β receptors	**Stimulate β receptors** Isoproterenol (Isuprel)	**Block β receptors** Propranolol (Inderal) and others (block β_1 and β_2) Atenolol (Tenormin) and others (block β_1) Butoxamine (blocks β_2)

[1] Clonidine stimulates α_2 receptors in the periphery, but along with other α_2 agonists that cross the blood-brain barrier, it also stimulates α_2 receptors in the brain that decrease sympathetic output. Therefore, the overall effect is decreased sympathetic discharge.

near the neurons, muscle cells, or gland cells on which they act. They bind to receptors on these cells, thus initiating their characteristic actions, and they are then removed from the area by reuptake or metabolism. Each of these steps can be stimulated or inhibited, with predictable consequences. In noradrenergic endings, certain drugs also cause the formation of compounds that replace norepinephrine in the granules, and these weak or inactive "false transmitters" are released instead of norepinephrine by the action potentials reaching the endings.

Some of the drugs and toxins that affect the activity of the autonomic nervous system and the mechanisms by which they produce their effects are listed in Table 13–3. Compounds with muscarinic actions include congeners of acetylcholine and drugs that inhibit acetylcholinesterase. Among the latter are the insecticide parathion and diisopropyl fluorophosphate (DFP), a component of the so-called nerve gases, which kill by producing massive inhibition of acetylcholinesterase. Atropine and scopolamine block muscarinic receptors. The drugs that block the effects of norepinephrine on visceral effectors are referred to as adrenergic blocking agents, peripheral sympathetic blocking agents, adrenolytic agents, or sympatholytic agents.

14

Central Regulation of Visceral Function

INTRODUCTION

The levels of autonomic integration within the central nervous system are arranged, like their somatic counterparts, in a hierarchy. Simple reflexes such as contraction of the full bladder are integrated in the spinal cord (see Chapter 12). More complex reflexes are the subject of this chapter. Those that regulate respiration and blood pressure are integrated in the medulla oblongata. Those that control pupillary responses to light and accommodation are integrated in the midbrain. The complex autonomic mechanisms that maintain the chemical constancy and temperature of the internal environment are integrated in the hypothalamus. The hypothalamus also functions with the limbic system as a unit that regulates emotional and instinctual behavior, and these aspects of hypothalamic function are discussed in the next chapter.

MEDULLA OBLONGATA

Control of Respiration, Heart Rate, & Blood Pressure

The medullary centers for the autonomic reflex control of the circulation, heart, and lungs are called the **vital centers** because damage to them is usually fatal. The afferent fibers to these centers originate in a number of instances in highly specialized visceral receptors. The specialized receptors include not only those of the carotid and aortic sinuses and bodies but also receptor cells that are located in the medulla itself. The motor responses are graded and delicately adjusted and include somatic as well as visceral components. The details of the reflexes themselves are discussed in the chapters on the regulation of the circulation and respiration.

Other Medullary Autonomic Reflexes

Swallowing, coughing, sneezing, gagging, and vomiting are also reflex responses integrated in the medulla oblongata. The swallowing reflex is initiated by the voluntary act of propelling the oral contents toward the back of the pharynx (see Chapter 26). Coughing is initiated by irritation of the lining of the trachea and extrapulmonary bronchi. The glottis closes, and strong contraction of the respiratory muscles builds up intrapulmonary pressure, whereupon the glottis suddenly opens, causing an explosive discharge of air (see Chapter 36). Sneezing is a somewhat similar response to irritation of the nasal epithelium. It is initiated by stimulation of pain fibers in the trigeminal nerves.

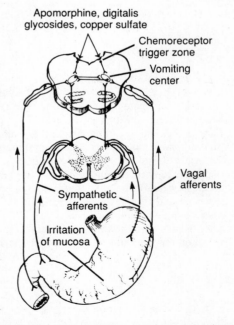

Figure 14–1. Afferent pathways for the vomiting reflex, showing the chemoreceptor trigger zone in the medulla. (Redrawn and reproduced, with permission, from *Research in the Service of Medicine.* Vol 44. Searle & Co, 1956.)

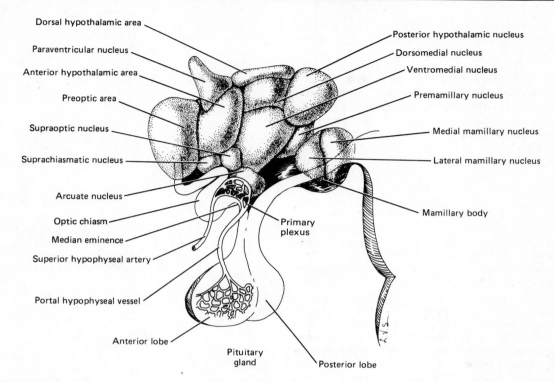

Figure 14–2. Human hypothalamus, with a superimposed diagrammatic representation of the portal-hypophyseal vessels.

Vomiting

Vomiting is another example of the way visceral reflexes integrated in the medulla include coordinated and carefully timed somatic as well as visceral components. Vomiting starts with salivation and the sensation of nausea. Reverse peristalsis empties material from the upper part of the small intestine into the stomach. The glottis closes, preventing aspiration of vomitus into the trachea. The breath is held in midinspiration. The muscles of the abdominal wall contract, and because the chest is held in a fixed position, the contraction increases intra-abdominal pressure. The lower esophageal sphincter and the esophagus relax, and the gastric contents are ejected.

The "vomiting center" in the reticular formation of the medulla at the level of the olivary nuclei controls these activities (Fig 14–1).

Afferents

Irritation of the mucosa of the upper gastrointestinal tract causes vomiting. Impulses are relayed from the mucosa to the vomiting center over visceral afferent pathways in the sympathetic nerves and vagi. Other afferents presumably reach the vomiting center from the diencephalon and limbic system, because emetic responses to emotionally charged stimuli also occur. Thus, we speak of "nauseating smells" and "sickening sights."

Chemoreceptor cells in the medulla initiate vomit-ing when they are stimulated by certain circulating chemical agents. The **chemoreceptor trigger zone** in which these cells are located (Fig 14–1) is in or near the **area postrema,** a V-shaped band of tissue on the lateral walls of the fourth ventricle near the obex. This structure is one of the circumventricular organs (see Chapter 32) and is more permeable to many substances than the underlying medulla. Lesions of the area postrema have little effect on the vomiting response to gastrointestinal irritation but abolish the vomiting that follows injection of apomorphine and a number of other emetic drugs. Such lesions also decrease vomiting in uremia and radiation sickness, both of which may be associated with endogenous production of circulating emetic substances. A chemoreceptor mechanism stimulated by circulating toxins would explain vomiting in many clinical disorders.

HYPOTHALAMUS

ANATOMIC CONSIDERATIONS

The hypothalamus (Figure 14–2) is the portion of the anterior end of the diencephalon that lies below the hypothalamic sulcus and in front of the inter-peduncular nuclei. It is divided into a variety of nu-clei and nuclear areas.

Relation to the Pituitary Gland

There are neural connections between the hypo-thalamus and the posterior lobe of the pituitary gland and vascular connections between the hypothalamus and the anterior lobe. Embryologically, the posterior pituitary arises as an evagination of the floor of the third ventricle. It is made up in large part of the end-ings of axons that arise from cell bodies in the supra-optic and paraventricular nuclei and pass to the poste-rior pituitary via the **hypothalamohypophyseal tract.** Most of the supraoptic fibers end in the poste-rior lobe itself, whereas some of the paraventricular fibers end in the median eminence. The anterior and intermediate lobes of the pituitary arise in the embryo from Rathke's pouch, an evagination from the roof of the pharynx (see Fig 22–1). Sympathetic nerve fibers reach the anterior lobe from its capsule, and parasym-pathetic fibers reach it from the petrosal nerves, but few if any nerve fibers pass to it from the hypothala-mus. However, the **portal hypophyseal vessels** form a direct vascular link between the hypothalamus and the anterior pituitary. Arterial twigs from the carotid arteries and circle of Willis form a network of fenes-trated capillaries called the **primary plexus** on the ventral surface of the hypothalamus (Fig 14–3). Cap-illary loops also penetrate the median eminence. The capillaries drain into the sinusoidal portal hypophy-seal vessels that carry blood down the pituitary stalk to the capillaries of the anterior pituitary. This system begins and ends in capillaries without going through the heart and is therefore a true portal system. In birds and some mammals, including humans, there is no other anterior hypophyseal arterial supply except capsular vessels and anastomotic connections from the capillaries of the posterior pituitary. In other mammals, some blood reaches the anterior lobe through a separate set of anterior hypophyseal arter-ies, but most still comes through the portal vessels. The **median eminence** is generally defined as the portion of the ventral hypothalamus from which the portal vessels arise. This region is "outside the blood-brain barrier" (see Chapter 32).

Afferent & Efferent Connections of the Hypothalamus

The principal afferent and efferent neural path-ways to and from the hypothalamus are listed in Table 14–1. Most of the fibers are unmyelinated. Many connect the hypothalamus to the limbic sys-tem. There are also important connections between the hypothalamus and nuclei in the midbrain tegmen-tum, pons, and hindbrain.

Norepinephrine-secreting neurons with their cell bodies in the hindbrain end in many different parts of the hypothalamus (see Fig 15–6). Paraventricular

Figure 14–3. Secretion of hypothalamic hormones. The hormones of the posterior lobe (PL) are released into the general circulation from the endings of supraoptic and paraventricular neurons, whereas hypophyseotropic hor-mones are secreted into the portal hypophyseal circulation from the endings of arcuate and other hypothalamic neu-rons. AL, anterior lobe; MB, mamillary bodies; OC, optic chi-asm.

neurons that probably secrete oxytocin and vasopres-sin project in turn to the hindbrain and the spinal cord. Neurons that secrete epinephrine have their cell bodies in the hindbrain and end in the ventral hypo-thalamus. There is an intrahypothalamic system of dopamine-secreting neurons which have their cell bodies in the arcuate nucleus and end on or near the capillaries that form the portal vessels in the median eminence. Serotonin-secreting neurons project to the hypothalamus from the raphe nuclei.

HYPOTHALAMIC FUNCTION

The major functions of the hypothalamus are sum-marized in Table 14–2. Some are fairly clear-cut vis-ceral reflexes, and others include complex behavioral and emotional reactions; however, all involve a par-ticular response to a particular stimulus. It is impor-tant to keep this pattern of stimulus-integration-response in mind in considering hypothalamic function.

RELATION TO AUTONOMIC FUNCTION

Many years ago, Sherrington called the hypothal-

Table 14–1. Principal pathways to and from the hypothalamus.

Tract	Type[1]	Description
Medial forebrain bundle	A, E	Connects limbic lobe and midbrain via lateral hypothalamus, where fibers enter and leave it; includes direct amygdalohypothalamic fibers, which are sometimes referred to as a separate pathway.
Fornix	A, E	Connects hippocampus to hypothalamus, mostly mamillary bodies.
Stria terminalis	A	Connects amygdala to hypothalamus, especially ventromedial region.
Mamillary peduncle	A	Connects brain stem to lateral mamillary nuclei.
Ventral noradrenergic bundle	A	Axons of noradrenergic neurones projecting from nucleus of tractus solitarius and ventrolateral medulla to paraventricular nuclei and other parts of hypothalamus.
Dorsal noradrenergic bundle	A	Axons of noradrenergic neurons projecting from locus ceruleus to dorsal hypothalamus.
Serotonergic neurons	A	Axons of serotonin-secreting neurons projecting from dorsal and other raphe nuclei to hypothalamus.
Adrenergic neurons	A	Axons of epinephrine-secreting neurons from medulla to ventral hypothalamus.
Retinohypothalamic fibers	A	Optic nerve fibers to suprachiasmatic nuclei from optic chiasm.
Thalamohypothalamic and pallidohypothalamic fibers	A	Connects thalamus and lenticular nucleus to hypothalamus.
Periventricular system (including dorsal longitudinal fasciculus of Schütz)	A, E	Interconnects hypothalamus and midbrain; efferent projections to spinal cord, afferent from sensory pathways.
Mamillothalamic tract of Vicq d'Azyr	E	Connects mamillary nuclei to anterior thalamic nuclei.
Mamillotegmental tract	E	Connects hypothalamus to reticular portions of midbrain.
Hypothalamohypophyseal tract (supraopticohypophyseal and paraventriculohypophyseal tracts)	E	Axons of neurons in supraoptic and paraventricular nuclei that end in median eminence, pituitary stalk, and posterior pituitary.
Neurons containing vasopressin, oxytocin	E	Run from paraventricular nucleus to nucleus of tractus solitarius, other brain stem nuclei, intermediolateral column of spinal cord; also from paraventricular nucleus to central nucleus of amygdala.
Neurons containing hypophyseotropic hormones	E	Run from various hypothalamic nuclei to median eminence.

[1]A, principally afferent; E, principally efferent.

amus "the head ganglion of the autonomic system." Stimulation of the hypothalamus produces autonomic responses, but the hypothalamus does not seem to be concerned with the regulation of visceral function per se. Rather, the autonomic responses triggered in the hypothalamus are part of more complex phenomena such as rage and other emotions.

Effects of Stimulation

Stimulation of the superior anterior hypothalamus occasionally causes contraction of the urinary bladder, a parasympathetic response. However, hypothalamic stimulation causes very few other parasympathetic responses. Thus, there is very little evidence that a localized "parasympathetic center" exists.

Stimulation of some parts of the hypothalamus can cause cardiac arrhythmias, and there is reason to believe that these are due to simultaneous activation of vagal and sympathetic nerves to the heart.

Sympathetic Responses

Stimulation of various parts of the hypothalamus, especially the lateral areas, produces a rise in blood pressure, pupillary dilation, piloerection, and other signs of diffuse noradrenergic discharge. The stimuli that trigger this pattern of responses in the intact animal are not regulatory impulses from the viscera but emotional stimuli, especially rage and fear. Noradrenergic responses are also triggered as part of the reactions that conserve heat (see below).

Low-voltage electrical stimulation of the middorsal portion of the hypothalamus causes vasodilation in muscle. Associated vasoconstriction in the skin and elsewhere maintains blood pressure at a fairly constant level. This observation and other evidence support the conclusion that the hypothalamus is a way station on the so-called cholinergic sympathetic vasodilator system which originates in the cerebral cortex. It may be this system that is responsible for the dilation of muscle blood vessels at the start of exercise (see Chapter 31).

Stimulation of the dorsomedial nuclei and posterior hypothalamic areas produces increased secretion of epinephrine and norepinephrine from the adrenal medulla. Increased adrenal medullary secretion is one of the physical changes associated with rage and

Table 14–2. Summary of principal hypothalamic regulatory mechanisms.

Function	Afferents From	Integrating Areas
Temperature regulation	Cutaneous cold receptors; temperature-sensitive cells in hypothalamus.	Anterior hypothalamus, response to heat; posterior hypothalamus, response to cold.
Neuroendocrine control of: Catecholamines	Limbic areas concerned with emotion.	Dorsomedial and posterior hypothalamus.
Vasopressin	Osmoreceptors, "volume receptors," others.	Supraoptic and paraventricular nuclei.
Oxytocin	Touch receptors in breast, uterus, genitalia.	Supraoptic and paraventricular nuclei.
Thyroid-stimulating hormone (thyrotropin, TSH) via TRH	Temperature receptors in infants, perhaps others.	Paraventricular nuclei and neighboring areas.
Adrenocorticotropic hormone (ACTH) and β-lipotropin (β-LPH) via CRH	Limbic system (emotional stimuli); reticular formation ("systemic" stimuli); hypothalamic or anterior pituitary cells sensitive to circulating blood cortisol level; suprachiasmatic nuclei (diurnal rhythm).	Paraventricular nuclei.
Follicle- stimulating hormone (FSH) and luteinizing hormone (LH) via GnRH	Hypothalamic cells sensitive to estrogens; eyes, touch receptors in skin and genitalia of reflex ovulating species.	Preoptic area, other areas.
Prolactin via PIH and PRH	Touch receptors in breasts, other unknown receptors.	Arcuate nucleus, other areas (hypothalamus inhibits secretion).
Growth hormone via somatostatin and GRH	Unknown receptors.	Periventricular nucleus, arcuate nucleus.
"Appetitive" behavior Thirst	Osmoreceptors, subfornical organ, and other circumventricular organs.	Lateral superior hypothalamus.
Hunger	"Glucostat" cells sensitive to rate of glucose utilization.	Ventromedial satiety center, lateral hunger center.
Sexual behavior	Cells sensitive to circulating estrogen and androgen, others.	Anterior ventral hypothalamus plus, in the male, piriform cortex.
Defensive reactions (fear, rage)	Sense organs and neocortex, paths unknown.	Diffuse, in limbic system and hypothalamus.
Control of various endocrine and activity rhythms	Retina via retinohypothalamic fibers.	Suprachiasmatic nuclei.

fear and may occur when the cholinergic sympathetic vasodilator system is activated. It has been claimed that there are separate hypothalamic centers for the control of epinephrine and norepinephrine secretion. Differential secretion of one or the other of these adrenal medullary catecholamines does occur in certain situations (see Chapter 20), but the selective increases are small.

RELATION TO SLEEP

Lesions of the posterior hypothalamus cause prolonged sleep, and stimulation of the dorsal hypothalamus in conscious animals causes them to go to sleep. However, posterior hypothalamic lesions that cause coma involve fibers of the RAS as they pass to the thalamus and cortex. Therefore, the sleep-producing effects of posterior hypothalamic lesions can be ex-

plained by damage to the RAS (see Chapter 11) rather than destruction of a hypothetical hypothalamic "wakefulness center." The stimulating frequency that produces sleep is approximately 8/s. As noted in Chapter 11, stimulation of the thalamus, the orbital surface of the frontal lobe, and some parts of the brain stem at a frequency of 8/s produces sleep. Thus, the reported effects of hypothalamic stimulation are probably due to stimulation of a diffuse system, the low-frequency stimulation of which produces sleep. On the basis of these considerations, it seems unlikely that the hypothalamus plays any direct or unique role in the regulation of sleep.

RELATION TO CYCLIC PHENOMENA

The suprachiasmatic nuclei are the dominant pacemakers for many circadian rhythms in the body.

These include the rhythms in the secretion of ACTH (see Chapter 20) and melatonin (see Chapter 24), as well as sleep-wake cycles, the body temperature rhythm, and activity patterns in laboratory animals. The nuclei receive an important input from the eyes via the retinohypothalamic fibers, as well as an input from the lateral geniculate nuclei, and their normal function is to synchronize, ie, to entrain the various body rhythms to the 24-hour light-dark cycle.

It is interesting that exposure to bright light can either advance, delay, or have no effect on the sleep-wake cycle in humans depending on the time of day when it is applied. During the usual daytime it has no effect, but just after dark it delays the onset of the sleep period, and just before dawn, it accelerates the onset of the sleep period. In experimental animals, exposure to light turns on immediate-early genes (see Chapter 1) in the suprachiasmatic nuclei, but only at times during the circadian cycle when light is capable of influencing entrainment. Stimulation during the day is ineffective. The suprachiasmatic nuclei also receive a prominent serotonergic innervation from the raphe nuclei, but the exact way the nuclei act to entrain rhythms is still unknown.

HUNGER

Feeding & Satiety

Food intake is generally regulated with great precision. Body weight is determined by the balance between caloric intake and energy expenditure. These are regulated on different time scales; ie, food intake is regulated on a meal-to-meal basis and energy output on a day-to-day basis. Nevertheless, the weight of a normal human is relatively constant over a long period. If animals are made obese by force-feeding and then permitted to eat as they wish, their spontaneous food intake decreases until their weight falls to the control level (Fig 14–4). Conversely, if animals are starved and then permitted to eat freely, their spontaneous food intake increases until they regain the lost weight. It appears that many factors are involved in the regulation of appetite, and much remains to be learned about the process. However, the hypothalamus is clearly involved.

Role of the Hypothalamus

Hypothalamic regulation of the appetite for food depends primarily upon the interaction of 2 areas: a lateral "**feeding center**" in the bed nucleus of the medial forebrain bundle at its junction with the pallidohypothalamic fibers, and a medial "**satiety center**" in the ventromedial nucleus. Stimulation of the feeding center evokes eating behavior in conscious animals, and its destruction causes severe, fatal anorexia in otherwise healthy animals (Fig 14–5). Stimulation of the ventromedial nucleus causes cessation of eating, whereas lesions in this region cause

hyperphagia and, if the food supply is abundant, the syndrome of **hypothalamic obesity** (Fig 14–6). Destruction of the feeding center in rats with lesions of the satiety center causes anorexia, which indicates that the satiety center functions by inhibiting the feeding center. It appears that the feeding center is chronically active and that its activity is transiently inhibited by activity in the satiety center after the ingestion of food. However, it is not certain that the feeding center and the satiety center simply control the desire for food. For example, rats with ventromedial lesions gain weight for a while, but their food intake then levels off. After their intake reaches a plateau, their appetite mechanism operates to maintain their new, higher weight. One theory that has been advanced to explain these observations is that it is the set point for body weight rather than food intake per se which is regulated by the hypothalamus.

Afferent Mechanisms

There is considerable debate about the signals that are sensed by the satiety and feeding center to regulate food intake at each meal. The activity of the satiety center is probably governed in part by the level of glucose utilization of cells within the center. These cells have therefore been called **glucostats.** It has been postulated that when their glucose utilization is low—and consequently when the arteriovenous blood glucose difference across them is low—their activity is decreased. Under these conditions, the activity of the feeding center is unchecked and the individual is hungry. When utilization is high, the activ-

Figure 14–4. Effects of changes in food intake and ventromedial hypothalamic lesions on spontaneous food intake and body weight. **A:** Rats were force-fed for weeks 3–6, then permitted free access to food. **B:** Rats were partially starved for weeks 3–6, then permitted free access to food. **C:** Bilateral ventromedial hypothalamic lesions were produced at 3 weeks and the rats allowed free access to food throughout. (Reproduced, with permission, from Stricker EM: Hyperphagia. N Engl J Med 1978;**298**:1010.)

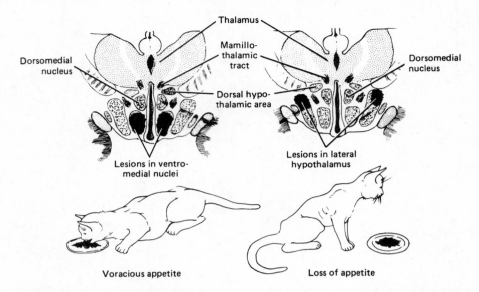

Figure 14–5. Diagrammatic summary of the effects of hypothalamic lesions on feeding. (Redrawn from an original by Netter FH in: *Ciba Clinical Symposia.* Copyright © 1956, Ciba Pharmaceutical Co. Reproduced with permission.)

ity of the glucostats is increased, the feeding center is inhibited, and the individual feels sated. This **glucostatic hypothesis** of appetite regulation is supported by an appreciable body of experimental data. Other factors undoubtedly affect appetite, but the glucostatic hypothesis has the merit of explaining the increased appetite in diabetes, in which the blood sugar is high but the glucose utilization of the cells is low because of the insulin deficiency. The objection has been raised that most neural tissue does not require insulin to metabolize glucose. However, the region of the ventromedial nucleus has been shown to be different from the rest of the brain in that its rate of glucose utilization does vary with the amount of insulin in the circulation.

Relatively large amounts of radioactive glucose are taken up by cells of the ventromedial nuclei. This affinity for glucose probably explains why injections of gold thioglucose produce obesity in mice. The ventromedial cells presumably take up the toxic gold-substituted glucose molecules, which destroy the cells and thus prevent inhibition of the lateral feeding center. In animals that have received gold thioglucose injections and have become obese, demonstrable lesions can be found in the ventromedial nuclei.

The transmitters in the neural circuits that appear to be involved in the regulation of appetite include serotonin and catecholamines. Activation of serotoninergic pathways reduces food intake, and centrally acting opiate agonists increase food intake. In the medial hypothalamus, activation of α_2-adrenergic receptors increases appetite, and in the lateral hypothalamus, activation of β-adrenergic and dopaminergic

receptors decreases appetite. Amphetamine and related drugs used in weight loss programs reduce appetite by acting on the lateral hypothalamus.

Role of Gastrointestinal Hormones

Food in the gastrointestinal tract may cause the secretion of gastrointestinal or other hormones that in-

Figure 14–6. Hypothalamic obesity. The animal on the right, in which bilateral lesions were placed in the ventromedial nuclei 4 months previously, weighs 1080 g. The control animal on the left weights 520 g. (Reproduced, with permission, from Stevenson JAF, in: *The Hypothalamus.* Haymaker W, Anderson E, Nauta WJH [editors]. Thomas, 1969.)

hibit further food intake. **Cholecystokinin (CCK;** see Chapter 26) and **calcitonin** (see Chapter 21) have been shown to decrease appetite. Much of the anorectic effect of circulating CCK is apparently due to an action on visceral receptors, since it is markedly reduced by subdiaphragmatic vagotomy. However, injection of CCK into the hypothalamus also decreases food intake. It is possible that peripheral and central sites are both involved. Peripheral CCK receptors **(CCK-A receptors)** differ from CCK receptors in the brain **(CCK-B receptors),** and it has recently been shown that selective antagonists of CCK-A receptors and those of CCK-B receptors both inhibit satiety. CCK-A antagonists but not CCK-B antagonists inhibit the anorectic effect of peripherally injected CCK. However, CCK-B antagonists are 100 times more potent in inhibiting satiety than are CCK-A antagonists. Thus the central CCK receptors seem to be more important.

Other Factors Regulating Food Intake

There is evidence that the size of body fat depots triggers either neural or humoral signals that are relayed to the brain and that appetite is controlled in this fashion **(lipostatic hypothesis).** Adipsin, a serine protease secreted into the circulation by adipocytes, could be such a signal. Several other circulating factors have been postulated to play a role, but their status is uncertain. In addition, there is some evidence that protein intake is regulated independently of body weight. A cold environment stimulates and a hot environment depresses appetite.

Distention of the gastrointestinal tract inhibits appetite, and contractions of an empty stomach **(hunger contractions)** stimulate appetite, but denervation of the stomach and intestines does not affect the amount of food eaten. In animals with esophageal fistulas, chewing and swallowing food causes some satiety even though the food never reaches the stomach. Especially in humans, cultural factors, environment, and past experiences related to the sight, smell, and taste of food also affect food intake.

It now appears that defective neural control of brown fat may contribute to the obesity of animals with ventromedial hypothalamic lesions produced electrolytically or by injection of gold thioglucose. **Brown fat,** a special form of body fat that has an extensive sympathetic innervation and produces more heat and less ATP than other tissues (see Chapter 17), is normally located between and around the scapulas and in other parts of the body. It is stimulated by cold and by food intake. In animals with ventromedial lesions, the response to cold is normal but food fails to produce the normal increase in sympathetic discharge to brown fat, and this abnormality has the effect of maintaining a high efficiency of food utilization and promoting obesity. In addition, the increased heat generated by food intake in normal animals may contribute to satiety, and this signal is reduced or absent in animals with ventromedial lesions.

Long-term Regulation of Appetite

The net effect of all the appetite-regulating mechanisms in normal adult animals and humans is an adjustment of food intake to the point where caloric intake balances energy expenditures, with the result that body weight is maintained. Children are notorious for their uneven food intake, their appetite for certain foods, and their unwillingness to eat others. However, over time they balance food intake with energy expenditure for immediate needs and growth and they grow and develop at a normal pace. Humans gain weight with advancing age, but this is normally a slow, carefully regulated process. One investigator calculated that the average woman gained 11 kg between the ages of 25 and 65 and that this was produced by an average excess food intake of only 350 mg/day. Considering that the total food intake of the woman over the 40-year period was more than 18 metric tons, the error in food intake over energy expenditure was less than 0.03%.

Obesity is discussed with other aspects of nutrition in Chapter 17. In the present context, however, it is worth noting that there is a strong genetic component in the determination of ultimate body weight, and studies of twins reared in different families and environments indicate that environmental factors account for less than 30% of body weight.

In hyperthyroidism and diabetes mellitus, hunger increases as energy output increases. After exercise, food intake is increased. During recovery from debilitating illness, food intake is increased in a catch-up fashion until lost weight is regained. How these long-term adjustments are brought about is not known.

THIRST

Another appetitive mechanism under hypothalamic control is thirst. Appropriately placed hypothalamic lesions diminish or abolish fluid intake, in some instances without any change in food intake, and electrical stimulation of the hypothalamus causes drinking.

Drinking is regulated by plasma osmolality and ECF volume in much the same fashion as vasopressin secretion (see below). Water intake is increased by increased effective osmotic pressure of the plasma (Fig 14–7), by decreases in ECF volume, and by psychologic and other factors. Osmolality acts via **osmoreceptors,** receptors that sense the osmolality of the body fluids. These osmoreceptors are located in the anterior hypothalamus.

Decreases in ECF volume also stimulate thirst by a pathway independent of that mediating thirst in response to increased plasma osmolality (Fig 14–8). Thus, hemorrhage causes increased drinking even

Figure 14–7. Relation of plasma osmolality to thirst in healthy adult humans during infusion of hypertonic saline. The intensity of thirst is measured on a special analog scale. (Reproduced, with permission, from Thompson CJ et al: The osmotic thresholds for thirst and vasopressin release are similar in healthy man. *Clin Sci Lond* 1986;**71**:651.)

though there is no change in the osmolality of the plasma. The effect of ECF volume depletion on thirst is mediated in part via the renin-angiotensin system (see Chapter 24). Renin secretion is increased by hypovolemia, and there is a resultant increase in circulating angiotensin II. The angiotensin II acts on the **subfornical organ,** a specialized receptor area in the diencephalon (see Fig 32–6), to stimulate the neural areas concerned with thirst. There is some evidence that it acts on the **organum vasculosum** of the **lam-**

ina terminalis (OVLT) as well. These areas are highly permeable and are 2 of the circumventricular organs located "outside the blood-brain barrier" (see Chapter 32). However, drugs which block the action of angiotensin II do not completely block the thirst response to hypovolemia, and it appears that the baroreceptors in the heart and blood vessels are also involved.

Whenever the sensation of thirst is obtunded, either by direct damage to the diencephalon or by depressed or altered states of consciousness, patients stop drinking adequate amounts of fluid. Dehydration results if appropriate measures are not instituted to maintain water balance. If the protein intake is high, the products of protein metabolism cause an osmotic diuresis (see Chapter 38), and the amounts of water required to maintain hydration are large. Most cases of hypernatremia are actually due to simple dehydration in patients with psychoses or cerebral disease who do not or cannot increase their water intake when their thirst mechanism is stimulated.

Other Factors Regulating Water Intake

A number of other well-established factors contribute to the regulation of water intake. Psychologic and social factors are important. Dryness of the pharyngeal mucous membrane causes a sensation of thirst. Patients in whom fluid intake must be restricted sometimes get appreciable relief of thirst by sucking ice chips or a wet cloth.

Dehydrated dogs, cats, camels, and some other animals rapidly drink just enough water to make up their water deficit. They stop drinking before the water is absorbed (while their plasma is still hypertonic), so some kind of pharyngeal or gastrointestinal "metering" must be involved. There is some evidence that humans have a similar metering ability, although it is not well developed.

CONTROL OF POSTERIOR PITUITARY SECRETION

Vasopressin & Oxytocin

In most mammals, the hormones secreted by the posterior pituitary gland are **arginine vasopressin (AVP)** and **oxytocin.** In hippopotamuses and most pigs, arginine in the vasopressin molecule is replaced by lysine to form **lysine vasopressin.** The posterior pituitaries of some species of pigs and marsupials contain a mixture of arginine and lysine vasopressin. The posterior lobe hormones are nonapeptides (considering each half cystine as a single amino acid) with a disulfide ring at one end (Fig 14–9).

Biosynthesis, Intraneuronal Transport, & Secretion

The hormones of the posterior pituitary gland are synthesized in the cell bodies of the magnocellular

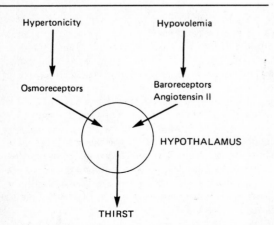

Figure 14–8. Diagrammatic representation of the way in which changes in plasma osmolality and changes in ECF volume affect thirst by separate pathways.

Figure 14–9. Arginine vasopressin and oxytocin.

neurons in the supraoptic and paraventricular nuclei and transported down the axons of these neurons to their endings in the posterior lobe, where they are secreted in response to electrical activity in the endings. Some of the neurons make oxytocin and others make vasopressin, and oxytocin-containing and vasopressin-containing cells are found in both nuclei.

Oxytocin and vasopressin are typical **neural hormones,** ie, hormones secreted into the circulation by nerve cells. This type of neural regulation is compared with other types in Fig 14–10. The term **neurosecretion** was originally coined to describe the secretion of hormones by neurons, but the term is somewhat misleading, because it appears that all neurons secrete chemical messengers (see Chapter 1).

Like other peptide hormones, the posterior lobe hormones are synthesized as part of larger precursor molecules. Vasopressin and oxytocin each have a characteristic **neurophysin** associated with them in the granules in the neurons that secrete them, neurophysin I in the case of oxytocin and neurophysin II

in the case of vasopressin. The neurophysins were originally thought to be binding polypeptides, but it now appears that they are simply parts of the precursor molecules. The precursor for arginine vasopressin, **prepropressophysin,** contains a leader sequence of 19 amino acid residues followed by arginine vasopressin, vasopressin neurophysin, and a glycopeptide (Fig 14–11). **Prepro-oxyphysin,** the precursor for oxytocin, is a similar but smaller molecule that lacks the glycopeptide.

The precursor molecules are synthesized in the ribosomes of the cell bodies of the neurons. They have their leader sequences removed in the endoplasmic reticulum, are packaged into secretory granules in the Golgi apparatus, and are transported down the axons by axoplasmic flow to the endings in the posterior pituitary. The secretory granules, called **Herring bodies,** are easy to stain in tissue sections, and they have been extensively studied. Cleavage of the precursor molecules occurs as they are being transported, and the storage granules in the endings contain free vasopressin or oxytocin and the corresponding neurophysin. In the case of vasopressin, the glycopeptide is also present. All these products are secreted, but the functions of the components other than the established posterior pituitary hormones are unknown.

Electrical Activity of Magnocellular Neurons

The oxytocin-secreting and vasopressin-secreting neurons also generate and conduct action potentials, and action potentials reaching their endings trigger release of hormone from them by Ca^{2+}-dependent exocytosis. At least in anesthetized rats, these neurons are silent at rest or discharge at low, irregular rates (0.1–3 spikes per second). However, their response to stimulation varies (Fig 14–12). Stimulation of the nipples causes a synchronous, high-frequency discharge of the oxytocin neurons after an appreciable

Figure 14–10. Neural control mechanisms. In the 2 situations on the left, neurotransmitters act at nerve endings on muscle; in the 2 in the middle, neurotransmitters regulate the secretion of endocrine glands; and in the 2 on the right, neurons secrete hormones into the hypophyseal portal or general circulation.

①	Signal Peptide	19 aa		①	Signal Peptide	19 aa
②	Vasopressin	9 aa		②	Oxytocin	9 aa
③	Neurophysin II	95 aa		③	Neurophysin I	93 aa
④	Glycopeptide	39 aa				

-Gly-Lys-Arg-   -Arg-   -Gly-Lys-Arg-   -Arg/His

Figure 14–11. Structure of bovine prepropressophysin *(left)* and prepro-oxyphysin *(right)*. Gly in the 10 position of both peptides is necessary for amidation of the Gly residue in position 9. (Reproduced, with permission, from Richter D: Molecular events in expression of vasopressin and oxytocin and their cognate receptors. *Am J Physiol* 1988;**255**:F207.)

Figure 14–12. Responses of magnocellular neurons to stimulation. The tracings show individual extracellularly recorded action potentials, discharge rates, and intramammary duct pressure. **A:** Response of an oxytocin-secreting neuron. HFD, high-frequency discharge; ME, milk ejection. Stimulation of nipples started before the onset of recording. **B:** Responses of a vasopressin-secreting neuron, showing no change in the slow firing rate in response to stimulation of nipples and a prompt increase in the firing rate when 5 mL of blood was drawn, followed by typical phasic discharge. (Tracings from Wakerly JB: Hypothalamic neurosecretory function: Insights from electrophysiological studies of the magnocellular nuclei. *IBRO News* 1985;**4**:15.)

latency. This discharge causes release of a pulse of oxytocin and consequent milk ejection (see below). On the other hand, stimulation of the vasopressin-secreting neurons by a stimulus such as hemorrhage causes an initial steady increase in firing rate followed by a prolonged pattern of phasic discharge in which periods of a high-frequency discharge alternate with periods of electrical quiescence (**phasic bursting**). These phasic bursts are generally not synchronous in different vasopressin-secreting neurons. They are well suited to maintain a prolonged increase in the output of vasopressin, as opposed to the synchronous, relatively short, high-frequency discharge of oxytocin-secreting neurons in response to stimulation of the nipples.

Vasopressin & Oxytocin in Other Locations

Vasopressin-secreting neurons are found in the suprachiasmatic nuclei, and vasopressin and oxytocin are also found in the endings of neurons that project from the paraventricular nuclei to the brain stem and spinal cord. These neurons appear to be involved in cardiovascular control. In addition, vasopressin and oxytocin are synthesized in the gonads and the adrenal cortex and there is oxytocin in the thymus. The functions of the peptides in these organs are unknown.

Effects of Vasopressin

Because one of its principal physiologic effects is the retention of water by the kidney, vasopressin is often called the **antidiuretic hormone (ADH).** It increases the permeability of the collecting ducts of the kidney, so that water enters the hypertonic interstitium of the renal pyramids (see Chapter 38). The urine becomes concentrated, and its volume decreases. The overall effect is therefore retention of water in excess of solute; consequently, the effective osmotic pressure of the body fluids is decreased. In the absence of vasopressin, the urine is hypotonic to plasma, urine volume is increased, and there is a net water loss. Consequently, the osmolality of the body fluids rises.

In large doses, vasopressin elevates arterial blood pressure by an action on the smooth muscle of the arterioles, but the amount of endogenous vasopressin in the circulation of normal individuals does not normally affect blood pressure. However, vasopressin is a potent vasoconstrictor in vitro, and the reason that it does not increase blood pressure when small doses are injected in vivo is that it also acts on the brain to cause a decrease in cardiac output. The site of this action is the area postrema, one of the circumventricular organs (see Chapter 32). Hemorrhage is a potent stimulus to vasopressin secretion. There is controversy about whether the blood pressure fall after hemorrhage is significantly greater or more prolonged when the posterior pituitary is absent than when it is present, but the fall is more marked in animals that have been treated with synthetic peptides which block the pressor action of vasopressin. Consequently, it appears that vasopressin does play a role in blood pressure homeostasis. The vasculature may be hyperresponsive to vasopressin in some forms of hypertension.

Vasopressin also causes glycogenolysis in the liver and, as noted above, is a putative neurotransmitter in the brain. In addition, it increases ACTH secretion from the corticotropes of the anterior pituitary gland.

Vasopressin Receptors & Mechanism of Action

There are at least 3 kinds of vasopressin receptors: V_{1A}, V_{1B}, and V_2. All are G protein-coupled. The V_{1A} and V_2 receptors have been cloned and are typical serpentine receptors (see Chapter 1). The V_{1A} and V_{1B} receptors act through phosphatidylinositol hydrolysis to increase the intracellular Ca^{2+}. The V_2 receptors act through G_s to increase cyclic GMP levels.

The V_{1A} receptors mediate the vasoconstrictor action of vasopressin. They are found in blood vessels, including renal vessels, in glomerular mesangial cells, and in the liver and brain.

The V_{1B} receptors are the receptors in the anterior pituitary gland that mediate vasopressin-induced increases in ACTH secretion.

The antidiuretic effect of vasopressin is mediated by V_2 receptors that are found in the nephrons on the blood side of the tubular cells in the thick ascending limb of Henle and the collecting duct (see Chapter 38). They activate adenylate cyclase, and in the collecting ducts, the resulting increase in cyclic AMP produces a striking increase in the permeability of the membrane on the luminal side of the cells to water, urea, and some other solutes. These cells contain endosomes with protein water channels in their membrane, and when intracellular cyclic AMP increases, the endosomes fuse with the luminal cell membrane, increasing its area and inserting many water channels. When the effect of vasopressin is over, the inserted channel-rich membrane is endocytosed and stored again in the endosomes. V_2 receptors are also found in blood vessels, where their function is uncertain.

Synthetic Agonists & Antagonists

Synthetic peptides that have selective actions and are more active than naturally occurring vasopressin and oxytocin have been produced by altering the amino acid residues. For example, 1-deamino-8-D-arginine vasopressin (DDAVP, desmopressin) has very high antidiuretic activity with little pressor activity, making it valuable in the treatment of vasopressin deficiency (see below). Antagonists that selectively block the pressor or antidiuretic activity of vasopressin have also been synthesized.

Metabolism

Circulating vasopressin is rapidly inactivated, principally in the liver and kidneys. It has a **biologic half-life** (time required for inactivation of half a given amount) of approximately 18 minutes in humans. Its effects on the kidney develop rapidly but are of short duration.

Control of Vasopressin Secretion: Osmotic Stimuli

Vasopressin is stored in the posterior pituitary and released into the bloodstream by impulses in the nerve fibers that contain the hormone. The factors affecting its secretion are summarized in Table 14–3. When the effective osmotic pressure of the plasma is increased above the normal 285 mosm/kg, the rate of discharge of these neurons increases and vasopressin secretion is increased (Fig 14–13). At 285 mosm/kg, plasma vasopressin is at or near the limits of detection by available assays, but there is probably a fur-

Table 14–3. Summary of stimuli affecting vasopressin secretion.

Vasopressin Secretion Increased	Vasopressin Secretion Decreased
Increased effective osmotic pressure of plasma	Decreased effective osmotic pressure of plasma
Decreased extracellular fluid volume	Increased extracellular fluid volume
Pain, emotion, "stress," exercise	Alcohol
Nausea and vomiting	Butorphanol, oxilorphan
Standing	
Morphine, nicotine, barbiturates	
Chlorpropamide, clofibrate, carbamazepine	
Angiotensin II	

Figure 14–13. Relation between plasma osmolality and plasma vasopressin in healthy adult humans during infusion of hypertonic saline. LD, limit of detection. (Reproduced, with permission, from Thompson CJ et al: The osmotic thresholds for thirst and vasopressin are similar in healthy man. *Clin Sci Lond* 1986;**71**:651.)

ther decrease when plasma osmolality is below this level. Vasopressin secretion is regulated by osmoreceptors located in the anterior hypothalamus. They are outside the blood-brain barrier and appear to be located in the circumventricular organs, primarily the organum vasculosum of the lamina terminalis (see Chapter 32). The osmotic threshold for thirst (Fig 14–7) is the same as or slightly greater than the threshold for increased vasopressin secretion (Fig 14–13), and it is still uncertain whether the same osmoreceptors mediate both effects.

Vasopressin secretion is thus controlled by a delicate feedback mechanism that operates continuously to defend the osmolality of the plasma. Significant changes in secretion occur when osmolality is changed as little as 1%. In this way, the osmolality of the plasma in normal individuals is maintained very close to 285 mosm/L.

Volume Effects

ECF volume also affects vasopressin secretion. Vasopressin secretion is increased when ECF volume is low and decreased when ECF volume is high (Table 14–3). There is an inverse relationship between the rate of vasopressin secretion and the rate of discharge in afferents from stretch receptors in the low- and high-pressure portions of the vascular system. The low-pressure receptors are those in the great veins, right and left atria, and pulmonary vessels; the high-pressure receptors are those in the carotid

sinuses and aortic arch (see Chapter 31). The exponential increases in plasma vasopressin produced by decreases in blood pressure are documented in Fig 14–14. However, the low-pressure receptors monitor the fullness of the vascular system, and moderate decreases in blood volume that decrease central venous pressure without lowering arterial pressure can also increase plasma vasopressin.

Thus, the low-pressure receptors are the primary mediators of volume effects on vasopressin secretion. Angiotensin II reinforces the response to hypovolemia and hypotension by acting on the brain to increase vasopressin secretion.

Hypovolemia and hypotension produced by conditions such as hemorrhage release large amounts of vasopressin, and in the presence of hypovolemia, the osmotic response curve is shifted to the left (Fig 14–15). Its slope is also increased. The result is water retention and reduced plasma osmolality. This includes hyponatremia, since Na+ is the most abundant osmotically active component of the plasma.

Other Stimuli Affecting Vasopressin Secretion

A variety of stimuli in addition to osmotic pressure changes and ECF volume aberrations increase vasopressin secretion. These include pain, nausea, surgical stress, some emotions, and a number of drugs (Table 14–3), including morphine, nicotine,

Figure 14–14. Relation of mean arterial blood pressure to plasma vasopressin in healthy adult humans in whom a progressive decline in blood pressure was induced by infusion of graded doses of the ganglionic blocking drug trimethaphan. The relation is exponential rather than linear. (Drawn from data in Baylis PH: Osmoregulation and control of vasopressin secretion in healthy humans. *Am J Physiol* 1987;**253**:R671.)

Figure 14–15. Effect of hypovolemia and hypervolemia on the relation between plasma vasopressin (pAVP) and plasma osmolality (pOsm). Seven blood samples were drawn at various times from 10 normal men when hypovolemia was induced by water deprivation (open circles, dashed line) and again when hypervolemia was induced by infusion of hypertonic saline (solid circles, solid line). Linear regression analysis defined the relationship pAVP = 0.52 (pOsm − 283.5) for water deprivation and pAVP = 0.38 (pOsm − 285.6) for hypertonic saline. LD, limit of detection. Note the steeper curve as well as the shift of the intercept to the left during hypovolemia. (Courtesy of CJ Thompson.)

and large doses of barbiturates. Nausea is associated with particularly large increases in vasopressin secretion. Alcohol and the opiate antagonists butorphanol and oxilorphan decrease vasopressin secretion.

Clinical Implications

In various clinical conditions, volume and other nonosmotic stimuli bias the osmotic control of vasopressin secretion. For example, patients who have had surgery may have elevated levels of plasma vasopressin because of pain and hypovolemia, and this may cause them to develop a low plasma osmolality and dilutional hyponatremia. In patients with hypersecretion of vasopressin, a high fluid intake can cause water intoxication. Water retention occurs in addition to salt retention in patients with edema due to congestive heart failure, cirrhosis of the liver, or nephrosis. The cause of excess vasopressin secretion in diseases of the heart, liver, and kidneys remains a matter of debate.

In the **syndrome of "inappropriate" hypersecretion of antidiuretic hormone (SIADH),** vasopressin is responsible not only for dilutional hyponatremia but also for loss of salt in the urine when

water retention is sufficient to expand the ECF volume, reducing aldosterone secretion (see Chapter 20). This occurs in patients with cerebral disease ("cerebral salt wasting") and pulmonary disease ("pulmonary salt wasting"). Hypersecretion of vasopressin in patients with pulmonary diseases such as lung cancer may be due in part to the interruption of inhibitory impulses in vagal afferents from the stretch receptors in the atria and great veins. However, a significant number of lung tumors and some other cancers secrete vasopressin. Patients with inappropriate hypersecretion of vasopressin have been successfully treated with demeclocycline, an antibiotic that reduces the renal response to vasopressin. If the increased amounts of vasopressin do not come from a tumor, secretion can be reduced by treatment with butorphanol and oxilorphan.

Diabetes insipidus is the syndrome that results when vasopressin deficiency develops because of disease processes in the supraoptic and paraventricular nuclei, the hypothalamohypophyseal tract, or the posterior pituitary gland. It has been estimated that 30% of the clinical cases are due to neoplastic lesions of the hypothalamus, either primary or metastatic; 30% are posttraumatic; 30% are idiopathic; and the remaining 10% are due to vascular lesions, infections, or systemic diseases such as sarcoidosis which affect the hypothalamus. The disease that develops after surgical removal of the posterior lobe of the pituitary may be temporary if only the distal ends of the supraoptic and paraventricular fibers are damaged, because the fibers recover, make new vascular connections, and begin to secrete vasopressin again. The symptoms of diabetes insipidus are passage of large amounts of dilute urine (**polyuria**) and the drinking of large amounts of fluid (**polydipsia**), provided the thirst mechanism is intact. It is the polydipsia that keeps these patients healthy. If their sense of thirst is depressed for any reason and their intake of dilute fluid decreases, they develop dehydration that can be fatal. In patients who have an incomplete rather than total vasopressin deficiency, treatment with drugs that increase vasopressin secretion, such as clofibrate, has proved to be of value. Chlorpropamide is also of value because it increases the renal response to vasopressin.

Another cause of diabetes insipidus is inability of the kidneys to respond to vasopressin. In this condition, which is known as **nephrogenic diabetes insipidus,** vasopressin fails to increase renal cyclic AMP.

The amelioration of diabetes insipidus produced by the development of concomitant anterior pituitary insufficiency is discussed in Chapter 22.

Effects of Oxytocin

Oxytocin acts primarily on the breasts and uterus, although it may also be involved in luteolysis (see Chapter 23). A G protein-coupled serpentine oxytocin receptor has been cloned from human myometrium, and a similar or identical receptor is found in mammary tissue and the ovary. It triggers increases in intracellular Ca^{2+} levels.

In mammals, oxytocin causes contraction of the **myoepithelial cells,** smooth-muscle-like cells that line the ducts of the breast. This squeezes the milk out of the alveoli of the lactating breast into the large ducts (sinuses) and thence out of the nipple (**milk ejection**). Many hormones acting in concert are responsible for breast growth and the secretion of milk into the ducts (see Chapter 23), but milk ejection in most species requires oxytocin.

The Milk Ejection Reflex

Milk ejection is normally initiated by a neuroendocrine reflex. The receptors involved are the touch receptors, which are plentiful in the breast—especially around the nipple. Impulses generated in these receptors are relayed from the somatic touch pathways to the supraoptic and paraventricular nuclei. Discharge of the oxytocin-containing neurons causes secretion of oxytocin from the posterior pituitary (Fig 14–12). The infant suckling at the breast stimulates the touch receptors, the nuclei are stimulated, oxytocin is released, and the milk is expressed into the sinuses, ready to flow into the mouth of the waiting infant. In lactating women, genital stimulation and emotional stimuli also produce oxytocin secretion, sometimes causing milk to spurt from the breasts.

Other Actions of Oxytocin

Oxytocin causes contraction of the smooth muscle of the uterus. The sensitivity of the uterine musculature to oxytocin is enhanced by estrogen and inhibited by progesterone. In late pregnancy, the uterus becomes very sensitive to oxytocin coincident with a marked increase in the number of oxytocin receptors and oxytocin receptor mRNA (see Chapter 23). Oxytocin secretion is increased during labor. After dilation of the cervix, descent of the fetus down the birth canal initiates impulses in the afferent nerves that are relayed to the supraoptic and paraventricular nuclei, causing secretion of sufficient oxytocin to enhance labor (see Fig 23–40). Plasma oxytocin is normal at the onset of labor, but it is possible that the marked increase in oxytocin receptors at this time causes normal oxytocin levels to initiate contractions, setting up a positive feedback (see Chapter 23).

Oxytocin may also act on the nonpregnant uterus to facilitate sperm transport. The passage of sperm up the female genital tract to the uterine tubes, where fertilization normally takes place, depends not only on the motile powers of the sperm but also, at least in some species, on uterine contractions. The genital stimulation involved in coitus releases oxytocin, but it has not been proved that it is oxytocin which initiates the rather specialized uterine contractions that transport the sperm. The secretion of oxytocin is in-creased by stressful stimuli and, like that of vasopressin, is inhibited by alcohol.

Circulating oxytocin increases at the time of ejaculation in males, and it is possible that this increase causes increased contraction of the smooth muscle of the vas deferens, propelling sperm toward the urethra.

CONTROL OF ANTERIOR PITUITARY SECRETION

Anterior Pituitary Hormones

The anterior pituitary secretes 6 hormones: **adrenocorticotropic hormone (corticotropin, ACTH), thyroid-stimulating hormone (thyrotropin, TSH), growth hormone, follicle-stimulating hormone (FSH), luteinizing hormone (LH),** and **prolactin (luteotropic hormone, LTH).** An additional polypeptide, β-lipotropin (β-LPH), is secreted with ACTH, but its physiologic role is unknown. The actions of these hormones are summarized in Fig 14–16, and synonyms and abbreviations for them are listed in Table 22–1. The hormones are discussed in detail in the chapters on the endocrine system. The hypothalamus plays an important stimulatory role in regulating the secretion of ACTH, β-LPH, TSH, growth hormone, FSH, and LH. It also regulates prolactin secretion, but its effect is predominantly inhibitory rather than stimulatory.

Nature of Hypothalamic Control

Anterior pituitary secretion is controlled by chemical agents carried in the portal hypophyseal vessels from the hypothalamus to the pituitary. These substances used to be called releasing and inhibiting factors, but now they are commonly called **hypophyseotropic hormones.** The latter term seems appropriate since they are secreted into the bloodstream and act at a distance from their site of origin. They do not escape into the general circulation to any degree, but they are in high concentration in portal hypophyseal blood. Consequently, the term hormone is used in this book.

Hypophyseotropic Hormones

There are 6 established hypothalamic releasing and inhibiting hormones (Fig 14–17): **corticotropin-releasing hormone (CRH); thyrotropin-releasing hormone (TRH); growth hormone-releasing hormone (GRH); growth hormone-inhibiting hormone (GIH;** also called **somatostatin); luteinizing hormone-releasing hormone (LHRH),** also known as **gonadotropin-releasing hormone (GnRH);** and **prolactin-inhibiting hormone (PIH).** In addition, hypothalamic extracts contain prolactin-releasing activity, and a **prolactin-releasing hormone (PRH)** has been postulated to exist. TRH, VIP, and several peptides found in the hypothalamus stimulate prolac-

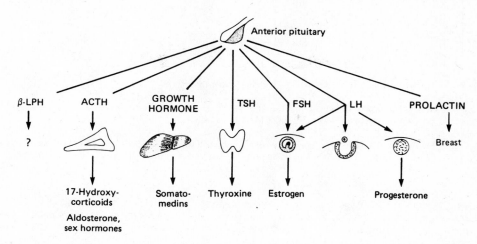

Figure 14–16. Anterior pituitary hormones. In women, FSH and LH act in sequence on the ovary to produce growth of the ovarian follicle, ovulation, and formation and maintenance of the corpus luteum. In men, FSH and LH control the functions of the testes (see Chapter 23). Prolactin stimulates lactation.

tin secretion, but it is uncertain whether one or more of these peptides or an as yet unisolated substance is the physiologic PRH. GnRH stimulates the secretion of FSH as well as that of LH, and it seems unlikely that there is a separate follicle-stimulating hormone-releasing hormone (FRH). Consequently, the term GnRH seems more appropriate than LHRH.

The structures of the 6 hypophyseotropic hormones are shown in Fig 14–18. The structures of the preprohormones for TRH, GnRH, somatostatin, CRH, and GRH are known. PreproTRH contains 6 copies of TRH (see Fig 1–18). Several other preprohormones may contain other hormonally active peptides in addition to the hypophyseotropic hormones.

The area from which the hypothalamic releasing and inhibiting hormones are secreted is the median eminence of the hypothalamus. This region contains few nerve cell bodies, but there are many nerve endings in close proximity to the capillary loops from which the portal vessels originate.

The locations of the cell bodies of the neurons that project to the external layer of the median eminence and secrete the hypophyseotropic hormones are shown in Fig 14–19, which also shows the location of the neurons secreting oxytocin and vasopressin. The GnRH-secreting neurons are primarily in the medial preoptic area, the somatostatin-secreting neurons are in the periventricular nuclei, the TRH-secreting and CRH-secreting neurons are in the medial parts of the paraventricular nuclei, and the GRH-secreting and dopamine-secreting neurons are in the arcuate nuclei.

Most, if not all, of the hypophyseotropic hormones affect the secretion of more than one anterior pituitary hormone (Fig 14–17). The FSH-stimulating activity of GnRH has been mentioned above. TRH stimulates the secretion of prolactin as well as TSH. Somatostatin inhibits the secretion of TSH as well as growth hormone. It does not normally inhibit the se-

Figure 14–17. Effects of hypophyseotropic hormones on the secretion of anterior pituitary hormones.

TRH	(pyro)Glu-His-Pro-NH₂
GnRH	(pyro)Glu-His-Trp-Ser-Tyr-Gly-Leu-Arg-Pro-Gly-NH₂

	┌S————————————————S┐
Somatostatin	Ala-Gly-Cys-Lys-Asn-Phe-Phe-Trp-Lys-Thr-Phe-Thr-Ser-Cys

CRH	Ser-Glu-Glu-Pro-Pro-Ile-Ser-Leu-Asp-Leu-Thr-Phe-His-Leu-Leu-Arg-Glu-Val-Leu-Glu-Met-Ala-Arg-Ala-Glu-Gln-Leu-Ala-Gln-Gln-Ala-His-Ser-Asn-Arg-Lys-Leu-Met-Glu-Ile-Ile-NH₂
GRH	Tyr-Ala-Asp-Ala-Ile-Phe-Thr-Asn-Ser-Tyr-Arg-Lys-Val-Leu-Gly-Gln-Leu-Ser-Ala-Arg-Lys-Leu-Leu-Gln-Asp-Ile-Met-Ser-Arg-Gln-Gln-Gly-Glu-Ser-Asn-Gln-Glu-Arg-Gly-Ala-Arg-Ala-Arg-Leu-NH₂
PIH	Dopamine

Figure 14–18. Structure of hypophyseotropic hormones in humans. The structure of somatostatin that is shown is the tetradecapeptide (somatostatin 14, SS14). In addition, preprosomatostatin is the source of an N-terminal extended polypeptide containing 28 amino acid residues (SS28) and a polypeptide containing 12 amino acid residues [SS28 (1–12)] that are found in tissues.

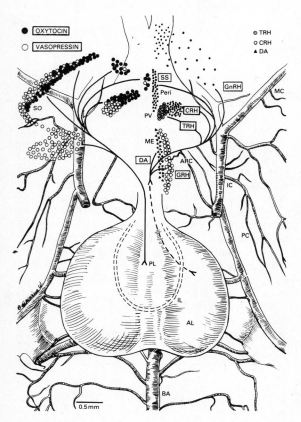

Figure 14–19. Location of cell bodies of hypophyseotropic hormone-secreting neurons projected on a ventral view of the hypothalamus and pituitary of the rat. AL, anterior lobe; ARC, arcuate nucleus; BA, basilar artery; IC, internal carotid; IL, intermediate lobe; MC, middle cerebral; ME, median eminence; PC, posterior cerebral; Peri, periventricular nucleus; PL, posterior lobe; PV, paraventricular nucleus; SO, supraoptic nucleus. The names of the hormones are enclosed in the boxes. (Courtesy of LW Swanson and ET Cunningham Jr.)

cretion of the other anterior pituitary hormones, but it inhibits the abnormally elevated secretion of ACTH in patients with Nelson's syndrome. CRH stimulates the secretion of ACTH and β-LPH (see Chapter 22).

Hypophyseotropic hormones function as neurotransmitters in other parts of the brain, the retina, and the autonomic nervous system (see Chapter 4). In addition, somatostatin is found in the pancreas (see Chapter 19), GRH is secreted by pancreatic tumors, and somatostatin and TRH are found in the gastrointestinal tract (see Chapter 26).

Significance & Clinical Implications

Research delineating the multiple neuroendocrine regulatory functions of the hypothalamus is important because it explains how endocrine secretion is made appropriate to the demands of a changing environment. The nervous system receives information about changes in the internal and external environment from the sense organs. It brings about adjustments to these changes through effector mechanisms that include not only somatic movement but also changes in the rate at which hormones are secreted. There are many examples of the operation of the neuroendocrine effector pathways. In birds and many mammals, the increasing number of daylight hours in the spring stimulates gonadotropin secretion, activating the quiescent gonads and starting the breeding season. In some avian species, the sight of a member of the opposite sex in a mating dance apparently produces the gonadotropin secretion necessary to cause ovulation. It has been claimed that Eskimo women cease to ovulate during the long winter night and that regular menstrual cycles and sexual vigor return in the spring. In more temperate climates, menstruation in women is a year-round phenomenon, but its regularity can be markedly affected by somatic and emotional stimuli. The disappearance of regular periods when young girls move away from home (sometimes called "boarding school amenorrhea") and the inhibition of menstruation that can be produced by fear of pregnancy are examples of the many ways in which psychic phenomena affect endocrine secretion.

The manifestations of hypothalamic disease are neurologic defects, endocrine changes, and metabolic abnormalities such as hyperphagia and hyperthermia. The relative frequencies of the signs and symptoms of hypothalamic disease in one large series of cases are shown in Table 14–4. The possibility of hypothalamic pathology should be kept in mind in evaluating all patients with pituitary dysfunction, especially those with isolated deficiencies of single pituitary tropic hormones. Many of these deficiencies are proving to be due to hypothalamic rather than primary pituitary disease.

TEMPERATURE REGULATION

In the body, heat is produced by muscular exercise, assimilation of food, and all the vital processes that contribute to the basal metabolic rate (see Chapter 17). It is lost from the body by radiation, conduction, and vaporization of water in the respiratory passages and on the skin. Small amounts of heat are also removed in the urine and feces. The balance between heat production and heat loss determines the body temperature. Because the speed of chemical reactions varies with the temperature and because the enzyme systems of the body have narrow temperature ranges in which their function is optimal, normal body function depends upon a relatively constant body temperature.

Table 14–4. Symptoms and signs in 60 autopsied patients with hypothalamic disease.[1]

Symptoms and Signs	Percentage of Cases
Endocrine and metabolic findings	
Precocious puberty	40
Hypogonadism	32
Diabetes insipidus	35
Obesity	25
Abnormalities of temperature regulation	22
Emaciation	18
Bulimia	8
Anorexia	7
Neurologic findings	
Eye signs	78
Pyramidal and sensory deficits	75
Headache	65
Extrapyramidal signs	62
Vomiting	40
Psychic disturbances, rage attacks, etc	35
Somnolence	30
Convulsions	15

[1]Data from Bauer HG: Endocrine and other clinical manifestations of hypothalamic disease. *J Clin Endocrinol* 1954;**14**:13. See also Kahana L et al: Endocrine manifestations of intracranial extrasellar lesions. *J Clin Endocrinol* 1962;**22**:304.

Invertebrates generally cannot adjust their body temperatures and so are at the mercy of the environment. In vertebrates, mechanisms for maintaining body temperature by adjusting heat production and heat loss have evolved. In reptiles, amphibia, and fish, the adjusting mechanisms are relatively rudimentary, and these species are called "cold-blooded" (**poikilothermic**) because their body temperature fluctuates over a considerable range. In birds and mammals, the "warm-blooded" (**homeothermic**) animals, a group of reflex responses that are primarily integrated in the hypothalamus operate to maintain body temperature within a narrow range in spite of wide fluctuations in environmental temperature. The hibernating mammals are a partial exception. While awake, they are homeothermic, but during hibernation, their body temperature falls.

Normal Body Temperature

In homeothermic animals, the actual temperature at which the body is maintained varies from species to species and, to a lesser degree, from individual to individual. In humans, the traditional normal value for the oral temperature is 37 °C (98.6 °F), but in one large series of normal young adults, the morning oral temperature averaged 36.7 °C, with a standard deviation of 0.2 °C. Therefore, 95% of all young adults would be expected to have a morning oral temperature of 36.3–37.1 °C (97.3–98.8 °F) (mean ± 1.96 standard deviations; see Appendix). Various parts of the body are at different temperatures, and the magnitude of the temperature difference between the parts varies with the environmental temperature (Fig 14–20). The extremities are generally cooler than the rest of the body. The temperature of the scrotum is carefully regulated at 32 °C. The rectal temperature is representative of the temperature at the core of the body and varies least with changes in environmental temperature. The oral temperature is normally 0.5 °C lower than the rectal temperature, but it is affected by many factors, including ingestion of hot or cold fluids, gum-chewing, smoking, and mouth breathing.

The normal human core temperature undergoes a regular circadian fluctuation of 0.5–0.7 °C. In individuals who sleep at night and are awake during the day (even when hospitalized at bed rest), it is lowest at about 6:00 AM and highest in the evenings (Fig 14–21). It is lowest during sleep, is slightly higher in the awake but relaxed state, and rises with activity. In women, there is an additional monthly cycle of temperature variation characterized by a rise in basal temperature at the time of ovulation (see Chapter 23 and Fig 23–29). Temperature regulation is less precise in young children, and they may normally have a temperature that is 0.5 °C or so above the established norm for adults.

During exercise, the heat produced by muscular contraction accumulates in the body, and the rectal temperature normally rises as high as 40 °C (104 °F).

Figure 14–20. Temperatures of various parts of the body of a naked subject at various ambient temperatures in a calorimeter. (Redrawn and reproduced, with permission, from Hardy JD, DuBois EF: Basal metabolism, radiation, convection and vaporization at temperatures of 22–35 °C. *J Nutr* 1938;**15**:477.)

This rise is due in part to the inability of the heat-dissipating mechanisms to handle the greatly increased amount of heat produced, but there is evidence that in addition there is an elevation of the body temperature at which the heat-dissipating mechanisms are activated during exercise. Body temperature also rises slightly during emotional excitement, probably owing to unconscious tensing of the muscles (Fig 14–22). It is chronically elevated by as much as 0.5 °C when the metabolic rate is high, as in hyperthyroidism, and lowered when the metabolic rate is low, as in hypothyroidism (Fig 14–21). Some apparently normal adults chronically have a temperature

Figure 14–21. Typical temperature chart of a hospitalized patient who does not have a febrile disease. Note the slight rise in temperature, due to excitement and apprehension, at the time of admission to the hospital, and the regular circadian temperature cycle.

Figure 14–22. Ranges in rectal and oral temperatures seen in normal humans. (Reproduced, with permission, from DuBois EF: *Fever and the Regulation of Body Temperature.* Thomas, 1948.)

above the normal range (constitutional hyperthermia).

Heat Production

Heat production and energy balance are discussed in Chapter 17. A variety of basic chemical reactions contribute to body heat production at all times. Ingestion of food increases heat production because of the specific dynamic action of the food (see Chapter 17), but the major source of heat is the contraction of skeletal muscle (Table 14–5). Heat production can be varied by endocrine mechanisms in the absence of food intake or muscular exertion. Epinephrine and norepinephrine produce a rapid but short-lived increase in heat production; thyroid hormones produce a slowly developing but prolonged increase. Furthermore, sympathetic discharge is decreased during fasting and increased by feeding (see Chapter 13).

A source of considerable heat, particularly in infants, is **brown fat** (see above and Chapter 17). This fat has a high rate of metabolism, and its thermogenic function has been likened to that of an electric blanket.

Heat Loss

The processes by which heat is lost from the body when the environmental temperature is below body temperature are listed in Table 14–5. **Conduction** is

Table 14–5. Body heat production and heat loss.

Body heat is produced by:	
Basic metabolic processes	
Food intake (specific dynamic action)	
Muscular activity	

Body heat is lost by:	Percentage of heat lost at 21°C
Radiation and conduction	70
Vaporization of sweat	27
Respiration	2
Urination and defecation	1

heat exchange between objects or substances at different temperatures that are in contact with one another. A basic characteristic of matter is that its molecules are in motion, with the amount of motion proportionate to the temperature. These molecules collide with the molecules in cooler objects, transferring thermal energy to them. The amount of heat transferred is proportionate to the temperature difference between the objects in contact (**thermal gradient**). Conduction is aided by **convection,** the movement of molecules away from the area of contact. Thus, for example, an object in contact with air at a different temperature changes the specific gravity of the air, and since warm air rises and cool air falls, a new supply of air is brought into contact with the object. Of course, convection is greatly aided if the object moves about in the medium or the medium moves past the object, eg, if a subject swims through water or a fan blows air through a room. **Radiation** is transfer of heat by infrared electromagnetic radiation from one object to another at a different temperature with which it is not in contact. When an individual is in a cold environment, heat is lost by conduction to the surrounding air and by radiation to cool objects in the vicinity. Conversely, of course, heat is transferred to an individual and the heat load is increased by these processes when the environmental temperature is above body temperature. Note that because of radiation, an individual can feel chilly in a room with cold walls even though the room is relatively warm. On a cold but sunny day, the heat of the sun reflected off bright objects exerts an appreciable warming effect. It is the heat reflected from the snow, for example, that makes it possible to ski in fairly light clothes even though the air temperature is below freezing.

Since conduction occurs from the surface of one object to the surface of another, the temperature of the skin determines to a large extent the degree to which the body heat is lost or gained. The amount of heat reaching the skin from the deep tissues can be varied by changing the blood flow to the skin. When the cutaneous vessels are dilated, warm blood wells into the skin, whereas in the maximally vasoconstricted state, heat is held centrally in the body. The rate at which heat is transferred from the deep tissues to the skin is called the **tissue conductance.** Birds have a layer of feathers next to the skin, and most mammals have a significant layer of hair or fur. Heat is conducted from the skin to the air trapped in this layer and from the trapped air to the exterior. When the thickness of the trapped layer is increased by fluffing the feathers or erection of the hairs (**horripilation**), heat transfer across the layer is reduced and heat losses (or, in a hot environment, heat gains) are decreased. "Goose pimples" are the result of horripilation in humans; they are the visible manifestation of cold-induced contraction of the piloerector muscles attached to the rather meager hair supply. Humans usually supplement this layer of hair with

one or more layers of clothes. Heat is conducted from the skin to the layer of air trapped by the clothes, from the inside of the clothes to the outside, and from the outside of the clothes to the exterior. The magnitude of the heat transfer across the clothing, a function of its texture and thickness, is the most important determinant of how warm or cool the clothes feel, but other factors, especially the size of the trapped layer of warm air, are important also. Dark clothes absorb radiated heat, and light-colored clothes reflect it back to the exterior.

The other major process transferring heat from the body in humans and other animals that sweat is vaporization of water on the skin and mucous membranes of the mouth and respiratory passages. Vaporization of 1 g of water removes about 0.6 kcal of heat. A certain amount of water is vaporized at all times. This **insensible water loss** amounts to 50 mL/h in humans. When sweat secretion is increased, the degree to which the sweat vaporizes depends upon the humidity of the environment. It is common knowledge that one feels hotter on a humid day. This is due in part to the decreased vaporization of sweat, but even under conditions in which vaporization of sweat is complete, an individual in a humid environment feels warmer than an individual in a dry environment. The reason for this difference is not known, but it seems related to the fact that in the humid environment sweat spreads over a greater area of skin before it evaporates. During muscular exertion in a hot environment, sweat secretion reaches values as high as 1600 mL/h, and in a dry atmosphere, most of this sweat is vaporized. Heat loss by vaporization of water therefore varies from 30 to over 900 kcal/h.

Some mammals lose heat by **panting.** This rapid, shallow breathing greatly increases the amount of water vaporization in the mouth and respiratory passages and therefore the amount of heat lost. Because the breathing is shallow, it produces relatively little change in the composition of alveolar air (see Chapter 34).

The relative contribution of each of the processes that transfer heat away from the body (Table 14–5) varies with the environmental temperature. At 21 °C, vaporization is a minor component in humans at rest. As the environmental temperature approaches body temperature, radiation losses decline and vaporization losses increase.

Temperature-Regulating Mechanisms

The reflex and semireflex thermoregulatory responses are listed in Table 14–6. They include autonomic, somatic, endocrine, and behavioral changes. One group of responses increases heat loss and decreases heat production; the other decreases heat loss and increases heat production. In general, exposure to heat stimulates the former group of responses and inhibits the latter, whereas exposure to cold does the opposite.

Table 14–6. Temperature-regulating mechanisms.

Mechanisms activated by cold:	
Shivering	
Hunger	Increase
Increased voluntary activity	heat
Increased secretion of norepinephrine and epinephrine	production
Cutaneous vasoconstriction	Decrease
Curling up	heat
Horripilation	loss
Mechanisms activated by heat:	
Cutaneous vasodilation	Increase
Sweating	heat
Increased respiration	loss
Anorexia	Decrease
Apathy and inertia	heat
	production

Curling up "in a ball" is a common reaction to cold in animals and has a counterpart in the position some people assume on climbing into a cold bed. Curling up decreases the body surface exposed to the environment. Shivering is an involuntary response of the skeletal muscles, but cold also causes a semiconscious general increase in motor activity. Examples include foot stamping and dancing up and down on a cold day. Increased catecholamine secretion is an important endocrine response to cold; adrenal medullectomized rats die faster than normal controls when exposed to cold. TSH secretion is increased by cold and decreased by heat in laboratory animals, but the change in TSH secretion produced by cold in adult humans is small and of questionable significance. It is common knowledge that activity is decreased in hot weather—the "it's too hot to move" reaction.

Thermoregulatory adjustments involve local responses as well as more general reflex responses. When cutaneous blood vessels are cooled, they become more sensitive to catecholamines and the arterioles and venules constrict. This local effect of cold directs blood away from the skin. Another heat-conserving mechanism that is important in animals living in cold water is heat transfer from arterial to venous blood in the limbs. The deep veins (**venae comitantes**) run alongside the arteries supplying the limbs, and heat is transferred from the warm arterial blood going to the limbs to the cold venous blood coming from the extremities (**counter current exchange;** see Chapter 38). This keeps the tips of the extremities cold but conserves body heat.

The reflex responses activated by cold are controlled from the posterior hypothalamus. Those activated by warmth are controlled primarily from the anterior hypothalamus, although some thermoregulation against heat still occurs after decerebration at the level of the rostral midbrain. Stimulation of the ante-rior hypothalamus causes cutaneous vasodilation and sweating, and lesions in this region cause hyperthermia, with rectal temperatures sometimes reaching 43 °C (109.4 °F). Posterior hypothalamic stimulation causes shivering, and the body temperature of animals with posterior hypothalamic lesions falls toward that of the environment.

There is some evidence in primates and humans that serotonin is a synaptic mediator in the centers controlling the mechanisms activated by cold and that norepinephrine plays a similar role in those activated by heat. However, there are marked species variations in the temperature responses to these amines. Peptides may also be involved, but the details of the central synaptic connections concerned with thermoregulation are still unknown.

Afferents

The signals that activate the hypothalamic temperature-regulating centers come from 2 sources: temperature-sensitive cells in the anterior hypothalamus, and cutaneous temperature receptors, especially cold receptors. Present evidence indicates that the stimuli which activate the defense against high temperatures in humans come mainly from the temperature-sensitive cells in the hypothalamus. This conclusion grew out of research in which the temperature of the back of the nasal cavity and the interior of the ear, near the hypothalamus, was correlated with thermoregulatory responses to changing environmental temperatures (Fig 14–23). The response of body heat production to cooling is modified by interactions between cutaneous and central stimuli. Heat production is increased when head temperature falls below a given threshold value, but the threshold for the response is lower and its magnitude decreased when the skin temperature is increased.

Fever

Fever is perhaps the oldest and most universally known hallmark of disease. It occurs not only in mammals but also in birds, reptiles, amphibia, and fish. When it occurs in homeothermic animals, the thermoregulatory mechanisms behave as if they were adjusted to maintain body temperature at a higher than normal level, ie, "as if the thermostat had been reset" to a new point above 37 °C. The temperature receptors then signal that the actual temperature is below the new set point, and the temperature-raising mechanisms are activated. This usually produces chilly sensations due to cutaneous vasoconstriction and occasionally enough shivering to produce a shaking chill. However, the nature of the response depends on the ambient temperature. The temperature rise in experimental animals injected with a pyrogen is due mostly to increased heat production if they are in a cold environment and mostly to decreased heat loss if they are in a warm environment.

The pathogenesis of fever is summarized in Fig

Figure 14–23. Quantitative relations in humans between the temperature of the interior of the head (internal temperature) and cutaneous blood flow (squares, scale at the right) and sweating (circles and triangles, scale at the left). The arrow points to the sharp threshold at which these parameters start to rise. In this subject, the threshold was at 36.9 °C. (Reproduced, with permission, from Benzinger TH: Receptor organs and quantitative mechanisms of human temperature control in a warm environment. *Fed Proc* 1960;**19**:32.)

Figure 14–24. Pathogenesis of fever.

14–24. Toxins from bacteria such as endotoxin act on monocytes, macrophages, and Kupffer cells to produce **interleukin-1 (IL-1;** see Chapter 27), a polypeptide also known as **endogenous pyrogen (EP).** IL-6 also causes fever. IL-1 has widespread effects in the body (see Chapter 27). It either enters the brain directly or, more probably, acts on the OVLT, one of the circumventricular organs. In any case, it activates the preoptic area of the hypothalamus. Diverse other agents, including the steroid etiocholanolone, a metabolite of testosterone (see Chapter 23), also cause the production of IL-1.

The fever produced by IL-1 may be due to local release of prostaglandins. Injection of prostaglandins into the hypothalamus produces fever. In addition, the antipyretic effect of aspirin is exerted directly on the hypothalamus, and aspirin inhibits prostaglandin synthesis. However, there is some evidence that the prostaglandins are not involved, and the question of their role in the production of fever is a matter of much debate.

The benefit of fever to the organism is uncertain. It is presumably beneficial, because it has evolved and persisted as a response to infections and other diseases. Many microorganisms grow best within a relatively narrow temperature range, and a rise in temperature inhibits their growth. In addition, antibody production is increased when body temperature is elevated. Before the advent of antibiotics, fevers were artificially induced for the treatment of neurosyphilis and proved to be beneficial. Hyperthermia benefits individuals infected with anthrax, pneumococcal pneumonia, leprosy, and various fungal, rickettsial, and viral diseases. Hyperthermia also slows the growth of some tumors. However, very high temperatures are harmful. When the rectal temperature is over 41 °C (106 °F) for prolonged periods, some permanent brain damage results. When it is over 43 °C, heat stroke develops, and death is common.

Hypothermia

In hibernating mammals, body temperature drops to low levels without causing any ill effects that are demonstrable upon subsequent arousal. This observation led to experiments on induced hypothermia. When the skin or the blood is cooled enough to lower the body temperature in nonhibernating animals and in humans, metabolic and physiologic processes slow down. Respiration and heart rate are very slow, blood pressure is low, and consciousness is lost. At rectal temperatures of about 28 °C, the ability to spontaneously return the temperature to normal is lost, but the individual continues to survive and, if rewarmed with external heat, returns to a normal state. If care is taken to prevent the formation of ice crystals in the tissues, the body temperature of experimental animals can be lowered to subfreezing levels without producing any damage that is detectable after subsequent rewarming.

Humans tolerate body temperatures of 21–24 °C (70–75 °F) without permanent ill effects, and induced hypothermia has been used extensively in surgery. In hypothermic patients, the circulation can be stopped for relatively long periods because the O_2 needs of the tissues are greatly reduced. Blood pressure is low, and bleeding is minimal. It is possible under hypothermia to stop and open the heart and to perform other procedures, especially brain operations, that would have been impossible without cooling.

Neural Basis of Instinctual Behavior & Emotions

15

INTRODUCTION

Emotions have both mental and physical components. They involve **cognition,** an awareness of the sensation and usually its cause; **affect,** the feeling itself; **conation,** the urge to take action; and **physical changes** such as hypertension, tachycardia, and sweating. The hypothalamus and limbic systems are intimately concerned with emotional expression and with the genesis of emotions.

This chapter reviews the physiologic basis of emotion, sexual behavior, fear, rage, and motivation. It also considers the relation of major neurotransmitter systems in the brain to these processes.

ANATOMIC CONSIDERATIONS

The term **limbic lobe** or **limbic system** is applied to the part of the brain that consists of a rim of cortical tissue around the hilus of the cerebral hemisphere and a group of associated deep structures—the amygdala, the hippocampus, and the septal nuclei (Figs 15–1 and 15–2). The region was formerly called the rhinencephalon because of its relation to olfaction, but only a small part of it is actually concerned with smell.

Histology

The limbic cortex is phylogenetically the oldest part of the cerebral cortex. Histologically, it is made up of a primitive type of cortical tissue called **allocortex,** surrounding the hilus of the hemisphere,

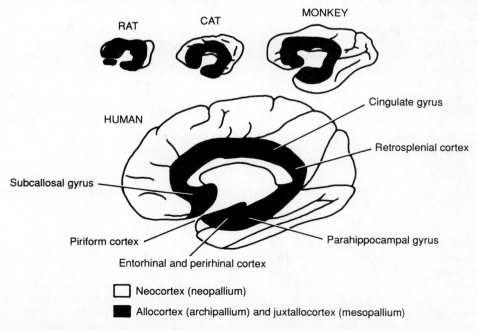

Figure 15–1. Relation of the limbic cortex to the neocortex in rats, cats, monkeys, and humans.

and a second ring of a transitional type of cortex called **juxtallocortex** between the allocortex and the rest of the cerebral hemisphere. The cortical tissue of the remaining nonlimbic portions of the hemisphere is called **neocortex** and is the most highly developed type. The actual extent of the allocortical and juxtallocortical areas has changed little as mammals have evolved, but these regions have been overshadowed by the immense growth of the neocortex, which reaches its greatest development in humans (Fig 15–1).

Afferent & Efferent Connections

The major connections of the limbic system are shown in Fig 15–2. The fornix connects the hippocampus to the mamillary bodies, which are in turn connected to the anterior nuclei of the thalamus by the mamillothalamic tract. The anterior nuclei of the thalamus project to the cingulate cortex, and from the cingulate cortex there are connections to the hippocampus, completing a complex closed circuit. This circuit was originally described by Papez and has been called the Papez circuit.

Correlations Between Structure & Function

One characteristic of the limbic system is the paucity of the connections between it and the neocortex. Nauta has aptly stated that "the neocortex sits astride the limbic system like a rider on a horse without reins." Actually, there are a few connections, and from a functional point of view, neocortical activity does modify emotional behavior and vice versa. However, one of the characteristics of emotion is that it cannot be turned on and off at will.

Another characteristic of limbic circuits is their prolonged after-discharge following stimulation. This may explain in part the fact that emotional responses are generally prolonged rather than evanescent and outlast the stimuli that initiate them.

LIMBIC FUNCTIONS

Stimulation and ablation experiments indicate that in addition to its role in olfaction (see Chapter 10), the limbic system is concerned with feeding behavior. Along with the hypothalamus, it is also concerned with sexual behavior, the emotions of rage and fear, and motivation. The relation of the limbic system to memory is discussed in Chapter 16.

Autonomic Responses & Feeding Behavior

Limbic stimulation produces autonomic effects, particularly changes in blood pressure and respiration. These responses are elicited from many limbic structures, and there is little evidence of localization of autonomic responses. This suggests that the autonomic effects are part of more complex phenomena, particularly emotional and behavioral responses.

Stimulation of the amygdaloid nuclei causes movements such as chewing and licking and other activities related to feeding. Lesions in the amygdala cause moderate hyperphagia, with indiscriminate ingestion of all kinds of food, probably because of the inability to categorize objects as edible or inedible (see Chapter 16).

SEXUAL BEHAVIOR

Mating is a basic but complex phenomenon in which many parts of the nervous system are involved. Copulation itself is made up of a series of reflexes integrated in spinal and lower brain stem centers, but the behavioral components that accompany it, the urge to copulate, and the coordinated sequence of events in the male and female that lead to pregnancy are regulated to a large degree in the limbic system and hypothalamus. Learning plays a part in the development of mating behavior, particularly in humans and other primates, but in nonprimate mammals, courtship and successful mating can occur with no previous sexual experience. The basic responses are therefore innate and are undoubtedly present in all mammals. However, in humans, the sexual functions have become extensively encephalized and conditioned by social and psychic factors. The basic physiologic mechanisms of sexual behavior in animals are therefore considered first and then compared with the responses in humans.

Relation to Endocrine Function

In nonprimate mammals, removal of the gonads leads eventually to decreased or absent sexual activity in both the male and the female—although the loss is slow to develop in the males of some species. Injections of gonadal hormones in castrate animals revive sexual activity. Testosterone in the male and estrogen in the female have the most marked effect. Large doses of progesterone are also effective in the female, while in the presence of smaller doses of progesterone, the dose of estrogen necessary to produce sexual activity is lowered. In primates, on the other hand, progesterone has no facilitatory effect and is even somewhat inhibitory. Large doses of testosterone and other androgens in castrate females initiate female behavior, and large doses of estrogens in castrate males trigger male mating responses. It is unsettled why responses appropriate to the sex of the animal occur when the hormones of the opposite sex are injected.

Effects of Hormones in Humans

In adult women, ovariectomy does not necessarily reduce libido (defined in this context as sexual interest and drive) or sexual ability. Postmenopausal women continue to have sexual relations, often without much change in frequency from their premeno-

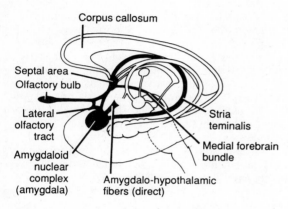

Figure 15–2. Principal connections of the limbic system. **Top:** Hippocampal system. **Bottom:** Olfactory and amygdaloid connections. (Reproduced, with permission, from de-Groot J: *Correlative Neuroanatomy,* 21st ed. Appleton & Lange, 1991.)

pausal pattern. This persistence is probably due to secretion of steroids from the adrenal cortex that are converted to circulating estrogens but may also be due to the greater degree of encephalization of sexual functions in humans and their relative emancipation from instinctual and hormonal control. Treatment with sex hormones increases sexual interest and drive in humans. Testosterone, for example, increases libido in males, and so does estrogen used to treat diseases such as carcinoma of the prostate. The behavioral pattern that was present before treatment is stimulated but not redirected. Thus, administration of testosterone to homosexuals intensifies their homosexual drive but does not convert it to a heterosexual drive.

Neural Control in the Male

In male animals, removal of the neocortex generally inhibits sexual behavior. Partial cortical ablations also produce some inhibition, the degree of the inhibition being independent of the coexisting motor deficit and most marked when the lesions are in the frontal lobes. On the other hand, cats and monkeys with bilateral limbic lesions localized to the piriform cortex overlying the amygdala (Fig 15–3) develop a marked intensification of sexual activity. These animals not only mount adult females; they also mount immature females and other males and attempt to copulate with animals of other species and with inanimate objects. Despite some claims to the contrary, such behavior is clearly abnormal in the species studied. The behavior is dependent upon the presence of testosterone but is not due to any increase in its secretion.

The hypothalamus is also involved in the control of sexual activity in males. Stimulation along the medial forebrain bundle and in neighboring hypothalamic areas causes penile erection with considerable emotional display in monkeys. In castrated rats, intrahypothalamic implants of testosterone restore the complete pattern of sexual behavior, and in intact rats, appropriately placed anterior hypothalamic lesions abolish interest in sex.

The extent to which the findings in male animals with periamygdaloid lesions are applicable to men is,

Figure 15–3. Site of lesions producing hypersexuality in male cats. The incidence of hypersexuality was proportionate to the intensity of the stippling, and when the black area was destroyed, hypersexuality was always present. Olf, olfactory. (Reproduced, with permission, from Green J et al: Rhinencephalic lesions and behavior in cats. *J Comp Neurol* 1957;**108:**505.)

of course, difficult to determine, but there are reports of hypersexuality in men with bilateral lesions in or near the amygdaloid nuclei.

Sexual Behavior in the Female

In mammals, the sexual activity of the male is more or less continuous, but in many species, the sexual activity of the female is cyclic. Most of the time, the female avoids the male and repulses his sexual advances. Periodically, however, there is an abrupt change in behavior and the female seeks out the male, attempting to mate. These short episodes of **heat** or **estrus** are so characteristic that the sexual cycle in mammalian species that do not menstruate is named the **estrous cycle.**

This change in female sexual behavior is brought on by a rise in the circulating blood estrogen level. Some animals, notably the rabbit and the ferret, come into heat and remains estrous until pregnancy or pseudopregnancy results. In these species, ovulation is due to a neuroendocrine reflex. Stimulation of the genitalia and other sensory stimuli at the time of copulation provoke release from the pituitary of the LH that makes the ovarian follicles rupture. In many other species, spontaneous ovulation occurs at regular intervals, and the periods of heat coincide with its occurrence. This is true in monkeys and apes. In captivity, these species mate at any time; but in the wild state, the females accept the male more frequently at the time of ovulation. In women, sexual activity occurs throughout the menstrual cycle, but careful studies indicate that, as in other primates, there is more spontaneous female-initiated sexual activity at about the time of ovulation.

In monkeys, the sex drive of the male is greater when he is exposed to a female at the time of ovulation than when he is exposed to a female at another time of her cycle. The "message" sent by the female to the male in this situation is olfactory, and the substances responsible are certain fatty acids in the vaginal secretions. These fatty acids are also found in increased amounts in human vaginal secretions at about midcycle. Substances produced by an animal that act at a distance to produce behavioral or other physiologic changes in another animal of the same species have been called **pheromones.** The sex attractants of certain insects are particularly well known examples, but it appears from the evidence cited above that pheromones also operate in the regulation of sexual behavior in primates.

Neural Control in the Female

In female animals, removal of the neocortex and the limbic cortex abolishes active seeking out of the male ("enticement reactions") during estrus, but other aspects of heat are unaffected. Amygdaloid and periamygdaloid lesions do not produce hypersexuality as they do in the male. However, discrete anterior hypothalamic lesions abolish behavioral heat (Fig

15–4) without affecting the regular pituitary-ovarian cycle (see Chapter 23).

Implantation of minute amounts of estrogen in the anterior hypothalamus causes heat in ovariectomized rats (see Fig 23–32). Implantation into other parts of the brain and outside the brain does not have this effect. Apparently, therefore, some element in the hypothalamus is sensitive to circulating estrogen and is stimulated by the hormone to initiate estrous behavior.

Effects of Sex Hormones in Infancy on Adult Behavior

In female experimental animals, exposure to sex steroids in utero or during early postnatal development causes marked abnormalities of sexual behavior when the animals reach adulthood. Female rats treated with a single relatively small dose of androgen before the fifth day of life do not have normal heat periods when they mature. They do not show the cyclic release of pituitary gonadotropins characteristic of the adult female, but rather the tonic, steady secretion characteristic of the adult male; their brains have been "masculinized" by the single brief exposure to androgens. They also show increased male sexual behavior. Conversely, male rats castrated at birth develop the female pattern of cyclic gonadotropin secretion and show considerable female sexual behavior when given doses of ovarian hormones that do not have this effect in intact males. Thus, the development of a "female hypothalamus" depends simply on the absence of androgens in early life rather than on exposure to female hormones.

Rats are particularly immature at birth, and animals of other species in which the young are more

Figure 15–4. Sites of hypothalamic lesions that blocked behavioral heat without affecting ovarian cycles in ewes. MI, massa intermedia; MB, mamillary body; OC, optic chiasm; PIT, pituitary. (From data of MT Clegg and WF Ganong.)

fully developed at birth do not show these changes when exposed to androgens during the postnatal period. However, these animals develop genital abnormalities when exposed to androgens in utero (see Chapter 23). Female monkeys exposed to androgens in utero do not lose the female pattern of gonadotropin secretion but do develop abnormalities of sexual behavior in adulthood.

Exposure of human females to androgens in utero does not change the cyclic pattern of gonadotropin secretion in adulthood (see Chapter 23). However, there is evidence that masculinizing effects on behavior do occur.

Maternal Behavior

Maternal behavior is depressed by lesions of the cingulate and retrosplenial portions of the limbic cortex in animals. Hormones do not appear to be necessary for its occurrence, but prolactin, which is secreted in large amounts during pregnancy and lactation, facilitates it.

FEAR & RAGE

Fear and rage are in some ways closely related emotions. The external manifestations of the **fear, fleeing,** or **avoidance reaction** in animals are autonomic responses such as sweating and pupillary dilation, cowering, and turning the head from side to side to seek escape. The **rage, fighting,** or **attack reaction** is associated in the cat with hissing, spitting, growling, piloerection, pupillary dilation, and well-directed biting and clawing. Both reactions—and sometimes mixtures of the two—can be produced by hypothalamic stimulation. When an animal is threatened, it usually attempts to flee. If cornered, it fights. Thus, fear and rage reactions are probably related instinctual protective responses to threats in the environment.

Fear

The fear reaction can be produced in conscious animals by stimulation of the hypothalamus and the amygdaloid nuclei. Conversely, the fear reaction and its autonomic and endocrine manifestations are absent in situations in which they would normally be evoked when the amygdalae are destroyed. A dramatic example is the reaction of monkeys to snakes. Monkeys are normally terrified by snakes. After bilateral temporal lobectomy, monkeys approach snakes without fear, pick them up, and even eat them.

Rage & Placidity

Most animals, including humans, maintain a balance between rage and its opposite, the emotional state that for lack of a better name is referred to here as placidity. Major irritations make normal individuals "lose their temper," but minor stimuli are ignored.

In animals with certain brain lesions, this balance is altered. Some lesions produce a state in which the most minor stimuli evoke violent episodes of rage; others produce a state in which the most traumatic and anger-provoking stimuli fail to ruffle the animal's abnormal calm.

Rage responses to minor stimuli are observed after removal of the neocortex and after destruction of the ventromedial hypothalamic nuclei and septal nuclei in animals with intact cerebral cortices. On the other hand, bilateral destruction of the amygdaloid nuclei in monkeys causes a state of abnormal placidity. Stimulation of some parts of the amygdala in cats produces rage. The placidity produced by amygdaloid lesions in animals is converted into rage by subsequent destruction of the ventromedial nuclei of the hypothalamus.

Rage can also be produced by stimulation of an area extending back through the lateral hypothalamus to the central gray area of the midbrain, and the rage response usually produced by amygdaloid stimulation is abolished by ipsilateral lesions in the lateral hypothalamus or rostral midbrain.

Gonadal hormones appear to affect aggressive behavior. In male animals, aggression is decreased by castration and increased by androgens. It is also conditioned by social factors; it is more prominent in males that live with females and increases when a stranger is introduced into an animal's territory.

"Sham Rage"

It was originally thought that rage attacks in animals with diencephalic and forebrain lesions represented only the physical, motor manifestations of anger, and the reaction was therefore called "sham rage." This now appears to be incorrect. Although rage attacks in animals with diencephalic lesions are induced by minor stimuli, they are usually directed with great accuracy at the source of the irritation. Furthermore, hypothalamic stimulation that produces the fear-rage reaction is apparently unpleasant to animals, because they become conditioned against the place where the experiments are conducted and try to avoid the experimental sessions. They can easily be taught to press a lever or perform some other act to avoid a hypothalamic stimulus that produces the manifestations of fear or rage. It is difficult if not impossible to form conditioned reflex responses (see Chapter 16) by stimulation of purely motor systems, and it is also difficult if the unconditioned stimulus does not evoke either a pleasant or unpleasant feeling. The fact that hypothalamic stimulation is a potent unconditioned stimulus for the formation of conditioned avoidance responses and the fact that the avoidance responses are extremely persistent indicate that the stimulus is unpleasant. There is therefore little doubt that rage attacks include the mental as well as the physical manifestations of rage, and the term "sham rage" should be dropped.

Significance & Clinical Correlates

It is tempting on the basis of the evidence cited above to speculate that there are 2 intimately related mechanisms in the hypothalamus and limbic system, one promoting placidity and the other rage. If this is true, the emotional state is probably determined by afferent impulses that adjust the balance between them. An arrangement of this sort would be analogous to the systems controlling feeding and body temperature.

Although emotional responses are much more complex and subtle in humans than in animals, the neural substrates are probably the same. It is doubtful if placidity would be recognized as a clinical syndrome in our culture, but rage attacks in response to trivial stimuli have been observed many times in patients with brain damage. They are a complication of pituitary surgery when there is inadvertent damage to the base of the brain. They also follow a number of diseases of the nervous system, especially epidemic influenza and encephalitis, which destroy neurons in the limbic system and hypothalamus. Stimulation of the amygdaloid nuclei and parts of the hypothalamus in conscious humans produces sensations of anger and fear. In Japan, bilateral amygdaloid lesions have been made in agitated, aggressive mental patients. The patients are said to have become placid and manageable, and it is of some interest that they were reported not to have developed hypersexuality or memory loss (see Chapter 16).

MOTIVATION

If an animal is placed in a box with a pedal or bar that can be pressed, the animal sooner or later accidentally presses it. Olds and his associates showed that if the bar is connected in such a way that each press delivers a stimulus to an electrode implanted in certain parts of the brain, the animal returns to the bar and presses it again and again. Pressing the bar soon comes to occupy most of the animal's time. Some animals go without food and water and others endure strong adversive stimuli to press the bar for brain stimulation, and some will continue until they fall over exhausted. Rats press the bar 5000–12,000 times per hour, and monkeys have been clocked at 17,000 bar presses per hour. On the other hand, when the electrode is in certain other areas, the animals avoid pressing the bar, and stimulation of these areas is a potent unconditioned stimulus for the development of conditioned avoidance responses.

The points where stimulation leads to repeated bar pressing are located in a medial band of tissue passing from the frontal cortex through the hypothalamus to the midbrain tegmentum (Fig 15–5). The highest rates are obtained from points in the medial forebrain bundle, with high rates also obtained from points in the tegmentum and septal nuclei. The points where

Figure 15–5. Areas where stimulation leads to repeated bar pressing projected on a parasagittal view of the rat brain. The regions where high rates of self-stimulation are produced are indicated by the rectangles. (Modified from Routtenberg A: The reward system of the brain. *Sci Am* [Nov] 1978;**239**:154.)

stimulation is avoided are in the lateral portion of the posterior hypothalamus, the dorsal midbrain, and the entorhinal cortex. The latter points are sometimes close to points where bar pressing is repeated, but they are part of a separate system. The areas where bar pressing is repeated are much more extensive than those where it is avoided. It has been calculated that in rats repeated pressing is obtained from 35% of the brain, avoidance from 5%, and indifferent responses (neither repetition nor avoidance) from 60%.

It is obvious that some effect of the stimulation causes the animals to stimulate themselves again and again, but what the animals feel is, of course, unknown. There are a number of reports of bar-pressing experiments in humans with chronically implanted electrodes. Most of the subjects were schizophrenics or epileptics, but a few were patients with visceral neoplasms and intractable pain. Like animals, humans press the bar repeatedly. They generally report that the sensations evoked are pleasurable, using phrases like "relief of tension" and "a quiet, relaxed feeling" to describe the experience. However, they rarely report "joy" or "ecstasy," and some persons with the highest self-stimulation rates cannot tell why they keep pushing the bar. When the electrodes are in the areas where stimulation is avoided, patients report sensations ranging from vague fear to terror. It is probably wise, therefore, to avoid vivid terms and call the brain systems involved the **reward** or **approach system** and the **punishment** or **avoidance system.**

Studies of the kind described above provide physiologic evidence that behavior is motivated not only by reduction or prevention of an unpleasant effect but also by primary rewards such as those produced by stimulation of the approach system of the brain. The implications of this fact are great in terms of the classic drive-reduction theory of motivation, in terms of the disruption and facilitation of ongoing behavior, and in terms of normal and abnormal emotional responses.

Stimulation of the reward system provides a potent motivation for learning mazes or performing other tasks. Drugs that block postsynaptic dopaminergic receptors reduce the rate of self-stimulation, whereas dopamine agonists increase it. Cocaine appears to produce its "high" by acting on the reward system, particularly on the ventral tegmental area and the nucleus accumbens. Cocaine binds to and inhibits the dopamine transporter (see Chapter 4). Therefore, dopamine reuptake is reduced and the extracellular dopamine level is increased. Cocaine also inhibits reuptake of serotonin and norepinephrine, and there is reason to believe that increased stimulation of serotonergic receptors augments the euphoria produced by cocaine. Amphetamine, another drug of abuse, releases dopamine, and there is evidence that nicotine and alcohol increase the amount of dopamine in the nucleus accumbens. Thus, a variety of habit-forming substances that produce addictive behavior may act by increasing dopaminergic activity in the reward system of the brain.

BRAIN CHEMISTRY, BEHAVIOR, & SYNAPTIC TRANSMISSION IN THE CENTRAL NERVOUS SYSTEM

Drugs that modify human behavior include **hallucinogenic agents,** drugs that produce hallucinations and other manifestations of the psychoses; **tranquilizers,** drugs that allay anxiety and various psychiatric symptoms; and **psychic energizers,** antidepressant drugs that elevate mood and increase interest and drive. These and many other drugs act by modifying transmission at synaptic junctions in the brain. The chemistry of the known and suspected synaptic transmitters in the peripheral as well as the central nervous system is discussed in Chapter 4. This chapter is concerned primarily with their relation to instinctual behavior and emotions.

Monoaminergic Systems in the Brain

There are 4 large monoaminergic systems in the brain that have in common the presence of their cell bodies in relatively few locations with multiple branched axons projecting to almost all parts of the nervous system (Fig 15–6). These are the sero-

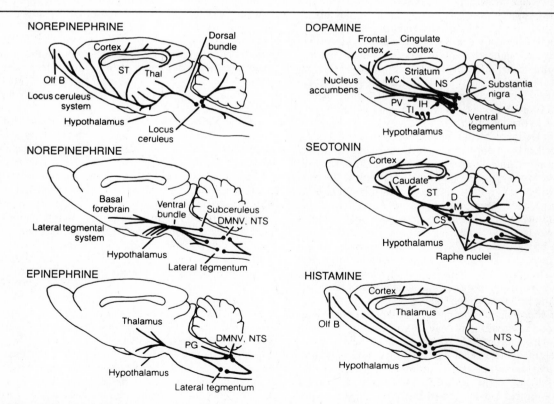

Figure 15–6. Monoaminergic pathways in rat brain. The pathways in humans appear to be similar. The 2 principal noradrenergic systems (locus ceruleus and lateral tegmental) are shown separately. Olf B, olfactory bulb; Thal, thalamus; ST, stria terminalis; DMNV, dorsal motor nucleus of vagus; NTS, nucleus of tractus solitarius; PG, periaqueductal gray; NS, nigrostriatal system; MC, mesocortical system; PV, periventricular system; IH, incertohypothalamic system; TI, tuberoinfundibular system; D, M, and CS, dorsal, medial, and central superior raphe nuclei.

tonergic, noradrenergic, adrenergic, and histaminergic systems. Dopaminergic neurons have their cell bodies in more locations, but their axons also project to many different areas.

Serotonin

Serotonin-containing neurons have their cell bodies in the midline raphe nuclei of the brain stem and project to portions of the hypothalamus, the limbic system, the neocortex, and the spinal cord (Fig 15–6).

The hallucinogenic agent lysergic acid diethylamide (LSD) is a serotonin agonist that produces its effects by activating $5HT_2$ receptors (see Chapter 4) in the brain. The transient hallucinations and other mental aberrations produced by this drug were discovered when the chemist who synthesized it inhaled some by accident. Its discovery called attention to the correlation between behavior and variations in brain serotonin content. Psilocin, a substance found in certain mushrooms, and N,N-dimethyltryptamine (DMT) are also hallucinogenic and, like serotonin, are derivatives of tryptamine. 2,5-Dimethoxy-4-methylamphetamine (DOM) and mescaline and its congeners, the other true hallucinogens, are phenylethylamines rather than indolamines. However, all these hallucinogens appear to exert their effects by binding to $5HT_2$ receptors. 3,4-Methylenedioxymethamphetamine, a drug known as MDMA or **ecstasy**, has been a popular drug of abuse. It produces euphoria, but this is followed by difficulty in concentrating, depression, and, in monkeys, insomnia. The drug causes release of serotonin followed by serotonin depletion; the euphoria may be due to the release and the later symptoms to the depletion.

Selective depletion of brain serotonin can be produced by administering *p*-chlorophenylalanine, a compound that blocks conversion of tryptophan to 5-hydroxytryptophan (see Fig 4–21). This is the rate-limiting step in serotonin biosynthesis. In animals, *p*-chlorophenylalanine produces prolonged wakefulness, suggesting that serotonin plays a role in sleep. However, wakefulness is not produced in humans, and no clear-cut psychic changes are produced even by large doses. Serotonergic neurons discharge rapidly in the awake state, slowly during drowsiness, more slowly with bursts during sleep, and not at all during REM sleep (see Chapter 11).

Other functions for brain serotonin have been proposed. Serotonin may play an excitatory role in the regulation of prolactin secretion (see Chapter 23). There is evidence that descending serotonergic fiber systems inhibit transmission in pain pathways in the dorsal horns. In addition, there is a prominent serotonergic innervation of the suprachiasmatic nuclei of the hypothalamus, and serotonin may be involved in the regulation of circadian rhythms (see Chapter 14). Pathophysiologically, there is evidence that discharge in serotonergic neurons in the dorsal

raphe nucleus causes migraine, and antimigraine drugs inhibit the firing of dorsal raphe neurons.

Norepinephrine

The cell bodies of the norepinephrine-containing neurons in the brain are located in the locus ceruleus and other nuclei in the pons and medulla. From the locus ceruleus, the axons of the noradrenergic neurons form a **locus ceruleus system.** They descend into the spinal cord, enter the cerebellum, and ascend to innervate the paraventricular, supraoptic, and periventricular nuclei of the hypothalamus, the thalamus, the basal telencephalon, and the entire neocortex (Fig 15–6). From cell bodies in the dorsal motor nucleus of the vagus, the nucleus of the tractus solitarius, and areas in the dorsal and lateral tegmentum, the axons of the noradrenergic neurons form a **lateral tegmental system** that projects the spinal chord, the brain stem, all of the hypothalamus, and the basal telencephalon.

Drugs that increase extracellular norepinephrine in the brain elevate mood, and drugs that decrease extracellular norepinephrine cause depression. Manic-depressive illness runs in families, and at least in some of these families, there is a genetic abnormality close to or in the gene on chromosome 11 that codes for tyrosine hydroxylase, the rate-limiting enzyme in catecholamine biosynthesis (see Chapter 4). On the other hand, it has not been possible to prove that extracellular norepinephrine is deficient in depressed patients, and there is evidence that some noradrenergic pathways are stimulated in depression while others are inhibited.

In the cerebellum, noradrenergic neurons inhibit Purkinje cells, and there is evidence that these inhibitory effects are mediated via β receptors and cyclic AMP.

The norepinephrine-containing neurons in the hypothalamus are involved in regulation of the secretion of vasopressin and oxytocin, and they adjust the secretion of the hypophyseotropic hormones that regulate the secretion of anterior pituitary hormones (see Chapter 14). Norepinephrine and serotonin both appear to be involved in the control of food intake and the regulation of body temperature.

Epinephrine

There is a system of phenylethanolamine-N-methyltransferase (PNMT)-containing neurons with cell bodies in the medulla that project to the hypothalamus. Those neurons secrete epinephrine, but their function is uncertain. Epinephrine-secreting neurons also project to the thalamus, periaqueductal gray, and spinal cord. There are appreciable quantities of tyramine in the central nervous system, but no function has been assigned to this agent.

Histamine

As noted in Chapter 4, histaminergic neurons have

their cell bodies in the tuberomamillary nucleus in the ventral portion of the posterior hypothalamus. The axons of these neurons project to all parts of the brain (Fig 15–6). The function of this diffuse histaminergic system is unknown, but there is evidence linking brain histamine to arousal, sexual behavior, blood pressure, drinking, pain thresholds, and regulation of the secretion of several anterior pituitary hormones.

Dopamine

Many dopaminergic neurons have their cell bodies in the midbrain (Fig 15–6). They project from the substantia nigra to the striatum (**nigrostriatal system**) and from other portions of the midbrain to the olfactory tubercle, nucleus accumbens, related limbic areas, and the frontal, cingulate, entorhinal, and perirhinal cortex (**mesocortical system**). A separate intrahypothalamic system of dopaminergic neurons (**tuberoinfundibular system**) projects from cell bodies in the arcuate nucleus to the external layer of the median eminence of the hypothalamus (Fig 15–7). Some dopaminergic neurons also project from the arcuate nucleus to the intermediate and posterior lobes

Figure 15–7. Dopaminergic neurons in the hypothalamus of the rat (transverse section). α-Methylnorepinephrine was injected before sacrifice to intensify fluorescence. Cell bodies can be seen above in the arcuate nucleus on either side of the third ventricle, and the dense nerve terminals can be seen below in the external layer of the median eminence. (Reproduced, with permission, from Hökfelt T, Fuxe K: On the morphology and the neuroendocrine role of the hypothalamic catecholamine neurons. In: *Brain-Endocrine Interaction.* Knigge K, Scott D, Weindl A [editors]. Karger, 1972.)

of the pituitary (**tuberohypophyseal system**). There is in addition an **incertohypothalamic system,** with cell bodies in the zona incerta and endings in the septum and dorsal hypothalamus, and a **periventricular system** made up of short dopaminergic neurons around the third and fourth ventricles. There are dopaminergic neurons in the **olfactory bulb** and the **retina.** Recent studies with positron emission tomography (PET) scanning (see Chapter 32) in normal humans show that there is a steady loss of dopamine receptors in the basal ganglia with age. The loss is greater in men than in women.

There appears to be a genetic defect in schizophrenia, possibly as a result of an abnormality in chromosome 5, and this abnormality may well involve the dopaminergic system. PET scanning of schizophrenics has shown that their brains contain elevated levels of D_2 dopamine receptors even if they have never received drugs. Amphetamine, which stimulates the secretion of norepinephrine and dopamine, produces a psychosis that resembles schizophrenia when administered in large doses. On the other hand, the major tranquilizers are effective in the relief of the symptoms of schizophrenia, and their antipsychotic activity generally parallels their ability to block D_2 receptors.

The nigrostriatal system is related to motor function. Parkinson's disease, the motor disorder caused by degeneration of this system, is discussed in Chapter 12.

Dopamine is the prolactin-inhibiting hormone secreted in the hypothalamus. Its role in the regulation of prolactin secretion is discussed in Chapter 23.

Acetylcholine

Acetylcholine is distributed throughout the central nervous system, with high concentrations in the cerebral cortex, thalamus, and various nuclei in the basal forebrain. The distribution of choline acetyltransferase and acetylcholinesterase parallels that of acetylcholine. Most of the acetylcholinesterase is in neurons, but some is found in glia. Pseudocholinesterase is found in many parts of the central nervous system.

The development of antibodies specific for choline acetyltransferase has permitted mapping the cholinergic pathways in the brain by immunocytochemical techniques. In addition to motor neurons and preganglionic autonomic neurons, which were previously known to be cholinergic, there are a modest number of cholinergic interneurons in the cerebral cortex. However, many of the cholinergic neurons are subcortical. There are cholinergic neurons that project from nuclei in the basal forebrain to the cerebral cortex, hippocampus, and related areas; and there are cholinergic neurons in the brain stem that project to the thalamus. In monkeys, cholinergic neurons in the medial septal nucleus and the nucleus of the vertical limb of the diagonal band on each side project to the hippocampus and are involved in memory. Cholinergic neurons in the nucleus of the horizontal limb of the diagonal band project to the olfactory bulb. There is also a large projection from the nucleus basalis of Meynert and adjacent nuclei to the amygdala and the entire neocortex, and these projections are probably involved in motivation, perception, and cognition. There is extensive cell loss in this projection in Alzheimer's disease (see Chapter 16). Cholinergic neurons in the pontopeduncular and lateral tegmental nuclei project through the pontomesencephalic reticular formation to the thalamus and are presumably involved in the attention and arousal functions of the ascending reticular system (see Chapter 11).

Acetylcholine has been linked directly or indirectly to a variety of brain functions. Injection of acetylcholine derivatives into the hypothalamus and parts of the limbic system causes drinking. In blinded rats, acetylcholinesterase activity is decreased in the superior colliculi and elevated in the occipital cortex. Cortical levels of acetylcholinesterase are greater in rats raised in a complex environment than in rats raised in isolation, but the significance of this type of correlation is uncertain. Acetylcholine is an excitatory transmitter in the basal ganglia, whereas dopamine is an inhibitory transmitter in these structures. In parkinsonism, the loss of dopamine alters the cholinergic-dopaminergic balance, and anticholinergic drugs are of benefit along with L-dopa in the treatment of the disease. The PGO spike system responsible for REM sleep is cholinergic.

Opioid Peptides

There are 3 types of opioid-peptide-secreting neurons in the brain, each producing one of the 3 opioid peptide precursor molecules (see Chapter 4). Proenkephalin-producing neurons are scattered throughout the brain, whereas pro-opiomelanocortin-producing neurons have their cell bodies in the arcuate nuclei and project to the thalamus and parts of the brain stem (Fig 15–8). Prodynorphin-producing neurons are located primarily in the hypothalamus, limbic system, and brain stem. The peptides they secrete are involved in various functions including, presumably, the phenomenon of tolerance and addiction produced by morphine, but the details are uncertain.

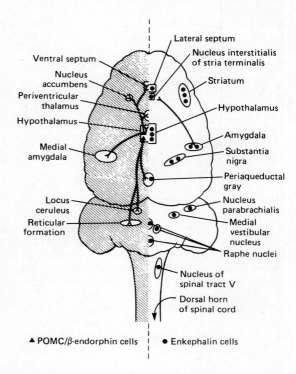

▲ POMC/β-endorphin cells · Enkephalin cells

Figure 15–8. Distribution of β-endorphin neurons **(left)** and enkephalin neurons **(right)** in the brain. (Reproduced, with permission, from Barchas JD et al: Behavioral neurochemistry: Neuroregulatory and behavioral states. *Science* 1978;**200:**964. Copyright © 1978 by the American Association for the Advancement of Science.)

Other Transmitters

The relation of GABA to anxiety and the anxiolytic actions of benzodiazepines are discussed in Chapter 4. Peptides other than enkephalins and β-endorphin are probably involved in emotional and other behavioral responses, but the details remain to be determined.

"Higher Functions of the Nervous System": Conditioned Reflexes, Learning, & Related Phenomena

16

INTRODUCTION

In previous chapters, somatic and visceral inputs to the brain and output from it have been described. The organization of the nervous system is summarized in Fig 16–1. Topics that have been discussed include specific inputs and outputs; the functions of the reticular core in maintaining an alert, awake state; and the functions of the limbic-hypothalamic circuit in maintenance of homeostatic equilibriums and regulation of instinctual and emotional behavior. There remain the phenomena called, for lack of a better or more precise term, the "higher functions of the nervous system": learning, memory, judgment, language, and the other functions of the mind. These phenomena are the subject of this chapter.

METHODS

Many of the phenomena of the mind are difficult to study because they are characteristic of humans and are much less developed, if at all, in other species. In addition, it is worth remembering that the brain of the rhesus monkey is only one-fourth the size of the brain of the chimpanzee, our nearest primate relative, and the chimpanzee brain is in turn one-fourth the size of the human brain. However, there are similarities between them, and much has been learned from the study of laboratory primates, particularly monkeys. In addition, valuable data have been obtained in humans. One method is stimulation of the exposed cerebral cortex in conscious humans undergoing neurosurgical procedures under local anesthesia, or in a few instances, stimulation of implanted electrodes. A second method is correlation of clinical observations with the site and extent of brain abnormalities at autopsy or, better, with observations of the living brain by PET or MRI at the time of clinical testing. Combining PET, MRI, and related methods (see Chapters 8, 32, and Appendix) with techniques that allow subtraction of nonspecific background

changes when a subject is performing a particular task has permitted remarkable resolution and localization and has led to major advances in understanding the "higher functions" of the normal human brain.

It should be emphasized that some forms of learning and memory are not unique to primates or even vertebrates and probably occur in all parts of the nervous system and in all animals with nervous systems. Studies of the molecular and cellular events underlying habituation, sensitization, and classical conditioning (see Chapter 4 and below) in species such as the sea snail *Aplysia* have been very fruitful, and indeed it is now possible to begin to talk about a "molecular biology of learning and memory."

LEARNING & MEMORY

Learning may be defined as the ability to alter behavior on the basis of experience, and memory is the ability to recall past events at the conscious or unconscious level. The 2 are obviously closely related and should be considered together.

Forms of Learning

Habituation is a simple form of learning in which a neutral stimulus is repeated many times. The first time it is applied, it is novel and evokes a reaction (the orienting reflex or "what is it?" response). However, it evokes less and less electrical response as it is repeated. Eventually, the subject becomes habituated to the stimulus and ignores it. **Sensitization** is in a sense the opposite reaction. A repeated stimulus produces a greater response if it is coupled one or more times with an unpleasant or a pleasant stimulus. An example of sensitization in a monkey is shown in Fig 16–2. It is common knowledge that similar intensification of the **arousal value** of stimuli occurs in humans. The mother who sleeps through many kinds of noise but wakes promptly when her baby cries is an example.

Habituation and sensitization are examples of

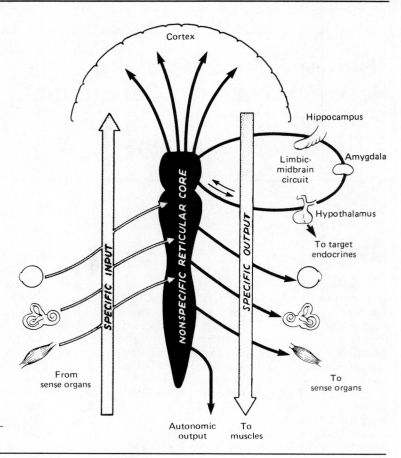

Figure 16–1. Organization of the nervous system.

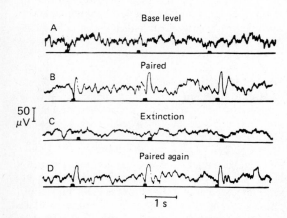

Figure 16–2. Records of the electrical activity of the hippocampus in a monkey exposed repeatedly to a tone stimulus (black signal marks). **A:** Control. **B:** After the tone was paired with a reward of food. **C:** After extinction of this response. **D:** After the tone was again paired with food. (Reproduced, with permission, from Hearst E et al: Some electrophysiological correlates of conditioning in the monkey. *Electroencephalogr Clin Neurophysiol* 1960;**12:**137.)

nonassociative learning. In nonassociative learning, the organism learns about a single stimulus. In associative learning, the organism learns about the relation of one stimulus to another. A classic example of associative learning is a conditioned reflex.

Conditioned Reflexes

A conditioned reflex is a reflex response to a stimulus that previously elicited little or no response, acquired by repeatedly pairing the stimulus with another stimulus that normally does produce the response. In Pavlov's classic experiments, the salivation normally induced by placing meat in the mouth of a dog was studied. A bell was rung just before the meat was placed in the dog's mouth, and this was repeated a number of times until the animal would salivate when the bell was rung even though no meat was placed in its mouth. In this experiment, the meat placed in the mouth was the **unconditioned stimulus (US),** the stimulus that normally produces a particular innate response. The **conditioned stimulus (CS)** was the bell-ringing. After the CS and US had been paired a sufficient number of times, the CS produced the response originally evoked only by the US. The CS had to precede the US. Another example is shown in Fig 16–3. This is so-called classical conditioning. An immense number of somatic, visceral, and other neural changes can be made to occur as conditioned reflex responses. Conditioning of visceral responses is often called **biofeedback.** The changes that can be produced include alterations in heart rate and blood pressure, and conditioned decreases in blood pressure have been advocated for the treatment of hypertension. However, the depressor responses that are produced in this fashion are small.

If the CS is presented repeatedly without the US, the conditioned reflex eventually dies out. This process is called **extinction** or **internal inhibition.** If the animal is disturbed by an external stimulus immediately after the CS is applied, the conditioned response may not occur **(external inhibition).** However, if the conditioned reflex is **reinforced** from time to time by again pairing the CS and US, the conditioned reflex persists indefinitely.

As noted in Chapter 15, conditioned reflexes are difficult to form unless the US is associated with a pleasant or unpleasant affect. Stimulation of the brain reward system is a powerful US pleasant (pleasant or **positive reinforcement**), and so is stimulation of the avoidance system or a painful shock to the skin (unpleasant or **negative reinforcement**).

Operant conditioning is a form of conditioning in which the animal is taught to perform some task ("operate on the environment") in order to obtain a reward or avoid punishment. The US is the pleasant or unpleasant event, and the CS is a light or some other signal that alerts the animal to perform the task. Conditioned motor responses that permit an animal to

avoid an unpleasant event are called **conditioned avoidance reflexes.** For example, an animal is taught that by pressing a bar it can prevent an electric shock to the feet. Another example is **food aversion conditioning.** An animal exposed to the taste of a food develops a strong aversion to the food if the tasting is coupled with injection of a drug that produces nausea or illness. Similar aversion responses occur in humans. These conditioned responses are very strong, can sometimes be learned with a single pairing of the CS and the US, and, unlike other conditioned responses, will develop when the CS and US are separated by an hour or more. The survival value of food aversion conditioning is obvious in terms of avoiding poisons, and it is not surprising that the brain is probably genetically "programmed" to facilitate the development of food aversion responses.

Intercortical Transfer of Learning

If a cat or monkey is conditioned to respond to a visual stimulus with one eye covered and then tested with the blindfold transferred to the other eye, it performs the conditioned response. This is true even if the optic chiasm has been cut, making the visual input from each eye go only to the ipsilateral cortex. If, in addition to the optic chiasm, the anterior and posterior commissures and the corpus callosum are sectioned ("split-brain animal"), no transfer of learning occurs. Similar results have been obtained in humans in whom the corpus callosum is congenitally absent or in whom it has been sectioned surgically in an effort to control epileptic seizures. This demonstrates that the neural coding necessary for "remembering with one eye what has been learned with the other" has been transferred somehow to the opposite cortex via the commissures. There is evidence for similar transfer of information acquired through other sensory pathways. Studies of humans with incomplete section of the corpus callosum indicate that visual information is transferred in the posterior part of this structure, whereas the anterior part is concerned with transfer of auditory and somatesthetic information. In some cases, callosal lesions block information transfer in one direction but not in the other.

Memory

It is obvious that there are various types of memory and that they range in complexity from the primitive types that underlie habituation and sensitization to the complex conscious memories of humans. The types are conveniently divided into **nondeclarative,** or **reflexive memory** and **declarative memory** (Table 16–1). Nondeclarative memory includes classical conditioning, skills, and habits and is largely and often totally unconscious. Declarative memory, on the other hand, involves conscious recall of events that have occurred and of facts. These memory types are processed in different ways and in different parts of the brain (see below). However, they are related in

Figure 16–3. Conditioned reflex in a rabbit. At the first arrow a tone is sounded (CS), and at second arrow a gentle puff of air (US) is blown on the rabbit's eye. At first the nictitating membrane extends only in response to the puff of air **(top trace).** After repeated pairing, however, the nictitating membrane starts to contract in response to the tone, before the puff of air **(bottom trace).** (Modified from Alkon DL: Memory storage and neural systems. *Sci Am* [July] 1989;**261**:42.)

Table 16–1. Forms of memory.[1]

Nondeclarative (implicit, reflexive)
Habituation
Sensitization
Simple classical conditioning
Skills and habits
Priming
Declarative (explicit, recognition)
Facts
Events

[1]Modified from Squire LR, Zola-Morgan S: The medial temporal lobe memory system. *Science* 1991;**253**:1380.

that declarative memory can be converted into reflexive memory by constant repetition. Thus, for example, athletes develop moves and responses that with training become "instinctive." In a similar fashion, many aspects of behavior as complex as driving a car or playing a piano become habit responses.

Declarative memory in humans is divided into **immediate recall** of ongoing events; **recent memory** of events occurring seconds to hours or even days before; and **remote memory** of events from the distant past. Recent memory encompasses the time during which memory of ongoing events is being consolidated and converted to remote, long-term memory. During this time, it is vulnerable and subject to erasure, whereas remote memories are remarkably resistant and persist in the presence of severe brain damage.

Clinical evidence for the vulnerability of recent memory is the observations that there is frequently a loss of memory of the events immediately preceding brain concussion or electroshock therapy (**retrograde amnesia**). In humans, this amnesia encompasses longer periods than it does in experimental animals—sometimes many days—but remote and reflexive memory are not affected. In animals, acquisition of learned responses—or at least their retrieval—is prevented if, within 5 minutes after each training session, the animals are anesthetized, given electroshock treatment, subjected to hypothermia, or treated with antibiotics that inhibit protein synthesis. Such treatment 4 hours after the training sessions has no effect on acquisition.

Once long-term memories have been established, they can be recalled or accessed by a large number of different associations. For example, the memory of a vivid scene can be evoked not only by a similar scene but also by a sound or smell associated with the scene and by words such as scene, vivid, and view. Thus, there must be multiple routes or keys to each stored memory. In addition, many memories have an emotional component or "color"; ie, in simplest terms, memories can be pleasant or unpleasant.

Biologic Bases of Learning & Memory

The biochemical events involved in habituation and sensitization in *Aplysia* and other invertebrates

have been worked out in considerable detail. As described in Chapter 4, habituation is due to a decrease in Ca^{2+} in the sensory endings that mediate the response to a particular stimulus, and sensitization is due to prolongation of the action potential in these endings with a resultant increase in intracellular Ca^{2+} that facilitates release of neurotransmitter by exocytosis. Posttetanic potentiation, a learninglike facilitation of transmission, is also due to accumulation of Ca^{2+} in presynaptic terminals, and long-term potentiation (LTP) in the mammalian brain, a much more prolonged facilitation of the effectiveness of synaptic transmission, is triggered by accumulation of Ca^{2+} in postsynaptic neurons (see Chapter 4).

Classical conditioning also occurs in *Aplysia,* and in mammals, in the isolated spinal cord. In *Aplysia,* the US acts presynaptically on the endings of neurons activated by the CS. This leaves free Ca^{2+} in the cell, leading to a long-term change in the adenylate cyclase molecule, so that when this enzyme is activated by the CS, more cyclic AMP is produced. This in turn closes K^+ channels and prolongs action potentials by the mechanism described in Chapter 4. The key point in this case is the temporal association, with the US coming soon after the CS.

In *Aplysia,* there are morphologic correlates to learning and memory. For example, 40% of the relevant sensory terminals normally contain active zones, whereas in habituated animals, 10% have active zones, and in sensitized animals, 65% have active zones. Long-term memory leads to activation of genes that produce increases in synaptic contacts.

Encoding Nondeclarative Memory in Mammals

Without much doubt, molecular events similar to those occurring in *Aplysia* underlie some aspects of nondeclarative memory in mammals. However, additional events involving various part of the CNS also contribute. Some investigators argue that the striatum is involved, and it is known that some reflexive tasks are disrupted by lesions of the basal ganglia. There is other evidence that the cerebellum is involved. For example, the vestibulo-ocular reflex (VOR), the reflex that maintains visual fixation while the head is moving (see Chapter 9), can be adjusted to new eye positions, and this plasticity is abolished by lesions of the flocculus. In addition, conditioning of an eye blink reflex by using a puff of air on the corneas as the US and a tone as the CS is prevented by lesions of the interpositus nucleus. In this case, it appears that impulses set up by the US act via the inferior lobe and climbing fibers to the cerebellar cortex to alter the Purkinje cell response to the tone arriving via the pontine nuclei and mossy fibers. Climbing-fiber-mediated modification of mossy-fiber-driven Purkinje cell discharge is also responsible for plastic changes in the VOR and learned muscle movements.

Arcuate fasciculus

Broca's area

Angular gyrus

Wernicke's area

Figure 16–8. Location of some of the areas that in the categorical hemisphere are concerned with language functions.

Wernicke's area appears to process information from words that are read in such a way that they can be converted into the auditory forms of the words in Wernicke's area.

Abnormalities of language functions that are not due to defects of vision or hearing or to motor paralysis are called **aphasias.** They are due to lesions in the categorical hemisphere. The most common cause is embolism or thrombosis of a cerebral blood vessel. Many different classifications of the aphasias have been published, and their nomenclature is chaotic. However, a convenient classification divides them into **fluent** and **nonfluent aphasias.** In nonfluent aphasia, the lesion is in Broca's area (Table 16–2). Speech is slow, and words are hard to come by. Patients with severe damage to this area are limited to 2 or 3 words with which to express the whole range of meaning and emotion. Sometimes the words retained are those which were being spoken at the time of the injury or vascular accident that caused the aphasia.

Table 16–2. Aphasias. Characteristic responses of patients with lesions in various areas when shown a picture of a chair.[1]

Type of Aphasia and Site of Lesion	Characteristic Naming Errors
Nonfluent (Broca's area)	"Tssair"
Fluent (Wernicke's area)	"Stool" or "choss" (neologism)
Fluent (areas 40, 41, and 42; conduction aphasia)	"Flair . . . no, swair . . . tair."
Anomic (angular gyrus)	"I know what it is . . . I have a lot of them."

[1]Modified from Goodglass H: Disorders of naming following brain injury. *Am Sci* 1980;**68**:647.

In one form of fluent aphasia, the lesion is in Wernicke's area. In this condition, speech itself is normal and sometimes the patients talk excessively. However, what they say is full of jargon and neologisms that make little sense. The patient also fails to comprehend the meaning of spoken or written words, so other aspects of the use of language are compromised.

Another form of fluent aphasia is a condition in which patients can speak relatively well and have good auditory comprehension but cannot put parts of words together or conjure up words. This is called **conduction aphasia** because it was thought to be due to lesions of the arcuate fasciculus connecting Wernicke's and Broca's areas. However, it now appears that it is due to lesions in and around the auditory cortex (areas 40, 41, and 42).

When there is a lesion damaging the angular gyrus in the categorical hemisphere without affecting Wernicke's or Broca's areas, there is no difficulty with speech or the understanding of auditory information, but there is trouble understanding written language or pictures, because visual information is not processed and transmitted to Wernicke's area. The result is a condition called **anomic aphasia.**

More selective speech defects have now been described. For example, lesions limited to the left temporal pole (area 38) cause inability to retrieve names of places and persons but preserves the ability to retrieve common nouns, ie, the names of nonunique objects. The ability to retrieve verbs and adjectives is also intact.

The isolated lesions that cause the selective defects described above occur in some patients, but brain destruction is often more general. Consequently, more than one form of aphasia is often present. Frequently, the aphasia is general **(global),** involving both receptive and expressive functions. In this situation, speech is scant as well as nonfluent. Writing is abnormal in all aphasias in which speech is abnormal, but the neural circuits involved are not known. In addition, deaf subjects lose their ability to communicate in sign language if they develop a lesion in the categorical hemisphere.

Although aphasias are produced by lesions of the categorical hemisphere, lesions in the representational hemisphere also have effects. For example, they may impair the ability to tell a story or make a joke. They may also impair a subject's ability to get the point of a joke and, more broadly, to comprehend the meaning of differences in inflection and the "color" of speech. This is one more example of the way the hemispheres are specialized rather than simply being dominant and nondominant.

Localization of Other Functions

In recent years, the availability of MRI and PET scanning has sparked a recrudescence of interest in detailed localization and functional analysis of defects caused by selective lesions of the cerebral cortex in patients with strokes and head injuries. These analyses permit insights, or at least glimpses, into the ways serial processing of sensory information produces cognition, reasoning, comprehension, and language. For example, patients with frontoparietal damage can have **acalculia,** a selective impairment of mathematical ability. One patient with a left parietal lesion had difficulty with the second half but not the first half of words. Some patients with parieto-occipital lesions write only with consonants and omit vowels. Bilateral lesions of the undersurface of the occipital and temporal lobes (Fig 16–9) cause **prosopagnosia,** a unique agnosia in which patients are unable to recognize faces. Patients with this abnormality can recognize forms and reproduce them. They can recognize people by their voices, and they show autonomic responses when they see familiar as opposed to unfamiliar faces. However, they cannot identify the familiar people they see. Additional studies of defects of this type on patients and laboratory primates should add greatly to our understanding of the functions of the neocortex.

Experimental Neurosis

Animals can be conditioned to respond to one stimulus and not to another even when the 2 stimuli are very much alike. However, when the stimuli are so nearly identical that they cannot be distinguished, the animal becomes upset, whines, fails to cooperate, and tries to escape. Pavlov called these symptoms the **experimental neurosis.** One may quarrel about

Figure 16–9. Basal view of the human brain, showing the facial recognition areas where bilateral lesions cause prosopagnosia. (Modified and reproduced, with permission, from Geschwind N: Specializations of the human brain. *Sci Am* [Sept] 1979;**241**:180. Copyright © 1979 by Scientific American. All rights reserved.)

whether this reaction is a true neurosis in the psychiatric sense, but the term is convenient.

Animals subjected to frontal lobectomy are still capable of discriminating between like stimuli up to a point; but when they can no longer discriminate, their failure does not upset them. Because of these results in animals, **prefrontal lobotomy** and various other procedures aimed at cutting the connections between the frontal lobes and the deeper portions of the brain were used to treat various mental diseases in humans.

In some mental patients, tensions resulting from real or imagined failures of performance and the tensions caused by delusions, compulsions, and phobias are so great as to be incapacitating. Successful lobotomy reduces the tension. The delusions and other symptoms are still there, but they no longer bother the patient. A similar lack of concern over severe pain led to the use of lobotomy in treating patients with intractable pain (see Chapter 7). Unfortunately, this lack of concern often extends to other aspects of the environment, including relations with associates, social amenities, and even toilet habits. Furthermore, the relief from the suffering associated with pain usually lasts less than a year.

The effects that lobotomy can have on the personality were well described 125 years ago by the physician who cared for a man named Phineas P. Gage. Gage was a construction foreman who was packing blasting powder into a hole with a tamping iron. The powder exploded, driving the tamping iron through his face and out the top of his skull, transecting his frontal lobes. After the accident, he became, in the words of his physician, ". . . fitful, irreverent, indulging at times in the grossest profanity (which was not previously his custom), manifesting but little deference to his fellows, impatient of restraint or advice when it conflicts with his desires, at times pertinaciously obstinate yet capricious and vacillating, devising many plans for future operation which are no sooner arranged than they are abandoned in turn for others appearing more feasible. . . . His mind was radically changed, so that his friends and acquaintances said he was no longer Gage."*

This description is classic, but it cannot be said to be typical. The effects of lobotomy in humans are highly variable from patient to patient, and complications are frequent. For these reasons and because the desirable effects of lobotomy can generally be achieved with tranquilizers and other drugs, lobotomies are now performed rarely, if ever, for the treatment of mental disease.

Temporal Lobectomy

The effects of bilateral temporal lobectomy were first described by Klüver and Bucy. Because of re-

*From Harlow JM: Recovery from the passage of an iron bar through the head. *Mass Med Soc Publications* 1868;**2**:329.

moval of limbic structures, temporal lobectomized monkeys ("Klüver-Bucy animals") are docile and hyperphagic and the males are hypersexual. The animals also demonstrate visual agnosia and a remarkable increase in oral activity. The monkeys repeatedly pick up all moveable objects in their environment. They manipulate each object in a compulsive way; mouth, lick, and bite it; and then, unless it is edible, discard it. However, discarded objects are

picked up again in a few minutes as if the animal had never seen them before and subjected to the same manipulation and oral exploration. It has been suggested that the cause of this pattern of behavior may be an inability to identify objects.

In humans, anxiety is associated with a bilateral increase in blood flow to a discrete portion of the anterior end of each temporal lobe.

REFERENCES

Section III: Functions of the Nervous System

Arnold AP, Gorski R: Gonadal steroid induction of structural sex differences in the central nervous system. *Annu Rev Neurosci* 1984;**7**:413.

Bonica JJ et al (editors): *The Management of Pain,* 2nd ed, 2 vols. Lea & Febiger, 1990.

Bradbury M: *The Concept of a Blood-Brain Barrier.* Wiley, 1979.

Bronisch FW: *The Clinically Important Reflexes* Grune & Stratton, 1952.

Burt DR, Kamatachi GL: GABA$_A$ receptor subtypes: From pharmacology to molecular biology. *FASEB J* 1991;**5**:2916.

Damasio AR: Aphasia. *N Engl J Med* 1992;**326**:531.

Davis M: The role of the amygdala in fear and anxiety. *Annu Rev Neurosci* 1992;**15**:353.

Edelman G: *The Remembered Present: A Biological Theory of Consciousness.* Basic Books, 1990.

Frégnac Y, Imbert M: Development of neuronal selectivity in primary visual cortex of cat. *Physiol Rev* 1984;**64**:325.

Ganong WF: The brain as an endocrine organ. *Acta Physiol Lat Am* 1983;**32**:31.

Gash DM, Boer GJ: *Vasopressin: Principles and Properties.* Plenum, 1987.

Georgopoulos AP: Higher order motor control. *Annu Rev Neurosci* 1991;**14**:361.

Gerfen CR: The neostrialal mosaic: Multiple levels of compartmental organization in the basal ganglia. *Annu Rev Neurosci* 1992;**15**:285.

Getchell TB et al (editors): *Smell and Taste in Health and Disease.* Raven Press, 1991.

Gillin JC, Byerley WF: The diagnosis and management of insomnia. *N Engl J Med* 1990;**322**:239.

Goldman-Rakic PS: Topography of cognition: parallel distributed networks in primate association cortex. *Annu Rev Neurosci* 1988;**11**:137.

Grossman SP: *Thirst and Sodium Appetite: Physiological Basis.* Academic Press, 1990.

Hall ZW: *An Introduction to Molecular Neurobiology.* Sinauer, 1992.

Harris RBS: Role of set-point theory in regulation of body weight. *FASEB J* 1990;**4**:3310.

Herzog DB, Copeland PM: Eating disorders. *N Engl J Med* 1985;**313**:295.

Himms-Hagen J: Neural control of brown adipose tissue thermogenesis, hypertrophy, and atrophy. *Front Neuroendocrinol* 1991;**12**:38.

Hudspeth AJ: How the ear's works work. *Nature* 1989;**341**:399.

Hunt CC: Mammalian muscle spindle: Peripheral mechanisms. *Physiol Rev* 1990;**70**:643.

Jacobs BL: How hallucinogenic drugs work. *Am Sci* 1987;**75**:386.

Jones BE: Paradoxical sleep and its chemical/structural substrates in the brain. *Neuroscience* 1991;**401**:637.

Imura H. Nakao K, Itoh H: The natriuretic peptide system in the brain: Implication in the central control of cardiovascular and neuroendocrine functions. *Front Neuroendocrinol* 1992;**13**:217.

Kandel ER, Schwartz JH, Jessell TM (editors): *Principles of Neural Science,* 3rd ed. Elsevier, 1991.

Katzman R: Alzheimer's disease. *N Engl J Med* 1986;**314**:964.

Koob GF, Bloom FE: Cellular and molecular mechanisms of drug dependence. *Science* 1988;**242**:715.

Kopin IT: Catecholamine metabolism: Basic metabolic aspects and clinical significance. *Pharmacol Rev* 1985;**37**:333.

Lam DM-K, Shatz CJ (editors): *Development of the Visual System.* MIT Press, 1991.

Leibovic KN: *Science of Vision.* Elsevier, 1990.

Le Magnen J: *Neurobiology of Feeding and Nutrition.* Academic Press, 1991.

Lisberger SG: The neural basis for learning simple motor skills. *Science* 1988;**242**:728.

Livingstone MS: Art, illusion, and the visual system. *Sci Am* (Jan) 1988;**258**:78.

Lopes da Silva FH et al: Anatomic organization and physiology of the limbic cortex. *Physiol Rev* 1990;**70**:453.

Maggio JE: Tachykinins. *Annu Rev Neurosci* 1988;**11**:13.

McLauglin SK, McKinnon PJ, Margolskee RF: Gustducin is a taste-cell-specific G protein closely related to transducins. *Nature* 1992;357:563.

Melzack R: The tragedy of needless pain. *Sci Am* (Feb) 1990;**262**:27.

Mendelson WB: *Human Sleep: Research and Clinical Care.* Plenum, 1987.

Moore RY, Bloom F: Central catecholamine neuronal systems: Anatomy and physiology of dopaminergic systems. *Annu Rev Neurosci* 1978;**1**:129.

Nathans J: The genes for color vision. *Sci Am* (Feb) 1989;**260**:42.

Nicholls OG, Martin AR, Wallace BG: *From Neuron to Brain,* 3rd ed. Sinauer, 1992.

Pavlov IP: *Conditioned Reflexes.* Oxford Univ Press, 1928.

Pickels JO: *An Introduction to the Physiology of Hearing,* 2nd ed. Academic Press, 1988.

Pulvirenti L, Koob GF: The neural substrates of drug addiction and dependence. *Funct Neurol* 1990;**5:**109.

Regan DM: *Human Brain Electrophysiology.* Elsevier, 1988.

Reis DJ, Doba N: The central nervous system and neurogenic hypertension. *Prog Cardiovasc Dis* 1974;**17:**51.

Rowland LP et al (editors): *Molecular Genetics in Diseases of the Brain, Nerve and Muscle.* Oxford Univ Press, 1989.

Scheuer ML, Pedley TA: The evaluation and treatment of seizures. *N Engl J Med* 1990;**323:**1486.

Schönbaum E, Lomax P (editors): *Thermoregulation.* 2 vols. Pergamon Press, 1990.

Sherrington CS: *The Integrative Action of the Nervous System.* Cambridge Univ Press, 1947.

Snyder SH: *Drugs and the Brain.* Freeman, 1986.

Sokoloff L: *Metabolic Probes of Central Nervous System Activity in Experimental Animals and Man.* Sinauer, 1984.

Squire LR et al (editors): *Memory: Organization and Locus of Change.* Oxford Univ Press, 1991.

Starke K, Göthert M, Kilbinger H: Modulation of neurotransmitter release by presynaptic autoreceptors. *Physiol Rev* 1989;**69:**864.

Stokes PE, Sikes CR: Hypothalamic-pituitary-adrenal axis in psychiatric disorders. *Annu Rev Med* 1991;**42:**519.

Stux G, Pomeranz B: *Basics of Acupuncture,* 2nd ed. Springer-Verlag, 1991.

Swanson LW, Sawchenko PE: Hypothalamic integration: Organization of the paraventricular and supraoptic nuclei. *Annu Rev Neurosci* 1983;**6:**269.

Thach WT, Goodkin HG, Keating JG: Cerebellum and adaptive coordination of movement. *Annu Rev Neurosci* 1992;**15:**403.

Tsukahara N: Synaptic plasticity in the mammalian central nervous system. *Annu Rev Neurosci* 1981;**4:**351.

Weiner RW, Ganong WF: Role of brain monoamines and histamine in regulation of anterior pituitary secretion. *Physiol Rev* 1978;**58:**905.

Wetterberg L (editor): *Genetics of Neuropsychiatric Disease.* Stockton Press, 1989.

Wolpaw JR, Schmidt JT, Vaughan TM (editors): Activity-driven CNS changes in learning and development. *Ann NY Acad Sci* 1992;**627:**1.

Yankner BA, Mesulam M-M: β-amyloid and the pathogenesis of Alzheimer's disease. *N Engl J Med* 1991;**325:**1849.

Zivin JA, Choi DW: Stroke therapy. *Sci Am* (July) 1991;**265:**56.

Section IV.
Endocrinology, Metabolism, & Reproductive Function

Energy Balance, Metabolism, & Nutrition

17

INTRODUCTION

The endocrine system, like the nervous system, adjusts and correlates the activities of the various body systems, making them appropriate to the changing demands of the external and internal environment. Endocrine integration is brought about by **hormones,** chemical messengers produced by ductless glands that are transported in the circulation to target cells. Some of the hormones are amines, and others are amino acids, polypeptides, proteins, or steroids.

The hormones regulate metabolic processes. The term **metabolism,** meaning literally "change," is used to refer to all the chemical and energy transformations that occur in the body.

The animal organism oxidizes carbohydrates, proteins, and fats, producing principally CO_2, H_2O, and the energy necessary for life processes. CO_2, H_2O, and energy are also produced when food is burned outside the body. However, in the body, oxidation is not a one-step, semiexplosive reaction but a complex, slow, stepwise process called **catabolism,** which liberates energy in small, usable amounts. Energy can be stored in the body in the form of special energy-rich phosphate compounds and in the form of proteins, fats, and complex carbohydrates synthesized from simpler molecules. Formation of these substances by processes that take up rather than liberate energy is called **anabolism.** This chapter sets the stage for consideration of endocrine function by providing a brief summary of the production and utilization of energy and the metabolism of carbohydrates, proteins, and fats.

ENERGY METABOLISM

Metabolic Rate

The amount of energy liberated by the catabolism of food in the body is the same as the amount liberated when food is burned outside the body. The energy liberated by catabolic processes in the body is used for maintaining body functions, digesting and metabolizing food, thermoregulation, and physical activity. It appears as external work, heat, and energy storage:

$$\text{Energy output} = \text{External work} + \text{Energy storage} + \text{Heat}$$

The amount of energy liberated per unit of time is the **metabolic rate.** Isotonic muscle contractions perform work at a peak efficiency approximating 50%:

$$\text{Efficiency} = \frac{\text{Work done}}{\text{Total energy expended}}$$

Essentially all of the energy of isometric contractions appears as heat, because little or no external work (the force multiplied by the distance that the force moves a mass) is done (see Chapter 3). Energy is stored by forming energy-rich compounds. The amount of energy storage varies, but in fasting individuals it is zero or negative. Therefore, in an individual who is not moving and who has not eaten recently, essentially all of the energy output appears as heat. When food is burned outside the body, all of the energy liberated also appears as heat.

Calories

The standard unit of heat energy is the **calorie (cal),** defined as the amount of heat energy necessary to raise the temperature of 1 g of water 1 degree, from 15 to 16 °C. This unit is also called the gram calorie, small calorie, or standard calorie. The unit commonly used in physiology and medicine is the **Calorie (kilocalorie, kcal),** which equals 1000 cal.

Calorimetry

The energy released by combustion of foodstuffs outside the body can be measured directly (**direct calorimetry**) by oxidizing the compounds in an apparatus such as a **bomb calorimeter,** a metal vessel surrounded by water inside an insulated container. The food is ignited by an electric spark. The change in the temperature of the water is a measure of the calories produced. Similar measurements of the energy released by combustion of compounds in living animals and humans are much more complex, but calorimeters have been constructed that can physically accommodate human beings. The heat produced by their bodies is measured by the change in temperature of the water in the walls of the calorimeter.

The caloric values of the common foodstuffs, as measured in a bomb calorimeter, are found to be 4.1 kcal/g of carbohydrate, 9.3 kcal/g of fat, and 5.3 kcal/g of protein. In the body, similar values are obtained for carbohydrate and fat, but the oxidation of protein is incomplete, the end products of protein catabolism being urea and related nitrogenous compounds in addition to CO_2 and H_2O (see below). Therefore, the caloric value of protein in the body is only 4.1 kcal/g.

Indirect Calorimetry

Energy production can also be calculated, at least in theory, by measuring the products of the energy-producing biologic oxidations—ie, CO_2, H_2O, and the end products of protein catabolism produced—or by measuring the O_2 consumed. This is **indirect calorimetry.** It is difficult to measure all of the end products, but measurement of O_2 consumption is relatively easy. Since O_2 is not stored and since its consumption, except when an O_2 debt is being incurred or repaid, always keeps pace with immediate needs, the amount of O_2 consumed per unit of time is proportionate to the energy liberated.

One problem with using O_2 consumption as a measure of energy output is that the amount of energy released per mole of O_2 consumed varies slightly with the type of compound being oxidized. The approximate energy liberation per liter of O_2 consumed is 4.82 kcal, and for many purposes this value is accurate enough. More accurate measurements require data on the foods being oxidized. Such data can be obtained from an analysis of the respiratory quotient and the nitrogen excretion.

Respiratory Quotient (RQ)

The **respiratory quotient (RQ)** is the ratio in the steady state of the volume of CO_2 produced to the volume of O_2 consumed per unit of time. It should be distinguished from the **respiratory exchange ratio (R),** which is the ratio of CO_2 to O_2 at any given time whether or not equilibrium has been reached. R is affected by factors other than metabolism. RQ and R can be calculated for reactions outside the body, for individual organs and tissues, and for the whole body. The RQ of carbohydrate is 1.00 and that of fat is about 0.70. This is because H and O are present in carbohydrate in the same proportions as in water,

whereas in the various fats, extra O_2 is necessary for the formation of H_2O.

Carbohydrate:

$$C_6H_{12}O_6 + 6O_2 \rightarrow 6CO_2 + 6H_2O$$
(glucose)

$$RQ = 6/6 = 1.00$$

Fat:

$$2C_{51}H_{96}O_6 + 145O_2 \rightarrow 102CO_2 + 98H_2O$$
(tripalmitin)

$$RQ = 102/145 = 0.703$$

Determining the RQ of protein in the body is a complex process, but an average value of 0.82 has been calculated. The approximate amounts of carbohydrate, protein, and fat being oxidized in the body at any given time can be calculated from the RQ and the urinary nitrogen excretion. RQ and R for the whole body differ in various conditions. For example, during hyperventilation, R rises because CO_2 is being blown off. During severe exercise, R may reach 2.00 because CO_2 is being blown off and lactic acid from anaerobic glycolysis is being converted to CO_2 (see below). After exercise, R may fall for a while to 0.50 or less. In metabolic acidosis, R rises because respiratory compensation for the acidosis causes the amount of CO_2 expired to rise (see Chapter 39). In severe acidosis, R may be greater than 1.00. In metabolic alkalosis, R falls.

The O_2 consumption and CO_2 production of an organ can be calculated at equilibrium by multiplying its blood flow per unit of time by the arteriovenous differences for O_2 and CO_2 across the organ, and the RQ can then be calculated. Data on the RQ of individual organs are of considerable interest in drawing inferences about the metabolic processes occurring in them. For example, the RQ of the brain is regularly 0.97–0.99, indicating that its principal but not its only fuel is carbohydrate. During secretion of gastric juice, the stomach has a negative R because it takes up more CO_2 from the arterial blood than it puts into the venous blood (see Chapter 26).

Measuring the Metabolic Rate

In determining the metabolic rate, O_2 consumption is usually measured with some form of oxygen-filled spirometer and a CO_2-absorbing system. Such a device is illustrated in Fig 17–1. The spirometer bell is connected to a pen that writes on a rotating drum as the bell moves up and down. The slope of a line joining the ends of each of the spirometer excursions is proportionate to the O_2 consumption. The amount of O_2 (in milliliters) consumed per unit of time is corrected to standard temperature and pressure (see

Figure 17–1. Diagram of a modified Benedict apparatus, a recording spirometer used for measuring human O_2 consumption, and the record obtained with it. The slope of the line AB is proportionate to the O_2 consumption. V, V: one-way check valves.

Chapter 34) and then converted to energy production by multiplying by 4.82 kcal/L of O_2 consumed.

Factors Affecting the Metabolic Rate

The metabolic rate is affected by many factors (Table 17–1). The most important is muscular exertion. O_2 consumption is elevated not only during exertion but also for as long afterward as is necessary to repay the O_2 debt (see Chapter 3). Recently ingested foods also increase the metabolic rate because of their **specific dynamic action (SDA).** The SDA of a food is the obligatory energy expenditure that occurs during its assimilation into the body. An amount of protein sufficient to provide 100 kcal increases the metabolic rate a total of 30 kcal; a similar amount of carbohydrate increases it 6 kcal; and a similar amount of fat, 4 kcal. This means, of course, that the calories available from the 3 foods are in effect reduced by this amount; the energy used in their assimilation must come from the food itself or from the body energy stores. The cause of the SDA, which may last up to 6 hours, is uncertain.

Another factor that stimulates metabolism is the environmental temperature. The curve relating the metabolic rate to the environmental temperature is U-shaped. When the environmental temperature is lower than body temperature, heat-conserving mechanisms such as shivering are activated and the metabolic rate rises. When the temperature is high enough to raise the body temperature, there is a general acceleration of metabolic processes, and the metabolic rate also rises.

The metabolic rate determined at rest in a room at a comfortable temperature 12–14 hours after the last meal is called the **basal metabolic rate (BMR).** Actually, the rate is not truly "basal"; the metabolic rate during sleep is lower than the "basal" rate. What the term basal actually denotes is a set of widely known and accepted standard conditions.

The BMR of a man of average size is about 2000 kcal/d. Large animals have higher absolute BMRs, but the ratio of BMR to body weight in small animals is much greater. One variable that correlates well with the metabolic rate in different species is the body surface area. This would be expected, since heat exchange occurs at the body surface. However, the line relating heat production to body weight is actually steeper than this; body surface area correlates to body weight raised to the 0.67 power, whereas heat production correlates to body weight raised to the 0.75 power (Fig 17–2). It has been argued that this difference represents the greater heat production of larger animals supporting their masses against gravity. Nevertheless, BMRs in humans are regularly related to body surface area.

The relationship between weight and height and body surface area in humans can be expressed by the following formula:

Table 17–1. Factors affecting the metabolic rate.

Muscular exertion during or just before measurement
Recent ingestion of food
High or low environmental temperature
Height, weight, and surface area
Sex
Age
Emotional state
Body temperature
Circulating levels of thyroid hormones
Circulating epinephrine and norepinephrine levels

Figure 17–2. Correlation between metabolic rate and body weight, plotted on logarithmic scales. The slope of the solid line is 0.75. The dashed line represents the way surface area increases with weight for geometrically similar shapes and has a slope of 0.67. (Modified from Kleiber M and reproduced, with permission, from McMahon TA: Size and shape in biology. *Science* 1973;**179**:1201. Copyright © 1973 by the American Association for the Advancement of Science.)

$$S = 0.007184 \times W^{0.425} \times H^{0.725}$$

where

$S =$ **Surface area in m^2**
$W =$ **Body weight in kg**
$H =$ **Height in cm**

Nomograms constructed from this formula are available for easy estimation of body surface area. A normal BMR in an adult male is about 40 kcal/m^2/h. For convenience, the BMR is usually expressed as a percent increase or decrease above or below a set of generally used standard normal values. Thus, a value of 65 means that the individual's BMR is 65% above the standard for that age and sex.

In females, the BMR at all ages is slightly lower than in males. O$_2$ consumption is increased in pregnancy because of the additional needs of the fetus. Basal metabolism is high in children and declines with age. Anxiety and tension elevate the BMR because they cause increased epinephrine secretion and increased tensing of the muscles, even when the individual is quiet. On the other hand, apathetic, depressed patients may have low BMRs. An increase in body temperature speeds up chemical reactions, and the BMR rises approximately 14% for each Celsius degree of fever (7% per Fahrenheit degree). The stimulatory effects of catecholamines and thyroid hormones on the BMR are discussed in the chapters on those hormones. During prolonged starvation, the BMR falls. Sympathetic function is also depressed, and the decrease in circulating catecholamines may contribute to the decrease in BMR. In addition, the circulating levels of biologically active thyroid hormones fall (see Chapter 18). The decrease in metabolic rate is the explanation of why, when an individual reduces, weight loss is initially rapid and then slows down. Conversely, sympathetic function and BMR increase after feeding.

Energy Balance

The first law of thermodynamics, the principle which states that energy is neither created nor destroyed when it is converted from one form to another, applies to living organisms as well as inanimate systems. One may therefore speak of an **energy balance** between caloric intake and energy output. If the caloric content of the food ingested is less than the energy output—ie, if the balance is negative—endogenous stores are utilized. Glycogen, body protein, and fat are catabolized, and the individual loses weight. If the caloric value of the food intake exceeds energy loss due to heat and work and the food is properly digested and absorbed—ie, if the balance is positive—energy is stored, and the individual gains weight.

Except in humans and some hibernating and domesticated animals, the appetite mechanism regulates food intake with such precision that obesity is rare. Appetite is regulated as if it were determined by caloric needs, although the link between caloric consumption and appetite is probably indirect. The mechanisms regulating food intake are discussed in Chapter 14.

To balance basal output so that the energy-consuming tasks essential for life can be performed, the average adult must take in about 2000 kcal/d. Caloric requirements above the basal level depend upon the individual's activity. The average sedentary student (or professor) needs another 500 kcal, whereas a lumberjack needs up to 3000 additional kilocalories per day. Children require a lower absolute intake because of their smaller size, but a larger relative intake is essential to promote growth.

INTERMEDIARY METABOLISM

General Considerations

The end products of the digestive processes discussed in Chapters 25 and 26 are for the most part amino acids, fat derivatives, and the hexoses fructose, galactose, and glucose. These compounds are absorbed and metabolized in the body by various routes. The details of their metabolism are the concern of biochemistry and are not considered here. However, an outline of carbohydrate, protein, and fat metabolism is included for completeness and because some knowledge of the pathways involved is essential to an understanding of the action of thyroid, pancreatic, and adrenal hormones.

General Plan of Metabolism

The short-chain fragments produced by hexose, amino acid, and fat catabolism are very similar. From this **common metabolic pool** of intermediates, carbohydrates, proteins, and fats can be synthesized, or the fragments can enter the citric acid cycle, a sort of final common pathway of catabolism, in which they are broken down to hydrogen atoms and CO_2. The hydrogen atoms are oxidized to form water by a chain of flavoprotein and cytochrome enzymes.

Energy Transfer

The energy liberated by catabolism is not used directly by cells but is applied instead to the formation of bonds between phosphoric acid residues and certain organic compounds. Because the energy of bond formation in some of these phosphates is particularly high, relatively large amounts of energy (10–12 kcal/mol) are released when the bond is hydrolyzed. Compounds containing such bonds are called **high-energy phosphate compounds.** Not all organic phosphates are of the high-energy type. Many, like glucose 6-phosphate, are low-energy phosphates that on hydrolysis liberate 2–3 kcal/mol. Some of the intermediates formed in carbohydrate metabolism are high-energy phosphates, but the most important high-energy phosphate compound is **adenosine triphosphate (ATP).** This ubiquitous molecule (Fig 17–3) is the energy storehouse of the body. Upon hydrolysis to adenosine diphosphate (ADP), it liberates energy directly to such processes as muscle contraction, active transport, and the synthesis of many chemical compounds. Loss of another phosphate to form adenosine monophosphate (AMP) releases more energy. Another energy-rich phosphate compound found in muscle is **creatine phosphate (phosphorylcreatine, CrP; Fig 17–21).** Other important phosphorylated

compounds, at least some of which can serve as energy donors, include the triphosphate derivatives of pyrimidine or purine bases other than adenine (Fig 17–22). These include the guanine derivative, guanosine triphosphate (GTP); the cytosine derivative, cytidine triphosphate (CTP); the uracil derivative, uridine triphosphate (UTP); and the hypoxanthine derivative, inosine triphosphate (ITP). Many catabolic reactions are associated with the formation of energy-rich phosphates.

Another group of high-energy compounds are the thioesters, the acyl derivatives of mercaptans. **Coenzyme A (CoA)** is a widely distributed mercaptan containing adenine, ribose, pantothenic acid, and thioethanolamine (Fig 17–4). Reduced CoA (usually abbreviated HS-CoA) reacts with acyl groups (R — CO —) to form R — CO — S-CoA derivatives. A prime example is the reaction of HS-CoA with acetic acid to form acetylcoenzyme A (acetyl-CoA), a compound of pivotal importance in intermediary metabolism. Because acetyl-CoA has a much higher energy content than acetic acid, it combines readily with substances in reactions that would otherwise require outside energy. Acetyl-CoA is therefore often called "active acetate." From the point of view of energetics, formation of 1 mol of any acyl-CoA compound is equivalent to the formation of 1 mol of ATP.

Biologic Oxidations

Oxidation is the combination of a substance with

Figure 17–3. Energy-rich adenosine derivatives. (Reproduced, with permission, from Murray RK et al: *Harper's Biochemistry,* 23rd ed. Appleton & Lange, 1993.)

Figure 17–4. *Top:* Formula of reduced CoA (HS-CoA). *Bottom:* Formula for reaction of CoA with biologically important compounds to form thioesters. R, rest of molecule.

O_2, or loss of hydrogen, or loss of electrons. The corresponding reverse processes are called **reduction.** Biologic oxidations are catalyzed by enzymes, a particular protein enzyme being responsible in most cases for one particular reaction. Cofactors (simple ions) or coenzymes (organic, nonprotein substances) are accessory substances that usually act as carriers for products of the reaction. Unlike the enzymes, the coenzymes may catalyze a variety of reactions.

A number of coenzymes serve as hydrogen acceptors. One common form of biologic oxidation is removal of hydrogen from an $R-OH$ group, forming $R=O$. In such dehydrogenation reactions, nicotinamide adenine dinucleotide (NAD^+) and nicotinamide adenine dinucleotide phosphate ($NADP^+$) pick up hydrogen, forming dihydronicotinamide adenine dinucleotide (NADH) and dihydronicotinamide adenine dinucleotide phosphate (NADPH) (Fig 17–5). The hydrogen is then transferred to the flavoprotein-cytochrome system, reoxidizing the NAD^+ and $NADP^+$. Flavin adenine nucleotide (FAD) can accept hydrogens in a similar fashion, forming its hydro (FADH) and dihydro ($FADH_2$) derivatives.

The flavoprotein-cytochrome system is a chain of enzymes that transfers hydrogen to oxygen, forming water. This process occurs in the mitochondria. Each enzyme in the chain is reduced and then reoxidized as the hydrogen is passed down the line. Each of the enzymes is a protein with an attached nonprotein prosthetic group. The final enzyme in the chain is cytochrome oxidase, which transfers hydrogens to O_2 forming H_2O. It is also known as cytochrome aa_3, because it contains the activities of 2 enzymes formerly thought to be separate. It contains 2 atoms of Fe and 2 of Cu.

Oxidative Phosphorylation

Production of ATP associated with oxidation by the flavoprotein-cytochrome system is called **oxidative phosphorylation.** ATP can also be formed in other situations when the release of free energy by a chemical reaction is great. This is called **oxidation at the substrate level.** Oxidative phosphorylation is chemiosmotic and involves the transfer of protons across an insulating membrane (the inner membrane that forms the cristae of the mitochondria), the transfer being driven by oxidation in the respiratory chain (Fig 17–6). This creates an electrochemical potential difference across the membrane, and the transport of protons from the intracristal space back into the matrix space (see Chapter 1) drives a reversible ATPase in the membrane (**ATP synthase,** F-ATPase) in the direction that converts ADP and inorganic phosphate (Pi) to ATP (Fig 17–7). The process depends on an adequate supply of ADP and consequently is in part under a form of feedback control; the more rapid the utilization of ATP in the tissues, the greater the accumulation of ADP and, consequently, the more rapid the rate of oxidative phosphorylation. Other factors regulating the rate of generation of ATP include the rate of delivery of fats, lactate, and glucose derivatives to the interior of the mitochondria and the availability of O_2.

In addition to its function in energy transfer, ATP is the precursor of cyclic AMP. The function of this compound is discussed in Chapter 1.

Figure 17–5. **Top:** Formula of the oxidized form of nicotinamide adenine dinucleotide (NAD^+). Nicotinamide adenine dinucleotide phosphate ($NADP^+$) has an additional phosphate group at the location marked by the asterisk. **Bottom:** Reaction by which NAD^+ and $NADP^+$ become reduced to form NADH and NADPH. R, remainder of molecule; R', hydrogen donor.

Figure 17–6. Summary of oxidative phosphorylation in mitochondria. Pyruvate, fats, and other reduced carbon chains provide hydrogen to NAD$^+$ and FAD, and these pass down the flavoprotein-cytochrome chain to cytochrome aa$_3$ (cyto aa$_3$). The free energy of NADH and FADH is used to transport protons out of the mitochondria, producing an H$^+$ electrochemical gradient. H$^+$ moves back into the mitochondria, driving ATP synthase to convert ADP and Pi into ATP. (Reproduced, with permission, from Balaban RS: Regulation of oxidative phosphorylation in the mammalian cell. *Am J Physiol* 1990;**258**:C377.)

CARBOHYDRATE METABOLISM

Dietary carbohydrates are for the most part polymers of hexoses, of which the most important are glucose, galactose, and fructose (Fig 17–8). Most of the monosaccharides occurring in the body are the D isomers. The principal product of carbohydrate digestion and the principal circulating sugar is glucose. The normal fasting level of glucose in peripheral venous blood is 70–110 mg/dL (3.9–6.1 mmol/L). In arterial blood, the glucose level is 15–30 mg/dL higher than in venous blood.

Once it enters the cells, glucose is normally phosphorylated to form glucose 6-phosphate. The enzyme

that catalyzes this reaction is **hexokinase.** In the liver, there is in addition an enzyme called **glucokinase,** which has greater specificity for glucose and which, unlike hexokinase, is increased by insulin and decreased in starvation and diabetes. The glucose 6-phosphate is either polymerized into glycogen or catabolized. The steps involved are outlined in Fig 17–9. The process of glycogen formation is called **glycogenesis,** and glycogen breakdown is called **glycogenolysis.** Glycogen, the storage form of glucose, is present in most body tissues, but the major supplies are in the liver and skeletal muscle. The breakdown of glucose to pyruvate or lactate (or both) is called **glycolysis.** Glucose catabolism proceeds via

Figure 17–7. Simplified diagram of transport of protons across the inner and outer lamellas of the inner mitochondrial membrane by the electron transport system (flavoprotein-cytochrome system), with return movement of protons down the proton gradient, generating ATP.

```
   H−C=O            H−C=O            CH₂OH
      |                |                |
   H−C−OH           H−C−OH           C=O
      |                |                |
  HO−C−H           HO−C−H           HO−C−H
      |                |                |
   H−C−OH           HO−C−H           H−C−OH
      |                |                |
   H−C−OH           H−C−OH           H−C−OH
      |                |                |
     CH₂OH            CH₂OH            CH₂OH

  D-Glucose        D-Galactose      D-Fructose
```

Figure 17–8. Structure of principal dietary hexoses. The naturally occurring D isomers are shown.

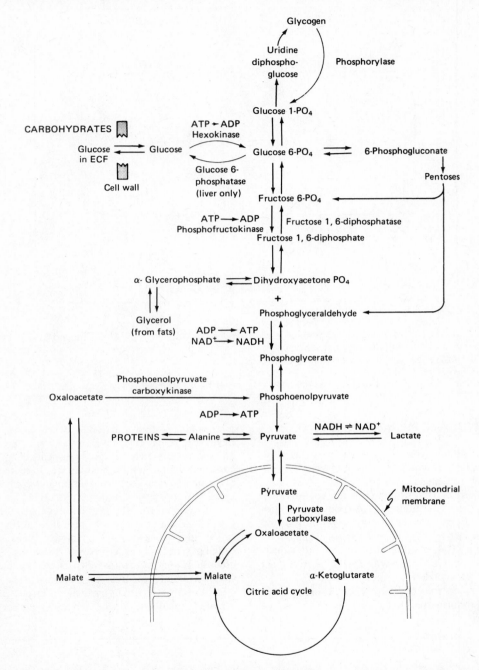

Figure 17–9. Outline of the metabolism of carbohydrate in cells, showing some of the principal enzymes involved.

cleavage to trioses, or via oxidation and decarboxylation to pentoses. The pathway to pyruvate through the trioses is the **Embden-Meyerhof pathway,** and that through 6-phosphogluconate and the pentoses is the **direct oxidative pathway (hexose monophosphate shunt)** (Fig 17–9). Pyruvate is converted to acetyl-CoA. Interconversions between carbohydrate,

protein, and fat (see below) include conversion of the glycerol from fats to dihydroxyacetone phosphate and conversion of a number of amino acids with carbon skeletons resembling intermediates in the Embden-Meyerhof pathway and citric acid cycle to these intermediates by deamination. In this way, and by conversion of lactate to glucose, nonglucose molecules can be converted to glucose (**gluconeogenesis**). Glucose can be converted to fats through acetyl-CoA, but since the conversion of pyruvate to acetyl-CoA, unlike most reactions in glycolysis, is irreversible (Fig 17–10), fats are not converted to glucose via this pathway. There is therefore very little net conversion of fats to carbohydrate in the body because, except for the quantitatively unimportant production from glycerol, there is no pathway for conversion.

Citric Acid Cycle

The **citric acid cycle** (Krebs cycle, tricarboxylic acid cycle) is a sequence of reactions in which acetyl-CoA is metabolized to CO_2 and H atoms. Acetyl-CoA is first condensed with the anion of a 4-carbon acid, oxaloacetate, to form citrate and HS-CoA. In a series of 7 subsequent reactions, 2 CO_2 molecules are split off, regenerating oxaloacetate (Fig 17–10). Four pairs of H atoms are transferred to the flavoprotein-cyto-

chrome chain, producing 12 ATP and 4 H_2O, of which 2 H_2O are used in the cycle. The citric acid cycle is the common pathway for oxidation to CO_2 and H_2O of carbohydrate, fat, and some amino acids. The major entry into it is through acetyl-CoA, but a number of amino acids can be converted to citric acid cycle intermediates by deamination. The citric acid cycle requires O_2 and does not function under anaerobic conditions.

Glycolysis to pyruvate occurs outside the mitochondria. Pyruvate then enters the mitochondria and is metabolized. Oxidative phosphorylation occurs only in the mitochondria. Within the mitochondria, the enzymes involved in this process are arranged in orderly sequences along the shelves and inner walls.

Energy Production

The net production of energy-rich phosphate compounds during the metabolism of glucose and glycogen to pyruvate depends on whether metabolism oc-

Figure 17–10. Citric acid cycle. The numbers (6C, 5C, etc) indicate the number of carbon atoms in each of the intermediates. The conversion of pyruvate to acetyl-CoA and each turn of the cycle provide 4 NADH and one $FADH_2$ for oxidation via the flavoprotein-cytochrome chain plus formation of one GTP that is readily converted to ATP. (Modified and reproduced, with permission, from Alberts BM et al: *Molecular Biology of the Cell,* 2nd ed. Garland, 1989.)

curs via the Embden-Meyerhof pathway or the hexose monophosphate shunt. By oxidation at the substrate level, the conversion of 1 mol of phosphoglyceraldehyde to phosphoglycerate generates 1 mol of ATP, and the conversion of 1 mol of phosphoenolpyruvate to pyruvate another. Since 1 mol of glucose 6-phosphate produces, via the Embden-Meyerhof pathway, 2 mol of phosphoglyceraldehyde, 4 mol of ATP are generated per mole of glucose metabolized to pyruvate. All these reactions occur in the absence of O_2 and consequently represent anaerobic production of energy. However, 1 mol of ATP is used in forming fructose 1,6-diphosphate from fructose 6-phosphate and 1 mol in phosphorylating glucose when it enters the cell. Consequently, when pyruvate is formed anaerobically from glycogen, there is a *net* production of 3 mol of ATP per mole of glucose 6-phosphate; however, when pyruvate is formed from 1 mol of blood glucose, the net gain is only 2 mol of ATP.

A supply of NAD^+ is necessary for the conversion of phosphoglyceraldehyde to phosphoglycerate. Under anaerobic conditions (anaerobic glycolysis), a block of glycolysis at the phosphoglyceraldehyde conversion step might be expected to develop as soon as the available NAD^+ is converted to NADH. However, pyruvate can accept hydrogen from NADH, forming NAD^+ and lactate.

$$\textbf{Pyruvate + NADH} \rightleftharpoons \textbf{Lactate + NAD}^+$$

In this way, glucose metabolism and energy production can continue for a while without O_2. The lactate that accumulates is converted back to pyruvate when the O_2 supply is restored, NADH transferring its hydrogen to the flavoprotein-cytochrome chain.

During aerobic glycolysis, the net production of ATP is 19 times as great as the 2 ATPs formed under anaerobic conditions. Six ATPs are formed by oxidation, via the flavoprotein-cytochrome chain, of the 2 NADH produced when 2 mol of phosphoglyceraldehyde are converted to phosphoglycerate (Fig 17-9), 6 ATPs are formed from the 2 NADH produced when 2 mol of pyruvate are converted to acetyl-CoA, and 24 ATPs are formed during the subsequent 2 turns of the citric acid cycle. Of these, 18 are formed by oxidation of 6 NADH, 4 by oxidation of 2 $FADH_2$, and 2 by oxidation at the substrate level when succinyl-CoA is converted to succinate. This reaction actually produces GTP, but the GTP is converted to ATP. Thus, the net production of ATP per mol of blood glucose metabolized aerobically via the Embden-Meyerhof pathway and citric acid cycle is 2 + [2 × 3] + [2 × 3] + [2 × 12] = 38.

Glucose oxidation via the hexosemonophosphate shunt generates large amounts of NADPH. A supply of this reduced coenzyme is essential for many metabolic processes. The pentoses formed in the process are building blocks for nucleotides (see below). The amount of ATP generated depends upon the amount of NADPH converted to NADH and then oxidized.

"Directional-Flow Valves"

Metabolism is regulated by a variety of hormones and other factors. To bring about any net change in a particular metabolic process, regulatory factors obviously must drive a chemical reaction in one direction. Most of the reactions in intermediary metabolism are freely reversible, but there are a number of "directional-flow valves," ie, reactions that proceed in one direction under the influence of one enzyme or transport mechanism and in the opposite direction under the influence of another. Five examples in the intermediary metabolism of carbohydrate are shown in Fig 17-11. The different pathways for fatty acid synthesis and catabolism (see below) are another example. Regulatory factors exert their influence on metabolism by acting directly or indirectly at these "directional-flow valves."

Phosphorylase

Glycogen breakdown is regulated by several hor-

Figure 17-11. Five examples of "directional-flow valves" in carbohydrate metabolism: reactions that proceed in one direction by one mechanism and in the other direction by a different mechanism. The double line in example 5 represents the mitochondrial membrane. Pyruvate is converted to malate in mitochondria, and the malate diffuses out of the mitochondria to the cytosol, where it is converted to phosphoenolpyruvate.

Figure 17–12. Glycogen formation and breakdown. The activation of phosphorylase a is summarized in Fig 17–13.

mones. Glycogen is synthesized from glucose 1-phosphate via uridine diphosphoglucose (UDPG; Fig 17–12), with the enzyme **glycogen synthase** catalyzing the final step. Glycogen is a branched glucose polymer with 2 types of glucoside linkage. Cleavage of the 1:4α linkage in the polymer chain is catalyzed by **phosphorylase,** while cleavage of the 1:6α linkages at branching points is catalyzed by another enzyme.

Phosphorylase is activated in part by the action of epinephrine on β-adrenergic receptors in the liver. This in turn initiates a sequence of reactions that provides a classic example of hormonal action via cyclic AMP (Fig 17–13). Protein kinase A is activated by cyclic AMP and catalyzes the transfer of a phosphate group to phosphorylase kinase, converting it to its active form. The phosphorylase kinase in turn catalyzes the phosphorylation and consequent activation of phosphorylase. Inactive phosphorylase is known as phosphorylase b (dephosphophosphorylase), and activated phosphorylase as phosphorylase a (phosphophosphorylase).

Activation of protein kinase A by cyclic AMP not only increases glycogen breakdown but also inhibits glycogen synthesis. Glycogen synthase (Fig 17–12) is active in its dephosphorylated form and inactive when phosphorylated, and it is phosphorylated along with phosphorylase kinase when protein kinase is activated.

Glycogen is also broken down by the action of catecholamines on α-adrenergic receptors in the liver.

Figure 17–13. Cascade of reactions by which epinephrine activates phosphorylase. Glucagon has a similar action in liver but not in skeletal muscle.

This breakdown is mediated by intracellular Ca^{2+} and involves an activation of phosphorylase kinase that is independent of cyclic AMP. Large doses of vasopressin and angiotensin II can also cause glycogenolysis by this mechanism, but it is doubtful if these hormones have any normal physiologic role in glucose homeostasis.

Because the liver contains the enzyme **glucose 6-phosphatase,** much of the glucose 6-phosphate that is formed in this organ can be converted to glucose and enter the bloodstream, raising the blood glucose level. The kidneys can also contribute to the elevation. Other tissues do not contain this enzyme, so in them a large proportion of the glucose 6-phosphate is catabolized via the Embden-Meyerhof pathway and hexosemonophosphate shunt pathway. Increased glucose catabolism in skeletal muscle causes a rise in the blood lactate level (see Chapter 3).

By stimulating adenylate cyclase, epinephrine causes activation of the phosphorylase in liver and skeletal muscle. The consequences of this activation are a rise in the blood glucose and lactate levels. Glucagon has a similar action, but it exerts its effect only on the phosphorylase in the liver. Consequently, glucagon causes a rise in blood glucose without any change in blood lactate.

McArdle's Disease

In the clinical condition known as **McArdle's disease** or **myophosphorylase deficiency glycogenosis,** glycogen accumulates in skeletal muscles because of a deficiency of muscle phosphorylase. Patients with this disease develop muscle pain and stiffness on exertion, and they have a greatly reduced exercise tolerance; they cannot break down their muscle glycogen to provide the energy for muscle contraction (see Chapter 3), and the glucose reaching their muscles from the bloodstream is sufficient only for the demands of very mild exercise. They respond with a normal rise in blood glucose when given glucagon or epinephrine, which indicates that their hepatic phosphorylase is normal.

The "Hepatic Glucostat"

There is a net uptake of glucose by the liver when the blood glucose is high and a net discharge when it is low. The liver thus functions as a sort of "glucostat," maintaining a constant circulating glucose level. This function is not automatic; glucose uptake and glucose discharge are affected by the actions of numerous hormones. Endocrine regulation of the blood glucose level and of carbohydrate metabolism in general is discussed in Chapter 19.

Renal Handling of Glucose

In the kidneys, glucose is freely filtered; but at normal blood glucose levels, all but a very small amount is reabsorbed in the proximal tubules (see Chapter 38). When the amount filtered increases,

reabsorption increases, but there is a limit to the amount of glucose the proximal tubules can reabsorb. When the tubular maximum for glucose (Tm_G) is exceeded, appreciable amounts of glucose appear in the urine (**glycosuria**). The **renal threshold** for glucose, the arterial blood level at which glycosuria appears, is reached when the venous blood glucose concentration is usually about 180 mg/dL, but it may be higher if the glomerular filtration rate is low.

Glycosuria

Glycosuria occurs when the blood glucose is elevated because of relative insulin deficiency (diabetes mellitus) or because of excessive glycogenolysis after physical or emotional trauma. In some individuals, the glucose transport mechanism in the renal tubules is congenitally defective, so that glycosuria is present at normal blood glucose levels. This condition is called **renal glycosuria. Alimentary glycosuria,** glycosuria after ingestion of a high-carbohydrate meal, has been claimed to occur in normal individuals, but many of these people actually have mild diabetes mellitus. The maximal rate of glucose absorption from the intestine is about 120 g/h.

Factors Determining the Blood Glucose Level

The blood glucose level at any given time is determined by the balance between the amount of glucose entering the bloodstream and the amount leaving it. The principal determinants are therefore the dietary intake; the rate of entry into the cells of muscle, adipose tissue, and other organs; and the glucostatic activity of the liver (Fig 17–14). Five percent of ingested glucose is promptly converted into glycogen in the liver, and 30–40% is converted into fat. The remainder is metabolized in muscle and other tissues. During fasting, liver glycogen is broken down and the liver adds glucose to the bloodstream. With more prolonged fasting, glycogen is depleted and there is increased gluconeogenesis from amino acids and glycerol in the liver. There is a modest decline in

Figure 17–14. Blood glucose homeostasis, illustrating the glucostatic function of the liver.

blood glucose to about 65 mg/dL in men and about 40 mg/dL in premenopausal women, but gluconeogenesis prevents the occurrence of more severe hypoglycemia, even during prolonged starvation. The cause of the lower fasting blood glucose concentrations in women is unknown, but it is interesting that values similar to those seen in women are seen in fasting prepuberal boys.

Carbohydrate Homeostasis in Exercise

In a 70-kg man, carbohydrate reserves total about 2500 kcal, stored in 400 g of muscle glycogen, 100 g of liver glycogen, and 20 g of glucose in extracellular fluid. In contrast, 112,000 kcal (about 80% of body fuel supplies) are stored in fat and the remainder in protein. Resting muscle utilizes fatty acids for its metabolism, and so does muscle after exercise. In the fasting human at rest, the brain accounts for 70–80% of the glucose utilized, and red blood cells account for most of the rest.

During exercise, the caloric needs of muscle are initially met by glycogenolysis in muscle and increased uptake of glucose. Blood glucose initially rises with increased hepatic glycogenolysis but may fall with strenuous, prolonged exercise. There is an increase in gluconeogenesis (Fig 17–15). Plasma insulin falls, and plasma glucagon and epinephrine rise.

After exercise, liver glycogen is replenished by additional gluconeogenesis and a decrease in hepatic glucose output. Insulin levels rise sharply, especially in hepatic portal blood. The insulin entering the liver presumably promotes glycogen storage (see Chapter 19).

Figure 17–15. Splanchnic (hepatic) glucose output, showing output due to glycogenolysis (open bars) and output presumably due to gluconeogenesis (in brackets). The values for gluconeogenesis are measured values for splanchnic uptake of the various gluconeogenetic precursors. (Reproduced, with permission, from Felig P, Wahren J: Fuel homeostasis in exercise. *N Engl J Med* 1975; **293**:1078.)

Metabolism of Hexoses Other Than Glucose

Other hexoses that are absorbed from the intestine include galactose, which is liberated by the digestion of lactose and converted to glucose in the body; and fructose, part of which is ingested and part produced by hydrolysis of sucrose. After phosphorylation, galactose is converted to uridine diphosphogalactose. The uridine diphosphogalactose is converted to uridine diphosphoglucose, which functions in glycogen synthesis (Fig 17–12). The latter reaction is reversible, and this is the way the galactose necessary for formation of glycolipids and mucoproteins is formed when dietary galactose intake is inadequate. The utilization of galactose, like that of glucose, is dependent upon insulin (see Chapter 19). In the inborn error of metabolism known as **galactosemia,** there is a congenital deficiency of phosphogalactose uridyl transferase, the enzyme responsible for the reaction between galactose 1-phosphate and uridine diphosphoglucose, so that ingested galactose accumulates in the circulation. Serious disturbances of growth and development result. Treatment with galactose-free diets improves this condition without leading to galactose deficiency because the enzymes necessary for the formation of uridine diphosphogalactose from uridine diphosphoglucose are present.

Fructose is converted in part to fructose 6-phosphate and then metabolized via fructose 1,6-diphosphate (Fig 17–9). The enzyme catalyzing the formation of fructose 6-phosphate is hexokinase, the same enzyme that catalyzes the conversion of glucose to glucose 6-phosphate. However, much more fructose is converted to fructose 1-phosphate in a reaction catalyzed by fructokinase. Most of the fructose 1-phosphate is then split into dihydroxyacetone phosphate and glyceraldehyde. The glyceraldehyde is phosphorylated, and it and the dihydroxyacetone phosphate enter the pathways for glucose metabolism. Since the reactions proceeding through phosphorylation of fructose in the 1 position can occur at a normal rate in the absence of insulin, it has been recommended that fructose be given to diabetics to replenish their carbohydrate stores. However, most of the fructose is metabolized in the intestines and liver, so its value in replenishing carbohydrate elsewhere in the body is limited.

Fructose 6-phosphate can also be phosphorylated in the 2 position, forming fructose 2,6-diphosphate. This compound is an important regulator of hepatic gluconeogenesis. When the fructose 2,6-diphosphate level is high, conversion of fructose 6-phosphate to fructose 1,6-diphosphate is facilitated, and thus breakdown of glucose to pyruvate is increased. A decreased level of fructose 2,6-diphosphate facilitates the reverse reaction and consequently aids gluconeogenesis. One of the actions of the protein kinase produced by the action of glucagon is to decrease hepatic fructose 2,6-diphosphate (see Chapter 19).

PROTEIN METABOLISM

Proteins

Proteins are made up of large numbers of amino acids (Fig 17–16) linked into chains by **peptide bonds** joining the amino group of one amino acid to the carboxyl group of the next. In addition, some proteins contain carbohydrates (glycoproteins) and lipids (lipoproteins). Smaller chains of amino acids are called **peptides** or **polypeptides.** The boundaries between peptides, polypeptides, and proteins are not well defined, but in this book, chains containing 2–10 amino acid residues are called peptides, chains containing more than 10 but fewer than 100 amino acid residues are called polypeptides, and chains containing 100 or more amino acid residues are called proteins. The term "oligopeptide," which is employed by others to refer to small peptides, is not used.

The order of the amino acids in the peptide chains is called the **primary structure** of a protein. The chains are twisted and folded in complex ways, and the term **secondary structure** of a protein refers to the spatial arrangement produced by the twisting and folding. A common secondary structure is a regular coil with 3.7 amino acid residues per turn (α-helix). Another common secondary structure is a β-sheet. An antiparallel β-sheet is formed when extended polypeptide chains fold back and forth on each other and hydrogen bonding occurs between the peptide bonds on neighboring chains. Parallel β-sheets between polypeptide chains also occur. In the so-called ribbon models of proteins (see Fig 27–6), the α-helices are shown as coils and the β-sheets as parallel arrows.

The **tertiary structure** of a protein is the arrangement of the twisted chains into layers, crystals, or fibers. Many protein molecules are made of subunits (eg, hemoglobin; see Chapter 27), and the term **quaternary structure** is used to refer to the arrangement of the subunits.

Amino Acids

The amino acids that are found in proteins are shown in Table 17–2. These amino acids are identified by the 3-letter abbreviations or the single-letter abbreviations shown in the table. Various other important amino acids such as ornithine, 5-hydroxytryptophan, L-dopa, taurine, and thyroxine (T_4) occur in the body but are not found in proteins. In higher animals, the L isomers of the amino acids are the only naturally occurring forms. The L isomers of hormones such as thyroxine are much more active than the D isomers. The amino acids are acidic, neutral, or basic in reaction, depending upon the relative proportions of free acidic ($-COOH$) or basic ($-NH_2$) groups in the molecule.

Some of the amino acids are **nutritionally essential amino acids,** which must be obtained in the diet, whereas others can be synthesized in vivo at rates sufficient to meet metabolic needs (see below).

The Amino Acid Pool

Small amounts of proteins are absorbed from the gastrointestinal tract in infants, and even smaller amounts in adults, but most ingested proteins are digested and their constituent amino acids absorbed. The body's own proteins are being continuously hydrolyzed to amino acids and resynthesized. The turnover rate of endogenous proteins averages 80–100 g/d, being highest in the intestinal mucosa and practically nil in collagen. The amino acids formed by endogenous protein breakdown are identical to those derived from ingested protein. With the latter, they form a common **amino acid pool** that supplies the needs of the body (Fig 17–17). In the kidney, most of the filtered amino acids are reabsorbed. During growth, the equilibrium between amino acids and body proteins shifts toward the latter, so that synthesis exceeds breakdown. At all ages, a small amount of protein is lost as hair. In women, small amounts are lost in the menstrual flow. Some small proteins are lost in the urine, and there are unreabsorbed protein digestive secretions in the stools. These losses are made up by synthesis from the amino acid pool.

Specific Metabolic Functions of Amino Acids

Thyroxine (T_4), catecholamines, histamine, serotonin, melatonin, and intermediates in the urea cycle are formed from specific amino acids. Methionine, cystine, and cysteine provide the sulfur contained in

Amino acid   Polypeptide chain

Figure 17–16. Amino acid structure and formation of peptide bonds. The dotted lines show how the peptide bonds are formed, with the production of H_2O. R, remainder of the amino acid. For example, in glycine, R=H; in glutamate, R = $-(CH_2)_2$ $-COOH$.

Table 17–2. Amino acids found in proteins. Those in bold type are the nutritionally essential amino acids. The generally accepted 3-letter and 1-letter abbreviations for the amino acids are shown in parentheses.

Neutral Amino Acids
  Amino acids with aliphatic side chains
    Glycine (Gly, G)
    Alanine (Ala, A)
    Valine (Val, V)
    Leucine (Leu, L)
    Isoleucine (Ile, I)
  Hydroxyl-substituted amino acids
    Serine (Ser, S)
    Threonine (Thr, T)
  Sulfur-containing amino acids
    Cysteine (Cys, C)
    Methionine (Met, M)
  Amino acids with aromatic ring side chains
    Phenylalanine (Phe, F)
    Tyrosine (Tyr, Y)
    Tryptophan (Trp, W)
	Amino Acids with Acidic Side Chains, or Their Amides
	  Aspartic acid (Asp, D)
	  Asparagine (Asn, N)
	  Glutamine (Gln, Q)
	  Glutamic acid (Glu, E)
	  γ-Carboxyglutamic acid[1] (Gla)
	Amino Acids with Side Chains Containing Basic Groups
	  Arginine[2] (Arg, R)
	  Lysine (Lys, K)
	  Hydroxylysine[1] (Hyl)
	  Histidine[2] (His, H)
	Imino Acids (contain imino group but no amino group)
	  Proline (Pro, P)
	  4-Hydroxyproline[1] (Hyp)
	  3-Hydroxyproline[1]

[1]There are no tRNAs for these 4 amino acids; they are formed by posttranslational modification of the corresponding unmodified amino acid in peptide linkage. There are tRNAs for the remaining 20 amino acids, and they are incorporated into peptides and proteins under direct genetic control.
[2]Arginine and histidine are sometimes called "semi-essential"; they are not necessary for maintenance of nitrogen balance but are needed for normal growth.

proteins, CoA, taurine, and other biologically important compounds. Methionine is converted into S-adenosylmethionine, which is the active methylating agent in the synthesis of compounds such as epinephrine, acetylcholine, and creatine. It is a major donor of biologically labile methyl groups, but methyl groups can also be synthesized from a derivative of formic acid bound to folic acid derivatives if the diet contains adequate amounts of folic acid and cyanocobalamin.

Urinary Sulfates

Sulfur-containing amino acids are the source of the sulfates in the urine. A few unoxidized sulfur-containing compounds are excreted (urinary **neutral sulfur**), but most of the urinary excretion is in the form of **sulfate** (SO_4^{2-}) accompanied by corresponding amounts of cation (Na^+, K^+, NH_4^+, or H^+). The **ethereal sulfates** in the urine are organic sulfate esters ($R - O - SO_3H$) formed in the liver from endogenous and exogenous phenols, including estrogens and other steroids, indoles, and drugs.

Deamination, Amination, & Transamination

Interconversions between amino acids and the products of carbohydrate and fat catabolism at the level of the common metabolic pool and the citric acid cycle involve transfer, removal, or formation of amino groups. **Transamination** reactions, conversion of one amino acid to the corresponding keto acid with simultaneous conversion of another keto acid to an amino acid, occur in many tissues:

Alanine + α-Ketoglutarate $\rightleftharpoons$ Pyruvate + Glutamate

The **transaminases** involved are also present in the circulation. When damage to many active cells occurs as a result of a pathologic process, serum transaminase levels rise. An example is the rise in **serum glutamic-oxaloacetic transaminase (SGOT)** following myocardial infarction.

Oxidative deamination of amino acids occurs in the liver. An imino acid is formed by dehydrogenation, and this compound is hydrolyzed to the corresponding keto acid, with production of NH_4^+:

Amino acid + NAD^+ $\rightarrow$ Imino acid + NADH

Imino acid + H_2O $\rightarrow$ Keto acid + NH_4^+

NH_4^+ is in equilibrium with NH_3. Amino acids can also take up NH_4^+, forming the corresponding amide. An example is the binding of NH_4^+ in the brain by glutamine (Fig 17–18). The reverse reaction occurs in the kidney, with conversion of NH_4^+ to NH_3 and secretion of NH_3 into the urine. The NH_3 reacts with H^+ in the

Figure 17–17. Amino acid metabolism.

Figure 17–18. Release and uptake of NH_4^+ by interconversion of glutamine and glutamate. NH_4^+ is in equilibrium with NH_3. The reaction goes predominantly to the right in the kidney, and NH_3 is secreted into the urine. The reaction goes predominantly to the left in the brain, removing NH_3, which is toxic to nerve cells.

urine to form NH_4^+, thus permitting more H^+ to be secreted into the urine (see Chapter 38).

Interconversions between the amino acid pool and the common metabolic pool are summarized in Fig 17–19. Leucine, isoleucine, phenylalanine, and tyrosine are said to be **ketogenic** because they are converted to the ketone body acetoacetate (see below). Alanine and many other amino acids are **glucogenic** or **gluconeogenic;** ie, they give rise to compounds that can readily be converted to glucose.

Urea Formation

Most of the NH_4^+ formed by deamination of amino acids in the liver is converted to urea, and the urea is excreted in the urine. The NH_4^+ forms carbamoyl phosphate, and in the mitochondria, it is transferred to ornithine, forming citrulline. The enzyme involved is ornithine carbamoyltransferase. Citrulline is converted to arginine, after which urea is split off and ornithine is regenerated (urea cycle; Fig 17–20). Most of the urea is formed in the liver, and in severe

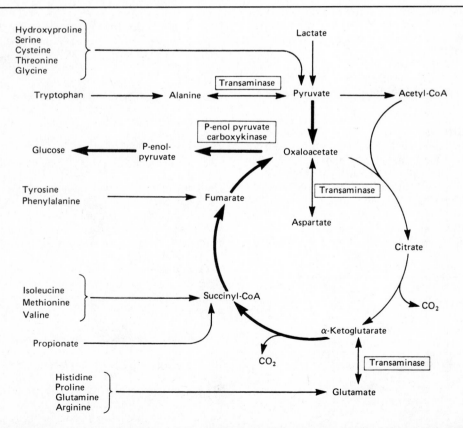

Figure 17–19. Involvement of the citric acid cycle in transamination and gluconeogenesis. The bold arrows indicate the main pathway of gluconeogenesis. (Reproduced, with permission, from Murray RK et al: *Harper's Biochemistry,* 23rd ed. Appleton & Lange, 1993.)

Figure 17–20. Urea cycle.

liver disease the blood urea nitrogen (BUN) falls and blood NH₃ rises. Congenital deficiency of ornithine carbamoyltransferase can also lead to NH₃ intoxication, even in individuals who are heterozygous for this deficiency.

Creatine & Creatinine

Creatine is synthesized in the liver from methionine, glycine, and arginine. In skeletal muscle, it is phosphorylated to form **phosphorylcreatine** (Fig 17–21), which is an important energy store for ATP synthesis (see Chapter 3). The ATP formed by glycolysis and oxidative phosphorylation reacts with creatine to form ADP and large amounts of phosphorylcreatine. During exercise, the reaction is reversed, maintaining the supply of ATP, which is the immediate source of the energy for muscle contraction.

The creatinine in the urine is formed from phosphorylcreatine. Creatine is not converted directly to creatinine. The rate of creatinine excretion is relatively constant from day to day. Indeed, creatinine output is sometimes measured as a check on the accuracy of the urine collections in metabolic studies; an average daily creatinine output is calculated, and the values for the daily output of other substances are corrected to what they would have been at this creatinine output.

Creatinuria occurs normally in children, in women during and after pregnancy, and occasionally in nonpregnant women. There is very little, if any, creatine in the urine of normal men, but appreciable quantities are excreted in any condition associated with extensive muscle breakdown. Thus, creatinuria occurs in starvation, thyrotoxicosis, poorly controlled diabetes mellitus, and the various primary and secondary diseases of muscle (**myopathies**).

Purines & Pyrimidines

The physiologically important **purines** and **pyrimidines** are shown in Fig 17–22. **Nucleosides**— purines or pyrimidines combined with ribose—are components not only of a variety of coenzymes and related substances (NAD⁺, NADP⁺, ATP, UDPG, etc) but of RNA and DNA as well (Table 17–3). The structure and function of DNA and RNA and their role in protein synthesis are discussed in Chapter 1.

Nucleic acids in the diet are digested and their constituent purines and pyrimidines absorbed, but most of the purines and pyrimidines are synthesized from amino acids, principally in the liver. The nucleotides and RNA and DNA are then synthesized. RNA is in dynamic equilibrium with the amino acid pool,

Figure 17–21. Creatine, phosphorylcreatine, and creatinine.

Adenine: 6-Aminopurine
Guanine: 1-Amino-6-oxypurine
Hypoxanthine: 6-Oxypurine
Xanthine: 2,6-Dioxypurine

Purine nucleus

Cytosine: 4-Amino-2-oxypyrimidine
Uracil: 2,4-Dioxypyrimidine
Thymine: 5-Methyl-2,4-dioxypyrimidine

Pyrimidine nucleus

Figure 17–22. Principal physiologically important purines and pyrimidines. Oxypurines and oxypyrimidines may form enol derivatives (hydroxypurines and hydroxypyrimidines) by migration of hydrogen to the oxygen substituents.

Table 17–3. Purine- and pyrimidine-containing compounds.

Purine or pyrimidine + Ribose or 2-deoxyribose = Nucleoside

Nucleoside + Phosphoric acid residue =
Nucleotide (mononucleotide)

Many nucleotides forming double-helical structure
of 2 polynucleotide chains = Nucleic acid

Nucleic acid + One or more simple basic proteins =
Nucleoprotein

Ribonucleic acids (RNA) contain ribose

Deoxyribonucleic acids (DNA) contain 2-deoxyribose

but DNA, once formed, is metabolically stable throughout life.

The purines and pyrimidines released by the breakdown of nucleotides may be reused or catabolized. Minor amounts are excreted unchanged in the urine. The pyrimidines are catabolized to CO_2 and NH_3, and the purines are converted to uric acid.

Protein Degradation

Like protein synthesis, protein degradation is a carefully regulated, complex process. Conjugation of proteins to the 74-amino-acid polypeptide **ubiquitin** tickets them for degradation. This polypeptide is highly conserved and is present in species ranging from bacteria to humans. It has been argued that a free NH_2-terminal group is essential to ubiquitin binding, but the conformations that are responsible are currently a matter of debate and active investigation. The rates at which individual proteins are metabolized vary, and the body appears to have mechanisms by which abnormal proteins are recognized and degraded more rapidly than normal body constituents. For example, abnormal hemoglobins are metabolized rapidly in individuals with congenital hemoglobinopathies (see Chapter 27). The rate of protein degradation is decreased during hypertrophy in exercised skeletal muscle and increased during atrophy in denervated or unused skeletal muscle. In addition, the rate of protein degradation is a factor in the determination of organ size (eg, the rate of degradation of liver protein is markedly reduced during the compensatory hypertrophy that follows partial hepatectomy).

Uric Acid

Uric acid is formed by the breakdown of purines and by direct synthesis from 5-phosphoribosyl pyrophosphate (5-PRPP) and glutamine (Fig 17–23). In humans, uric acid is excreted in the urine; but in other mammals, uric acid is further oxidized to allantoin before excretion. The normal blood uric acid level in humans is approximately 4 mg/dL (0.24 mmol/L). In the kidney, uric acid is filtered, reabsorbed, and se-

Figure 17–23. Uric acid metabolism.

creted. Normally, 98% of the filtered uric acid is reabsorbed and the remaining 2% makes up approximately 20% of the amount excreted. The remaining 80% comes from the tubular secretion. The uric acid excretion on a purine-free diet is about 0.5 g/24 h and on a regular diet about 1 g/24 h.

"Primary" & "Secondary" Gout

Gout is a disease characterized by recurrent attacks of arthritis; urate deposits in the joints, kidneys, and other tissues; and elevated blood and urine uric acid levels. The joint most commonly affected initially is the metatarsophalangeal joint of the great toe. There are 2 forms of "primary" gout: one in which, because of various enzyme abnormalities, uric acid production is increased and another in which there is a selective deficit in renal tubular transport of uric acid. In "secondary" gout, the uric acid levels in the body fluids are elevated as a result of decreased excretion or increased production secondary to some other disease process. For example, excretion is decreased in patients treated with thiazide diuretics (see Chapter 38) and those with renal disease. Production is increased in leukemia and pneumonia because of increased breakdown of uric acid-rich white blood cells.

The treatment of gout is aimed at relieving the acute arthritis with drugs such as colchicine or nonsteroidal anti-inflammatory agents and decreasing the uric acid level in the blood. Colchicine does not affect uric acid metabolism, and it apparently relieves gouty attacks by inhibiting the phagocytosis of uric acid crystals by leukocytes, a process that in some way produces the joint symptoms. Phenylbutazone and probenecid inhibit uric acid reabsorption in the renal tubules. Allopurinol, which inhibits xanthine oxidase (Fig 17–23), is one of the drugs used to decrease uric acid production.

Nitrogen Balance

A moderate daily protein intake is necessary to replace protein and amino acid losses. This requirement is not for protein itself but for the constituent amino acids, and the need can be met by feeding pure amino acids. Loss of protein and its derivatives in the stools is normally very small. Consequently, the amount of nitrogen in the urine is a reliable indicator of the amount of irreversible protein and amino acid breakdown. When the amount of nitrogen in the urine is equal to the nitrogen content of the protein in the diet, the individual is said to be in **nitrogen balance.** If protein intake is increased in a normal individual, the extra amino acids are deaminated and urea excretion increases, maintaining nitrogen balance. However, in conditions in which secretion of the catabolic hormones of the adrenal cortex is elevated or that of insulin is decreased, and during starvation and forced immobilization, nitrogen losses exceed intake and the nitrogen balance is **negative.** During growth or recovery from severe illnesses, or following administration of anabolic steroids such as testosterone, nitrogen intake exceeds excretion and nitrogen balance is **positive.**

When any one of the nutritionally essential amino acids necessary for synthesis of a particular protein is unavailable, the protein is not synthesized. The other amino acids that would have gone into the protein are deaminated, like other excess amino acids, and their nitrogen is excreted as urea. This is probably why nitrogen balance becomes negative whenever a single essential amino acid is omitted from the diet.

Response to Starvation

When an individual eats a diet that is low in protein but calorically adequate, excretion of urea and inorganic and ethereal sulfates declines. Uric acid excretion falls by 50%. Creatinine excretion is not affected. The creatinine and about half of the uric acid in the urine must therefore be the result of "wear-and-tear" processes that are unaffected by the protein intake. Total nitrogen excretion fails to fall below 3.6 g/d during protein starvation when the diet is calorically adequate because of the negative nitrogen balance produced by essential amino acid deficiencies.

On a diet that is inadequate in calories as well, urea nitrogen excretion averages about 10 g/d as proteins are catabolized for energy. Small amounts of glucose counteract this catabolism to a marked degree (**protein-sparing effect** of glucose). This protein-sparing effect is probably due for the most part to the increased insulin secretion produced by the glucose. The insulin in turn inhibits the breakdown of protein in muscle. Intravenous injection of relatively small amounts of amino acids also exerts a considerable protein-sparing effect.

Fats also spare nitrogen. During prolonged starvation, keto acids derived from fats (see below) are used by the brain and other tissues. These substances share cofactors for metabolism in muscle with 3 branched-chain amino acids, leucine, isoleucine, and valine, and to the extent that the fat-derived keto acids are utilized, these amino acids are apparently spared. Infusion of the non-nitrogen-containing analogs of these amino acids produces protein sparing and decreases urea and ammonia formation in patients with renal and hepatic failure.

Most of the protein burned during total starvation comes from the liver, spleen, and muscles and relatively little from the heart and brain. The blood glucose falls somewhat after liver glycogen is depleted (see above), but is maintained above levels that produce hypoglycemic symptoms by gluconeogenesis. Ketosis is present, and neutral fat is rapidly catabolized. When fat stores are used up, protein catabolism increases even further, and death soon follows. An average 70-kg man has 0.1 kg of glycogen in his liver, 0.4 kg of glycogen in his muscles, and 12 kg of fat. The glycogen is enough fuel for about 1 day of starvation. In hospitalized obese patients given nothing except water and vitamins, weight loss was observed to be about 1 kg/d for the first 10 days. It then declined and stabilized at about 0.3 kg/d. The patients did quite well for a time, although postural hypotension and attacks of acute gouty arthritis were troublesome complications in some instances. In the Irish prisoners who starved themselves to death several years ago, the average time from the start of the fast to death was about 60 days.

There has been considerable interest in reducing diets in which the sole source of calories is liquid protein or amino acid mixtures. Such diets have been hypothesized to promote ketosis, which in turn decreases appetite while sparing body protein. However, it is uncertain whether these diets are more successful than others in reducing body weight, and there has been a relatively large number of unexplained sudden deaths in patients on these diets. Until the cause of these deaths is elucidated, it seems unwise to use the diets.

FAT METABOLISM

Lipids

The biologically important lipids are the fatty acids and their derivatives, the neutral fats (triglycerides), the phospholipids and related compounds, and the sterols. The triglycerides are made up of 3 fatty acids bound to glycerol (Table 17–4). Naturally occurring fatty acids contain an even number of carbon atoms. They may be saturated (no double bonds) or

Table 17–4. Lipids.

Typical fatty acids:

Palmitic acid:
$$CH_3(CH_2)_{14}-\overset{\overset{\displaystyle O}{\|}}{C}-OH$$

Stearic acid:
$$CH_3(CH_2)_{16}-\overset{\overset{\displaystyle O}{\|}}{C}-OH$$

Oleic acid:
$$CH_3(CH_2)_7CH=CH(CH_2)_7-\overset{\overset{\displaystyle O}{\|}}{C}-OH$$
(Unsaturated)

Triglycerides (triacylglycerols): Esters of glycerol and 3 fatty acids.

$$
\begin{array}{l}
CH_2-O-\overset{\overset{\displaystyle O}{\|}}{C}-R \\
CH-O-\overset{\overset{\displaystyle O}{\|}}{C}-R \;+\; 3H_2O \rightleftharpoons \\
CH_2-O-\overset{\overset{\displaystyle O}{\|}}{C}-R \\
\text{Triglyceride}
\end{array}
\qquad
\begin{array}{l}
CH_2OH \\
CHOH \;+\; 3HO-\overset{\overset{\displaystyle O}{\|}}{C}-R \\
CH_2OH \\
\text{Glycerol}
\end{array}
$$

R = Aliphatic chain of various lengths and degrees of saturation.

Phospholipids:
  A. Esters of glycerol, 2 fatty acids, and
    1. Phosphate = phosphatidic acid
    2. Phosphate plus inositol = phosphatidylinositol
    3. Phosphate plus choline = phosphatidylcholine (lecithin)
    4. Phosphate plus ethanolamine = phosphatidylethanolamine (cephalin)
    5. Phosphate plus serine = phosphatidylserine
  B. Other phosphate-containing derivatives of glycerol
  C. Sphingomyelins: Esters of fatty acid, phosphate, choline, and the amino alcohol sphingosine.

Cerebrosides: Compounds containing galactose, fatty acid, and sphingosine.

Sterols: Cholesterol and its derivatives, including steroid hormones, bile acids, and various vitamins.

unsaturated (dehydrogenated, with various numbers of double bonds). The phospholipids are constituents of cell membranes. The sterols include the various steroid hormones and cholesterol.

Fatty Acid Oxidation & Synthesis

In the body, fatty acids are broken down to acetyl-CoA, which enters the citric acid cycle. Fatty acid oxidation begins with activation of the fatty acid (Fig 17–24), and this reaction occurs both inside and outside the mitochondria. Active long-chain fatty acids formed outside the mitochondria must be linked to **carnitine,** a lysine derivative, to cross the mitochondrial membrane; thus, carnitine stimulates fat oxidation. The rest of the oxidation steps occur in the mitochondria, where 2-carbon fragments are serially split off the fatty acid (β-oxidation). The energy yield of this process is large. For example, catabolism of 1 mol of a 6-carbon fatty acid through the citric acid cycle to CO_2 and H_2O generates 44 mol of ATP, compared with the 38 mol generated by catabolism of 1 mol of the 6-carbon carbohydrate glucose.

Defective β-oxidation of fatty acids can be produced by carnitine deficiency or other abnormalities. The general pattern that is produced is **hypoketonemic hypoglycemia** because adequate amounts of CoA are not made available to the liver to make ketone bodies (see below) and insufficient ATP is generated to support gluconeogenesis. The clinical manifestations of this abnormality are acute episodes of vomiting and lethargy, which may progress to coma. The mortality is high.

Many tissues can synthesize fatty acids from acetyl-CoA. Some synthesis of long-chain fatty acids from short-chain fatty acids occurs in the mitochondria by simple reversal of the reactions shown in Fig 17–24. There is in addition a similar chain elongation system outside the mitochondria. However, most of the synthesis of fatty acids occurs de novo from acetyl-CoA via a different pathway located principally outside the mitochondria, in the microsomes. The steps in this pathway are summarized in Fig 17–25.

For unknown reasons, fatty acid synthesis stops in practically all cells when the chain is 16 carbon atoms long. Only small amounts of 12- and 14-carbon fatty acids are formed, and none with more than 16 carbons. Particularly in fat depots, the fatty acids are combined with glycerol to form neutral fats. This combination takes place in the mitochondria.

Ketone Bodies

In many tissues, acetyl-CoA units condense to form acetoacetyl-CoA (Fig 17–26). In the liver, which (unlike other tissues) contains a deacylase, free acetoacetate is formed. This β-keto acid is converted to β-hydroxybutyrate and acetone, and because these compounds are metabolized with difficulty in the liver, they diffuse into the circulation. Acetoacetate is also formed in the liver via the formation of β-hydroxy-β-methylglutaryl-CoA (Fig 17–26), and this pathway is quantitatively more important than deacylation. Acetoacetate, β-hydroxybutyrate, and acetone are called **ketone bodies.** Tissues other than liver transfer CoA from succinyl-CoA to acetoacetate and metabolize the "active" acetoacetate to CO_2 and H_2O via the citric acid cycle. There are also other pathways whereby ketone bodies are metabolized.

Fatty acid

"Active" fatty acid

$$R-CH_2CH_2-\overset{O}{\overset{\|}{C}}-OH + HS\cdot CoA \xrightarrow[ATP \to ADP]{Mg^{2+}} H_2O + R-CH_2CH_2-\overset{O}{\overset{\|}{C}}-S-CoA$$

Oxidized flavoprotein

↓

Reduced flavoprotein

$$R-\overset{OH}{\underset{H}{\overset{|}{C}}}-CH_2-\overset{O}{\overset{\|}{C}}-S-CoA \longleftarrow H_2O + R-CH=CH-\overset{O}{\overset{\|}{C}}-S-CoA$$

β-Hydroxy fatty acid –CoA   α,β-Unsaturated fatty acid –CoA

$NAD^+ \to NADH + H^+$

$$R-\overset{O}{\overset{\|}{C}}-CH_2-\overset{O}{\overset{\|}{C}}-S-CoA + HS\cdot CoA \longrightarrow R-\overset{O}{\overset{\|}{C}}-S-CoA + CH_3-\overset{O}{\overset{\|}{C}}-S-CoA$$

β-Keto fatty acid –CoA   "Active" fatty acid + Acetyl·CoA

R = Rest of fatty acid chain.

Figure 17–24. Fatty acid oxidation. This process, splitting off 2 carbon fragments at a time, is repeated to the end of the chain.

$$CH_3-\overset{O}{\overset{\|}{C}}-S-CoA + CO_2 \xrightarrow[ATP \to ADP]{Mn^{2+} \atop Biotin} CH_2\overset{COOH}{\underset{C-S-CoA \atop \|}{\overset{}{\diagdown}}}$$

Acetyl·CoA   Malonyl·CoA

$$CH_2\overset{COOH}{\underset{\overset{\|}{O}}{\overset{}{\diagdown}}}_{C-S-CoA} + \boxed{ACP}-SH \longrightarrow CH_2\overset{COOH}{\underset{\overset{\|}{O}}{\overset{}{\diagdown}}}_{C-S-\boxed{ACP}}$$

Malonyl·ACP

$$CH_2\overset{COOH}{\underset{\overset{\|}{O}}{\overset{}{\diagdown}}}_{C-S-\boxed{ACP}} + CH_3-\overset{O}{\overset{\|}{C}}-S-\boxed{ACP} \longrightarrow CH_3-\overset{O}{\overset{\|}{C}}-CH_2-\overset{O}{\overset{\|}{C}}-S-\boxed{ACP} + HS-\boxed{ACP} + CO_2$$

$$CH_3-\overset{O}{\overset{\|}{C}}-CH_2-\overset{O}{\overset{\|}{C}}-S-\boxed{ACP} + 2H^+ \xrightarrow{2H^+ + 2NADPH \to 2NADP^+} CH_3CH_2CH_2-\overset{O}{\overset{\|}{C}}-S-\boxed{ACP} + H_2O$$

Butyryl·ACP

Figure 17–25. Fatty acid synthesis via the pathway found in microsomes. ACP = acyl carrier protein, a protein that is part of an enzyme complex to which the acyl residues are attached as fatty acid synthesis proceeds. The process shown here repeats itself, adding one acetyl-ACP unit to the butyryl-ACP to form a 6-carbon fatty acid-ACP derivative, then another acetyl-ACP to form an 8-carbon unit, etc.

$$CH_3-\overset{O}{\overset{\|}{C}}-S-CoA \; + \; CH_3-\overset{O}{\overset{\|}{C}}-S-CoA \; \underset{}{\overset{\beta\text{-Ketothiolase}}{\rightleftharpoons}} \; CH_3-\overset{O}{\overset{\|}{C}}-CH_2-\overset{O}{\overset{\|}{C}}-S-CoA \; + \; HS\text{-}CoA$$

2 Acetyl-CoA $\qquad\qquad\qquad\qquad$ Acetoacetyl-CoA

$$CH_3-\overset{O}{\overset{\|}{C}}-CH_2-\overset{O}{\overset{\|}{C}}-S-CoA \; + \; H_2O \; \underset{\text{(liver only)}}{\overset{\text{Deacylase}}{\longrightarrow}} \; CH_3-\overset{O}{\overset{\|}{C}}-CH_2-\overset{O}{\overset{\|}{C}}-O^- \; + \; H^+ \; + \; HS\text{-}CoA$$

Acetoacetyl-CoA $\qquad\qquad\qquad\qquad$ Acetoacetate

$$\text{Acetyl-CoA} \; + \; \text{Acetoacetyl-CoA} \; \rightarrow \; CH_3-\overset{OH}{\underset{\overset{|}{CH_2-COO^-}}{\overset{|}{C}}}-CH_2-\overset{O}{\overset{\|}{C}}-S-CoA \; + \; H^+$$

β-Hydroxy-β-methylglutaryl-CoA
(HMG-CoA)

$$\text{HMG-CoA} \; \rightarrow \; \text{Acetoacetate} \; + \; H^+ \; + \; \text{Acetyl-CoA}$$

Acetoacetate

$$CH_3-\overset{O}{\overset{\|}{C}}-CH_2-\overset{O}{\overset{\|}{C}}-O^- \; + \; H^+ \quad\overset{\text{Tissues except liver}}{\longrightarrow}\quad CO_2 \; + \; ATP$$

$$-CO_2$$

$$+2H \; \big\updownarrow \; -2H \qquad\qquad CH_3-\overset{O}{\overset{\|}{C}}-CH_3$$

Acetone

$$CH_3-CHOH-CH_2-\overset{O}{\overset{\|}{C}}-O^- \; + \; H^+$$

β-Hydroxybutyrate

Figure 17–26. Formation and metabolism of ketone bodies. Note that there are 2 pathways for the formation of acetoacetate.

The ketones are an important source of energy in some conditions. Acetone is discharged in the urine and expired air.

The normal blood ketone level in humans is low (about 1 mg/dL) and less than 1 mg is excreted per 24 hours, because the ketones are normally metabolized as rapidly as they are formed. However, if the entry of acetyl-CoA into the citric acid cycle is depressed because of a decreased supply of the products of glucose metabolism, or if the entry does not increase when the supply of acetyl-CoA increases, acetyl-CoA accumulates, the rate of condensation to acetoacetyl-CoA increases, and more acetoacetate is formed in the liver. The ability of the tissues to oxidize the ketones is soon exceeded, and they accumulate in the bloodstream (**ketosis**). Two of the 3 ketone bodies, acetoacetate and β-hydroxybutyrate, are anions of the moderately strong acids acetoacetic acid and β-hydroxybutyric acid, and they are buffered, reducing the decline in pH that would otherwise occur. However, the buffering capacity can be exceeded, and the

metabolic acidosis that develops in conditions such as diabetic ketosis can be severe and even fatal.

Three conditions lead to deficient intracellular glucose supplies: starvation, diabetes mellitus, and a high-fat, low-carbohydrate diet. In diabetes, glucose entry into cells is impaired. When most of the caloric intake is supplied by fat, carbohydrate deficiency develops because there is no major pathway for converting fat to carbohydrate. The liver cells also become filled with fat, which damages them and displaces any glycogen that is formed. In all of these conditions, ketosis develops primarily because the supply of ketones is overabundant.

The acetone odor on the breath of children who have been vomiting is due to the ketosis of starvation. Parenteral administration of relatively small amounts of glucose abolishes the ketosis, and it is for this reason that carbohydrate is said to be **antiketogenic.**

Cellular Lipids

The lipids in cells are of 2 main types: **structural**

lipids, which are an inherent part of the membranes and other parts of cells; and **neutral fat,** stored in the adipose cells of the fat depots. Neutral fat is mobilized during starvation, but structural lipid is preserved. The fat depots obviously vary in size, but in nonobese individuals they make up about 15% of body weight in men and 21% in women. They are not the inert lumps they were once thought to be, but active dynamic tissues undergoing continuous breakdown and resynthesis. In the depots, glucose is metabolized to fatty acids, and neutral fats are synthesized. Neutral fat is also broken down, and free fatty acids are released into the circulation.

Brown Fat

A third, special type of lipid is **brown fat,** which makes up a small percentage of total body fat. Brown fat, which is somewhat more abundant in infants but is present in adults as well, is located between the scapulas, at the nape of the neck, along the great vessels in the thorax and abdomen, and in other scattered locations in the body. In brown fat depots, the fat cells as well as the blood vessels have an extensive sympathetic innervation. This is in contrast to white fat depots, in which there may be innervation of some fat cells but the principal sympathetic innervation is solely on blood vessels. In addition, ordinary lipocytes have only a single large droplet of white fat, whereas brown fat cells contain several small droplets of fat. Brown fat cells also contain many mitochondria. In these mitochondria, there is the usual inward proton conductance that generates ATP (oxidative phosphorylation; see above), but there is in addition a second proton conductance that does not generate ATP. This "short-circuit" conductance, which is associated with a polypeptide of molecular weight 32,000 in the membrane, causes uncoupling of metabolism and generation of ATP, so that more heat is produced (Fig 17–27). Stimulation of the sympathetic innervation to brown fat releases norepinephrine, which acts via β-adrenergic receptors to increase lipolysis, and increased fatty acid oxidation in

the mitochondria increases heat production. Thus, variations in the activity in nerves to brown fat produce variations in the efficiency with which food is utilized and energy produced; ie, they provide a mechanism for varying the weight gained per unit of food ingested.

There is evidence that brown fat functions in this way in 2 situations: (1) In animals and presumably in humans adapted to cold, the rate of heat production in brown fat is increased, and there is in addition a marked increase in blood flow. (2) Nerve discharge to brown fat is also increased after eating, so that heat production is increased. Note that there are 2 components to the heat production after eating: the prompt specific dynamic action (SDA; see above) due to assimilation of foodstuff, and a second, somewhat slower increase in heat produced by brown fat. The possible role of abnormalities in brown fat metabolism in the genesis of obesity is discussed below, and the relation of brown fat to food intake is discussed in Chapter 14.

Plasma Lipids & Lipid Transport

The major lipids in plasma do not circulate in the free form. **Free fatty acids** (variously called FFA, UFA, or NEFA) are bound to albumin, whereas cholesterol, triglycerides, and phospholipids are transported in the form of lipoprotein complexes. There are 6 families of lipoproteins (Table 17–5), which are graded in size and lipid content. The density of these lipoproteins (and consequently the speed at which they sediment in the ultracentrifuge) is inversely proportionate to their lipid content. In general, the lipoproteins consist of a hydrophobic core of triglycerides and cholesteryl esters surrounded by phospholipids and protein (Fig 17–28). Their interrelations and their roles in lipid transport are summarized in Fig 17–29.

The protein constituents of the lipoproteins are called **apoproteins.** The major apoproteins are called apo E, apo C, and apo B (Fig 17–29). There are 2 forms of apo B, a low-molecular-weight form called apo B–48, which is characteristic of the exogenous system that transports exogenous ingested lipids (see below), and a high-molecular-weight form called apo B–100, which is characteristic of the endogenous system that transports lipids that come from the liver.

Chylomicrons are formed in the intestinal mucosa during the absorption of the products of fat digestion (see Chapter 25). They are very large lipoprotein complexes that enter the circulation via the lymphatic ducts. After meals, there are so many of these particles in the blood that the plasma may have a milky appearance **(lipemia).** The chylomicrons are cleared from the circulation by the action of **lipoprotein lipase,** which is located on the surface of the endothelium of the capillaries. This enzyme catalyzes the breakdown of the triglyceride in the chylomicrons to FFA and glycerol, which then enter adipose cells and

Figure 17–27. Proton transport across the mitochondrial membrane in brown fat. Protons are transported outward by the electron transport system, as in other mitochondria, but in addition to the inward proton movement that generates ATP, there is an inward proton "leak" that does not generate ATP. Consequently, metabolism of fat and generation of ATP are uncoupled.

Table 17-5. The principal lipoproteins. The plasma lipids include these components, free fatty acids from adipose tissue, which circulate bound to albumin, and chylomicron remnants.

Lipoprotein	Size (nm)	Composition (%)					Origin
		Protein	Free Cholesterol	Cholesteryl Esters	Triglyceride	Phospholipid	
Chylomicrons	75–1000	2	2	3	90	3	Intestine
Chylomicron remnants	30–80	...	...	...	...	...	Capillaries
Very low density lipoproteins (VLDL)	30–80	8	4	16	55	17	Liver and intestine
Intermediate-density lipoproteins (IDL)	25–40	10	5	25	40	20	VLDL
Low-density lipoproteins (LDL)	20	20	7	46	6	21	IDL
High-density lipoproteins (HDL)	7.5–10	50	4	16	5	25	Liver and intestine

are reesterified. Alternatively, the FFA remain in the circulation bound to albumin. Lipoprotein lipase, which requires heparin as a cofactor, also removes triglycerides from circulating **very low density lipoproteins (VLDL;** see below). Chylomicrons and VLDL contain apoprotein C, a complex of proteins,

Figure 17–28. Diagrammatic representation of the structure of low-density lipoprotein (LDL), the LDL receptor, and the binding of the LDL to the receptor via apo B–100. (Courtesy of the National Institutes of Health.)

which leaves them in the capillaries. One component of the complex, apolipoprotein C-II, activates lipoprotein lipase.

Chylomicrons depleted of their triglyceride remain in the circulation as cholesterol-rich lipoproteins called **chylomicron remnants,** which are 30–80 nm in diameter. The remnants are carried to the liver, where they bind to chylomicron remnant and LDL receptors, are immediately internalized by receptor-mediated endocytosis (see Chapter 1), and are degraded in lysosomes.

The chylomicrons and their remnants constitute a transport system for ingested exogenous lipids (Fig 17–29). There is also an endogenous system made up of VLDL, **intermediate-density lipoproteins (IDL), low-density lipoproteins (LDL),** and **high-density lipoproteins (HDL),** which transports triglycerides and cholesterol throughout the body. VLDL are formed in the liver and transport triglycerides formed from fatty acids and carbohydrates in the liver to extrahepatic tissues. After their triglyceride is largely removed by the action of lipoprotein lipase, they become IDL. The IDL give up phospholipids and, through the action of the plasma enzyme **lecithin-cholesterol acyltransferase (LCAT;** Fig 17–29), pick up cholesteryl esters formed from cholesterol in the HDL. Some IDL are taken up by the liver. The remaining IDL then lose more triglyceride and protein, probably in the sinusoids of the liver, and become LDL. During this conversion, they lose apo E, but apo B–100 remains.

LDL provide cholesterol to the tissues. The cholesterol is an essential constituent in cell membranes and is used by gland cells to make steroid hormones. In the liver and most extrahepatic tissues, LDL are taken up by receptor-mediated endocytosis in coated pits (see Chapter 1). The receptors recognize the apo B–100 component of the LDL (Fig 17–28). They also bind apo E but do not bind apo B–48.

Figure 17–29. Lipoprotein systems for transporting lipids in humans. In the exogenous system, chylomicrons rich in triglycerides of dietary origin are converted to chylomicron remnants rich in cholesteryl esters by the action of lipoprotein lipase. In the endogenous system, VLDL rich in triglycerides are secreted by the liver and converted to IDL and then to LDL rich in cholesteryl esters. LCAT, lecithin-cholesterol acyltransferase. The letters on the chylomicrons, chylomicron remnants, VLDL, IDL, and LDL identify the principal apoproteins found in them. One-third of the LDL is taken up by macrophages and other cells by alternative mechanisms. (Slightly modified and reproduced, with permission, from Goldstein JL, Kita T, Brown MS: Defective lipoprotein receptors and atherosclerosis. *N Engl J Med* 1983;**309:**288.)

The human LDL receptor is a large, complex molecule made up of a cysteine-rich region of 292 amino acid residues that binds LDL; a region of about 400 amino acid residues that is homologous to the precursor for epidermal growth factor; a 58-amino-acid region that is rich in serine and threonine and is the site of glycosylation; a stretch of 22 hydrophobic amino acid residues that spans the cell membrane; and a portion of 50 amino acid residues that projects into the cytoplasm (Fig 17–28). The gene for this protein contains 18 exons, 13 of which encode protein sequences homologous to sequences in other proteins. Thus, it appears that the LDL receptor is a mosaic protein formed by exons which code for parts of other proteins.

In the process of receptor-mediated endocytosis, each coated pit is pinched off to form a coated vesicle and then an endosome. Protein pumps in the membranes of the endosomes lower the pH in this organelle. In the case of the LDL receptor, but not the chylomicron remnant receptor, this triggers release of the LDL receptors, which recycle to the cell membrane (Fig 17–30). The endosome then fuses with a lysosome, where cholesterol formed from the cholesteryl esters by the acid lipase in the lysosomes becomes available to meet the cell's needs (Fig

Figure 17–30. Cellular uptake and metabolism of cholesterol. LDL bind to receptors and are internalized by receptor-mediated endocytosis. Receptors are freed and recycle to the membrane. The cholesteryl esters enter lysosomes, where free cholesterol is released and is used for cellular processes. Cholesterol also (1) inhibits HMG-CoA reductase, (2) is processed in part to other cholesteryl esters by the enzyme acetyl-CoA:cholesterol acyltransferase (ACAT), and (3) inhibits the formation of LDL receptors. (Courtesy of MS Brown.)

17–30). The cholesterol in the cells also inhibits intracellular synthesis of cholesterol by inhibiting HMG-CoA reductase (see below), stimulates esterification of any excess cholesterol that is released, and inhibits synthesis of new LDL receptors. All of these reactions provide feedback control of the amount of cholesterol in the cell.

LDL are also taken up by a lower-affinity system in the macrophages and some other cells. When overloaded by a high plasma level of LDL, the macrophages become full of cholesteryl esters and make up the "foam cells" that appear early in atherosclerotic lesions and may play a part in their genesis. There is some evidence that macrophages preferentially take up LDL that have been modified by oxidation. The LDL receptor on macrophages and related cells is called the **scavenger receptor.** It is different from the receptor on other cells and has a greater affinity for altered LDL.

It is apparent that in the steady state, cholesterol leaves as well as enters cells. The cholesterol leaving the cells is absorbed into HDL, lipoproteins that are synthesized in the liver and the intestine. Some of the HDL contain apo E and bind to LDL receptors on other cells, thus transporting cholesterol from one cell to another. In addition, HDL provides, through LCAT, the cholesteryl esters that are transferred to IDL and thence back again to LDL.

Apo E is synthesized by astrocytes in the brain and by cells in the spleen, lung, adrenal, ovary, and kidney, as well as the liver. Its concentration is greatly increased in injured nerves, where it appears to play a role in nerve regeneration by adjusting the cholesterol concentration during the repair process.

Free Fatty Acid Metabolism

Free fatty acids (FFA) are provided to fat cells and other tissues by chylomicrons and VLDL (see above). They are also synthesized in the fat depots in which they are stored. They circulate bound to albumin and are a major source of energy for many organs. They are used extensively in the heart, but probably all tissues, including the brain, can oxidize FFA to CO_2 and H_2O.

The supply of FFA to the tissues is regulated by 2 lipases. As noted above, lipoprotein lipase on the surface of the endothelium of the capillaries hydrolyzes the triglycerides in chylomicrons and VLDL, providing FFA and glycerol, which are reassembled into new triglycerides in the fat cells. The intracellular **hormone-sensitive lipase** of adipose tissue catalyzes the breakdown of stored triglycerides into glycerol and fatty acids, with the latter entering the circulation.

The hormone-sensitive lipase is converted from an inactive to an active form by cyclic AMP via protein kinase A (Fig 17–31). The adenylate cyclase in adipose cells is in turn activated by glucagon. It is also activated by the catecholamines norepinephrine and

Figure 17–31. Mechanism by which catecholamines increase the activity of the hormone-sensitive lipase in adipose tissue.

epinephrine via a β_1-adrenergic receptor and possibly in some species, a β_3 receptor. ACTH, TSH, LH, serotonin, and vasopressin increase lipolysis via cyclic AMP, but the physiologic role of these substances in the regulation of lipolysis is uncertain. Growth hormone, glucocorticoids, and thyroid hormone also increase the activity of the hormone-sensitive lipase, but they do it by a slower process that requires synthesis of new protein. Growth hormone appears to produce a protein that increases the ability of catecholamines to activate cyclic AMP, whereas cortisol produces a protein that increases the action of cyclic AMP. On the other hand, insulin and prostaglandin E decrease the activity of the hormone-sensitive lipase, possibly by inhibiting the formation of cyclic AMP.

Given the hormonal effects described in the preceding paragraph, it is not surprising that the activity of the hormone-sensitive lipase is increased by fasting and stress and decreased by feeding and insulin. Conversely, feeding increases and fasting and stress decrease the activity of lipoprotein lipase.

Cholesterol Metabolism

Cholesterol is the precursor of the steroid hormones and bile acids and is an essential constituent of cell membranes (see Chapter 1). It is found only in animals. Related sterols occur in plants, but plant sterols are not normally absorbed from the gastrointestinal tract. Most of the dietary cholesterol is contained in egg yolks and animal fat.

Cholesterol is absorbed from the intestine and incorporated into the chylomicrons formed in the mucosa. After the chylomicrons discharge their triglyceride in adipose tissue, the chylomicron remnants bring cholesterol to the liver. The liver and other tissues also synthesize cholesterol. Some of the cholesterol in the liver is excreted in the bile, both in the free form and as bile acids. Some of the biliary cholesterol is reabsorbed from the intestine. Most of the cholesterol in the liver is incorporated into VLDL, and all of it circulates in lipoprotein complexes (see above).

The biosynthesis of cholesterol from acetate is summarized in Fig 17–32. Cholesterol feeds back to inhibit its own synthesis by inhibiting **HMG-CoA reductase,** the enzyme that converts β-hydroxy-β-methylglutaryl-CoA to mevalonic acid. Thus, when dietary cholesterol intake is high, hepatic cholesterol synthesis is decreased, and vice versa. However, the feedback compensation is incomplete, because a diet that is low in cholesterol and saturated fats leads to a modest decline in circulating blood cholesterol.

The plasma cholesterol level is decreased by thyroid hormones, which increase LDL receptors, and by estrogens, which increase HDL and lower LDL, probably by increasing the number of LDL receptors in the liver and thus accelerating hepatic clearance of LDL. Plasma cholesterol is elevated by biliary obstruction and in untreated diabetes mellitus. If bile acid reabsorption in the intestine is decreased by resins such as cholestipol, more cholesterol is diverted to bile acid formation. However, the drop in plasma cholesterol is relatively small because there is a compensatory increase in cholesterol synthesis. Large doses of the vitamin niacin decrease LDL and increase HDL. Compactin, mevinolin, and their derivatives inhibit HMG-CoA reductase and are of considerable value in lowering plasma cholesterol.

Relation to Atherosclerosis

A subject of great interest is the role of cholesterol in the etiology and course of **atherosclerosis.** This extremely widespread disease predisposes to myocardial infarction, cerebral thrombosis, and other serious illnesses. It is characterized by infiltration of cholesterol and appearance of foam cells in certain lesions of the arterial walls, distorting the vessels and making them rigid. Laymen often refer to this condition as arteriosclerosis, but technically, arteriosclerosis is a more general term that means loss of elasticity or hardening of the arteries from any cause. In individuals with elevated plasma cholesterol levels, there is an increased incidence of atherosclerosis and its complications. Of the cholesterol in the body, 93% is in cells and only 7% in the plasma, but it is the increases in the 7% in plasma that predispose to atherosclerosis.

Familial hypercholesterolemia can be produced by a variety of mutations in the LDL receptor. In individuals with this condition and in experimental animals, lowering plasma cholesterol slows the progression of atherosclerosis. In addition, it now seems clear that in humans with plasma cholesterol levels in the upper part of the normal range, lowering the plasma cholesterol by diet, supplemented, if necessary, with drugs, decreases the incidence of atherosclerosis, myocardial infarctions, and strokes. The plasma concentrations of LDL and HDL, as well as of cholesterol, must be considered when evaluating patients. Individuals with elevated LDL have a higher than normal incidence of the disease and its complications, whereas individuals with elevated HDL have a lower incidence. It is interesting that women, who have a lower incidence of myocardial infarction than men, have higher HDL levels. In addition, HDL levels are increased in individuals who exercise and those who drink 1–2 alcoholic drinks per day, whereas they are decreased in individuals who smoke, are obese, or live sedentary lives. Moderate drinking decreases the incidence of myocardial infarction, and obesity and smoking are known risk factors that increase it. There is evidence that elevated levels of IDL and chylomicron remnants also predispose to atherosclerosis, whereas elevated chylomicron and VLDL levels do not.

Essential Fatty Acids

Animals fed a fat-free diet fail to grow, develop skin and kidney lesions, and become infertile. Adding linolenic, linoleic, and arachidonic acids to the diet cures all the deficiency symptoms. These 3 acids are polyunsaturated fatty acids and because of their

Figure 17–32. Biosynthesis of cholesterol. Six mevalonic acid molecules condense to form squalene, which is then hydroxylated and converted to cholesterol. The dashed arrow indicates feedback inhibition by cholesterol of HMG-CoA reductase, the enzyme that catalyzes mevalonic acid formation.

action are called **essential fatty acids.** Similar deficiency symptoms have not been unequivocally demonstrated in humans, but there is reason to believe that some unsaturated fats are essential dietary constituents, especially in children. Dehydrogenation of fats is known to occur in the body, but there does not appear to be any synthesis of carbon chains with the arrangement of double bonds found in the essential fatty acids.

Eicosanoids

One of the reasons, and possibly the only reason, that essential fatty acids are necessary for health is that they are the precursors of prostaglandins, prostacyclin, thromboxanes, lipoxins, and leukotrienes (Fig 17–33). These substances are called **eicosanoids,** reflecting their origin from the 20-carbon (eicosa-) polyunsaturated fatty acid **arachidonic acid** and the 20-carbon derivatives of linoleic and linolenic acids.

The **prostaglandins** are a series of 20-carbon unsaturated fatty acids containing a cyclopentane ring. They were first isolated from semen but have now been shown to be synthesized in most and possibly in all organs in the body. The structures of some of them are shown in Fig 17–34. The prostaglandins are divided into groups—PGE and PGF, for example—on the basis of the configuration of the cyclopentane ring. The number of double bonds in the side chains is indicated by subscript numbers; for example, the E series prostaglandin shown in Fig 17–34 is PGE_2. Its precursor is PGG_2, which is formed from arachidonic acid in a reaction catalyzed by cyclooxygenase. PGG_2 is also converted via 2 different pathways to **prostacyclin** and to **thromboxane A_2.**

Arachidonic acid is also converted to 5-hydro-peroxyeicosatetraenoic acid (5-HPETE), which is then converted to the **leukotrienes.** Four of the leukotrienes are **aminolipids** that contain amino acids; leukotriene C_4 (LTC_4) contains the tripeptide glutathione, LTD_4 contains glycine and cysteine, LTE_4 contains cysteine, and LTF_4 contains cysteine and glutamic acid. In addition, arachidonic acid is converted to **lipoxins** (Figs 17–33 and 17–34) and a few other compounds.

The leukotrienes, thromboxanes, lipoxins, and prostaglandins have been called local hormones. They have short half-lives and are inactivated in many different tissues. They undoubtedly act mainly in the tissues and sites in which they are produced.

The leukotrienes are mediators of allergic responses and inflammation. Their release is provoked when specific allergens combine with IgE antibodies on the surfaces of mast cells. They produce bronchoconstriction, constrict arterioles, increase vascular permeability, and attract neutrophils and eosinophils to inflammatory sites. The leukotriene receptors that have been isolated are coupled to G proteins. Diseases in which they may be involved include asthma, psoriasis, adult respiratory distress syndrome, allergic rhinitis, rheumatoid arthritis, Crohn's disease, and ulcerative colitis.

Thromboxane A_2 is synthesized by platelets and promotes vasoconstriction and platelet aggregation. The thromboxane A_2 receptor is a typical serpentine receptor coupled to a G protein (see Chapter 1), and it acts via phosphatidylinisitol to open Ca^{2+}-activated Cl^- channels. Thromboxane B_2 is mainly a metabolite of thromboxane A_2. Prostacyclin, on the other hand, inhibits platelet aggregation and is a vasodilator. It is produced by endothelial and smooth muscle cells in the walls of blood vessels. Release of thromboxane A_2 by platelets at the immediate site of injury to a blood vessel promotes clot formation, whereas prostacyclin formation in the neighboring portions of the blood vessel wall keeps the clot localized and maintains the patency of the rest of the vessel. Obviously, the effects of inhibiting cyclooxygenase in the platelets and the vessel walls would cancel each other. However, small doses of aspirin inhibit platelet cyclooxygenase without inhibiting vessel wall cyclooxygenase, and doses in the range of 300 mg/d have been used with some success to prevent formation of clots in cerebral and myocardial vessels.

Lipoxin A dilates the microvasculature, and lipoxin A and lipoxin B both inhibit the cytotoxic effects of natural killer cells (see Chapter 27). However, their physiologic role is still uncertain.

The prostaglandins have attracted great attention because they have profound effects when administered in very small doses. In addition, they have a bewildering array of actions. Prostaglandins may play a role in regulating the capacity of red blood cells to undergo deformation in passing through capillaries. Some of them decrease gastric acid secretion and in-

Figure 17–33. Principal biologically active substances produced from arachidonic acid.

Figure 17–34. Biosynthesis of prostaglandins and thromboxanes **(top),** lipoxins **(middle),** and leukotrienes **(bottom)** from arachidonic acid. Leukotriene C_4 (LTC_4) is converted to leukotriene D_4 by removal of glutamic acid, and leukotriene D_4 is converted to leukotriene E_4 by further removal of glycine. Leukotriene E_4 can take up glutamic acid to form leukotriene F_4. 5-HPETE, 5-hydroperoxyeicosatetraenoic acid.

hibit the formation of peptic ulcers in experimental animals. They cause luteolysis and may play a role in the regulation of the female reproductive cycle (see Chapter 23). They also have the capacity to modify pituitary responses to hypothalamic hormones. They induce abortion when injected intra-amniotically in midtrimester pregnancy, and they induce labor when injected near term. They can mimic the effects of TSH, ACTH, and other hormones. They stimulate renin secretion. Some of them are antilipolytic and may play a role along with glucagon, catecholamines, and other hormones in the regulation of FFA release. They appear to be released from inflamed tissue and cause fever when injected into the third ventricle (see Chapter 14). They are found in the eye and produce miosis. They exist in the brain and may alter the release or effects of neurotransmitters.

It is difficult to find a common theme in these multifarious actions, but it may be significant that cyclic AMP is involved in most if not all of them. It may be that the various prostaglandins act to adjust the generation of cyclic AMP in response to various stimuli.

Anti-inflammatory steroids such as cortisol inhibit the release of arachidonic acid from phospholipid stores and thus block formation of leukotrienes as well as prostacyclin, PGE_2 and related prostaglandins, and thromboxanes. Nonsteroidal anti-inflammatory drugs such as aspirin and indomethacin inhibit cyclooxygenase, leaving the leukotriene pathway intact. Aspirin irreversibly inactivates cyclooxygenase by acetylating it, and the effect of aspirin only wears off as new enzyme is synthesized by the cells. Much effort is being devoted to a search for drugs that will produce even more selective alterations by acting farther along the biosynthetic pathways.

Obesity

Obesity in humans is a health problem of appreciable magnitude, since it is associated with an increased incidence of various diseases, including cardiovascular and gallbladder disease and diabetes. In addition, obese individuals have an increased mortality rate. According to a commonly used definition, obesity is said to be present when more than 20% of body weight is due to fat in men and more than 25% in women. Normal values of fat are 12–18% for men and 18–24% for women. Normal height and weight tables are also used extensively, but a value that correlates better to body fat is the **Quetelet index,** which is the body weight (in kg) divided by the square of the height (in m). The normal value for this index is 20–25 kg/m². Lean body mass rises to a plateau in the third decade of life, then in men declines at an accelerating rate with advancing age. In women, the decline is small until age 50–55 but rapid thereafter. Consequently, if food intake is not reduced with advancing age, obesity will result. In addition, the basal metabolic rate declines with age.

There is a strong genetic component to obesity in humans, but environmental factors contribute as well. For instance, in the USA, there is a significantly lower incidence of obesity in women from higher socioeconomic groups than in those from lower ones. The relation of obesity to diabetes is discussed in Chapter 19.

The first law of thermodynamics applies to obese as well as thin individuals. When the energy provided by food intake exceeds energy output, the excess is stored primarily as fat. The metabolic rate per unit of fat-free mass is similar in obese and nonobese subjects. However, in formerly obese individuals who have lost weight, the number of calories needed to maintain weight is smaller than in persons who have never been obese, and the metabolic rate of formerly obese individuals is restored to normal by eating more food than is necessary to maintain weight. In addition, the incidence of obesity is much greater in families with congenitally low metabolic rates than it is in controls. Thus it may be that some sort of homeostatic mechanism drives certain individuals to overeat in order to raise their metabolic rates to normal.

The relation of neural control of brown fat to obesity is discussed above.

NUTRITION

The aim of the science of nutrition is the determination of the kinds and amounts of foods that promote health and well-being. This includes not only the problems of undernutrition but those of overnutrition, taste, and availability. However, certain substances are essential constituents of any human diet. Many of these compounds have been mentioned in previous sections of this chapter, and a brief summary of the essential and desirable dietary components is presented below.

Essential Dietary Components

An optimal diet includes, in addition to sufficient water (see Chapter 38), adequate calories, protein, fat, minerals, and vitamins (Table 17–6).

Caloric Intake & Distribution

As noted above, the caloric value of the dietary intake must be approximately equal to the energy expended as heat and work if body weight is to be maintained. When the caloric intake is insufficient, body stores of protein and fat are catabolized, and when the intake is excessive, obesity results. In addition to the 2000 kcal/d necessary to meet basal needs, 500–2500 kcal/d (or more) are required to meet the energy demands of daily activities.

The distribution of the calories among carbohydrate, protein, and fat foodstuffs is determined partly by physiologic factors and partly by taste and eco-

Table 17–6. Recommended daily dietary allowances.[1] (Revised 1980.) Designed for the maintenance of good nutrition of practically all healthy people in the USA.

Category	Age (years)	Weight (kg)	Weight (lb)	Height (cm)	Height (in)	Protein (g)	Fat-Soluble Vitamins Vitamin A (µg RE)[2]	Vitamin D (µg)[3]	Vitamin E (mg α-TE)[4]	Water-Soluble Vitamins Vitamin C (mg)	Thiamine (mg)	Riboflavin (mg)	Niacin (mg NE)[5]	Vitamin B$_6$ (mg)	Folacin[6] (µg)	Vitamin B$_{12}$ (µg)	Minerals Calcium (mg)	Phosphorus (mg)	Magnesium (mg)	Iron (mg)	Zinc (mg)	Iodine (µg)
Infants	0.0–0.5	6	13	60	24	kg × 2.2	420	10	3	35	0.3	0.4	6	0.3	30	0.5[7]	360	240	50	10	3	40
	0.5–1.0	9	20	71	28	kg × 2.0	400	10	4	35	0.5	0.6	8	0.6	45	1.5	540	360	70	15	5	50
Children	1–3	13	29	90	35	23	400	10	5	45	0.7	0.8	9	0.9	100	2.0	800	800	150	15	10	70
	4–6	20	44	112	44	30	500	10	6	45	0.9	1.0	11	1.3	200	2.5	800	800	200	10	10	90
	7–10	28	62	132	52	34	700	10	7	45	1.2	1.4	16	1.6	300	3.0	800	800	250	10	10	120
Males	11–14	45	99	157	62	45	1000	10	8	50	1.4	1.6	18	1.8	400	3.0	1200	1200	350	18	15	150
	15–18	66	145	176	69	56	1000	10	10	60	1.4	1.7	18	2.0	400	3.0	1200	1200	400	18	15	150
	19–22	70	154	177	70	56	1000	7.5	10	60	1.5	1.7	19	2.2	400	3.0	800	800	350	10	15	150
	23–50	70	154	178	70	56	1000	5	10	60	1.4	1.6	18	2.2	400	3.0	800	800	350	10	15	150
	51+	70	154	178	70	56	1000	5	10	60	1.2	1.4	16	2.2	400	3.0	800	800	350	10	15	150
Females	11–14	46	101	157	62	46	800	10	8	50	1.1	1.3	15	1.8	400	3.0	1200	1200	300	18	15	150
	15–18	55	120	163	64	46	800	10	8	60	1.1	1.3	14	2.0	400	3.0	1200	1200	300	18	15	150
	19–22	55	120	163	64	44	800	7.5	8	60	1.1	1.3	14	2.0	400	3.0	800	800	300	18	15	150
	23–50	55	120	163	64	44	800	5	8	60	1.0	1.2	13	2.0	400	3.0	800	800	300	18	15	150
	51+	55	120	163	64	44	800	5	8	60	1.0	1.2	13	2.0	400	3.0	800	800	300	10	15	150
Pregnant						+30	+200	+5	+2	+20	+0.4	+0.3	+2	+0.6	+400	+1.0	+400	+400	+150	8	+5	+25
Lactating						+20	+400	+5	+3	+40	+0.5	+0.5	+5	+0.5	+100	+1.0	+400	+400	+150	8	+10	+50

Reference: *Recommended Dietary Allowances*, 9th ed. Food and Nutrition Board, National Research Council—National Academy of Sciences, 1980.

[1] The allowances are intended to provide for individual variations among most normal persons as they live in the United States under usual environmental stresses. Diets should be based on a variety of common foods in order to provide other nutrients for which human requirements have been less well defined.

[2] Retinol equivalents. 1 retinol equivalent = 1 µg of retinol or 6 µg of β-carotene.

[3] As cholecalciferol. 10 µg of cholecalciferol = 400 IU of vitamin D.

[4] α-Tocopherol equivalents. 1 mg of α-tocopherol = 1 α-TE.

[5] 1 NE (niacin equivalent) is equal to 1 mg of niacin or 60 mg of dietary tryptophan.

[6] The folacin allowances refer to dietary sources as determined by *Lactobacillus casei* assay after treatment with enzymes (conjugases) to make polyglutamyl forms of the vitamin available to the test organism.

[7] The recommended dietary allowance for vitamin B$_{12}$ in infants is based on the average concentration of the vitamin in human milk. The allowances after weaning are based on energy intake (as recommended by the American Academy of Pediatrics) and consideration of other factors, such as intestinal absorption.

[8] The increased requirement during pregnancy cannot be met by the iron content of habitual American diets or by the existing iron stores of many women; therefore the use of 30–60 mg of supplemental iron is recommended. Iron needs during lactation are not substantially different from those of nonpregnant women, but continued supplementation of the mother for 2–3 months after parturition is advisable in order to replenish stores depleted by pregnancy.

nomic considerations. A daily protein intake of at least 1 g/kg body weight to supply the 8 nutritionally essential amino acids and other amino acids is now regarded as desirable. The source of the protein is also important. **Grade I proteins,** the animal proteins of meat, fish, and eggs, contain amino acids in approximately the proportions required for protein synthesis and other uses. Some of the plant proteins are also grade I, but most are **grade II** because they supply different proportions of amino acids and some lack one or more of the essential amino acids. Protein needs can be met with a mixture of grade II proteins, but the intake must be large because of the amino acid wastage.

Fat is the most compact form of food, since it supplies 9.3 kcal/g. However, it is also the most expensive. Indeed, there is a reasonably good positive correlation between fat intake and standard of living, and in the past, western diets have contained moderately large amounts (100 g/d or more). The evidence indicating that a high unsaturated/saturated fat ratio in the diet is of value in the prevention of atherosclerosis and the current interest in preventing obesity may change this. In Central and South American Indian communities where corn (carbohydrate) is the dietary staple, adults live without ill effects for years on a very low fat intake. Therefore, provided that the needs for essential fatty acids are met, a low fat intake does not seem to be harmful, and a diet low in saturated fats is desirable.

Carbohydrate is the cheapest source of calories and provides 50% or more of the calories in most diets. In the average middle-class American diet, approximately 50% of the calories come from carbohydrate, 15% from protein, and 35% from fat. When calculating dietary needs, it is usual to meet the protein requirement first and then split the remaining calories between fat and carbohydrate, depending upon taste, income, and other factors. For example, a 70-kg man who is moderately active needs about 2800 kcal/d. He should eat at least 65 g of protein daily, supplying 267 (65 × 4.1) kcal. Some of this should be grade I protein. A reasonable figure for fat intake is 50–60 g. The rest of the caloric requirement can be met by supplying carbohydrate.

Mineral Requirements

A number of minerals must be ingested daily for the maintenance of health. Besides those for which recommended daily dietary allowances have been set (Table 17–6), a variety of different trace elements should be included. Trace elements are defined as elements found in tissues in minute amounts. Those believed to be essential for life, at least in experimental animals, are listed in Table 17–7. In humans, iron deficiency causes anemia (see Chapter 26). Cobalt is part of the vitamin B_{12} molecule, and vitamin B_{12} deficiency leads to megaloblastic anemia (see Chapter 26). Iodine deficiency causes thyroid disorders (see

Table 17–7. Trace elements believed essential for life.[1]

Arsenic	Manganese
Chromium	Molybdenum
Cobalt	Nickel
Copper	Selenium
Fluorine	Silicon
Iodine	Vanadium
Iron	Zinc

[1]Data from Mertz W: The essential trace elements. *Science* 1981;**213**:1332.

Chapter 18). Zinc deficiency causes skin ulcers, depressed immune responses, and hypogonadal dwarfism. Copper deficiency causes anemia, changes in ossification, and possibly elevated plasma cholesterol. Chromium deficiency causes insulin resistance. Fluorine deficiency increases the incidence of dental caries. Trace-element deficiencies are rare, because any diet that is adequate in other respects easily supplies the needed minerals.

Conversely, some minerals can be toxic when present in the body in excess. For example, iron overload causes hemochromatosis (see Chapter 25), copper excess causes brain damage (Wilson's disease; see Chapter 12), and aluminum poisoning in patients with renal failure who are receiving dialysis treatment causes a rapidly progressive dementia that resembles Alzheimer's disease (see Chapter 16).

Sodium and potassium are also essential minerals, but listing them is academic, because it is very difficult to prepare a sodium-free or potassium-free diet. A low-salt diet is well tolerated for prolonged periods because of the compensatory mechanisms that conserve sodium, and salt depletion is a problem only when sodium is lost in excessive amounts in the stools, sweat, or urine.

Vitamins

Vitamins were discovered when it was observed that diets adequate in calories, essential amino acids, fats, and minerals failed to maintain health. The term **vitamin** has now come to refer to any organic dietary constituent necessary for life, health, and growth that does not function by supplying energy.

Because there are minor differences in metabolism between mammalian species, some substances are vitamins in one species and not in another. The sources and functions of the major vitamins in humans are listed in Table 17–8 and the recommended daily dietary allowances in Table 17–6. Most vitamins have important functions in intermediary metabolism or the special metabolism of the various organ systems. Vitamin A and vitamin D are bound to transport proteins in the circulation. Those that are water-soluble (vitamin B complex, vitamin C) are easily absorbed, but the fat-soluble vitamins (vitamins A, D, E, and K) are poorly absorbed in the absence of bile or pancreatic lipase. Some dietary fat intake is necessary for

Table 17–8. Vitamins essential or probably essential to human nutrition. Choline is not listed because it is synthesized in the body in adequate amounts except in special circumstances.

Vitamin	Action	Deficiency Symptoms	Sources	Chemistry
A (A_1, A_2)	Constituents of visual pigments (see Chapter 8); necessary for fetal development and for cell development throughout life.	Night blindness, dry skin	Yellow vegetables and fruit	Vitamin A_1 alcohol (retinol)
B complex Thiamine (vitamin B_1)	Cofactor in decarboxylations.	Beriberi, neuritis	Liver, unrefined cereal grains	
Riboflavin (vitamin B_2)	Constituent of flavoproteins.	Glossitis, cheilosis	Liver, milk	
Niacin	Constituent of NAD^+ and $NADP^+$	Pellagra	Yeast, lean meat, liver	Can be synthesized in body from tryptophan.
Pyridoxine (vitamin B_6)	Forms prosthetic group of certain decarboxylases and transaminases. Converted in body into pyridoxal phosphate and pyridoxamine phosphate.	Convulsions, hyperirritability	Yeast, wheat, corn, liver	
Pantothenic acid	Constituent of CoA.	Dermatitis, enteritis, alopecia, adrenal insufficiency	Eggs, liver, yeast	
Biotin	Catalyzes CO_2 "fixation" (in fatty acid synthesis, etc).	Dermatitis, enteritis	Egg yolk, liver, tomatoes	
Folates (folic acid) and related compounds	Coenzymes for "1-carbon" transfer; involved in methylating reactions.	Sprue, anemia	Leafy green vegetables	Folic acid
Cyanocobalamin (vitamin B_{12})	Coenzyme in amino acid metabolism. Stimulates erythropoiesis.	Pernicious anemia (see Chapter 26)	Liver, meat, eggs, milk	Complex of 4 substituted pyrrole rings around a cobalt atom (see Chapter 26).

(continued)

Table 17–8 (cont'd). Vitamins essential or probably essential to human nutrition. Choline is not listed because it is synthesized in the body in adequate amounts except in special circumstances.

Vitamin	Action	Deficiency Symptoms	Sources	Chemistry
C	Necessary for hydroxylation of proline and lysine in collagen synthesis.	Scurvy	Citrus fruits, leafy green vegetables	Ascorbic acid (synthesized in most mammals except guinea pigs and primates, including humans).
D group	Increase intestinal absorption of calcium and phosphate (see Chapter 21).	Rickets	Fish liver	Family of sterols (see Chapter 21).
E group	Antioxidants; cofactors in electron transport in cytochrome chain?	Muscular dystrophy and fetal death in animals	Milk, eggs, meat, leafy vegetables	a-Tocopherol; β- and γ-tocopherol also active.
K group	Catalyze γ carboxylation of glutamic acid residues on various proteins concerned with blood clotting.	Hemorrhagic phenomena	Leafy green vegetables	Vitamin K_3; a large number of similar compounds have biologic activity.

their absorption, and in obstructive jaundice or in pancreatic disease, deficiencies of the fat-soluble vitamins can develop even if their intake is adequate (see Chapter 26).

The diseases caused by deficiency of each of these vitamins are listed in Table 17–8. It is worth remembering, however, particularly in view of the advertising campaigns for vitamin pills and supplements, that very large doses of the fat-soluble vitamins are definitely toxic. **Hypervitaminosis A** is characterized by anorexia, headache, hepatosplenomegaly, irritability, scaly dermatitis, patchy loss of hair, bone pain, and hyperostosis. Acute vitamin A intoxication was first described by Arctic explorers, who developed headache, diarrhea, and dizziness after eating polar bear liver. The liver of this animal is particularly rich in vitamin A. **Hypervitaminosis D** is associated with weight loss and calcification of many soft tissues. Excessive vitamin D intake eventually causes renal failure. **Hypervitaminosis K** is characterized by gastrointestinal disturbances and anemia. Large doses of water-soluble vitamins have been thought to be less likely to cause problems because they can be rapidly cleared from the body. However, it has now been demonstrated that ingestion of megadoses of pyridoxine (vitamin B_6) can produce peripheral neuropathy.

The Thyroid Gland

18

INTRODUCTION

The thyroid gland maintains the level of metabolism in the tissues that is optimal for their normal function. Thyroid hormone stimulates the O_2 consumption of most of the cells in the body, helps regulate lipid and carbohydrate metabolism, and is necessary for normal growth and maturation. The thyroid gland is not essential for life, but its absence causes mental and physical slowing, poor resistance to cold, and, in children, mental retardation and dwarfism. Conversely, excess thyroid secretion leads to body wasting, nervousness, tachycardia, tremor, and excess heat production. Thyroid function is controlled by the thyroid-stimulating hormone (TSH, thyrotropin) of the anterior pituitary. The secretion of this tropic hormone is in turn regulated in part by a direct inhibitory feedback of high circulating thyroid hormone levels on the pituitary and in part via neural mechanisms operating through the hypothalamus. In this way, changes in the internal and external environment bring about appropriate adjustments in the rate of thyroid secretion.

In mammals, the thyroid gland also secretes calcitonin, a calcium-lowering hormone. This hormone is discussed in Chapter 21.

ANATOMIC CONSIDERATIONS

Thyroid tissue is present in all vertebrates. In mammals, the thyroid originates from an evagination of the floor of the pharynx, and a **thyroglossal duct** marking the path of the thyroid from the tongue to the neck sometimes persists in the adult. The 2 lobes of the human thyroid are connected by a bridge of tissue, the **thyroid isthmus,** and there is sometimes a **pyramidal lobe** arising from the isthmus in front of the larynx (Fig 18–1). The gland is well vascularized, and the thyroid has one of the highest rates of blood flow per gram of tissue of any organ in the body.

The thyroid is made up of multiple **acini (follicles).** Each spherical follicle is surrounded by a single layer of cells and filled with pink-staining proteinaceous material called **colloid.** When the gland is inactive, the colloid is abundant, the follicles are large, and the cells lining them are flat. When the gland is active, the follicles are small, the cells are cuboid or columnar, and the edge of the colloid is scalloped, forming many small "reabsorption lacunae" (Fig 18–2).

Microvilli project into the colloid from the apexes of the thyroid cells, and canaliculi extend into them. There is a prominent endoplasmic reticulum, a feature common to most glandular cells, and secretory droplets of thyroglobulin are seen (Fig 18–3). The individual thyroid cells rest on a basal lamina that separates them from the adjacent capillaries. The endothelial cells are attenuated at places, forming gaps (fenestrations; see Chapter 30) in the walls of the capillaries (Fig 18–3). This fenestration of the capillary walls occurs in most endocrine organs.

Figure 18–1. The human thyroid.

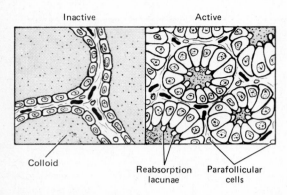

Inactive   Active

Colloid

Reabsorption   Parafollicular
lacunae   cells

Figure 18–2. Thyroid histology. Note the small, punched-out "reabsorption lacunae" in the colloid next to the cells in the active gland.

FORMATION & SECRETION OF THYROID HORMONES

Chemistry

The principal hormones secreted by the thyroid are **thyroxine (T_4)** and **triiodothyronine (T_3)**. T_3 is also formed in the peripheral tissues by deiodination of T_4 (see below). Both hormones are iodine-containing amino acids (Fig 18–4). Small amounts of reverse triiodothyronine (3,3′,5′-triiodothyronine, RT_3), monoiodotyrosine, and other compounds are also found in thyroid venous blood. T_3 is more active than T_4, whereas RT_3 is inactive. The naturally occurring forms of T_4 and its congeners with an asymmetric carbon atom are the L isomers. D-Thyroxine has only a small fraction of the activity of the L form.

Thyroglobulin

T_4 and T_3 are synthesized in the colloid by iodination and condensation of tyrosine molecules bound in peptide linkage in **thyroglobulin.** This glycoprotein is made up of 2 subunits and has a molecular weight of 660,000. It contains 10% carbohydrate by weight. It also contains 123 tyrosine residues, but only 4–8 of these are normally incorporated into thyroid hormones. Thyroglobulin is synthesized in the thyroid cells and secreted into the colloid by exocytosis of granules that also contain thyroid peroxidase (see below). The hormones remain bound to thyroglobulin until secreted. When they are secreted, colloid is ingested by the thyroid cells, the peptide bonds are hydrolyzed, and free T_4 and T_3 are discharged into the capillaries. The thyroid cells thus have 3 functions: They collect and transport iodine; they synthesize thyroglobulin and secrete it into the colloid; and they remove the thyroid hormones from thyroglobulin and secrete them into the circulation.

Thyroglobulin enters the blood as well as the colloid. The normal serum thyroglobulin concentration

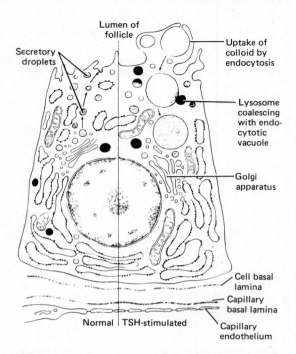

Lumen of
follicle

Secretory
droplets

Uptake of
colloid by
endocytosis

Lysosome
coalescing
with endo-
cytotic
vacuole

Golgi
apparatus

Cell basal
lamina

Capillary
basal lamina

Normal | TSH-stimulated

Capillary
endothelium

Figure 18–3. Thyroid cell. *Left:* Normal pattern. *Right:* After marked TSH stimulation. The arrows on the left show the secretion of thyroglobulin into the colloid. On the right, endocytosis of the colloid and merging of a colloid-containing vacuole with a lysosome are shown. The cell rests on a capillary with gaps (fenestrations) in the endothelial wall. (Modified and reproduced, with permission, from Fawcett DW, Long JA, Jones AL: The ultrastructure of endocrine glands. *Recent Prog Horm Res* 1969;**25:**315.)

in humans is about 6 ng/mL, and this level is increased in hyperthyroidism and some forms of thyroid cancer. However, the function, if any, of circulating thyroglobulin is unknown.

Iodine Metabolism

Iodine is a raw material essential for thyroid hormone synthesis. Ingested iodine is converted to iodide and absorbed. The fate of the absorbed I⁻ is summarized in Fig 18–5. The minimum daily iodine intake that will maintain normal thyroid function is 150 μg in adults (see Table 17–6), but in the USA the average dietary intake is approximately 500 μg/d. The normal plasma I⁻ level is about 0.3 μg/dL, and I⁻ is distributed in a "space" of approximately 25 L (35% of body weight). The principal organs that take up the I⁻ are the thyroid, which uses it to make thyroid hormones, and the kidneys, which excrete it in the urine. About 120 μg/d enter the thyroid at normal rates of thyroid hormone synthesis and secretion. The thyroid secretes 80 μg/d as iodine in T_3 and T_4. Forty micrograms of I⁻ per day diffuse into the ECF. The secreted T_3 and T_4 are metabolized in the liver and

Figure 18–4. Thyroid hormones. The numbers in the rings in the T_4 formula indicate the numbering of positions in the molecule. RT_3 is 3,3′,5′-triodothyronine.

3,5,3′,5′-Tetraiodothyronine (thyroxine, T_4)

3,5,3′-Triiodothyronine (T_3)

other tissues, with the release of 60 μg of I^- per day into the ECF. Some thyroid hormone derivatives are excreted in the bile, and some of the iodine in them is reabsorbed (enterohepatic circulation), but there is a net loss of I^- in the stool of approximately 20 μg/d. The total amount of I^- entering the ECF is thus 500 + 40 + 60, or 600 μg/d; 20% of this I^- enters the thyroid, whereas 80% is excreted in the urine.

Iodide Trapping

The thyroid concentrates iodide by actively transporting it from the circulation to the colloid. The transport mechanism is frequently called the "iodide-trapping mechanism" or "iodide pump." The thyroid cell is about 50 mV negative relative to the interstitial area and the colloid; ie, it has a resting membrane potential of –50 mV. Iodide is pumped into the cell at its

base against this electrical gradient, then diffuses down the electrical gradient into the colloid. Iodine uptake can be studied by administering radioactive iodine in **tracer doses,** amounts so small that they do not significantly increase the amount of iodide in the body. In the gland, iodide is rapidly oxidized and bound to tyrosine. Despite the binding, however, the ratio of thyroid to plasma serum free iodide (**T/S ratio**) is normally greater than 1. If binding to tyrosine residues is blocked by antithyroid drugs such as propylthiouracil (see below), iodide accumulates in the thyroid and the T/S ratio is markedly increased. Perchlorate and a number of other anions decrease iodide transport by competitive inhibition. The active transport mechanism of iodide is stimulated by TSH. It also depends on Na^+-K^+ ATPase and consequently is inhibited by ouabain (see Chapter 1).

It is worth noting that the relation of thyroid function to iodide is unique; as discussed in more detail below, iodide is essential for normal thyroid function, but iodide deficiency and iodide excess both inhibit this function.

The salivary glands, the gastric mucosa, the placenta, the ciliary body of the eye, the choroid plexus, and the mammary glands also transport iodide against a concentration gradient, but their uptake is not affected by TSH. The mammary glands also bind the iodine; diiodotyrosine is formed in mammary tissue, but T_4 and T_3 are not. The physiologic significance of all these extrathyroidal iodide-concentrating mechanisms is obscure. It has been claimed that traces of T_4 may be formed by nonthyroidal tissues, but if such formation does occur, the amount is insufficient to prevent the development of the full picture of hypothyroidism after surgical thyroidectomy.

Thyroid Hormone Synthesis

In the thyroid gland, iodide is oxidized to iodine and bound in a matter of seconds to the 3 position of tyrosine molecules attached to thyroglobulin (Fig 18–6). The enzyme responsible for the oxidation and binding of iodide is **thyroid peroxidase,** with hydrogen peroxide accepting the electrons. Monoiodotyrosine (MIT) is next iodinated in the 5 position to form diiodotyrosine (DIT). Two DIT molecules then undergo an oxidative condensation to form T_4 with the elimination of the alanine side chain from the molecule that forms the outer ring. There are 2 theories of how this **coupling reaction** occurs. One holds that the coupling occurs with both DIT molecules attached to thyroglobulin (intramolecular coupling). The other holds that the DIT that forms the outer ring is first detached from thyroglobulin (intermolecular coupling). In any case, thyroid peroxidase is probably involved in coupling as well as iodination. T_3 is probably formed by condensation of MIT with DIT. A small amount of RT_3 is also formed, probably by condensation of DIT with MIT. In the normal human thyroid, the average distribution of iodinated compounds

Figure 18–5. Iodine metabolism.

Figure 18–6. Outline of thyroid hormone biosynthesis. Iodination of tyrosine and the condensation reaction take place at the apical border of the thyroid cells while the molecules are bound in peptide linkage in thyroglobulin.

is 23% MIT, 33% DIT, 35% T_4, and 7% T_3. Only traces of RT_3 and other components are present.

MIT and DIT residues that are not coupled are deiodinated in the thyroid, and deiodination of these residues normally provides about twice as much iodide for hormone synthesis as the iodide pump. In patients with congenital absence of the thyroid deiodinase, MIT and DIT appear in the urine and there are symptoms of iodine deficiency (see below).

Secretion

The human thyroid secretes about 80 μg (103 nmol) of T_4, 4 μg (7 nmol) of T_3, and 2 μg (3.5 nmol) of RT_3 per day (Fig 18–7). The thyroid cells ingest colloid by endocytosis (see Chapter 1). This chewing away at the edge of the colloid produces the reabsorption lacunae seen in active glands (Fig 18–2). In the cells, the globules of colloid merge with lysosomes (Fig 18–3). The peptide bonds between the iodinated residues and the thyroglobulin are broken by the proteases in the lysosomes, and T_4, T_3, diiodotyrosine, and monoiodotyrosine are liberated into the cytoplasm. The iodinated tyrosines are deiodinated by a

microsomal **iodotyrosine dehalogenase,** but this enzyme does not attack iodinated thyronines, and T_4 and T_3 pass on the circulation. The iodine liberated by deiodination of the tyrosines is reutilized.

TRANSPORT & METABOLISM OF THYROID HORMONES

Protein Binding

The normal total **plasma T_4** level is approximately 8 μg/dL (103 nmol/L), and the **plasma T_3** level is approximately 0.15 μg/dL (2.3 nmol/L). Large amounts of both are bound to plasma proteins. Both are now measured by radioimmunoassay, and their measurement has largely replaced use of **protein-bound iodine (PBI)** as an index of the circulating level of thyroid hormones. The normal PBI level is approximately 6 μg/dL.

The free thyroid hormones in plasma are in equilibrium with the protein-bound thyroid hormones in plasma and in tissues (Fig 18–8). Free thyroid hormones are added to the circulating pool by the thy-

Figure 18–7. Secretion and interconversion of thyroid hormones in normal adult humans. Figures are in micrograms per day. Note that most of the T_3 and RT_3 are formed from T_4 deiodination in the tissues, and only small amounts are secreted by the thyroid.

roid. It is the free thyroid hormones in plasma that are physiologically active and that inhibit pituitary secretion of TSH.

Many other hormones are bound to plasma proteins, and there is an equilibrium between their free active forms and their bound inactive forms in the circulation. The function of protein-binding appears to be maintenance of a large pool of readily available free hormone. In addition, at least for T_3, hormone binding prevents excess uptake by the first cells encountered and promotes uniform tissue distribution.

Capacity & Affinity of Plasma Proteins for Thyroid Hormones

The plasma proteins that bind thyroid hormones are **albumin;** a prealbumin called **thyroxine-binding prealbumin (TBPA);** and a globulin with an electrophoretic mobility between those of α_1- and α_2-globulin, **thyroxine-binding globulin (TBG).** Of the 3 proteins, albumin has the largest **capacity** to bind T_4—ie, it can bind the most T_4 before becoming saturated—and TBG the smallest. However, the **affinities** of the proteins for T_4—ie, the avidity with which

they bind T_4 under physiologic conditions—are such that most of the circulating T_4 is bound to TBG (Table 18–1), with over a third of the binding sites on the protein occupied. Smaller amounts of T_4 are bound to TBPA and albumin. The half-life of TBPA is 2 days, that of TBG is 5 days, and that of albumin 13 days.

Normally, 99.98% of the T_4 in plasma is bound; the free T_4 level is only about 2 ng/dL. There is very little T_4 in the urine. Its biologic half-life is long (about 6–7 days), and its volume of distribution is less than that of ECF (10 L, or about 15% of body weight). All of these properties are characteristic of a substance that is strongly bound to protein.

T_3 is not bound to quite as great an extent; of the 0.15 µg/dL normally found in plasma, 0.2% (0.3 ng/dL) is free. The remaining 99.8% is protein-bound, 46% to TBG and most of the remainder to albumin, with very little binding to TBPA (Table 18–1). The lesser binding of T_3 correlates with the facts that T_3 has a shorter half-life than T_4 and that its action on the issues is much more rapid. RT_3 also binds to TBG.

Free T_4 and T_3 can be measured by equilibrium dialysis, but this is a laborious technique. Tissue uptake of the free hormones is proportionate to their concentration in plasma, and consequently red blood cell uptake is sometimes measured as an index of their level. Similar results are obtained when a resin is used to take up the free thyroid hormones. For example, radioactive T_4 is added to plasma, and after equilibration, resin is added and the percentage of the radioactive T_4 taken up by the resin is determined. Multiplication of the resin uptake by the total T_4 concentration provides the **free T_4 index (FT$_4$I),** an index of the amount of free T_4 in the specimen. In a similar fashion, a **free T_3 index (FT$_3$I)** can be calculated. It is important to emphasize that the FT$_4$I and FT$_3$I are not direct measures of free T_4 and T_3 but indices of their concentration.

Fluctuations in Binding

When there is a sudden, sustained increase in the

Figure 18–8. Distribution of T_4 in the body. The distribution of T_3 is similar. The dashed arrow indicates inhibition of TSH secretion by increases in the free T_4 level of ECF. Approximate concentrations in human blood are shown in parentheses.

Table 18–1. Binding of thyroid hormones to plasma proteins in normal adult humans.

Protein	Plasma Concentration (mg/dL)	Amount of Circulating Hormone Bound (%)	
		T_4	T_3
Thyroxine-binding globulin (TBG)	2	67	46
Thyroxine-binding prealbumin (TBPA)	15	20	1
Albumin	3500	13	53

concentration of thyroid-binding proteins in the plasma, the concentration of free thyroid hormones falls. This change is temporary, however, because the decrease in the concentration of free thyroid hormones in the circulation stimulates TSH secretion, which in turn causes an increase in the production of free thyroid hormones. A new equilibrium is eventually reached at which the total quantity of thyroid hormones in the blood is elevated but the concentration of free hormones, the rate of their metabolism, and the rate of TSH secretion are normal. Corresponding changes in the opposite direction occur when the concentration of thyroid-binding proteins is reduced. Consequently, patients with elevated or decreased concentrations of binding proteins, particularly TBG, are neither hyper- nor hypothyroid; ie, they are **euthyroid.**

TBG levels are elevated in estrogen-treated patients and during pregnancy, as well as after treatment with various drugs (Table 18–2). They are depressed by glucocorticoids, androgens, the antiestrogen danazol, and the cancer chemotherapeutic agent L-asparaginase. A number of other drugs, including salicylates, the anticonvulsant phenytoin, and the cancer chemotherapeutic agents mitotane (o,p'-DDD) and 5-fluorouracil inhibit binding of T_4 and T_3 to TBG and consequently produce changes similar to those produced by a decrease in TBG concentration. Changes in total plasma T_4 and T_3 can also be produced by changes in plasma concentrations of albumin and prealbumin.

Metabolism of Thyroid Hormones

T_4 and T_3 are deiodinated in the liver, the kidneys, and many other tissues. One-third of the circulating T_4 is normally converted to T_3 in adult humans, and 45% is converted to RT_3. As shown in Fig 18–7, only about 13% of the circulating T_3 is secreted by the thyroid and 87% is formed by deiodination of T_4; similarly, only 5% of the circulating RT_3 is secreted by the thyroid and 95% is formed by deiodination of T_4. Two different enzymes are involved, 5'-deiodinase catalyzing the formation of T_3 and 5-deiodinase catalyzing the formation of RT_3 (Fig 18–9). T_3 and RT_3 are then converted to various diiodothyronines. 5'-deiodinase (type I iodothyronine deiodinase), which

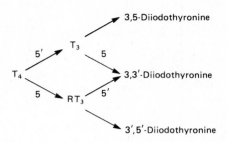

Figure 18–9. Conversion of T_4 to T_3 and RT_3 and thence to diiodothyronines in liver, kidney, and other tissues. 5' = 5'-deiodinase; 5 = 5-deiodinase.

has now been cloned, contains selenium, and conversion of T_4 to T_3 is impaired in selenium deficiency. Type II iodothyronine deiodinase is a related 5'-deiodinase localized to the anterior pituitary gland. 5'-Deiodinase also catalyzes the conversion of RT_3 to 3,3'-diiodothyronine, whereas 5-deiodinase catalyzes the conversion of T_3 to 3,3'-diiodothyronine. 3,5-Diiodothyronine and 3',5'-diiodothyronine are also formed. Much more RT_3 and much less T_3 are formed during fetal life, and the ratio shifts to that of adults about 6 weeks after birth.

In the liver, T_4 and T_3 are conjugated to form sulfates and glucuronides. These conjugates enter the bile and pass into the intestine. The thyroid conjugates are hydrolyzed, and some are reabsorbed (enterohepatic circulation), but some are excreted in the stool. In addition, some T_4 and T_3 pass directly from the circulation to the intestinal lumen. The iodide lost by these routes amounts to about 4% of the total daily iodide loss.

Fluctuations in Deiodination

Various drugs inhibit the 5'-deiodinase that converts T_4 to T_3 and RT_3 to 3,3'-diiodothyronine, producing a fall in plasma T_3 and a rise in plasma RT_3. A wide variety of nonthyroidal illnesses also depress 5'-deiodinase. These include burns, trauma, advanced cancer, cirrhosis, renal failure, myocardial infarction, and febrile states. The low T_3 state produced by these

Table 18–2. Effect of variations in the concentrations of thyroid hormone-binding proteins in the plasma on various parameters of thyroid function after equilibrium has been reached.

Condition	Concentrations of Binding Proteins	Total Plasma T_4, T_3, RT_3	Free Plasma T_4, T_3, RT_3	Plasma TSH	Clinical State
Hyperthyroidism	Normal	High	High	Low	Hyperthyroid
Hypothyroidism	Normal	Low	Low	High	Hypothryoid
Estrogens, methadone, heroin, major tranquilizers, clofibrate	High	High	Normal	Normal	Euthyroid
Glucocorticoids, androgens, danazol, L-asparaginase	Low	Low	Normal	Normal	Euthyroid

conditions disappears with recovery. It is difficult to decide whether individuals with the low T_3 state produced by drugs and illness have mild hypothyroidism, and thyroid hormone measurements are difficult to interpret if thyroid disease is suspected, partly because many of the drugs have other effects on thyroid hormones and partly because the drugs and the illnesses produce other changes that make clinical hypothyroidism hard to diagnose.

Diet also has a clear-cut effect on conversion of T_4 to T_3. In fasted individuals, plasma T_3 is reduced 10–20% in 24 hours and about 50% in 3–7 days, with a corresponding rise in RT_3 (Fig 18–10). Free and bound T_4 levels remain normal. During more prolonged starvation, RT_3 returns to normal but T_3 remains depressed. At the same time, the BMR falls and urinary nitrogen excretion, an index of protein breakdown, is decreased. Thus, the decline in T_3 conserves calories and protein. Conversely, overfeeding increases T_3 and reduces RT_3.

EFFECTS OF THYROID HORMONES

Most of the widespread effects of thyroid hormones in the body are secondary to stimulation of O_2 consumption **(calorigenic action),** although the hormones also affect growth and development in mammals, help regulate lipid metabolism, and increase the

absorption of carbohydrates from the intestine. They also increase the dissociation of oxygen from hemoglobin by increasing red cell 2,3-diphosphoglycerate (DPG) (see Chapter 35).

Mechanism of Action

Thyroid hormones enter cells, and T_3 binds to receptors in the nuclei (see Fig 1–31). T_4 can also bind, but not as avidly. In addition, as noted above, much of the circulating T_4 is converted to T_3, and in that sense T_4 is a prohormone to T_3. The thyroid hormone-receptor complex then binds to DNA and increases the expression of specific genes. The resultant mRNAs trigger the production of various enzymes that alter cell function.

In most of its actions, T_3 acts more rapidly and is 3–5 times more potent than T_4 (Fig 18–11). This is because it is less tightly bound to plasma proteins but binds more avidly to thyroid hormone receptors. RT_3 is inert.

Calorigenic Action

T_4 and T_3 increase the O_2 consumption of almost all metabolically active tissues. The exceptions are the adult brain, testes, uterus, lymph nodes, spleen, and anterior pituitary. T_4 actually depresses the O_2 consumption of the anterior pituitary, presumably because it inhibits TSH secretion. The increase in metabolic rate produced by a single dose of T_4 becomes measurable after a latency period of several hours and lasts 6 days or more. The magnitude of the calorigenic effect depends on the level of catecholamine secretion and on the metabolic rate before injection. If the initial rate is low, the rise is great; but if it is high, the rise is small. This is true not only in euthyroid subjects but in athyreotic T_4-treated patients as well. The cause of the decreased effect at higher metabolic rates is obscure.

Figure 18–10. Effect of starvation on plasma levels of T_4, T_3, and RT_3 in humans. Similar changes occur in wasting diseases. The scale for T_3 and RT_3 is on the left and the scale for T_4 is on the right. (Reproduced, with permission, from Burger AG: New aspects of the peripheral action of thyroid hormones. *Triangle, Sandoz J Med Sci* 1983;**22:**175. Copyright © Sandoz Ltd., Basel, Switzerland.)

Figure 18–11. Calorigenic responses of thyroidectomized rats to subcutaneous injections of T_4 and T_3. (Redrawn and reproduced, with permission, from Barker SB: Peripheral actions of thyroid hormones. *Fed Proc* 1962;**21:**635.)

Thyroid hormones increase the activity of the membrane-bound Na^+-K^+ ATPase in many tissues, and it has been argued that it is the increase in energy consumption associated with the increase in Na^+ transport which is responsible for the increase in metabolic rate. However, inhibition of the increase in Na^+-K^+ ATPase activity with ouabain does not completely abolish the calorigenic effect of thyroid hormones. Mitochondrial protein synthesis is increased, but the role of the mitochondria in the response is obscure and the old theory that thyroid hormones increase energy consumption by uncoupling oxidation of substrate from phosphorylation in the mitochondria has now been largely abandoned.

Effects Secondary to Calorigenesis

When the metabolic rate is increased by T_4 and T_3 in adults, nitrogen excretion is increased; if food intake is not increased, endogenous protein and fat stores are catabolized and weight is lost. In hypothyroid children, small doses of thyroid hormones cause a positive nitrogen balance because they stimulate growth, but large doses cause protein catabolism similar to that produced in the adult. The potassium liberated during protein catabolism appears in the urine, and there is an increase in urinary hexosamine and uric acid excretion.

The skin normally contains a variety of proteins combined with polysaccharides, hyaluronic acid, and chondroitin sulfuric acid. In hypothyroidism, these complexes accumulate, promoting water retention and the characteristic puffiness of the skin (myxedema). When thyroid hormones are administered, the proteins are mobilized, and diuresis continues until the myxedema is cleared.

Large doses of thyroid hormones cause enough extra heat production to lead to a slight rise in body temperature (see Chapter 14), which in turn activates heat-dissipating mechanisms. Peripheral resistance decreases because of cutaneous vasodilation, but cardiac output is increased by the combined action of thyroid hormones and catecholamines on the heart, so that pulse pressure and cardiac rate are increased and circulation time is shortened.

When the metabolic rate is increased, the need for all vitamins is increased and vitamin deficiency syndromes may be precipitated. Thyroid hormones are necessary for hepatic conversion of carotene to vitamin A, and the accumulation of carotene in the bloodstream (**carotenemia**) in hypothyroidism is responsible for the yellowish tint of the skin. Carotenemia can be distinguished from jaundice because in the former condition the scleras are not yellow.

Milk secretion is decreased in hypothyroidism and stimulated by thyroid hormones, a fact sometimes put to practical use in the dairy industry. Thyroid hormones do not stimulate the metabolism of the uterus but are essential for normal menstrual cycles and fertility.

Effects on the Nervous System

In hypothyroidism, mentation is slow and the CSF protein level elevated. Thyroid hormones reverse these changes, and large doses cause rapid mentation, irritability, and restlessness. This is somewhat surprising, because it is generally agreed that cerebral blood flow and the glucose and O_2 consumption of the brain are normal in adult hypo- and hyperthyroidism. However, thyroid hormones enter the brain in adults and are found in gray matter in numerous different locations. In addition, the brain converts T_4 to T_3, and there is a sharp increase in brain 5'-deiodinase activity after thyroidectomy that is reversed within 4 hours by a single intravenous dose of T_3. Some of the effects of thyroid hormones on the brain are probably secondary to increased responsiveness to catecholamines, with consequent increased activation of the reticular activating system (see Chapter 11). There is disagreement about whether thyroid hormones affect brain O_2 consumption in young animals. However, thyroid hormones have marked effects on brain development. In hypothyroid infants, synapses develop abnormally, myelination is defective, and mental development is seriously retarded. The mental changes are irreversible if replacement therapy is not begun soon after birth.

Thyroid hormones also exert effects on the peripheral nervous system. The reaction time of stretch reflexes (see Chapter 6) is shortened in hyperthyroidism and prolonged in hypothyroidism. Measurement of the reaction time of the ankle jerk (Achilles reflex) has attracted considerable attention as a clinical test for evaluating thyroid function, but the reaction time is also affected by certain other diseases.

Effects on the Heart

Thyroid hormones increase the number and affinity of β-adrenergic receptors in the heart and consequently increase its sensitivity to the inotropic and chronotropic effects of catecholamines. They also affect the type of myosin found in cardiac muscle. The heart contains 2 myosin heavy chain (MHC) isoforms, α- and β-MHC. They are encoded by 2 highly homologous genes located in tandem in humans on the short arm of chromosome 17 (Fig 18–12). The α-MHC gene is expressed only in the heart, whereas the β-MHC gene is expressed in the heart and skeletal muscle. Each myosin molecule consists of 2 heavy chains and 2 pairs of light chains (see Chapter 3). The myosin containing β-MHC has less ATPase activity than the myosin containing α-MHC. α-MHC predominates in the ventricles in adults, and its level is increased by treatment with thyroid hormone. However, expression of the α-MHC gene is depressed and that of the β-MHC gene is enhanced in hypothyroidism.

Effects on Skeletal Muscle

Muscle weakness occurs in most patients with hy-

Cardiac myosin heavy chain

Figure 18–12. Cardiac myosin heavy chain genes, showing their location in tandem on the short arm of chromosome 17 in humans and the very similar proteins that are produced. (Reproduced, with permission, from Ridgway EC III: Molecular biology of thyroid hormone action: TSH subunit and cardiac myosin ATPase genes. *Thyroid Today* [July] 1989;**9**:1. The Boots Co [USA] Inc.)

perthyroidism **(thyrotoxic myopathy),** and when the hyperthyroidism is severe and prolonged, the myopathy may be severe. The muscle weakness may be due in part to increased protein catabolism. Thyroid hormones affect the expression of the MHC genes in skeletal as well as cardiac muscle. However, the effects produced are complex and their relation to the myopathy is not established. Hypothyroidism is also associated with muscle weakness, cramps, and stiffness.

Relation to Catecholamines

The actions of thyroid hormones and the catecholamines norepinephrine and epinephrine are intimately interrelated. Epinephrine increases the metabolic rate, stimulates the nervous system, and produces cardiovascular effects similar to those of thyroid hormones, although the duration of these actions is brief. Norepinephrine has generally similar actions. Thyroid hormones increase the number and affinity of β-adrenergic receptors in the heart and possibly in some other tissues, and the effects of thyroid hormones on the heart resemble those of β-adrenergic stimulation. The toxicity of the catecholamines is markedly increased in rats treated with T_4. Although plasma catecholamines are normal in hyperthyroidism, the cardiovascular effects, tremulousness, and sweating produced by thyroid hormones can be reduced or abolished by sympathectomy. They can also be reduced by drugs such as propranolol that block β-adrenergic receptors. Indeed, propranolol and other β blockers are used extensively in the treat-

ment of thyrotoxicosis and in the treatment of the severe exacerbations of hyperthyroidism called **thyroid storms.** However, even though β blockers are weak inhibitors of extrathyroidal conversion of T_4 to T_3, and consequently may produce a small fall in plasma T_3, they have little effect on the other actions of thyroid hormones.

Effects on Carbohydrate Metabolism

Thyroid hormones increase the rate of absorption of carbohydrate from the gastrointestinal tract, an action that is probably independent of their calorigenic action. In hyperthyroidism, therefore, the blood glucose level rises rapidly after a carbohydrate meal, sometimes exceeding the renal threshold. However, it falls again at a high rate.

Effects on Cholesterol Metabolism

Thyroid hormones lower circulating cholesterol levels. The plasma cholesterol level drops before the metabolic rate rises, which indicates that this action is independent of the stimulation of O_2 consumption. As noted in Chapter 17, the decrease in plasma cholesterol concentration is due to increased formation of LDL receptors in the liver, resulting in increased removal of cholesterol from the circulation. Despite considerable effort, it is not yet possible to produce a clinically useful thyroid hormone analog that lowers plasma cholesterol without increasing metabolism.

Effects on Growth & Development

Thyroid hormones are essential for normal growth

and skeletal maturation (see Chapter 22). In hypothyroid children, bone growth is slowed and epiphyseal closure delayed. In the absence of thyroid hormones, growth hormone secretion is also depressed, and thyroid hormones potentiate the effect of growth hormone on the tissues.

Another example of the role of thyroid hormones in growth and maturation is their effect on amphibian metamorphosis. Tadpoles treated with T_4 and T_3 metamorphose early into dwarf frogs, whereas hypothyroid tadpoles never become frogs. The effects of T_4 on metamorphosis are probably independent of its effects on O_2 consumption, even though it raises the O_2 consumption of metamorphosing tadpole skin in vitro. The hormone also exerts a calorigenic effect on tadpoles in vivo.

REGULATION OF THYROID SECRETION

Thyroid function is regulated primarily by variations in the circulating level of pituitary TSH. TSH secretion is increased by the hypophyseotropic hormone TRH (see Chapter 14) and inhibited in a negative feedback fashion by circulating free T_4 and T_3. TSH secretion is also inhibited by stress, and in experimental animals it is increased by cold and decreased by warmth.

Chemistry & Metabolism of TSH

Human TSH is a glycoprotein that contains 211 amino acid residues, plus hexoses, hexosamines, and sialic acid. It is made up of 2 subunits, designated α and β. The α subunit is encoded by a gene on chromosome 6, and the β subunit by a gene on chromosome 1. The α and β subunits become noncovalently linked in the thyrotropes. TSH-α is identical in structure to the α subunit of LH, FSH, and hCG-α (see Chapters 22 and 23). The functional specificity of TSH is conferred by the β unit. The structure of TSH varies from species to species, but other mammalian TSHs are biologically active in humans.

Recent research has done much to clarify the mechanisms involved in the glycosylation of TSH. Glycosylation of both subunits begins with attachment in the rough endoplasmic reticulum of a 14-sugar unit to asparagine residues with modification of these carbohydrate side chains in the Golgi apparatus. The initial glycosylation is necessary for proper combination of the α and the β subunits, and the final pattern of carbohydrate side chains is necessary for full biologic activity. In addition, deglycosylated TSH is removed more rapidly from the circulation; thus, its effectiveness is decreased. TRH increases the biologic activity of TSH by altering its glycosylation. In patients with TRH deficiency, there is a drop in circulating TSH bioactivity, sometimes with little if any decrease in circulating TSH immunoactivity.

The biologic half-life of human TSH is about 60 minutes. TSH is degraded for the most part in the kidneys and to a lesser extent in the liver. Secretion is pulsatile, and mean output starts to rise at about 9:00 PM, peaks at midnight, and then declines during the day. The normal secretion rate is about 110 µg/d. The average plasma level is about 2 µU/mL (Fig 18–13).

The α subunit in hCG is the same as that in TSH, and large amounts of hCG can activate thyroid receptors. In some patients with benign or malignant tumors of placental origin, plasma hCG levels can rise so high that they produce mild hyperthyroidism.

Effects of TSH on the Thyroid

When the pituitary is removed, thyroid function is depressed and the gland atrophies; when TSH is administered, thyroid function is stimulated. Within a few minutes after the injection of TSH, there are increases in iodide binding; synthesis of T_3, T_4, and iodotyrosines; secretion of thyroglobulin into the colloid; and endocytosis of colloid. Iodide trapping is increased in a few hours; blood flow increases; and, with chronic TSH treatment, the cells hypertrophy and the weight of the gland increases.

Whenever TSH stimulation is prolonged, the thyroid becomes detectably enlarged. Enlargement of the thyroid is called **goiter**.

TSH Receptors

The TSH receptor is a typical serpentine receptor that activates adenylate cyclase through G_s (see

Figure 18–13. Relation between plasma TSH, measured by a highly sensitive immunoenzymatic assay, and the free T_4 index (FT$_4$I). The shaded area identifies the normal range. (Reproduced, with permission, from Spencer CA: Clinical utility of sensitive TSH assays. *Thyroid Today* [Apr] 1986;**9**:1. The Boots Co [USA] Inc.)

Chapter 1), and, like the FSH and LH receptors, it has an extended, glycosylated extracellular domain.

Other Factors Affecting Thyroid Growth

In addition to TSH receptors, the thyroid cells contain receptors for IGF-I, EGF, and other growth factors. IGF-I and EGF promote growth whereas γ interferon and tumor necrosis factor α inhibit growth. The exact physiologic role of these factors in the thyroid has not been established.

Control Mechanisms

The mechanisms regulating thyroid secretion are summarized in Fig 18–14. The negative feedback effect of thyroid hormones on TSH secretion is exerted in part at the hypothalamic level, but it is also due in large part to an action on the pituitary, since T_4 and T_3 block the increase in TSH secretion produced by TRH. The relation between plasma TSH and the FT_4I index is shown in Fig 18–13. Infusion of T_4 as well as T_3 reduces the circulating level of TSH, which declines measurably within 1 hour. In experimental animals, there is an initial rise in pituitary TSH content before the decline, indicating that thyroid hormones inhibit secretion before they inhibit synthesis. The effects on secretion and synthesis of TSH both appear to depend on protein synthesis, even though the former is relatively rapid.

The day-to-day maintenance of thyroid secretion depends on the feedback interplay of thyroid hormones with TSH and TRH (Fig 18–14). The adjustments that appear to be mediated via TRH include the

increased secretion of thyroid hormones produced by cold and, presumably, the decrease produced by heat. It is worth noting that although cold produces clearcut increases in circulating TSH in experimental animals and human infants, the rise produced by cold in adult humans is negligible. Consequently, in adults, increased heat production due to increased thyroid hormone secretion (**thyroid hormone thermogenesis**) plays little if any role in the response to cold. Stress has an inhibitory effect on TRH secretion. Dopamine and somatostatin act at the pituitary level to inhibit TSH secretion, but it is not known whether they play a physiologic role in the regulation of TSH secretion. Glucocorticoids also inhibit TSH secretion.

The amount of thyroid hormone necessary to maintain normal cellular function in thyroidectomized individuals used to be defined as the amount necessary to normalize the BMR, but it is now defined as the amount necessary to return plasma TSH to normal. Indeed, with the accuracy and sensitivity of modern assays for TSH and the marked inverse correlation between plasma free thyroid hormone levels and plasma TSH, measurement of TSH is now widely regarded as one of the best tests of thyroid function. The amount of T_4 that normalizes plasma TSH in athyreotic individuals averages 112 μg of T_4 by mouth per day in adults. About 80% of this dose is absorbed from the gastrointestinal tract. It produces a slightly greater than normal FT_4I but a normal FT_3I, indicating that in humans, unlike some experimental animals, it is circulating T_3 rather than T_4 that is the principal feedback regulator of TSH secretion.

CLINICAL CORRELATES

The signs, symptoms, and complications of hypothyroidism and hyperthyroidism in humans are predictable consequences of the physiologic effects of thyroid hormones discussed above.

Hypothyroidism

The syndrome of adult hypothyroidism is generally called **myxedema,** although this term is also used to refer specifically to the skin changes in this syndrome. Hypothyroidism may be the end result of a number of diseases of the thyroid gland, or it may be secondary to pituitary failure (pituitary hypothyroidism) or hypothalamic failure (hypothalamic hypothyroidism). In the latter 2 conditions, unlike the first, the thyroid responds to a test dose of TSH, and at least in theory, hypothalamic hypothyroidism can be distinguished from pituitary hypothyroidism by the presence in the former of a rise in plasma TSH following a test dose of TRH. The TSH response to TRH is usually normal in hypothalamic hypothyroidism, while it is increased in hypothyroidism caused by thyroid disease and decreased in hyperthyroidism

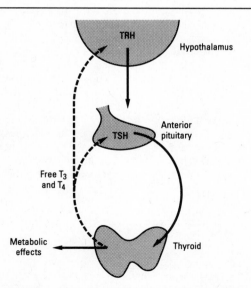

Figure 18–14. Feedback control of thyroid secretion. The dashed arrows indicate inhibitory effects, and the solid arrows indicate stimulatory effects. Compare with Figs 20–21, 22–11, 23–22, and 23–36.

because of the feedback of thyroid hormones on the pituitary gland.

In completely athyreotic humans, the BMR falls to about –40. The hair is coarse and sparse, the skin is dry and yellowish (carotenemia), and cold is poorly tolerated. The voice is husky and slow, the basis of the aphorism that "myxedema is the one disease that can be diagnosed over the telephone." Mentation is slow and memory is poor, and in some patients there are severe mental symptoms ("myxedema madness"). Plasma cholesterol is elevated.

Cretinism

Children who are hypothyroid from birth are called cretins. They are dwarfed and mentally retarded and have potbellies and enlarged, protruding tongues (Fig 18–15). Before the use of iodized salt became widespread, the most common cause of cretinism was maternal iodine deficiency. Various congenital abnormalities of the hypothalamo-pituitary-thyroid axis that cause goiter can also cause congenital hypothyroidism with cretinism. T_4 crosses the placenta, and unless the mother is hypothyroid,

growth and development are normal until birth. However, treatment must be started soon after birth if mental retardation is to be prevented.

Hyperthyroidism

Hyperthyroidism (thyrotoxicosis) is characterized by nervousness; weight loss; hyperphagia; heat intolerance; increased pulse pressure; a fine tremor of the outstretched fingers; a warm, soft skin; sweating; and a BMR from + 10 to as high as + 100. It may be caused by a variety of thyroid disorders (Table 18–3), including, in rare instances, TSH-secreting pituitary tumors. However, the most common form is **Graves' disease (exophthalmic goiter)**. In Graves' disease, the thyroid is diffusely enlarged and hyperplastic and there is a protrusion of the eyeballs called **exophthalmos** (Fig 18–16). It now seems clear that Graves' disease is an autoimmune disease (see Chapter 27) in which T lymphocytes activated by antigens in the thyroid gland stimulate B lymphocytes to produce circulating antibodies against the antigens. Some of the antibodies damage the thyroid, producing **Hashimoto's thyroiditis,** and this condition can progress to hypothyroidism. However, other antibodies are formed against components of the TSH receptors. These antibodies, which are called **thyroid-stimulating immunoglobulins (TSI),** activate the receptors, producing hyperthyroidism. One theory of the genesis of Graves' disease holds that there is a defect in suppressor T lymphocytes that permits helper T lymphocytes to stimulate B lymphocytes to produce the thyroid autoantibodies. However, the mechanism that initiates the immunologic events is still unknown. There is marked stimulation of the secretion of thyroid hormones, and the high circulating T_4 and T_3 levels inhibit TSH secretion (Fig 18–17), so the circulating TSH level is depressed. Increasing levels of thyroid hormones tend to cause increased formation of the antibodies, whereas decreased levels produced by treatment of the disease tend to lower but not completely prevent formation of the antibodies.

The exophthalmos in Graves' disease is due to swelling of the extraocular muscles and, to a lesser extent, the connective tissue within the rigid bony walls of the orbits. This pushes the eyeballs forward. Thyroid hormones are not directly responsible for the

Figure 18–15. Fraternal twins, age 8 years. The boy has congenital hypothyroidism. (Reproduced, with permission, from Wilkins L in: *Clinical Endocrinology I.* Astwood EB, Cassidy CE [editors]. Grune & Stratton, 1960.)

Table 18–3. Causes of hyperthyroidism.

Thyroid overactivity
Graves' disease
Solitary toxic adenoma
Toxic multinodular goiter
TSH-secreting pituitary tumor
Thyroiditis
Extrathyroidal
Administration of T_3 or T_4 (factitious or iatrogenic
hyperthyroidism)
Eptopic thyroid tissue

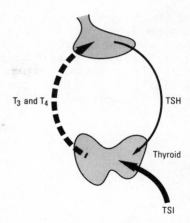

Figure 18–17. Thyroid and pituitary function in Graves' disease. Thyroid-stimulating immunoglobulins (TSI) act on the thyroid to produce a marked increase in the secretion of T_4 and T_3, and secretion of TSH from the pituitary is markedly inhibited.

Figure 18–16. Graves' disease. Note the goiter and the exophthalmos. (Courtesy of PH Forsham.)

exophthalmos, and although exophthalmos is usually associated with hyperthyroidism, it may occur before, during, or after the thyrotoxic state. In a significant proportion of patients with thyrotoxicosis it is made worse by thyroidectomy. In addition, it is occasionally seen in euthyroid patients with Hashimoto's thyroiditis, in patients with myxedema, and in patients with no known thyroid disease. The cause of the exophthalmos is almost certainly an autoimmune attack on the extraocular muscles and orbital connective tissue by cytotoxic antibodies. The antibodies are formed to antigens common to the eye muscles, the thyroid, and some other tissues. The condition responds to glucocorticoids or other agents that produce immunosuppression.

Thyrotoxicosis places a considerable load on the cardiovascular system, and in some patients with hyperthyroidism, most or even all of the symptoms are cardiovascular. Thyrotoxic heart disease is curable but the diagnosis is often missed. It is important to remember that heart failure develops whenever the cardiac output, even though it is elevated, is inadequate to maintain tissue perfusion. The drop in peripheral resistance caused by cutaneous vasodilation

in thyrotoxicosis is equivalent to opening a large arteriovenous fistula; if the compensatory increase in cardiac output is not great enough, "high-output failure" results.

Iodine Deficiency

When the dietary iodine intake falls below 10 μg/d, thyroid hormone synthesis is inadequate and secretion declines. As a result of increased TSH secretion, the thyroid hypertrophies, producing an **iodine deficiency goiter** that may become very large. Such "endemic goiters" have been known since ancient times. Before the practice of adding iodide to table salt became widespread, they were very common in Central Europe and the area around the Great Lakes in the USA, the inland "goiter belts" where iodine has been leached out of the soil by rain so that food grown in the soil is iodine-deficient.

Radioactive Iodine Uptake

Iodine uptake is an index of thyroid function that can be easily measured by using tracer doses of radioactive isotopes of iodine that have no known deleterious effect on the thyroid. The tracer is administered orally and the thyroid uptake determined by placing a gamma ray counter over the neck. An area such as the thigh is also counted, and counts in this region are subtracted from the neck counts to correct for nonthyroidal radioactivity in the neck. The isotope of iodine that is most commonly used is ^{123}I because it has a half-life of only 0.55 day, compared with ^{131}I, which has a half-life of 8.1 days, and ^{125}I, which has a half-life of 60 days. Diagnostic use of radioactive iodine uptake has decreased because of the general availability of methods for measuring T_4, T_3, and TSH in plasma. In addition, the use of iodized salt is widespread, and this causes uptake to be low because the iodide pool is so large that the tracer is excessively diluted. However, an analysis of radioactive iodine uptake is helpful in understanding the physiology of the thyroid gland. The uptake in a normal subject is plotted in Fig 18–18. In hyperthyroidism, iodide is rapidly incorporated into T_4 and T_3, and these

hormones are released at an accelerated rate. Therefore, the amount of radioactivity in the thyroid rises sharply, but it then levels off and may start to decline within 24 hours, at a time when the uptake in normal subjects is still rising. In hypothyroidism, the uptake is low.

Large amounts of radioactive iodine destroy thyroid tissue because the radiation kills the cells. Radioiodine therapy is useful in some cases of thyroid cancer and can be used to treat benign thyroid disease, although, especially in young patients, the dangers of possible radiation carcinogenesis and germ cell mutations must be kept in mind.

Radioactive isotopes of iodine are major products of nuclear fission, and if fission products are released into the atmosphere as a result of an accident at a nuclear power plant or explosion of a nuclear bomb, the isotopes spread for considerable distances in the atmosphere because they are more volatile than the other products. Treatment with potassium iodide is regularly instituted in fallout areas to enlarge the iodide pool and depress thyroid uptake to low levels.

Antithyroid Drugs

Most of the drugs that inhibit thyroid function act either by interfering with the iodide-trapping mechanism or by blocking the organic binding of iodine. In either case, TSH secretion is stimulated by the decline in circulating thyroid hormones, and goiter is produced. A number of monovalent anions compete with iodide for active transport into the thyroid and therefore inhibit uptake to the point where the T/S ratio approaches unity; this inhibition of iodide transport can be overcome by administration of extra iodide. The anions include chlorate, pertechnetate, periodate, bi-iodate, nitrate, and perchlorate. Thiocyanate, another monovalent anion, inhibits iodide transport but is not itself concentrated within the gland. The activity of perchlorate is about 10 times that of thiocyanate.

The **thiocarbamides,** a group of compounds related to thiourea, inhibit the iodination of monoiodotyrosine (organic binding of iodide) and block the coupling reaction. The 2 used clinically are propylthiouracil and methimazole (Fig 18–19). Iodination of tyrosine is inhibited because propylthiouracil and methimazole compete with tyrosine residues for iodine and become iodinated. In addition, propylthiouracil but not methimazole inhibits the conversion of T_4 to T_3 in extrathyroidal tissues. Both drugs may also ameliorate thyrotoxicosis by suppressing the immune system and thereby depressing TSI formation. They may also inhibit the biosynthesis or alter the structure of thyroglobulin.

The effect of thiocarbamides on coupling is produced by lower doses than the doses that affect iodination. The thiocarbamides do not block iodide trapping. Because of the increase in TSH secretion, the initial uptake of radioactive iodine is actually in-

Figure 18–18. Distribution of radioactive iodine in individuals on a relatively low-iodine diet. Percentages are plotted against time after an oral dose of radioactive iodine. In hyperthyroidism, plasma radioactivity falls rapidly and then rises again as a result of release of labeled T_4 and T_3 from the thyroid. (Modified from Ingbar SH, Woeber KA in: *Textbook of Endocrinology,* 4th ed. Williams RH [editor]. Saunders, 1968.)

Figure 18–19. Antithyroid thiocarbamides.

Figure 18–20. The naturally occurring goitrogen in vegetables of the family Brassicaceae.

creased during thiocarbamide therapy, and the T/S ratio may reach 250. However, since binding is inhibited, the iodine is not retained; 24 hours after administration of the isotope, uptake is subnormal.

Another substance that inhibits thyroid function under certain conditions is iodide itself. The position of iodide in thyroid physiology is thus unique in that while some iodide is needed for normal thyroid function, too little iodide and too much both cause abnormal thyroid function. In normal individuals, large doses of iodides act directly on the thyroid to produce a mild and transient inhibition of organic binding of iodide and hence of hormone synthesis. This inhibition is known as the **Wolff-Chaikoff effect.** The Wolff-Chaikoff effect is greater and more prolonged when iodide transport is increased, and this is why patients with thyrotoxicosis are more responsive to iodide than normal individuals. Susceptibility to the effect is also increased when there is a defect in the organic binding mechanism, and consequently the inhibition is increased in individuals with partially destroyed or removed thyroids or thyroiditis. There are at least 2 additional mechanisms by which excess I⁻ inhibits thyroid function. It reduces the effect of TSH on the gland by reducing the cyclic AMP response to this hormone, and it inhibits proteolysis of thyroglobulin. There is no direct effect on the I⁻ trapping mechanism, but the total I⁻ uptake is low because of the inhibition of organic binding and, to a lesser extent, because the amount of circulating iodide is so great that added tracer is immensely diluted. In thyrotoxicosis, iodides cause colloid to accumulate, and the vascularity of the hyperplastic gland is decreased, making iodide treatment of considerable value in preparing thyrotoxic patients for surgery.

Naturally Occurring Goitrogens

Thiocyanates are sometimes ingested with food, and there are relatively large amounts of naturally occurring goitrogens in some foods. Vegetables of the Brassicaceae family, particularly rutabagas, cabbage, and turnips, contain **progoitrin** and a substance that converts this compound into **goitrin,** an active antithyroid agent (Fig 18–20). The progoitrin activator in

vegetables is heat-labile, but because there are activators in the intestine (presumably of bacterial origin), goitrin is formed even if the vegetables are cooked. The goitrin intake on a normal mixed diet is usually not great enough to be harmful, but in vegetarians and food faddists, "cabbage goiters" do occur. Other as yet unidentified plant goitrogens probably exist and may be responsible for the occasional small "goiter epidemics" reported from various parts of the world.

Use of Thyroid Hormones in Nonthyroidal Diseases

When the pituitary-thyroid axis is normal, doses of exogenous thyroid hormone that provide less than the amount secreted endogenously have no significant effect on metabolism, because there is a compensatory decline in endogenous secretion resulting from inhibition of TSH secretion. In euthyroid humans, the oral dose of T_4 that merely suppresses thyroid function is 100–125 µg/d. Suppression of TSH secretion due to larger doses of T_4 or pituitary disease leads eventually to thyroid atrophy. An atrophic gland initially responds sluggishly to TSH, and if the TSH suppression has been prolonged, it may take some time for normal thyroid responsiveness to return. The adrenal cortex and some other endocrine glands respond in an analogous fashion; when they are deprived of the support of their tropic hormones for some time, they become atrophic and only sluggishly responsive to their tropic hormone until the hormone has had some time to act on the gland.

In patients with so-called metabolic insufficiency—fatigued individuals who feel weak and have a slightly low BMR with normal plasma T_4 and T_3 levels—thyroid therapy has been proved to be of no more value than placebo medication. Use of thyroid hormones to promote weight loss is of value only if the patient pays the price of some nervousness and heat intolerance and curbs the appetite so that there is no compensatory increase in caloric intake.

19

Endocrine Functions of the Pancreas & the Regulation of Carbohydrate Metabolism

INTRODUCTION

At least 4 peptides with hormonal activity are secreted by the islets of Langerhans in the pancreas. Two of these hormones, **insulin** and **glucagon,** have important functions in the regulation of the intermediary metabolism of carbohydrates, proteins, and fats. The third hormone, **somatostatin,** plays a role in the regulation of islet cell secretion, and the physiologic function of the fourth, **pancreatic polypeptide,** is unsettled. Glucagon, somatostatin, and possibly pancreatic polypeptide are also secreted by cells in the mucosa of the gastrointestinal tract.

Insulin is anabolic, increasing the storage of glucose, fatty acids, and amino acids. Glucagon is catabolic, mobilizing glucose, fatty acids, and amino acids from stores into the bloodstream. The 2 hormones are thus reciprocal in their overall action and are reciprocally secreted in most circumstances. Insulin excess causes hypoglycemia, which leads to convulsions and coma. Insulin deficiency, either absolute or relative, causes diabetes mellitus, a complex and debilitating disease that if untreated is eventually fatal. Glucagon deficiency can cause hypoglycemia, and glucagon excess makes diabetes worse. Excess pancreatic production of somatostatin causes hyperglycemia and other manifestations of diabetes.

A variety of other hormones also have important roles in the regulation of carbohydrate metabolism.

ISLET CELL STRUCTURE

The islets of Langerhans (Fig 19–1) are ovoid, 76- × 175-μm collections of cells scattered throughout the pancreas, although they are more plentiful in the tail than in the body and head. They make up 1–2% of the weight of the pancreas. In humans, there are 1–2 million islets. Each has a copious blood supply; and blood from the islets, like that from the gastrointestinal tract but unlike that from any other endocrine organs, drains into the hepatic portal vein.

The cells in the islets can be divided into types on the basis of their staining properties and morphology. There are at least 4 distinct cell types in humans: A, B, D, and F cells. A, B, and D cells are also called α, β, and δ cells. However, this leads to confusion in view of the use of Greek letters to refer to other structures in the body, particularly adrenergic receptors (see Chapter 4). The A cells secrete glucagon, the B cells secrete insulin, the D cells secrete somatostatin, and the F cells secrete pancreatic polypeptide. The B cells, which are the most common and account for 60–75% of the cells in the islets, are generally located in the center of each islet. They tend to be surrounded by the A cells, which make up 20% of the total, and the less common D and F cells. The islets in the tail, the body, and the anterior and superior part of the head of the human pancreas have many A cells and few if any F cells in the outer rim, whereas in rats and probably in humans, the islets in the posterior part of the head of the pancreas have a relatively large number of F cells and few A cells. The A-cell-rich (glucagon-rich) islets arise embryologically from the dorsal pancreatic bud, and the F-cell-rich (pancreatic polypeptide-rich) islets arise from the ventral pancreatic bud. These buds arise separately from the duodenum.

The B cell granules are packets of insulin in the cell cytoplasm. Each packet is contained in a membrane-lined vesicle (Fig 19–2), and, characteristically, there is a clear space (halo) between the wall of the vesicle and the packet. The shape of the packets varies from species to species; in humans, some are round whereas others are rectangular. In the B cells, the insulin molecule forms polymers and also complexes with zinc. The differences in the shape of the packets are probably due to differences in the size of polymers or zinc aggregates of insulin. The A granules, which contain glucagon, are relatively uniform from species to species (Fig 19–3). The D cells also contain large numbers of relatively homogeneous granules.

Figure 19–1. Islet of Langerhans in the rat pancreas. Darkly stained cells are B cells. Surrounding pancreatic acinar tissue is light-colored. (× 400; courtesy of LL Bennett.)

STRUCTURE, BIOSYNTHESIS, & SECRETION OF INSULIN

Structure & Species Specificity

Insulin is a polypeptide containing 2 chains of amino acids linked by disulfide bridges (Table 19–1). There are minor differences in the amino acid composition of the molecule from species to species. The differences are generally not sufficient to affect the biologic activity of a particular insulin in heterolo-

Figure 19–2. Electron micrograph of 2 adjoining B cells in the human pancreas. The B granules are the membrane-lined vesicles containing crystals that vary in shape from rhombic to round. (× 26,000; courtesy of A Like. Reproduced, with permission, from Fawcett DW: *Bloom and Fawcett, A Textbook of Histology,* 11th ed. Saunders, 1986.)

gous species but are sufficient to make the insulin antigenic. If insulin of one species is injected for a prolonged period into another species, the anti-insulin antibodies formed inhibit the injected insulin. Almost all humans who have received commercial beef insulin for more than 2 months have antibodies against beef insulin, but the titer is usually low. Pork insulin differs from human insulin by only one amino acid residue and has low antigenicity. Human insulin produced in bacteria by recombinant DNA technology is also used.

Biosynthesis & Secretion

Insulin is synthesized in the endoplasmic reticulum of the B cells. It is then transported to the Golgi apparatus, where it is packaged in membrane-bound granules. These granules move to the cell wall by a process that appears to involve microtubules, and their membranes fuse with the membrane of the cell, expelling the insulin to the exterior by exocytosis (see Chapter 1). The insulin then crosses the basal laminas of the B cell and a neighboring capillary and the fenestrated endothelium of the capillary (Fig 19–3) to reach the bloodstream. The fenestrations are discussed in detail in Chapter 30.

Like other polypeptide hormones and related proteins that enter the endoplasmic reticulum, insulin is synthesized as part of a larger preprohormone (see Chapter 1). The gene for insulin is located on the short arm of chromosome 11 in humans. It has 2 introns and 3 exons (Fig 19–4). **Preproinsulin** has a 23-amino-acid signal peptide (leader sequence) removed as it enters the endoplasmic reticulum. The remainder of the molecule is then folded, and the disulfide bonds are formed to make **proinsulin.** The peptide segment connecting the A and B chains is normally detached in the granules before secretion. Without the connection, proper folding of the molecule for the formation of the disulfide bridges would be difficult. The polypeptide that remains in addition to insulin after the connection is severed is called the **connecting peptide (C peptide).** Normally, 90–97% of the product released from the B cells is insulin along with equimolar amounts of C peptide. The rest is mostly proinsulin. C peptide can be measured by radioimmunoassay, and its level provides an index of B cell function in patients receiving exogenous insulin.

Tissue kallikrein (see Chapter 31) may play a role in the conversion of proinsulin to insulin. This endopeptidase is found in pancreatic islets, and its distribution parallels that of insulin.

FATE OF SECRETED INSULIN

Insulin & Insulinlike Activity in Blood

Plasma contains a variety of substances with insulinlike activity in addition to insulin (Table 19–2).

Figure 19–3. A and B cells, showing the relation of the cells to blood vessels. Note that insulin from the B cell and glucagon from the A cell are secreted by exocytosis and cross the basal lamina of the cell and the basal lamina of the capillary before entering the lumen of the fenestrated capillary. (Reproduced, with permission, from Junqueira IC, Carneiro J, Kelley RO: *Basic Histology,* 7th ed. Appleton & Lange, 1992.)

The activity that is not suppressed by antibodies has been called **nonsuppressible insulinlike activity (NSILA).** Most, if not all, of this activity persists after pancreatectomy and is due to the insulinlike growth factors **IGF-I** and **IGF-II** (see Chapter 22). These IGFs are polypeptides. Small amounts are free in the plasma (low-molecular-weight fraction), but large amounts are bound to proteins (high-molecular-weight fraction).

One may well ask why pancreatectomy causes diabetes mellitus (see below) when the NSILA persists in the plasma. However, the insulinlike activities of IGF-I and IGF-II are weak compared with that of insulin.

Metabolism

The half-life of insulin in the circulation in humans is about 5 minutes. Insulin binds to insulin receptors and is internalized (see below); it is destroyed in the endosomes formed by the endocytotic process. The principal enzyme involved is **insulin protease,** an enzyme on the cell membrane that is internalized with the insulin. It should be noted that insulin receptors are found in many cells in the body in addition to

Table 19–1. Structure of human insulin (molecular weight 5808) and (below) variations in this structure in other mammalian species. In the rat, the islet cells secrete 2 slightly different insulins, and in certain fish 4 different chains are found.

	Variations From Human Amino Acid Sequence	
Species	**A Chain Position** 8 9 10	**B Chain Position** 30
Pig, dog, sperm whale	Thr-Ser-Ile	Ala
Rabbit	Thr-Ser-Ile	Ser
Cattle, goat	Ala-Ser-Val	Ala
Sheep	Ala-Gly-Val	Ala
Horse	Thr-Gly-Ile	Ala
Sei whale	Ala-Ser-Thr	Ala

Figure 19–4. Biosynthesis of insulin. The 3 exons of the insulin gene (top) are separated by 2 introns (INT 1 and INT 2). Exons 1 and 2 code for an untranslated part of the mRNA, exon 2 codes for the signal peptide (P) and the B chain (B), exons 2 and 3 code for the C peptide (C), and exon 3 codes for the A chain (A) plus an untranslated part of the mRNA; bp, base pairs; nt, nucleotides. The signal peptide guides the polypeptide chain into the endoplasmic reticulum and is then removed. The molecule is next folded, with formation of the disulfide bonds. The C peptide is separated by one or more converting enzymes in the secretory granule.

Table 19–2. Substances with insulinlike activity in human plasma.

Insulin
Proinsulin
Nonsuppressible insulinlike activity (NSILA)
    Low-molecular-weight fraction
      IGF-I
      IGF-II
    High-molecular-weight fraction (mostly IGF bound to protein)

the classic "insulin-sensitive" cells in which the hormone increases glucose uptake. Indeed, almost all tissues in the body can metabolize insulin. However, 80% of secreted insulin is normally degraded in the liver and kidneys.

EFFECTS OF INSULIN

The physiologic effects of insulin are far-reaching and complex. They are conveniently divided into rapid, intermediate, and delayed actions, as listed in Table 19–3. The best known is the hypoglycemic effect, but there are additional effects on amino acid and electrolyte transport, many enzymes, and growth (see below). The net effect of the hormone is storage of carbohydrate, protein, and fat. Therefore, insulin is appropriately called the "hormone of abundance."

Hypoglycemic Action of Insulin

Glucose enters all cells, but in muscles, fat, and a variety of other tissues (Table 19–4), insulin facilitates its entry into the cells by increasing the number of glucose transporters in the cell membranes. The rate of phosphorylation of the glucose, once it has entered the cells, is regulated by other hormones. Growth hormone and cortisol both inhibit phosphorylation in certain tissues. However, the process is normally so rapid that it is a rate-limiting step in glucose metabolism only when the rate of glucose entry is high.

Because of the rapid phosphorylation, the intracellular free glucose concentration is normally low and

Table 19–3. Principal actions of insulin.[1]

Rapid (seconds)
    Increased transport of glucose, amino acids, and K+ into insulin-sensitive cells

Intermediate (minutes)
    Stimulation of protein synthesis
    Inhibition of protein degradation
    Activation of glycogen synthase and glycolytic enzymes
    Inhibition of phosphorylase and gluconeogenic enzymes

Delayed (hours)
    Increase in mRNAs for lipogenic and other enzymes, c-*fos*

[1]Courtesy of ID Goldfine.

Table 19-4. Effect of insulin on glucose uptake in tissues in which it has been investigated.

Tissues in which insulin facilitates glucose uptake
  Skeletal muscle
  Cardiac muscle
  Smooth muscle
  Adipose tissue
  Leukocytes
  Crystalline lens of the eye
  Pituitary
  Fibroblasts
  Mammary gland
  Aorta
  A cells of pancreatic islets
Tissues in which insulin does not facilitate glucose uptake
  Brain (except probably part of hypothalamus)
  Kidney tubules
  Intestinal mucosa
  Red blood cells

the concentration gradient for glucose is directed inward. The lipid cell membrane is impermeable to glucose. Some glucose is transported into cells in insulin-sensitive tissues in the absence of insulin, but insulin greatly increases the rate of glucose transport. The transport shows the saturation kinetics of a system in which there is limited availability of a specific carrier. It is thus an example of **facilitated diffusion.** The process is highly specific, being limited for the most part to glucose and a few other sugars with the same configuration in the first 3 carbon atoms; furthermore, these sugars compete with one another for transport.

Glucose Transporters

The glucose transporters that are responsible for facilitated diffusion of glucose across cell membranes are a family of closely related proteins that cross the cell membrane 12 times (Fig 19–5). They

Figure 19–5. Orientation of glucose transport protein in the plasma membrane. There are 12 membrane-spanning domains, indicated by the striped areas, and there is one glycosylation site (CHO) on the outside of the membrane. The amino and the carboxy terminals are both located in the cytoplasm inside the cell. (Reproduced, with permission, from Kasanicki MA, Pilch PF: Regulation of glucose-transporter function. *Diabetes Care* 1990;**13:**219. Copyright © 1990 by the American Diabetes Association, Inc.)

differ from and have no homology with the sodium-dependent glucose transporter (SGLT 1, Na$^+$-glucose cotransporter) responsible for the secondary active transport of glucose out of the intestine (see Chapter 25) and renal tubules (see Chapter 38). Particularly in transmembrane helical segments 3, 5, 7, and 11, the amino acids of the facilitative transporters appear to surround channels that glucose can enter. Presumably conformation then changes, and glucose is released inside the cell.

Five different glucose transporters, called in order of discovery GLUT 1 through GLUT 5, have been characterized (Table 19–5). They contain 492–524 amino acids, and their affinity for glucose varies. Each transporter appears to have evolved for special tasks. GLUT 4 is the transporter in muscle and adipose tissue that is stimulated by insulin. A pool of GLUT 4 molecules is maintained in the cytoplasm of insulin-sensitive cells, and when these cells are exposed to insulin, the transporters move rapidly into the cell membrane, apparently by exocytosis. When the insulin stimulation stops, they return to the cytoplasm, probably by endocytosis, and are ready for the next exposure to insulin. The other GLUT transporters appear to stay in the cell membrane.

Insulin Preparations

The maximal decline in blood glucose occurs 30 minutes after intravenous injection of crystalline insulin. After subcutaneous administration, the maximal fall occurs in 2–3 hours (Table 19–6). A wide variety of insulin preparations are now available commercially. These include insulins that have been complexed with protamine and other polypeptides to delay absorption. In general, however, they fall into 3 categories: rapid-, intermediate-, and long-acting. The main intermediate-acting insulins are NPH and lente insulins (Table 19–6), and the main long-acting insulin is protamine zinc insulin.

Relation to Potassium

Insulin causes K$^+$ to enter cells, with a resultant lowering of the extracellular K$^+$ concentration. Infusions of insulin and glucose significantly lower the plasma K$^+$ level in normal individuals and are very effective for the temporary relief of hyperkalemia in patients with renal failure. Hypokalemia often develops when patients with diabetic acidosis are treated with insulin. The reason for the intracellular migration of K$^+$ is still uncertain. However, insulin increases the activity of Na$^+$-K$^+$ ATPase in cell membranes, so that more K$^+$ is pumped into cells.

Other Actions

The hypoglycemic and other effects of insulin are summarized in temporal terms in Table 19–3, and the net effects on various tissues are summarized in Table 19–7. The action on glycogen synthase fosters glycogen storage, and the actions on glycolytic en-

Table 19–5. Glucose transporters in mammals.[1]

	Function	K_m (mM)[2]	Major sites of expression
Active transport			
SGLT 1 (Na⁺-glucose cotransporter)	Secondary active transport of glucose	0.1–10	Small intestine, renal tubules
Facilitated diffusion			
GLUT 1	Basal glucose uptake	1–2	Placenta, brain, red cells, kidneys, colon, many other organs
GLUT 2	B cell glucose sensor; transport out of intestinal and renal epithelial cells	12–20	B cells of islets, liver, epithelial cells of small intestine, kidneys
GLUT 3	Basal glucose uptake	<1	Brain, placenta, kidneys, many other organs
GLUT 4	Insulin-stimulated glucose uptake	5	Skeletal and cardiac muscle, adipose tissue, other tissues (see Table 19–4)
GLUT 5	Dietary absorption	1–2	Jejunum

[1]Modified from Bell GI: Molecular defects in diabetes mellitus. *Diabetes* 1991;**40**:413.
[2]The K_m is the glucose concentration at which transport is half maximal.

zymes favor glucose metabolism to 2 carbon fragments (see Fig 17–9), with resulting promotion of lipogenesis. Stimulation of protein synthesis from amino acids entering the cells and inhibition of protein degradation foster growth. The effects on RNAs are largely related to increasing enzyme stores.

The anabolic effect of insulin is aided by the protein-sparing action of adequate intracellular glucose supplies. Failure to grow is a symptom of diabetes in children, and insulin stimulates the growth of immature hypophysectomized rats to almost the same degree as growth hormone. Maximum insulin-induced growth is present, however, only when the protein-sparing action of glucose is fostered by feeding a high-carbohydrate diet.

MECHANISM OF ACTION

Insulin Receptors

The insulin receptor is a complex protein with a molecular weight of approximately 340,000. Insulin receptors are found on many different cells in the body, including cells in which insulin does not increase glucose uptake. The receptor is a tetramer made up of 2 α and 2 β glycoprotein subunits, all of which are synthesized on a single mRNA and then proteolytically separated and bound to each other by disulfide bonds (Fig 19–6). The gene for the insular receptor has 22 exons and is located on chromosome 19. The α subunits bind insulin and are extracellular, whereas the β subunits span the membrane. The intracellular ends of the β subunits have tyrosine kinase activity. The α and β subunits are both glycosylated,

Table 19–6. Characteristics of the blood-glucose-lowering effects of crystalline zinc insulin and several modified insulins.

Type of Insulin	Hours After Subcutaneous Administration	
	Peak Action	Duration of Action
Crystalline zinc insulin (CZI)	2–3	5–7
NPH insulin (neutral protamine-Hagedorn insulin)	10–20	24–28
Lente insulin	10–20	24–28
Protamine zinc insulin (PZI)	16–24	36 +

Table 19–7. Effects of insulin on various tissues.

Adipose tissue
1. Increased glucose entry
2. Increased fatty acid synthesis
3. Increased glycerol phosphate synthesis
4. Increased triglyceride deposition
5. Activation of lipoprotein lipase
6. Inhibition of hormone-sensitive lipase
7. Increased K⁺ uptake

Muscle
1. Increased glucose entry
2. Increased glycogen synthesis
3. Increased amino acid uptake
4. Increased protein synthesis in ribosomes
5. Decreased protein catabolism
6. Decreased release of gluconeogenic amino acids
7. Increased ketone uptake
8. Increased K⁺ uptake

Liver
1. Decreased ketogenesis
2. Increased protein synthesis
3. Increased lipid synthesis
4. Decreased glucose output due to decreased gluconeogenesis and increased glycogen synthesis

General
1. Increased cell growth

Figure 19–6. Diagrammatic representation of the structure of the insulin receptor. The receptor is a tetrameric protein made up of 2 α and 2 β subunits joined by disulfide (− S − S −) bonds. Insulin (INS) binds to the α subunits, and this triggers autophosphorylation of the tyrosine kinase portions of the β subunits inside the cell. The autophosphorylation in turn triggers the rest of the multiple and extensive effects of insulin. (Modified from Andersen AS: Reception and transmission. *Nature* 1989;**337:**12.)

Figure 19–7. Comparison of the structure of the insulin receptor with the structures of the receptors for IGF-I, IGF-II, nerve growth factor (NGF), and LDL. The diagonally striped boxes identify intracellular tyrosine kinase domains. Note the marked similarity between the insulin receptor and the IGF-I receptor; also note the 15 repeat sequences in the extracellular portion of the IGF-II receptor. (Modified from Morgan DA et al: Insulinlike growth factor II receptor as a multifunctional binding protein. *Nature* 1987;**329:**301.)

with sugar residues extending into the interstitial fluid. Binding of insulin triggers the tyrosine kinase activity of the β subunits, producing autophosphorylation of the β subunits on tyrosine residues. This autophosphorylation is necessary for insulin to exert its biologic effects. At present, it is uncertain how phosphorylation of the insulin receptors triggers the actions of the hormone, but the phosphorylated receptor acts by phosphorylating and dephosphorylating other proteins inside the cell, particularly on serine and threonine residues. This action may be direct or mediated by one or more second messengers in insulin-sensitive cells. There is evidence that an inositol and a structurally distinct diacylglycerol may function as such messengers, but it is too soon to draw firm conclusions.

It is interesting to compare the insulin receptor with other related receptors. It is very similar to the receptor for IGF-I but different from the receptor for IGF-II (Fig 19–7). Several other receptors for growth factors and receptors for various oncogenes are also tyrosine kinases. However, the amino acid composition of these receptors is quite different.

When insulin binds to its receptors, they aggregate in patches and are taken into the cell by receptor-mediated endocytosis (see Chapter 1). Eventually, the insulin-receptor complexes enter lysosomes, where the receptors are presumably broken down or recycled. The half-life of the insulin receptor is about 7 hours.

The number or the affinity, or both, of insulin receptors is affected by insulin and other hormones, exercise, food, and other factors. Exposure to increased amounts of insulin decreases receptor concentration (down regulation), and exposure to decreased insulin levels increases the affinity of the receptors. The number of receptors per cell is increased in starvation and decreased in obesity and acromegaly. The affinity of the receptors is increased in adrenal insufficiency and decreased by excess glucocorticoids (see below).

CONSEQUENCES OF INSULIN DEFICIENCY

The far-reaching physiologic effects of insulin are highlighted by a consideration of the extensive and serious consequences of insulin deficiency.

Diabetes Mellitus

In humans, insulin deficiency is a common and serious pathologic condition. In animals, it can be produced by pancreatectomy; by administration of alloxan, streptozocin, or other toxins that in appropriate doses cause selective destruction of the B cells of the pancreatic islets; by administration of drugs that inhibit insulin secretion; and by administration of antiinsulin antibodies. Strains of mice, rats, hamsters, guinea pigs, miniature swine, and monkeys that have a high incidence of spontaneous diabetes mellitus have also been described.

The constellation of abnormalities caused by insulin deficiency is called **diabetes mellitus.** Greek and Roman physicians used the term "diabetes" to refer to conditions in which the cardinal finding was a large urine volume, and 2 types were distinguished: "diabetes mellitus," in which the urine tasted sweet; and "diabetes insipidus," in which the urine was tasteless. Today, the term diabetes insipidus is reserved for the condition produced by lesions of the supraoptic-posterior pituitary system (see Chapter 14), and the unmodified word diabetes is generally used as a synonym for diabetes mellitus.

Diabetes is characterized by polyuria, polydipsia, weight loss in spite of polyphagia (increased appetite), hyperglycemia, glycosuria, ketosis, acidosis, and coma. There are widespread biochemical abnormalities, but the fundamental defects to which most of the abnormalities can be traced are (1) reduced entry of glucose into various "peripheral" tissues and (2) increased liberation of glucose into the circulation from the liver (increased **hepatic glucogenesis**). There is therefore an extracellular glucose excess and, in many cells, an intracellular glucose deficiency, a situation that has been called "starvation in the midst of plenty." There is also a decrease in the entry of amino acids into muscle and an increase in lipolysis.

It is now clear that there is an absolute or relative hypersecretion of glucagon in diabetes. This is true even when the pancreas is removed, because glucagon is secreted by the gastrointestinal tract as well as the pancreas. Somatostatin inhibits the secretion of insulin and glucagon (see Chapter 14), and when it is infused in pancreatectomized animals, the blood glucose falls toward normal. Consequently, it appears that the extracellular glucose excess in diabetes is due in part to hyperglucagonemia. However, some hyperglycemia persists even when glucagon secretion is reduced to zero.

Glucose Tolerance

In diabetes, glucose piles up in the bloodstream, especially after meals. If a glucose load is given to a diabetic, the blood glucose rises higher and returns to the baseline more slowly than it normally does. The response to a standard oral test dose of glucose, the **oral glucose tolerance test,** is used in the clinical diagnosis of diabetes (Fig 19–8).

Impaired glucose tolerance in diabetes is due in part to reduced entry of glucose into cells **(decreased peripheral utilization).** In the absence of insulin, the entry of glucose into skeletal, cardiac, and smooth muscle and other tissues is decreased (Fig 19–9). Glucose uptake by the liver is also reduced, but the effect is indirect (see below). Intestinal absorption of glucose is unaffected, as is its reabsorption from the urine by the cells of the proximal tubules of the kidneys. Glucose uptake by most of the brain and the red blood cells is also normal.

The second and the major cause of hyperglycemia in diabetes is derangement of the glucostatic function of the liver (see Chapter 17). The liver takes up glu-

Figure 19–8. Oral glucose tolerance test. Adults were given 75 g of glucose in 300 mL of water. In normal individuals, the fasting venous plasma glucose is less than 115 mg/dL, the 2-hour value is less than 140 mg/dL, and no value is greater than 200 mg/dL. Diabetes mellitus is present if the 2-hour value and one other value are greater than 200 mg/dL. Impaired glucose tolerance is diagnosed when the values are above the upper limits of normal but below the values diagnostic of diabetes.

cose from the bloodstream and stores it as glycogen, but because the liver contains glucose 6-phosphatase it also discharges glucose into the bloodstream. Indeed, Claude Bernard spoke of the liver as an endocrine gland that secreted glucose. Insulin facilitates glycogen synthesis and inhibits hepatic glucose output. When the blood glucose is high, insulin secretion is normally increased and hepatic glucogenesis is decreased. This effect is missing in diabetes. Glucagon also contributes to the hyperglycemia, and glucose

Figure 19–9. Disordered blood glucose homeostasis in insulin deficiency. Compare with Fig 17–14. The heavy arrows indicate reactions that are accentuated. The rectangles across arrows indicate reactions that are blocked.

output by the liver is facilitated by catecholamines, cortisol, and growth hormone when the stress of illness is severe.

Effects of Hyperglycemia

Hyperglycemia by itself can cause symptoms resulting from the hyperosmolality of the blood. In addition, there is glycosuria because the renal capacity for glucose reabsorption is exceeded. Excretion of the osmotically active glucose molecules entails the loss of large amounts of water (osmotic diuresis; see Chapter 38). The resultant dehydration activates the mechanisms regulating water intake, leading to polydipsia. There is an appreciable urinary loss of Na^+ and K^+ as well. For every gram of glucose excreted, 4.1 kcal are lost from the body. Increasing oral caloric intake to cover this loss simply raises the blood glucose further and increases the glycosuria, so mobilization of endogenous protein and fat stores and weight loss are not prevented.

When plasma glucose is episodically elevated over time, small amounts of hemoglobin A are nonenzymatically glycosylated to form HbA_{IC} (see Chapter 27). Careful control of the diabetes with insulin reduces this level, and consequently HbA_{IC} concentration is measured clinically as an integrated index of diabetic control for the 4–6-week period before the measurement. Albumin is also glycosylated, and glycosylated albumin turns over more rapidly, so it provides an index of more rapid fluctuations in diabetic control.

Effects of Intracellular Glucose Deficiency

The plethora of glucose outside the cells in diabetes contrasts with the intracellular deficit. Glucose catabolism is normally a major source of energy for cellular processes, and in diabetes, energy requirements can be met only by drawing on protein and fat reserves. Mechanisms are activated that greatly increase the catabolism of protein and fat, and one of the consequences of increased fat catabolism is ketosis.

Deficient glucose utilization in the cells of the hypothalamic ventromedial nuclei is probably the cause of the hyperphagia in diabetes. When the activity of the satiety center is decreased in response to decreased glucose utilization in its cells, the lateral appetite center operates unopposed, and food intake is increased (see Chapter 14).

Glycogen depletion is a common consequence of intracellular glucose deficit, and the glycogen content of liver and skeletal muscle in diabetic animals is usually reduced.

Changes in Protein Metabolism

In diabetes, the rate at which amino acids are catabolized to CO_2 and H_2O is increased. In addition, more amino acids are converted to glucose in the liver. This shift is indicated in Fig 19–10, which also shows the other principal abnormalities of intermediary metabolism in the liver.

Some idea of the rate of gluconeogenesis in fasting diabetic animals is obtained by measuring the ratio of glucose (dextrose) to nitrogen in the urine (**D/N ratio**). In fasting animals, liver glycogen is depleted and glycerol is converted to glucose at a very limited rate, so that the only important source of plasma glucose is protein (see Chapter 17). It can be calculated that the amount of carbon in the protein represented by 1 g of urinary nitrogen is sufficient to form 8.3 g of glucose. Consequently, the D/N ratio of approximately 3 seen in diabetes indicates the conversion to glucose of about 33% of the carbon of the protein metabolized.

The increased gluconeogenesis has many causes. Glucagon stimulates gluconeogenesis, and hyperglucagonemia is generally present in diabetes. Adrenal glucocorticoids also contribute to increased gluconeogenesis when they are elevated in severely ill diabetics. There is an increased supply of amino acids for gluconeogenesis because, in the absence of insulin, less protein synthesis occurs in muscle and blood amino acid levels rise. Alanine is particularly easily converted to glucose. In addition, the activity of the enzymes that catalyze the conversion of pyruvate and other 2-carbon metabolic fragments to glucose is increased. These include phosphoenolpyruvate carboxykinase, which facilitates conversion of oxaloacetate to phosphoenolpyruvate (see Chapter 17). They also include fructose 1,6-diphosphatase, which catalyzes the conversion of fructose diphosphate to fructose 6-phosphate, and glucose 6-phosphatase, which controls the entry of glucose into the circulation from the liver. Increased acetyl-CoA increases pyruvate carboxylase activity, and insulin deficiency increases the supply of acetyl-CoA because lipogenesis is decreased. Pyruvate carboxylase catalyzes the conversion of pyruvate to oxaloacetate (see Fig 17–9).

Negative Nitrogen Balance

In diabetes, the net effect of accelerated protein catabolism to CO_2 and H_2O and to glucose, plus diminished protein synthesis, is a markedly negative nitrogen balance, protein depletion, and wasting. Protein depletion from any cause is associated with poor "resistance" to infections, and the sugar-rich body fluids are undoubtedly good culture media for microorganisms. This is probably why diabetics are particularly prone to bacterial infections.

Fat Metabolism in Diabetes

The principal abnormalities of fat metabolism in diabetes are acceleration of lipid catabolism, with increased formation of ketone bodies, and decreased synthesis of fatty acids and triglycerides. The manifestations of the disordered lipid metabolism are so prominent that diabetes has been called "more a disease of lipid than of carbohydrate metabolism."

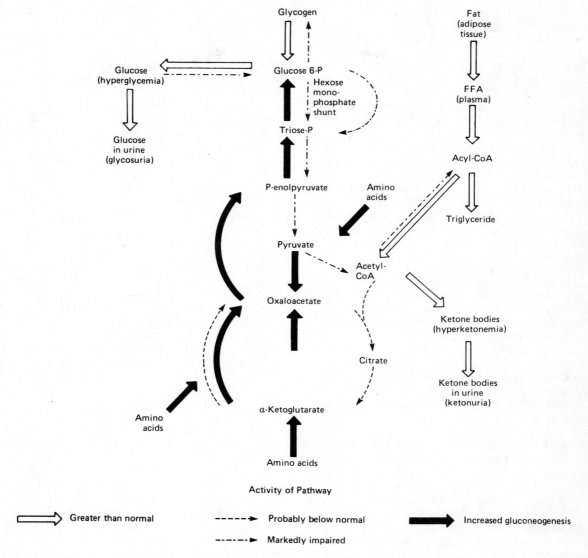

Figure 19–10. Metabolic abnormalities in the liver in uncontrolled diabetes. (Reproduced, with permission, from Murray RK et al: *Harper's Biochemistry,* 23rd ed. Appleton & Lange, 1993.)

Fifty percent of an ingested glucose load is normally burned to CO_2 and H_2O; 5% is converted to glycogen; and 30–40% is converted to fat in the fat depots. In diabetes, less than 5% is converted to fat even though the amount burned to CO_2 and H_2O is also decreased and the amount converted to glycogen is not increased. Therefore, glucose accumulates in the bloodstream and spills over into the urine.

The role of lipoprotein lipase and hormone-sensitive lipase in the regulation of the metabolism of fat depots is discussed in Chapter 17. In diabetes, there is decreased conversion of glucose to fatty acids in the depots because of the intracellular glucose defi-ciency. Insulin inhibits the hormone-sensitive lipase in adipose tissue, and, in the absence of this hormone, the plasma level of **free fatty acids** (NEFA, UFA, FFA) is more than doubled. The increased glucagon also contributes to the mobilization of FFA. Thus, the FFA level parallels the blood glucose level in diabetes and in some ways is a better indicator of the severity of the diabetic state. In the liver and other tissues, the fatty acids are catabolized to acetyl-CoA. Some of the acetyl-CoA is burned along with amino acid residues to yield CO_2 and H_2O in the citric acid cycle. However, the supply exceeds the capacity of the tissues to catabolize the acetyl-CoA.

The events occurring in the liver in diabetes are summarized in Fig 19–10. In addition to the previously mentioned increase in gluconeogenesis and marked outpouring of glucose into the circulation, there is a marked impairment of the conversion of acetyl-CoA to malonyl-CoA and thence to fatty acids. This is due to a deficiency of acetyl carboxylase, the enzyme that catalyzes the conversion. The excess acetyl-CoA is converted to ketone bodies (see below).

In uncontrolled diabetes, there is an increase in the plasma concentration of triglycerides and chylomicrons as well as FFA, and the plasma is often lipemic. The rise in these constituents is due mainly to decreased removal of triglycerides into the fat depots. The decreased activity of lipoprotein lipase contributes to this decreased removal.

Ketosis

When there is excess acetyl-CoA in the body, some of it is converted to acetoacetyl-CoA and then, in the liver, to acetoacetate. Acetoacetate and its derivatives, acetone and β-hydroxybutyrate, enter the circulation in large quantities (see Chapter 17).

These circulating ketone bodies are an important source of energy in fasting. Half of the metabolic rate in fasted normal dogs is said to be due to metabolism of ketones. The rate of ketone utilization in diabetics is also appreciable. It has been calculated that the maximal rate at which fat can be catabolized without significant ketosis is 2.5 g/kg body weight/d in diabetic humans. In untreated diabetes, production is much greater than this, and ketone bodies pile up in the bloodstream. There is some evidence that in severe diabetes the rate of ketone utilization may also decline, making the ketosis worse, and insulin is said to increase ketone uptake in muscle.

Acidosis

Most of the hydrogen ions liberated from acetoacetate and β-hydroxybutyrate are buffered, but severe metabolic acidosis still develops. The low plasma pH stimulates the respiratory center, producing the rapid, deep respiration described by Kussmaul as "air hunger" and named, for him, **Kussmaul breathing.** The urine becomes acidic. However, when the ability of the kidneys to replace the plasma cations accompanying the organic anions with H^+ and NH_4^+ is exceeded, Na^+ and K^+ are lost in the urine. The electrolyte and water losses lead to dehydration, hypovolemia, and hypotension. Finally, the acidosis and dehydration depress consciousness to the point of coma. Diabetic acidosis is a medical emergency. Now that the infections which used to complicate the disease can be controlled with antibiotics, acidosis is the commonest cause of early death in clinical diabetes.

In severe acidosis, total body sodium is markedly depleted, and when sodium loss exceeds water loss, plasma Na^+ is not infrequently low. Total body potassium is also low, but the plasma K^+ is usually normal, partly because ECF volume is reduced and partly because K^+ moves from cells to ECF when the ECF H^+ concentration is high. Another factor tending to maintain the plasma K^+ is the lack of insulin-induced entry of K^+ into cells.

The degree to which ketoacidosis complicates experimental diabetes varies in different species. The size of the body fat stores is also a factor conditioning the response to diabetes. Before the isolation of insulin by Banting and Best in 1921, the principal treatment of human diabetes was a starvation diet (**Allen regimen**). This not only lowered the blood glucose level but also reduced depot fat stores to the point where there was little fat to mobilize.

Coma

Coma in diabetes can be due to acidosis and dehydration. However, the blood glucose can be elevated to such a degree that independent of plasma pH, the hyperosmolarity of the plasma causes unconsciousness (**hyperosmolar coma**). Accumulation of lactate in the blood (**lactic acidosis**) may also complicate diabetic ketoacidosis if the tissues become hypoxic (see Chapter 33), and lactic acidosis may itself cause coma. Brain edema is seen in a significant number of patients with diabetic acidosis, and it can cause coma. The reason for the brain edema is unknown, but it is a serious complication with a bad prognosis.

Cholesterol Metabolism

In diabetes, the plasma cholesterol level is usually elevated, and this plays a role in the accelerated development of the atherosclerotic vascular disease that is a major long-term complication of diabetes in humans. The rise in plasma cholesterol level is due to an increase in the plasma concentration of VLDL and LDL (see Chapter 17), which may be due to increased hepatic production of VLDL or decreased removal of VLDL and LDL from the circulation.

Summary

Because of the complexities of the metabolic abnormalities in diabetes, a summary is in order. One of the key features of insulin deficiency (Fig 19–11) is decreased entry of glucose into many tissues (decreased peripheral utilization). There is also increased net release of glucose from the liver (increased production), due in part to glucagon excess. The resultant hyperglycemia leads to glycosuria and a dehydrating osmotic diuresis. Dehydration leads to polydipsia. In the face of intracellular glucose deficiency, appetite is stimulated, glucose is formed from protein (gluconeogenesis), and energy supplies are maintained by metabolism of proteins and fats. Weight loss, debilitating protein deficiency, and inanition are the result.

Fat catabolism is increased, and the system is

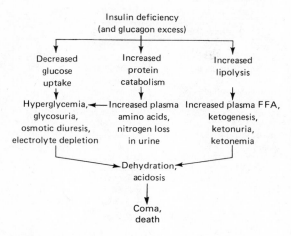

Figure 19–11. Effects of insulin deficiency. (Courtesy of RJ Havel.)

flooded with triglycerides and FFA. Fat synthesis is inhibited, and the overloaded catabolic pathways cannot handle the excess acetyl-CoA that is formed. In the liver, the acetyl-CoA is converted to ketone bodies. The ketones are for the most part organic acids, which accumulate in the circulation (ketosis) because their production rate exceeds the ability of the body to utilize them. Therefore, metabolic acidosis develops as the ketones accumulate. Na^+ and K^+ depletion are added to the dehydration because these plasma cations are excreted with the organic anions not covered by the H^+ and NH_4^+ secreted by the kidney. Finally, the acidotic, hypovolemic, hypotensive, depleted animal or patient becomes comatose because of the toxic effects of acidosis, dehydration, and hyperosmolarity on the nervous system and dies if treatment is not instituted.

All of these abnormalities are corrected by administration of insulin. Although emergency treatment of acidosis also includes administration of alkali to combat the acidosis and parenteral water, Na^+, and K^+ to replenish body stores, only insulin repairs the fundamental defects in a way that permits a return to normal.

INSULIN EXCESS

Symptoms

All the known consequences of insulin excess are manifestations, directly or indirectly, of the effects of hypoglycemia on the nervous system. Except in individuals who have been fasting for some time, glucose is the only fuel used in appreciable quantities by the brain. The carbohydrate reserves in neural tissue are very limited, and normal function depends upon a continuous glucose supply. As the blood glucose level falls, the cortex and the other brain areas with high metabolic rates are affected first, followed by the more slowly respiring vegetative centers in the diencephalon and hindbrain (see Chapter 32). Thus, early cortical symptoms of confusion, weakness, dizziness, and hunger are followed by convulsions and coma. If the hypoglycemia is prolonged, irreversible changes develop in the same cortical-diencephalic-medullary sequence, and death results from depression of the respiratory center. Therefore, prompt treatment with glucose is in order. Although a dramatic disappearance of symptoms is the usual response, abnormalities ranging from some intellectual dulling to permanent coma may persist if the hypoglycemia was severe or prolonged. Hypoglycemia is a potent stimulus to sympathetic discharge and increased secretion of catecholamines, particularly epinephrine. The tremors, palpitations, and nervousness in hypoglycemia are probably due to sympathetic overactivity. Hypoglycemia is sometimes said to be present when the blood glucose concentration is below 45 mg/dL, but the concentration at which symptoms of hypoglycemia appear is variable.

Compensatory Mechanisms

Hypoglycemia triggers increased secretion of 5 counterregulatory hormones that antagonize the fall in the blood glucose level: epinephrine, norepinephrine, glucagon, growth hormone, and cortisol. Epinephrine, norepinephrine, glucagon, and cortisol increase the hepatic output of glucose, the first 3 by increasing glycogenolysis. Growth hormone decreases the utilization of glucose in various peripheral tissues, and cortisol has a similar action (see below). The keys to counterregulation appear to be epinephrine and glucagon: if the plasma concentration of either increases, the decline in the blood glucose level is reversed; but if both fail to increase, there is little if any compensatory rise in the blood glucose level. The actions of the other hormones are supplementary.

REGULATION OF INSULIN SECRETION

The normal concentration of insulin measured by radioimmunoassay in the peripheral venous plasma of fasting normal humans is 0–70 μU/mL (0–502 pmol/L). The amount of insulin secreted in the basal state is about 1 U/h, with a 5–10-fold increase following ingestion of food. Therefore, the average amount secreted per day in a normal human is about 40 U (287 m.mol).

Effect of the Blood Glucose Level

The major control of insulin secretion is exerted by a feedback effect of blood glucose directly on the B cells of the pancreas. Glucose enters B cells via the GLUT 2 transporter, which does not depend on insulin for activation. The glucose is metabolized, generating ATP. The ATP closes ATP-sensitive K^+ channels, and the resultant decrease in K^+ efflux depolarizes the cell membrane. This opens voltage-sensitive Ca^{2+} channels, and Ca^{2+} enters the cells and activates Ca^{2+}-dependent protein kinases, which in turn trigger the release of insulin by exocytosis. The response is terminated when a separate group of voltage-gated K^+ channels activated by the membrane depolarization permits enough K^+ efflux to restore the membrane potential to its previous level. If the extracellular glucose level remains high, the cycle repeats itself and more insulin is released. The relation of insulin output to blood glucose is shown in Fig 19–12. Mannose also stimulates insulin secretion (Table 19–8). So does fructose, which is converted to glucose intracellularly. Mannoheptulose and 2-deoxyglucose, which prevent glucose from being metabolized, inhibit insulin secretion.

In experimental animals (Fig 19–13) and humans, the action of glucose on insulin secretion is biphasic; there is a rapid increase in secretion followed by a more slowly developing prolonged response. The initial response is due to release of preformed insulin by increased cytoplasmic Ca^{2+}. The prolonged response is due to release of newly synthesized insulin, since it is blocked by inhibitors of protein synthesis. If glucose perfusion is resumed after a rest period, the response to glucose is increased.

The feedback control of blood glucose on insulin secretion normally operates with great precision, so that blood glucose and blood insulin levels parallel each other with remarkable consistency.

Table 19–8. Factors affecting insulin secretion.

Stimulators	Inhibitors
Glucose	Somatostatin
Mannose	2-Deoxyglucose
Amino acids (leucine, arginine, others)	Mannoheptulose
	α-Adrenergic stimulators (norepinephrine, epinephrine)
Intestinal hormones (GIP, gastrin, secretin, CCK, glucagon, others?)	β-Adrenergic blockers (propranolol)
β-Keto acids	Galanin
Acetylcholine	Diazoxide
Glucagon	Thiazide diuretics
Cyclic AMP and various cyclic AMP-generating substances	K^+ depletion
β-Adrenergic stimulators	Phenytoin
Theophylline	Alloxan
Sulfonylureas	Microtubule inhibitors
	Insulin

Protein & Fat Derivatives

Arginine, leucine, and certain other amino acids stimulate insulin secretion, as do β-keto acids such as acetoacetate. Like glucose, these compounds generate ATP when metabolized, and apparently this closes ATP-sensitive K^+ channels in the B cells. It is worth noting that insulin stimulates the incorporation of amino acids into proteins and combats the fat catabolism that produces the β-keto acids.

Cyclic AMP & Insulin Secretion

Stimuli that increase cyclic AMP in B cells in-

Figure 19–12. Mean rates of insulin and glucagon delivery from an artificial pancreas at various blood glucose levels. The device was programmed to establish and maintain normal blood glucose in insulin-requiring diabetic humans, and the values for hormone output approximate the output of the normal human pancreas. The shape of the insulin curve also resembles the insulin response of incubated B cells to graded concentrations of glucose. (Reproduced, with permission, from Marliss EB et al: Normalization of glycemia in diabetics during meals with insulin and glucagon delivery by the artificial pancreas. *Diabetes* 1977;**26:**663.)

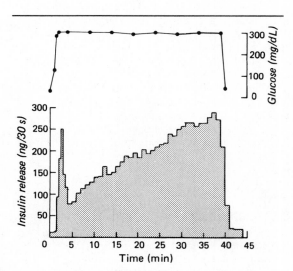

Figure 19–13. Insulin secretion from perfused rat pancreas in response to sustained glucose infusion. Values are means of 3 preparations. The top record shows the glucose concentration in the effluent perfusion mixture. (Reproduced, with permission, from Curry DL, Bennett LL, Grodsky GM: Dynamics of insulin secretion by the perfused rat pancreas. *Endocrinology* 1968;**83:**572.)

crease insulin secretion, probably by increasing intracellular Ca^{2+}. These include β-adrenergic agonists, glucagon, and phosphodiesterase inhibitors such as theophylline.

Catecholamines have a dual effect on insulin secretion; they inhibit insulin secretion via α_2-adrenergic receptors and stimulate insulin secretion via β-adrenergic receptors. The net effect of epinephrine and norepinephrine is usually inhibition. However, if catecholamines are infused after administration of α-adrenergic blocking drugs, the inhibition is converted to stimulation.

Effect of Autonomic Nerves

Branches of the right vagus nerve innervate the pancreatic islets, and stimulation of the right vagus causes increased insulin secretion via M_4 receptors (see Table 4–2). Atropine blocks the response, and acetylcholine stimulates insulin secretion. The effect of acetylcholine, like that of glucose, is due to increased cytoplasmic Ca^{2+}, but acetylcholine activates phospholipase C, with the released IP_3 releasing the Ca^{2+} from the endoplasmic reticulum.

Stimulation of the sympathetic nerves to the pancreas inhibits insulin secretion. The inhibition is produced by released norepinephrine acting on α_2-adrenergic receptors, and the inhibitory response is converted to an excitatory one by infusion of α-adrenergic blockers. On the other hand, the polypeptide galanin is found in some of the autonomic nerves innervating the islets, and galanin inhibits insulin secretion by activating the K^+ channels that are inhibited by ATP. Thus, the autonomic innervation of the pancreas is involved in the regulation of insulin secretion. The effects of glucose do not require intact innervation, since they occur in the transplanted pancreas, but there is some evidence that the nerve fibers maintain normal islet sensitivity to glucose.

Intestinal Hormones

Orally administered glucose exerts a greater insulin-stimulating effect than intravenously administered glucose, and orally administered amino acids also produce a greater insulin response than intravenous amino acids. These observations led to exploration of the possibility that a substance secreted by the gastrointestinal mucosa stimulated insulin secretion. Glucagon, secretin, cholecystokinin (CCK), gastrin, and gastric inhibitory peptide (GIP) all have such an action (see Chapter 26), and CCK potentiates the insulin-stimulating effects of amino acids. However, it appears that GIP is the physiologic "gut factor" that normally stimulates insulin secretion, since this factor produces stimulation when administered in doses that produce blood GIP levels comparable to those produced by oral glucose. The blood glucose level following oral glucose administration is lower when CCK secretion is increased than when it is not, but this is due to delayed gastric emptying rather than to

increased insulin secretion. GIP probably acts by increasing cyclic AMP in the B cells. The possible role of pancreatic somatostatin and glucagon in the regulation of insulin secretion is discussed below.

Oral Hypoglycemic Agents

Tolbutamide and other sulfonylurea derivatives such as acetohexamide, tolazamide, glipizide, and glyburide are orally active hypoglycemic agents that increase insulin release by closing the ATP-sensitive K^+ channels on B cells. There is some evidence that they also increase tissue sensitivity to insulin, but the clinical significance of any such effect is doubtful since they lower blood glucose only in patients with some remaining B cells and are ineffective after pancreatectomy or in insulin-dependent diabetes.

Phenformin and other biguanides that have been used as oral hypoglycemic agents do not affect insulin secretion, but they do increase glucose utilization, possibly because they inhibit oxidative metabolism of glucose and consequently increase anaerobic glycolysis within the cells. They also decrease glucose absorption from the gastrointestinal tract and may have other actions that lower blood glucose. However, they cause lactic acidosis in some patients, and because of the seriousness of this complication, they have been removed from the market in the USA.

Effects of K+ Depletion

K^+ depletion decreases insulin secretion, and K^+-depleted patients, eg, patients with primary hyperaldosteronism (see Chapter 20), develop diabetic glucose tolerance curves. These curves are restored to normal by K^+ repletion. The thiazide diuretics, which cause loss of K^+ as well as Na^+ in the urine (see Table 38–11), decrease glucose tolerance and make diabetes worse. They apparently exert this effect primarily because of their K^+-depleting effects, although some of them also cause pancreatic islet cell damage.

Exercise

The entry of glucose into skeletal muscle is increased during exercise in the absence of insulin. The increased muscle uptake may be due in part to relative O_2 deficiency, because glucose entry into cells is increased under anaerobic conditions. However, exercise also increases the affinity of the insulin receptors in muscle. Exercise can precipitate hypoglycemia in diabetics taking insulin not only because of the effect on insulin receptors and the production of relative O_2 deficiency but also because absorption of injected insulin is more rapid during exercise. Patients with diabetes should take in extra calories or reduce their insulin dosage when they exercise.

Long-Term Changes in B Cell Responses

The magnitude of the insulin response to a given stimulus is determined in part by the secretory history of the B cells. Individuals fed a high-carbohydrate diet for several weeks not only have higher fasting plasma insulin levels but also show a greater secretory response to a glucose load than individuals fed an isocaloric low-carbohydrate diet.

Although the B cells respond to stimulation with hypertrophy like other endocrine cells, they become exhausted and stop secreting (**B cell exhaustion**) when the stimulation is marked or prolonged. With such stimulation, they become vacuolated and hyalinized (hydropic and hyaline degeneration). If the stimulation stops soon after the cells stop secreting, they can recover; but if it is continued, they eventually die and disappear. In spite of some claims to the contrary, there is no convincing evidence that such "exhaustion atrophy" occurs in any other endocrine gland.

The pancreatic reserve is large, and it is difficult to produce B cell exhaustion in normal animals; but if the pancreatic reserve is reduced by partial pancreatic reserve is reduced by partial pancreatectomy or small doses of alloxan, exhaustion of the remaining B cells can be produced by any procedure that chronically raises the blood glucose level. This is the cause of the diabetes produced in animals with limited pancreatic reserves by anterior pituitary extracts, growth hormone, thyroid hormones, or the prolonged continuous infusion of glucose alone. The diabetes precipitated by hormones in animals is at first reversible, but with prolonged treatment it becomes permanent. It is usually named for the agent producing it, eg, "hypophyseal diabetes," "thyroid diabetes." Permanent diabetes persisting after treatment has been discontinued is indicated by the prefix meta-, eg, "metahypophyseal diabetes" or "metathyroid diabetes." When insulin is administered along with the diabetogenic hormones, the B cells are protected, probably because the blood glucose is lowered, and diabetes does not develop. Adrenal glucocorticoids raise blood glucose, but it has proved difficult to produce permanent diabetes in animals with adrenal steroid treatment alone.

GLUCAGON

Chemistry

Human glucagon, a linear polypeptide with a molecular weight of 3485, is produced by the A cells of the pancreatic islets. It contains 29 amino acid residues (see Table 26–2). All mammalian glucagons appear to have the same structure. Glucagon and glicentin, a larger molecule with some glucagon activity, are also found in the gastrointestinal mucosa (see Chapter 26). The C-terminal portion of glicentin consists of glucagon extended at its C terminal by an octapeptide. Human preproglucagon is a 179-amino-acid polypeptide that contains glicentin near its N terminal, followed by glucagon and then by 2 glucagonlike peptides of unknown function.

It has been demonstrated that glicentin as well as glucagon occurs in the A cells of the pancreatic islets, with glicentin in the peripheral portion of the A cell granules and glucagon in the center. However, the relation between these 2 molecules in the pancreas is still unknown.

Action

Glucagon is glycogenolytic, gluconeogenic, lipolytic, and ketogenic (Fig 19–14). When it binds to receptors on liver cells, it acts via G_s to activate adenylate cyclase and increase intracellular cyclic AMP. This leads via protein kinase A to activation of phosphorylase and therefore to increased breakdown of glycogen and an increase in blood glucose. The steps involved in this classic example of mediation of a hormone effect via cyclic AMP are discussed in Chapter 17. However, glucagon acts on different glucagon receptors located on the same hepatic cells to activate phospholipase C, and the resulting increase in cytoplasmic Ca^{2+} also stimulates glycogenolysis. Protein kinase A also decreases the metabolism of glucose 6-phosphate (Fig 19–15) by inhibiting the conversion of phosphoenolpyruvate to pyruvate. It also decreases the concentration of fructose 2,6-diphosphate, and this in turn inhibits the conversion of fructose 6-phosphate to fructose 1,6-diphosphate. The resultant buildup of glucose 6-phosphate leads to increased release of glucose.

Glucagon does not cause glycogenolysis in muscle. It increases gluconeogenesis from available amino acids in the liver and elevates the metabolic rate. It increases ketone body formation by decreasing malonyl-CoA in the liver (see Chapter 17). Its lipolytic activity, which leads in turn to increased ketogenesis, is discussed in Chapter 17. The calorigenic action of glucagon is not due to the hyperglycemia per se but probably to the increased hepatic deamination of amino acids.

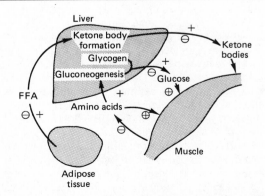

Figure 19–14. Bihormonal control of human substrate metabolism. ⊕, increased by insulin; ⊖, decreased by insulin; +, increased by glucagon. (Modified from Gerich J et al: Prevention of human diabetic ketoacidosis by somatostatin. *N Engl J Med* 1975;**295**:985.)

Figure 19–15. Mechanisms by which glucagon increases glucose output from the liver. Solid arrows indicate facilitation, and dashed arrows indicate inhibition. (Modified from Foster DW, McGarry JD: The metabolic derangements and treatment of diabetic ketoacidosis. *N Engl J Med* 1983;**309**:159.)

Large doses of exogenous glucagon exert a positively inotropic effect on the heart (see Chapter 29) without producing increased myocardial excitability, presumably because they increase myocardial cyclic AMP. Use of this hormone in the treatment of heart disease has been advocated, but there is no evidence for a physiologic role of glucagon in the regulation of cardiac function. Glucagon also stimulates the secretion of growth hormone, insulin, and pancreatic somatostatin.

Glucagon receptors have been identified in a variety of tissues. The molecular weight of the receptor protein is approximately 190,000.

Metabolism

Glucagon has a half-life in the circulation of 5–10 minutes. It is degraded by many tissues but particularly by the liver. Since glucagon is secreted into the portal vein and reaches the liver before it reaches the peripheral circulation, peripheral blood levels are relatively low. The rise in peripheral blood glucagon levels produced by excitatory stimuli (see below) is exaggerated in patients with cirrhosis, presumably because of decreased hepatic degradation of the hormone.

Regulation of Secretion

The principal factors known to affect glucagon secretion are summarized in Table 19–9. Secretion is increased by hypoglycemia and decreased by a rise in plasma glucose. The B cells contain GABA, and

there is evidence that coincident with the increased insulin secretion produced by hyperglycemia, GABA is released and acts on the A cells to inhibit glucagon secretion by activating $GABA_A$ receptors. The $GABA_A$ receptors are Cl^- channels, and the resulting Cl^- influx hyperpolarizes the A cells.

Secretion is also increased by stimulation of the sympathetic nerves to the pancreas, and this sympathetic effect is mediated via β-adrenergic receptors and cyclic AMP. It appears that the A cells are like the B cells in that stimulation of β-adrenergic receptors increases secretion and stimulation of α-adrenergic receptors inhibits secretion (see above). However, the pancreatic response to sympathetic stimulation in the absence of blocking drugs is increased secretion of glucagon, so the effect of β-receptors predominates in the glucagon-secreting cells. The stimulatory effects of various stresses and possibly of exercise and infection are mediated at least in part via the sympathetic nervous system. Vagal stimulation also increases glucagon secretion.

A protein meal and infusion of various amino acids increase glucagon secretion. It seems appropriate that the glucogenic amino acids are particularly potent in this regard, since these are the amino acids that are converted to glucose in the liver under the influence of glucagon. The increase in glucagon secretion following a protein meal is also valuable, since the amino acids stimulate insulin secretion, and the secreted glucagon prevents the development of hypoglycemia while the insulin promotes storage of the absorbed carbohydrates, fat, and lipids. Glucagon secretion increases during starvation. It reaches a peak on the third day of a fast, at the time of maximal gluconeogenesis. Thereafter, the plasma glucagon level declines as fatty acids and ketones become the major sources of energy.

During exercise, there is a balanced increase in glucose utilization and glucose production. Plasma glucagon increases and plasma insulin falls. The increased glucose uptake is due to hypoxia, increased affinity of insulin receptors, increased muscle blood flow permitting insulin to gain access to the muscles, and other factors that in effect amplify the action of

Table 19–9. Factors affecting glucagon secretion.

Stimulators	Inhibitors
Amino acids (particularly the glucogenic amino acids: alanine, serine, glycine, cysteine, and threonine)	Glucose
	Somatostatin
	Secretin
	FFA
CCK, gastrin	Ketones
Cortisol	Insulin
Exercise	Phenytoin
Infections	α-Adrenergic stimulators
Other stresses	GABA
β-Adrenergic stimulators	
Theophylline	
Acetylcholine	

insulin. The increase in glucose production by the liver is made possible by the decrease in plasma insulin and is brought about primarily by the increase in plasma glucagon.

The glucagon response to oral administration of amino acids is greater than the response to intravenous infusion of amino acids, suggesting that a glucagon-stimulating factor is secreted from the gastrointestinal mucosa. CCK and gastrin increase glucagon secretion, whereas secretin inhibits it. Since CCK and gastrin secretion are both increased by a protein meal, either hormone could be the gastrointestinal mediator of the glucagon response. The inhibition produced by somatostatin is discussed below.

Glucagon secretion is also inhibited by FFA and ketones. However, this inhibition can be overridden, since plasma glucagon levels are high in diabetic ketoacidosis.

Insulin-Glucagon Molar Ratios

As noted above, insulin is glycogenic, antigluconeogenetic, antilipolytic, and antiketotic in its actions. It thus favors storage of absorbed nutrients and is a "hormone of energy storage." Glucagon, on the other hand, is glycogenolytic, gluconeogenetic, lipolytic, and ketogenic. It mobilizes energy stores and is a "hormone of energy release." Because of their opposite effects, the blood levels of both hormones must be considered in any given situation. A good example is exercise, as discussed above. It is convenient to think in terms of the molar ratios of these hormones. The insulin-glucagon molar ratio in any given blood specimen can readily be calculated from the blood levels of the hormones, as determined by immunoassays, and their molecular weights.

The insulin-glucagon molar ratios fluctuate markedly because the secretion of glucagon and insulin are both modified by the conditions that preceded the application of any given stimulus (Table 19–10). Thus, for example, the insulin-glucagon molar ratio on a balanced diet is approximately 2.3. An infusion of arginine increases the secretion of both hormones and raises the ratio to 3.0. After 3 days of starvation, the

Table 19–10. Insulin-glucagon molar ratios (I/G) in blood in various conditions.[1]

Condition	Hepatic Glucose Storage (S) or Production (P)[2]	I/G
Glucose availability		
Large carbohydrate meal	4 + (S)	70
IV glucose	2 + (S)	25
Small meal	1 + (S)	7
Glucose need		
Overnight fast	1 + (P)	2.3
Low-carbohydrate diet	2 + (P)	1.8
Starvation	4 + (P)	0.4

[1]Courtesy of RH Unger.
[2]1+ to 4+ indicate relative magnitude.

ratio falls to 0.4, and an infusion of arginine in this state lowers the ratio to 0.3. Conversely, the ratio is 25 in individuals receiving a constant infusion of glucose and rises to 170 on ingestion of a protein meal during the infusion. The rise is due to the fact that insulin secretion rises sharply, while the usual glucagon response to a protein meal is abolished. Thus, when energy is needed during starvation, the insulin-glucagon molar ratio is low, favoring glycogen breakdown and gluconeogenesis; conversely, when the need for energy mobilization is low, the ratio is high, favoring the deposition of glycogen, protein, and fat.

OTHER ISLET CELL HORMONES

In addition to insulin and glucagon, the pancreatic islets secrete somatostatin and pancreatic polypeptide. These substances are released from the pancreas. In addition, somatostatin may be involved in regulatory processes within the islets that adjust the pattern of hormones secreted in response to various stimuli.

Somatostatin

Somatostatin 14 and its N-terminal-extended form somatostatin 28 are found in the D cells of pancreatic islets. Both forms inhibit the secretion of insulin, glucagon, and pancreatic polypeptide and may act locally within the pancreatic islets in a paracrine fashion. Furthermore, patients with somatostatin-secreting pancreatic tumors develop hyperglycemia and other manifestations of diabetes that disappear when the tumor is removed. They also develop dyspepsia due to slow gastric emptying and decreased gastric acid secretion, and gallstones, which are precipitated by decreased gallbladder contraction due to inhibition of CCK secretion. The secretion of pancreatic somatostatin is increased by several of the same stimuli that increase insulin secretion, ie, glucose and amino acids, particularly arginine and leucine. Its secretion is also increased by CCK. Somatostatin is released from the pancreas and the gastrointestinal tract into the peripheral blood.

Pancreatic Polypeptide

Human pancreatic polypeptide is a linear peptide containing 36 amino acid residues. It is closely related to 2 other 36-amino-acid polypeptides: **polypeptide YY (PYY)**, which is found in the intestine and may be a gastrointestinal hormone, and **neuropeptide Y**, which is found in the brain and the autonomic nervous system (see Chapter 4). Pancreatic polypeptide can be measured by radioimmunoassay in peripheral blood. At least in part, its secretion is under cholinergic control; plasma levels fall after administration of atropine. Its secretion is increased by a meal containing protein and by fasting, exercise, and acute hypoglycemia. Its secretion is decreased by somatostatin and intravenous glucose. Infusions of leucine, arginine, and alanine do not affect it, so the stimulatory effect of a protein meal may be mediated indirectly. A closely related but different polypeptide, neuropeptide Y, is found in the brain (see Chapter 4). Pancreatic polypeptide slows the absorption of food in humans, and it may smooth out the peaks and valleys of absorption. However, its exact physiologic function is unknown.

Organization of the Pancreatic Islets

The presence in the pancreatic islets of hormones that affect the secretion of other islet hormones suggests that the islets function as secretory units in the regulation of nutrient homeostasis. Somatostatin inhibits the secretion of insulin, glucagon, and pancreatic polypeptide (Fig 19–16); insulin inhibits the secretion of glucagon; and glucagon stimulates the secretion of insulin and somatostatin. As noted above, A and D cells and pancreatic polypeptide-secreting cells are generally located around the periphery of the islets, with the B cells in the center. There are clearly 2 types of islets, glucagon-rich islets and pancreatic polypeptide-rich islets, but the functional significance of this separation is not known. The islet cell hormones released into the ECF probably diffuse to other islet cells and influence their function (paracrine communication; see Chapter 1). It has been demonstrated that there are gap junctions between A, B, and D cells and that these permit the passage of ions and other small molecules from one cell to another, which could coordinate their secretory functions.

ENDOCRINE REGULATION OF CARBOHYDRATE METABOLISM

Many hormones in addition to insulin, IGF-I, IGF-II, glucagon, and somatostatin have important roles in the regulation of carbohydrate metabolism. They include epinephrine, thyroid hormones, glucocorticoids, and growth hormone. The other functions of these hormones are considered elsewhere, but it seems wise to summarize their effects on carbohydrate metabolism in the context of the present chapter. The "directional flow valves" in metabolism at

which these hormones act are discussed in Chapter 17.

Catecholamines

The activation of phosphorylase in liver by catecholamines is discussed in Chapters 17 and 20. Activation occurs via β-adrenergic receptors, which increase intracellular cyclic AMP, and α-adrenergic receptors, which increase intracellular Ca^{2+}. Hepatic glucose output is increased, producing hyperglycemia. In muscle, the phosphorylase is also activated via cyclic AMP and presumably via Ca^{2+}, but the glucose 6-phosphate formed can be catabolized only to pyruvate because of the absence of glucose 6-phosphatase. For reasons that are not entirely clear, large amounts of pyruvate are converted to lactate, which diffuses from the muscle into the circulation (Fig 19–17). The lactate is oxidized in the liver to pyruvate and converted to glycogen. Therefore, the response to an injection of epinephrine is an initial glycogenolysis followed by a rise in hepatic glycogen content. Lactate oxidation may be responsible for the calorigenic effect of epinephrine (see Chapter 20). Epinephrine and norepinephrine also liberate FFA into the circulation, and epinephrine decreases peripheral utilization of glucose.

Adrenal medullary tumors (pheochromocytomas) that secrete epinephrine cause hyperglycemia, glycosuria, and an elevated metabolic rate. However, the hyperglycemic effect of epinephrine is generally too transient to produce permanent diabetes, and the metabolic abnormalities disappear as soon as the tumors are removed.

Thyroid Hormones

Thyroid hormones make experimental diabetes worse; thyrotoxicosis aggravates clinical diabetes; and metathyroid diabetes can be produced in animals with decreased pancreatic reserve. The principal

Figure 19–16. Effects of islet cell hormones on the secretion of other islet cell hormones. Solid arrows indicate stimulation; dashed arrows indicate inhibition.

Figure 19–17. Effect of epinephrine on tissue glycogen, blood glucose, and blood lactate levels in fed rats. (Reproduced, with permission, from Ruch TC, Patton HD [editors]: *Physiology and Biophysics,* 20th ed. Vol. 3. Saunders, 1973.)

diabetogenic effect of thyroid hormones is to increase absorption of glucose from the intestine, but the hormones also cause (probably by potentiating the effects of catecholamines) some degree of hepatic glycogen depletion. Glycogen-depleted liver cells are easily damaged. When the liver is damaged, the glucose tolerance curve is diabetic because the liver takes up less of the absorbed glucose. Thyroid hormones may also accelerate the degradation of insulin. All these actions have a hyperglycemic effect and, if the pancreatic reserve is low, may lead to B cell exhaustion.

Adrenal Glucocorticoids

Glucocorticoids from the adrenal cortex (see Chapter 20) elevate blood glucose and produce a diabetic type of glucose tolerance curve. In humans, this effect may occur only in individuals with a genetic predisposition to diabetes. Glucose tolerance is reduced in 80% of patients with Cushing's syndrome (see Chapter 20), and 20% of these patients have frank diabetes. The glucocorticoids are necessary for glucagon to exert its gluconeogenetic action during fasting. They are gluconeogenetic themselves, but their role is mainly permissive. In adrenal insufficiency, the blood glucose is normal as long as food intake is maintained, but fasting precipitates hypoglycemia and collapse. The blood-glucose-lowering effect of insulin is greatly enhanced in patients with adrenal insufficiency. In animals with experimental diabetes, adrenalectomy markedly ameliorates the diabetes.

The complex actions of the glucocorticoids that affect carbohydrate metabolism are summarized in Table 19–11. The major diabetogenic effects are an increase in protein catabolism with increased gluconeogenesis in the liver; increased hepatic glycogenesis and ketogenesis; and a decrease in peripheral glucose utilization relative to the blood insulin level that may be due to inhibition of glucose phosphorylation (see below).

Growth Hormone

The diabetogenic effect of anterior pituitary extracts is due in part to the ACTH and TSH they contain, but it is also due to growth hormone. There is some species variation in the diabetogenicity of growth hormone. However, human growth hormone makes clinical diabetes worse, and 25% of patients with growth-hormone-secreting tumors of the anterior pituitary have diabetes. Hypophysectomy ameliorates diabetes and increases sensitivity to insulin even more than adrenalectomy, whereas growth hormone treatment decreases insulin responsiveness.

Growth hormone mobilizes FFA from adipose tissue, thus favoring ketogenesis. It decreases glucose uptake into some tissues ("anti-insulin action"), increases hepatic glucose output, and may decrease tissue binding of insulin. Indeed, it has been suggested

Table 19–11. Actions of glucocorticoids that affect carbohydrate metabolism.[1]

1. Increased protein catabolism in periphery, resulting in increased concentration of amino acids in plasma
2. Increased hepatic uptake ("trapping") of amino acids
3. Increased deamination and transamination of amino acids
4. Increased hepatic conversion of oxaloacetate to phosphopyruvate
5. Increased hepatic fructose diphosphatase activity, facilitating dephosphorylation of fructose 1,6-diphosphate
6. Increased hepatic glucose 6-phosphatase activity, releasing more glucose into the circulation
7. Decrease in glucose utilization peripherally and in the liver, possibly due to inhibition of phosphorylation
8. Increased blood lactate and pyruvate
9. Decreased hepatic lipogenesis
10. Increased plasma FFA levels and increased ketone formation (when pancreatic reserve is low)
11. Increased formation of the active form of glycogen synthase

[1]Items 1–6 represent reactions leading to an elevated blood glucose level due to increased gluconeogenesis.

that the ketosis and decreased glucose tolerance produced by starvation are due to hypersecretion of growth hormone. Growth hormone does not stimulate insulin secretion directly, but the hyperglycemia it produces secondarily stimulates the pancreas and may eventually exhaust the B cells.

There is evidence that growth hormone decreases the number of insulin receptors and glucocorticoids decrease their affinity. However, the decrease in glucose utilization produced by these hormones is due more to inhibition of glucose phosphorylation than to inhibition of glucose entry into cells. Entry into cells is usually the rate-limiting step in glucose metabolism, but when adequate insulin is present, a decline in phosphorylation can decrease utilization.

HYPOGLYCEMIA & DIABETES MELLITUS IN HUMANS

Hypoglycemia

"Insulin reactions" are common in juvenile diabetics, and occasional hypoglycemic episodes are the price of good diabetic control in most diabetics. The increase in glucose uptake of skeletal muscle and the increased absorption of injected insulin during exercise have been mentioned above. Diabetics taking insulin should be taught to anticipate this effect by adjusting their diet or insulin dose when they exercise.

Symptomatic hypoglycemia also occurs in nondiabetics, and a review of some of the more important causes serves to emphasize the variables affecting blood glucose homeostasis. Chronic mild hypoglycemia can cause incoordination and slurred speech, and the condition can be mistaken for drunkenness. Mental aberrations and convulsions in the absence of frank coma also occur. When the level of insulin secretion is chronically elevated by an insulin-secreting tumor (**insulinoma**) or as a result of hyperplasia of the B cells, symptoms are most common in the morning. This is because a night of fasting has depleted hepatic glycogen reserves. However, symptoms can develop at any time, and in such patients, the diagnosis may be missed. Some cases of insulinoma have been erroneously diagnosed as epilepsy or psychosis. Hypoglycemia also occurs in some patients with large malignant tumors that do not involve the pancreatic islets, and the hypoglycemia in these cases is apparently due to excess secretion of IGF-II. A third cause of hypoglycemia is an autoimmune disorder in which spontaneously produced antibodies to insulin bind appreciable amounts of the hormone, then release free insulin at irregular intervals.

In liver disease, the glucose tolerance curve is diabetic but the fasting blood glucose level is low (Fig 19–18). In **functional hypoglycemia,** the blood glucose rise is normal after a test dose of glucose, but the subsequent fall overshoots to hypoglycemic levels, producing symptoms 3–4 hours after meals. This pattern is sometimes seen in individuals who later develop diabetes. Patients with this syndrome should be distinguished from patients with similar symptoms due to psychologic or other problems who do not have hypoglycemia when blood is drawn during the symptomatic episode. Most patients with functional hypoglycemia are tense, conscientious people with symptoms of autonomic overactivity. It has been postulated that the overshoot of the blood glucose is due to insulin secretion stimulated by impulses in the right vagus, but cholinergic blocking agents do not routinely correct the abnormality. In some thyrotoxic patients and in patients who have had gastrectomies or other operations that speed the passage of food into the intestine, glucose absorption is abnormally rapid. The blood glucose rises to a high, early peak, but it then falls rapidly to hypoglycemic levels because the wave of hyperglycemia evokes a greater than normal rise in insulin secretion. Symptoms characteristically occur about 2 hours after meals.

Infants born to diabetic mothers often have high birth weights and large organs (**macrosomia).** This condition is caused by excess circulating insulin in the fetus, which in turn is caused in part by stimulation of the fetal pancreas by glucose and amino acids from the blood of the mother. Free insulin in maternal blood is destroyed by proteases in the placenta, but antibody-bound insulin is protected, so it reaches the fetus. Therefore fetal macrosomia also occurs in women who develop antibodies against various animal insulins, then continue to receive the animal insulin during pregnancy.

Infants with GLUT 1 deficiency have now been described. They have defective transport of glucose across the blood-brain barrier, low CSF glucose in the presence of normal blood glucose, seizures, and developmental delay.

Diabetes Mellitus

Diabetes mellitus is common in humans, although it also occurs in other species. In the USA, it has been calculated that almost 5% of the population has the disease. Not only does it sometimes cause acidosis and coma, but in long-standing diabetes, other complications develop. These complications are associated with prolonged hyperglycemia and may be due in part to accumulation of sorbitol in the tissues or to the formation of glycosylated end products that cross-link matrix proteins. They include proliferative scarring of the retina (**diabetic retinopathy);** renal disease (**diabetic nephropathy);** loss of nerve function, particularly in the autonomic nervous system (**diabetic neuropathy);** and accelerated atherosclerosis. The atherosclerosis leads to circulatory insufficiency in the legs, with chronic ulceration and gangrene, and to an increased incidence of stroke and myocardial infarction.

The cause of clinical diabetes is always a deficiency of the action of insulin at the tissue level, but the deficiency may be due to a variety of abnormalities. Thus, diabetes, like hypertension (see Chapter 33), is a syndrome with multiple causes. There are 2 common forms, type I and type II diabetes (Table 19–12), and a collection of other less frequent forms due to conditions such as excess secretion of diabetogenic hormones or congenitally defective insulin receptors. To date, 21 insulin receptor abnormalities have been described. **Type II diabetes** usually develops after age 40 and is not associated with

Figure 19–18. Typical glucose tolerance curves after an oral glucose load in liver disease and in conditions causing excessively rapid absorption of glucose from the intestine. The horizontal line is the approximate blood glucose level at which hypoglycemic symptoms may appear.

Table 19–12. Types of human diabetes mellitus.

Type	Other Names
Type I	Insulin-dependent diabetes (IDDM). Juvenile diabetes. Ketosis-prone diabetes.
Type II	Non-insulin-dependent diabetes (NIDDM). Maturity-onset diabetes. Ketosis-resistant diabetes.
Diabetes associated with other conditions	Examples include diabetes due to pancreatoectomy or pancreatic disease; diabetes due to mutant forms of insulin or insulin receptors; and diabetes in patients with Cushing's syndrome, acromegaly, or other endocrine diseases.

total loss of the ability to secrete insulin. Consequently, it is also called **maturity-onset diabetes** and **non-insulin-dependent diabetes mellitus (NIDDM).** It has an insidious onset, is rarely associated with ketosis, and is usually associated with normal B cell morphology and insulin content if the B cells have not become exhausted.

One hallmark of this disease is impaired insulin secretion. Another is insulin resistance, especially in skeletal muscle, but there is still much debate about which comes first. There is an exaggerated, prolonged insulin response to glucose. Recent evidence indicates that this is secondary to an inadequate early B cell response which does not suppress glucose output from the liver to a normal degree. The resulting hyperglycemia causes the large, late insulin response. Patients with NIDDM also have a deficiency of GLUT 4 glucose transporters in insulin-sensitive tissues. There is also a strong genetic component in this form of the disease. When one identical twin develops maturity-onset diabetes, the other almost invariably does as well—ie, the **concordance rate** is close to 100%.

In **type I diabetes,** on the other hand, the concordance rate is about 50%, indicating that an environmental factor as well as a genetic predisposition is in-

volved. This disease usually develops before age 40, although it can occur at any age, and it is characterized by loss of B cells with eventual absence of insulin in the circulation. Consequently, it is also called **juvenile diabetes** and **insulin-dependent diabetes mellitus (IDDM).** It is not associated with obesity and is commonly complicated by ketosis and acidosis. About 10% of diabetics have IDDM. There is general agreement that it is an autoimmune disease, with anti-B-cell antibodies causing destruction of B cells. However, a genetic factor predisposes to the development of the antibodies. There is considerable evidence that aspartic acid in position 57 of the β-chains of human HLA-DQ, one of the class II MHC molecules (see Chapter 27), protects against the formation of the autoantibodies that cause IDDM, but the inheritance of protection is probably more complicated than this. Some of the antibodies are against GAD (see Chapter 4), although the function of this enzyme in β cells is unknown. Immunosuppression with drugs such as cyclosporin produces amelioration of type I diabetes if given early in the course of the disease before all the B cells are lost.

Insulin Requirements

The dose of insulin required to control diabetes mellitus varies from patient to patient and from time to time in the same patient. Because insulin secretion normally increases after each meal, efforts have been made to duplicate the rise in diabetics by the use of programmable implanted insulin pumps. Considerable effort has also been devoted to transplanting islet tissue that will respond to changes in plasma glucose and amino acids in the same way as islet tissue in normal individuals. It is interesting, however, that in patients with long-term pancreatic transplants whose blood glucoses are normal, insulin resistance is still increased.

In general, the factors affecting insulin requirements are those that influence glucose tolerance. Thus, insulin requirements rise when patients gain weight, when they secrete or receive increased amounts of glucocorticoids and other hormones that are diabetogenic, during pregnancy, and in the presence of infection and fever. They fall when patients lose weight and during exercise.

The Adrenal Medulla & Adrenal Cortex

20

INTRODUCTION

There are 2 endocrine organs in the adrenal gland, one surrounding the other. The main secretions of the inner **adrenal medulla** (Fig 20–1) are the catecholamines **epinephrine, norepinephrine,** and **dopamine;** the outer **adrenal cortex** secretes steroid hormones.

The adrenal medulla is in effect a sympathetic ganglion in which the postganglionic neurons have lost their axons and become secretory cells. The cells secrete when stimulated by the preganglionic nerve fibers that reach the gland via the splanchnic nerves. Adrenal medullary hormones are not essential for life, but they help to prepare the individual to deal with emergencies.

On the other hand, the adrenal cortex is essential for life. It secretes **glucocorticoids,** steroids with widespread effects on the metabolism of carbohydrate and protein; a **mineralocorticoid** essential to the maintenance of sodium balance and ECF volume; and **sex hormones** that exert minor effects on reproductive function. Of these, the mineralocorticoids and the glucocorticoids are necessary for survival. Adrenocortical secretion is controlled primarily by ACTH from the anterior pituitary, but mineralocorticoid secretion is also subject to independent control by circulating factors, of which the most important is angiotensin II, a peptide formed in the bloodstream. The formation of angiotensin is in turn dependent on renin secreted by the kidney.

ADRENAL MORPHOLOGY

The adrenal medulla, which constitutes 28% of the mass of the adrenal gland, is made up of interlacing cords of densely innervated granule-containing cells

that abut on venous sinuses. Two cell types can be distinguished morphologically: an epinephrine-secreting type that has larger, less dense granules; and a norepinephrine-secreting type in which smaller, very dense granules fail to fill the vesicles in which they are contained (Fig 20–2). Ninety percent of the cells are the epinephrine-secreting type, and 10% are the norepinephrine-secreting type. The type of cell that secretes dopamine is unknown. **Paraganglia,** small groups of cells resembling those in the adrenal medulla, are found near the thoracic and abdominal sympathetic ganglia (Fig 20–1).

Figure 20–1. Human adrenal glands. Adrenocortical tissue is stippled; adrenal medullary tissue is black. Note the location of the adrenals at the superior pole of each kidney. Also shown are extra-adrenal sites at which cortical and medullary tissue is sometimes found. (Reproduced, with permission, from Forsham PH: The adrenal cortex. In: *Textbook of Endocrinology,* 4th ed. Williams RH [editor]. Saunders, 1968.)

In adult mammals, the adrenal cortex is divided into 3 zones of variable distinctness (Fig 20–3). The outer **zona glomerulosa** is made up of whorls of cells that are continuous with the columns of cells which form the **zona fasciculata.** These columns are separated by venous sinuses. The inner portion of the zona fasciculata merges into the **zona reticularis,** where the cell columns become interlaced in a network. The zona glomerulosa makes up 15% of the mass of the adrenal gland, the zona fasciculata 50%, and the zona reticularis 7%. The adrenal cells contain abundant lipid, especially in the outer portion of the zona fasciculata. All 3 cortical zones secrete corticosterone (see below), but the active enzymatic mechanism for aldosterone biosynthesis is limited to the zona glomerulosa, whereas the enzymatic mechanism for forming cortisol and sex hormones is found in the 2 inner zones.

Arterial blood reaches the adrenal from many small branches of the phrenic and renal arteries and the aorta. From a plexus in the capsule, blood flows to the sinusoids of the medulla. The medulla is also supplied by a few arterioles that pass directly to it from the capsule. In most species, including humans, there is a single large adrenal vein. The blood flow through the adrenal is large, as it is in most endocrine glands.

During fetal life, the human adrenal is large and under pituitary control, but the 3 zones of the permanent cortex represent only 20% of the gland. The remaining 80% is the large **fetal adrenal cortex,** which undergoes rapid degeneration at the time of birth. A major function of this fetal adrenal is secretion of sulfate conjugates of androgens that are converted in the placenta to estrogens (see Chapter 23). There is no structure comparable to the human fetal adrenal in laboratory animals.

An important function of the zona glomerulosa, in addition to aldosterone biosynthesis, is the formation of new cortical cells. Like other tissues of neural origin, the adrenal medulla does not regenerate; but when the inner 2 zones of the cortex are removed, a new zona fasciculata and zona reticularis regenerate from glomerular cells attached to the capsule. Small capsular remnants will regrow large pieces of adrenocortical tissue. Immediately after hypophysectomy, the zona fasciculata and zona reticularis begin to atrophy, whereas the zona glomerulosa is unchanged (Fig 20–3), because of the action of angiotensin II on this zone. The ability to secrete aldosterone and conserve Na^+ is normal for some time, but in long-standing hypopituitarism, aldosterone deficiency may develop, apparently because of the absence of a pituitary factor that maintains the responsiveness of the zona glomerulosa (see below). Injections of ACTH and stimuli that cause endogenous ACTH secretion produce hypertrophy of the zona fasciculata and zona reticularis but do not increase the size of the zona glomerulosa.

Figure 20–2. Norepinephrine-secreting adrenal medullary cell. The granules are released by exocytosis, and the granule contents enter the bloodstream (arrow). (Modified from Poirier J, Dumas JLR: *Review of Medical Histology.* Saunders, 1977.)

The cells of the adrenal cortex contain large amounts of smooth endoplasmic reticulum, which is involved in the steroid-forming process. Other steps in steroid biosynthesis occur in the mitochondria. The structure of steroid-secreting cells is very similar throughout the body. The typical features of such cells are shown in Fig 20–4.

Figure 20–3. Effect of hypophysectomy on the morphology of the adrenal cortex of the dog. Note that the atrophy does not involve the zona glomerulosa. The morphology of the human adrenal is similar.

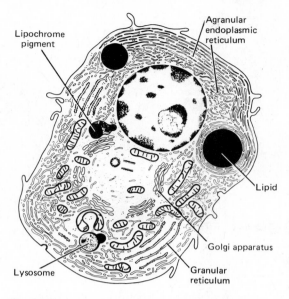

Figure 20–4. Diagrammatic representation of the cytologic features of steroid-secreting cells. Note the abundant agranular endoplasmic reticulum, the pleomorphic mitochondria, and the lipid droplets. (Reproduced, with permission, from Fawcett DW, Long JA, Jones AL: The ultrastructure of endocrine glands. *Recent Prog Horm Res* 1969;**25:**315.)

ADRENAL MEDULLA

STRUCTURE & FUNCTION OF MEDULLARY HORMONES

Catecholamines

Norepinephrine, epinephrine, and dopamine are secreted by the adrenal medulla. Cats and some other species secrete mainly norepinephrine, but in dogs and humans, most of the catecholamine output in the adrenal vein is epinephrine. Norepinephrine also enters the circulation from noradrenergic nerve endings.

The structures of norepinephrine, epinephrine, and dopamine and the pathways for their biosynthesis and metabolism are shown in Figs 4–18, 4–19, and 4–20. Norepinephrine is formed by hydroxylation and decarboxylation of tyrosine, and epinephrine by methylation of norepinephrine. Phenylethanolamine-N-methyltransferase (PNMT), the enzyme that catalyzes the formation of epinephrine from norepinephrine, is found in appreciable quantities only in the brain and the adrenal medulla. Adrenal medullary PNMT is induced by glucocorticoids, and although relatively large amounts are required, the glucocorti-

coid concentration is high in the blood draining from the cortex to the medulla. After hypophysectomy, the glucocorticoid concentration of this blood falls and epinephrine synthesis is decreased.

In plasma, about 95% of the dopamine and 70% of the norepinephrine and epinephrine are conjugated to sulfate. Sulfate conjugates are inactive, and their function is unsettled. In recumbent humans, the normal plasma level of free norepinephrine is about 300 pg/mL (1.8 nmol/L). There is a 50–100% increase upon standing (Fig 20–5). Plasma norepinephrine is generally unchanged after adrenalectomy, but the free epinephrine level, which is normally about 30 pg/mL (0.16 nmol/L), falls to essentially zero. The epinephrine found in tissues other than the adrenal medulla and the brain is for the most part absorbed from the bloodstream rather than synthesized in situ. Interestingly, low levels of epinephrine reappear in the blood some time after bilateral adrenalectomy, and these levels are regulated like those secreted by the adrenal medulla. However, their source is unknown.

The plasma free dopamine level is about 35 pg/mL (0.23 nmol/L), and there are appreciable quantities of dopamine in the urine. Half the plasma dopamine comes from the adrenal medulla, whereas the remaining half presumably comes from the sympathetic ganglia or other components of the autonomic nervous system.

The catecholamines have a half-life of about 2 minutes in the circulation. For the most part, they are methoxylated and then oxidized to 3-methoxyl-4-hydroxymandelic acid (vanillylmandelic acid, VMA). About 50% of the secreted catecholamines appear in the urine as free or conjugated metanephrine and normetanephrine, and 35% as VMA. Only small amounts of free norepinephrine and epinephrine are excreted. In normal humans, about 30 μg of norepinephrine, 6 μg of epinephrine, and 700 μg of VMA are excreted per day.

Other Substances Secreted by the Adrenal Medulla

In the medulla, norepinephrine and epinephrine are stored in granules with ATP. The granules also contain chromogranin A (see Chapter 4). Secretion is initiated by acetylcholine released from the preganglionic neurons that innervate the secretory cells. The acetylcholine increases the permeability of the cells, and the Ca^{2+} that enters the cells from the ECF triggers exocytosis (see Chapter 1). In this fashion, the catecholamines, ATP, and proteins in the granules are all released together.

Epinephrine-containing cells of the medulla also contain and secrete opioid peptides (see Chapter 4). The precursor molecule is preproenkephalin (see Table 4–4), a protein containing a leader sequence of 25 amino acid residues and 242 additional amino acid residues that include 4 met-enkephalins, a heptapep-

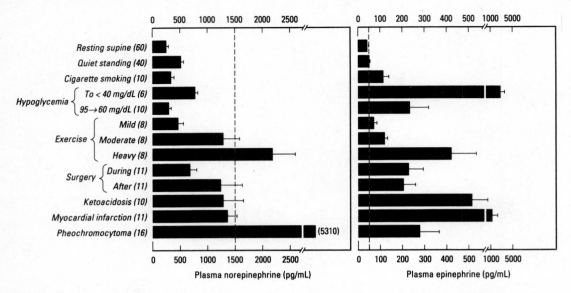

Figure 20–5. Norepinephrine and epinephrine levels in human venous blood in various physiologic and pathologic states. Note that the horizontal scales are different. The numbers in parentheses are the numbers of subjects tested. In each case, the vertical dashed line identifies the threshold plasma concentration at which detectable physiologic changes are observed. (Modified and reproduced, with permission, from Cryer PE: Physiology and pathophysiology of the human sympathoadrenal neuroendocrine system. *N Engl J Med* 1980;**303:**436.)

tide which contains met-enkephalin, an octapeptide which contains met-enkephalin, and one leu-enkephalin (see Fig 1–18). Most of the circulating met-enkephalin comes from the adrenal medulla. However, the circulating opioid peptides do not cross the blood-brain barrier to any degree, and their function in the blood is presently unknown.

Effects of Epinephrine & Norepinephrine

In addition to mimicking the effects of noradrenergic nervous discharge, norepinephrine and epinephrine exert metabolic effects that include glycogenolysis in liver and skeletal muscle, mobilization of FFA, increased plasma lactate, and stimulation of the metabolic rate. The effects of norepinephrine and epinephrine are brought about by actions on 2 classes of receptors, α- and β-adrenergic receptors. Alpha receptors are subdivided into 2 groups, α_1 and α_2 receptors, and β receptors into β_1, β_2, and β_3 receptors, as outlined in Chapter 4. Table 13–2 lists the receptors in each of the visceral organs.

Norepinephrine and epinephrine both increase the force and rate of contraction of the isolated heart. These responses are mediated by β_1 receptors. The catecholamines also increase myocardial excitability, causing extrasystoles and, occasionally, more serious cardiac arrhythmias. Norepinephrine produces vasoconstriction in most if not all organs via α_1 receptors, but epinephrine dilates the blood vessels in skeletal muscle and the liver via β_2 receptors. This usually overbalances the vasoconstriction produced by epi-

nephrine elsewhere, and the total peripheral resistance drops. When norepinephrine is infused slowly in normal animals or humans, the systolic and diastolic blood pressures rise. The hypertension stimulates the carotid and aortic baroreceptors, producing reflex bradycardia that overrides the direct cardioacceleratory effect of norepinephrine. Consequently, cardiac output per minute falls. Epinephrine causes a widening of the pulse pressure, but because baroreceptor stimulation is insufficient to obscure the direct effect of the hormone on the heart, cardiac rate and output increase. These changes are summarized in Fig 20–6.

Catecholamines increase alertness (see Chapter 11). Epinephrine and norepinephrine are equally potent in this regard, although in humans epinephrine usually evokes more anxiety and fear.

The catecholamines have several different actions that affect blood glucose. Epinephrine and norepinephrine both cause glycogenolysis. They produce this effect via β-adrenergic receptors that increase cyclic AMP, with activation of phosphorylase, and via α-adrenergic receptors that increase intracellular Ca^{2+} (see Chapter 17). In addition, the catecholamines increase the secretion of insulin and glucagon via β-adrenergic mechanisms and inhibit the secretion of these hormones via α-adrenergic mechanisms.

Norepinephrine and epinephrine also produce a prompt rise in the metabolic rate that is independent of the liver and a smaller, delayed rise that is abolished by hepatectomy and coincides with the rise in

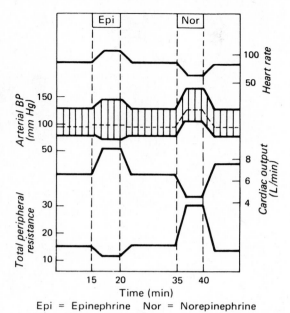

Epi = Epinephrine   Nor = Norepinephrine

Figure 20–6. Circulatory changes produced in humans by the slow intravenous infusion of epinephrine and norepinephrine. (Modified and reproduced, with permission, from Barcroft H, Swan HJC: *Sympathetic Control of Human Blood Vessels.* Arnold, 1953.)

blood lactate concentration. The calorigenic action does not occur in the absence of the thyroid and the adrenal cortex. The cause of the initial rise in metabolic rate is not clearly understood. It may be due to cutaneous vasoconstriction, which decreases heat loss and leads to a rise in body temperature, or to increased muscular activity, or both. The second rise is probably due to oxidation of lactate in the liver.

When injected, epinephrine and norepinephrine cause an initial rise in plasma K^+ because of release of K^+ from the liver and then a prolonged fall in plasma K^+ because of an increased entry of K^+ into skeletal muscle that is mediated by β_2-adrenergic receptors. There is some evidence that activation of α receptors opposes this effect. Thus, the catecholamines may play a significant role in regulating the ratio between extracellular and intracellular K^+.

The increase in plasma norepinephrine and epinephrine that are needed to produce the various effects listed above have been determined by infusion of catecholamines in resting humans. In general, the threshold for the cardiovascular and the metabolic effects of norepinephrine is about 1500 pg/mL, ie, about 5 times the resting value (Fig 20–5). Epinephrine, on the other hand, produces tachycardia when the plasma level is about 50 pg/mL, ie, about twice the resting value. The threshold for increased systolic blood pressure and lipolysis is about 75 pg/mL; the

threshold for hyperglycemia, increased plasma lactate, and decreased diastolic blood pressure is about 150 pg/mL; and the threshold for the α-mediated decrease in insulin secretion is about 400 pg/mL. Plasma epinephrine often exceeds those thresholds. On the other hand, plasma norepinephrine rarely exceeds the threshold for its cardiovascular and metabolic effects, and most of its effects are due to its local release from postganglionic sympathetic neurons. Most adrenal medullary tumors (**pheochromocytomas**) secrete norepinephrine, and episodic or sustained hypertension is the most prominent finding. Pheochromocytomas that secrete epinephrine cause less hypertension and frequently produce episodic hyperglycemia, glycosuria, and other metabolic effects.

Effects of Dopamine

The physiologic function of the dopamine in the circulation is unknown. However, injected dopamine produces renal vasodilation, probably by acting on a specific dopaminergic receptor. It also produces vasodilation in the mesentery. Elsewhere, it produces vasoconstriction, probably by releasing norepinephrine, and it has a positively inotropic effect on the heart by an action on β_1-adrenergic receptors. The net effect of moderate doses of dopamine is an increase in systolic pressure and no change in diastolic pressure. Because of these actions, dopamine is useful in the treatment of traumatic and cardiogenic shock (see Chapter 33).

REGULATION OF ADRENAL MEDULLARY SECRETION

Neural Control

Certain drugs act directly on the adrenal medulla, but physiologic stimuli affect medullary secretion through the nervous system. Catecholamine secretion is low in basal states, but the secretion of epinephrine and, to a lesser extent, that of norepinephrine is reduced even further during sleep.

Increased adrenal medullary secretion is part of the diffuse sympathetic discharge provoked in emergency situations, which Cannon called the "emergency function of the sympathoadrenal system." The ways in which this discharge prepares the individual for flight or fight are described in Chapter 13, and the increases in plasma catecholamines under various conditions are shown in Fig 20–5. Some of the secreted norepinephrine is taken up by noradrenergic neurons, providing them with more transmitter and thus reinforcing their function.

The metabolic effects of circulating catecholamines are probably important, especially in certain situations. The calorigenic action of catecholamines in animals exposed to cold is an example. Animals with denervated adrenal glands shiver sooner and more vigorously than normal controls when exposed to cold. The glycogenolysis produced by epinephrine and norepinephrine in hypoglycemic animals is another example. Hypoglycemia is a potent stimulus to catecholamine secretion. As noted in Chapter 19, glucagon can substitute for catecholamines as a counterregulatory hormone, and vice versa; but if the secretion of both is blocked, insulin tolerance is markedly reduced.

Selective Secretion

When adrenal medullary secretion is increased, the ratio of norepinephrine to epinephrine in the adrenal effluent is generally unchanged. However, asphyxia and hypoxia increase this ratio. The fact that the output of norepinephrine can be increased selectively has unfortunately led to the teleologic speculation that the adrenal medulla secretes epinephrine or norepinephrine depending upon which hormone best equips the animal to meet the emergency it faces. The fallacy of such speculation is illustrated by the response to hemorrhage, in which the predominant catecholamine secreted is not norepinephrine but epinephrine, which lowers the peripheral resistance. Norepinephrine secretion is increased by emotional stresses with which the individual is familiar, whereas epinephrine secretion rises in situations in which the individual does not know what to expect.

ADRENAL CORTEX

STRUCTURE & BIOSYNTHESIS OF ADRENOCORTICAL HORMONES

Classification & Structure

The hormones of the adrenal cortex are derivatives of cholesterol. Like cholesterol, bile acids, vitamin D, and ovarian and testicular steroids, they contain the **cyclopentanoperhydrophenanthrene nucleus** (Fig 20–7). Gonadal and adrenocortical steroids are of 3 types: C_{21} steroids, which have a 2-carbon side chain at position 17; C_{19} steroids, which have a keto or hydroxyl group at position 17; and C_{18} steroids, which, in addition to a 17-keto or hydroxyl group, have no angular methyl group attached to position 10. The adrenal cortex secretes primarily C_{21} and C_{19} steroids. Most of the C_{19} steroids have a keto group at position 17 and are therefore called **17-ketosteroids.** The C_{21} steroids that have a hydroxyl group at the 17 position in addition to the side chain are often called 17-hydroxycorticoids or 17-hydroxycorticosteroids.

The C_{19} steroids have androgenic activity. The C_{21} steroids are classified, using Selye's terminology, as mineralocorticoids or glucocorticoids. All secreted

Figure 20–7. Basic structure of adrenocortical and gonadal steroids. The letters in the formula for cholesterol identify the 4 basic rings, and the numbers identify the positions in the molecule. As shown here, the angular methyl groups (positions 18 and 19) are usually indicated simply by straight lines. (Modified from Sperloff L, Glass RH, Kase NG: *Clinical Gynecologic Endocrinology & Infertility,* 3rd ed. Williams & Wilkins, 1989.)

C_{21} steroids have both mineralocorticoid and glucocorticoid activity; **mineralocorticoids** are those in which effects on Na^+ and K^+ excretion predominate, and **glucocorticoids** those in which effects on glucose and protein metabolism predominate.

The details of steroid nomenclature and isomerism can be found in texts on steroid nomenclature, but it is pertinent to mention that the Greek letter Δ indicates a double bond and that the groups which lie above the plane of each of the steroid rings are indicated by the Greek letter β and a solid line (—OH), whereas those which lie below are indicated by α and a dotted line (- - -OH). Thus, the C_{21} steroids secreted by the adrenal have a Δ^4-3-keto configuration in the A ring. In most naturally occurring adrenal steroids, 17-hydroxy groups are in the α configuration, whereas 3-, 11-, and 21-hydroxy groups are in the β configuration. The 18-aldehyde configuration on naturally occurring aldosterone is the D form. L-Aldosterone is physiologically inactive.

Secreted Steroids

Innumerable steroids have been isolated from adrenal tissue, but the only steroids normally secreted in physiologically significant amounts are the mineralocorticoid **aldosterone,** the glucocorticoids **cortisol** and **corticosterone,** and the androgens **dehydroepiandrosterone (DHEA)** and **androstenedione.** The structures of these steroids are shown in Figs 20–8 and 20–9. **Deoxycorticosterone** is a mineralocorticoid that is normally secreted in about the same amount as aldosterone (Table 20–1) but has only 3% of the mineralocorticoid activity of aldosterone. Its effect on mineral metabolism is usually negligible, but in diseases in which its secretion is increased, its effect can be appreciable. The adrenals may also secrete small amounts of estrogen, although most of the estrogens that are not formed in the ovaries are produced in the circulation from adrenal androstenedione. Almost all the dehydroepiandrosterone is secreted conjugated with sulfate, although most if not all of the other steroids are secreted in the free, unconjugated form.

The secretion rate for individual steroids can be determined by injecting a very small dose of isotopically labeled steroid and determining the degree to which the radioactive steroid excreted in the urine is diluted by unlabeled secreted hormone. This technique is used to measure the output of many different hormones.

Species Differences

In all species from amphibia to humans, the major C_{21} steroid hormones secreted by adrenocortical tissue appear to be aldosterone, cortisol, and corticosterone, although the ratio of cortisol to corticosterone varies. Birds, mice, and rats secrete corticosterone almost exclusively; dogs secrete approximately equal amounts of the 2 glucocorticoids; and cats, sheep, monkeys, and humans secrete predominantly cortisol. In humans, the ratio of secreted cortisol to corticosterone is approximately 7:1.

Synthetic Steroids

As with many other naturally occurring substances, the activity of adrenocortical steroids can be increased by altering their structure. A number of synthetic steroids are now available that have many times the activity of cortisol. The relative glucocorticoid and mineralocorticoid potencies of the natural steroids are compared to those of the synthetic steroids 9α-fluorocortisol, prednisolone, and dexamethasone in Table 20–2. The potency of dexamethasone is due to its high affinity for glucocorticoid re-

Figure 20–8. Outline of hormone biosynthesis in the zona fasciculata and zona reticularis of the adrenal cortex. The major secretory products are underlined. The enzymes for the reactions are shown on the left and at the top of the chart. When a particular enzyme is deficient, hormone production is blocked at the points indicated by the shaded bars.

Figure 20–9. Hormone synthesis in the zona glomerulosa. The zona glomerulosa lacks 17α-hydroxylase activity, and only the zona glomerulosa can convert corticosterone to aldosterone because it is the only zone that normally contains aldosterone synthase. AII, angiotensin II.

Table 20–1. Principal adrenocortical hormones in adult humans.[1]

Name	Synonyms	Average Plasma Concentration (Free and Bound)[1] (μg/dL)	Average Amount Secreted (mg/24 h)
Cortisol	Compound F, hydrocortisone	13.9	10
Corticosterone	Compound B	0.4	3
Aldosterone		0.006	0.15
Deoxycorticosterone	DOC	0.006	0.20
Dehydroepiandrosterone	DEA, DHEA	175.0	20

[1] All plasma concentration values except DEA are morning values after overnight recumbency.

Table 20–2. Relative potencies of corticosteroids compared to cortisol.[1]

Steroid	Glucocorticoid Activity	Mineralocorticoid Activity
Cortisol	1.0	1.0
Corticosterone	0.3	15
Aldosterone	0.3	3000
Deoxycorticosterone	0.2	100
Cortisone	0.7	1.0
Prednisolone	4	0.8
9α-Fluorocortisol	10	125
Dexamethasone	25	~0

[1] Values are approximations based on liver glycogen deposition or anti-inflammatory assays for glucocorticoid activity, and effect on urinary Na^+/K^+ or maintenance of adrenalectomized animals for mineralocorticoid activity. The last 3 steroids listed are synthetic compounds that do not occur naturally. (Data from various sources.)

ceptors and its long half-life (see below). Prednisolone also has a long half-life.

Steroid Biosynthesis

The major paths by which the naturally occurring adrenocortical hormones are synthesized in the body are summarized in Figs 20–8 and 20–9. The precursor of all steroids is cholesterol. Some of the cholesterol is synthesized from acetate, but most of it is taken up from LDL in the circulation (see Chapter 17). LDL receptors are especially abundant in adrenocortical cells. The cholesterol is esterified and stored in lipid droplets. **Cholesterol ester hydrolase** catalyzes the formation of free cholesterol in the lipid droplets (Fig 20–10). The cholesterol is transported to mitochondria by a sterol carrier protein. In the mitochondria, it is converted to pregnenolone in a reaction catalyzed by **cholesterol desmolase.** This enzyme is a mitochondrial cytochrome P450, which is also known as **side-chain cleavage enzyme** or **P450scc.**

The pregnenolone then moves to the smooth endoplasmic reticulum, where some of it is dehydrogenated to form progesterone in a reaction catalyzed by **3β-hydroxysteroid dehydrogenase.** This enzyme has a molecular weight of 46,000 and is not a P450. In the zona fasciculata and the zona reticularis, some of the pregnenolone and some of the progesterone are hydroxylated in the smooth endoplasmic reticulum to 17α-hydroxypregnenolone. The enzyme that catalyzes these reactions, **17α-hydroxylase,** is another P450, **P450c17.** Some of the 17α-hydroxypregnenolone is converted to 17α-hydroxyprogesterone. P450c17 is also the **17,20-lyase** that catalyzes the cleavage of the 17,20 bond to convert 17α-hydroxypregnenolone and 17α-hydroxyprogesterone to dehydroepinandrosterone and androstenedione, respectively (Fig 20–8).

Hydroxylation of progesterone and 17α-

Figure 20–10. Mechanism of action of ACTH on cortisol-secreting cells in the inner 2 zones of the adrenal cortex. When ACTH binds to its receptor (R), adenylate cyclase (AC) is activated via G_s. The resulting increase in cyclic AMP activates protein kinase A, and the kinase phosphorylates cholesterol ester hydrolase (CEH), increasing its activity. Consequently, more free cholesterol is formed and converted to pregnenolone in the mitochondria. Note that in the subsequent steps in steroid biosynthesis, products are shuttled between the mitochondria and the smooth endoplasmic reticulum (SER). Corticosterone is also synthesized and secreted.

hydroxyprogesterone also occurs in the smooth endoplasmic reticulum, where it is catalyzed by **21β-hydroxylase,** a cytochrome that is also called **P450c21.** The end products, 11-deoxycorticosterone and 11-deoxycortisol, move back to the mitochondria, where they are 11-hydroxylated to form corticosterone and cortisol, respectively. The enzyme that catalyzes these reactions, **11β-hydroxylase,** is a fourth cytochrome, **P450c11.** Thus, there are 4 P450 cytochromes involved in the production of glucocorticoids and sex hormones in the adrenal cortex. Two of these, P450s11 and P450c11, are located in the mitochondria, and 2, P450c17 and P450c21, are located in the smooth endoplasmic reticulum.

In the zona fasciculata and zona reticularis, the corticosterone and the cortisol that are formed by the action of P450c11 diffuse into the circulation. However, in the zona glomerulosa, corticosterone is converted to 18-hydroxycorticosterone and then to aldosterone (Fig 20–9). The enzyme that catalyzes this reaction is called **aldosterone synthase.** The gene encoding the enzyme is 95% identical to the gene for P450c11, and both are located in the same part of chromosome 8, but the synthase is normally found only in the zona glomerulosa. The zona glomerulosa also lacks 17α-hydroxylase. This is why the zona glomerulosa makes aldosterone but fails to make 17-hydroxysteroids or sex hormones.

Much of the dehydroepiandrosterone formed in the inner 2 zones is converted to dehydroepiandrosterone sulfate by **adrenal sulfokinase.** Androsterone is converted to testosterone and estradiol.

Action of ACTH

ACTH binds to high-affinity receptors on the plasma membrane of adrenocortical cells. This activates adenylate cyclase via G_s (see Chapter 1), and the resultant increase in intracellular cyclic AMP activates protein kinase A. Protein kinase A phosphorylates cholesterol ester hydrolase, increasing its activity, and conversion of cholesterol esters to free cholesterol is increased (Fig 20–10). This, in turn, leads to a prompt increase in the formation of pregnenolone and its derivatives. Over longer periods, ACTH also increases the synthesis of all 5 P450s involved in the formation of adrenocortical hormones.

Action of Angiotensin II

Angiotensin II binds to receptors in the zona glomerulosa which act via G_p to activate phospholipase C (see Chapter 1). The resulting increase in protein kinase C fosters the conversion of cholesterol to pregnenolone (Fig 20–9) and facilitates the formation of 18-hydroxycorticosterone, which, in turn, facilitates the production of aldosterone.

Enzyme Deficiencies

The consequences of inhibiting any of the enzyme systems involved in steroid biosynthesis can be predicted from Figs 20–8 and 20–9. Single enzyme defects can occur as congenital "inborn errors of metabolism." Congenital defects in the enzymes lead to deficient cortisol secretion and the syndrome of **congenital adrenal hyperplasia.** The hyperplasia is due to increased ACTH secretion. The disease is severe when cholesterol desmolase (P450scc) is deficient, so that conversion of cholesterol to pregnenolone is compromised. The gene coding for this enzyme is on chromosome 15. The defect, which fortunately is rare, causes diffuse adrenal insufficiency and, in most cases, death soon after birth. Since androgens are not formed, female genitalia develop regardless of genetic sex (see Chapter 23). In 3β-hydroxysteroid dehydrogenase deficiency, another rare condition, DHEA secretion is increased. This steroid is a weak androgen that can cause some masculinization in females with the disease, but it is not adequate to produce full masculinization of the genitalia in genetic males. Consequently, hypospadias is common. In fully developed 17α-hydroxylase (P450c17) deficiency, a third rare condition due to deficiency of an enzyme encoded by a gene on chromosome 10, no sex hormones are produced, so there are female external genitalia. However, the pathway leading to corticosterone and aldosterone is intact, and elevated levels of 11-deoxycorticosterone and other mineralocorticoids produce hypertension and hypokalemia.

Cortisol is deficient, but this is compensated by the glucocorticoid activity of corticosterone. Variants of this syndrome have been reported in which the lack of 17,20-lyase activity prevents the production of normal amounts of sex hormones but 17α-hydroxylase activity is adequate to produce cortisol.

Unlike the 3 defects discussed in the preceding paragraph, 21β-hydroxylase (P450c21) and 11β-hydroxylase (P450c11) deficiencies are common; the former accounts for 90–95% of the cases of congenital adrenal hyperplasia, and the latter accounts for all but 1% of the remaining cases. Both are characterized by **virilization**, because the increase in ACTH secretion causes steroids to pile up behind the blockades and to be shunted to the production of androgens. The characteristic pattern that develops in females in the absence of treatment is the **adrenogenital syndrome** (Fig 20–11). P450c21 is on the short arm of chromosome 6, closely linked to the HLA major histocompatibility complex (see Chapter 19), and various defects can produce degrees of 21β-hydroxylase deficiency that range from mild to severe. Enough glucocorticoids and mineralocorticoids are usually formed to sustain life. In severe cases, the genitalia of genetic females are masculinized (**female pseudohermaphroditism;** see Chapter 23). However, masculinization may not be marked until later in life, and mild cases can be detected only by laboratory tests. About 30% of the patients with 21β-hydroxylase deficiency lose sodium to an excessive degree (**salt-losing form** of congenital virilizing adrenal hyperplasia). The sodium loss is due both to the mineralocorticoid deficiency and to the aldosterone-

antagonizing effect of some of the steroids being secreted in excess amounts. In 11β-hydroxylase deficiency, there is virilization plus excess secretion of 11-deoxycortisol and deoxycorticosterone; since the latter is an active mineralocorticoid, salt and water retention and hypertension result (**hypertensive form** of congenital virilizing adrenal hyperplasia).

Glucocorticoid treatment is indicated in all of the virilizing forms of congenital adrenal hyperplasia because it repairs the glucocorticoid deficit and inhibits ACTH secretion (see below), reducing the abnormal secretion of androgens and other steroids. However, doses that just restore glucocorticoid activity to normal do not completely eliminate excess androgen production because the intrinsic abnormality in steroid biosynthesis is still present and some endogenous substrate is diverted to the androgen pathway. Some clinicians get around this problem by giving supernormal amounts of glucocorticoids. Others suggest combining glucocorticoid treatment with drugs that prevent the effects of excess sex hormones.

TRANSPORT, METABOLISM & EXCRETION OF ADRENOCORTICAL HORMONES

Glucocorticoid Binding

Cortisol is bound in the circulation to an α globulin called **transcortin** or **corticosteroid-binding globulin (CBG).** There is also a minor degree of binding to albumin (see Table 23–5). Corticosterone is similarly bound, but to a lesser degree. The half-life of cortisol in the circulation is therefore longer (about 60–90 minutes) than that of corticosterone (50 minutes). Bound steroids appear to be physiologically inactive. Because of protein binding, there is relatively little free cortisol and corticosterone in the urine.

The equilibrium between cortisol and its binding protein and the implications of binding in terms of tissue supplies and ACTH secretion are summarized in Fig 20–12. The bound cortisol functions as a circulating reservoir of hormone that keeps a supply of free cortisol available to the tissues. The relationship is similar to that of T_4 and its binding protein (see Fig 18–8). At normal levels of total plasma cortisol (13.5 μg/dL, or 375 nmol/L), there is very little free cortisol in the plasma, but the binding sites on CBG become saturated when the total plasma cortisol exceeds 20 μg/dL. At higher plasma levels, there is some increased binding to albumin but the main increase is in the unbound fraction.

CBG is synthesized in the liver, and its production is increased by estrogen. CBG levels are elevated during pregnancy and depressed in cirrhosis, nephrosis, and multiple myeloma. When the CBG level rises, more cortisol is bound, and initially there is a drop in the free cortisol level. This stimulates ACTH

ADRENOGENITAL SYNDROME

Baldness
Receding hairline
Hirsutism
Androgenic flush
Small breasts
Male escutcheon
Heavy arms and legs
Enlarged clitoris

Figure 20–11. Typical findings in the adrenogenital syndrome in a postpubertal woman. (Reproduced, with permission, from Forsham PH, Di Raimondo VC: *Traumatic Medicine and Surgery for the Attorney.* Butterworth, 1960.)

Figure 20–12. The interrelationships of free and bound cortisol. The dashed arrow indicates that cortisol inhibits ACTH secretion. The value for free cortisol is an approximation; in most studies, it is calculated by subtracting the protein-bound cortisol from the total plasma cortisol.

secretion, and more cortisol is secreted until a new equilibrium is reached at which the bound cortisol is elevated but the free cortisol is normal. Changes in the opposite direction occur when the CBG level falls. This explains why pregnant women have high total plasma cortisol levels without symptoms of glucocorticoid excess and, conversely, why some patients with nephrosis have low total plasma cortisol without symptoms of glucocorticoid deficiency.

Metabolism & Excretion of Glucocorticoids

Cortisol is metabolized in the liver, which is the principal site of glucocorticoid catabolism. Most of the cortisol is reduced to dihydrocortisol and then to tetrahydrocortisol, which is conjugated to glucuronic acid (Fig 20–13). The glucuronyl transferase system responsible for this conversion also catalyzes the formation of the glucuronides of bilirubin (see Chapter 26) and a number of hormones and drugs. There is competitive inhibition between these substrates for the enzyme system.

In the liver and other tissues containing the enzyme 11β-hydroxysteroid dihydrogenase (see below), cortisol is converted to cortisone. Cortisone is an active glucocorticoid and is well known because of its extensive use in medicine, but it is not secreted in appreciable quantities by the adrenal glands. Little if any of the cortisone formed in the liver enters the circulation, because it is promptly reduced and conjugated to form tetrahydrocortisone glucuronide. The tetrahydroglucuronide derivatives ("conjugates") of cortisol and corticosterone are freely soluble. They enter the circulation, where they do not become bound to protein. They are rapidly excreted in the urine, in part by tubular secretion.

About 10% of the secreted cortisol is converted in the liver to the 17-ketosteroid derivatives of cortisol

Figure 20–13. Outline of hepatic metabolism of cortisol. The enzyme that catalyzes the conversion of cortisol to cortisone is 11β-hydroxysteroid dehydrogenase.

and cortisone. The ketosteroids are conjugated for the most part to sulfate and then excreted in the urine. Other metabolites, including 20-hydroxy derivatives, are formed. There is an enterohepatic circulation of glucocorticoids, and about 15% of the secreted cortisol is excreted in the stool. The metabolism of corticosterone is similar to that of cortisol, except that it does not form a 17-ketosteroid derivative.

Variations in the Rate of Hepatic Metabolism

The rate of hepatic inactivation of glucocorticoids is depressed in liver disease and, interestingly, during surgery and other stresses. Thus, in stressed humans, the plasma free cortisol level rises higher than it does with maximal ACTH stimulation in the absence of stress.

Aldosterone

Aldosterone is bound to protein to only a slight extent, and its half-life is short (about 20 minutes). The amount secreted is small (Table 20–1), and the total plasma aldosterone level in humans is normally about 0.006 μg/dL (0.17 nmol/L), compared with a cortisol level (bound and free) of about 13.5 μg/dL (375 nmol/L). Much of the aldosterone is converted in the liver to the tetrahydroglucuronide derivative, but some is changed in the liver and in the kidneys to an 18-glucuronide. This glucuronide, which is unlike the breakdown products of other steroids, is converted to free aldosterone by hydrolysis at pH 1.0, and it is therefore often referred to as the "acid-labile conjugate." Less than 1% of the secreted aldosterone appears in the urine in the free form. Another 5% is in the form of the acid-labile conjugate, and up to 40% is in the form of the tetrahydroglucuronide.

17-Ketosteroids

The major adrenal androgen is the 17-ketosteroid dehydroepiandrosterone, although androstenedione is also secreted. The 11-hydroxy derivative of androstenedione and the 17-ketosteroids formed from cortisol and cortisone by side chain cleavage in the liver are the only 17-ketosteroids that have an = O or an − OH group in the 11 position ("11-oxy-17-ketosteroids"). Testosterone is also converted to a 17-ketosteroid. Since the daily 17-ketosteroid excretion in normal adults is 15 mg in men and 10 mg in women, about two-thirds of the urinary ketosteroids in men are secreted by the adrenal or formed from cortisol in the liver, and about one-third is of testicular origin.

Etiocholanolone, one of the metabolites of the adrenal androgens and testosterone (see Fig 23–18), can cause fever when it is unconjugated (see Chapter 14). Certain individuals have episodic bouts of fever due to periodic accumulation in the blood of unconjugated etiocholanolone ("etiocholanolone fever").

EFFECTS OF ADRENAL ANDROGENS & ESTROGENS

Androgens

Androgens are the hormones that exert masculinizing effects, and they promote protein anabolism and growth (see Chapter 23). Testosterone from the testes is the most active androgen, and the adrenal androgens have less than 20% of its activity. Secretion of the adrenal androgens is controlled by ACTH and possibly by a pituitary adrenal androgen-stimulating hormone (see Chapter 23) but not by gonadotropins. The normal plasma concentration of dehydroepiandrosterone in adult men and women is about 300 μg/dL, and all but 0.1% of this is conjugated to sulfate. After menopause, the plasma concentration in women falls to about 90 μg/dL. The secretion of adrenal androgens is as great in castrated males and females as it is in normal males, so it is clear that these hormones exert very little masculinizing effect when secreted in normal amounts. However, they can produce appreciable masculinization when secreted in excessive amounts. In adult males, excess adrenal androgens merely accentuate existing characteristics; but in prepuberal boys, they can cause precocious development of the secondary sex characteristics without testicular growth (**precocious pseudopuberty**). In females they cause female pseudohermaphroditism and the adrenogenital syndrome (Fig 20–11).

Estrogens

The adrenal androgen androstenedione is converted to testosterone and to estrogens (aromatized) in the circulation. It is an important source of estrogens in men and postmenopausal women (see Chapter 23).

PHYSIOLOGIC EFFECTS OF GLUCOCORTICOIDS

Adrenal Insufficiency

In untreated adrenalectomized animals and humans, there is Na+ loss with circulatory insufficiency, hypotension, and, eventually, fatal shock due to the mineralocorticoid deficiency. Because glucocorticoids are absent, water, carbohydrate, protein, and fat metabolism are abnormal, and collapse and death follow exposure to even minor noxious stimuli. Small amounts of glucocorticoids correct the metabolic abnormalities, in part directly and in part by permitting other reactions to occur. It is important to separate these physiologic actions of glucocorticoids from the quite different effects produced by large amounts of the hormones.

Mechanism of Action

The multiple effects of glucocorticoids are triggered by binding to glucocorticoid receptors, which promote the transcription of certain segments of DNA (see Chapter 1). This, in turn, leads via the appropriate mRNAs (see Fig 1–31) to synthesis of enzymes that alter cell function.

Effects on Intermediary Metabolism

The actions of glucocorticoids on the intermediary metabolism of carbohydrate, protein, and fat are discussed in Chapter 19. They include increased protein catabolism and increased hepatic glycogenesis and gluconeogenesis (Fig 20–14). Glucose 6-phosphatase activity is increased, and the blood glucose level rises. Glucocorticoids exert an anti-insulin action in peripheral tissues and make diabetes worse. However, the brain and the heart are spared, so the increase in blood glucose provides extra glucose to these vital organs. In diabetics, glucocorticoids raise plasma lipid levels and increase ketone body formation, but in normal individuals, the increase in insulin secretion provoked by the rise in blood glucose obscures these actions. In adrenal insufficiency, the blood glucose level is normal as long as an adequate caloric intake is maintained, but fasting causes hypoglycemia that can be fatal. The adrenal cortex is not essential for the ketogenic response to fasting.

Permissive Action

Small amounts of glucocorticoids must be present

Figure 20–14. Actions of glucocorticoids in promoting gluconeogenesis. –, inhibition; +, stimulation. (Modified and reproduced, with permission, from Baxter JD, Forsham PH: Tissue effects of glucocorticoids. *Am J Med* 1972;**52**:573.)

for a number of metabolic reactions to occur, although the glucocorticoids do not produce the reactions by themselves. This effect is called their **permissive action.** Permissive effects include the requirement for glucocorticoid to be present for glucagon and catecholamines to exert their calorigenic effects (see above and Chapter 19), for catecholamines to exert their lipolytic effects, and for catecholamines to produce pressor responses and bronchodilation.

Effects on ACTH Secretion

Glucocorticoids inhibit ACTH secretion, and ACTH secretion is increased in adrenalectomized animals. The consequences of the feedback action of cortisol on ACTH secretion are discussed below in the section on regulation of glucocorticoid secretion.

Vascular Reactivity

In adrenally insufficient animals, vascular smooth muscle becomes unresponsive to norepinephrine and epinephrine. The capillaries dilate and, terminally, become permeable to colloidal dyes. Failure to respond to the norepinephrine liberated at noradrenergic nerve endings probably impairs vascular compensation for the hypovolemia of adrenal insufficiency and promotes vascular collapse. Glucocorticoids restore vascular reactivity.

Effects on the Nervous System

Changes in the nervous system in adrenal insufficiency that are reversed only by glucocorticoids include the appearance of electroencephalographic waves slower than the normal α rhythm, personality changes, and increased sensitivity to olfactory and gustatory stimuli. The personality changes, which are mild, include irritability, apprehension, and inability to concentrate.

Effects on Water Metabolism

Adrenal insufficiency is characterized by inability to excrete a water load, and only glucocorticoids repair this deficit. The load is eventually excreted, but the excretion is so slow that there is danger of water intoxication. In patients with adrenal insufficiency who have not received glucocorticoids, glucose infusions may cause high fever ("glucose fever") followed by collapse and death. Presumably, the glucose is metabolized, the water dilutes the plasma, and the resultant osmotic gradient between the plasma and the cells causes the cells of the thermoregulatory centers in the hypothalamus to swell to such an extent that their function is disrupted. It should be noted that water intoxication can occur in adrenal insufficiency even without an absolute excess of water in the body; the important factor is retention of water in excess of Na+.

The cause of defective water excretion in adrenal insufficiency is unsettled. Plasma vasopressin and neurophysin levels are elevated in adrenal insufficiency and reduced by glucocorticoid treatment. The glomerular filtration rate is low, and this probably contributes to the deficiency in water excretion. The selective effect of glucocorticoids on the abnormal water excretion is consistent with this possibility, because even though the mineralocorticoids improve filtration by restoring plasma volume, the glucocorticoids raise the glomerular filtration rate to a much greater degree.

Effects on the Blood Cells & Lymphatic Organs

Glucocorticoids decrease the number of circulating eosinophils by increasing their sequestration in the spleen and lungs. Glucocorticoids also lower the number of basophils in the circulation and increase the number of neutrophils, platelets, and red blood cells (Table 20–3).

Glucocorticoids decrease the circulating lymphocyte count and the size of the lymph nodes and thymus by inhibiting lymphocyte mitotic activity. It appears that their primary action is to inhibit production of IL-2 by T lymphocytes, thus effectively stopping lymphocyte proliferation (see Chapter 27). Production of IL-1 by monocytes and macrophages is also inhibited.

Resistance to "Stress"

When an animal or human is exposed to any of an immense variety of noxious or potentially noxious stimuli, there is an increased secretion of ACTH and, consequently, a rise in the circulating glucocorticoid level. This rise is essential for survival. Hypophysectomized animals, or adrenalectomized animals treated with maintenance doses of glucocorticoids, die when exposed to the same noxious stimuli.

Selye defined noxious stimuli that increase ACTH secretion as "stressors," and it is fashionable today to lump these stimuli together under the term "stress." This word is a short, emotionally charged word for something that otherwise takes many words to say, and it is a convenient term to use as long as it is understood that, in the context of this book, it denotes only those stimuli that have been proved to increase ACTH secretion in normal animals and humans.

Table 20–3. Typical effects of cortisol on the white and red blood cell counts in humans (cells/μL).

Cell	Normal	Cortisol-Treated
White blood cells		
Total	9000	10,000
PMNs	5760	8330
Lymphocytes	2370	1080
Eosinophils	270	20
Basophils	60	30
Monocytes	450	540
Red blood cells	5 million	5.2 million

The reason an elevated circulating glucocorticoid level is essential for resisting stress remains for the most part unknown. Most of the stressful stimuli that increase ACTH secretion also activate the sympathetic nervous system, and part of the function of circulating glucocorticoids may be maintenance of vascular reactivity to catecholamines. Glucocorticoids are also necessary for the catecholamines to exert their full FFA-mobilizing action, and the FFAs are an important emergency energy supply. However, sympathectomized animals tolerate a variety of stresses with relative impunity. Another theory holds that glucocorticoids prevent other stress-induced changes from becoming excessive. At present, all that can be said is that stress causes increases in plasma glucocorticoids to high "pharmacologic" levels that in the short run are life-saving but in the long run are definitely harmful and disruptive.

PHARMACOLOGIC & PATHOLOGIC EFFECTS OF GLUCOCORTICOIDS

Cushing's Syndrome

The clinical picture produced by prolonged increases in plasma glucocorticoids is called **Cushing's syndrome** (Fig 20–15). It may be caused not only by the administration of large amounts of exogenous hormones but also by glucocorticoid-producing adrenocortical tumors or adrenal hyperplasia, which is apparently caused in some instances by circulating adrenal-stimulating immunoglobulins. Alternatively, it can be caused by hypersecretion of ACTH. Most patients with hypersecretion of ACTH from the pituitary have small pituitary tumors (**microadenomas**) that are often so small that the sella turcica is not enlarged. However, some cases of excess ACTH secretion may be due to hypersecretion of corticotropin-releasing hormone (CRH). Cushing's syndrome is also produced by tumors of nonendocrine tissues that secrete substances with CRH activity or, more commonly, ACTH (**ectopic ACTH syndrome**). Tumors of nonendocrine tissues have also been reported to secrete substances with actions like those of almost every known hormone.

Patients with Cushing's syndrome are protein-depleted as a result of excess protein catabolism. The skin and subcutaneous tissues are therefore thin, and the muscles are poorly developed. Wounds heal poorly, and minor injuries cause bruises and ecchymoses. The hair is thin and scraggly. Many patients with the disease have some increase in facial hair and acne, but this is caused by the increased secretion of adrenal androgens that often accompanies the increase in glucocorticoid secretion.

Body fat is redistributed in a characteristic way. The extremities are thin, but fat collects in the abdominal wall, face, and upper back, where it produces a "buffalo hump." As the thin skin of the abdo-

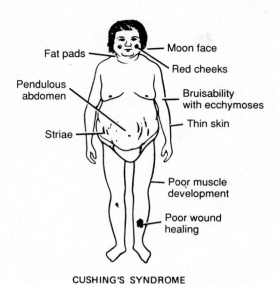

Fat pads
Moon face
Red cheeks
Pendulous abdomen
Bruisability with ecchymoses
Thin skin
Striae
Poor muscle development
Poor wound healing

CUSHING'S SYNDROME

Figure 20–15. Typical findings in Cushing's syndrome. (Reproduced, with permission, from Forsham PH, Di Raimondo VC: *Traumatic Medicine and Surgery for the Attorney.* Butterworth, 1960.)

men is stretched by the increased subcutaneous fat depots, the subdermal tissues rupture to form prominent reddish-purple **striae.** These scars are seen normally whenever there is rapid stretching of skin (eg, around the breasts of girls at puberty or in the abdominal skin during pregnancy), but in normal individuals, the striae are usually inconspicuous and lack the intense purplish color.

Many of the amino acids liberated from catabolized proteins are converted into glucose in the liver, and the resultant hyperglycemia and decreased peripheral utilization of glucose may be sufficient to precipitate insulin-resistant diabetes mellitus, especially in patients genetically predisposed to diabetes. Hyperlipemia and ketosis are associated with the diabetes, but acidosis is usually not severe.

The glucocorticoids are present in such large amounts in Cushing's syndrome that they exert a significant mineralocorticoid action. Deoxycorticosterone secretion is also elevated in cases due to ACTH hypersecretion. The salt and water retention plus the facial obesity cause the characteristic plethoric, rounded "moon-faced" appearance, and there may be significant K^+ depletion and weakness. About 85% of patients with Cushing's syndrome are hypertensive. The hypertension may be due to increased deoxycorticosterone secretion, increased secretion of angiotensinogen (see Chapter 24), or a direct glucocorticoid effect on blood vessels.

Excess glucocorticoids lead to bone dissolution by decreasing bone formation and increasing bone resorption. This leads to **osteoporosis,** a loss of bone

mass that leads eventually to collapse of vertebral bodies and other fractures. The mechanisms by which glucocorticoids produce their effects on bone are discussed in Chapter 21.

Glucocorticoids in excess accelerate the basic electroencephalographic rhythms and produce mental aberrations ranging from increased appetite, insomnia, and euphoria to frank toxic psychoses. As noted above, glucocorticoid deficiency is also associated with mental symptoms, but the symptoms produced by glucocorticoid excess are more severe.

Anti-Inflammatory & Antiallergic Effects of Glucocorticoids

In large doses, glucocorticoids inhibit the inflammatory response to tissue injury. The glucocorticoids also suppress manifestations of allergic disease that are due to the release of histamine from tissues. Both of these effects require high levels of circulating glucocorticoids and cannot be produced without producing the other manifestations of glucocorticoid excess. Furthermore, large doses of exogenous glucocorticoids inhibit ACTH secretion to the point that severe adrenal insufficiency can be a dangerous problem when therapy is stopped.

Glucocorticoids inhibit fibroblastic activity, decrease local swelling, and block the systemic effects of bacterial toxins. The decreased local inflammatory reaction is due, at least in part, to inhibition of phospholipase A_2, with a consequent reduction in the release of arachidonic acid from tissue phospholipids. Therefore, the formation of leukotrienes, thromboxanes, prostaglandins, and prostacyclin (see Fig 17–33) is reduced. The leukotrienes are powerful mediators of inflammation, and high concentrations of leukotriene B_4 have been found in synovial fluid in rheumatoid arthritis. By stabilizing lysosomal membranes, glucocorticoids inhibit the breakdown of lysosomes that takes place in inflamed tissue. Despite their protein-catabolic effects, they slow the degrading effect of collagenase on joint tissues in rheumatoid arthritis, and this is the basis of their effectiveness following intra-articular administration. The glucocorticoids also inhibit the release of IL-1 (endogenous pyrogen) (see above). Large doses of glucocorticoids in humans initially elevate and later depress antibody levels. The inhibition of fibroblastic activity prevents the walling off of infections and also prevents such phenomena as keloid formation and the development of adhesions after abdominal surgery.

The actions of glucocorticoids in patients with bacterial infections are dramatic but dangerous. For example, in pneumococcal pneumonia or active tuberculosis, the febrile reaction, the toxicity, and the lung symptoms disappear; but unless antibiotics are given at the same time, the bacteria spread throughout the body. It is important to remember that the symptoms are the warning that disease is present;

when these symptoms are masked by treatment with glucocorticoids, there may be serious and even fatal delays in diagnosis and the institution of treatment with antimicrobial drugs.

When certain types of antibodies combine with their antigens, they provoke the release of histamine from mast cells in various tissues, and this in turn causes many of the symptoms of allergy. Glucocorticoids do not affect the combination of antigen with antibody and have no influence on the effects of histamine once it is released, but they prevent histamine release. Glucocorticoids therefore relieve the symptoms of asthma and delayed hypersensitivity reactions and may be of benefit in such immunoglobulin-mediated diseases as serum sickness.

Other Effects

Large doses of glucocorticoids inhibit growth, decrease growth hormone secretion (see Chapter 22), induce PNMT, and decrease TSH secretion. During fetal life, glucocorticoids accelerate the maturation of surfactant in the lungs (see Chapter 34).

REGULATION OF GLUCOCORTICOID SECRETION

Role of ACTH

Both basal secretion of glucocorticoids and the increased secretion provoked by stress are dependent upon ACTH from the anterior pituitary. Angiotensin II also stimulates the adrenal cortex, but its effect is mainly on aldosterone secretion. Large doses of a number of other naturally occurring substances, including vasopressin, serotonin, and VIP, are capable of stimulating the adrenal directly, but there is no evidence that these agents play any role in the physiologic regulation of glucocorticoid secretion.

Chemistry & Metabolism of ACTH

ACTH is a single-chain polypeptide containing 39 amino acids. Its origin from pro-opiomelanocortin (POMC) in the pituitary is discussed in Chapter 22. The first 23 amino acids in the chain, which are the same in all species that have been examined, constitute the active "core" of the molecule, and a synthetic peptide containing these 23 amino acids has been shown to have the full activity of the 39-amino-acid peptide. Amino acids 24–39 are therefore a "tail" that stabilizes the molecule and varies slightly in amino acid composition from species to species (Fig 20–16). Each of the ACTHs that have been isolated is active in all species but generally antigenic in heterologous species.

ACTH is inactivated in blood in vitro more slowly than in vivo; its half-life in the circulation in humans is about 10 minutes. A large part of an injected dose of ACTH is found in the kidneys, but neither nephrectomy nor evisceration appreciably enhances its

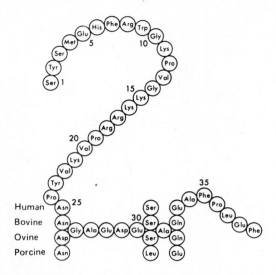

Figure 20–16. Structure of ACTH in various species. As shown, the amino acid composition varies only at positions 25, 31, and 33. (Reproduced, with permission, from Li CH: Adrenocorticotropin 45: Revised amino acid sequences for sheep and bovine hormones. *Biochem Biophys Res Commun* 1972;**49**:835.)

in vivo activity, and the site of its inactivation is not known.

Effect of ACTH on the Adrenal

After hypophysectomy, glucocorticoid synthesis and output decline within 1 hour to very low levels, although some hormone is still secreted. Within a short time after an injection of ACTH (in dogs, less than 2 minutes), glucocorticoid output rises (Fig 20–17). With low doses of ACTH, there is a linear relationship between the log of the dose and the increase in glucocorticoid secretion. However, the maximal rate at which glucocorticoids can be secreted is rapidly reached; and in dogs, doses larger than 10 mU only prolong the period of maximal secretion. A similar "ceiling on output" exists in rats and humans. The effects of ACTH on adrenal morphology and the mechanism by which it increases steroid secretion are discussed above.

Adrenal Responsiveness

ACTH not only produces prompt increases in glucocorticoid secretion but also increases the sensitivity of the adrenal to subsequent doses of ACTH. Conversely, single doses of ACTH do not increase glucocorticoid secretion in chronically hypophysectomized animals and patients with hypopituitarism, and repeated injections or prolonged infusions of ACTH are necessary to restore normal adrenal responses to ACTH. Decreased responsiveness is also produced by doses of glucocorticoids that inhibit

Figure 20–17. Changes in cortisol output in the hypophysectomized dog following the intravenous (IV) administration of various doses of ACTH. (Reproduced, with permission, from Ganong WF: The central nervous system and the synthesis and release of ACTH. In: *Advances in Neuroendocrinology.* Nalbandov A [editor]. Univ of Illinois Press, 1963.)

ACTH secretion. The decreased adrenal responsiveness to ACTH is detectable within 24 hours after hypophysectomy and increases progressively with time (Fig 20–18). It is marked when the adrenal is atrophic but develops before there are visible changes in adrenal size or morphology. There is evidence that the sensitivity of the inner 2 zones and the zona glomerulosa to ACTH is increased by a portion of the N-terminal region of the POMC molecule (see Chapter 22) that is secreted along with ACTH.

Circadian Rhythm

ACTH is secreted in irregular bursts throughout the day, and plasma cortisol tends to rise and fall in response to these bursts (Fig 20–19). In humans, the bursts are most frequent in the early morning, and about 75% of the daily production of cortisol occurs between 4 and 10 AM. The bursts are least frequent in the evening. This **diurnal (circadian) rhythm** in ACTH secretion is present in patients with adrenal insufficiency receiving constant doses of glucocorticoids. It is not due to the stress of getting up in the morning, traumatic as that may be, because the increased ACTH secretion occurs before waking up. If the "day" is lengthened experimentally to more than 24 hours—ie, if the individual is isolated and the day's activities are spread over more than 24 hours—the adrenal cycle also lengthens, but the increase in ACTH secretion still occurs during the period of sleep. The biologic clock responsible for the diurnal ACTH rhythm is located in the suprachiasmatic nuclei of the hypothalamus (see Chapter 14).

The Response to Stress

The morning plasma ACTH concentration in a healthy resting human is about 25 pg/mL (5.5 pmol/L). ACTH and cortisol values in various abnormal conditions are summarized in Fig 20–20. During severe stress, the amount of ACTH secreted exceeds the amount necessary to produce maximal glucocorticoid output. However, prolonged exposure to ACTH in conditions such as the ectopic ACTH syndrome increases the adrenal maximum.

Figure 20–18. Loss of ACTH responsiveness when ACTH secretion is decreased in humans. The 1–24-amino-acid sequence of ACTH was infused intravenously (IV) in a dose of 250 μg over 8 hours. N, normal subjects; DX, dexamethasone 0.75 mg every 8 hours for 3 days; CST, long-term corticosteroid therapy; HI, anterior pituitary insufficiency. (Reproduced, with permission, from Kolanowski J et al: Adrenocortical response upon repeated stimulation with corticotrophin in patients lacking endogenous corticotrophin secretion. *Acta Endocrinol [Kbh]* 1977;**85**:595.)

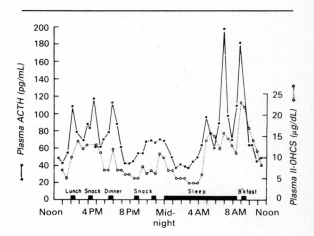

Figure 20–19. Fluctuations in plasma ACTH and glucocorticoids throughout the day in a normal girl (age 16). The ACTH was measured by immunoassay and the glucocorticoids as 11-oxysteroids (11-OHCS). Note the greater ACTH and glucocorticoid rises in the morning, before awakening. (Reproduced, with permission, from Krieger DT et al: Characterization of the normal temporal pattern of plasma corticosteroid levels. *J Clin Endocrinol Metab* 1971;**32**:266.)

Condition	Plasma ACTH (pg/mL)	Plasma cortisol (µg/dL)
	5 50 500 5000	0 12 25 50 100
Normal, morning	⊢—⊣	⊢—⊣
Normal, evening	⊢⊣	⊢⊣
Normal, dexamethasone	⊢	⊢
Normal, metyrapone	⊢—⊣	⊢—⊣
Normal, stress	⊢—⊣	⊢—⊣
Addison's disease	⊢—⊣	⊢
Hypopituitarism	⊣	⊢
Congenital adrenal hyperplasia	⊢—⊣	⊢—⊣
Cushing's, hyperplasia	⊢—⊣	⊢—⊣
Cushing's, dexamethasone	⊢—⊣	⊢—⊣
Cushing's, postadrenalectomy	⊢——⊣	⊢
Cushing's, ectopic ACTH syndrome	⊢—⊣	⊢——⊣
Cushing's, adrenal tumor	⊢	⊢—⊣
	0 5 50 500 5000	0 12 25 50 100

Figure 20–20. Plasma concentrations of ACTH and cortisol in various clinical states. (Reproduced, with permission, from Liddle G: The adrenal cortex. In: *Textbook of Endocrinology,* 5th ed. Williams RH [editor]. Saunders, 1974.)

Increases in ACTH secretion to meet emergency situations are mediated almost exclusively through the hypothalamus via release of CRH. This polypeptide is secreted in the median eminence and transported in the portal-hypophyseal vessels to the anterior pituitary, where it stimulates ACTH secretion (see Chapter 14). If the median eminence is destroyed, some basal glucocorticoid secretion continues and the adrenals do not atrophy, but increased secretion in response to many different stresses is blocked. Afferent nerve pathways from many parts of the brain converge on the median eminence. Fibers from the amygdaloid nuclei mediate responses to emotional stresses, and fear, anxiety, and apprehension cause marked increases in ACTH secretion. Input from the suprachiasmatic nuclei also provides the drive for the diurnal rhythm. Impulses ascending to the hypothalamus via the nociceptive pathways and the reticular formation trigger ACTH secretion in response to injury (Fig 20–21). There is an inhibitory input from the baroreceptors via the nucleus of the tractus solitarius. Contrary to previously proposed theories, circulating epinephrine and norepinephrine do not increase ACTH secretion in humans, and adrenocortical and adrenal medullary secretion are independently regulated by the hypothalamus.

Intracerebroventricular injection of CRH produces hyperglycemia, increased cardiac output, diminished reproductive function, diminished gastrointestinal function, and other changes that are also seen upon exposure to stress. However, it is not possible as yet to say that endogenously secreted CRH is responsible for these changes during stress.

Glucocorticoid Feedback

Free glucocorticoids inhibit ACTH secretion, and the degree of pituitary inhibition is proportionate to the circulating glucocorticoid level. The inhibitory effect is exerted at both the pituitary and the hypothalamic levels. The inhibition is due to an action on DNA, and maximal inhibition takes several hours to develop, although there is evidence for an additional, more rapid "fastfeedback." The ACTH-inhibiting activity of the various steroids parallels their glucocorticoid potency. A drop in resting corticoid levels stimulates ACTH secretion, and in chronic adrenal insufficiency there is a marked increase in the rate of ACTH synthesis and secretion.

It was once thought that stress stimulated ACTH

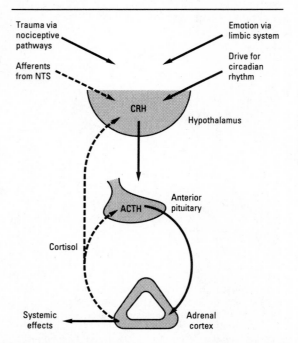

Figure 20–21. Feedback control of the secretion of cortisol and other glucocorticoids. The dashed arrows indicate inhibitory effects, and the solid arrows indicate stimulatory effects. NTS, nucleus tractus solitarius. Compare with Figs 18–14, 22–11, 23–22, and 23–36.

secretion by initially lowering the plasma corticoid level, but this has been shown to be incorrect. In fact, an untreated adrenalectomized animal whose circulating corticoid level is zero responds to stress with a much greater than normal rise in plasma ACTH. Thus, it seems clear that the rate of ACTH secretion is determined by 2 opposing forces: the sum of the neural and possibly other stimuli converging through the median eminence to increase ACTH secretion, and the magnitude of the braking action of glucocorticoids on ACTH secretion, which is proportionate to their level in the circulating blood (Fig 20–21).

The dangers involved when prolonged treatment with anti-inflammatory doses of glucocorticoids is stopped deserve emphasis. Not only is the adrenal atrophic and unresponsive after such treatment; even if its responsiveness is restored by injecting ACTH, the pituitary may be unable to secrete normal amounts of ACTH for as long as a month. The cause of the deficiency is presumably diminished ACTH synthesis. Thereafter, there is a slow rise in ACTH to supranormal levels. These in turn stimulate the adrenal, and glucocorticoid output rises, with feedback inhibition gradually reducing the elevated ACTH levels to normal (Fig 20–22). The complications of sudden cessation of steroid therapy can usually be avoided by slowly decreasing the steroid dose over a long period.

EFFECTS OF MINERALOCORTICOIDS

Actions

Aldosterone and other steroids with mineralocorticoid activity increase the reabsorption of Na^+ from the urine, sweat, saliva, and gastric juice.

Sodium ions diffuse out of the urine (or saliva, sweat, or gastric juice) into the surrounding epithelial cells and are actively transported from these cells into the interstitial fluid. The amount of Na^+ removed from these fluids is proportionate to the rate of active transport of Na^+. Thus, mineralocorticoids cause retention of Na^+ in the ECF. In the kidney, they act primarily on the epithelium of the cortical collecting ducts. They may also increase the K^+ and decrease the Na^+ levels in muscle and brain cells. Under the influence of aldosterone, increased amounts of Na^+ are in effect exchanged for K^+ and H^+ in the renal tubules, producing a K^+ diuresis (Fig 20–23) and an increase in urine acidity.

Mechanism of Action

Aldosterone binds to the mineralocorticoid receptor, one of a family of intracellular steroid receptors (see Fig 1–32). This exposes the DNA-binding domain of the receptor, and the subsequent binding to DNA promotes the formation of mRNAs that in turn foster the formation of new proteins (see Fig 1–31). The principal new protein that is produced is Na^+-K^+ ATPase, thus providing the target cells with more Na^+ pumps. Aldosterone also appears to increase the abundance of amiloride-inhibitable Na^+ channels in the membranes of target cells and to increase the H^+ pump activity in cortical collecting ducts. In addition, there may be a small number of Na^+-H^+ antiports (see

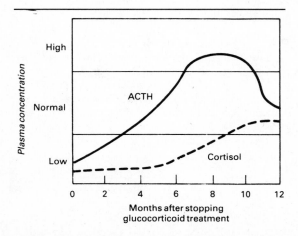

Figure 20–22. Pattern of plasma ACTH and cortisol values in patients recovering from prior long-term daily treatment with large doses of glucocorticoids. (Courtesy of R Ney.)

Figure 20–23. Effect of aldosterone (5 µg as a single dose injected into the aorta) on electrolyte excretion in an adrenalectomized dog.

Fig 38–21) in the distal as well as the proximal portions of the nephrons, and the activity of these distal antiports may be increased by aldosterone.

The process of new protein synthesis takes 10–30 minutes, and this is why it takes 10–30 minutes for aldosterone to start exerting its effect on Na^+ excretion, even when it is injected directly into the renal artery (Fig 20–23).

Relation of Mineralocorticoid to Glucocorticoid Receptors

It is intriguing that in vitro, the mineralocorticoid receptor has an appreciably higher affinity for glucocorticoids than the glucocorticoid receptor does. This raises the question of why glucocorticoids do not bind to the mineralocorticoid receptors in the kidneys and other locations and produce mineralocorticoid effects. The answer is that the kidneys and other mineralocorticoid-sensitive tissues also contain an enzyme, **11β-hydroxysteroid dehydrogenase,** which catalyzes the conversion of cortisol to cortisone (Fig 20–13) and corticosterone to its 11-oxy derivative, but leaves aldosterone untouched. The 11-oxy derivatives do not bind to the receptor. The target tissues that lack this enzyme are the heart and part of the brain.

This arrangement has significant clinical implications. Obviously, if the 11β-hydroxysteroid dehydrogenase were congenitally absent or inhibited, one would expect cortisol to have marked mineralocorticoid effects because its concentration in plasma is so much greater than the concentration of aldosterone. Congenital absence of the enzyme has been reported, and the condition that is produced is the clinical picture of hyperaldosteronism (see below). In addition, licorice produces Na^+ retention and hypertension, and glycyrrhetinic acid, which is found in licorice, inhibits the enzyme.

Other Mineralocorticoids

Aldosterone is the principal mineralocorticoid secreted by the adrenal, although corticosterone is secreted in sufficient amounts to exert a minor mineralocorticoid effect (Tables 20–1 and 20–2). Deoxycorticosterone, which is secreted in appreciable amounts only in abnormal situations, has about 3% of the activity of aldosterone. It is used clinically as a mineralocorticoid because its synthetic acetate (**desoxycorticosterone acetate, DOCA**) is cheaper and more readily available than aldosterone. Large amounts of progesterone and some other steroids cause natriuresis, but there is little evidence that they play any normal role in the control of Na^+ excretion.

Effect of Adrenalectomy

In adrenal insufficiency, Na^+ is lost in the urine. K^+ is retained, and the plasma K^+ rises. When adrenal insufficiency develops rapidly, the amount of Na^+ lost from the extracellular fluid exceeds the amount

excreted in the urine, indicating that Na^+ also must be entering the cells. When the posterior pituitary is intact, salt loss exceeds water loss, and the plasma Na^+ falls (Table 20–4). However, the plasma volume also is reduced, resulting in hypotension, circulatory insufficiency, and, eventually, fatal shock. These changes can be prevented to a degree by increasing the dietary NaCl intake. Rats survive indefinitely on extra salt alone, but in dogs and most humans, the amount of supplementary salt needed is so large that it is almost impossible to prevent eventual collapse and death unless mineralocorticoid treatment is also instituted.

Secondary Effects of Excess Mineralocorticoids

K^+ depletion due to prolonged K^+ diuresis is a prominent feature of prolonged mineralocorticoid treatment (Table 20–4). When the K^+ loss is marked, intracellular K^+ is replaced by Na^+. This shift can be prevented by the administration of supplemental K^+ along with the mineralocorticoid. There is no net increase in body sodium if salt intake is restricted to the level of output, but on a normal diet, total body sodium rises. However, the plasma Na^+ is elevated only slightly if at all, because water is retained with the osmotically active sodium ions. Consequently, ECF volume is expanded and the blood pressure rises. When the ECF expansion passes a certain point, Na^+ excretion is usually increased in spite of the continued action of mineralocorticoids on the renal tubules. This **escape phenomenon** (Fig 20–24) is probably due to increased secretion of ANP (see Chapter 24). Because of increased excretion of Na^+ when the ECF volume is expanded, mineralocorticoids do not produce edema in normal individuals and patients with hyperaldosteronism. However, escape may not occur in certain disease states, and in these situations, continued expansion of ECF volume leads to edema (see Chapter 38).

Primary Hyperaldosteronism

The effects of chronic mineralocorticoid excess are seen in patients with aldosterone-secreting tumors of the adrenal cortex that produce the syndrome of **primary hyperaldosteronism** (Conn's syndrome). These patients are severely K^+-depleted. They are hypertensive, and their ECF volume is expanded, but

Table 20–4. Typical plasma electrolyte levels in normal humans and patients with adrenocortical diseases.

State	Plasma Electrolytes (meq/L)			
	Na^+	K^+	Cl^-	HCO_3^-
Normal	142	4.5	105	25
Adrenal insufficiency	120	6.7	85	25
Primary hyper-aldosteronism	145	2.4	96	41

Figure 20–24. "Escape" from the sodium-retaining effect of desoxycorticosterone acetate (DOCA) in an adrenalectomized patient. ECF, extracellular fluid volume; TBV, total blood volume; RCV, red cell volume; PV, plasma volume. (Courtesy of EG Biglieri.)

they are not edematous or markedly hypernatremic, because the escape phenomenon discussed in the preceding paragraph causes the excretion of extra ingested salt.

Prolonged K^+ depletion damages the kidneys, resulting in a loss of concentrating ability and polyuria **(hypokalemic nephropathy)**. The K^+ depletion also causes muscle weakness and metabolic alkalosis (see Chapter 38), and the alkalosis lowers the plasma Ca^{2+} level to the point where latent or even frank tetany is present (see Chapter 21). The K^+ depletion causes a minor but detectable decrease in glucose tolerance that is corrected by K^+ repletion (see Chapter 19).

REGULATION OF ALDOSTERONE SECRETION

Stimuli

The principal stimuli that increase aldosterone secretion are summarized in Table 20–5. Some of them also increase glucocorticoid secretion; others selectively affect the output of aldosterone. The primary regulatory factors involved are ACTH from the pituitary, renin from the kidney via angiotensin II, and a

direct stimulatory effect of a rise in plasma K^+ concentration—or a drop in plasma Na^+ concentration—on the adrenal cortex.

Effect of ACTH

When first administered, ACTH stimulates the output of aldosterone as well as that of glucocorticoids and sex hormones. Although the amount of ACTH required to increase aldosterone output is somewhat greater than the amount that stimulates maximal glucocorticoid secretion (Fig 20–25), it is

Table 20–5. Stimuli that increase aldosterone secretion.

Glucocorticoid secretion also increased
  Surgery
  Anxiety
  Physical trauma
  Hemorrhage
Glucocorticoid secretion unaffected
  High potassium intake
  Low sodium intake
  Constriction of inferior vena cava in thorax
  Standing
  Secondary hyperaldosteronism (in some cases of congestive heart failure, cirrhosis, and nephrosis)

Figure 20–25. Changes in adrenal venous output of steroids produced by ACTH in nephrectomized hypophysectomized dogs.

Figure 20–26. Changes in adrenal venous output of steroids produced by angiotensin II in nephrectomized hypophysectomized dogs.

well within the range of endogenous ACTH secretion. The effect is transient, and even if ACTH secretion remains elevated, aldosterone output begins to decline in 1 or 2 days. On the other hand, the output of the mineralocorticoid deoxycorticosterone remains elevated. The decline in aldosterone output is partly due to decreased renin secretion secondary to hypervolemia (see below), but it is possible that some other factor also decreases the conversion of corticosterone to aldosterone. After hypophysectomy, the basal rate of aldosterone secretion is normal. The increase normally produced by surgical and other stresses is absent, but the increase produced by dietary salt restriction is unaffected for some time. Later on, atrophy of the zona glomerulosa complicates the picture in long-standing hypopituitarism, and this may lead to salt loss and hypoaldosteronism (see above).

Effects of Angiotensin II & Renin

The octapeptide angiotensin II is formed in the body from angiotensin I, which is liberated by the action of renin on a circulating α_2 globulin (see Chapter 24). Injections of angiotensin II stimulate adrenocortical secretion and, in small doses, affect primarily the secretion of aldosterone (Fig 20–26). The sites of action of angiotensin II are both early and late in the steroid biosynthetic pathway. The early action is on the conversion of cholesterol to pregnenolone, and the late action is on the conversion of corticosterone to aldosterone. Angiotensin II does not increase the secretion of deoxycorticosterone, which is controlled by ACTH.

Renin is secreted from the juxtaglomerular cells that surround the renal afferent arterioles as they enter the glomeruli (see Chapter 24). Aldosterone secretion is regulated via the renin-angiotensin system

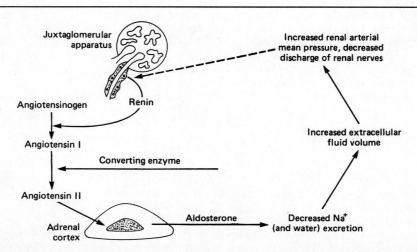

Figure 20–27. Feedback mechanism regulating aldosterone secretion. The dashed arrow indicates inhibition.

in a feedback fashion (Fig 20–27). A drop in ECF volume or intra-arterial vascular volume leads to a reflex increase in renal nerve discharge and decreases renal arterial pressure. Both changes increase renin secretion, and the angiotensin II formed by the action of the renin increases the rate of secretion of aldosterone. The aldosterone causes Na$^+$ and water retention, expanding ECF volume and shutting off the stimulus that initiated increased renin secretion.

Hemorrhage stimulates ACTH and renin secretion. Like hemorrhage, standing and constriction of the thoracic inferior vena cava decrease intra-arterial vascular volume. Dietary sodium restriction also increases aldosterone secretion via the renin-angiotensin system (Fig 20–28). Such restriction reduces ECF volume, but aldosterone and renin secretion is increased before there is any consistent decrease in blood pressure. Consequently, the initial increase in renin secretion produced by dietary sodium restriction is probably due to reflex increases in the activity of the renal nerves. Sodium depletion increases the binding of angiotensin II to the adrenal cortex, first by increasing the affinity of angiotensin II receptors and then by increasing their number. At the same time, there is a decrease in angiotensin II binding to blood vessels.

Effects of Electrolytes

It seems unlikely that changes in plasma Na$^+$ are a major factor in the regulation of aldosterone secretion, since, at least when it occurs acutely as an isolated change, the plasma Na$^+$ concentration must drop about 20 meq/L to produce stimulation. Changes of this magnitude are unusual. However, the plasma K$^+$ level need increase only 1 meq/L or less to stimulate aldosterone secretion, and transient increases of this magnitude may occur after a meal, particularly if it is rich in K$^+$. Like angiotensin II, K$^+$ stimulates the conversion of cholesterol to pregnenolone and the conversion of corticosterone to aldosterone. It appears to act by increasing intracellular Ca^{2+}. The sensitivity of the zona glomerulosa to angiotensin II and consequently to a low-sodium diet is decreased by a low-potassium diet. However, in most clinical situations, K$^+$ depletion is produced by an aldosterone-producing tumor or is secondary to hyperaldosteronism from some other cause.

In normal individuals, there is an increase in plasma aldosterone concentration during the portion of the day that the individual is carrying on activities in the upright position. This increase is due to a decrease in the rate of removal of aldosterone from the circulation by the liver and an increase in aldosterone secretion due to a postural increase in renin secretion. Individuals who are confined to bed show a circadian rhythm of aldosterone and renin secretion, with the highest values in the early morning before awakening.

Figure 20–28. Effect of low-, normal-, and high-sodium diets on sodium metabolism and plasma renin activity, aldosterone, vasopressin, and ANP in normal humans. (Data from Sagnella GA et al: Plasma atrial natriuretic peptide: Its relationship to changes in sodium intake, plasma renin activity, and aldosterone in man. *Clin Sci* 1987;**72**:25.)

Other Factors

ANP inhibits renin secretion and decreases the responsiveness of the zona glomerulosa to angiotensin II (see Chapter 24). From time to time, there have been reports that other factors affect aldosterone secretion, and there is evidence that a pituitary hormone other than ACTH helps maintain the responsiveness of the zona glomerulosa. However, there is as yet little evidence that these factors play a significant role in the day-to-day regulation of renin secretion.

The mechanisms by which ACTH, angiotensin II, and K+ stimulate aldosterone secretion are summarized in Table 20–6.

Clinical Correlates

Circulating renin and angiotensin II levels are elevated in many patients with congestive heart failure, nephrosis, or cirrhosis of the liver, and this in turn increases aldosterone secretion (**secondary hyperaldosteronism**). Increased renin secretion is also found in individuals with the salt-losing form of the adrenogenital syndrome (see above), presumably because their ECF volume is low. In patients with elevated renin secretion due to renal artery constriction, aldosterone secretion is increased; in those in whom renin secretion is not elevated, aldosterone secretion is normal. The relationship of aldosterone to hypertension is discussed in Chapter 33.

ROLE OF MINERALOCORTICOIDS IN THE REGULATION OF SALT BALANCE

Variation in aldosterone secretion is only one of many factors affecting Na+ excretion. Other major factors include the glomerular filtration rate, ANP, the presence or absence of osmotic diuresis, and changes in tubular reabsorption of Na+ independent of aldosterone. It takes some time for aldosterone to act. Thus, when one rises from the supine to the standing position, aldosterone secretion is increased and there is Na+ retention; but the decrease in Na+ excretion develops too rapidly to be explained solely by increased aldosterone secretion. The primary function of the aldosterone-secreting mechanism is probably the defense of intravascular volume, but it is only one of the homeostatic mechanisms involved.

SUMMARY OF THE EFFECTS OF ADRENOCORTICAL HYPER- & HYPOFUNCTION IN HUMANS

Recapitulating the manifestations of excess and deficiency of the adrenocortical hormones in humans is a convenient way to summarize the multiple and complex actions of these steroids. A characteristic

clinical syndrome is associated with excess secretion of each of the types of hormones. Excess androgen secretion causes masculinization (**adrenogenital syndrome**) and precocious pseudopuberty or female pseudohermaphroditism. Feminizing estrogen-secreting adrenal tumors are sometimes seen. Excess glucocorticoid secretion produces a moon-faced, plethoric appearance, with trunk obesity, purple abdominal striae, hypertension, osteoporosis, protein depletion, mental abnormalities, and, frequently, diabetes mellitus (**Cushing's syndrome**). Excess mineralocorticoid secretion leads to K+ depletion and Na+ retention, usually without edema but with weakness, hypertension, tetany, polyuria, and hypokalemic alkalosis (**Conn's syndrome**).

If hormone treatment is stopped in a surgically adrenalectomized patient, Na+ excretion is increased and Na+ enters the cells. The plasma volume declines, the blood pressure falls, and fatal shock develops in a few days if treatment is not reinstituted. These changes can be prevented by mineralocorticoids. Many adrenalectomized patients, provided they eat regularly, can be maintained for prolonged periods on mineralocorticoid replacement alone, but their existence is a precarious one. Fasting causes fatal hypoglycemia, and any stress causes collapse. Water is retained, and there is always the danger of water intoxication. Circulating ACTH levels are elevated. The diffuse tanning of the skin and the spotty pigmentation characteristic of chronic glucocorticoid

Table 20–6. Second messengers involved in the regulation of aldosterone secretion.

Secretogague	Intracellular mediator
ACTH	Cyclic AMP, protein kinase A
Angiotensin II	Diacylglycerol, protein kinase C
K+	Ca2+

Figure 20–29. Pigmentation in Addison's disease. **A,** tan and vitiligo; **B,** pigmentation of scars from lesions that occurred after the development of the disease; **C,** pigmentation of skin creases; **D,** darkening of areolas; **E,** pigmentation of pressure points; **F,** pigmentation of the gums. (Reproduced, with permission, from Forsham PH, Di Raimondo VC: *Traumatic Medicine and Surgery for the Attorney.* Butterworth, 1960.)

deficiency (Fig 20–29) are due, at least in part, to the MSH activity of the ACTH in the blood (see Chapter 22). Minor menstrual abnormalities occur in women, but the deficiency of adrenal sex hormones usually has little effect in the presence of normal testes or ovaries.

Adrenocortical insufficiency is called **Addison's disease.** It can be caused by adrenocortical atrophy due to autoimmune disease or by destruction of the adrenal glands in diseases such as tuberculosis or cancer. Total adrenal insufficiency is rapidly fatal, but in Addison's disease, destruction of the adrenals is usually incomplete. Therefore, patients with this disease have time to develop marked pigmentation and a decrease in cardiac size that is apparently secondary to chronic hypotension and a decrease in cardiac work. They often get along fairly well until some minor stress precipitates a collapse **(addisonian crisis)** requiring emergency medical treatment.

Cases of isolated aldosterone deficiency have also been reported in patients with renal disease and a low circulating renin level **(hyporeninemic hypoaldosteronism).** In addition, **pseudohypoaldosteronism** is produced when there is resistance to the action of aldosterone. Patients with these syndromes have marked hyperkalemia, salt wasting, and hypotension, and they may develop metabolic acidosis.

Hormonal Control of Calcium Metabolism & the Physiology of Bone

INTRODUCTION

Three hormones are primarily concerned with the regulation of calcium metabolism. **1,25-Dihydroxycholecalciferol** is a steroid hormone formed from vitamin D by successive hydroxylations in the liver and kidneys. Its primary action is to increase calcium absorption from the intestine. **Parathyroid hormone** is secreted by the parathyroid glands. Its main action is to mobilize calcium from bone and increase urinary phosphate excretion. **Calcitonin,** a calcium-lowering hormone that in mammals is secreted primarily by cells in the thyroid gland, inhibits bone resorption. Although the role of calcitonin seems to be relatively minor, all 3 hormones probably operate in concert to maintain the constancy of the Ca^{2+} level in the body fluids. Glucocorticoids, growth hormone, estrogens, and various growth factors also affect calcium metabolism.

CALCIUM & PHOSPHORUS METABOLISM

Calcium

The adult human body contains about 1100 g (27.5 mol) of calcium (1.5% of body weight). Ninety-nine percent of the calcium is in the skeleton. The plasma calcium, normally about 10 mg/dL (5 meq/L, 2.5 mmol/L), is partly bound to protein and partly diffusible (Table 21–1). The distribution of calcium inside cells is discussed in Chapter 1.

It is the free, ionized calcium in the body fluids that is a vital second messenger (see Chapter 1) and is necessary for blood coagulation, muscle contraction, and nerve function. A decrease in extracellular Ca^{2+} at the myoneural junction inhibits transmission (see Chapter 4), but this effect is overbalanced by the excitatory effect of a low Ca^{2+} level on the nerve and muscle cells (see Chapter 2). The result is **hypocalcemic tetany,** which is due to increased activity of the motor nerve fibers. This condition is characterized by extensive spasms of skeletal muscle, involving especially the muscles of the extremities and the larynx.

Laryngospasm becomes so severe that the airway is obstructed and fatal asphyxia is produced. Ca^{2+} plays an important role in clotting (see Chapter 27) and many different aspects of cell function (see Chapter 1); in vivo, however, the level of plasma Ca^{2+} at which fatal tetany occurs is still above the level at which clotting defects would occur.

Since the extent of Ca^{2+} binding by plasma proteins is proportionate to the plasma protein level, it is important to know the plasma protein level when evaluating the total plasma calcium. Plasma ionized calcium can be measured by use of a calcium-sensitive electrode. Other electrolytes and pH affect the Ca^{2+} level. Thus, for example, symptoms of tetany appear at much higher total calcium levels if the patient hyperventilates, increasing plasma pH. Plasma proteins are more ionized when the pH is high, providing more protein anion to bind with Ca^{2+}.

The calcium in bone is of 2 types: a readily exchangeable reservoir and a much larger pool of stable calcium that is only slowly exchangeable. There are 2 independent but interacting homeostatic systems affecting the calcium in bone. One is the system that regulates plasma Ca^{2+}, and in the operation of this system, about 500 mmol of Ca^{2+} per day moves into and out of the readily exchangeable pool in the bone (Fig 21–1). The other system is the one concerned with bone remodeling by the constant interplay of bone resorption and deposition (see below), which, in the adult, accounts for 95% of bone formation. However, the Ca^{2+} interchange between plasma and this stable pool of bone calcium is only about 7.5 mmol/d.

A large amount of calcium is filtered in the kidneys, but 98–99% of the filtered calcium is reabsorbed. About 60% of the reabsorption occurs in the proximal tubules and the remainder in the ascending limb of the loop of Henle and the distal tubule. Distal tubular reabsorption is regulated by parathyroid hormone.

The absorption of Ca^{2+} from the gastrointestinal tract is discussed in Chapter 25. Ca^{2+} is actively transported out of the intestine by a system in the brush border of the epithelial cells that involves a calcium-

Table 21–1. Distribution (mmol/L) of calcium in normal human plasma.

Total diffusible	1.34
Ionized (Ca^{2+})	1.18
Complexed to HCO_3^-, citrate, etc	0.16
Total nondiffusible (protein-bound)	1.16
Bound to albumin	0.92
Bound to globulin	0.24
Total plasma calcium	2.50

dependent ATPase, and this process is regulated by 1,25-dihydroxycholecalciferol (see below). There is also some absorption by passive diffusion. When Ca^{2+} intake is high, the active transport mechanism becomes saturated. In addition, 1,25-dihydroxycholecalciferol levels fall in the presence of increased plasma Ca^{2+}. Consequently, Ca^{2+} absorption undergoes adaptation; ie, it is high when the calcium intake is low and decreased when the calcium intake is high. Calcium absorption is also decreased by substances that form insoluble salts with Ca^{2+} (eg, phosphates and oxalates) or by alkalis, which favor formation of insoluble calcium soaps. A high-protein diet increases absorption in adults.

Phosphorus

Phosphorus metabolism is not as finely regulated as calcium metabolism, but phosphorus is found in ATP, cyclic AMP, 2,3-diphosphoglycerate, and other vital compounds in the body. Total body phosphorus is 500–800 g (16.1–25.8 mol), 85–90% of which is in the skeleton. Total plasma phosphorus is about 12 mg/dL, with two-thirds of this total in organic compounds and the remaining inorganic phosphorus (Pi) mostly in PO_4^{3-}, HPO_4^{2-}, and $H_2PO_4^-$.

The amount of phosphorus normally entering bone is about 3 mg (97 μmol)/kg/d, with an equal amount leaving via reabsorption.

The Pi in the plasma is filtered in the glomeruli, and 85–90% of the filtered Pi is reabsorbed. Active transport in the proximal tubule accounts for most of the reabsorption, and this active transport process is powerfully inhibited by parathyroid hormone (see below).

Pi is absorbed in the duodenum and small intestine by both active transport and passive diffusion. However, unlike the absorption of Ca^{2+}, the absorption of Pi is linearly proportionate to dietary intake. Many stimuli that increase Ca^{2+} absorption, including 1,25-dihydroxycholecalciferol, also increases Pi absorption.

BONE PHYSIOLOGY

Structure of Bone

Bone is a special form of connective tissue made up of a collagenous matrix that has been impregnated with mineral salts, especially phosphates of calcium. Because of its high calcium and phosphate content, it plays a vital role in calcium homeostasis (see below). It protects vital organs, and the rigidity that it provides permits locomotion and the support of loads

Figure 21–1. Calcium metabolism in an adult human ingesting 25 mmol (1000 mg) of calcium per day.

against gravity. It is a living tissue; bone is constantly being resorbed and new bone formed (see below), permitting it to respond to the stresses and strains that are put upon it. It is well vascularized and has a total blood flow of 200–400 mL/min in adult humans.

The protein in bone matrix is mostly type I collagen, which is also the major structural protein in tendons and skin. This collagen, which weight for weight is as strong as steel, is made up of a triple helix of 3 polypeptides bound tightly together. Two of these are identical α_1 polypeptides encoded by one gene, and one is an α_2 polypeptide encoded by a different gene. Collagens represent a family of structurally related proteins that maintain the integrity of many different organs. To date, 15 different types have been identified, and these are encoded by more than 20 different genes.

Adequate amounts of both protein and minerals must be available for the maintenance of normal bone structure. Mineral in bone is mostly in the form of **hydroxyapatites,** which have the general formula $Ca^{2+}_{10-x}(H_3O^+)_{2x} \cdot (PO_4^{3-})_6(OH^-)_2$. These salts form crystals that measure 20 by 3–7 nm. Sodium and small amounts of magnesium and carbonate are also present in bone.

Most bones are made up of an outer layer of compact bone surrounding trabecular bone and, in many instances, a bone marrow cavity (Fig 21–2). **Trabecular** or **spongy bone** is made up of bone spicules separated by spaces. **Compact bone** is much denser and is less active metabolically. About 75% of the bone in the body is compact, and 25% is trabecular. In spongy bone, nutrients diffuse from bone ECF into the trabeculae, but in compact bone, nutrients are provided via **Haversian canals** (Fig 21–3), which contain blood vessels. Around each Haversian canal, collagen is arranged in concentric layers, forming cylinders called **osteons** or **Haversian systems.**

The cells in bone that are concerned primarily with bone formation and resorption are osteoblasts, osteocytes, and osteoclasts. **Osteoblasts** are the bone-forming cells that secrete collagen, forming a matrix around themselves which then calcifies. When surrounded by calcified matrix, they are called **osteocytes** (Fig 21–3). They send processes into the canaliculi that ramify throughout the bone and, via tight junctions, connect to the processes of other osteocytes.

Osteoclasts are multinuclear cells that erode and resorb previously formed bone. They are derived from hematopoietic stem cells via monocytes (see Chapter 27). They attach to bone by membrane extensions that surround an isolated area between the bone and a portion of the osteoclast. Proton pumps, which are H⁺-dependent ATPases, then move from en-

Figure 21–2. Structure of a typical long bone before *(left)* and after *(right)* epiphyseal closure. (Modified from Junqueira LC, Carneiro J, Kelley RO: *Basic Histology,* 7th ed. Appleton & Lange, 1992.)

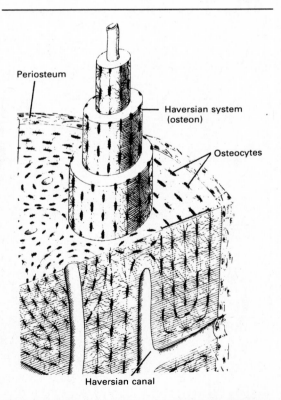

Figure 21–3. Schematic drawing of compact bone. (Modified from Junqueira LC, Carneiro J, Kelley RO: *Basic Histology,* 7th ed. Appleton & Lange, 1992.)

dosomes into the cell membrane apposed to the isolated area, and they acidify the area to approximately pH 4. Similar proton pumps are found in the endosomes and lysosomes of all eukaryotic cells (see Chapter 1), but in only a few other instances do they move into the cell membrane. Note in this regard that the sealed-off space formed by the osteoclast resembles a large lysosome. The acid pH dissolves hydroxyapatite, and acid proteases secreted by the cell dissolve collagen.

Osteoblasts, on the other hand, develop from pluripotent mesenchymal fibroblasts that are also the progenitor stem cells for chondrocytes and other types of cells. Osteoblasts remain in contact with one another via tight junctions between long protoplasmic processes that run through canaliculi in the bone.

Bone Growth

The bones of the skull are formed by ossification of membranes (**intramembranous** bone formation). The long bones are first modeled in cartilage and then transformed into bone by ossification that begins in the shaft and in the ends of the bone (**enchondral** bone formation). Osteoblasts form a network of collagen fibers. This matrix then calcifies. In the meantime, the cartilage in the center of the shaft is invaded by osteoclasts that erode it away.

During growth, specialized areas at the ends of each long bone (**epiphyses**) are separated from the shaft of the bone by a plate of actively proliferating cartilage, the **epiphyseal plate** (Fig 21–2). The bone increases in length as this plate lays down new bone on the end of the shaft. The width of the epiphyseal plate is proportionate to the rate of growth. The width is affected by a number of hormones but most markedly by the pituitary growth hormone and IGF-I (see Chapter 22). Changes in the width of the tibial epiphyseal plate are used as the end point in the "tibia test," a bioassay that has been used to measure growth hormone.

Linear bone growth can occur as long as the epiphyses are separated from the shaft of the bone, but such growth ceases after the epiphyses unite with the shaft (**epiphyseal closure**). The epiphyses of the various bones close in an orderly temporal sequence, the last epiphyses closing after puberty. The normal age at which each of the epiphyses closes is known, and the "bone age" of an individual can be determined by x-raying the skeleton and noting which epiphyses are open and which are closed.

Bone Formation & Resorption

Throughout life, bone is being constantly resorbed and new bone being formed. The calcium in bone turns over at a rate of 100% per year in infants and 18% per year in adults. Bone remodeling is a local process carried out in small areas sometimes called "units," in which osteoclasts first absorb bone and then osteoblasts lay down new bone in the same gen-

eral area. This cycle takes about 100 days. About 5% of the bone mass is being remodeled by about 2 million bone-remodeling units in the human skeleton at any one time. The renewal rate for bone is about 4% per year for compact bone and 20% per year for trabecular bone. The remodeling is related in part to the stresses and strains imposed on the skeleton by gravity and other factors and is regulated by hormones in the systemic circulation and by growth factors, most of which appear to act locally. The principal growth factors found in bone are listed in Table 21–2. IGF-I stimulates the synthesis of bone collagen and acts in conjunction with growth hormone to cause proliferation of cartilage in epiphyseal plate (see Chapter 22). IL-1β activates osteoclasts. The actions of the other growth factors are complex, sometimes biphasic, and incompletely understood. However, it seems clear that the factors are involved, probably in a paracrine fashion, in bone growth and the remodeling cycle. Prostaglandins may be involved as well, but their effects on bone are also complex.

The details of the process responsible for calcification of newly formed bone matrix are also unclear despite intensive study. Whether calcium phosphate precipitates out of a solution depends upon the product of concentrations of Ca^{2+} and PO_4^{3-}. At a certain value for this product (the **solubility product**), the solution is saturated. Whenever $[Ca^{2+}] \cdot [PO_4^{3-}]$ exceeds the solubility product, calcium phosphate precipitates. Associated with the osteoblasts in bone is an alkaline phosphatase that hydrolyzes phosphate esters. The phosphate liberated by ester hydrolysis increases the concentration of phosphate in the vicinity of the osteoblasts to a point where the solubility product is exceeded and calcium phosphate precipitates. This mechanism may be involved in the calcification process, but it is almost certainly only part of the story.

A number of the bone proteins in addition to collagen have been isolated and characterized. Matrix

Table 21-2. Growth factors that are found in bone or that act on bone cells in culture.[1]

Fibroblast growth factors (FGF)
  Acidic
  Basic
Transforming growth factors (TGF)
  TGFα
  TGFβ$_1$
  TGFβ$_2$
  TGFβ$_{1.2}$
Insulinlike growth factors (IGF)
  IGF-I
  IGF-II
β$_2$-Microglobulin (β$_2$m)
Platelet-derived growth factor (PDGF)
Interleukin-1β (IL-1β)
Gamma interferon (IFN-γ)

[1] Modified from Canalis E: Bone–related growth factors. *Triangle, Sandoz J Med Sci* 1988;**27**:11.

Gla protein (MGP) and bone Gla protein (BGP, osteocalcin) both contain γ-carboxylated glutamic acid residues (Gla), and γ-carboxylation is catalyzed by vitamin K (see Chapter 17). The Gla residues bind Ca²⁺, but vitamin K deficiency only causes skeletal abnormalities in the fetus. Two additional proteins, osteonectin and osteopontin, are synthesized by osteoblasts. Synthesis of osteocalcin and osteopontin increases as calcification begins. However, the exact functions of all these proteins in bone remain to be determined.

Uptake of Other Minerals

Lead and some other toxic elements are taken up and released by bone in a manner similar to that in which calcium is turned over. The rapid bone uptake of these elements is sometimes called a "detoxifying mechanism" because it serves to remove them from the body fluids, thus ameliorating the toxic manifestations. The radioactive elements radium and plutonium (and the radioactive isotope of strontium that is a by-product of atomic bomb explosions) are also taken up by bone. In this situation, the uptake is definitely harmful, because radiation from these elements may cause malignant degeneration of bone cells and the formation of osteogenic sarcomas. Fluoride taken up by bone causes new bone to form. It is also incorporated into the enamel of the teeth, and small amounts of fluoride increase resistance to dental caries. However, large amounts cause discoloration of the enamel (**mottled enamel**).

Bone Disease

The diseases produced by selective abnormalities of the cells and processes discussed above illustrate the interplay of factors that maintain normal bone function. Various mutations in the genes that code for collagen make bones brittle, causing **osteogenesis imperfecta** (brittle bones disease). Other mutations in collagen genes cause the various chondrodysplasias and at least one form of Ehlers-Danlos syndrome. It is possible that some cases of osteoporosis (see below), osteoarthritis, and aortic aneurysm are also caused by mutated collagen genes. In **osteopetrosis,** another rare and often severe disease, the osteoclasts are defective and are unable to resorb bone in their usual fashion. The result is a steady increase in bone density, neurologic defects due to narrowing and distortion of foramina through which nerves normally pass, and hematologic abnormalities due to crowding out of the marrow cavities. The condition in which the amount of calcium accretion per unit bone matrix is deficient is called **rickets** in children and **osteomalacia** in adults. In **osteoporosis,** matrix and mineral are both lost, and there is a loss of bone mass (Fig 21–4) and strength, with an increased incidence of fractures. This disease, in a sense, is the opposite of osteopetrosis in that it is characterized over time by a net excess of bone resorption over

Figure 21–4. Normal trabecular bone *(left)* compared with trabecular bone from a patient with osteoporosis *(right).*

bone formation. It has multiple causes but is associated mostly with advancing age and the menopause. Involutional osteoporosis is a common disease and has become a major public health problem in the USA and Europe as the number of elderly people in the population has increased.

All normal humans gain bone early in life, during growth. After a plateau, they begin to lose bone as they grow older (Fig 21–5). When this loss is accelerated or exaggerated, as it is in osteoporosis, it leads to an increased incidence of fractures, particularly fractures of the distal forearm (Colles' fracture), vertebral bodies, and hip. All these areas have a high content of trabecular bone, and since trabecular bone is more active metabolically, it is lost more rapidly. Fractures of the vertebras with compression cause kyphosis, with the production of the typical widow's hump that is common in elderly women with osteoporosis. Fractures of the hip in elderly individuals are associated with a mortality rate of 12–20%, and half of those who survive require prolonged, expensive care.

Adult women have less bone mass than adult men, and after menopause they lose it more rapidly than men of comparable age do. Consequently, they are more prone to development of serious osteoporosis. There are estrogen receptors on human osteoblasts, and the cause of the bone loss after menopause is primarily estrogen deficiency. Estrogen treatment arrests the progress of the disease but apparently does not reverse it. It appears that the tendency for estrogen to cause cancer of the endometrium is avoided if small doses are used, but there is some evidence that even small doses increase the incidence of cancer of the breast. Increased intake of calcium, particularly from natural sources such as milk, and moderate exercise may also help prevent or slow the progress of the disease, although their effects are not great. Etidronate, an oral diphosphonate compound that inhibits osteoclastic activity, increases the mineral con-

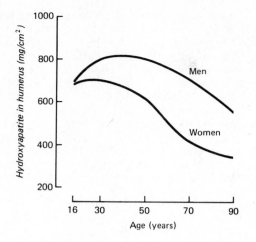

Figure 21–5. Changes in bone mass from adolescence to old age in normal men and women. Curves are drawn from values of hydroxyapatite (mg/cm of the radius) determined by radiodensitometry.

tent of bone when administered in a cyclic fashion and decreases the rate of new vertebral fractures. Fluoride stimulates osteoblasts, making bone more dense, and may also be beneficial in the treatment of the disease.

In patients who are immobilized for any reason,

and during space flight (see Chapter 33), bone resorption exceeds bone formation and disuse osteoporosis develops. The plasma calcium level is not markedly elevated, but plasma concentrations of parathyroid hormone and 1,25-dihydroxycholecalciferol fall and large amounts of calcium are lost in the urine. Osteoporosis also occurs in patients with excess glucocorticoid secretion (Cushing's syndrome; see Chapter 20).

VITAMIN D & THE HYDROXYCHOLECALCIFEROLS

Chemistry

The active transport of Ca^{2+} and PO_4^{3-} from the intestine is increased by a metabolite of **vitamin D.** The term "vitamin D" is used to refer to a group of closely related sterols produced by the action of ultraviolet light on certain provitamins (Fig 21–6). Vitamin D_3, which is also called cholecalciferol, is produced in the skin of mammals from 7-dehydrocholesterol by the action of sunlight. The reaction involves the rapid formation of previtamin D_3, which is then converted more slowly to vitamin D_3 (cholecalciferol). Vitamin D_3 and its hydroxylated derivatives are transported in the plasma bound to a specific globulin vitamin D-binding protein (DBP). The affinity of previtamin D_3 for DBP is low, but the

Figure 21–6. Formation and hydroxylation of vitamin D_3. 25-Hydroxylation takes place in the liver, and the other hydroxylations occur primarily in the kidneys. The formulas of 7-dehydrocholesterol, vitamin D_3, and 1,25-dihydroxycholecalciferol are also shown.

affinity of vitamin D_3 is high, so DBP moves vitamin D_3 from the skin into the circulation. Vitamin D_3 is also ingested in the diet.

In the liver, vitamin D_3 is converted to **25-hydroxycholecalciferol** (calcidiol, 25-OHD$_3$). The 25-hydroxycholecalciferol is converted in the proximal tubules of the kidneys to the more active metabolite **1,25-dihydroxycholecalciferol,** which is also called calcitriol or 1,25-(OH)$_2$D$_3$. The enzyme involved is mitochondrial 1α-hydroxylase, a P450 mixed-function oxidase that is located in the cells of the proximal tubules. 1,25-Dihydroxycholecalciferol is also made in the placenta, in keratinocytes in the skin, and in macrophages. In patients with sarcoidosis, pulmonary alveolar macrophages also produce 1,25-dihydroxycholecalciferol, apparently upon stimulation by γ-interferon. The normal plasma level of 25-hydroxycholecalciferol is about 30 ng/mL, and that of 1,25-dihydroxycholecalciferol is about 0.03 ng/mL (approximately 100 pmol/L). The less active metabolite 24,25-dihydroxycholecalciferol is also formed in the kidneys (Fig 21–6).

Vitamin D_3 and its derivatives are **secosteroids;** ie, they are steroids in which one of the rings has been opened. In this case, it is the B ring (Fig 21–6). 1,25-Dihydroxycholecalciferol is a hormone because it is produced in the body and transported in the bloodstream to produce effects in target cells. Like other steroids, most of it circulates bound to its binding protein, which in this case is DBP.

Mechanism of Action

Since 1,25-dihydroxycholecalciferol is a steroid, it is not surprising that it acts via a receptor which is one of the superfamily of receptors by which steroids, thyroid hormones, and a number of other substances trigger changes in gene expression (see Chapter 1). Binding of the steroid to the receptor exposes a DNA-binding region, and the result in this case is increased transcription of some mRNAs and inhibition of the transcription of others.

1,25-Dihydroxycholecalciferol receptors are found in many tissues other than the intestine, kidneys, and bone. Among them are the skin, lymphocytes, monocytes, skeletal and cardiac muscle, breast, and anterior pituitary gland.

Actions

The mRNAs that are produced in response to 1,25-dihydroxycholecalciferol dictate the formation of a family of **calbindin-D** proteins. These are members of the troponin C superfamily of Ca^{2+}-binding proteins that also includes calmodulin (see Chapter 1). Calbindin-Ds are found in human intestine, brain, and kidneys and in many different tissues in rats. In the intestinal epithelium and many other tissues, 2 calbindins are induced: calbindin-D_{9K}, which has a molecular weight of 9000 and binds 2 Ca^{2+}; and calbindin-D_{28K}, which has a molecular weight of

28,000 and normally binds 4 Ca^{2+} even though it has 6 Ca^{2+}-binding sites. Calbindin-Ds are not vitamin D-dependent, despite their name. In the intestine, increases in calbindin-D_{9K} and calbindin-D_{28K} are correlated with increased Ca^{2+} transport, but the precise way they facilitate Ca^{2+} movement across the intestinal epithelium is still uncertain.

In addition to increasing Ca^{2+} absorption from the intestine, 1,25-dihydroxycholecalciferol facilitates Ca^{2+} reabsorption in the kidneys. It also acts on bone, where it mobilizes Ca^{2+} and PO_4^{3-}, probably by an initial action to release osteoblast factors that activate osteoclasts and subsequently by increasing the number of osteoclasts.

The poor intestinal absorption of Ca^{2+} in vitamin D deficiency often leads to hypocalcemia, and because of the calcium deficiency, the protein of new bone fails to mineralize. The result in children and young animals is the disease called **rickets.** Rickets used to be most commonly due to inadequate intake of the provitamins that are converted to vitamin D_3 by sunlight, or to inadequate exposure to sun, but it can also be due to failure of the liver to produce 25-hydroxycholecalciferol, failure of the kidneys to make 1,25-dihydroxycholecalciferol, or defects in target cell receptors that bind 1,25-dihydroxycholecalciferol.

Evidence is accumulating that 1,25-dihydroxycholecalciferol stimulates the differentiation of immune cells and keratinocytes in the skin. It is interesting in this regard that there is an increased incidence of infections in rickets and that 1,25-dihydroxycholecalciferol has shown promise in the treatment of psoriasis. However, the details of the other actions of 1,25-dihydroxycholecalciferol remain to be determined.

Regulation of Synthesis

The formation of 25-hydroxycholecalciferol does not appear to be stringently regulated, but the formation of 1,25-dihydroxycholecalciferol in the kidneys, which is catalyzed by 1α-hydroxylase, is regulated in a feedback fashion by plasma Ca^{2+} and PO_4^{3-} (Fig 21–7). The hormone acts on intestine and bone to increase plasma Ca^{2+} and PO_4^{3-} levels. Its formation is facilitated by PTH, and when the plasma Ca^{2+} level is low, PTH secretion is increased. When the plasma Ca^{2+} level is high, little 1,25-dihydroxycholecalciferol is produced, and the kidneys produce the relatively inactive metabolite 24,25-dihydroxycholecalciferol instead. This effect of Ca^{2+} on production of 1,25-dihydroxycholecalciferol is the mechanism that brings about adaptation of Ca^{2+} absorption from the intestine (see above). The production of 1,25-dihydroxycholecalciferol is also increased by low and inhibited by high plasma PO_4^{3-} levels, by a direct inhibitory effect of PO_4^{3-} on 1α-hydroxylase. Additional control of 1,25-dihydroxycholecalciferol formation is exerted by a direct negative feedback effect

Figure 21–7. Feedback control of the formation of 1,25-dihydroxycholecalciferol (1,25-[OH]$_2$D$_3$) from 25-hydroxy-cholecalciferol (25-OHD$_3$) in the kidneys. Solid arrows indicate stimulation, and dashed arrows indicate inhibition.

Figure 21–8. The human parathyroid glands, viewed from behind.

of the metabolite on 1α-hydroxylase, a positive feedback action on the formation of 24,25-dihydroxy-cholecalciferol, and a direct action on the parathyroid gland to inhibit the production of the mRNA for PTH. Prolactin increases the activity of 1α-hydroxylase, and circulating 1,25-dihydroxycholecalciferol is increased during lactation. Estrogen increases total circulating 1,25-dihydroxycholecalciferol, but this is probably due to an increase in the secretion of its binding protein without any steady-state change in the free hormone. Hyperthyroidism is associated with decreased circulating 1,25-dihydroxycholecalciferol and an increased incidence of osteoporosis. 1,25-Dihydroxycholecalciferol production is depressed by metabolic acidosis. Growth hormone, hCS, and calcitonin also stimulate 1,25-dihydroxycholecalciferol formation.

THE PARATHYROID GLANDS

Anatomy

In humans, there are usually 4 parathyroid glands: 2 embedded in the superior poles of the thyroid and 2 in its inferior poles (Fig 21–8). However, the location of the individual parathyroids and their number vary considerably. Parathyroid tissue is sometimes found in the mediastinum.

Each parathyroid gland is a richly vascularized disk, about $3 \times 6 \times 2$ mm, containing 2 distinct types of cells. The abundant **chief cells,** which contain a prominent Golgi apparatus plus endoplasmic reticulum and secretory granules (Fig 21–9), synthesize and secrete PTH. The less abundant and larger **oxyphil cells** contain oxyphil granules and large numbers of mitochondria in their cytoplasm. In humans, few are seen before puberty, and thereafter they increase in number with age. Their function is unknown, although some have argued that they are degenerated chief cells.

Synthesis & Metabolism of PTH

Human **parathyroid hormones (PTH)** is a linear polypeptide with a molecular weight of 9500 that contains 84 amino acid residues (Fig 21–10). Its structure is very similar to that of bovine and porcine PTH. It is synthesized as part of a larger molecule containing 115 amino acid residues **(prepro-PTH).** Upon entry of prepro-PTH into the endoplasmic re-

Figure 21–9. Section of human parathyroid. (Reduced 50% from × 960.) Small cells are chief cells; large stippled cells (especially prominent in the lower left of picture) are oxyphil cells. (Reproduced, with permission, from Fawcett DW: *Bloom and Fawcett, A Textbook of Histology,* 11th ed. Saunders, 1986.)

HUMAN– ◯   BOVINE– ⬭   PORCINE– ◿

Figure 21–10. Parathyroid hormone. The structure of human PTH is shown. The symbols above and below the human structure show where amino acid residues are different in bovine and porcine PTH. (Reproduced, with permission, from Keutmann HT et al: Complete amino acid sequence of human parathyroid hormone. *Biochemistry* 1978;**17**:5723. Copyright © by the American Chemical Society.)

ticulum, a leader sequence containing 25 amino acid residues is removed from the N terminal to form the 90-amino-acid polypeptide **pro-PTH.** Six additional amino acid residues are removed from the N terminal of pro-PTH in the Golgi apparatus, and the 84-amino-acid polypeptide PTH is the main secretory product of the chief cells.

The normal plasma level of intact PTH is 10–55 pg/mL. The half-life of PTH is less than 20 minutes, and the secreted polypeptide is rapidly cleaved by the Kupffer cells in the liver into 2 polypeptides, a biologically inactive C-terminal fragment with a molecular weight of 7000 and a biologically active N-terminal fragment with a molecular weight of 2500. It is interesting in this regard that a synthetic polypeptide containing amino acid residues 1–34 of bovine PTH has all the known biologic effects of the full molecule. Further metabolism occurs more slowly. There is some evidence that the 2500- and 7000-molecular-weight fragments are also formed in the chief cells, with secretion of small amounts of each. The fragments may also be produced in the kidneys.

Actions

PTH acts directly on bone to increase bone resorption and mobilize Ca^{2+}. In addition to increasing the plasma Ca^{2+} and depressing the plasma phosphate, PTH increases phosphate excretion in the urine. This **phosphaturic action** is due to a decrease in reabsorption of phosphate in the proximal tubules. PTH also increases reabsorption of Ca^{2+} in the distal tubules, although Ca^{2+} excretion is usually increased in hyperparathyroidism because the increase in the amount filtered overwhelms the effect on reabsorption. PTH also increases the formation of 1,25-dihydroxycholecalciferol, the physiologically active metabolite of vitamin D (see above). It increases Ca^{2+} absorption from the intestine, but this action is apparently due solely to the stimulation of 1,25-dihydroxycholecalciferol production.

Mechanism of Action

The actions of PTH on the bones and the kidneys involve activation of adenylate cyclase by way of a membrane receptor and G_s, with consequent increased formation of cyclic AMP in the affected cells. How cyclic AMP affects calcium in bone is unsettled. The hormone increases the permeability of osteocytes and osteoblasts to the Ca^{2+} in bone fluid, and Ca^{2+} enters these cells. Presumably, the osteoblasts then pump the Ca^{2+} into the ECF. The pump is stimulated by 1,25-dihydroxycholecalciferol. This is why the action of PTH is facilitated by 1,25-dihydroxycholecalciferol without any effect of the vitamin derivative on cyclic AMP. Starting about 12 hours after injection, PTH also promotes increased osteoclastic activity and triggers the formation of more osteoclasts. The cyclic AMP content of osteoblasts is also increased. However, with sustained increases in PTH, bone resorption always exceeds bone formation.

In the disease called **pseudohypoparathyroidism,** the signs and symptoms of hypoparathyroidism develop but the circulating level of PTH is normal or elevated. Since the tissues fail to respond to the hormone, the disease is sometimes classified as a receptor disease. There are 2 forms, both associated with an abnormal response to PTH. In the more common form, there is a congenital 50% reduction of the activity of G_s, one of the nucleotide regulatory proteins associated with adenylate cyclase in cell membranes (see Chapter 1), and PTH fails to produce a normal increase in cyclic AMP concentration. In a different, less common form, the cyclic AMP response is normal but the phosphaturic action of the hormone is defective.

Regulation of Secretion

It has been proved by perfusion experiments that the circulating level of ionized calcium acts directly on the parathyroid glands in a feedback fashion to regulate the secretion of PTH. When the plasma Ca^{2+}

level is high, secretion is inhibited and the Ca^{2+} is deposited in the bones. When it is low, secretion is increased and Ca^{2+} is mobilized from the bones. 1,25-Dihydroxycholecalciferol acts directly on the parathyroid glands to decrease prepro-PTH mRNA. Increased plasma phosphate stimulates PTH secretion by lowering plasma Ca^{2+} and inhibiting the formation of 1,25-dihydroxycholecalciferol. A decrease in plasma Mg^{2+} stimulates parathyroid secretion, and Mg^{2+} may exert a direct inhibitory effect on the gland. Secretion of PTH is also increased by β-adrenergic discharge and cyclic AMP.

Another factor regulating plasma Ca^{2+} is the readily exchangeable bone calcium pool. However, this pool can maintain the total plasma calcium at only about 7 mg/dL in the absence of PTH, and maintenance of the normal level of 10 mg/dL is due to the activity of the parathyroid glands.

Effects of Parathyroidectomy

PTH is essential for life. After parathyroidectomy, there is a steady decline in the plasma calcium level. Signs of neuromuscular hyperexcitability appear, followed by full-blown hypocalcemic tetany (see above). Plasma phosphate levels usually rise as the plasma calcium level falls after parathyroidectomy, but the rise does not always occur.

In humans, tetany is most often due to inadvertent parathyroidectomy during thyroid surgery. Symptoms usually develop 2–3 days postoperatively but may not appear for several weeks or more. In rats on a low-calcium diet, tetany develops much more rapidly, and death occurs 6–10 hours after parathyroidectomy. Injections of PTH correct the chemical abnormalities, and the symptoms disappear. Injections of calcium salts give temporary relief, and 1-hydroxylated derivatives of vitamin D restore plasma Ca^{2+} to normal. 25-Hydroxycholecalciferol has a direct stimulatory effect on Ca^{2+} absorption from the intestine when administered in large doses, and it is sometimes used in the treatment of hypoparathyroidism even though there is little 1-hydroxylation of vitamin D metabolites in the absence of parathyroid hormone.

The signs of tetany in humans include **Chvostek's sign,** a quick contraction of the ipsilateral facial muscles elicited by tapping over the facial nerve at the angle of the jaw; and **Trousseau's sign,** a spasm of the muscles of the upper extremity that causes flexion of the wrist and thumb with extension of the fingers (Fig 21–11). In individuals with mild tetany in whom spasm is not evident, Trousseau's sign can sometimes be produced by occluding the circulation for a few minutes with a blood pressure cuff.

Parathyroid Hormone Excess

Hyperparathyroidism due to injections of large doses of parathyroid extract in animals or hypersecretion of a functioning parathyroid tumor in humans is

Figure 21–11. Position of the hand in hypocalcemic tetany (Trousseau's sign).

characterized by hypercalcemia, hypophosphatemia, demineralization of the bones, hypercalciuria, and the formation of calcium-containing kidney stones. The bone disease caused by hyperparathyroidism (**osteitis cystica**) is characterized by multiple bone cysts.

Secondary Hyperparathyroidism

In conditions such as chronic renal disease and rickets, in which the plasma Ca^{2+} level is chronically low, feedback stimulation of the parathyroid glands causes compensatory parathyroid hypertrophy and **secondary hyperparathyroidism.** The plasma Ca^{2+} level is low in chronic renal disease primarily because the diseased kidneys lose the ability to form 1,25-dihydroxycholecalciferol. However, phosphate retention with resulting hyperphosphatemia also contributes to the decrease in plasma Ca^{2+}.

CALCITONIN

Origin

In dogs, perfusion of the thyroproparathyroid region with solutions containing high concentrations of Ca^{2+} leads to a fall in peripheral plasma calcium, and after damage to this region, Ca^{2+} infusions cause a greater increase in plasma Ca^{2+} than they do in control animals. These and other observations led to the discovery that a Ca^{2+}-lowering as well as a Ca^{2+}-elevating hormone was secreted by structures in the neck. The Ca^{2+}-lowering hormone has been named **calcitonin.** In nonmammalian vertebrates, the source of calcitonin is the **ultimobranchial bodies,** a pair of glands derived embryologically from the fifth branchial arches. In mammals, these bodies have for the most part become incorporated into the thyroid gland, where the ultimobranchial tissue is distributed around the follicles as the **parafollicular cells** (Figs 18–2 and 21–12). Since these cells (also known as clear cells or C cells) secrete the hormone, it is sometimes

Parafollicular cells   Follicular cell

Colloid

Capillary

Figure 21–12. Parafollicular cells in the thyroid. (Modified from Poirier J, Dumas JLR: *Review of Medical Histology.* Saunders, 1977.)

called **thyrocalcitonin.** However, total thyroidectomy does not reduce the circulating level of the hormone to zero, and the hormone has been found by radioimmunoassay in various other tissues. The significance of extrathyroidal calcitonin is unsettled, but at least it seems appropriate to refer to it by its original name, calcitonin.

Structure

Human calcitonin has a molecular weight of 3500 and contains 32 amino acid residues (Fig 21–13). Preprocalcitonin is the source not only of calcitonin but also of an additional circulating peptide, **katacalcin,** the physiologic significance of which is unsettled. Much of the mRNA transcribed from the calcitonin gene is processed to a different mRNA in the nervous system, so that calcitonin-gene-related peptide (CGRP) is formed rather than calcitonin (see Chapter 4). CGRP is also found in the intermediate lobe of the pituitary, the adrenal medulla, the pancreas, and the circulating blood. Some is also found in the thyroid. When injected systemically, CGRP produces vasodilation, but its physiologic function is

unknown. It is now known that 2 CGRPs that differ by only one amino acid residue, CGRPα and CGRPβ, are formed in humans. These polypeptides are encoded by 2 different genes (see Chapter 4). However, it is not known whether their effects differ.

The calcitonins of the other species that have been studied also contain 32 amino acid residues, but the amino acid composition varies considerably. Salmon calcitonin is of interest because it is more than 20 times more active in humans than human calcitonin is.

Secretion & Metabolism

Secretion of calcitonin is increased when the thyroid gland is perfused with solutions containing a high Ca^{2+} concentration. Measurement of circulating calcitonin by immunoassay indicates that it is not secreted until the plasma calcium level reaches approximately 9.5 mg/dL and that above this calcium level, plasma calcitonin is directly proportionate to plasma calcium (Fig 21–14). β-Adrenergic agonists, dopamine, and estrogens also stimulate calcitonin secretion. Gastrin, CCK, glucagon, and secretin have all been reported to stimulate calcitonin secretion, gastrin being the most potent stimulus (see Chapter 26). The plasma calcitonin level is elevated in Zollinger-Ellison syndrome (see Chapter 26) and in pernicious anemia, in which the plasma gastrin level is also elevated. However, the dose of gastrin needed to stimulate calcitonin secretion produces an increase in plasma gastrin concentration greater than that produced by food, so it is premature to conclude that calcium in the intestine initiates secretion of a calcium-lowering hormone before the calcium is absorbed.

Human calcitonin has a half-life of less than 10 minutes.

Actions

Serpentine receptors for calcitonin are found in bones and the kidneys. Calcitonin lowers the circulating calcium and phosphate levels. It exerts its calcium-lowering effect by inhibiting bone resorption. This action is direct, and calcitonin inhibits the activity of osteoclasts in vitro. It also increases Ca^{2+} excretion in the urine.

The exact physiologic role of calcitonin is uncertain. The calcitonin content of the human thyroid is low, and after thyroidectomy, bone density and plasma Ca^{2+} level are normal as long as the parathyroid glands are intact. Indeed, there are only transient

```
       ┌─S ─────────── S ─┐
Cys-Gly-Asn-Leu-Ser-Thr-Cys-Met-Leu-Gly-Thr-Tyr-Thr-Gln-Asp-Phe-Asn-
  1   2   3   4   5   6   7   8   9  10  11  12  13  14  15  16  17

Lys-Phe-His-Thr-Phe-Pro-Gln-Thr-Ala-Ile-Gly-Val-Gly-Ala-Pro-NH2
 18  19  20  21  22  23  24  25  26  27  28  29  30  31  32
```

Figure 21–13. Human calcitonin.

Figure 21–14. Concentration of calcitonin (CT) and parathyroid hormone (PTH) as a function of serum calcium level in pigs given EDTA or calcium infusions to decrease or increase serum calcium. (Reproduced, with permission, from Arnaud CD et al, in: *Calcitonin: Proceedings of the Second International Symposium.* Taylor S [editor]. Heinemann, 1969.)

abnormalities of Ca^{2+} metabolism when a Ca^{2+} load or parathyroid extract is injected. This may be explained in part by secretion of calcitonin from tissues other than the thyroid. However, there is general agreement that the hormone has little long-term effect on the plasma Ca^{2+} level in adult animals and humans. In addition, patients with medullary carcinoma of the thyroid have a very high circulating calcitonin level but no symptoms directly attributable to the hormone, and their bones are essentially normal. No syndrome due to calcitonin deficiency has been described. More hormone is secreted in young individuals, and it may play a role in skeletal development. It may protect against postprandial hypercalcemia (see above). In addition, it may protect the bones of the mother from excess calcium loss during pregnancy. Bone formation in the infant and lactation are major drains on calcium stores, and plasma concentrations of 1,25-dihydroxycholecalciferol are elevated in pregnancy. This would cause bone loss in the mother if bone resorption were not simultaneously inhibited by an increase in the plasma calcitonin level.

Clinical Correlates

Calcitonin is useful in the treatment of Paget's disease, a condition in which increased osteoclastic activity triggers compensatory formation of disorganized new bone. It also has beneficial effects in severe hypercalcemia, but the hormone must be injected and its effect generally wears off.

Summary

The actions of the 3 principal hormones that regulate the plasma concentration of Ca^{2+} are summarized in Fig 21–15. PTH increases plasma Ca^{2+} by mobilizing this ion from bone. It increases Ca^{2+} reabsorption in the kidney, but this may be offset by the increase in filtered Ca^{2+}. It also increases the formation of 1,25-dihydroxycholecalciferol. 1,25-Dihydroxycholecalciferol increases Ca^{2+} absorption from the intestine, mobilizes the ion from bone, and increases Ca^{2+} reabsorption in the kidneys. Calcitonin inhibits bone resorption and increases the amount of Ca^{2+} in the urine.

EFFECTS OF OTHER HORMONES & HUMORAL AGENTS ON CALCIUM METABOLISM

Calcium metabolism is affected by various hormones in addition to 1,25-dihydroxycholecalciferol, PTH, calcitonin, and the factors listed in Table 21–2. The glucocorticoids lower plasma Ca^{2+} levels by inhibiting osteoclast formation and activity, but over long periods they cause osteoporosis by decreasing bone formation and increasing bone resorption (see Chapter 20). They decrease bone formation by inhibiting cellular replication and protein synthesis in bone. They also decrease the absorption of Ca^{2+} and PO_4^{3-} from the intestine by an anti-vitamin D action and increase the renal excretion of these ions. This is why they depress the hypercalcemia of vitamin D intoxication. The decrease in plasma Ca^{2+} concentration increases the secretion of PTH, and bone resorption is facilitated. **Growth hormone** increases calcium excretion in the urine, but it also increases intestinal absorption of calcium, and this effect may be greater than the effect on excretion, with a resultant positive calcium balance. In addition, IGF-I generated by the action of growth hormone stimulates

Figure 21–15. Major actions of hormones involved in the control of plasma Ca^{2+}. PTH, parathyroid hormone; 1,25-$(OH)_2D_3$, 1,25-dihydroxycholecalciferol; CT, calcitonin. See text for details.

protein synthesis in bone. As noted above, **thyroid hormones** may cause hypercalcemia, hypercalciuria, and, in some instances, osteoporosis. **Estrogens** prevent osteoporosis, probably by a direct effect on osteoblasts (see above). **Insulin** increases bone formation, and there is significant bone loss in untreated diabetes.

Hypercalcemia is a common metabolic complication of cancer. In patients with bone metastases, local erosion of the bone contributes to the hypercalcemia. There is evidence that this erosion is produced by prostaglandins such as PGE from the tumor.

Hypercalcemia is also seen in cancer patients in the absence of bone metastases. The cause in many of these cases is the production by the tumors of **parathyroid hormone-like protein (PLP).** This protein is larger than PTH, having 140 amino acid residues, and is encoded by a gene on human chromosome 12, whereas PTH is encoded by a gene on chromosome 11. However, PLP and PTH have marked homology at their N-terminal ends, with 8 of the first 13 amino acid residues in the same positions, and they both bind to the PTH receptor. In normal individuals PLP is produced in large quantities in the breast, and large amounts are found in milk. However, the plasma level does not increase during lactation, and the function of PLP in milk is unknown.

The Pituitary Gland

22

INTRODUCTION

The anterior, intermediate, and posterior lobes of the pituitary gland are actually 3 more or less separate endocrine organs that, at least in some species, contain 14 or more hormonally active substances. The intermediate lobe is rudimentary in humans. The 6 established hormones that are secreted by the anterior pituitary are **thyroid-stimulating hormone (TSH, thyrotropin), adrenocorticotropic hormone (ACTH), luteinizing hormone (LH), follicle-stimulating hormone (FSH), prolactin,** and **growth hormone** (Table 22–1). ACTH, prolactin, and growth hormone are simple polypeptides or proteins, whereas TSH, LH, and FSH are glycoproteins. Prolactin acts on the breast (see Fig 14–16). The remaining 5 are, at least in part, **tropic hormones;** ie, they stimulate secretion of hormonally active substances by other endocrine glands or, in the case of growth hormone, the liver and other tissues (see below). In addition, the anterior lobe of the pituitary secretes β-lipotropin (β-LPH). This linear polypeptide contains 91 amino acid residues, and although its physiologic role is uncertain, it contains the amino acid sequences of endorphins and enkephalins, peptides that bind to opiate receptors (see Chapter 4). In addition, the anterior and intermediate lobes contain other hormonally active derivatives of the pro-opiomelanocortin molecule (see below). The hormones tropic to a particular endocrine gland are discussed in the chapter on that gland: TSH in Chapter 18; ACTH in Chapter 20; and the gonadotropins FSH and LH in Chapter 23, along with prolactin. The hormones secreted by the posterior pituitary in mammals (**oxytocin** and **vasopressin**) and the neural regulation of anterior and posterior pituitary secretion are discussed in Chapter 14. Growth hormone, β-lipotropin, and the melanocyte-stimulating hormones of the intermediate lobe of the pituitary, α-MSH and β-MSH, are the subject of this chapter, along with a number of general considerations about the pituitary.

MORPHOLOGY

Gross Anatomy

The anatomy of the pituitary gland is summarized in Fig 22–1 and discussed in detail in Chapter 14. The posterior pituitary is made up largely of the endings on blood vessels of axons from the supraoptic and paraventricular nuclei of the hypothalamus, whereas the anterior pituitary has a special vascular connection with the brain, ie, the portal hypophyseal vessels. The intermediate lobe is formed in the embryo from the dorsal half of **Rathke's pouch,** an evagination of the roof of the pharynx, but is closely adherent to the posterior lobe in the adult. It is separated from the anterior lobe by the remains of Rathke's pouch, the **residual cleft.**

Histology

In the posterior lobe, the endings of the supraoptic and paraventricular axons can be observed in close relation to blood vessels. In some species, the endings are rodlike palisaded structures. There are also neuroglial cells and **pituicytes,** stellate cells containing fat globules that were once thought to secrete posterior lobe hormones but are now believed to be modified astroglia.

The intermediate lobe is rudimentary in humans and a few other mammalian species. In these species, most of its cells are agranular, although there are often a few basophilic elements that resemble anterior lobe cells. Along the residual cleft are small thyroid-like follicles, some containing a little colloid. The function of the colloid, if any, is unknown.

The anterior pituitary is made up of interlacing cell cords and an extensive network of sinusoidal capillaries. The endothelium of the capillaries is fenestrated, like that in other endocrine organs. The cells contain granules of stored hormone that are extruded from the cells by exocytosis. The granules have not been seen entering the capillaries, and they presumably break down in the pericapillary space.

Cell Types in the Anterior Pituitary

Human anterior pituitary cells have traditionally been divided on the basis of their staining reactions into agranular chromophobes and granular chromophils. The chromophilic cells are subdivided into

Table 22–1. Pituitary hormones in mammals.[1]

Name and Source	Principal Actions
Anterior lobe	
Thyroid-stimulating hormone (TSH, thyrotropin)	Stimulates thyroid secretion and growth of thyroid gland.
Adrenocorticotropic hormone (ACTH, corticotropin)	Stimulates secretion and growth of zona fasciculata and zona reticularis of adrenal cortex.
Growth hormone (GH, somatotropin, STH)	Accelerates body growth; stimulates secretion of IGF-I.
Follicle-stimulating hormone (FSH)	Stimulates ovarian follicle growth in female and spermatogenesis in male.
Luteinizing hormone (LH, interstitial cell-stimulating hormone, ICSH)	Stimulates ovulation and luteinization of ovarian follicles in female and testosterone secretion in male.
Prolactin (luteotropic hormone, LTH, luteotropin, lactogenic hormone, mammotropin)	Stimulates secretion of milk and maternal behavior. Maintains corpus luteum in female rodents but apparently not in other species.
β-Lipotropin (β-LPH)	?
γ-Melanocyte-stimulating hormone (γ-MSH)	May maintain adrenal sensitivity.
Intermediate lobe	
α- and β-melanocyte-stimulating hormones (α- and β-MSH; referred to collectively as melanotropin or intermedin)	Expands melanophores in fish, amphibians, and reptiles; stimulates melanin synthesis in melanocytes in humans.
γ-Lipotropin (γ-LPH), corticotropinlike intermediate lobe peptide (CLIP), other fragments of pro-opio-melanocortin	?
Posterior lobe	
Vasopressin (antidiuretic hormone, ADH)	Promotes water retention.
Oxytocin	Causes milk ejection.

[1] In addition, a variety of gastrointestinal and other polypeptides are found in one or more lobes of the pituitary gland. These include CCK, gastrin, renin, angiotensin II, and calcitonin-gene-related peptide (CGRP).

acidophils (about 40% of the cells), which stain with acidic dyes, and basophils (about 10% of the cells), which stain with basic dyes. The cells secreting growth hormone and prolactin are acidophilic, whereas those secreting the glycoprotein hormones TSH, LH, and FSH are basophilic and those secreting ACTH are variously classified as basophilic or chromophobic.

With the advent of immunocytochemistry and electron microscopy, 5 types of secretory cells were distinguished (Fig 22–2): somatotropes, which se-

crete growth hormone; lactotropes (also called mammotropes), which secrete prolactin; thyrotropes, which secrete TSH; gonadotropes, which appear in most instances to secrete both LH and FSH; and corticotropes, which secrete both ACTH and β-LPH. Note that in this classification, 2 or more polypeptide hormones were secreted by several types of cells. In addition, investigators described subtypes of the 5 primary types. However, the classification became clearly inadequate when it was demonstrated that many other cells contained at least 2 hormones. For example, it has now been proved that in addition to acidophils that secrete growth hormone and acidophils that secrete prolactin, there are cells called **mammosomatotropes** that secrete both hormones. These may be precursor cells that differentiate into somatotropes or mammotropes, but this has not been proved. In addition, it seems clear that there are 2 types of morphologically distinct corticotropes. Finally, there is immunocytochemical evidence for various other combinations of hormones in single cells. Examples include **corticothyrotropes,** which contain TSH and ACTH, and **corticogonadotropes,** which contain gonadotropins and ACTH. The physiologic significance of these other types of multihormonal cells remains uncertain.

The anterior pituitary also contains folliculostel-

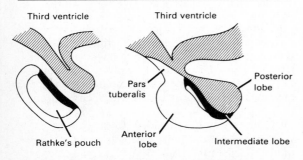

Figure 22–1. Diagrammatic outline of the formation of the pituitary and the various parts of the organ in the adult.

Figure 22–2. Ultrastructure of the anterior pituitary, showing the characteristics of 5 types of secretory cells. Note that in this illustration, 3 morphologically different subtypes of gonadotropes (A, B, and C) have been identified. STELL, folliculostellate cell. (Reproduced, with permission, from Steger RW, Peluso JJ: Anterior pituitary. In: *Handbook of Endocrinology.* Gass GH, Kaplan HM [editors]. CRC Press, 1982. Copyright © CRC Press, Inc., Boca Raton, FL.)

late cells, which send processes between established secretory cells (Fig 22–2). Current evidence indicates these cells contain and secrete the cytokine IL-6 (see Chapter 27), but their physiologic role remains unsettled.

Two-Unit Structure of FSH, LH, & TSH

The 3 pituitary glycoprotein hormones, FSH, LH, and TSH, are each made up of 2 subunits. The subunits, which have been designated α and β, have some activity but must be combined for maximal physiologic activity. In addition, the placental glycoprotein gonadotropin human chorionic gonadotropin (hCG) has α and β subunits (see Chapter 23). All of the α subunits of these hormones are products of a single gene and have the same amino acid composition, although their carbohydrate residues vary. The β subunits, which are produced by separate genes and differ in structure, confer hormonal specificity. The α units are remarkably interchangeable, and hybrid molecules can be created. However, the physiologic and evolutionary significance of the unique 2-unit structure of these glycoprotein hormones remains to be determined.

INTERMEDIATE LOBE HORMONES

Pro-opiomelanocortin

Intermediate lobe cells and corticotropes of the anterior lobe both synthesize a large precursor protein that is cleaved to form a family of hormones. After removal of the signal peptide, this prohormone is known as **pro-opiomelanocortin (POMC).** This molecule is also synthesized in the hypothalamus and other parts of the nervous system, the lungs, the gastrointestinal tract, and the placenta (see Chapter 4). Its structure is shown in Fig 22–3. In the corticotropes, it is hydrolyzed to ACTH and β-LPH plus a small amount of β-endorphin, and these substances are secreted. It also appears that γ-MSH is secreted. In the intermediate lobe, POMC is further hydrolyzed to α-MSH, corticotropinlike intermediate-lobe peptide (CLIP), γ-LPH, and appreciable quantities of β-endorphin, and these substances are presumably secreted. The functions, if any, of CLIP and γ-LPH are unknown, whereas β-endorphin is an opioid peptide (see Chapter 4) that has the 5 amino acid residues of met-enkephalin at its N-terminal end. The intermediate lobe in humans is rudimentary, and it appears that neither α-MSH nor β-MSH is secreted in adults.

Melanocyte-Stimulating Hormones

In the skins of fish, reptiles, and amphibians, there are cells called **melanophores.** These contain melanin granules. Melanins are synthesized from tyrosine via dopa (see Chapter 4) and dopaquinone. They are of 2 types: eumelanins, which are responsible for black or brown pigmentation, and pheomelanins, which contain more sulfhydryl groups and are responsible for yellow and red pigmentation. Melanins are synthesized in cellular organelles called **melanosomes.** Other cells called **iridophores** contain reflecting platelets. When the granules aggregate around the nuclei of melanophores and the reflecting platelets disperse to the periphery of the iridophores, the color of the skin lightens; when the melanin granules disperse and the reflecting platelets aggregate, the skin darkens. Birds and mammals do not have pigment cells of this type, but they have cells that contain melanin, and their pituitaries contain 3 polypeptides called **melanotropins** or **melanocyte-stimulating hormones (MSHs),** which when injected into lower vertebrates disperse melanophore granules and aggregate iridophore platelets.

α-MSH is made up of amino acid residues 1–13 of the ACTH molecule (Fig 22–3). β-MSH is made up of the 17 amino acid residues at the C-terminal end of γ-LPH. It contains a 7-amino-acid sequence that is identical to amino acid residues 4–10 in α-MSH and ACTH. Consequently, it is not surprising that ACTH has considerable MSH activity. The effects of γ-MSH resemble those of the other MSHs. It is present in highest concentration in the intermediate lobe, but it is also present in the anterior lobe, and its blood level is increased in patients with adrenal insufficiency and decreased in patients with high levels of circulating glucocorticoids.

Control of Skin Coloration

Changes of skin coloration in fish, reptiles, and

Figure 22–3. Schematic representation of the prepro-opiomelanocortin molecule formed in pituitary cells, neurons, and other tissues. The numbers in parentheses identify the amino acid sequences in each of the polypeptide fragments. For convenience, the amino acid sequences are numbered from the N terminal of ACTH and read toward the C-terminal portion of the parent molecule, whereas the amino acid sequences in the other portion of the molecule read to the left to −131, the N terminal of the parent molecule. The locations of Lys-Arg and other pairs of basic amino acids residues are also indicated; these are the sites of proteolytic cleavage in the formation of the smaller fragments of the parent molecule. AL, anterior lobe; IL, intermediate lobe.

amphibians are probably mediated in part via neuroendocrine reflex mechanisms. The receptors for these reflexes are in the retina, and the changes are brought about by MSH. On a dark background, MSH secretion is not inhibited and the animal is dark. On a white background, release of MSH is inhibited and the animal's color lightens.

In mammals, there are no melanophores containing pigment granules that disperse and aggregate as they do in fish, reptiles, and birds, but there are **melanocytes,** which contain melanosomes and synthesize melanins. The melanocytes then transfer the melanosomes to skin cells (keratinocytes) in hair follicles and other portions of the skin. This accounts for the pigmentation of hair and skin. Treatment with MSHs accelerates melanin synthesis and causes readily detectable darkening of the skin of humans in 24 hours. However, the physiologic function of MSHs in humans is unknown.

Pigment Abnormalities in Humans

The pigmentary changes in several endocrine diseases are probably due to changes in circulating ACTH, since ACTH has MSH activity. For example, abnormal pallor is a hallmark of hypopituitarism. Hyperpigmentation occurs in patients with adrenal insufficiency due to primary adrenal disease. Indeed, the presence of hyperpigmentation in association with adrenal insufficiency rules out the possibility

that the insufficiency is secondary to pituitary disease, because the pituitary must be intact for pigmentation to occur.

Albinos have a congenital inability to synthesize melanin. Albinism occurs in humans and many other mammalian species. It can be due to a variety of different genetic defects in the pathways for melanin synthesis. **Piebaldism** is characterized by patches of skin that lack melanin as a result of congenital defects in the migration of pigment cell precursors from the neural crest during embryonic development. Not only the condition but also the precise pattern of the loss is passed from one generation to the next. **Vitiligo** is due to a similar patchy loss of melanin, but the loss develops after birth and is progressive.

GROWTH HORMONE

Biosynthesis & Chemistry

The long arm of human chromosome 17 contains the growth hormone-hCS cluster that contains 5 genes: one, hGH-N (for normal), codes for the most abundant form of growth hormone; a second, hGH-V (for variant), codes for the variant form of growth hormone (see below); 2 code for hCS (see Chapter 23); and the fifth is probably an hCS pseudogene.

"Normal" human growth hormone, the product of hGH-N, accounts for 75% of the circulating hGH.

Because of its molecular weight of 22,000, it is also known as 22K hGH. Figure 22–4 shows its structure and compares it with the structure of hCS, which also has 191 amino acid residues and differs by only 29 residues. The hGH-V gene, which is expressed primarily in the placenta, produces a 191-amino-acid growth hormone that differs from normal human growth hormone by 13 amino acids scattered throughout the polypeptide chain. Appreciable amounts appear in the circulation only in pregnancy. The mRNA produced by the hGH-N gene undergoes alternative splicing to produce a growth hormone identical to 22K hGH except that amino acid residues 32–46 are deleted. This smaller form, 20K hGH, is also biologically active and makes up about 10% of the circulating growth hormone. The physiologic significance of having the 2 principal forms, 22K and 20K, is unknown. Desamino forms and various other forms also circulate, so there are many different forms of growth hormone in the body.

Species Specificity

Not surprisingly, there is considerable variation in the structure of growth hormone from one species to another. Porcine and simian growth hormones have only a transient effect in the guinea pig, probably because they stimulate the rapid formation of anti-growth-hormone antibodies. In monkeys and humans, bovine and porcine growth hormones do not even have a significant transient effect on growth, although monkey and human growth hormones are fully active in both monkeys and humans. Human growth hormone has intrinsic lactogenic activity.

Plasma Levels, Binding, & Metabolism

The basal growth hormone level measured by radioimmunoassay in adult humans is normally less than 3 ng/mL. This is mostly 22K hGH, since the affinity of 20K hGH for most antibodies is about 30% of the affinity of 22K hGH.

Growth hormone is bound to 2 proteins in plasma. One has a high affinity and is a large fragment of the extracellular domain of the growth hormone receptor (see below). Its concentration is an index of the number of growth hormone receptors in the tissues. The other has a low affinity. About half the growth hormone activity is bound, providing a reservoir of the hormone to compensate for wide fluctuations in secretion. The high-affinity binding protein is present at low levels in African pygmies and absent in Laron dwarfs (see below).

Growth hormone is metabolized rapidly, probably at least in part in the liver. The half-life of circulating growth hormone in humans is 6–20 minutes, and the daily growth hormone output has been calculated to be 0.2–1.0 mg/d in adults.

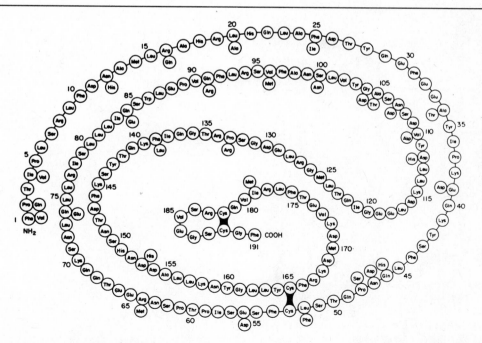

Figure 22–4. Structure of the principal human growth hormone (continuous chain). The black bars indicate disulfide bridges. The 29 residues alongside the chain identify residues that differ in human chorionic somatomammotropin (hCS; see Chapter 23). All the other residues in hCS are the same, and hCS also has 191 amino acid residues. (Reproduced, with permission, from Parsons JA [editor]: *Peptide Hormones.* University Park Press, 1976.)

Growth Hormone Receptors

The growth hormone receptor is a 620-amino-acid protein with a large extracellular portion, a transmembrane domain, and a large cytoplasmic portion. It is a member of the cytokine receptor superfamily, which also includes the prolactin receptor (see Chapters 1 and 23). It has recently been demonstrated that hGH has 2 binding sites for the receptor, and that it binds 2 receptors, producing dimerization. However, it is not yet known how this brings about the intracellular changes produced by growth hormone.

An important action is stimulation of the secretion of the somatomedin **IGF-I** (see below), and IGF-I synergizes with growth hormone in producing many of its effects (see below).

Effects on Growth

In young animals in which the epiphyses have not yet fused to the long bones (see Chapter 21), growth is inhibited by hypophysectomy (Fig 22–5) and stimulated by growth hormone. Chondrogenesis is accelerated, and as the cartilaginous epiphyseal plates widen, they lay down more bone matrix at the ends of long bones (Fig 22–6). In this way, stature is increased, and prolonged treatment leads to gigantism. The increase in width of the tibial epiphyseal plate in hypophysectomized rats has been used as a bioassay for growth hormone ("tibia test").

Figure 22–6. Effect of growth hormone treatment for 4 days on the proximal tibial epiphysis of the hypophysectomized rat. Note the increased width of the unstained cartilage plate in the tibia of the treated animal on the right, compared with the control on the left. (Reproduced, with permission, from Evans HM et al: Bioassay of pituitary growth hormone. *Endocrinology* 1943;**32**:14.)

When the epiphyses are closed, linear growth is no longer possible, and growth hormone produces the pattern of bone and soft tissue deformities known in humans as **acromegaly** (Fig 22–7). The sizes of most of the viscera are increased. Endocrine organs may also be affected by growth hormone, and the hormone is said to synergize with ACTH in increasing adrenal size and with androgens in increasing the size

Figure 22–5. Effect of hypophysectomy on growth of the immature rhesus monkey. Both monkeys were the same size and weight 2 years previously, when the one on the left was hypophysectomized. (Reproduced, with permission, from Knobil E, in: *Growth in Living Systems*. Zarrow MX [editor]. Basic Books, 1961.)

Figure 22–7. Typical findings in acromegaly.

of accessory reproductive organs. The protein content of the body is increased, and the fat content is decreased.

Effects on Protein & Electrolyte Metabolism

Growth hormone is a protein anabolic hormone and produces a positive nitrogen and phosphorus balance, a rise in plasma phosphorus, and a fall in the blood urea nitrogen and amino acid levels. In adults with growth hormone deficiency, recombinantly produced human growth hormone produces an increase in lean body mass and a decrease in body fat, along with an increase in metabolic rate and a fall in plasma cholesterol. Gastrointestinal absorption of Ca^{2+} is increased. Na^+ and K^+ excretion is reduced by an action independent of the adrenal glands, probably because these electrolytes are diverted from the kidneys to the growing tissues. Excretion of the amino acid 4-hydroxyproline is increased during growth and in acromegaly, but it is also increased in a number of other diseases. Much of the excreted hydroxyproline comes from collagen, and hydroxyproline excretion is increased in diseases associated with increased collagen destruction. However, it is also increased when synthesis of soluble collagen is increased and growth hormone stimulates the synthesis of soluble collagen.

Effects on Carbohydrate & Fat Metabolism

The actions of growth hormone on carbohydrate metabolism are discussed in Chapter 19. Growth hormone is diabetogenic because it increases hepatic glucose output and exerts an anti-insulin effect in muscle. It is ketogenic because it increases circulating FFA levels. The increase in plasma FFA, which takes several hours to develop, provides a ready source of energy for the tissues during hypoglycemia, fasting, and stressful stimuli. Growth hormone does not stimulate B cells of the pancreas directly, but it increases the ability of the pancreas to respond to insulinogenic stimuli such as arginine and glucose. This is an additional way growth hormone promotes growth, since insulin has a protein anabolic effect (see Chapter 19).

Somatomedins

The effects of growth hormone on growth, cartilage, and protein metabolism depend on an interaction between growth hormone and **somatomedins**, which are polypeptide growth factors secreted by the liver and other tissues in response to stimulation by growth hormone. The first of these factors isolated was called sulfation factor because it stimulated the incorporation of sulfate into cartilage. However, it also stimulated collagen formation, and its name was changed to somatomedin. It then became clear that there are a variety of different somatomedins and that they are related to an increasingly long list of **growth factors** that affect many different tissues and organs. These include nerve growth factor (NGF; see Chapter 2), epidermal growth factor (EGF), ovarian growth factor (OGF), fibroblast growth factor (FGF), and platelet-derived growth factor (PDGF). Several of the growth factors are related to the proteins produced by oncogenes (see Chapter 1).

The principal (and in humans probably the only) circulating somatomedins are insulinlike growth factor I (IGF-I, somatomedin C) and insulinlike growth factor II (IGF-II). These factors are closely related to insulin, except that their C chains are not separated (Fig 22–8) and they have an extension of the A chain called the D domain. The hormone relaxin (see Chapter 23) is also a member of this family; there are 2 related forms of relaxin in humans, and both resemble IGF-II. In humans a variant form of IGF-I lacking 3 N-terminal amino acid residues has been found in the brain, and there are several variant forms of human IGF-II (Fig 22–9). The mRNAs for IGF-I and IGF-II are found in the liver, in cartilage, and in many other tissues, indicating that they are synthesized in these tissues.

The properties of IGF-I, IGF-II, and insulin are compared in Table 22–2. Both are tightly bound to proteins in the plasma, and at least for IGF-I, this prolongs the IGF half-life in the circulation. The contribution of the IGFs to the insulinlike activity in blood is discussed in Chapter 19. Secretion of IGF-I is stimulated by growth hormone, and it has pronounced

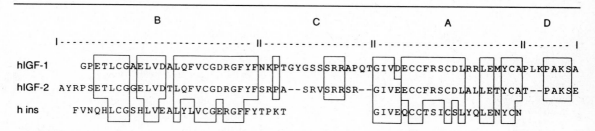

Figure 22–8. Structure of human IGF-I, IGF-II, and insulin (ins), with amino acid residues aligned to show the homologies identified by the boxed areas. Single-letter codes are used for amino acid residues. Note that in insulin the C peptide is removed, whereas in IGF-I and IGF-II the A and B chains (domains) remain connected by the C peptide and there are D domains. (Modified and reproduced, with permission, from Sara VR, Hall K: Insulin-like growth factors and their binding proteins. *Physiol Rev* 1990;**70**:591.)

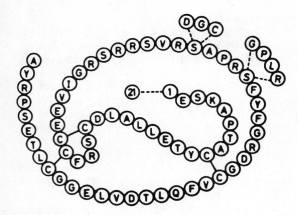

Figure 22–9. Primary structure of human IGF-II and 3 identified variants: a 21-amino-acid-residue extension of the C terminal, a tetrapeptide substitution at Ser-29, and a tripeptide substitution of Ser-33. Single-letter codes are used for amino acid residues. (Reproduced, with permission, from Sara VR, Hall K: Insulin-like growth factors and their binding proteins. *Physiol Rev* 1990;**70:**591.)

growth-stimulating activity. IGF-II is much less affected by growth hormone and plays a role in the growth of the fetus before birth. In adults, the gene for IGF-II is expressed only in the choroid plexus and meninges.

The secretion of somatomedins is affected by various factors in addition to growth hormone. Gluco-

Table 22–2. Comparison of insulin and the insulinlike growth factors.[1]

	Insulin	IGF-I	IGF-II
Other names	. . .	Somatomedin C	Multiplication-stimulating activity (MSA)
Number of amino acids	51	70	67
Source	Pancreatic B cells	Liver and other tissues	Diverse tissues
Level regulated by	Glucose	Growth hormone, nutritional status	Unknown
Plasma levels	0.3–2 ng/mL	ng/mL range	ng/mL range
Plasma-binding protein	No	Yes	Yes
Major physiologic role	Control of metabolism	Skeletal and cartilage growth	Growth during fetal development

[1]Modified and reproduced, with permission, from Murray RK et al: *Harper's Biochemistry,* 23rd ed. Appleton & Lange, 1993.

corticoids and protein deficiency reduce plasma somatomedin activity, and large doses of estrogens inhibit the production of IGF-I. The secretion of somatomedins is reduced in untreated diabetes and restored to normal by insulin treatment.

Mechanism of Action of Growth Hormone

Ideas about the mechanism of action of growth hormone have undergone a series of changes as new information has become available. Growth hormone was originally thought to produce growth by a direct action on tissues, and later it was believed to act solely through somatomedins. However, if growth hormone is injected into one proximal tibial epiphysis, a unilateral increase in cartilage width is produced, and cartilage, like other tissues, makes IGF-I. A current hypothesis to explain these results holds that growth hormone acts on cartilage to convert stem cells into cells that respond to IGF-I and then locally produced and circulating IGF-I makes the cartilage grow. However, the role of circulating IGF-I remains important, since infusion of IGF-I to hypophysectomized rats restores bone and body growth.

Figure 22–10 is a summary of current views of the other actions of growth hormone and IGF-I. However, growth hormone probably combines with circulating and locally produced IGF-I in various proportions to produce at least some of these effects.

Hypothalamic Control of Growth Hormone Secretion

The secretion of growth hormone is controlled via the hypothalamus. The hypothalamus secretes growth hormone-releasing hormone (GRH) and the growth hormone-inhibiting hormone somatostatin into the portal hypophyseal blood, and hypothalamic lesions or section of the pituitary stalk inhibits growth hormone secretion. If growth hormone were secreted in a constant and continuous fashion and only in children, the hypothalamic control of its secretion might be surprising. However, there are large quantities of growth hormone in the pituitaries of adults as well as children, and in both children and adults the rate of growth hormone secretion undergoes marked and rapid spontaneous fluctuations as well as increases or decreases in response to specific stimuli. Fluctuations in somatostatin secretion are the major factor controlling secretion, and the effect of GRH is generally more tonic.

Growth hormone secretion is under feedback control, like the secretion of other anterior pituitary hormones. Growth hormone increases circulating IGF-I, and IGF-I in turn exerts a direct inhibitory action on growth hormone secretion from the pituitary. It also stimulates somatostatin secretion (Fig 22–11).

Stimuli Affecting Growth Hormone Secretion

As noted above, the basal plasma growth hormone

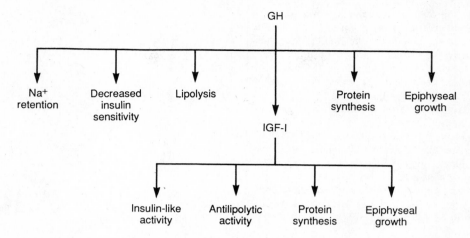

Figure 22–10. Current view of actions mediated by growth hormone (GH) and IGF-I. (Courtesy of R Clark and N Gesundheit.)

concentration ranges from 0 to 3 ng/mL in normal adults. The values in newborn infants are higher, but average plasma growth hormone levels throughout the rest of childhood are not significantly greater than they are in adults. On the other hand, growth hormone is secreted in irregular "spikes" throughout the day, and these spikes increase in amplitude during puberty so that the daily mean plasma growth hormone is elevated at this time.

The stimuli that increase growth hormone secretion are summarized in Table 22–3. Most of them fall into 3 general categories: (1) conditions such as hypoglycemia and fasting in which there is an actual or threatened decrease in the substrate for energy production in the cells; (2) conditions in which there are increased amounts of certain amino acids in the plasma; and (3) stressful stimuli. It has been claimed that the response to glucagon is useful as a test of the growth-hormone-secreting mechanism in patients with endocrine diseases. Although the growth hormone responses to the other stimuli are relatively reproducible, they may vary from individual to individ-

Figure 22–11. Feedback control of growth hormone secretion. The dashed arrows indicate inhibitory effects and the solid arrows stimulatory effects. Note that IGF-I stimulates the secretion of somatostatin (SS) from the hypothalamus and acts directly on the pituitary to inhibit growth hormone (GH) secretion. Compare with Figs 18–14, 20–21, 23–22, and 23–36.

Table 22–3. Stimuli that affect growth hormone secretion in humans.

Stimuli that increase secretion
  Deficiency of energy substrate
    Hypoglycemia
    2-Deoxyglucose
    Exercise
    Fasting
  Increase in circulating levels of certain amino acids
    Protein meal
    Infusion of arginine and some other amino acids
  Glucagon
  Stressful stimuli
    Pyrogen
    Lysine vasopressin
    Various psychologic stresses
  Going to sleep
  L-Dopa and α-adrenergic agonists that penetrate the brain
  Apomorphine and other dopamine receptor agonists
  Estrogens and androgens
Stimuli that decrease secretion
  REM sleep
  Glucose
  Cortisol
  FFA
  Medroxyprogesterone
  Growth hormone

ual. A spike in growth hormone secretion occurs with considerable regularity upon going to sleep, but the significance of the association between growth hormone and sleep is an enigma. Growth hormone secretion is increased in subjects deprived of REM sleep (see Chapter 11) and inhibited during normal REM sleep.

Glucose infusions lower plasma growth hormone levels and inhibit the response to exercise. The increase produced by 2-deoxyglucose is presumably due to intracellular glucose deficiency, since this compound blocks the catabolism of glucose 6-phosphate. Sex hormones, particularly estrogens, increase growth hormone responses to provocative stimuli such as arginine and insulin. Growth hormone secretion is inhibited by cortisol, FFA, and medroxyprogesterone.

Growth hormone secretion is increased by L-dopa, which increases the release of dopamine and norepinephrine in the brain, and by the dopamine receptor agonist apomorphine. Central release of norepinephrine is known to increase GRH secretion, but it is uncertain whether the effect mediated by dopamine receptors is central or peripheral.

PHYSIOLOGY OF GROWTH

Growth is a complex phenomenon that is affected not only by growth hormone and somatomedins but also by thyroid hormones, androgens, estrogens, glucocorticoids, and insulin. It is also affected, of course, by genetic factors, and it depends on adequate nutrition. It is normally accompanied by an orderly sequence of maturational changes, and it involves accretion of protein and increase in length and size, not just an increase in weight, which may be due to the formation of fat or retention of salt and water.

Role of Nutrition

The food supply is the most important extrinsic factor affecting growth. The diet must be adequate not only in protein content but also in essential vitamins and minerals (see Chapter 17) and in calories, so that ingested protein is not burned for energy. However, the age at which a dietary deficiency occurs appears to be an important consideration. For example, once the pubertal growth spurt has commenced, considerable linear growth continues even if caloric intake is reduced. Injury and disease stunt growth because they increase protein catabolism.

Growth Periods

Patterns of growth vary somewhat from species to species. Rats continue to grow, although at a declining rate, throughout life. In humans, there are 2 periods of rapid growth (Fig 22–12), the first in infancy and the second in late puberty just before growth stops. The first period of accelerated growth is partly

Figure 22–12. Rate of growth in boys and girls from birth to age 20.

a continuation of the fetal growth period. The second growth spurt at the time of puberty is due to growth hormone, androgens, and estrogens, and the subsequent cessation of growth is due in large part to the action of androgens and estrogens. Since girls mature earlier than boys, this growth spurt appears earlier in girls. Of course, in both sexes the rate of growth of individual tissues varies (Fig 22–13).

Following illnesses in children, there is a period of "catch-up growth" during which the growth rate may be as much as 400% above normal. The acceleration continues until the previous growth curve is reached, at which point growth slows to normal. The mechanisms that bring about and control "catch-up growth" are not known.

Hormonal Effects

The contributions of hormones to growth after birth are shown diagrammatically in Fig 22–14. In laboratory animals and in humans, growth in utero is independent of fetal growth hormone. Rats hypophysectomized at birth continue to grow for about 30 days, and then growth stops. Interestingly, the brain continues to grow in these rats after growth of the skull stops, and the resulting compression of the brain may lead to death.

As noted above, plasma growth hormone is elevated in newborns. Subsequently, average resting levels fall, but the spikes of growth hormone secretion are larger during puberty so the mean plasma level over 24 hours is increased. One of the factors

Figure 22–13. Growth of different tissues at various ages as percentage of size at age 20. The curves are composites that include both boys and girls.

stimulating IGF-I secretion is growth hormone, and plasma IGF-I levels rise during childhood, reaching a peak at 13–17 years of age. In contrast, IGF-II levels are constant throughout postnatal growth.

The growth spurt that occurs at the time of puberty (Fig 22–12) is due in part to the protein anabolic effect of androgens, and the secretion of adrenal androgens increases at this time in both sexes. However, it is also due to an interaction between sex steroids, growth hormone, and IGF-I. Treatment with estrogens and androgens increases the growth hormone responses to stimuli such as insulin and arginine. Sex steroids also increase plasma IGF-I but fail to produce this increase in individuals with growth hor-

mone deficiency. Thus, it appears that the sex hormones produce an increase in the amplitude of the spikes in growth hormone secretion that increases IGF-I secretion, and this in turn causes growth.

Although androgens and estrogens initially stimulate growth, they ultimately terminate growth by causing the epiphyses to fuse to the long bones (epiphyseal closure). Once the epiphyses have closed, linear growth ceases (see Chapter 21). This is why pituitary dwarfs treated with testosterone first grow a few inches and then stop. It is also why patients with sexual precocity are apt to be dwarfed. On the other hand, men who were castrated before puberty tend to be tall because the epiphyses remain open and some growth continues past the normal age of puberty.

When growth hormone is administered to hypophysectomized animals, the animals do not grow as rapidly as they do when treated with growth hormone plus thyroid hormones. Thyroid hormones alone have no effect on growth in this situation. Their action is therefore permissive to that of growth hormone, possibly via potentiation of the actions of somatomedins. Thyroid hormones also appear to be necessary for a completely normal rate of growth hormone secretion; basal growth hormone levels are normal in hypothyroidism, but the response to hypoglycemia is frequently subnormal in hypothyroid children. Thyroid hormones have widespread effects on the ossification of cartilage, the growth of teeth, the contours of the face, and the proportions of the body. Cretins are therefore dwarfed and have infantile features (Fig 22–15). Patients who are dwarfed because of panhypopituitarism have features consistent with their chronologic age until puberty, but since they do not mature sexually, they have juvenile features in adulthood.

The effect of insulin on growth is discussed in Chapter 19. Diabetic animals fail to grow, and insulin causes growth in hypophysectomized animals. However, the growth is appreciable only when large amounts of carbohydrate and protein are supplied with the insulin.

Adrenocortical hormones other than androgens exert a permissive action on growth in the sense that adrenalectomized animals fail to grow unless their blood pressures and circulations are maintained by replacement therapy. On the other hand, glucocorticoids are potent inhibitors of growth because of their direct action on cells, and treatment of children with pharmacologic doses of steroids slows or stops growth for as long as the treatment is continued.

Dwarfism

Short stature can be due to GRH deficiency, growth hormone deficiency, deficient secretion of IGF-I, or other causes. Isolated growth hormone deficiency is often due to GRH deficiency, and in these instances, the growth hormone response to GRH is normal. However, some patients with isolated growth

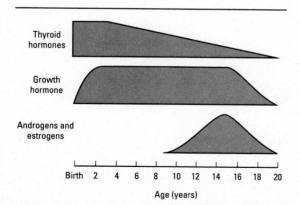

Figure 22–14. Relative importance of hormones in human growth at various ages. (Courtesy of DA Fisher.)

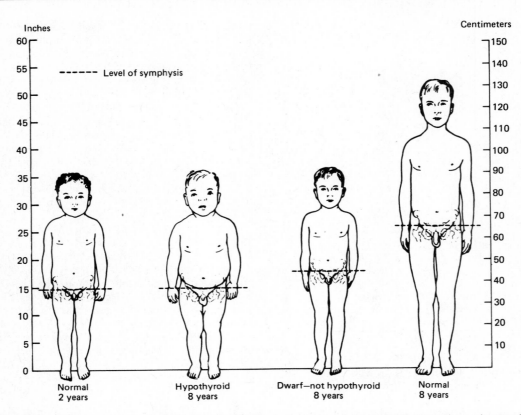

Figure 22–15. Normal and abnormal growth. Hypothyroid dwarfs (cretins) retain their infantile proportions, whereas dwarfs of the constitutional type and, to a lesser extent, of the hypopituitary type have proportions characteristic of their chronologic age. (Reproduced, with permission, from Wilkins L: *The Diagnosis and Treatment of Endocrine Disorders in Childhood and Adolescence,* 3rd ed. Thomas, 1966.)

hormone deficiency have abnormalities of the growth hormone-secreting cells. In another group of dwarfed children, the plasma growth hormone concentration is normal or elevated but there is end-organ unresponsiveness to growth hormone and consequently a deficiency of circulating growth factors. In this syndrome, which is known as **growth hormone insensitivity syndrome** or **Laron dwarfism,** there are congenital abnormalities of the growth hormone receptor, so the level of high-affinity growth hormone-binding protein in the plasma is low and IGF-I is not secreted in adequate amounts. African pygmies have normal plasma growth hormone levels, but there is a modest reduction in the level of plasma high-affinity growth hormone-binding protein. The plasma IGF-I concentration fails to increase at the time of puberty. However, there is less growth than in nonpygmy controls throughout the prepuberal period. Thus the explanation for the short stature of pygmies is still unsettled.

As noted above, short stature is characteristic of cretinism and occurs in patients with precocious puberty. It is also part of the syndrome of **gonadal dys-** **genesis** seen in patients who have an XO chromosomal pattern instead of an XX or XY pattern (see Chapter 23). Various bone and metabolic diseases also cause stunted growth, and in many cases there is no known cause ("constitutional delayed growth").

PITUITARY INSUFFICIENCY

Changes in Other Endocrine Glands

The widespread changes that develop when the pituitary is removed surgically or destroyed by disease in humans or animals are predictable in terms of the known hormonal functions of the gland. The adrenal cortex atrophies, and the secretion of adrenal glucocorticoids and sex hormones falls to low levels, although some secretion persists. Stress-induced rises in aldosterone secretion are absent, but basal aldosterone secretion and increases induced by salt depletion are normal, at least for some time. Since there is no mineralocorticoid deficiency, salt loss and hypovolemic shock do not develop, but the inability to increase glucocorticoid secretion makes patients with pituitary insufficiency sensitive to stress. The development of salt loss in long-standing hypopituitarism is discussed in Chapter 20. Growth is inhibited (see above). Thyroid function is depressed to low levels, and cold is tolerated poorly. The gonads atrophy, sexual cycles stop, and some of the secondary sex characteristics disappear.

Insulin Sensitivity

Hypophysectomized animals have a tendency to become hypoglycemic, especially when fasted. In some species, but not in humans, fatal hypoglycemic reactions are fairly common. Hypophysectomy ameliorates diabetes mellitus (see Chapter 19) and markedly increases the hypoglycemic effect of insulin. This is due in part to the deficiency of adrenocortical hormones, but hypophysectomized animals are more sensitive to insulin than adrenalectomized animals because they also lack the anti-insulin effect of growth hormone.

Water Metabolism

Although selective destruction of the supraoptic-posterior pituitary mechanism causes diabetes insipidus (see Chapter 14), removal of both the anterior and posterior pituitary usually causes no more than a transient polyuria. In the past, there was speculation that the anterior pituitary secreted a "diuretic hormone," but the amelioration of the diabetes insipidus is explained on the basis of a decrease in the osmotic load presented for excretion. Osmotically active particles hold water in the renal tubules (see Chapter 38). Because of the ACTH deficiency, the rate of protein catabolism is decreased in hypophysectomized animals. Because of the TSH deficiency, the metabolic rate is low. Consequently, fewer osmotically active products of catabolism are filtered and urine volume declines, even in the absence of vasopressin. Growth hormone deficiency contributes to the depression of the glomerular filtration rate in hypophysectomized animals, and growth hormone increases the glomerular filtration rate and renal plasma flow in humans. Finally, because of the glucocorticoid deficiency, there is the same defective excretion of a water load that is seen in adrenalectomized animals. The "diuretic" activity of the anterior pituitary can thus be explained in terms of the actions of ACTH, TSH, and growth hormone.

Other Defects

The deficiency of ACTH and other pituitary hormones with MSH activity may be responsible for the pallor of the skin in patients with hypopituitarism. There may be some loss of protein in adults, but it should be emphasized that wasting is not a feature of hypopituitarism in humans, and most patients with pituitary insufficiency are well nourished (Fig 22–16). It used to be thought that cachexia was part of the clinical picture, but it is now generally ac-

Figure 22–16. Typical picture of hypopituitarism in adults. Note the well-nourished appearance and pallor. (Reproduced, with permission, from Daughaday WH: The adenohypophysis. In: *Textbook of Endocrinology*, 5th ed. Williams RH [editor]. Saunders, 1974.)

cepted that emaciated patients described in the older literature had anorexia nervosa rather than hypopituitarism.

Causes of Pituitary Insufficiency in Humans

Tumors of the anterior pituitary are generally classified on the basis of staining characteristics as chromophobe, acidophil, or basophil tumors. Some of the apparently nonfunctioning tumors are chromophobe tumors, and they produce hypopituitarism by destroying normal pituitary tissue. Up to 70% of patients with chromophobe tumors have high circulating levels of prolactin. This indicates that the tumors secrete prolactin, although pressure on the hypothalamus or pituitary stalk interrupting the transport of dopamine from the hypothalamus to the anterior pituitary could also contribute. Suprasellar cysts, remnants of Rathke's pouch that enlarge and compress the pituitary, are another cause of hypopituitarism. In women who have an episode of shock in the course of pregnancy, the pituitary may become infarcted, with the subsequent development of postpartum necrosis. The blood supply to the anterior lobe is vulnerable because it descends on the pituitary stalk through the rigid diaphragma sellae, and during pregnancy the pituitary is enlarged. Pituitary infarction is usually extremely rare in men, but it was fairly common in soldiers who contracted hemorrhagic fever in Korea. In hemorrhagic fever, there was a diffuse vasculitis that probably caused pituitary enlargement as a result of edema, and the patients who developed pituitary infarction were those who went into shock in the course of their disease.

Partial Pituitary Insufficiency

The anterior pituitary has a large reserve, and much of it can be destroyed without producing detectable endocrine abnormalities. It now appears that with progressive loss of pituitary tissue, growth hormone secretion is the first function to be impaired. In dogs and rats, there is a depression of gonadotropin secretion when 70–90% of the anterior pituitary is removed or destroyed. When 90–95% is destroyed, thyroid function is also impaired, but almost 100% of the anterior pituitary must be destroyed before a marked degree of adrenal insufficiency is seen (Table 22–4). The same relationships hold in humans. Isolated pituitary tropin deficiencies occur, but they are often due to hypothalamic rather than pituitary disorders. For example, an appreciable number of patients with isolated TSH deficiency show an increase in TSH secretion when TRH is injected. Similarly, some patients with gonadotropin deficiency respond to GnRH.

In rats and in some other species, there appears to be some localization within the pituitary of the cells secreting the various hormones. It is not known if similar localization occurs in humans.

PITUITARY HYPERFUNCTION IN HUMANS

Acromegaly

Tumors of the somatotropes of the anterior pituitary secrete large amounts of growth hormone, leading in children to **gigantism** and in adults to **acromegaly.** Hypersecretion of growth hormone is accompanied by hypersecretion of prolactin in 20–40% of patients with acromegaly because they have tumors made up of mammosomatotropes. In addition, secretion of glycoprotein hormone α subunits is said to occur in up to 37% of patients. Acromegaly can be caused by extrapituitary as well as intrapituitary growth hormone-secreting tumors and by hypothalamic tumors that secrete GRH, but these are rare.

The principal findings in acromegaly are those related to the local effects of the tumor (enlargement of the sella turcica, headache, visual disturbances) and those due to growth hormone secretion. In adults, there is enlargement of the hands and feet (**acral**

Table 22–4. Effects of removing various amounts of pituitary tissue on endocrine function in male dogs.[1]

Number of Dogs	Gonadal Atrophy	Thyroid Atrophy	Adrenal Atrophy	Mean % Anterior Pituitary Remaining
6	0	0	0	27
3	+	0	0	11
2	+	+	0	5
19	+	+	+	2

[1](+) indicates that the finding was present and (0) that it was absent.

parts; hence the term acromegaly) and a protrusion of the lower jaw (**prognathism**) (Fig 22–7). Overgrowth of the malar, frontal, and basal bones combines with prognathism to produce the coarse facial features called **acromegalic facies.** Body hair is increased in amount. The skeletal changes predispose to osteoarthritis. About 25% of patients have abnormal glucose tolerance tests, and 4% develop lactation in the absence of pregnancy.

Cushing's Syndrome

The clinical picture of Cushing's syndrome and its various causes are described in Chapter 20. Many patients with bilaterally hyperplastic adrenals have small ACTH-secreting pituitary tumors (microadenomas) that are difficult to detect. However, a significant percentage of the patients who have bilaterally hyperplastic adrenals removed develop rapidly growing ACTH-secreting pituitary tumors (**Nelson's syndrome).** These tumors cause hyperpigmentation of the skin and neurologic signs due to pressure on structures in the sellar region. Most of them are made up of chromophobic cells rather than basophils, and some are malignant. Blood ACTH levels are extremely high, and the intrinsic MSH activity of the ACTH probably accounts for the cutaneous pigmentation. It is difficult to say whether these patients had undetected tumors to start with or developed neoplastic changes in the pituitary when the feedback check on ACTH secretion was removed.

Other Hormone-Secreting Tumors

Animals sometimes develop TSH-secreting tumors after thyroidectomy and gonadotropin-secreting tumors after gonadectomy. TSH-secreting tumors are rare in humans. On the other hand, prolactin-secreting tumors are common (see Chapter 23). In addition, it has been discovered that many apparently nonsecretory pituitary tumors in women secrete gonadotropins, their α or β subunits, or both. Secretion of those substances does not cause clinical abnormalities, but elevated concentrations in the blood can be detected by suitable laboratory tests.

The Gonads: Development & Function of the Reproductive System

INTRODUCTION

Modern genetics and experimental embryology make it clear that, in most species of mammals, the multiple differences between the male and the female depend primarily on a single chromosome (the Y chromosome) and a single pair of endocrine structures, the testes in the male and the ovaries in the female. The differentiation of the primitive gonads into testes or ovaries in utero is genetically determined in humans, but the formation of male genitalia depends upon the presence of a functional, secreting testis. There is evidence that male sexual behavior and, in some species, the male pattern of gonadotropin secretion are due to the action of male hormones on the brain in early development. After birth, the gonads remain quiescent until adolescence, when they are activated by gonadotropins from the anterior pituitary. Hormones secreted by the gonads at this time cause the appearance of features typical of the adult male or female and the onset of the sexual cycle in the female. In males, the gonads remain more or less active from puberty onward. In human females, ovarian function regresses after a number of years and sexual cycles cease (the menopause).

In both sexes, the gonads have a dual function: the production of germ cells (**gametogenesis**) and the secretion of **sex hormones.** The **androgens** are the steroid sex hormones that are masculinizing in their action; the **estrogens** are those that are feminizing. Both types of hormones are normally secreted in both sexes. The testes secrete large amounts of androgens, principally **testosterone,** but they also secrete small amounts of estrogens. The ovaries secrete large amounts of estrogens and small amounts of androgens. Androgens are secreted from the adrenal cortex in both sexes, and soma of the androgens are converted to estrogens in fat and other tissues. The ovaries also secrete **progesterone,** a steroid that has special functions in preparing the uterus for pregnancy.

During pregnancy, the ovaries secrete the peptide hormone **relaxin,** which loosens the ligaments of the pubic symphysis and softens the cervix, facilitating delivery of the fetus. In both sexes, the gonads secrete other polypeptides, including **inhibin,** a polypeptide that inhibits FSH secretion.

The secretory and gametogenic functions of the gonads are both dependent upon the secretion of the anterior pituitary gonadotropins, FSH and LH. The sex hormones and inhibin feed back to inhibit gonadotropin secretion. In males, gonadotropin secretion is noncyclic; but in postpubertal females an orderly, sequential secretion of gonadotropins is necessary for the occurrence of menstruation, pregnancy, and lactation.

SEX DIFFERENTIATION & DEVELOPMENT

CHROMOSOMAL SEX

The Sex Chromosomes

Sex is determined genetically by 2 chromosomes, called the **sex chromosomes** to distinguish them from the **somatic chromosomes (autosomes).** In humans and many other mammals, the sex chromosomes are called X and Y chromosomes. The Y chromosome is necessary and sufficient for the production of testes, and the testis-determining gene product is in all probability encoded by a gene called SRY (for sex-determining region of the Y chromosome). This gene is located on the short arm of the human Y chromosome. Male cells with the diploid number of chromosomes contain an X and a Y chromosome (XY pattern), whereas female cells contain two X chromosomes (XX pattern). As a consequence of meiosis during gametogenesis, each normal ovum contains a single X chromosome, but half the normal sperms contain an X chromosome and half a Y chromosome (Fig 23–1). When a sperm containing a Y chromosome fertilizes an ovum, an XY pattern results and the zygote develops into a **genetic male.** When fertilization occurs with an X-containing sperm, an XX pattern and a **genetic female** result. Cell division and the chemical nature of chromosomes are discussed in Chapter 1.

Human Chromosomes

Human chromosomes can be studied in detail. Human cells are grown in tissue culture; treated with the drug colchicine, which arrests mitosis at the metaphase; exposed to a hypotonic solution that makes the chromosomes swell and disperse; and then "squashed" onto slides. Fluorescent and other staining techniques make it possible to identify the individual chromosomes and study them in detail (Fig 23–2). There are 46 chromosomes: in males, 22 pairs of autosomes plus an X chromosome and a small Y chromosome; in females, 22 pairs of autosomes plus two X chromosomes. The individual chromosomes are usually arranged in an arbitrary pattern (**karyotype**). The individual autosome pairs are identified by the numbers 1–22 on the basis of their morphologic characteristics. The human Y chromosome is smaller than the X chromosome, and it has been hypothesized that sperms containing the Y chromosome are lighter and able to "swim" faster up the female genital tract, thus reaching the ovum more rapidly. This supposedly accounts for the fact that the number of males born is slightly greater than the number of females.

Figure 23–2. Karyotype of chromosomes from a normal male. The chromosomes have been stained with Giemsa's stain, which produces a characteristic banding pattern. This banding and the patterns produced by other techniques make it possible to identify individual chromosomes with great accuracy. (Reproduced, with permission, from Conte FA, Grumbach MM: Abnormalities of sexual differentiation. In: *Basic & Clinical Endocrinology,* 3rd ed. Greenspan FS, Forsham PH [editors]. Appleton & Lange, 1991.)

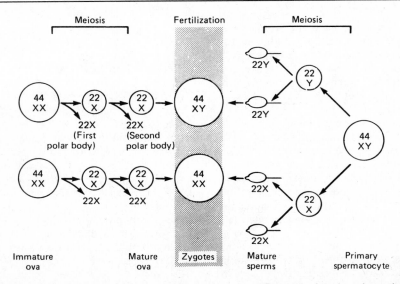

Figure 23–1. Basis of genetic sex determination. In the 2-stage meiotic division in the female, only one body survives as the mature ovum. In the male, the meiotic division results in the formation of 4 sperms, 2 containing the X and 2 the Y chromosome. Fertilization thus produces a 44XY (male) zygote or a 44XX (female) zygote.

Sex Chromatin

Soon after cell division has started during embryonic development, one or the other of the two X chromosomes of the somatic cells in normal females becomes functionally inactive. In abnormal individuals with more than 2 X chromosomes, only one remains active. The process that is normally responsible for inactivation appears to be initiated in an X-inactivation center in the chromosome, and probably inactivation by methylation of DNA spreads in both directions from the center. In any case, the choice of which X chromosome remains active is random, so normally one X chromosome remains active in approximately half of the cells and the other X chromosome is active in the other half. The selection persists through subsequent divisions of these cells, and consequently some of the somatic cells in adult females contain an active X chromosome of paternal origin and some contain an active X chromosome of maternal origin.

In normal cells, the inactive X chromosome condenses and can be seen in various types of cells, usually near the nuclear membrane, as the **Barr body** (Fig 23–3). Thus, there is a Barr body for each X chromosome in excess of one in the cell. The inactive X chromosome is also visible as a small "drumstick" of chromatin projecting from the nuclei of 1–15% of the polymorpho-nuclear leukocytes in females but not in males (Fig 23–3).

The Barr body has also been called sex chromatin. Another form of sex chromatin is the **F body,** a portion of the Y chromosome that can be stained by modern fluorescence techniques. Thus, the number of Y chromosomes in a cell can be estimated by the number of F bodies it contains.

EMBRYOLOGY OF THE HUMAN REPRODUCTIVE SYSTEM

Development of the Gonads

On each side of the embryo, a primitive gonad arises from the genital ridge, a condensation of tissue near the adrenal gland. The gonad develops a **cortex** and a **medulla** (Fig 23–4). Until the sixth week of development, these structures are identical in both sexes. In genetic males, the medulla develops during the seventh and eighth weeks into a testis, and the cortex regresses (Fig 23–5). The testis-determining gene responsible for this development is located on the short arm of the Y chromosome (see above). The mechanism of action of the product of SRY is unsettled. It used to be thought that a different protein, the H-Y antigen, was responsible for development of testes, but the H-Y antigen is produced by a gene on the long arm of the Y chromosome. Leydig cells appear next, and testosterone and müllerian inhibiting substance (see below) are secreted. In genetic females, the cortex develops into an ovary and the medulla regresses. The embryonic ovary does not secrete hormones. Hormonal treatment of the mother has no effect on gonadal (as opposed to ductal and genital) differentiation in humans, although it does in some experimental animals.

Embryology of the Genitalia

In the seventh week of gestation, the embryo has both male and female primordial genital ducts (Fig 23–6). In a normal female fetus, the müllerian duct system then develops into uterine tubes (oviducts) and a uterus. In the normal male fetus, the wolffian duct system on each side develops into the epididymis and vas deferens. The external genitalia are similarly bipotential until the eighth week (Fig 23–7). Thereafter, the urogenital slit disappears and male genitalia form, or, alternatively, it remains open and female genitalia form.

When there are functional testes in the embryo, male internal and external genitalia develop. The Leydig cells of the fetal testis secrete testosterone, and the Sertoli cells secrete **müllerian inhibiting substance** (MIS; also called müllerian regression factor, or MRF). MIS is a 535-amino-acid dimeric glycoprotein that is related to inhibin (see below). In their effects on the internal as opposed to the external genitalia, MIS and testosterone act unilaterally. MIS

Figure 23–3. Barr body (arrows) in the epidermal spinous cell layer *(left)* and nuclear appendage ("drumstick") in white blood cells *(right).* (Reproduced, with permission, from Grumbach MM, Barr ML: Cytologic tests of chromosomal sex in relation to sex anomalies in man. *Recent Prog Horm Res* 1958;**14**:255.)

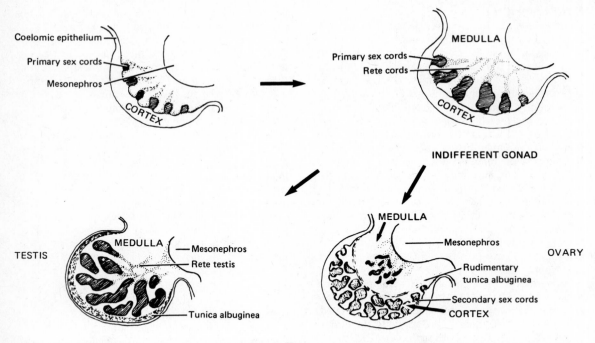

Figure 23–4. Diagrammatic representation of the development of an ovary from the cortex or a testis from the medulla of a bipotential primordial human gonad. (Reproduced, with permission, from Grumbach M in: *Clinical Endocrinology I.* Astwood EB [editor]. Grune & Stratton, 1960.)

Figure 23–5. Diagrammatic summary of normal sex determination, differentiation, and development in humans. MIS, müllerian inhibiting substance; T, testosterone or other androgen.

Figure 23–6. Embryonic differentiation of male and female internal genitalia (genital ducts) from wolffian (male) and müllerian (female) primordia. (After Corning HK, Wilkins L. Redrawn and reproduced, with permission, from Van Wyk J, Grumbach M in: *Textbook of Endocrinology,* 5th ed. Williams RH [editor]. Saunders, 1974.)

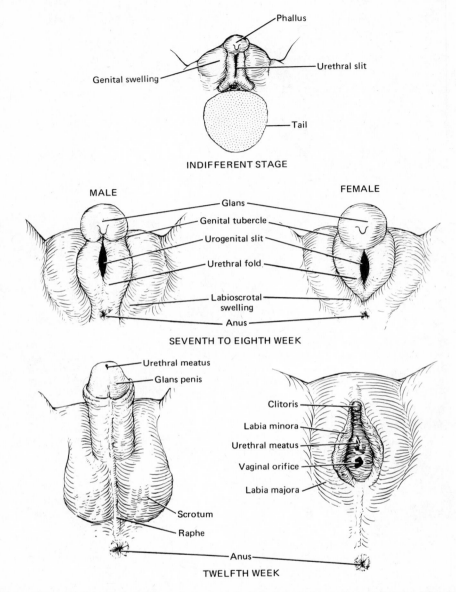

Figure 23–7. Differentiation of male and female external genitalia from indifferent primordial structures in the embryo.

causes regression of the müllerian ducts on the side on which it is secreted, and testosterone fosters the development of the vas deferens and related structures from the wolffian ducts. Testosterone induces the formation of male external genitalia (Fig 23–5). MIS is also produced by the granulosa cells of the ovarian follicles, where its function is unknown. It now appears that MIS requires a high androgen to estrogen ratio to exert its effects on the müllerian ducts, and this is present in the male but not in the female fetus, so the ducts persist in the female despite the MIS in the ovaries.

Development of the Brain

At least in some species, the development of the brain as well as the external genitalia is affected by androgens early in life. In rats, a brief exposure to androgens during the first few days of life causes the male pattern of sexual behavior and the male pattern of hypothalamic control of gonadotropin secretion to develop after puberty. In the absence of androgens, female patterns develop (see Chapter 15). In monkeys, similar effects on sexual behavior are produced by exposure to androgens in utero, but the pattern of gonadotropin secretion remains cyclic. Early exposure of female human fetuses to androgens also appears to cause subtle but significant masculinizing effects on behavior. However, women with adrenogenital syndrome due to congenital adrenocortical enzyme deficiency (see Chapter 20) develop normal menstrual cycles when treated with cortisol. Thus, the human, like the monkey, appears to retain the cyclic pattern of gonadotropin secretion despite the exposure to androgens in utero.

ABERRANT SEXUAL DIFFERENTIATION

Chromosomal Abnormalities

From the preceding discussion, it might be expected that abnormalities of sexual development could be caused by genetic or hormonal abnormalities as well as by other nonspecific teratogenic influences, and this is indeed the case. The major classes of abnormalities are listed in Table 23–1.

An established defect in gametogenesis is **nondisjunction,** a phenomenon in which a pair of chromosomes fail to separate, so that both go to one of the daughter cells during meiosis. Four of the abnormal zygotes that can form as a result of nondisjunction of one of the X chromosomes during oogenesis are shown in Fig 23–8. In individuals with the XO chromosomal pattern, the gonads are rudimentary or absent, so that female external genitalia develop. Stature is short, other congenital abnormalities are often present, and no maturation occurs at puberty. This syndrome is now called **gonadal dysgenesis** or, alternatively, **ovarian agenesis** or **Turner's syndrome.** Individuals with the XXY pattern, the most common sex chromosome disorder, have the genitalia of a normal male. Testosterone secretion at puberty is often great enough for the development of male characteristics. However, the seminiferous tubules are abnor-

Figure 23–8. Summary of 4 possible defects produced by maternal nondisjunction of the sex chromosomes at the time of meiosis. The YO combination is believed to be lethal, and the fetus dies in utero.

mal, and there is a higher than normal incidence of mental retardation. This syndrome is known as **seminiferous tubule dysgenesis** or **Klinefelter's syndrome.** The XXX ("superfemale") pattern is second in frequency only to the XXY pattern and may be even more common in the general population, since it does not seem to be associated with any characteristic abnormalities. The YO combination is probably lethal.

Meiosis is a 2-stage process, and although nondisjunction usually occurs during the first meiotic division, it can occur in the second, producing more complex chromosomal abnormalities. In addition, nondisjunction or simple loss of a sex chromosome can occur during the early mitotic divisions after fertilization. The result of faulty mitoses in the early zygote is the production of a mosaic, an individual with 2 or more populations of cells with different chromosome complements. **True hermaphroditism,** the condition in which the individual has both ovaries and testes, is probably due to XX/XY mosaicism and related mosaic patterns, although other genetic aberrations are possible.

Chromosomal abnormalities also include transposition of parts of chromosomes to other chromosomes. Rarely, genetic males are found to have the XX karyotype because the short arm of their father's Y chromosome was transposed to their father's X

Table 23–1. Classification of the major disorders of sex differentiation in humans. Many of these syndromes can have great variation in degree and, consequently, in manifestations.

Chromosomal disorders
  Gonadal dysgenesis (XO and variants)
  "Superfemales" (XXX)
  Seminiferous tubule dysgenesis (XXY and variants)
  True hermaphroditism
Developmental disorders
  Female pseudohermaphroditism
    Congenital virilizing adrenal hyperplasia of fetus
    Maternal androgen excess
      Virilizing ovarian tumor
      Iatrogenic: Treatment with androgens or certain synthetic progestational drugs
  Male pseudohermaphroditism
    Androgen resistance
    Defective testicular development
  Congenital 17α-hydroxylase deficiency
  Congenital adrenal hyperplasia due to blockade of pregnenolone formation
  Various nonhormonal anomalies

chromosome during meiosis and they received that X chromosome along with their mother's. Similarly, deletion of the small portion of the Y chromosome containing SRY produces females with the XY karyotype.

Sex chromosome abnormalities are of course not the only abnormalities associated with disease states; nondisjunction of several different autosomal chromosomes is known to occur. For example, nondisjunction of chromosome 21 (Fig 23–2) produces **trisomy 21,** the chromosomal abnormality associated with **Down's syndrome** (mongolism). In most instances, nondisjunction occurs in the ovary rather than the testis and in the first rather than the second meiotic division.

There are many other chromosomal abnormalities as well as numerous diseases due to defects in single genes. These conditions are generally diagnosed in utero by analysis of fetal cells in a sample of amniotic fluid collected by inserting a needle in the uterus (**amniocentesis**) or, earlier in pregnancy, by examining fetal cells obtained by a biopsy of chorionic villi (**chorionic villus sampling).**

Hormonal Abnormalities

Development of the male external genitalia occurs normally in genetic males in response to androgen secreted by the embryonic testes, but male genital development may also occur in genetic females exposed to androgens from some other source during the eighth to the thirteenth weeks of gestation. The syndrome that results is **female pseudohermaphroditism.** A pseudohermaphrodite is an individual with the genetic constitution and gonads of one sex and the genitalia of the other. After the thirteenth week, the genitalia are fully formed, but exposure to androgens can cause hypertrophy of the clitoris. Female pseudohermaphroditism may be due to congenital virilizing adrenal hyperplasia (see Chapter 20), or it may be caused by androgens administered to the mother. Conversely, development of female external genitalia in genetic males (**male pseudohermaphroditism**) occurs when the embryonic testes are defective. Because the testes also secrete MIS, genetic males with defective testes have female internal genitalia.

Another cause of male pseudohermaphroditism is **androgen resistance,** in which, as a result of various congenital abnormalities, male hormones cannot exert their full effects on the tissues. One form of androgen resistance is a **5α-reductase deficiency,** in which the enzyme responsible for the formation of dihydrotestosterone, the active form of testosterone, is decreased. The consequences of this deficiency are discussed in the section on the male reproductive system. In other forms of androgen resistance, androgen receptors are absent or abnormal. When the loss of receptor function is complete, the **testicular feminizing syndrome** results. In this condition, MIS is present and testosterone is secreted at normal or even elevated rates. The external genitalia are female, but the vagina ends blindly because there are no female internal genitalia. Individuals with this syndrome develop enlarged breasts at puberty and usually are considered to be normal women until they are diagnosed when they seek medical advice because of lack of menstruation. Androgen resistance can also occur in the presence of normal androgen receptors when there are congenital defects in the molecular events that occur after the receptor binding (**receptor-positive androgen resistance).**

It is worth noting that genetic males with congenital blockade of the formation of pregnenolone are pseudohermaphrodites because testicular as well as adrenal androgens are normally formed from pregnenolone. Male pseudohermaphroditism also occurs when there is a congenital deficiency of 17α-hydroxylase (see Chapter 20).

PUBERTY

As noted above, there is a burst of testosterone secretion in male fetuses before birth. In the neonatal period there is another burst, whose function is unknown, but thereafter the Leydig cells become quiescent. There follows in all mammals a period in which the gonads of both sexes are quiescent until they are activated by gonadotropins from the pituitary to bring about the final maturation of the reproductive system. This period of final maturation is known as **adolescence.** It is often also called **puberty,** although puberty, strictly defined, is the period when the endocrine and gametogenic functions of the gonads have first developed to the point where reproduction is possible. In girls, the first event is **thelarche,** the development of breasts, followed by **pubarche,** the development of axillary and pubic hair, and then by **menarche,** the first menstrual period. The initial periods are generally anovulatory, and regular ovulation appears about a year later. In contrast to the situation in adulthood, removal of the gonads during the period from soon after birth to puberty causes little or no increase in gonadotropin secretion, so gonadotropin secretion is not being held in check by the gonadal hormones. In children between the ages of 7 and 10, a slow increase in estrogen and androgen secretion precedes the more rapid rise in the early teens (Fig 23–9).

The age at the time of puberty is variable. In Europe and the USA, it has been declining at the rate of 1–3 months per decade for more than 175 years. In the USA in recent years, puberty generally occurs between the ages of 8 and 13 in girls and 9 and 14 in boys.

Another event that occurs in humans at the time of puberty is an increase in the secretion of adrenal androgens (Fig 23–10). The onset of this increase is

Figure 23–10. Change in serum dehydroepiandrosterone sulfate (DHEAS) with age. There are no significant differences between males and females. The middle line is the mean, and the dashed lines identify ±1.96 standard deviations. (Reproduced, with permission, from Smith MR et al: A radioimmunoassay for the estimation of serum dehydroepiandrosterone sulfate in normal and pathological sera. Clin Chim Acta 1975;65:5.)

Figure 23–9. Changes in plasma hormone concentrations during puberty in boys *(top)* and girls *(bottom).* Stage 1 of puberty is preadolescence in both sexes. In boys, stage 2 is characterized by beginning enlargement of the testes, stage 3 by penile enlargement, stage 4 by growth of the glans penis, and stage 5 by adult genitalia. In girls, stage 2 is characterized by breast buds, stage 3 by elevation and enlargement of the breasts, stage 4 by projection of the areolas, and stage 5 by adult breasts. (Reproduced, with permission, from Grumbach MM: Onset of puberty. In: *Puberty: Biologic and Psychosocial Components.* Berenberg SR [editor]. HE Stenfoert Kroese BV, 1975.)

called **adrenarche.** It occurs at age 8–10 years, without any change in the secretion of cortisol or ACTH. It may be due to a change in the enzyme systems in the adrenal so that more pregnenolone is diverted to the androgen pathway (see Chapter 20). On the other hand, there is some evidence that it is due to increased secretion of an as yet unisolated **adrenal androgen-stimulating hormone (AASH)** from the pituitary gland.

Control of the Onset of Puberty

The gonads of children can be stimulated by gonadotropins; their pituitaries contain gonadotropins; and their hypothalami contain GnRH (see Chapter 14). However, their gonadotropins are not secreted. In immature monkeys, normal menstrual cycles can be brought on by pulsatile injection of GnRH, and they persist as long as the pulsatile injection is contin-

ued. In addition, GnRH is secreted in a pulsatile fashion in utero. Thus, it seems clear that during the period from birth to puberty, a neural mechanism is operating to prevent the normal pulsatile release of GnRH. The nature of the mechanism inhibiting the GnRH pulse generator is unknown.

PRECOCIOUS & DELAYED PUBERTY

Sexual Precocity

The major causes of precocious sexual development in humans are listed in Table 23–2. Early development of secondary sexual characteristics without gametogenesis is caused by abnormal exposure of immature males to androgen or females to estrogen. This syndrome should be called **precocious pseudopuberty** to distinguish it from **true precocious puberty** due to an early but otherwise normal pubertal pattern of gonadotropin secretion from the pituitary (Fig 23–11).

In one large series of cases, precocious puberty was the most frequent endocrine symptom of hypothalamic disease (see Chapter 14). It is interesting that in experimental animals and humans, lesions of the ventral hypothalamus near the infundibulum cause precocious puberty. The effect of the lesions may be due to interruption of neural pathways that produce inhibition of the GnRH pulse generator, or to chronic stimulation of GnRH secretion originating in irritative foci around the lesion. Pineal tumors are sometimes associated with precocious puberty, but there is evidence that these tumors are associated

Table 23-2. Classification of the causes of precocious sexual development in humans.

True precocious puberty
  Constitutional
  Cerebral: Disorders involving posterior hypo-
      thalamus
    Tumors
    Infections
    Developmental abnormalities
  Gonadotropin-independent precocity
Precocious pseudopuberty (no spermatogenesis
  or ovarian development)
  Adrenal
    Congenital virilizing adrenal hyperplasia (without
        treatment in males; following cortisone treat-
        ment in females)
    Androgen-secreting tumors (in males)
    Estrogen-secreting tumors (in females)
  Gonadal
    Leydig cell tumors of testis
    Granulosa cell tumors of ovary
  Miscellaneous

Figure 23–11. Constitutional precocious puberty in a 3½-year-old girl. The patient developed pubic hair and started to menstruate at the age of 17 months. (Reproduced, with permission, from Jolly H: *Sexual Precocity.* Thomas, 1955.)

with precocious puberty only when there is secondary damage to the hypothalamus. Precocity due to this and other forms of hypothalamic damage probably occurs with equal frequency in both sexes, although the constitutional form of precocious puberty is more common in girls. In addition, precocious gametogenesis and steroidogenesis can occur without the pubertal pattern of gonadotropin secretion (gonadotropin-independent precocity).

Delayed or Absent Puberty

The normal variation in the age at which adolescent changes occur is so wide that puberty cannot be considered to be pathologically delayed until the menarche has failed to occur by the age of 17 or testicular development by the age of 20. Failure of maturation due to panhypopituitarism is associated with dwarfing and evidence of other endocrine abnormalities. Patients with the XO chromosomal pattern and gonadal dysgenesis are also dwarfed. In some individuals, puberty is delayed even though the gonads are present and other endocrine functions are normal. In males, this clinical picture is called **eunuchoidism.** In females, it is called **primary amenorrhea** (see below).

MENOPAUSE

The human ovaries become unresponsive to gonadotropins with advancing age, and their function declines, so that sexual cycles disappear (**menopause**). This unresponsiveness is associated with and probably caused by a decline in the number of primordial follicles, which becomes precipitous at the time of menopause (Fig 23–12). The ovaries no longer secrete progesterone and 17β-estradiol in appreciable quantities, and estrogen is formed only in small amounts by aromatization of androstenedione in the circulation (see Chapter 20). The uterus and the vagina gradually become atrophic. As the negative feedback effect of estrogens and progesterone is reduced, secretion of FSH and LH is increased, and plasma FSH and LH increase to high levels. Old female mice and rats have long periods of diestrus and increased levels of gonadotropin secretion, but a clear-cut "menopause" has apparently not been described in animals.

In women, the menses usually become irregular and cease between the ages of 45 and 55. The average age at onset of the menopause has been increasing since the turn of the century and is currently 52 years. Sensations of warmth spreading from the trunk to the face ("hot flashes") and various psychic symptoms are common after ovarian function has ceased. The hot flashes, which also occur when early menopause is produced by bilateral ovariectomy, are prevented by estrogen treatment. Their cause is unknown. However, they coincide with surges of LH

Figure 23–12. Number of primordial follicles per ovary in women at various ages. □, premenopausal women (regular menses); ■, perimenopausal women (irregular menses for at least 1 year); ▲, postmenopausal women (no menses for at least 1 year). Note that the vertical scale is a log scale and that the values are from one rather than 2 ovaries. (Redrawn by PM Wise and reproduced, with permission, from Richardson SJ, Senikas V, Nelson JF: Follicular depletion during the menopausal transition: evidence for accelerated loss and ultimate exhaustion. *J Clin Endocrinol Metab* 1987;**65:**1231.)

secretion. LH is secreted in episodic bursts at intervals of 30–60 minutes or more (**circhoral secretion**), and in the absence of gonadal hormones these bursts are large. Each hot flash begins with the start of a burst. However, LH itself is not responsible for the symptoms, because they can continue after removal of the pituitary. Instead, it appears that some estrogen-sensitive event in the hypothalamus initiates both the release of LH and the episode of flushing.

Although the function of the testes tends to decline slowly with advancing age, the evidence is clear that there is no "male menopause" (**climacteric**) similar to that occurring in women.

PITUITARY GONADOTROPINS & PROLACTIN

Chemistry

FSH and LH are each made up of an α and a β subunit whose nature is discussed in Chapter 22. They are glycoproteins that contain the hexoses mannose and galactose, the hexosamines N-acetylgalactosamine and N-acetylglycosamine, and the methylpentose fucose. They also contain sialic acid. The carbohydrate in the gonadotropin molecules increases their potency by markedly slowing their metabolism. The half-life of human FSH is about 170 minutes; the half-life of LH is about 60 minutes.

Human pituitary prolactin contains 199 amino acid residues and 3 disulfide bridges (Fig 23–13) and has considerable structural similarity to human growth hormone and hCS. The half-life of prolactin, like that of growth hormone, is about 20 minutes. Structurally similar prolactins are secreted by the endometrium and by the placenta (see below).

Receptors

The receptors for FSH and LH are serpentine receptors coupled to adenylate cyclase through G_s (see Chapter 1). In addition, each has an extended, glycosylated extracellular domain.

The human prolactin receptor resembles the growth hormone receptor and is one of the superfamily of receptors that includes the growth hormone receptor and receptors for many cytokines and hematopoietic growth factors (see Chapters 1, 22, 24, and 27). When activated, its cytoplasmic domain may have tyrosine kinase activity, but the exact mechanism of signal transduction is unknown.

Actions

The testes and ovaries become atrophic when the pituitary is removed or destroyed. The actions of prolactin and the gonadotropins FSH and LH, as well as those of the gonadotropin secreted by the placenta, are described in detail in succeeding sections of this chapter. In brief, FSH helps maintain the spermatogenic epithelium by stimulating Sertoli cells in the male and is responsible for the early growth of ovarian follicles in the female. LH is tropic to the Leydig cells and, in females, is responsible for the final maturation of the ovarian follicles and estrogen secretion from them. It is also responsible for ovulation, the initial formation of the corpus luteum, and secretion of progesterone.

Prolactin causes milk secretion from the breast after estrogen and progesterone priming. Its effect on the breast involves increased action of mRNA and increased production of casein and lactalbumin. However, the action of the hormone is not exerted on the cell nucleus and is prevented by inhibitors of microtubules. Prolactin also inhibits the effects of gonadotropins, possibly by an action at the level of the ovary. Its role in preventing ovulation in lactating women is discussed below. The function of prolactin in males (if any) is unknown. Prolactin maintains the corpus luteum in rodents but is not luteotropic in humans; the human corpus luteum is maintained by LH. An action of prolactin that has been used as the basis for bioassay of this hormone is stimulation of the growth and "secretion" of the crop sacs in pigeons and other birds. The paired crop sacs are outpouchings of the esophagus which form, by desquamation of their inner cell layers, a nutritious material

Figure 23–13. Structure of human prolactin.

("milk") that the birds feed to their young. However, prolactin, FSH, and LH are now regularly measured by radioimmunoassay.

Regulation of Prolactin Secretion

The normal plasma prolactin concentration is approximately 5 ng/mL in men and 8 ng/mL in women. Secretion is tonically inhibited by the hypothalamus, and section of the pituitary stalk leads to an increase in circulating prolactin. Thus, the effect of hypothalamic prolactin-inhibiting hormone (PIH) is usually greater than the effect of the putative prolactin-releasing hormone (see Chapter 14). It now seems clear that PIH is dopamine secreted by the tuberoinfundibular dopaminergic neurons into the portal hypophyseal vessels. In humans, prolactin secretion is increased by exercise, surgical and psychologic stresses, and stimulation of the nipple (Table 23–3). The plasma prolactin level rises during sleep, the rise starting after the onset of sleep and persisting throughout the sleep period. Secretion is increased during pregnancy, reaching a peak at the time of parturition. After delivery, the plasma concentration falls to nonpregnant levels in about 8 days. Suckling produces a prompt increase in secretion, but the magnitude of this rise gradually declines after a woman has as been nursing for more than 3 months. With prolonged lactation, milk secretion occurs with prolactin levels that are in the normal range.

L-Dopa decreases prolactin secretion by increas-

Table 23–3. Factors affecting the secretion of human prolactin and growth hormone.[1]

Factor	Prolactin[2]	Growth Hormone[2]
Sleep	I+	I+
Nursing	I++	N
Breast stimulation in nonlactating women	I	N
Stress	I+	I+
Hypoglycemia	I	I+
Strenuous exercise	I	I
Sexual intercourse in women	I	N
Pregnancy	I++	N
Estrogens	I	I
Hypothyroidism	I	N
TRH	I+	N
Phenothiazines, butyrophenones	I+	N
Opiates	I	I
Glucose	N	D
Somatostatin	N	D+
L-Dopa	D+	I+
Apomorphine	D+	I+
Bromocriptine and related ergot derivatives	D+	I

[1] Modified from Frantz A: Prolactin. *N Engl J Med* 1978; **298**:201.
[2] I, moderate increase; I+, marked increase; I++, very marked increase; N, no change; D, moderate decrease; D+, marked decrease.

ing the formation of dopamine, and bromocriptine and other dopamine agonists inhibit secretion because they stimulate dopamine receptors. Chlorpromazine and related drugs that block dopamine receptors increase prolactin secretion. TRH stimulates the secretion of prolactin in addition to TSH, but it seems likely that there is an additional prolactin-releasing hormone (PRH) in hypothalamic tissue that is different from TRH. Estrogens also produce a slowly developing increase in prolactin secretion.

It has now been established that prolactin facilitates the secretion of dopamine in the median eminence. Thus, prolactin acts in the hypothalamus in a negative feedback fashion to inhibit its own secretion.

Hyperprolactinemia

Up to 70% of the patients with chromophobe adenomas of the anterior pituitary have elevated plasma prolactin levels. In some instances, the elevation may be due to damage to the pituitary stalk, but in most cases, the tumor cells are actually secreting the hormone. The hyperprolactinemia may cause galactorrhea, but in many individuals there are no demonstrable abnormalities. Indeed, most women with galactorrhea have normal prolactin levels; definite elevations are found in less than a third of patients with this condition.

Another interesting observation is that 15–20% of women with secondary amenorrhea have elevated prolactin levels and that when prolactin secretion is reduced, normal menstrual cycles and fertility return. It appears that the prolactin may produce amenorrhea by blocking the action of gonadotropins on the ovaries, but definitive proof of this hypothesis must await further research. The hypogonadism produced by prolactinomas is associated with osteoporosis due to estrogen deficiency.

In men, hyperprolactinemia is associated with impotence and hypogonadism that disappear when prolactin secretion is reduced.

THE MALE REPRODUCTIVE SYSTEM

STRUCTURE

The testes are made up of loops of convoluted **seminiferous tubules,** along the walls of which the spermatozoa are formed from the primitive germ cells **(spermatogenesis).** Both ends of each loop drain into a network of ducts in the head of the **epididymis.** From there, spermatozoa pass through the tail of the epididymis into the **vas deferens.** They enter through the **ejaculatory ducts** into the urethra in the body of the **prostate** at the time of ejaculation (Fig 23–14). Between the tubules in the testes are nests of cells containing lipid granules, the **interstitial cells of Leydig,** which secrete testosterone into the bloodstream (Figs 23–15 and 23–16). The spermatic arteries to the testes are tortuous, and blood in them runs parallel but in the opposite direction to blood in the pampiniform plexus of spermatic veins. This anatomic arrangement may permit countercurrent exchange of heat and testosterone. The principles of countercurrent exchange are considered in detail in relation to the kidney in Chapter 38.

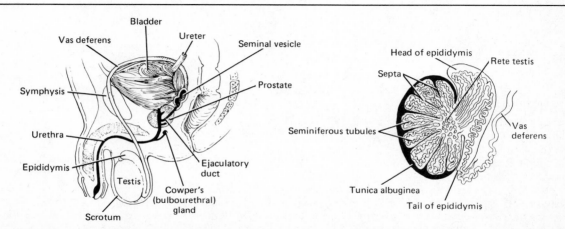

Figure 23–14. *Left:* Male reproductive system. *Right:* Duct system of the testis.

Figure 23–15. Section of human testis.

GAMETOGENESIS & EJACULATION

Blood-Testis Barrier

The walls of the seminiferous tubules are lined by primitive germ cells (see below) and by **Sertoli cells,** large, complex glycogen-containing cells that stretch from the basal lamina of the tubule to the lumen (Fig 23–16). Tight junctions between adjacent Sertoli cells near the basal lamina form a **blood-testis barrier** that prevents many large molecules from passing from the interstitial tissue and the part of the tubule near the basal lamina (basal compartment) to the region near the tubular lumen (adluminal compartment) and the lumen. However, steroids penetrate this barrier with ease, and there is evidence that some proteins pass from the Sertoli cells to the Leydig cells and vice versa in a paracrine fashion. In addition, maturing germ cells must pass through the barrier as they move to the lumen. This appears to occur without disruption of the barrier by progressive breakdown of the tight junctions above the germ cells, with concomitant formation of new tight junctions below them.

The fluid in the lumen of the seminiferous tubules is quite different from plasma; it contains very little protein and glucose but is rich in androgens, estro-

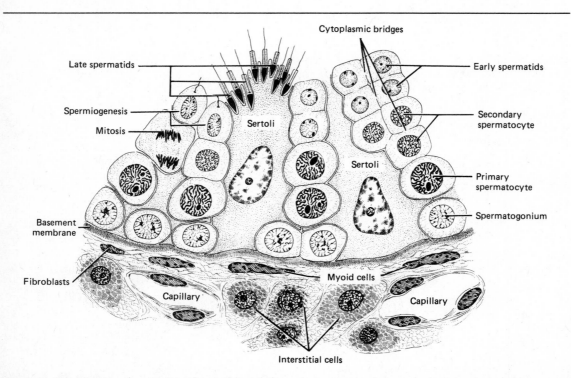

Figure 23–16. Seminiferous epithelium. Note that maturing germ cells remain connected by cytoplasmic bridges through the early spermatid stage and that these cells are closely invested by Sertoli cell cytoplasm as they move from the basal lamina to the lumen. (Reproduced, with permission, from Junqueira LC, Carneiro J, Kelley RO: *Basic Histology.* 7th ed. Appleton & Lange, 1992.)

gens, K+, inositol, and glutamic and aspartic acids. Maintenance of its composition presumably depends on the blood-testis barrier. The barrier also protects the germ cells from blood-borne noxious agents, prevents antigenic products of germ cell division and maturation from entering the circulation and generating an autoimmune response, and may help establish an osmotic gradient that facilitates movement of fluid into the tubular lumen.

Spermatogenesis

The **spermatogonia,** the primitive germ cells next to the basal lamina of the seminiferous tubules, mature into **primary spermatocytes** (Fig 23–16). This process begins during adolescence. The primary spermatocytes undergo meiotic division, reducing the number of chromosomes. In this 2-stage process, they divide into **secondary spermatocytes** and then into **spermatids,** which contain the haploid number of 23 chromosomes. The spermatids mature into **spermatozoa (sperms).** As a single spermatogonium divides and matures, its descendants remain tied together by cytoplasmic bridges until the late spermatid stage. This apparently ensures synchrony of the differentiation of each clone of germ cells. The estimated number of spermatids formed from a single spermatogonium is 512. In humans, it takes an average of 74 days to form a mature sperm from a primitive germ cell by this orderly process of spermatogenesis. Each sperm is an intricate motile cell, rich in DNA, with a head that is made up mostly of chromosomal material (Fig 23–17). Covering the head like a cap is the **acrosome,** a lysosome-like organelle rich in enzymes involved in sperm penetration of the ovum and other events involved in fertilization. The motile tail of the sperm is wrapped in its proximal portion by a sheath holding numerous mitochondria.

The spermatids mature into spermatozoa in deep folds of the cytoplasm of the Sertoli cells (Fig 23–16). Mature spermatozoa are released from the Sertoli cells and become free in the lumens of the tubules. The Sertoli cells secrete **androgen-binding protein (ABP), inhibin,** and **MIS.** They do not synthesize androgens, but they contain **aromatase,** the enzyme responsible for conversion of androgens to estrogens, and they can produce estrogens. ABP probably functions to maintain a high, stable supply of androgen in the tubular fluid. Inhibin inhibits FSH secretion (see below). MIS causes regression of the müllerian ducts in males during fetal life (see above).

FSH and androgens maintain the gametogenic function of the testis. After hypophysectomy, injection of LH produces a high local concentration of androgen in the testes, and this maintains spermatogenesis. There is some evidence that testosterone passes from the spermatic veins to the spermatic arteries as they parallel each other in the scrotum, thus helping to maintain a high androgen concentration in the testes. The stages from spermatogonia to spermatids appear to be androgen-independent. However, the maturation from spermatids to spermatozoa depends on androgen acting on the Sertoli cells in which the developing spermatozoa are embedded. FSH acts on the Sertoli cells to facilitate the last stages of spermatid maturation. In addition, it promotes the production of ABP.

Spermatozoa leaving the testes are not fully mobile. They continue their maturation and acquire motility during their passage through the epididymis. There is some evidence that sperm motility is further improved by relaxin, which is probably produced by the prostate. The capacity of sperm to produce fertilization is further enhanced if they spend some additional time in the female reproductive tract. The initial process occurring in the female reproductive tract is called **capacitation.** It makes the spermatozoa better able to adhere to the ovum. The subsequent events that occur when sperms attach to the ovum are discussed below, in the section on fertilization.

Effect of Temperature

Spermatogenesis requires a temperature consider-

Figure 23–17. Human spermatozoan.

ably lower than that of the interior of the body. The testes are normally maintained at a temperature of about 32 °C. They are kept cool by air circulating around the scrotum and probably by heat exchange in a countercurrent fashion between the spermatic arteries and veins. When the testes are retained in the abdomen or when, in experimental animals, they are held close to the body by tight cloth binders, degeneration of the tubular walls and sterility result. Hot baths (43–45 °C for 30 minutes per day) and insulated athletic supporters reduce the sperm count in humans, in some cases by 90%. However, the reductions produced in this manner are not consistent enough to make the procedures reliable forms of male contraception. In addition, there is evidence suggesting a seasonal effect in men, with sperm counts being greater in the winter regardless of the temperature to which the scrotum is exposed.

Semen

The fluid that is ejaculated at the time of orgasm, the **semen,** contains sperms and the secretions of the seminal vesicles, prostate, Cowper's glands, and, probably, the urethral glands (Table 23–4). An average volume per ejaculate is 2.5–3.5 mL after several days of continence. The volume of semen and the sperm count decrease rapidly with repeated ejaculation. Even though it takes only one sperm to fertilize the ovum, there are normally about 100 million sperms per milliliter of semen. Fifty percent of men

Table 23-4. Composition of human semen.

Color: White, opalescent	
Specific gravity: 1.028	
pH: 7.35–7.50	
Sperm count: Average about 100 million/mL, with fewer than 20% abnormal forms	
Other components:	
Fructose (1.5– 6.5 mg/mL) Phosphorylcholine Ergothioneine Ascorbic acid Flavins Prostaglandins	From seminal vesicles (contribute 60% of total volume)
Spermine Citric acid Cholesterol, phospholipids Fibrinolysin, fibrinogenase Zinc Acid phosphatase	From prostate (contributes 20% of total volume)
Phosphate Bicarbonate	Buffers
Hyaluronidase	

with counts of 20–40 million/mL and essentially all of those with counts under 20 million/mL are sterile. The **prostaglandins** in semen, which actually come from the seminal vesicles, are in high concentration, but the function of these fatty acid derivatives in semen is not known. Their structure and their multiple actions in other parts of the body are discussed in Chapter 17.

Human sperms move at a speed of about 3 mm/min through the female genital tract. Sperms reach the uterine tubes 30–60 minutes after copulation. In some species, contractions of the female organs facilitate the transport of the sperms to the uterine tubes, but it is not known if such contractions occur in humans.

Ejaculation

Ejaculation is a 2-part spinal reflex that involves **emission,** the movement of the semen into the urethra; and **ejaculation** proper, the propulsion of the semen out of the urethra at the time of orgasm. The afferent pathways are mostly fibers from touch receptors in the glans penis that reach the spinal cord through the internal pudendal nerves. Emission is a sympathetic response, integrated in the upper lumbar segments of the spinal cord and effected by contraction of the smooth muscle of the vasa deferentia and seminal vesicles in response to stimuli in the hypogastric nerves. The semen is propelled out of the urethra by contraction of the bulbocavernosus muscle, a skeletal muscle. The spinal reflex centers for this part of the reflex are in the upper sacral and lowest lumbar segments of the spinal cord, and the motor pathways traverse the first to third sacral roots and the internal pudendal nerves.

Erection

Erection is initiated by dilation of the arterioles of the penis. As the erectile tissue of the penis fills with blood, the veins are compressed, blocking outflow and adding to the turgor of the organ. The integrating centers in the lumbar segments of the spinal cord are activated by impulses in afferents from the genitalia and descending tracts that mediate erection in response to erotic psychic stimuli. The efferent parasympathetic fibers are in the pelvic splanchnic nerves **(nervi erigentes).** The fibers presumably contain acetylcholine and VIP as cotransmitters (see Chapter 4), and release of both produces the vasodilation. In any case, the role of VIP seems to be established. Injection of VIP into the erectile tissue produces erection, and injection of an antibody to VIP causes erection to subside. Normally erection is terminated by sympathetic vasoconstrictor impulses to the arterioles.

Vasectomy

Bilateral ligation of the vas deferens (vasectomy) has proved to be a relatively safe and convenient contraceptive procedure. However, it has proved difficult to restore the patency of the vas in those wishing to restore fertility, and the current success rate for such operations, as measured by the subsequent production of pregnancy, is about 50%. Half of the men who have been vasectomized develop antibodies against spermatozoa, and in monkeys, the presence of such antibodies is associated with a higher incidence of infertility after restoration of the patency of the vas. However, there do not appear to be any other adverse effects of the antisperm antibodies.

ENDOCRINE FUNCTION OF THE TESTES

Chemistry & Biosynthesis of Testosterone

Testosterone, the principal hormone of the testes, is a C_{19} steroid (see Chapter 20) with an $-$ OH group in the 17 position (Fig 23–18). It is synthesized from cholesterol in the Leydig cells and is also formed from androstenedione secreted by the adrenal cortex.

According to current concepts, the biosynthetic pathways in all endocrine organs that form steroid hormones are similar, the organs differing only in the enzyme systems they contain. In the Leydig cells, the 11- and 21-hydroxylases found in the adrenal cortex (see Fig 20–8) are absent, but 17α-hydroxylase is present. Pregnenolone is therefore hydroxylated in the 17 position and then subjected to side chain cleavage to form dehydroepiandrosterone. Androstenedione is also formed via progesterone and 17-hydroxyprogesterone, but this pathway is less prominent in humans. Dehydroepiandrosterone and androstenedione are then converted to testosterone.

The secretion of testosterone is under the control of LH, and the mechanism by which LH stimulates the Leydig cells involves increased formation of cyclic AMP via the serpentine LH receptor and G_s. Cyclic AMP increases the formation of cholesterol from cholesteryl esters and the conversion of cholesterol to pregnenolone via the activation of protein kinase A.

Figure 23–18. Biosynthesis and metabolism of testosterone. The formulas of the precursor steroids are shown in Fig 20–8. Although the main secretory product of the Leydig cells is testosterone, some of the precursors also enter the circulation. Note that dihydrotestosterone is formed in some target tissues and that the three 17-ketosteroids produced from testosterone in the liver are isomers of one another.

Secretion

The testosterone secretion rate is 4–9 mg/d (13.9–31.2 nmol/d) in normal adult males. Small amounts of testosterone are also secreted in females, probably from the ovary but possibly from the adrenal as well.

Transport & Metabolism

Ninety-eight percent of the testosterone in plasma is bound to protein: 65% is bound to a β-globulin called **gonadal steroid-binding globulin (GBG)** or **sex steroid-binding globulin,** and 33% to albumin (Table 23–5). GBG also binds estradiol. The plasma testosterone level (free and bound) is approximately 525 ng/dL (18.2 nmol/L) in adult men (Figure 23–19) and 30 ng/dL (1.0 nmol/L) in adult women. It declines somewhat with age in males.

A small amount of circulating testosterone is converted to estrogen somewhere in the body (see below), but most of the testosterone is converted into 17-ketosteroids and excreted in the urine (Fig 23–18). About two-thirds of the urinary 17-ketosteroids are of adrenal origin, and one-third are of testicular origin. Although most of the 17-ketosteroids are weak androgens (they have 20% or less the potency of testosterone), it is worth emphasizing that not all 17-ketosteroids are androgens and not all androgens are 17-ketosteroids. Etiocholanolone, for example, has no androgenic activity, and testosterone itself is not a 17-ketosteroid.

Actions

In addition to their actions during development (see above), testosterone and other androgens exert an inhibitory feedback effect on pituitary LH secretion; develop and maintain the male secondary sex characteristics; and exert an important protein-anabolic, growth-promoting effect. Along with FSH, testosterone is responsible for the maintenance of gametogenesis (see above).

Secondary Sex Characteristics

The widespread changes in hair distribution, body configuration, and genital size that develop in boys at puberty—the male **secondary sex characteristics**—are summarized in Table 23–6. The prostate and seminal vesicles enlarge, and the seminal vesicles begin

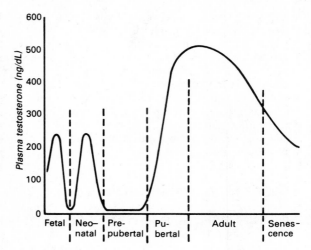

Figure 23–19. Plasma testosterone at various ages in human males.

to secrete fructose. This sugar appears to function as the main nutritional supply for the spermatozoa. The psychic effects of testosterone are difficult to define in humans, but in experimental animals, androgens provoke boisterous and aggressive play. The effects of androgens and estrogens on sexual behavior are considered in detail in Chapter 15. Although body hair is increased by androgens, scalp hair is decreased (Fig 23–20). Hereditary baldness often fails to develop unless androgens are present.

Anabolic Effects

Androgens increase the synthesis and decrease the breakdown of protein, leading to an increase in the rate of growth. They also cause the epiphyses to fuse to the long bones, thus eventually stopping growth. Their role in the adolescent growth spurt is discussed in Chapter 22. Secondary to their anabolic effect,

Table 23–5. Distribution of gonadal steroids and cortisol in plasma.[1]

| Steroid | % Free | % Bound to | | |
		CBG	GBG	Albumin
Testosterone	2	0	65	33
Androstenedione	7	0	8	85
Estradiol	2	0	38	60
Progesterone	2	18	0	80
Cortisol	4	90	0	6

[1]CBG, corticosteroid-binding globulin; GBG, gonadal steroid-binding globulin. (Courtesy of S Munroe.)

Table 23–6. Body changes at puberty in boys (male secondary sex characteristics).

External genitalia: Penis increases in length and width. Scrotum becomes pigmented and rugose.
Internal genitalia: Seminal vesicles enlarge and secrete and begin to form fructose. Prostate and bulbourethral glands enlarge and secrete.
Voice: Larynx enlarges, vocal cords increase in length and thickness, and voice becomes deeper.
Hair growth: Beard appears. Hairline on scalp recedes anterolaterally. Pubic hair grows with male (triangle with apex up) pattern. Hair appears in axillas, on chest, and around anus; general body hair increases.
Mental: More aggressive, active attitude. Interest in opposite sex develops.
Body conformation: Shoulders broaden, muscles enlarge.
Skin: Sebaceous gland secretion thickens and increases (predisposing to acne).

Figure 23–20. Hairline in children and adults. The hairline of the woman is like that of the child, whereas that of the man is indented in the lateral frontal region. (Reproduced, with permission, from Greulich WW et al: Somatic and endocrine studies of pubertal and adolescent boys. *Monogr Soc Res Child Dev* 1942;**7**:1.)

they cause moderate sodium, potassium, water, calcium, sulfate, and phosphate retention; and they also increase the size of the kidneys. Doses of exogenous testosterone that exert significant anabolic effects are also masculinizing and increase libido, which limits the usefulness of the hormone as an anabolic agent in patients with wasting diseases. Attempts to develop synthetic steroids in which the anabolic action is divorced from the androgenic action have not been particularly successful.

Mechanism of Action

Like other steroids (see Chapter 1), testosterone binds to an intracellular receptor, and the receptor-steroid complex then binds to DNA in the nucleus, facilitating transcription of various genes. In addition, testosterone is converted to dihydrotestosterone (DHT) by 5α-reductase in some target cells (Fig 23–18), and DHT binds to the same intracellular receptor as testosterone. DHT also circulates, with a plasma level that is about 10% of the testosterone level. Testosterone-receptor complexes are less stable than DHT-receptor complexes in target cells, and they transform less well to the DNA-binding state. Thus, DHT formation is a way of amplifying the action of testosterone in target tissues.

Testosterone-receptor complexes are responsible for the maturation of wolffian duct structures and consequently for the formation of male internal genitalia during development, but DHT-receptor complexes are needed to form male external genitalia. DHT-receptor complexes are also primarily responsible for enlargement of the prostate and probably of the penis at the time of puberty as well as the facial hair, the acne, and the temporal recession of the hairline. On the other hand, the increase in muscle mass and the development of male sex drive and libido depend primarily on testosterone rather than DHT.

Congenital 5α-reductase deficiency, which is common in certain parts of the Dominican Republic, produces an interesting form of male pseudohermaphroditism. Individuals with this syndrome are born with male internal genitalia including testes, but they have female external genitalia and are usually raised as girls. However, when they reach puberty, LH secretion and circulating testosterone levels are increased. Consequently, they develop male body contours and male libido. At this point, they usually change their gender identities and "become boys." They develop little facial hair or baldness, but their clitorises enlarge ("penis-at-12 syndrome") to the point that some of them can have intercourse with women. This enlargement probably occurs because with the high LH, there is enough testosterone to overcome the need for DHT amplification in the genitalia, and possibly because high levels of testosterone induce an increase in 5α-reductase. It has now been demonstrated that there are 2 different 5α-reductases: 5α-reductase 1, which is normal in this syndrome, and 5α-reductase 2, which is absent because of a defect or deletion in the gene that encodes it.

Testicular Production of Estrogens

Seventy percent of the estradiol in the plasma of adult men is formed by aromatization of circulating testosterone and androstenedione. Small amounts may be secreted by the adrenal cortex, but most of the remaining 30% is secreted by the testes. Some of the estradiol in testicular venous blood comes from the Leydig cells, but some is also produced by aromatization of androgens in Sertoli cells. In men, the plasma estradiol level is approximately 2 ng/dL (70 pmol/L) and the total production rate is 0.05 mg/d (0.18 μmol/d). In contrast to the situation in women, there is a moderate increase in estrogen production with advancing age in men.

CONTROL OF TESTICULAR FUNCTION

FSH is tropic to the Sertoli cells, and FSH and androgens maintain the gametogenic function of the testes. FSH also stimulates the secretion of androgen-binding protein and inhibin. Inhibin feeds back to inhibit FSH secretion. LH is tropic to the Leydig cells and stimulates secretion of testosterone, which in turn feeds back to inhibit LH secretion. Hypothalamic lesions in animals and hypothalamic disease in humans lead to atrophy of the testes and loss of their function.

Inhibin

Testosterone reduces plasma LH, but except in large doses, it has no effect on plasma FSH. Plasma FSH is elevated in patients who have atrophy of the seminiferous tubules but normal levels of testosterone and LH secretion. These observations led to the search for **inhibin,** a factor of testicular origin that inhibits FSH secretion. There are 2 inhibins in extracts of testes and, in women, in antral fluid from ovarian follicles. They are formed from 3 polypeptide subunits: a glycosylated α subunit with a molecular weight of 18,000, and 2 nonglycosylated β subunits, β_A and β_B, each with a molecular weight of 14,000. The subunits are formed from precursor proteins (Fig 23–21). The α subunit combines with β_A to form a heterodimer and with β_B to form another heterodimer, with the subunits linked by disulfide bonds. Both $\alpha\beta_A$ (inhibin A) and $\alpha\beta_B$ (inhibin B) inhibit FSH secretion by a direct action on the pituitary. Inhibins are produced by Sertoli cells in males and granulosa cells in females.

The heterodimer $\beta_A\beta_B$ and the homodimer $\beta_A\beta_A$ also form from the inhibin precursors, and $\beta_B\beta_B$ may form as well. They stimulate FSH secretion. Consequently, they have been called **activins.** Furthermore, the β dimers have marked homology to transforming growth factor-β (TGFβ), a dimeric polypeptide with a molecular weight of 25,000; and TGFβ also stimulates FSH secretion. The physiologic roles of these various stimulators in the regulation of gonadotropin

secretion are not yet known. However, activins are also present in the brain and the bone marrow, and in the marrow they appear to be involved in the development of white blood cells.

Steroid Feedback

A current "working hypothesis" of the way the functions of the testes are regulated is shown in Fig 23–22. Castration is followed by a rise in the pituitary content and secretion of FSH and LH, and hypothalamic lesions prevent this rise. Testosterone inhibits LH secretion by acting directly on the anterior pituitary and by inhibiting the secretion of GnRH from the hypothalamus. Inhibin acts directly on the anterior pituitary to inhibit FSH secretion.

In response to LH, some of the testosterone secreted from the Leydig cells bathes the seminiferous epithelium and provides the high local concentration of androgen to the Sertoli cells that is necessary for normal spermatogenesis. Systemically administered testosterone does not raise the androgen level in the testes to as great a degree, and it inhibits LH secretion. Consequently, the net effect of systemically administered testosterone is generally a decrease in sperm count. Testosterone therapy has been suggested as a means of male contraception. However, the dose of testosterone needed to suppress spermatogenesis causes sodium and water retention. The possible use of inhibin as a male contraceptive is now being explored.

Figure 23–21. Inhibin precursor proteins and the various inhibins and activins that are formed from the C-terminal regions of these precursors. SS, disulfide bonds.

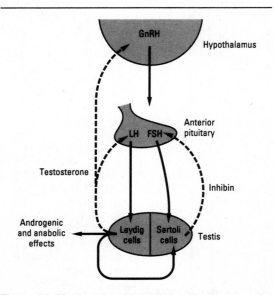

Figure 23–22. Postulated interrelationships between the hypothalamus, anterior pituitary, and testes. Solid arrows indicate excitatory effects; dashed arrows indicate inhibitory effects. Compare with Figs 18–14, 20–21, 22–11, and 23–36.

ABNORMALITIES OF TESTICULAR FUNCTION

Cryptorchidism

The testes develop in the abdominal cavity and normally migrate to the scrotum during fetal development. Testicular descent is incomplete on one or, less commonly, both sides in 10% of newborn males, the testes remaining in the abdominal cavity or inguinal canal. Spontaneous descent of these testes is the rule, however, and the proportion of boys with undescended testes (**cryptorchidism**) falls to 2% at age 1 year and 0.3% after puberty. Gonadotropic hormone treatment speeds descent in some cases, or the defect can be corrected surgically. Treatment should be instituted, probably before puberty, because there is a higher incidence of malignant tumors in undescended than in scrotal testes and because after puberty the higher temperature in the abdomen eventually causes irreversible damage to the spermatogenic epithelium.

Male Hypogonadism

The clinical picture of male hypogonadism depends upon whether testicular deficiency develops before or after puberty and whether the gametogenic or the endocrine function is compromised. The causes of testicular deficiency include hypothalamic and pituitary disease as well as a variety of primary testicular and chromosomal disorders. Loss or failure of maturation of the gametogenic function causes sterility. If the endocrine function is lost in adulthood, the secondary sex characteristics regress slowly because it takes very little androgen to maintain them once they are established. The growth of the larynx during adolescence is permanent, and the voice remains deep. Men castrated in adulthood suffer some loss of libido, although the ability to copulate persists for some time. They occasionally have hot flashes and are generally more irritable, passive, and depressed than men with intact testes. When the Leydig cell deficiency dates from childhood, the clinical picture is that of **eunuchoidism.** Eunuchoid individuals over the age of 20 are characteristically tall, although not so tall as hyperpituitary giants, because their epiphyses remain open and some growth continues past the normal age of puberty. They have narrow shoulders and small muscles, a body configuration resembling that of the adult female. The genitalia are small and the voice high-pitched. Pubic and axillary hair does appear, because of adrenocortical androgen secretion; but the hair is sparse, and the pubic hair has the female "triangle with the base up" distribution rather than the "triangle with the base down" pattern (male escutcheon) seen in normal males.

Androgen-Secreting Tumors

"Hyperfunction" of the testes in the absence of tumor formation is not a recognized entity. Androgen-secreting Leydig cell tumors are rare and cause detectable endocrine symptoms only in prepuberal boys, who develop precocious pseudopuberty (Table 23–2).

THE FEMALE REPRODUCTIVE SYSTEM

THE MENSTRUAL CYCLE

The reproductive system of the female (Fig 23–23), unlike that of the male, shows regular cyclic changes that teleologically may be regarded as periodic preparations for fertilization and pregnancy. In primates, the cycle is a **menstrual** cycle, and its most conspicuous feature is the periodic vaginal bleeding that occurs with the shedding of the uterine mucosa

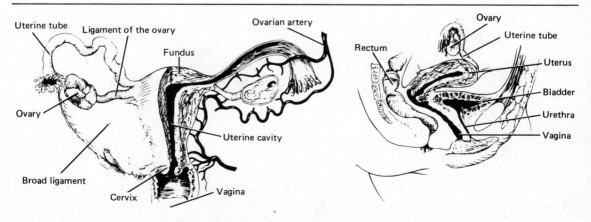

Figure 23–23. The female reproductive system.

(menstruation). The length of the cycle is notoriously variable in women, but an average figure is 28 days from the start of one menstrual period to the start of the next. By common usage, the days of the cycle are identified by number, starting with the first day of menstruation.

Ovarian Cycle

From the time of birth, there are many **primordial follicles** under the ovarian capsule. Each contains an immature ovum (Fig 23–24). At the start of each cycle, several of these follicles enlarge, and a cavity forms around the ovum **(antrum formation).** This cavity is filled with follicular fluid. In humans, one of the follicles in one ovary starts to grow rapidly on about the sixth day and becomes the **dominant follicle,** while the others regress, forming **atretic follicles.** It is uncertain how one follicle is selected to be the dominant follicle in this **follicular phase** of the menstrual cycle, but it seems to be related to the ability of the follicle to secrete the estrogen inside it that is needed for final maturation. However, when women are given highly purified human pituitary gonadotropin preparations by injection, many follicles develop simultaneously.

The structure of a maturing ovarian **(graafian)** follicle is shown in Fig 23–24. The cells of the **theca interna** of the follicle are the primary source of circulating estrogens. However, the follicular fluid has a high estrogen content, and much of this estrogen comes from the granulosa cells (see below).

At about the 14th day of the cycle, the distended follicle ruptures, and the ovum is extruded into the abdominal cavity. This is the process of **ovulation.** The ovum is picked up by the fimbriated ends of the uterine tubes (oviducts). It is transported to the uterus and, unless fertilization occurs, on out through the vagina.

The follicle that ruptures at the time of ovulation promptly fills with blood, forming what is sometimes called a **corpus hemorrhagicum.** Minor bleeding from the follicle into the abdominal cavity may cause peritoneal irritation and fleeting lower abdominal pain ("mittelschmerz"). The granulosa and theca cells of the follicle lining promptly begin to proliferate, and the clotted blood is rapidly replaced with yellowish, lipid-rich **luteal cells,** forming the **corpus luteum.** This initiates the **luteal phase** of the menstrual cycle, during which the luteal cells secrete estrogens and progesterone. If pregnancy occurs, the corpus luteum persists, and there are usually no more periods until after delivery. If there is no pregnancy, the corpus luteum begins to degenerate about 4 days before the next menses (24th day of the cycle) and is eventually replaced by scar tissue, forming a **corpus albicans.**

The ovarian cycle in other mammals is similar, except that in many species more than one follicle ovu-

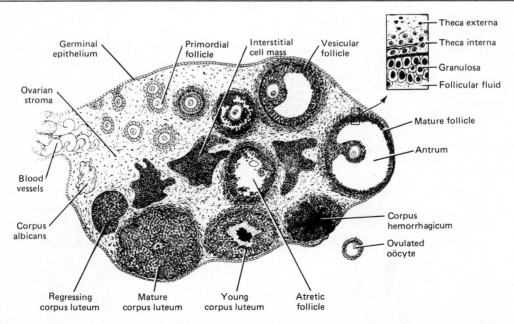

Figure 23–24. Diagram of a mammalian ovary, showing the sequential development of a follicle, formation of a corpus luteum, and, in the center, follicular atresia. A section of the wall of a mature follicle is enlarged at the upper right. The interstitial cell mass is not prominent in primates. (Reproduced, with permission, from Gorbman A, Bern H: *Textbook of Comparative Endocrinology.* Wiley, 1962.)

lates, and multiple births are the rule. Corpora lutea form in some submammalian species but not in others.

In humans, no new ova are formed after birth. During fetal development, the ovaries contain over 7 million primordial follicles. However, many undergo atresia (involution) before birth, and others are lost after birth. At the time of birth, there are 2 million ova, but 50% of these are atretic. The million that are normal undergo the first part of the first meiotic division at about this time and enter a stage of arrest in prophase in which those that survive persist until adulthood. There is continuing atresia during development, and the number of ova in both of the ovaries at the time of puberty is less than 300,000 (Fig 23–12). Only one of these ova per cycle (or about 500 in the course of a normal reproductive life) normally reaches maturity; the remainder degenerate. Just before ovulation, the first meiotic division is completed. One of the daughter cells, the **secondary oocyte,** receives most of the cytoplasm, while the other, the **first polar body,** fragments and disappears. The secondary oocyte immediately begins the second meiotic division, but this division stops at metaphase and is completed only when a sperm penetrates the oocyte. At that time, the **second polar body** is cast off and the fertilized ovum proceeds to form a new individual. The arrest in metaphase is due, at least in some species, to formation in the ovum of the protein **pp39**^{mos}, which is encoded by the **c-mos** protooncogene. When fertilization occurs, the pp39^{mos} is destroyed within 30 minutes by **calpain,** a calcium-dependent cysteine protease.

Uterine Cycle

At the end of menstruation, all but the deep layers of the endometrium have sloughed. Under the influence of estrogens from the developing follicle, the endometrium increases rapidly in thickness from the 5th to the 14th days of the menstrual cycle. As the thickness increases, the uterine glands are drawn out so that they lengthen (Fig 23–25), but they do not become convoluted or secrete to any degree. These endometrial changes are called proliferative, and this part of the menstrual cycle is sometimes called the **proliferative phase.** It is also called the preovulatory or follicular phase of the cycle. After ovulation, the endometrium becomes more highly vascularized and slightly edematous under the influence of estrogen and progesterone from the corpus luteum. The glands become coiled and tortuous (Fig 23–25), and they begin to secrete a clear fluid. Consequently, this phase of the cycle is called the **secretory** or **luteal phase.** Late in the luteal phase, the endometrium, like the anterior pituitary, produces prolactin, but the function of this endometrial prolactin is unknown.

The endometrium is supplied by 2 types of arteries. The superficial two-thirds of the endometrium that is shed during menstruation, the **stratum functionale,** is supplied by long, coiled **spiral arteries** (Fig 23–26), whereas the deep layer that is not shed, the **stratum basale,** is supplied by short, straight **basilar arteries.**

When the corpus luteum regresses, hormonal support for the endometrium is withdrawn. The endometrium becomes thinner, which adds to the coiling of the spiral arteries. Foci of necrosis appear in the endometrium, and these coalesce. There is in addition spasm and then necrosis of the walls of the spiral arteries, leading to spotty hemorrhages that become confluent and produce the menstrual flow.

The vasospasm is probably produced by locally released prostaglandins. There are large quantities of prostaglandins in the secretory endometrium and in menstrual blood, and infusions of $PGF_{2\alpha}$ produce endometrial necrosis and bleeding. One theory of the onset of menstruation holds that in necrotic endometrial cells, lysosomal membranes break down, with the release of enzymes that foster the formation of prostaglandins from cellular phospholipids.

From the point of view of endometrial function, the proliferative phase of the menstrual cycle represents restoration of the epithelium from the preceding menstruation, and the secretory phase represents

Figure 23–25. Changes in the endometrium during the menstrual cycle.

Myometrium

Figure 23–26. Spiral artery of endometrium. Drawing of a spiral artery *(left)* and 2 uterine glands **(right)** from the endometrium of a rhesus monkey; early progestational phase. The uterine cavity is at the top. (Reproduced, with permission, from Daron GH: The arterial pattern of the tunica mucosa of the uterus in the *Macacus rhesus. Am J Anat* 1936;**58:**349.)

preparation of the uterus for implantation of the fertilized ovum. The length of the secretory phase is remarkably constant at about 14 days, and the variations seen in the length of the menstrual cycle are due for the most part to variations in the length of the proliferative phase. When fertilization fails to occur during the secretory phase, the endometrium is shed and a new cycle starts.

Normal Menstruation

Menstrual blood is predominantly arterial, with only 25% of the blood being of venous origin. It contains tissue debris, prostaglandins, and relatively large amounts of fibrinolysin from endometrial tissue. The fibrinolysin lyses clots, so that menstrual blood does not normally contain clots unless the flow is excessive.

The usual duration of the menstrual flow is 3–5 days, but flows as short as 1 day and as long as 8 days can occur in normal women. The amount of blood lost may range normally from slight spotting to 80 mL; the average amount lost is 30 mL. Loss of more than 80 mL is abnormal. Obviously, the amount of flow can be affected by various factors, including the thickness of the endometrium, medication, and diseases that affect the clotting mechanism. After menstruation, a new endometrium regenerates from the stratum basale.

Anovulatory Cycles

In some instances, ovulation fails to occur during the menstrual cycle. Such anovulatory cycles are common for the first 12–18 months after menarche and again before the onset of the menopause. When ovulation does not occur, no corpus luteum is formed and the effects of progesterone on the endometrium are absent. Estrogens continue to cause growth, however, and the proliferative endometrium becomes thick enough to break down and begins to slough. The time it takes for bleeding to occur is variable, but it usually occurs in less than 28 days from the last menstrual period. The flow is also variable and ranges from scanty to relatively profuse.

Cyclic Changes in the Uterine Cervix

Although it is continuous with the body of the uterus, the cervix of the uterus is different in a number of ways. The mucosa of the uterine cervix does not undergo cyclic desquamation, but there are regular changes in the cervical mucus. Estrogen makes the mucus thinner and more alkaline, changes that promote the survival and transport of sperm. Progesterone makes it thick, tenacious, and cellular. The mucus is thinnest at the time of ovulation, and its elasticity, or **spinnbarkeit,** increases so that by midcycle, a drop can be stretched into a long, thin thread that may be 8–12 cm or more in length. In addition, it dries in an arborizing, fernlike pattern (Fig 23–27) when a thin layer is spread on a slide. After ovulation and during pregnancy, it becomes thick and fails to form the fern pattern.

Vaginal Cycle

Under the influence of estrogens, the vaginal epithelium becomes cornified, and cornified epithelial cells can be identified in the vaginal smear. Under the influence of progesterone, a thick mucus is secreted, and the epithelium proliferates and becomes infiltrated with leukocytes. The cyclic changes in the vaginal smear in rats are relatively marked (Fig 23–28). The changes in humans and other species are similar but not so clear-cut.

Cyclic Changes in the Breasts

Although lactation normally does not occur until the end of pregnancy, there are cyclic changes in the breasts during the menstrual cycle. Estrogens cause proliferation of mammary ducts, whereas progesterone causes growth of lobules and alveoli. The breast

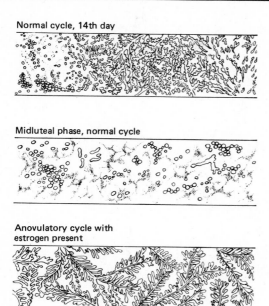

Normal cycle, 14th day

Midluteal phase, normal cycle

Anovulatory cycle with estrogen present

Figure 23–27. Patterns formed when cervical mucus is smeared on a slide, permitted to dry, and examined under the microscope. Progesterone makes the mucus thick and cellular. In the smear from a patient who failed to ovulate *(bottom),* there is no progesterone to inhibit the estrogen-induced fern pattern.

Vaginal epithelium   Cells in vaginal lumen

Diestrus

Estrus

Figure 23–28. Changes in the vaginal wall *(left)* and type of cells found in the vaginal smear *(right)* during the estrous cycle in rats. The vaginal smear pattern in women is similar but not so clear-cut. (Modified and reproduced, with permission, from Turner CD, Bagnara JT: *General Endocrinology,* 6th ed. Saunders, 1976.)

swelling, tenderness, and pain experienced by many women during the 10 days preceding menstruation are probably due to distention of the ducts, hyperemia, and edema of the interstitial tissue of the breast. All these changes regress, along with the symptoms, during menstruation.

Changes During Intercourse

During sexual excitement in women, fluid is secreted onto the vaginal walls, probably because of release of VIP from vaginal nerves. A lubricating mucus is also secreted by the vestibular glands. The upper part of the vagina is sensitive to stretch, while tactile stimulation from the labia minora and clitoris adds to the sexual excitement. These stimuli are reinforced by tactile stimuli from the breasts and, as in men, by visual, auditory, and olfactory stimuli, which may build to the crescendo known as orgasm. During orgasm, there are autonomically mediated rhythmic contractions of the vaginal walls. Impulses also travel via the pudendal nerves and produce rhythmic contraction of the bulbocavernosus and ischiocavernosus muscles. The vaginal contractions may aid sperm transport but are not essential for it, since fertilization of the ovum is not dependent on orgasm.

Indicators of Ovulation

It is often important in clinical practice to know that ovulation has occurred and to know when during the cycle it occurred. The finding of a secretory pattern in a biopsy of the endometrium indicates that a functioning corpus luteum is present. Less reliably, finding thick, cellular cervical mucus that does not form a fern pattern in a woman who has regular menses is evidence of the same thing. A convenient and reasonably reliable indicator of the time of ovulation is a change—usually a rise—in the basal body temperature (Fig 23–29). Women interested in obtaining an accurate temperature chart should use a thermometer with wide gradations and take their temperatures (oral or rectal) in the morning before getting out of bed. The cause of the temperature change at the time of ovulation is probably the increase in progesterone secretion, since progesterone is thermogenic (see Chapter 14).

Ovulation normally occurs 36–38 hours after the start of the LH surge at midcycle (Fig 23–29). The ovum lives for approximately 72 hours after it is extruded from the follicle, but it is fertilizable for less than half this time. Sperms apparently survive in the female genital tract for no more than 72 hours. Consequently, the "fertile period" during a 28-day cycle is about 120 hours. Unfortunately for those interested in the "rhythm method" of contraception, the time of ovulation is rather variable even from one menstrual cycle to another in the same woman. Before the 9th

Figure 23–29. Typical basal body temperature and plasma hormone concentrations during a normal 28-day human menstrual cycle. M, menstruation. (Modified from Midgley AR in: *Human Reproduction.* Hafez ESE, Evans TN [editors]. Harper & Row, 1973; and McLachlan RI et al: Circulating immunoreactive inhibin levels during the normal human menstrual cycle. *J Clin Endocrinol Metab* 1987;**65:**954.)

and after the 20th day, there is little chance of conception; but there are documented cases of pregnancy resulting from isolated coitus on every day of the cycle.

The Estrous Cycle

Mammals other than primates do not menstruate, and their sexual cycle is called an **estrous cycle.** It is named for the conspicuous period of "heat" (**estrus**) at the time of ovulation, normally the only time during which the sexual interest of the female is aroused (see Chapter 15). In spontaneously ovulating species with estrous cycles, such as the rat, there is no episodic vaginal bleeding but the underlying endocrine events are essentially the same as those in the menstrual cycle. In other species, ovulation is induced by copulation (reflex ovulation).

OVARIAN HORMONES

Chemistry, Biosynthesis, & Metabolism of Estrogens

The naturally occurring estrogens are **17β-estradiol, estrone,** and **estriol** (Fig 23–30). They are C_{18} steroids (see Fig 20–7); ie, they do not have an angular methyl group attached to the 10 position or a Δ^4-3-keto configuration in the A ring. They are secreted by

Figure 23–30. Biosynthesis and metabolism of estrogens. The formulas of the precursor steroids are shown in Fig 20–8.

the theca interna and granulosa cells of the ovarian follicles, the corpus luteum, and the placenta. The biosynthetic pathway involves their formation from androgens. They are also formed by aromatization of androstenedione in the circulation. **Aromatase** is the enzyme that catalyzes the conversion of androstenedione to estrone and the conversion of testosterone to estradiol (Fig 23–30).

Theca interna cells have many LH receptors, and LH acts via cyclic AMP to increase conversion of cholesterol to androstenedione. Some of the andro-

stenedione is converted to estradiol, which enters the circulation. The theca interna cells also supply androstenedione to the granulosa cells. The granulosa cells make estradiol when provided with androgens (Fig 23–31), and it appears that the estradiol they form in primates is secreted into the follicular fluid. Granulosa cells have many FSH receptors, and FSH facilitates their secretion of estradiol by acting via cyclic AMP to increase their aromatase activity. Mature granulosa cells also acquire LH receptors, and LH also stimulates estradiol production.

Figure 23–31. Interactions between theca and granulosa cells in estradiol synthesis and secretion.

The stromal tissues of the ovary also have the potential to make androgens and estrogens. However, they probably do so in insignificant amounts in normal premenopausal women. 17β-Estradiol, the major secreted estrogen, is in equilibrium in the circulation with estrone. Estrone is further metabolized to estriol, probably for the most part in the liver. Estradiol is the most potent estrogen of the three, and estriol the least.

Two percent of the circulating estradiol is free, and the remainder is bound to protein: 60% to albumin and 38% to the same gonadal steroid-binding globulin (GBG) that binds testosterone (Table 23–5).

In the liver, estrogens are oxidized or converted to glucuronide and sulfate conjugates. Appreciable amounts are secreted in the bile and reabsorbed into the bloodstream (enterohepatic circulation). There are at least 10 different metabolites of estradiol in human urine.

Secretion

The concentration of estradiol in the plasma during the menstrual cycle is shown in Fig 23–29. Almost all of this estrogen comes from the ovary, and there are 2 peaks of secretion: one just before ovulation and one during the midluteal phase. The estradiol secretion rate is 36 μg/d (133 μmol/d) in the early follicular phase, 380 μg/d just before ovulation, and 250 μg/d during the midluteal phase (Table 23–7). After menopause, estrogen secretion declines to low levels.

As noted above, the estradiol production rate in men is about 50 μg/d (180 μmol/d).

Effects on the Female Genitalia

Estrogens facilitate the growth of the ovarian follicles and increase the motility of the uterine tubes. Their role in the cyclic changes in the endometrium, cervix, and vagina is discussed above. They increase uterine blood flow and have important effects on the smooth muscle of the uterus. In immature and castrate females, the uterus is small and the myometrium atrophic and inactive. Estrogens increase the amount of uterine muscle and its content of contractile proteins. Under the influence of estrogens, the muscle becomes more active and excitable, and action potentials in the individual fibers become more frequent (see Chapter 3). The "estrogen-dominated" uterus is also more sensitive to oxytocin.

Chronic treatment with estrogens causes the endometrium to hypertrophy. When estrogen therapy is discontinued, there is sloughing with **withdrawal bleeding.** Some "breakthrough" bleeding may occur during treatment when estrogens are given for long periods.

Effects on Endocrine Organs

Estrogens decrease FSH secretion. Under some circumstances, they inhibit LH secretion (negative feedback); in other circumstances, they increase LH secretion (positive feedback; see below). Estrogens also increase the size of the pituitary. Women are sometimes given large doses of estrogens for 4–6 days to prevent conception after coitus during the fertile period (postcoital or "morning-after" contraception). However, in this instance, pregnancy is probably prevented by interference with implantation of the fertilized ovum rather than changes in gonadotropin secretion.

Estrogens cause increased secretion of angiotensinogen (see Chapter 24) and thyroid-binding globulin (see Chapter 18). They exert an important protein anabolic effect in chickens and cattle, possibly by stimulating the secretion of androgens from the adrenal, and estrogen treatment has been used commercially to increase the weight of domestic animals. Estrogens have been reported to exert anabolic effects, and they cause epiphyseal closure in humans.

Behavioral Effects

The estrogens are responsible for estrous behavior in animals, and they increase libido in humans. They apparently exert this action by a direct effect on certain neurons in the hypothalamus (Fig 23–32). The relation of estrogens, progesterone, and androgens to sexual behavior is discussed in Chapter 15.

Effects on the Breasts

Estrogens produce duct growth in the breasts and are largely responsible for breast enlargement at puberty in girls; they have been called the growth hormones of the breast. Breast enlargement that occurs when estrogen-containing skin creams are applied locally is due primarily to systemic absorption of the estrogen, although a slight local effect is also produced. Estrogens are responsible for the pigmenta-

Table 23–7. Twenty-four hour production rates of sex steroids in women at different stages of the mentrual cycle.[1]

Sex Steroids	Early Follicular	Preovulatory	Midluteal
Progesterone (mg)	1.0	4.0	25.0
17-Hydroxyprogesterone (mg)	0.5	4.0	4.0
Dehydroepiandrosterone (mg)	7.0	7.0	7.0
Androstenedione (mg)	2.6	4.7	3.4
Testosterone (μg)	200.0	240.0	170.0
Estrone (μg)	50.0	350.0	250.0
Estradiol (μg)	36.0	380.0	250.0

[1]Reproduced, with permission, from Yen SSC, Jaffe RB: *Reproductive Endocrinology,* 3rd ed. Saunders, 1991.

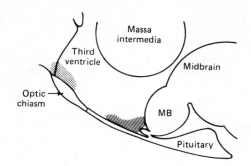

Figure 23–32. Loci where implantations of estrogen in the hypothalamus affect ovarian weight and sexual behavior in rats, projected on a sagittal section of the hypothalamus. The implants that stimulate sex behavior are located in the suprachiasmatic area above the optic chiasm (dotted area), whereas ovarian atrophy is produced by implants in the arcuate nucleus and surrounding ventral hypothalamus (striped area). MB, mamillary body.

tion of the areolas, although pigmentation usually becomes more intense during the first pregnancy than it does at puberty. The role of the estrogens in the overall control of breast growth and lactation is discussed below.

Female Secondary Sex Characteristics

The body changes that develop in girls at puberty—in addition to enlargement of breasts, uterus, and vagina—are due in part to estrogens, which are the "feminizing hormones," and in part simply to the absence of testicular androgens. Women have narrow shoulders and broad hips, thighs that converge, and arms that diverge (wide **carrying angle**). This body configuration, plus the female distribution of fat in the breasts and buttocks, is seen also in castrate males. In women, the larynx retains its prepuberal proportions and the voice stays high-pitched. There is less body hair and more scalp hair, and the pubic hair generally has a characteristic flat-topped pattern (female escutcheon). Growth of pubic and axillary hair in the female is due primarily to androgens rather than estrogens, although estrogen treatment may cause some hair growth. The androgens come from the adrenal cortex and, to a lesser extent, from the ovaries.

Other Actions

Normal women retain salt and water and gain weight just before menstruation. Estrogens cause some degree of salt and water retention. However, aldosterone secretion is slightly elevated in the luteal phase, and this also contributes to the premenstrual fluid retention.

Estrogens are said to make sebaceous gland secretions more fluid and thus to counter the effect of testosterone and inhibit formation of **comedones**

("blackheads") and acne. The liver palms, spider angiomas, and slight breast enlargement seen in advanced liver disease are due to increased circulating estrogens. The increase is due not only to alterations in the hepatic metabolism of estrogens but also to increased conversion of androgens to estrogens.

Estrogens have a significant plasma cholesterol-lowering action (see Chapter 17), and they inhibit atherogenesis. They contribute to the low incidence of myocardial infarction and other complications of atherosclerotic vascular disease in premenopausal women. However, large doses of orally active estrogens appear to promote thrombosis because they reach the liver in high concentrations in the portal blood and alter hepatic production of clotting factors.

Mechanism of Action

Like other steroids (see Chapter 1), estrogens combine with an intracellular protein receptor, and the complex binds to DNA, promoting formation of mRNAs that in turn direct the formation of new proteins which modify cell function. Almost all estrogen actions are produced in this fashion, although there also appear to be poorly understood rapid actions on cell membranes in some situations.

Synthetic Estrogens

The ethinyl derivative of estradiol (Fig 23–33) is a potent estrogen and—unlike the naturally occurring estrogens—is relatively active when given by mouth, because it is resistant to hepatic metabolism. The activity of the naturally occurring hormones is low when they are administered by mouth because the portal venous drainage of the intestine carries them to the liver, where they are inactivated before they can reach the general circulation. Some nonsteroidal substances and a few compounds found in plants have estrogenic activity. The plant estrogens are rarely a problem in human nutrition, but they may cause undesirable effects in farm animals. Diethylstilbestrol (Fig 23–33) and a number of related compounds are estrogenic, possibly because they are converted to a steroidlike ring structure in the body.

Chemistry, Biosynthesis, & Metabolism of Progesterone

Progesterone is a C_{21} steroid (Fig 23–34) secreted by the corpus luteum, the placenta, and (in small amounts) the follicle. It is an important intermediate in steroid biosynthesis in all tissues that secrete steroid hormones, and small amounts apparently enter the circulation from the testes and adrenal cortex. 17α-Hydroxyprogesterone is apparently secreted along with estrogens from the ovarian follicle, and its secretion parallels that of 17β-estradiol. About 2% of the circulating progesterone is free (Table 23–5), whereas 80% is bound to albumin and 18% is bound to corticosteroid-binding globulin (see Chapter 20). Progesterone has a short half-life and is converted in

Figure 23–33. Synthetic estrogens.

Figure 23–34. Biosynthesis of progesterone and major pathway for its metabolism. Other metabolites are also formed.

the liver to pregnanediol, which is conjugated to glucuronic acid and excreted in the urine (Fig 23–34).

Secretion

In men, the plasma progesterone level is approximately 0.3 ng/mL (1 nmol/L). In women, the level is approximately 0.9 ng/mL (3 nmol/L) during the follicular phase of the menstrual cycle (Fig 23–29). The difference is due to secretion of small amounts of progesterone by cells in the ovarian follicles; theca cells provide pregnenolone to the granulosa cells, which convert it to progesterone. Late in the follicular phase, progesterone secretion begins to increase. During the luteal phase, the corpus luteum produces large quantities of progesterone, and ovarian secretion increases about 20-fold (Table 23–7). The result is an increase in plasma progesterone to a peak value of approximately 18 ng/mL (60 nmol/L).

The stimulating effect of LH on progesterone secretion by the corpus luteum is due to activation of adenylate cyclase and involves a subsequent step that is dependent on protein synthesis.

Actions

The principal target organs of progesterone are the uterus, the breasts, and the brain. Progesterone is responsible for the progestational changes in the endometrium and the cyclic changes in the cervix and vagina described above. It has an antiestrogenic effect on the myometrial cells, decreasing their excitability, their sensitivity to oxytocin, and their spontaneous electrical activity while increasing their membrane potential. It also decreases the number of estrogen receptors in the endometrium and increases the rate of conversion of 17β-estradiol to less active estrogens.

In the breast, progesterone stimulates the development of lobules and alveoli. It induces differentiation of estrogen-prepared ductal tissue and supports the secretory function of the breast during lactation.

The feedback effects of progesterone are complex and are exerted at both the hypothalamic and the pituitary levels. Large doses of progesterone inhibit LH secretion and potentiate the inhibitory effect of estrogens. Progesterone injections can prevent ovulation in humans.

Progesterone is thermogenic and is probably responsible for the rise in basal body temperature at the time of ovulation. It stimulates respiration, and the alveolar P_{CO_2} (P_{ACO_2}; see Chapter 34) in women during the luteal phase of the menstrual cycle is lower than that in men. In pregnancy, the P_{ACO_2} falls as progesterone secretion rises.

Large doses of progesterone produce natriuresis, probably by blocking the action of aldosterone on the kidney. The hormone does not have a significant anabolic effect.

The effects of progesterone, like those of other steroids, are brought about by an action on DNA to initiate synthesis of new mRNA. As noted in Chapter 1, the progesterone receptor is bound to a heat shock protein in the absence of the steroid, and progesterone binding releases the heat shock protein, exposing the DNA-binding domain of the receptor. The synthetic steroid **mifepristone (RU-486)** binds to the receptor but does not release the heat shock protein, and it blocks the binding of progesterone. Since the maintenance of early pregnancy depends on the stimulatory effect of progesterone on endometrial growth and its inhibition of uterine contractility, mifepristone causes abortion. In some countries, mifepristone combined with a prostaglandin is used to produce elective abortions.

Substances that mimic the action of progesterone are sometimes called **progestational agents, gestagens,** or **progestins.** They are used along with synthetic estrogens as oral contraceptive agents (see below).

Relaxin

Relaxin is a polypeptide hormone that relaxes the pubic symphysis and other pelvic joints and softens and dilates the uterine cervix during pregnancy. Thus, it facilitates delivery. It also inhibits uterine contractions and may play a role in the development of the mammary glands. Human relaxin is encoded by 2 genes on chromosome 9 which produce 2 structurally related polypeptides. Both are related to insulin (see Chapters 19 and 22). However, only one of the genes is active in the ovary. In nonpregnant women, relaxin is found in the corpus luteum and the endometrium during the secretory but not the proliferative phase of the menstrual cycle. In men, it is found in semen, and appears to be produced in the prostate.

CONTROL OF OVARIAN FUNCTION

FSH from the pituitary is responsible for the early maturation of the ovarian follicles, and FSH and LH together are responsible for their final maturation. A burst of LH secretion (Fig 23–29) is responsible for ovulation and the initial formation of the corpus luteum. There is also a smaller midcycle burst of FSH secretion, whose significance is uncertain. LH stimulates the secretion of estrogen and progesterone from the corpus luteum.

Hypothalamic Components

The hypothalamus occupies a key position in the control of gonadotropin secretion. Hypothalamic control is exerted by GnRH secreted into the portal hypophyseal vessels (see Chapter 14). GnRH stimulates the secretion of FSH as well as LH, and it is unlikely that there is an additional separate FRH.

GnRH is normally secreted in episodic bursts, and these bursts produce the circhoral peaks of LH secretion (Fig 23–35). They are essential for normal secretion of gonadotropins. If GnRH is administered by

Figure 23–35. Episodic secretion of LH (○) and FSH (●) during the follicular stage **(top)** and the luteal stage **(bottom)** of the menstrual cycle. The numbers above each graph indicate the numbers of LH pulses per 12 hours and the plasma estradiol (E_2) and progesterone (P) concentrations at these 2 times of the cycle. (Reproduced, with permission, from Marshall JC, Kelch RO: Gonadotropin-releasing hormone: role of pulsatile secretion in the regulation of reproduction. *N Engl J Med* 1986;**315:**1459.)

constant infusion, the GnRH receptors in the anterior pituitary down-regulate (see Chapter 1) and LH secretion declines to zero. However, if GnRH is administered episodically at a rate of one pulse per hour, LH secretion is stimulated. This is true even when endogenous GnRH secretion has been prevented by a lesion of the ventral hypothalamus.

It is now clear not only that episodic secretion of GnRH is a general phenomenon but also that fluctuations in the frequency and amplitude of the GnRH bursts are important in generating the other hormonal changes that are responsible for the menstrual cycle. Frequency is increased by estrogens and decreased by progesterone and testosterone. The frequency increases late in the follicular phase of the cycle, culminating in the LH surge. During the proliferative phase, the frequency decreases as a result of the action of progesterone (Fig 23–35), but when estrogen and progesterone secretion decrease at the end of the cycle, the frequency once again increases.

At the time of the LH surge, the sensitivity of the gonadotropes to GnRH is greatly increased because of their exposure to GnRH pulses of the frequency that exist at this time. This self-priming effect of

GnRH is important in producing a maximum LH response.

The nature and the exact location of the GnRH pulse generator in the hypothalamus are still unsettled. However, it is known in a general way that norepinephrine and possibly epinephrine in the hypothalamus increase GnRH pulse frequencies. Conversely, opioid peptides such as the enkephalins and β-endorphin reduce the frequency of GnRH pulses.

The down regulation of pituitary receptors and the consequent decrease in LH secretion produced by constantly elevated levels of GnRH has led to the use of long-acting GnRH analogs to inhibit LH secretion in precocious puberty and in cancer of the prostate (see below).

Feedback Effects

Changes in plasma LH, FSH, sex steroids, and inhibin during the menstrual cycle are shown in Fig 23–29, and their feedback relations are diagrammed in Fig 23–36. During the early part of the follicular phase, inhibin is low and FSH is modestly elevated, fostering follicular growth. LH secretion is held in check by the negative feedback effect of the rising plasma estrogen level. At 36–48 hours before ovulation, the estrogen feedback effect becomes positive, and this initiates the burst of LH secretion (LH surge)

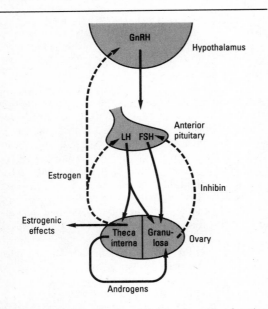

Figure 23–36. Feedback regulation of ovarian function. The cells of the theca interna provide androgens to the granulosa cells, and theca cells also produce the circulating estrogens that inhibit the secretion of GnRH, LH, and FSH. Inhibin from the granulosa cells inhibits FSH secretion. LH regulates the thecal cells, whereas the granulosa cells are regulated by both LH and FSH. The dashed arrows indicate inhibitory effects and the solid arrows stimulatory effects. Compare with Figs 18–14, 20–21, 22–11, and 23–22.

that produces ovulation. Ovulation occurs about 9 hours after the LH peak. FSH secretion also peaks, despite a small rise in inhibin, probably because of the strong stimulation of gonadotropes by GnRH. During the luteal phase, the secretion of LH and FSH is low because of the elevated levels of estrogen, progesterone, and inhibin.

It should be emphasized that a moderate, constant level of circulating estrogen exerts a negative feedback effect on LH secretion, whereas an elevated estrogen level exerts a positive feedback effect and stimulates LH secretion. It has been demonstrated that in monkeys there is also a minimum time that estrogens must be elevated to produce positive feedback. When circulating estrogen was increased about 300% for 24 hours, only negative feedback was seen; but when it was increased about 300% for 36 hours or more, a brief decline in secretion was followed by a burst of LH secretion that resembled the midcycle surge. When circulating levels of progesterone were high, the positive feedback effect of estrogen was inhibited. There is evidence that in primates, both the negative and the positive feedback effects of estrogen are exerted in the mediobasal hypothalamus, but exactly how negative feedback is switched to positive feedback and then back to negative feedback in the luteal phase remains unknown.

Control of the Cycle

In an important sense, regression of the corpus luteum (**luteolysis**) starting 3–4 days before menses is the key to the menstrual cycle. There is evidence that prostaglandins may play a role in this process, possibly by inhibiting the effect of LH on cyclic AMP, but their exact function is unknown. In some domestic animals, oxytocin appears to be secreted by the corpus luteum and to exert a local luteolytic effect, possibly by causing release of prostaglandins. Once luteolysis begins, the estrogen and progesterone levels fall and the secretion of FSH and LH increases. A new crop of follicles develops, and then a single dominant follicle matures as a result of the action of FSH and LH. Near midcycle, there is a rise in estrogen secretion from the follicle. This rise augments the responsiveness of the pituitary to GnRH and triggers a burst of LH secretion. The resulting ovulation is followed by formation of a corpus luteum. There is a drop in estrogen secretion, but progesterone and estrogen levels then rise together, along with inhibin. The elevated levels inhibit FSH and LH secretion for a while, but luteolysis again occurs and a new cycle starts.

Reflex Ovulation

Female cats, rabbits, mink, and some other animals have long periods of estrus, during which they ovulate only after copulation. Such **reflex ovulation** is brought about by afferent impulses from the genitalia and the eyes, ears, and nose that converge on the ventral hypothalamus and provoke an ovulation-inducing release of LH from the pituitary. In species such as rats, monkeys, and humans, ovulation is a spontaneous periodic phenomenon, but neural mechanisms are also involved. Ovulation can be prevented in rats by administering pentobarbital or various other neurally active drugs 12 hours before the expected time of ovulation. In women, menstrual cycles may be markedly influenced by emotional stimuli.

Contraception

Methods commonly used to prevent conception are listed in Table 23–8, along with their failure rates. Once conception has occurred, abortion can be produced by progesterone antagonists such as mifepristone (see above). Male contraception is discussed in a preceding section of this chapter.

Implantation of foreign bodies in the uterus causes changes in the duration of the sexual cycle in a number of mammalian species. In humans, such foreign bodies do not alter the menstrual cycle, but they act as effective contraceptive devices. Intrauterine implantation of pieces of metal or plastic (**intrauterine devices, IUDs**) has been used in programs aimed at controlling population growth. The mechanism by which these devices exert their contraceptive effect is unsettled, but there is some evidence that they speed the passage of the fertilized ovum through the uterus, preventing its implantation in the endometrium. They also disturb the orderly sequential changes in the endometrium during the menstrual cycle, and this may be a factor. Their usefulness is limited by their tendency to cause intrauterine infections.

Women undergoing long-term treatment with relatively large doses of estrogen do not ovulate, probably because they have depressed FSH levels and multiple irregular bursts of LH secretion rather than a single midcycle peak. Women treated with similar doses of estrogen plus a progestational agent do not ovulate because the secretion of both gonadotropins is suppressed. In addition, the progestin makes the cervical mucus thick and unfavorable to sperm migration, and it may also interfere with implantation. For contraception, an orally active estrogen such as ethinylestradiol (Fig 23–33) is often combined with a synthetic progestin such as norethindrone (Fig 23–37). The pills are administered for 21 days, then withdrawn for 5–7 days to permit menstrual flow, and started again. Norethindrone has an ethinyl group on position 17 of the steroid nucleus, so it is resistant to hepatic metabolism and consequently is effective by mouth. In addition to being a progestin, it is partly metabolized to ethinyl estradiol, and for this reason it also has estrogenic activity. It is now clear that small as well as large doses of estrogen are effective (Table 23–8); the use of small dose reduces the risk of thromboses or other complications. Progestins alone can be used for contraception, although they are more effective when combined with estrogens.

Implants made up primarily of progestins are now seeing increased use in some parts of the world. These are inserted under the skin and can prevent pregnancy for up to 5 years. They often produce amenorrhea, but otherwise they appear to be effective and well tolerated.

ABNORMALITIES OF OVARIAN FUNCTION

Menstrual Abnormalities

Some women who are infertile have **anovulatory cycles;** they fail to ovulate but have menstrual periods at fairly regular intervals. As noted above, anovulatory cycles are the rule for the first 1–2 years after menarche and again before the menopause. **Amenorrhea** is the absence of menstrual periods. If menstrual bleeding has never occurred, the condition is called **primary amenorrhea.** Some women with primary amenorrhea have small breasts and other signs of failure to mature sexually. Cessation of cycles in a woman with previously normal periods is called sec-

Table 23–8. Relative effectiveness of frequently used contraceptive methods.[1]

Method	Failures per 100 Woman-Years
Vasectomy	0.02
Tubal ligation and similar procedures	0.13
Oral contraceptive	
>50 μg estrogen and progestin	0.32
<50 μg estrogen and progestin	0.27
Progestin only	1.2
IUD	
Copper 7	1.5
Loop D	1.3
Diaphragm	1.9
Condom	3.6
Withdrawal	6.7
Spermicide	11.9
Rhythm	15.5

[1]Data from Vessey M, Lawless M, Yeates D: Efficacy of different contraceptive methods. *Lancet* 1982;1:841. Reproduced with permission.

Figure 23–37. Norethindrone, a synthetic progestational agent.

ondary amenorrhea. The commonest cause of secondary amenorrhea is pregnancy, and the old clinical maxim that "secondary amenorrhea should be considered to be due to pregnancy until proven otherwise" has considerable merit. Other causes of amenorrhea include emotional stimuli and changes in the environment, hypothalamic diseases, pituitary disorders, primary ovarian disorders, and various systemic diseases. There is evidence that in some women with hypothalamic amenorrhea, the frequency of GnRH pulses is slowed as a result of excess opioid activity in the hypothalamus. In encouraging preliminary studies, the frequency of GnRH pulses has been increased by administration of the orally active opioid blocker naltrexone.

The terms **hypomenorrhea** and **menorrhagia** refer to scanty and abnormally profuse flow, respectively, during regular periods. **Metrorrhagia** is bleeding from the uterus between periods, and **oligomenorrhea** is reduced frequency of periods. **Dysmenorrhea** is painful menstruation. The severe menstrual cramps that are common in young women quite often disappear after the first pregnancy. Most of the symptoms of dysmenorrhea are due to accumulation of prostaglandins in the uterus, and symptomatic relief has been obtained by treatment with inhibitors of prostaglandin synthesis (see Chapter 17).

Some women develop symptoms such as irritability, bloating, edema, emotional lability, decreased ability to concentrate, depression, headache, and constipation during the last 7–10 days of their menstrual cycles. These symptoms of the **premenstrual syndrome** have been attributed to salt and water retention. However, it seems unlikely that this or any of the other hormonal alterations that occur in the late luteal phase are responsible because the time course and severity of the symptoms is not modified if the luteal phase is terminated early and menstruation produced by administration of mifepristone. Therefore, at present the cause of the premenstrual syndrome must be listed as unknown.

The Polycystic Ovary Syndrome

An interesting cause of infertility and amenorrhea is the **polycystic ovary syndrome** (Stein-Leventhal syndrome), a condition characterized by thickening of the ovarian capsule and formation of multiple follicular cysts. Plasma testosterone, estradiol, estrone, and LH are elevated in this syndrome, whereas plasma FSH and inhibin are low. There has been much debate about this syndrome, and the causes are probably multiple. Some cases may be due to an abnormality of the hypothalamic pulse generator in which GnRH pulses are too frequent. This would favor LH secretion and diminish FSH secretion (see above).

Ovarian Tumors

Androgen-secreting tumors can cause masculinization, and estrogen-secreting ovarian tumors in childhood can cause precocious sexual development (Table 23–2).

PREGNANCY

Fertilization & Implantation

In humans, **fertilization** of the ovum by the sperm usually occurs in the midportion of the uterine tube. Millions of sperms are deposited in the vagina during intercourse. There is evidence that the human ovum secretes an attractant or chemotactic factor that attracts sperms to the egg. Eventually, 50–100 sperms reach the ovum. Many of these contact the **zona pellucida,** a membranous structure that surrounds the ovum. The sperms then bind to the zona pellucida by a reaction between sperm receptors in the zona and a specific egg-binding protein on the sperm plasma membrane. Binding is followed by the **acrosomal reaction,** which appears to be triggered by the sperm receptor. This reaction is the breakdown of the acrosome, the lysosomelike organelle on the head of the sperm (Fig 23–17), with the release of various enzymes including the trypsinlike protease **acrosin.** Acrosin facilitates the penetration of the sperm through the zona pellucida. When one sperm reaches the membrane of the ovum, it fuses to it, setting off a reduction in the membrane potential of the ovum that prevents polyspermy, the fertilization of the ovum by more than one sperm. This transient potential change is followed by a structural change in the zona pellucida that provides protection against polyspermy on a more long-term basis. Fusion of the cell membranes of the sperm and the ovum activates the cells, and embryonic development begins.

The developing embryo, now called a **blastocyst,** moves down the tube into the uterus. This journey takes about 3 days, during which the blastocyst reaches the 8- or 16-cell stage. Once in contact with the endometrium, the blastocyst becomes surrounded by an outer layer of **syncytiotrophoblast,** a multinucleate mass with no discernible cell boundaries, and an inner layer of **cytotrophoblast** made up of individual cells. The syncytiotrophoblast erodes the endometrium, and the blastocyst burrows into it (**implantation**). The implantation site is usually on the dorsal wall of the uterus. A placenta then develops, and the trophoblast remains associated with it.

It should be noted that the fetus and the mother are 2 genetically distinct individuals, and the fetus is in effect a transplant of foreign tissue in the mother. However, the transplant is tolerated, and the rejection reaction that is characteristically produced when other foreign tissues are transplanted (see Chapter 27) fails to occur. The reason the "fetal graft" is not rejected is unknown, although it may be that the major histocompatibility antigen expressed by the trophoblast in areas where it is in contact with maternal tissue is not variable in structure and hence not immunogenic.

The vexing clinical problem of **infertility** often requires extensive investigation before a cause is found. In 30–40% of cases the problem is in the man, in 50–60% the problem is in the woman, and in the remaining 5–10% no cause can be found. **In vitro fertilization,** ie, removing mature ova, fertilizing them with capacitated sperm, and implanting one or more of them in the uterus at the 4-cell stage, has a 5–10% chance of producing a live birth and hence is of value in some of these cases.

Endocrine Changes

In all mammals, the corpus luteum in the ovary at the time of fertilization fails to regress and instead enlarges in response to stimulation by gonadotropic hormones secreted by the placenta. The placental gonadotropin in humans is called **human chorionic gonadotropin (hCG).** The enlarged **corpus luteum of pregnancy** secretes estrogens, progesterone, and relaxin. The relaxin helps maintain pregnancy by inhibiting myometrial contractions. In most species, removal of the ovaries at any time during pregnancy precipitates abortion. In humans, however, the placenta produces sufficient estrogen and progesterone from maternal and fetal precursors to take over the function of the corpus luteum after the sixth week of pregnancy. Ovariectomy before the sixth week leads to abortion but thereafter has no effect on the pregnancy. The function of the corpus luteum begins to decline after 8 weeks of pregnancy, but it persists throughout pregnancy. hCG secretion decreases after an initial marked rise, but estrogen and progesterone secretion increase until just before parturition (Fig 23–38).

hCG

hCG is a glycoprotein that contains galactose and hexosamine. It is produced by the syncytiotrophoblast. Like the pituitary glycoprotein hormones, it is made up of α and β subunits. hCG-α is identical to the α subunit of LH, FSH, and TSH. The molecular weight of hCG-α is 18,000, and that of hCG-β is 28,000. hCG is primarily luteinizing and luteotropic and has little FSH activity. It can be measured by radioimmunoassay and detected in the blood as early as 6 days after conception. Its presence in the urine in early pregnancy is the basis of the various laboratory tests for pregnancy, and it can sometimes be detected in the urine as early as 14 days after conception. It appears to act on the same receptor as LH.

hCG is not absolutely specific for pregnancy.

Figure 23–38. Hormone levels during normal pregnancy. (Data from various authors.)

Small amounts are secreted by a variety of gastrointestinal and other tumors in both sexes, and hCG has been measured in individuals with suspected tumors as a "tumor marker." It also appears that the fetal liver and kidney normally produce small amounts of hCG.

hCS

The syncytiotrophoblast also secretes large amounts of a protein hormone that is lactogenic and has a small amount of growth-stimulating activity. This hormone has been called **chorionic growth hormone-prolactin** (CGP) and **human placental lactogen** (hPL), but it is now generally called **human chorionic somatomammotropin** (hCS). The structure of hCS is very similar to that of human growth hormone (see Fig 22–4), and it appears that these 2 hormones and prolactin evolved from a common progenitor hormone. Large quantities of hCS are found in maternal blood, but very little reaches the fetus. Secretion of growth hormone from the maternal pituitary is not increased during pregnancy and may actually be decreased by hCS. However, hCS has most of the actions of growth hormone and apparently functions as a "maternal growth hormone of pregnancy" to bring about the nitrogen, potassium, and calcium retention, lypolysis, and decreased glucose utilization seen in this state. These latter 2 actions divert glucose to the fetus. The amount of hCS secreted is proportionate to the size of the placenta, which normally weighs about one-sixth as much as the fetus, and low hCS levels are a sign of placental insufficiency.

Other Placental Hormones

In addition to hCG, hCS, progesterone, and estrogens, the human placenta apparently secretes relaxin (Fig 23–38). Human placental fragments probably produce POMC. In culture, they release CRH, β-endorphin, αMSH, and dynorphin A, all of which appear to be identical to their hypothalamic counterparts. They also secrete GnRH and inhibin, and since GnRH stimulates and inhibin inhibits hCG secretion, locally produced GnRH and inhibin may act in a paracrine fashion to regulate hCG secretion. The placenta also secretes prolactin in a number of variant forms.

Finally, the placenta also secretes the α subunits of hCG, and the plasma concentration of free α subunits rises throughout pregnancy. These α subunits acquire a carbohydrate composition that makes them unable to combine with β subunits, and their prominence suggests that they have a function of their own. It is interesting in this regard that the secretion of the prolactin produced by the endometrium also appears to increase throughout pregnancy, and it may be that the circulating α subunits stimulate endometrial prolactin secretion.

The cytotrophoblast of the human chorion contains prorenin (see Chapter 24), a material that appears to be identical to GnRH, and inhibin. There are large amounts of prorenin in amniotic fluid, but its function in this location is unknown.

Fetoplacental Unit

The fetus and the placenta interact in the formation of steroid hormones. The placenta synthesizes pregnenolone and progesterone from cholesterol. Some of the progesterone enters the fetal circulation and provides the substrate for the formation of cortisol and corticosterone in the fetal adrenal glands (Fig 23–39). Some of the pregnenolone enters the fetus and, along with pregnenolone synthesized in the fetal

liver, is the substrate for the formation of dehydroepiandrosterone sulfate (DHEAS) and 16-hydroxydehydroepiandrosterone sulfate (16-OHDHEAS) in the fetal adrenal. Some 16-hydroxylation also occurs in the fetal liver. DHEAS and 16-OHDHEAS are transported back to the placenta, where DHEAS forms estradiol and 16-OHDHEAS forms estriol. The principal estrogen formed is estriol, and since fetal 16-OHDHEAS is the principal substrate for the estrogens, the urinary estriol excretion of the mother can be monitored as an index of the state of the fetus.

Parturition

The duration of pregnancy in humans averages 270 days from fertilization (284 days from the first day of the menstrual period preceding conception). Irregular uterine contractions increase in frequency in the last month of pregnancy.

The difference between the body of the uterus and the cervix becomes evident at the time of delivery. The cervix, which is firm in the nonpregnant state and throughout pregnancy until near the time of delivery, softens and dilates, while the body of the uterus contracts and expels the fetus.

The number of oxytocin receptors in the myometrium and the decidua (the endometrium of pregnancy) increases more than 100-fold during pregnancy and reaches a peak during early labor. Estrogens increase the number of oxytocin receptors, and uterine distention late in pregnancy may also increase their formation. In early labor, the oxytocin concentration in maternal plasma is not elevated from the prelabor value of about 25 pg/mL, but the marked increase in oxytocin receptors may cause the uterus to respond to normal plasma oxytocin concentrations. Once labor is started, the uterine contractions dilate the cervix, and this dilation in turn sets up signals in afferent nerves that increase oxytocin secretion (Fig 23–40). The plasma oxytocin level rises, and more oxytocin becomes available to act on the uterus. Thus, a positive feedback loop is established that aids delivery and terminates on expulsion of the products of conception. Oxytocin increases uterine contractions in 2 ways: (1) It acts directly on uterine smooth muscle cells to make them contract, and (2) it stimulates the formation of prostaglandins in the decidua, and the prostaglandins enhance the oxytocin-induced contractions. There is evidence that both processes are needed for normal labor to occur; labor is prolonged by inhibitors of prostaglandin synthesis and by inhibition of oxytocin secretion. Some women with diabetes insipidus have been reported to have normal deliveries, but this is probably because some oxytocin-secreting neurons were spared by the disease process that caused the defect in vasopressin secretion.

During labor, spinal reflexes and voluntary contractions of the abdominal muscles ("bearing down") also aid in delivery. However, it appears that delivery

Figure 23–39. Interactions between the placenta and the fetal adrenal cortex in the production of steroids.

Figure 23–40. Role of oxytocin in parturition.

can occur without bearing down and without a reflex increase in secretion of oxytocin from the posterior pituitary gland, since paraplegic women can go into labor and deliver.

The fetus may also play a role in triggering the onset of labor. It is interesting in this regard that the plasma oxytocin concentration in the human fetus rises in very early labor, before the maternal concentration rises. However, it is uncertain whether fetal oxytocin can reach the decidua and the myometrium.

In some species, an increase in fetal plasma glucocorticoid concentration may be involved in the initiation of labor, but it is uncertain whether such a mechanism plays any role in humans.

LACTATION

Development of the Breasts

Many hormones are necessary for full mammary development. In general, estrogens are primarily responsible for proliferation of the mammary ducts and progesterone for the development of the lobules. In rats, some prolactin is also needed for development of the glands at puberty, but it is not known if prolactin is necessary in humans. In hypophysectomized rats, glucocorticoids, insulin, and growth hormone are necessary for mammary development in response to other hormones, but they do not by themselves cause growth of the breasts (Fig 23–41). During pregnancy, prolactin levels increase steadily until term, and under the influence of this hormone plus the high levels of estrogens and progesterone, full lobulo-alveolar development of the breasts takes place.

Secretion & Ejection of Milk

The composition of human and cows' milk is

Figure 23–41. Hormonal control of breast development and lactation in rats. Estrogens (E) plus some progesterone (P) and some prolactin in the presence of glucocorticoids (C), insulin (I), and growth hormone (GH) cause duct proliferation and growth at puberty *(right).* During pregnancy, all of these hormones bring about full alveolar development and some milk secretion *(below).* After delivery, increased secretion of prolactin and a decline in estrogen and progesterone levels bring about copious secretion and, in the presence of oxytocin, ejection of milk *(left).* Chorionic mammotropin is the lactogenic hormone presumably secreted by the placenta in rats and is analogous to hCS. It supplements the action of prolactin.

shown in Table 23–9. In estrogen- and progesterone-primed rodents, injections of prolactin cause the formation of milk droplets and their secretion into the ducts. Oxytocin causes contraction of the myoepithelial cells lining the duct walls, with consequent ejection of the milk through the nipple (Fig 23–41). The reflex release of oxytocin initiated by touching the nipples and areolas (milk ejection reflex) is discussed in Chapter 14. Oxytocin is not essential for milk ejection in some species, but it is in humans.

The other hormonal relations in humans are generally similar to those in rats, although normal breast growth and lactation can occur in dwarfs with congenital growth hormone deficiency.

Initiation of Lactation After Delivery

The breasts enlarge during pregnancy in response to high circulating levels of estrogens, progesterone, prolactin, and possibly hCG. Some milk is secreted into the ducts as early as the fifth month, but the amounts are small compared with the surge of milk secretion that follows delivery. In most animals, milk is secreted within an hour after delivery, but in women it takes 1–3 days for the milk to "come in."

After expulsion of the placenta at parturition, there is an abrupt decline in circulating estrogens and progesterone. The drop in circulating estrogen initiates lactation. Prolactin and estrogen synergize in producing breast growth, but estrogen antagonizes the milk-producing effect of prolactin on the breast. Indeed, in women who do not wish to nurse their babies, estrogens may be administered to stop lactation.

Table 23–9. Composition of colostrum and milk.[1]
(Units are weight per deciliter.)

Component	Human Colostrum	Human Milk	Cows' Milk
Water, g	. . .	88	88
Lactose, g	5.3	6.8	5.0
Protein, g	2.7	1.2	3.3
Casein:lactalbumin ratio	. . .	1:2	3:1
Fat, g	2.9	3.8	3.7
Linoleic acid	. . .	8.3% of fat	1.6% of fat
Sodium, mg	92	15	58
Potassium, mg	55	55	138
Chloride, mg	117	43	103
Calcium, mg	31	33	125
Magnesium, mg	4	4	12
Phosphorus, mg	14	15	100
Iron, mg	0.09[2]	0.15[2]	0.10[2]
Vit A, μg	89	53	34
Vit D, μg	. . .	0.03[2]	0.06[2]
Thiamine, μg	15	16	42
Riboflavin, μg	30	43	157
Nicotinic acid, μg	75	172	85
Ascorbic acid, mg	4.4[2]	4.3[2]	1.6[2]

[1]Reproduced, with permission, from Findlay ALR: Lactation, *Res Reprod* (Nov) 1974; **6(6).**
[2]Poor source.

Suckling not only evokes reflex oxytocin release and milk ejection; it also maintains and augments the secretion of milk because of the stimulation of prolactin secretion produced by suckling (see above).

Effect of Lactation on Menstrual Cycles

Women who do not nurse their infants usually have their first menstrual period 6 weeks after delivery. However, women who nurse regularly have amenorrhea for 25–30 weeks. Nursing stimulates prolactin secretion, and there is evidence that prolactin inhibits GnRH secretion, inhibits the action of GnRH on the pituitary, and antagonizes the action of gonadotropins on the ovaries. Ovulation is inhibited, and the ovaries are inactive, so estrogen and progesterone output falls to low levels. Consequently, only 5–10% of women become pregnant again during the suckling period, and nursing has long been known to be an important if only partly effective method of birth control. Furthermore, almost 50% of the cycles in the first 6 months after resumption of menses are anovulatory.

Chiari-Frommel Syndrome

An interesting although rare condition is persistence of lactation (**galactorrhea**) and amenorrhea in women who do not nurse after delivery. This condition, called the **Chiari-Frommel syndrome,** may be associated with some genital atrophy and is due to persistent prolactin secretion without the secretion of the FSH and LH necessary to produce maturation of new follicles and ovulation. A similar pattern of galactorrhea and amenorrhea with high circulating prolactin levels is seen in nonpregnant women with chromophobe pituitary tumors and in women in whom the pituitary stalk has been sectioned in treatment of cancer.

Gynecomastia

Breast development in the male is called **gynecomastia.** It may be unilateral but is more commonly bilateral. It is seen in mild, transient form in 70% of normal boys at the time of puberty. It is a complication of estrogen therapy and occurs in patients with estrogen-secreting tumors. It also occurs in a wide variety of seemingly unrelated conditions, including eunuchoidism due to primary testicular disease, hypothyroidism, hyperthyroidism, and cirrhosis of the liver. Digitalis can produce it, apparently because cardiac glycosides are weakly estrogenic. It has been seen in malnourished prisoners of war, but only after they were liberated and eating an adequate diet. A feature common to many cases of gynecomastia is an increase in the plasma estrogen:androgen ratio due either to increased circulating estrogens or decreased circulating androgens. However, there are many other theories about the causes of gynecomastia, and the causes are probably multiple.

Hormones & Cancer

About 35% of carcinomas of the breast in women of childbearing age are **estrogen-dependent;** their continued growth depends upon the presence of estrogens in the circulation. The tumors are not cured by decreasing estrogen secretion, but symptoms are dramatically relieved, and the tumor regresses for months or years before recurring. Women with estrogen-dependent tumors often have a remission when their ovaries are removed. The incidence of a favorable response is greater when the tumor contains estrogen receptors and greatest when it contains both estrogen and progesterone receptors, because estrogen stimulates the formation of progesterone receptors and their presence indicates that estrogen is not only binding to but acting on the tumor cells. However, a few women with neither type of receptor still respond to this type of endocrine therapy. When the disease recurs, another remission follows bilateral adrenalectomy. Since ovarian and adrenal estrogen secretion are both inhibited by hypophysectomy, this operation has been performed in cancer patients. The number of remissions produced by hypophysectomy is at least as great as the number produced by castration. There also is some evidence that growth hormone and prolactin stimulate the growth of breast carcinomas, and hypophysectomy removes these stimuli. Hypophysectomy induces a significant incidence of remissions in carcinoma of the male breast, an uncommon but serious disease.

Some carcinomas of the prostate are **androgen-dependent** and regress temporarily after the removal of the testes or treatment with GnRH agonists in doses that are sufficient to produce down-regulation of the GnRH receptors on gonadotropes and decrease LH secretion. The formation of pituitary tumors after removal of the target endocrine glands controlled by pituitary tropic hormones is discussed in Chapter 22.

24

Other Endocrine Organs

INTRODUCTION

The organs with endocrine functions include numerous structures in addition to the posterior, intermediate, and anterior lobes of the pituitary; the thyroid; the parathyroids; the pancreas; the adrenal cortex; the adrenal medulla; and the gonads. Hormones that stimulate or inhibit the secretion of anterior pituitary hormones are secreted by the hypothalamus (see Chapter 14), and a number of hormones are secreted by the mucosa of the gastrointestinal tract (see Chapter 26). Many different cells produce cytokines, interleukins, and growth factors (see Chapters 1, 22, and 27). The kidneys produce 3 hormones: 1,25-dihydroxycholecalciferol (see Chapter 21), renin, and erythropoietin. Atrial natriuretic peptide, a substance secreted by the heart, increases excretion of sodium by the kidneys, and an additional natriuretic hormone inhibits Na^+-K^+ ATPase. The pineal gland secretes melatonin, and this indole may have an endocrine function. The endocrine functions of the kidneys, heart, and pineal gland are considered in this chapter.

THE RENIN-ANGIOTENSIN SYSTEM

Renin & Angiotensin

The rise in blood pressure produced by injection of kidney extracts is due to **renin,** an acid protease secreted by the kidneys into the bloodstream. This glycoprotein hormone has a molecular weight of 37,326 in humans. The molecule is made up of 2 lobes, or domains, between which the active site of the enzyme is located in a deep cleft. Two aspartic acid residues, one at position 32 and one at position 215, are juxtaposed at the mouth of the cleft and are essential for activity. Thus, renin is an aspartyl protease.

Like other hormones, renin is synthesized as a large preprohormone. Human **preprorenin** contains 406 amino acid residues. The **prorenin** that remains after removal of a leader sequence of 23 amino acid residues from the N terminal contains 383 amino acid residues, and after removal of the pro sequence from the N terminal of prorenin, active **renin** contains 340 amino acid residues. Prorenin has relatively little biologic activity.

Some prorenin is converted to renin in the kidneys, and some is secreted. There is very little conversion to active renin in the circulation. Prorenin can be converted to renin by tissue kallikrein (see Chapter 31) and cathepsin B, but the details of the process by which renin is formed from prorenin in vivo are still unsettled.

Renin has a half-life in the circulation of 80 minutes or less. It splits the decapeptide **angiotensin I** from the N terminal of a glycoprotein in the α_2 globulin fraction of the proteins in the circulating plasma (Fig 24–1). The α_2 globulin, which contains about 13% carbohydrate, is made up of 453 amino acid residues. It is synthesized in the liver with a 32-amino-acid signal sequence that is removed in the endoplasmic reticulum, and it is called **angiotensinogen** or **renin substrate.** Its circulating level is increased by glucocorticoids, thyroid hormones, and estrogens. **Angiotensin-converting enzyme (ACE)** is a dipeptidyl-carboxypeptidase that splits off histidyl-leucine from the physiologically inactive angiotensin I, forming the octapeptide **angiotensin II** (Fig 24–2). The same enzyme inactivates bradykinin (see Chapter 31). Most of the converting enzyme that forms angiotensin II in the circulation is located in endothelial cells. Much of the conversion occurs as the blood passes through the lungs, but there is also conversion in many other parts of the body.

Angiotensin II is destroyed rapidly, its half-life in humans being 1–2 minutes. The enzymes that destroy angiotensin II are lumped together under the term **angiotensinase.** They include an aminopeptidase that removes the Asp residue from the N terminal of the peptide. The resulting heptapeptide, unlike the other peptide fragments, has physiologic activity and is sometimes called **angiotensin III** (see below). In addition aminopeptidase can act on angiotensin I to produce (des-Asp^1)angiotensin I, and this compound can be converted directly to angiotensin III by the action of angiotensin-converting enzyme. Angiotensinase activity is found in red blood cells and many tissues. In addition, angiotensin II appears to be removed from the circulation by some sort of trapping mechanism in the vascular beds of tissues other than the lungs.

Figure 24–1. Formation and metabolism of circulating angiotensins. ACE, angiotensin-converting enzyme; AP, aminopeptidase.

Renin is usually measured by incubating the sample to be assayed and measuring by immunoassay the amount of angiotensin I generated. This measures the **plasma renin activity (PRA)** of the sample. Deficiency of angiotensinogen as well as renin can cause low PRA values, and to avoid this problem, exogenous angiotensinogen is often added, so that **plasma renin concentration (PRC)** rather than PRA is measured. The normal PRA in supine subjects eating a normal amount of sodium is approximately 1 ng of angiotensin I generated per milliliter per hour. The plasma angiotensin II concentration in such subjects is about 25 pg/mL (approximately 25 pmol/L).

Actions of Angiotensins

Angiotensin I appears to function solely as the precursor of angiotensin II and does not have any other established action.

Angiotensin II—previously called hypertensin or angiotonin—produces arteriolar constriction and a rise in systolic and diastolic blood pressure. It is one of the most potent vasoconstrictors known, being 4–8 times as active as norepinephrine on a weight basis in normal individuals. However, its pressor activity is decreased in sodium-depleted individuals and in patients with cirrhosis and some other diseases. In these conditions, circulating angiotensin II is increased, and this down regulates the angiotensin receptors in vascular smooth muscle. Consequently, there is less response to injected angiotensin II.

Angiotensin II also acts directly on the adrenal cortex to increase the secretion of aldosterone, and the renin-angiotensin system is a major regulator of aldosterone secretion (see Chapter 20). Additional actions of angiotensin II include facilitation of the release of norepinephrine by a direct action on post-

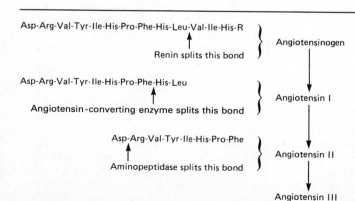

Figure 24–2. Structure of the N-terminal end of angiotensinogen and angiotensins I, II, and III in humans. R, remainder of protein. After removal of a 24-amino-acid leader sequence, angiotensinogen contains 453 amino acid residues. The structure of angiotensin II in dogs, rats, and many other mammals is the same as that in humans. Bovine and ovine angiotensin II have valine instead of isoleucine in position 5.

ganglionic sympathetic neurons, contraction of mesangial cells with a resultant decrease in glomerular filtration rate (see Chapter 38), and additional direct effects on the kidneys.

Angiotensin II also acts on the brain to increase blood pressure, increase water intake (see Chapter 14), and increase the secretion of vasopressin and ACTH. It does not penetrate the blood-brain barrier, but it triggers these responses by acting on the circumventricular organs, 4 small structures in the brain that are outside the blood-brain barrier (see Chapter 32). One of these structures, the area postrema, is primarily responsible for the pressor effect, whereas 2 of the others, the subfornical organ (SFO) and the organum vasculosum of the lamina terminalis (OVLT), are responsible for the increase in water intake (dipsogenic effect). It is not certain which of the circumventricular organs are responsible for the increases in vasopressin and ACTH secretion.

Angiotensin III [(des-Asp^1)angiotensin II] has about 40% of the pressor activity of angiotensin II but 100% of the aldosterone-stimulating activity. It has been suggested that angiotensin III is the natural aldosterone-stimulating peptide whereas angiotensin II is the blood-pressure-regulating peptide. However, this appears not to be the case, and instead angiotensin III is simply a breakdown product with some biologic activity.

Tissue Renin-Angiotensin Systems

Evidence is accumulating that in addition to the system that generates circulating angiotensin II, many different tissues contain independent renin-angiotensin systems that generate angiotensin II, apparently for local use. Components of the renin-angiotensin system are found in the walls of blood vessels and in the uterus, the placenta, and the fetal membranes. There is a high concentration of prorenin in amniotic fluid. In addition, there are tissue renin-angiotensin systems in the adrenal cortex, testis, ovary, anterior and intermediate lobes of the pituitary, pineal, and brain. The submaxillary glands of male mice have a particularly high content of renin. The functions of these tissue renin-angiotensin systems are unsettled. However, tissue renin contributes very little to the circulating renin pool since plasma renin activity falls to near zero after the kidneys are removed.

Angiotensin II Receptors

There are at least 2 types of angiotensin II receptors (Fig 24-1). The AT$_1$ receptor, which has been cloned, is a serpentine receptor coupled by a G protein to phospholipase C, and angiotensin II increases cytosolic free Ca^{2+}. AT$_1$ receptors are found in blood vessel walls, the adrenal cortex, the brain, and many other organs, and they mediate most of the known effects of angiotensin II. AT$_2$ receptors have an as yet unknown second messenger that is not IP$_3$ or cyclic AMP and their functions are unknown. AT$_2$ receptors are more plentiful in fetal and neonatal life, but they persist in the brain and other organs in adults. It appears likely that there are AT$_3$ receptors as well.

Although both appear to be AT$_1$ receptors, the receptors in the arterioles and the adrenal cortex are regulated in opposite ways: an excess of angiotensin II down regulates the vascular receptors, but it up regulates the adrenocortical receptors, making the gland more sensitive to the aldosterone-stimulating effect of the peptide.

The Juxtaglomerular Apparatus

The renin in kidney extracts and the bloodstream is produced by the **juxtaglomerular cells (JG cells).** These epithelioid cells are located in the media of the afferent arterioles as they enter the glomeruli. The membrane-lined secretory granules in them have been shown to contain renin. Renin is also found in agranular **lacis cells** that are located in the junction between the afferent and efferent arterioles, but its significance in this location is unknown.

At the point where the afferent arteriole enters the glomerulus and the efferent arteriole leaves it, the tubule of the nephron touches the arterioles of the glomerulus from which it arose. At this point, which marks the start of the distal convolution, there is a modified region of tubular epithelium called the macula densa (Fig 24-3). The macula densa is in close proximity to the JG cells. The lacis cells, the JG cells, and the macula densa constitute the **juxtaglomerular apparatus.**

Regulation of Renin Secretion

Renin secretion is increased by stimuli that decrease ECF volume and blood pressure or increase sympathetic output (Table 24-1). At least 5 different regulatory factors appear to be involved (Fig 24-4), and the rate of renin secretion at any given time is due to the summed activity of these factors. One is an intrarenal baroreceptor mechanism that causes renin secretion to increase when the intra-arteriolar pressure at the level of JG cells is decreased and to decrease when this pressure is increased. Another sensor involved in the regulation of renin secretion is the macula densa. There is some debate about what is being sensed, but the bulk of the evidence supports the view that renin secretion is inversely proportionate to the rate of transport of Na$^+$ and Cl$^-$ across this portion of the tubule. The rate of transport depends not only on the mechanisms in the macula densa cells but also on the amount of electrolyte reaching the macula densa. Therefore, decreased delivery of Na$^+$ and Cl$^-$ to the distal tubules is associated with increased renin secretion. Prostaglandins, especially prostacyclin (see Chapter 17), stimulate renin secretion, apparently by a direct action on the juxtaglomerular cells. There is some evidence that the effects of the macula densa on renin secretion are mediated by prostaglandins generated in the renal cortex. Renin secretion

Figure 24–3. **Left:** Diagram of glomerulus, showing the juxtaglomerular apparatus. (Modified from Ham AW: *Histology,* 6th ed. Lippincott, 1969). **Right:** Phase contrast photomicrograph of afferent arteriole in an unstained, freeze-dried preparation of the kidney of a mouse. Note the red blood cell in the lumen of the arteriole and the granulated juxtaglomerular cells in the wall. (Courtesy of C Peil.)

also varies inversely with the plasma K^+ level, but the effect of K^+ appears to be mediated by the changes it produces in Na^+ and Cl^- delivery to the macula densa.

Angiotensin II feeds back to inhibit renin secretion by a direct action on the JG cells. Vasopressin also inhibits renin secretion in vitro and in vivo, although there is some debate about whether its in vivo effect is direct or indirect.

Finally, increased activity of the sympathetic nervous system increases renin secretion. The increase is mediated both by way of increased circulating catecholamines and by way of the renal sympathetic nerves. The sympathetic effects on renin secretion are mediated mostly via β_1-adrenergic receptors and cyclic AMP generated in the JG cells.

Pharmacologic Manipulation of the Renin-Angiotensin System

It is now possible to inhibit the secretion or the effects of renin in a variety of ways. Inhibitors of

prostaglandin synthesis such as **indomethacin** and β-adrenergic blocking drugs such as **propranolol** reduce renin secretion. The peptide **pepstatin** and newly developed renin inhibitors such as **enalkiren** prevent renin from generating angiotensin I, and angiotensin-converting enzyme inhibitors (ACE inhibitors) such as **captopril** and **enalapril** prevent conversion of angiotensin I to angiotensin II. **Saralasin** and several other analogs of angiotensin II are competitive inhibitors of the action of angiotensin II on both AT_1 and AT_2 receptors. Losartan (DuP-753) selectively blocks AT_1 receptors, and PD-123177 and several other drugs selectively block AT_2 receptors. These drugs are proving to be important tools in exploring the physiology of the renin-angiotensin system.

Role of Renin in Clinical Hypertension

Constriction of one renal artery causes a prompt increase in renin secretion and the development of sustained hypertension (**renal** or **Goldblatt hypertension**). Removal of the ischemic kidney or the arterial constriction cures the hypertension if it has not persisted too long. In general, the hypertension produced by constricting one renal artery with the other kidney intact (one-clip 2-kidney Goldblatt hypertension; see Table 33–5) is associated with increased circulating renin. The clinical counterpart of this condition is **renal hypertension** due to atheromatous narrowing of one renal artery or other abnormalities of the renal circulation. However, plasma renin activity is usually normal in one-clip one-kidney Goldblatt

Table 24–1. Stimuli that increase renin secretion.

Sodium depletion
Diuretics
Hypotension
Hemorrhage
Upright posture
Dehydration
Constriction of renal artery or aorta
Cardiac failure
Cirrhosis
Various psychologic stimuli

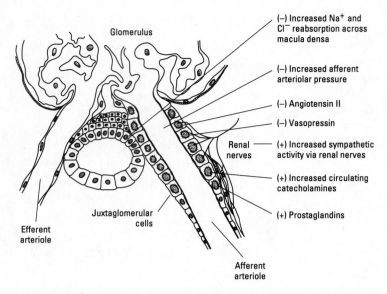

Figure 24–4. Factors that inhibit (–) and stimulate (+) renin secretion.

hypertension. The explanation of the hypertension in this situation is unsettled. However, many patients with hypertension respond to treatment with ACE inhibitors even when their renal circulation appears to be normal and they have normal or even low plasma renin activity. Preliminary data indicate that they also respond to renin inhibitors and losartan, so the fall in blood pressure cannot be explained by increased circulating bradykinin. This raises the possibility that these patients are abnormally sensitive to angiotensin II or that the tissue renin-angiotensin system in blood vessel walls is also involved in the pathogenesis of hypertension.

An interesting syndrome occurs in patients with idiopathic hypertrophy and hyperplasia of their juxtaglomerular apparatuses **(Bartter's syndrome).** These patients have persistent hypokalemia, elevated aldosterone secretion, and high circulating levels of angiotensin II. However, their vascular AT_1 receptors appear to be down regulated and their blood pressure is normal. The role of renin in a feedback mechanism that helps maintain the constancy of ECF volume through regulation of aldosterone secretion has been described in Chapter 20. A high level of renin secretion is responsible for the elevated aldosterone secretion **(secondary hyperaldosteronism)** seen in some normotensive patients with cirrhosis and nephrosis.

ERYTHROPOIETIN

Structure & Function

When an individual is bled or becomes hypoxic, hemoglobin synthesis is enhanced, and production and release of red blood cells from the bone marrow **(erythropoiesis)** are increased (see Chapter 27). Conversely, when the red cell volume is increased above normal by transfusion, the erythropoietic activity of the bone marrow decreases. These adjustments are brought about by changes in the circulating level of **erythropoietin,** a circulating glycoprotein that contains 165 amino acid residues and 4 oligosaccharide chains that are necessary for its activity in vivo. Its blood level is markedly increased in anemia (Fig 24–5). The gene for the hormone has been cloned, and recombinant erythropoietin produced in animal cells is available for clinical use. The recombinant erythropoietin is of value in the treatment of the anemia associated with renal failure and is also used to stimulate red cell production in individuals who are banking a supply of their own blood in preparation for autologous transfusions during elective surgery (see Chapter 27).

Erythropoietin increases the number of erythropoietin-sensitive committed stem cells in the bone marrow that are converted to red blood cell precursors and subsequently to mature erythrocytes (see Fig 27–2). The receptor for erythropoietin is a linear protein with a single transmembrane domain that is a member of the cytokine receptor superfamily (see Chapter 1). The mechanism by which it produces its intracellular effects is unsettled, but it is probably a tyrosine kinase. However, it appears that when eryth-

Figure 24–5. Plasma erythropoietin levels in normal blood donors (▲) and patients with various forms of anemia (■). (Reproduced, with permission, from Erslev AJ: Erythropoietin. *N Engl J Med* 1991;**324**:1339.)

ropoietin levels are low, erythroid stem cells show DNA cleavage followed by programmed cell death **(apoptosis)**. Apoptosis of various types of cells in now known to be part of normal development in many different tissues. In the erythroid stem cells, erythropoietin reduces the DNA cleavage and causes the cells to survive.

The principal site of inactivation of erythropoietin is the liver, and the hormone has a half-life in the circulation of about 5 hours. However, the increase in circulating red cells that it triggers takes 2–3 days to appear, since red cell maturation is a relatively slow process. Loss of even a small portion of the sialic acid residues in the carbohydrate moieties that are part of the erythropoietin molecule shortens its half-life to 5 minutes, making it biologically ineffective.

Sources

In adults, about 85% of the erythropoietin comes from the kidneys and 15% from the liver. Both these organs contain the mRNA for erythropoietin. Erythropoietin can also be extracted from the spleen and salivary glands, but these tissues do not contain the mRNA and consequently do not appear to manufacture the hormone. During fetal and neonatal life, the major site of erythropoiesis is the liver, and it is also the major site of erythropoietin production before erythropoiesis is taken over by the bone marrow and erythropoietin production by the kidneys. When renal mass is reduced in adults by renal disease or nephrectomy, the liver cannot compensate and anemia develops.

Erythropoietin is produced by endothelial cells of the peritubular capillaries in the renal cortex. In the liver, both Kupffer cells and hepatocytes have been claimed to produce erythropoietin.

Regulation of Secretion

The usual stimulus for erythropoietin secretion is hypoxia, but secretion of the hormone can also be stimulated by cobalt salts and androgens. Recent evidence suggests that the O_2 sensor regulating erythropoietin secretion in the kidneys and the liver is a heme protein that in the dioxy form stimulates and in the oxy form inhibits transcription of the erythropoietin gene to form erythropoietin mRNA. Secretion of hormone is facilitated by the alkalosis that develops at high altitudes. Like renin secretion, erythropoietin secretion is facilitated by catecholamines via a β-adrenergic mechanism, although the renin-angiotensin system is totally separate from the erythropoietin system. Erythropoietin production is also stimulated by adenosine and inhibited by the adenosine antagonist theophylline.

THE ENDOCRINE FUNCTION OF THE HEART: ATRIAL NATRIURETIC PEPTIDE

Structure

The existence of various **natriuretic hormones** has been postulated for some time. One type is secreted by the heart. The muscle cells in the atria contain secretory granules (Fig 24–6) that increase in

Figure 24–6. ANP granules (g) interspersed between mitochondria (m) in rat atrial muscle cell. G, Golgi complex; M, myofilaments; N, nucleus. The arrow identifies the cell membrane. The granules in human atrial cells are similar. (× 17,640. Courtesy of M Cantin.)

number when sodium chloride intake is increased and extracellular fluid expanded, and extracts of atrial tissue cause natriuresis.

The first natriuretic hormone isolated from the heart was **atrial natriuretic peptide (ANP),** a polypeptide with a characteristic 17-amino-acid ring formed by a disulfide bond between 2 cysteines. The circulating form of this polypeptide has 28 amino acid residues (Fig 24–7). It is formed from a large

precursor molecule that contains 151 amino acid residues, including a 24-amino-acid signal peptide. ANP was subsequently isolated from the brain, where it exists in 2 forms that are smaller than circulating ANP, and from other tissues. Next, brain natriuretic peptide (BNP), which has different amino acid residues in its 17-member ring as well as in its amino and carboxy terminals, was isolated from porcine brain. Interestingly, it does not appear to be present in human brain but is found in the human heart. A third polypeptide, CNP, is present in the human brain but apparently not in the human heart. ANP, BNP, and CNP are encoded by different genes.

Actions

ANP causes natriuresis, which may be due to an increase in glomerular filtration rate. There are ANP receptors on the mesangial cells in the glomeruli, and the relaxation of these cells produced by ANP presumably increases the effective surface area available for filtration (see Chapter 38). Alternatively, ANP could act on the tubules to promote sodium excretion. However, the exact mechanism of action of ANP on the kidneys is currently unsettled, and additional research is needed. ANP also lowers blood pressure, decreases the responsiveness of vascular smooth muscle to many vasoconstrictor substances, decreases the responsiveness of the zona glomerulosa to stimuli that normally increase aldosterone secretion, and inhibits the secretion of vasopressin. It is worth noting that these actions are generally the opposite of those of angiotensin II, and, indeed, the actions of ANP are most readily demonstrated against a back-

Figure 24–7. Human ANP, BNP, and CNP. **Top:** Single-letter codes for amino acid residues aligned to show common sequences (shaded). **Bottom:** Shape of molecules. Note that one cysteine is the C-terminal amino acid residue in CNP, so there is no C-terminal extension from the 17-member ring. (Modified from Imura H, Nakao K, Itoh H: The natriuretic peptide system in the brain: Implication in the central control of cardiovascular and neuroendocrine functions. *Front Neuroendocrinol* 1992;**13:**217.)

ground of stimulation by angiotensin II. In addition, ANP inhibits renin secretion and consequently lowers circulating angiotensin II levels. The actions of BNP and CNP, to the extent that they have been studied, resemble those of ANP. All 3 increase cyclic GMP in cells and appear to exert their effects via this nucleotide.

In the brain, ANP is present in neurons, and an ANP-containing neural pathway projects from the anterior, medial part of the hypothalamus to the areas in the lower brain stem that are concerned with neural regulation of the cardiovascular system. In general, the effects of ANP in the brain are opposite to those of angiotensin II, and ANP-containing neural circuits appear to be involved in lowering blood pressure and promoting natriuresis. The details of the CNP-containing neural circuits and, in species that have them, the BNP-containing circuits are unsettled.

ANP Receptors

Three different ANP receptors, ANPR-A, ANPR-B, and ANPR-C have been isolated and characterized (Fig 24–8). The ANPR-A and ANPR-B receptors both span the cell membrane and have cytoplasmic domains that are guanylate cyclases. ANP has the greatest affinity for the ANPR-A receptor, and CNP has the greatest affinity for the ANPR-B receptor. The third receptor, ANPR-C, binds all 3 natriuretic peptides but has a markedly truncated cytoplasmic domain. There is some evidence that it acts via G proteins to activate phospholipase C and inhibit adenylate cyclase. However, it has also been argued that this receptor does not trigger any intracellular change and is instead a clearance receptor which removes natriuretic peptides from the bloodstream and then releases them at a later time, helping to maintain a steady blood level of the hormones.

Secretion and Metabolism

The concentration of ANP in plasma is about 5 fmol/mL in normal humans ingesting moderate amounts of sodium. ANP secretion is increased when the ECF volume is increased by infusion of isotonic saline or ingestion of a high-sodium diet. It is also increased by immersion in water up to the neck (Fig 24–9), a procedure that counteracts the effect of gravity on the circulation and increases central venous and consequently atrial pressure. Note that immersion also decreases the secretion of renin and aldosterone. Conversely, there is a small but measurable decrease in plasma ANP in association with a decrease in central venous pressure on rising from the supine to the standing position. Strips of atrial muscle release ANP when stretched in vitro. Thus, it appears likely that the atria respond directly to stretch in vivo

Figure 24–9. Effect of immersion in water up to the neck for 3 hours on plasma concentrations of ANP, PRA, and aldosterone. (Modified and reproduced, with permission, from Epstein M et al: Increases in circulating atrial natriuretic factor during immersion-induced central hypervolaemia in normal humans. *Hypertension* 1986;**4[suppl 2]:**593.)

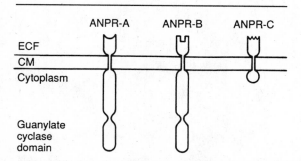

Figure 24–8. Diagrammatic representation of ANP receptors. The ANPR-A and ANPR-B receptor molecules have intracellular guanylate cyclase domains, whereas the clearance receptor, ANPR-C, has only a small cytoplasmic domain. CM, cell membrane.

and that the rate of ANP secretion is proportionate to the degree to which the atria are stretched by increases in central venous pressure.

Circulating ANP has a short half life. It is metabolized by neutral endopeptidase (NEP), which is inhibited by thiorphan. Therefore, administration of thiorphan increases circulating ANP.

Na⁺-K⁺ ATPase-Inhibiting Factor

There is another natriuretic factor in blood. This factor produces natriuresis by inhibiting Na^+-K^+ ATPase and raises rather than lowers blood pressure. Current evidence indicates that it is the digitalis-like steroid **ouabain** and that it comes from the adrenal glands. However, its physiologic significance is not yet known.

PINEAL

The **pineal** (epiphysis), believed by Descartes to be the seat of the soul, has at one time or another been regarded as having a wide variety of functions. It is now known to secrete melatonin, and there is speculation that it may function as a timing device to keep internal events synchronized with the light-dark cycle in the environment.

Anatomy

The pineal arises from the roof of the third ventricle under the posterior end of the corpus callosum and is connected by a stalk to the posterior commissure and habenular commissure. There are nerve fibers in the stalk, but they apparently do not reach the gland. The pineal stroma contains neuroglia and parenchymal cells with features suggesting that they have a secretory function (Fig 24–10). Like other endocrine glands, the pineal has highly permeable fenestrated capillaries. In young animals and infants, the pineal is large, and the cells tend to be arranged in alveoli. It begins to involute before puberty, and, in humans, small concretions of calcium phosphate and carbonate (**pineal sand**) appear in the tissue. Because the concretions are radiopaque, the normal pineal is often visible on x-ray films of the skull in adults. Displacement of a calcified pineal from its normal position indicates the presence of a space-occupying lesion such as a tumor in the brain.

Melatonin

The amphibian pineal contains an indole, N-acetyl-5-methroxytryptamine, named **melatonin** because it lightens the skin of tadpoles by an action on melanophores. However, it does not appear to play a physiologic role in the regulation of skin color. Melatonin and the enzymes responsible for its synthesis from serotonin by N-acetylation and O-methylation

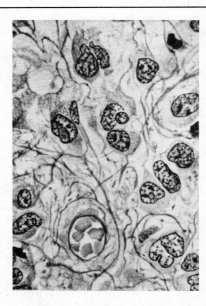

Figure 24–10. *Left:* Sagittal section of human brain stem showing the pineal and its innervation. Retino-hypothalamic fibers synapse in the suprachiasmatic nuclei (SCN), and there are connections from the SCN to the intermediolateral gray column in the spinal cord. Preganglionic neurons pass from the spinal cord to the superior cervical ganglion, and the postganglionic neurons project from this ganglion to the pineal in the nervi conarii. *Right:* Histology of pineal. Drawing of hematoxylin and eosin-stained section. (Reproduced, with permission, from Fawcett DW: *Bloom and Fawcett, A Textbook of Histology,* 11th ed. Saunders, 1986.)

(Fig 24–11) are also present in mammalian tissue, although some melatonin is synthesized in other parts of the body as well. Melatonin is synthesized by pineal parenchymal cells and secreted by them into the blood and the cerebrospinal fluid.

In humans and all other species studied to date, melatonin synthesis and secretion are increased during the dark period of the day and maintained at a low level during the daylight hours (Fig 24–12). This remarkable diurnal variation in secretion is brought about by norepinephrine secreted by the postganglionic sympathetic nerves (nervi conarii) that innervate the pineal (Fig 24–10). The norepinephrine acts via β-adrenergic receptors in the pineal to increase intracellular cyclic AMP, and the cyclic AMP in turn produces a marked increase in N-acetyltransferase activ-

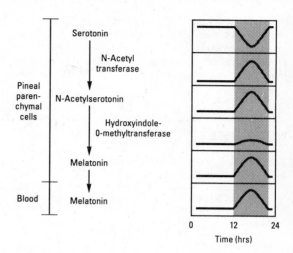

Figure 24–12. Diurnal rhythms of various compounds in the pineal and melatonin in blood. The shaded area represents the hours of darkness during the 24-hour day. (Modified from Reiter RJ: The pineal gland: An important link to the environment. *News Physiol Sci* 1986;**1:**20.)

ity. This results in increased melatonin synthesis and secretion.

The discharge of the sympathetic nerves to the pineal is entrained to the light-dark cycle in the environment via the retinohypothalamic nerve fibers to the suprachiasmatic nuclei (Fig 24–10). From the hypothalamus, descending pathways converge on the intermediolateral gray column of the thoracic spinal cord and end on the preganglionic sympathetic neurons that in turn innervate the superior cervical ganglion, the site of origin of the postganglionic neurons to the pineal.

Circulating melatonin is rapidly metabolized in the liver by 6-hydroxylation followed by conjugation, and over 90% of the melatonin that appears in the urine is in the form of 6-hydroxy conjugate. The pathway by which the brain metabolizes melatonin is unsettled but may involve cleavage of the indole nucleus.

Function of the Pineal

Injected melatonin has some effects on the gonads, but these effects vary markedly from species to species and also depend on the time of injection. In some situations, melatonin inhibits gonadal function, and in others, its effect is facilitatory. This variability has led to the hypothesis that it is not the melatonin per se but the diurnal change in melatonin secretion that functions as some sort of timing signal which coordinates internal events with the light-dark cycle in the environment. This may be correct, especially in seasonally breeding animals, which respond to changes in day length. However, the events that are regulated in humans have not been identified.

Figure 24–11. Formation and metabolism of melatonin. HIOMT, hydroxyindole-O-methyltransferase. For details of the synthesis and metabolism of serotonin, see Fig 4–21.

It has been argued that the pineal normally inhibits the onset of puberty in humans, because pineal tumors are sometimes associated with sexual precocity. However, as noted in Chapter 23, there is evidence that pineal tumors produce precocity only when they produce hypothalamic damage, and except for a small rise during breast development in girls, there are no significant changes in melatonin excretion during puberty. Thus, the functions of melatonin and the pineal in humans remain obscure.

REFERENCES

Section IV: Endocrinology, Metabolism, & Reproductive Function

Baulieu EE: Contragestion and other clinical applications of RU 486, on antiprogesterone at the receptor. *Science* 1989;**245**:1351.

Baumann G: Growth hormone heterogenicity: Genes, isohormones, variants, and binding proteins. *Endocr Rev* 1991;**12**:224.

Bender AE, Brookes LJ (editors): *Body Weight Control: The Physiology, Clinical Treatment and Prevention of Obesity.* Churchill Livingstone, 1987.

Bulleti C, Gurpide E (editors): The primate endometrium. *Ann NY Acad Sci* 1991;**622**:1.

Burger H, de Kretser D (editors): *The Testis,* 2nd ed. Academic Press, 1989.

Bremner WJ: Inhibin: From hypothesis to clinical application. *N Engl J Med* 1989;**321**:826.

Brody S, Uelard K (editors): *Endocrine Disorders in Pregnancy.* Appleton & Lange, 1989.

Brown MS, Goldstein JL: A receptor-mediated pathway for cholesterol homeostasis. *Science* 1986;**232**:34.

Burrow GN, Oppenheimer JH, Volpé R: *Thyroid Function and Disease.* Saunders, 1989.

Cantin M, Genest J: The heart and the atrial natriuretic factor. *Endocr Rev* 1985;**6**:107.

Carlson HE: Gynecomastia. *N Engl J Med* 1980;**303**:795.

Coe FL, Favus MJ (editors): *Disorders of the Bone and Mineral Metabolism.* Raven, 1991.

Cohen MP: *Diabetes and Protein Glycosylation.* Springer-Verlag, 1986.

Conn PM, Crowley WJ Jr: Gonadotropin-releasing hormone and its analogs. *N Engl J Med* 1991;**324**:93.

Cooper DS: Antithyroid drugs. *N Engl J Med* 1984;**311**:1353.

Cryer PE, Gerich JE: Glucose counter-regulation, hypoglycemia, and intensive insulin therapy in diabetes mellitus. *N Engl J Med* 1985;**313**:232.

Daughaday WH, Rotwein P: Insulin-like growth factors I and II. Peptide, messenger ribonucleic acid and gene structures, serum and tissue concentrations. *Endocr Rev* 1989;**10**:68.

DeGroot LJ (editor): *Endocrinology,* 2nd ed. Saunders, 1988.

Deluca H: The vitamin D story: A collaborative effort of basic science and clinical medicine. *FASEB J* 1988;**2**:224.

Dwyer JT: Nutritional consequences of vegetarianism. *Annu Rev Nutr* 1991;**11**:61.

Eckel RH: Lipoprotein lipase: A multifunctional enzyme relevant to common metabolic diseases. *N Engl J Med* 1989;**320**:1060.

Eggo MC, Burrow GN (editors): *Thyroglobulin: The Prothyroid Hormone,* Raven, 1985.

Eisan LJ, Longo M: Glucose transporters. *Annu Rev Med* 1992;**43**:377.

Falkner F, Tanner JM (editors): *Human Growth,* 2nd ed. 3 vols. Plenum, 1986.

Frawley LS, Boockfor FR: Mammosomatotropes: Presence and functions in normal and neoplastic pituitary tissue. *Endocr Rev* 1991;**12**:337.

Ganong WF, Martini L: Neuroendocrinology, 1990. *Front Neuroendocrinol* 1990;**11**:1.

Gerich JE: Oral hypoglycemic agents. *N Engl J Med* 1989;**321**:1231.

Goldstein JL, Brown MS: Regulation of the mevalonate pathway. *Nature* 1990;**343**:425.

Goodman DS: Vitamin A and retinoids in health and disease. *N Engl J Med* 1984;**310**:1023.

Gosden RG: *Biology of Menopause.* Academic Press, 1985.

Griffin JE: Androgen resistance—the clinical and molecular spectrum. *N Engl J Med* 1992;**326**:611.

Gross M, Kumar R: Physiology and biochemistry of vitamin D-dependent calcium binding proteins. *Am J Physiol* 1990;**259**:F195.

Hackenthal E et al: Morphology, physiology and molecular biology of renin secretion. *Physiol Rev* 1990;**70**:1067.

Heaning VJ, Tsukamato K: Enzymatic control of pigmentation in mammals. *FASEB J* 1991;**5**:2902.

Himms-Hagen J: Neural control of brown adipose tissue thermogenesis, hypertrophy, and atrophy. *Front Neuroendocrinol* 1991;**12**:38.

Howell RR, Morriss FH Jr, Pickering LK (editors): *Human Milk in Infant Nutrition and Health.* Thomas. 1986.

Kelly PA et al: The prolactin/growth hormone receptor family. *Endocr Rev* 1991;**12**:235.

Klabinski A, Zervas NT: Diagnosis and management of hormone-secreting pituitary adenomas. *N Engl J Med* 1991;**324**:822.

Knobil E: A hypothalamic pulse generator governs mammalian reproduction. *News Physiol Sci* 1987;**2**:42.

Knobil E, Neill JD (editors): *The Physiology of Reproduction.* 2 vols. Raven, 1987.

Krane RJ, Goldstein I, Saenz de Tejada I: Impotence. *N Engl J Med* 1989;**321**:1648.

Krantz SB: Erythropoietin. *Blood* 1991;**77**:419.

Lewis RA, Austen KF, Soberman RJ: Leukotrienes and other products of the 5-lipoxygenase pathway. *N Engl J Med* 1990;**323**:645.

Lynch KR, Peach MJ: Molecular biology of angiotensinogen. *Hypertension* 1991;**17**:263.

Lyon MF: the quest for the X-inactivation centre. *Trends Genet* 1991;**7**:69.

Maack T: Receptors for atrial natriuretic factor. *Annu Rev Physiol* 1992;**54**:11.

Mahley RW: Apolipoprotein E: cholesterol transport protein with expanding role in cell biology. *Science* 1988;**240**:622.

Martin JB, Reichlin S: *Clinical Neuroendocrinology*, 2nd ed. Davis, 1987.

Melmed S: Acromegaly. *N Engl J Med* 1990;**322**:966.

Mertz W: The essential trace elements. *Science* 1981; **213**:1332.

Mishell DR Jr: Contraception. *N Engl J Med* 1989;**320**:777.

Monder C: Corticosteroid receptors and the organ-specific functions of 11β-hydroxysteroid dehydrogenase. *FASEB J* 1991;**5**:347.

Muller DE, Flier JS: Insulin resistance mechanisms, syndromes and implications. *N Engl J Med* 1991; **325**:938.

Myant NB: *Cholesterol Metabolism, LDL, and the LDL Receptor*. Academic Press, 1990.

Petersen OH: Control of insulin secretion in pancreatic β-cells. *News Physiol Sci* 1990;**5**:254.

Pickup J, Williams G (editors): *Textbook of Diabetes*. 2 vols. Blackwell, 1991.

Prockop D: Mutations in collagen genes as a cause of connective tissue disease. *N Engl J Med* 1992; **326**:540.

Ramirez F, Di Liberto M: Complex and diversified regulatory programs control the expression of vertebrate collagen genes. *FASEB J* 1990;**4**:1616.

Rodger JC, Drake BL: The enigma of the fetal graft. *Am Sci* 1987;**75**:51.

Samuelsson B et al: Leukotrienes and lipoxins: structures, biosynthesis, and biological effects. *Science* 1987;**237**:1171.

Seibel MM: A new era in reproductive technology: In vitro fertilization, gamete intrafallopian transfer, and donated gametes and embryos. *N Engl J Med* 1988; **318**:828.

Selden RF et al: Regulation of insulin-gene expression. Implications for gene therapy. *N Engl J Med* 1987; **317**:1067.

Shils ME, Young VR (editors): *Modern Nutrition in Health and Disease*, 7th ed. Lea & Febiger, 1987.

Simonson DC, Defronzo RA: Indirect calorimetry: Methodological and interpretative problems. *Am J Physiol* 1990;**258**:E399.

Smith RD et al: Pharmacology of nonpeptide angiotensin II receptor antagonists. *Annu Rev Pharmacol Toxicol* 1992;**32**:135.

Stoner HB: Metabolism after trauma and in sepsis. *Circ Shock* 1986;**19**:75.

Taylor AL, Fishman LM: Corticotropin-releasing hormone. *N Engl J Med* 1988;**319**:213.

Therman E: *Human Chromosomes: Structure, Behavior, Effects*, 2nd ed. Springer-Verlag, 1986.

Ungar A, Phillips JH: Regulation of adrenal medulla. *Physiol Rev* 1983;**63**:787.

Unger RA: Diabetic hyperglycemia: Link to impaired glucose transport in pancreatic β cells. *Science* 1991;**251**:1200.

Vaitukaitis JL: Premenstrual syndrome. *N Engl J Med* 1984;**311**:1371.

Van Middlesworth L: Nuclear reactor accidents and the thyroid. *Thyroid Today* (Apr) 1987;**10**:1.

Volpé R: *Autoimmune Diseases of the Endocrine System*. CRC Press, 1990.

Walker WH et al: The human placental lactogen genes: Structure, function, evolution, and transcriptional regulation. *Endocr Rev* 1991;**12**:316.

Wall JR: Pathogenesis and management of thyroid associated ophthalmopathy: An update. *Thyroid Today* (Jan) 1991;**14**:1.

Wasserman DH, Cherrington AD: Hepatic fuel metabolism during muscular work: Role and regulation. *Am J Physiol* 1991;**210**:E811.

Wasserman PM. The biology and chemistry of fertilization. *Science* 1987;**235**:553.

White PC, New MI, DuPont B: Congenital adrenal hyperplasia. *N Engl J Med* 1987;**316**:1519.

Wilson JD: Androgen abuse by athletes. *Endocr Rev* 1989;**9**:181.

Wilson JD, Foster DW (editors): *Williams Textbook of Endocrinology*, 8th ed. Saunders, 1992.

Woeber KA: Thyrotoxicosis and the heart. *N Engl J Med* 1992;**327**:94.

Woodruff TK, Mayo KE: Regulation of inhibin synthesis in the rat ovary *Annu Rev Physiol* 1990;**52**:807.

Yanase T, Simpson ER, Waterman MR: 17α-hydroxylase/17,20-lyase deficiency: From clinical investigation to molecular definition. *Endocr Rev* 1991;**12**:91.

Yen SSC, Jaffe KB: *Reproductive Endocrinology*, 3rd ed. Saunders, 1991.

Zinman B: The physiologic replacement of insulin: An elusive goal. *N Engl J Med* 1989;**321**:363.

Section V.
Gastrointestinal Function

Digestion & Absorption

25

INTRODUCTION

The gastrointestinal system is the portal through which nutritive substances, vitamins, minerals, and fluids enter the body. Proteins, fats, and complex carbohydrates are broken down into absorbable units (**digested**), principally in the small intestine. The products of digestion and the vitamins, minerals, and water cross the mucosa and enter the lymph or the blood (**absorption**). The digestive and absorptive processes are the subject of this chapter. The details of the functions of the various parts of the gastrointestinal system are considered in Chapter 26.

Digestion of the major foodstuffs is an orderly process involving the action of a large number of **digestive enzymes** (Table 25–1). Enzymes from the salivary and lingual glands attack carbohydrates and fats; enzymes from the stomach attack proteins and fats; and enzymes from the exocrine portion of the pancreas attack carbohydrates, proteins, lipids, DNA, and RNA. Other enzymes that complete the digestive process are found in the luminal membranes and the cytoplasm of the cells that line the small intestine. The action of the enzymes is aided by the hydrochloric acid secreted by the stomach and the bile secreted by the liver.

The mucosal cells in the small intestine are called enterocytes. In the small intestine they have a **brush border** made up of numerous microvilli lining their apical surface (see Fig 26–25). This border is rich in enzymes. It is lined on its luminal side by a layer that is rich in neutral and amino sugars, the **glycocalyx.** The membranes of the mucosal cells contain glycoprotein enzymes that hydrolyze carbohydrates and peptides, and the glycocalyx is made up in part of the carbohydrate portions of these glycoproteins that extend into the intestinal lumen. Next to the brush border and glycocalyx is an **unstirred layer** similar to the layer adjacent to other biologic membranes (see Chapter 1). Solutes must diffuse across this layer to reach the mucosal cells. The mucous coat overlying the cells also constitutes a significant barrier to diffusion.

Substances pass from the lumen of the gastrointestinal tract to the extracellular fluid and thence to the lymph and blood by diffusion, facilitated diffusion, solvent drag, active transport, secondary active transport (coupled transport), and endocytosis. Most substances must pass from the intestinal lumen into the mucosal cells and then out of the mucosal cells to the extracellular fluid, and the processes responsible for movement across the luminal cell membrane are often quite different from those responsible for movement across the basal and lateral cell membranes to the extracellular fluid. The dynamics of transport in all parts of the body are considered in Chapter 1.

CARBOHYDRATES

Digestion

The principal dietary carbohydrates are polysaccharides, disaccharides, and monosaccharides. Starches (glucose polymers) and their derivatives are the only polysaccharides that are digested to any degree in the human gastrointestinal tract. In glycogen, the glucose molecules are mostly in long chains (glucose molecules in 1:4α linkage), but there is some chain-branching (produced by 1:6α linkages; see Fig 17–12). Amylopectin, which constitutes 80–90% of dietary starch, is similar but less branched, whereas amylose is a straight chain with only 1:4α linkages. Glycogen is found in animals, whereas amylose and amylopectin are of plant origin. The disaccharides **lactose** (milk sugar) and **sucrose** (table sugar) are also ingested, along with the monosaccharides fructose and glucose.

Starch is attacked by ptyalin, the α-amylase in the saliva. However, the optimal pH for this enzyme is 6.7, and its action is inhibited by the acid gastric juice when food enters the stomach. In the small intestine, the potent pancreatic α-amylases also act on the ingested polysaccharides. Both the salivary and the pancreatic α-amylase also acts on the ingested polysaccharides. Both the salivary and the pancreatic α-amylases hydrolyze 1:4α linkages but spare 1:6α linkages, terminal 1:4α linkages, and the 1:4α linkages next to branching points. Consequently, the end products of α-amylase digestion are oligosaccharides: the disaccharide **maltose;** the trisaccharide

Table 25-1. Principal digestive enzymes. The corresponding proenzymes are shown in parentheses.

Source	Enzyme	Activator	Substrate	Catalytic Function or Products
Salivary glands	Salivary α-amylase	Cl⁻	Starch	Hydrolyzes 1:4α linkages, producing α-limit dextrins, maltotriose, and maltose
Lingual glands	Lingual lipase		Triglycerides	Fatty acids plus 1,2-diacylglycerols
Stomach	Pepsins (pepsinogens)	HCl	Proteins and polypeptides	Cleave peptide bonds adjacent to aromatic amino acids
	Gastric lipase		Triglycerides	Fatty acids and glycerol
Exocrine pancreas	Trypsin (trypsinogen)	Enteropeptidase	Proteins and polypeptides	Cleave peptide bonds adjacent to arginine or lysine
	Chymotrypsins (chymotrypsinogens)	Trypsin	Proteins and polypeptides	Cleaves peptide bonds adjacent to arginine or lysine
	Elastase (proelastase)	Trypsin	Elastin, some other proteins	Cleaves bonds adjacent to aliphatic amino acids
	Carboxypeptidase A (procarboxypeptidase A)	Trypsin	Proteins and polypeptides	Cleaves carboxy-terminal amino acids that have aromatic or branched aliphatic side chains
	Carboxypeptidase B (procarboxypeptidase B)	Trypsin	Proteins and polypeptides	Cleaves carboxy-terminal amino acids that have basic side chains
	Colipase (procolipase)	Trypsin	Fat droplets	Binds to bile salt-triglyceride-water interface, making anchor for lipase
	Pancreatic lipase	. . .	Triglycerides	Monoglycerides and fatty acids
	Cholesteryl ester hydrolase	. . .	Cholesteryl esters	Cholesterol
	Pancreatic α-amylase	Cl⁻	Starch	Same as salivary α-amylase
	Ribonuclease	. . .	RNA	Nucleotides
	Deoxyribonuclease	. . .	DNA	Nucleotides
	Phospholipase A₂ (prophospholipase A₂)	Trypsin	Phospholipids	Fatty acids, lysophospholipids
Intestinal mucosa	Enteropeptidase	. . .	Trypsinogen	Trypsin
	Aminopeptidases	. . .	Polypeptides	Cleave N-terminal amino acid from peptide
	Dipeptidases	. . .	Dipeptides	Two amino acids
	Glucoamylase	. . .	Maltose, maltotriose	Glucose
	Lactase	. . .	Lactose	Galactose and glucose
	Sucrase[1]	. . .	Sucrose	Fructose and glucose
	α-Limit dextrinase[1]	. . .	α-Limit dextrins	Glucose
	Nuclease and related enzymes	. . .	Nucleic acids	Pentoses and purine and pyrimidine bases
Cytoplasm of mucosal cells	Various peptidases	. . .	Di-, tri-, and tetrapeptides	Amino acids

[1]Sucrase and α-limit dextrinase are separate polypeptide chains that are subunits of a single protein.

maltotriose; some slightly larger polymers with glucose in 1:4α linkage; and α-limit dextrins, branched polymers containing an average of about 8 glucose molecules (Fig 25–1).

The oligosaccharidases responsible for the further digestion of the starch derivatives are located in the outer portion of the brush border, the membrane of the microvilli of the small intestine. α-Limit dex-trinase hydrolyzes the α-limit dextrins, and glucoamylase splits glucose from maltose, maltotriose, and other polymers of glucose in 1:4α linkage. Most of the glucose molecules that are formed enter adjacent mucosal cells, although some remain in the intestinal lumen and are absorbed farther along. Ingested disaccharides are hydrolyzed by lactase or sucrase on the luminal surface of mucosal cells (Fig 25–2). De-

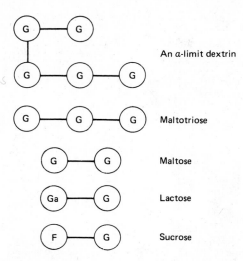

Figure 25–1. Principal end products of carbohydrate digestion in the intestinal lumen. Each circle represents a hexose molecule. G, glucose; F, fructose; Ga, galactose.

ficiency of one or more of these disaccharidases may cause diarrhea, bloating, and flatulence after ingestion of sugar. The diarrhea is due to the increased number of osmotically active oligosaccharide molecules that remain in the intestinal lumen, causing the volume of the intestinal contents to increase. In the colon, bacteria break down some of the oligosaccha-

rides, further increasing the number of osmotically active particles. The bloating and flatulence are due to the production of gas (CO_2 and H_2) from disaccharide residues in the lower small intestine and colon. The problem of milk intolerance can be relieved by administration of commercial lactase preparations, but this is expensive. Yogurt is better tolerated than milk in intolerant individuals because it contains its own bacterial lactase.

Lactase is of interest because, in most mammals and in many races of humans, intestinal lactase activity is high at birth, declines to low levels during childhood, and remains low in adulthood. The low lactase levels are associated with intolerance to milk (lactose intolerance). Most Europeans and their American descendants retain their intestinal lactase activity in adulthood; the incidence of lactase deficiency in northern and western Europeans is only about 15%. However, the incidence in blacks, American Indians, Orientals, and Mediterranean populations is 70–90%.

Absorption

Hexoses and pentoses are rapidly absorbed across the wall of the small intestine (Table 25–2). Essentially all of the hexoses are removed before the remains of a meal reach the terminal part of the ileum. The sugar molecules pass from the mucosal cells to the blood in the capillaries draining into the portal vein.

The transport of some sugars is uniquely affected

Figure 25–2. Outline of carbohydrate digestion and absorption. Some of the monosaccharides are also released into the intestinal lumen. (Modified from Gray GM: Carbohydrate digestion and absorption. *N Engl J Med* 1975;**292:**1225.)

Table 25–2. Normal transport of substances by the intestine and location of maximum absorption or secretion.[1]

Absorption of:	Small Intestine			Colon
	Upper[2]	Mid	Lower	
Sugars (glucose, galactose, etc)	++	+++	++	0
Amino acids	++	+++	++	0
Water-soluble and fat-soluble vitamins except vitamin B_{12}	+++	++	0	0
Betaine, dimethylglycine, sarcosine	+	++	++	?
Antibodies in newborns	+	++	+++	?
Pyrimidines (thymine and uracil)	+	+	?	?
Fatty acid absorption and conversion to triglyceride	+++	++	+	0
Bile salts	+	+	+++	
Vitamin B_{12}	0	+	+++	0
Na^+	+++	++	+++	+++
K^+	+	+	+	Sec
Ca^{2+}	+++	++	+	?
Fe^{2+}	+++	++	+	?
Cl^-	+++	++	+	+
SO_4^{2-}	++	+	0	?

[1]Amount of absorption is graded + to +++. Sec: secreted when luminal $K^+ < 25$mm.

[2]Upper small intestine refers primarily to jejunum, although the duodenum is similar in most cases studied (with the notable exception that the duodenum secretes HCO_3^- and shows little net absorption or secretion of NaCl.

by the amount of Na^+ in the intestinal lumen; a high concentration of Na^+ on the mucosal surface of the cells facilitates and a low concentration inhibits sugar influx into the epithelial cells. This is because glucose and Na^+ share the same **cotransporter,** or **symport.** This molecule, the **sodium-dependent glucose transporter** (SGLT 1, Na^+-glucose cotransporter) resembles the glucose transporters responsible for facilitated diffusion (see Chapter 19) in that it crosses the cell membrane 12 times. However, there is no homology to the GLUT series of transporters. The same molecule is responsible for glucose transport out of the renal tubules (see Chapter 38).

Since intracellular Na^+ is low in intestinal and renal cells, as it is in other cells, Na^+ moves into the cell along its concentration gradient. Glucose moves with the Na^+ and is released in the cell (Fig 25–3). The Na^+ is transported into the lateral intercellular spaces, and the glucose is transported by GLUT 2 into the interstitium and thence to the capillaries. Thus, glucose transport is an example of secondary active transport (see Chapter 1); the energy for glucose transport is provided indirectly, by the active transport of Na^+ out of the cell. This maintains the concentration gradient across the luminal border of the cell, so that more Na^+ and consequently more glucose enter. When the Na^+/glucose cotransporter is absent or defective, the resulting glucose/galactose malabsorption causes severe diarrhea that may be fatal. The use of glucose and its polymers to retain Na^+ in diarrheal disease is discussed below.

The glucose mechanism also transports galactose. Fructose utilizes a different carrier, and its absorption is independent of Na^+ or the transport of glucose and galactose; it is transported instead by facilitated diffusion. Some fructose is converted to glucose in the mucosal cells. Pentoses are absorbed by simple diffusion.

Insulin has little effect on intestinal transport of sugars. In this respect, intestinal absorption resembles glucose reabsorption in the proximal convoluted tubules of the kidneys (see Chapter 38); neither process requires phosphorylation, and both are essentially normal in diabetes but are depressed by the drug phlorhizin. The maximal rate of glucose absorption from the intestine is about 120 g/h.

Figure 25–3. Mechanism for glucose transport across intestinal epithelium. Glucose transport into the intestinal cell is coupled to Na^+ transport, utilizing a common carrier protein. Na^+ is then actively transported out of the cell, and glucose enters the interstitium by simple diffusion (SD) or facilitated diffusion (FD). From there, it diffuses into the blood. (Modified from Gray GM: Carbohydrate digestion and absorption. *N Engl J Med* 1975;**292**:1225.)

PROTEINS & NUCLEIC ACIDS

Protein Digestion

Protein digestion begins in the stomach, where pepsins cleave some of the peptide linkages. Like many of the other enzymes concerned with protein digestion, pepsins are secreted in the form of inactive precursors (**proenzymes**) and activated in the gastrointestinal tract. The pepsin precursors are called pepsinogens and are activated by gastric hydrochloric acid. Human gastric mucosa contains a number of related pepsinogens, which can be divided into 2 immunohistochemically distinct groups, pepsinogen I and pepsinogen II. Pepsinogen I is found only in acid-secreting regions, whereas pepsinogen II is also found in the pyloric region. Maximal acid secretion correlates with pepsinogen I levels, and patients with congenitally elevated circulating pepsinogen I levels have a 5-fold greater incidence of peptic ulcers than individuals with normal levels (see Chapter 26).

Pepsins hydrolyze the bonds between aromatic amino acids such as phenylalanine or tyrosine and a second amino acid, so the products of peptic digestion are polypeptides of very diverse sizes. A **gelatinase** that liquefies gelatin is also found in the stomach. **Chymosin,** a milk-clotting gastric enzyme also known as **rennin,** is found in the stomachs of young animals but is probably absent in humans.

Because pepsins have a pH optimum of 1.6–3.2, their action is terminated when the gastric contents are mixed with the alkaline pancreatic juice in the duodenum and jejunum. The pH of the intestinal contents in the duodenal cap is 2.0–4.0, but in the rest of the duodenum, it is about 6.5.

In the small intestine, the polypeptides formed by digestion in the stomach are further digested by the powerful proteolytic enzymes of the pancreas and intestinal mucosa. Trypsin, the chymotrypsins, and elastase act at interior peptide bonds in the peptide molecules and are called **endopeptidases.** The formation of the active endopeptidases from their inactive precursors is discussed in Chapter 26. The carboxypeptidases of the pancreas and the aminopeptidases of the brush border are **exopeptidases** that hydrolyze the amino acids at the carboxy and amino ends of the polypeptides. Some free amino acids are liberated in the intestinal lumen, but others are liberated at the cell surface by the aminopeptidases and dipeptidases in the brush border of the mucosal cells. Some di- and tripeptides are actively transported into the intestinal cells and hydrolyzed by intracellular peptidases, with the amino acids entering the bloodstream. Thus, the final digestion to amino acids occurs in 3 locations: the intestinal lumen, the brush border, and the cytoplasm of the mucosal cells.

Absorption

After ingestion of a protein meal, there is a sharp transient rise in the amino nitrogen content of the portal blood. L Amino acids are absorbed more rapidly than the corresponding D isomers. The D amino acids are apparently absorbed solely by passive diffusion, whereas most L amino acids are actively transported out of the intestinal lumen. Transport is effected by various different carriers, many of which cotransport amino acids and Na^+ in a manner similar to the cotransport of glucose and Na^+ (Fig 25–3). A neutral amino acid carrier, a phenylalanine and methionine carrier, and an amino acid carrier in the brush border are Na^+-dependent. The Na^+-independent carriers in the brush border include a carrier for basic amino acids and a carrier for neutral amino acids that prefers hydrophobic side chains. There are other carriers in the basolateral membrane of the enterocytes. A separate system transports di- and tripeptides into the mucosal cells. The transported amino acids and those produced by intracellular hydrolysis of di- and tripeptides accumulate in the mucosal cells, and from these cells they enter the ECF by simple diffusion and facilitated diffusion. The only small peptides known to enter portal blood are those from gelatin that contain proline and hydroxyproline and those from certain meats that contain carnosine and anserine.

Absorption of amino acids is rapid in the duodenum and jejunum but slow in the ileum. Approximately 50% of the digested protein comes from ingested food, 25% from proteins in digestive juices, and 25% from desquamated mucosal cells. Only 2–5% of the protein in the small intestine escapes digestion and absorption. Some of the ingested protein enters the colon and is eventually digested by bacterial action. The protein in the stools is not of dietary origin but comes from bacteria and cellular debris. There is evidence that the peptidase activities of the brush border and the mucosal cell cytoplasm are increased by resection of part of the ileum and that they are independently altered in starvation. Thus, these enzymes appear to be subject to homeostatic regulation. In humans, a congenital defect in the mechanism that transports neutral amino acids in the intestine and renal tubules causes **Hartnup disease.** A congenital defect in the transport of basic amino acids causes **cystinuria.**

In infants, moderate amounts of undigested proteins are also absorbed. The protein antibodies in maternal colostrum that contribute to passive immunity against infections enter the circulation from the intestine, although this transfer is relatively minor in humans. Absorption is by endocytosis and subsequent exocytosis. Protein absorption declines with age, but adults still absorb small quantities. Foreign proteins that enter the circulation provoke the formation of antibodies, and the antigen-antibody reaction occurring upon subsequent entry of more of the same protein may cause allergic symptoms. Thus, absorption of proteins from the intestine may explain the occurrence of allergic symptoms after eating certain foods.

However, true food allergies are probably rare, and the diagnosis is difficult to establish.

Absorption of protein antigens, particularly bacterial and viral proteins, takes place in large **microfold cells** or **M cells,** specialized intestinal epithelial cells that overlie aggregates of lymphoid tissue (Peyer's patches). These cells pass the antigens to the lymphoid cells, and lymphoblasts are activated. The activated lymphoblasts enter the circulation, but they later return to the intestinal mucosa and other epithelia, where they secrete IgA in response to subsequent exposures to the same antigen. This **secretory immunity** is an important defense mechanism; it is discussed in more detail in Chapter 27.

Nucleic Acids

Nucleic acids are split into nucleotides in the intestine by the pancreatic nucleases, and the nucleotides are split into the nucleosides and phosphoric acid by enzymes that appear to be located on the luminal surfaces of the mucosal cells. The nucleosides are then split into their constituent sugars and purine and pyrimidine bases. The bases are absorbed by active transport.

LIPIDS

Fat Digestion

A lingual lipase is secreted by Ebner's glands on the dorsal surface of the tongue, and the stomach also secretes a lipase (Table 25–1). The gastric lipase is of little importance except in pancreatic insufficiency, but lingual lipase is active in the stomach and can digest as much as 30% of dietary triglyceride.

Most fat digestion begins in the duodenum, pancreatic lipase being the most important enzyme involved. This enzyme hydrolyzes the 1- and 3-bonds of the triglycerides (triacylglycerols) with relative ease but acts on the 2-bonds at a very low rate, so the principal products of its action are free fatty acids and 2-monoglycerides (2-monoacylglycerols). It acts on fats that have been emulsified. However, it cannot act on fat droplets covered by emulsifying agents without **colipase,** a protein with a molecular weight of about 11,000 that binds to the surface of the fat droplets, displacing the emulsifying agents and anchoring lipase to the droplet. Colipase is secreted into the pancreatic juice in an inactive proform (Table 25–1) and is activated in the intestinal lumen by trypsin.

Most of the dietary cholesterol is in the form of cholesteryl esters, and cholesteryl ester hydrolyzes these esters in the intestinal lumen.

Fats are finely emulsified in the small intestine by the detergent action of bile salts, lecithin, and monoglycerides; bile salts alone are relatively poor emulsifying agents, but in the presence of phospholipids and monoglycerides, particles 200–5000 nm in diameter

are formed. The structure of the bile salts is discussed in Chapter 26.

When the concentration of bile salts in the intestine is high, as it is after contraction of the gallbladder, lipids and bile salts interact spontaneously to form **micelles** (Figs 25–4 and 26–20). These cylindrical aggregates take up lipids (see Chapter 26), and although their lipid concentration varies, they generally contain fatty acids, monoglycerides, and cholesterol in their hydrophobic centers. Micellar formation further solubilizes the lipids and provides a mechanism for their transport to the enterocytes. Thus, the micelles move down their concentration gradient through the unstirred layer to the brush border of the mucosal cells. The lipids diffuse out of the micelles, and a saturated aqueous solution of the lipids is maintained in contact with the brush border of the mucosal cells (Fig 25–4). The lipids enter the cells by passive diffusion and are rapidly esterified inside the cells, maintaining a favorable concentration gradient from the lumen into the cells (Fig 25–5). Unlike the ileal mucosa, the rate of uptake of bile salts by the jejunal mucosa is low, and for the most part the bile salts remain in the intestinal lumen, where they are available for the formation of new micelles. Thus, the bile salt micelles solubilize lipids, transport them across the unstirred layer, and keep a saturated solution of lipids in contact with the mucosal cells.

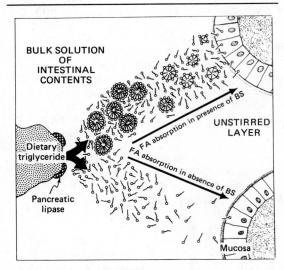

Figure 25–4. Lipid digestion and passage to intestinal mucosa. Fatty acids (FA) are liberated by the action of pancreatic lipase on dietary triglycerides and, in the presence of bile salts (BS), form micelles (the circular structures), which diffuse through the unstirred layer to the mucosal surface. (Reproduced, with permission, from Thomson ABR: Intestinal absorption of lipids: Influence of the unstirred water layer and bile acid micelle. In: *Disturbances in Lipid and Lipoprotein Metabolism.* Dietschy JM, Gotto AM Jr, Ontko JA [editors]. American Physiological Society, 1978.)

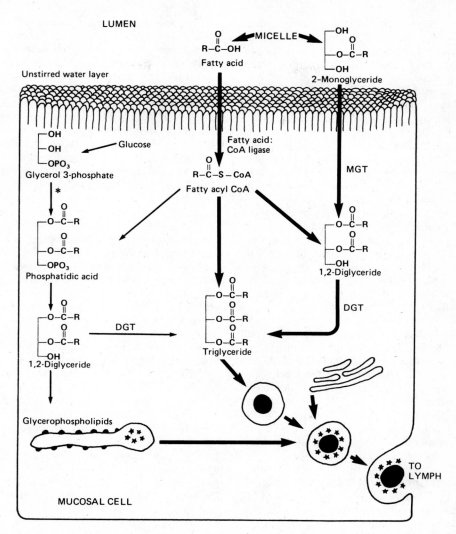

Figure 25–5. Lipid absorption. Triglycerides are formed in the mucosal cells from monoglycerides and fatty acids. Some of the glycerides also come from glucose via phosphatidic acid. The triglycerides are then converted to chylomicrons and released by exocytosis. From the extracellular space, they enter the lymph. Heavy arrows indicate major pathways. *, reaction inhibited by monoglyceride; MGT, monoacylglycerol acyltransferase; DGT, diacylglycerol acyltransferase. (Modified and reproduced, with permission, from Johnston JM: Esterification reactions in the intestinal mucosa and lipid absorption. In: *Disturbances in Lipid and Lipoprotein Metabolism.* Dietschy JM, Gotto AM Jr, Ontko JA [editors]. American Physiological Society, 1978.)

Pancreatectomized animals and patients with diseases that destroy the exocrine portion of the pancreas have fatty, bulky, clay-colored stools (**steatorrhea**) because of the impaired digestion and absorption of fat. The steatorrhea is due mostly to the lipase deficiency, but in addition, in the absence of the bicarbonate that is secreted from the pancreas, the relatively acidic milieu in the duodenum precipitates some bile salts. Acid also inhibits pancreatic lipase, and this is why patients with excess secretion of gastric acid due to a gastrin-secreting tumor (gastrinoma; see Chapter 26) and a consequent low duodenal pH may develop steatorrhea. Another cause of steatorrhea is defective reabsorption of bile salts in the distal ileum (see Chapter 26).

Fat Absorption

Monoglycerides, cholesterol, and fatty acids from the micelles enter the mucosal cells by passive diffusion. The subsequent fate of the fatty acids depends on their size. Fatty acids containing less than 10–12 carbon atoms pass from the mucosal cells directly into the portal blood, where they are transported as free (unesterified) fatty acids. The fatty acids containing more than 10–12 carbon atoms are reesterified to triglycerides in the mucosal cells. In addition, some of the absorbed cholesterol is esterified. The triglycerides and cholesteryl esters are then coated with a layer of protein, cholesterol, and phospholipid to form chylomicrons, which leave the cell and enter the lymphatics (Fig 25–5).

In mucosal cells, most of the triglyceride is formed by the acylation of the absorbed 2-monoglycerides, primarily in the smooth endoplasmic reticulum. However, some of the triglyceride is formed from glycerophosphate, which in turn is a product of glucose catabolism. Glycerophosphate is also converted into glycerophospholipids that participate in chylomicron formation. The acylation of glycerophosphate and the formation of lipoproteins occur in the rough endoplasmic reticulum. Carbohydrate moieties are added to the proteins in the Golgi apparatus, and the finished chylomicrons are extruded by exocytosis from the basal or lateral aspects of the cell.

Fat absorption is greatest in the upper parts of the small intestine, but appreciable amounts are also absorbed in the ileum (Fig 25–6). On a moderate fat intake, 95% or more of the ingested fat is absorbed. The stools contain 5% fat, but much of the fecal fat is probably derived from cellular debris and microorganisms rather than from the diet. The processes involved in fat absorption are not fully mature at birth, and infants fail to absorb 10–15% of ingested fat. Thus, they are more susceptible to the ill effects of disease processes that reduce fat absorption.

Absorption of Cholesterol & Other Sterols

Cholesterol is readily absorbed from the small intestine if bile, fatty acids, and pancreatic juice are present. Closely related sterols of plant origin are poorly absorbed. Almost all the absorbed cholesterol is incorporated into chylomicrons that enter the circulation via the lymphatics, as noted above. Nonabsorbable plant sterols such as those found in soybeans reduce the absorption of cholesterol, probably by competing with cholesterol for esterification with fatty acids.

ABSORPTION OF WATER & ELECTROLYTES

Water, Sodium, & Potassium

Overall water balance in the gastrointestinal tract is summarized in Table 25–3. The intestines are presented each day with about 2000 mL of ingested fluid plus 7000 mL of secretions from the mucosa of the gastrointestinal tract and associated glands, Ninety-eight percent of this fluid is reabsorbed, with a daily fluid loss of only 200 mL in the stools.

It should be noted that the figures for intestinal reabsorption are net rather than gross. Only small amounts of water move across the gastric mucosa, but water moves in both directions across the mucosa of the small and large intestines in response to osmotic gradients. Some Na^+ diffuses into or out of the small intestine depending on the concentration gradient. Because the luminal membranes of all enterocytes in the small intestine and colon are permeable to Na^+ and their basolateral membranes contain Na^+-K^+ ATPase, Na^+ is also actively absorbed throughout the small and large intestines.

In the small intestine, active transport of Na^+ is important in bringing about absorption of glucose, some amino acids, and other substances (see above). Conversely, the presence of glucose in the intestinal lumen facilitates the reabsorption of Na^+. This is the physiologic basis for the treatment of Na^+ and water loss in diarrhea by oral administration of solutions containing NaCl and glucose. Cereals containing carbohydrates are also useful in the treatment of diarrhea. This type of treatment has even proved to be beneficial in the treatment of cholera, a disease associated with very severe and, if untreated, frequently fatal diarrhea.

Cl^- is normally secreted into the lumen of the small intestines by Cl^- channels that are activated by cyclic AMP. Enterocytes also take up Na^+, K^+, and Cl^- by a $1Na^+$-$1K^+$-$2Cl^-$ cotransporter in their basolateral membranes (Fig 25–7). The cholera vibrio stays in the intestinal lumen, but it produces a toxin that binds adenosine diphosphate ribose to the subunit of G_s, inhibiting its GTPase activity. Therefore, it produces prolonged stimulation of adenylate cyclase and a consequent marked increase in intracellular cyclic AMP. Some strains of diarrhea-producing *Escherichia coli* produce a similar toxin. The accumulation of cyclic AMP increases Cl^- secretion from the intestinal glands and inhibits the function of the mucosal carrier for Na^+, reducing NaCl absorption. The resultant increase in electrolyte and water content of the intestinal contents causes the diarrhea. However, Na^+-K^+ ATPase and the Na^+/glucose cotransporter are unaffected, so coupled reabsorption of glucose and Na^+ bypasses the defect.

Water moves into or out of the intestine until the osmotic pressure of the intestinal contents equals that of the plasma. The osmolality of the duodenal contents may be hypertonic or hypotonic, depending on the meal ingested, but by the time the meal enters the jejunum, its osmolality is close to that of plasma. This osmolality is maintained throughout the rest of the small intestine; the osmotically active particles produced by digestion are removed by absorption, and water moves passively out of the gut along the osmotic gradient thus generated. In the colon, Na^+ is pumped out and water moves passively with it, again along the osmotic gradient. **Saline cathartics** such as magnesium sulfate are poorly absorbed salts that retain their osmotic equivalent of water in the intestine,

Figure 25–6. Fat absorption, based on measurement after a fat meal in humans. (Redrawn and reproduced, with permission, from Davenport HW: *Physiology of the Digestive Tract,* 2nd ed. Year Book, 1966.)

Figure 25–7. Movement of ions across enterocytes in the small intestine. PP, paracellular path. (Reproduced, with permission, from Field M, Roa MC, Chang EB: Intestinal electrolyte transport and diarrheal disease. *N Engl J Med* 1989:**321**:800.)

thus increasing intestinal volume and consequently exerting a laxative effect.

There is some secretion of K^+ into the intestinal lumen, especially as a component of mucus, but for the most part, the movement of K^+ across the gastrointestinal mucosa is due to diffusion. On the other hand, there are K^+ channels in the luminal as well as the basolateral membrane of the enterocytes of the colon, so K^+ is secreted into the colon. In addition, K^+ moves passively down its electrochemical gradient. The accumulation of K^+ in the colon is partially offset to a degree by H^+-K^+ ATPase in the luminal membrane of cells in the distal colon, with resulting active transport of K^+ into the cells. Nevertheless, loss of ileal or colonic fluids in chronic diarrhea can lead to severe hypokalemia.

When the dietary intake of K^+ is high for a prolonged period, aldosterone secretion is increased and more K^+ is secreted into the colon. This is due in part to the appearance of more Na^+-K^+ ATPase pumps in the basolateral membranes of the cells, with a consequent increase in intracellular K^+ and K^+ diffusion across the luminal membranes of the cells.

Chloride & Bicarbonate

In the ileum and the colon, it appears that Cl^- is actively reabsorbed in a one-for-one exchange for HCO_3^-. This tends to make the intestinal contents more alkaline. However, the physiologic significance of this exchange is uncertain.

ABSORPTION OF VITAMINS & MINERALS

Vitamins

Absorption of water-soluble vitamins is rapid, but absorption of the fat-soluble vitamins A, D, E, and K is deficient if fat absorption is depressed because of lack of pancreatic enzymes or if bile is excluded from the intestine by obstruction of the bile duct. Most vitamins are absorbed in the upper small intestine, but vitamin B_{12} is absorbed in the ileum. This vitamin binds to intrinsic factor, a protein secreted by the stomach, and the complex is absorbed across the ileal mucosa (see Chapter 26).

Calcium

From 30 to 80% of ingested calcium is absorbed. Active transport of Ca^{2+} out of the intestinal lumen occurs primarily in the upper small intestine, and there is also some absorption by passive diffusion. Active transport is facilitated by 1,25-dihydroxy-

Table 25–3. Daily net water turnover (mL) in the gastrointestinal tract[1]

Ingested		2000
Endogenous secretions		7000
Salivary glands	1500	
Stomach	2500	
Bile	500	
Pancreas	1500	
Intestine	1000	
	7000	
Total input		9000
Reabsorbed		8800
Jejunum	5500	
Ileum	2000	
Colon	1300	
	8800	
Balance in stool		200

[1]Data from Moore EW: *Physiology of Intestinal Water and Electrolyte Absorption.* American Gastroenterological Society, 1976.

cholecalciferol, the metabolite of vitamin D that is produced in the kidneys. The metabolite induces the synthesis of a Ca^{2+}-binding protein in the mucosal cells (see Chapter 21). The rate of production of 1,25-dihydroxycholecalciferol is increased when the plasma calcium level is decreased and reduced when the plasma calcium level is elevated (see Chapter 21). Consequently, Ca^{2+} absorption is adjusted to body needs; absorption is increased in the presence of Ca^{2+} deficiency and decreased in the presence of Ca^{2+} excess. Ca^{2+} absorption is also facilitated by protein. It is inhibited by phosphates and oxalates because these anions form insoluble salts with Ca^{2+} in the intestine. Magnesium absorption is facilitated by protein.

Iron

In adults, the amount of iron lost from the body is relatively small. The losses are generally unregulated, and total body stores of iron are regulated by changes in the rate at which it is absorbed from the intestine. Men lose about 0.6 mg/d, whereas women have a variable, larger loss averaging about twice this value because of the additional iron lost in the blood shed during menstruation. The average daily iron intake in the USA and Europe is about 20 mg, but the amount absorbed is equal only to the losses; if it were any greater, iron overload would develop. Thus, the amount of iron absorbed ranges normally from about 3 to 6% of the amount ingested.

Iron is more readily absorbed in the ferrous state (Fe^{2+}), but most of the dietary iron is in the ferric form (Fe^{3+}). No more than a trace of iron is absorbed in the stomach, but the gastric secretions dissolve the iron and permit it to form soluble complexes with ascorbic acid and other substances that aid its reduction to the Fe^{2+} form. The importance of this function in humans is indicated by the fact that iron deficiency anemia is a troublesome and relatively frequent complication of partial gastrectomy. Heme is also absorbed, and the Fe^{2+} that it contains is released in the mucosal cells. Various dietary factors affect the availability of iron for absorption; for example, the phytic acid found in cereals reacts with iron to form insoluble compounds in the intestine. So do phosphates and oxalates. Pancreatic juice inhibits iron absorption.

Most of the iron is absorbed in the upper part of the small intestine (Table 25–2). Other mucosal cells can transport iron, but the duodenum and adjacent jejunum contain most of the iron suitable for absorption. The mucosal cells contain an intracellular iron carrier. Some iron is supplied to the mitochondria from the carrier, but the remainder is partitioned between **apoferritin** in the mucosal cells and **transferrin,** the iron-transporting polypeptide in the plasma (Fig 25–8). Apoferritin, which is also found in many other tissues, combines with iron to form **ferritin.** The iron bound to ferritin in intestinal cells is lost with the cells when they are shed into the intestinal

Figure 25–8. Diagrammatic summary of the control of iron absorption by the intestinal mucosa. C, intracellular iron carrier; F, ferritin; M, mitochondria; TF, plasma transferrin. The width of the black arrows is proportionate to the movement of iron from one compartment to another. Mucosal cells are shed into the intestinal lumen, and the more ferritin they contain, the more they contribute to iron loss in the stools. Note that the amount of TF in the plasma is inversely proportionate to the amount of iron in the body and that the greater the saturation of TF, the more iron goes into ferritin in the mucosal cells. (Modified from Stanbury JB et al: *The Metabolic Basis of Inherited Disease,* 5th ed. McGraw-Hill, 1983.)

lumen at the end of their life cycle and passed in the stool.

Apoferritin is a globular protein made up of 24 subunits. Iron forms a micelle of ferric hydroxyphosphate, and in ferritin, the subunits surround this micelle. The ferritin molecule can contain as many as 4500 atoms of iron. Ferritin is readily visible under the electron microscope and has been used as a tracer in studies of phagocytosis and related phenomena. It is the principal storage form of iron in tissues. Ferritin molecules in lysosomal membranes may aggregate in deposits that contain as much as 50% iron.

These deposits are called **hemosiderin.** Seventy percent of the iron in the body is in hemoglobin, 3% in myoglobin, and the remainder in ferritin. Ferritin is also found in plasma, but most iron is transported bound to transferrin. This polypeptide has 2 iron-binding sites. Normally, transferrin is about 35% saturated with iron, and the normal plasma iron level is about 130 μg/dL (23 μmol/L) in men and 110 μg/dL (19 μmol/L) in women.

Iron absorption is increased when body iron stores are depleted or when erythropoiesis is increased, and decreased under the opposite conditions. In iron deficiency, the amount of transferrin in the plasma is increased and its percent saturation with iron is decreased. Consequently, more iron moves from the intracellular iron carrier to transferrin and less binds to apoferritin (Fig 25–8). Ferritin stores in the mucosa are decreased, and consequently, less iron is lost when the mucosal cells are shed. Conversely, in the presence of iron overload, the amount of circulating transferrin is decreased and its saturation is increased, so more iron is diverted to apoferritin, ferritin stores increase, and more iron is lost when the mucosal cells are shed. Iron fosters apoferritin synthesis by binding to a sequence at the 5′ untranslated portion of the ferritin mRNA, making the mRNA more active. This is one of the few instances of regulation of this type at the level of translation rather than transcription.

The normal operation of the factors that maintain iron balance is essential for health. Iron deficiency causes anemia. If more iron is absorbed than is excreted, iron overload results. However, normal individuals can maintain a normal rate of absorption even when the ingested load is 5 or 10 times more than needed. Hemosiderin accumulates in the tissues when the overload is prolonged and severe, producing **hemosiderosis.** Large amounts of hemosiderin can damage the tissues, producing **hemochromatosis.** This syndrome is characterized by pigmentation of the skin, pancreatic damage with diabetes ("bronze diabetes"), cirrhosis of the liver, a high incidence of hepatic carcinoma, and gonadal atrophy. Hemochromatosis can be produced not only by prolonged excessive iron intake but also by a number of other conditions. In idiopathic hemochromatosis, a congenital disorder due to an autosomal recessive gene, the mucosal regulatory mechanism behaves as if the iron deficiency were present and absorbs iron at a high rate in the face of elevated rather than depleted body iron stores.

Regulation of Gastrointestinal Function

INTRODUCTION

The digestive and absorptive functions of the gastrointestinal system outlined in the previous chapter depend upon a variety of mechanisms that soften the food, propel it through the gastrointestinal tract, and mix it with bile from the gallbladder and digestive enzymes secreted by the salivary glands and pancreas. Some of these mechanisms depend upon intrinsic properties of the intestinal smooth muscle. Others involve the operation of reflexes involving the neurons intrinsic to the gut, reflexes involving the CNS, paracrine effects of chemical messengers, and gastrointestinal hormones. The hormones are humoral agents secreted by parts of the mucosa and transported in the circulation to influence the functions of the stomach, the intestines, the pancreas, and the gallbladder.

GENERAL CONSIDERATIONS

Organization

The organization of the structures that make up the wall of the gastrointestinal tract from the posterior pharynx to the anus is shown in Fig 26–1. There is some local variation, but in general there are 3 layers of smooth muscle, 2 longitudinal and 1 circular. The wall is lined by mucosa and, except in the case of the esophagus and distal rectum, is covered by serosa. The serosa continues onto the mesentery, which contains the nerves, lymphatics, and blood vessels supplying the tract.

The Enteric Nervous System

There are 2 major networks of nerve fibers that are intrinsic to the gastrointestinal tract: the **myenteric plexus** (Auerbach's plexus), between the outer longitudinal and middle circular muscle layers, and the **submucous plexus** (Meissner's plexus), between the middle circular layer and the mucosa (Fig 26–1). The plexuses are interconnected, and they contain motor neurons that innervate smooth muscle; secretory neurons that regulate endocrine and exocrine secretion in the mucosa; sensory neurons that respond to stretch,

tonicity, glucose, or amino acids; and interneurons. Collectively, these neurons constitute the **enteric nervous system.** This system contains 80–100 million neurons in humans and is included by many physiologists with the sympathetic and parasympathetic divisions (see Chapter 13) as a third division of the autonomic nervous system. The substances secreted by the neurons in the enteric nervous system include acetylcholine, serotonin, GABA, and a large number of polypeptides. The main polypeptides and examples of their actions are listed in Table 26–1. Some of the substances secreted by the neurons act as synaptic transmitters; others diffuse in the ECF and act in a paracrine fashion; and still others enter the bloodstream, becoming hormones. In some instances, single neurons contain more than one of the sub-

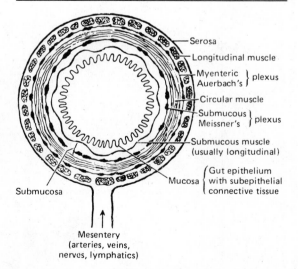

Figure 26–1. Diagrammatic representation of the layers of the wall of the stomach, small intestine, and colon. The structure of the esophagus and the distal rectum is similar, except that they have no serosa or mesentery. (Reproduced, with permission, from Bell GH, Emslie-Smith D, Paterson CR: *Textbook of Physiology and Biochemistry,* 9th ed. Churchill Livingstone, 1976.)

Table 26–1. Polypeptides secreted by neurons in the enteric nervous system.[1]

Polypeptide	Principal effect
Motor neurons	
VIP	Relaxation in front of peristaltic wave, relaxation of sphincters
Substance P	Contraction of intestinal smooth muscle
CCK	Inhibition of gastric emptying
Somatostatin	Inhibition of intestinal motility
Neurotensin	Relaxation of circular smooth muscle
Encephalins	Relaxation of intestinal smooth muscle, pyloric contraction
Galanin	Contraction of intestinal smooth muscle
Secretory neurons	
GRP	Release of gastrin
CGRP	Release of somatostatin
Somatostatin	Inhibition of acid and intestinal secretion
VIP	Stimulation of intestinal secretion
Substance P	Inhibition of acid secretion, stimulation of pepsin secretion
Sensory neurons	
CGRP	
Substance P	

[1]Modified from Debas HT, Mulvihill SJ: Neuroendocrine design of the gut. *Am J Surg* 1991;**161**:243.

stances, and they are secreted as cotransmitters (see Chapter 4).

The specific functions of the enteric nervous system are considered in subsequent parts of this chapter. In a general way, however, the submucous plexus is concerned with control of exocrine and endocrine secretion by cells in the gastrointestinal tract, and the myenteric plexus is concerned with the control of peristaltic activity.

Extrinsic Innervation

The intestine receives a dual extrinsic innervation from the autonomic nervous system, with parasympathetic cholinergic activity generally increasing the activity of intestinal smooth muscle and sympathetic noradrenergic activity generally decreasing it while causing sphincters to contract. The preganglionic parasympathetic fibers consist of about 2000 vagal efferents and other efferents in the sacral nerves. They generally end on cholinergic nerve cells of the myenteric and submucous plexuses. The sympathetic fibers are postganglionic, but many of them end on postganglionic cholinergic neurons, where they inhibit acetylcholine secretion. Others appear to end directly on intestinal smooth muscle cells. The electrical properties of intestinal smooth muscle are discussed in Chapter 3. Still others innervate blood vessels, where they produce vasoconstriction. It appears that the intestinal blood vessels have a dual innervation; they have an extrinsic noradrenergic innervation and an intrinsic innervation by peptidergic fibers of the enteric nervous system. VIP and serotonin are among the mediators in the intrinsic innerva-

tion, which seems among other things to be responsible for the hyperemia that accompanies digestion of food. It is unsettled whether there is an additional cholinergic innervation of the blood vessels.

Peristalsis

Peristalsis is a reflex response that is initiated when the gut wall is stretched by the contents of the lumen, and it occurs in all parts of the gastrointestinal tract from the esophagus to the rectum. The stretch usually initiates a circular contraction behind the stimulus and an area of relaxation in front of it. The wave of contraction then moves in an oral to caudal direction, propelling the contents of the lumen forward at rates that vary from 2 to 25 cm/s. Peristaltic activity can be increased or decreased by the autonomic input to the gut, but its occurrence is independent of the extrinsic innervation. Indeed, progression of the contents is not blocked by removal and resuture of a segment of intestine in its original position and is blocked only if the segment is reversed before it is sewn back into place. Peristalsis is an excellent example of the integrated activity of the enteric nervous system. The initiating stretch of the wall is sensed by neurons containing calcitonin gene-related polypeptide (CGRP), the wave of contraction behind the stimulus is produced by acetylcholine and substance P, and the relaxation in front of it is produced by vasoactive intestinal polypeptide (VIP).

Gastrointestinal Circulation

The blood flow to the stomach, intestines, pancreas, and liver is arranged in a series of parallel circuits, with all the blood from the intestines and pancreas draining via the portal vein to the liver. The physiology of this important portion of the circulation is discussed in Chapter 32.

GASTROINTESTINAL HORMONES

Biologically active polypeptides that are secreted in the mucosa act in a paracrine fashion, but they also enter the circulation. Experiments with them and measurement of their concentrations in blood by radioimmunoassay have identified the roles these **gastrointestinal hormones** play in the regulation of gastrointestinal secretion and motility. When large doses of the hormones are given, their actions overlap. However, their physiologic effects appear to be relatively discrete. On the basis of structural similarity (Table 26–2) and, to a degree, similarity of function, many of the hormones fall into one of 2 families: the gastrin family, the primary members of which are gastrin and cholecystokinin (CCK); and the secretin family, the primary members of which are secretin, glucagon, glicentin (GLI), VIP, and gastric inhibitory polypeptide (GIP).

Table 26–2. Structures of hormonally active polypeptides secreted by cells in the human gastrointestinal tract.[1]

Gastrin Family		Secretin Family				Other Polypeptides			
CCK 39	Gastrin 34	GIP	Glucagon	Secretin	VIP	Motilin	Substance P	GRP	Somatostatin 14
Tyr		Tyr	His	His	His	Phe	Arg	Val	Ala
Ile		Ala	Ser	Ser	Ser	Val	Pro	Pro	Gly
Gln		Glu	Gln	Asp	Asp	Pro	Lys	Leu	Cys
Gln		Gly	Gly	Gly	Ala	Ile	Pro	Pro	Lys
Ala		Thr	Thr	Thr	Val	Phe	Gln	Ala	Asn
Arg	(pyro)Glu	Phe	Phe	Phe	Phe	Thr	Gln	Gly	Phe
Lys	Leu	Ile	Thr	Thr	Thr	Tyr	Phe	Gly	Phe
Ala	Gly	Ser	Ser	Ser	Asp	Gly	Phe	Gly	Trp
Pro	Pro	Asp	Asp	Glu	Asn	Glu	Gly	Thr	Lys
Ser	Gln	Tyr	Tyr	Leu	Tyr	Leu	Leu	Val	Thr
Gly	Gly	Ser	Ser	Ser	Thr	Gln	Met-NH₂	Leu	Phe
Arg	Pro	Ile	Lys	Arg	Arg	Arg		Thr	Thr
Met	Pro	Ala	Tyr	Leu	Leu	Met		Lys	Ser
Ser	His	Met	Leu	Arg	Arg	Gln		Met	Cys
Ile	Leu	Asp	Asp	Glu	Lys	Glu		Tyr	
Val	Val	Lys	Ser	Gly	Gln	Lys		Pro	
Lys	Ala	Ile	Arg	Ala	Met	Glu		Arg	
Asn	Asp	His	Arg	Arg	Ala	Arg		Gly	
Leu	Pro	Gln	Ala	Leu	Val	Asn		Asn	
Gln	Ser	Gln	Gln	Gln	Lys	Lys		His	
Asn	Lys	Asp	Asp	Arg	Lys	Gly		Trp	
Leu	Lys	Phe	Phe	Leu	Tyr	Gln		Ala	
Asp	Gln	Val	Val	Leu	Leu			Val	
Pro	Gly	Asn	Gln	Gln	Asn			Gly	
Ser	Pro	Trp	Trp	Gly	Ser			His	
His	Trp	Leu	Leu	Leu	Ile			Leu	
Arg	Leu	Leu	Met	Leu	Leu			Met-NH₂	
Ile	Glu	Ala	Asn	Val-NH₂	Asn-NH₂				
Ser	Glu	Gln	Thr						
Asp	Glu	Lys							
Arg	Glu	Gly							
Asp	Glu	Lys							
Tys	Ala	Lys							
Met	Tys	Asn							
Gly	Gly	Asp							
Trp	Trp	Trp							
Met	Met	Lys							
Asp	Asp	His							
Phe-NH₂	Phe-NH₂	Asn							
		Ile							
		Thr							
		Gln							

[1]Homologous amino acid residues are enclosed by the lines that generally cross from one polypeptide to another. Arrows indicate points of cleavage to form smaller variants. Tys, tyrosine sulfate. All gastrins occur in unsulfated (gastrin I) and sulfated (gastrin II) forms. Glicentin, an additional member of the secretin family which is not shown, is a C-terminally extended relative of glucagon.

Gastrin

Gastrin is produced by cells called G cells in the lateral walls of the glands in the antral portion of the gastric mucosa (Fig 26–2). G cells are flask-shaped, with a broad base containing many gastrin granules and a narrow apex that reaches the mucosal surface. Microvilli project from the apical end into the lumen. Receptors mediating gastrin responses to changes in gastric contents are present on the microvilli. Other cells in the gastrointestinal tract that secrete hormones have a similar morphology. G cells and other enteroendocrine cells also contain amines related to norepinephrine or serotonin and appear to be of neural crest origin. Because they take up amine precursors and decarboxylate them, they are sometimes called APUD cells (for *a*mine *p*recursor *u*ptake and *d*ecarboxylation). There are APUD cells in many different parts of the body in addition to the gastrointestinal tract. A second type of gastrin-producing cell, the TG cell, is found throughout the stomach and small intestine. It contains G 34 and the C-terminal tetrapeptide of gastrin but lacks G 17 (see below).

Gastrin is also found in the pancreatic islets in fetal life. Gastrin-secreting tumors, called **gastrinomas** occur in the pancreas, but it is uncertain whether any gastrin is present in the pancreas in normal adults. In addition, gastrin is found in the anterior and intermediate lobes of the pituitary gland, in the hypothalamus and medulla oblongata, and in the vagus and sciatic nerves.

Gastrin is typical of a number of polypeptide hormones in that it shows both **macroheterogeneity** and **microheterogeneity.** Macroheterogeneity refers to the occurrence in tissues and body fluids of peptide chains of various lengths; microheterogeneity refers to differences in molecular structure due to derivatization of single amino acid residues. Pre-

progastrin is processed into fragments of various sizes. Three main forms of gastrin contain 34, 17, and 14 amino acid residues. All have the same C-terminal configuration (Table 26–2). These forms are also known as G 34, G 17, and G 14 gastrins, respectively. Another form is the C-terminal tetrapeptide, and there is also a large form that is extended at the N terminal and contains more than 45 amino acid residues. One form of derivatization is sulfation of the tyrosine that is the sixth amino acid residue from the C terminal. There are approximately equal amounts of nonsulfated and sulfated forms in blood and tissues, and they are equally active. Another derivatization is amidation of the C-terminal phenylalanine.

What is the physiologic significance of this marked heterogeneity? There are some differences in activity between the various components, and the proportions of the components also differ in the various tissues in which gastrin is found. This suggests that different forms are tailored for different actions. However, all that can be concluded at present is that G 17 is the principal form with respect to gastric acid secretion. The C-terminal tetrapeptide has all the activities of gastrin but only 10% of the strength of G 17.

G 14 and G 17 have half-lives of 2–3 minutes in the circulation, whereas G 34 has a half-life of 15 minutes. Gastrins are inactivated primarily in the kidney and small intestine.

In large doses, gastrin has a variety of actions, but its principal physiologic actions are stimulation of gastric acid and pepsin secretion and stimulation of the growth of the mucosa of the stomach and small and large intestines (**trophic action**). Stimulation of gastric motility is probably a physiologic action as well. Gastrin also causes contraction of the musculature that closes the gastroesophageal junction (see below), but this effect is of questionable physiologic significance. It stimulates insulin and glucagon secretion, but only after a protein meal and not after a carbohydrate meal does circulating endogenous gastrin reach the level necessary to stimulate the B cells. The functions of gastrin in the pituitary gland, brain, and peripheral nerves are unknown.

Gastrin secretion is affected by the contents of the stomach, the rate of discharge of the vagus nerves, and blood-borne factors (Table 26–3). Atropine does not inhibit the gastrin response to a test meal in humans because the transmitter secreted by the postganglionic vagal fibers that innervate the G cells is gastrin-releasing polypeptide (GRP; see below) rather than acetylcholine. Gastrin secretion is also increased by the presence of the products of protein digestion in the stomach, particularly amino acids, which act directly on the G cells. Phenylalanine and tryptophan are particularly effective.

Acid in the antrum inhibits gastrin secretion. The effect of acid is the basis of a negative feedback loop regulating gastrin secretion. Increased secretion of

Figure 26–2. Distribution of gastrointestinal peptides along the gastrointestinal tract. The thickness of each bar is proportionate to the concentration of the peptide in the mucosa.

Table 26–3. Stimuli that affect gastrin secretion.[1]

Stimuli that increase gastrin secretion
  Luminal
    Peptides and amino acids
    Distention
  Neural
    Increased vagal discharge, probably noncholinergic
  Blood-borne
    Calcium
    Epinephrine
Stimuli that inhibit gastrin secretion
  Luminal
    Acid
  Blood-borne
    Secretin, GIP, VIP, glucagon, calcitonin

[1]Modified from Walsh JH, Grossman MI: Gastrin. *N Engl J Med* 1975;**292**:1324.

the hormone increases acid secretion, but the acid then feeds back to inhibit further gastrin secretion.

The role of gastrin in the pathophysiology of duodenal ulcers is discussed below. In conditions such as pernicious anemia in which the acid-secreting cells of the stomach are damaged, gastrin secretion is chronically elevated.

Cholecystokinin-Pancreozymin

It was formerly thought that a hormone called cholecystokinin produced contraction of the gallbladder whereas a separate hormone called pancreozymin increased the secretion of pancreatic juice rich in enzymes. It is now clear that a single hormone secreted by cells in the mucosa of the upper small intestine has both activities, and the hormone has therefore been named **cholecystokinin-pancreozymin.** It is also called **CCK-PZ** or, most commonly, **CCK.**

Like gastrin, CCK shows both macroheterogeneity and microheterogeneity. Prepro-CCK is processed into many fragments. A large CCK contains 58 amino acid residues (CCK 58). There are, in addition, CCK peptides containing 39 amino acid residues (CCK 39) and 33 amino acid residues (CCK 33), several forms that have 12 (CCK 12) or slightly more amino acid residues, and a form containing 8 amino acid residues (CCK 8). All of these forms have the same 5 amino acids at the C terminal as gastrin (Table 26–2). The C-terminal tetrapeptide (CCK 4) also exists in tissues. The C terminal is amidated, and the tyrosine that is the seventh amino acid residue from the C terminal is sulfated. Unlike gastrin, the nonsulfated form of CCK has not been found in tissues. However, it is known that derivatization of other amino acid residues in CCK can occur. The half-life of circulating CCK is about 5 minutes, but little is known about its metabolism.

In addition to its secretion by endocrine cells, the I cells in the upper intestine, CCK is found in nerves in the distal ileum and colon. It is also found in neurons in the brain, especially the cerebral cortex, and

in nerves in many parts of the body (see Chapter 4). The CCK secreted in the duodenum and jejunum is probably mostly CCK 8 and CCK 12, although CCK 58 is also present in the intestine and circulating blood in some species. The enteric and pancreatic nerves contain primarily CCK 4. CCK 58 and CCK 8 are found in the brain.

In addition to causing contraction of the gallbladder and secretion of a pancreatic juice rich in enzymes, CCK augments the action of secretin in producing secretion of an alkaline pancreatic juice. It also inhibits gastric emptying, exerts a trophic effect on the pancreas, increases the secretion of enterokinase, and may enhance the motility of the small intestine and colon. There is some evidence that, along with secretin, it augments the contraction of the pyloric sphincter, thus preventing the reflux of duodenal contents into the stomach. Gastrin and CCK stimulate glucagon secretion, and since the secretion of both gastrointestinal hormones is increased by a protein meal, either or both may be the "gut factor" that stimulates glucagon secretion (see Chapter 19). The action of CCK on pancreatic acinar cells is produced by activation of phospholipase C, with consequent increases in intracellular IP_3 and DAG.

The secretion of CCK is increased by contact of the intestinal mucosa with the products of digestion, particularly peptides and amino acids, and also by the presence in the duodenum of fatty acids containing more than 10 carbon atoms. Since the bile and pancreatic juice that enter the duodenum in response to CCK further the digestion of protein and fat and the products of this digestion stimulate further CCK secretion, a sort of positive feedback operates in the control of the secretion of this hormone. The positive feedback is terminated when the products of digestion move on to the lower portions of the gastrointestinal tract.

Secretin

Secretin occupies a unique position in the history of physiology. In 1902, Bayliss and Starling first demonstrated that the excitatory effect of duodenal stimulation on pancreatic secretion was due to a blood-borne factor. Their research led to the identification of secretin. They also suggested that many chemical agents might be secreted by cells in the body and pass in the circulation to affect organs some distance away, and Starling introduced the term **hormone** to categorize such "chemical messengers." Modern endocrinology is the proof of the correctness of this hypothesis.

Secretin is secreted by S cells that are located deep in the glands of the mucosa of the upper portion of the small intestine. The structure of secretin (Table 26–2) is different from that of CCK and gastrin but very similar to that of glucagon, GLI, VIP, and GIP. Only one form of secretin has been isolated, and the fragments of the molecule that have been tested to date are inactive. Its half-life is about 5 minutes, but little is known about its metabolism.

Secretin increases the secretion of bicarbonate by the duct cells of the pancreas and biliary tract. It thus causes the secretion of a watery, alkaline pancreatic juice. Its action on pancreatic duct cells is mediated via cyclic AMP. It also augments the action of CCK in producing pancreatic secretion of digestive enzymes. It decreases gastric acid secretion and may cause contraction of the pyloric sphincter.

The secretion of secretin is increased by the products of protein digestion and by acid bathing the mucosa of the upper small intestine. The release of secretin by acid is another example of feedback control: secretin causes alkaline pancreatic juice to flood into the duodenum, neutralizing the acid from the stomach and thus stopping further secretion of the hormone.

GIP

GIP contains 43 amino acid residues (Table 26–2) and is produced by K cells in the mucosa of the duodenum and jejunum. Its secretion is stimulated by glucose and fat in the duodenum, and it inhibits gastric secretion and motility. It also stimulates insulin secretion. Gastrin, CCK, secretin, and glucagon also have this effect, but the evidence summarized in Chapter 19 indicates that GIP is the physiologic B-cell-stimulating hormone of the gastrointestinal tract. For this reason, it is sometimes called **glucose-dependent insulinotropic polypeptide.**

The integrated action of gastrin, CCK, secretin, and GIP in facilitating digestion and utilization of absorbed nutrients is summarized in Fig 26–3.

VIP

VIP contains 28 amino acid residues (Table 26–2). It is found in nerves in the gastrointestinal tract. Prepro-VIP contains both VIP and a closely related polypeptide (**PHM-27** in humans, PHI-27 in other species). VIP is also found in blood, in which it has a half-life of about 2 minutes. In the intestine, it markedly stimulates intestinal secretion of electrolytes and hence of water. Its other actions include relaxation of intestinal smooth muscle, including sphincters; dilation of peripheral blood vessels; and inhibition of gastric acid secretion. It is also found in the brain and many autonomic nerves (see Chapter 4), where it often occurs in the same neurons as acetylcholine. It potentiates the action of acetylcholine in salivary glands. However, VIP and acetylcholine do not coexist in neurons that innervate other parts of the gastrointestinal tract. VIP-secreting tumors (VIPomas) have been described in patients with severe diarrhea. The relation of GIP and VIP to **enterogastrone,** a putative hormone that inhibits gastric acid secretion and motility, is unknown.

Figure 26–3. Integrated action of gastrointestinal hormones in regulating digestion and utilization of absorbed nutrients. The dashed arrows indicate inhibition.

Other Gastrointestinal Hormones

Motilin, a polypeptide containing 22 amino acid residues (Table 26–2), is secreted by cells in the duodenal mucosa. Some of these are **enterochromaffin cells (EC cells),** in which motilin is colocalized with serotonin. Others are called M cells, but this is confusing, since the term M cells is also used to refer to the cells that pass antigens to gut-associated lymphoid tissue (see Chapter 25). Motilin causes contraction of intestinal smooth muscle and appears to be a regulator of interdigestive motility, preparing the intestine for the next meal. **Neurotensin,** a 13-amino-acid polypeptide, is produced by cells that are abundant in the mucosa of the ileum. Its release is stimulated by fatty acids, and it inhibits gastrointestinal motility and increases ileal blood flow. **Substance P** (Table 26–2) is found in endocrine cells in the gastrointestinal tract, but it has not been proved to enter the circulation. It increases the motility of the small intestine. **GRP** contains 27 amino acid residues, and the 10 amino acid residues at its C terminal are almost identical to amphibian bombesin. It is present in the vagal nerve endings that terminate on G cells and is the neurotransmitter producing vagally mediated increases in gastrin secretion. **Somatostatin,** the growth-hormone-inhibiting hormone originally isolated from the hypothalamus, is secreted into the circulation by D cells in the pancreatic islets (see Chapter 19) and by similar D cells in the gastrointestinal mucosa. As noted in Chapter 14, it exists in tissues in 2 forms, somatostatin 14 (Table 26–2) and somatostatin 28, and both are secreted. In addition, SS28 (1–12) is found in tissues. Somatostatin inhibits the secretion of gastrin, VIP, GIP, secretin, and motilin. Like several other gastrointestinal hormones, somatostatin is secreted in larger amounts into the gastric lumen than into the bloodstream. Its secretion is stimulated by acid in the lumen, and it may act in a "paracrine" fashion via the gastric juice to mediate the inhibition of gastrin secretion produced by acid. It also inhibits pancreatic exocrine secretion; gastric acid secretion and motility; gallbladder contraction; and the absorption of glucose, amino acids, and triglycerides. **Glucagon** is secreted by the A cells in the mucosa of the stomach and duodenum in addition to the A cells in the pancreatic islets. As noted in Chapter 19, glucagon from the gastrointestinal tract appears to play a role in the hyperglycemia of diabetes. Preproglucagon, which is made up of 179 amino acid residues, is processed in the mucosa of the lower small intestine and colon to a 69-amino-acid polypeptide instead of glucagon. This polypeptide, which is also secreted, is known as **glicentin (glucagonlike immunoreactivity, GLI).** Its function is unknown. Glicentin is found along with glucagon in the granules of A cells in the pancreatic islets (see Chapter 19).

Serotonin is present in and secreted by endocrine enterochromaffin cells as well as nerve endings, so in the gastrointestinal tract it is both a hormone and a neurotransmitter. Enkephalins may also be gastrointestinal hormones. Substance P, somatostatin, enkephalins, CCK, and neurotensin are present in gland cells or nerve fibers in the gastrointestinal tract. All these substances, along with gastrin and glucagon, are also found in the brain (see Chapter 4). The opioid peptide dynorphin is found in the posterior pituitary and the duodenum. TRH and ACTH are found in the gastrointestinal tract, but their functions in this location are unknown. TRH is also found in the pancreatic islets, where it is probably located in B cells. A substance called **urogastrone,** which was originally isolated from urine and found to aid the healing of ulcers, is now known to be epidermal growth factor (see Chapter 22).

The possible role of **pancreatic polypeptide** in the regulation of gastrointestinal function is discussed in Chapter 19.

Caerulein

An interesting decapeptide isolated from the skin of the Australian frog *Hyla caerulea* has the same 5 C-terminal and 2 N-terminal amino acid residues as gastrin and also has a sulfated tyrosyl residue. This material, which has been named **caerulein,** has all the properties of gastrin and CCK in mammals and has been used in humans to contract the gallbladder during cholecystography (see below). The skin of various species of frogs also contains TRH and many other biologically active peptides (see Chapter 14).

MOUTH & ESOPHAGUS

In the mouth, food is mixed with saliva and propelled into the esophagus. Peristaltic waves in the esophagus move the food into the stomach.

Mastication

Chewing (**mastication**) breaks up large food particles and mixes the food with the secretions of the salivary glands. This wetting and homogenizing action aids subsequent digestion. Large food particles can be digested, but they cause strong and often painful contractions of the esophageal musculature. Edentulous patients are generally restricted to a soft diet and have considerable difficulty eating dry food.

Salivary Glands & Saliva

In the salivary glands, the secretory (**zymogen**) granules containing the salivary enzymes are discharged from the acinar cells into the ducts (Fig 26–4). The characteristics of each of the 3 pairs of salivary glands in humans are summarized in Table 26–4.

About 1500 mL of saliva are secreted per day. The pH of saliva from resting glands is slightly less than 7.0, but during active secretion, it approaches 8.0. Saliva contains 2 digestive enzymes: **lingual lipase,** secreted by glands on the tongue, and **ptyalin (salivary**

Figure 26–4. Structure of submandibular gland (also known as submaxillary gland). Note that the cells in the mucous acini have flattened basal nuclei, whereas the cells in the serous acini have round nuclei and collections of zymogen secretory granules at their apexes. The intercalated ducts drain into the striated ducts, where the cells are specialized for ion transport. (Modified and reproduced, with permission, from Junqueira LC, Carneiro J, Kelley RO: *Basic Histology,* 7th ed. Appleton & Lange, 1992.)

Table 26–4. Characteristics of each of the pairs of salivary glands in humans.

Gland	Histologic Type[1]	Secretion	Percentage of Total Saliva in Humans (1.5 L/d)
Parotid	Serous	Water	20
Submandibular (submaxillary)	Mixed	Moderately viscous	70
Sublingual	Mucous	Viscous	5

[1]Serous cells secrete ptyalin; mucous cells secrete mucin. The remaining 5% of salivary volume is contributed by lingual and other minor glands in the oral cavity.

α-amylase), secreted by the salivary glands. The functions of these enzymes are discussed in Chapter 25. Saliva also contains **mucins,** glycoproteins that lubricate the food and protect the oral mucosa. It also contains IgA (see Chapter 27), the first immunologic defense against bacteria and viruses; lysozyme, which attacks the walls of bacteria; lactoferrin, which binds iron and is bacteriostatic; and proline-rich proteins that protect tooth enamel and bind toxic tannins.

Saliva performs a number of important functions. It facilitates swallowing, keeps the mouth moist, serves as a solvent for the molecules that stimulate the taste buds, aids speech by facilitating movements of the lips and tongue, and keeps the mouth and teeth clean. The saliva may also have some antibacterial action, and patients with deficient salivation (**xerostomia**) have a higher than normal incidence of dental caries. The buffers in saliva help maintain the oral pH at about 7.0. They also help neutralize gastric acid and relieve heartburn when gastric juice is regurgitated into the esophagus.

Ionic Composition of Saliva

There is considerable variation in the ionic composition of saliva from species to species and from gland to gland. In general, however, saliva secreted in the acini is probably isotonic, with concentrations of Na^+, K^+, Cl^-, and HCO_3^- that are close to those in plasma. The excretory ducts and probably the intercalated ducts that drain into them modify the composition of the saliva by extracting Na^+ and Cl^- and adding K^+ and HCO_3^-. The ducts are relatively impermeable to water, and saliva becomes hypotonic in the duct system. Therefore, at low salivary flows, the saliva that reaches the mouth is hypotonic, alkaline, and rich in K^+ but relatively depleted of Na^+ and Cl^-. When salivary flow is rapid, there is less time for ionic composition to change in the ducts. Consequently, although still hypotonic in humans, saliva is closer to isotonic, with higher concentrations of Na^+ and Cl^-. Aldosterone increases the K^+ concentration and reduces the Na^+ concentration of saliva in an action analogous to its action on the kidneys (see Chapters 20 and 38), and a high salivary Na^+/K^+ ratio is seen when aldosterone is deficient in Addison's disease.

Control of Salivary Secretion

Salivary secretion is under neural control. Stimulation of the parasympathetic nerve supply causes profuse secretion of watery saliva with a relatively low content of organic material. Associated with this secretion is a pronounced vasodilation in the gland, which appears to be due to the local release of VIP. This polypeptide is a cotransmitter with acetylcholine in some of the postganglionic parasympathetic neurons. Atropine and other cholinergic blocking agents reduce salivary secretion. Stimulation of the sympathetic nerve supply causes vasoconstriction and, in humans, secretion of small amounts of saliva rich in organic constituents from the submandibular glands.

Food in the mouth causes reflex secretion of saliva, and so does stimulation of the vagal afferent fibers at the gastric end of the esophagus. Salivary secretion is easily conditioned, as shown in Pavlov's original experiments (see Chapter 16). In humans, the sight, smell, and even thought of food causes salivary secretion ("makes the mouth water").

Like the thyroid gland, the salivary glands and the gastric mucosa concentrate iodide from the plasma, the salivary/plasma iodide ratio sometimes reaching 60. The physiologic significance of this iodide-trapping phenomenon is uncertain. The salivary glands have also been reported to contain somatostatin and glucagon, and in male mice, they contain renin and nerve growth factor. The functions of most of these factors in the salivary glands are unknown, but it has been suggested that glucagon secreted from the salivary glands contributes to the hyperglycemia in pancreatectomized animals (see Chapter 19).

Deglutition

Swallowing (deglutition) is a reflex response that is triggered by afferent impulses in the trigeminal, glossopharyngeal, and vagus nerves. These impulses are integrated in the nucleus of the tractus solitarius and the nucleus ambiguus. The efferent fibers pass to the pharyngeal musculature and the tongue via the trigeminal, facial, and hypoglossal nerves. Swallowing is initiated by the voluntary action of collecting the oral contents on the tongue and propelling them backward into the pharynx. This starts a wave of involuntary contraction in the pharyngeal muscles that pushes the material into the esophagus. Inhibition of respiration and glottic closure are part of the reflex response. Swallowing is difficult if not impossible when the mouth is open, as anyone who has spent time in the dentist's chair feeling saliva collect in the throat is well aware. A normal adult swallows frequently while eating, but swallowing also continues between meals. The total number of swallows per day is about 600: 200 while eating and drinking, 350 while awake without food, and 50 while sleeping.

At the pharyngoesophageal junction, there is a 3-cm segment of esophagus in which the resting wall tension is high. This segment relaxes reflexly upon swallowing, permitting the swallowed material to enter the body of the esophagus. A peristaltic ring contraction of the esophageal muscle forms behind the material, which is then swept down the esophagus at a speed of approximately 4 cm/s. When humans are in an upright position, liquids and semisolid foods generally fall by gravity to the lower esophagus ahead of the peristaltic wave. Unlike the rest of the esophagus, the musculature of the gastroesophageal junction (**lower esophageal sphincter, LES**) is tonically active but relaxes upon swallowing. The relaxation appears to be mediated via neurons that secrete VIP. The tonic activity of the lower esophageal sphincter between meals prevents reflux of gastric contents into the esophagus. Large doses of gastrin increase the tone of the lower esophageal sphincter, but this effect is not produced by doses of gastrin that produce circulating gastrin levels comparable to those that occur after a meal.

Motor Disorders of the Esophagus

Achalasia is a condition in which food accumulates in the esophagus and the organ becomes massively dilated. It is due to increased resting lower esophageal sphincter tension, incomplete relaxation of this sphincter on swallowing, and weak esophageal peristalsis. The number of VIP-containing neurons in the lower esophagus is decreased, and the VIP content is low, suggesting that the condition is caused by VIP deficiency. It can be treated by pneumatic dilation of the sphincter or incision of the esophageal muscle (myotomy).

The opposite condition is lower esophageal sphincter incompetence, which permits reflux of acid gastric contents into the esophagus. This causes heartburn and esophagitis and can lead to ulceration and stricture due to scarring. The condition can be treated by inhibition of acid secretion by using H_2 receptor blockers or omeprazole (see below), or surgically by making a fold of gastric tissue (**fundoplication**).

Aerophagia & Intestinal Gas

Nervous persons who hyperventilate sometimes swallow large amounts of air, and some air is un-avoidably swallowed in the process of eating and drinking (**aerophagia**). Some of the swallowed air is regurgitated (belching), and some of the gases it contains are absorbed, but much of it passes on to the colon. Here, some of the oxygen is absorbed, and hydrogen, hydrogen sulfide, carbon dioxide, and methane formed by the colonic bacteria from carbohydrates and other substances are added to it. It is then expelled as **flatus.** The smell is largely due to sulfides. The volume of gas normally found in the human gastrointestinal tract is about 200 mL, and the daily production is 500–1500 mL. In some individuals, gas in the intestines causes cramps, **borborygmi** (rumbling noises), and abdominal discomfort.

STOMACH

Food is stored in the stomach; mixed with acid, mucus, and pepsin; and released at a controlled, steady rate into the duodenum.

Anatomic Considerations

The gross anatomy of the stomach is shown in Fig 26–5. The gastric mucosa contains many deep glands. In the pyloric and cardiac regions, the glands secrete mucus. In the body of the stomach, including the fundus, the glands contain **parietal (oxyntic) cells,** which secrete hydrochloric acid and intrinsic factor, and **chief (zymogen, peptic) cells,** which secrete pepsinogens (Fig 26–6). These secretions mix with mucus secreted by the cells in the necks of the glands. Several of the glands open on a common chamber (**gastric pit**) that opens in turn on the surface of the mucosa. Mucus is also secreted along with

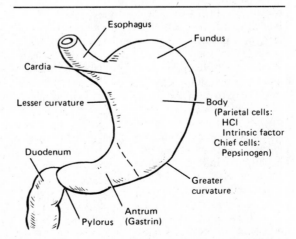

Figure 26–5. Anatomy of the stomach. The principal secretions are listed in parentheses under the labels indicating the locations where they are produced. In addition, mucus is secreted in all parts of the stomach. The dashed line marks the border between the body and the antrum.

Figure 26–6. Diagram of glands in the mucosa of the body of the human stomach. (Reproduced, with permission, from Bell GH, Davidson N, Scarborough G: *Textbook of Physiology and Biochemistry,* 6th ed. Livingstone, 1965.)

HCO_3^- by mucus cells on the surface of the epithelium between glands.

The stomach has a very rich blood and lymphatic supply. Its parasympathetic nerve supply comes from the vagi and its sympathetic supply from the celiac plexus.

Gastric Secretion

The cells of the gastric glands secrete about 2500 mL of **gastric juice** daily. This juice contains a variety of substances (Table 26–5). The gastric enzymes are discussed in Chapter 25. The **hydrochloric acid** secreted by the glands in the body of the stomach kills many ingested bacteria, aids protein digestion, provides the necessary pH for pepsin to start protein digestion, and stimulates the flow of bile and pancreatic juice. It is concentrated enough to cause tissue

Table 26–5. Contents of normal gastric juice (fasting state).

Cations: Na^+, K^+, Mg^{2+}, H^+ (pH approximately 1.0)
Anions: Cl^-, HPO_4^{2-}, SO_4^{2-}
Pepsins
Gelatinase
Mucus
Intrinsic factor
Water

damage, but in normal individuals the gastric mucosa does not become irritated or digested, in part because the gastric juice also contains **mucus.** Mucus, which is secreted by the neck and surface mucous cells in the body and fundus and similar cells elsewhere in the stomach, is made up of glycoproteins. Each mucus glycoprotein contains 4 subunits joined by disulfide bridges. The mucus forms a flexible gel that coats the mucosa.

The surface mucosa cells also secrete HCO_3^-. The HCO_3^- and the mucus form an unstirred layer that has a pH of about 7.0. This unstirred layer plus the surface membranes of the mucosal cells and the tight junctions between them constitute the mucosal bicarbonate barrier that protects the mucosal surface from damage by gastric acid. Substances that tend to disrupt the barrier and cause gastric irritation include ethanol, vinegar, bile salts, and aspirin and other nonsteroidal anti-inflammatory drugs. Prostaglandins stimulate mucus secretion, and aspirin and related drugs inhibit prostaglandin synthesis.

The electrolyte content of the gastric juice varies with the rate of secretion. At low secretory rates, the Na^+ concentration is high and the H^+ concentration is low, but as acid secretion increases, the Na^+ concentration falls.

Pepsinogen Secretion

The chief cells that secrete pepsinogens, the precursors of the pepsins in gastric juice (see Chapter 25), contain zymogen granules. The secretory process is similar to that involved in the secretion of ptyalin by the salivary glands and trypsinogen and the other pancreatic enzymes by the pancreas. Pepsinogen activity can be detected in the plasma and in the urine, where it is called **uropepsinogen.**

Hydrochloric Acid Secretion

It is difficult to obtain the products of parietal cell secretion free of contamination by other gastric secretions, but the purest specimens that have been analyzed are essentially isotonic. Their H^+ concentration is equivalent to approximately 0.17 N HCl, with pHs as low as 0.87. Therefore, parietal cell secretion may well be an isotonic solution of essentially pure HCl that contains 150 meq of Cl^- and 150 meq of H^+ per liter. Yet the pH of the cytoplasm of the parietal cells, like that of other cells, is 7.0–7.2, and the comparable concentrations per liter of plasma are about 100 meq of Cl^- and 0.00004 meq of H^+.

It is H^+-K^+ ATPase in the apical membrane of the parietal cells that pumps H^+ against a concentration gradient of this magnitude (Fig 26–7). The parietal cells are polarized, with an apical membrane facing the lumen of the gastric glands and a basolateral membrane in contact with the interstitial fluid. Canaliculi extend from the apical surface into the cell (Fig 26–8). At rest, the cells also contain abundant tubulovesicular structures with H^+-K^+ ATPase mole-

Interstitial fluid | Parietal cell | Gastric lumen

Figure 26–7. HCl secretion by parietal cells in the stomach. Active transport by ATPase is indicated by arrows in circles. H^+ is secreted into the lumen of the canaliculi in exchange for K^+ by H^+-K^+ ATPase. HCO_3^- is exchanged for Cl^- in the interstitial fluid by an antiport, and Na^+-K^+ ATPase keeps intracellular Na^+ low. Dashed arrows indicate diffusion. Compare with Fig 38–21.

Figure 26–8. Composite diagram of a parietal cell, showing the resting state *(lower left)* and the active state *(upper right)*. The resting cell has intracellular canaliculi (IC), which open on the apical membrane of the cell, and many tubulovesicular structures (TV) in the cytoplasm. When the cell is activated, the TVs fuse with the cell membrane and microvilli (MV) project into the canaliculi, so the cell membrane in contact with gastric lumen is greatly increased. M, mitochondrion; G, Golgi apparatus. (Reproduced, with permission, from Junqueira LC, Carneiro J, Kelley RO: *Basic Histology,* 7th ed. Appleton & Lange, 1992.)

cules in their walls. The K^+ inside the vesicles is reduced to low levels, and the ATPase molecules are inactive. When the parietal cells are stimulated, the tubulovesicular structures move to the apical membrane and fuse with it, thus inserting many more H^+-K^+ ATPase molecules into the membrane. This increases their surface area and forms numerous microvilli, which project into the canaliculi. The ATPase molecules are now exposed to the K^+ in the ECF, and H^+-K^+ exchange begins.

Pumping H^+ out of the parietal cells in exchange for K^+ requires appreciable energy, and this is provided by hydrolysis of ATP. Inhibition of ATP generation prevents H^+ secretion. Cl^- is also extruded down its electrochemical gradient through channels that are activated by cyclic AMP in the apical membrane; the concentration gradient is inward, but the outwardly directed electrical gradient is much greater. K^+ is also extruded. The H^+ that is extruded comes from H_2CO_3, and H_2CO_3 in turn is formed by the hydration of CO_2 (Fig 26–7). This latter reaction is catalyzed by carbonic anhydrase, and the parietal cells are particularly rich in this enzyme. The HCO_3^- formed by dissociation of H_2CO_3 is extruded by an antiport in the basolateral membrane of the parietal cells that exchanges HCO_3^- for another anion, and since Cl^- is the most abundant anion in interstitial fluid, the exchange is mainly for Cl^-. Because of the efflux of HCO_3^- into the blood, the stomach has a negative respiratory quotient (RQ)—ie, the amount of CO_2 in arterial blood is greater than the amount in gastric venous blood. When gastric acid secretion is elevated after a meal, sufficient H^+ may be secreted to raise the pH of systemic blood and make the urine alkaline. This is the probable explanation of the high pH of the urine excreted after a meal, the so-called **postprandial alkaline tide.**

Acid secretion is stimulated by histamine via H_2 receptors, by acetylcholine via M_1 muscarinic receptors, and by gastrin via gastrin receptors in the membranes of the parietal cells (Fig 26–9). The H_2 receptors increase intracellular cyclic AMP via G_s, whereas the muscarinic receptors and the gastrin receptors exert their effects by increasing intracellular free Ca^{2+}. The intracellular events interact so that activation of one receptor type potentiates the response of another to stimulation. The histamine comes from cells in the mucosa that resemble mast cells, the acetylcholine comes from the endings of postganglionic cholinergic neurons innervating the parietal cells, and the gastrin reaches the cells via the circulation. Prostaglandins, particularly those of the E series, inhibit acid secretion by activating G_i, and this may explain in part the increased incidence of ulcers in patients taking anti-inflammatory drugs that inhibit prostaglandin synthesis.

Gastric Motility & Emptying

When food enters the stomach, the organ relaxes

Figure 26–9. Regulation of gastric acid secretion by the parietal cell. Acid secretion is increased by acetylcholine acting on M_1 muscarinic receptors to increase intracellular Ca^{2+} and by gastrin acting on gastrin receptors to increase intracellular Ca^{2+}. Histamine binds to H_2 receptors, and via G_s, this increases adenylate cyclase (AC) activity and intracellular cyclic AMP. PGE_2 acts via G_i to decrease adenylate cyclase activity and intracellular cyclic AMP. Cyclic AMP and Ca^{2+} act via protein kinases to increase the transport of H^+ into the gastric lumen by H^+-K^+ ATPase.

by the reflex process of **receptive relaxation.** This relaxation of the gastric musculature is triggered by movement of the pharynx and esophagus. It is followed by peristaltic contractions that mix the food and squirt it into the duodenum at a controlled rate. The peristaltic waves are most marked in the distal half of the stomach. When well developed, they occur at a rate of 3/min.

The pyloric sphincter has a limited function in the control of gastric emptying. Gastric emptying is normal if the pylorus is held open or even if it is surgically resected. The antrum, pylorus, and upper duodenum apparently function as a unit. Contraction of the antrum is followed by sequential contraction of the pyloric region and the duodenum. In the antrum, contraction ahead of the advancing gastric contents prevents solid masses from entering the duodenum. The gastric contents are thus squirted a bit at a time into the small intestine. Normally, regurgitation from the duodenum does not occur, because the contraction of the pyloric segment tends to persist slightly longer than that of the duodenum. The prevention of regurgitation may also be due to the stimulating action of CCK and secretin on the pyloric sphincter.

Gastric Slow Wave

Peristaltic contractions in the stomach are coordinated by the **gastric slow wave,** a wave of depolarization of smooth muscle cells proceeding from the circular muscle of the fundus of the stomach to the pylorus approximately every 20 seconds. This wave is also called the **basic electric rhythm (BER).** It is the pacemaker for antral peristalsis; the wave and, consequently, peristalsis become irregular and chaotic after vagotomy or transection of the stomach wall. Thus, the slow wave plays a major role in the control of gastric emptying. A similarly aborally directed slow wave coordinates smooth muscle contractions in the small intestine and the colon (see below).

Hunger Contractions

The musculature of the stomach is rarely inactive. Soon after the stomach is emptied, mild peristaltic contractions begin. They gradually increase in intensity over a period of hours. The more intense contractions can be felt and may even be mildly painful. These **hunger contractions** are associated with the sensation of hunger and were once thought to be an important regulator of appetite. However, it has now been demonstrated that food intake is normal in animals after denervation of the stomach and intestines (see Chapter 14).

Regulation of Gastric Secretion

Gastric motility and secretion are regulated by neural and humoral mechanisms. The neural components are local autonomic reflexes, involving cholinergic neurons, and impulses from the central nervous system by way of the vagus nerves. The humoral components are the hormones discussed above. Vagal stimulation increases gastrin secretion by release of gastrin-releasing peptide (see above). Other vagal fibers release acetylcholine, which acts directly on the cells in the glands in the body and the fundus to increase acid and pepsin secretion. Stimulation of the vagus nerve in the chest or neck increases acid and pepsin secretion, but vagotomy does not abolish the secretory response to local stimuli.

For convenience, the physiologic regulation of gastric secretion is usually discussed in terms of cephalic, gastric, and intestinal influences, although these overlap. The **cephalic** influences are vagally mediated responses induced by activity in the central nervous system. The **gastric** influences are primarily local reflex responses and responses to gastrin. The **intestinal** influences are the reflex and hormonal feedback effects on gastric secretion initiated from the mucosa of the small intestine.

Cephalic Influences

The presence of food in the mouth reflexly stimulate gastric secretion. The efferent fibers for this reflex are in the vagus nerves. Vagally mediated increases in gastric secretion are easily conditioned. In humans, for example, the sight, smell, and thought of food increase gastric secretion. These increases are due to alimentary conditioned reflexes that become established early in life.

The neural mechanisms involved in the formation of conditioned reflexes are discussed in Chapter 16. Stimulation of the anterior hypothalamus and parts of the adjacent orbital frontal cortex increases vagal efferent activity and gastric secretion. Cephalic influences are responsible for one-third to one-half the acid secreted in response to a normal meal.

Emotional Responses

Psychic states have effects on gastric secretion and motility that are mediated principally via the vagi. Among his famous observations on Alexis St. Martin, the Canadian with a permanent gastric fistula resulting from a gunshot wound, William Beaumont noted that anger and hostility were associated with turgor, hyperemia, and hypersecretion of the gastric mucosa. Similar observations have been made on other patients with gastric fistulas. Fear and depression decrease gastric secretion and blood flow and inhibit gastric motility.

Gastric Influences

Food in the stomach accelerates the increase in gastric secretion produced by the sight and smell of food and the presence of food in the mouth (Fig 26–10). Receptors in the wall of the stomach and the mucosa respond to stretch and chemical stimuli, mainly amino acids and related products of digestion. The fibers from the receptors enter Meissner's plexus, where the cell bodies of the receptor neurons are located. They synapse on postganglionic parasympathetic neurons that end on parietal cells and

Figure 26–10. Human gastric acid secretion after a steak meal. (Reproduced, with permission, from Brooks FP: Integrative lecture: Response of the GI tract to a meal. *Undergraduate Teaching Project.* American Gastroenterological Association, 1974.)

stimulate acid secretion. Thus, the acid responses are produced by local reflexes in which the reflex arc is totally within the wall of the stomach. The postganglionic neurons in the local reflex arc are the same ones innervated by the descending vagal preganglionic neurons from the brain that mediate the cephalic phase of secretion. The products of protein digestion also bring about increased secretion of gastrin, and this augments the flow of acid.

Intestinal Influences

Although there are gastrin-containing cells in the mucosa of the small intestine as well as in the stomach, instillation of amino acids directly into the duodenum does not increase circulating gastrin. Fats, carbohydrates, and acid in the duodenum inhibit gastric acid and pepsin secretion and gastric motility via GIP and secretin and possibly via other hormones. Gastric acid secretion is increased following removal of large parts of the small intestine. The hypersecretion, which is roughly proportionate in degree to the amount of intestine removed, may be due in part to removal of the source of hormones that inhibit acid secretion.

Other Influences

Hypoglycemia acts via the brain and vagal efferents to stimulate acid and pepsin secretion. Other stimulants include **alcohol** and **caffeine,** both of which act directly on the mucosa. The beneficial effects of moderate amounts of alcohol on appetite and digestion, as a result of this stimulatory effect on gastric secretion, have been known since ancient times.

Regulation of Gastric Motility & Emptying

The rate at which the stomach empties into the duodenum depends on the type of food ingested. Food rich in carbohydrate leaves the stomach in a few hours. Protein-rich food leaves more slowly, and emptying is slowest after a meal containing fat (Fig 26–11). The rate of emptying also depends on the osmotic pressure of the material entering the duodenum. Hyperosmolality of the duodenal contents is sensed by "duodenal osmoreceptors" that initiate a decrease in gastric emptying which is probably neural in origin.

Products of protein digestion and hydrogen ions bathing the duodenal mucosa initiate a neurally mediated decrease in gastric motility, the **enterogastric reflex.** Distention of the duodenum also initiates this reflex. GIP and CCK inhibit gastric motility. Cutting the vagus nerves slows gastric emptying. In humans, vagotomy may cause relatively severe gastric atony and distention. Excitement is said to hasten gastric emptying, and fear to slow it.

Since fats are particularly effective in inhibiting gastric emptying, some people drink milk, cream, or even olive oil before a cocktail party. The fat keeps the alcohol in the stomach for a long time, where its

(300-mL Liquid meals)

Standard meal (inert pectin)

Protein

Protein + lipid

Mean ± S.E.

Volume of test meal emptied (mL)

Time after feeding (min)

Figure 26–11. Effect of protein and fat on the rate of emptying of the human stomach. Subjects were fed 300-mL liquid meals. (Reproduced, with permission, from Brooks FP: Integrative lecture: Response of the GI tract to a meal. *Undergraduate Teaching Project.* American Gastroenterological Association, 1974.)

absorption is slower than in the small intestine, and the intoxicant enters the small intestine in a slow, steady stream so that—theoretically, at least—a sudden rise of the blood alcohol to a high level and consequent embarrassing intoxication are avoided.

Peptic Ulcer

Gastric and duodenal ulceration in humans is related in some poorly understood way to a breakdown of the barrier that normally prevents irritation and autodigestion of the mucosa by the gastric secretions. It appears that infection with a bacterium, *Helicobacter pylori,* contributes to the breakdown. In addition, aspirin and other nonsteroidal anti-inflammatory drugs (NSAIDs) contribute, and these are widely used. In patients with ulcers of the duodenum and prepyloric portion of the stomach, gastric acid is always present ("no acid, no ulcer") and its secretion is often elevated (Fig 26–10). As noted in Chapter 25, there is a correlation between pepsinogen I levels, maximum acid secretion, and the incidence of peptic ulcer. Resting gastrin levels are not elevated in most patients with gastric and duodenal ulcers, but their gastrin responses to feeding are greater than normal and their parietal cells are hyperresponsive to gastrin. Another illustration of the importance of acid hypersecretion in the genesis of duodenal and prepyloric ulcers is provided by the **Zollinger-Ellison syndrome.** This syndrome is seen in patients with gastrinomas (tumors that secrete gastrin). These tumors can occur in the stomach and duodenum, but most of them are found in the pancreas. The gastrin causes prolonged hypersecretion of acid and severe ulcers are produced.

More extensive discussions of the problem of peptic ulceration can be found in textbooks of pathophysiology. However, it is pertinent to note that the procedures used in the treatment of ulcers are aimed at inhibiting acid secretion and enhancing mucosal resistance to acid. A variety of different antacids, most of which contain aluminum hydroxide, magnesium hydroxide, or calcium carbonate, are available. Blockade of the gastric H_2 receptors with the H_2 receptor-blocking drugs cimetidine, ranitidine, nizatidine, and famotidine produces effective reduction of acid responses because it removes the potentiating effect of stimulation of this receptor on the responses to other stimuli (see above); indeed, on the basis of its ability to inhibit acid secretion, cimetidine was at one point the most commonly prescribed drug in the USA. M_1 muscarinic receptors can be blocked with atropine or the new, more specific anticholinergic drug pirenzepine. Gastric receptors are sometimes blocked with proglumide. The gastric H^+-K^+ ATPase can be inhibited by omeprazole, and this drug has proved to be effective over long periods. Sucralfate, a basic aluminum salt of sucrose octasulfate, increases the resistance of the mucosa to acid, probably by forming adherent protein and other complexes at the ulcer site, and it is also of value in the treatment of ulcers.

The effectiveness of H_2 receptor blockers and omeprazole has made surgical treatment of ulcers rare. However, severe duodenal and prepyloric ulcers are sometimes treated by vagotomy combined with removal of the gastrin-secreting antral mucosa.

Other Functions of the Stomach

The stomach has a number of functions in addition to the storage of food and the control of its release into the duodenum. The hydrochloric acid kills many ingested bacteria. The parietal cells in the gastric mucosa also secrete **intrinsic factor,** a substance necessary for the absorption of **cyanocobalamin** (vitamin B_{12}) from the small intestine.

Cyanocobalamin (Fig 26–12) is a complex cobalt-containing vitamin that is necessary for normal erythropoiesis. Inadequate absorption of this vitamin causes an anemia characterized by the appearance in the bloodstream of large primitive red cell precursors called **megaloblasts.** A complete remission of the deficiency syndrome occurs when cyanocobalamin is injected parenterally but not when it is administered by mouth unless the intrinsic factor secreted by the gastric mucosa is present. Anemia due to an inadequate dietary intake of cyanocobalamin is very rare, apparently because the minimum daily requirements are quite low and the vitamin is found in most foods of animal origin. The deficiency states seen clinically are therefore due primarily to defective cyanocobalamin absorption. Inadequate absorption may be due to primary intestinal diseases such as sprue or to intrinsic factor deficiency following gastrectomy. In pa-

Figure 26–12. Cyanocobalamin (vitamin B_{12}). Empiric formula: $C_{63}H_{88}O_{14}N_{14}PCo$.

tients with the specific megaloblastic anemia called **pernicious anemia,** there is intrinsic factor deficiency due to an idiopathic atrophy of much of the gastric mucosa.

Intrinsic factor is a glycoprotein with a molecular weight of 45,000. Cyanocobalamin becomes firmly bound to it in the intestine. The intrinsic factor-cyanocobalamin complex then becomes bound to specific receptors in the ileum, and the cyanocobalamin is transferred across the intestinal epithelium. Trypsin is required for this process to be efficient, and absorption is sometimes decreased in patients with pancreatic insufficiency. The details of the absorptive process are uncertain, but it has been suggested that the role of intrinsic factor in the process is to stimulate endocytosis. This would explain how such a large molecule can be absorbed. A syndrome has been described in which there is congenital absence of the ileal receptors for the cyanocobalamin-intrinsic factor complex.

In totally gastrectomized patients, the intrinsic factor deficiency must be circumvented by parenteral injection of cyanocobalamin. Protein digestion is normal in the absence of pepsin, and nutrition can be maintained. However, these patients are prone to develop iron deficiency anemia (see Chapter 25) and

other abnormalities, and they must eat frequent small meals. Because of rapid absorption of glucose from the intestine and the resultant hyperglycemia and abrupt rise in insulin secretion, gastrectomized patients sometimes develop hypoglycemic symptoms about 2 hours after meals (see Chapter 19). Weakness, dizziness, and sweating after meals, due in part to hypoglycemia, are part of the picture of the "dumping syndrome," a distressing syndrome that develops in patients in whom portions of the stomach have been removed or the jejunum has been anastomosed to the stomach. Another cause of the symptoms is rapid entry of hypertonic meals into the intestine; this provokes the movement of so much water into the gut that significant hypovolemia and hypotension are produced.

EXOCRINE PORTION OF THE PANCREAS

The pancreatic juice contains enzymes that are of major importance in the digestion (see Table 25–1). Its secretion is controlled in part by a reflex mechanism and in part by the gastrointestinal hormones secretin and CCK.

Anatomic Considerations

The portion of the pancreas that secretes pancreatic juice is a compound alveolar gland resembling the salivary glands. Granules containing the digestive enzymes (**zymogen granules**) are formed in the cell and discharged by exocytosis (see Chapter 1) from the apexes of the cells into the lumens of the pancreatic ducts (Fig 26–13). The small duct radicles coalesce into a single duct (pancreatic duct of Wirsung), which usually joins the common bile duct to form the ampulla of Vater (Fig 26–14). The ampulla opens through the duodenal papilla, and its orifice is encircled by the sphincter of Oddi. In some individuals, there is an accessory pancreatic duct (duct of Santorini) that enters the duodenum more proximally. There is said to be a high incidence of nonpatency of the accessory pancreatic duct in patients with gastric acid hypersecretion and duodenal ulcer. It is possible that failure of the alkaline pancreatic secretion to enter the upper part of the duodenum via this duct contributes to the development of the ulcer.

Composition of Pancreatic Juice

The pancreatic juice is alkaline (Table 26–6) and has a high HCO_3^- content (approximately 113 meq/L versus 24 meq/L in plasma). About 1500 mL of pancreatic juice is secreted per day. Bile and intestinal juices are also neutral or alkaline, and these 3 secretions neutralize the gastric acid, raising the pH of the duodenal contents to 6.0–7.0. By the time the chyme reaches the jejunum, its reaction is nearly neutral, but the intestinal contents are rarely alkaline.

The powerful protein-splitting enzymes of the

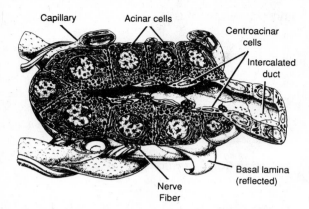

Figure 26–13. Acinar cells surrounding a terminal branch of the pancreatic ducts. Note the abundant rough endoplasmic reticulum and the zymogen granules concentrated at the apexes of the cells. (Reproduced, with permission, from Krstić RV: *Die Gewebe des Menschen und der Säugetiere.* Springer-Verlag, 1978.)

pancreatic juice are secreted as inactive proenzymes. Trypsinogen is converted to the active enzyme trypsin by the brush border enzyme **enteropeptidase (enterokinase)** when the pancreatic juice enters the duodenum. The secretion of enteropeptidase is increased by CCK. Enteropeptidase contains 41% polysaccharide, and this high polysaccharide content apparently prevents it from being digested itself before it can exert its effect. Trypsin converts chymotrypsinogens into chymotrypsins and other proenzymes into active enzymes (Fig 26–15). Trypsin can also activate trypsinogen; therefore, once some trypsin is formed, there is an autocatalytic chain reaction. Enteropeptidase deficiency occurs as a congenital abnormality and leads to protein malnutrition.

The potential danger of the release into the pancreas of a small amount of trypsin is apparent; the resulting chain reaction would produce active enzymes that could digest the pancreas. It is therefore not surprising that the pancreas normally contains a trypsin inhibitor.

Another enzyme activated by trypsin is phospholipase A_2. This enzyme splits a fatty acid off lecithin, forming lysolecithin. Lysolecithin damages cell membranes. It has been hypothesized that in **acute pancreatitis,** a severe and sometimes fatal disease, phospholipase A_2 is activated in the pancreatic ducts, with the formation of lysolecithin from the lecithin that is a normal constituent of bile. This causes disruption of pancreatic tissue and necrosis of surrounding fat.

A small amount of pancreatic α-amylase normally leaks into the circulation, but in acute pancreatitis, the circulating level of this enzyme rises markedly. Measurement of plasma amylase is therefore of value in diagnosing the disease.

Regulation of the Secretion of Pancreatic Juice

Secretion of pancreatic juice is primarily under hormonal control. Secretin acts on the pancreatic ducts to cause copious secretion of a very alkaline pancreatic juice that is rich in HCO_3^- and poor in enzymes. The effect on duct cells is due to an increase

Table 26–6. Composition of normal human pancreatic juice.

Cations: Na^+, K^+, Ca^{2+}, Mg^{2+} (pH approximately 8.0)
Anions: HCO_3^-, Cl^-, SO_4^{2-}, HPO_4^{2-}
Digestive enzymes (see Table 25–1; 95% of protein in juice)
Other proteins

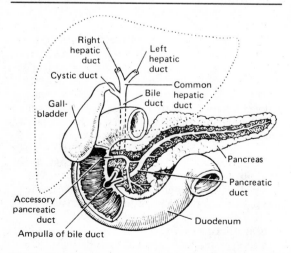

Figure 26–14. Connections of the ducts of the gallbladder, liver, and pancreas. (Reproduced, with permission, from Bell GH, Emslie-Smith D, Paterson CR: *Textbook of Physiology and Biochemistry,* 9th ed. Churchill Livingstone, 1976.)

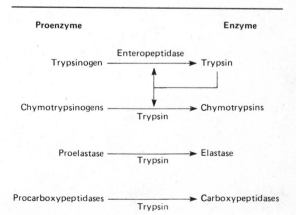

Figure 26–15. Activation of the pancreatic proteases in the duodenal lumen.

in intracellular cyclic AMP. Secretin also stimulates bile secretion. CCK acts on the acinar cells to cause release of zymogen granules and production of pancreatic juice rich in enzymes. Its effect is mediated by phospholipase C (see Chapter 1).

The response to intravenous secretin is shown in Fig 26–16. Note that as the volume of pancreatic secretion increases, its Cl^- concentration falls and its HCO_3^- concentration increases. Although HCO_3^- is secreted in the small ducts, it is reabsorbed in the large ducts in exchange for Cl^-. The magnitude of the exchange is inversely proportionate to the rate of flow.

Like CCK, acetylcholine acts on acinar cells to cause discharge of zymogen granules, and stimulation of the vagi causes secretion of a small amount of pancreatic juice rich in enzymes. Acetylcholine also exerts its effects by activating phospholipase C. There is evidence for vagally mediated conditioned reflex secretion of pancreatic juice in response to the sight or smell of food.

Figure 26–16. Effect of a single dose of secretin on the composition and volume of the pancreatic juice in humans. (Redrawn and reproduced, with permission, from Janowitz HD: Pancreatic secretion. *Physiol Physicians* [Nov] 1964;**2**:11.)

LIVER & BILIARY SYSTEM

Bile is secreted by the cells of the liver into the bile duct, which drains into the duodenum. Between meals, the duodenal orifice of this duct is closed and bile flows into the gallbladder, where it is stored. When food enters the mouth, the sphincter around the orifice relaxes; when the gastric contents enter the duodenum, the hormone CCK from the intestinal mucosa causes the gallbladder to contract.

Anatomic Considerations

The liver is organized in lobules, within which blood flows past hepatic cells via sinusoids from branches of the portal vein to the central vein of each lobule (Fig 26–17). The endothelium of the sinusoids has large fenestrations, and plasma is in close contact with liver cells (Fig 26–18). There is usually only one layer of hepatocytes between sinusoids, so the total area of contact between liver cells and plasma is very large. Hepatic artery blood also enters the sinusoids. The central veins coalesce to form the hepatic veins, which drain into the inferior vena cava. Hepatic blood flow is discussed in more detail in Chapter 32. The average transit time for blood across the liver lobule from the portal venule to the central hepatic vein is about 8.4 seconds. Numerous macrophages (**Kupffer cells**) are anchored to the endothelium of the sinusoids and project into the lumen. The functions of these phagocytic cells are discussed in Chapter 27.

Each liver cell is also apposed to several bile canaliculi (Fig 26–17). The canaliculi drain into intralobular bile ducts, and these coalesce via interlobular bile ducts to form the right and left hepatic ducts. These ducts join outside the liver to form the common hepatic duct. The cystic duct drains the gallbladder. The hepatic duct unites with the cystic duct to form the common bile duct. The common bile duct enters the duodenum at the duodenal papilla. Its orifice is surrounded by the sphincter of Oddi, and it usually unites with the main pancreatic duct just before entering the duodenum (Fig 26–14).

The walls of the extrahepatic biliary ducts and the gallbladder contain fibrous tissue and smooth muscle. The mucous membrane contains mucous glands and is lined by a layer of columnar cells. In the gallbladder, the mucous membrane is extensively folded; this increases its surface area and gives the interior of the gallbladder a honeycombed appearance. In primates, the mucous membranes of the cystic duct are also folded to form the so-called spiral valves.

Functions of the Liver

The liver, the largest gland in the body, has many complex functions, including those listed in Table 26–7. Discussion of these functions in one place requires that they be taken out of their proper context at considerable cost in terms of clarity and integration

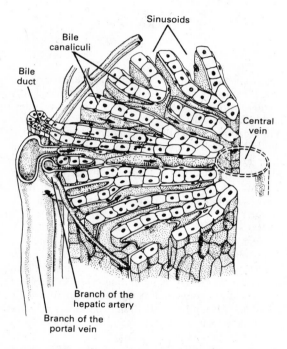

Figure 26–17. ***Top:*** Organization of the liver. CV, central vein. PS, portal space containing branches of bile duct, portal vein, and hepatic artery. ***Bottom:*** Arrangement of plates of liver cells, sinusoids, and bile ducts in a liver lobule, showing centripetal flow of blood in sinusoids to central vein and centrifugal flow of bile in bile canaliculi to bile ducts. (Reproduced, with permission, from Fawcett DW: *Bloom and Fawcett, A Textbook of Histology,* 11th ed. Saunders, 1986.)

Figure 26–18. Liver cell, showing relation of hepatocytes to bile canaliculi and sinusoids. Note the wide openings between the endothelial cells next to the hepatocyte. (Reproduced, with permission, from Fawcett DW: *Bloom and Fawcett, A Textbook of Histology,* 11th ed. Saunders, 1986.)

of function. Therefore, in this book they are discussed individually in the chapters on the systems of which each is a part. References to the relevant pages can be found in the index under the heading "liver functions."

Composition of Bile

Bile is made up of bile salts, bile pigments, and other substances dissolved in an alkaline electrolyte solution that resembles pancreatic juice (Table 26–8).

About 500 mL is secreted per day. Some of the components of the bile are reabsorbed in the intestine and then excreted again by the liver (**enterohepatic circulation**).

The glucuronides of the **bile pigments,** bilirubin and biliverdin, are responsible for the golden-yellow color of bile. The formation of these breakdown products of hemoglobin is discussed in detail in Chapter 27, and their excretion is discussed below.

The **bile salts** are sodium and potassium salts of **bile acids** conjugated to glycine or taurine, a derivative of cystine. The bile acids are synthesized from cholesterol. The 4 found in humans are listed in Fig 26–19. In common with vitamin D, cholesterol, a variety of steroid hormones, and the digitalis glycosides, the bile acids contain the cyclopentano-

Table 26–7. Principal functions of the liver.

Formation of bile
Carbohydrate storage and release
Formation of urea
Cholesterol metabolism
Manufacture of plasma proteins
Many functions related to metabolism of fat
Inactivation of some polypeptide hormones
Reduction and conjugation of adrenocortical and gonadal steroid hormones
Synthesis of 25-hydroxycholecalciferol
Detoxification of many drugs and toxins

Table 26–8. Composition of human hepatic duct bile.

Water	97.0%
Bile salts	0.7%
Bile pigments	0.2%
Cholesterol	0.06%
Inorganic salts	0.7%
Fatty acids	0.15%
Lecithin	0.1%
Fat	0.1%
Alkaline phosphatase	. . .

Figure 26–20. Cross section of disk-shaped bile acid-lipid mixed micelle with free fatty acids and cholesterol in its hydrophobic interior. The surface of each bile acid that faces outward is hydrophilic because of the polar peptide bond and the carboxyl and OH groups (shaded).

perhydrophenanthrene nucleus (see Chapter 20). The 2 principal (primary) bile acids formed in the liver are cholic acid and chenodeoxycholic acid. In the colon, bacteria convert cholic acid to deoxycholic acid and chenodeoxycholic acid to lithocholic acid. Since they are formed by bacterial action, deoxycholic acid and lithocholic acid are called secondary bile acids. Conjugation of the acids to glycine or taurine occurs in the liver, and the conjugates—eg, **glycocholic acid** and **taurocholic acid**—form sodium and potassium salts in the alkaline hepatic bile.

The bile salts have a number of important actions. They reduce surface tension and, in conjunction with phospholipids and monoglycerides, are responsible for the emulsification of fat preparatory to its digestion and absorption in the small intestine (see Chapter 25). In addition, they are **amphipathic;** ie, they have both hydrophilic and hydrophobic domains; one surface of the molecule is hydrophilic because the polar peptide bond and the carboxyl and hydroxyl groups are on that surface, whereas the other surface is hydrophobic. Therefore, the bile salts tend to form cylindrical disks called **micelles,** with the hydrophilic surfaces facing out and a hydrophobic core (Fig

26–20). Above a certain concentration called the **critical micelle concentration,** all bile salts added to a solution form micelles. Lipids collect in the micelles, with cholesterol in the hydrophobic center and amphipathic phospholipids and monoglycerides lined up with their hydrophilic heads on the outside and their hydrophobic tails in the center. The micelles play an important role in keeping lipids in solution and transporting them to the brush border of the intestinal epithelial cells, where they are absorbed (see Chapter 25).

From 90 to 95% of the bile salts are absorbed from the small intestine. Some are absorbed by nonionic diffusion, but most are absorbed from the terminal ileum by an extremely efficient active transport process (Fig 26–21). The remaining 5–10% enter the colon and are converted to the salts of deoxycholic acid and lithocholic acid. Lithocholate is relatively insoluble and is mostly excreted in the stools; only 1% is absorbed. However, deoxycholate is absorbed. The absorbed bile salts are transported back to the liver in the portal vein and reexcreted in the bile (enterohepatic circulation). Those lost in the stool are replaced by synthesis in the liver; the normal rate of bile salt synthesis is 0.2–0.4 g/d. The total bile salt pool of approximately 3.5 g recycles repeatedly via the enterohepatic circulation; it has been calculated that the entire pool recycles twice per meal and 6–8 times per day. When bile is excluded from the intestine, up to 50% of ingested fat appears in the feces. There is also severe malabsorption of fat-soluble vitamins. When bile salt reabsorption is prevented by resection of the terminal ileum or by disease in this portion of the small intestine, fat in the stools is also increased because when the enterohepatic circulation is interrupted, the liver cannot increase the rate of bile salt production to a sufficient degree to compensate for the loss. The other effects of resection of the terminal ileum are discussed in the section on malabsorption.

OH

CH$_3$

12

COOH

CH$_3$

3   7

HO   OH

Cholic acid

	Group at Position			Percent in
	3	7	12	Human Bile
Cholic acid	OH	OH	OH	50
Chenodeoxycholic acid	OH	OH	H	30
Deoxycholic acid	OH	H	OH	15
Lithocholic acid	OH	H	H	5

Figure 26–19. Human bile acids. The numbers in the formula for cholic acid refer to the positions in the steroid ring.

Figure 26–21. Enterohepatic circulation of bile salts. The solid lines entering the portal system represent bile salts of hepatic origin, whereas the dashed lines represent bile salts resulting from bacterial action. (Courtesy of M Tyor. Modified from Way LW [editor]: *Current Surgical Diagnosis & Treatment,* 9th ed. Appleton & Lange, 1991.)

Figure 26–22. Metabolism of bilirubin in the liver. P, intracellular binding proteins; UDPGA, uridine diphosphoglucuronic acid; UDP, uridine diphosphate.

Bilirubin Metabolism & Excretion

Most of the bilirubin in the body is formed in the tissues by the breakdown of hemoglobin (see Chapter 27). The bilirubin is bound to albumin in the circulation. Some of it is tightly bound, but most of it can dissociate in the liver, and free bilirubin enters liver cells, where it is bound to cytoplasmic proteins (Fig 26–22). It is next conjugated to glucuronic acid in a reaction catalyzed by the enzyme **glucuronyl transferase.** This enzyme is located primarily in the smooth endoplasmic reticulum. Each bilirubin molecule reacts with 2 uridine diphosphoglucuronic acid (UDPGA) molecules to form bilirubin diglucuronide. This glucuronide, which is more water-soluble than the free bilirubin, is then transported against a concentration gradient by a presumably active process into the bile canaliculi. A small amount of the bilirubin glucuronide escapes into the blood, where it is bound less tightly to albumin than free bilirubin, and is excreted in the urine. Thus, the total plasma bilirubin normally includes free bilirubin plus a small amount of conjugated bilirubin. Most of the bilirubin glucuronide passes via the bile ducts to the intestine.

The intestinal mucosa is relatively impermeable to conjugated bilirubin but is permeable to unconjugated bilirubin and to urobilinogens, a series of colorless derivatives of bilirubin formed by the action of bacteria in the intestine. Consequently, some of the bile pigments and urobilinogens are reabsorbed in the portal circulation. Some of the reabsorbed substances are again excreted by the liver (enterohepatic circulation), but small amounts of urobilinogens enter the general circulation and are excreted in the urine.

Jaundice

When free or conjugated bilirubin accumulates in the blood, the skin, the scleras, and mucous membranes turn yellow. This yellowness is known as **jaundice** (icterus) and is usually detectable when the total plasma bilirubin is greater than 2 mg/dL (34 μmol/L). Hyperbilirubinemia may be due to (1) excess production (hemolytic anemia, etc); (2) decreased uptake of bilirubin into hepatic cells; (3) disturbed intracellular protein binding or conjugation; (4) disturbed secretion of conjugated bilirubin into the bile canaliculi; or (5) intrahepatic or extrahepatic bile duct obstruction. When it is due to one of the first 3 processes, the free bilirubin rises. When it is due to disturbed secretion of conjugated bilirubin or bile duct obstruction, bilirubin glucuronide regurgitates into the blood, and it is predominantly the conjugated bilirubin in the plasma that is elevated.

Other Substances Conjugated by Glucuronyl Transferase

The glucuronyl transferase system in the smooth endoplasmic reticulum catalyzes the formation of the glucuronides of a variety of substances in addition to bilirubin. The list includes steroids (see Chapters 20 and 23) and various drugs. These other compounds can compete with bilirubin for the enzyme system when they are present in appreciable amounts. In addition, several barbiturates, antihistamines, anticonvulsants, and other compounds cause marked proliferation of the smooth endoplasmic reticulum in the hepatic cells, with a concurrent increase in hepatic glucuronyl transferase activity. Phenobarbital has been used successfully for the treatment of a congenital disease in which there is a relative deficiency of glucuronyl transferase (type 2 UDP glucuronyl transferase deficiency).

Other Substances Excreted in the Bile

Cholesterol and alkaline phosphatase are excreted in the bile. In patients with jaundice due to intra- or extrahepatic obstruction of the bile duct, the blood levels of these 2 substances usually rise; a much smaller rise is generally seen when the jaundice is due to nonobstructive hepatocellular disease. Adrenocortical and other steroid hormones and a number of drugs are excreted in the bile and subsequently reabsorbed (enterohepatic circulation). The dye sulfobromophthalein (Bromsulphalein, BSP) is removed from the bloodstream by the liver cells and excreted in the bile. Its rate of removal from the circulation has been used as a test of hepatic function. The rate depends not only on the functional capacity of the liver cells but also on the hepatic blood flow.

Functions of the Gallbladder

In normal individuals, bile flows into the gallbladder when the sphincter of Oddi is closed. In the gallbladder, the bile is concentrated by absorption of water. The degree of this concentration is shown by the increase in the concentration of solids (Table 26–9); liver bile is 97% water, whereas the average water content of gallbladder bile is 89%. When the bile duct and cystic duct are clamped, the intrabiliary pressure rises to about 320 mm of bile in 30 minutes, and bile secretion stops. However, when the bile duct is clamped and the cystic duct is left open, water is reabsorbed in the gallbladder, and the intrabiliary pressure rises only to about 100 mm of bile in several hours. Acidification of the bile is another function of the gallbladder (Table 26–9).

Regulation of Biliary Secretion

When food enters the mouth, the resistance of the sphincter of Oddi decreases. Fatty acids and amino acids in the duodenum release CCK, which causes gallbladder contraction. Acid and Ca^{2+} also stimulate the secretion of CCK. Substances that cause contraction of the gallbladder are called **cholagogues.**

Table 26–9. Comparison of human hepatic duct bile and gallbladder bile.

	Hepatic Duct Bile	Gallbladder Bile
Percentage of solids	2–4	10–12
Bile salts (mmol/L)	10–20	50–200
pH	7.8–8.6	7.0–7.4

The production of bile is increased by stimulation of the vagus nerves and by the hormone secretin, which increases the water and HCO_3^- content of bile. Substances that increase the secretion of bile are known as **choleretics.** Bile salts themselves are among the most important physiologic choleretics. The bile salts reabsorbed from the intestine actually inhibit the synthesis of new bile acids, but they themselves are promptly secreted, and they markedly increase bile flow.

Effects of Cholecystectomy

The periodic discharge of bile from the gallbladder aids digestion but is not essential for it. Cholecystectomized patients maintain good health and nutrition with a constant slow discharge of bile into the duodenum, although eventually the bile duct becomes somewhat dilated, and more bile tends to enter the duodenum after meals than at other times. Cholecystectomized patients can even tolerate fried foods, although they generally must avoid foods that are particularly high in fat content.

Cholecystography

Certain radiopaque iodine-containing dyes such as tetraiodophenolphthalein are excreted in the bile and concentrated with the bile in the gallbladder. When the gallbladder contains an adequate concentration of such a dye, it can be visualized by x-rays (**cholecystography**). To determine if the gallbladder is capable of normal contraction, a fatty meal can be fed. CCK is released and the gallbladder contracts, the size of its x-ray shadow normally decreasing to less than one-third of its original size within 30 minutes. When administered intravenously, technetium 99-labeled derivatives of iminidoacetic acid are excreted in the bile and provide excellent gamma camera images of the gallbladder and bile ducts (**radionuclide scan**). Ultrasonography of the right upper quadrant of the abdomen is also valuable in detecting gallstones.

Gallstones

Cholelithiasis, ie, the presence of gallstones, is a common condition. Its incidence increases with age, so that in the USA, for example, 20% of the women and 5% of the men between the ages of 50 and 65 have gallstones. The stones are of 2 types: calcium bilirubinate stones and cholesterol stones. In the USA and Europe, 85% of the stones are cholesterol stones.

Three factors appear to be involved in the formation of cholesterol stones. One is bile stasis; stones form in the bile that is sequestrated in the gallbladder rather than the bile that is flowing in the bile ducts. A second is supersaturation of the bile with cholesterol. Cholesterol is very insoluble in bile, and it is maintained in solution in micelles only at certain concentrations of bile salts and lecithin (Fig 26–23). At concentrations above line ABC in Fig 26–23, the bile is supersaturated and contains small crystals of choles-

Figure 26–23. Cholesterol solubility in bile as a function of the proportions of lecithin, bile salts, and cholesterol. In bile that has a composition described by any point below line ABC (eg, point P), cholesterol is solely in micellar solution; points above line ABC describe bile in which there are cholesterol crystals as well. (Reproduced, with permission, from Small DM: Gallstones. *N Engl J Med* 1968;**279**:588.)

terol in addition to micelles. However, many normal individuals who do not develop gallstones also have supersaturated bile. The third factor is a mix of nucleation factors that favors formation of stones from the supersaturated bile. Outside the body, bile from patients with cholelithiasis forms stones in 2–3 days, whereas it takes more than 2 weeks for stones to form in bile from normal individuals. The exact nature of the nucleation factors is unsettled, although glycoproteins in gallbladder mucus have been implicated. In addition, it is unsettled whether stones form as a result of excess production of components that favor nucleation or decreased production of antinucleation components that prevent stones from forming in normal individuals.

SMALL INTESTINE

In the small intestine, the intestinal contents are mixed with the secretions of the mucosal cells and with pancreatic juice and bile. Digestion, which begins in the mouth and stomach, is completed in the lumen and mucosal cells of the small intestine, and the products of digestion are absorbed, along with most of the vitamins and fluid. The small intestine is presented with about 9 L of fluid per day—2 L from dietary sources and 7 L of gastrointestinal secretions (see Table 25–3); however, only 1–2 L passes into the colon.

Anatomic Considerations

The general arrangement of the muscular layers, nerve plexuses, and mucosa in the small intestine is shown in Fig 26–1. The first portion of the duodenum is sometimes called the duodenal cap or bulb. It is the region struck by the acid gastric contents squirted through the pylorus and is a common site of peptic ulceration. At the ligament of Treitz, the duodenum becomes the jejunum. Arbitrarily, the upper 40% of the small intestine below the duodenum is called the jejunum and the lower 60% the ileum, although there is no sharp anatomic boundary between the two. The ileocecal valve marks the point where the ileum ends in the colon.

The small intestine is shorter during life than it is in cadavers; it relaxes and elongates after death. In living humans, the distance from the pylorus to the ileocecal valve has been reported to be 285 cm (Table 26–10), but at autopsy, this distance is about 700 cm.

The mucosa of the small intestine contains **solitary lymph nodules** and, especially in the ileum, **aggregated lymphatic nodules** (Peyer's patches) along the antimesenteric border. Throughout the small intestine there are simple tubular **intestinal glands** (crypts of Lieberkühn). In the duodenum there are in addition the small, coiled acinotubular **duodenal glands** (Brunner's glands). The **enterochromaffin cells** in the mucosa of the intestine secrete serotonin (see Chapter 4). These and other endocrine cells are often located deep in the intestinal glands. There are many valvelike folds (**valvulae conniventes**) in the mucous membrane.

Throughout the length of the small intestine, the mucous membrane is covered by **villi** (Fig 26–24). There are 20–40 villi per square millimeter of mucosa. Each intestinal villus is a fingerlike projection, 0.5–1 mm long, covered by a single layer of columnar epithelium and containing a network of capillaries and a lymphatic vessel (**lacteal**). Fine extensions of the smooth muscle of the submucosa run longitudinally up each villus to its tip. The free edges of the cells of the epithelium of the villi are divided into minute **microvilli** (Fig 26–25), which form a **brush border.** The cells are connected to each other by tight junctions. The outer layer of the cell membrane of the mucosal cells contains many of the enzymes involved

Table 26–10. Mean lengths of various segments of the gastrointestinal tract as measured by intubation in living humans.[1]

Segment	Length (cm)
Pharynx, esophagus, and stomach	65
Duodenum	25
Jejunum and ileum	260
Colon	110

[1]Data from Hirsch JE, Ahrens EH Jr, Blankenhorn DH: Measurement of human intestinal length in vivo and some causes of variation. *Gastroenterology* 1956;**31**:274.

Villi

Muscularis mucosae

Lamina propria

Glands
(crypts of
Lieberkühn)

Figure 26–24. Structure of the small intestine. The epithelium lining the glands continues over the villi. Mucosal cells are formed in the glands and migrate up to the tips of the villi, where they are shed into the intestinal lumen. (Modified from Junqueira LC, Carneiro J, Kelley RO: *Basic Histology,* 7th ed. Appleton & Lange, 1992.)

Microvilli
(brush border)

Tight junction

Desmosome

Microtubules

Lysosomes

Granular reticulum

Mitochondria

Smooth-surfaced reticulum

Unattached ribosomes

Golgi material

Intercellular space

Nucleus

Basal lamina

Lamina propria

Figure 26–25. Diagram of enterocytes from human small intestine. Note the microvilli, the tight connections of cells at the mucosal edge (desmosomes), and the space between the cells at base (intercellular space). (Reproduced, with permission, from Trier JS: Structure of the mucosa of the small intestine as it relates to intestinal function. *Fed Proc* 1967;**26:**1391.)

in the digestive processes initiated by salivary, gastric, and pancreatic enzymes. Enzymes found in this membrane include various disaccharidases, peptidases, and enzymes involved in the breakdown of nucleic acids (see Chapter 25). The brush border is lined on its luminal side by an amorphous layer called the glycocalyx, which is rich in neutral and amino sugars and may serve a protective function.

The absorptive surface of the small intestine is increased about 600-fold by the valvulae conniventes, villi, and microvilli. It has been estimated that the inner surface area of a mucosal cylinder the size of the small intestine would be about 3300 cm^2, that the valvulae increase the surface area to 10,000 cm^2, that the villi increase it to 100,000 cm^2, and that the microvilli increase it to 2 million cm^2.

The enterocytes in the small intestine are formed from mitotically active undifferentiated cells in the crypts of Lieberkuhn. They migrate up to the tips of the villi, where they are sloughed into the intestinal lumen in large numbers (Fig 26–24). The average life of these cells is 2–5 days, depending on the species. The number of cells shed per day has been calculated to be about 17 billion in humans, and the amount of protein "secreted" in this fashion is about 30 g/d. Mucosal cells are also rapidly sloughed and replaced in the stomach. The crypts are also the site of cyclic AMP-mediated secretion of water and electrolytes (see Chapter 25).

Intestinal Mucus

Mucus is secreted by surface epithelial cells throughout the gastrointestinal tract, by Brunner's glands in the duodenum, and by characteristic goblet cells in the mucosa of the small and large intestine. Secretion actually occurs in the form of mucins, which are complex, high-molecular-weight glycoproteins. The secreted mucins hydrate and gel, forming a blanket of mucus that covers and protects the intestinal epithelium. The mucus blanket also lubricates, binds some bacteria, and holds immunoglobulins in place so they can bind to pathogens. The goblet cells have a narrow base and an expanded apical portion facing the intestinal lumen. The expanded portion harbors numerous mucin-containing secretory granules. Mucin secretion is accelerated by cholinergic stimulation and by chemical and physical irritation. Somewhat different mucins are secreted by different goblet cells. The composition of mucins is altered in individuals with tumors of the small intestine and in patients with ulcerative colitis.

Intestinal Motility

The contractions of the small intestine are coordinated by the **small bowel slow wave,** a wave of smooth muscle depolarization that moves caudally from the circular smooth muscle of the duodenum. The frequency of the slow waves decreases from about 12/min in the jejunum to about 9/min in the ileum.

The movements of the small intestine mix and churn the intestinal contents (**chyme**) and propel it toward the large intestine. There are 2 types of movement, segmentation contractions and peristaltic waves. Both occur in the absence of extrinsic innervation but require an intact myenteric nerve plexus. The innervation of the intestine is discussed above. **Segmentation contractions** are ringlike contractions that appear at fairly regular intervals along the gut, then disappear and are replaced by another set of ring contractions in the segments between the previous contractions (Fig 26–26). They move the chyme to and fro and increase its exposure to the mucosal surface.

Peristaltic waves are discussed above. Very intense peristaltic waves called **peristaltic rushes** are not seen in normal individuals, but do occur when the intestine is obstructed. Weak antiperistalsis is sometimes seen in the colon, but most waves pass regularly in an oral-caudal direction.

Regulation of Intestinal Secretion

Brunner's glands in the duodenum secrete a thick alkaline mucus that probably helps protect the duodenal mucosa from the gastric acid. There is also an appreciable secretion of HCO_3^- that is independent of Brunner's glands. Decreased duodenal HCO_3^- secretion may play a role in the genesis of duodenal ulcers.

The intestinal glands secrete an isotonic fluid. Most of the enzymes usually found in this secretion are in desquamated mucosal cells; cell-free intestinal juice probably contains few if any enzymes. Gastrointestinal hormones such as VIP (see above) stimulate the secretion of intestinal juice. Vagal stimulation increases the secretion of Brunner's glands but probably has no effect on that of the intestinal glands.

The Malabsorption Syndrome

The digestive and absorptive functions of the small intestine are essential for life. Removal of short segments of the jejunum or ileum generally does not cause severe symptoms, and there is compensatory hypertrophy and hyperplasia of the remaining mucosa, with gradual return of the absorptive function toward normal (**intestinal adaptation**). This adaptation is partly due to a direct effect of nutrients in the intestinal lumen on the mucosa and partly due to circulating factors such as gastrointestinal hormones. However, when more than 50% of the small intestine is resected or bypassed, the absorption of nutrients and vitamins is so compromised that it is very difficult to prevent malnutrition and wasting. Various disease processes also impair absorption (Table 26–11). The pattern of deficiencies that results is sometimes called the **malabsorption syndrome.** This pattern varies somewhat with the cause, but it can include deficient absorption of amino acids, with marked body wasting and, eventually, hypoproteinemia and edema. Carbohydrate and fat absorption is also depressed. Because of the defective fat absorption, the fat-soluble vitamins (vitamins A, D, E, and K) are not absorbed in adequate amounts. The amount of fat and protein in the stools is increased, and the stools become bulky, pale, foul-smelling, and greasy (**steatorrhea**).

The increased gastric acid secretion produced by intestinal resection has been mentioned above. Resection of the ileum prevents the absorption of bile acids, and this leads in turn to deficient fat absorption. It also causes diarrhea because the unabsorbed bile salts enter the colon, where they increase adenylate cyclase activity, thus increasing intestinal secretion (see Chapter 25). They also increase the entry of small molecules into the lumen. Because the capacity of the jejunum to adapt is lower than that of the ileum, distal small bowel resection causes a greater degree of malabsorption than removal of a comparable length of proximal small bowel. Other complications of intestinal resection or bypass include hypocalcemia, arthritis, hyperuricemia, and possibly fatty infiltration of the liver, followed by cirrhosis. Operations in which segments of the small intestine are bypassed have been recommended for the treatment of obesity. In addition, partial ileal bypass has

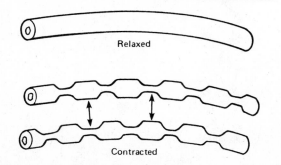

Figure 26–26. Diagram of segmentation contractions of the intestine. Arrows indicate how areas of relaxation become areas of constriction and vice versa.

Table 26–11. Disease processes associated with malabsorption.

Abnormalities of digestion in the intestinal lumen
  Inadequate lipolysis (eg, due to pancreatic insufficiency or excess secretion of gastric acid)
  Decreased conjugated bile salts (eg, due to ileal resection or bacterial overgrowth)
Abnormalities of mucosal cell transport
  Nonspecific, due to tropical sprue, celiac disease, etc
  Specific, due to deficiency of various disaccharidases, etc
Abnormalities of fat transport in intestinal lymphatics

been investigated for the prevention of atherosclerosis and its complications because it lowers plasma cholesterol and increases plasma LDL. However, bypass operations should not be undertaken lightly in view of their complications and dangers. Another surgical approach to obesity is reduction of the volume the stomach can hold, but procedures to accomplish this also have complications, including vomiting and unexplained neurologic disorders.

The defective intestinal function in tropical sprue (Table 26–11) is apparently due in part to folic acid deficiency. However, the intestinal changes in experimentally produced folic acid deficiency are less intense than those in tropical sprue, and patients with tropical sprue respond to treatment with antibiotics such as tetracycline. In celiac disease, the absorption defect is caused by congenital absence of the enzyme gluten hydrolase from the mucosal cells. This results in the formation of the toxic polypeptide gliadin from gluten, a protein in wheat, rye, barley, and oats. Gliadin disrupts the formation of microvilli. When these grains are omitted from the diet, bowel function is generally restored to normal.

Adynamic Ileus

When the intestines are traumatized, there is a decrease in intestinal motility caused by direct inhibition of smooth muscle. When the peritoneum is irritated, there is reflex inhibition due to increased discharge of noradrenergic fibers in the splanchnic nerves. Both types of inhibition operate to cause **paralytic (adynamic) ileus** after abdominal operations. Because of the diffuse decrease in peristaltic activity in the small intestine, its contents are not propelled into the colon, and it becomes irregularly distended by pockets of gas and fluid. Intestinal peristalsis returns in 6–8 hours, followed by gastric peristalsis, but colonic activity takes 2–3 days to return. Adynamic ileus can be relieved by passing a tube through the nose down to the small intestine and aspirating the fluid and gas for a few days until peristalsis returns.

Mechanical Obstruction of the Small Intestine

Localized mechanical obstruction of the small intestine causes severe cramping pain (**intestinal colic**), whereas adynamic ileus is often painless. The segment of intestine above the point of mechanical obstruction dilates and becomes filled with fluid and gas. The pressure in the segment rises, and the blood vessels in its wall are compressed, causing local ischemia. Activity in visceral afferent nerve fibers from the distended segment causes sweating, a drop in blood pressure, and severe vomiting, with resultant metabolic alkalosis and dehydration. If the obstruction is not relieved, the condition is fatal.

COLON

The main function of the colon is absorption of water, Na^+, and other minerals. By removal of about 90% of the fluid, it converts the 1000–2000 mL of isotonic chyme that enters it each day from the ileum to about 200–250 mL of semisolid feces.

Anatomic Considerations

The diameter of the colon is greater than that of the small intestine. Its length is about 100 cm in living adults and about 150 cm at autopsy. The fibers of its external muscular layer are collected into 3 longitudinal bands, the **teniae coli.** Because these bands are shorter than the rest of the colon, the wall of the colon forms outpouchings (**haustra**) between the teniae (Fig 26–27). There are no villi on the mucosa. The colonic glands are short inward projections of the mucosa that secrete mucus. Solitary lymph follicles are present, especially in the cecum and appendix.

Motility & Secretion of the Colon

The portion of the ileum containing the ileocecal valve projects slightly into the cecum, so that increases in colonic pressure squeeze it shut whereas increases in ileal pressure open it. Therefore, it effectively prevents reflux of colonic contents into the ileum. It is normally closed. Each time a peristaltic wave reaches it, it opens briefly, permitting some of the ileal chyme to squirt into the cecum. If the valve is resected in experimental animals, the chyme enters the colon so rapidly that absorption in the small intestine is reduced; however, a significant reduction does not occur in humans. When food leaves the stomach,

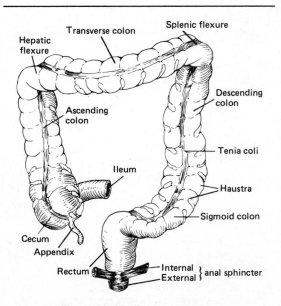

Figure 26–27. The human colon.

the cecum relaxes and the passage of chyme through the ileocecal valve increases (**gastroileal reflex**). This is presumably a vagal reflex, although there is some argument about whether vagal stimulation affects the ileocecal valve. Sympathetic stimulation increases the contraction of the valve.

The movements of the colon include segmentation contractions and peristaltic waves like those occurring in the small intestine. Segmentation contractions mix the contents of the colon and, by exposing more of the contents to the mucosa, facilitate absorption. Peristaltic waves propel the contents toward the rectum, although weak antiperistalsis is sometimes seen. A third type of contraction that occurs only in the colon is the **mass action contraction**, in which there is simultaneous contraction of the smooth muscle over large confluent areas. These contractions move material from one portion of the colon to another. They also move material into the rectum, and rectal distention initiates the defecation reflex (see below).

The movements of the colon are coordinated by a slow wave of the colon. The frequency of this wave, unlike the wave in the small intestine, increases along the colon, from about 2/min at the ileocecal valve to 6/min at the sigmoid.

Transit Time in the Small Intestine & Colon

The first part of a test meal reaches the cecum in about 4 hours, and all of the undigested portions have entered the colon in 8 or 9 hours. On the average, the first remnants of the meal reach the hepatic flexure in 6 hours, the splenic flexure in 9 hours, and the pelvic colon in 12 hours. From the pelvic colon to the anus, transport is much slower. As much as 25% of the residue of a test meal may still be in the rectum in 72 hours. When small colored beads are fed with a meal, an average of 70% of them are recovered in the stool in 72 hours, but total recovery requires more than a week. Transit time, pressure fluctuations, and changes in pH in the gastrointestinal tract can be observed by monitoring the progress of a small pill that contains sensors and a miniature radio transmitter.

Absorption in the Colon

The absorptive capacity of the mucosa of the large intestine is great. Na^+ is actively transported out of the colon, and water follows along the osmotic gradient thus generated. Normally, there is net secretion of K^+ and HCO_3^- into the colon (see Chapter 25). The absorptive capacity of the colon makes rectal instillation a practical route for drug administration, especially in children. Many compounds, including anesthetics, sedatives, tranquilizers, and steroids, are absorbed rapidly by this route. Some of the water in an enema is absorbed, and if the volume of an enema is large, absorption may be rapid enough to cause water intoxication. Coma and death due to water in-

toxication have been reported following tapwater enemas in children with megacolon.

Feces

The stools contain inorganic material, undigested plant fibers, bacteria, and water. Their composition (Table 26–12) is relatively unaffected by variations in diet because a large fraction of the fecal mass is of nondietary origin. This is why appreciable amounts of feces continue to be passed during prolonged starvation.

Intestinal Bacteria

The chyme in the jejunum normally contains few if any bacteria. There are more microorganisms in the ileum, but it is only the colon that regularly contains large numbers of bacteria. The reason for the relative sterility of the jejunal contents is unsettled, although gastric acid and the comparatively rapid transit of the chyme through this region may inhibit bacterial growth.

The microorganisms present in the colon include not only bacilli such as *Escherichia coli* and *Enterobacter aerogenes* but also pleomorphic organisms such as *Bacteroides fragilis,* cocci of various types, and organisms such as gas gangrene bacilli, which can cause serious disease in tissues outside the colon. Great masses of bacteria are passed in the stool. At birth, the colon is sterile, but the intestinal bacterial flora becomes established early in life.

The effects of the intestinal bacteria on their host are complex, some being definitely beneficial and others possibly harmful.

Antibiotics improve growth rates in a variety of species, including humans; and small amounts of antibiotics are frequently added to the diets of domestic animals. Animals raised under sanitary but not germ-free conditions grow faster than controls. They assimilate food better and do not require certain amino acids that are essential dietary constituents in other

Table 26–12. Approximate composition of feces on an average diet.

Component	Percentage of Total Weight
Water	75
Solids	25
	Percentage of Total Solids
Cellulose and other indigestible fiber	Variable
Bacteria	30
Inorganic material (mostly calcium and phosphates)	15
Fat and fat derivatives	5
Also desquamated mucosal cells, mucus, and small amounts of digestive enzymes	

animals. They also have larger litters and a lower neonatal death rate.

The reason for the improved growth is unsettled. Nutritionally important substances such as ascorbic acid, cyanocobalamin, and choline are utilized by some intestinal bacteria. On the other hand, some enteric microorganisms synthesize vitamin K and a number of B complex vitamins, and the folic acid produced by bacteria can be shown to be absorbed in significant amounts.

The brown color of the stools is due to pigments formed from the bile pigments by the intestinal bacteria. When bile fails to enter the intestine, the stools become white (**acholic stools**). Bacteria produce some of the gases in the flatus. Organic acids formed from carbohydrates by bacteria are responsible for the slightly acid reaction of the stools (pH 5.0–7.0). Intestinal bacteria also appear to play a role in cholesterol metabolism, since the poorly absorbed antibiotic neomycin that modifies the intestinal flora lowers LDL and the plasma cholesterol level.

A number of amines, including such potentially toxic substances as histamine and tyramine, are formed in the colon by bacterial enzymes that decarboxylate amino acids. At one time, the symptoms seen in constipated patients were blamed on absorption of these amines, but this theory of "autointoxication" has been discredited. Other amines formed by the intestinal bacteria, especially indole and skatole, contribute to the odor of the feces, as do sulfides.

Ammonia is also produced in the colon and absorbed. When the liver is diseased, ammonia is not removed from the blood, and hyperammonemia can cause neurologic symptoms (**hepatic encephalopathy**). Elevated blood ammonia is also one of the causes of hepatic encephalopathy that occurs in patients in whom the portal vein has been anastomosed to the vena cava in order to reduce portal pressure and stop bleeding from dilated esophageal varices. There is improvement in encephalopathy if the colon is removed or isolated by anastomosing the ileum to the sigmoid. However, the improvement is transient because colonic bacteria soon colonize the ileum, and the procedure is not used therapeutically.

Secretory Immunity in the Intestine

It is now well established that cells in the gut-associated lymphoid tissue are stimulated to proliferate and differentiate by intestinal antigens that reach them via the M cells. The lymphoid cells enter the circulation but return to the intestinal wall, where they differentiate further into plasma cells that secrete immunoglobulins, particularly IgA and IgM (see Chapter 27). These antibodies coat the mucosa and do much to prevent pathogenic organisms from penetrating the intestinal wall.

When normal animals with the usual intestinal bacterial flora are exposed to ionizing radiation, the body defenses that prevent intestinal bacteria from invading the rest of the body break down, and a major cause of death in **radiation poisoning** is overwhelming sepsis. Germ-free animals have extremely hypoplastic lymphoid tissue and poorly developed immune mechanisms, probably because these mechanisms have never been challenged. However, they are much more resistant to radiation than animals with the usual intestinal flora because they have no intestinal bacteria to cause sepsis.

Blind Loop Syndrome

Overgrowth of bacteria within the intestinal lumen can cause definite harmful effects. Such overgrowth occurs when there is stasis of the contents of the small intestine, and it causes macrocytic anemia, malabsorption of cyanocobalamin (vitamin B_{12}), and steatorrhea. Because the condition is prominent in patients with surgically created blind loops of small intestine, it has acquired the name **blind loop syndrome;** however, it can occur in any condition that promotes massive bacterial contamination of the small intestine. The cause of the anemia appears to be the cyanocobalamin deficiency, and the deficiency in cyanocobalamin absorption is probably due to uptake of ingested cyanocobalamin by the bacteria. The steatorrhea is probably due to excessive hydrolysis of conjugated bile salts by the bacteria. The important role of bile salts in fat digestion is discussed in Chapter 25.

Dietary Fiber

Adequate nutrition in herbivorous animals depends upon the action of gastrointestinal microorganisms that break down cellulose and related plant carbohydrates. In humans, there is no appreciable digestion of these vegetable products. Cellulose, hemicellulose, and lignin in the diet are important components of the **dietary fiber,** which by definition is all ingested food that reaches the large intestine in an essentially unchanged state. Various gums, algal polysaccharides, and pectic substances also contribute to dietary fiber.

If the amount of dietary fiber is low, the diet is said to lack **bulk.** Since the amount of material in the colon is small, the colon is inactive and bowel movements are infrequent. So-called bulk laxatives work by providing a larger volume of indigestible material to the colon. It has been claimed that some individuals with chronic constipation have a greater than normal capacity to break down cellulose and related products, thus reducing the residue in their colons.

There has been a recent upsurge of interest in dietary fiber because of epidemiologic evidence indicating that groups of people who live on a diet which contains large amounts of vegetable fiber have a low incidence of diverticulitis, cancer of the colon, diabetes mellitus, and coronary artery disease. However, the relationship between dietary fiber and the incidence of disease is still unsettled and needs further study.

Defecation

Distention of the rectum with feces initiates reflex contractions of its musculature and the desire to defecate. In humans, the sympathetic nerve supply to the internal (involuntary) anal sphincter is excitatory, whereas the parasympathetic supply is inhibitory. This sphincter relaxes when the rectum is distended. The nerve supply to the external anal sphincter, a skeletal muscle, comes from the pudendal nerve. The sphincter is maintained in a state of tonic contraction, and moderate distention of the rectum increases the force of its contraction (Fig 26–28). The urge to defecate first occurs when rectal pressure increases to about 18 mm Hg. When this pressure reaches 55 mm Hg, the external as well as the internal sphincter relaxes and the contents of the rectum are expelled. This is why there is reflex evacuation of the rectum in chronic spinal animals and humans. Before the pressure that relaxes the external anal sphincter is reached, voluntary defecation can be initiated by voluntarily relaxing the external sphincter and contracting the abdominal muscles (straining), thus aiding the reflex emptying of the distended rectum. Defecation is therefore a spinal reflex that can be voluntarily inhibited by keeping the external sphincter contracted or facilitated by relaxing the sphincter and contracting the abdominal muscles.

Distention of the stomach by food initiates contractions of the rectum and, frequently, a desire to defecate. The response is called the **gastrocolic reflex,** although there is some evidence that it is due to an action of gastrin on the colon and is not neurally mediated. Because of the response, defecation after meals is the rule in children. In adults, habit and cultural factors play a large role in determining when defecation occurs.

Effects of Colectomy

Humans can survive after total removal of the colon if fluid and electrolyte balance is maintained. When total colectomy is performed, the ileum is brought out through the abdominal wall (**ileostomy)** and the chyme expelled from the ileum is collected in a plastic bag fastened around the opening. If the diet is carefully regulated, the volume of ileal discharge decreases and its consistency increases over time. Care of an ileostomy used to be a time-consuming, difficult job, but when ileostomies are constructed by modern techniques, they can be relatively trouble-free and patients with them can lead essentially normal lives.

Figure 26–28. Responses to distention of the rectum by pressures less than 55 mm Hg. Distention produces passive tension due to stretching of the wall of the rectum, and additional active tension when the smooth muscle in the wall contracts. (Reproduced, with permission, from Davenport HW: *A Digest of Digestion,* 2nd ed. Year Book, 1978.)

Constipation

In bowel-conscious America, the amount of misinformation and undue apprehension about constipation probably exceeds that about any other health topic. Patients with persistent constipation, and particularly those with a recent change in bowel habits, should of course be examined carefully to rule out underlying organic disease. However, many normal humans defecate only once every 2–3 days, even though others defecate once a day and some as often as 3 times a day. Furthermore, the only symptoms caused by constipation are slight anorexia and mild abdominal discomfort and distention. These symptoms are not due to absorption of "toxic substances," because they are promptly relieved by evacuating the rectum and can be reproduced by distending the rectum with inert material. Other symptoms attributed by the lay public to constipation are due to anxiety or other causes.

Megacolon

The lack of harmful effects of infrequent bowel movements is emphasized by the relative absence of symptoms other than abdominal distention, anorexia, and lassitude in children with **aganglionic megacolon** (Hirschsprung's disease). This disease is due to congenital absence of the ganglion cells in both the myenteric and submucous plexuses of a segment of the distal colon, probably as a result of failure of the normal cranial-to-caudal migration of neuroblasts during development. There are no VIP-containing nerve cells or fibers, and the involved bowel is hyperresponsive to exogenous VIP. The substance P content is also low. The absence of peristalsis causes feces to pass the aganglionic region with difficulty, and children with the disease may defecate as infrequently as once every 3 weeks. The condition can be relieved if the aganglionic portion of the colon is resected and the portion of the colon above it anastomosed to the rectum.

Diarrhea

Diarrhea has many causes, some of which have been mentioned in particular contexts in preceding sections of this chapter. Severe diarrhea is debilitating and can be fatal, especially in infants. Large amounts of Na^+, K^+, and water are washed out of the colon and the small intestine in the diarrheal stools, causing dehydration, hypovolemia, and, eventually, shock and cardiovascular collapse. A more insidious complication of chronic diarrhea, if fluid balance is maintained, is severe hypokalemia.

REFERENCES

Section V: Gastrointestinal Function

Boyer JL, Graf J, Meier PJ: Hepatocyte transport systems regulating pH$_i$, cell volume, and bile secretion. *Annu Rev Physiol* 1992;**54:**415.

Brooks FP (editor): *Peptic Ulcer Disease.* Churchill Livingstone, 1985.

Castell DO, Johnson LF (editors): *Esophageal Function in Health and Disease.* Elsevier, 1983.

Cohen S, Soloway RD (editors): *Gallstones.* Churchill Livingstone, 1985.

Diamond JM, Karasov WH: Trophic control of the intestinal mucosa. *Nature* 1983;**304:**18.

Dobbins WO: Gut immunophysiology: A gastroenterologist's view with emphasis on pathophysiology. *Am J Physiol* 1982;**242:**G1.

Field M, Rao MC, Chang EB: Intestinal electrolyte transport and diarrheal disease. *N Engl J Med* 1989;**321:**800.

Finch CA, Huebers H: Perspectives in iron metabolism. *N Engl J Med* 1982;**306:**1520.

Fraser CL, Arieff AI: Hepatic encephalopathy. *N Engl J Med* 1985;**313:**865.

Furness JB, Costa M: *The Enteric Nervous System.* Churchill Livingstone, 1987.

Gallavan RH Jr, Parks DA, Jacobson ED: Pathophysiology of gastrointestinal circulation. *Handbook of Physiology,* Section 6, *The Gastrointestinal System,* vol I, part 1:1713. Schultz S (editor). Am Physiol Soc, 1989.

Granger DM, Barrowman J, Kvietys PR: *Clinical Gastrointestinal Physiology.* Saunders, 1985.

Hirsch DJ, Hayslett JP: Adaptation to potassium. *News Physiol Sci* 1986;**1:**54.

Howat HT, Searles H (editors): *The Exocrine Pancreas.* Saunders, 1979.

Johnson LR: Regulation of the mucosal gastrin receptor. *Proc Soc Exp Biol Med* 1983;**173:**167.

Johnson LR (editor): *Gastrointestinal Physiology,* 4th ed. Mosby-Year Book, 1991.

Klaassen CD, Watkins JB III: Mechanisms of bile formation, hepatic uptake and biliary excretion. *Pharmacol Rev* 1984;**361:**1.

Kretchmer N: Lactose and lactase. *Sci Am* (Oct) 1972; **227:**70.

Levitt MD, Bond JH: Volume, composition and source of intestinal gas. *Gastroenterology* 1970;**59:**921.

Mackowiak PA: The normal microbial flora. *N Engl J Med* 1982;**307:**83.

Mayer EA, Sun XP, Willenbucher RF: Contraction coupling in colonic smooth muscle. *Annu Rev Physiol* 1992;**54:**395.

McArthur KE, Jensen RT, Gardner JD: Treatment of acid-peptic diseases by inhibition of gastric H^+, K^+-ATPase. *Annu Rev Med* 1986;**37:**47.

Mendeloff AI: Dietary fiber and human health. *N Engl J Med* 1977;**297:**811.

Pollack JM: Endocrine cells of the gut. *Handbook of Physiology,* Section 6, *The Gastrointestinal System,* vol II:79. Schultz S (editor). Am Physiol Soc, 1989.

Rabon EC, Reuben MA: The mechanism and structure of the gastric H, K-ATPase. *Annu Rev Physiol* 1990; **52:**321.

Reichlin S: Somatostatin. *N Engl J Med* 1983;**309:**18.

Sanders KM: Colonic electrical activity: Concerto for two pacemakers. *News Physiol Sci* 1989;**4:**176.

Shepherd AP, Granger DN (editors): *Physiology of the Intestinal Circulation.* Raven, 1984.

Sleisenger MH, Brandborg LL: *Malabsorption.* Saunders, 1977.

Soll AH: Pathogenesis of peptic ulcer and implications for therapy. *N Engl J Med* 1990;**322:**909.

Specian RD, Oliver MG: Functional biology of intestinal goblet cells. *Am J Med* 1991;**260:**C183.

Thompson JC et al (editors): *Gastrointestinal Endocrinology.* McGraw-Hill, 1987.

Weisbrodt NW: Gastrointestinal motility. *Annu Rev Physiol* 1981;**43:**7.

Williams JA: Regulatory mechanisms in pancreas and salivary acini. *Annu Rev Physiol* 1984;**46:**361.

Williamson RCN: Intestinal adaptation. *N Engl J Med* 1978;**298:**1393.

Wolfe MM, Jensen RT: Zollinger-Ellison syndrome: Current concepts in diagnosis and management. *N Engl J Med* 1987;**317:**1200.

Young JA, van Lennep EW: *The Morphology of Salivary Glands.* Academic Press, 1978.

Zakim D, Boyer TD (editors): *Hepatology: A Textbook of Liver Disease.* Saunders, 1982.

Section VI.
Circulation

Circulating Body Fluids

27

INTRODUCTION

The **circulatory system** is the transport system that supplies O_2 and substances absorbed from the gastrointestinal tract to the tissues, returns CO_2 to the lungs and other products of metabolism to the kidneys, functions in the regulation of body temperature, and distributes hormones and other agents that regulate cell function. The blood, the carrier of these substances, is pumped through a closed system of blood vessels by the heart, which in mammals is really 2 pumps in series with each other. From the left ventricle, blood is pumped through the arteries and arterioles to the capillaries, where it equilibrates with the interstitial fluid. The capillaries drain through venules into the veins and back to the right atrium. This is the **major (systemic) circulation.** From the right atrium, blood flows to the right ventricle, which pumps it through the vessels of the lungs—the **lesser (pulmonary) circulation**—and the left atrium to the left ventricle. In the pulmonary capillaries, the blood equilibrates with the O_2 and CO_2 in the alveolar air. Some tissue fluids enter another system of closed vessels, the lymphatics, which drain lymph via the thoracic duct and the right lymphatic duct into the venous system (the **lymphatic circulation**). The circulation is controlled by multiple regulatory systems that function in general to maintain adequate capillary blood flow—when possible in all organs, but particularly in the heart and brain. This chapter is concerned with blood and lymph and with the multiple functions of the cells they contain.

BLOOD

The cellular elements of the blood—white blood cells, red blood cells, and platelets—are suspended in the plasma. The normal total circulating blood volume is about 8% of the body weight (5600 mL in a 70-kg man). About 55% of this volume is plasma.

BONE MARROW

In the adult, red blood cells, many white blood cells, and platelets are formed in the bone marrow. In the fetus, blood cells are also formed in the liver and spleen, and in adults such **extramedullary hematopoiesis** may occur in diseases in which the bone marrow becomes destroyed or fibrosed. In children, blood cells are actively produced in the marrow cavities of all the bones. By age 20, the marrow in the cavities of the long bones, except for the upper humerus and femur, has become inactive (Fig 27–1). Active cellular marrow is called **red marrow;** inactive marrow that is infiltrated with fat is called **yellow marrow.**

The bone marrow is actually one of the largest organs in the body, approaching the size and weight of the liver. It is also one of the most active. Normally, 75% of the cells in the marrow belong to the white blood cell-producing myeloid series, and only 25% are maturing red cells, even though there are over 500 times as many red cells in the circulation as there are white cells. This difference in the marrow probably reflects the fact that the average life span of white cells is short, whereas that of red cells is long.

Figure 27–1. Changes in red bone marrow cellularity with age. 100% equals the degree of cellularity at birth. (Reproduced, with permission, from Whitby LEH, Britton CJC: *Disorders of the Blood,* 10th ed. Churchill Livingstone, 1969.)

The bone marrow contains multipotent uncommitted stem cells that differentiate into committed stem cells, and the committed stem cells differentiate into the various differentiated cell types found in marrow and blood (Fig 27–2). The multipotent uncommitted stem cells are few in number but are capable of completely replacing the bone marrow when injected into a host whose own bone marrow has been completely destroyed. It appears likely that the bone marrow contains separate pools of committed stem cells for megakaryocytes, lymphocytes, erythrocytes, eosinophils, and basophils, whereas neutrophils and monocytes arise from a common precursor.

WHITE BLOOD CELLS

There are normally 4000–11,000 white blood cells per microliter of human blood (Table 27–1). Of these, the **granulocytes (polymorphonuclear leukocytes, PMNs)** are the most numerous. Young granulocytes have horseshoe-shaped nuclei that become multilobed as the cells grow older (Fig 27–2). Most of them contain neutrophilic granules (**neutrophils**), but a few contain granules that stain with acid dyes (**eosinophils**), and some have basophilic granules (**basophils**). The other 2 cell types found normally in peripheral blood are **lymphocytes**, which have large, round nuclei and scanty cytoplasm; and **monocytes**, which have abundant agranular cytoplasm and kidney-shaped nuclei (Fig 27–2). Acting together, these cells provide the body with powerful defenses against tumors and viral, bacterial, and parasitic infections.

Granulocytes

All granulocytes have cytoplasmic granules that contain biologically active substances involved in inflammatory and allergic reactions. The basophils, which resemble but are not identical to mast cells, contain histamine and heparin. They release histamine and other inflammatory mediators when activated by a histamine-releasing factor secreted by T lymphocytes (see below) and are essential for immediate-type hypersensitivity reactions. These range from mild urticaria and rhinitis to severe anaphylactic shock. The eosinophils attack parasites that are too large to be engulfed by phagocytosis, and they produce leukotriene C_4 and PAF (see below), both of which are involved in allergic diseases. They enter tissues and are especially abundant in the mucosa of the respiratory, lower urinary, and gastrointestinal tracts, where they are probably involved in mucosal immunity. The circulating eosinophil level is often elevated in patients with allergic diseases. The neutrophils seek out, ingest, and kill bacteria and have been called the body's first line of defense against bacterial infections.

The average half-life of neutrophils in the circulation is 6 hours. To maintain the normal circulating blood level, it is therefore necessary to produce over 100 billion neutrophils per day. Many of the neutrophils enter the tissues. They first adhere to the endothelium by binding to neutrophil adhesion molecules of the integrin family. Then they insinuate themselves through the walls of the capillaries between endothelial cells by a process called **diapedesis.** Many of those that leave the circulation enter the gastrointestinal tract and are lost from the body.

Invasion of the body by bacteria triggers the **inflammatory response.** The bone marrow is stimulated to produce and release large numbers of neutrophils. Bacterial products interact with plasma factors and cells to produce agents that attract neutrophils to the infected area (**chemotaxis**). The chemotactic agents include a component of the complement system (C5a), leukotrienes, and polypeptides from lymphocytes, mast cells, and basophils. Other plasma factors act on the bacteria to make them "tasty" to the phagocytes (**opsonization**). The principal opsonins that coat the bacteria are immunoglobulins of a particular class (IgG) and complement proteins (see below). The coated bacteria then bind to the neutrophil cell membrane, triggering ingestion of the bacteria by endocytosis (**phagocytosis**). In addition, the neutrophil granules move to the cell membrane and discharge their contents into the extracellular medium and the phagocytic vacuole (**degranulation**), while the cell membrane-bound enzyme **NADPH oxidase** is activated, with the production of toxic oxygen metabolites. The combination of the toxic oxygen metabolites and the proteolytic enzymes from the granules makes the neutrophil a very effective killing machine.

Activation of NADPH oxidase is associated with a sharp increase in O_2 uptake and metabolism in the neutrophil (the **respiratory burst**) and generation of O_2^- by the following reaction:

$$NADPH + H^+ + 2O_2 \rightarrow NADP^+ + 2H^+ + 2O_2^-$$

O_2^- is a **free radical** formed by the addition of one electron to O_2, and this is sometimes indicated by adding a dot to make the symbol $O_2^{\cdot-}$. Two O_2^- react to form H_2O_2.

$$2O_2^- + 2^+ \rightarrow H_2O_2 + O_2$$

O_2^- and H_2O_2 are both oxidants that are effective bactericidal agents. In addition, neutrophils discharge the enzyme **myeloperoxidase,** which catalyzes the conversion of Cl^-, Br^-, I^-, and SCN^- to the corresponding hypothalous acids (HOCl, HOBr, etc). These acids are also potent oxidants. Since Cl^- is present in greatest abundance in body fluids, the principal product is HOCl.

In addition to myeloperoxidase, neutrophil granules contain an elastase, 2 metalloproteinases that at-

Figure 27–2. Development of various formed elements of the blood from bone marrow cells. Cells below the horizontal line are found in normal peripheral blood. The principal sites of action of erythropoietin (erythro) and the various colony-stimulating factors (CSF) that stimulate the differentiation of the components are indicated. G, granulocyte; M, macrophage; IL, interleukin; see Tables 27–2 and 27–4.

Table 27-1. Normal values for the cellular elements in human blood.

Cell	Cells/μL (average)	Approximate Normal Range	Percentage of Total White Cells
Total WBC	9000	4000–11,000	...
Granulocytes Neutrophils	5400	3000–6000	50–70
Eosinophils	275	150–300	1–4
Basophils	35	0–100	0.4
Lymphocytes	2750	1500–4000	20–40
Monocytes	540	300–600	2–8
Erythrocytes Females	4.8×10^6	...	...
Males	5.4×10^6	...	...
Platelets	300,000	200,000–500,000	...

Figure 27–3. Tissue macrophages in the pituitary gland. The macrophages, which are identified by arrows, have been stained immunocytochemically with a monoclonal antibody that is specific for these cells. AL, anterior lobe; IL, intermediate lobe. (Courtesy of S Gordon.)

tack collagen, and a variety of other proteases that help destroy invading organisms. These enzymes act in a cooperative fashion with the O_2^-, H_2O_2, and HOCl formed by the action of the NADPH oxidase and myeloperoxidase to produce a killing zone around the activated neutrophil. This zone is effective in killing invading organisms, but in certain diseases, eg, rheumatoid arthritis, the neutrophils may also cause local destruction of host tissue.

The movements of the cell in phagocytosis, as well as migration to the site of infection, involve microtubules and microfilaments (see Chapter 1). Proper function of the microfilaments involves the interaction of the actin they contain with myosin-I on the inside of the cell membrane (see Chapter 1). Neutrophils also release thromboxanes that are vasoconstrictors and platelet-aggregating agents, leukotrienes that increase vascular permeability and attract other neutrophils to the site, and other prostaglandins that exert a moderate anti-inflammatory effect.

Monocytes

Monocytes enter the blood from the bone marrow and circulate for about 72 hours. They then enter the tissues and become **tissue macrophages** (Fig 27–3). Their lifespan in the tissues is unknown, but bone marrow transplantation data in humans suggest that they persist for about 3 months. It appears that they do not reenter the circulation. Some of them end up as the multinucleated giant cells seen in the chronic inflammatory diseases such as tuberculosis. The tissue macrophages include the Kupffer cells of the liver, pulmonary alveolar macrophages (see Chapter 34), osteoclasts, and microglia in the brain, and all come from the circulation. In the past, they have been called the **reticuloendothelial system,** but the general term **tissue macrophage system** seems more appropriate.

The macrophages become activated by lymphokines from T lymphocytes. The activated macrophages migrate in response to chemotactic stimuli and engulf and kill bacteria by processes generally similar to those occurring in neutrophils. They play a key role in immunity (see below). They also secrete up to 100 different substances, including factors that affect lymphocytes and other cells, prostaglandins of the E series, and clot-promoting factors.

Granulocyte & Macrophage Colony-Stimulating Factors

The production of red and white blood cells is regulated with great precision in healthy individuals, and the production of granulocytes is rapidly and dramatically increased in infections. The proliferation and maturation of the cells that enter the blood from the marrow are regulated by glycoprotein growth factors or hormones that cause cells in one or more of the committed cell lines to proliferate and mature (Fig 27–2, Table 27–2). The regulation of erythrocyte production by **erythropoietin** is discussed in Chapter 24. Three additional factors are called **colony-stimulating factors (CSFs),** because they cause appropriate single stem cells to proliferate in soft agar, forming colonies in this culture medium. The factors stimulating the production of committed stem cells include **granulocyte-macrophage CSF (GM-CSF), granulocyte CSF (G-CSF),** and **macrophage CSF (M-CSF).** Interleukins **IL-1** and **IL-5** followed by **IL-3** (Table 27–2) act in sequence to convert multipotent uncommitted stem cells to committed stem cells (Fig 27–2). IL-3 is also known as **multi-CSF.** Each of the CSFs has a predominant action, but all the CSFs and interleukins also have other overlapping actions. In addition, they activate and sustain mature blood cells. It is interesting in this regard that

Table 27–2. Factors regulating hematopoiesis.

Name	Cellular Sources	Cell Types Produced in Increased Numbers[1]
Erythropoietin	Kidney cells, Kupffer cells	rbc
G-CSF	Monocytes, fibro-blasts, endothelial cells	n
M-CSF	Monocytes, fibro-blasts, endothelial cells	m
GM-CSF	T cells, monocytes, endothelial cells, fi-broblasts	n, m, e, meg, rbc
IL-1	Macrophages, endo-thelial cells, fibro-blasts	n, m, e, b, meg, rbc
IL-3	T cells	n, m, e, b, meg, rbc
IL-4	T cells	b
IL-5	T cells	e
IL-6	Macrophages, endo-thelial cells, fibro-blasts	n, m, e, b, meg, rbc

[1]n, neutrophils; m, monocytes; e, eosinophils; b, basophils; meg, megakaryocytes; rbc, red blood cells.

the genes for many of these factors are located together on the long arm of chromosome 5 and may have originated by duplication of an ancestral gene.

As noted in Chapter 24, erythropoietin is produced in part by kidney cells and is a circulating hormone. The other factors are produced by macrophages, activated T cells, fibroblasts, and endothelial cells. For the most part, they act locally in the bone marrow. The cellular interactions that lead to their production and that of other lymphokines (see below) are summarized in Fig 27–4.

Disorders of Phagocytic Function

More than 15 primary defects in neutrophil function have been described, along with at least 30 other conditions in which there is a secondary depression of the function of the neutrophils. Patients with these diseases are prone to infections that are relatively mild when only the neutrophil system is involved but severe when the monocyte-tissue macrophage system is also involved. In one syndrome (neutrophil hypomotility), actin in the neutrophils does not polymerize normally, and the neutrophils move slowly. In another, there is a congenital deficiency of leukocyte integrins. In a more serious disease (chronic granulomatous disease), there is a failure to generate O_2^- in both the neutrophils and monocytes and consequent inability to kill many phagocytosed bacteria. In severe congenital glucose 6-phosphate dehydrogenase deficiency, there are multiple infections because of failure to generate the NADPH necessary for O_2^- production. In congenital myeloperoxidase deficiency,

Figure 27–4. Production of colony-stimulating growth factors. See Tables 27–2 and 27–4. (Reproduced, with permission, from Groopman JE, Molina JM, Scadden DT: Hematopoietic growth factors: Biology and clinical applications. *N Engl J Med* 1989;**321**:1449. Reprinted by permission of the *New England Journal of Medicine*.)

microbial killing power is reduced because hypohalite ions are not formed. However, it is not absent, because other bactericidal mechanisms are intact.

Lymphocytes

After birth, some lymphocytes are formed in the bone marrow, but most are formed in the lymph nodes, thymus, and spleen from precursor cells that originally came from the bone marrow. Lymphocytes enter the bloodstream for the most part via the lymphatics. It has been calculated that in humans, 3.5×10^{10} lymphocytes per day enter the circulation via the thoracic duct alone; however, this count includes cells that reenter the lymphatics and thus traverse the thoracic duct more than once. The effects of adrenocortical hormones on the lymphoid organs, the circulating lymphocytes, and the granulocytes are discussed in Chapter 20.

IMMUNE MECHANISMS

Lymphocytes are key constituents of the immune system. In mammals, this system has the remarkable ability to produce antibodies against many millions of different foreign agents that may invade the body.

In addition, the immune system "remembers," and a second exposure to a foreign substance produces a more rapid and greater response. There are 2 types of immune defense systems: humoral and cellular. Both react to **antigens.** The most potent antigens are complex proteins and polypeptides, but antibodies can be formed against complex carbohydrate moieties. Antibodies against nucleic acids and lipids can be formed if these are presented as nucleoproteins and lipoproteins, and antibodies to smaller molecules can be produced experimentally when the molecules are bound to protein. **Humoral immunity** is immunity due to circulating antibodies in the γ globulin fraction of the plasma proteins. It is a major defense against bacterial infections. **Cellular immunity** is responsible for delayed allergic reactions and rejection of transplants of foreign tissue. It constitutes a major defense against infections due to viruses, fungi, and a few bacteria such as the tubercle bacillus. It also helps defend against tumors. There have been spectacular advances in immunology in recent years, and the field is now large and complex. Only the fundamentals are presented in this book.

Development of the Immune System

During fetal development, lymphocyte precursors come from the bone marrow. Those that populate the thymus (Fig 27–5) become transformed by the environment in this organ into the lymphocytes responsible for cellular immunity (**T lymphocytes**). In birds, the precursors that populate the bursa of Fabricius, a lymphoid structure near the cloaca, become transformed into the lymphocytes responsible for humoral immunity (**B lymphocytes**). There is no bursa in mammals, and it appears that the transformation to B lymphocytes occurs in **bursal equivalents,** ie, the fetal liver and, after birth, the bone marrow. After residence in the thymus or liver, many of the T and B lymphocytes migrate to the lymph nodes and bone marrow. T and B lymphocytes are morphologically indistinguishable but can be identified by special techniques. B cells differentiate into **plasma cells** and **memory B cells.** Four different varieties of T cells have been identified: **helper/inducer T cells** (Fig 27–5), **suppressor T cells, cytotoxic T cells** (which are also known as **effector T cells** or **killer cells),** and **memory T cells.** The first 2 types are involved in the regulation of antibody production by B cell derivatives, whereas the cytotoxic T cells destroy transplanted and other foreign cells. Cytotoxic and suppressor T cells have on their surface the glycoprotein CD8, which can be detected by monoclonal antibodies, so they are frequently called T_8 cells. Helper/inducer T cells have on their surface the glycoprotein CD4 and are therefore called T_4 cells. CD8 is a coreceptor for MHC class I molecules (see below), and CD4 is a coreceptor for MHC class II molecules.

Memory B and T cells are cells that have been exposed to an antigen and are readily converted to effector cells by a later encounter with the same antigen. Unlike other lymphocytes, they persist in the body for months or even years.

Lymphocytes, macrophages, and other cells involved in immune responses communicate in part by hormonelike chemical messengers called **interleukins** and **cytokines.** These are discussed in detail below.

Most of the processing of the lymphocyte precursors in the thymus and bursal equivalents and the migration to lymphoid nodes and other tissues occurs during fetal and neonatal life. However, there is also a slow, continuous production of new lymphocytes from stem cells with processing in the bone marrow in adults.

Major Histocompatibility Complex

The genes of the **major histocompatibility complex (MHC),** which are located on the short arm of human chromosome 6, encode glycoproteins that are located on the surface of all cells and function in antibody processing and distinguishing self from nonself. They are divided into 2 classes on the basis of

Figure 27–5. Development of the immune system.

tissue distribution and function. Class I antigens are composed of a 45-kilodalton heavy chain associated noncovalently with β_2 microglobulin encoded by a gene outside the MHC (Fig 27–6). They are found on all nucleated cells and must be presented with antigen to activate T_8 cells. The known class I loci in humans are called HLA-A, -B, and -C. Class II antigens are heterodimers made up of a 29–34K α chain associated noncovalently with a 25–28K β chain. They are found on macrophages, B cells, and activated T cells. They must be presented with antigen to activate T_4 cells. Three families of loci code for class II antigens: HLA-DR, -DQ, and -DP.

Humoral Immunity

When viruses, bacteria, or other foreign proteins and related substances enter the body, they are ingested by macrophages and partially digested. Peptide fragments then combine with MHC and move to the cell surface, where they are exposed (Fig 27–7).

Figure 27–6. Structure of human histocompatibility antigen HLA-A2. The antigen-binding pocket is at the top and is formed by the α_1 and the α_2 parts of the molecule. The α_3 portion and the associated β_2 microglobulin are close to the membrane. The extension of the C terminal from α_3 that provides the transmembrane domain and the small cytoplasmic portion of the molecule has been omitted. (Reproduced, with permission, from Bjorkman PJ et al: Structure of the human histocompatibility antigen HLA-A2. *Nature* 1987;**329**:506.)

The macrophages then contact lymphocytes. T_4 cells are activated when their T cell receptors bind simultaneously to the antigen and a class II MHC protein on the surface of the macrophage (Fig 27–8). The T_4 cells then contact B cells, activating them and causing them to proliferate and transform into **memory B cells** and **plasma cells.** The plasma cells secrete large quantities of antibodies into the general circulation. The antibodies circulate in the globulin fraction of the plasma (see below) and, like antibodies elsewhere, are called **immunoglobulins.** The immunoglobulins are actually the secreted form of receptors on the B cell membrane, and B cells bind free antigens in blood and lymph. They can produce antibodies to complex carbohydrates without coming into contact with T cells. However, for efficient production of antibodies to proteins or polypeptides, B cells require contact with helper/inducer T_4 cells.

Antigens can also be processed and presented to T_4 cells by various types of cells in the body in addition to macrophages. The other types of **antigen-presenting cells** include the B cells themselves, the Langerhans cells of the skin, and specialized cells called dendritic cells in the lymph nodes and spleen.

The number of different antigens recognized by lymphocytes in the body is extremely large. It is now clear that the recognition ability is innate and develops without exposure to the antigen. Stem cells differentiate into many million different T and B lymphocytes, each with the ability to respond to a particular antigen. When the antigen first enters the body, it is processed by antigen-binding cells and binds to the appropriate lymphocytes by the process described above. These cells are stimulated to divide, forming **clones** of cells that respond to this antigen **(clonal selection).**

Immunoglobulins

Antibodies protect their host by binding to and neutralizing some protein toxins, by blocking the attachment of some viruses to cells, by opsonizing bacteria (see above), by activating complement, and by activating NK cells (see below). Five general types of immunoglobulin antibodies are produced by the lymphocyte-plasma cell system (Table 27–3). The basic component of each is a symmetric unit containing 4 polypeptide chains (Fig 27–9). The 2 long chains are called **heavy chains,** whereas the 2 short chains are called **light chains.** There are 2 types of light chains, κ and λ, and 5 types of heavy chains. The chains are joined by disulfide bridges that permit mobility, and there are intrachain disulfide bridges as well. In addition, the heavy chains are flexible in a region called the hinge. Each heavy chain has a variable (V) segment in which the amino acid sequence is highly variable, a diversity (D) segment in which the amino acid segment is also highly variable, a joining (J) segment in which it is moderately variable, and a constant (C) segment in which the sequence is constant. Each light

Figure 27–7. Activation of the immune system by viruses and other foreign substances. Macrophages and other antigen-presenting cells ingest the foreign material and express part of it (antigen) plus class I and II proteins of the major histocompatibility complex (MHC) on their surfaces. When antigen and MHC-II proteins are presented to T_4 lymphocytes, the T_4 lymphocytes in turn contact B lymphocytes that have been exposed to antigen and MHC-II proteins, and the B cells proliferate and secrete immunoglobulins. Antigen and MHC-I proteins activate T_8 lymphocytes to proliferate and form cytotoxic and suppressor T cells. Macrophages that have ingested foreign material secrete interleukin-1 (IL-1), which also stimulates T_4 cells and produces systemic effects. Activated T_4 cells produce interleukin-2 (IL-2), which causes proliferation of T_8 cells and B cells.

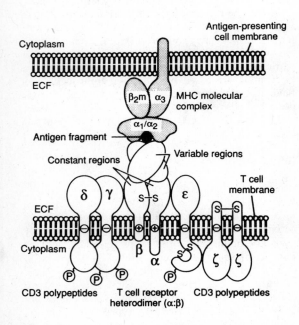

Figure 27–8. Interaction between antigen-presenting cell *(top)* and T lymphocyte *(bottom)*. The MHC proteins and their peptide antigen fragment bind to the α and β units that combine to form the T cell receptor. The associated CD3 polypeptides are part of the receptor complex and are probably involved in signal transduction. (Modified and reproduced, with permission, from Krensky AM et al: T-lymphocyte-antigen interactions in transplant rejection. *N Engl J Med* 1990;**322:**510. Reprinted by permission of the *New England Journal of Medicine.*)

chain has a V, a J, and a C segment. The V segments form part of the antibody-binding sites (Fab portion of the molecule; see Fig 27–9). The Fc portion of the molecule is the effector portion, which mediates the reactions initiated by antibodies.

Two of the classes of immunoglobulins contain additional polypeptide components (Table 27–3). In IgMs, 5 of the basic immunoglobulin units join around a polypeptide called the J chain to form a pentamer. In IgAs, the **secretory immunoglobulins,** the immunoglobulin units form dimers and trimers around a J chain and a polypeptide that comes from epithelial cells, the secretory component (SC).

In the intestine, bacterial and viral antigens are taken up by M cells (see Chapter 25) and passed on to underlying aggregates of lymphoid tissue, where they stimulate lymphoblasts. These lymphoblasts then enter the circulation via the lymphatic ducts, but after "maturing" in the circulation, they move to diffuse lymphoid tissue underlying the intestinal mucosa and epithelia in the lungs, breast, genitourinary tract, and female reproductive tract. There they secrete large amounts of IgAs when exposed again to the antigen that initially stimulated them. The epithelial cells produce the SC, which acts as a receptor for and binds the IgA. The resulting secretory immunoglobulin passes through the epithelial cell and is secreted by exocytosis. This system of **secretory immunity** is an important and effective defense mechanism.

Cellular Immunity

Cellular immunity is mediated by T_8 cells. These cells are activated when they are presented with antigens and MHC-I proteins on the surfaces of antigen-presenting cells. When also exposed to interleukin-2 (see below), they proliferate and differentiate into cytotoxic T cells (Figs 27–5 and 27–7). The cytotoxic

Table 27–3. Human immunoglobulins. In all instances, the light chains are κ or λ.

Immuno- globulin	Function	Heavy Chain	Additional Chain	Structure	Plasma Concentration (µg/mL)
IgG	Complement fixation.	$\gamma_1, \gamma_2,$ γ_3, γ_4		Monomer.	12,100
IgA	Localized protection in external se- cretions (tears, intestinal secre- tions, etc).	α_1, α_2	J, SC	Monomer; dimer with J or SC chain; trimer with J chain.	2600
IgM	Complement fixation.	μ	J	Pentamer with J chain.	930
IgD	Anitgen recognition by B cells.	δ		Monomer.	23
IgE	Reagin activity; releases hista- mine from basophils and mast cells.	ε		Monomer.	0.5

T cells attack and destroy cells that have the antigen which activated them. They kill by inserting pore-forming molecules (**perforins**) in the membranes of their target cells in the same fashion as the complement system does. They may also act by inducing within the cells an as yet undefined change that leads to death. Suppressor T cells, which develop more slowly than cytotoxic T cells, help terminate the immune response by dampening the immune responses of T and B cells. This includes turning off the helper/inducer cells.

T Cell Receptors

Receptors on T cells must recognize MHC anti-

Figure 27–9. Typical immunoglobulin G molecule. Fab, portion of the molecule that is concerned with antigen binding; Fc, effector portion of the molecule. The constant regions are shaded, and the variable regions are clear. The constant segment of the heavy chain is subdivided into C_H1, C_H2, and C_H3. The lines indicating the presence of intrasegmental disulfide bonds are omitted on the right side, and the J_H, D, V_H, J_L, and V_L segments are labeled on that side.

gens combined with fragments of a wide variety of different antigens. T cell receptors are made up of 2 polypeptide units designated α and β. The α and β units form heterodimers that recognize the MHC proteins and the antigen fragments with which they are combined (Fig 27–8). Associated with the α and β chains are other polypeptides that assemble with the α and β units intracellularly and are transported with them to the cell surface. These **CD3 polypeptides** are designated γ, δ, ε, and ζ, and their function is unsettled, but they probably play roles in signal transduction.

The α and β units contain variable as well as constant regions, and like the immunoglobulins, their structure is highly variable. Indeed, it has been calculated that the variable regions permit the generation of up to 10^{15} distinct T cell receptors.

Adhesion Receptors

In a real sense, the T cell receptor and the MHC-I and MHC-II proteins are adhesion receptors that bind cells together. They have an immunoglobulin-like structure and are part of a superfamily of **immunoglobulin adhesion receptors** that also include CD4, CD8, and several **cell adhesion molecules (CAMs)** that bind tissues together and guide developing nerve fibers in the nervous system. Another family of adhesion molecules that play a role in the immune and nervous systems is the **integrin family,** and a third is the **selectin family** of lectinlike receptors.

Genetic Basis of Diversity in the Immune System

The genetic mechanism for the production of the immensely large number of different configurations of immunoglobulins in the body store of lymphocytes is a fascinating biologic problem. Diversity is brought about in part by the fact that there are 2 kinds of light chains and 5 kinds of heavy chains. As noted above, there are areas of great variability (**hypervariable regions**) in each chain. The variable portion of the heavy chains consists of the V, D, and J segments.

In the gene family responsible for this region, there are several hundred different coding regions for the V segment, about 20 for the D segment, and 4 for the J segment. During B cell development, one V coding region, one D coding region, and one J coding region are selected at random and recombined to form the gene that produces that particular variable portion. There is similar variable recombination in the coding regions responsible for the 2 variable segments (V and J) in the light chain. In addition, the J segments are variable because the gene segments join in an imprecise and variable fashion (junctional site diversity) and nucleotides are sometimes added (junctional insertion diversity). Finally, somatic mutation adds to the variability. It has been calculated that these mechanisms permit the production of 10^8–10^{10} different immunoglobulin molecules.

Similar gene rearrangement and joining mechanisms operate to produce the diversity in T cell receptors. In humans, the α subunit has a V region encoded by one of about 50 different genes and a J region encoded by one of another 50 different genes. The β subunits have a V region encoded by one of about 50 genes, a D region encoded by one of 2 genes, and a J region encoded by one of 13 genes.

Monoclonal Antibodies

It is now possible to obtain large quantities of the immunoglobulin produced by a single plasma cell by fusing the cell with a tumor cell, producing an antibody "factory." In practice, animals are immunized with a particular antigen or cell preparation. They are then sacrificed, and the antibody-producing cells are extracted from their spleens and fused to myeloma cells. Myelomas are B lymphocyte tumors that readily fuse with plasma cells to form antibody-producing **hybridomas,** which grow and reproduce very well. The fused cells are separated by standard techniques, and each starts a line, a **clone,** of cells descended from a single cell. The hybridomas have the potential to produce the heavy and light chains of the spleen cell and the heavy and light chains of the myeloma cell, but hybridomas tend to lose chromosomes, and clones can be selected that form only the antibody of the spleen cell (monoclonal antibody). Alternatively, the spleen cells can be fused to mutant myeloma cells that do not produce an immunoglobulin. The value of this technology is very great not only in terms of producing large amounts of pure antibodies for research use but also in terms of using the antibodies to treat disease.

The Complement System

The effects of circulating antibodies and cellular immunity are mediated in part by a system of plasma enzymes called the **complement system.** The enzymes are identified by the numbers C1–C9. C1 is made up of 3 subunits, C1q, C1r, and C1s, so that there are actually 11 proteins in the system. C1 binds to immunoglobulins that have bound antigen, and this triggers a sequence of events that activates other components of the system; activation of complement in this fashion is called activation via the **classic pathway.** One consequence of activating the system is insertion of pore-forming molecules (perforins) in the membranes of antibody-sensitized cells. Ions move through these pores, and the cells become lysed. Another consequence of activation is formation of the fragments C3a and C5a from C3 and C5, respectively. These fragments release histamine from granulocytes, mast cells, and platelets. The histamine dilates blood vessels and increases capillary permeability. C5a and a complex formed by C5b, C6, and C7 are chemotactic and attract leukocytes to the site of the antigen-antibody reaction. C3b is the complement component responsible for opsonization of bacteria, the step that precedes phagocytosis by neutrophils. IgG can also serve as an opsonin.

Complement-mediated cell lysis can also occur in the absence of antibodies via the **alternative,** or **properdin, pathway.** The key to this pathway is a circulating protein, **factor I,** that recognizes repetitive sugar structures such as polyglucose or polyfructose in cell membranes. These sequences are found in bacteria and viruses but not in mammalian cells. Interaction of factor I with the surface of invading cells triggers reactions that activate C3 and C5. **Properdin** is another circulating protein that stabilizes the activating enzyme complex.

Recognition of Self

An intriguing question is why T and B cells do not form antibodies against and destroy the cells and organs of the individual in which they develop. It appears that many of the antiself T cells are eliminated in the thymus during their early development (**clonal abortion**). For the B cell, the major mechanism preventing destruction of self is **clonal anergy;** when B cells are exposed to high concentrations of antigen in the absence of lymphokines, they enter a prolonged hyporesponsive state. Clonal anergy also occurs in T cells. In addition, suppressor T cells may serve as an ancillary mechanism to keep the development of antiself antibodies in check. Finally, in a few tissues, eg, the lens of the eye, self antigens are more or less isolated from the cells of the immune system.

Natural Killer Cells

A third type of lymphocyte found in the body is the **natural killer cell (NK cell),** which is sometimes called the non-T, non-B lymphocyte. The natural killer cells are large lymphocytes that make up 10–15% of the circulating mononuclear cells. They kill cells without any apparent prior sensitization and without the involvement of major histocompatibility antigens. They destroy cells that have undergone malignant transformation and so help prevent the establishment of cancerous tumors. In addition, they attack viruses and have Fc receptors that permit them to kill antibody-coated viruses. Their activity is increased by IL-2. Current evidence indicates that they represent an important first line of defense against viral infections, combating the spread of disease while the more specific T and B cell responses are activated. In addition, they may represent a primitive immune system from which the T and B cell system evolved.

Lymphokines & Cytokines

Lymphocytes, macrophages, and in some instances endothelial cells, neurons, glial cells, and other types of cells secrete many hormonelike chemical messengers that affect the immune response. The messengers secreted by lymphocytes are often called lymphokines. However, since they are produced by other cells as well, it seems more appropriate to call them cytokines. This field is growing very rapidly, and most of the cytokines are initially named for other actions, eg, B cell-differentiating factor, B cell-stimulating factor 2. There is a convention that once the amino acid sequence of a factor in humans is known, its name is changed to **interleukin.** Thus, for example, the name of B cell-differentiating factor was changed to interleukin-4. However, nomenclature in this field remains somewhat confused and uncertain. In a field that is moving this fast, it is very difficult to construct a comprehensive list of factors, but many of the principal cytokines are listed in Table 27–4.

Interleukin-1 (IL-1) exists in 2 forms (α and β) with little sequence homology but identical activities.

Their molecular weights are about 17,000. In addition, there are 2 IL-1 receptors, and a naturally occurring IL-1 antagonist has been isolated and cloned. IL-1 effects are very widespread. They include activation of T_4 and B lymphocytes during activation of the immune system. In addition, IL-1 exerts effects on other blood cells, the brain, the liver, and the vascular wall (Table 27–5). Cachectin has overlapping, similar systemic effects. Thus, many of the effects are those that are seen in humans at the start of an infection.

Interleukin-2 (IL-2), a polypeptide with a molecular weight of approximately 15,000, is a growth factor produced by T_4 cells that increases the synthesis and maturation of T_8 and B cells (Fig 27–7).

The receptors for human IL-2, IL-4, IL-6, and IL-7 have been cloned, and along with the human GM-CSF receptor, the erythropoietin receptor, and growth hormone and prolactin receptors, they are part of a superfamily of cytokine receptors (see Chapter 1).

α and β interferons have antiviral effects, and α interferon fosters the appearance of MCH-I antigens. α and β interferons are quite different from γ interferon, which is a potent activator of macrophages and an inducer of MHC-II antigens. α interferon also increases the activity of NK cells. This action has led to the hope that α interferon may be of value in the treatment of cancer.

Tissue Transplantation

It is the T lymphocyte system that is responsible for the rejection of transplanted tissue. When tissues

Table 27–4. Principal cytokines.

Cytokine	Sources	Principal Actions
Interleukin-1α, Interleukin-1β (IL-1α, IL-1β)	Macrophages, T lymphocytes, keratinocytes, glial cells, etc	Multiple; see Table 27–5; α and β act on same receptor.
Interleukin-2 (IL-2)	T4 lymphocytes	Proliferation of T_8 and B lymphocytes.
Interleukin-3 (IL-3), G-CSF, M-CSF, GM-CSF	See Table 27–2	See Table 27–2
Interleukin-4 (IL-4; B-cell differentiating factor)	T_4 lymphocytes	Differentiation of B lymphocytes, basophils.
Interleukin-5 (IL-5)	T lymphocytes	B cell activation, eosinophil differentiation.
Interleukin-6 (IL-6; B-cell stimulatory factor 2, interferon-β2)	Fibroblasts, tumor cells, macrophages	Increased synthesis and secretion of immunoglobulins by B lymphocytes.
Interleukin-7 (IL-7)	Bone marrow cells	Proliferation of T and B lymphocytes
Tumor necrosis factor (TNF; cachectin)	Macrophages	Multiple (include fever, lysis of bone, hemorrhagic necrosis in tumors)
Tumor necrosis factor β (TNF-β; lymphotoxin)	T cells	Generally similar to those of cachectin; acts on same receptor as cachectin.
α-Interferon (α-IFN)	Leukocytes	Antiviral; antiproliferative; induces MHC-I antigens on lymphocytes; increases NK cell activity.
β-Interferon (β-IFN)	Fibroblasts	Antiviral; antiproliferative.
γ-Interferon (γ-IFN)	T lymphocytes, NK cells	Activates macrophages; induces MHC-II antigens on macrophages.
Platelet-derived growth factor (PDGF)	Platelets, macrophages, endothelial cells	Mitogen for vacular smooth muscle; fosters wound healing.
Platelet-activating factor (PAF)	Neutrophils, monocytes, platelets	Platelet aggregation; inflammatory processes, including activation of neutrophils and neutrophil chemotaxis; bronchoconstriction; increased capillary permeability.

Table 27-5. Systemic effects of recombinant IL-1.[1]

Central nervous system effects
Fever
Increased slow-wave sleep
Increased secretion of CRH
Anorexia
Metobolic effects
Increased synthesis of hepatic proteins
Increased sodium excretion
Decreased plasma zinc and iron
Decreased cytochrome P450
Lactic acidosis
Hematologic effects
Increased circulating neutrophils
Decreased circulating lymphocytes
Inhibition of lipoprotein lipase
Increased secretion of colony-stimulating factors
Increased nonspecific resistance
Vascular wall effects
Increased leukocyte adherence
Increased prostaglandin synthesis
Increased release of platelet-activating factor
Increased capillary permeability
Hypertension

[1]Modified from Dinarello CA: Biology of interleukin-1. *FASEB J* 1988;**2**:108.

such as skin and kidneys are transplanted from a donor to a recipient of the same species, the transplants "take" and function for a while but then become necrotic and are "rejected" because the recipient develops an immune response to the transplanted tissue. This is generally true even if the donor and recipient are close relatives, and the only transplants that are never rejected are those from an identical twin.

A number of treatments have been developed to overcome the rejection of transplanted organs in humans. The goal of treatment is to stop rejection without leaving the patient vulnerable to massive infections. One approach is to kill T lymphocytes by killing all rapidly dividing cells with drugs such as azathioprine, a purine antimetabolite, but this makes patients susceptible to infections and cancer. Another is to administer glucocorticoids, which inhibit cytotoxic T cell proliferation by inhibiting production of IL-2 by T_4 cells but cause osteoporosis, mental changes, and the other stigmas of Cushing's syndrome (see Chapter 20). A third is to suppress T cells with antibodies against these cells, using **antilymphocyte globulin** or monoclonal antibodies against T lymphocytes. A fourth approach is treatment with **cyclosporin A, FK-506,** or **rapamycin.** Cyclosporin A is a cyclic peptide extracted from fungi that, despite some problems with kidney damage and hypertension, has been mainly responsible for the recent major advances in organ transplantation. FK-506 is a newer compound that is a macrolide antibiotic of fungal origin, and rapamycin is similar to FK-506. Cyclosporin A and FK-506 prevent T cells from transcribing the IL-2 gene, whereas rapamycin does not

affect IL-2 production but prevents activation of the IL-2 receptor from producing T cell differentiation and proliferation. All 3 bind to immunosuppressant-binding proteins (**immunophilins**) in the cytoplasm of lymphocytes, but it is uncertain how this binding leads to the changes related to IL-2.

Other Clinical Correlates

As knowledge about immune mechanisms has increased, diseases due to defects at various stages in the development of the immune system have now been identified. Failure of the lymphocyte precursors to develop or to migrate to the thymus, liver, and spleen causes absence of cellular and humoral immunity, with marked susceptibility to infections. In individuals with congenital absence of the thymus (DiGeorge's syndrome), cellular immunity is absent but humoral immunity is present. Conversely, in Bruton's X-linked agammaglobulinemia, the B lymphocytes fail to develop. Patients with this condition have many bacterial infections but are relatively resistant to viral and fungal diseases.

Malignant transformation can also occur at various stages of development. Most if not all cases of chronic lymphocytic leukemia are due to uncontrolled proliferation of B lymphocytes, whereas multiple myeloma is due to malignant proliferation of clones of mature plasma cells. Some cases of acute lymphocytic leukemia are T lymphocyte cancers.

Acquired immune deficiency syndrome (AIDS), a disease that is currently a major problem, is unique in that **HIV (human immunodeficiency virus),** the retrovirus that causes many cases of it, binds to CD4 and produces a decrease in the number of circulating T_4 cells. The loss of helper lymphocytes leads in turn to failure of proliferation of T_8 and B cells, with eventual loss of immune function and death from infections due to normally nonpathogenic bacteria or cancer.

Normally, the immune system does an excellent job of distinguishing between self and nonself. However, antibodies are sometimes formed against various constituents of the host's body. **Autoimmune diseases** produced in this fashion are relatively common and can be chronic, debilitating, and even fatal. Examples include insulin-dependent diabetes (antibodies against pancreatic islet B cells), myasthenia gravis (antibodies against nicotinic cholinergic receptors), and probably multiple sclerosis (antibodies against myelin basic protein). In some instances, the antibodies are against receptors and are capable of activating receptors; for example, antibodies against TSH receptors increase thyroid activity and cause Graves' disease. Some of these conditions are associated with mutations in MHC class II proteins. Others may be due to the production of antibodies against invading organisms that cross-react with normal body constituents (**molecular mimicry**). However, much is still uncertain about the pathogenesis of autoimmune disease.

PLATELETS

The platelets are small, granulated bodies 2–4 μm in diameter (Fig 27–2). There are about 300,000/μL of circulating blood, and they normally have a half-life of about 4 days. The **megakaryocytes,** giant cells in the bone marrow, form platelets by pinching off bits of cytoplasm and extruding them into the circulation. Platelet production is regulated by the colony-stimulating factors that control the production of megakaryocytes (Table 27–2). Between 60 and 75% of the platelets that have been extruded from the bone marrow are in the circulating blood, but the remainder are mostly in the spleen. Splenectomy causes an increase in the platelet count (**thrombocytosis**).

Platelets have a ring of microtubules around their periphery and an extensively invaginated membrane with an intricate canalicular system in contact with the ECF. Their membranes contain receptors for collagen, vessel wall von Willebrand factor (see below), and fibrinogen. Their cytoplasm contains actin, myosin, glycogen, lysosomes, and 2 types of granules: (1) dense granules, which contain the nonprotein substances that are secreted in response to platelet activation, including serotonin and ADP and other adenine nucleotides, and (2) α-granules, which contain secreted proteins other than the hydrolases in lysosomes. These proteins include clotting factors and **platelet-derived growth factor (PDGF).** PDGF is also produced by macrophages and endothelial cells. It is a dimer made up of A and B subunit polypeptides. Homodimers (AA and BB), as well as the heterodimer (AB), are produced. PDGF stimulates wound healing and is a potent mitogen for vascular smooth muscle. Platelets as well as blood vessel walls contain von Willebrand factor, and in addition to its role in adhesion, this factor regulates circulating levels of factor VIII (see below).

When a blood vessel wall is injured, platelets adhere to the exposed collagen, laminin, and von Willebrand factor in the wall via integrins (see Chapter 1). This process of **platelet adhesion,** unlike aggregation, does not require platelet metabolic activity. However, binding to collagen initiates **platelet activation.** Activation can also be produced by ADP and thrombin. The activated platelets change shape, put out pseudopodia, discharge their granules, and stick to other platelets (**platelet aggregation**). Aggregation is also fostered by **platelet-activating factor (PAF),** a cytokine secreted by neutrophils and monocytes as well as platelets (Table 27–4). This compound is an ether phospholipid, 1-alkyl-2-acetylglyceryl-3-phosphorylcholine, which is produced from membrane lipids. Its receptor, which has now been cloned, is a typical serpentine receptor coupled to a G protein. It activates phospholipase C, and the diacylglycerol that is formed causes the platelets to release the contents of their granules. The increased cytoplasmic Ca^{2+} and the diacylglycerol activate phospholipase A_2, which causes release of arachidonic acid from membrane phospholipids, and the arachidonic acid is converted to thromboxane A_2 (see Chapter 17). Thromboxane A_2 facilitates further Ca^{2+} influx and phosphatidylinositol breakdown. Thrombin also acts by fostering Ca^{2+} influx and hydrolysis of phosphatidylinositol. Aspirin causes a moderate inhibition of thromboxane formation, and there is evidence that when taken in low doses, it is of value in preventing myocardial infarctions, unstable angina, transient ischemic attacks, and stroke by altering the balance between platelet thromboxane A_2 and vessel wall prostacyclin (see Chapter 31).

When the platelet count is low, clot retraction is deficient and there is poor constriction of ruptured vessels. The resulting clinical syndrome (**thrombocytopenic purpura**) is characterized by easy bruisability and multiple subcutaneous hemorrhages. Purpura may also occur when the platelet count is normal, and in some of these cases, the circulating platelets are abnormal (**thrombasthenic purpura**). Individuals with thrombocytosis are predisposed to thrombotic events.

RED BLOOD CELLS

The red blood cells (**erythrocytes**) carry hemoglobin in the circulation. They are biconcave disks (Fig 27–10) that are manufactured in the bone marrow. In mammals, they lose their nuclei before entering the circulation. In humans, they survive in the circulation for an average of 120 days. The average normal red blood cell count is 5.4 million/μL in men and 4.8 million/μL in women. Each human red blood cell is about 7.5 μm in diameter and 2 μm thick, and each contains approximately 29 pg of hemoglobin (Table 27–6). There are thus about 3×10^{13} red blood cells and about 900 g of hemoglobin in the circulating blood of an adult man (Fig 27–11).

Erythropoiesis

The formation of red blood cells (**erythropoiesis**) is subject to a feedback control. It is inhibited by a rise in the circulating red cell level to supernormal values and is stimulated by anemia. It is also stimulated by hypoxia, and an increase in the number of circulating red cells is a prominent feature of acclimatization to altitude (see Chapter 37). The control of erythropoiesis by the circulating glycoprotein hormone **erythropoietin** is discussed in Chapter 24. The role of IL-1, IL-3, IL-6, and GM-CSF in develop of the relevant erythroid stem cells is sh 27–2.

Figure 27–10. Human red blood cells and fibrin fibrils. Blood was placed on a polyvinyl chloride surface, then fixed and photographed with a scanning electron microscope. Reduced from × 2590. (Courtesy of NF Rodman.)

Figure 27–11. Red cell formation and destruction. RBC, red blood cells.

Red Cell Fragility

Red blood cells, like other cells, shrink in solutions with an osmotic pressure greater than that of normal plasma. In solutions with a lower osmotic pressure they swell, becoming spherical rather than disk-shaped, and eventually lose their hemoglobin (**hemolysis**). The hemoglobin of hemolyzed red cells dissolves in the plasma, coloring it red. A 0.9% sodium chloride solution is isotonic with plasma. When **osmotic fragility** is normal, red cells begin to hemolyze when suspended in 0.5% saline; 50% lysis occurs in 0.40–0.42% saline, and lysis is complete in

0.35% saline. In **hereditary spherocytosis** (congenital hemolytic icterus), the cells are spherocytic in normal plasma and hemolyze more readily than normal cells in hypotonic sodium chloride solutions (**abnormal red cell fragility**). Spherocytes are also removed by the spleen (see below). Consequently, hereditary spherocytosis is one of the most common causes of **hereditary hemolytic anemia.** The spherocytosis is caused by various abnormalities of the protein network that maintains the shape and flexibility of the red cell membrane. The membrane skeleton is attached to transmembrane proteins by the protein **ankyrin,** which is also found in association with the cytoskeleton of neurons and other cells. Another important protein in the red cell cytoskeleton is **spectrin,** and spectrin and ankyrin defects have both been reported as causes of spherocytosis.

Red cells can also be lysed by drugs and infec-

Table 27–6. Characteristics of human red cells.[1,2]

		Male	Female
Hematocrit (Hct) (%)		47	42
Red blood cells (RBC) ($10^6/\mu L$)		5.4	4.8
Hemoglobin (Hb) (g/dL)		16	14
Mean corpuscular volume (MCV) (fL)	$= \dfrac{Hct \times 10}{RBC\ (10^6/\mu L)}$	87	87
Mean corpuscular hemoglobin (MCH) (pg)	$= \dfrac{Hb \times 10}{RBC\ (10^6/\mu L)}$	29	29
Mean corpuscular hemoglobin concentration (MCHC) (g/dL)	$= \dfrac{Hb \times 100}{Hct}$	34	34
Mean cell diameter (MCD) (μm)	$= \dfrac{\text{Mean diameter of}}{500\ \text{cells in smear}}$	7.5	7.5

[1]Values are from Wintrobe M: *Clinical Hematology*, 6th ed. Lea & Febiger, 1967.
[2]Cells with MCVs > 95 are called macrocytes; cells with VCVs < 80 are called microcytes; cells with MCHs < 25 are called hypochromic.

tions. The susceptibility of red cells to hemolysis by these agents is increased by deficiency of the enzyme glucose 6-phosphate dehydrogenase (G6PD), which catalyzes the initial step in the oxidation of glucose via the hexose monophosphate pathway (see Chapter 17). This pathway generates NADPH, which is needed in some way for the maintenance of normal red cell fragility. Severe G6PD deficiency also inhibits the killing of bacteria by granulocytes and predisposes to severe infections (see above).

Role of the Spleen

The spleen is an important blood filter that removes spherocytes and other abnormal red cells. It also contains many platelets and plays a significant role in the immune system. The circulation of the spleen has 2 components: a fast component, mainly nutritive in function, in which blood stays within blood vessels; and a slow component in which blood leaves arterioles and percolates through large numbers of phagocytes and lymphocytes before entering the splenic sinuses and passing back into the general circulation. The phagocytes remove bacteria and initiate immune responses. Abnormal red cells are removed if they are not as flexible as normal red cells and consequently are unable to squeeze through the slits between the endothelial cells that line the splenic sinuses. In the absence of the spleen, bacterial infections are more common and more severe. In addition, malaria has a higher mortality rate because the deformed red cells that contain the malaria parasite are not removed.

Hemoglobin

The red, oxygen-carrying pigment in the red blood cells of vertebrates is **hemoglobin,** a protein with a molecular weight of 64,450. Hemoglobin is a globular molecule made up of 4 subunits (Fig 27–12). Each subunit contains a **heme** moiety conjugated to a polypeptide. Heme is an iron-containing porphyrin derivative (Fig 27–13). The polypeptides are referred to collectively as the **globin** portion of the hemoglobin molecule. There are 2 pairs of polypeptides in each hemoglobin molecule, 2 of the subunits containing one type of polypeptide and 2 containing another. In normal adult human hemoglobin (**hemoglobin A),** the 2 types of polypeptide are called the α chains, each of which contains 141 amino acid residues, and the β chains, each of which contains 146 amino acid residues. Thus, hemoglobin A is designated $\alpha_2\beta_2$. Not all the hemoglobin in the blood of normal adults is hemoglobin A. About 2.5% of the hemoglobin is hemoglobin A_2, in which β chains are replaced by δ chains ($\alpha_2\delta_2$). The δ chains also contain 146 amino acid residues, but 10 individual residues differ from those in the β chains.

There are small amounts of 3 hemoglobin A derivatives closely associated with hemoglobin A that probably represent glycosylated hemoglobins. One of

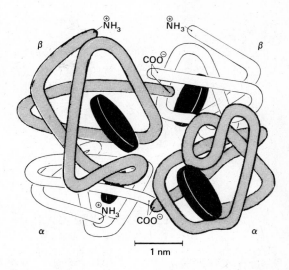

Figure 27–12. Diagrammatic representation of a molecule of hemoglobin A, showing the 4 subunits. There are 2 α and 2 β polypeptide chains, each containing a heme moiety. These moieties are represented by the disks. (Reproduced, with permission, from Harper HA et al: *Physiologische Chemie.* Springer-Verlag, 1975.)

these, **hemoglobin A_{IC} (HbA$_{IC}$),** has a glucose attached to the terminal valine in each β chain and is of special interest because the quantity in the blood increases in poorly controlled diabetes mellitus (see Chapter 19).

Reactions of Hemoglobin

Hemoglobin binds O_2 to form **oxyhemoglobin,** O_2 attaching to the Fe^{2+} in the heme. The affinity of hemoglobin for O_2 is affected by pH, temperature, and the concentration in the red cells of 2,3-diphosphoglycerate (2,3-DPG). 2,3-DPG and H^+ compete with O_2 for binding to deoxygenated hemoglobin, decreasing the affinity of hemoglobin for O_2 by shifting the positions of the 4 peptide chains (quaternary structure). The details of the oxygenation and deoxygenation of hemoglobin and the physiologic role of these reactions in O_2 transport are discussed in Chapter 35.

When blood is exposed to various drugs and other oxidizing agents in vitro or in vivo, the ferrous iron (Fe^{2+}) in the molecule is converted to ferric iron (Fe^{3+}), forming **methemoglobin.** Methemoglobin is dark-colored, and when it is present in large quantities in the circulation, it causes a dusky discoloration of the skin resembling cyanosis (see Chapter 37). Some oxidation of hemoglobin to methemoglobin occurs normally, but an enzyme system in the red cells, the NADH-methemoglobin reductase system, converts methemoglobin back to hemoglobin. Congenital absence of this system is one cause of hereditary methemoglobinemia.

Figure 27–13. Reaction of heme with O_2. The abbreviations M, V, and P stand for the groups shown on the molecule on the left.

Carbon monoxide reacts with hemoglobin to form **carbonmonoxyhemoglobin (carboxyhemoglobin).** The affinity of hemoglobin for O_2 is much lower than its affinity for carbon monoxide, which consequently displaces O_2 on hemoglobin, reducing the oxygen-carrying capacity of blood (see Chapter 37).

Heme is also part of the structure of **myoglobin,** an oxygen-binding pigment found in red (slow) muscles (see Chapter 35) and the respiratory chain enzyme **cytochrome c** (see Chapter 17). Porphyrins other than that found in heme play a role in the pathogenesis of a number of metabolic diseases (congenital and acquired porphyria, etc).

Hemoglobin in the Fetus

The blood of the human fetus normally contains **fetal hemoglobin (hemoglobin F).** Its structure is similar to that of hemoglobin A except that the β chains are replaced by γ chains; ie, hemoglobin F is $\alpha_2\gamma_2$. The γ chains also contain 146 amino acid residues but have 37 that differ from those in the β chain. Fetal hemoglobin is normally replaced by adult hemoglobin soon after birth (Fig 27–14). In certain individuals, it fails to disappear and persists throughout life. In the body, its O_2 content at a given P_{O_2} is greater than that of adult hemoglobin because it binds 2,3-DPG less avidly. This facilitates movement of O_2 from the maternal to the fetal circulation (see Chapter 32). In young embryos there are, in addition, ζ and ε chains, forming Gower 1 hemoglobin ($\zeta_2\varepsilon_2$) and Gower 2 hemoglobin ($\alpha_2\varepsilon_2$). Switching from one form of hemoglobin to another during development seems to be regulated by a locus control region (LCR) in the DNA upstream from the human gene for ε hemoglobin, but the details of the switching remain to be determined.

Abnormalities of Hemoglobin Production

The amino acid sequences in the polypeptide chains of hemoglobin are determined by globin genes.

There are 2 major types of inherited disorders of hemoglobin in humans: the **hemoglobinopathies,** in which abnormal polypeptide chains are produced, and the **thalassemias** and related disorders, in which the chains are normal in structure but produced in decreased amounts or absent because of defects in the globin genes. The α and β thalassemias are defined by decreased or absent α and β polypeptides, respectively.

Mutant genes that cause the production of abnormal hemoglobins are widespread, and many abnormal hemoglobins have been described in humans. They are usually identified by letter—hemoglobin C,

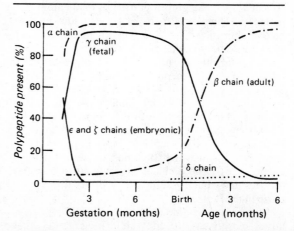

Figure 27–14. Development of human hemoglobin chains.

E, I, J, S, etc. In most instances, the abnormal hemoglobins differ from normal hemoglobin A in the structure of the polypeptide chains. For example, in hemoglobin S, the α chains are normal but the β chains are abnormal, because among the 146 amino acid residues in each β polypeptide chain, one glutamic acid residue has been replaced by a valine residue (Table 27–7).

When an abnormal gene inherited from one parent dictates formation of an abnormal hemoglobin—ie, when the individual is heterozygous—half the circulating hemoglobin is abnormal and half is normal. When identical abnormal genes are inherited from both parents, the individual is homozygous and all of the hemoglobin is abnormal. It is theoretically possible to inherit 2 different abnormal hemoglobins, one from the father and one from the mother. Studies of the inheritance and geographic distribution of abnormal hemoglobins have made it possible in some cases to decide where the mutant gene originated and approximately how long ago the mutation occurred. In general, harmful mutations tend to die out, but mutant genes that confer traits with survival value persist and spread in the population.

Many of the abnormal hemoglobins are harmless. However, some have abnormal O_2 equilibria. Others cause anemia. For example, hemoglobin S is very insoluble at low O_2 tensions, and this causes the red cells to become sickle-shaped. These abnormal sickle cells hemolyze, producing the severe anemia known as **sickle cell anemia.** Heterozygous individuals have the **sickle cell trait** and rarely have severe symptoms, but homozygous individuals develop the full-blown disease. The sickle cell gene is an example of a gene that has persisted and spread in the population. It originated in the black population in Africa, and it confers resistance to one type of malaria. This is an important benefit in Africa, and in some parts of Africa 40% of the population have the sickle cell trait. In the American black population its incidence is about 10%.

Figure 27–15. Bilirubin. The abbreviations M, V, and P stand for the groups shown on the molecule on the left in Fig 27–13.

Synthesis of Hemoglobin

The average normal hemoglobin content of blood is 16 g/dL in men and 14 g/dL in women, all of it in red cells. In the body of a 70-kg man, there is about 900 g of hemoglobin, and 0.3 g of hemoglobin is destroyed and 0.3 g synthesized every hour (Fig 27–11). The heme portion of the hemoglobin molecule is synthesized from glycine and succinyl-CoA.

Catabolism of Hemoglobin

When old red blood cells are destroyed in the tissue macrophage system, the globin portion of the hemoglobin molecule is split off, and the heme is converted to **biliverdin.** In humans, most of the biliverdin is converted to **bilirubin** (Fig 27–15) and excreted in the bile (see Chapter 26). The iron from the heme is reused for hemoglobin synthesis.

Exposure of the skin to white light converts bilirubin to Lumirubin, which has a shorter half-life than bilirubin. **Phototherapy** (exposure to light) is of value in treating infants with jaundice due to hemolysis. Iron is essential for hemoglobin synthesis; if blood is lost from the body and the iron deficiency is

Table 27-7. Partial amino acid composition of normal human β chain, and some hemoglobins with abnormal β chains.[1]

Hemoglobin	Positions on β Polypeptide Chain of Hemoglobin									
	1	2	3	6	7	26	63	67	121	146
A (normal)	Val	His	Leu	Glu	Glu	Glu	His	Val	Glu	His
S (sickle cell)				Val						
C				Lys						
G_San Jose					Gly					
E						Lys				
M_Saskatoon							Tyr			
M_Milwaukee								Glu		
O_Arabia									Lys	

[1]Other hemoglobins have abnormal α chains. Abnormal hemoglobins that are very similar electrophoretically but differ slightly in composition are indicated by the same letter and a subscript indicating the geographic location where they were first discovered; hence, M_Saskatoon and M_Milwaukee.

not corrected, **iron deficiency anemia** results. The metabolism of iron is discussed in Chapter 25.

BLOOD TYPES

The membranes of human red cells contain a variety of **blood group antigens,** which are also called **agglutinogens.** The most important and best known of these are the A and B antigens, but there are many more.

The ABO System

The A and B antigens are inherited as mendelian dominants, and individuals are divided into 4 major **blood types** on this basis. Type A individuals have the A antigen, type B have the B_1, type AB have both, and type O have neither. These antigens are found in many tissues in addition to blood: these include salivary glands, saliva, pancreas, kidney, liver, lungs, testes, semen, and amniotic fluid.

The A and B antigens are actually complex oligosaccharides that differ in their terminal sugar. On red cells they are mostly glycosphingolipids, whereas in other tissues they are glycoproteins. An H gene codes for a fucose transferase that puts a fucose on the end of these glycolipids or glycoproteins, forming the H antigen that is usually present in individuals of all blood types (Fig 27–16). Individuals who are type A have a gene which codes for a transferase that catalyzes placement of a terminal N-acetylgalactosamine on the H antigen, whereas individuals who are type B have a gene which codes for a transferase that places a terminal galactose. Individuals who are type AB have both transferases. Individuals who are type O have neither, so the H antigen persists. It now appears

that O individuals have a single-base deletion in their corresponding gene. This creates and open reading frame, and consequently they produce a protein that has no transferase activity.

Subgroups of blood types A and B have been described, the most important being A_1 and A_2. However, the difference between A_1 and A_2 appears to be quantitative; each A_1 cell has about 1,000,000 copies of the A antigen on its surface, and each A_2 cell has about 250,000.

Antibodies against red cell agglutinogens are called **agglutinins.** It appears that antigens very similar to A and B are common in intestinal bacteria and possibly foods to which newborn individuals are exposed, and the infants rapidly develop antibodies against those not present in their own cells. Thus, type A individuals develop anti-B antibodies, type B individuals develop anti-A antibodies, type O individuals develop both, and type AB individuals develop neither (Table 27–8). When the plasma of a type A individual is mixed with type B red cells, the anti-A antibodies cause the type B red cells to clump (agglutinate), as shown in Figure 27–17. The other agglutination reactions produced by mismatched plasma and red cells are summarized in Table 27–8. **Blood typing** is performed by mixing an individual's red blood cells with antisera containing the various agglutinins on a slide and seeing whether agglutination occurs.

Transfusion Reactions

Dangerous **hemolytic transfusion reactions** occur when blood is transfused into an individual with an incompatible blood type, ie, an individual who has agglutinins against the red cells in the transfusion. The plasma in the transfusion is usually so diluted in the recipient that it rarely causes agglutination even when the titer of agglutinins against the recipient's cells is high. However, when the recipient's plasma has agglutinins against the donor's red cells, the cells agglutinate and hemolyze. Free hemoglobin is liberated into the plasma. The severity of the resulting transfusion reaction may vary from an asymptomatic minor rise in the plasma bilirubin level to severe jaundice and renal tubular damage (caused in some way by the products liberated from hemolyzed cells), with anuria and death.

Incompatibilities in the ABO blood group system

F = fucose
G = N-acetylgalactosamine
glucose = glucose
galactose = galactose
C = ceramide
lipid bilayer = lipid bilayer

Figure 27–16. Antigens of the ABO system on the surface of red blood cells. (Modified from Donegan E, Bossom EL: Blood banking and immunohematology. In: *Basic and Clinical Immunology,* 7th ed. Stites DP, Terr AL (editors), p. 284. Appleton & Lange, 1991.)

Table 27–8. Summary of ABO system.

Blood Type	Agglutinins in Plasma	Frequency in USA (%)	Plasma Agglutinates Red Cells of Type:
O	Anti-A, anti-B	45	A, B, AB
A	Anti-B	41	B, AB
B	Anti-A	10	A, AB
AB	None	4	None

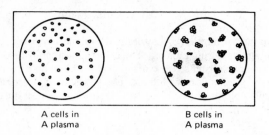

Figure 27–17. Red cell agglutination in incompatible plasma.

are summarized in Table 27–8. Persons with type AB blood are "universal recipients" because they have no circulating agglutinins and can be given blood of any type without developing a transfusion reaction due to ABO incompatibility. Type O individuals are "universal donors" because they lack A and B antigens, and type O blood can be given to anyone without producing a transfusion reaction due to ABO incompatibility. This does not mean, however, that blood should ever be transfused without being cross-matched except in the most extreme emergencies, since the possibility of reactions or sensitization due to incompatibilities in systems other than ABO systems always exists. In cross-matching, donor red cells are mixed with recipient plasma on a slide and checked for agglutination. It is advisable to check the action of the donor's plasma on the recipient cells even though, as noted above, this is rarely a source of trouble.

A procedure that has recently become popular is to withdraw the patient's own blood in advance of elective surgery and then infuse this blood back (**autologous transfusion**) if a transfusion is needed during the surgery. With iron treatment, 1000–1500 mL can be withdrawn over a 3-week period. The popularity of banking one's own blood is due primarily to fear of transmission of AIDS by heterologous transfusions, but of course another advantage is elimination of the risk of transfusion reactions.

Inheritance of A & B Antigens

The A and B antigens are inherited as mendelian allelomorphs, A and B being dominants. For example, an individual with type B blood may have inherited a B antigen from each parent or a B antigen from one parent and an O from the other; thus, an individual whose **phenotype** is B may have the **genotype** BB (**homozygous**) or BO (**heterozygous**).

When the blood types of the parents are known, the possible genotypes of their children can be stated. When both parents are type B, they could have children with genotype BB (B antigen from both parents), BO (B antigen from one parent, O from the other heterozygous parent), or OO (O antigen from

both parents, both being heterozygous). When the blood types of a mother and her child are known, typing can prove that a man could not be the father, although it cannot prove that he is the father. The predictive value is increased if the blood typing of the parties concerned includes identification of antigens other than the ABO agglutinogens. With the addition of HLA typing and other molecular biologic techniques, the exclusion rate rises to close to 100%.

Other Agglutinogens

In addition to the ABO system of antigens in human red cells, there are systems such as the Rh, MNSs, Lutheran, Kell, Kidd, and many others. There are over 500 billion possible known blood group phenotypes, and because undiscovered antigens undoubtedly exist, it has been calculated that the number of phenotypes is actually in the trillions.

The number of blood groups in animals is as large as it is in humans. An interesting question is why this degree of polymorphism developed and has persisted through evolution. Certain diseases are more common in individuals with one blood type or another, but the differences are not great. It seems likely that the actual function of the blood group antigens is cell recognition, but the exact significance of a recognition code of this complexity is unknown.

The Rh Group

Aside from the antigens of the ABO system, those of the Rh system are of the greatest clinical importance. The "Rh factor," named for the rhesus monkey because it was first studied using the blood of this animal, is a system composed primarily of the C, D, and E antigens, although it actually contains many more. Unlike the ABO antigens, the system has not been detected in tissues other than red cells. D is by far the most antigenic component, and the term "Rh-positive" as it is generally used means that the individual has agglutinogen D. The D protein is not glycosylated, and its function is unknown. The "Rh-negative" individual has no D antigen and forms the anti-D agglutinin when injected with D-positive cells. The Rh typing serum used in routine blood typing is anti-D serum. Eighty-five percent of Caucasians are D-positive and 15% are D-negative; over 99% of Asians are D-positive. Unlike the antibodies of the ABO system, anti-D antibodies do not develop without exposure of a D-negative individual to D-positive red cells by transfusion or entrance of fetal blood into the maternal circulation. However, D-negative individuals who have received a transfusion of D-positive blood (even years previously) can have appreciable anti-D titers and thus may develop transfusion reactions when transfused again with D-positive blood.

Hemolytic Disease of the Newborn

Another complication due to "Rh incompatibility"

arises when an Rh-negative mother carries an Rh-positive fetus. Small amounts of fetal blood leak into the maternal circulation at the time of delivery, and some mothers develop significant titers of anti-Rh agglutinins during the postpartum period. During the next pregnancy, the mother's agglutinins cross the placenta to the fetus. In addition, there are some cases of fetal-maternal hemorrhage during pregnancy, and sensitization can occur during pregnancy. In any case, when anti-Rh agglutinins cross the placenta to an Rh-positive fetus, they can cause hemolysis and various forms of **hemolytic disease of the newborn (erythroblastosis fetalis).** If hemolysis in the fetus is severe, the infant may die in utero or may develop anemia, severe jaundice, and edema (**hydrops fetalis**). **Kernicterus,** a neurologic syndrome in which the bile pigments are deposited in the basal ganglia, may also develop, especially if birth is complicated by a period of hypoxia. Bile pigments do not enter the brain in the adult, but in the fetus and newborn infant the blood-brain barrier is not developed (see Chapter 32).

About 50% of Rh-negative individuals are sensitized (develop an anti-Rh titer) by transfusion of Rh-positive blood. Since sensitization of Rh-negative mothers by carrying an Rh-positive fetus generally occurs at birth, the first child is usually normal. However, hemolytic disease occurs in about 17% of the Rh-positive fetuses born to Rh-negative mothers who have previously been pregnant one or more times with Rh-positive fetuses. Fortunately, it is usually possible to prevent sensitization from occurring the first time by administering a single dose of anti-Rh antibodies in the form of Rh immune globulin during the postpartum period. Such passive immunization does not harm the mother and has been demonstrated to prevent active antibody formation by the mother. In obstetric clinics, the institution of such treatment on a routine basis to unsensitized Rh-negative women who have delivered an Rh-positive baby has reduced the overall incidence of hemolytic disease by more than 90%. Treatment with a small dose during pregnancy will also prevent sensitization due to fetal-maternal hemorrhage before delivery.

PLASMA

The fluid portion of the blood, the **plasma,** is a remarkable solution containing an immense number of ions, inorganic molecules, and organic molecules that are in transit to various parts of the body or aid in the transport of other substances. The normal plasma volume is about 5% of body weight, or roughly 3500 mL in a 70-kg man. Plasma clots on standing, remaining fluid only if an anticoagulant is added. If whole blood is allowed to clot and the clot is removed, the remaining fluid is called **serum.** Serum has essentially the same composition as plasma ex-

cept that its fibrinogen and clotting factors II, V, and VIII (Table 27–9) have been removed and it has a higher serotonin content because of the breakdown of platelets during clotting. The normal plasma levels of various substances are discussed in the chapters on the systems with which the substances are concerned and summarized on the inside back cover of this book.

Plasma Proteins

The plasma proteins consist of **albumin, globulin,** and **fibrinogen** fractions. The globulin fraction is subdivided into numerous components. One classification divides it into α_1, α_2, β_1, β_2, and γ globulins and fibrinogen. Protein fractions can be separated and characterized by their relative speed of sedimentation in the ultracentrifuge or their rate of migration in an electrical field (electrophoresis). The molecular weights and configurations of several of the plasma proteins are shown in Fig 27–18.

The capillary walls are relatively impermeable to the proteins in plasma, and the proteins therefore exert an osmotic force of about 25 mm Hg across the capillary wall (**oncotic pressure;** see Chapter 1) that tends to pull water into the blood. The plasma proteins are also responsible for 15% of the buffering capacity of the blood (see Chapter 35), because of the weak ionization of their substituent COOH and NH_2 groups. At the normal plasma pH of 7.40, the proteins are mostly in the anionic form (see Chapter 1) and constitute a significant part of the anionic complement of plasma. Plasma proteins such as antibodies and the proteins concerned with blood clotting have

Table 27–9. System for naming blood-clotting factors.

Factor[1]	Names
I	Fibrogen
II	Prothrombin
III	Thromboplastin
IV	Calcium
V	Proaccelerin, labile factor, accelerator globulin
VII	Proconvertin, SPCA, stable factor
VIII	Antihemophilic factor (AHF), antihemophilic factor A, antihemophilic globulin (AHG)
IX	Plasma thromboplastic component (PTC), Christmas factor, antihemophilic factor B
X	Stuart-Prower factor
XI	Plasma thromboplastin antecedent (PTA), antihemophilic factor C
XII	Hageman factor, glass factor
XIII	Fibrin-stabilizing factor, Laki-Lorand factor
HMW-K	High-molecular-weight kininogen, Fitzgerald factor
Pre-K	Prekallikrein, Fletcher factor
Ka	Kallikrein
PL	Platelet phospholipid

[1]Factor VI is not a separate entity and has been dropped.

Figure 27–18. Relative dimensions and molecular weights of some of the protein molecules in the blood. (Modified and reproduced, with permission, from Murray RK et al: *Harper's Biochemistry,* 23rd ed. Appleton & Lange, 1993.)

specific functions. Some of the proteins function in the transport of thyroid, adrenocortical, gonadal, and other hormones. Binding keeps these hormones from being rapidly filtered through the glomeruli and provides a stable reservoir of hormones on which the tissues can draw. In addition, albumin serves as a carrier for metals, ions, fatty acids, amino acids, bilirubin, enzymes, and drugs.

Origin of Plasma Proteins

Circulating antibodies in the γ globulin fraction of the plasma proteins are manufactured in the plasma cells (see above). The albumin fraction and the proteins concerned with blood clotting (fibrinogen, prothrombin, etc) are manufactured in the liver.

Data on the turnover of albumin provide an indication of the role played by synthesis in the maintenance of normal albumin levels. In normal adult humans, the plasma albumin level is 3.5–5.0 g/dL, and the total exchangeable albumin pool is 4.0–5.0 g/kg body weight; 38–45% of this albumin is intravascular, and much of the rest of it is in the skin. Six to 10% of the exchangeable pool is degraded per day, and the degraded albumin is replaced by hepatic synthesis of 200–400 mg/kg/d. The albumin is probably transported to the extravascular areas by vesicular transport across the walls of the capillaries (see Chapter 1). Albumin synthesis is carefully regulated. It is decreased during fasting and increased in conditions such as nephrosis in which there is excessive albumin loss.

Hypoproteinemia

Plasma protein levels are maintained during starvation until body protein stores are markedly de-

pleted; but in prolonged starvation and in the malabsorption syndrome due to intestinal diseases such as sprue, plasma protein levels are low (**hypoproteinemia).** They are also low in liver disease, because hepatic protein synthesis is depressed; and in nephrosis, because large amounts of albumin are lost in the urine. Because of the decrease in the plasma oncotic pressure, edema tends to develop (see Chapter 30). Rarely, there may be congenital absence of one or another plasma protein fraction. An example of congenital protein deficiency is the congenital form of **afibrinogenemia,** characterized by defective blood clotting.

HEMOSTASIS

When a small blood vessel is transected or damaged, the injury initiates a series of events (Fig 27–19) that leads to the formation of a clot (**hemostasis).** This seals off the damaged region and prevents further blood loss. The initial event is constriction of the vessel and formation of a temporary **hemostatic plug** of platelets that is triggered when platelets bind to collagen (see below). This is followed by conversion of the plug into the definitive clot.

The in vivo action of the clotting mechanism is balanced by limiting reactions that normally prevent clots from developing in uninjured vessels and maintain the blood in a fluid state. It is worth emphasizing that a balance between many complex, interrelated systems must be maintained to prevent hemorrhage

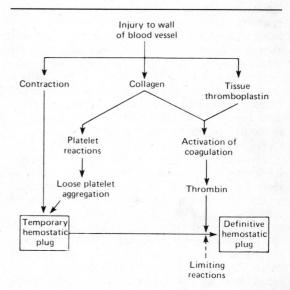

Figure 27–19. Summary of reactions involved in hemostasis. The dashed arrow indicates inhibition. (Modified and reproduced, with permission, from Deykin D: Thrombogenesis, *N Engl J Med* 1967;**276:**622.)

while preventing intravascular coagulation. The factors involved include the endothelium of the blood vessels and the collagen underlying it, vascular tone, the platelets, the clotting and fibrinolytic systems, and the flow characteristics of blood within the blood vessels.

Local Vasoconstriction

The constriction of an injured arteriole or small artery may be so marked that its lumen is obliterated. The vasoconstriction is due to serotonin and other vasoconstrictors liberated from platelets that adhere to the walls of the damaged vessels. It is claimed that at least for a time after being divided transversely, arteries as large as the radial artery constrict and stop bleeding. However, this is no excuse for delay in ligating the damaged vessel. Furthermore, arterial walls cut longitudinally or irregularly do not constrict in such a way that the lumen of the artery is occluded, and bleeding continues.

The Clotting Mechanism

The loose aggregation of platelets in the temporary plug is bound together and converted into the definitive clot by **fibrin.** The clotting mechanism responsible for the formation of fibrin involves a cascade of reactions in which inactive enzymes are activated, and the activated enzymes in turn activate other inactive enzymes. The complexity of the system has in the past been compounded by variations in nomenclature, but acceptance of a numbering system for most of the various clotting factors (Table 27–9) has simplified the situation.

The fundamental reaction in the clotting of blood is conversion of the soluble plasma protein fibrinogen to insoluble fibrin (Fig 27–20). The process involves the release of 2 pairs of polypeptides from each fibrinogen molecule. The remaining portion, **fibrin monomer,** then polymerizes with other monomer molecules to form **fibrin.** The fibrin is initially a loose mesh of interlacing strands. It is converted by the formation of covalent cross-linkages to a dense, tight aggregate. This latter reaction is catalyzed by factor III, the fibrin-stabilizing factor, and requires Ca^{2+}.

The conversion of fibrinogen to fibrin is catalyzed by thrombin. Thrombin is a serine protease that is formed from its circulating precursor, prothrombin, by the action of activated factor X. Factor X can be activated by reactions in either of 2 systems, an intrinsic and an extrinsic system (Fig 27–20).

The initial reaction in the **intrinsic system** is conversion of inactive factor XII to active factor XII (XIIa). This activation, which is catalyzed by high-molecular-weight kininogen and kallikrein (see Chapter 31), can be brought about in vitro by exposing the blood to electronegatively charged wettable surfaces such as glass and collagen fibers. Activation in vivo occurs when blood is exposed to the collagen

Figure 27–20. The clotting mechanism. a, active form of clotting factor; HMW, high-molecular-weight; PL, platelet phospholipid; TPL, tissue thromboplastin. (Reproduced, with permission, from Wessler S, Gitel SN: Warfarin: From bedside to bench. *N Engl J Med* 1984;**311:**645. Reprinted by permission of the *New England Journal of Medicine.*)

fibers underlying the endothelium in the blood vessels. Active factor XII then activates factor XI, and active factor XI activates factor IX. Activated factor IX forms a complex with factor VIII, activating factor X. Phospholipids from aggregated platelets (PL) and Ca^{2+} are necessary for full activation of factor X. The **extrinsic system** is triggered by the release of tissue thromboplastin, a protein-phospholipid mixture that activates factor VII. The tissue thromboplastin and factor VII activate factors IX and X. In the presence of PL, Ca^{2+}, and factor V, activated factor X catalyzes the conversion of prothrombin to thrombin.

Anticlotting Mechanisms

The tendency of blood to clot is balanced in vivo by limiting reactions that tend to prevent clotting inside the blood vessels and to break down any clots that do form. These reactions include removal of some activated clotting factors from the circulation by the liver and reduction in the supply of clotting factors to the degree that they are used up during clotting. Another reaction is the interaction between the platelet-aggregating effect of thromboxane A_2 and the antiaggregating effect of prostacyclin, which causes clots to form at the site when a blood vessel is injured but keeps the vessel lumen free of clot (see Chapter 31).

Antithrombin III is a circulating protease inhibitor that binds to the serine proteases in the coagulation system, blocking their activity as clotting factors. This binding is facilitated by **heparin,** a naturally occurring anticoagulant that is a mixture of sulfated polysaccharides with molecular weights averaging 15,000–18,000. The clotting factors that are inhibited are the active forms of factors IX, X, XI, and XII. Note that "antithrombin" is a misnomer; it is actually a different substance, heparin cofactor II, that inhibits thrombin.

The endothelium of the blood vessels also plays an active role in preventing the extension of clots to normal blood vessels. All endothelial cells except those in the cerebral microcirculation produce **thrombomodulin,** a thrombin-binding protein that converts thrombin into **protein C activator.** This activates protein C (Fig 27–21), a naturally occurring anticoagulant protein that inactivates factors V and VIII and inactivates an inhibitor of tissue plasminogen activator, increasing the formation of plasmin.

Plasmin (fibrinolysin) is the active component of the **fibrinolytic system** (Fig 27–21). This enzyme lyses fibrin and fibrinogen, with the production of fibrinogen degradation products (FDP) that inhibit thrombin. Plasmin is formed from its inactive precursor, plasminogen, by the action of thrombin and a **tissue plasminogen activator (TPA).** Human plasminogen consists of a 560-amino-acid heavy chain and a 241-amino-acid light chain (Fig 27–22). The heavy chain, with glutamate at its N terminal, is folded into 5 loop structures called kringles because of their resemblance to a type of German cookie. The kringles are lysine-binding sites by which the molecule binds to fibrin and other clot proteins. It is converted to active plasmin when TPA hydrolyzes the bond between Arg 560 and Val 561.

Human TPA is now produced by recombinant DNA techniques and is available for clinical use. It lyses clots in the coronary arteries if given to patients soon after the onset of myocardial infarction. Streptokinase, a bacterial enzyme, and urokinase, an enzyme produced by kidney cells, are also fibrinolytic and are used in the treatment of early myocardial infarction (see Chapter 32).

Anticoagulants

As noted above, heparin is a naturally occurring anticoagulant that facilitates the action of antithrombin III. It is also a cofactor for the lipoprotein lipase (clearing factor; see Chapter 17). The highly basic protein protamine forms an irreversible complex with heparin and is used clinically to neutralize heparin.

Heparin and histamine are found in the granules of the **mast cells,** wandering cells found in large numbers in tissues that are rich in connective tissue. In rodents but not in humans, mast cells also contain se-

Figure 27–21. The fibrinolytic system and its regulation by protein C. (Modified from Wessler S, Gitel SN: Warfarin: From bedside to bench. *N Engl J Med* 1984;**311**:645.)

Figure 27–22. Structure of human plasminogen. Note the Glu at the N terminal, the Asn at the C terminal, and 5 uniquely shaped loop structures (kringles). Hydrolysis by TPA at the arrow separates the C-terminal light chain from the N-terminal heavy chain but leaves the disulfide bonds intact. This activates the molecule. (Modified and reproduced, with permission, from Bachman F, in: *Thrombosis and Hemostasis.* Verstraete M et al [editors], p. 231. Leuven University Press, 1987.)

rotonin. Histamine is also found in the granules of circulating basophils, along with heparinlike polysulfated chondroitins. Mast cells have IgE receptors on their surfaces and discharge the contents of their granules when IgE-coated antigens bind to the receptors. The cells also bring about the secretion of leukotrienes from other cells (see Chapter 17) and mediate some aspects of allergic and inflammatory reactions. In addition to mediating allergic reactions, the mast cells may aid in defense against parasitic infections.

In vivo, a plasma Ca^{2+} level low enough to interfere with blood clotting is incompatible with life, but clotting can be prevented in vitro if Ca^{2+} is removed from the blood by the addition of substances such as oxalates, which form insoluble salts with Ca^{2+}, or **chelating agents,** which bind Ca^{2+}. Coumarin derivatives such as **dicumarol** and **warfarin** are also effective anticoagulants. They inhibit the action of vitamin K, and this vitamin is a necessary cofactor for the enzyme that catalyzes the conversion of glutamic acid residues to γ-carboxyglutamic acid residues. Six of the proteins involved in clotting require conversion of a number of glutamic acid residues to γ-carboxyglutamic acid residues before being released into the circulation, and hence, all 6 are vitamin K-dependent. These proteins are factors II (prothrombin), VII, IX, and X, protein C, and protein S, which facilitates the inactivation of factors Va and VIIIa by facilitating the action of activated protein C.

Abnormalities of Hemostasis

In addition to clotting abnormalities due to platelet disorders (see above), hemorrhagic diseases can be produced by selective deficiencies of most of the clotting factors (Table 27–10). Hemophilia A is of interest because it is relatively common. The disease has been treated with factor VIII-rich preparations made from plasma, but unfortunately this has led to transmission of HIV to a significant number of patients. However, highly purified preparations treated to inactivate viruses are available. So is factor VIII produced by recombinant DNA techniques. In addition to promoting platelet adherence (see above), **von Willebrand factor** forms a complex with factor VIII and regulates its plasma levels. Congenital deficiency of von Willebrand factor also causes a bleeding disorder (von Willebrand's disease).

The absorption of the K vitamins, along with that of the other fat-soluble vitamins, is depressed in obstructive jaundice because of the lack of bile in the intestine and consequent depression of fat absorption (see Chapter 26). The resulting clotting factor deficiencies may cause the development of a significant bleeding tendency.

Formation of clots inside blood vessels is called **thrombosis** to distinguish it from the normal extravascular clotting of blood. Thromboses are a major medical problem. They are particularly prone to

Table 27–10. Examples of diseases due to deficiency of clotting factors.

Deficiency of Factor	Clinical Syndrome	Cause
I	Afibrinogenemia	Depletion during pregnancy with premature separation of placenta; also congenital (rare).
II	Hypoprothrombinemia (hemorrhagic tendency in liver disease)	Decreased hepatic synthesis, usually secondary to vitamin K deficiency.
V	Parahemophilia	Congenital.
VII	Hypoconvertinemia	Congenital.
VIII	Hemophilia A (classic hemophilia)	Congenital defect due to various abnormalities of the gene on X chromosome that codes for factor VIII; disease is therefore inherited as sex-linked characteristic.
IX	Hemophilia B (Christmas disease)	Congenital.
X	Stuart-Prower factor deficiency	Congenital.
XI	PTA deficiency	Congenital.
XII	Hageman trait	Congenital.

occur where blood flow is sluggish, eg, in the veins of the legs after operations and delivery, because the slow flow permits activated clotting factors to accumulate instead of being washed away. They also occur in vessels such as the coronary and cerebral arteries at sites where the intima is damaged by atherosclerotic plaques, and over areas of damage to the endocardium. They frequently occlude the arterial supply to the organs in which they form, and bits of thrombus (**emboli**) sometimes break off and travel in the bloodstream to distant sites, damaging other organs. Examples are obstruction of the pulmonary artery or its branches (**pulmonary embolism**) by thrombi from the leg veins, and embolism of cerebral or leg vessels by bits of clot breaking off from a thrombus in the left ventricle (**mural thrombus**) overlying a myocardial infarct.

Congenital absence of protein C leads to uncontrolled intravascular coagulation and, in general, death in infancy. If this condition is diagnosed and treatment is instituted with blood concentrates rich in protein C, the coagulation defect disappears.

Table 27–11. Probable approximate protein content of lymph in humans[1]

Source of Lymph	Protein Content (g/dL)
Choroid plexus	0
Ciliary body	0
Skeletal muscle	2
Skin	2
Lung	4
Gastrointestinal tract	4.1
Heart	4.4
Liver	6.2

[1] Data largely from JN Diana.

LYMPH

Lymph is tissue fluid that enters the lymphatic vessels. It drains into the venous blood via the thoracic and right lymphatic ducts. It contains clotting factors and clots on standing in vitro. In most locations, it also contains proteins that traverse capillary walls and return to the blood via the lymph. Its protein content is generally lower than that of plasma, which contains about 7 g/dL, but lymph protein varies with the region from which the lymph drains (Table 27–11). Water-insoluble fats are absorbed from the intestine into the lymphatics, and the lymph in the thoracic duct after a meal is milky because of its high fat content (see Chapter 25). Lymphocytes enter the circulation principally through the lymphatics, and there are appreciable numbers of lymphocytes in thoracic duct lymph.

28

Origin of the Heartbeat & the Electrical Activity of the Heart

INTRODUCTION

The parts of the heart normally beat in an orderly sequence: Contraction of the atria (**atrial systole**) is followed by contraction of the ventricles (**ventricular systole,**) and during **diastole** all 4 chambers are relaxed. The heartbeat originates in a specialized **cardiac conduction system** and spreads via this system to all parts of the myocardium. The structures that make up the conduction system (Fig 28–1) are the **sinoatrial node** (SA node), the **internodal atrial pathways,** the **atrioventricular node** (AV node), the **bundle of His** and its branches, and the **Purkinje system.** The various parts of the conduction system and, under abnormal conditions, parts of the myocardium are capable of spontaneous discharge. However, the SA node normally discharges most rapidly, depolarization spreading from it to the other regions before they discharge spontaneously. The SA node is therefore the normal **cardiac pacemaker,** its rate of discharge determining the rate at which the heart beats. Impulses generated in the SA node pass through the atrial pathways to the AV node, through this node to the bundle of His, and through the branches of the bundle of His via the Purkinje system to the ventricular muscle.

ORIGIN & SPREAD OF CARDIAC EXCITATION

Anatomic Considerations

In the human heart, the SA node is located at the junction of the superior vena cava with the right atrium. The AV node is located in the right posterior portion of the interatrial septum (Fig 28–1). There are 3 bundles of atrial fibers that contain Purkinje type fibers and connect the SA node to the AV node: the anterior internodal tract of Bachman, the middle internodal tract of Wenckebach, and the posterior internodal tract of Thorel. However, there is debate about the role of these bundles versus the regular atrial myocytes in conduction from the SA node to the AV node. The AV node is continuous with the bundle of His, which gives off a left bundle branch at the top of

the interventricular septum and continues as the right bundle branch. The left bundle branch divides into an anterior fascicle and a posterior fascicle. The branches and fascicles run subendocardially down either side of the septum and come into contact with the Purkinje system, whose fibers spread to all parts of the ventricular myocardium.

The histology of cardiac muscle is described in Chapter 3. The conduction system is composed for the most part of modified cardiac muscle that has fewer striations and indistinct boundaries. The SA node and, to a lesser extent, the AV node, also contain small round cells with few organelles, which are connected by gap junctions. These are probably the actual pacemaker cells, and therefore they are called **P cells.** The atrial muscle fibers are separated from those of the ventricles by a fibrous tissue ring, and normally the only conducting tissue between the atria and ventricles is the bundle of His.

The SA node develops from structures on the right side of the embryo and the AV node from structures on the left. This is why in the adult the right vagus is distributed mainly to the SA node and the left vagus mainly to the AV node. Similarly, the sympathetic innervation on the right side is distributed primarily to the SA node and the sympathetic innervation on the left side primarily to the AV node. On each side, most sympathetic fibers come from the stellate ganglion. Cholinergic and noradrenergic fibers are epicardial, whereas the vagal fibers are endocardial. It should also be noted that connections exist for reciprocal inhibitory effects of the sympathetic and parasympathetic innervation of the heart on each other. Thus, acetylcholine acts presynaptically to reduce norepinephrine release from the sympathetic nerves, and conversely, neuropeptide Y released from noradrenergic endings may inhibit the release of acetylcholine.

Properties of Cardiac Muscle

The electrical responses of cardiac muscle and nodal tissue and the ionic fluxes that underlie them are discussed in detail in Chapter 3. As shown in Fig 3–14, myocardial fibers have a resting membrane potential of approximately –90 mV. The relation of the

Figure 28–1. Conducting system of the heart. Typical transmembrane action potentials for the SA and AV nodes, other parts of the conduction system, and the atrial and ventricular muscles are shown along with the correlation to the extracellularly recorded electrical activity, ie, the electrocardiogram (ECG). The action potentials and ECG are plotted on the same time axis but with different zero points on the vertical scale. LAF, left anterior fascicle.

action potential to the contractile response is shown in Fig 3–13. The individual fibers are separated by membranes, but depolarization spreads radially through them as if they were a synctium, because of the presence of gap junctions. The transmembrane action potential of single cardiac muscle cells is characterized by rapid depolarization, a plateau, and a slow repolarization process (see Fig 3–14). The initial depolarization is due to Na^+ influx through rapidly opening Na^+ channels (the Na^+ current, I_{Na}). Ca^{2+} influx through more slowly opening Ca^{2+} channels (the Ca^{2+} current, I_{Ca}) produces the plateau phase, and repolarization is due to K^+ efflux through 2 types of K^+ channels, producing the inward rectifier current I_{KI} and delayed rectifier current I_K. Recorded extracellularly, the summed electrical activity of all the cardiac muscle fibers is the ECG. The timing of the discharge of the individual units relative to the ECG is shown in Fig 28–1.

Pacemaker Potentials

Rhythmically discharging cells have a membrane potential that, after each impulse, declines to the firing level. Thus, this **prepotential** or **pacemaker potential** (Fig 28–2) triggers the next impulse. At the peak of each impulse, I_K begins and brings about repolarization. I_K then declines, and as K^+ efflux decreases, the membrane begins to depolarize, forming

Figure 28–2. Diagram of the membrane potential of pacemaker tissue. The current responsible for each part of the potential is shown under or beside the component. L, long-lasting; T, transient. Note that the resting membrane potential of pacemaker tissue is somewhat lower than that of atrial and ventricular muscle.

the first part of the prepotential. Ca^{2+} channels then open. These are of 2 subtypes in the heart, the **T** (for transient) **channels** and the **L** (for long-lasting) **channels.** The calcium current (I_{Ca}) due to opening of T channels completes the prepotential, and I_{Ca} due to opening of L channels produces the impulse.

The actions potentials in the SA and AV nodes are largely due to Ca^{2+}, with little contribution by Na^+ influx. Consequently there is no sharp, rapid depolarizing spike before the plateau, as there is in other parts of the conduction system and the atrial and ventricular fibers (Fig 28–1). In addition, prepotentials are normally prominent only in the SA and AV nodes. However, there are "latent pacemakers" in other portions of the conduction system that can take over when the SA and AV nodes are depressed or conduction from them is blocked. Atrial and ventricular muscle fibers do not have prepotentials, and they discharge spontaneously only when injured or abnormal.

When the cholinergic vagal fibers to nodal tissue are stimulated, the membrane becomes hyperpolarized and the slope of the prepotentials is decreased (Fig 28–3) because the acetylcholine released at the nerve endings increases the K^+ conductance of nodal tissue. This action is mediated by M_2 muscarinic receptors, which, via a G protein, open a special set of K^+ channels. The resulting I_{KAch} counters the decay of I_K. In addition, activation of the M_2 receptors decreases cyclic AMP in the cells, and this slows the opening of the Ca^{2+} channels. The result is a decrease in firing rate. Strong vagal stimulation may abolish spontaneous discharge for some time.

Conversely, stimulation of the sympathetic cardiac nerves makes the membrane potential fall more rapidly, and the rate of spontaneous discharge increases (Fig 28–3). Norepinephrine secreted by the sympathetic endings binds to β_1 receptors, and the resulting increase in intracellular cyclic AMP facilitates the opening of L channels, increasing I_{Ca} and the rapidity of the depolarization phase of the impulse. Interestingly, the membrane hyperpolarizes by a mechanism that is still unsettled. This activates an inward current called I_f, which rapidly depolarizes the membrane to the firing level. In addition, activation of β_1 receptors on cardiac muscle fibers increases intracellular Ca^{2+}, and consequently the strength of this contraction is increased.

Because of the "sidedness" of the cardiac innervation (see above), stimulation of the right vagus slows the heart by inhibiting the SA node, whereas stimulation of the left vagus mainly slows AV conduction. Similarly, stimulation of the right stellate ganglion accelerates the heart, whereas stimulation of the left stellate ganglion shortens the AV nodal conduction time and refractoriness.

The rate of discharge of the SA node and other nodal tissue is influenced by temperature and by drugs. The discharge frequency is increased when the temperature rises, and this may contribute to the tachycardia associated with fever. Digitalis depresses nodal tissue and exerts an effect like that of vagal stimulation, particularly on the AV node.

Spread of Cardiac Excitation

The speeds of conduction in the various types of cardiac tissue are shown in Table 28–1. It is worth noting that there are marked differences in the rate of conduction in the various parts of the conducting system.

Depolarization initiated in the SA node spreads radially through the atria, then converges on the AV node. Atrial depolarization is complete in about 0.1 second. Because conduction in the AV node is slow, there is a delay of about 0.1 second (**AV nodal delay**) before excitation spreads to the ventricles. This delay is shortened by stimulation of the sympathetic nerves to the heart and lengthened by stimulation of the vagi. From the top of the septum, the wave of depolarization spreads in the rapidly conducting Purkinje fibers to all parts of the ventricles in the 0.08–0.1 second. In humans, depolarization of the ventricular muscle starts at the left side of the interventricular septum and moves first to the right across the midportion of the septum. The wave of depolarization then spreads down the septum to the apex of the heart. It returns along the ventricular walls to the AV groove, proceeding from the endocardial to the epicardial surface (Fig 28–4). The last parts of the heart to be depolarized are the posterobasal portion of the left ventricle,

Figure 28–3. Effect of sympathetic (noradrenergic) and vagal (cholinergic) stimulation on the membrane potential of the SA node.

Table 28-1. Conduction speeds in cardiac tissue.

Tissue	Conduction Rate (m/s)
SA node	0.05
Atrial pathways	1
AV node	0.05
Bundle of His	1
Purkinje system	4
Ventricular muscle	1

SA node — Atrial activation

Septal activation from left to right

R   L

Activation of anteroseptal region of the ventricular myocardium

Activation of major portion of ventricular myocardium from endocardial surfaces

Late activation of posterobasal portion of the left ventricle and the pulmonary conus

Figure 28–4. Normal spread of electrical activity in the heart. (Reproduced, with permission, from Goldman MJ: *Principles of Clinical Electrocardiography,* 12th ed. Lange, 1986.)

the pulmonary conus, and the uppermost portion of the septum.

THE ELECTROCARDIOGRAM

Because the body fluids are good conductors (ie, because the body is a **volume conductor**), fluctuations in potential that represent the algebraic sum of the action potentials of myocardial fibers can be recorded extracellularly. The record of these potential fluctuations during the cardiac cycle is the **electrocardiogram (ECG).** Most electrocardiograph ma-

chines record these fluctuations on a moving strip of paper.

The ECG may be recorded by using an **active** or **exploring** electrode connected to an indifferent electrode at zero potential (**unipolar** recording), or by using 2 active electrodes (**bipolar** recording). In a volume conductor, the sum of the potentials at the points of an equilateral triangle with a current source in the center is zero at all times. A triangle with the heart at its center (**Einthoven's triangle**) can be approximated by placing electrodes on both arms and on the left leg. These are 3 **standard limb leads** used in electrocardiography. If these electrodes are connected to a common terminal, an indifferent electrode that stays near zero potential is obtained. Depolarization moving toward an active electrode in a volume conductor produces a positive deflection, whereas depolarization moving in the opposite direction produces a negative deflection.

The names of the various waves and segments of the ECG in humans are shown in Fig 28–5. By convention, an upward deflection is written when the active electrode becomes positive relative to the indifferent electrode, and a downward deflection is written when the active electrode becomes negative. The P wave is produced by atrial depolarization, the QRS complex by ventricular depolarization, and the ST segment and T wave by ventricular repolarization. The manifestations of atrial repolarization are not normally seen because they are obscured by the QRS complex. The U wave is an inconstant finding, believed to be due to slow repolarization of the papillary muscles. The intervals between the various waves of the ECG and the events in the heart that occur during these intervals are shown in Table 28–2.

Bipolar Leads

Bipolar leads were used before unipolar leads

Figure 28–5. Waves of the ECG.

Table 28-2. ECG intervals.

Intervals	Normal Duration(s)		Events in the Heart During Interval
	Average	Range	
PR interval[1]	0.18[2]	0.12–0.20	Atrial depolarization and conduction through AV node
QRS duration	0.08	to 0.10	Ventricular depolarization and atrial repolarization
QT interval	0.40	to 0.43	Ventricular depolarization plus ventricular repolarization
ST interval (QT minus QRS)	0.32	. . .	Ventricular repolarization

[1]Measured from the beginning of the P wave to the beginning of the QRS complex.
[2]Shortens as heart rate increases from average of 0.18 at a rate of 70 to 0.14 at a rate of 130.

were developed. The **standard limb leads,** leads I, II, and III, each record the differences in potential between 2 limbs. Since current flows only in the body fluids, the records obtained are those that would be obtained if the electrodes were at the points of attachment of the limbs, no matter where on the limbs the electrodes are placed. In lead I, the electrodes are connected so that an upward deflection is inscribed when the left arm becomes positive relative to the right (left arm positive). In lead II, the electrodes are on the right arm and left leg, with the leg positive; and in lead III, the electrodes are on the left arm and left leg, with the leg positive.

Unipolar (V) Leads

An additional 9 unipolar leads, ie, leads that record the potential difference between an exploring electrode and an indifferent electrode, are commonly used in clinical electrocardiography. There are 6 unipolar chest leads (precordial leads) designated V_1–V_6 (Fig 28–6) and 3 unipolar limb leads: VR (right arm), VL (left arm), and VF (left foot). **Augmented** limb leads, designated by the letter a (aVR, aVL, aVF) are generally used. The augmented limb leads are recordings between one limb and the other 2 limbs. This increases the size of the potentials by 50% without any change in configuration from the nonaugmented record.

An esophageal electrode is sometimes used to study atrial activity. The electrode is inserted in a catheter and swallowed. Unipolar leads can also be placed at the tips of catheters and inserted into the heart.

Normal ECG

The ECG of a normal individual is shown in Fig 28–7. The sequence in which the parts of the heart are depolarized (Fig 28–4) and the position of the heart relative to the electrodes are the important considerations in interpreting the configurations of the waves in each lead. The atria are located posteriorly in the chest. The ventricles form the base and anterior surface of the heart, and the right ventricle is anterolateral to the left. Thus, aVR "looks at" the cavities of the ventricles. Atrial depolarization, ventricular de-

polarization, and ventricular repolarization move away from the exploring electrode, and the P wave, QRS complex, and T wave are therefore all negative (downward) deflections; aVL and aVF look at the ventricles, and the deflections are therefore predominantly positive or biphasic. There is no Q wave in V_1 and V_2, and the initial portion of the QRS complex is a small upward deflection because ventricular depolarization first moves across the midportion of the septum from left to right toward the exploring electrode. The wave of excitation then moves down the septum and into the left ventricle away from the electrode, producing a large S wave. Finally, it moves back along the ventricular wall toward the electrode, producing the return to the isoelectric line. Conversely, in the left ventricular leads (V_4–V_6) there may be an initial small Q wave (left to right septal depolarization), and there is a large R wave (septal and left ventricular depolarization) followed in V_4 and V_5 by a moderate S wave (late depolarization of the ventricular walls moving back toward the AV junction).

There is considerable variation in the position of

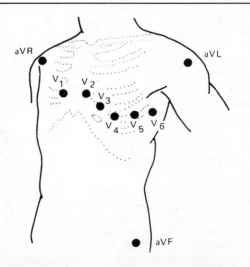

Figure 28–6. Unipolar electrocardiographic leads.

Figure 28–7. Normal ECG. (Reproduced, with permission, from Goldman MJ: *Principles of Clinical Electrocardiography*, 12th ed. Lange, 1986.)

the normal heart, and the position affects the configuration of the electrocardiographic complexes in the various leads.

Bipolar Limb Leads & the Cardiac Vector

Because the standard limb leads are records of the potential differences between 2 points, the deflection in each lead at any instant indicates the magnitude and direction in the axis of the lead of the electromotive force generated in the heart (**cardiac vector** or **axis**). The vector at any given moment in the 2 dimensions of the frontal plane can be calculated from any 2 standard limb leads (Fig 28–8) if it is assumed that the 3 electrode locations form the points of an equilateral triangle (Einthoven's triangle) and that the heart lies in the center of the triangle. These assumptions are not completely warranted, but calculated vectors are useful approximations. An approximate **mean QRS vector** ("electrical axis of the heart") is often plotted by using the average QRS de-

Figure 28–8. Cardiac vector. **Left:** Einthoven's triangle. Perpendiculars dropped from the midpoints of the sides of the equilateral triangle intersect at the center of electrical activity. RA, right arm; LA, left arm; LL, left leg. **Center:** Calculation of mean QRS vector. In each lead, distances equal to the height of the R wave minus the height of the largest negative deflection in the QRS complex are measured off from the midpoint of the side of the triangle representing that lead. An arrow drawn from the center of electrical activity to the point of intersection of perpendiculars extended from the distances measured off on the sides represents the magnitude and direction of the mean QRS vector. **Right:** Reference axes for determining the direction of the vector.

flection in each lead as shown in Fig 28–8. This is a **mean** vector as opposed to an **instantaneous** vector, and the average QRS deflections should be measured by integrating the QRS complexes. However, they can be approximated by measuring the net differences between the positive and negative peaks of the QRS. The normal direction of the mean QRS vector is generally said to be –30 to +110 degrees on the coordinate system shown in Fig 28–8. **Left** or **right axis deviation** is said to be present if the calculated axis falls to the left of –30 degrees or to the right of +110 degrees, respectively. Right axis deviation suggests right ventricular hypertrophy, and left axis deviation may be due to left ventricular hypertrophy. However, simple differences in heart position may produce axis values in the "abnormal" range, and there are better and more reliable electrocardiographic criteria for ventricular hypertrophy.

Vectorcardiography

If the tops of the arrows representing all of the **instantaneous** instantaneous cardiac vectors in the frontal plane during the cardiac cycle are connected, from first to last, the line connecting them forms a series of 3 loops: one for the P wave, one for the QRS complex, and one for the T wave. This can be done electronically and the loops, called **vectorcardiograms,** projected on the face of a cathode ray oscilloscope.

His Bundle Electrogram

In patients with heart block, the electrical events in the AV node, bundle of His, and Purkinje system are frequently studied with a catheter containing ring electrodes at its tip that is passed through a vein to the right side of the heart and manipulated into a position close to the tricuspid valve. Three or more standard electrocardiographic leads are recorded simultaneously. The record of the electrical activity obtained with the catheter (Fig 28–9) is the **His bundle electrogram (HBE).** It normally shows an A deflection when the AV node is activated, an H spike during transmission through the His bundle, and a V deflection during ventricular depolarization. With the HBE and the standard electrocardiographic leads, it is possible to accurately time 3 intervals: (1) the PA interval, the time from the first appearance of atrial depolarization to the A wave in the HBE, which represents conduction time from the SA node to the AV node; (2) the AH interval, from the A wave to the start of the H spike, which represents the AV nodal conduction time; and (3) the HV interval, the time from the start of the H spike to the start of the QRS deflection in the ECG, which represents conduction in the bundle of His and the bundle branches. The approximate normal values for these intervals in adults are PA, 27 ms; AH, 92 ms; and HV, 43 ms. These values illustrate the relative slowness of conduction in the AV node (Table 28–1).

Figure 28–9. Normal His bundle electrogram (HBE) with simultaneously recorded ECG.

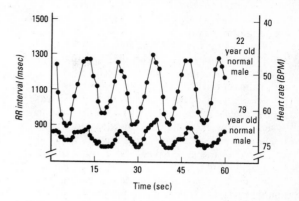

Figure 28–10. Sinus arrhythmia in a young man and an old man. The subjects breathed 5 times per minute, and with each inspiration the R–R interval declined, indicating an increase in heart rate. Note the marked reduction in the magnitude of the arrhythmia in the older man. These records were obtained after β-adrenergic blockade but would have been generally similar in its absence. (Reproduced, with permission, from Pfiefer MA et al: Differential changes of autonomic nervous system function with age in man. *Am J Med* 1983;**75**:49.)

Monitoring

The ECG is often recorded continuously in hospital coronary care units, with alarms arranged to sound at the onset of life-threatening arrhythmias. Using a small portable tape recorder (**Holter monitor**), it is also possible to record the ECG in ambulatory individuals as they go about their normal activities. The recording is later played back at high speed and analyzed. Long-term continuous records can be obtained, and transtelephonic ECG devices are also available so that records can be forwarded to a doctor's office or a laboratory by telephone. Long-term recordings have proved valuable in the diagnosis of arrhythmias and in planning the treatment of patients recovering from myocardial infarctions.

CARDIAC ARRHYTHMIAS

Normal Cardiac Rate

In the normal human heart, each beat originates in the SA node (**normal sinus rhythm, NSR**). The heart beats about 70 times a minute at rest. The rate is slowed (**bradycardia**) during sleep and accelerated (**tachycardia**) by emotion, exercise, fever, and many other stimuli. The control of heart rate is discussed in Chapter 31. In healthy young individuals breathing at a normal rate, the heart rate varies with the phases of respiration: it accelerates during inspiration and decelerates during expiration, especially if the depth of breathing is increased. This **sinus arrhythmia** (Fig 28–10) is a normal phenomenon and is due primarily to fluctuations in parasympathetic output to the heart. During inspiration, impulses in the vagi from the stretch receptors in the lungs inhibit the cardioinhibitory area in the medulla oblongata. The tonic vagal discharge that keeps the heart rate slow decreases, and the heart rate rises.

Disease processes affecting the sinus node lead to marked bradycardia accompanied by dizziness and syncope (**sick sinus syndrome**). When the condition causes severe symptoms, the treatment is implantation of a pacemaker. Sinus node dysfunction accounts for over half of the pacemaker implants in the USA.

Abnormal Pacemakers

The AV node and other portions of the conduction system can in abnormal situations become the cardiac pacemaker. In addition, diseased atrial and ventricular muscle fibers can have their membrane potentials reduced and discharge repetitively.

A simple but enlightening experiment that demonstrates the hierarchy of pacemakers in the heart can be performed in cold-blooded animals such as frogs and turtles. In these animals, the heartbeat originates in a separate cardiac chamber, the **sinus venosus,** instead of in an SA node. There is no coronary circulation, the heart receiving O_2 by diffusion from the blood in the cardiac chambers. If a ligature (Stannius ligature I) is tied around the junction of the sinus venosus and right atrium, conduction from the sinus to the rest of the heart is prevented. The atria and ventricles stop for a moment, then resume beating at a lower rate as a focus in the atrium becomes the pacemaker for the portion of the heart below the ligature. If a second ligature (Stannius ligature II) is tied between the atria and ventricles, the ventricles stop, then resume beating at an even lower rate than the atria. With both ligatures in place, there are 3 separate regions of the heart beating at 3 different rates.

It is not possible to perform the Stannius experiment in mammals because the SA node is embedded in the atrial wall, and tying a ligature between the atria and the ventricles interrupts the coronary circulation, usually causing immediate ventricular fibrillation. However, experiments on laboratory mammals involving cooling or crushing of the SA and AV nodes and the "experiments of nature" in humans with diseased nodal tissue show that the same pacemaker hierarchy prevails.

When conduction from the atria to the ventricles is completely interrupted, **complete (third-degree) heart block** results, and the ventricles beat at a low rate **(idioventricular rhythm)** independently of the atria (Fig 28–11). The block may be due to disease in the AV node **(AV nodal block)** or in the conducting system below the node **(infranodal block)**. In patients with AV nodal block, the remaining nodal tissue becomes the pacemaker, and the rate of the idioventricular rhythm is approximately 45 beats/min. In patients with infranodal block due to disease in the bundle of His, the ventricular pacemaker is located more peripherally in the conduction system, and the ventricular rate is lower; it averages 35 beats/min, but in individual cases it can be as low as 15 beats/min. In such individuals, there may also be periods of asystole lasting a minute or more. The resultant cerebral ischemia causes dizziness and fainting **(Stokes-Adams syndrome)**. Causes of third-degree heart block include septal myocardial infarction and damage to the bundle of His during surgical correction of congenital interventricular septal defects. In infranodal block with Stokes-Adams syndrome, implantation of a permanent cardiac pacemaker is indicated in order to drive the ventricles at a more rapid, regular rate.

When conduction between the atria and ventricles is slowed but not completely interrupted, **incomplete heart block** is present. In the form called **first-degree heart block,** all the atrial impulses reach the ventricles, but the PR interval is abnormally long. In the form called **second-degree heart block,** not all atrial impulses are conducted to the ventricles. There may be, for example, a ventricular beat following every second or every third atrial beat (2:1 block, 3:1 block, etc). In another form of incomplete heart block, there are repeated sequences of beats in which the PR interval lengthens progressively until a ventricular beat is dropped **(Wenckebach phenomenon).** The PR interval of the cardiac cycle that follows each dropped beat is usually normal or only slightly prolonged (Fig 28–11).

Sometimes one branch of the bundle of His is interrupted, causing **right** or **left bundle branch block.** In bundle branch block, excitation passes normally down the bundle on the intact side and then sweeps back through the muscle to activate the ventricle on the blocked side. The ventricular rate is therefore normal, but the QRS complexes are prolonged and deformed (Fig 28–11). Block can also occur in the anterior or posterior fascicle of the left bundle branch, producing the condition called **hemiblock** or **fascicular block.** Left anterior hemiblock produces abnormal left axis deviation in the ECG, whereas left posterior hemiblock produces abnormal right axis deviation. It is not uncommon to

PR = 0.16 s

Normal complex

PR = 0.38 s

First-degree heart block

Second-degree heart block
(2:1 heart block)

Second-degree heart block
(Wenckebach phenomenon)

Complete heart block. Atrial rate, 107; ventricular rate, 43

2 V leads in left
bundle branch block

Figure 28–11. Heart block.

find combinations of fascicle and branch blocks. The His bundle electrogram permits detailed analysis of the site of block when there is a defect in the conduction system.

Ectopic Foci of Excitation

Normally, myocardial cells do not discharge spontaneously, and the possibility of spontaneous discharge of the His bundle and Purkinje system is low because the normal pacemaker discharge of the SA node is more rapid than their rate of spontaneous discharge. However, in abnormal conditions, the His-Purkinje fibers or the myocardial fibers may discharge spontaneously. In these conditions, **increased automaticity** of the heart is said to be present. If an irritable **ectopic focus** discharges once, the result is a beat that occurs before the expected next normal beat and transiently interrupts the cardiac rhythm (atrial, nodal, or ventricular **extrasystole** or **premature beat**). If the focus discharges repetitively at a rate higher than that of the SA node, it produces rapid, regular tachycardia (atrial, ventricular, or nodal **paroxysmal tachycardia** or **atrial flutter**).

Reentry

A more common cause of paroxysmal arrhythmias is a defect in conduction that permits a wave of excitation to propagate continuously within a closed circuit (**circus movement**). For example, if there is a transient block on one side of a portion of the conducting system, the impulse can go down the other side. If the block then wears off, the impulse may conduct in a retrograde direction back to the origin and then descend again, establishing a circus movement. An example of this in a ring of tissue is shown in Fig 28–12. If the reentry is in the AV node, the reentrant activity depolarizes the atrium, and the resulting atrial beat is called an echo beat. In addition, the reentrant activity in the node propagates back down to the ventricle, producing paroxysmal nodal tachycardia. Circus movements can also become es-

tablished in the atrial or ventricular muscle fibers. In individuals with an abnormal bundle of muscle or nodal tissue connecting the atria to the ventricles (bundle of Kent), the circus activity can pass in one direction through the AV node and in the other direction through the bundle, thus involving both the atria and the ventricles.

Atrial Arrhythmias

Excitation spreading from an independently discharging focus in the atria stimulates the AV node prematurely and is conducted to the ventricles. The P waves of atrial extrasystoles are abnormal, but the QRST configurations are normal (Fig 28–13). The

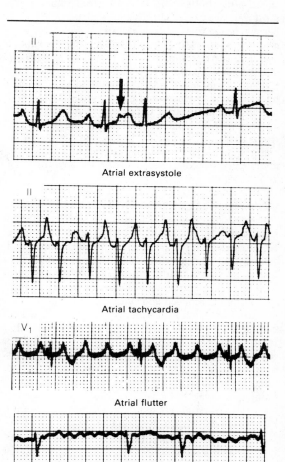

Atrial extrasystole

Atrial tachycardia

Atrial flutter

Atrial fibrillation

Figure 28–13. Atrial arrhythmias. The illustration shows atrial premature beat (arrow), atrial tachycardia, atrial flutter with 4:1 AV block, and atrial fibrillation with a totally irregular ventricular rate. (Tracings reproduced, with permission, from Goldschlager N, Goldman MJ: *Principles of Clinical Electrocardiography.* 13th ed. Appleton & Lange, 1989.)

Figure 28–12. Depolarization of a ring of cardiac muscle. Normally, the impulse spreads in both directions in the ring **(left)** and the tissue immediately behind each branch of the impulse is refractory. When there is a transient block on one side **(center;** shaded area), the impulse goes around the ring, and if the transient block has now worn off **(right),** the impulse passes this area and continues to circle indefinitely (circus movement).

excitation depolarizes the SA node, which must repolarize and then depolarize to the firing level before it can initiate the next normal beat. Consequently, there is a pause between the extrasystole and the next normal beat that is usually equal in length to the interval between the normal beats preceding the extrasystole, and the rhythm is "reset" (see below).

Atrial tachycardia occurs when an atrial focus discharges regularly or there is reentrant activity producing atrial rates up to 220/min. Sometimes, especially in digitalized patients, some degree of atrioventricular block is associated with the tachycardia **(paroxysmal atrial tachycardia with block).** When the atrial rate is 200–350/min, the condition is called atrial flutter (Fig 28–13). Atrial flutter is almost always associated with 2:1 or greater AV block, because the normal AV node has a long refractory period and in adults it cannot conduct more than about 230 impulses per minute.

In **atrial fibrillation,** the atria beat very rapidly (300–500/min) in a completely irregular and disorganized fashion. The cause of atrial fibrillation is still a matter of debate, but it is probably due to multiple, concurrently circulating reentrant excitation wave fronts in the atrial muscle. Because the AV node discharges at irregular intervals, the ventricles beat at a completely irregular rate, usually 80–160/min (Fig 28–13).

Consequences of Atrial Arrhythmias

Occasional atrial extrasystoles occur from time to time in most normal humans and have no pathologic significance. In paroxysmal atrial tachycardia and flutter, the ventricular rate may be so high that diastole is too short for adequate filling of the ventricles with blood between contractions. Consequently, cardiac output is reduced and symptoms of heart failure appear. The relationship between cardiac rate and cardiac output is discussed in detail in Chapter 29. Heart failure may also complicate atrial fibrillation when the ventricular rate is high. Acetylcholine liberated at vagal endings depresses conduction in the atrial musculature and AV node. This is why stimulating reflex vagal discharge by pressing on the eyeball **(oculocardiac reflex)** or massaging the carotid sinus often converts tachycardia and sometimes converts atrial flutter to normal sinus rhythm. Alternatively, vagal stimulation increases the degree of AV block, abruptly lowering the ventricular rate. Digitalis also depresses AV conduction and is used to lower a rapid ventricular rate in atrial fibrillation.

Ventricular Arrhythmias

Premature beats that originate in an ectopic ventricular focus usually have bizarrely shaped prolonged QRS complexes (Fig 28–14) because of the slow spread of the impulse from the focus through the ventricular muscle to the rest of the ventricle. They are usually incapable of exciting the bundle of His,

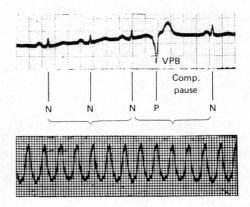

Figure 28–14. *Top:* Ventricular premature beats (VPB). The lines under the tracing illustrate the compensatory pause and show that the duration of the premature beat plus the preceding normal beat is equal to the duration of 2 normal beats. ***Bottom:*** Ventricular tachycardia.

and retrograde conduction to the atria therefore does not occur. In the meantime, the next succeeding normal SA nodal impulse depolarizes the atria. The P wave is usually buried in the QRS of the extrasystole. If the normal impulse reaches the ventricles, they are still in the refractory period following depolarization from the ectopic focus. However, the second succeeding impulse from the SA node produces a normal beat. Thus, ventricular premature beats are followed by a **compensatory pause** that is longer than the pause after an atrial extrasystole. Furthermore, ventricular premature beats do not interrupt the regular discharge of the SA node, whereas atrial premature beats interrupt and "reset" the normal rhythm. These differences make it possible to distinguish between atrial and ventricular premature beats by palpation of the pulse or auscultation of the heart at the bedside. If one foot is tapped in time with the regular normal rhythm, the beats after an atrial premature beat occur early with respect to the previous regular beats. However, the first normal beat after a ventricular premature beat occurs on the second foot tap after the premature beat, because the duration of a ventricular premature beat plus the preceding normal beat is equal to the duration of 2 normal beats (Fig 28–14).

Atrial and ventricular premature beats are not strong enough to produce a pulse at the wrist if they occur early in diastole, when the ventricles have not had time to fill with blood and the ventricular musculature is still in its relatively refractory period. They may not even open the aortic and pulmonary valves, in which case there is, in addition, no second heart sound (see Chapter 29).

Paroxysmal ventricular tachycardia (Fig 28–14) is in effect a series of rapid, regular ventricular depolarizations usually due to a circus movement involving the ventricles. Tachycardias originating above the ventricles (supraventricular tachycardias such as paroxysmal nodal tachycardia) can be distinguished from paroxysmal ventricular tachycardia by use of the HBE; in supraventricular tachycardias, there is a His bundle (H) deflection, whereas in ventricular tachycardias, there is none. Ventricular premature beats are common and, in the absence of ischemic heart disease, usually benign. Ventricular tachycardia is more serious because cardiac output is decreased, and ventricular fibrillation is an occasional complication of ventricular tachycardia.

In **ventricular fibrillation** (Fig 28–15), the ventricular muscle fibers contract in a totally irregular and ineffective way because of the very rapid discharge of multiple ventricular ectopic foci or a circus movement. The fibrillating ventricles, like the fibrillating atria, look like a quivering "bag of worms." Ventricular fibrillation can be produced by an electric shock or an extrasystole during a critical interval, the **vulnerable period.** The vulnerable period coincides in time with the midportion of the T wave—ie, it occurs at a time when some of the ventricular myocardium is depolarized, some is incompletely repolarized, and some is completely repolarized. These are excellent conditions in which to establish reentry and a circus movement. The fibrillating ventricles cannot pump blood effectively, and circulation of the blood stops. Therefore, in the absence of emergency treatment, ventricular fibrillation that lasts more than a few minutes is fatal. The most frequent cause of sudden death in patients with myocardial infarcts is ventricular fibrillation.

Although ventricular fibrillation may be produced by electrocution, it can often be stopped and converted to normal sinus rhythm by means of electrical shocks. Electronic defibrillators are now available not only in hospitals but also in emergency vehicles and should be used as rapidly as possible. In addition, defibrillators can now be implanted surgically in patients who are at high risk for attacks of ventricular fibrillation.

In patients who are fibrillating or whose hearts have stopped, cardiac output and perfusion of the coronaries can be maintained by **cardiac massage.** Effective massage can be carried out without opening the chest. The person conducting external massage places the heel of one hand on the lower sternum

above the xiphoid process and the heel of the other hand on top of the first (Fig 28–16). Pressure is applied straight down, depressing the sternum 4 or 5 cm toward the spine. This procedure is repeated 60 times per minute. Manually squeezing the ventricles is also effective if the chest is already open, but emergency thoracotomies should not be performed. If breathing has also stopped, cardiac massage should alternate with mouth-to-mouth resuscitation (see Chapter 37).

Accelerated AV Conduction

An interesting condition seen in some otherwise normal individuals who are prone to attacks of paroxysmal atrial arrhythmias is **accelerated AV conduction (Wolff-Parkinson-White syndrome).** Normally, the only conducting pathway between the atria and the ventricles is the AV node. Individuals with Wolff-Parkinson-White syndrome have an additional aberrant muscular or nodal tissue connection (**bundle of Kent**) between the atria and ventricles. This conducts more rapidly than the slow-conducting AV node, and one ventricle is excited early. The manifestations of its activation merge with the normal QRS pattern, producing a short PR interval and a prolonged QRS deflection slurred on the upstroke (Fig 28–17), with a normal interval between the start of the P wave and the end of the QRS complex ("PJ in-

Figure 28–16. Technique of external, closed-chest cardiac massage. The black circle on the diagram of the heart shows the area where force should be applied. Circles on the supine figure at the apex of the heart and just to the right of the upper portion of the sternum show where electrodes should be applied for defibrillation. (Reproduced, with permission, from Schroeder SA et al [editors]: *Current Medical Diagnosis & Treatment, 1990.* Appleton & Lange, 1990.)

Figure 28–15. Ventricular fibrillation triggered by a ventricular premature beat (VPB) in the vulnerable period in a patient with a myocardial infarct. The patient was immediately defibrillated electrically and made an uneventful recovery.

Figure 28–17. Accelerated AV conduction. ***Top:*** Normal sinus beat. ***Middle:*** Short PR interval; wide, slurred QRS complex; normal PJ interval (Wolff-Parkinson-White syndrome). ***Bottom:*** Short PR interval, normal QRS complex (Lown-Ganong-Levine syndrome). (Reproduced, with permission, from Goldschlager N, Goldman MJ: *Principles of Clinical Electrocardiography,* 13th ed. Appleton & Lange, 1989.)

terval"). The paroxysmal atrial tachycardias seen in this syndrome often follow an atrial premature beat. This beat conducts normally down the AV node but finds the aberrant bundle refractory, since the bundle has a longer refractory period than the AV node. However, when ventricular activation spreads to the aberrant bundle, it is no longer refractory, and the impulse is transmitted retrograde to the atrium. A circus movement is thus established. Less commonly, an atrial premature beat finds the AV node refractory but reaches the ventricles via the bundle of Kent, setting up a circus movement in which the impulse passes from the ventricles to the atria via the AV node.

Attacks of paroxysmal supraventricular tachycardia, usually nodal tachycardia, are seen in individuals with short PR intervals and normal QRS complexes **(Lown-Ganong-Levine syndrome).** In this condition, depolarization presumably passes from the atria to the ventricles via an aberrant bundle that bypasses the AV node but enters the intraventricular conducting system distal to the node.

Antiarrhythmic Drugs

Many different drugs have now been developed that are of value in the treatment of arrhythmias because they slow conduction in the conduction system and the myocardium. This depresses ectopic activity and reduces the discrepancy between normal and re-entrant paths so that reentry does not occur. Some block Na^+ channels (class I antiarrhythmic drugs). Others block β-adrenergic receptors or otherwise reduce adrenergic activity in the heart (class II drugs), prolong the refractory period of cardiac muscle (class III drugs), or block Ca^{2+} channels (class IV drugs). These drugs are relatively effective in clinical situa-

tions. However, it has now become clear that in a small percentage of patients any of them can be **proarrhythmic,** rather than antiarrhythmic; ie, they can also cause various arrhythmias. Some of these can be life-threatening. Therefore, the drugs should be used only under careful medical supervision.

Many class IV drugs are also of value in the treatment of angina pectoris and myocardial infarction because they inhibit contraction of vascular smooth muscle, producing vasodilation and decreasing the afterload on the heart by lowering blood pressure (see Chapter 29).

ELECTROCARDIOGRAPHIC FINDINGS IN OTHER CARDIAC & SYSTEMIC DISEASES

Myocardial Infarction

When the blood supply to part of the myocardium is interrupted, there are profound changes in the myocardium that lead to irreversible changes and death of muscle cells **(myocardial infarction;** see Chapter 32). The ECG is very useful for diagnosing and locating areas of infarction. The underlying electrical events and the resulting electrocardiographic changes are complex, and only a brief review can be presented here.

The 3 major abnormalities that cause electrocardiographic changes in acute myocardial infarction are summarized in Table 28–3. The first change, abnormally rapid repolarization after discharge of the infarcted muscle fibers as a result of accelerated opening of K^+ channels, develops seconds after occlusion of a coronary artery in experimental animals. It lasts only a few minutes, but before it is over the resting membrane potential of the infarcted fibers declines because of the loss of intracellular K^+. Starting about 30 minutes later, the infarcted fibers also begin to depolarize more slowly than the surrounding normal fibers.

All 3 of these changes cause current flow that produces elevation of the ST segment in electrocardiographic leads recorded with electrodes over the in-

Table 28–3. Summary of the 3 major abnormalities of membrane polarization associated with acute myocardial infarction.

Defect in Infarcted Cells	Current Flow	Resultant ECG Change in Leads over Infarct
Rapid repolorization	Out of infarct	ST segment elevation
Delayed depolarization	Out of infarct	ST segment elevation
Decreased resting membrane potential	Into infarct	TQ segment depression (manifested as ST segment elevation)

farcted area (Fig 28–18). Because of the rapid repolarization in the infarct, the membrane potential of the area is greater than it is in the normal area during the latter part of repolarization, making the normal region negative relative to the infarct. Extracellularly, current therefore flows out of the infarct into the normal area (since, by convention, current flow is from positive to negative). This current flows toward electrodes over the injured area, causing increased positivity between the S and T waves of the ECG. Similarly, the delayed depolarization of the infarcted cells causes the infarcted area to be positive relative

Figure 28–18. Diagrammatic illustration of serial electrocardiographic patterns in anterior infarction. **A:** Normal tracing. **B:** Very early pattern (hours after infarction): ST segment elevation in I, aVL, and V_{3-6}; reciprocal ST depression in II, III, and aVF. **C:** Later pattern (many hours to a few days): Q waves have appeared in I, aVL, and V_{5-6}. QS complexes are present in V_{3-4}. This indicates that the major transmural infarction is underlying the area recorded by V_{3-4}; ST segment changes persist but are of lesser degree, and the T waves are beginning to invert in the leads in which the ST segments are elevated. **D:** Late established pattern (many days to weeks): The Q waves and QS complexes persist; the ST segments are isoelectric; the T waves are symmetric and deeply inverted in leads that had ST elevation and tall in leads that had ST depression. This pattern may persist for the remainder of the patient's life. **E:** Very late pattern: This may occur many months to years after the infarction. The abnormal Q waves and QS complexes persist. The T waves have gradually returned to normal. (Reproduced, with permission, from Goldschlager N, Goldman MJ: *Principles of Clinical Electrocardiography,* 13th ed. Appleton & Lange, 1989.)

to the healthy tissue (Table 28–3) during the early part of repolarization, and the result is also ST segment elevation. The remaining change, the decline in resting membrane potential during diastole, causes a current flow into the infarct during ventricular diastole. The result of this current flow is a depression of the TQ segment of the ECG. However, the electronic arrangement in electrocardiographic recorders is such that a TQ segment depression is recorded as an ST segment elevation. Thus, the hallmark of acute myocardial infarction is elevation of the ST segments in the leads overlying the area of infarction (Fig 28–18). Leads on the opposite side of the heart show ST segment depression.

After some days or weeks, the ST segment abnormalities subside. The dead muscle and scar tissue become electrically silent. The infarcted area is therefore negative relative to the normal myocardium during systole, and it fails to contribute its share of positivity to the electrocardiographic complexes. The manifestations of this negativity are multiple and subtle. Common changes include the appearance of a Q wave in some of the leads in which it was not previously present and an increase in the size of the normal Q wave in some of the other leads, although so-called non-Q-wave infarcts are also seen. These infarcts tend to be less severe, but there is a high incidence of subsequent reinfarction. Another finding in infarction of the anterior left ventricle is "failure of progression of the R wave"; the R wave fails to become successively larger in the precordial leads as the electrode is moved from right to left over the left ventricle. If the septum is infarcted, the conduction system may be damaged, causing bundle branch block or other forms of heart block.

Myocardial infarctions are often complicated by serious ventricular arrhythmias, with the threat of ventricular fibrillation and death. In experimental animals and presumably in humans, ventricular arrhythmias occur during 3 periods. During the first 30 minutes of an infarction, arrhythmias due to reentry are common. There follows a period relatively free from arrhythmias, but, starting 12 hours after infarction, arrhythmias occur as a result of increased automaticity. Arrhythmias occurring 3 days to several weeks after infarction are once again usually due to reentry.

Normal tracing (plasma K⁺ 4–5.5 meq/L). PR interval = 0.16 s; QRS interval = 0.06 s; QT interval = 0.4 s (normal for an assumed heart rate of 60).

Hyperkalemia (plasma K⁺ ±7.0 meq/L). The PR and QRS intervals are within normal limits. Very tall, slender peaked T waves are now present.

Hyperkalemia (plasma K⁺ ±8.5 meq/L). There is no evidence of atrial activity; the QRS complex is broad and slurred and the QRS interval has widened to 0.2 s. The T waves remain tall and slender. Further elevation of the plasma K⁺ level may result in ventricular tachycardia and ventricular fibrillation.

Hypokalemia (plasma K⁺ ±3.5 meq/L). PR interval = 0.2 s; QRS interval = 0.06 s; ST segment depression. A prominent U wave is now present immediately following the T. The actual QT interval remains 0.4 s. If the U wave is erroneously considered a part of the T, a falsely prolonged QT interval of 0.6 s will be measured.

Hypokalemia (plasma K⁺ ±2.5 meq/L). The PR interval is lengthened to 0.32 s; the ST segment is depressed; the T wave is inverted; a prominent U wave is seen. The true QT interval remains normal.

Figure 28–19. Correlation of plasma K⁺ level and the ECG, assuming that the plasma calcium level is normal. The diagrammed complexes are left ventricular epicardial leads. (Reproduced, with permission, from Goldman MJ: *Principles of Clinical Electrocardiography*, 12th ed. Lange, 1986.)

It is worth noting in this regard that infarcts that damage the epicardial portions of the myocardium interrupt sympathetic nerve fibers, producing denervation supersensitivity to catecholamines (see Chapter 4) in the area beyond the infarct. Alternatively, endocardial lesions can selectively interrupt vagal fibers (see above), leaving the actions of sympathetic fibers unopposed. Increased local sympathetic activity can cause arrhythmias. Interruption of the sympathetic outflow to the heart from the left stellate ganglion, which innervates the AV node, has been reported to be of value in the treatment of ventricular arrhythmias and prevention of sudden death after myocardial infarction.

Effects of Changes in the Ionic Composition of the Blood

Changes in ECF Na^+ and K^+ concentration would be expected to affect the potentials of the myocardial fibers, because the electrical activity of the heart depends upon the distribution of these ions across the muscle cell membranes. Clinically, a fall in the plasma level of Na^+ may be associated with low-voltage electrocardiographic complexes, but changes in the plasma K^+ level produce severe cardiac abnormalities. Hyperkalemia is a very dangerous and potentially lethal condition because of its effects on the heart. As the plasma K^+ level rises, the first change in the ECG is the appearance of tall peaked T waves, a manifestation of altered repolarization (Fig 28–19). At higher K^+ levels, paralysis of the atria and prolongation of the QRS complexes occur. Ventricular arrhythmias may develop. The resting membrane potential of the muscle fibers decreases as the extracellular K^+ concentration increases. The fibers eventually become unexcitable, and the heart stops in diastole. Conversely, a decrease in the plasma K^+ level causes prolongation of the PR interval, prominent U waves, and, occasionally, late T wave inversion in the precordial leads. If the T and U waves merge, the apparent QT interval is often prolonged; if the T and U waves are separated, the true QT interval is seen to be of normal duration. Hypokalemia is a serious condition, but it is not as rapidly fatal as hyperkalemia.

Increases in extracellular Ca^{2+} concentration enhance myocardial contractility. When large amounts of Ca^{2+} are infused into experimental animals, the heart relaxes less during diastole and eventually stops in systole (**calcium rigor**). However, in clinical conditions associated with hypercalcemia, the plasma calcium level is rarely if ever high enough to affect the heart. Hypocalcemia causes prolongation of the ST segment and consequently of the QT interval, a change that is also produced by phenothiazines and tricyclic antidepressant drugs and by various diseases of the central nervous system.

Changes in plasma calcium and potassium levels produce relatively marked changes in the sensitivity of the heart to digitalis. Hypercalcemia enhances digitalis toxicity, and hyperkalemia counteracts it. Mg^{2+}, which depresses the myocardium, also counteracts digitalis toxicity.

29

The Heart as a Pump

INTRODUCTION

The orderly depolarization process described in the previous chapter triggers a wave of contraction that spreads through the myocardium. In single muscle fibers, contraction starts just after depolarization and lasts until about 50 ms after repolarization is completed (see Fig 3–13). Atrial systole starts after the P wave of the ECG; ventricular systole starts near the end of the R wave and ends just after the T wave. The contraction produces sequential changes in pressures and flows in the heart chambers and blood vessels. It should be noted that the term **systolic pressure** in the vascular system refers to the peak pressure reached during systole, not the mean pressure; similarly, the **diastolic pressure** refers to the lowest pressure during diastole.

MECHANICAL EVENTS OF THE CARDIAC CYCLE

Events in Late Diastole

Late in diastole, the mitral and tricuspid valves between the atria and ventricles are open and the aortic and pulmonary valves are closed. Blood flows into the heart throughout diastole, filling the atria and ventricles. The rate of filling declines as the ventricles become distended, and—especially when the heart rate is low—the cusps of the atrioventricular

(AV) valves drift toward the closed position (Fig 29–1). The pressure in the ventricles remains low.

Atrial Systole

Contraction of the atria propels some additional blood into the ventricles, but about 70% of the ventricular filling occurs passively during diastole. Contraction of the atrial muscle that surrounds the orifices of the superior and inferior vena cava and pulmonary veins narrows their orifices, and the inertia of the blood moving toward the heart tends to keep blood in it; however, there is some regurgitation of blood into the veins during atrial systole.

Ventricular Systole

At the start of ventricular systole, the mitral and tricuspid (AV) valves close. Ventricular muscle initially shortens relatively little, but intraventricular pressure rises sharply as the myocardium presses on the blood in the ventricle (Fig 29–2). This period of **isovolumetric (isovolumic, isometric) ventricular contraction** lasts about 0.05 second, until the pressures in the left and right ventricles exceed the pressures in the aorta (80 mm Hg; 10.6 kPa) and pulmonary artery (10 mm Hg) and the aortic and pulmonary valves open. During isovolumetric contraction, the AV valves bulge into the atria, causing a small but sharp rise in atrial pressure (Fig 29–3).

When the aortic and pulmonary valves open, the phase of **ventricular ejection** begins. Ejection is

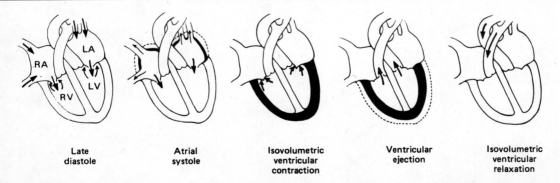

| Late diastole | Atrial systole | Isovolumetric ventricular contraction | Ventricular ejection | Isovolumetric ventricular relaxation |

Figure 29–1. Blood flow in the heart and great vessels during the cardiac cycle. The portions of the heart contracting in each phase are indicated in black. RA and LA, right and left atria; RV and LV, right and left ventricles.

Figure 29–2. Pressure-volume loop of the cardiac ventricles at rest and during exercise.

rapid at first, slowing down as systole progresses. The intraventricular pressure rises to a maximum and then declines somewhat before ventricular systole ends. Peak left ventricular pressure is about 120 mm Hg, and peak right ventricular pressure is 25 mm Hg or less. Late in systole, the aortic pressure actually exceeds the ventricular, but for a short period momentum keeps the blood moving forward. The AV valves are pulled down by the contractions of the ventricular muscle, and atrial pressure drops. The amount of blood ejected by each ventricle per stroke at rest is 70–90 mL. The **end-diastolic ventricular volume** is about 130 mL. Thus, the **ejection fraction,** the percentage of the ventricular volume ejected with each stroke, is about 65%, and about 50 mL of blood remains in each ventricle at the end of systole (**end-systolic ventricular volume**). The ejection fraction is a relatively good index of ventricular function, and as such is calculated in a variety of clinical situations.

Early Diastole

Once the ventricular muscle is fully contracted, the already falling ventricular pressures drop more rapidly. This is the period of **protodiastole.** It lasts about 0.04 second. It ends when the momentum of the ejected blood is overcome and the aortic and pulmonary valves close, setting up transient vibrations in the blood and blood vessel walls. After the valves are closed, pressure continues to drop rapidly during the period of **isovolumetric ventricular relaxation.** Isovolumetric relaxation ends when the ventricular pressure falls below the atrial pressure and the AV

valves open, permitting the ventricles to fill. Filling is rapid at first, then slows as the next cardiac contraction approaches. Atrial pressure continues to rise after the end of ventricular systole until the AV valves open, then drops and slowly rises again until the next atrial systole.

Pericardium

The heart is separated from the rest of the thoracic viscera by the pericardium. The myocardium itself is covered by the fibrous epicardium. The pericardial sac normally contains 5–30 mL of clear fluid, which lubricates the heart and permits it to contract with minimal friction.

Timing

Although events on the 2 sides of the heart are similar, they are somewhat asynchronous. Right atrial systole precedes left atrial systole, and contraction of the right ventricle starts after that of the left (see Chapter 28). However, since pulmonary arterial pressure is lower than aortic pressure, right ventricular ejection begins before left ventricular ejection. During expiration, the pulmonary and aortic valves close at the same time; but during inspiration, the aortic valve closes slightly before the pulmonary. The slower closure of the pulmonary valve is due to lower impedance of the pulmonary vascular tree. When measured over a period of minutes, the outputs of the 2 ventricles are, of course, equal, but transient differences in output during the respiratory cycle occur in normal individuals.

Length of Systole & Diastole

Cardiac muscle has the unique property of contracting and repolarizing faster when the heart rate is high (see Chapter 3), and the duration of systole decreases from 0.3 second at a heart rate of 65 to 0.16 second at a rate of 200 (Table 29–1). The shortening is due mainly to a decrease in the duration of systolic ejection. However, the duration of systole is much more fixed than that of diastole, and when the heart rate is increased, diastole is shortened to a much greater degree. For example, at a heart rate of 65, the duration of diastole is 0.62 second, whereas at a heart rate of 200, it is only 0.14 second. This fact has important physiologic and clinical implications. It is during diastole that the heart muscle rests, and coronary blood flow to the subendocardial portions of the left ventricle occurs only during diastole (see Chapter 32). Furthermore, most of the ventricular filling occurs in diastole. At heart rates up to about 180 beats/min, filling is adequate as long as there is ample venous return, and cardiac output per minute is increased by an increase in rate. However, at very high heart rates, filling may be compromised to such a degree that cardiac output per minute falls and symptoms of heart failure develop.

Because it has a prolonged action potential, car-

Figure 29–3. Events of the cardiac cycle at a heart rate of 75 beats/min. The phases of the cardiac cycle identified by the numbers at the bottom are as follows: 1, atrial systole; 2, isovolumetric ventricular contraction; 3, ventricular ejection; 4, isovolumetric ventricular relaxation; 5, ventricular filling. Note that late in systole, aortic pressure actually exceeds left ventricular pressure. However, the momentum of the blood keeps it flowing out of the ventricle for a short period. The pressure relationships in the right ventricle and pulmonary artery are similar. Abbreviations: Atr. syst., atrial systole; Ventric. syst., ventricular systole.

Table 29–1. Variation in length of action potential and associated phenomena with cardiac rate.[1]

	Heart Rate 75/min	Heart Rate 200/min	Skeletal Muscle
Duration, each cardiac cycle	0.80	0.30	. . .
Duration of systole	0.27	0.16	. . .
Duration of action potential	0.25	0.15	0.005
Duration of absolute refractory period	0.20	0.13	0.004
Duration of relative refractory period	0.05	0.02	0.003
Duration of diastole	0.53	0.14	. . .

[1]All values are in seconds. (Courtesy of AC Barger and GS Richardson.)

diac muscle is in its refractory period and will not contract in response to a second stimulus until near the end of the initial contraction (see Fig 3–13). Therefore, cardiac muscle cannot be tetanized like skeletal muscle. The highest rate at which the ventricles can contract is theoretically about 400/min, but in adults the AV node will not conduct more than about 230 impulses/min because of its long refractory period. A ventricular rate of more than 230 is seen only in paroxysmal ventricular tachycardia (see Chapter 28).

Exact measurement of the duration of isometric ventricular contraction is difficult in clinical situations, but it is relatively easy to measure the duration of **total electromechanical systole (QS$_2$),** the **pre-ejection period (PIP),** and the **left ventricular ejection time (LVET)** by recording the ECG, phonocardiogram, and carotid pulse simultaneously. QS$_2$ is the period from the onset of the QRS complex to the closure of the aortic valves, as determined by the onset of the second heart sound. LVET is the period from the beginning of the carotid pressure rise to the dicrotic notch (see below). PEP is the difference between QS$_2$ and LVET and represents the time for the electrical as well as the mechanical events that precede systolic ejection. The ratio PEP/LVET is normally about 0.35, and it increases without a change in QS$_2$ when left ventricular performance is compromised in a variety of cardiac diseases.

Arterial Pulse

The blood forced into the aorta during systole not only moves the blood in the vessels forward but also sets up a pressure wave that travels along the arteries. The pressure wave expands the arterial walls as it travels, and the expansion is palpable as the **pulse.** The rate at which the wave travels, which is independent of and much higher than the velocity of blood flow, is about 4 m/s in the aorta, 8 m/s in the large arteries, and 16 m/s in the small arteries of young adults. Consequently, the pulse is felt in the radial artery at the wrist about 0.1 second after the peak of systolic ejection into the aorta (Fig 29–3). With advancing age, the arteries become more rigid, and the pulse wave moves faster.

The strength of the pulse is determined by the pulse pressure and bears little relation to the mean pressure. The pulse is weak ("thready") in shock. It is strong when stroke volume is large, eg, during exercise or after the administration of histamine. When the pulse pressure is high, the pulse waves may be large enough to be felt or even heard by the individual (palpitation, "pounding heart"). When the aortic valve is incompetent (aortic insufficiency), the pulse is particularly strong, and the force of systolic ejection may be sufficient to make the head nod with each heartbeat. The pulse in aortic insufficiency is called a **collapsing, Corrigan,** or **water-hammer pulse.** A water hammer is an evacuated glass tube half-filled with water that was a popular toy in the 19th century. When held in the hand and inverted, it delivers a short, hard knock.

The **dicrotic notch,** a small oscillation on the falling phase of the pulse wave caused by vibrations set up when the aortic valve snaps shut (Fig 29–3), is visible if the pressure wave is recorded but is not palpable at the wrist. There is also a dicrotic notch on the pulmonary artery pressure curve produced by the closure of the pulmonary valves.

Atrial Pressure Changes & the Jugular Pulse

Atrial pressure rises during atrial systole and continues to rise during isovolumetric ventricular contraction when the AV valves bulge into the atria. When the AV valves are pulled down by the contracting ventricular muscle, pressure falls rapidly and then rises as blood flows into the atria until the AV valves open early in diastole. The return of the AV valves to their relaxed position also contributes to this pressure rise by reducing atrial capacity. The atrial pressure changes are transmitted to the great veins, producing 3 characteristic waves in the jugular pulse (Figs 29–3 and 29–4). The **a wave** is due to atrial systole. As noted above, some blood regurgitates into the great veins when the atria contract, even though the orifices of the great veins are constricted. In addition, venous inflow stops, and the resultant rise in venous pressure contributes to the a wave. The **c wave** is the transmitted manifestation of the rise in atrial pressure produced by the bulging of the tricuspid valve into the atria during isometric ventricular contraction. The **v wave** mirrors the rise in atrial pressure before the tricuspid valve opens during diastole. The jugular pulse waves are superimposed on the respiratory fluctuations in venous pressure. Venous pressure falls during inspiration as a result of the increased negative intrathoracic pressure and rises again during expiration.

Figure 29–4. Jugular venous pressure record (JVP), showing a, c, and v waves, compared with the simultaneously recorded ECG and carotid pressure record. (Redrawn and reproduced, with permission, from Wood PW: *Diseases of the Heart.* Lippincott, 1956.)

Careful bedside inspection of the pulsations of the jugular veins may give clinical information of some importance. For example, in tricuspid insufficiency there is a giant c wave with each ventricular systole. In complete heart block, when the atria and ventricles are beating at different rates, the a waves that are not synchronous with the radial pulse can be made out, and there is a giant a wave ("cannon wave") whenever the atria contract while the tricuspid valve is closed. It may also be possible to distinguish atrial from ventricular extrasystoles by inspection of the jugular pulse, because atrial premature beats produce an a wave, whereas ventricular premature beats do not.

Heart Sounds

Two sounds are normally heard through a stethoscope during each cardiac cycle. The first is a low, slightly prolonged "lub" **(first sound),** caused by vibrations set up by the sudden closure of the mitral and tricuspid valves at the start of ventricular systole (Fig 29–3). The second is a shorter, high-pitched "dup" **(second sound),** caused by vibrations associated with closure of the aortic and pulmonary valves just after the end of ventricular systole. A soft, low-pitched **third sound** is heard about one-third of the way through diastole in many normal young individuals. It coincides with the period of rapid ventricular filling and is probably due to vibrations set up by the inrush of blood. A **fourth sound** can sometimes be heard immediately before the first sound when atrial pressure is high or the ventricle is stiff in conditions

such as ventricular hypertrophy. It is due to ventricular filling and is rarely heard in normal adults.

The first sound has a duration of about 0.15 second and a frequency of 25–45 Hz. It is soft when the heart rate is low, because the ventricles are well filled with blood and the leaflets of the AV valves float together before systole. The second sound lasts about 0.12 second, with a frequency of 50 Hz. It is loud and sharp when the diastolic pressure in the aorta or pulmonary artery is elevated, causing the respective valves to shut briskly at the end of systole. The interval between aortic and pulmonary valve closure during inspiration is frequently long enough for the second sound to be reduplicated (physiologic splitting of the second sound). Splitting also occurs in various diseases. The third sound has a duration of 0.1 second.

Murmurs

Murmurs and **bruits** are abnormal sounds heard in various parts of the vascular system. As discussed in detail in Chapter 30, blood flow is laminar and nonturbulent up to a critical velocity; above this velocity, and beyond an obstruction, blood flow is turbulent. Laminar flow is silent, but turbulent flow creates sounds. Blood flow speeds up when an artery or a heart valve is narrowed.

Examples of vascular sounds outside the heart are the bruit heard over a large, highly vascular goiter and the murmurs heard over an aneurysmal dilatation of one of the large arteries, an arteriovenous (A-V) fistula, or a patent ductus arteriosus.

The major, but certainly not the only, cause of cardiac murmurs is disease of the heart valves. When a valve is narrowed **(stenosis),** blood flow through it in the normal direction is accelerated and turbulent. When a valve is incompetent, blood flows backward through it **(regurgitation** or **insufficiency),** again through a narrow orifice that accelerates flow. The timing (systolic or diastolic) of a murmur due to stenosis or insufficiency of any particular valve (Table 29–2) can be predicted from a knowledge of the mechanical events of the cardiac cycle. Murmurs due to disease of a particular valve can generally be heard best when the stethoscope is over that particular valve; thus, murmurs due to disorders of the aortic and pulmonic valves are usually heard best at the base of the heart, and murmurs due to mitral disease

Table 29 – 2. Heart murmurs.

Valve	Abnormality	Timing of Murmur
Aortic or pulmonary	Stenosis	Systolic
	Insufficiency	Diastolic
Mitral or tricuspid	Stenosis	Diastolic
	Insufficiency	Systolic

are usually heard best at the apex. There are other aspects of the duration, character, accentuation, and transmission of the sound that help to locate its origin in one valve or the other. One of the loudest murmurs is that produced when blood flows backward in diastole through a hole in a cusp of the aortic valve. Most murmurs can be heard only with the aid of the stethoscope, but this high-pitched musical diastolic murmur is sometimes audible to the unaided ear several feet from the patient.

In patients with congenital interventricular septal defects, flow from the left to the right ventricle causes a systolic murmur. Soft murmurs may also be heard in patients with interatrial septal defects, although they are not a constant finding.

Soft systolic murmurs are common in individuals, especially children, who have no cardiac disease. Systolic murmurs are also common in anemic patients as a result of the low viscosity of the blood and the rapid flow (see Chapter 30).

Echocardiography

Wall movement and other aspects of cardiac function can be evaluated by **echocardiography,** a noninvasive technique that does not involve injections or insertion of a catheter. In echocardiography, pulses of ultrasonic waves, commonly at a frequency of 2.25 MHz, are emitted from a transducer that also functions as a receiver to detect waves reflected back from various parts of the heart. Reflections occur wherever acoustic impedance changes, and a recording of the echoes displayed against time on an oscilloscope provides a record of the movements of the ventricular wall, septum, and valves during the cardiac cycle. When combined with Doppler techniques, echocardiography can be used to measure velocity and volume of flow through valves. It has considerable clinical usefulness, particularly in evaluating and planning therapy in patients with valvular lesions.

CARDIAC OUTPUT

Methods of Measurement

In experimental animals, cardiac output can be measured with an electromagnetic flow meter placed on the ascending aorta. Two methods of measuring output that are applicable to humans, in addition to Doppler combined with echocardiography, are the **direct Fick method** and the **indicator dilution method.**

The **Fick principle** states that the amount of a substance taken up by an organ (or by the whole body) per unit of time is equal to the arterial level of the substance minus the venous level (**A-V difference**) times the blood flow. This principle can be applied, of course, only in situations in which the arterial blood is the sole source of the substance taken up.

The principle can be used to determine cardiac output by measuring the amount of O_2 consumed by the body in a given period and dividing this value by the A-V difference across the lungs. Because arterial blood has the same content in all parts of the body, the arterial O_2 content can be measured in a sample obtained from any convenient artery. A sample of venous blood in the pulmonary artery is obtained by means of a cardiac catheter. Right atrial blood has been used in the past, but mixing of this blood may be incomplete, so that the sample is not representative of the whole body. An example of the calculation of cardiac output using a typical set of values is as follows:

$$\text{Output of left ventricle} = \frac{O_2 \text{ consumption (mL/min)}}{[A_{O_2}] - [V_{O_2}]}$$

$$= \frac{250 \text{ mL/min}}{\begin{array}{c}190 \text{ mL/L arterial blood} - \\ 140 \text{ mL/L venous blood in} \\ \text{pulmonary artery}\end{array}}$$

$$= \frac{250 \text{ mL/min}}{50 \text{ mL/L}}$$

$$= 5 \text{ L/min}$$

It has now become commonplace to insert a long catheter through a forearm vein and to guide its tip into the heart with the aid of a fluoroscope. The technique was initially developed by Forssmann, who catheterized himself but was summarily fired from his job when he sought permission to explore the use of the catheter in others for diagnostic purposes. However, the procedure is now known to be essentially harmless. Catheters can be inserted not only into the right atrium but also through the atrium and the right ventricle into the small branches of the pulmonary artery. Catheters can also be inserted in peripheral arteries and guided in a retrograde direction to the heart and into coronary or other arteries.

In the indicator dilution technique, a known amount of a dye such as indocyanine green or a radioactive isotope is injected into an arm vein and the concentration of the indicator in serial samples of arterial blood is determined. The output of the heart is equal to the amount of indicator injected divided by its average concentration in arterial blood after a single circulation through the heart (Fig 29–5). The indicator must, of course, be a substance that stays in the bloodstream during the test and has no harmful or hemodynamic effects. In practice, the log of the indicator concentration in the serial arterial samples is plotted against time as the concentration rises, falls, and then rises again as the indicator recirculates. The initial decline in concentration, linear on a semilog plot, is extrapolated to the abscissa, giving the time for first passage of the indicator through the circulation. The cardiac output for that period is calculated (Fig 29–5) and then converted to output per minute.

$$F = \frac{E}{\int_{0}^{\infty} Cdt}$$

F = flow
E = amount of indicator injected
C = instantaneous concentration of indicator in arterial blood

In the **rest** example above,

Flow in 39 s
(time of first passage) = $\dfrac{5 \text{ mg injection}}{1.6 \text{ mg/L}}$
(avg concentration)

Flow = 3.1 L in 39 s

Flow (cardiac output)/min = $3.1 \times \dfrac{60}{39} = 4.7$ L

For the **exercise** example,

Flow in 9 s = $\dfrac{5 \text{ mg}}{1.51 \text{ mg/L}} = 3.3$ L

Flow/min = $3.3 \times \dfrac{60}{9} = 22.0$ L

Figure 29–5. Determination of cardiac output by indicator (dye) dilution. (Data and graph from Asmussen E, Nielsen M: The cardiac output in rest and work determined by the acetylene and the dye injection methods. *Acta Physiol Scand* 1952;**27**:217.)

Similar estimates of ventricular output have been made by injecting a suitable radioactive substance (radionuclide) and monitoring radioactivity over the heart during the first pass of the radionuclide through it.

One of the most popular indicator dilution techniques in recent years has been **thermodilution**, in which the indicator used is cold saline. The saline is injected into the right atrium through one side of a double-lumen catheter, and the temperature change in the blood is recorded in the pulmonary artery, using a thermistor in the other, longer side of the catheter. The temperature change is inversely proportionate to the amount of blood flowing through the pulmonary artery, ie, to the extent that the cold saline is diluted by blood. This technique has 2 important advantages: (1) the saline is completely innocuous; and (2) the cold is dissipated in the tissues so recirculation is not a problem, and it is easy to make repeated determinations.

Cardiac Output in Various Conditions

The amount of blood pumped out of each ventricle per beat, the **stroke volume**, is about 70 mL in a resting man of average size in the supine position (70 mL from the left ventricle and 70 mL from the right, with the 2 ventricular pumps in series). The output of the heart per unit time is the **cardiac output.** In a resting, supine man, it averages about 5.0 L/min (70 mL × 72 beats/min). There is a correlation between resting cardiac output and body surface area. The output per minute per square meter of body surface (the **cardiac index**) averages 3.2 L. The effects of various conditions on cardiac output are summarized in Table 29–3.

Factors Controlling Cardiac Output

Variations in cardiac output can be produced by changes in cardiac rate or stroke volume (Fig 29–6). The cardiac rate is controlled primarily by the cardiac innervation, sympathetic stimulation increasing the rate and parasympathetic stimulation decreasing it (see Chapter 28). The stroke volume is also determined in part by neural input, sympathetic stimuli making the myocardial muscle fibers contract with greater strength at any given length and parasympathetic stimuli having the opposite effect. When the strength of contraction increases without an increase in fiber length, more of the blood that normally remains in the ventricles is expelled; ie, the ejection fraction increases and the end-systolic ventricular blood volume falls. The cardiac accelerator action of

Table 29–3. Effect of various conditions on cardiac output.

	Condition or Factor[1]
No change	Sleep Moderate changes in environmental temperature
Increase	Anxiety and excitement (50–100%) Eating (30%) Exercise (up to 700%) High environmental temperature Pregnancy Epinephrine
Decrease	Sitting or standing from lying position (20–30%) Rapid arrhythmias Heart disease

[1]Approximate percent changes are shown in parentheses.

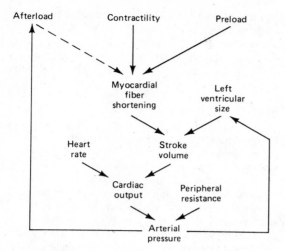

Figure 29–6. Interactions between the components that regulate cardiac output and arterial pressure. Solid lines indicate increases, and the dashed line indicates a decrease. (Modified from Braunwald E: Regulation of the circulation. *N Engl J Med* 1974;**290**:1124.)

the catecholamines liberated by sympathetic stimulation is referred to as their **chronotropic action,** whereas their effect on the strength of cardiac contraction is called their **inotropic action.** Factors that increase the strength of cardiac contraction are said to be positively inotropic; those that decrease it are said to be negatively inotropic.

The force of contraction of cardiac muscle is dependent upon its preloading and its afterloading. These factors are illustrated in Fig 29–7, in which a muscle strip is stretched by a load (the **preload**) that

Figure 29–7. Model for isotonic contraction of afterloaded muscles. *A:* Rest. *B:* Partial contraction of the contractile element of the muscle (CE), with stretching of the series elastic element (SE) but no shortening. *C:* Complete contraction, with shortening. (Reproduced, with permission, from Sonnenblick EH in: *The Myocardial Cell: Structure, Function and Modification.* Briller SA, Conn HL [editors]. Univ Pennsylvania Press, 1966.)

rests on a platform. The initial phase of the contraction is isometric; the elastic component in series with the contractile element is stretched, and tension increases until it is sufficient to lift the load. The tension at which the load is lifted is the **afterload.** The muscle then contracts isotonically without developing further tension. In vivo, the preload is the degree to which the myocardium is stretched before it contracts and the afterload is the resistance against which blood is expelled.

Relation of Tension to Length in Cardiac Muscle

The length-tension relationship in cardiac muscle (see Fig 3–15) is similar to that in skeletal muscle (see Fig 3–9); as the muscle is stretched, the developed tension increases to a maximum and then declines as stretch becomes more extreme. Starling pointed this out when he stated that the "energy of contraction is proportional to the initial length of the cardiac muscle fiber." This pronouncement has come to be known as **Starling's law of the heart** or the **Frank-Starling law.** For the heart, the length of the muscle fibers (ie, the extent of the preload) is proportionate to the end-diastolic volume. The relation between ventricular stroke volume and end-diastolic volume is called the Frank-Starling curve.

Regulation of cardiac output as a result of changes in cardiac muscle fiber length is sometimes called **heterometric regulation,** whereas regulation due to changes in contractility independent of length is sometimes called **homometric regulation.**

Heart-Lung Preparation

The effect of changes in peripheral resistance can be demonstrated in the **heart-lung preparation** (Fig 29–8). In this preparation, the heart and lungs of an anesthetized experimental animal, usually a dog, are cannulated in such a way that blood flows from the aorta through a system of tubing and reservoirs to the right atrium, and from there through the animal's heart and lungs back to the aorta. Deprived of its blood supply, the rest of the animal dies, so that the heart is functionally denervated and the heart rate varies little if at all. By decreasing the caliber of the outflow tubing, the resistance against which the heart pumps (**peripheral resistance**) can be increased. When the peripheral resistance is increased, the heart puts out less blood than it receives for several beats. Blood accumulates in the ventricles, and the size of the heart increases. The distended heart beats more forcefully, and output returns to its previous level.

The effect of variations in venous return can also be demonstrated in the heart-lung preparation. When the reservoir emptying into the right atrium in Fig 29–8 is raised, venous pressure and hence venous return are increased. The myocardial fibers are stretched, and cardiac output is increased.

Figure 29–8. Heart-lung preparation.

Factors Affecting End-Diastolic Volume

The factors that normally operate to regulate end-diastolic volume, ie, the degree to which cardiac muscle is stretched, are listed in Table 29–4. An increase in intrapericardial pressure limits the extent to which the ventricle can fill. So does a decrease in ventricular compliance, ie, an increase in ventricular stiffness produced by myocardial infarction, infiltrative disease, and other abnormalities. Atrial contractions aid ventricular filling. The other factors affect the amount of blood returning to the heart and hence the degree of cardiac filling during diastole. An increase in total blood volume increases venous return. Constriction of the veins reduces the size of the venous reservoirs, decreasing venous pooling and thus increasing venous return. An increase in the normal

Table 29–4. Factors that normally increase or decrease the length of ventricular cardiac muscle fibers.

Increase
Stronger atrial contractions
Increased total blood volume
Increased venous tone
Increased pumping action of skeletal muscle
Increased negative intrathoracic pressure
Decrease
Standing
Increased intrapericardial pressure
Decreased ventricular compliance

negative intrathoracic pressure increases the pressure gradient along which blood flows to the heart, whereas a decrease impedes venous return. Standing decreases venous return, and muscular activity increases it as a result of the pumping action of skeletal muscle.

Myocardial Contractility

The contractility of the myocardium exerts a major influence on stroke volume. When the sympathetic nerves to the heart are stimulated, the whole length-tension curve shifts upward and to the left (Fig 29–9). The positively inotropic effect of the norepinephrine liberated at the nerve endings is augmented by circulating norepinephrine, and epinephrine has a similar effect. There is a negatively inotropic effect of vagal stimulation on the atrial muscle and a small negatively inotropic effect on the ventricular muscle.

Changes in cardiac rate and rhythm also affect myocardial contractility (force-frequency relation, Fig 29–9). Ventricular extrasystoles condition the myocardium in such a way that the next succeeding contraction is stronger than the preceding normal contraction. This **postextrasystolic potentiation** is independent of ventricular filling, since it occurs in isolated cardiac muscle, and is due to increased availability of intracellular Ca^{2+}. A sustained increment in contractility can be produced by delivering paired electrical stimuli to the heart in such a way that the second stimulus is delivered shortly after the refractory period of the first. It has also been shown that myocardial contractility increases as the heart rate increases, although this effect is relatively small.

The catecholamines exert their inotropic effect via an action on cardiac β_1 adrenergic receptors and G_s, with resultant activation of adenylate cyclase and increased intracellular cyclic AMP. Xanthines such as caffeine and theophylline that inhibit the breakdown of cyclic AMP are positively inotropic. Glucagon, which increases the formation of cyclic AMP, is positively inotropic, and it has been recommended for use in the treatment of some heart diseases. The positively inotropic effect of digitalis and related drugs (Fig 29–9) is due to their inhibitory effect on the Na^+-K^+ ATPase in the myocardium. The inhibition causes an increase in intracellular Na^+, which in turn increases the availability of Ca^{2+} in the cell. The key role of Ca^{2+} in the initiation of contraction in skeletal and cardiac muscle is discussed in Chapter 3. Hypercapnia, hypoxia, acidosis, and drugs such as quinidine, procainamide, and barbiturates depress myocardial contractility. The contractility of the myocardium is also reduced in heart failure (intrinsic depression). The cause of this depression is not known. If part of the myocardium becomes fibrotic and nonfunctional as a result of a myocardial infarction, total ventricular performance is reduced.

Figure 29–9. Effect of changes in myocardial contractility on the Frank-Starling curve. The curve shifts downward and to the right as contractility is decreased. The major factors influencing contractility are summarized on the right. The dashed lines indicate portions of the ventricular function curves where maximum contractility has been exceeded; ie, they identify points on the "falling limb" of the Frank-Starling curve. EDV, end-diastolic volume. (Reproduced, with permission, from Braunwald E, Ross J, Sonnenblick EH: Mechanisms of contraction of the normal and failing heart. *N Engl J Med* 1967;**277**:794. Courtesy of Little, Brown, Inc.)

Integrated Control of Cardiac Output

In intact animals and humans, the mechanisms listed above operate in an integrated way to maintain cardiac output. During muscular exercise, there is increased sympathetic discharge, so that myocardial contractility is increased and the heart rate rises. The increase in heart rate is particularly prominent in normal individuals, and there is only a modest increase in stroke volume (Table 29–5). However, patients with transplanted hearts are able to increase their cardiac output during exercise in the absence of cardiac innervation through the operation of the Frank-Starling mechanism (Fig 29–10). Circulating catecholamines also contribute. The increase seen in these patients is not as rapid, and their maximal increase is smaller than in normal individuals, but is still appre-

ciable. It is important to note in this regard that the heart is the servant, rather than the master; it normally adjusts its function so that it expels all the blood that returns in the veins. If venous return increases and there is no change in sympathetic tone, venous pressure rises, diastolic inflow is greater, ventricular end-diastolic pressure increases, and the heart muscle contracts more forcefully. During muscular exercise, venous return is increased by the pumping action of the muscles and the increase in respiration (see Chapter 33). In addition, because of vasodilation in the contracting muscles, peripheral resistance and consequently afterload are decreased. The end result in both normal and transplanted hearts is thus a prompt and marked increase in cardiac output.

One of the differences between untrained individ-

Table 29–5. Changes in cardiac function with exercise. Note that stroke volume levels off, then falls somewhat (as a result of the shortening of diastole) when the heart rate rises to high values.[1]

Work (kg·m/min)	O₂ Usage (mL/min)	Pulse Rate (per min)	Cardiac Output (L/min)	Stroke Volume (mL)	A-V O₂ Difference (mL/dL)
Rest	267	64	6.4	100	4.3
288	910	104	13.1	126	7.0
540	1430	122	15.2	125	9.4
900	2143	161	17.8	110	12.3
1260	3007	173	20.9	120	14.5

[1]Reproduced, with permission, from Asmussen E, Nielsen M: The cardiac output in rest and work determined by the acetylene and the dye injection methods. *Acta Physiol Scand* 1952;**27**:217.

Figure 29–10. Cardiac responses to moderate supine exercise in normal humans and patients with cardiac transplants. (Reproduced, with permission, from Kent KM, Cooper T: The denervated heart. *N Engl J Med* 1974;**291:** 1017.)

uals and trained athletes is that the athletes have lower heart rates, greater end-systolic ventricular volumes, and greater stroke volumes at rest. Therefore, they can potentially achieve a given increase in cardiac output by further increases in stroke volume without increasing their heart rate to as great a degree as an untrained individual.

Oxygen Consumption of the Heart

The basal O_2 consumption of the myocardium, which can be determined by stopping the heart while artificially maintaining the coronary circulation, is about 2 mL/100 g/min. This value is considerably higher than that of resting skeletal muscle. O_2 consumption of the beating heart is about 9 mL/100 g/min at rest. Increases occur during exercise and in a number of different states. Cardiac venous O_2 tension is low, and little additional O_2 can be extracted from the blood in the coronaries, so increases in O_2 consumption require increases in coronary blood flow. The regulation of coronary flow is discussed in Chapter 32.

The O_2 consumption of the heart is determined primarily by the intramyocardial tension, the contractile state of the myocardium, and the heart rate. Ventricular work per beat correlates with O_2 consumption. The work is the product of stroke volume and mean arterial pressure in the pulmonary artery (for the right ventricle) or the aorta (for the left ventricle). Since aortic pressure is 7 times greater than pulmonary artery pressure, the stroke work of the left ventricle is approximately 7 times the stroke work of the right ventricle. In theory, a 25% increase in stroke volume without a change in arterial pressure should produce the same increase in O_2 consumption as a 25% increase in arterial pressure without a change in stroke volume. However, for reasons that are incompletely understood, pressure work produces a greater increase in O_2 consumption than volume work. In other words, an increase in afterload causes a greater increase in cardiac O_2 consumption than an increase in preload does. This is why angina pectoris due to deficient delivery of O_2 to the myocardium is more common in aortic stenosis than in aortic insufficiency. In aortic stenosis, intraventricular pressure must be increased to force blood through the stenotic valve, whereas in aortic insufficiency, regurgitation of blood produces an increase in stroke volume with little change in aortic impedance.

It is worth noting that the increase in O_2 consumption produced by increased stroke volume when the myocardial fibers are stretched is an example of the operation of the law of Laplace. This law, which is discussed in detail in Chapter 30, states that the tension developed in the wall of a hollow viscus is proportionate, among other things, to the radius of the viscus, and the radius of a dilated heart is increased. O_2 consumption per unit time increases when the heart rate is increased by sympathetic stimulation because of the increased number of beats and the increased velocity and strength of each contraction. However, this is somewhat offset by the decrease in end-systolic volume and hence in the radius of the heart.

Dynamics of Blood & Lymph Flow   30

INTRODUCTION

The blood vessels are a closed system of conduits that carry blood from the heart to the tissues and back to the heart. Some of the interstitial fluid enters the lymphatics and passes via these vessels to the vascular system. Blood flows through the vessels primarily because of the forward motion imparted to it by the pumping of the heart, although in the case of the systemic circulation, diastolic recoil of the walls of the arteries, compression of the veins by skeletal muscles during exercise, and the negative pressure in the thorax during inspiration also move the blood forward. The resistance to flow depends to a minor degree upon the viscosity of the blood but mostly upon the diameter of the vessels, principally the arterioles. The blood flow to each tissue is regulated by local chemical and general neural and humoral mechanisms that dilate or constrict the vessels of the tissue. All of the blood flows through the lungs, but the systemic circulation is made up of numerous different circuits in parallel (Fig 30–1), an arrangement that permits wide variations in regional blood flow without changing total systemic flow.

This chapter is concerned with the general principles that apply to all parts of the circulation and with pressure and flow in the systemic circulation. The homeostatic mechanisms operating to adjust flow are the subject of Chapter 31. The special characteristics of pulmonary and renal circulation are discussed in Chapters 34 and 38 and the unique features of the circulation to other organs in Chapter 32.

ANATOMIC CONSIDERATIONS

Arteries & Arterioles

The characteristics of the various types of blood vessels are listed in Table 30–1. The walls of all arteries are made up of an outer layer of connective tissue, the adventitia; a middle layer of smooth muscle, the media; and an inner layer, the intima, made up of the endothelium and underlying connective tissue. The walls of the aorta and other arteries of large diameter contain a relatively large amount of elastic tissue. They are stretched during systole and recoil on the blood during diastole. The walls of the arterioles contain less elastic tissue but much more smooth muscle. The muscle is innervated by noradrenergic nerve fibers, which are constrictor in function, and in some instances by cholinergic fibers, which dilate the vessels. The arterioles are the major site of the resistance to blood flow, and small changes in their caliber cause large changes in the total peripheral resistance.

Capillaries

The arterioles divide into smaller muscle-walled vessels, sometimes called **metarterioles,** and these in turn feed into capillaries (Fig 30–2). In some of the vascular beds that have been studied in detail, a metarteriole is connected directly with a venule by a capillary **thoroughfare vessel,** and the true capillaries are an anastomosing network of side branches of this thoroughfare vessel. The openings of the true capillaries are surrounded on the upstream side by minute smooth muscle **precapillary sphincters.** It is unsettled whether the metarterioles are innervated, and it appears that the precapillary sphincters are not. However, they can of course respond to local or circulating vasoconstrictor substances. The true capil-

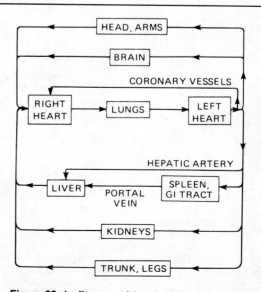

Figure 30–1. Diagram of the circulation in the adult.

Table 30-1. Characteristics of various types of blood vessels in humans.

Vessel	Lumen Diameter	Wall Thickness	All Vessels of Each Type	
			Approximate Total Cross-Sectional Area (cm²)	Percentage of Blood Volume Contained[1]
Aorta	2.5 cm	2 mm	4.5	2
Artery	0.4 cm	1 mm	20	8
Arteriole	30 μm	20 μm	400	1
Capillary	5 μm	1 μm	4500	5
Venule	20 μm	2 μm	4000	
Vein	0.5 cm	0.5 mm	40	54
Vena cava	3 cm	1.5 mm	18	

[1] In systemic vessels. There is an additional 12% in the heart and 18% in the pulmonary circulation.

laries are about 5 μm in diameter at the arterial end and 9 μm in diameter at the venous end. When the sphincters are dilated, the diameter of the capillaries is just sufficient to permit red blood cells to squeeze through in "single file." As they pass through the capillaries, the red cells become thimble- or parachute-shaped, with the concavity pointing in the direction of flow. This configuration appears to be due simply to the pressure in the center of the vessel whether or not the edges of the red blood cell are in contact with the capillary walls.

The total area of all the capillary walls in the body exceeds 6300 m² in the adult. The walls, which are

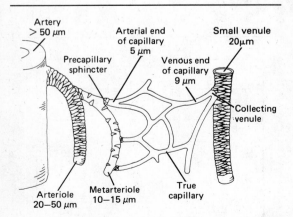

Figure 30–2. The microcirculation. Arterioles give rise to metarterioles, which give rise to capillaries. The capillaries drain via short collecting venules to the venules. The walls of the arteries, arterioles, and small venules contain relatively large amounts of smooth muscle. There are scattered smooth muscle cells in the walls of the metarterioles, and the openings of the capillaries are guarded by muscular precapillary sphincters. The diameters of the various vessels are also shown. (Courtesy of JN Diana.)

about 1 μm thick, are made up of a single layer of endothelial cells. The structure of the walls varies from organ to organ. In many beds, including those in skeletal, cardiac, and smooth muscle, the junctions between the endothelial cells (Fig 30–3) permit the passage of molecules up to 10 nm in diameter. It also appears that plasma and its dissolved proteins are taken up by endocytosis, transported across the endothelial cells, and discharged by exocytosis (**vesicular transport;** see Chapter 1). However, this process can account for only a small portion of the transport across the endothelium. In the brain, the capillaries resemble the capillaries in muscle, but the junctions between endothelial cells are tighter; they permit the passage of small molecules only. In most endocrine glands, the intestinal villi, and parts of the kidneys, the cytoplasm of the endothelial cells is attenuated to form gaps called **fenestrations.** These fenestrations are 20–100 nm in diameter. They permit the passage of relatively large molecules and make the capillaries porous. Except in the renal glomeruli, they appear to be closed by a thin membrane. However, in a number of different tissues, the membrane can be shown by a rapid freeze-fracture technique to be discontinuous, consisting of a central hub joined by spokes of membrane to the edges of the fenestration (Fig 30–4). In the liver, where the sinusoidal capillaries are extremely porous, the endothelium is discontinuous and there are large gaps between endothelial cells that are not closed by membranes (see Fig 26–18). Some of the gaps are 600 nm in diameter, and others may be as large as 3000 nm. The permeabilities of capillaries in various parts of the body, expressed in terms of their hydraulic conductivity, are summarized in Table 30–2.

Lymphatics

The lymphatics drain from the lungs and from the rest of the body tissues via a system of vessels that

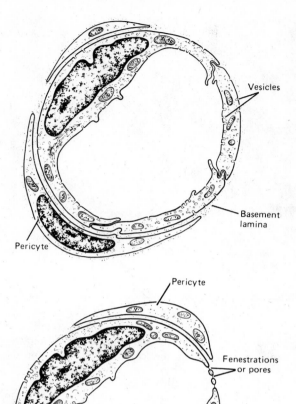

Figure 30–3. Cross-sections of capillaries. **Top:** Continuous type of capillary found in muscle. **Bottom:** Fenestrated type of capillary. (Reproduced, with permission, from Fawcett DW: *Bloom and Fawcett, Textbook of Histology,* 11th ed. Saunders, 1986.)

coalesce and eventually enter the right and left subclavian veins at their junctions with the respective internal jugular veins. The lymph vessels contain valves and regularly traverse lymph nodes along their course. The ultrastructure of the small lymph vessels differs from that of the capillaries in several details: There are no visible fenestrations in the lymphatic endothelium: there is very little if any basal lamina under the endothelium; and the junctions between endothelial cells are open, with no tight intercellular connections.

Arteriovenous Anastomoses

In the fingers, palms, and ear lobes of humans and the paws, ears, and some other tissues of animals, there are short channels that connect arterioles to venules, bypassing the capillaries. These **arteriovenous (A-V) anastomoses,** or **shunts,** have thick, muscular walls and are abundantly innervated, presumably by vasoconstrictor nerve fibers.

Venules & Veins

The walls of the venules are only slightly thicker than those of the capillaries. The walls of the veins are also thin and easily distended. They contain relatively little smooth muscle, but considerable venoconstriction is produced by activity in the noradrenergic nerves to the veins and by chemical agents such as norepinephrine. Anyone who has had trouble making venipunctures has observed the marked local venospasm produced in superficial forearm veins by injury. Variations in venous tone are important in circulatory adjustments.

The intima of the limb veins is folded at intervals to form **venous valves** that prevent retrograde flow. The way these valves function was first demonstrated by William Harvey (Fig 30–5). There are no valves in the very small veins, the great veins, or the veins from the brain and viscera.

Angiogenesis

When tissues grow, blood vessels must proliferate if the tissue is to maintain a normal blood supply. Therefore, angiogenesis, the formation of new blood vessels, is important during fetal life and growth to adulthood. It is also important in adulthood for processes such as wound-healing, formation of the corpus luteum after ovulation, and formation of new endometrium after menstruation. Abnormally, it is important in tumor growth, and if tumors do not develop a blood supply, they do not grow. Numerous growth factors have angiogenic activity (Table 30–3), and other factors inhibit angiogenesis. Interaction with the extracellular matrix is known to be involved. There is a problem in that angiogenesis depends in large part on secreted factors, and the precursors of the FGFs and PDECGF lack signal sequences. Therefore, if they are secreted, it is not known how they get out of cells. Several of the angiogenic factors bind heparin, and heparin in the presence of adrenocortical steroids is said to inhibit angiogenesis, but its exact role is unknown. Consequently, it is not possible at present to put together a comprehensive hypothesis of how these factors lead to new blood vessel formation in living animals.

BIOPHYSICAL CONSIDERATIONS

Flow, Pressure, & Resistance

Blood always flows, of course, from areas of high

Figure 30–4. Fenestrations in capillaries in pancreatic islets. **Left:** Arrows identify fenestrations in a grazing section of the endothelium (× 15,000). ECS, extracellular space. **Right:** Fenestrations are seen in a quick-frozen, deep-etched face view of the capillary endothelium (× 64,000). Note the central material and the spokes attaching it to the rim of each fenestration. (Reproduced, with permission, from Orci L: The insulin cell: Its cellular environment and how it processes (pro)-insulin. *Diabetes Metab Rev* 1986;**2**:71.)

pressure to areas of low pressure, except in certain situations when momentum transiently sustains flow (see Fig 29–3). The relationship between mean flow, mean pressure, and resistance in the blood vessels is analogous in a general way to the relationship between the current, electromotive force, and resistance in an electrical circuit expressed in Ohm's law:

$$\text{Current (I)} = \frac{\text{Electromotive force (E)}}{\text{Resistance (R)}}$$

$$\text{Flow (F)} = \frac{\text{Pressure (P)}}{\text{Resistance (R)}}$$

Flow in any portion of the vascular system is equal to the **effective perfusion pressure** in that portion divided by the **resistance.** The effective perfusion pressure is the mean intraluminal pressure at the arterial end minus the mean pressure at the venous end. The

Table 30–2. Hydraulic conductivity of capillaries in various parts of the body.[1]

Organ	Conductivity[2]	Type of Endothelium
Brain (excluding circumventricular organs)	3	Continuous
Skin	100	
Skeletal muscle	250	
Lung	340	
Heart	860	
Gastrointestinal tract (intestinal mucosa)	13,000	Fenestrated
Glomerulus in kidney	15,000	

[1] Data courtesy of JN Diana.
[2] Units of conductivity are 10^{-13} cm^3 s^{-1} dyne^{-1}.

units of resistance (pressure divided by flow) are dyne • s/cm^5. To avoid dealing with such complex units, resistance in the cardiovascular system is sometimes expressed in **R units,** which are obtained by dividing pressure in mm Hg by flow in mL/s (see also Table 32–1). Thus, for example, when the mean aortic pressure is 90 mm Hg and the left ventricular output is 90 mL/s, the total peripheral resistance is

$$\frac{90 \text{ mm Hg}}{90 \text{ mL/s}} = 1 \text{ R unit}$$

Methods for Measuring Blood Flow

Blood flow can be measured by cannulating a blood vessel, but this has obvious limitations. Various devices have been developed to measure flow in a blood vessel without opening it. **Electromagnetic flow meters** depend on the principle that a voltage is generated in a conductor moving through a magnetic field and that the magnitude of the voltage is proportionate to the speed of movement. Since blood is a conductor, a magnet is placed around the vessel, and the voltage, which is proportionate to the volume flow, is measured with an appropriately placed electrode on the surface of the vessel. Blood flow velocity can be measured with **Doppler flow meters.** Ultrasonic waves are sent into a vessel diagonally from one crystal, and the waves reflected from the red and white blood cells are picked up by a second, downstream crystal. The frequency of the reflected waves is higher by an amount that is proportionate to the rate of flow toward the second crystal because of the Doppler effect.

Indirect methods for measuring the blood flow of various organs in humans include adaptations of the Fick and indicator dilution techniques described in

Figure 30–5. Illustration showing the function of the venous valves in the human forearm, from William Harvey's book *Exercitatio Anatomica de Motu Cordis et Sanguinis in Animalibus,* which was published in 1628. A is a tourniquet above the elbow, obstructing the venous return. If the vein (Fig 2) is milked from G to H, it fills from above only to the valve, O. If it is milked from K to O with another finger (Fig 3), it bulges above O, but O to H remains empty. If the vein is occluded at L (Fig 4) and milked with another finger (M) from L to the valve N, it remains empty from L to N. If the finger at L is removed, the vein fills from below.

Chapter 29. An ingenious example of the use of the Fick principle to measure flow in an organ is the Kety N_2O method for measuring cerebral blood flow (see Chapter 32). Another example is determination of the renal blood flow by measuring the clearance of para-aminohippuric acid (see Chapter 38). A considerable amount of data on blood flow in the extremities has been obtained by **plethysmography** (Fig 30–6). The forearm, for example, is sealed in a watertight chamber **(plethysmograph).** Changes in the volume of the forearm, reflecting changes in the amount of blood and interstitial fluid it contains, displace the water, and this displacement is measured with a volume recorder. When the venous drainage of the forearm is occluded, the rate of increase in the volume of the forearm is a function of the arterial blood flow **(venous occlusion plethysmography).**

Applicability of Physical Principles to Flow in Blood Vessels

Physical principles and equations that are applicable to the description of the behavior of perfect fluids in rigid tubes have often been used indiscriminately to explain the behavior of blood in blood vessels. Blood vessels are not rigid tubes, and the blood is not a perfect fluid but a 2-phase system of liquid and cells. Therefore, the behavior of the circulation deviates, sometimes markedly, from that predicted by these principles. However, the physical principles are of value when used as an aid to understanding what goes on in the body rather than as an end in themselves or as a test of the memorizing ability of students.

Table 30–3. Factors that may be involved in angiogenesis.

Stimulators
  Acidic fibroblast growth factor (FGF)
  Basic FGF
  Angiogenin
  Platelet-derived endothelial cell growth factor
    (PDECGF)
  Vascular endothelial growth factor (VEGF)
  Vascular permeability factor (VPF)
Inhibitors
  Heparin
  Platelet factor 4
  16K fragment of prolactin

Figure 30–6. Plethysmography.

Laminar Flow

The flow of blood in the blood vessels, like the flow of liquids in narrow rigid tubes, is normally **laminar (streamline).** Within the blood vessels, an infinitely thin layer of blood in contact with the wall of the vessel does not move. The next layer within the vessel has a low velocity, the next a higher velocity, and so forth, velocity being greatest in the center of the stream (Fig 30–7). Laminar flow occurs at velocities up to a certain **critical velocity.** At or above this velocity, flow is turbulent. Streamline flow is silent, but turbulent flow creates sounds.

The probability of turbulence is also related to the diameter of the vessel and the viscosity of the blood. This probability can be expressed by the ratio of inertial to viscous forces as follows:

$$Re = \frac{\rho D \dot{V}}{\eta}$$

where Re is the Reynolds number, named for the man who described the relationship; ρ is the density of the fluid; D is the diameter of the tube under consideration; V is the velocity of the flow; and η is the viscosity of the fluid. The higher the value of Re, the greater the probability of turbulence. When D is in cm, V is in cm/s^{-1}, and η is in poises, flow is usually not turbulent if Re is less than 2000. When it is more than 3000, turbulence is almost always present. Constriction of an artery increases the velocity of blood flow through the constriction, producing turbulence, and consequently sound, beyond the constriction (Fig 30–8). Examples are bruits heard over arteries constricted by atherosclerotic plaques and the sound of Korotkow heard when measuring blood pressure (see below).

In humans, the critical velocity is sometimes exceeded in the ascending aorta at the peak of systolic ejection, but it is usually exceeded only when an artery is constricted. Turbulence occurs more frequently in anemia because the viscosity of the blood

Figure 30–8. **Top:** Effect of constriction (C) on the profile of velocities in a blood vessel. The arrows indicate direction of velocity components, and their length is proportionate to their magnitude. **Bottom:** Range of velocities at each point along the vessel. In the area of turbulence, there are many different anterograde (A) and some retrograde (R) velocities. (Modified and reproduced, with permission, from Richards KE: Doppler echocardiography in diagnosis and quantification of vascular disease. *Mod Concepts Cardiovasc Dis* 1987;**56**:43. By permission of the American Heart Association, Inc.)

is lower. This may be the explanation of the systolic murmurs that are common in anemia.

Average Velocity

When considering flow in a system of tubes, it is important to distinguish between velocity, which is displacement per unit time (eg, cm/s), and flow, which is volume per unit time (eg, cm^3/s). Velocity (V) is proportionate to flow (Q) divided by the area of the conduit (A):

$$\dot{V} = \frac{Q}{A}$$

The average velocity of fluid movement at any point in a system of tubes is inversely proportionate to the *total* cross-sectional area at that point. Therefore, the average velocity of the blood is high in the aorta, declines steadily in the smaller vessels, and is lowest in the capillaries, which have 1000 times the *total* cross-sectional area of the aorta (Table 30–1). The average velocity of blood flow increases again as the blood enters the veins and is relatively high in the vena cava, although not so high as in the aorta. Clinically, the velocity of the circulation can be measured by injecting a bile salt preparation into an arm vein and timing the first appearance of the bitter taste it produces (Fig 30–9). The average normal arm-to-tongue **circulation time** is 15 seconds.

Poiseuille-Hagen Formula

The relation between the flow in a long narrow tube, the viscosity of the fluid, and the radius of the

Figure 30–7. Diagram of the velocities of concentric laminas of a viscous fluid flowing in a tube, illustrating the parabolic distribution of velocities (streamline flow).

Site of injection
(antecubital vein)

Figure 30–9. Pathway traversed by the injected material when the arm-to-tongue circulation time is measured.

tube is expressed mathematically in the **Poiseuille-Hagen formula:**

$$F = (P_A - P_B) \times \left(\frac{\pi}{8}\right) \times \left(\frac{1}{\eta}\right) \times \left(\frac{r^4}{L}\right)$$

where

F = flow
$P_A - P_B$ = pressure difference between the 2 ends of the tube
η = viscosity
r = radius of tube
L = length of tube

Since flow is equal to pressure difference divided by resistance (R),

$$R = \frac{8\eta L}{\pi r^4}$$

Since flow varies directly and resistance inversely with the fourth power of the radius, blood flow and resistance in vivo are markedly affected by small changes in the caliber of the vessels. Thus, for example, flow through a vessel is doubled by an increase of only 19% in its radius; and when the radius is doubled, resistance is reduced to 6% of its previous value. This is why organ blood flow is so effectively regulated by small changes in the caliber of the arterioles and why variations in arteriolar diameter have such a pronounced effect on systemic arterial pressure.

Viscosity & Resistance

The resistance to blood flow is determined not only by the radius of the blood vessels (**vascular hindrance**) but also by the viscosity of the blood. Plasma is about 1.8 times as viscous as water, whereas whole blood is 3–4 times as viscous as water. Thus viscosity depends for the most part on the **hematocrit,** ie, the percentage of the volume of blood occupied by red blood cells. The effect of viscosity in vivo deviates from that predicted by the Poiseuille-Hagen formula. In large vessels, increases in hematocrit cause appreciable increases in viscosity. However, in vessels smaller than 100 µm in diameter, ie, in arterioles, capillaries, and venules, the viscosity change per unit change in hematocrit is much less than it is in large-bore vessels. This is due to a difference in the nature of flow through the small vessels. Therefore, the net change in viscosity per unit change in hematocrit is considerably smaller in the body than it is in vitro (Fig 30–10). This is why hematocrit changes have relatively little effect on the peripheral resistance except when the changes are large. In severe polycythemia, the increase in resistance does increase the work of the heart. Conversely, in anemia, peripheral resistance is decreased, in part because of the decline in viscosity. Of course, the decrease in hemoglobin decreases the O_2-carrying ability of the blood, but the improved blood flow due to the decrease in viscosity partially compensates for this.

Viscosity is also affected by the composition of the plasma and the resistance of the cells to deforma-

Figure 30–10. Effect of changes in hematocrit on the relative viscosity of blood measured in a glass viscometer and in the hind leg of a dog. In each case, the middle line represents the mean and the upper and lower lines the standard deviation. (Reproduced, with permission, from Whittaker SRF, Winton FR: The apparent viscosity of blood flowing in the isolated hind limb of the dog, and its variation with corpuscular concentration. *J Physiol [Lond]* 1933;**78**:338.)

tion. Clinically significant increases in viscosity are seen in diseases in which plasma proteins such as the immunoglobulins are markedly elevated and in diseases such as hereditary spherocytosis, in which the red blood cells are abnormally rigid.

In the vessels, red cells tend to accumulate in the center of the flowing stream. Consequently, the blood along the side of the vessels has a low hematocrit, and branches leaving a large vessel at right angles may receive a disproportionate amount of red-cell-poor blood. This phenomenon, which has been called **plasma skimming,** may be the reason why the hematocrit of capillary blood is regularly about 25% lower than the whole-body hematocrit.

Critical Closing Pressure

In rigid tubes the relation between pressure and flow of homogeneous fluids is linear, but in blood vessels in vivo it is not. When the pressure in a small blood vessel is reduced, a point is reached at which there is no flow of blood even though the pressure is not zero (Fig 30–11). This is in part a manifestation of the fact that it takes some pressure to force red cells through capillaries which have smaller diameters than the red cells. Also, the vessels are surrounded by tissues that exert a small but definite pressure on the vessels, and when the intraluminal pressure falls below the tissue pressure, the vessels collapse. In inactive tissues, for example, the pressure in many capillaries is low because the precapillary sphincters and metarterioles are constricted, and many of these capillaries are collapsed. The pressure at which flow ceases is called the **critical closing pressure.**

Law of Laplace

The relation between distending pressure and tension is shown diagrammatically in Fig 30–12. It is perhaps surprising that structures as thin-walled and delicate as the capillaries are not more prone to rupture. The principal reason for their relative invulnerability is their small diameter. The protective effect of small size in this case is an example of the operation of the **law of Laplace,** an important physical principle with several other applications in physiology. This law states that the distending pressure (P) in a distensible hollow object is equal at equilibrium to the tension in the wall (T) divided by the 2 principal radii of curvature of the object (r_1 and r_2):

$$P = T\left(\frac{1}{r_1} + \frac{1}{r_2}\right)$$

In the equation, P is actually the **transmural pressure,** the pressure on one side of the wall minus that on the other. T is expressed in dynes/cm and r_1 and r_2 in cm, so P is expressed in dynes/cm^2. In a sphere, $r_1 = r_2$ so

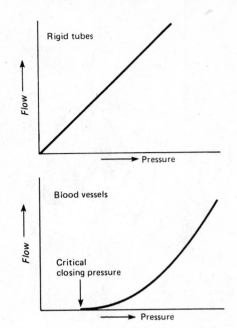

Figure 30–11. Relation of pressure to flow in a rigid-walled system **(top)** and the vascular system **(bottom).**

$$P = \frac{2T}{r}$$

In a cylinder such as a blood vessel, one radius is infinite, so

$$P = \frac{T}{r}$$

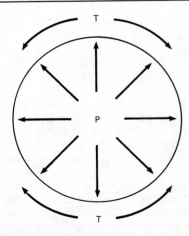

Figure 30–12. Relation between distending pressure (P) and wall tension (T) in a hollow viscus.

Consequently, the smaller the radius of a blood vessel, the lower the tension in the wall necessary to balance the distending pressure. In the human aorta, for example, the tension at normal pressures is about 170,000 dynes/cm, and in the vena cava it is about 21,000 dynes/cm; but in the capillaries, it is approximately 16 dynes/cm.

The law of Laplace also makes clear a disadvantage faced by dilated hearts. When the radius of a cardiac chamber is increased, a greater tension must be developed in the myocardium to produce any given pressure; consequently, a dilated heart must do more work than a nondilated heart. In the lungs, the radii of curvature of the alveoli become smaller during expiration, and these structures would tend to collapse because of the pull of surface tension if the tension were not reduced by the surface-tension-lowering agent, surfactant (see Chapter 34). Another example of the operation of this law is seen in the urinary bladder (see Chapter 38).

Resistance & Capacitance Vessels

When a segment of aorta is filled and then distended with increasing volumes of fluid, the pressure in the segment initially rises in a linear fashion. When the same experiment is carried out on a segment of the vena cava or another large distensible vein, the pressure does not rise rapidly until large volumes of fluid are injected. In vivo, the veins are an important blood reservoir. Normally they are partially collapsed and oval in cross section. A large amount of blood can be added to the venous system before the veins become distended to the point where further increments in volume produce a large rise in venous pressure. The veins are therefore called **capacitance vessels.** The small arteries and arterioles are referred to as **resistance vessels** because they are the principal site of the peripheral resistance (see below).

At rest, at least 50% of the circulating blood volume is in the systemic veins. Twelve percent is in the heart cavities, and 18% is in the low-pressure pulmonary circulation. Only 2% is in the aorta, 8% in the arteries, 1% in the arterioles, and 5% in the capillaries (Table 30–1). When extra blood is administered by transfusion, less than 1% of it is distributed in the arterial system (the **"high-pressure system"**), and all the rest is found in the systemic veins, pulmonary circulation, and heart chambers other than the left ventricle (the **"low-pressure system"**).

ARTERIAL & ARTERIOLAR CIRCULATION

The pressure and velocities of the blood in the various parts of the systemic circulation are summarized in Fig 30–13. The general relationships in the pulmonary circulation are similar, but the pressure in the pulmonary artery is 25/10 mm Hg or less.

Figure 30–13. Diagram of the changes in pressure and velocity as blood flows through the systemic circulation. TA, total cross-sectional area of the vessels, which increases from 4.5 cm² in the aorta to 4500 cm² in the capillaries (Table 30–1). RR, relative resistance, which is highest in the arterioles.

Velocity & Flow of Blood

Although the mean velocity of the blood in the proximal portion of the aorta is 40 cm/s, the flow is phasic, and velocity ranges from 120 cm/s during systole to a negative value at the time of the transient backflow before the aortic valve closes in diastole. In the distal portions of the aorta and in the large arteries, velocity is also greater in systole than it is in diastole. However, the vessels are elastic, and forward flow is continuous because of the recoil during diastole of the vessel walls that have been stretched during systole (Fig 30–14). This recoil effect is sometimes called the **Windkessel effect,** and the vessels are called Windkessel vessels; Windkessel is the German word for an elastic reservoir. Pulsatile flow appears, in some poorly understood way, to maintain optimal function of the tissues. If an organ is perfused with a pump that delivers a nonpulsatile flow, there is a gradual rise in vascular resistance, and tissue perfusion fails.

Arterial Pressure

The pressure in the aorta and in the brachial and other large arteries in a young adult human rises to a peak value (**systolic pressure**) of about 120 mm Hg during each heart cycle and falls to a minimum value (**diastolic pressure**) of about 70 mm Hg. The arterial pressure is conventionally written as systolic pressure

Figure 30–14. Changes in blood flow during the cardiac cycle in the dog. Diastole is followed by systole starting at 0.1 and again at 0.5 second. Flow patterns in humans are similar. Ao, aorta; PA, pulmonary artery; PV, pulmonary vein; IVC, inferior vena cava; RA, renal artery. (Reproduced, with permission, from Milnor WR: Pulsatile blood flow. *N Engl J Med* 1972;**287:**27.)

Figure 30–15. Brachial artery pressure curve of a normal young human, showing the relation of systolic and diastolic pressure to mean pressure. The shaded area above the mean pressure line is equal to the shaded area below it.

over diastolic pressure—eg, 120/70 mm Hg. One millimeter of Hg equals 0.133 kPa, so in SI units (see appendix) this value is 16.0/9.3 kPa. The **pulse pressure,** the difference between the systolic and diastolic pressures, is normally about 50 mm Hg. The **mean pressure** is the average pressure throughout the cardiac cycle. Because systole is shorter than diastole, the mean pressure is slightly less than the value halfway between systolic and diastolic pressure. It can actually be determined only by integrating the area of the pressure curve (Fig 30–15); however, as an approximation, mean pressure equals the diastolic pressure plus one-third of the pulse pressure.

The pressure falls very slightly in the large and medium-sized arteries because their resistance to flow is small, but it falls rapidly in the small arteries and arterioles, which are the main sites of the peripheral resistance against which the heart pumps. The mean pressure at the end of the arterioles is 30–38 mm Hg. Pulse pressure also declines rapidly to about 5 mm Hg at the ends of the arterioles (Fig 30–13). The magnitude of the pressure drop along the arterioles varies considerably depending upon whether they are constricted or dilated.

Effect of Gravity

The pressures in Fig 30–13 are those in blood ves-

sels at heart level. The pressure in any vessel below heart level is increased and that in any vessel above heart level is decreased by the effect of gravity. The magnitude of the gravitational effect—the product of the density of the blood, the acceleration due to gravity (980 cm/s/s), and the vertical distance above or below the heart—is 0.77 mm Hg/cm at the density of normal blood. Thus, in an adult human in the upright position, when the mean arterial pressure at heart level is 100 mm Hg, the mean pressure in a large artery in the head (50 cm above the heart) is 62 mm Hg $(100 - [0.77 \times 50])$ and the pressure in a large artery in the foot (105 cm below the heart) is 180 mm Hg $(100 + [0.77 \times 105])$ (Fig 30–16). The effect of gravity on venous pressure is similar (see below).

Methods of Measuring Blood Pressure

If a cannula is inserted into an artery, the arterial pressure can be measured directly with a mercury manometer or a suitably calibrated strain gauge and an oscillograph arranged to write directly on a moving strip of paper. When an artery is tied off beyond the point at which the cannula is inserted, an **end pressure** is recorded. Flow in the artery is interrupted, and all the kinetic energy of flow is converted into pressure energy. If, alternatively, a T tube is inserted into a vessel and the pressure is measured in the side arm of the tube, the recorded **side pressure** under conditions where pressure drop due to resistance is negligible is lower than the end pressure by the kinetic energy of flow. This is because in a tube or a blood vessel the total energy—the sum of the kinetic energy of flow and the pressure energy—is constant (**Bernoulli's principle**).

It is worth noting that the pressure drop in any segment of the arterial system is due both to resistance and to conversion of potential into kinetic energy. The pressure drop due to energy lost in overcoming resistance is irreversible, since the energy is dissipated as heat; but the pressure drop due to conversion of potential to kinetic energy as a vessel narrows is reversed when the vessel widens out again (Fig 30–17).

Figure 30–16. Effects of gravity on arterial and venous pressure. The scale on the right indicates the increment (or decrement) in mean pressure in a large artery at each level. The mean pressure in all large arteries is approximately 100 mm Hg when they are at the level of the left ventricle. The scale on the left indicates the increment in venous pressure at each level due to gravity. The manometers on the left of the figure indicate the height to which a column of blood in a tube would rise if connected to an ankle vein (A), the femoral vein (B), or the right atrium (C), with the subject in the standing position. The approximate pressures in these locations in the recumbent position—ie, when the ankle, thigh, and right atrium are at the same level—are (A), 10 mm Hg; (B), 7.5 mm Hg; and (C), 4.6 mm Hg.

Figure 30–17. Bernoulli's principle. When fluid flows through the narrow portion of the tube, the kinetic energy of flow is increased as the velocity increases, and the pressure energy is reduced. Consequently, the measured pressure (P) is lower than it would have been at that point if the tube had not been narrowed. The dashed line indicates what the pressure drop due to frictional forces would have been if the tube had been of uniform diameter.

Bernoulli's principle also has a significant application in pathophysiology. According to the principle, the greater the velocity of flow in a vessel, the lower the lateral pressure distending its walls. When a vessel is narrowed, the velocity of flow in the narrowed portion increases and the distending pressure decreases. Therefore, when a vessel is narrowed by a pathologic process such as an arteriosclerotic plaque, the lateral pressure at the constriction is decreased and the narrowing tends to maintain itself.

Auscultatory Method

The arterial blood pressure in humans is routinely measured by the **auscultatory method.** An inflatable cuff (**Riva-Rocci cuff**) attached to a mercury manometer (**sphygmomanometer**) is wrapped around the arm and a stethoscope is placed over the brachial artery at the elbow (Fig 30–18). The cuff is rapidly inflated until the pressure in it is well above the expected systolic pressure in the brachial artery. The artery is occluded by the cuff, and no sound is heard with the stethoscope. The pressure in the cuff is then

lowered slowly. At the point at which systolic pressure in the artery just exceeds the cuff pressure, a spurt of blood passes through with each heartbeat and, synchronously with each beat, a tapping sound is heard below the cuff. The cuff pressure at which the sounds are first heard is the systolic pressure. As the cuff pressure is lowered further, the sounds become louder, then dull and muffled; finally, in most individuals, they disappear. These are the **sounds of Korotkow.** When direct and indirect blood pressure measurements are made simultaneously, the diastolic pressure in resting adults correlates best with the pressure at which the sound disappears. However, in

Figure 30–18. Determination of blood pressure by the auscultatory method. (Reproduced, with permission, from Schottelius BA, Schottelius D: *Textbook of Physiology,* 18th ed. Mosby, 1978.)

adults after exercise and in children, the diastolic pressure correlates best with the pressure at which the sounds become muffled. This is also true in diseases such as hyperthyroidism and aortic insufficiency.

The sounds of Korotkow are produced by turbulent flow in the brachial artery. The streamline flow in the unconstricted artery is silent, but when the artery is narrowed, the velocity of flow through the constriction exceeds the **critical velocity** and turbulent flow results (Fig 30–8). At cuff pressures just below the systolic pressure, flow through the artery occurs only at the peak of systole, and the intermittent turbulence produces a tapping sound. As long as the pressure in the cuff is above the diastolic pressure in the artery, flow is interrupted at least during part of diastole, and the intermittent sounds have a staccato quality. When the cuff pressure is near the arterial diastolic pressure, the vessel is still constricted, but the turbulent flow is continuous. Continuous sounds have a muffled rather than a staccato quality.

The auscultatory method is accurate when used properly, but a number of precautions must be observed. The cuff must be at heart level to obtain a pressure that is uninfluenced by gravity. The blood pressure in the thighs can be measured with the cuff around the thigh and the stethoscope over the popliteal artery, but there is more tissue between the cuff and the artery in the leg than there is in the arm, and some of the cuff pressure is dissipated. Therefore, pressures obtained by using the standard arm cuff are falsely high. The same thing is true when brachial arterial pressures are measured in individuals with obese arms, because the blanket of fat dissipates some of the cuff pressure. In both situations, accurate pressures can be obtained by using a cuff that is wider than the standard arm cuff. If the cuff is left inflated for some time, the discomfort may cause generalized reflex vasoconstriction, raising the blood pressure. It is always wise to compare the blood pressure in both arms when examining an individual for the first time. Persistent major differences between the pressure on the 2 sides indicate the presence of vascular obstruction.

Palpation Method

The systolic pressure can be determined by inflating an arm cuff and then letting the pressure fall and determining the pressure at which the radial pulse first becomes palpable. Because of the difficulty in determining exactly when the first beat is felt, pressures obtained by this **palpation method** are usually 2–5 mm Hg lower than those measured by the auscultatory method.

It is wise to form a habit of palpating the radial pulse while inflating the blood pressure cuff during measurement of the blood pressure by the auscultatory method. When the cuff pressure is lowered, the sounds of Korotkow sometimes disappear at pressures well above diastolic pressure, then reappear at lower pressures ("auscultatory gap"). If the cuff is initially inflated until the radial pulse disappears, the examiner can be sure that the cuff pressure is above systolic pressure, and falsely low pressure values will be avoided.

Normal Arterial Blood Pressure

The blood pressure in the brachial artery in young adults in the sitting or lying position at rest is approximately 120/70 mm Hg. Since the arterial pressure is the product of the cardiac output and the peripheral resistance, it is affected by conditions that affect either or both of these factors. Emotion, for example, increases the cardiac output, and it may be difficult to obtain a truly resting blood pressure in an excited or tense individual. In general, increases in cardiac output increase the systolic pressure, whereas increases in peripheral resistance increase the diastolic pressure. There is a good deal of controversy about where to draw the line between normal and elevated blood pressure levels (**hypertension**), particularly in older patients. However, the evidence seems incontrovertible than in apparently healthy humans both the systolic and the diastolic pressures rise with age (Fig 30–19). The systolic pressure increase is greater than the diastolic. An important cause of the systolic pressure rise is decreased distensibility of the arteries as their walls become increasingly rigid. At the same level of cardiac output, the systolic pressure is higher in old subjects than in young ones because there is less increase in the volume of the arterial system during systole to accommodate the same amount of blood.

CAPILLARY CIRCULATION

At any one time, only 5% of the circulating blood is in the capillaries, but this 5% is in a sense the most important part of the blood volume because it is across the systemic capillary walls that O_2 and nutrients enter the interstitial fluid and CO_2 and waste products enter the bloodstream. The exchange across the capillary walls is essential to the survival of the tissues.

Methods of Study

It is difficult to obtain accurate measurements of capillary pressures and flows. The capillaries in the mesentery of experimental animals and the fingernail beds of humans are readily visible under the dissecting microscope, and observations on flow patterns under various conditions have been made in these and in some other tissues. Capillary pressure has been estimated by determining the amount of external pressure necessary to occlude the capillaries or the amount of pressure necessary to make saline start to flow through a micropipette inserted so that its tip faces the arteriolar end of the capillary.

Figure 30–19. Arterial pressure as a function of age in the general population. The size of the squares and the circles is proportionate to the number of subjects studied in that age group. (Reproduced, with permission, from Hamilton M et al: The aetiology of essential hypertension. 1. The arterial pressure in the general population. *Clin Sci* 1954;**13**:11.)

Capillary Pressure & Flow

Capillary pressures vary considerably, but typical values in human nail bed capillaries are 32 mm Hg at the arteriolar end and 15 mm Hg at the venous end. The pulse pressure is approximately 5 mm Hg at the arteriolar end and zero at the venous end. The capillaries are short, but blood moves slowly (about 0.07 cm/s) because the total cross-sectional area of the capillary bed is large. Transit time from the arteriolar to the venular end of an average-sized capillary is 1–2 seconds.

Equilibration With Interstitial Fluid

As noted above, the capillary wall is a thin membrane made up of endothelial cells. Substances pass through the junctions between endothelial cells and through fenestrations, when they are present. Some also pass through the cells by vesicular transport or, in the case of lipid-soluble substances, through the cytoplasm.

The factors other than vesicular transport that are responsible for transport across the capillary wall are diffusion and filtration (see Chapter 1). Diffusion is quantitatively much more important in terms of the exchange of nutrients and waste materials between blood and tissue. O_2 and glucose are in higher concentration in the bloodstream than in the interstitial fluid and diffuse into the interstitial fluid, whereas CO_2 diffuses in the opposite direction.

The rate of filtration at any point along a capillary depends upon a balance of forces sometimes called the **Starling forces** after the physiologist who first described their operation in detail. One of these forces is the **hydrostatic pressure gradient** (the hydrostatic pressure in the capillary minus the hydrostatic pressure of the interstitial fluid) at that point. The interstitial fluid pressure varies from one organ to another, and there is considerable evidence that it is subatmospheric—about –2 mm Hg—in subcutaneous tissue. It is positive in the liver and kidneys and is as high as 6 mm Hg in the brain. The other force is the **osmotic pressure gradient** across the capillary wall (colloid osmotic pressure of plasma minus colloid osmotic pressure of interstitial fluid). This component is directed inward.

Thus:

$$\text{Fluid movement} = k[(P_c + \pi_i) - (P_i + \pi_c)]$$

where

$$
\begin{aligned}
k &= \textbf{capillary filtration coefficient} \\
P_c &= \textbf{capillary hydrostatic pressure} \\
P_i &= \textbf{interstitial hydrostatic pressure} \\
\pi_c &= \textbf{capillary colloid osmotic pressure} \\
\pi_i &= \textbf{interstitial colloid osmotic pressure}
\end{aligned}
$$

π_i is usually negligible, so the osmotic pressure gradient ($\pi_c - \pi_i$) usually equals the oncotic pressure. The capillary filtration coefficient takes into account, and is proportionate to, the permeability of the capillary wall and the area available for filtration. The magnitude of the Starling forces along a typical muscle capillary is shown in Fig 30–20. Fluid moves into the interstitial space at the arteriolar end of the capillary, where the filtration pressure across its wall exceeds the oncotic pressure, and into the capillary at the venular end, where the oncotic pressure exceeds the filtration pressure. In other capillaries, the balance of Starling forces is different and, for example, fluid moves out of almost the entire length of the capillaries in the renal glomeruli. On the other hand, fluid moves into the capillaries through almost their entire length in the intestines.

It is worth noting that small molecules often equilibrate with the tissues near the arteriolar end of each capillary. In this situation, total diffusion can be increased by increasing blood flow; ie, exchange is **flow-limited** (Fig 30–21). Conversely, transfer of substances that do not reach equilibrium with the tis-

Figure 30–20. Schematic representation of pressure gradients across the wall of a muscle capillary. The numbers at the arteriolar and venular ends of the capillary are the hydrostatic pressures in mm Hg at these locations. The arrows indicate the approximate magnitude and direction of fluid movement. In this example, the pressure differential at the arteriolar end of the capillary is 11 mm Hg ([37 – 1] – 25) outward; at the opposite end, it is 9 mm Hg (25 – [17 – 1]) inward.

sues during their passage through the capillaries is said to be **diffusion-limited.**

It has been estimated that about 24 L of fluid is filtered through the capillaries per day. This is about 0.3% of the cardiac output. About 85% of the filtered fluid is reabsorbed into the capillaries, and the remainder returns to the circulation via the lymphatics.

Active & Inactive Capillaries

In resting tissues, most of the capillaries are collapsed, and blood flows for the most part through the thoroughfare vessels from the arterioles to the ven-

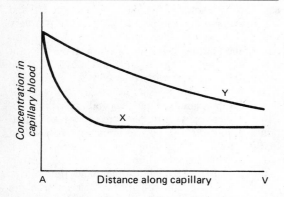

Figure 30–21. Flow-limited and diffusion-limited exchange across capillary walls. A and V indicate the arteriolar and venular ends of the capillary. Substance X equilibrates with the tissues (movement into the tissues equals movement out) well before the blood leaves the capillary, whereas substance Y does not equilibrate. If other factors stay constant, the amount of X entering the tissues can only be increased by increasing blood flow; ie, it is flow-limited. The movement of Y is diffusion-limited.

ules. In active tissues, the metarterioles and the precapillary sphincters dilate. The intracapillary pressure rises, overcoming the critical closing pressure of the vessels, and blood flows through all the capillaries. Relaxation of the smooth muscle of the metarterioles and precapillary sphincters is due to the action of vasodilator metabolites formed in active tissue (see Chapter 31) and possibly also to a decrease in the activity of the sympathetic vasoconstrictor nerves that innervate the smooth muscle.

After noxious stimulation, substance P released by the axon reflex (see Chapter 32) increases capillary permeability. Bradykinin and histamine also increase capillary permeability. When capillaries are stimulated mechanically, they empty (white reaction; see Chapter 32), but this is probably due to contraction of the precapillary sphincters.

LYMPHATIC CIRCULATION & INTERSTITIAL FLUID VOLUME

Lymphatic Circulation

Fluid efflux normally exceeds influx across the capillary walls, but the extra fluid enters the lymphatics and drains through them back into the blood. This keeps the interstitial fluid pressure from rising and promotes the turnover of tissue fluid. The normal 24-hour lymph flow is 2–4 L. The composition of lymph is discussed in Chapter 27.

Lymph flow is due to movements of skeletal muscle, the negative intrathoracic pressure during inspiration, the suction effect of high-velocity flow of blood in the veins in which the lymphatics terminate, and rhythmic contractions of the walls of the large lymph ducts. Since lymph vessels have valves that prevent backflow, skeletal muscle contractions push the lymph toward the heart. Pulsations of arteries near lymphatics may have a similar effect. However, the contractions of the walls of the lymphatic ducts are important, and the rate of these contractions increases in direct proportion to the volume of lymph in the vessels. There is evidence that the contractions are the principal factor propelling the lymph.

Agents that increase lymph flow are called **lymphagogues.** They include a variety of agents that increase capillary permeability. Agents that cause contraction of smooth muscle also increase lymph flow from the intestines.

Other Functions of the Lymphatic System

Appreciable quantities of protein enter the interstitial fluid in the liver and intestine, and smaller quantities enter from the blood in other tissues. The walls of the lymphatics are permeable to macromolecules, and the proteins are returned to the bloodstream via the lymphatics. The amount of protein returned in this fashion in 1 day is equal to 25–50% of the total circulating plasma protein. In the kidneys, formation of a maximally concentrated urine depends upon an intact lymphatic circulation; removal of reabsorbed water from the medullary pyramids is essential for the efficient operation of the countercurrent mechanism (see Chapter 38), and water enters the vasa recta only if an appreciable osmotic gradient is maintained between the medullary interstitium and the vasa recta blood by drainage of protein-containing interstitial fluid into the renal lymphatics. The transport of absorbed long-chain fatty acids and cholesterol from the intestine via the lymphatics has been discussed in Chapter 25.

Interstitial Fluid Volume

The amount of fluid in the interstitial spaces depends upon the capillary pressure, the interstitial fluid pressure, the oncotic pressure, the capillary filtration coefficient, the number of active capillaries, the lymph flow, and the total ECF volume. The ratio of precapillary to postcapillary venular resistance is also important. Precapillary constriction lowers filtration pressure, whereas postcapillary constriction raises it. Changes in any of these variables lead to changes in the volume of interstitial fluid. Factors promoting an increase in this volume are summarized in Table 30–4. **Edema** is the accumulation of interstitial fluid in abnormally large amounts.

In active tissues, capillary pressure rises, often to the point where it exceeds the oncotic pressure throughout the length of the capillary. In addition, osmotically active metabolites may temporarily accumulate in the interstitial fluid because they cannot be washed away as rapidly as they are formed. To the extent that they accumulate, they exert an osmotic effect that decreases the magnitude of the osmotic gradient due to the oncotic pressure. The amount of fluid leaving the capillaries is therefore markedly increased and the amount entering them reduced. Lymph flow is increased, decreasing the degree to which the fluid would otherwise accumulate, but exercising muscle, for example, still increases in volume by as much as 25%.

Table 30-4. Causes of increased interstitial fluid volume and edema.

Increased filtration pressure
  Arteriolar dilation
  Venular constriction
  Increased venous pressure (heart failure, incompetent valves, venous obstruction, increased total ECF volume, effect of gravity, etc)
Decreased osmotic pressure gradient across capillary
  Decreased plasma protein level
  Accumulation of osmotically active substances in interstitial space
Increased capillary permeability
  Substance P
  Histamine and related substances
  Kinins, etc
Inadequate lymph flow

Interstitial fluid tends to accumulate in dependent parts because of the effect of gravity. In the upright position, the capillaries in the legs are protected from the high arterial pressure by the arterioles, but the high venous pressure is transmitted to them through the venules. Skeletal muscle contractions keep the venous pressure low by pumping blood toward the heart (see below) when the individual moves about; but if one stands still for long periods, fluid accumulates and edema eventually develops. The ankles also swell during long trips when travelers sit for prolonged periods with their feet in a dependent position. Venous obstruction may contribute to the edema in these situations.

Whenever there is abnormal retention of salt in the body, water is also retained. The salt and water are distributed throughout the ECF, and since the interstitial fluid volume is therefore increased, there is a predisposition to edema. Salt and water retention is a factor in the edema seen in heart failure, nephrosis, and cirrhosis, but there are also variations in the mechanisms that govern fluid movement across the capillary walls in these diseases. In congestive heart failure, for example, there is usually an elevation in venous pressure, and capillary pressure is consequently elevated. In cirrhosis of the liver, oncotic pressure is low because hepatic synthesis of plasma proteins is depressed; and in nephrosis, oncotic pressure is low because large amounts of protein are lost in the urine.

Another cause of edema is inadequate lymphatic drainage. A complication of **radical mastectomy,** an operation for cancer of the breast in which the axillary lymph nodes are also removed, is edema of the arm due to interruption of its lymph drainage. In filariasis, parasitic worms migrate into the lymphatics and obstruct them. Fluid accumulation plus tissue reaction lead in time to massive swelling, usually of the legs or scrotum (**elephantiasis**). The extent of the reaction is perhaps most graphically illustrated by the remarkable account of the man with elephantiasis whose scrotum was so edematous that he had to place it in a wheelbarrow and wheel it along with him when he walked.

VENOUS CIRCULATION

Blood flows through the blood vessels, including the veins, primarily because of the pumping action of the heart, although venous flow is aided by the heartbeat, the increase in the negative intrathoracic pressure during each inspiration, and contractions of skeletal muscles that compress the veins (**muscle pump**).

Venous Pressure & Flow

The pressure in the venules is 12–18 mm Hg. It falls steadily in the larger veins to about 5.5 mm Hg in the great veins outside the thorax. The pressure in the great veins at their entrance into the right atrium (**central venous pressure**) averages 4.6 mm Hg but fluctuates with respiration and heart action.

Peripheral venous pressure, like arterial pressure, is affected by gravity. It is increased by 0.77 mm Hg for each cm below the right atrium and decreased a like amount for each cm above the right atrium the pressure is measured (Fig 30–16).

When blood flows from the venules to the large veins, its average velocity increases as the total cross-sectional area of the vessels decreases. In the great veins, the velocity of blood is about one-fourth that in the aorta, averaging about 10 cm/s.

Thoracic Pump

During inspiration, the intrapleural pressure falls from –2.5 to –6 mm Hg. This negative pressure is transmitted to the great veins and, to a lesser extent, the atria, so that central venous pressure fluctuates from about 6 mm Hg during expiration to approximately 2 mm Hg during quiet inspiration. The drop in venous pressure during inspiration aids venous return. When the diaphragm descends during inspiration, intra-abdominal pressure rises, and this also squeezes blood toward the heart because backflow into the leg veins is prevented by the venous valves.

Effects of Heartbeat

The variations in atrial pressure are transmitted to the great veins to produce the **a, c,** and **v waves** of the venous pressure-pulse curve (see Chapter 29). Atrial pressure drops sharply during the ejection phase of ventricular systole because the atrioventricular valves are pulled downward, increasing the capacity of the atria. This action sucks blood into the atria from the great veins. The sucking of the blood into the atria during systole contributes appreciably to the venous return, especially at rapid heart rates.

Close to the heart, venous flow becomes pulsatile. When the heart rate is slow, 2 periods of peak flow are detectable, one during ventricular systole, due to pulling down of the atrioventricular valves, and one in early diastole, during the period of rapid ventricular filling (Fig 30–14).

Muscle Pump

In the limbs, the veins are surrounded by skeletal muscles, and contraction of these muscles during activity compresses the veins. Pulsations of nearby arteries may also compress veins. Since the venous valves prevent reverse flow, the blood moves toward the heart. During quiet standing, when the full effect of gravity is manifest, venous pressure at the ankle is 85–90 mm Hg (Fig 30–16). Pooling of blood in the leg veins reduces venous return, with the result that cardiac output is reduced, sometimes to the point where fainting occurs. Rhythmic contractions of the leg muscles while the person is standing serve to lower the venous pressure in the legs to less than 30

mm Hg by propelling blood toward the heart. This heartward movement of the blood is decreased in patients with **varicose veins,** whose valves are incompetent, and such patients may have venous stasis and ankle edema. However, even when the valves are incompetent, muscle contractions will continue to produce a basic heartward movement of the blood because the resistance of the larger veins in the direction of the heart is less than the resistance of the small vessels away from the heart.

Venous Pressure in the Head

In the upright position, the venous pressure in the parts of the body above the heart is decreased by the force of gravity. The neck veins collapse above the point where the venous pressure is close to zero, and the pressure all along the collapsed segments is close to zero rather than subatmospheric. However, the dural sinuses have rigid walls and cannot collapse. The pressure in them in the standing or sitting position is therefore subatmospheric. The magnitude of the negative pressure is proportionate to the vertical distance above the top of the collapsed neck veins and in the superior sagittal sinus may be as much as –10 mm Hg. This fact must be kept in mind by neurosurgeons. Neurosurgical procedures are sometimes performed with the patient seated. If one of the sinuses is opened during such a procedure, it sucks air, causing **air embolism.**

Air Embolism

Because air, unlike fluid, is compressible, its presence in the circulation has serious consequences. The forward movement of the blood depends upon the fact that blood is incompressible. Large amounts of air fill the heart and effectively stop the circulation, causing sudden death, because most of the air is compressed by the contracting ventricles rather than propelled into the arteries. Small amounts of air are swept through the heart with the blood, but the bubbles lodge in the small blood vessels. The surface capillarity of the bubbles markedly increases the resistance to blood flow, and flow is reduced or abolished. Blockage of small vessels in the brain leads to serious and even fatal neurologic abnormalities. In experimental animals, the amount of air that produces fatal air embolism varies considerably, depending in part upon the rate at which it enters the veins. Sometimes as much as 100 mL can be injected without ill effects, whereas at other times as little as 5 mL is lethal.

Measuring Venous Pressure

Central venous pressure can be measured directly by inserting a catheter into the thoracic great veins. **Peripheral venous pressure** correlates well with central venous pressure in most conditions. To measure peripheral venous pressure, a needle attached to a manometer containing sterile saline is inserted into an arm vein. The peripheral vein should be at the level of the right atrium (a point 10 cm or half the chest diameter from the back in the supine position). The values obtained in mm of saline can be converted into mm Hg by dividing by 13.6 (the density of mercury). The amount by which peripheral venous pressure exceeds central venous pressure increases with the distance from the heart along the veins. The mean pressure in the antecubital vein is normally 7.1 mm Hg, compared with a mean pressure of 4.6 mm Hg in the central veins.

A fairly accurate estimate of central venous pressure can be made without any equipment by simply noting the height to which the external jugular veins are distended when the subject lies with the head slightly above the heart. The vertical distance between the right atrium and the place the vein collapses (the place where the pressure in it is zero) is the venous pressure in mm of blood.

Central venous pressure is decreased during negative pressure breathing and shock. It is increased by positive pressure breathing, straining, expansion of the blood volume, and heart failure. In advanced congestive heart failure or obstruction of the superior vena cava, the pressure in the antecubital vein may reach values of 20 mm Hg or more.

31 Cardiovascular Regulatory Mechanisms

INTRODUCTION

In humans and other mammals, multiple cardiovascular regulatory mechanisms have evolved. These mechanisms increase the blood supply to active tissues and increase or decrease heat loss from the body by redistributing the blood. In the face of challenges such as hemorrhage, they maintain the blood flow to the heart and brain. When the challenge faced is severe, flow to these vital organs is maintained at the expense of the circulation to the rest of the body.

Circulatory adjustments are effected by altering the output of the pump (the heart), changing the diameter of the resistance vessels (primarily the arterioles), or altering the amount of blood pooled in the capacitance vessels (the veins). Regulation of cardiac output is discussed in Chapter 29. The caliber of the arterioles is increased in active tissues by locally produced vasodilator metabolites, is affected by substances secreted by the endothelium, and is regulated systematically by circulating vasoactive substances and the nerves that innervate the arterioles. The caliber of the capacitance vessels is also affected by circulating vasoactive substances and by vasomotor nerves. The systemic regulatory mechanisms synergize with the local mechanisms and adjust vascular responses throughout the body.

The terms **vasoconstriction** and **vasodilation** are generally used to refer to constriction and dilation of the resistance vessels. Changes in the caliber of the veins are referred to specifically as **venoconstriction** or **venodilation**.

LOCAL REGULATORY MECHANISMS

Autoregulation

The capacity of tissues to regulate their own blood flow is referred to as **autoregulation.** Most vascular beds have an intrinsic capacity to compensate for moderate changes in perfusion pressure by changes in vascular resistance, so that blood flow remains relatively constant. This capacity is well developed in the kidneys (see Chapter 38), but it has also been observed in the mesentery, skeletal muscle, brain, liver, and myocardium. It is probably due in part to the intrinsic contractile response of smooth muscle to stretch (**myogenic theory of autoregulation**). As the pressure rises, the blood vessels are distended and the vascular smooth muscle fibers that surround the vessels contract. If it is postulated that the muscle responds to the tension in the vessel wall, this theory could explain the greater degree of contraction at higher pressures; the wall tension is proportionate to the distending pressure times the radius of the vessel (law of Laplace; see Chapter 30), and the maintenance of a given wall tension as the pressure rises would require a decrease in radius. Vasodilator substances tend to accumulate in active tissues, and these "metabolites" also contribute to autoregulation (**metabolic theory of autoregulation**). When blood flow decreases, they accumulate and the vessels dilate; when blood flow increases, they tend to be washed away. It has also been argued that as blood flow increases, the accumulation of interstitial fluid compresses the capillaries and venules; but the bulk of the available evidence is against this **tissue pressure hypothesis of autoregulation.**

Vasodilator Metabolites

The metabolic changes that produce vasodilation include, in most tissues, decreases in O_2 tension and pH. These changes cause relaxation of the arterioles and precapillary sphincters. Increases in CO_2 tension and osmolality also dilate the vessels. The direct dilator action of CO_2 is most pronounced in the skin and brain. The neurally mediated vasoconstrictor effects of systemic as opposed to local hypoxia and hypercapnia are discussed below. A rise in temperature exerts a direct vasodilator effect, and the temperature rise in active tissues (due to the heat of metabolism) may contribute to the vasodilation. K^+ is another substance that accumulates locally, has demonstrated dilator activity, and probably plays a role in the dilation that occurs in skeletal muscle. Lactate may also contribute to the dilation. In injured tissues, histamine released from damaged cells increases capillary permeability. Thus, it is probably responsible for some of the swelling in areas of inflammation. Adenosine may play a vasodilator role in cardiac muscle but not in skeletal muscle. It also inhibits the release of norepinephrine.

Localized Vasoconstriction

Injured arteries and arterioles constrict strongly. The constriction appears to be due in part to the local liberation of serotonin from platelets that stick to the vessel wall in the injured area (see Chapter 27).

A drop in tissue temperature causes vasoconstriction, and this local response to cold plays a part in temperature regulation (see Chapter 14).

SUBSTANCES SECRETED BY THE ENDOTHELIUM

Prostacyclin & Thromboxane A_2

Prostacyclin is produced by endothelial cells and thromboxane A_2 by platelets from their common precursor arachidonic acid via the cyclooxygenase pathway (see Fig 17–34). Thromboxane A_2 promotes platelet aggregation and vasoconstriction, whereas prostacyclin inhibits platelet aggregation and promotes vasodilation. The balance between platelet thromboxane A_2 and prostacyclin fosters localized platelet aggregation and consequent clot formation (see Chapter 27) while preventing excessive extension of the clot and maintaining blood flow around it.

The thromboxane A_2-prostacyclin balance can be shifted toward prostacyclin by administration of low doses of aspirin. Aspirin produces irreversible inhibition of cyclooxygenase by acetylating a serine residue in its active site. Obviously, this reduces production of both thromboxane A_2 and prostacyclin. However, endothelial cells produce new cyclooxygenase in a matter of hours whereas platelets cannot manufacture the enzyme, and the level only rises as new platelets enter the circulation. This is a slow process because platelets have a half-life of about 4 days. Therefore, administration of small amounts of aspirin for prolonged periods reduces clot formation and has been shown to be of value in preventing myocardial infarctions, unstable angina, transient ischemic attacks, and stroke.

Endothelium-Derived Relaxing Factor

Several years ago, a chance observation led to the discovery that the endothelium also plays a key role in vasodilation. Many different stimuli act on the endothelial cells to produce **endothelium-derived relaxing factor (EDRF)**, a substance that has now been identified as nitric oxide (NO). NO is synthesized from arginine by the calmodulin-dependent enzyme NO synthetase (Fig 31–1). It activates the soluble guanylate cyclase in cells (see Chapter 1), producing cyclic GMP, which in turn mediates relaxation of vascular smooth muscle. NO is inactivated by hemoglobin. It is released by a variety of different compounds. Acetylcholine, for example, is a vasoconstrictor when applied to vessels from which the endothelium has been stripped off, but it is a relatively potent vasodilator when applied to intact blood

Figure 31–1. Synthesis of NO from arginine and its action via guanylate cyclase to increase intracellular cyclic GMP (cGMP). The reaction is catalyzed by NO synthetase and requires NADPH and Ca^{2+}, although the exact details of the synthesis are still unsettled.

vessels because it releases NO. Adenosine, ANP, and histamine via H_2 receptors produce relaxation of vascular smooth muscle that is independent of the endothelium. However, bradykinin, VIP, substance P, and some other polypeptides act via the endothelium, and various vasoconstrictors that act directly on vascular smooth muscle would produce much greater constriction if they did not simultaneously cause the release of EDRF. When flow to a tissue is suddenly increased by arteriolar dilation, the large arteries to the tissue also dilate. This flow-induced dilation is probably due to local release of EDRF. Products of platelet aggregation also cause release of EDRF, and the resulting vasodilation probably helps keep blood vessels with an intact endothelium patent. This is in contrast to injured blood vessels, where the endothelium is damaged at the site of injury and platelets therefore aggregate and produce vasoconstriction (see Chapter 27).

Further evidence for a physiologic role of EDRF is the observation that when various derivatives of orginine that inhibit NO synthetase are administered to experimental animals, there is a prompt rise in blood pressure. This suggests that tonic release of EDRF is necessary to maintain normal blood pressure.

There is now some evidence that EDRF deficiency can cause clinical hypertension, and EDRF may also be involved in the pathogenesis of atherosclerosis. It is interesting in this regard that some patients with heart transplants develop an accelerated form of atherosclerosis in the vessels of the transplant, and there is reason to believe that this is triggered by endothelial damage. Nitroglycerin and other nitrovasodilators that are of great value in the treatment of angina act by stimulating guanylate cyclase in the same manner as EDRF. So does the blood pressure-lowering drug nitroprusside. It has also been suggested that deficient EDRF secretion may be responsible for some cases of impotence, since intracavernosal injection of nitrovasodilators has been reported to cause erections.

Other Functions of EDRF

It has become almost commonplace to discover a compound that plays an important role in cardiovascular regulation and then learn that it is produced in other systems and has additional diverse functions. This is true, for example, of angiotensin II (see Chapter 24) and the endothelins (see below). EDRF is no exception. NO is produced by macrophages, and it appears to be necessary for the cytotoxic action of these cells, including their ability to kill cancer cells. In addition, NO is produced in the cerebellum and other parts of the brain, and evidence is accumulating that, acting via cyclic GMP, it plays a role in brain function.

Endothelins

Endothelial cells also produce the most potent vasoconstrictor agent yet isolated, a polypeptide that was initially called endothelin. Subsequent research demonstrated that there is a family of at least 4 similar 21-amino-acid polypeptides now called **endothelin-1** (ET-1), **endothelin-2** (ET-2), **endothelin-3** (ET-3), and **vasoactive intestinal contractor** (VIC). They are encoded by at least 3 different genes. All have 2 disulfide bridges, which give their molecules a unique shape (Fig 31–2). ET-1 is produced by endothelial cells, and in humans its gene produces a 38-amino-acid proendothelin-1. This is cleaved in a somewhat unusual way, hydrolysis of a Trp-Val bond, to form ET-1. The putative enzyme involved has been called **endothelin-converting enzyme.** It is interesting that the structure of the endothelins is very similar to that of the sarafotoxins, polypeptides found in the venom of a snake, the Israeli burrowing asp (Fig 31–2). Two different endothelin receptors have been cloned, both of which are serpentine receptors coupled to G proteins (see Chapter 1). The ET_A receptor, which is specific for ET-1, is found in many tissues and is presumably the vascular receptor. The ET_B receptor responds to all 3 endothelins, and its function is unknown.

When injected intravenously, ET-1 produces a transient fall in blood pressure followed by a marked, prolonged pressor response that lasts for more than 1 hour, even though the half-life of endothelins in the blood is less than 1 minute. The many other actions of endothelins are summarized in Table 31–1. ET-1 is released by thrombin and epinephrine, but the release is slow. The exact physiologic role of the endothelin produced by endothelial cells has not been determined, but there is some evidence that it is released in

Figure 31–2. Endothelins. **Top:** Structure of endothelin-1 (ET-1). **Bottom:** Sequence homology of other endothelins and one of the sarafotoxin peptides, S6b. Single-letter codes are used for constituent amino acids. (Reproduced, with permission, from Simonson MS, Dunn MJ: Cellular signaling by peptides of the endothelin gene family. *FASEB J* 1990;**4**:2989.)

	1	2	3	4	5	6	7	8	9	10	11	12	13	14	15	16	17	18	19	20	21
ET-1	C	S	C	S	S	L	M	D	K	E	C	V	Y	F	C	H	L	D	I	I	W
ET-2	C	S	C	S	S	W	L	D	K	E	C	V	Y	F	C	H	L	D	I	I	W
ET-3	C	T	C	F	T	Y	K	D	K	E	C	V	Y	Y	C	H	L	D	I	I	W
Mouse/Rat ET-2	C	S	C	N	D	W	L	D	K	E	C	V	Y	F	C	H	L	D	I	I	W
S6b	C	S	C	K	D	M	T	D	K	E	C	L	Y	F	C	H	Q	D	V	I	W

response to stretching of blood vessels, a stimulus known to produce local vasoconstriction. ET-1 also activates phospholipase A_2, increasing the production of prostacyclin and thromboxane A_2. ET-1 is present in the circulating blood, and the blood level is elevated in hypertension.

Table 31–1. Biologic actions of endothelins.[1]

Hemodynamic effects
Causes initial depressor response followed by sustained pressor effect (regional differences in vasoconstriction exist)
Cardiac effects
Evokes positive inotropic and chronotropic effects on myocardium
Stimulates intense vasoconstriction of coronary arteries
Neuroendocrine effects
Increases plasma levels of ANF, renin, aldosterone, and catecholamines
Modulates synaptic transmission
Renal effects
Increases renal vascular resistance
Decreases glomerular filtration rate, renal blood flow, the glomerular ultrafiltration coefficient
Increases Na^+ reabsorption through hemodynamic actions
Decreases Na^+ reabsorption through inhibition of Na^+-K^+-ATPase
Smooth muscle effects
Contracts vascular smooth muscle; veins possibly more sensitive than arteries
Contracts numerous nonvascular smooth muscles
Promitogenic effects
Stimulates mitogenesis in vascular smooth muscle cells, 3T3 fibroblasts, and glomerular mesangial cells
Gene expression
Up-regulates mRNA expression of VL30 gene and the c-*fos* and c-*myc* protoncogenes

[1]Reproduced, with permission, from Simonson MS, Dunn MJ: cellular signaling by peptides of the endothelin gene family. *FASEB J* 1990;**4**:2989.

Other Functions of Endothelins

ET-1 is found in the brain and the kidneys as well as in endothelial cells. Like VIC, ET-2 is found primarily in the intestine, and ET-3 is found in the intestine and adrenal glands. The actions listed in Table 31–1 suggest functions for endothelins in these various locations. In the brain, endothelin is found in dorsal root ganglia, ventral horn cells, the cortex, the hypothalamus, and cerebellar Purkinje cells. There are endothelin receptors on glial cells (see Chapter 38), and the polypeptide presumably produces mesangial cell-mediated decreases in the glomerular filtration rate.

SYSTEMIC REGULATION BY HORMONES

Prostacyclin, EDRF, and ET-1 appear to act primarily as paracrine mediators of local hormones in regulating cardiovascular function. On the other hand, many circulating hormones affect the vascular system. The vasodilator hormones include kinins, VIP, and ANP. Circulating vasoconstrictor hormones include vasopressin, norepinephrine, epinephrine, and angiotensin II.

Kinins

Two related vasodilator peptides called **kinins** are found in the body. One is the nonapeptide **bradykinin,** and the other is the decapeptide **lysylbradykinin,** also known as **kalidin** (Fig 31–3). Lysylbradykinin can be converted to bradykinin by aminopeptidase. Both peptides are metabolized to inactive fragments by **kininase I,** a carboxypeptidase that removes the C-terminal Arg. In addition, the dipeptidylcarboxypeptidase **kininase II** inactivates bradykinin and lysylbradykinin by removing Phe-Arg from the C terminal. Kininase II is the same enzyme as **angiotensin-converting enzyme** (see Chapter 24), which removes His-Leu from the C-terminal end of angiotensin I.

Bradykinin and lysylbradykinin are formed from 2 precursor proteins, **high-molecular-weight kininogen** and **low-molecular-weight kininogen** (Fig 31–4). Proteases called **kallikreins** release the pep-

Figure 31–4. Formation of kinins from high-molecular-weight (HMW) and low-molecular-weight (LMW) kininogens.

tides from these precursors. There are 2 kallikreins: **plasma kallikrein,** which circulates in an inactive form, and **tissue kallikrein,** which appears to be located primarily on the apical membranes of cells concerned with transcellular electrolyte transport. Tissue kallikrein acts on high-molecular-weight kininogen and low-molecular-weight kininogen to form lysylbradykinin. When activated, plasma kallikrein acts on high-molecular-weight kininogen to form bradykinin.

Inactive plasma kallikrein (**prekallikrein**) is activated by prekallikrein activators, which are proteolytic fragments of the active form of clotting factor XII (Hageman factor). The formation of these activators from active factor XII is catalyzed by plasmin (see Chapter 27) and in a positive feedback fashion by plasma kallikrein itself. Plasma kallikrein and high-molecular-weight kininogen also catalyze the activation of factor XII (see Fig 27–20).

There are important interactions between the kinin system and the renin-angiotensin system. As noted above, kininase II and angiotensin-converting enzyme are the same enzyme. In addition, plasma kallikrein can convert prorenin to active renin in vitro, and it may have the same action in vivo.

The actions of the kinins resemble those of histamine. They cause contraction of visceral smooth muscle, but they relax vascular smooth muscle via EDRF, lowering blood pressure. They also increase capillary permeability, attract leukocytes, and cause pain upon injection under the skin. They appear to be formed during active secretion in sweat glands, salivary glands, and the exocrine portion of the pancreas (see Chapter 26), and they are probably responsible for the increase in blood flow when these tissues are actively secreting their products. They are present in the kidneys, where their function is uncertain.

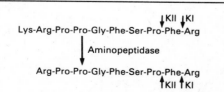

Figure 31–3. Kinins. Lysylbradykinin *(top)* can be converted to bradykinin *(bottom)* by aminopeptidase. Both peptides are inactivated by kininase I (KI) or kininase II (KII) at the sites indicated by the arrows.

Two subtypes of bradykinin receptors, β1 and β2, have been identified.

Natriuretic Hormones

The atrial natriuretic peptide (ANP) secreted by the heart (see Chapter 24) antagonizes the action of various vasoconstrictor agents and lowers blood pressure, but its exact role in the regulation of the circulation is still unsettled. The natriuretic Na^+-K^+ ATPase inhibitor, which is now thought to be endogenously produced ouabain, apparently raises rather than lowers blood pressure.

Circulating Vasoconstrictors

Vasopressin is a potent vasoconstrictor, but when it is injected in normal individuals, there is a compensating decrease in cardiac output, so that there is little change in blood pressure. Its role in blood pressure regulation is discussed in Chapter 14.

Norepinephrine has a generalized vasoconstrictor action, whereas epinephrine dilates the vessels in skeletal muscle and the liver. The relative unimportance of circulating norepinephrine, as opposed to norepinephrine released from vasomotor nerves, is pointed out in Chapter 20, in which the cardiovascular actions of catecholamines are discussed in detail.

The octapeptide angiotensin II has a generalized vasoconstrictor action. It is formed from angiotensin I liberated by the action of renin from the kidney on circulating angiotensinogen (see Chapter 24). Its formation is increased because renin secretion is increased when the blood pressure falls or ECF volume is reduced, and it helps maintain blood pressure. Angiotensin II also increases water intake and stimulates aldosterone secretion, and increased formation of angiotensin II is part of a homeostatic mechanism that operates to maintain ECF volume (see Chapter 20). In addition, there are renin-angiotensin systems in many different organs, and one of these is in vascular smooth muscle. Angiotensin II produced in blood vessel walls may be important in some forms of clinical hypertension.

SYSTEMIC REGULATION BY THE NERVOUS SYSTEM

Neural Regulatory Mechanisms

Although the arterioles and the other resistance vessels are most densely innervated, all blood vessels except capillaries and venules contain smooth muscle and receive motor nerve fibers from the sympathetic division of the autonomic nervous system. The fibers to the resistance vessels regulate tissue blood flow and arterial pressure. The fibers to the venous capacitance vessels vary the volume of blood "stored" in the veins. The innervation of most veins is sparse, but the splanchnic veins are well innervated. Venoconstriction is produced by stimuli that also activate the

vasoconstrictor nerves to the arterioles. The resultant decrease in venous capacity increases venous return, shifting blood to the arterial side of the circulation.

Innervation of the Blood Vessels

Noradrenergic fibers end on vessels in all parts of the body (Fig 31–5). The noradrenergic fibers are vasoconstrictor in function. In addition to their vasoconstrictor innervation, the resistance vessels of the skeletal muscles are innervated by vasodilator fibers, which, although they travel with the sympathetic nerves, are cholinergic (the **sympathetic vasodilator system**). There is some evidence that blood vessels in the heart, lungs, kidneys, and uterus also receive a cholinergic innervation. Bundles of noradrenergic and cholinergic fibers form a plexus on the adventitia of the arterioles. Fibers with multiple varicosities extend from this plexus to the media and end primarily on the outer surface of the smooth muscle of the media without penetrating it. Transmitter reaches the inner portions of the media by diffusion, and current spreads from one smooth muscle cell to another via gap junctions.

There is no tonic discharge in the vasodilator fibers, but the vasoconstrictor fibers to most vascular beds have some tonic activity. When the sympathetic nerves are cut (**sympathectomy**), the blood vessels

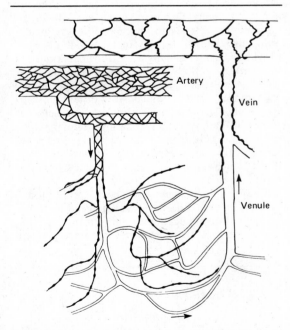

Figure 31–5. Relation of noradrenergic nerve fibers to blood vessels in the mesentery of the rat. Arrows indicate direction of flow. (Reproduced, with permission, from Furness JB, Marshall JM: Correlation of the directly observed responses of mesenteric vessels of the rat to nerve stimulation and noradrenaline with the distribution of adrenergic nerves. *J Physiol* 1974;**75:**239.)

dilate. In most tissues, vasodilation is produced by decreasing the rate of tonic discharge in the vasoconstrictor nerves, although in skeletal muscles it can also be produced by activating the sympathetic vasodilator system (Table 31–2).

Nerves containing polypeptides are found on many blood vessels. The cholinergic nerves also contain VIP, which produces vasodilation; and postganglionic sympathetic nerves also contain neuropeptide Y, which is a vasoconstrictor. Substance P and CGRPα, which also produce vasodilation, are found in sensory nerves near blood vessels.

Afferent impulses in sensory nerves from the skin are relayed antidromically down branches of the sensory nerves that innervate blood vessels, and these impulses cause release of substance P from the nerve endings. Substance P causes vasodilation and increased capillary permeability. This local neural mechanism is called the **axon reflex** (see Fig 32–16). Other cardiovascular reflexes are integrated in the central nervous system.

Cardiac Innervation

Impulses in the noradrenergic sympathetic nerves to the heart increase the cardiac rate (chronotropic effect) and the force of cardiac contraction (inotropic effect). Impulses in the cholinergic vagal cardiac fibers decrease the heart rate. There is a moderate amount of tonic discharge in the cardiac sympathetic nerves at rest, but there is a good deal of tonic vagal discharge (**vagal tone**) in humans and other large animals. When the vagi are cut in experimental animals, the heart rate rises, and after the administration of parasympatholytic drugs such as atropine, the heart rate in humans increases from its normal resting value of 70 to 150–180 beats/min because the sympathetic tone is unopposed. In humans in whom both noradrenergic and cholinergic systems are blocked, the heart rate is approximately 100.

Table 31–2. Summary of factors affecting the caliber of the arterioles.

Constriction	Dilation
Increased noradrenergic discharge	Decreased noradrenergic discharge
Circulating catecholamines (except epinephrine in skeletal muscle and liver)	Circulating epinephrine in skeletal muscle and liver
Circulating angiotensin II	Circulating ANP
Circulating AVP	Activation of cholinergic dilators in skeletal muscle
Locally released serotonin	Histamine
Decreased local temperature	Kinins
Endothelin-I	Substance P (axon reflex)
Neuropeptide Y	CGRPα
Circulating Na^+-K^+ ATPase inhibitor	VIP
	EDRF
	Decreased O_2 tension
	Increased CO_2 tension
	Decreased pH
	Lactate, K^+, adenosine, etc
	Increased local temperature

Vasomotor Control

The sympathetic nerves that constrict arterioles and veins and increase heart rate and stroke volume discharge in a tonic fashion, and blood pressure is adjusted by variations in the rate of this tonic discharge. Spinal reflex activity affects blood pressure, but the main control of blood pressure is exerted by groups of neurons in the medulla oblongata that are sometimes called collectively the **vasomotor area** or **vasomotor center.** Neurons that mediate increased sympathetic discharge to blood vessels and the heart (Fig 31–6) project directly to sympathetic preganglionic neurons in the intermediolateral gray column (IML) of the spinal cord. On each side, the cell bodies of these neurons are located near the pial surface of the medulla in the rostral ventrolateral reticular nucleus and are known as the C1 group of cells. Their axons course dorsally and medially and then descend in the lateral column of the spinal cord to the IML. They contain PNMT (see Chapter 4), but it appears that the excitatory transmitter they secrete is not epinephrine and may instead be glutamate. A second group of neurons with cell bodies in the caudal ventrolateral medulla mediate inhibition of vasomotor discharge. These neurons are in the same area as the A1 noradrenergic neurons, but it is uncertain whether they secrete norepinephrine and whether they project directly to the IML.

Impulses reaching the medulla also affect the heart rate via vagal discharge to the heart. The neurons from which the vagal fibers arise are in the dorsal motor nucleus of the vagus, the nucleus of the tractus solitarius, and the nucleus ambiguus.

When vasoconstrictor tone is increased, there is increased arteriolar constriction and a rise in blood pressure. Venoconstriction and a decrease in the stores of blood in the venous reservoirs usually accompany these changes, although changes in the capacitance vessels do not always parallel changes in the resistance vessels. Heart rate and stroke volume are increased because of activity in the sympathetic nerves to the heart, and cardiac output is increased. There is usually an associated decrease in the tonic activity of vagal fibers to the heart. Conversely, a decrease in the discharge rate of vasoconstrictor fibers causes vasodilation, a fall in blood pressure, and an increase in the storage of blood in the venous reservoirs. There is usually a concomitant decrease in heart rate, but this is mostly due to stimulation of the vagal innervation of the heart.

Afferents to the Vasomotor Area

The afferents that converge on the vasomotor area are summarized in Table 31–3. They include not only the very important fibers from arterial and venous baroreceptors but also fibers from other parts of the nervous system and from the carotid and aortic chemoreceptors. In addition, some stimuli act directly on the vasomotor area.

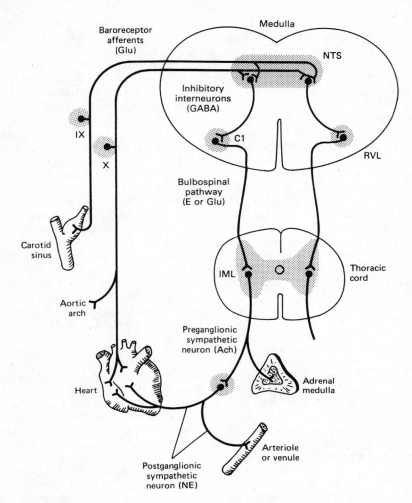

Figure 31–6. Basic pathways involved in the medullary control of blood pressure. The vagal efferent pathways that slow the heart are not shown. The putative neurotransmitters in the pathways are indicated in parentheses. Glu, glutamate; GABA, γ-aminobutyric acid; E, epinephrine; Ach, acetylcholine; NE, norepinephrine; IML, intermediolateral gray column; NTS, nucleus of the tractus solitarius; RVL, rostral ventrolateral reticular nucleus; IX and X, glossopharyngeal and vagus nerves. (Modified and reproduced, with permission, from Reis DJ et al: Role of adrenaline neurons of ventrolateral medulla [the C1 group] in the tonic and phasic control of arterial pressure. *Clin Exp Hypertens* [A] 1984;**6**:221. By courtesy of Marcel Dekker, Inc.)

There are descending tracts to the vasomotor area from the cerebral cortex (particularly the limbic cortex) that relay in the hypothalamus and possibly also in the mesencephalon. These fibers are responsible for the blood pressure rise and tachycardia produced by emotions such as sexual excitement and anger. The connections between the hypothalamus and the vasomotor area are reciprocal, with afferents from the brain stem closing the loop.

Inflation of the lungs causes vasodilation and a decrease in blood pressure. This response is mediated via vagal afferents from the lungs that inhibit vasomotor discharge. Pain usually causes a rise in blood pressure via afferent impulses in the reticular formation converging on the vasomotor area. However,

Table 31–3. Factors affecting the activity of the vasomotor area in the medulla.

Direct stimulation
    CO_2
    Hypoxia
Excitatory inputs
    From cortex via hypothalamus
    From pain pathways and muscles
    From carotid and aortic chemoreceptors
Inhibitory inputs
    From cortex via hypothalamus
    From lungs
    From carotid, aortic, and cardiopulmonary baroreceptors

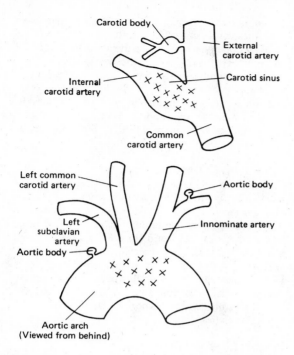

Figure 31–7. Baroreceptor areas in carotid sinus and aortic arch. The Xs identify sites where receptors are located.

Figure 31–8. Discharges (vertical lines) in a single afferent nerve fiber from the carotid sinus at various arterial pressures, plotted against changes in aortic pressure with time. (Reproduced, with permission, from Berne RM, Levy MN: *Cardiovascular Physiology*, 3rd ed. Mosby, 1977.)

prolonged severe pain may cause vasodilation and fainting.

Baroreceptors

The **baroreceptors** are stretch receptors in the walls of the heart and blood vessels. The **carotid sinus** and **aortic arch** receptors monitor the arterial circulation. Receptors are also located in the walls of the right and left atria at the entrance of the superior and inferior venae cavae and the pulmonary veins, in the wall of the left ventricle, and in the pulmonary circulation. These receptors in the low-pressure part of the circulation and the left ventricle are referred to collectively as the cardiopulmonary receptors. The baroreceptors are stimulated by distention of the structures in which they are located, and so they discharge at an increased rate when the pressure in these structures rises. Their afferent fibers pass via the glossopharyngeal and vagus nerves to the medulla. Most of them end in the nucleus of the tractus solitarius (NTS), and the excitatory transmitter they secrete is probably glutamate. An inhibitory pathway made up of one or more neurons projects from the NTS to the C1 cells. The neurotransmitter that inhibits the C1 cells appears to be GABA, but this may be produced by excitation of inhibitory interneurons in the C1 area rather than GABA-secreting neurons that project directly from the NTS to the C1 neurons. In any case, it is important to remember that impulses generated in the baroreceptors *inhibit* the tonic discharge of the vasoconstrictor nerves and *excite* the vagal innervation of the heart, producing vasodilation, venodilation, a drop in blood pressure, bradycardia, and a decrease in cardiac output.

Carotid Sinus & Aortic Arch

The carotid sinus is a small dilation of the internal carotid artery just above the bifurcation of the common carotid into external and internal carotid branches (Fig 31–7). Baroreceptors are located in this dilation. They are also found in the wall of the arch of the aorta. The receptors are located in the adventitia of the vessels. They are extensively branched, knobby, coiled, and intertwined ends of myelinated nerve fibers that resemble Golgi tendon organs (see Fig 6–5). Similar receptors have been found in various other parts of the large arteries of the thorax and neck in some species. The afferent nerve fibers from the carotid sinus and carotid body form a distinct branch of the glossopharyngeal nerve, the **carotid sinus nerve,** but the fibers from the aortic arch form a separate distinct branch of the vagus only in the rabbit. The carotid sinus nerves and vagal fibers from the aortic arch are commonly called the **buffer nerves.**

Buffer Nerve Activity

At normal blood pressure levels, the fibers of the buffer nerves discharge at a slow rate (Fig 31–8). When the pressure in the sinus and aortic arch rises, the discharge rate increases; and when the pressure falls, the rate declines. The compensatory response produced by increased discharge is a fall in blood pressure, because activity in the baroreceptor afferents inhibits the tonic discharge in the vasoconstrictor nerves.

When one carotid sinus of a monkey is isolated and perfused and the other baroreceptors are denervated, there is no discharge in the afferent fibers from the perfused sinus and no drop in the animal's arterial pressure or heart rate when the perfusion pressure is below 30 mm Hg. At perfusion pressures of 70–110 mm Hg, there is an essentially linear relation between the perfusion pressure and the fall in blood pressure and heart rate produced in the monkey. At perfusion pressures above 150 mm Hg there is no further increase in response (Fig 31–9), presumably because the rate of baroreceptor discharge and the degree of inhibition of the vasomotor center are maximal.

The carotid receptors respond both to sustained pressure and to pulse pressure. A decline in carotid pulse pressure without any change in mean pressure decreases the rate of baroreceptor discharge and provokes a rise in blood pressure and tachycardia. The receptors also respond to changes in pressure as well as steady pressure; when the pressure is fluctuating, they sometimes discharge during the rises and are silent during the falls (Fig 31–8) at mean pressures at which if there were no fluctuations, there would be a steady discharge.

The aortic receptors have not been studied in such great detail, but there is no reason to believe that their responses differ significantly from those of the receptors in the carotid sinus.

From the foregoing discussion, it is apparent that the baroreceptors on the arterial side of the circulation, their afferent connections to the vasomotor and cardioinhibitory areas, and the efferent pathways from these areas constitute a reflex feedback mechanism that operates to stabilize the blood pressure and heart rate. Any drop in systemic arterial pressure decreases the inhibitory discharge in the buffer nerves, and there is a compensatory rise in blood pressure and cardiac output. Any rise in pressure produces dilation of the arterioles and decreases cardiac output until the blood pressure returns to its previous normal level.

Baroreceptor Resetting

In chronic hypertension, the baroreceptor reflex mechanism is "reset" to maintain an elevated rather than a normal blood pressure. In perfusion studies on hypertensive experimental animals, raising the pressure in the isolated carotid sinus lowers the elevated systemic pressure, and decreasing the perfusion pressure raises the systemic pressure (Fig 31–9). Little is known about how and why this occurs, but resetting occurs rapidly in experimental animals. It is also rapidly reversible, both in experimental animals and in clinical situations.

Effect of Carotid Clamping & Buffer Nerve Section

Bilateral clamping of the carotid arteries proximal to the carotid sinuses elevates the blood pressure and heart rate because the procedure lowers the pressure in the sinuses. Cutting the carotid sinus nerves on each side has the same effect. The pressor response following these 2 procedures is moderate, because the aortic baroreceptors are still functioning normally, and they buffer the rise. If baroreceptor afferents in the vagi are also interrupted, blood pressure rises to 300/200 mm Hg or higher and is unstable. Bilateral lesions of the nucleus of the tractus solitarius, the site of termination of the baroreceptor afferents, cause severe hypertension that can be fatal. These forms of experimental hypertension are called **neurogenic hypertension.**

Atrial Stretch Receptors

The stretch receptors in the atria are of 2 types: those that discharge primarily during atrial systole (type A), and those that discharge primarily late in diastole, at the time of peak atrial filling (type B). The discharge of type B baroreceptors is increased when venous return is increased and decreased by positive-pressure breathing, indicating that these baroreceptors respond primarily to distention of the atrial walls. The reflex circulatory adjustments initiated by increased discharge from most if not all of these receptors include vasodilation and a fall in blood pressure. However, the heart rate is increased rather than decreased. The atrial baroreceptors are an important part of a reflex mechanism that responds to increases in central venous pressure and regulates the ECF volume.

Figure 31–9. Solid line: Fall in system blood pressure produced by raising the pressure in the isolated carotid sinus of a normal monkey to various values. **Dashed line:** Response in a hypertensive monkey, demonstrating baroreceptor resetting.

Role of Baroreceptors in Endocrine Defense of ECF Volume

When the ECF volume falls, central venous pressure declines, and the decreased firing of the atrial stretch receptors leads to increased secretion of vasopressin. Sympathetic activity is increased, and this leads to increased secretion of renin. The increase in renin secretion increases the secretion of aldosterone. When loss of volume is more severe, arterial pressure falls and decreased discharge of the carotid and aortic baroreceptors also contributes to the increases in hormone secretion. Other factors contributing to the increase in vasopressin and renin secretion are discussed in Chapters 20, 24, and 39. The net result is a retention of water and sodium that helps restore the ECF volume.

Bainbridge Reflex

Rapid infusion of blood or saline in anesthetized animals sometimes produces a rise in heart rate if the initial heart rate is low. This effect was described by Bainbridge in 1915, and since then it has been known as the **Bainbridge reflex.** It appears to be a true reflex rather than a response to local stretch, since it is abolished by bilateral vagotomy, and infusion of fluids in animals with transplanted hearts increases the rate of the recipient's atrial remnant but fails to affect the rate of the transplanted heart. The receptors may be the tachycardia-producing atrial receptors mentioned above. The reflex competes with the baroreceptor-mediated decrease in heart rate produced by volume expansion and is diminished or absent when the initial heart rate is high. There has been much debate about its significance, and its physiologic role remains unsettled.

Left Ventricular Receptors

When the left ventricle is distended in experimental animals, there is a fall in systemic arterial pressure and heart rate. It takes considerable ventricular distention to produce this response, and its physiologic significance is uncertain. However, left ventricular stretch receptors may play a role in the maintenance of the vagal tone that keeps the heart rate low at rest.

In experimental animals, injections of serotonin, veratridine, capsaicin, phenyldiguanide, and some other drugs into the coronary arteries supplying the left ventricle cause apnea followed by rapid breathing, hypotension, and bradycardia (the **coronary chemoreflex** or **Bezold-Jarisch reflex**). The receptors are probably C fiber endings, and the afferents are vagal. The response is not produced by injections into the blood supply of the atria or the right ventricle. Its physiologic role is uncertain, but in patients with myocardial infarcts, substances released from the infarcted tissue may stimulate ventricular receptors, contributing to the hypotension that is not infrequently a stubborn complication of this disease. A similar response is produced by stimulation of C fiber endings in the lungs (**pulmonary chemoreflex**; see Chapter 36).

Pulmonary Receptors

Injections of serotonin, capsaicin, veratridine, and related drugs into the pulmonary artery activates C fiber endings close to capillaries in the lungs and produces apnea followed by rapid breathing, hypotension, and bradycardia (**pulmonary chemoreflex**). This response, which is discussed in Chapter 36, is blocked by vagotomy and is essentially the same as the coronary chemoreflex caused by injection of drugs into the arterial supply of the left ventricle. However, the response is too rapid to be caused by the drugs reaching the left ventricular receptors.

Clinical Testing & Stimulation

The changes in pulse rate and blood pressure that occur in humans on standing up or lying down (see Chapter 33) are due for the most part to baroreceptor reflexes. The function of the receptors can be tested by monitoring changes in heart rate as a function of increasing arterial pressure during infusion of the α-adrenergic agonist phenylephrine. A normal response is shown in Fig 31–10; from a systolic pressure of about 120–150 mm Hg, there is a linear relation between pressure and lowering of the heart rate (greater R-R interval).

The function of the receptors can also be tested by monitoring the changes in pulse and blood pressure that occur in response to brief periods of straining (forced expiration against a closed glottis: the **Valsalva maneuver**). The blood pressure rises at the

Figure 31–10. Baroreflex-mediated lowering of the heart rate during infusion of phenylephrine in a human subject. Note that the values for the R-R interval of the EKG, which are plotted on the vertical axis, are inversely proportionate to the heart rate. (Reproduced, with permission, from Kotrly K et al: Effects of fentanyl-diazepam-nitrous oxide anaesthesia on arterial baroreflex control of heart rate in man *Br J Anaesth* 1986;**58:**406.)

onset of straining (Fig 31–11) because the increase in intrathoracic pressure is added to the pressure of the blood in the aorta. It then falls because the high intrathoracic pressure compresses the veins, decreasing venous return and cardiac output. The decreases in arterial pressure and pulse pressure inhibit the baroreceptors, causing tachycardia and a rise in peripheral resistance. When the glottis is opened and the intrathoracic pressure returns to normal, cardiac output is restored but the peripheral vessels are constricted. The blood pressure therefore rises above normal, and this stimulates the baroreceptors, causing bradycardia and a drop in pressure to normal levels.

In sympathectomized patients, heart rate changes still occur because the baroreceptors and the vagi are intact. However, in patients with autonomic insufficiency, a syndrome in which there is widespread disruption of autonomic function, the heart rate changes are absent. For reasons that are still obscure, patients with primary hyperaldosteronism also fail to show the heart rate changes and the blood pressure rise when the intrathoracic pressure returns to normal. Their response to the Valsalva maneuver returns to normal after removal of the aldosterone-secreting tumor.

Effects of Chemoreceptor Stimulation on the Vasomotor Area

Afferents from the chemoreceptors in the carotid and aortic bodies exert their main effect on respiration, and their function is discussed in Chapter 36. However, they also converge on the vasomotor area. The cardiovascular response to chemoreceptor stimulation consists of peripheral vasoconstriction and bradycardia. However, hypoxia also produces hyperpnea and increased catecholamine secretion from the adrenal medulla, both of which produce tachycardia and an increase in cardiac output. Hemorrhage that produces hypotension leads to chemoreceptor stimu-

lation. This is due to decreased blood flow to the chemoreceptors and consequent stagnant anoxia of these organs (see Chapter 37). In hypotensive animals, baroreceptor discharge is low (see below), and section of the glossopharyngeal and vagus nerves leads to a fall rather than a rise in blood pressure, because the chemoreceptor drive to the vasomotor area is removed. Chemoreceptor discharge may also contribute to the production of **Mayer waves.** These should not be confused with **Traube-Hering waves,** which are fluctuations in blood pressure synchronized with respiration. The Mayer waves are slow regular oscillations in arterial pressure that occur at the rate of about one per 20–40 seconds during hypotension. Under these conditions, hypoxia stimulates the chemoreceptors. The stimulation raises the blood pressure, which improves the blood flow in the receptor organs and eliminates the stimulus to the chemoreceptors, so that the pressure falls and a new cycle is initiated. However, Mayer waves are reduced but not abolished by chemoreceptor denervation and are sometimes present in spinal animals, so oscillation in spinal vasopressor reflexes is also involved.

Direct Effects on the Vasomotor Area

Hypoxia and hypercapnia both stimulate the vasomotor area directly and presumably act on the C1 cells, although the direct effect of hypoxia is small. When intracranial pressure is increased, the blood supply to the vasomotor area is compromised, and the local hypoxia and hypercapnia increase its discharge. The resultant rise in systemic arterial pressure (**Cushing reflex**) tends to restore the blood flow to the medulla. The rise in blood pressure causes a reflex decrease in heart rate via the arterial baroreceptors (see below), and this is why bradycardia rather than tachycardia is characteristically seen in patients with increased intracranial pressure.

A rise in arterial P_{CO_2} stimulates the vasomotor

Figure 31–11. Diagram of the response to straining (the Valsalva maneuver) in a normal man, recorded with a needle in the brachial artery. (Courtesy of M McIlroy.)

area, but the direct peripheral effect of hypercapnia is vasodilation. Therefore, the peripheral and central actions tend to cancel each other. Moderate hyperventilation, which significantly lowers the CO_2 tension of the blood, causes cutaneous and cerebral vasoconstriction in humans, but there is little change in blood pressure. Exposure to high concentrations of CO_2 is associated with marked cutaneous and cerebral vasodilation, but there is vasoconstriction elsewhere and usually a slow rise in blood pressure.

Somatosympathetic Reflex

Pain causes increased arterial pressure, and activity in afferents from exercising muscles probably exerts a similar pressor effect via the C1 neurons in the rostral ventrolateral medulla. The pressor response to stimulation of somatic afferent nerves is called the somatosympathetic reflex. The pathways involved are shown in Fig 31–12.

Sympathetic Vasodilator System

The cholinergic sympathetic vasodilator fibers are part of a regulatory system that originates in the cerebral cortex, relays in the hypothalamus and mesencephalon, and passes through the medulla without interruption to the intermediolateral gray column of the spinal cord (Fig 31–13). The preganglionic neurons which are part of this system activate postganglionic neurons to blood vessels in skeletal muscle that are anatomically sympathetic but secrete acetylcholine. Stimulation of this system produces vasodilation in skeletal muscle, but the resultant increase in blood flow is associated with a decrease rather than an increase in muscle O_2 consumption. This suggests that the blood is being diverted through thoroughfare

Figure 31–12. The somatosympathetic reflex. Impulses from receptors in skin and muscle relay to the C1 cells in the rostral ventrolateral reticular nucleus (RVL). The axons of the C1 cells project directly to the intermediolateral gray column (IML), where they end on preganglionic sympathetic neurons. ICP, inferior cerebellar peduncle; ION, inferior olivary nucleus; MLF, medial longitudinal fasciculus; NA, nucleus ambiguus; NTS, nucleus of the tractus solitarius; P, pyramid; STN, spinal trigeminal nucleus. (Reproduced, with permission, from Reis DJ, Ruggerio DA, Morrison SF: The C₁ area of the rostral ventrolateral medulla oblongata: A critical brainstem region for control of resting and reflex integration of arterial pressure. *Am J Hypertens* 1989;**2**:363S.)

Figure 31–13. The sympathetic vasodilator pathways. (Reproduced, with permission, from Lindgren P: The mesencephalic and vascular system. *Acta Physiol Scand Suppl* 1955;**35**:121.)

Table 31–4. Factors affecting heart rate.[1]

Heart rate **accelerated** by:
    Decreased activity of baroreceptors in the arteries,
      left ventricle, and pulmonary circulation
    Increased activity of atrial stretch receptors
    Inspiration
    Excitement
    Anger
    Most painful stimuli
    Hypoxia
    Exercise
    Norepinephrine
    Epinephrine
    Thyroid hormones
    Fever
    Bainbridge reflex
Heart rate **slowed** by:
    Increased activity of baroreceptors in the arteries,
      left ventricle, and pulmonary circulation
    Expiration
    Fear
    Grief
    Stimulation of pain fibers in trigeminal nerve
    Increased intracranial pressure

[1]Norepinephrine has a direct chronotropic effect on the heart, but in the intact animal its pressor action stimulates the baroreceptors, leading to enough reflex increase in vagal tone to overcome the direct effect and produce bradycardia.

channels rather than capillaries. Adrenal medullary secretion of norepinephrine and epinephrine is apparently increased when this system is stimulated, the epinephrine probably reinforcing the dilation of muscle blood vessels. In cats and dogs, the system discharges in response to emotional stimuli such as fear, apprehension, and rage. Its role in humans is uncertain, but it has been suggested that the sympathetic vasodilator system is responsible for fainting in emotional situations. In addition, there is direct evidence for cholinergically mediated vasodilation in muscle at or even before the start of muscular exercise (see Chapter 33). However, it has been argued that the vasodilation before the start of exercise is not a constant or marked phenomenon.

Control of Heart Rate

The sympathetic and parasympathetic nerves to the heart and baroreceptor-mediated reflex changes in heart rate have been considered in detail in preceding sections of this chapter. However, Table 31–4 is a convenient summary of conditions that affect the heart rate. In general, stimuli that increase the heart rate also increase the blood pressure, whereas those that decrease the heart rate lower the blood pressure. However, there are exceptions such as the production of hypotension and tachycardia by stimulation of atrial stretch receptors (see above). Another is the production of hypertension and bradycardia by increased intracranial pressure. The 2 occur together in this latter condition because, as noted above, there is hypercapnic stimulation of the vasomotor center and reflex bradycardia.

Circulation Through
Special Regions

32

INTRODUCTION

The distribution of the cardiac output to various parts of the body at rest in a normal man is shown in Table 32–1. The general principles described in preceding chapters apply to the circulation of all these regions, but the vascular supplies of most organs have additional special features. The portal circulation of the anterior pituitary is discussed in Chapter 14, the renal circulation in Chapter 38, and the pulmonary circulation in Chapter 34. The circulation of skeletal muscle is discussed with the physiology of exercise in Chapter 33. This chapter is concerned with the circulation of the brain, the heart, the splanchnic area, the skin, the placenta, and the fetus.

CEREBRAL CIRCULATION

ANATOMIC CONSIDERATIONS

Vessels

The principal arterial inflow to the brain in humans is via 4 arteries: 2 internal carotids and 2 vertebrals. The vertebral arteries unite to form the basilar artery; and the circle of Willis, formed by the carotids and the basilar artery, is the origin of the 6 large vessels supplying the cerebral cortex. In some animals the vertebrals are large and the internal carotids small, but in humans a relatively small fraction of the total arterial flow is carried by the vertebral arteries. Substances injected into one carotid artery are distributed almost exclusively to the cerebral hemisphere on that side. There is normally no crossing over, probably because the pressure is equal on both sides. Even when it is not, the anastomotic channels in the circle do not permit a very large flow. Occlusion of one carotid artery, particularly in older pa-

tients, often causes serious symptoms of cerebral ischemia. There are precapillary anastomoses between the cerebral arterioles in humans and some other species, but flow through these channels is generally insufficient to maintain the circulation and prevent infarction when a cerebral artery is occluded.

Venous drainage from the brain by way of the deep veins and dural sinuses empties principally into the internal jugular veins in humans, although a small amount of venous blood drains through the ophthalmic and pterygoid venous plexuses, through emissary veins to the scalp, and down the system of paravertebral veins in the spinal canal. In other species, the internal jugular veins are small, and the venous blood from the brain mixes with blood from other structures.

The cerebral vessels have a number of unique anatomic features. In the choroid plexuses there are gaps between the endothelial cells of the capillary wall, but the choroid epithelial cells that separate them from the cerebrospinal fluid (CSF) are connected to one another by tight junctions. The capillaries in the brain substance resemble nonfenestrated capillaries in muscle and other parts of the body (see Chapter 30). However, there are tight junctions between the endothelial cells that do not permit the passage of substances which pass through the junctions between endothelial cells in other tissues. In addition, there are relatively few vesicles in the endothelial cytoplasm, and presumably there is little vesicular transport. The brain capillaries are surrounded by the endfeet of astrocytes (Fig 32–1). These endfeet are closely applied to the basal lamina of the capillaries, but they do not cover the entire capillary wall, and there are gaps of about 20 nm between endfeet. The protoplasm of astrocytes is also found around synapses, where it appears to isolate the synapses in the brain from one another.

Innervation

Three systems of nerves innervate the cerebral blood vessels. Postganglionic sympathetic neurons have their cell bodies in the superior cervical ganglia, and their endings contain norepinephrine and neuropeptide Y. Cholinergic neurons that probably originate in the sphenopalatine ganglia also innervate the

Table 32–1. Resting blood flow and O_2 consumption of various organs in a 63-kg adult man with a mean arterial blood pressure of 90 mm Hg and an O_2 consumption of 250 mL/min.[1]

Region	Mass (kg)	Blood Flow		Arteriovenous Oxygen Difference (mL/L)	Oxygen Consumption		Resistance (R units)[2]		Percentage of Total	
		mL/ min	mL/100 g/min		mL/ min	mL/100 g/min	Absolute	per kg	Cardiac Output	Oxygen Consumption
Liver	2.6	1500	57.7	34	51	2.0	3.6	9.4	27.8	20.4
Kidneys	0.3	1260	420.0	14	18	6.0	4.3	1.3	23.3	7.2
Brain	1.4	750	54.0	62	46	3.3	7.2	10.1	13.9	18.4
Skin	3.6	462	12.8	25	12	0.3	11.7	42.1	8.6	4.8
Skeletal muscle	31.0	840	2.7	60	50	0.2	6.4	198.4	15.6	20.0
Heart muscle	0.3	250	84.0	114	29	9.7	21.4	6.4	4.7	11.6
Rest of body	23.8	336	1.4	129	44	0.2	16.1	383.2	6.2	17.6
Whole body	63.0	5400	8.6	46	250	0.4	1.0	63.0	100.0	100.0

[1] Reproduced, with permission, from Bard P (editor): *Medical Physiology,* 11th ed. Mosby, 1961.
[2] R units are pressure (mm Hg) divided by blood flow (mL/s).

cerebral vessels, and the postganglionic cholinergic neurons on the blood vessels contain acetylcholine, VIP, and PHM-27 (see Chapter 26). Sensory nerves on the blood vessels have their cell bodies in the trigeminal ganglia and contain substance P, another polypeptide closely related to substance P and produced by the same gene, and CGRP. Substance P, CGRP, VIP, and PHM-27 cause vasodilation, whereas neuropeptide Y is a vasoconstrictor. Touching or pulling on the cerebral vessels causes pain.

Figure 32–1. Membrane of end-feet *(1)* of fibrous astrocytes around a cerebral capillary *(2)*. On the left, the processes of an astrocyte *(3)* also envelop parts of the dendrites of neighboring neurons *(4)*. (Reproduced, with permission, from Krstić RV: *Die Gewebe des Menschen und der Säugetiere.* Springer-Verlag, 1978.)

CEREBROSPINAL FLUID

Formation & Absorption

CSF fills the ventricles and subarachnoid space. In humans, the volume of CSF is about 150 mL and the rate of CSF production is about 550 mL/d. Thus the CSF turns over about 3.7 times a day. In experiments on animals, it has been estimated that 50–70% of the CSF is formed in the choroid plexuses and the remainder is formed around blood vessels and along ventricular walls. Presumably, the situation in humans is similar. The CSF in the ventricles flows through the foramens of Magendie and Luschka to the subarachnoid space, which is composed of closed projections of the CSF space and is absorbed through the arachoroid villi, into veins, primarily the cerebral venous sinuses. Bulk flow through these villi is about 500 mL/d, with additional small amounts of CSF being absorbed by diffusion into cerebral blood vessels.

The composition of CSF (Table 32–2) is essentially the same as that of brain ECF, which in living humans makes up 15% of the brain volume. There appears to be free communication between the brain interstitial fluid and CSF, although the diffusion distances from parts of the brain to the CSF are appreciable. Consequently, equilibration may take some time to occur, and local areas of the brain may have extracellular microenvironments that are transiently different from CSF. Brain interstitial fluid is in effect an ultrafiltrate of plasma (Table 32–2) with its composition modified by transport processes in the endothelial cells of the cerebral capillaries and the choroid epithelium. There is active transport and facilitated diffusion both into and out of the brain. There are some differences in transport between the endothelial cells and the choroid epithelium (Fig 32–2), but, as a result of mixing, these differences do not produce

Table 32-2. Concentration of various substances in human CSF and plasma.

Substance		CSF	Plasma	Ratio CSF/Plasma
Na$^+$	(meq/kg H$_2$O)	147.0	150.0	0.98
K$^+$	(meq/kg H$_2$O)	2.9	4.6	0.62
Mg^{2+}	(meq/kg H$_2$O)	2.2	1.6	1.39
Ca^{2+}	(meq/kg H$_2$O)	2.3	4.7	0.49
Cl$^-$	(meq/kg H$_2$O)	113.0	99.0	1.14
HCO$_3$$^-$	(meq/L)	25.1	24.8	1.01
P$_{CO_2}$	(mm Hg)	50.2	39.5	1.28
pH		7.33	7.40	...
Osmolality	(mosm/kg H$_2$O)	289.0	289.0	1.00
Protein	(mg/dL)	20.0	6000.0	0.003
Glucose	(mg/dL)	64.0	100.0	0.64
Inorganic P	(mg/dL)	3.4	4.7	0.73
Urea	(mg/dL)	12.0	15.0	0.80
Creatinine	(mg/dL)	1.5	1.2	1.25
Uric acid	(mg/dL)	1.5	5.0	0.30
Lactic acid	(mg/dL)	18.0	21.0	0.86
Cholesterol	(mg/dL)	0.2	175.0	0.001

persistent local differences in the composition of brain fluid.

Lumbar CSF pressure is normally 70–180 mm CSF. Up to pressures well above this range, the rate of CSF formation is independent of intraventricular pressure. However, absorption, which takes place largely by bulk flow, is proportionate to the pressure (Fig 32–3). At a pressure of 112 mm CSF, which is the average normal CSF pressure, filtration and absorption are equal. Below a pressure of approximately 68 mm CSF, absorption stops. Large amounts of fluid accumulate when the reabsorptive capacity of the arachnoid villi is decreased (**external hydrocephalus, communicating hydrocephalus**). Fluid also accumulates proximal to the block and distends the ventricles when the foramens of Luschka and Magendie are blocked or there is obstruction within the ventricular system (**internal hydrocephalus, noncommunicating hydrocephalus**).

Protective Function

The meninges and the CSF protect the brain. The dura is attached firmly to bone. There is normally no "subdural space," the arachnoid being held to the dura by the surface tension of the thin layer of fluid between the 2 membranes. As shown in Figure 32–4, the brain itself is supported within the arachnoid by the blood vessels and nerve roots and by the multiple fine fibrous **arachnoid trabeculae.** The brain weighs about 1400 g in air, but in its "water bath" of CSF it has a net weight of only 50 g. The buoyancy of the brain in the CSF permits its relatively flimsy attachments to suspend it very effectively. When the head

receives a blow, the arachnoid slides on the dura and the brain moves, but its motion is gently checked by the CSF cushion and by the arachnoid trabeculae.

The pain produced by spinal fluid deficiency illustrates the importance of CSF in supporting the brain. Removal of CSF during lumbar puncture can cause a severe headache after the fluid is removed, because the brain hangs on the vessels and nerve roots, and traction on them stimulates pain fibers. The pain can be relieved by intrathecal injection of sterile isotonic saline.

Head Injuries

Without the protection of the spinal fluid and the meninges, the brain would probably be unable to withstand even the minor traumas of everyday living; but with the protection afforded, it takes a fairly severe blow to produce cerebral damage. The brain is damaged most commonly when the skull is fractured and bone is driven into neural tissue (depressed skull fracture), when the brain moves far enough to tear the delicate bridging veins from the cortex to the bone, or when the brain is accelerated by a blow on the head and is driven against the skull or the tentorium at a point opposite where the blow was struck (**contrecoup injury**).

THE BLOOD-BRAIN BARRIER

Many years ago, it was demonstrated that when acidic dyes such as trypan blue were injected into living animals, all the tissues were stained except most

Figure 32–3. CSF formation and absorption in humans at various CSF pressures. Note that at 112 mm CSF formation and absorption are equal, and at 68 mm CSF absorption is zero. (Modified and reproduced, with permission, from Cutler RWP et al: Formation and absorption of cerebrospinal fluid in man. *Brain* 1968;**91:**707.)

Figure 32–2. ***Top:*** Transport across cerebral capillaries. ***Bottom:*** Transport across cells of the choroid plexus. (Modified from Fishman RA: *Cerebrospinal Fluid in Diseases of the Nervous System,* 2nd ed. Saunders, 1992.)

of the brain and spinal cord. To explain the failure of the neural tissue to stain, the existence of a **blood-brain barrier** was postulated. Subsequent research has established that only water, CO_2, and O_2 enter the brain with ease, and the exchange of other substances between blood and brain is slow. The site of the barrier is the endothelium of cerebral capillaries and the choroid plexus epithelium. Some physiologists use the term blood-brain barrier to refer to the barrier in the capillary walls and the term **blood-CSF barrier** to refer to the barrier in the choroid epithelium. However, the barriers are similar, and it seems more ap-

Figure 32–4. Investing membranes of the brain, showing their relation to the skull and to brain tissue. (Reproduced, with permission, from Wheater PR et al, *Functional Histology.* Churchill Livingstone, 1979.)

propriate to use the term blood-brain barrier to refer to net exchange across both barriers.

The general features of exchange across capillary walls between the plasma and the interstitial fluid are described in Chapter 30. There is considerable variation in capillary permeability from organ to organ in the body. However, the exchange across the cerebral vessels is so different from that in other capillary beds—and the rate of exchange of many physiologically important substances is so low—that it seems justifiable to speak specifically of a blood-brain barrier.

Penetration of Substances Into the Brain

In general, the rapidity with which substances penetrate brain tissue is inversely related to their molecular size and directly related to their lipid solubility. Water-soluble polar compounds generally cross slowly. There is an H^+ gradient between brain ECF and blood; the pH of brain ECF is 7.33, whereas that of blood is 7.40 (Table 32–2). There are bidirectional transport systems in the endothelium of brain capillaries for K^+, organic acids, amino acids, and glucose. Water, CO_2, and O_2 cross the blood-brain barrier readily, whereas glucose crosses more slowly. Na^+, K^+, Mg^{2+}, Cl^-, HCO_3^-, and HPO_4^{2-} in plasma require 3–30 times as long to equilibrate with spinal fluid as they do with other portions of the interstitial fluid. The relatively slow penetration of urea into the brain and the CSF is illustrated in Fig 32–5. Bile salts and catecholamines do not enter the adult brain in more than minute amounts. Proteins cross the barrier to a very limited extent, and it is because acidic dyes are bound to protein that they fail to stain neural tissue. It is worth noting that no substance is completely excluded from the brain and that the important consideration is the rate of transfer of the substance. Certain compounds cross the blood-brain barrier slowly, whereas closely related compounds enter at a more rapid rate. For example, the amines dopamine and serotonin penetrate to a very limited degree, but their corresponding acids, L-dopa and 5-hydroxytryptophan, enter with relative ease (see Chapter 15).

Development of the Blood-Brain Barrier

The cerebral capillaries are much more permeable at birth than in adulthood, and the blood-brain barrier develops during the early years of life. In severely jaundiced infants, bile pigments penetrate into the nervous system and, in the presence of asphyxia, damage the basal ganglia (kernicterus). However, in jaundiced adults, the nervous system is unstained and not directly affected.

Circumventricular Organs

When an acidic dye is injected into an animal, 4 small areas in or near the brain stain like the tissues outside the brain. These areas are (1) the **posterior pituitary** (neurohypophysis) and the adjacent ventral part of the median eminence of the hypothalamus, (2) the **area postrema,** (3) the **organum vasculosum of the lamina terminalis** (OVLT, supraoptic crest), and (4) the **subfornical organ** (intercolumnar tubercle).

These areas are referred to collectively as the **circumventricular organs** (Fig 32–6). All have fenestrated capillaries, and because of their permeability

Figure 32–5. Penetration of urea into muscle, brain, spinal cord, and CSF. Urea was administered by constant infusion. (Modified and reproduced, with permission, from Kleeman CR, Davson H, Levin E: Urea transport in the central nervous system. *Am J Physiol* 1962;**203:**739.)

Figure 32–6. Circumventricular organs. The neurohypophysis (NH), organum vasculosum of the lamina terminalis (OVLT, supraoptic crest), subfornical organ (SFO), and area postrema (AP) are shown projected on a sagittal section of the human brain. SCO, subcommissural organ; PI, pineal.

they are said to be "outside the blood-brain barrier." Some of them function as **neurohemal organs,** ie, areas in which substances secreted by neurons enter the circulation; for example, oxytocin and vasopressin enter the general circulation in the posterior pituitary, and hypothalamic hypophyseotropic hormones enter the portal hypophyseal circulation in the median eminence. Other circumventricular organs function as chemoreceptor zones, ie, areas in which substances in the circulating blood can act to trigger changes in brain function without penetrating the blood-brain barrier. The area postrema is a chemoreceptor zone that initiates vomiting in response to chemical changes in the plasma (see Chapter 14). It is also concerned with cardiovascular control, and in many species circulating angiotensin II acts on the area postrema to trigger an increase in blood pressure. Angiotensin II also acts on the subfornical organ and possibly on the OVLT to increase water intake. In addition, it appears that the OVLT is the site of the osmoreceptor controlling vasopressin secretion (see Chapter 14), and there is evidence that circulating IL-1 produces fever by acting on this circumventricular organ.

The subcommissural organ (Fig 32–6) is closely associated with the pineal and histologically resembles the circumventricular organs. However, it does not have fenestrated capillaries, is less permeable, and has no established function. The pineal and the anterior pituitary have fenestrated capillaries and are outside the blood-brain barrier, but both are endocrine glands and are not part of the brain.

Function of the Blood-Brain Barrier

The blood-brain barrier probably maintains the constancy of the environment of the neurons in the central nervous system. These neurons are so dependent upon the concentration of K^+, Ca^{2+}, Mg^{2+}, H^+, and other ions in the fluid bathing them that even minor variations have far-reaching consequences. The constancy of the composition of the ECF in all parts of the body is maintained by multiple homeostatic mechanisms (see Chapters 1 and 39), but because of the sensitivity of the cortical neurons to ionic change it is not surprising that an additional defense has evolved to protect them. Similarly, a blood-testis barrier protects the composition of the fluid surrounding the germinal epithelium and a placental barrier protects the composition of the body fluids of the fetus.

Other suggested functions for the blood-brain barrier are protection of the brain from endogenous and exogenous toxins in the blood and prevention of the escape of neurotransmitters into the general circulation.

Clinical Implications

The physician must know the permeability of the blood-brain barrier to drugs in order to treat diseases of the nervous system intelligently. For example, among the antibiotics, penicillin and chlortetracycline enter the brain to a very limited degree. Sulfadiazine and erythromycin, on the other hand, enter quite readily.

Another important clinical consideration is the fact that the blood-brain barrier tends to break down in areas of the brain that are irradiated, infected, or the site of tumors. The breakdown helps in identifying the location of tumors; substances such as radioactive iodine-labeled albumin penetrate normal brain tissue very slowly, but they enter tumor tissue, making the tumor stand out as an island of radioactivity in the surrounding normal brain. The blood-brain barrier can also be temporarily disrupted by sudden marked increases in blood pressure or by intravenous injection of hypertonic fluids.

CEREBRAL BLOOD FLOW

Kety Method

According to the **Fick principle** (see Chapter 29), the blood flow of any organ can be measured by determining the amount of a given substance (Q_x) removed from the bloodstream by the organ per unit of time and dividing that value by the difference between the concentration of the substance in arterial blood and the concentration in the venous blood from the organ ($[A_x] - [V_x]$). Thus:

$$\text{Cerebral blood flow (CBF)} = \frac{Q_x}{[A_x] - [V_x]}$$

When an individual inhales small subanesthetic amounts of nitrous oxide (N_2O), this gas is taken up by the brain, and the brain N_2O equilibrates with the N_2O in blood in 9–11 minutes. After equilibration, the N_2O concentration in cerebral venous blood is equal to the concentration in the brain because the brain-blood partition coefficient for N_2O is 1. Therefore, the level in cerebral venous blood after equilibration divided by the mean arteriovenous N_2O difference during equilibration equals the cerebral blood flow (CBF) per unit of brain. This is the **Kety method** for measuring cerebral blood flow. The average cerebral blood flow in young adults is 54 mL/100 g/min. The average adult brain weighs about 1400 g, so the flow for the whole brain is about 756 mL/min. Note that the Kety method provides an average value for perfused areas of brain during the 10-minute equilibration period; that it gives no information about regional differences in blood flow; and that since it depends on N_2O uptake, it measures flow to perfused parts of the brain only. If the blood flow to a portion of the brain is occluded, there is no change in the measured flow, because the nonperfused area does not take up any N_2O.

Blood Flow in Various Parts of the Brain

Regional blood flow in the brain can be determined by measuring the distribution of an inert radioactive gas in frozen sections of the brain and comparing the values obtained with the level of the gas in the blood. It can also be determined in living animals and humans by monitoring the distribution of ^{133}Xe, ^{123}I-labeled iodoamphetamine (^{123}I-IMP), or other tracers. The arrival and clearance of the tracer are monitored by a battery of 254 scintillation detectors placed over the head. The output from the detectors is processed in a computer and displayed on a color television screen in such a way that the color corresponding to the location of each detector is proportionate to the flow it is detecting. Resolution can be improved with computerized tomographic reconstruction of the type used for computed tomography (CT) and magnetic resonance imaging (MRI) (see appendix); this technique is called single photon emission computed tomography (SPECT).

Since blood flow is tightly coupled to brain metabolism, local uptake of 2-deoxyglucose is also a good index of blood flow (see below and Chapter 8). If the 2-deoxyglucose is labeled with short-half-life positron emitters such as ^{18}F, [11]O, and ^{15}O, its concentration in any part of the brain can be monitored by **positron emission tomography (PET)** scanning through the intact skull in living subjects.

Values for regional blood flow obtained by using the frozen-section method are shown in Table 32–3. Blood flow in the cerebral cortex and cerebellar cortex is large, but the part of the brain with the largest blood flow per gram is the inferior colliculus. The blood flow in gray matter is about 6 times that in white matter. Using the ^{133}Xe method, the average hemispheric blood flow in resting humans is 48 mL/100 g/min, with the flow in gray matter averag-ing 69 mL/100 g/min compared with 28 mL/100 g/min in white matter.

A striking feature of cerebral function is the marked fluctuation in regional blood flow with fluctuation in activity. In subjects who are awake but at rest, blood flow is greatest in the premotor and frontal regions. This is the part of the brain that is believed to be concerned with decoding and analyzing afferent input and with intellectual activity. During voluntary clenching of the right hand, flow is increased in the hand area of the left motor cortex and the corresponding sensory areas in the postcentral gyrus. Especially when the movements being performed are sequential, the flow is also increased in the supplementary motor area. When subjects talk, there is a bilateral increase in blood flow in the face, tongue, and mouth-sensory and motor areas and the upper premotor cortex in the categorical (usually the left) hemisphere. When the speech is stereotyped, Broca's and Wernicke's areas do not show increased flow, but when the speech is creative, ie, when it involves ideas, there are flow increases in both these areas. Reading produces widespread increases in blood flow. Problem solving, reasoning, and motor ideation without movement produce increases in selected areas of the premotor and frontal cortex (Fig 32–7). In right-handed individuals, blood flow to the left hemisphere is greater when a verbal task is being performed and blood flow to the right hemisphere is greater when a spatial task is being performed.

SPECT and PET scanning have been applied to the study of various diseases. Epileptic foci are hyperemic during seizures, whereas flow is reduced in other parts of the brain. Between seizures, flow is sometimes reduced in the foci that generate the seizures. There is decreased temporal lobe flow in patients with memory deficits and decreased parieto-occipital flow in patients with symptoms of agnosia (see Chapter 16). In senile dementia, there are patchy decreases in flow, particularly in the temporoparietal cortex, and the magnitude of the decreases is proportionate to the degree of dementia. In Huntington's disease, there is a bilateral reduction in the blood flow to the caudate nucleus, and this alteration in flow occurs so early in the disease that it is probably of potential value in detecting asymptomatic individuals at risk for the disease. In manic depressives but not in patients with unipolar depression, there is a general decrease in cortical blood flow when the patients are depressed. In schizophrenia there is some evidence for decreased blood flow in the frontal lobes, temporal lobes, and basal ganglia.

Cerebral Vascular Resistance

The cerebral vascular resistance (CVR) is equal to the cerebral perfusion pressure divided by the cerebral blood flow. The CVR in the supine position can be calculated on the basis of the mean brachial artery pressure (ignoring the relatively low cerebral venous

Table 32–3. Blood flow of representative areas of the brain of unanesthetized cats.[1]

Area	Mean Blood Flow (mL/g/min)
Inferior colliculus	1.80
Sensorimotor cortex	1.38
Auditory cortex	1.30
Visual cortex	1.25
Medial geniculate body	1.22
Lateral geniculate body	1.21
Superior colliculus	1.15
Caudate nucleus	1.10
Thalamus	1.03
Association cortex	0.88
Cerebellar nuclei	0.87
Cerebellar white matter	0.24
Cerebral white matter	0.23
Spinal cord white matter	0.14

[1] Data from Landau WM et al: The local circulation of the living brain: Values in the unanesthetized cat. *Trans Am Neurol Assoc* 1955;**80**:125.

Figure 32–7. Areas showing increases in regional blood flow in left and right cerebral hemispheres of humans carrying out 2 kinds of thinking. **Top:** Subjects performing a mental subtraction task. **Bottom:** Subjects imagining that they are walking a specific route in familiar surroundings. The numbers beside each area indicate the percent increase. The left hemisphere was analyzed in 6 subjects and the right in 5. Note that the areas activated are different for the 2 kinds of thinking. However, the areas activated during each kind of thinking are remarkably similar in all subjects. (Redrawn and reproduced, with permission, from Roland PE: Change in brain blood and oxidase metabolism during mental activity. *News Physiol Sci* 1987;**2:**120).

pressure) without introducing a very large error. Calculated in this way, the normal CVR is approximately 10 "R units" per kilogram of brain (7.2 R units for the whole brain) (Table 32–1).

REGULATION OF CEREBRAL CIRCULATION

Normal Flow

The cerebral circulation is regulated in such a way that a constant total cerebral blood flow is generally maintained under varying conditions. For example, despite extensive shifts in the pattern of flow, total cerebral blood flow is not increased by strenuous mental activity. The factors affecting the total cerebral blood flow are the arterial pressure at brain level, the venous pressure at brain level, the intracranial pressure, the viscosity of the blood, and the degree of active constriction or dilation of the cerebral arterioles (Fig 32–8). The caliber of the arterioles is controlled by local vasodilator metabolites that include the products of metabolism and other factors, and by autoregulation, locally acting substances produced by

the endothelium, circulating peptides such as angiotensin II, and vasomotor nerves.

Role of Intracranial Pressure

In adults, the brain, spinal cord, and spinal fluid are encased, along with the cerebral vessels, in a rigid bony enclosure. The cranial cavity normally contains a brain weighing approximately 1400 g, 75 mL of blood, and 75 mL of spinal fluid. Because brain tissue and spinal fluid are essentially incompressible, the volume of blood, spinal fluid, and brain in the cranium at any time must be relatively constant (**Monro-Kellie doctrine**). More importantly, the cerebral vessels are compressed whenever the intracranial pressure rises. Any change in venous pressure promptly causes a similar change in intracranial pressure. Thus, a rise in venous pressure decreases cerebral blood flow both by decreasing the effective perfusion pressure and by compressing the cerebral vessels. This relationship helps to compensate for changes in arterial blood pressure at the level of the head. For example, if the body is accelerated upward (positive *g*), blood moves toward the feet and arterial pressure at the level of the head decreases. However,

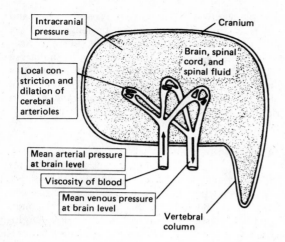

Figure 32–8. Diagrammatic summary of the factors affecting cerebral blood flow.

venous pressure also falls and intracranial pressure falls, so that the pressure on the vessels decreases and blood flow is much less severely compromised than it would otherwise be. Conversely, during acceleration downward, force acting toward the head (negative g) increases arterial pressure at head level, but intracranial pressure also rises, so that the vessels are supported and do not rupture. The cerebral vessels are protected during the straining associated with defecation or delivery in the same way.

Effect of Intracranial Pressure Changes on Systemic Blood Pressure

When intracranial pressure is elevated to more than 450 mm of water (33 mm Hg) over a short period, cerebral blood flow is significantly reduced. The resultant ischemia stimulates the vasomotor area (see Chapter 31), and systemic blood pressure rises. Stimulation of vagal outflow produces bradycardia, and respiration is slowed. The blood pressure rise, which was described by Cushing and is sometimes called the **Cushing reflex,** helps to maintain the cerebral blood flow. Over a considerable range, the rise in systemic blood pressure is proportionate to the rise in intracranial pressure, although eventually a point is reached where the intracranial pressure exceeds the arterial pressure and cerebral circulation ceases.

Effects of Brain Metabolism on Cerebral Vessels

As noted above, there are marked local fluctuations in blood flow with alterations in neuronal activity that are produced by vasodilator metabolites such as K^+, H^+, and adenosine. The arterioles in the brain, like those in other parts of the body, are directly affected by local changes in CO_2 and O_2 tension. A rise

in P_{CO_2} which increases the local H^+ concentration, exerts a particularly potent dilator effect on the cerebral vessels. A fall in P_{CO_2} has a constrictor effect, and cerebral vasoconstriction is an important factor in the production of the cerebral symptoms seen when the arterial P_{CO_2} falls during hyperventilation. A low P_{O_2} in brain tissue is associated with vasodilation and a high P_{O_2} with mild vasoconstriction.

Autoregulation

Autoregulation is prominent in the brain (Fig 32–9). This process, by which the flow to many tissues is maintained at relatively constant levels despite variations in perfusion pressure, is discussed in Chapter 31. In the brain, autoregulation maintains a normal cerebral blood flow at arterial pressures of 65–140 mm Hg.

Role of Vasomotor Nerves

The innervation of the cerebral blood vessels by postganglionic sympathetic, postganglionic parasympathetic, and sensory nerves is described above. The role of these nerves remains a matter of debate. Stimulation of the cervical sympathetic nerves does cause some constriction of pial vessels in animals, and it has been argued that noradrenergic discharge occurs when the blood pressure is markedly elevated. This reduces the resultant passive increase in blood flow and helps protect the blood-brain barrier from the disruption that could otherwise occur (see above).

Vasomotor discharges also affect autoregulation. With sympathetic stimulation, the constant-flow, or plateau, part of the pressure-flow curve is extended to the right (Fig 32–9); ie, greater increases in pressure can occur without an increase in flow. On the other hand, the vasodilator hydralazine and the ACE inhibitor captopril reduce the length of the plateau.

Figure 32–9. Autoregulation of cerebral blood flow (CBF) during steady-state conditions. The dotted line shows the alteration produced by sympathetic stimulation during autoregulation. (Modified from Busija DW, Heistad DD, Marcus ML: Effects of sympathetic nerves on cerebral blood vessels during acute moderate increases in arterial pressure in cats and dogs. *Circ Res* 1980;**46:**696.)

Integrated Operation of Regulatory Mechanisms

Several examples of the way the regulatory mechanisms discussed above maintain cerebral blood flow are discussed in Chapter 33. It is worth remembering that these mechanisms operate together to overcome such formidable challenges as the effect of gravity on the cerebral blood flow, not only in humans but also, for example, in giraffes when these animals stoop to drink water and when they raise their heads to nibble leaves from the tops of trees.

BRAIN METABOLISM & OXYGEN REQUIREMENTS

Uptake & Release of Substances by the Brain

If the cerebral blood flow is known, it is possible to calculate the consumption or production by the brain of O_2, CO_2, glucose, or any other substance present in the bloodstream by multiplying the cerebral blood flow by the difference between the concentration of the substance in arterial blood and its concentration in cerebral venous blood (Table 32–4). When calculated in this fashion, a negative value indicates that the brain is producing the substance.

Oxygen Consumption

O_2 consumption by the human brain (**cerebral metabolic rate for O_2, CMRO$_2$**) averages about 3.5 mL/100 g of brain/min (49 mL/min for the whole brain) in an adult. This figure represents approximately 20% of the total resting O_2 consumption (Table 32–1). The brain is extremely sensitive to hypoxia, and occlusion of its blood supply produces unconsciousness in as short a period as 10 seconds. The vegetative structures in the brain stem are more resistant to hypoxia than the cerebral cortex, and patients

may recover from accidents such as cardiac arrest and other conditions causing fairly prolonged hypoxia with normal vegetative functions but severe, permanent intellectual deficiencies. The basal ganglia use O_2 at a very high rate, and symptoms of Parkinson's disease as well as intellectual deficits can be produced by chronic hypoxia. The thalamus and the inferior colliculus are also very susceptible to hypoxic damage.

Energy Sources

Glucose is the major ultimate source of energy for the brain; under normal conditions, 90% of the energy needed to maintain ion gradients across cell membranes and transmit electrical impulses comes from this source. Glucose enters the brain via a glucose transporter found in high concentration in cerebral capillaries that appears to be identical to the type I glucose transporter found in erythrocytes (see Chapter 19). Congenital defects in this transporter may lead to low CSF glucose levels and seizures despite a normal blood glucose level.

Glucose is taken up from the blood in large amounts, and the RQ (respiratory quotient; see Chapter 17) of cerebral tissue is 0.95–0.99 in normal individuals. This does not mean that the total source of energy is always glucose. During prolonged starvation, there is appreciable utilization of other substances. Indeed, there is evidence that as much as 30% of the glucose taken up under normal conditions is converted to amino acids, lipids, and proteins and that substances other than glucose are metabolized for energy during convulsions. There may also be some utilization of amino acids from the circulation even though the amino acid arteriovenous difference across the brain is normally minute. Insulin is not required for most cerebral cells to utilize glucose.

Glucose uptake is increased in active neurons, and 2-deoxyglucose is taken up like glucose but is not metabolized. Therefore, the uptake of radioactive 2-deoxyglucose can be used to map the activity of neurons on a fine scale in experimental animals and in living humans by PET scanning. For example, illumination of one eye in monkeys causes an increase in the amount of 2-deoxyglucose, as determined by radioautography, in columns of cells in the visual cortex that are less than 1 mm apart (see Chapter 8). In general, glucose utilization at rest parallels blood flow (Table 32–3) and O_2 consumption. There is an average decrease of 30% in the uptake by all areas during slow-wave sleep.

Hypoglycemia

The symptoms of hypoglycemia are anxiety, sweating, confusion, and other mental changes progressing to coma and convulsions (see Chapter 19). The total glycogen content of the brain is about 1.6 mg/g in fasted animals, but the available glycogen and glucose are used up in 2 minutes if the blood sup-

Table 32–4. Utilization and production of substances by the adult human brain in vivo.[1]

Substance	Uptake (+) or output (−) per 100 g brain/min	Total/min
Substances utilized Oxygen	+3.5 mL	+49 mL
Glucose	+5.5 mg	+77 mg
Glutamate	+0.4 mg	+5.6 mg
Substances produced Carbon dioxide	−3.5 mL	−49 mL
Glutamine	−0.6 mg	−8.4 mg

Substances not utilized or produced in the fed state: lactate, pyruvate, total ketones, α-ketoglutarate.

[1]Modified from Sokoloff L: Metabolism of the central nervous system in vivo. In: *Handbook of Physiology.* Section 1, Vol III, p. 1843, American Physiological Society, 1960.

ply is totally occluded. Thus, the brain can withstand hypoglycemia for somewhat longer periods than it can withstand hypoxia, but glucose and O_2 are both needed for survival. The cortical regions are more sensitive to hypoglycemia than the vegetative centers in the brain stem, and sublethal exposures to hypoglycemia, like similar exposures to hypoxia, may cause irreversible cortical changes. In diabetic patients with chronic hyperglycemia, maximum glucose transport capacity across the blood-brain barrier is reduced, and these patients may develop symptoms of hypoglycemia when their blood glucose is reduced to values that are still in the normal rather than the hypoglycemic range for normal individuals.

Glutamate & Ammonia Removal

The brain's uptake of glutamate is approximately balanced by its output of glutamine. Glutamate entering the brain takes up ammonia and leaves as glutamine (see Chapter 17). The glutamate-glutamine conversion in the brain—the opposite of the reaction in the kidney that produces some of the ammonia entering the tubules—probably serves as a detoxifying mechanism to keep the brain free of ammonia. Ammonia is very toxic to nerve cells, and ammonia intoxication is believed to be a major cause of the bizarre neurologic symptoms in hepatic coma.

CORONARY CIRCULATION

Anatomic Considerations

The 2 coronary arteries that supply the myocardium arise from the sinuses behind the cusps of the aortic valve at the root of the aorta (Fig 32–10). Eddy currents keep the valves away from the orifices of the arteries, and they are patent throughout the cardiac cycle. The right coronary artery has a greater flow in 50% of individuals, the left has a greater flow in 20%, and the flow is equal in 30%. Most of the venous blood returns to the heart through the coronary sinus and anterior cardiac veins (Fig 32–11), which drain into the right atrium. In addition, there are other vessels that empty directly into the heart chambers. These include **arteriosinusoidal vessels,** sinusoidal capillary-like vessels that connect arterioles to the chambers; **thebesian veins** that connect capillaries to the chambers; and a few **arterioluminal vessels** that are small arteries draining directly into the chambers. There are a few anastomoses between the coronary arterioles and extracardiac arterioles, especially around the mouths of the great veins. Anastomoses between coronary arterioles in humans only pass particles less than 40 μm in diameter, but there is evi-

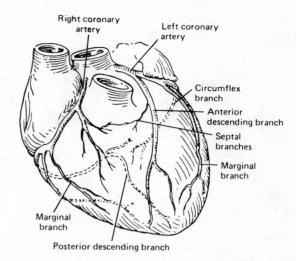

Figure 32–10. Coronary arteries and their principal branches in humans. (Reproduced, with permission, from Ross G: The cardiovascular system. In: *Essentials of Human Physiology.* Ross G [editor]. Copyright © 1978 by Year Book Medical Publishers, Inc., Chicago.)

dence that these channels enlarge and increase in number in patients with coronary artery disease.

Pressure Gradients & Flow in the Coronary Vessels

The heart is a muscle that, like skeletal muscle, compresses its blood vessels when it contracts. The pressure inside the left ventricle is slightly higher than in the aorta during systole (Table 32–5). Consequently, flow occurs in the arteries supplying the subendocardial portion of the left ventricle only during diastole, although the force is sufficiently dissipated in the more superficial portions of the left ventricular myocardium to permit some flow in this region throughout the cardiac cycle. Since diastole is shorter when the heart rate is high, left ventricular coronary

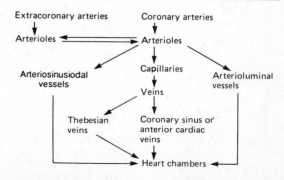

Figure 32–11. Diagram of the coronary circulation.

Table 32–5. Pressures in aorta and left and right ventricles in systole and diastole.

	Pressure (mm Hg) in			Pressure Differential (mm Hg) Between Aorta and	
	Aorta	Left Vent	Right Vent	Left Vent	Right Vent
Systole	120	121	25	–1	95
Diastole	80	0	0	80	80

flow is reduced during tachycardia. On the other hand, the pressure differential between the aorta and the right ventricle, and the differential between the aorta and the atria, are somewhat greater during systole than during diastole. Consequently, coronary flow in those parts of the heart is not appreciably reduced during systole. Flow in the right and left coronary arteries is shown in Fig 32–12. Because there is no blood flow during systole in the subendocardial portion of the left ventricle, this region is prone to ischemic damage and is the most common site of

Figure 32–12. Blood flow in the left and right coronary arteries during various phases of the cardiac cycle. Systole occurs between the 2 vertical dashed lines. (Reproduced, with permission, from Berne RM, Levy MN: *Physiology.* Mosby, 1983.)

myocardial infarction. Blood flow to the left ventricle is decreased in patients with stenotic aortic valves because in aortic stenosis the pressure in the left ventricle must be much higher than that in the aorta to eject the blood. Consequently, the coronary vessels are severely compressed during systole. Patients with this disease are particularly prone to develop symptoms of myocardial ischemia, in part because of this compression and in part because the myocardium requires more O_2 to expel blood through the stenotic aortic valve. Coronary flow is also decreased when the aortic diastolic pressure is low. The rise in venous pressure in conditions such as congestive heart failure reduces coronary flow because it decreases effective coronary perfusion pressure.

Coronary blood flow has been measured by inserting a catheter into the coronary sinus and applying the Kety method to the heart on the assumption that the N_2O content of coronary venous blood is typical of the entire myocardial effluent. Coronary flow at rest in humans is about 250 mL/min (5% of the cardiac output). A number of techniques utilizing **radionuclides,** radioactive tracers that can be detected with γ scintillation cameras over the chest, have been used to study regional blood flow in the heart and to detect areas of ischemia and infarct as well as to evaluate ventricular function. Radionuclides such as thallium 201 (^{201}Tl) are pumped into cardiac muscle cells by Na^+-K^+ ATPase and equilibrate with the intracellular K^+ pool. For the first 10–15 minutes after intravenous injection, ^{201}Tl distribution is directly proportionate to myocardial blood flow, and areas of ischemia can be detected by their low uptake. The uptake of this isotope is often determined soon after exercise and again several hours later to bring out areas in which exertion leads to compromised flow. Conversely, radiopharmaceuticals such as technetium 99m stannous pyrophosphate (^{99m}Tc-PYP) are selectively taken up by infarcted tissue by an incompletely understood mechanism and make infarcts stand out as "hot spots" on scintiscans of the chest. Coronary angiography can be combined with measurement of ^{133}Xe washout (see above) to provide detailed analysis of coronary blood flow. Radiopaque contrast medium is first injected into the coronary arteries, x-rays being used to outline their distribution. The angiographic camera is then replaced with a multiple-crystal scintillation camera and ^{133}Xe washout is measured. An example of normal flow distribution after injection in a left coronary artery is shown in Fig 32–13.

Variations in Coronary Flow

At rest, the heart extracts 70–80% of the O_2 from each unit of blood delivered to it (Table 32–1). O_2 consumption can be increased significantly only by increasing blood flow. Therefore, it is not surprising that blood flow increases when the metabolism of the myocardium is increased. The caliber of the coronary

Mean LV flow (65 mL/100 g/min)

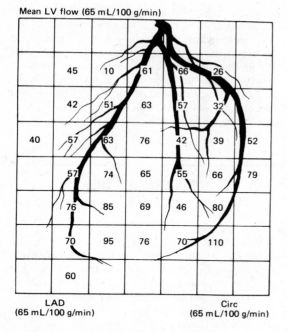

LAD
(65 mL/100 g/min)

Circ
(65 mL/100 g/min)

Figure 32–13. Normal human myocardial perfusion pattern following injection of ^{133}Xe into the left main coronary artery. The branches of the artery are shown in black, and the numbers in the squares are the flow values (in mL/100 g/min) for the regions under each scintillation detector. Circ, circumflex artery; LAD, left anterior descending artery; LV, left ventricle. (Reproduced, with permission, from Cannon PJ et al: Evaluation of myocardial circulation with radionuclides. *Cardiovasc Med* 1978;**2:**371.)

vessels, and consequently the rate of coronary blood flow, is influenced not only by pressure changes in the aorta but also by chemical and neural factors. The coronary circulation shows considerable autoregulation.

Chemical Factors

The close relationship between coronary blood flow and myocardial O_2 consumption indicates that one or more of the products of metabolism cause coronary vasodilation. Factors suspected of playing this role include O_2 lack and increased local concentrations of CO_2, H^+, K^+, lactate, prostaglandins, adenine nucleotides, and adenosine. More than one of these vasodilator metabolites could be involved. Asphyxia, hypoxia, and intracoronary injections of cyanide all increase coronary blood flow 200–300% in denervated as well as intact hearts, and the feature common to these 3 stimuli is hypoxia of the myocardial fibers. A similar increase in flow is produced in the area supplied by a coronary artery if the artery is occluded and then released. This **reactive hyperemia** is similar to that seen in the skin (see below).

Neural Factors

The coronary arterioles contain α adrenergic receptors, which mediate vasoconstriction, and β adrenergic receptors, which mediate vasodilation. Activity in the noradrenergic nerves to the heart and injections of norepinephrine cause coronary vasodilation. However, norepinephrine increases the heart rate and the force of cardiac contraction, and the vasodilation is due to production of vasodilator metabolites in the myocardium secondary to the increase in its activity. When the inotropic and chronotropic effects of noradrenergic discharge are blocked by a β adrenergic blocking drug, stimulation of the noradrenergic nerves or injection of norepinephrine in unanesthetized animals elicits coronary vasoconstriction. Thus, the direct effect of noradrenergic stimulation is constriction rather than dilation of the coronary vessels. Stimulation of vagal fibers to the heart dilates the coronaries.

When the systemic blood pressure falls, the overall effect of the reflex increase in noradrenergic discharge is increased coronary blood flow secondary to the metabolic changes in the myocardium at a time when the cutaneous, renal, and splanchnic vessels are constricted. In this way the circulation of the heart, like that of the brain, is preserved when flow to other organs is compromised.

Coronary Artery Disease

When flow through a coronary artery is reduced to the point that the myocardium it supplies becomes hypoxic, "P factor" accumulates and **angina pectoris** develops (see Chapter 7). If the myocardial ischemia is severe and prolonged, irreversible changes occur in the muscle, and the result is **myocardial infarction.** The cause of myocardial infarction is usually obstruction of at least 75% of the lumen of a coronary artery by a thrombus in a region narrowed by atherosclerotic plaques. Myocardial infarction is a common cause of death, particularly in developed countries. Other events that precipitate a thrombosis include spasm of a coronary artery, transient platelet aggregation in a severely sclerotic vessel, and rupture of or hemorrhage into an atherosclerotic plaque.

The electrical events produced by myocardial ischemia and infarction and the resulting changes produced in the ECG are discussed in Chapter 28. It is interesting that before the ischemic muscle cells die, they cease contracting, and this cessation of contraction has the homeostatic effect of prolonging the period before the onset of irreversible changes. The cause of the loss of contractile activity is uncertain. In dogs in which blood flow to a portion of myocardium has been interrupted, necrosis starts in 20 minutes but is not complete for 3–6 hours.

Damaged cells leak enzymes into the circulation, and the rises in the serum levels of enzymes and isoenzymes produced by myocardial infarction play an important ancillary role in the diagnosis of this disease. The enzymes most commonly measured are creatine kinase (CK) and lactate dehydrogenase (LDH). The MB isoenzyme of CK and fraction 1 of LDH are found in higher concentrations in heart muscle than in many other organs, and measurement of the serum concentration of the particular isoenzymes helps improve the diagnostic specificity of the enzyme measurement.

Nitrates such as nitroglycerin often produce prompt relief of the pain in angina pectoris. These substances dilate normal arterial vessels, and even though atherosclerotic coronary arteries are generally too rigid to dilate to any degree, large doses produce dilation of coronary arteries with eccentric stenoses. However, the main effect of the nitrates is to decrease venous return to the heart. This reduces stroke volume, and consequently myocardial O_2 consumption is reduced. The decreased venous return is secondary to dilation of peripheral veins, with pooling of blood in the periphery.

Discrete areas of narrowing in the coronary arteries can be detected by arteriography and can be bypassed by implantation of a graft (aortocoronary artery bypass graft). This procedure often produces marked and sustained relief of intractable angina, but it is unsettled whether it protects against myocardial infarction or produces significant prolongation of life when compared with modern medical treatment of coronary artery disease. In the treatment of myocardial infarction, emphasis is now placed on removing the obstruction and/or lysing the clot as soon as possible after the onset of pain in order to establish reperfusion and prevent irreversible changes. One technique is to insert a catheter, manipulate it into the affected coronary artery, and then use it to dilate the area of constriction above where the clot has formed (**coronary angioplasty**). However, this is associated with a high incidence of reocclusion at a later date, apparently because of a predilection to clot formation in the area where the endothelium is damaged. It is significant in this regard that injection of serotonin into the coronary arteries of individuals without coronary atherosclerosis causes vasodilation, but in individuals with atheromas and endothelial damage, the response is vasoconstriction.

Another approach is lysis of the intracoronary clot by intravenous or intracoronary injection of an agent that produces plasmin (see Chapter 27). One frequently used agent is **streptokinase**, a protein from hemolytic streptococci that converts plasminogen to plasmin (see Chapter 27). Another lytic agent that converts plasminogen to plasmin in the thrombus is **tissue plasminogen activator (TPA)**, a naturally occurring protease that is produced by means of recombinant DNA techniques. TPA has the advantage that it preferentially activates plasminogen bound to fibrin, thus preferentially producing fibrinolysis in the clot and limiting activation of plasminogen elsewhere in the body. However, careful trials have not shown any consistent increases in the clinical effectiveness of TPA over that of streptokinase, and streptokinase is much cheaper. In any case, if a lytic agent is injected within the first few hours after the onset of pain in developing myocardial infarction, coronary perfusion can be improved in a significant number of patients and the increased patency can be maintained with antiplatelet drugs, calcium channel blockers, and other medical therapy.

SPLANCHNIC CIRCULATION

The blood from the intestines, pancreas, and spleen drains via the portal vein to the liver and from the liver via the hepatic veins to the inferior vena cava. The viscera and the liver receive about 30% of the cardiac output via the celiac, superior mesenteric, and inferior mesenteric arteries (Fig 32–14). The liver receives about 1000 mL/min from the portal vein and 500 mL/min from the hepatic artery.

Intestinal Circulation

The intestines are supplied by a series of parallel circulations via the branches of the superior and inferior mesenteric arteries (Fig 32–14). There are extensive anastomoses between these vessels, but blockage of a large intestinal artery still leads to infarction of the bowel. The blood flow to the mucosa is greater

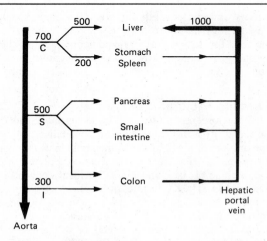

Figure 32–14. Splanchnic circulation. Note that most of the viscera are supplied by a series of parallel circuits, whereas the liver receives blood from the hepatic artery and the portal vein. The figures are average blood flows (mL/min). C, celiac axis; S, superior mesenteric artery; I, inferior mesenteric artery. (Modified from Sernka T, Jacobson E: *Gastrointestinal Physiology—The Essentials.* Williams & Wilkins, 1979.)

than that to the rest of the intestinal wall, and it responds to changes in metabolic activity. Thus, blood flow to the small intestine (and hence blood flow in the portal vein) doubles after a meal, and the increase lasts up to 3 hours. The intestinal circulation is capable of extensive autoregulation.

Hepatic Circulation

There are large fenestrations in the walls of hepatic sinusoids, and they are highly permeable. The way the intrahepatic branches of the hepatic artery and portal vein converge on the sinusoids and drain into the central lobular veins of the liver is shown in Figs 26–17 and 32–15. The functional unit of the liver is the acinus. Each acinus is at the end of a vascular stalk containing terminal branches of portal veins, hepatic arteries, and bile ducts. Blood flows from the center of this functional unit to the terminal branches of the hepatic veins at the periphery (Fig 32–15). This is why the central portion of the acinus, sometimes called zone 1, is well oxygenated, the intermediate zone (zone 2) is moderately well oxygenated, and the peripheral zone (zone 3) is least well oxygenated and most susceptible to anoxic injury. The hepatic veins drain into the inferior vena cava. The acini have been likened to grapes or berries, each on a vascular stem. There are about 100,000 acini in the human liver.

Portal venous pressure is normally about 10 mm Hg in humans, and hepatic venous pressure is approx-

imately 5 mm Hg. The mean pressure in the hepatic artery branches that converge on the sinusoids is about 90 mm Hg, but the pressure in the sinusoids is lower than the portal venous pressure, so there is a marked pressure drop along the hepatic arterioles. This pressure drop is adjusted so that there is an inverse relationship between hepatic arterial and portal venous blood flow. This inverse relationship may be maintained in part by the rate at which adenosine is removed from the region around the arterioles. According to this hypothesis, adenosine is produced by metabolism at a constant rate. When portal flow is reduced, it is washed away more slowly, and the local accumulation of adenosine dilates the terminal arterioles.

The intrahepatic portal vein radicles have smooth muscle in their walls that is innervated by noradrenergic vasoconstrictor nerve fibers reaching the liver via the third to eleventh thoracic ventral roots and the splanchnic nerves. The vasoconstrictor innervation of the hepatic artery comes from the hepatic sympathetic plexus. There are no known vasodilator fibers reaching the liver. At rest, circulation in the peripheral portions of the liver is sluggish, and only a portion of the organ is actively perfused. When systemic venous pressures rises, the portal vein radicles are dilated passively and the amount of blood in the liver increases. In congestive heart failure, this hepatic venous congestion may be extreme. Conversely, when there is diffuse noradrenergic discharge in response to a drop in systemic blood pressure, the intrahepatic portal radicles constrict, portal pressure rises, and blood flow through the liver is brisk, bypassing most of the organ. Most of the blood in the liver enters the systemic circulation. Constriction of the hepatic arterioles diverts blood from the liver, and constriction of the mesenteric arterioles reduces portal inflow. In severe shock, hepatic blood flow may be reduced to such a degree that there is patchy necrosis of the liver.

Reservoir Function of the Splanchnic Circulation

In dogs and other carnivores, there is a large amount of smooth muscle in the capsule of the spleen. The spleen traps blood, and rhythmic contractions of its capsule pump plasma into the lymphatics. The spleen therefore contains a reservoir of blood rich in cells. Noradrenergic nerve discharge and epinephrine make the spleen contract strongly, discharging the blood into the circulation. This function of the spleen is quantitatively unimportant in humans. However, the reservoir function of the whole visceral circulation is important. For example, 25–30% of the volume of the liver is accounted for by blood. Contraction of the capacitance vessels in the viscera can pump a liter of blood into the arterial circulation in less than a minute.

Other blood reservoirs that contain a large volume

Figure 32–15. Concept of the acinus as the functional unit of the liver. In each acinus, blood in the portal venule and hepatic arteriole enters the center of the acinus and flows outward to the hepatic venule. (Reproduced, with permission, from Lautt WW, Greenway CV: Conceptual review of the hepatic vascular bed. *Hepatology* 1987;**7**:952. Copyright © by The American Association for the Study of Liver Diseases.)

Terminal hepatic arteriole

Terminal hepatic venule

Terminal hepatic venule

Terminal bile duct

Terminal portal venule

of blood at rest are the skin and lungs. During vigorous exercise, constriction of the vessels in these organs and decreased blood "storage" in the liver and other portions of the splanchnic bed, the skin, and the lungs may increase the volume of actively circulating blood perfusing the muscles by as much as 30%.

CIRCULATION OF THE SKIN

The amount of the heat lost from the body is regulated to a large extent by varying the amount of blood flowing through the skin (see Chapter 14). The fingers, toes, palms, and earlobes contain well-innervated anastomotic connections between arterioles and venules (arteriovenous anastomoses; see Chapter 30). Blood flow in response to thermoregulatory stimuli can vary from 1 to as much as 150 mL/100 g of skin per minute, and it has been postulated that these variations are possible because blood can be shunted through the anastomoses. The subdermal capillary and venous plexus is a blood reservoir of some importance, and the skin is one of the few places where the reactions of blood vessels can be observed visually.

White Reaction

When a pointed object is drawn lightly over the skin, the stroke lines become pale (white reaction). The mechanical stimulus apparently initiates contraction of the precapillary sphincters, and blood drains out of the capillaries and small veins. The response appears in about 15 seconds.

Triple Response

When the skin is stroked more firmly with a pointed instrument, instead of the white reaction there is reddening at the site that appears in about 10 seconds (red reaction). This is followed in a few minutes by local swelling and diffuse, mottled reddening around the injury. The initial redness is due to capillary dilation, a direct response of the capillaries to pressure. The swelling (wheal) is local edema due to increased permeability of the capillaries and postcapillary venules, with consequent extravasation of fluid. The redness spreading out from the injury (flare) is due to arteriolar dilation. This three-part response, the red reaction, wheal, and flare, is called the triple response and is part of the normal reaction to injury (see Chapter 20). It is present after total sympathectomy. The flare is absent in locally anesthetized skin and in denervated skin after the sensory nerves have degenerated, but it is present immediately after nerve block or section above the site of the injury. This plus other evidence indicates that it is

due to an axon reflex, a response in which impulses initiated in sensory nerves by the injury are relayed antidromically down other branches of the sensory nerve fibers (Fig 32–16). This is the one situation in the body where there is substantial evidence for a physiologic effect due to antidromic conduction. The increase in vascular permeability responsible for wheal formation is produced in part by histamine or a histaminelike substance released from local mast cells and mediated via H_1 receptors. Serotonin released from mast cells also contributes to the swelling in rats, but human mast cells do not contain serotonin. The transmitter released at the central termination of the sensory C fiber neurons is substance P (see Chapter 4), and substance P and CGRP are present in all parts of the neurons. Both dilate arterioles, and, in addition, substance P causes extravasation of fluid. Furthermore, substance P antagonists reduce the extravasation. Thus, there is a neurogenic as well as a histaminergic and serotonergic component in the wheal.

Reactive Hyperemia

A response of the blood vessels that occurs in many organs but is visible in the skin is reactive hyperemia, an increase in the amount of blood in a region when its circulation is reestablished after a period of occlusion. When the blood supply to a limb is occluded, the cutaneous arterioles below the occlusion dilate; and when the circulation is reestablished, blood flowing into the dilated vessels makes the skin become fiery red. O_2 diffuses a short distance through the skin, and reactive hyperemia is prevented if the circulation of the limb is occluded in an atmosphere

Spinal cord   Sensory neuron

Endings in skin

Endings near arteriole

→ Orthodromic conduction
----→ Antidromic conduction
········→ Direction taken by impulses

Figure 32–16. Axon reflex. Hypothetical pathway involved in the flare response.

of 100% O_2. Therefore, the arteriolar dilation is apparently due to a local effect of hypoxia, and it is believed to be effected by the release of a chemical substance.

Generalized Responses

Noradrenergic nerve stimulation and circulating epinephrine and norepinephrine constrict cutaneous blood vessels. There are no known vasodilator nerve fibers to the cutaneous vessels, and vasodilation is brought about by a decrease in constrictor tone as well as the local production of bradykinin in sweat glands and vasodilator metabolites. Skin color and temperature also depend on the state of the capillaries and venules. A cold blue or gray skin is one in which the arterioles are constricted and the capillaries dilated; a warm red skin is one in which both are dilated.

Because painful stimuli cause diffuse noradrenergic discharge, a painful injury causes generalized cutaneous vasoconstriction in addition to the local triple response. When the body temperature rises during exercise, the cutaneous blood vessels dilate in spite of continuing noradrenergic discharge in other parts of the body. Dilation of cutaneous vessels in response to a rise in hypothalamic temperature (see Chapter 14) is a prepotent reflex response that overcomes other reflex activity. Cold causes cutaneous vasoconstriction; however, with severe cold, superficial vasodilation may supervene. This vasodilation is the cause of the ruddy complexion seen on a cold day.

Shock is more profound in patients with elevated temperatures because of the cutaneous vasodilation, and patients in shock should not be warmed to the point that their body temperature rises. This is sometimes a problem because well-meaning laymen have read in first-aid books that "injured patients should be kept warm," and they pile blankets on accident victims who are in shock.

PLACENTAL & FETAL CIRCULATION

Uterine Circulation

The blood flow of the uterus parallels the metabolic activity of the myometrium and endometrium and undergoes cyclic fluctuations that correlate well with the menstrual cycle in nonpregnant women. The function of the spiral and basilar arteries of the endometrium in menstruation is discussed in Chapter 23. During pregnancy, blood flow increases rapidly as the uterus increases in size (Fig 32–17). Metabolites with a vasodilator action are undoubtedly produced in the uterus, as they are in other active tissues; but in early pregnancy the arteriovenous O_2 difference

Figure 32–17. Changes in uterine blood flow and the amount of O_2 in uterine venous blood during pregnancy. (After Barcroft H. Modified and redrawn, with permission, from Keele CA, Neil E: *Samson Wright's Applied Physiology*, 12th ed. Oxford Univ Press, 1971.)

across the uterus is small, and it has been suggested that estrogens act on the blood vessels to increase uterine blood flow in excess of tissue O_2 needs. However, even though uterine blood flow increases 20-fold during pregnancy, the size of the conceptus increases much more, changing from a single cell to a fetus plus a placenta that weigh 4–5 kg at term in humans. Consequently, more O_2 is extracted from the uterine blood during the latter part of pregnancy, and the O_2 saturation of uterine blood falls. Just before parturition there is a sharp decline in uterine blood flow, but the significance of this is not clear.

Placenta

The placenta is the "fetal lung." Its maternal portion is in effect a large blood sinus. Into this "lake" project the villi of the fetal portion containing the small branches of the fetal umbilical arteries and vein (Fig 32–18). O_2 is taken up by the fetal blood and CO_2 is discharged into the maternal circulation across the walls of the villi in a fashion analogous to O_2 and CO_2 exchange in the lungs (see Chapter 34), but the cellular layers covering the villi are thicker and less permeable than the alveolar membranes in the lungs, and exchange is much less efficient. The placenta is also the route by which all nutritive materials enter the fetus and by which fetal wastes are discharged to the maternal blood.

Figure 32–18. Diagram of a section through the human placenta. AM, amnion; S, septa; UA, umbilical arteries; UV, umbilical vein; UTA, uterine artery; V, veins. (Reproduced, with permission, from Harrison RG: *Textbook of Human Embryology,* 2nd ed. Blackwell, 1964.)

Fetal Circulation

The arrangement of the circulation in the fetus is shown diagrammatically in Fig 32–19. Fifty-five percent of the fetal cardiac output goes through the placenta. The blood in the umbilical vein in humans is believed to be about 80% saturated with O_2, compared with 98% saturation in the arterial circulation of the adult. The **ductus venosus** (Fig 32–20) diverts some of this blood directly to the inferior vena cava, and the remainder mixes with the portal blood of the fetus. The portal and systemic venous blood of the fetus is only 26% saturated, and the saturation of the mixed blood in the inferior vena cava is approximately 67%. Most of the blood entering the heart through the inferior vena cava is diverted directly to the left atrium via the patent foramen ovale. Most of the blood from the superior vena cava enters the right ventricle and is expelled into the pulmonary artery. The resistance of the collapsed lungs is high, and the pressure in the pulmonary artery is several mm Hg higher than it is in the aorta, so that most of the blood in the pulmonary artery passes through the **ductus arteriosus** to the aorta. In this fashion, the relatively unsaturated blood from the right ventricle is diverted to the trunk and lower body of the fetus, while the head of the fetus receives the better-oxygenated blood from the left ventricle. From the aorta, some of the blood is pumped into the umbilical arteries and back to the placenta. The O_2 saturation of the blood in the lower aorta and umbilical arteries of the fetus is approximately 60%.

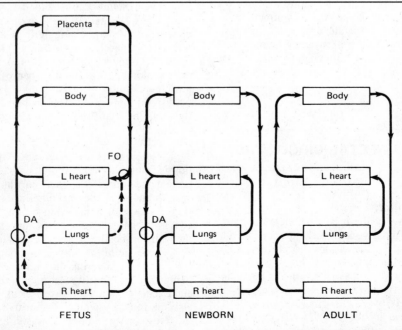

Figure 32–19. Diagram of the circulation in the fetus, the newborn infant, and the adult. DA, ductus arteriosus; FO, foramen ovale. (Redrawn and reproduced, with permission, from Born GVR et al: Changes in the heart and lungs at birth. *Cold Spring Harbor Symp Quant Biol* 1954;**19**:102.)

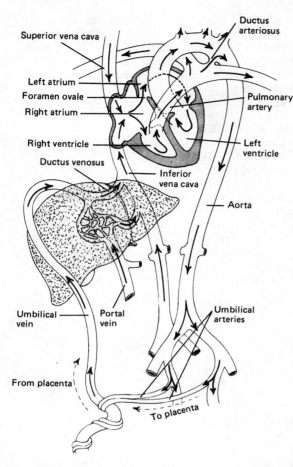

Figure 32–20. Circulation in the fetus. Most of the oxygenated blood reaching the heart via the umbilical vein and inferior vena cava is diverted through the foramen ovale and pumped out the aorta to the head, while the deoxygenated blood returned via the superior vena cava is mostly pumped through the pulmonary artery and ductus arteriosus to the feet and the umbilical arteries.

Fetal Respiration

The tissues of fetal and newborn mammals have a remarkable but poorly understood resistance to hypoxia. However, the O_2 saturation of the maternal blood in the placenta is so low that the fetus might suffer hypoxic damage if fetal red cells did not have a greater O_2 affinity than adult red cells (Fig 32–21). The fetal red cells contain fetal hemoglobin (hemoglobin F), while the adult cells contain adult hemoglobin (hemoglobin A). The cause of the difference in O_2 affinity between the two is that hemoglobin F binds 2,3-DPG less effectively than hemoglobin A does. The decrease in O_2 affinity due to the binding of 2,3-DPG is discussed in Chapter 35. The quantitative aspects of gas exchange across the placenta, based on experiments on the cow, are shown in Table 32–6.

Figure 32–21. Dissociation curves of hemoglobin in human maternal and fetal blood. (Modified from Longo LD: Carbon monoxide: Effects on oxygenation of the fetus in utero. *Science* 1976;**194**:523.)

In humans, hemoglobin A first appears in the fetal circulation at about the 20th week, when the bone marrow first begins to function (see Chapter 27). At birth, only 20% of the circulating hemoglobin is of the adult type. However, no more hemoglobin F is normally formed after birth, and by the age of 4 months 90% of the circulating hemoglobin is hemoglobin A.

Changes in Fetal Circulation & Respiration at Birth

Because of the patent ductus arteriosus and foramen ovale (Fig 32–20), the left and right heart pump in parallel in the fetus rather than in series as they do in the adult. At birth, the placental circulation is cut off and the peripheral resistance suddenly rises. The pressure in the aorta rises until it exceeds that in the pulmonary artery. Meanwhile, because the placental circulation has been cut off, the infant becomes increasingly asphyxic, Finally, the infant gasps several times, and the lungs expand. The markedly negative

Table 32-6. Summary of gaseous interchange across the placenta in the cow.[1]

Vessel	Hemoglobin % Saturation	Partial Pressure (mm Hg)	
		O_2	CO_2
Maternal artery	90	70	41
Uterine vein	70	41.5	46.5
Umbilical vein		11.5	48
Umbilical artery		5.5	50

[1]Data from Roos J, Romijn C: *J Physiol (Lond)* 1938;**92**:261.

intrapleural pressure (–30 to –50 mm Hg) during the gasps contributes to the expansion of the lungs, but other poorly understood factors are also involved. The sucking action of the first breath plus constriction of the umbilical veins squeezes as much as 100 mL of blood from the placenta (the "placental transfusion").

Once the lungs are expanded, the pulmonary vascular resistance falls to less than 20% of the in utero value, and pulmonary blood flow increases markedly. Blood returning from the lungs raises the pressure in the left atrium, closing the foramen ovale by pushing the valve that guards it against the interatrial septum. The ductus arteriosus constricts within a few minutes after birth but, at least in sheep, does not close completely for 24–48 hours. Eventually, the foramen ovale and the ductus arteriosus both fuse shut in normal infants, and by the end of the first few days of life the adult circulatory pattern is established. The mechanism responsible for obliteration of the ductus arteriosus, like that responsible for expansion of the lungs, is incompletely understood, although there is evidence that the rise in arterial P_{O_2} after birth contributes. Bradykinin has been shown to constrict the umbilical vessels and the ductus arteriosus while dilating the pulmonary vascular bed. Prostacyclin appears to have a role in maintaining the patency of the ductus arteriosus before birth, and rectal administration of one or 2 small doses of indomethacin, a drug that inhibits prostacyclin and prostaglandin synthesis (see Chapter 17), closes the ductus in many infants who would otherwise require surgical closure. On the other hand, closure of the ductus before birth causes pulmonary hypertension, and in experimental animals, administration of prostaglandin D_2 during the neonatal period increases pulmonary blood flow and decreases pulmonary artery pressure. There is some evidence for an increased incidence of pulmonary hypertension in the children of women given prostaglandin inhibitors to delay the onset of labor.

Cardiovascular Homeostasis in Health & Disease

33

INTRODUCTION

The compensatory adjustments of the cardiovascular system to the challenges faced by the circulation normally in everyday life and abnormally in disease illustrate the integrated operation of the cardiovascular regulatory mechanisms described in the preceding chapters. The adjustments to gravity, exercise, inflammation, shock, fainting, hypertension, and heart failure are considered in this chapter.

COMPENSATIONS FOR GRAVITATIONAL EFFECTS

In the standing position, as a result of the effect of gravity on the blood (see Chapter 30), the mean arterial blood pressure in the feet of a normal adult is 180–200 mm Hg and venous pressure is 85–90 mm Hg. The arterial pressure at head level is 60–75 mm Hg, and the venous pressure is zero. If the individual does not move, 300–500 mL of blood pools in the venous capacitance vessels of the lower extremities, fluid begins to accumulate in the interstitial spaces because of increased hydrostatic pressure in the capillaries, and stroke volume is decreased up to 40%. Symptoms of cerebral ischemia develop when the cerebral blood flow decreases to less than about 60% of the flow in the recumbent position. If there were no compensatory cardiovascular changes, the reduction in cardiac output due to pooling on standing would lead to a reduction of cerebral flow of this magnitude, and consciousness would be lost.

The major compensations on assuming the upright position are triggered by the drop in blood pressure in the carotid sinus and aortic arch. The heart rate increases, helping to maintain cardiac output. There is relatively little venoconstriction in the periphery, but there is a prompt increase in the circulating levels of renin and aldosterone. The arterioles constrict, helping to maintain blood pressure. The actual blood pressure change at heart level is variable, depending upon the balance between the degree of arteriolar constriction and the drop in cardiac output (Fig 33–1).

In the cerebral circulation there are additional

compensatory changes. The arterial pressure at head level drops 20–30 mm Hg, but jugular venous pressure falls 5–8 mm Hg, reducing the drop in perfusion pressure (arterial pressure minus venous pressure). Cerebral vascular resistance is reduced because intracranial pressure falls as venous pressure falls, decreasing the pressure on the cerebral vessels. The decline in cerebral blood flow increases the partial pressure of CO_2 (P_{CO_2}) and decreases the P_{O_2} and the

Figure 33–1. Effect on the cardiovascular system of rising from the supine to the upright position. Figures shown are average changes. Changes in abdominal and limb resistance and in blood pressure are variable from individual to individual. (Redrawn and reproduced, with permission, from Brobeck JR [editor]: *Best and Taylor's Physiological Basis of Medical Practice,* 9th ed. Williams & Wilkins, 1973.)

pH in brain tissue, further actively dilating the cerebral vessels. Because of the operation of these autoregulatory mechanisms, cerebral blood flow declines only 20% on standing. In addition, the amount of O_2 extracted from each unit of blood increases, and the net effect is that cerebral O_2 consumption is about the same in the supine and the upright positions.

Prolonged standing presents an additional problem because of increasing interstitial fluid volume in the lower extremities. As long as the individual moves about, the operation of the muscle pump (see Chapter 30) keeps the venous pressure below 30 mm Hg in the feet, and venous return is adequate. However, with prolonged quiet standing (eg, in military personnel standing at attention for long periods), fainting may result. In a sense, the fainting is a "homeostatic mechanism," because falling to the horizontal position promptly restores venous return, cardiac output, and cerebral blood flow to adequate levels.

The effects of gravity on the circulation in humans depend in part upon the blood volume. When the blood volume is low, these effects are marked; when it is high, they are minimal.

The compensatory mechanisms that operate on assumption of the erect posture are better developed in humans than in quadrupeds even though these animals have sensitive carotid sinus mechanisms. Quadrupeds tolerate tilting to the upright position poorly. Of course, giraffes are an exception. These long-legged animals do not develop ankle edema despite the very large increment in vascular pressure in their legs due to gravity because they have tight skin and fascia in the lower legs—in a sense a built-in antigravity suit (see below)—and a very effective muscle pump. Perfusion in the head is maintained by a high mean arterial pressure and valves that prevent backflow in the jugular vein. When giraffes lower their heads to drink, blood is pumped up the jugular vein to the chest, presumably by rhythmic contractions of the muscles of the jaws.

Postural Hypotension

In some individuals, sudden standing causes a fall in blood pressure, dizziness, dimness of vision, and even fainting. The causes of this **orthostatic (postural) hypotension** are multiple. It is common in patients receiving sympatholytic drugs. It also occurs in diseases such as diabetes and syphilis in which there is damage to the sympathetic nervous system, underscoring the importance of the sympathetic vasoconstrictor fibers in compensating for the effects of gravity on the circulation. Another cause of postural hypotension is **primary autonomic insufficiency.** Several forms of this condition are now recognized (Table 33–1), including one in which there appears to be a congenital deficiency of dopamine β-hydroxylase (see Chapter 4) and consequently little or no production of norepinephrine and epinephrine.

Table 33–1. Major forms of primary autonomic failure.[1]

Bradbury-Eggleston syndrome (idiopathic orthostatic hypotension)
  Onset late in life
  Sympathetic and parasympathetic failure
  Absent or minimal other neurologic involvement
  Plasma norepinephrine/dopamine ratio greater than 1
Shy-Drager syndrome (multiple system atrophy)
  Onset in mid to late life
  Sympathetic and parasympathetic failure
  Other neurologic involvement (extrapyramidal, cerebellar, etc.)
  Plasma norepinephrine/dopamine ratio greater than 1
Riley-Day syndrome (familial dysautonomia)
  Congenital onset and premature mortality
  Ashkenazi Jewish extraction
  Sympathetic and parasympathetic involvement
  Emotional liability
  Plasma norepinephrine/dopamine ratio greater than 1
Dopamine β-hydroxylase deficiency
  Congenital onset
  Sympathoadrenomedullary failure (orthostatic hypotension)
  Intact sweating
  Parasympathetic sparing
  Plasma norepinephrine/dopamine ratio much less than 1

[1]Reproduced, with permission, from Robertson D et al: Dopamine β-hydroxylase deficiency: A genetic disorder of cardiovascular regulation. *Hypertension* 1991;**18**:1. By permission of the American Heart Association, Inc.

Baroreceptor reflexes are also abnormal in patients with primary hyperaldosteronism. However, these patients generally do not have postural hypotension, because their blood volumes are expanded sufficiently to maintain cardiac output in spite of changes in position. Indeed, mineralocorticoids are used to treat patients with postural hypotension.

Effects of Acceleration

The effects of gravity on the circulation are multiplied during acceleration or deceleration in vehicles that in modern civilization range from elevators to rockets. Force acting on the body as a result of acceleration is commonly expressed in *g* units, 1 *g* being the force of gravity on the earth's surface. "Positive *g*" is force due to acceleration acting in the long axis of the body, from head to foot; "negative *g*" is force due to acceleration acting in the opposite direction. During exposure to positive *g*, blood is "thrown" into the lower part of the body. The cerebral circulation is protected by the fall in venous pressure and intracranial pressure. Cardiac output is maintained for a time because blood is drawn from the pulmonary venous reservoir and because the force of cardiac contraction is increased. At accelerations producing more than 5 *g*, however, vision fails ("blackout") in about 5 seconds and unconsciousness follows almost immediately thereafter. The effects of positive g are effectively cushioned by the use of antigravity "g suits," double-walled pressure suits containing water or compressed air and regulated in such a way that they compress the abdomen and legs with a force proportionate to the positive *g*. This decreases venous pooling and helps maintain venous return (Fig 33–1).

Negative *g* causes increased cardiac output, a rise in cerebral arterial pressure, intense congestion of the head and neck vessels, ecchymoses around the eyes, severe throbbing head pain, and, eventually, mental confusion ("redout"). In spite of the great rise in cerebral arterial pressure, the vessels in the brain do not rupture, because there is a corresponding increase in intracranial pressure and their walls are supported in the same way as during straining (see Chapter 32). The tolerance for *g* forces exerted across the body is much greater than it is for axial *g*. Humans tolerate 11 *g* acting in a back-to-chest direction for 3 minutes and 17 *g* acting in a chest-to-back direction for 4 minutes. Astronauts are therefore positioned to take the *g* forces of rocket flight in the chest-to-back direction. The tolerances in this position are sufficiently large to permit acceleration to orbital or escape velocity and deceleration back into the earth's atmosphere without ill effects.

Effects of Zero Gravity on the Cardiovascular System

From the data available to date, weightlessness for up to 366 days appears to have only transient adverse effects on the circulation. One would expect that the circulation in zero gravity would be essentially similar to that in the recumbent position at sea level, with possible minor variations due to changes in the position of viscera. Transient postural hypotension has been present after return to earth from orbital flights, and full readaptation to normal gravity after return from space flights to date has required 4–7 weeks. In the absence of the increases in cardiac output normally occasioned by the efforts of everyday living, there is some atrophy of the myocardium. There has been speculation that prolonged weightlessness during future trips to the planets may lead to more severe "disuse atrophy" of the heart and the cardiovascular and somatic reflex mechanisms (see Chapter 12) responsible for postural adjustments.

Other Effects of Zero Gravity

Muscular effort is much reduced when the objects to be moved are weightless, and the decrease in the extensive normal proprioceptive input due to the action of gravity on the body leads to flaccidity and atrophy of skeletal muscles. A program of regular exercises against resistance, eg, pushing against a wall or stretching a heavy rubber band, appears to decrease the loss of muscle. However, the compensation is incomplete.

Other changes produced by exposure to space flight include space motion sickness, a problem that has proved to be of greater magnitude than initially expected; loss of plasma volume due to headward shift of body fluids, with subsequent diuresis; loss of

muscle mass; steady loss of bone mineral, with increased Ca^{2+} excretion; loss of red-cell mass; and alterations in plasma lymphocytes. The loss of body Ca^{2+} is equivalent to 0.4% of the total body Ca^{2+} per month, and although there is some evidence that the loss tapers off during prolonged space flight, loss at this rate might create problems of appreciable magnitude if continued for more than 9–12 months. A high-calcium diet helps overcome this problem, but no totally effective treatment has as yet been developed. The psychologic problems associated with the isolation and monotony of prolonged space flight are also a matter of concern.

EXERCISE

Exercise is associated with very extensive alterations in the circulatory and respiratory systems. For convenience, the circulatory adjustments are considered in this chapter and the respiratory adjustments in Chapter 37. However, it should be emphasized that they occur together in an integrated fashion as part of the homeostatic responses that make moderate to severe exercise possible.

Muscle Blood Flow

The blood flow of resting skeletal muscle is low (2–4 mL/100 g/min). When a muscle contracts, it compresses the vessels in it if it develops more than 10% of its maximal tension (Fig 33–2); when it develops more than 70% of its maximal tension, blood flow is completely stopped. Between contractions, however, flow is so greatly increased that blood flow per unit of time in a rhythmically contracting muscle is increased as much as 30-fold. Blood flow some-

Figure 33–2. Blood flow through a portion of the calf muscles during rhythmic contraction. (Reproduced, with permission, from Barcroft H, Swan HJC: *Sympathetic Control of Human Blood Vessels.* Arnold, 1953.)

times increases at or even before the start of exercise, so the initial rise is probably a neurally mediated response. Impulses in the sympathetic vasodilator system (see Chapter 31) may be involved. The blood flow in resting muscle doubles after sympathectomy, so some decrease in tonic vasoconstrictor discharge may also be involved. However, once exercise has started, local mechanisms maintain the high blood flow, and there is no difference in flow in normal and sympathectomized animals.

Local mechanisms maintaining a high blood flow in exercising muscle include a fall in tissue P_{O_2}, a rise in tissue P_{CO_2} and accumulation of K^+ and other vasodilator metabolites (see Chapter 31). The temperature rises in active muscle, and this further dilates the vessels. Dilation of the arterioles and precapillary sphincters causes a 10–100-fold increase in the number of open capillaries. The average distance between the blood and the active cells—and the distance O_2 and metabolic products must diffuse—is thus greatly decreased. The dilation increases the cross-sectional area of the vascular bed, and the velocity of flow therefore decreases. The capillary pressure increases until it exceeds the oncotic pressure throughout the length of the capillaries. In addition, the accumulation of osmotically active metabolites more rapidly than they can be carried away decreases the osmotic gradient across the capillary walls. Therefore, fluid transudation into the interstitial spaces is tremendously increased. Lymph flow is also greatly increased, limiting the accumulation of interstitial fluid and in effect greatly increasing its turnover. The decreased pH and increased temperature shift the dissociation curve for hemoglobin to the right, so that more O_2 is given up by the blood. Increased concentration of 2,3-DPG in the red blood cells has been reported, and this would further decrease the O_2 affinity of hemoglobin (see Chapters 27 and 35). The net result is an up to 3-fold increase in the arteriovenous O_2 difference, and the transport of CO_2 out of the tissue is also facilitated. All of these changes combine to make it possible for the O_2 consumption of skeletal muscle to increase 100-fold during exercise. An even greater increase in energy output is possible for short periods during which the energy stores are replenished by anaerobic metabolism of glucose and the muscle incurs an O_2 debt (see Chapter 3). The overall changes in intermediary metabolism during exercise are discussed in Chapter 17.

K^+ dilates arterioles in exercising muscle, particularly during the early part of exercise. Muscle blood flow increases to a lesser degree during exercise in K^+-depleted individuals, and there is a greater tendency for severe disintegration of muscle (**exertional rhabdomyolysis**) to occur.

Systemic Circulatory Changes

The systemic cardiovascular response to exercise depends on whether the muscle contractions are pri-

marily isometric or primarily isotonic with the performance of external work. With the start of an isometric muscle contraction, the heart rate rises. This increase still occurs if the muscle contraction is prevented by local infusion of a neuromuscular blocking drug. It also occurs with just the thought of performing a muscle contraction, so it is probably the result of psychic stimuli acting on the medulla oblongata. The increase is largely due to decreased vagal tone, although increased discharge of the cardiac sympathetic nerves plays some role. Within a few seconds of the onset of an isometric muscle contraction, systolic and diastolic blood pressures rise sharply. Stroke volume changes relatively little, and blood flow to the steadily contracting muscles is reduced as a result of compression of their blood vessels.

The response to exercise involving isotonic muscle contraction is similar in that there is a prompt increase in heart rate but different in that there is a prompt increase in heart rate but different in that there is a marked increase in stroke volume. In addition, there is a net fall in total peripheral resistance (Fig 33–3) due to vasodilation in exercising muscles (Table 33–2). Consequently, systolic blood pressure

Figure 33–3. Effects of different levels of isotonic exercise on cardiovascular function. (Reproduced, with permission, from Berne RM, Levy MN: *Cardiovascular Physiology*, 5th ed. Mosby, 1986.)

Table 33–2. Cardiac output and regional blood flow in a sedentary man.[1]

	mL/min	
	Quiet Standing	Exercise
Cardiac output	5900	24,000
Blood flow to:		
Heart	250	1000
Brain	750	750
Active skeletal muscle	650	20,850
Inactive skeletal muscle	650	300
Skin	500	500
Kidney, liver, gastrointestinal tract, etc	3100	600

[1]Values are at rest and during isotonic exercise at maximal oxygen uptake. (From Mitchell JH, Blomqvist G: Maximal oxygen uptake. *N Engl J Med* 1971;**284:**1018.)

rises only moderately, whereas diastolic pressure usually remains unchanged or falls. The difference in response to isometric and isotonic exercise is explained in part by the fact that the active muscles are tonically contracted during isometric exercise and consequently contribute to increased total peripheral resistance. In addition, there is a general increase in muscle sympathetic nerve activity, apparently because of a signal from the contracted muscle, but since there is cholinergic sympathetic vasodilation in the inactive skeletal muscles, the significance of this increase is unclear.

Cardiac output is increased during isotonic exercise to values that may exceed 35 L/min, the amount being proportionate to the increase in O_2 consumption. The mechanisms responsible for this increase are discussed in Chapter 29. The increase is due to an increase in heart rate and stroke volume, the heart muscle contracting more forcibly and discharging more of the end-systolic volume of blood in the ventricles. These chronotropic and inotropic effects on the heart are both due to increased activity in the noradrenergic sympathetic nerves to the heart. The activity is apparently initiated by psychic stimuli. Decreased vagal tone also contributes to the increase in heart rate. The increase in heart rate is sustained by the autonomic changes plus the stimulatory effect of the increase in P_{CO_2} on the medulla. The maximal heart rate achieved during exercise decreases with age. In children, it rises to 200 or more beats per minute; in adults it rarely exceeds 195 beats per minute, and in elderly individuals the rise is even smaller.

There is a great increase in venous return, although the increase in venous return is not the primary cause of the increase in cardiac output. Venous return is increased by the great increase in the activity of the muscle and thoracic pumps; by mobilization of blood from the viscera; by increased pressure transmitted through the dilated arterioles to the veins; and by noradrenergically mediated venoconstriction,

which decreases the volume of blood in the veins. The amount of blood mobilized from the splanchnic area and other reservoirs may increase the amount of blood in the arterial portion of the circulation by as much as 30% during strenuous exercise.

After exercise, the blood pressure may transiently drop to subnormal levels, presumably because accumulated metabolites keep the muscle vessels dilated for a short period. However, the blood pressure soon returns to the preexercise level. The heart rate returns to normal more slowly.

Temperature Regulation

The quantitative aspects of heat dissipation during exercise are summarized in Fig 33–4. In many locations, the skin is supplied by branches of muscle arteries, so that some of the blood warmed in the muscles is transported directly to the skin, where some of the heat is radiated to the environment. There is a marked increase in ventilation (see Chapter 37), and some heat is lost in the expired air. The body temperature rises, and the hypothalamic centers that control heat-dissipating mechanisms are activated. The temperature increase is due in part to inability of the heat-dissipating mechanism to handle the great increase in heat production, but there is evidence that during exercise there is in addition a "resetting of the thermostat"—ie, an increase in the body temperature at which the heat-dissipating mechanisms are activated. Sweat secretion is greatly increased, and vaporization of this sweat is the major path for heat loss. The cuta-

Figure 33–4. Energy exchange in muscular exercise. The shaded area represents the excess of heat production over heat loss. The total energy output equals the heat production plus the work done. (Redrawn and reproduced, with permission, from Nielsen M: Die Regulation der Körpertemperatur bei Muskelarbeit. *Skand Arch Physiol* 1938: **79:**193.)

neous vessels also dilate. This dilation is primarily due to inhibition of vasoconstrictor tone, although local release of vasodilator polypeptides may also contribute (see Chapter 31).

Training

Both at rest and at any given level of exercise, trained athletes have a larger stroke volume and lower heart rate than untrained individuals (see Chapter 29), and they tend to have larger hearts. Training increases the maximal oxygen consumption ($V_{O_{2max}}$) that can be produced by exercise in an individual. $V_{O_{2max}}$ averages about 38 mL/kg/min in active healthy men and about 29 mL/kg/min in active healthy women. It is lower in sedentary individuals. $V_{O_{2max}}$ is the product of maximal cardiac output and maximal O_2 extraction by the tissues, and both increase with training.

The changes that occur in skeletal muscles with training include increases in the number of mitochondria and the enzymes involved in oxidative metabolism. There is an increase in the number of capillaries, with better distribution of blood to the muscle fibers. The net effect is more complete extraction of O_2 and consequently, for a given work load, less increase in lactate production. There is less increase in blood flow to muscles as well, and because of this, less increase in heart rate and cardiac output than in an untrained individual. This is one of the reasons that exercise is of benefit to patients with heart disease.

In view of the current vogue for jogging and other forms of exercise, it is worth noting that one of the clear-cut benefits of exercise regimens is psychologic; patients who exercise regularly "feel better." Regular physical exertion also increases the probability that an individual can remain active past the standard age of retirement. In addition, there is evidence that regular exercise decreases the incidence and severity of myocardial infarctions.

INFLAMMATION & WOUND HEALING

Inflammation is a complex localized response to foreign substances such as bacteria or in some instances to internally produced substances. It includes a sequence of reactions initially involving neutrophils, adhesion molecules, complement, and IgG. PAF, an agent with potent inflammatory effects (see Chapter 27), also plays a role. Later, monocytes and lymphocytes are involved. Cytokines released from these cells up-regulate relevant CAMs. Arterioles in the inflamed area dilate, and capillary permeability is increased (see Chapters 31 and 32). When the inflammation occurs in or just under the skin, it is characterized by redness, swelling, tenderness, and pain. The intensity of the response is affected by the circulating level of glucocorticoids, and the anti-inflammatory

effects of large doses of these hormones are discussed in Chapter 20.

When tissue is damaged, platelets adhere to exposed matrix via integrins that bind to collagen and laminin (Fig 33–5). Blood coagulation produces thrombin, which promotes platelet aggregation and granule release. The platelet granules generate an inflammatory response, leading to extravasation of white blood cells. Cytokines released by the white blood cells and platelets up-regulate integrins on macrophages, which migrate to the area of injury, and on fibroblasts and epithelial cells, which mediate wound healing and scar formation.

SHOCK

General Considerations

Shock is a syndrome about which there has been a great deal of confusion and controversy. Part of the difficulty lies in the loose use of the term by physiologists and physicians as well as laymen. Electric shock and spinal shock, for example, bear no relation to the condition produced by hemorrhage and related cardiovascular abnormalities. Shock in the restricted sense of "circulatory shock" is still a collection of different entities that share certain common features. However, the feature that is common to all the entities is inadequate tissue perfusion as a result of inadequate cardiac output. The cardiac output may be in-

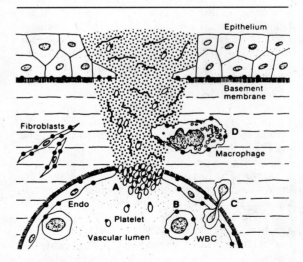

Figure 33–5. Role of CAMs in wound healing and coagulation. Platelets bind to areas of tissue damage and aggregate **(A)**. Cytokines and tissue factors activate endothelial cells and white blood cells and up-regulate their adhesion molecules **(B)**, fostering extravasation **(C)** and migration to the wound site **(D)**. Up-regulation of CAMs on fibroblasts and epithelial cells speeds the healing process. (Reproduced, with permission, from Albelda SM, Buck CA: Integrins and other cell adhesion molecules. *FASEB J* 1990;**4**:2868.)

adequate because the amount of fluid in the vascular system is inadequate to fill it (**hypovolemic shock**). Alternatively, it may be inadequate because the size of the vascular system is increased by vasodilation even though the blood volume is normal (**distributive, vasogenic,** or **low-resistance shock**). Shock may also be caused by inadequate pumping action of the heart as a result of myocardial abnormalities (**cardiogenic shock**), and by inadequate cardiac output as a result of obstruction of blood flow in the lungs or heart (**obstructive shock**). These forms of shock are listed in Table 33–3, along with examples of the disease processes that can cause them.

Hypovolemic Shock

Hypovolemic shock is also called "cold shock." It is characterized by hypotension; a rapid, thready pulse; a cold, pale, clammy skin; intense thirst; rapid respiration; and restlessness or, alternatively, torpor. None of these findings, however, are invariably present. The hypotension may be relative. A hypertensive patient whose blood pressure is regularly 240/140, for example, may be in severe shock when the blood pressure is 120/90.

Hypovolemic shock is commonly subdivided into categories on the basis of cause. The use of terms such as hemorrhagic shock, traumatic shock, surgical shock, and burn shock is of some benefit because, although there are similarities between these various forms of shock, there are important features that are unique to each.

Hemorrhagic Shock

It is useful to consider the effects of hemorrhage in some detail because they illustrate the features of a major form of hypovolemic shock and the multiple

Table 33–3. Types of shock, with examples of conditions or diseases that can cause each type.

Hypovolemic shock (decreased blood volume)
Hemorrhage
Trauma
Surgery
Burns
Distributive shock (marked vasodilation; also called vasogenic or low-resistance shock)
Fainting (neurogenic shock)
Anaphylaxis
Sepsis (also causes hypovolemia due to increased capillary permeability with loss of fluid into tissues)
Cardiogenic shock (inadequate output by a diseased heart)
Myocardial infarction
Congestive heart failure
Arrhythmias
Obstructive shock (obstruction of blood flow)
Tension pneumothorax
Pulmonary embolism
Cardiac tumor
Cardiac tamponade

compensatory reactions that come into play to defend ECF volume. The principal reactions are listed in Table 33–4.

The decline in blood volume produced by bleeding decreases venous return, and cardiac output falls. The heart rate is increased, and with severe hemorrhage, there is always a fall in blood pressure. With moderate hemorrhage (5–15 mL/kg body weight), pulse pressure is reduced, but mean arterial pressure may be normal. The blood pressure changes vary from individual to individual, even when exactly the same amount of blood is lost. The skin is cool and pale and may have a grayish tinge because of stasis in the capillaries and a small amount of cyanosis. Respiration is rapid, and in patients whose consciousness is not obtunded, intense thirst is a prominent symptom.

In hypovolemic and other forms of shock, the inadequate perfusion of the tissue leads to increased anaerobic glycolysis, with the production of large amounts of lactic acid. In severe cases, the blood lactate level rises from the normal value of about 1 mmol/L to 9 mmol/L or more. The resulting **lactic acidosis** depresses the myocardium, decreases peripheral vascular responsiveness to catecholamines, and may be severe enough to cause coma.

Rapid Compensatory Reactions

When blood volume is reduced and venous return is decreased, the arterial baroreceptors are stretched to a lesser degree, and sympathetic output is increased. Even if there is no drop in mean arterial pressure, the decrease in pulse pressure decreases the rate of discharge in the arterial baroreceptors, and reflex tachycardia and vasoconstriction result.

The vasoconstriction is generalized, sparing only the vessels of the brain and heart. The vasoconstrictor innervation of the cerebral arterioles is probably insignificant from a functional point of view, and the coronary vessels are dilated because of the increased myocardial metabolism secondary to the increase in heart rate (see Chapter 32). Vasoconstriction is most marked in the skin, where it accounts for the coolness and pallor, and in the kidneys and viscera.

Hemorrhage also evokes a widespread reflex

Table 33–4. Compensatory reactions activated by hemorrhage.

Vasoconstriction
Tachycardia
Venoconstriction
Increased thoracic pumping
Increased skeletal muscle pumping (in some cases)
Increased movement of interstitial fluid into capillaries
Increased secretion of norepinephrine and epinephrine
Increased secretion of vasopressin
Increased secretion of glucocorticoids
Increased secretion of renin and aldosterone
Increased secretion of erythropoietin
Increased plasma protein synthesis

venoconstriction that helps maintain the filling pressure of the heart, although the receptors that initiate the venoconstriction are unsettled. The intense vasoconstriction in the splanchnic area shifts blood from the visceral reservoir into the systemic circulation. Blood is also shifted out of the subcutaneous and pulmonary veins. Contraction of the spleen discharges more "stored" blood into the circulation, although the volume mobilized in this way in humans is small.

In the kidneys, both afferent and efferent arterioles are constricted, but the efferent vessels are constricted to a greater degree. The glomerular filtration rate is depressed, but renal plasma flow is decreased to a greater extent, so that the filtration fraction (glomerular filtration rate divided by renal plasma flow) increases. There may be shunting of the blood through the medullary portions of the kidneys, bypassing the cortical glomeruli. Very little urine is formed. Sodium retention is marked, and there is retention of the nitrogenous products of metabolism in the blood (**azotemia** or **uremia**). Especially when the hypotension is prolonged, there may be severe renal tubular damage (**acute renal failure**).

Hemorrhage is a potent stimulus to adrenal medullary secretion (see Chapter 20). Circulating norepinephrine is also increased because of the increased discharge of sympathetic noradrenergic neurons. The increase in circulating catecholamines probably contributes relatively little to the generalized vasoconstriction, but it may lead to stimulation of the reticular formation (see Chapter 11). Possibly because of such reticular stimulation, some patients in hemorrhagic shock are restless and apprehensive. Others are quiet and apathetic, and their sensorium is dulled, probably because of cerebral ischemia and acidosis. When restlessness is present, increased motor activity and increased respiratory movements increase the muscular and thoracic pumping of venous blood.

The loss of red cells decreases the O_2-carrying power of the blood, and the blood flow in the carotid and aortic bodies is reduced. The resultant anemia and stagnant hypoxia (see Chapter 37), as well as the acidosis, stimulate the chemoreceptors. Increased activity in chemoreceptor afferents is probably the main cause of respiratory stimulation in shock. Chemoreceptor activity also excites the vasomotor center, increasing vasoconstrictor discharge. In fact, in hemorrhaged dogs with arterial pressures of less than 70 mm Hg, cutting the nerves to the carotid baroreceptors and chemoreceptors may cause a further fall in blood pressure rather than a rise. This paradoxic result occurs because there is no baroreceptor discharge at pressures below 70 mm Hg and activity in fibers from the carotid chemoreceptors is driving the vasomotor center beyond the maximal rate produced by release of baroreceptor inhibition.

The increase in the level of circulating angiotensin II produced by the increase in plasma renin activity during hemorrhage causes thirst by an action on the subfornical organ (see Chapter 32), and ingestion of fluid helps restore the ECF volume. The increase in angiotensin II also helps to maintain blood pressure. The blood pressure fall produced by removal of a given volume of blood is greater in animals infused with drugs that block angiotensin II receptors than it is in controls. Vasopressin also raises blood pressure when administered in large doses in normal animals, but infusion of doses that produce the same plasma vasopressin levels produced by hemorrhage causes only a small increase in blood pressure, probably because of buffering by baroreceptor mechanisms. However, blood pressure falls when peptides that antagonize the effects of vasopressin are injected following hemorrhage. Thus, it appears that vasopressin also plays a significant role in maintaining blood pressure. The increases in circulating angiotensin II and ACTH levels increase aldosterone secretion, and the increased circulating levels of aldosterone and vasopressin cause retention of Na^+ and water, which helps reexpand the blood volume. However, aldosterone takes about 30 minutes to exert its effect, and the initial decline in urine volume and Na^+ excretion is certainly due for the most part to the hemodynamic alterations in the kidney.

When the arterioles constrict and the venous pressure falls because of the decrease in blood volume, there is a drop in capillary pressure. Fluid moves into the capillaries along most of their course, helping to maintain the circulating blood volume. This decreases interstitial fluid volume, and fluid moves out of the cells.

Long-Term Compensatory Reactions

After a moderate hemorrhage, the circulating plasma volume is restored in 12–72 hours (Fig 33–6). There is also a rapid entry of preformed albumin from extravascular stores, but most of the tissue fluids that are mobilized are protein-free. They dilute the plasma proteins and cells, but when whole blood is lost, the hematocrit may not fall for several hours after the onset of bleeding. After the initial influx of preformed albumin, the rest of the plasma protein losses are replaced, presumably by hepatic synthesis, over a period of 3–4 days. Erythropoietin appears in the circulation, and the reticulocyte count increases, reaching a peak in 10 days. The red cell mass is restored to normal in 4–8 weeks. However, a low hematocrit is remarkably well tolerated because of various compensatory mechanisms. One of these is an increase in the concentration of 2,3-DPG in the red blood cells, which causes hemoglobin to give more O_2 to the tissues (see Chapter 27). In long-standing anemia in otherwise healthy individuals, exertional dyspnea is not observed until the hemoglobin concentration is about 7.5 g/dL. Weakness becomes appreciable at about 6 g/dL; dyspnea at rest appears at about 3 g/dL; and the heart fails when the hemoglobin level falls to 2 g/dL.

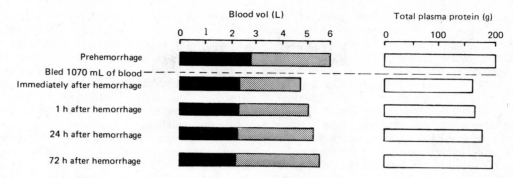

Figure 33–6. Changes in plasma volume (dotted bars), red cell volume (black bars), and total plasma protein following hemorrhage in a normal human subject. (From data in Keele CA, Neil E: *Samson Wright's Applied Physiology*, 11th ed. Oxford Univ Press, 1965.)

Refractory Shock

Depending largely upon the amount of blood lost, some patients die soon after hemorrhage and others recover as the compensatory mechanisms, aided by appropriate treatment, gradually restore the circulation to normal. In an intermediate group of patients, shock persists for hours and gradually progresses to a state in which there is no longer any response to vasopressor drugs and in which, even if the blood volume is returned to normal, cardiac output remains depressed. This is known as **refractory shock.** The condition is not unique to hemorrhagic shock, but occurs in other forms as well. It used to be called **irreversible shock,** and patients still do die despite vigorous treatment. However, more and more patients are saved as understanding of the pathophysiologic mechanisms increases and treatment is improved. Therefore, refractory shock seems to be a more appropriate term.

Spasm of the precapillary sphincters and venules, especially in the splanchnic region, is a prominent feature at this stage. The reduced capillary perfusion due to the constriction of the precapillary sphincters leads to hypoxic tissue damage. After 3–5 hours, the precapillary sphincters dilate while the venules remain constricted. Blood now enters the capillaries but stagnates in these vessels, so that the tissue hypoxia continues. Capillary hydrostatic pressure rises, and fluid leaves the vascular system in increasing amounts. Granulocytes adhere to the injured vessel walls and release free radicals, especially O_2^-, and this causes further tissue damage. An antibody to the protein complex that binds neutrophils to tissue has been reported to improve survival in severe shock. There is evidence that damage to the gastrointestinal mucosa permits entry of bacteria into the circulation.

Various positive feedback mechanisms contribute to the production of refractory shock. For example, severe cerebral ischemia leads eventually to depression of the vasomotor and cardiac areas of the brain,

with slowing of the heart rate and vasodilation. These both make the blood pressure drop further, with a further reduction in cerebral blood flow and further depression of the vasomotor and cardiac areas.

Another important example of this type of positive feedback is myocardial depression. In severe shock, the coronary blood flow is reduced because of the hypotension and tachycardia (see Chapter 32), even though the coronary vessels are dilated. The myocardial failure makes the shock and the acidosis worse, and this in turn leads to further depression of myocardial function. If the reduction is marked and prolonged, the myocardium may be damaged to the point where cardiac output cannot be restored to normal in spite of reexpansion of the blood volume.

A late complication of shock that can be fatal is pulmonary damage with the production of **adult respiratory distress syndrome** (shock lung; see Chapter 37). The primary cause is pulmonary microembolism. This is due to intravascular coagulation triggered by coagulants released by damaged cells and clumps of aggregated platelets and red cells washed out of hypoxic tissues.

Other Forms of Hypovolemic Shock

Traumatic shock develops when there is severe damage to muscle and bone. This is the type of shock seen in battle casualties and automobile accident victims. Frank bleeding into the injured areas is the principal cause of the shock, although some plasma also enters the tissues. The amount of blood which can be lost into an injury that appears relatively minor is remarkable; the thigh muscles can accommodate 1 L of extravasated blood, for example, with an increase in the diameter of the thigh of only 1 cm.

Breakdown of skeletal muscle (**rhabdomyolysis**) is a serious additional problem when shock is accompanied by extensive muscle crushing (**crush syndrome**). When pressure on the tissues is relieved and they are once again perfused with blood, free radicals are generated, and these cause further tissue destruction (**reperfusion-induced injury**). Many of the free radicals are generated by conversion of tissue xanthine dehydrogenase to xanthine oxidase, an enzyme that generates O_2 and H_2O_2. Adherent white blood cells in the tissues also generate free radicals. The xanthine oxidase inhibitor allopurinol and antibodies that prevent neutrophil adhesion reduce the severity of reperfusion injury. An additional cause of tissue damage in reperfused areas is marked intracellular Ca^{2+} excess, which is presumably due to exchange of excess intracellular Na^+ in damaged tissues for extracellular Ca^{2+}. Kidney damage is also common in the crush syndrome. It is due to accumulation of myoglobin and other products from reperfused tissue in kidneys in which glomerular filtration is already reduced by shock. The products damage and clog the tubules, frequently causing anuria, which may be fatal.

Surgical shock is due to the combination in various proportions of external hemorrhage, bleeding into injured tissues, and dehydration.

In **burn shock**, the most apparent abnormality is loss of plasma as exudate from the burned surfaces. Since the loss in this situation is plasma rather than whole blood, the hematocrit rises and **hemoconcentration** is a prominent finding. Burns also cause complex, poorly understood metabolic changes in addition to fluid loss. For example, there is a 50% rise in metabolic rate of nonthyroidal origin, and some burned patients develop hemolytic anemia. Because of these complications, plus the severity of the shock and the problems of sepsis and kidney damage, the mortality rate when third-degree burns cover more than 75% of the body is still close to 100%.

Hypovolemic shock is a complication of various metabolic and infectious diseases. For example, although the mechanism is different in each case, adrenal insufficiency, diabetic ketoacidosis, and severe diarrhea are all characterized by loss of Na^+ from the circulation. The resultant decline in plasma volume may be severe enough to precipitate cardiovascular collapse.

Distributive Shock

As noted above, distributive shock occurs when the blood volume is normal but the capacity of the circulation is increased by marked vasodilation. It is also called "warm shock" because the skin is not cold and clammy, as it is in hypovolemic shock. A good example is **anaphylactic shock,** a rapidly developing, severe allergic reaction that sometimes occurs when an individual who has previously been sensitized to an antigen is reexposed to it. The resultant antigen-antibody reaction releases large quantities of histamine, causing increased capillary permeability and widespread dilation of arterioles and capillaries.

Another common form of distributive shock is **septic shock.** In this condition, bacterial toxins cause vasodilation. In addition, sepsis depresses the myocardium and increases capillary permeability, so that plasma leaks into the tissues and blood volume falls. Consequently, septic shock is cardiogenic and hypovolemic as well as distributive.

In febrile patients, shock is likely to be more severe because the cutaneous blood vessels are often dilated (see Chapter 32), increasing the disparity between the capacity of the vascular system and the available circulating blood volume.

The gram-negative bacteria that often cause septic shock release **endotoxin,** the cell wall lipopolysaccharide of the organisms. It causes macrophages to produce increased amounts of **cachectin (tumor necrosis factor),** a polypeptide that inhibits lipoprotein lipase and has other effects which make shock worse. Mice that congenitally produce only small amounts of cachectin are resistant to endotoxin, and the glucocorticoid dexamethasone, which helps protect animals from endotoxic shock, inhibits the translation of cachectin mRNA in macrophages. Antibodies against endotoxin-producing bacteria reduce the mortality rate in septic shock. Endotoxin causes release of IL-1 and other cytokines, and an antibody against the IL-1 receptor has also been reported to be beneficial in treating septic shock in experimental animals.

A third type of distributive shock is **neurogenic shock,** in which sudden autonomic activity results in vasodilation and pooling of blood in the veins. Examples include fainting in response to strong emotion such as overwhelming fear or grief.

Fainting

Fainting, or **syncope,** is sudden, transient loss of consciousness. It can be due to metabolic or neurologic abnormalities, but more commonly it is due to peripheral vascular or cardiac abnormalities that cause inadequate cerebral blood flow. It is often benign and is most commonly due to abrupt vasodilation. This produces hypotension, generally in association with bradycardia. The term **vasovagal syncope** has been coined to denote this entity. **Postural syncope** is fainting due to pooling of blood in the dependent parts of the body on standing. **Micturition syncope,** fainting during urination, occurs in patients with orthostatic hypotension. It is due to the combination of the orthostasis and reflex bradycardia induced by voiding in these patients. Pressure on the carotid sinus, produced, for example, by a tight collar, can cause such marked bradycardia and vasodilation that fainting results (**carotid sinus syncope**). Rarely, vasodilation and bradycardia may be precipitated by swallowing (**deglutition syncope**). **Cough syncope** occurs when the increase in intrathoracic pressure during straining or coughing is sufficient to block venous return. **Effort syncope** is fainting on exertion as a result of inability to increase cardiac output to meet the increased demands of the tissues and is particularly common in patients with aortic or pulmonary stenosis.

Syncope can also be due to more serious abnormalities. About 25% of the attacks are of cardiac origin and are due to either transient obstruction of blood flow through the heart or sudden decreases in cardiac output owing to various cardiac arrhythmias. Fainting is the presenting symptom in 7% of patients with myocardial infarctions. Thus, all cases of syncope should be investigated to determine the cause.

Cardiogenic & Obstructive Shock

When there is a gross decline in cardiac output as a result of disease of the heart rather than inadequate blood volume, the condition that results is called cardiogenic shock. The symptoms are those of shock plus congestion of the lungs and viscera due to failure of the heart to put out all the venous blood returned to it. Consequently, the condition is sometimes called "congested shock." The incidence of this shock in patients with myocardial infarction is about 10%, and it usually occurs when there is extensive damage to the left ventricular myocardium. It has been suggested that release of chemical agents such as serotonin from the infarcted area activates ventricular receptors which trigger a reflex inhibition of vasomotor tone (Bezold-Jarisch reflex; see Chapter 31). This limits the compensatory vasoconstrictor responses and makes the shock worse.

The picture of congested shock is also seen in obstructive shock. When the obstruction is due to tension pneumothorax with kinking of the great veins (see Chapter 37) or bleeding into the pericardium with external pressure on the heart (**cardiac tamponade),** prompt surgical intervention is required to prevent death.

Treatment of Shock

The treatment of shock should be aimed at correcting the cause and helping the physiologic compensatory mechanisms to restore an adequate level of tissue perfusion. In hemorrhagic, traumatic, and surgical shock, for example, the primary cause of the shock is blood loss, and the treatment should include early and rapid transfusion of adequate amounts of compatible whole blood. Saline is of limited temporary value. The immediate goal is restoration of an adequate circulating blood volume, and since saline is distributed in ECF, only 25% of the amount administered stays in the vascular system. In burn shock and other conditions in which there is hemoconcentration, plasma is the treatment of choice to restore the fundamental defect, the loss of plasma. "Plasma expanders," solutions of sugars of high molecular weight and related substances that do not cross capillary walls, have some merit. Concentrated human serum albumin and other hypertonic solutions expand the blood volume by drawing fluid out of the interstitial spaces. They are valuable in emergency treatment but have the disadvantage of further dehydrating the tissues of an already dehydrated patient.

In anaphylactic shock, epinephrine has a highly beneficial and almost specific effect that must represent more than just constriction of the dilated vessels. In all types of shock, restoration of an adequate arterial pressure is important to maintain coronary blood flow. Vasopressor agents such as norepinephrine are of value for this purpose, but their use should be discontinued as soon as possible. Because it produces renal vasodilation and has a positively inotropic effect while producing vasoconstriction elsewhere (see Chapter 20), dopamine is of value in the treatment of traumatic and cardiogenic shock.

A number of measures sometimes used in the treatment of shock inhibit the operation of the physiologic compensatory mechanisms. Sedatives and other drugs that depress the central nervous system should be used as sparingly as possible, because they depress the discharge of the vasomotor areas. Alcohol is particularly unphysiologic because it depresses the central nervous system and, through a central action, dilates cutaneous vessels. Care should be taken to avoid overheating and resultant cutaneous vasodilation. Further compromising the circulation by permitting the patient to sit or stand is, of course, detrimental, and gravity should be put to work to help rather than hinder the compensatory mechanisms. Raising the foot of the bed 6–12 inches (15–30 cm) with "shock blocks" is a simple but valuable therapeutic maneuver that aids venous return from the lower half of the body and improves cerebral blood flow. However, the head-down position also causes the abdominal viscera to press on the diaphragm, making adequate ventilation more difficult to maintain and favoring the development of pulmonary complications. Consequently, it should not be used for prolonged periods.

Other, more specific treatments such as the use of antibodies that reduce reperfusion-induced injury are discussed in the preceding paragraphs of this section.

HYPERTENSION

Hypertension is a sustained elevation of the systemic arterial pressure. **Pulmonary hypertension** can also occur, but the pressure in the pulmonary artery (see Chapter 34) is relatively independent of that in the systemic arteries.

Experimental Hypertension

The arterial pressure is determined by the cardiac output and the peripheral resistance (pressure = flow × resistance; see Chapter 30). The peripheral resistance is determined by the viscosity of the blood and, more importantly, by the caliber of the resistance vessels. Hypertension can be produced by elevating the cardiac output, but sustained hypertension is usually due to increased peripheral resistance. Some of the procedures that have been reported to produce sustained hypertension in experimental animals are listed in Table 33–5. For the most part, the procedures involve manipulation of the kidneys, the nervous system, or the adrenals. There are in addition a number of strains of rats that develop hypertension either spontaneously (SHR rats) or when fed a high-sodium diet (Dahl strain rats).

The hypertension that follows constriction of the renal arterial blood supply or compression of the kidney is called **renal hypertension.** As noted in Chapter 24, some animals with renal hypertension have elevated plasma renin activity, whereas others do not. In general, one-clip 2-kidney Goldblatt hypertension (see Table 33–5) is renin-dependent, whereas one-clip one-kidney Goldblatt hypertension is not. An additional factor that probably contributes to renal hypertension is decreased ability of the constricted kidney to excrete Na^+.

Neurogenic hypertension is discussed in Chapter 31. Provided that salt intake is normal or high, deoxycorticosterone causes hypertension which may persist after treatment is stopped. The hypertension is more severe in unilaterally nephrectomized animals.

Hypertension in Humans

Hypertension is a very common abnormality in humans. It can be produced by many diseases (Table 33–6). It is also a prominent symptom of **toxemia of pregnancy,** a condition that may be caused by a pressor polypeptide secreted by the placenta.

Hypertension causes a number of serious disorders. When the resistance against which the left ventricle must pump (afterload) is elevated for a long period, the cardiac muscle hypertrophies. Left

Table 33–5. Procedures that produce sustained hypertension in experimental animals.

1. Interference with renal blood flow (renal hypertension)
   a. Constriction of one renal artery; other kidney removed (one-clip, one-kidney Goldblatt hypertension)
   b. Constriction of one renal artery; other kidney intact (one-clip, 2-kidney Goldblatt hypertension)
   c. Constriction of aorta or both renal arteries (2-clip, 2-kidney Goldblatt hypertension)
   d. Compression of kidney by rubber capsules, production of perinephritis, etc.
2. Interruptions of afferent input from arterial baroreceptors (neurogenic hypertension)
   a. Denervation of carotid sinuses and aortic arch
   b. Bilateral lesions of nucleus of tractus solitarius
3. Treatment with corticosteroids
   a. Deoxycorticosterone and salt
   b. Other mineralocorticoids
4. Partial adrenalectomy (adrenal regeneration hypertension)
5. Genetic
   a. Spontaneous hypertension in various strains of rats
   b. Salt-induced hypertension in genetically sensitive rats

Table 33–6. Principal causes of sustained diastolic hypertension in humans.[1]

1. Unknown (essential hypertension)
2. Adrenocortical diseases
   a. Hypersecretion of aldosterone (Conn's syndrome)*
   b. Hypersecretion of other mineralocorticoids (hypertensive form of congenital virilizing adrenal hyperplasia; 17 α-hydroxylase deficiency)
   c. Hypersecretion of glucocorticoids (Cushing's syndrome)*
3. Catecholamine-secreting tumors of adrenal medullary or paraganglionic origin (pheochromocytoma)*
4. Tumors of juxtaglomerular cells*
5. Narrowing of one or both renal arteries (renal hypertension)*
6. Renal disease
   a. Glomerulonephritis
   b. Pyelonephritis
   c. Polycystic disease
7. Narrowing (coarctation) of the aorta*
8. Severe polycythemia
9. Oral contraceptives*

[1]The conditions marked by asterisks are quite often curable. In all of them except Cushing's syndrome, sustained hypertension may be the only physical finding.

ventricular hypertrophy is associated with a poor prognosis. The total O_2 consumption of the heart, already increased by the work of expelling blood against a raised pressure (see Chapter 29), is increased further because there is more muscle. Therefore, any decrease in coronary blood flow has more serious consequences in hypertensive patients than it does in normal individuals, and degrees of coronary narrowing that do not produce symptoms when the size of the heart is normal may produce myocardial infarction when the heart is enlarged. There is an increased incidence of atherosclerosis in hypertension, and myocardial infarcts are common even when the heart is not enlarged. Starling's law (see Chapter 29) operates in hypertension, dilation of the heart stretching the muscle fibers and increasing their strength of contraction. However, the ability to compensate for the high peripheral resistance is eventually exceeded, and the heart fails. Hypertensive individuals are also predisposed to thromboses of cerebral vessels and cerebral hemorrhage. An additional complication is renal failure. However, the incidence of heart failure, strokes, and renal failure can be markedly reduced by active treatment of hypertension, even when the hypertension is relatively mild.

Malignant Hypertension

Chronic hypertension can enter an accelerated phase in which necrotic arteriolar lesions develop and there is a rapid downhill course with papilledema, cerebral symptoms, and progressive renal failure. This syndrome is known as **malignant hypertension,** and without treatment it is fatal in less than 2 years. It can be triggered by hypertension of any cause. However, its progression can be stopped, and it can be reversed by appropriate antihypertensive therapy.

Essential Hypertension

In 90% of patients with elevated blood pressure, the cause of the hypertension is unknown, and they are said to have **essential hypertension.** The remainder have blood pressure elevations due to a variety of different diseases (Table 33–6). Early in the course of essential hypertension, the blood pressure elevations are intermittent and there is an exaggerated pressor response to stimuli such as cold and excitement that produce only a moderate blood pressure elevation in normal individuals. This suggests that overactive autonomic reactions are responsible for the arteriolar spasm, and treatment with drugs that block sympathetic outflow markedly slows the progress of the disease. Later the blood pressure elevation becomes sustained. The baroreceptor mechanism is "reset" so that the blood pressure is maintained at the elevated level (see Chapter 31). Spasm of the arterioles leads to hypertrophy of their musculature, and there is some organic narrowing of the vessels. At this stage, even a normal rate of autonomic discharge is associated with an elevated pressure.

The progression of untreated essential hypertension is variable. Particularly in women, it is often benign, with elevated pressure being the only finding for many years. However, it may also progress rapidly into the malignant phase.

At present, essential hypertension is a treatable but not a curable disease. On the other hand, several of the other disease processes that cause hypertension in humans are not only treatable but also curable. It is therefore important to recognize these latter diseases and institute appropriate definitive treatment.

Evidence is accumulating that some humans, like Dahl salt-sensitive rats, have a relatively marked increase in blood pressure when fed a high-sodium diet, whereas others, like Dahl salt-resistant rats, have small increases in blood pressure. Since there is no easy test to distinguish salt-responsive from salt-resistant humans, the current emphasis on decreasing sodium intake in all humans makes some sense. However, the degree to which hypertension can be ameliorated or prevented will obviously vary with the salt sensitivity of the individual.

Other Causes of Hypertension

In humans, deoxycorticosterone and aldosterone both elevate the blood pressure, and hypertension is a prominent feature of primary hyperaldosteronism. It is also seen in patients who secrete excess deoxycorticosterone (see Chapter 20). The hypokalemia that these hormones produce damages the kidneys **(hypokalemic nephropathy),** and the hypertension may be partly renal in origin. Expansion of the blood volume because of Na^+ retention also plays a role.

Plasma renin activity is low in the forms of hypertension that are due to excess secretion of aldosterone and deoxycorticosterone. It has also been found to be low, for unknown reasons, in the presence of normal or low aldosterone and deoxycorticosterone secretion rates in 10–15% of patients with what otherwise appears to be essential hypertension **(low-renin hypertension).**

Hypertension is also seen in Cushing's syndrome, in which aldosterone secretion is usually normal. The cause of the hypertension in this syndrome is uncertain; it may be due to the increased secretion of deoxycorticosterone produced by increased circulating ACTH, increased secretion of angiotensinogen due to the increase in circulating glucocorticoids, a direct action of glucocorticoids on the arterioles, or a combination of these factors.

Pheochromocytomas—tumors of the adrenal medulla or of catecholamine-secreting tissue elsewhere in the body—also produce hypertension. The hypertension is often episodic, and particularly when the tumors secrete predominantly epinephrine, the blood glucose level and the metabolic rate are intermittently elevated. However, in some patients with pheochromocytomas, sustained hypertension is the only finding.

Renal hypertension due to narrowing of the renal arteries is discussed above and in Chapter 24. Various other types of renal disease are associated with hypertension, but it is not known whether the association is due to renin secretion or to nonhumoral mechanisms. **Coarctation of the aorta,** a congenital narrowing of a segment of the thoracic aorta, increases the resistance to flow, producing severe hypertension in the upper part of the body. The blood pressure in the lower part of the body is usually normal but may be elevated as a result of increased renin secretion.

Chronic treatment with oral contraceptives containing progestins and estrogens produces significant hypertension in some women. The hypertension is due, at least in part, to an increase in circulating levels of angiotensinogen, the production of which is stimulated by estrogens (see Chapter 24). There is some evidence that women developing "pill hypertension" are those predisposed to hypertension in any case. Certainly the occurrence of hypertension is not a reason for normotensive women to avoid oral contraceptives, but it would seem wise for them to have their blood pressures checked at 6-month intervals.

HEART FAILURE

Manifestations

The manifestations of heart failure range from sudden death (eg, in ventricular fibrillation or air embolism), through the shocklike state called cardiogenic shock, to chronic **congestive heart failure,** depending upon the degree of circulatory inadequacy and the rapidity with which it develops. The principal symptoms and signs of congestive failure are cardiac enlargement, weakness, edema (particularly of the dependent portions of the body), a prolonged circulation time, hepatic enlargement (**hepatomegaly**), a sensation of shortness of breath and suffocation (**dyspnea**), and distention of the neck veins. Dyspnea on exertion is a prominent symptom. In advanced cases, a common finding is dyspnea that is precipitated by lying flat and relieved by sitting up (**orthopnea**). The dyspnea may be paroxysmal and sometimes progresses to frank **pulmonary edema.** Some of these abnormalities are due to "backward failure"—failure of the left and right heart pumps to put out all the blood returned to them in the veins, with a resultant rise in venous pressure and congestion of the lungs and viscera. Others are due to "forward failure"— failure of the heart to maintain a cardiac output ade-

quate for normal perfusion of all the tissues (Table 33–7). The decline in cardiac output may be relative rather than absolute. In thyrotoxicosis and thiamin deficiency, for example, output may be elevated in absolute terms, but when it is inadequate relative to the needs of the tissues, heart failure is present ("high-output failure").

Pathogenesis

The initial pathologic event in heart failure is reduced myocardial contractility. At first, this is manifest only during exercise (Fig 33–7), and exertion is limited by dyspnea and fatigue. Later, resting cardiac output is also reduced. The ejection fraction falls from its normal value of about 0.6 to as low as 0.2.

The cause of the decreased contractility is unsettled and may vary from one disease to another. The most common diseases producing congestive heart failure are hypertension and coronary artery disease. The failing myocardium has decreased myosin ATPase activity. In rats and also in humans, there are changes in the isoforms of myosin in the failing heart, and abnormal actin and tropomyosin are also synthesized. Relaxation is slowed, possibly because of slowed uptake of Ca^{2+} by the sarcoplasmic reticulum. There is in addition a reduced density of β-adrenergic receptors in the myocardium, but this is probably due to down-regulation of the receptors produced by elevated circulating catecholamine levels (see below).

The decrease in cardiac output triggers increased sympathetic discharge via the baroreceptors (Fig 33–8). Venous pressure also starts to rise, and the viscera become congested. Circulating catecholamine

Table 33–7. Simplified summary of pathogenesis of major findings in congestive heart failure.

Abnormality	Cause
Ankle, sacral edema	"Backward failure" of right ventricle → increased venous pressure → fluid transudation.
Hepatomegaly	Increased venous pressure → increased resistance to portal flow.
Pulmonary congestion	"Backward failure" of left ventricle → increased pulmonary venous pressure → pulmonary venous distention and transudation of fluid into air spaces
Dyspnea on exertion	Failure of left ventricular output to rise during exercise → increased pulmonary venous pressure.
Paroxysmal dyspnea, pulmonary edema	Probably sudden failure of left heart output to keep up with right heart output → acute rise in pulmonary venous and capillary pressure → transudation of fluid into air spaces
Orthopnea	Normal pooling of blood in lungs in supine position added to already congested polmonary vascular system; increased venous return not put out by left ventricle. (Relieved by sitting up, raising head of bed, lying on extra pillows.)
Weakness, exercise intolerance	"Forward failure" of left ventricle; cardiac output inadequate to perfuse muscles; especially, failure of output to rise with exercise.
Cardiac dilation	Greater ventricular end-diastolic volume.

Figure 33–7. Decreased cardiac output in congestive heart failure. R, rest; E, maximal exercise. Note that with moderate failure, resting cardiac output is normal and only the response to exercise is reduced. As failure progresses, resting cardiac output is also reduced. (Modified and reproduced, with permission, from Zelis R et al: Vasoconstrictor mechanisms in congestive heart failure, Part I. *Mod Concepts Cardiovasc Dis* 1989;**58:**7. By permission of the American Heart Association, Inc.)

Figure 33–8. Sequence of events leading to congestive heart failure.

levels are increased, and there is widespread vasoconstriction. This in turn increases the afterload, causing a further decrease in cardiac output.

Constriction of the vascular bed in the kidneys causes retention of Na^+ and water. In addition, the decrease in effective arterial blood volume and the increase in sympathetic discharge increase renin secretion in many patients, and the resultant increase in aldosterone secretion adds to the salt and water retention. Vasopressin secretion is also increased. Escape fails to occur (see Chapter 20), and edema is produced. It is interesting in this regard that the plasma ANP level is generally increased rather than decreased in the patients who have congestive heart failure, but the failure to escape may be due to end-organ unresponsiveness to the atrial peptide.

Treatment of congestive heart failure is aimed at improving cardiac contractility, treating the symptoms, and decreasing the load on the heart. Classically, myocardial contractility has been improved by administration of digitalis and related positively inotropic cardiac glycosides (see Chapter 3). Diuretics are used to overcome sodium retention. The afterload is reduced by use of arterial vasodilators, and venodilators diminish preload. Reduction of the plasma angiotensin II level by administration of ACE inhibitors (see Chapter 24) reduces afterload, and use of ACE inhibitors has been so successful that they have now become one of the cornerstones of the treatment of heart failure. The decline in the angiotensin II level also decreases the plasma aldosterone level.

REFERENCES

Section VI: Circulation

Beutler E: Glucose-6-phosphate dehydrogenase deficiency. *N Engl J Med* 1991;**324:**169.

Bhoola KD, Figuerca CD, Worthy K: Bioregulation of kinins: Kallikreins, kininogens, and kininase. *Pharmacol Rev* 1992;**44:**4.

Birrer RB (editor): *Sports Medicine for the Primary Care Physician.* Appleton-Century-Crofts, 1984.

Clouse LH, Comp PC: The regulation of hemostasis: The protein C system. *N Engl J Med* 1986;**314:**129.

Cooper ASL, Plum F: Biochemistry and physiology of brain ammonia. *Physiol Rev* 1987;**67:**440.

Donald DH, Shepherd JT: Autonomic regulation of the peripheral circulation. *Annu Rev Physiol* 1980; **42:**429.

Elgsaeter A et al: The molecular basis of erythrocyte shape. *Science* 1986;**234:**1217.

Fauci AS: The human immunodeficiency virus: infectiv-

ity and mechanisms of pathogenesis. *Science* 1988; **299:**617.

Fenstermacher JD, Rapoport SF: Blood-brain barrier. In: *Handbook of Physiology.* Section 2, Vol IV, p. 969. American Physiological Society, 1984.

Fishman RA: *Cerebrospinal Fluids in Diseases of the Nervous System,* 2nd ed. Saunders, 1992.

Flye MW (editor): *Principles of Organ Transplantation.* Saunders, 1989.

Freda VJ, Pollack W, Gorman JG: Rh disease: How near the end? *Hosp Pract* (June) 1978;**13:**61.

French DL, Laskov R, Scharff MD: The role of somatic hypermutation in the generation of antibody diversity. *Science* 1989;**244:**1152.

Furie B, Furie BC: Molecular and cellular biology of blood coagulation. *N Engl J Med* 1992;**326:**800.

Furster V et al: The pathogenesis of coronary artery disease and the acute coronary syndrome. *N Engl J Med* 1992;**326:**242.

Galbo H: *Hormonal and Metabolic Adaptation to Exercise.* Thieme-Stratton, 1983.

Gazes PC: Angina pectoris: Classification and diagnosis. *Mod Concepts Cardiovasc Dis* 1989;**57**:19.

George JN, Shattil SJ: The clinical importance of acquired abnormalities of platelet function. *N Engl J Med* 1991;**24**:27.

Golub ES, Green DR: *Immunology: A Synthesis,* 2nd ed. Sinauer, 1991.

Guyton AC: Blood pressure control—special role of the kidneys and body fluids. *Science* 1991;**252**:1813.

Hallett MB (editor): *The Neutrophil: Cellular Biochemistry and Physiology.* CRC Press, 1989.

Halliwell B (editor): *Oxygen Radicals and Tissue Injury.* Upjohn, 1988.

Harpel PC: Protease inhibitors: A precarious balance. *N Engl J Med* 1983;**309**:725.

Harrison MH: Effects of thermal stress and exercise on blood volume in humans. *Physiol Rev* 1985;**65**:149.

Hoffman JI, Spaan JA: Pressure-flow relations in coronary circulation. *Physiol Rev* 1990;**70**:331.

Honig GR, Adams JG III: *Human Hemoglobin Genetics.* Springer-Verlag, 1986.

Honjo T, Alt FW, Rabbits TH (editors): *Immunoglobulin Genes.* Academic Press, 1989.

Jelliffe RW: *Fundamentals of Electrocardiography.* Springer-Verlag, 1990.

Johnston RB Jr: Monocytes and macrophages. *N Engl J Med* 1988;**318**:747.

Kaplan NM: *Clinical Hypertension,* 5th ed. Williams & Wilkins, 1990.

Katz AM: *Physiology of the Heart,* 2nd ed. Raven, 1992.

Kirkendall WM et al: Recommendations for human blood pressure determination by sphygmomanometers. *Hypertension* 1981;**3**:510A.

Klagsbrun M, D'Amore PA: Regulation of angiogenesis. *Annu Rev Physiol* 1991;**53**:217.

Lyons KP: *Cardiovascular Nuclear Medicine.* Appleton & Lange, 1988.

Metcalf D: Control of granulocytes and macrophages: Molecular and clinical aspects. *Science* 1991; **254**:529.

Mizel SB: The interleukins. *FASEB J* 1989;**2**:2379.

Morady F, Scheinman MM: Paroxysmal supraventricular tachycardia. *Mod Concepts Cardiovasc Dis* 1982; **51**:107.

Morgan JP: Abnormal intracellular modulation of calcium as a major cause of cardiac contractile dysfunction. *N Engl J Med* 1991;**325**:625.

Nicogossian AE, Leach-Huntoon (editors): *Space Physiology and Medicine,* 2nd ed. Lea & Febiger, 1989.

Odeh M: The role of reperfusion-induced injury in the pathogenesis of the crush syndrome. *N Engl J Med* 1991;**324**:1417.

Payan DG, McGillis JP, Goetzl EJ: Neuroimmunology. *Adv Immunol* 1986;**39**:299.

Phelps ME, Mazziota JC: Positron emission tomography: Human brain function and biochemistry. *Science* 1985;**228**:799.

Popp RL: Echocardiography. *N Engl J Med* 1990; **323**:101.

Putnam FW (editor): *The Plasma Proteins: Structure, Function and Genetic Control,* 2nd ed. Academic Press, 1975.

Reis DJ, Ruggiero DA, Morrison SF: The C1 of the rostral ventrolateral medulla oblongata: A critical brainstem region for control of resting and reflex integration of arterial pressure. *Am J Hypertension* 1989; **2**:363S.

Roland PE: Changes in brain blood flow and oxidative metabolism during mental activity. *News Physiol Sci* 1987;**2**:120.

Rosen FS, Cooper MD, Wedgewood RJP: The primary immunodeficiencies. *N Engl J Med* 1984;**311**:235.

Royer HD, Reinherz FC: T lymphocytes: Ontogeny, function, and relevance to clinical disorders. *N Engl J Med* 1987;**317**:1136.

Rubany GM (editor): *Endothelin.* Oxford Univ Press, 1992.

Schmid-Schönbein GW: Microlymphatics and lymph flow. *Physiol Rev* 1990;**70**:987.

Schrieber SL: Chemistry and biology of the immunophilins and their immunosuppressive ligands. *Science* 1991;**251**:283.

Sell S: *Basic Immunology: Immune Mechanisms in Health and Disease.* Elsevier, 1987.

Snyder AH, Bredt DS: Biological roles of nitric oxide. *Sci Am* (May) 1992;**266**:68.

Sporn MB, Roberts AB: Peptide growth factors are multifunctional. *Nature* 1988;**332**:217.

Stamatoyannopoulos G et al (editors): *The Molecular Basis of Blood Diseases.* Saunders, 1987.

Staub NC, Taylor AE (editors): *Edema.* Raven, 1984.

Stull JT et al: Vascular smooth muscle contractile elements: Cellular regulation. *Hypertension* 1991;**47**: 723.

Vane JR, Äungärd EE, Botting EM: Regulatory functions of the vascular endothelium. *N Engl J Med* 1990;**327**:27.

Vane J, Botting R: Inflammation and the mechanism of action of anti-inflammatory drugs. *FASEB J* 1987; **1**:89.

von Boehmer H, Kisielow P: How the immune system learns about self. *Sci Am* (Oct) 1991;**265**:74.

Waldman TA: Monoclonal antibodies in diagnosis and therapy. *Science* 1991;**252**:1657.

Weiss SJ: Tissue destruction by neutrophils. *N Engl J Med* 1989;**320**:365.

Weller PF: The immunobiology of eosinophils. *N Engl J Med* 1991;**324**:1110.

Section VII.
Respiration

Pulmonary Function

34

INTRODUCTION

Respiration, as the term is generally used, includes 2 processes: **external respiration,** the absorption of O_2 and removal of CO_2 from the body as a whole; and **internal respiration,** the utilization of O_2 and production of CO_2 by cells and the gaseous exchanges between the cells and their fluid medium. Details of the utilization of O_2 and the production of CO_2 by cells are considered in Chapter 17. This chapter is concerned with the functions of the respiratory system in external respiration, ie, the processes responsible for the uptake of O_2 and excretion of CO_2 in the lungs. Chapter 35 is concerned with the transport of O_2 and CO_2 to and from the tissues.

The respiratory system is made up of a gas-exchanging organ (the lungs) and a pump that ventilates the lungs. The pump consists of the chest wall; the respiratory muscles, which increase and decrease the size of the thoracic cavity; the areas in the brain that control the muscles; and the tracts and nerves that connect the brain to the muscles. At rest, a normal human breathes 12–15 times a minute. Five hundred milliliters of air per breath, or 6–8 L/min, is inspired and expired. This air mixes with the gas in the alveoli, and, by simple diffusion, O_2 enters the blood in the pulmonary capillaries while CO_2 enters the alveoli. In this manner, 250 mL of O_2 enters the body per minute and 200 mL of CO_2 is excreted.

Traces of other gases such as methane from the intestines are also found in expired air. Alcohol and acetone are expired when present in appreciable quantities in the body. Indeed, over 250 different volatile substances have been identified in human breath.

PROPERTIES OF GASES

Partial Pressures

Unlike liquids, gases expand to fill the volume available to them, and the volume occupied by a given number of gas molecules at a given temperature and pressure is (ideally) the same regardless of the composition of the gas.

$$P = \frac{nRT}{V} \quad \text{(from equation of state of ideal gas)}$$

where

P = Pressure
n = Number of moles
R = Gas constant
T = Absolute temperature
V = Volume

Therefore, the pressure exerted by any one gas in a mixture of gases (its **partial pressure**) is equal to the total pressure times the fraction of the total amount of gas it represents.

The composition of dry air is 20.98% O_2, 0.04% CO_2, 78.06% N_2, and 0.92% other inert constituents such as argon and helium. The barometric pressure (P_B) at sea level is 760 mm Hg (one atmosphere). The partial pressure (indicated by the symbol P) of O_2 in dry air is therefore 0.21 × 760, or 160 mm Hg at sea level. The partial pressure of N_2 and the other inert gases is 0.79 × 760, or 600 mm Hg; and the P_{CO_2} is 0.0004 × 760, or 0.3 mm Hg. The water vapor in the air in most climates reduces these percentages, and therefore the partial pressures, to a slight degree. Air equilibrated with water is saturated with water vapor, and inspired air is saturated by the time it reaches the lungs. The P_{H_2O} at body temperature (37 °C) is 47 mm Hg. Therefore, the partial pressures at sea level of the other gases in the air reaching the lungs are P_{O_2}, 149 mm Hg; P_{CO_2}, 0.3 mm Hg; and P_{N_2} (including the other inert gases), 564 mm Hg.

Gas diffuses from areas of high pressure to areas of low pressure, the rate of diffusion depending upon the concentration gradient and the nature of the barrier between the 2 areas. When a mixture of gases is in contact with and permitted to equilibrate with a liquid, each gas in the mixture dissolves in the liquid to an extent determined by its partial pressure and its solubility in the fluid. The partial pressure of a gas in a liquid is that pressure which in the gaseous phase in equilibrium with the liquid would produce the concentration of gas molecules found in the liquid.

Methods of Quantitating Respiratory Phenomena

Respiratory excursions can be recorded by devices that measure chest expansion, or by recording spirometers (see Fig 17–1), which also permit measurement of gas intake and output. Since gas volumes vary with temperature and pressure and since the amount of water vapor in them varies, it is important to correct respiratory measurements involving volume to a stated set of standard conditions. The 4 most commonly used standards and their abbreviations are shown in Table 34–1. Modern techniques for gas analysis make possible rapid, reliable measurements of the composition of gas mixtures and the gas content of body fluids. For example, O_2 and CO_2 electrodes, small probes sensitive to O_2 or CO_2, can be inserted into the airway or into blood vessels or tissues and the P_{O_2} and P_{CO_2} recorded continuously. CO_2, carbon monoxide (CO), and many anesthetic gases can be measured rapidly by infrared absorption spectroscopy. Gases can also be measured by gas chromatography or mass spectrometry.

ANATOMY OF THE LUNGS

Air Passages

After passing through the nasal passages and pharynx, where it is warmed and takes up water vapor, the inspired air passes down the trachea and through the bronchioles, respiratory bronchioles, and alveolar ducts to the alveoli (Fig 34–1).

Between the trachea and the alveolar sacs, the airways divide 23 times. The first 16 generations of passages form the conducting zone of the airways that transports gas from and to the exterior. They are made up of bronchi, bronchioles, and terminal bronchioles. The remaining 7 generations form the transitional and respiratory zones where gas exchange occurs and are made up of respiratory bronchioles, alveolar ducts, and alveoli. These multiple divisions greatly increase the total cross-sectional area of the airways, from 2.5 cm^2 in the trachea to 11,800 cm^2 in the alveoli. Consequently, the velocity of air flow in the small airways declines to very low values.

The alveoli are surrounded by pulmonary capillaries, and in most areas the structures between the air and the capillary blood across which O_2 and CO_2 diffuse are exceedingly thin (Fig 34–2). There are 300

million alveoli in humans, and the total area of the alveolar walls in contact with capillaries in both lungs is about 70 m^2.

The alveoli are lined by 2 types of epithelial cells. **Type I cells** are flat cells with large cytoplasmic extensions and are the primary lining cells. **Type II cells (granular pneumocytes)** are thicker and contain numerous lamellar inclusion bodies. These cells secrete surfactant (see below). There may be other special types of epithelial cells, and the lung also contains pulmonary alveolar macrophages (PAMs), lymphocytes, plasma cells, APUD cells (see Chapter 26), and mast cells. The mast cells contain heparin, various lipids, histamine, and polypeptides that participate in allergic reactions.

The Bronchi and Their Innervation

The trachea and bronchi have cartilage in their walls but relatively little smooth muscle. They are lined by a ciliated epithelium that contains mucous and serous glands. Cilia are present as far as the respiratory bronchioles, but glands are absent from the epithelium of the bronchioles and terminal bronchioles, and their walls do not contain cartilage. However, their walls contain more smooth muscle, of which the largest amount relative to the thickness of the wall is present in the terminal bronchioles. The walls of the bronchi and bronchioles are innervated by the autonomic nervous system. There are abundant muscarinic receptors, and cholinergic discharge causes bronchoconstriction. There are β_1 and β_2 adrenergic receptors in the bronchial epithelium and smooth muscle and in mast cells. Many are not innervated. However, some may be located on cholinergic endings and ganglia, where they inhibit acetylcholine release. In humans, the β_2 receptors predominate and inhaled or injected β agonists such as isoproterenol cause bronchodilation and decreased bronchial secretion.

There is in addition a noncholinergic, nonadrenergic innervation of the bronchioles that produces bronchodilation, and there is evidence that VIP is the mediator responsible for the dilation. The leukotrienes LTC_4, LTD_4, and LTE_4 (see Chapter 17) are potent bronchoconstrictors, particularly when administered by inhalation. Some of the nerves in the lungs contain substance P, and others contain CCK 8, CGRP, and other polypeptides, but the function of these polypeptides is not known.

Pulmonary Circulation

Almost all the blood in the body passes via the pulmonary artery to the pulmonary capillary bed, where it is oxygenated and returned to the left atrium via the pulmonary veins. The separate and much smaller bronchial arteries come from the aorta, and the bronchial veins drain into the azygos vein. The bronchial circulation nourishes the bronchi and pleura, but there are extensive anastomoses between

Table 34–1. Standard conditions to which measurements involving gas volumes are corrected.

STPD	0 °C, 760 mm Hg, dry (standard temperature and pressure, dry)
BTPS	Body temperature and pressure, saturated with water vapor
ATPD	Ambient temperature and pressure, dry
ATPS	Ambient temperature and pressure, saturated with water vapor

Figure 34–1. Structure of the lung. A, alveolus; AD, alveolar duct; RB, respiratory bronchiole; TB, terminal bronchiole. (Reproduced, with permission, from Staub NC: The pathophysiology of pulmonary edema. *Hum Pathol* 1970;**1**:419.)

Figure 34–2. Electron photomicrograph of human lung showing a pulmonary capillary adjacent to an alveolar sac. The capillary contains a neutrophil (GR). EP1, type I pulmonary epithelial cell; EN, endothelial cell, with its nucleus (N); J, junction between 2 endothelial cells. (Reproduced, with permission, from Weibel ER: Lung cell biology. In: *Handbook of Physiology,* Section 3, *The Respiratory System.* Vol I, p. 47. Fishman AP [editor]. American Physiological Society, 1985.)

the bronchial and pulmonary capillary beds. Lymphatic channels are more abundant in the lungs than in any other organ (Fig 34–1).

Further aspects of the pulmonary circulation are considered below.

MECHANICS OF RESPIRATION

Inspiration & Expiration

The lungs and the chest wall are elastic structures. Normally, there is no more than a thin layer of fluid between the lungs and the chest wall. The lungs slide easily on the chest wall but resist being pulled away from it in the same way that 2 moist pieces of glass slide on each other but resist separation. The pressure in the "space" between the lungs and chest wall (intrapleural pressure) is subatmospheric (Fig 34–3). The lungs are stretched when they are expanded at birth, and at the end of quiet expiration their tendency to recoil from the chest wall is just balanced by the tendency of the chest wall to recoil in the opposite direction. If the chest wall is opened, the lungs collapse; and if the lungs lose their elasticity, the chest expands and becomes barrel-shaped.

Inspiration is an active process. The contraction of the inspiratory muscles increases intrathoracic volume. During quiet breathing, the intrapleural pressure at the base of the lungs, which is about –2.5 mm Hg

(relative to atmospheric) at the start of inspiration, decreases to about –6 mm Hg. The lungs are pulled into a more expanded position. The pressure in the airway becomes slightly negative, and air flows into the lungs (Fig 34–3). At the end of inspiration, the lung recoil begins to pull the chest back to the expiratory position, where the recoil pressures of the lungs and chest wall balance. The pressure in the airway becomes slightly positive, and air flows out of the lungs. Expiration during quiet breathing is passive in the sense that no muscles which decrease intrathoracic volume contract. However, there is some contraction of the inspiratory muscles in the early part of expiration. This contraction exerts a braking action on the recoil forces and slows expiration.

Strong inspiratory efforts reduce intrapleural pressure to values as low as –30 mm Hg, producing correspondingly greater degrees of lung inflation. When ventilation is increased, the extent of lung deflation is also increased by active contraction of expiratory muscles that decrease intrathoracic volume.

Lung Volumes

The amount of air that moves into the lungs with each inspiration (or the amount that moves out with each expiration) is called the **tidal volume.** The air inspired with a maximal inspiratory effort in excess of the tidal volume is the **inspiratory reserve volume.** The volume expelled by an active expiratory ef-

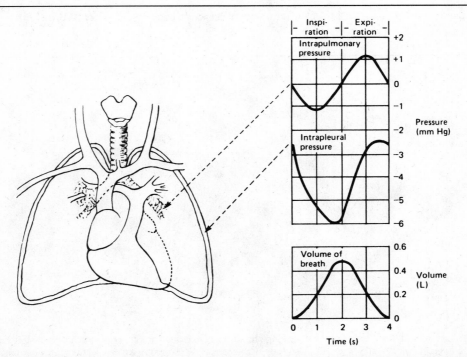

Figure 34–3. Changes in intrapleural (intrathoracic) and intrapulmonary pressure relative to atmospheric pressure during inspiration and expiration.

fort after passive expiration is the **expiratory reserve volume,** and the air left in the lungs after a maximal expiratory effort is the **residual volume.** Normal values for these lung volumes, and names applied to combinations of them, are shown in Fig 34–4. The space in the conducting zone of the airways occupied by gas that does not exchange with blood in the pulmonary vessels is the **respiratory dead space.** The **vital capacity,** the largest amount of air that can be expired after a maximal inspiratory effort, is frequently measured clinically as an index of pulmonary function. It gives useful information about the strength of the respiratory muscles and other aspects of pulmonary function. The fraction of the vital capacity expired in 1 second (**timed vital capacity;** also called forced expired volume in 1 second, or FEV 1″) gives additional information; the vital capacity may be normal but the timed vital capacity greatly reduced in diseases such as asthma, in which the resistance of the airways is increased owing to bronchial constriction. The amount of air inspired per minute (**pulmonary ventilation, respiratory minute volume**) is normally about 6 L (500 mL/breath × 12 breaths/min). The **maximal voluntary ventilation (MVV),** or, as it was formerly called, the **maximal breathing capacity,** is the largest volume of gas that can be moved into and out of the lungs in 1 min-

ute by voluntary effort. The normal MVV is 125–170 L/min.

Respiratory Muscles

Movement of the **diaphragm** accounts for 75% of the change in intrathoracic volume during quiet inspiration. Attached around the bottom of the thoracic cage, this muscle arches over the liver and moves downward like a piston when it contracts. The distance it moves ranges from 1.5 cm to as much as 7 cm with deep inspiration (Fig 34–5).

The diaphragm has 3 parts: the costal portion, made up of muscle fibers that arise from the ribs around the bottom of the thoracic cage; the crural portion, made up of fibers that arise from the ligaments along the vertebras; and the central tendon, into which the costal and the crural fibers insert. The crural fibers pass on either side of the esophagus. The central tendon is also the inferior part of the pericardium. The costal and crural portions are innervated by different parts of the phrenic nerve and can contract separately. For example, during vomiting and eructation, intra-abdominal pressure is increased by contraction of the costal fibers but the crural fibers remain relaxed, allowing material to pass from the stomach into the esophagus.

The other important **inspiratory muscles** are the

IRV = Inspiratory reserve volume   TV = Tidal volume
ERV = Expiratory reserve volume   RV = Residual volume

		Volume (L)		
		Men	Women	
Vital capacity	IRV	3.3	1.9	Inspiratory capacity
	TV	0.5	0.5	
	ERV	1.0	0.7	Functional residual capacity
	RV	1.2	1.1	
Total lung capacity		6.0	4.2	

Respiratory minute volume (rest): 6 L/min   Timed vital capacity: 83% of total in 1 s; 97% in 3 s
Alveolar ventilation (rest): 4.2 L/min   Work of quiet breathing: 0.5 kg-m/min
Maximal voluntary ventilation (BTPS): 125–170 L/min   Maximal work of breathing: 10 kg-m/breath

Figure 34–4. Lung volumes and some measurements related to the mechanics of breathing. The diagram at the upper right represents the excursions of a spirometer plotted against time. (Modified from Comroe JH Jr et al: *The Lung: Clinical Physiology and Pulmonary Function Tests,* 2nd ed. Year Book, 1962.)

Figure 34–5. X-ray of chest in full expiration *(left)* and full inspiration *(right).* The dashed white line on the right is an outline of the lungs in full expiration. (From Comroe JH Jr: *Physiology of Respiration,* 2nd ed. Copyright © 1974 by Year Book Medical Publishers, Inc., Chicago. Used by permission.)

external intercostal muscles, which run obliquely downward and forward from rib to rib. The ribs pivot as if hinged at the back, so that when the external intercostals contract they elevate the lower ribs. This pushes the sternum outward and increases the anteroposterior diameter of the chest. The transverse diameter is actually changed little if at all. Either the diaphragm or the external intercostal muscles alone can maintain adequate ventilation at rest. Transection of the spinal cord above the third cervical segment is fatal without artificial respiration, but transection below the fifth cervical segment is not, because it leaves the phrenic nerves that innervate the diaphragm intact; the phrenic nerves arise from cervical segments 3–5. Conversely, in patients with bilateral phrenic nerve palsy but intact innervation of their intercostal muscles, respiration is somewhat labored but adequate to maintain life. The scalene and sternocleidomastoid muscles in the neck are accessory inspiratory muscles that help to elevate the thoracic cage during deep labored respiration.

A decrease in intrathoracic volume and forced expiration result when the **expiratory muscles** contract. The internal intercostals have this action because they pass obliquely downward and posteriorly from rib to rib and therefore pull the rib cage downward when they contract. Contractions of the muscles of the anterior abdominal wall also aid expiration by pulling the rib cage downward and inward and by in-

creasing the intra-abdominal pressure, which pushes the diaphragm upward.

Glottis

The abductor muscles in the larynx contract early in inspiration, pulling the vocal cords apart and opening the glottis. During swallowing or gagging, there is reflex contraction of the adductor muscles that closes the glottis and prevents aspiration of food, fluid, or vomitus into the lungs. In unconscious or anesthetized patients, glottic closure may be incomplete and vomitus may enter the trachea, causing an inflammatory reaction in the lung (**aspiration pneumonia).**

The laryngeal muscles are supplied by the vagus nerves. When the abductors are paralyzed, there is inspiratory stridor. When the adductors are paralyzed, food and fluid enter the trachea, causing aspiration pneumonia and edema. Bilateral cervical vagotomy in animals causes the slow development of fatal pulmonary congestion and edema. The edema is due at least in part to aspiration, although some edema develops even if a tracheostomy is performed before the vagotomy.

Bronchial Tone

In general, the smooth muscle in the bronchial walls helps maintain an even distribution of ventilation. Stimulation of sensory receptors in the airways by irritants and chemicals such as sulfur dioxide produces reflex bronchoconstriction that is mediated via cholinergic pathways. Cool air also causes bronchoconstriction, and so does exercise, possibly because the increased respiration associated with it cools the airways. Leukotrienes cause bronchoconstriction, and an antagonist to LTD_4 reduces the bronchoconstriction produced by exercise. In addition, the bronchial muscles protect the bronchi during coughing. There is a circadian rhythm in bronchial tone, with maximal constriction at about 6:00 AM and maximal dilation at about 6:00 PM.

VIP relaxes bronchial smooth muscle, and a **nonadrenergic noncholinergic nervous system** that regulates the bronchi via VIP has been postulated to exist. VIP has been reported to be deficient or absent in a large number of patients with asthma.

Compliance of the Lungs & Chest Wall

The interaction between the recoil of the lungs and recoil of the chest can be demonstrated in living subjects. The nostrils are clipped shut, and the subject breathes through a spirometer that has a valve just beyond the mouthpiece. The mouthpiece contains a pressure-measuring device. After the subject inhales a given amount, the valve is shut, closing off the airway. The respiratory muscles are then relaxed while the pressure in the airway is recorded. The procedure is repeated after inhaling or actively exhaling various volumes. The curve of airway pressure obtained in

this way, plotted against volume, is the **relaxation pressure curve** of the total respiratory system (Fig 34–6). The pressure is zero at a lung volume that corresponds to the volume of gas in the lungs at the end of quiet expiration (**relaxation volume,** which equals the functional residual capacity). It is positive at greater volumes and negative at smaller volumes. The change in lung volume per unit change in airway pressure ($\Delta V/\Delta P$) is the stretchability (**compliance**) of the lungs and chest wall. It is normally measured in the pressure range where the relaxation pressure curve is steepest, and the normal value is approximately 0.2 L/cm H_2O. However, compliance depends on lung volume; an individual with only one lung has approximately half the ΔV for a given ΔP. Compliance is also slightly greater when measured during deflation than when measured during inflation. Consequently, it is more informative to examine the whole pressure-volume curve. The curve is shifted downward and to the right (compliance is decreased) by pulmonary congestion and interstitial pulmonary fibrosis (Fig 34–7). It is shifted upward and to the left (compliance is increased) in emphysema (see Chapter 37). It should be noted that compliance is a static measure of lung and chest recoil. The **resistance** of the lung and chest is the pressure difference required for a unit of air flow, and this measurement, which is dynamic rather than static, also takes into account the resistance to air flow in the airways.

Figure 34–6. Relation between intrapulmonary pressure and volume. The middle curve is the static pressure curve of values obtained when the lungs are inflated or deflated by various amounts and the intrapulmonary pressure (elastic recoil pressure) is measured with the airway closed. The relaxation volume is the point where the recoil of the chest and the recoil of the lungs balance. The slope of the curve is the compliance of the lungs and chest wall. The maximal inspiratory and expiratory curves are the airway pressures that can be developed during maximal inspiratory and expiratory efforts. The dashed lines, the horizontal distances between the inspiratory and expiratory curves and the relaxation pressure curve, represent the net pressure developed by the inspiratory and expiratory muscles.

Figure 34–7. Static expiratory pressure-volume curves of lungs in normal subjects and subjects with severe emphysema and pulmonary fibrosis. (Modified and reproduced, with permission, from Pride NB, Mackem PT: Lung mechanics in disease. In: *Handbook of Physiology,* Section 3, *The Respiratory System.* Vol III, part 2. Fishman AP [editor]. American Physiological Society, 1986.)

Figure 34–8. Pressure-volume relations in the lungs of a cat after removal from the body. ***Air:*** lungs inflated (Inf) and deflated (Def) with air. ***Saline:*** lungs inflated and deflated with saline. (Reproduced, with permission, from Morgan TE: Pulmonary surfactant. *N Engl J Med* 1971;**284:**1185.)

Alveolar Surface Tension

An important factor affecting the compliance of the lungs is the surface tension of the film of fluid that lines the alveoli. The magnitude of this component at various lung volumes can be measured by removing the lungs from the body and distending them alternately with saline and with air while measuring the intrapulmonary pressure. Because saline reduces the surface tension to nearly zero, the pressure-volume curve obtained with saline measures only the tissue elasticity (Fig 34–8), whereas the curve obtained with air measures both tissue elasticity and surface tension. The difference between the 2 curves, the elasticity due to surface tension, is much smaller at small than at large lung volumes. The surface tension is also much lower than the expected surface tension at a water-air interface of the same dimensions.

Surfactant

The low surface tension when the alveoli are small is due to the presence in the fluid lining the alveoli of **surfactant,** a lipid surface-tension-lowering agent. Surfactant is a mixture of dipalmitoylphosphatidylcholine (DPPC), other lipids, and proteins (Table 34–2). If the surface tension is not kept low when the alveoli become smaller during expiration, they collapse in accordance with the law of Laplace (see Chapter 30). In spherical structures like the alveoli, the distending pressure equals 2 times the tension divided by the radius (P 2T/r); if T is not reduced as

r is reduced, the tension overcomes the distending pressure. Surfactant also helps to prevent pulmonary edema. It has been calculated that if it were not present, the unopposed surface tension in the alveoli would produce a 20-mm-Hg force favoring transudation of fluid from the blood into the alveoli.

Phospholipids, which have a hydrophilic "head" and 2 parallel hydrophobic fatty acids "tails" (see Chapter 1), line up in the alveoli with their tails facing the alveolar lumen (Fig 34–9), and surface tension is inversely proportionate to their concentration per unit area. They move farther apart as the alveoli enlarge during inspiration, and surface tension increases, whereas it decreases when they move closer together during expiration.

Surfactant is produced by type II alveolar epithelial cells (Fig 34–9). Typical **lamellar bodies,** membrane-bound organelles containing whorls of phospholipid, are formed in these cells and secreted into the alveolar lumen by exocytosis. Tubes of lipid called **tubular myelin** form from the extruded bodies, and the tubular myelin in turn forms the phospholipid film. Some of the protein-lipid complexes in surfactant are taken up by endocytosis in type II alveolar cells and recycled.

Table 34–2. Approximate composition of surfactant.

Component	Percent Composition
Dipalmitoylphosphatidylcholine	62
Phosphatidylglycine	5
Other phospholipids	10
Neutral lipids	13
Proteins	8
Carbohydrate	2

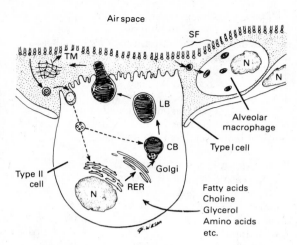

Figure 34–9. Formation and metabolism of surfactant. Lamellar bodies (LB) are formed in type II alveolar epithelial cells and secreted by exocytosis. The released lamellar body material is converted to tubular myelin (TM), and the TM is probably the source of the phospholipid surface film (SF). Some surfactant is taken up by alveolar macrophages, but more is taken up by endocytosis in type II epithelial cells. (Reproduced, with permission, from Wright JR: Metabolism and turnover of lung surfactant. *Am Rev Respir Dis* 1987;**135**:426.)

Formation of the phospholipid film is greatly facilitated by the proteins in surfactant. This material contains 3 major proteins, SP-A, SP-B, and SP-C. SP-A is a large glycoprotein and has a collagenlike domain within its structure. It probably has multiple functions, including regulation of the feedback uptake of surfactant by the type II alveolar epithelial cells that secrete it. SP-B and SP-C are smaller proteins, which facilitate formation of the monomolecular film of phospholipid.

Surfactant is important at birth. The fetus makes respiratory movements in utero, but the lungs remain collapsed until birth. After birth, the infant makes several strong inspiratory movements and the lungs expand. Surfactant keeps them from collapsing again. Surfactant deficiency is the cause of **hyaline membrane disease, (respiratory distress syndrome, RDS),** the serious pulmonary disease that develops in infants born before their surfactant system is functional. Surface tension in the lungs of these infants is high, and there are many areas in which the alveoli are collapsed (atelectasis). Administration of phospholipid alone by inhalation has little value in the treatment of RDS. However, encouraging results have been obtained with administration of bovine surfactant, which contains both phospholipids and proteins.

The size and number of inclusions in type II cells are increased by thyroid hormones, and RDS is more common and more severe in infants with low plasma levels of thyroid hormones than in those with normal plasma levels. Maturation of surfactant in the lungs is also accelerated by glucocorticoid hormones. There is an increase in fetal and maternal cortisol near term, and the lungs are rich in glucocorticoid receptors. In addition, insulin inhibits SP-A accumulation in cultured human fetal lung tissue, and fetal hyperinsulinism occurs in diabetic mothers. This may explain the increased incidence of RDS in infants born to diabetic mothers.

Patchy atelectasis is also associated with surfactant deficiency in patients who have undergone cardiac surgery during which a pump oxygenator was used and the pulmonary circulation was interrupted. In addition, surfactant deficiency may play a role in some of the abnormalities that develop following occlusion of a main bronchus, occlusion of one pulmonary artery, or long-term inhalation of 100% O_2. There is a decrease in surfactant in the lungs of cigarette smokers.

Work of Breathing

Work is performed by the respiratory muscles in stretching the elastic tissues of the chest wall and lungs (elastic work), moving inelastic tissues (viscous resistance), and moving air through the respiratory passages (Table 34–3). Since pressure times volume ($g/cm^2 \times cm^3 = g \times cm$) has the same dimensions as work (force × distance), the work of breathing can be calculated from the relaxation pressure curve (Figs 34–6 and 34–10). In Fig 34–10, the total elastic work required for inspiration is area ABCA. Note that the relaxation pressure curve of the total respiratory system differs from that of the lungs alone. The actual elastic work required to increase the volume of the lungs alone is area ABDEA. The amount of elastic work required to inflate the whole respiratory system is less than the amount required to inflate the lungs alone because part of the work comes from elastic energy stored in the thorax. The elastic energy lost from the thorax (area AFGBA) is equal to that gained by the lungs (area AEDCA).

The frictional resistance to air movement is relatively small during quiet breathing, but it does cause the intrapleural pressure changes to lead the lung volume changes during inspiration and expiration (Fig 34–3), producing a **hysteresis loop** rather than a straight line when pressure is plotted against volume (Fig 34–11). In this diagram, area AXBYA represents

Table 34–3. Components that make up the work of breathing during quiet inspiration, and the percent contribution of each.

Nonelastic work
Viscous resistance (7%)
Airway resistance (28%)
Elastic work (65%)

Figure 34–10. Relaxation pressure curve of the total respiratory system (P_{TR}) and its components, and the relaxation pressure curves of the lungs (P_L) and the chest (P_W). The transmural pressure is intrapulmonary pressure minus intrapleural pressure in the case of the lungs, intrapleural pressure minus outside (barometric) pressure in the case of the chest wall, and intrapulmonary pressure minus barometric pressure in the case of the total respiratory system. (Modified from Mines AH: *Respiratory Physiology,* 2nd ed. Raven, 1986.)

the work done to overcome airway resistance and lung viscosity. If the air flow becomes turbulent during rapid respiration, the energy required to move the air is greater than when the flow is laminar.

Estimates of the total work of quiet breathing range from 0.3 up to 0.8 kg-m/min. The value rises markedly during exercise, but the energy cost of breathing in normal individuals represents less than 3% of the total energy expenditure during exercise. The work of breathing is greatly increased in diseases

such as emphysema, asthma, and congestive heart failure with dyspnea and orthopnea. The respiratory muscles have length-tension relations like those of other skeletal and cardiac muscles, and when they are severely stretched, they contract with less strength. They can also become fatigued and fail (pump failure), leading to inadequate ventilation (see Chapter 37). For unknown reasons, aminophylline increases the force of contraction of the human diaphragm and is useful in the treatment of pump failure.

Differences in Ventilation & Blood Flow in Different Parts of the Lung

In the upright position, ventilation per unit lung volume is greater at the base of the lung than at the apex. The reason for this is that at the start of inspiration, intrapleural pressure is less negative at the base than at the apex (Fig 34–12), and since the intrapulmonary-intrapleural pressure difference is less than at the apex, the lung is less expanded. Conversely, at the apex, the lung is more expanded, ie, the percentage of maximum lung volume is greater. Because of the stiffness of the lung, the increase in lung volume per unit increase in pressure is less when the lung is initially more expanded, and ventilation is consequently greater at the base. Blood flow is also greater at the base than the apex (see below). The rel-

Figure 34–11. Diagrammatic representation of pressure and volume changes during quiet inspiration (line AXB) and expiration (line BZA). Line AYB is the compliance line.

Figure 34–12. Intrapleural pressures in the upright position and their effect on ventilation. Note that because intrapulmonary pressure is atmospheric, the more negative intrapleural pressure at the apex holds the lung in a more expanded position at the start of inspiration. Further increases in volume per unit increase in intrapleural pressure are less than at the base because the expanded lung is stiffer. (Reproduced, with permission, from West JB: *Ventilation/Blood Flow and Gas Exchange,* 3rd ed. Blackwell, 1977.)

ative change in blood flow from the apex to the base is greater than the relative change in ventilation, so the ventilation/perfusion ratio is low at the base and high at the apex.

The ventilation and perfusion differences from the apex to the base of the lung have usually been attributed to gravity; they tend to disappear in the supine position, and the weight of the lung would be expected to make the intrapleural pressure lower at the base in the upright position. However, in a recent space flight, the inequalities of ventilation and blood flow in humans were found to persist to a remarkable degree in weightlessness. Therefore, other as yet unknown factors apparently also play a role in producing the inequalities.

It should be noted that at very low lung volumes such as those after forced expiration, intrapleural pressure at the bases of the lungs can actually exceed the atmospheric pressure in the airways, and the small airways such as respiratory bronchioles collapse (**airway closure**). In older people and in those with chronic lung disease, some of the elastic recoil is lost, with a resulting decrease in intrapleural pressure. Consequently, airway closure may occur in the bases of the lungs in the upright position without forced expiration, at volumes as high as the functional residual capacity.

A clinical correlate of the presumed effect of gravity on ventilation is that arterial oxygenation is improved in unilateral lung diseases when patients lie on their sides so that the good lung is in the dependent position. For reasons that are not entirely clear, the situation is opposite in infants, who do better with the diseased lung in the dependent position.

Dead Space & Uneven Ventilation

Since gaseous exchange in the respiratory system occurs only in the terminal portions of the airways, the gas that occupies the rest of the respiratory system is not available for gas exchange with pulmonary capillary blood. Normally, the volume of this dead space is approximately equal to the body weight in pounds. Thus, in a man who weighs 150 lb (68 kg), only the first 350 mL of the 500 mL inspired with each breath at rest mixes with the air in the alveoli. Conversely, with each expiration, the first 150 mL expired is gas that occupied the dead space, and only the last 350 mL is gas from the alveoli. Consequently, the **alveolar ventilation,** ie, the amount of air reaching the alveoli per minute, is less than the respiratory minute volume. Note in addition that because of the dead space, rapid shallow breathing produces much less alveolar ventilation than slow deep breathing at the same respiratory minute volume (Table 34–4).

It is important to distinguish between the **anatomic dead space** (respiratory system volume exclusive of alveoli) and the **total (physiologic) dead space** (volume of gas not equilibrating with blood, ie, wasted ventilation). In healthy individuals, the 2 dead

Table 34–4. Effect of variations in respiratory rate and depth on alveolar ventilation.

Respiratory rate	30/min	10/min
Tidal volume	200 mL	600 mL
Minute volume	6 L	6 L
Alveolar ventilation	$(200 - 150) \times 30$ $= 1500$ mL	$(600 - 150) \times 10$ $= 4500$ mL

spaces are identical; but in disease states, there may be no exchange between the gas in some of the alveoli and the blood, and some of the alveoli may be overventilated. The volume of gas in nonperfused alveoli and any volume of air in the alveoli in excess of that necessary to arterialize the blood in the alveolar capillaries is part of the dead space (nonequilibrating) gas volume. The anatomic dead space can be measured by analysis of the single-breath N_2 curves (Fig 34–13). From midinspiration, the subject takes as deep a breath as possible of pure O_2, then exhales steadily while the N_2 content of the expired gas is continuously measured. The initial gas exhaled (phase I) is the gas that filled the dead space and that consequently contains no N_2. This is followed by a mixture of dead space and alveolar gas (phase II) and then by alveolar gas (phase III). The volume of the dead space is the volume of the gas expired from peak inspiration to the midportion of phase II (Fig 34–13).

Phase III of the single-breath N_2 curve terminates at the **closing volume (CV)** and is followed by phase IV, during which the N_2 content of the expired gas is increased. The CV is the lung volume above residual volume at which airways in the lower, dependent parts of the lungs begin to close off because of the lesser transmural pressure in these areas (see above). The gas in the upper portions of the lungs is richer in N_2 than the gas in the lower, dependent portions be-

Figure 34–13. Single-breath N_2 curve. From midinspiration, the subject takes a deep breath of pure O_2, then exhales steadily. The changes in the N_2 concentration of expired gas during expiration are shown, with the various phases of the curve indicated by roman numerals. DS, dead space; CV, closing volume; RV, residual volume. (Modified from Buist AS: New tests to assess lung function: The single-breath nitrogen test. *N Engl J Med* 1975;**293:**438.)

cause the alveoli in the upper portions are more distended at the start of the inspiration of O_2 (see above) and, consequently, the N_2 in them is less diluted with O_2. It is also worth noting that in most normal individuals, phase III has a slight positive slope even before phase IV is reached. This indicates that even during phase III there is a gradual increase in the proportion of the expired gas coming from the relatively N_2-rich upper portions of the lungs.

The pattern of ventilation in the lungs can be assessed by having the subject inhale a radioactive isotope of the inert gas xenon (^{133}Xe) while the chest is monitored with a battery of radiation detectors. Areas that show little radioactivity are poorly ventilated.

The total dead space can be calculated from the P_{CO_2} of expired air, the P_{CO_2} of arterial blood, and the tidal volume. The tidal volume (V_T) times the P_{CO_2} of the expired gas ($P_{E_{CO_2}}$) equals the arterial P_{CO_2} ($P_{a_{CO_2}}$) times the difference between the tidal volume and the dead space (V_D) plus the P_{CO_2} of inspired air ($P_{I_{CO_2}}$) times V_D:

$$P_{E_{CO_2}} \times V_T = P_{a_{CO_2}} \times (V_T - V_D) + P_{I_{CO_2}} \times V_D \text{ (Bohr's equation).}$$

However, the term $P_{I_{CO_2}} \times V_D$ is so small that it can be ignored and the equation solved for V_D. If, for example,

$$P_{E_{CO_2}} = 28 \text{ mm Hg}$$
$$P_{a_{CO_2}} = 40 \text{ mm Hg}$$
$$V_T = 500 \text{ mL}$$

then,

$$V_D = 150 \text{ mL}$$

The equation can also be used to measure the anatomic dead space if one replaces $P_{a_{CO_2}}$ with alveolar P_{CO_2} ($P_{A_{CO_2}}$), which is the P_{CO_2} of the last 10 mL of expired gas (see below). $P_{A_{CO_2}}$ is an average of gas from different alveoli in proportion to their ventilation regardless of whether they are perfused. This is in contrast to $P_{a_{CO_2}}$, which is gas equilibrated only with perfused alveoli and, consequently in individuals with unperfused alveoli, is greater than $P_{A_{CO_2}}$.

Although it is possible to stand under water and breathe through a tube that projects above the surface, it may be worth noting that such a tube is in effect an extension of the respiratory dead space. For each milliliter of tube volume, the depth of inspiration would have to be increased 1 mL to supply the same volume of air to the alveoli. Thus, if the volume of the tube were at all large, breathing would become very laborious. Additional effort is also required to expand the chest against the pressure of the surrounding water.

GAS EXCHANGE IN THE LUNGS

Composition of Alveolar Air

Oxygen continuously diffuses out of the gas in the alveoli (**alveolar air**) into the bloodstream, and CO_2 continuously diffuses into the alveoli from the blood. In the steady state, inspired air mixes with the alveolar gas, replacing the O_2 that has entered the blood and diluting the CO_2 that has entered the alveoli. Part of this mixture is expired. The O_2 content of the alveolar gas then falls and its CO_2 content rises until the next inspiration. Since the volume of gas in the alveoli is about 2 L at the end of expiration (functional residual capacity; Fig 34–4), each 350-mL increment of inspired and expired air changes the P_{O_2} and P_{CO_2} very little. Indeed, the composition of alveolar gas remains remarkably constant, not only at rest but in a variety of other conditions as well (see Chapter 36).

Sampling Alveolar Air

Theoretically, all but the first 150 mL expired with each expiration is alveolar air, but there is always some mixing at the interface between the dead-space gas and the alveolar air (Fig 34–13). A later portion of expired air is therefore the portion taken for analysis. Using modern apparatus with a suitable automatic valve, it is possible to collect the last 10 mL expired during quiet breathing. The composition of alveolar gas is compared with that of inspired and expired air in Fig 34–14.

Diffusion Across the Alveolocapillary Membrane

Gases diffuse from the alveoli to the blood in the pulmonary capillaries or vice versa across the thin alveolocapillary membrane made up of the pulmonary epithelium, the capillary endothelium, and their fused basement membranes (Fig 34–2). Whether or not substances passing from the alveoli to the capillary blood reach equilibrium in the 0.75 second that blood takes to traverse the pulmonary capillaries at rest depends on their reaction with substances in the blood. Thus, for example, the anesthetic gas nitrous oxide does not react, and N_2O reaches equilibrium in about 0.1 second (Fig 34–15). In this situation, the amount of N_2O entering the body is not limited by diffusion but by the amount of blood flowing through the pulmonary capillaries; ie, it is **perfusion-limited.** On the other hand, carbon monoxide is taken up by the hemoglobin in the red blood cells at such a high rate that the partial pressure of CO in the capillaries stays very low and equilibrium is not reached in the 0.75 second the blood is in the pulmonary capillaries. Therefore, the transfer of CO is not limited by perfusion at rest, and instead is **diffusion-limited.** O_2 is intermediate between N_2O and CO; it is taken up by hemoglobin, but much less avidly than CO, and it reaches equilibrium with capillary blood in about 0.3 second. Thus, its uptake is also perfusion-limited.

Inspired air

O$_2$	158.0
CO$_2$	0.3
H$_2$O	5.7
N$_2$	596.0

Expired gas

O$_2$	116.0
CO$_2$	32.0
H$_2$O	47.0
N$_2$	565.0

Alveoli

O$_2$	100.0
CO$_2$	40.0
H$_2$O	47.0
N$_2$	573.

Dead space

Physiologic shunt

Right heart

Left heart

Veins

O$_2$	40.0
CO$_2$	46.0
H$_2$O	47.0
N$_2$	573.0

Arteries

O$_2$	95.0
CO$_2$	40.0
H$_2$O	47.0
N$_2$	573.0

Capillaries

O$_2$	40.0−
CO$_2$	46.0+
H$_2$O	47.0
N$_2$	573.0

Tissues

Figure 34–14. Partial pressures of gases (mm Hg) in various parts of the respiratory system and in the circulatory system.

The **diffusion capacity** of the lung for a given gas is directly proportionate to the size of the alveolocapillary membrane and inversely proportionate to its thickness. The diffusion capacity for CO ($D_{L_{CO}}$) is measured as an index of diffusion capacity because its uptake is diffusion-limited. $D_{L_{CO}}$ is proportionate to the amount of CO entering the blood (VCO) divided by the partial pressure of CO in the al-

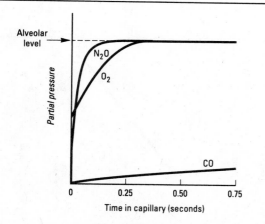

Figure 34–15. Uptake of various substances during the 0.75 second they are in transit through a pulmonary capillary. N$_2$O is not bound in blood, so its partial pressure in blood rises rapidly to its partial pressure in the alveoli. Conversely, CO is avidly taken up by red blood cells, so its partial pressure reaches only a fraction of its partial pressure in the alveoli. O$_2$ is intermediate between the two.

veoli minus the partial pressure of CO in the blood entering the pulmonary capillaries. Except in habitual cigarette smokers, this latter term is close to zero, so it can be ignored and the equation becomes

$$D_{L_{CO}} = \frac{\dot{V}_{CO}}{P_{A_{CO}}}$$

The normal value of $D_{L_{CO}}$ at rest is about 25 mL/min/mm Hg. It increases up to 3-fold during exercise because of capillary dilation and an increase in the number of active capillaries.

The P_{O_2} of alveolar air is normally 100 mm Hg (Fig 34–14), and the P_{O_2} of the blood entering the pulmonary capillaries is 40 mm Hg. The diffusion capacity for O$_2$, like that for CO at rest, is about 25 mL/min/mm Hg, and the P_{O_2} of blood is raised to 97 mm Hg, a value just under the alveolar P_{O_2}. This falls to 95 mm Hg in the aorta because of the physiologic shunt (see below). $D_{L_{O_2}}$ increases to 65 mL/min/mm Hg or more during exercise and is reduced in diseases such as sarcoidosis and beryllium poisoning (berylliosis) that cause fibrosis of the alveolar walls. Another cause of pulmonary fibrosis is excess secretion of PDGF (see Chapter 27) by alveolar macrophages, with resulting stimulation of neighboring mesenchymal cells.

The P_{CO_2} of venous blood is 46 mm Hg, whereas that of alveolar air is 40 mm Hg, and CO$_2$ diffuses from the blood into the alveoli along this gradient. The P_{CO_2} of blood leaving the lungs is 40 mm Hg. CO$_2$ passes through all biologic membranes with ease, and the pulmonary diffusion capacity for CO$_2$ is much greater than the capacity for O$_2$. It is for this reason that CO$_2$ retention is rarely a problem in patients with alveolar fibrosis even when the reduction in diffusion capacity for O$_2$ is severe.

PULMONARY CIRCULATION

Pulmonary Blood Vessels

The pulmonary vascular bed resembles the systemic (see Chapter 30), except that the walls of the pulmonary artery and its large branches are about 30% as thick as the wall of the aorta, and the small arterial vessels, unlike the systemic arterioles, are endothelial tubes with relatively little muscle in their walls. There is also some smooth muscle in the walls of the postcapillary vessels. The pulmonary capillaries are large, and there are multiple anastomoses, so that each alveolus sits in a capillary basket.

Pressure, Volume, & Flow

With 2 quantitatively minor exceptions, the blood put out by the left ventricle returns to the right atrium and is ejected by the right ventricle, making the pulmonary vasculature unique in that it accommodates a blood flow that is almost equal to that of all the other organs in the body. One of the exceptions is part of the bronchial blood flow. As noted above, there are extensive anastomoses between the bronchial and pulmonary capillaries, and although some of the bronchial blood enters the bronchial veins, some enters the pulmonary capillaries, bypassing the right ventricle. The other exception is blood that flows from the coronary arteries into the chambers of the left side of the heart (see Chapter 32). Because of the small **physiologic shunt** created by those 2 exceptions, the blood in systemic arteries has a P_{O_2} about 2 mm Hg lower than that of blood that has equilibrated with alveolar air, and the saturation of hemoglobin is 0.5% less (see Chapter 35.)

The entire pulmonary vascular system is a distensible low-pressure system. The pulmonary arterial pressure is about 24/9 mm Hg, and the mean pressure is about 15 mm Hg. The pressure in the left atrium is about 8 mm Hg during diastole, so that the pressure gradient in the pulmonary system is about 7 mm Hg, compared with a gradient of about 90 mm Hg in the systemic circulation (Fig 34–16). It is interesting that the pressure fall from the pulmonary artery to the capillaries is relatively small and that there is an appreciable pressure drop in the veins.

The volume of blood in the pulmonary vessels at any one time is about 1 L, of which less than 100 mL is in the capillaries. The mean velocity of the blood in the root of the pulmonary artery is the same as that in the aorta (about 40 cm/s). It falls off rapidly, then rises slightly again in the larger pulmonary veins. It takes a red cell about 0.75 second to traverse the pulmonary capillaries at rest, and 0.3 second or less during exercise.

Capillary Pressure

Pulmonary capillary pressure is about 10 mm Hg, whereas the oncotic pressure is 25 mm Hg, so that there is an inward-directed pressure gradient of about 15 mm Hg which keeps the alveoli free of fluid. When the pulmonary capillary pressure is more than 25 mm Hg—as it may be, for example, when there is "backward failure" of the left ventricle—pulmonary congestion and edema result. Patients with mitral stenosis also have a chronic, progressive rise in pulmonary capillary pressure and extensive fibrotic changes in the pulmonary vessels. Pulmonary edema is not as prominent a symptom in mitral stenosis as in frank congestive heart failure, probably because the fibrosis and constriction of the pulmonary arterial vessels "protect" the capillaries.

Effect of Gravity

Gravity has a relatively marked effect on the pulmonary circulation. In the upright position, the upper portions of the lungs are well above the level of the heart, and the bases are at or below it. Consequently, there is a relatively marked pressure gradient in the pulmonary artery from the top to the bottom of the lungs and a resulting linear increase in pulmonary blood flow from the apices to the bases of the lungs (Fig 34–17). The pressure in the capillaries at the top of the lungs is close to the atmospheric pressure in the alveoli. Pulmonary arterial pressure is normally just sufficient to maintain perfusion, but if it is reduced or

Figure 34–16. Blood pressures (mm Hg) in the pulmonary and systemic circulation. (Modified from Comroe JH Jr: *Physiology of Respiration,* 2nd ed. Copyright © 1974 by Year Book Medical Publishers, Inc., Chicago. Used by permission.)

Figure 34–17. Relative blood flow from the top to the bottom of the lung in the upright position. The values for lung blood flow are scaled so that if flow were uniform, the value would be 100 throughout.

if alveolar pressure is increased, some of the capillaries collapse. Under these circumstances, there is no gas exchange in the affected alveoli and they become part of the physiologic dead space.

In the middle portions of the lungs, the pulmonary arterial and capillary pressure exceeds alveolar pressure, but the pressure in the pulmonary venules may be lower than alveolar pressure during normal expiration, so they are collapsed. Under these circumstances, blood flow is determined by the pulmonary artery-alveolar pressure difference rather than the pulmonary artery-pulmonary vein difference. Beyond the constriction, blood "falls" into the pulmonary veins, which are compliant and take whatever amount of blood the constriction lets flow into them. This has been called the **waterfall effect,** but the term is inappropriate because there is no restriction to flow at the top of a real waterfall.

In the lower portions of the lungs, alveolar pressure is lower than the pressure in all parts of the pulmonary circulation and blood flow is determined by the arterial-venous pressure difference.

Ventilation/Perfusion Ratios

The ratio of pulmonary ventilation to pulmonary blood flow for the whole lung at rest is about 0.8 (4.2 L/min ventilation divided by 5.5 L/min blood flow). However, there are relatively marked differences in this **ventilation/perfusion ratio** in various parts of the normal lung as a result of the effect of gravity, and local changes in the ventilation/perfusion ratio are common in disease. If the ventilation to an alveolus is reduced relative to its perfusion, the P_{O_2} in the alveolus falls because less O_2 is delivered to it and the P_{CO_2} rises because less CO_2 is expired. Conversely, if perfusion is reduced relative to ventilation, the P_{CO_2} falls because less CO_2 is delivered and the P_{O_2} rises because less O_2 enters the blood. These effects are summarized in Fig 34–18.

As noted above, ventilation, as well as perfusion, in the upright position declines in a linear fashion from the bases to the apices of the lungs. However, the ventilation/perfusion ratios are high in the upper portions of the lungs. It is said that the high ventilation/perfusion ratios at the apices account for the predilection of tuberculosis for this area because the relatively high alveolar P_{O_2} that results provides a favorable environment for the growth of the tuberculosis bacteria.

When widespread, nonuniformity of ventilation and perfusion in the lungs can cause declines in systemic arterial P_{O_2} and P_{CO_2} retention. The consequences of nonuniformity in disease states are discussed in Chapter 37.

Pulmonary Reservoir

Because of their distensibility, the pulmonary veins are an important blood reservoir. When a normal individual lies down, the pulmonary blood vol-

Figure 34–18. Effects of decreasing or increasing the ventilation/perfusion ratio ($\dot{V}_A/\dot{Q}$) on the P_{CO_2} and P_{O_2} in an alveolus. The drawings above the curve represent an alveolus and a pulmonary capillary, and the black areas indicate sites of blockage. With complete obstruction of the airway to the alveolus, P_{CO_2} and P_{O_2} approximate the values in mixed venous blood ($\bar{V}$). With complete block of perfusion, P_{CO_2} and P_{O_2} approximate the values in inspired air. (Reproduced, with permission, from West JB: *Ventilation/Blood Flow and Gas Exchange,* 3rd ed. Blackwell, 1977.)

ume increases by up to 400 mL, and when the person stands up this blood is discharged into the general circulation. This shift is the cause of the decrease in vital capacity in the supine position and is responsible for the occurrence of orthopnea in heart failure (see Chapter 33).

Regulation of Pulmonary Blood Flow

Pulmonary arterioles are constricted by norepinephrine, epinephrine, angiotensin II, thromboxanes, and $PGF_{2\alpha}$; they are dilated by isoproterenol, acetylcholine, and PGI_2. Pulmonary venules are constricted by serotonin, histamine, and *Escherichia coli* endotoxin. Pulmonary vessels are plentifully supplied with sympathetic vasoconstrictor nerve fibers, and stimulation of the cervical sympathetic ganglia decreases pulmonary blood flow by as much as 30%.

Despite this innervation and the reactivity of the vessels, it appears that regulation of overall pulmonary blood flow is largely passive, and local adjustments of perfusion to ventilation are determined by local effects of O_2 or its lack. With exercise, cardiac output increases and pulmonary arterial pressure rises proportionately with little or no vasodilation. More red cells move through the lungs without any reduction in the O_2 saturation of the hemoglobin in them, and consequently, the total amount of O_2 delivered to the systemic circulation is increased. Capillaries dilate, and previously underperfused capillaries are "recruited" to carry blood. The net effect is a marked increase in pulmonary blood flow with few if any alterations in autonomic outflow to the pulmonary vessels.

Local alterations in pulmonary blood flow are produced in large part by local alterations in tissue O_2

content. ^{133}Xe can be used to survey blood flow by injecting a saline solution of the gas intravenously while monitoring the chest. The gas rapidly enters the alveoli that are perfused normally but fails to enter those that are not perfused. Another technique for locating poorly perfused areas is injection of macroaggregates of albumin labeled with radioactive iodine. These aggregates are large enough to block capillaries and small arterioles, and they lodge only in vessels in which blood was flowing when they reached the lungs. Although it seems paradoxic to study patients with defective pulmonary blood flow by producing vascular obstruction, the technique is safe because relatively few particles are injected. The particles block only a small number of pulmonary vessels and are rapidly removed by the body.

When a bronchus or a bronchiole is obstructed, hypoxia develops in the underventilated alveoli beyond the obstruction. The O_2 deficiency apparently acts directly on vascular smooth muscle in the area to produce constriction, shunting blood away from the hypoxic area. Accumulation of CO_2 leads to a drop in pH in the area, and a decline in pH also produces vasoconstriction in the lungs, as opposed to the vasodilation it produces in other tissues. Conversely, reduction of the blood flow to a portion of the lung lowers the alveolar P_{CO_2} in that area, and this leads to constriction of the bronchi supplying it, shifting ventilation away from the poorly perfused area.

Systemic hypoxia also causes the pulmonary arterioles to constrict, with a resultant increase in pulmonary arterial pressure.

Pulmonary Embolization

One of the normal functions of the lungs is to filter out small blood clots, and this occurs without any symptoms. When emboli block larger branches of the pulmonary artery, they provoke a rise in pulmonary arterial pressure and rapid, shallow respiration (**tachypnea**). The rise in pulmonary arterial pressure is apparently due to reflex vasoconstriction via the sympathetic nerve fibers, although there is controversy on this point and reflex vasoconstriction appears to be absent when large branches of the pulmonary artery are blocked. The tachypnea is a reflex response to activation of vagally innervated pulmonary receptors close to the vessel walls (see Table 36–2) that appear to be activated by serotonin released from platelets at the site of embolization.

OTHER FUNCTIONS OF THE RESPIRATORY SYSTEM

Lung Defense Mechanisms

The respiratory passages that lead from the exterior to the alveoli do more than serve as gas conduits. They humidify and cool or warm the inspired air so that even very hot or very cold air is at or near body temperature by the time it reaches the alveoli. Bronchial secretions contain secretory immunoglobulins (IgA; see Chapter 27) and other substances that help resist infections and maintain the integrity of the mucosa.

The pulmonary alveolar macrophages (PAMs, "dust cells") are important components of the pulmonary defense mechanisms. Like other macrophages (see Chapter 27), these cells come originally from the bone marrow. They are actively phagocytic and ingest inhaled bacteria and small particles. They also help process inhaled antigens for immunologic attack, and they secrete substances that attract granulocytes to the lungs as well as substances that stimulate granulocyte and monocyte formation in the bone marrow. Their role in the pathogenesis of emphysema is discussed in Chapter 37. When the macrophages ingest large amounts of the substances in cigarette smoke, they may also release lysosomal products into the extracellular space. This causes inflammation. Silica and asbestos particles also cause extracellular release of lysosomal enzymes.

Various mechanisms operate to prevent foreign matter from reaching the alveoli. The hairs in the nostrils strain out many particles larger than 10 μm in diameter. Most of the remaining particles of this size settle on mucous membranes in the nose and pharynx; because of their momentum, they do not follow the airstream as it curves downward into the lungs, and they impact on or near the **tonsils** and **adenoids,** large collections of immunologically active lymphoid tissue in the back of the pharynx. Particles 2–10 μm in diameter generally fall on the walls of the bronchi as the air flow slows in the smaller passages. There they initiate reflex bronchial constriction and coughing (see Chapter 14). They are also moved away from the lungs by the "ciliary escalator." The epithelium of the respiratory passages from the anterior third of the nose to the beginning of the respiratory bronchioles is ciliated, and the cilia, which are covered with mucus, beat in a coordinated fashion at a frequency of 1000–1500 cycles per minute. The ciliary mechanism is capable of moving particles away from the lungs at a rate of at least 16 mm/min. Particles less than 2 μm in diameter generally reach the alveoli, where they are ingested by the macrophages. The importance of these defense mechanisms is evident when one remembers that in modern cities, each liter of air may contain several million particles of dust and irritants.

When ciliary motility is defective, mucus transport is virtually absent. This leads to chronic sinusitis, recurrent lung infections, and bronchiectasis. Ciliary immotility may be produced by various air pollutants, or it may be congenital. One congenital form is **Kartagener's syndrome,** in which the axonemal dynein, the ATPase molecular motor that produces ciliary beating (see Chapter 1), is absent. Patients with this condition also are infertile because

Table 34–5. Biologically active substances metabolized by the lungs.

Synthesized and used in the lungs
Surfactant
Synthesized or stored and released into the blood
Prostaglandins
Histamine
Kallikrein
Partially removed from the blood
Prostaglandins
Bradykinin
Adenine nucleotides
Serotonin
Norepinephrine
Acetylcholine
Activated in the lungs
Angiotensin I → angiotensin II

they lack motile sperm, and they often have situs inversus, presumably because the cilia necessary for rotating the viscera are nonfunctional during embryonic development.

Metabolic & Endocrine Functions of the Lungs

In addition to their functions in gas exchange, the lungs have a number of metabolic functions. They manufacture surfactant for local use as noted above. They also contain a fibrinolytic system that lyses clots in the pulmonary vessels. They release a variety of substances that enter the systemic arterial blood (Table 34–5), and they remove other substances from the systemic venous blood that reaches them via the pulmonary artery. Prostaglandins are removed from the circulation, but they are also synthesized in the lungs and released into the blood when lung tissue is stretched.

The lungs also activate one hormone; the physiologically inactive decapeptide angiotensin I is converted to the pressor, aldosterone-stimulating octapeptide angiotensin II in the pulmonary circulation (see Chapter 24). Large amounts of the angiotensin-converting enzyme responsible for this activation are located on the surface of the endothelial cells of the pulmonary capillaries and particularly in small pits **(caveolae)** on the vascular surface of these cells. The converting enzyme also inactivates bradykinin. Circulation time through the pulmonary capillaries is less than 1 second, yet 70% of the angiotensin I reaching the lungs is converted to angiotensin II in a single trip through the capillaries. Four other peptidases are located on the surface of the pulmonary vascular surface, but their physiologic role is unsettled.

Removal of serotonin and norepinephrine reduces the amounts of these vasoactive substances reaching the systemic circulation. However, many other vasoactive hormones pass through the lungs without being metabolized. These include epinephrine, dopamine, oxytocin, vasopressin, and angiotensin II.

Gas Transport Between the Lungs & the Tissues

INTRODUCTION

The partial pressure gradients for O_2 and CO_2 plotted in graphic form in Fig 35–1 emphasize that they are the key to gas movement and that O_2 "flows downhill" from the air through the alveoli and blood into the tissues whereas CO_2 "flows downhill" from the tissues to the alveoli. However, the amount of both of these gases transported to and from the tissues would be grossly inadequate if it were not that about 99% of the O_2 which dissolves in the blood combines with the O_2-carrying protein hemoglobin and that about 94.5% of the CO_2 which dissolves enters into a series of reversible chemical reactions that convert it into other compounds. Thus, the presence of hemoglobin increases the O_2-carrying capacity of the blood 70-fold, and the reactions of CO_2 increase the blood CO_2 content 17-fold (Table 35–1). This chapter reviews the mechanisms involved in O_2 and CO_2 transport.

OXYGEN TRANSPORT

Oxygen Delivery to the Tissues

The O_2 delivery system in the body consists of the lungs and the cardiovascular system. O_2 delivery to a particular tissue depends on the amount of O_2 entering the lungs, the adequacy of pulmonary gas exchange, the blood flow to the tissue, and the capacity of the blood to carry O_2. The blood flow depends on the degree of constriction of the vascular bed in the tissue and the cardiac output. The amount of O_2 in the blood is determined by the amount of dissolved O_2, the amount of hemoglobin in the blood, and the affinity of the hemoglobin for O_2.

Reaction of Hemoglobin & Oxygen

The dynamics of the reaction of hemoglobin with O_2 make it a particularly suitable O_2 carrier. Hemoglobin (see Chapter 27) is a protein made up of 4 subunits, each of which contains a **heme** moiety attached to a polypeptide chain. Heme (see Fig 27–13) is a complex made up of a porphyrin and 1 atom of ferrous iron. Each of the 4 iron atoms can bind revers-

ibly one O_2 molecule. The iron stays in the ferrous state, so that the reaction is an **oxygenation,** not an oxidation. It has been customary to write the reaction of hemoglobin with O_2 as $Hb + O_2 \rightleftarrows HbO_2$. Since it contains 4 Hb units, the hemoglobin molecule can also be represented as Hb_4, and it actually reacts with 4 molecules of O_2 to form Hb_4O_8.

$$Hb_4 + O_2 \rightleftarrows Hb_4O_2$$
$$Hb_4O_2 + O_2 \rightleftarrows Hb_4O_4$$
$$Hb_4O_4 + O_2 \rightleftarrows Hb_4O_6$$
$$Hb_4O_6 + O_2 \rightleftarrows Hb_4O_8$$

The reaction is rapid, requiring less than 0.01 second. The deoxygenation (reduction) of Hb_4O_8 is also very rapid.

The quaternary structure of hemoglobin determines its affinity for O_2; by shifting the relationship of its 4 component polypeptide chains, the molecule fosters either O_2 uptake or O_2 delivery. The movement of the chains is associated with a change in the position of the heme moieties, which assume a relaxed or **R state** that favors O_2 binding or a tense or **T state** that decreases O_2 binding. The transition from one state to another involves breaking or forming bridges between the polypeptide chains, and it has been calculated that these shifts occur about 10^8 times in the life of a red blood cell.

The **oxygen-hemoglobin dissociation curve,** the curve relating percentage saturation of the O_2-carrying power of hemoglobin to the P_{O_2} (Fig 35–2), has a characteristic sigmoid shape. When hemoglobin takes up a small amount of O_2, the R state is favored and additional uptake of O_2 is facilitated. Combination of the first heme in the Hb molecule with O_2 increases the affinity of the second heme for O_2, and oxygenation of the second increases the affinity of the third, etc, so that the affinity of Hb for the fourth O_2 molecule is many times that for the first. When hemoglobin takes up O_2, the two β chains move closer together; when O_2 is given up, they move farther apart. This shift is essential for the shift in affinity for O_2 to occur.

When blood is equilibrated with 100% O_2 ($P_{O_2} = 760$ mm Hg), the normal hemoglobin becomes 100%

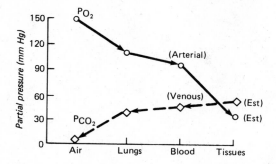

Figure 35–1. Summary of P_{O_2} and P_{CO_2} values in air, lungs, blood, and tissues, graphed to emphasize the fact that both O_2 and CO_2 diffuse "downhill" along gradients of decreasing partial pressure. (Redrawn and reproduced, with permission, from Kinney JM: Transport of carbon dioxide in blood. *Anesthesiology* 1960;**21**:615.)

Figure 35–2. Oxygen-hemoglobin dissociation curve. pH 7.40, temperature 38 °C. (Redrawn and reproduced, with permission, from Comroe JH Jr et al: *The Lung: Clinical Physiology and Pulmonary Function Tests,* 2nd ed. Year Book, 1962.)

saturated. When fully saturated, each gram of normal hemoglobin contains 1.39 mL of O_2. However, blood normally contains small amounts of inactive hemoglobin derivatives, and the measured value in vivo is lower. The traditional figure is 1.34 mL of O_2. The hemoglobin concentration in normal blood is about 15 g/dL (14 g/dL in women and 16 g/dL in men; see Chapter 27). Therefore, 1 dL of blood contains 20.1 mL (1.34 mL × 15) of O_2 bound to hemoglobin when the hemoglobin is 100% saturated. The amount of dissolved O_2 is a linear function of the P_{O_2} (0.003 mL/dL blood/mm Hg P_{O_2}).

In vivo, the hemoglobin in the blood at the ends of the pulmonary capillaries is about 97.5% saturated with O_2 (P_{O_2} = 97 mm Hg). Because of a slight admixture with venous blood that bypasses the lungs (physiologic shunt), the hemoglobin in systemic arterial blood is only 97% saturated. The arterial blood therefore contains a total of about 19.8 mL of O_2 per deciliter: 0.29 mL in solution and 19.5 mL bound to hemoglobin. In venous blood at rest, the hemoglobin is 75% saturated and the total O_2 content is about 15.2 mL/dL: 0.12 mL in solution and 15.1 mL bound to hemoglobin. Thus, at rest the tissues remove about 4.6 mL of O_2 from each deciliter of blood passing

through them (Table 35–1); 0.17 mL of this total represents O_2 that was in solution in the blood, and the remainder represents O_2 that was liberated from hemoglobin. In this way, 250 mL of O_2 per minute is transported from the blood to the tissues at rest.

Factors Affecting the Affinity of Hemoglobin for Oxygen

Three important conditions affect the oxygen-hemoglobin dissociation curve: the pH, the temperature, and the concentration of 2,3-diphosphoglycerate (**DPG; 2,3-DPG**). A rise in temperature or a fall in pH shifts the curve to the right (Fig 35–3). When the curve is shifted in this direction, a higher P_{O_2} is required for hemoglobin to bind a given amount of O_2. Conversely, a fall in temperature or a rise in pH shifts the curve to the left, and a lower P_{O_2} is required to bind a given amount of O_2. A convenient index of such shifts is the P_{50}, the P_{O_2} at which hemoglobin is half saturated with O_2. The higher the P_{50}, the lower the affinity of hemoglobin for O_2.

The decrease in O_2 affinity of hemoglobin when the pH of blood falls is called the **Bohr effect** and is closely related to the fact that deoxygenated hemoglobin (deoxyhemoglobin) binds H^+ more actively than does oxyhemoglobin. The pH of blood falls as its CO_2 content increases (see below), so that when the P_{CO_2} rises, the curve shifts to the right and the P_{50} rises. Most of the unsaturation of hemoglobin that occurs in the tissues is secondary to the decline in the P_{O_2}, but an extra 1–2% unsaturation is due to the rise in P_{CO_2} and consequent shift of the dissociation curve to the right.

2,3-DPG is very plentiful in red cells. It is formed

Table 35-1. Gas content of blood.

	mL/dL of Blood Containing 15 g of Hemoglobin			
Gas	**Arterial Blood** (P_{O_2} 95 mm Hg; P_{CO_2} 40 mm Hg; Hb 97% Saturated)		**Venous Blood** (P_{O_2} 40 mm Hg; P_{CO_2} 46 mm Hg; Hb 75% Saturated)	
	Dissolved	**Combined**	**Dissolved**	**Combined**
O_2	0.29	19.5	0.12	15.1
CO_2	2.62	46.4	2.98	49.7
N_2	0.98	0	0.98	0

Figure 35–3. Effect of temperature and pH on the oxygen-hemoglobin dissociation curve. Ordinates and abscissas are as in Fig 35–2. (Redrawn and reproduced, with permission, from Comroe JH Jr et al: *The Lung: Clinical Physiology and Pulmonary Function Tests,* 2nd ed. Year Book, 1962.)

(Fig 35–4) from 3-phosphoglyceraldehyde, which is a product of glycolysis via the Embden-Meyerhof pathway (see Chapter 17). It is a highly charged anion that binds to the β chains of deoxyhemoglobin. One mole of deoxyhemoglobin binds 1 mol of 2,3-DPG. In effect,

$$\text{HbO}_2 + \text{2,3-DPG} \rightleftharpoons \text{Hb–2,3-DPG} + \text{O}_2$$

In this equilibrium, an increase in the concentration of 2,3-DPG shifts the reaction to the right, causing more O_2 to be liberated. ATP binds to deoxyhemoglobin to a lesser extent, and some other organic phosphates bind to a minor degree.

Factors affecting the concentration of 2,3-DPG in the red cells include pH. Because acidosis inhibits red cell glycolysis, the 2,3-DPG concentration falls when the pH is low. Thyroid hormones, growth hormone, and androgens increase the concentration of 2,3-DPG and the P_{50}.

Exercise has been reported to produce an increase in 2,3-DPG within 60 minutes, although the rise may not occur in trained athletes. The P_{50} is also increased during exercise, because the temperature rises in active tissues and CO_2 and metabolites accumulate, lowering the pH. In addition, much more O_2 is removed from each unit of blood flowing through active tissues because the tissues P_{O_2} declines. Finally, at low P_{O_2} values, the oxygen-hemoglobin dissociation curve is steep, and large amounts of O_2 are liberated per unit drop in P_{O_2}.

Ascent to high altitude triggers a substantial rise in 2,3-DPG concentration in red cells, with a consequent increase in P_{50} and increase in the availability of O_2 to tissues. The rise in 2,3-DPG, which has a half-life of 6 hours, is secondary to the rise in blood pH (see Chapter 37). 2,3-DPG levels drop to normal upon return to sea level.

The greater affinity of fetal hemoglobin (hemoglobin F) than adult hemoglobin (hemoglobin A) for O_2 facilitates the movement of O_2 from the mother to the fetus (see Chapters 27 and 32). The cause of this greater affinity is the poor binding of 2,3-DPG by the γ polypeptide chains that replace β chains in fetal hemoglobin. Some abnormal hemoglobins in adults have low P_{50} values, and the resulting high O_2 affinity of the hemoglobin causes enough tissue hypoxia to stimulate increased red cell formation, with resulting polycythemia (see Chapter 24). It is interesting to speculate that these hemoglobins may not bind 2,3-DPG.

Red cell 2,3-DPG concentration is increased in anemia and in a variety of diseases in which there is chronic hypoxia. This facilitates the delivery of O_2 to the tissues by raising the P_{O_2} at which O_2 is released

Figure 35–4. Formation and catabolism of 2,3-DPG.

in peripheral capillaries. In bank blood that is stored, the 2,3-DPG level falls and the ability of this blood to release O_2 to the tissues is reduced. This decrease, which obviously limits the benefit of the blood if it is transfused into a hypoxic patient, is less if the blood is stored in citrate-phosphate-dextrose solution rather than the usual acid-citrate-dextrose solution.

Other aspects of the chemistry of hemoglobin are discussed in Chapter 27. Fetal hemoglobin and transplacental O_2 exchange are discussed in Chapter 32.

Myoglobin

Myoglobin is an iron-containing pigment found in skeletal muscle. It resembles hemoglobin but binds one rather than 4 mol of O_2 per mole. Its dissociation curve is a rectangular hyperbola rather than a sigmoid curve. Because its curve is to the left of the hemoglobin curve (Fig 35–5), it takes up O_2 from hemoglobin in the blood. It releases O_2 only at low P_{O_2} values, but the P_{O_2} in exercising muscle is close to zero. The myoglobin content is greatest in muscles specialized for sustained contraction. The muscle blood supply is compressed during such contractions, and myoglobin may provide O_2 when blood flow is cut off. There is also evidence that myoglobin facilitates the diffusion of O_2 from the blood to the mitochondria, where the oxidative reactions occur.

Blood Substitutes

The solubility of O_2 in plasma is limited. Certain perfluoro compounds dissolve much more O_2 and have been used to totally replace blood for short periods in experimental animals. Some of these compounds have been tested in patients, and although much more testing is needed, they may be of value as substitutes for blood in emergencies until blood can be obtained and cross-matched.

Figure 35–5. Dissociation curve of hemoglobin and myoglobin.

BUFFERS

Since CO_2 forms carbonic acid in the blood, a discussion of the buffer systems in the body is a necessary preliminary to a consideration of CO_2 transport.

The Henderson-Hasselbalch Equation

The general equation for a buffer system is

$$HA \rightleftharpoons H^+ + A^-$$

A^- represents any anion and HA the undissociated acid. If an acid stronger than HA is added to a solution containing this system, the equilibrium is shifted to the left. Hydrogen ions are "tied up" in the formation of more undissociated HA, so the increase in H^+ concentration is much less than it would otherwise be. Conversely, if a base is added to the solution, H^+ and OH^- react to form H_2O; but more HA dissociates, limiting the decrease in H^+ concentration. By the law of mass action, the product of the concentrations of the products in a chemical reaction divided by the product of the concentration of the reactants at equilibrium is a constant:

$$\frac{[H^+][A^-]}{[HA]} = K$$

If this equation is solved for H^+ and put in pH notation (pH is the negative log of $[H^+]$), the resulting equation is that originally derived by Henderson and Hasselbalch to describe the pH changes resulting from addition of H^+ or OH^- to any buffer system (**Henderson-Hasselbalch equation**):

$$pH = pK + \log \frac{[A^-]}{[HA]}$$

It is apparent from these equations that the buffering capacity of a system is greatest when the amount of free anion is equal to the amount of undissociated HA, ie, when $[A^-]/[HA] = 1$, so that log $[A^-]/[HA]$ 0 and pH-pK. This is why the most effective buffers in the body would be expected to be those with pKs close to the pH in which they operate. The pH of the blood is normally 7.4; that of the cells is probably about 7.2; and that of the urine varies from 4.5 to 8.0.

It should be noted that the equilibrium constant, K, applies only to infinitely dilute solutions in which interionic forces are negligible. In body fluids, it is more appropriate to use the apparent ionization constant, K'.

Buffers in Blood

In the blood, proteins—particularly the **plasma proteins**—are effective buffers because both their free carboxyl and their free amino groups dissociate:

$$RCOOH \rightleftharpoons RCOO^- + H^+$$

$$pH = pK'_{RCOOH} + \log \frac{[RCOO^-]}{[RCOOH]}$$

$$RNH_3^+ \rightleftharpoons RNH_2 + H^+$$

$$pH = pK'_{RNH_3^+} + \log \frac{[RNH_2]}{[RNH_3^+]}$$

Another important buffer system is provided by the dissociation of the imidazole groups of the histidine residues in **hemoglobin**:

In the pH 7.0–7.7 range, the free carboxyl and amino groups of hemoglobin contribute relatively little to its buffering capacity. However, the hemoglobin molecule contains 38 histidine residues, and on this basis—plus the fact that hemoglobin is present in large amounts—the hemoglobin in blood has 6 times the buffering capacity of the plasma proteins. In addition, the action of hemoglobin is unique because the imidazole groups of deoxyhemoglobin dissociate less than those of oxyhemoglobin, making Hb a weaker acid and therefore a better buffer than HbO_2. Titration curves for Hb and HbO_2 are shown in Fig 35–6.

The third major buffer system in blood is the **carbonic acid-bicarbonate** system:

$$H_2CO_3 \rightleftharpoons H^+ + HCO_3^-$$

The Henderson-Hasselbalch equation for this system is as follows:

$$pH = pK + \log \frac{[HCO_3^-]}{[H_2CO_3]}$$

The pK for this system in an ideal solution is low (about 3), and the amount of H_2CO_3 is small and hard to measure accurately. However, in the body, H_2CO_3 is in equilibrium with CO_2.

$$H_2CO_3 \rightleftharpoons CO_2 + H_2O$$

If the pK is changed to pK' (see above) and $[CO_2]$ is substituted for $[H_2CO_3]$, the pK' is 6.1.

$$pH = 6.10 + \log \frac{[HCO_3^-]}{[CO_2]}$$

The clinically relevant form of this equation is as follows:

$$pH = 6.10 + \log \frac{[HCO_3^-]}{0.0301\ P_{CO_2}}$$

since the amount of dissolved CO_2 is proportionate to the partial pressure of CO_2 and the solubility coefficient of CO_2 in mmol/L/mm Hg is 0.0301. $[HCO_3^-]$ cannot be measured directly, but pH and P_{CO_2} can be measured with suitable accuracy with pH and P_{CO_2} glass electrodes, and $[HCO_3^-]$ can then be calculated.

The pK' of this system is still low relative to the pH of the blood, but the system is one of the most effective buffer systems in the body because the amount of dissolved CO_2 is controlled by respiration. In addition, the plasma concentration of HCO_3^- is regulated by the kidneys. When H^+ is added to the blood, HCO_3^- declines as more H_2CO_3 is formed. If the extra H_2CO_3 were not converted to CO_2 and H_2O and the CO_2 excreted in the lungs, the H_2CO_3 concentration would rise. When enough H^+ had been added to halve the plasma HCO_3^-, the pH would have dropped from 7.4 to 6.0. However, not only is all the extra H_2CO_3 that is formed removed, but also the H^+ rise stimulates respiration and therefore produces a drop in P_{CO_2}, so that some additional H_2CO_3 is removed. The pH thus falls only to 7.2 or 7.3 (see Fig 39–5).

The reaction $CO_2 + H_2O \rightleftharpoons H_2CO_3$ proceeds slowly in either direction unless the enzyme **carbonic anhydrase** is present. There is no carbonic anhydrase in plasma, but there is an abundant supply in red blood cells. It is also found in high concentration in gastric acid-secreting cells (see Chapter 26) and in renal tubular cells (see Chapter 38). Carbonic an-

Figure 35–6. Titration curves of Hb and HbO_2. The arrow from a to c indicates the number of millimoles of H^+ that can be added without pH shift. The arrow from a to b indicates the pH shift on deoxygenation. (Modified from Davenport HW: *The ABC of Acid-Base Chemistry,* 6th ed. Univ of Chicago Press, 1974.)

hydrase is a protein with a molecular weight of 30,000 that contains an atom of zinc in each molecule. It is inhibited by cyanide, azide, and sulfide. The sulfonamides also inhibit this enzyme, and sulfonamide derivatives have been used clinically as diuretics because of their inhibitory effects on carbonic anhydrase in the kidney (see Chapter 38).

The system $H_2PO_4^- \rightleftharpoons H^+ + HPO_4^{2-}$ has a pK of 6.80. In the plasma, the phosphate concentration is too low for this system to be a quantitatively important buffer, but it is important intracellularly, and it frequently plays a significant role in the urine (see Chapter 38).

Buffering in Vivo

Buffering in vivo is of course not limited to the blood. The principal buffers in the blood, interstitial fluid, and intracellular fluid are listed in Table 35–2. The principal buffers in cerebrospinal fluid and urine are the bicarbonate and phosphate systems. In metabolic acidosis, only 15–20% of the acid load is buffered by the H_2CO_3-HCO_3^- system in the extracellular fluid, and most of the remainder is buffered in cells. In metabolic alkalosis, about 30–35% of the OH^- load is buffered in cells, whereas in respiratory acidosis and alkalosis (see Chapter 39), almost all the buffering is intracellular.

Summary

When a strong acid is added to the blood, the major buffer reactions are driven to the left. The blood levels of the 3 "buffer anions"—Hb^- (hemoglobin), $Prot^-$ (protein), and HCO_3^-—consequently drop. The anions of the added acid are filtered into the renal tubules. They are accompanied ("covered") by cations, particularly Na^+, because electrochemical neutrality is maintained. By processes that are discussed in Chapter 38, the tubules replace the Na^+ with H^+ and in so doing reabsorb equimolar amounts of Na^+ and HCO_3^-, thus conserving the cations, eliminating the acid, and restoring the supply of buffer anions to normal. When CO_2 is added to the blood, similar reactions occur, except that since it is H_2CO_3 that is formed, the plasma HCO_3^- rises rather than falls.

CARBON DIOXIDE TRANSPORT

Fate of Carbon Dioxide in Blood

The solubility of CO_2 in blood is about 20 times that of O_2, so that there is considerably more CO_2 than O_2 in simple solution at equal partial pressures. The CO_2 that diffuses into red blood cells is rapidly hydrated to H_2CO_3 because of the presence of carbonic anhydrase. The H_2CO_3 dissociates to H^+ and HCO_3^-, and the H^+ is buffered, primarily by hemoglobin, while the HCO_3^- enters the plasma. The decline in the O_2 saturation of the hemoglobin as the blood passes through the tissue capillaries improves its buffering capacity because deoxygenated hemoglobin binds more H^+ than oxyhemoglobin (see above). Some of the CO_2 in the red cells reacts with the amino groups of proteins, principally hemoglobin, to form **carbamino compounds:**

$$CO_2 + R-N\genfrac{}{}{0pt}{}{H}{H} \rightleftharpoons R-N\genfrac{}{}{0pt}{}{H}{COOH}$$

Since deoxygenated hemoglobin forms carbamino compounds much more readily than HbO_2 does, transport of CO_2 is facilitated in venous blood. About 11% of the CO_2 added to the blood in the systemic capillaries is carried to the lungs as carbamino-CO_2.

In the plasma, CO_2 reacts with plasma proteins to form small amounts of carbamino compounds, and small amounts of CO_2 are hydrated; but the hydration reaction is slow in the absence of carbonic anhydrase.

Chloride Shift

Since the rise in the HCO_3^- content of red cells is much greater than that in plasma as the blood passes through the capillaries, about 70% of the HCO_3^- formed in the red cells enters the plasma. The excess HCO_3^- leaves the red cells in exchange for Cl^- (Fig 35–7), a process mediated by **Band 3,** a major membrane protein. This exchange is called the **chloride shift.** The Cl^- content of the red cells in venous blood is therefore significantly greater than in arterial

Table 35–2. Principal buffers in body fluids.

Blood	$H_2CO_3 \rightleftharpoons H^+ + HCO_3^-$ $HProt \rightleftharpoons H^+ + Prot^-$ $HHb \rightleftharpoons H^+ + Hb^-$
Interstitial fluid	$H_2CO_3 \rightleftharpoons H^+ + HCO_3^-$
Intracellular fluid	$HProt \rightleftharpoons H^+ + Prot^-$ $H_2PO_4^- \rightleftharpoons H^+ + HPO_4^{2-}$

Figure 35–7. Summary of changes that occur in a red cell upon addition of CO_2 to blood. Note that for each CO_2 molecule that enters the red cell, there is an additional HCO_3^- or Cl^- ion in the cell.

Table 35–3. Fate of CO_2 in blood.

In plasma
1. Dissolved
2. Formation of carbamino compounds with plasma protein
3. Hydration, H^+ buffered, HCO_3^- in plasma

In red blood cells
1. Dissolved
2. Formation of carbamino-Hb
3. Hydration, H^+ buffered, 70% of HCO_3^- enters the plasma
4. Cl^- shifts into cells; mosm/L in cells increases

Figure 35–8. CO_2 dissociation curves. The arterial point *(a)* and the venous point *(v)* indicate the total CO_2 content found in arterial blood and venous blood of normal resting humans. (Modified and reproduced, with permission, from Schmidt RF, Thews G [editors]: *Human Physiology.* Springer-Verlag, 1983.)

blood. The chloride shift occurs rapidly and is essentially complete in 1 second.

Note that for each CO_2 molecule added to a red cell, there is an increase of one osmotically active particle—either an HCO_3^- or a Cl^-—in the red cell (Fig 35–7). Consequently, the red cells take up water and increase in size. For this reason, plus the fact that a small amount of fluid in the arterial blood returns via the lymphatics rather than the veins, the hematocrit of venous blood is normally 3% greater than that of the arterial blood. In the lungs, the Cl^- moves out of the cells and they shrink.

Summary of Carbon Dioxide Transport

For convenience, the various fates of CO_2 in the plasma and red cells are summarized in Table 35–3. The extent to which they increase the capacity of the blood to carry CO_2 is indicated by the difference between the lines indicating the dissolved CO_2 and the total CO_2 in the dissociation curves for CO_2 shown in Fig 35–8.

Of the approximately 49 mL of CO_2 in each deciliter of arterial blood (Table 35–1), 2.6 mL is dissolved, 2.6 mL is in carbamino compounds, and 43.8 mL is in HCO_3^-. In the tissues, 3.7 mL of CO_2 per deciliter of blood is added; 0.4 mL stays in solution, 0.8 mL forms carbamino compounds, and 2.5 mL forms HCO_3^-. The pH of the blood drops from 7.40 to 7.36. In the lungs, the processes are reversed, and the 3.7 mL of CO_2 is discharged into the alveoli. In this fashion, 200 mL of CO_2 per minute at rest and much larger amounts during exercise are transported from the tissues to the lungs and excreted. It is worth noting that this amount of CO_2 is equivalent in 24 hours to over 12,500 meq of H^+.

Regulation of Respiration

36

INTRODUCTION

Spontaneous respiration is produced by rhythmic discharge of motor neurons that innervate the respiratory muscles. This discharge is totally dependent on nerve impulses from the brain; breathing stops if the spinal cord is transected above the origin of the phrenic nerves.

The rhythmic discharges from the brain that produce spontaneous respiration are regulated by alternations in arterial P_{O_2}, P_{CO_2}, and H^+ concentration, and this chemical control of breathing is supplemented by a number of nonchemical influences.

NEURAL CONTROL OF BREATHING

Control Systems

Two separate neural mechanisms regulate respiration. One is responsible for voluntary control and the other for automatic control. The voluntary system is located in the cerebral cortex and sends impulses to the respiratory motor neurons via the corticospinal tracts. The automatic system is located in the pons and medulla, and the efferent output from this system to the respiratory motor neurons is located in the white matter of the spinal cord between the lateral and ventral corticospinal tracts. The nerve fibers mediating inspiration converge on the phrenic motor neurons located in the ventral horns from C3 to C5 and the internal intercostal motor neurons in the ventral horns throughout the thoracic cord. The fibers concerned with expiration converge primarily on the external intercostal motor neurons in the thoracic cord.

The motor neurons to the expiratory muscles are inhibited when those supplying the inspiratory muscles are active, and vice versa. Although spinal reflexes contribute to this **reciprocal innervation** (see Chapter 6), it is due primarily to activity in descending pathways. Impulses in these descending pathways excite agonists and inhibit antagonists. The one exception to the reciprocal inhibition is a small amount of activity in phrenic axons for a short period after inspiration. The function of this postinspiratory output appears to be to brake the lung's elastic recoil and make respiration smooth.

Medullary Systems

Rhythmic discharge of neurons in the medulla and pons produces automatic respiration; transection of the brain stem below the medulla (Section D in Fig 36–1) stops respiration, and automatic respiration is normal after transection rostral to the pons (Section A in Fig 36–1). Brain stem respiratory neurons are of 2 types: those that discharge during inspiration (I neurons) and those that discharge during expiration (E neurons). Many of these discharge at increasing frequencies during inspiration, in the case of I neurons, or during expiration, in the case of E neurons. Some discharge at decreasing frequencies, and some discharge at the same high rate during inspiration or expiration. However, expiration is passive during quiet breathing, and E neurons are quiet: they become active only when ventilation is increased.

The area in the medulla that is concerned with respiration has classically been called the **respiratory center,** but there are actually 2 groups of respiratory neurons (Fig 36–1). The **dorsal group** of neurons is located in and near the nucleus of the tractus solitarius. The **ventral group** is a long column of neurons that extends through the nucleus ambiguus and the nucleus retroambiguus in the ventrolateral medulla. The dorsal group is made up primarily of I neurons, some of which project monosynaptically to the phrenic motor neurons. They probably receive afferents from the airways and carotid and aortic bodies which terminate in the nucleus tractus solitarius. The ventral group has E neurons at its caudal end, I neurons in its midportion, and E neurons at its rostral end. Some of these neurons project to respiratory motor neurons. Those in the rostral end of the group appear to inhibit I neurons during expiration.

The main components of the **respiratory control pattern generator** responsible for automatic respiration are located in the medulla, since spontaneous respiration continues, albeit somewhat irregular and gasping, after transection of the brain stem at the inferior border of the pons (Section C in Fig 36–1). It has been shown that in an invertebrate, a simple 3-neuron network consisting of a facilitating dopaminergic neuron connected to an inspiratory and an expiratory neuron can discharge in a rhythmic fashion. The inspiratory and expiratory neurons have inhibitory connections to each other, and discharge of

Figure 36–1. Respiratory neurons in the brain stem. Dorsal view of brain stem; cerebellum removed. The effects of various lesions and brain stem transections are also shown. The spirometer tracings at the right indicate the depth and rate of breathing. DRG, dorsal group of respiratory neurons; VRG, ventral group of respiratory neurons; NPBL, nucleus parabrachialis (pneumotaxic center); 4th vent, fourth ventricle; IC, inferior colliculus; CP, middle cerebellar peduncle. The roman numerals identify cranial nerves. (Modified and reproduced, with permission, from Mitchell RA, Berger A: State of the art: Review of neural regulation of respiration. *Am Rev Respir Dis* 1975;**111**:206.

one initiates discharge of the other after a delay because after a period of inhibition, each fires during a period of postinhibitory rebound excitation. The dopaminergic input keeps the reciprocating discharge from dying out. However, despite intensive research, the mechanism responsible for rhythmic respiratory discharge in mammals remains unknown. No pacemaker cells have been identified, and neither the dorsal nor the ventral groups of respiratory neurons are obligatory parts of the generator since selective lesions of one or the other reduce the amplitude of respiration but do not abolish it.

Pontine & Vagal Influences

Although the rhythmic discharge of medullary neurons concerned with respiration is spontaneous, it is modified by neurons in the pons and afferents in the vagus from receptors in the airways and lungs. An area known as the **pneumotaxic center** in the medial parabrachial and Kölliker-Fuse nuclei of the dorsolateral pons contains I and E neurons as well as neurons that are active in both phases of respiration. When this area is damaged, respiration becomes slower and tidal volume greater, and when the vagi are also cut in anesthetized animals, there are prolonged inspiratory spasms that resemble breath holding (**apneusis; Section B in Fig 36–1**). The normal function of the pneumotaxic center is unknown, but it may play a role in switching between inspiration and expiration.

Stretching of the lungs during inspiration initiates impulses in afferent pulmonary vagal fibers. These impulses inhibit inspiratory discharge. This is why the depth of inspiration is increased after vagotomy (Fig 36–1) and apneusis develops if the vagi are cut after damage to the pneumotaxic center. As shown in Fig 36–2, the vagal feedback activity does not alter the rate of rise of the neural activity in respiratory motor neurons, but the activity is prolonged.

When the activity of the inspiratory neurons is increased in intact animals, the rate and the depth of breathing are increased. The depth of respiration is increased because the lungs are stretched to a greater degree before the amount of vagal and pneumotaxic center inhibitory activity is sufficient to overcome the more intense inspiratory neuron discharge. The respiratory rate is increased because the after-discharge in the vagal and possibly the pneumotaxic afferents to the medulla is rapidly overcome.

REGULATION OF RESPIRATORY ACTIVITY

A rise in the P_{CO_2} or H^+ concentration of arterial blood or a drop in its P_{O_2} increases the level of respiratory neuron activity in the medulla, and changes in the opposite direction have a slight inhibitory effect. The effects of variations in blood chemistry on ventilation are mediated via respiratory **chemoreceptors**—the carotid and aortic bodies and collections of cells in the medulla that are sensitive to changes in

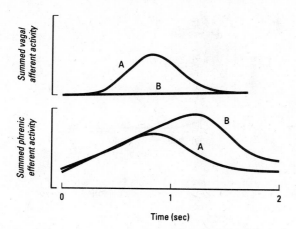

Figure 36–2. Superimposed records of 2 breaths: **(A)** with and **(B)** without feedback vagal afferent activity from stretch receptors in the lungs. Note that the rate of rise in phrenic nerve activity to the diaphragm is unaffected but the discharge is prolonged in the absence of vagal input. (Modified from von Euler C: Central pattern generation during breathing. *Trends Neurosci* 1980;**3**:275.)

the chemistry of the blood. They initiate impulses that stimulate the respiratory center. Superimposed on this basic **chemical control of respiration,** other afferents provide nonchemical controls that affect breathing in particular situations (Table 36–1).

CHEMICAL CONTROL OF BREATHING

The chemical regulatory mechanisms adjust ventilation in such a way that the alveolar P_{CO_2} is normally held constant, the effects of excess H^+ in the blood are combatted, and the P_{O_2} is raised when it falls to a potentially dangerous level. The respiratory minute volume is proportionate to the metabolic rate, but the link between metabolism and ventilation is CO_2, not O_2. The receptors in the carotid and aortic bodies are stimulated by a rise in the P_{CO_2} or H^+ concentration of

Table 36–1. Stimuli affecting the respiratory center.

Chemical control
  CO_2 (via CSF and brain interstitial fluid H^+
    concentration)

  $\left.\begin{array}{l} O_2 \\ H^+ \end{array}\right\}$ (via carotid and aortic bodies)

Nonchemical control
  Vagal afferents from receptors in the airways and
    lungs
  Afferents from the pons, hypothalamus, and limbic
    system
  Afferents from proprioceptors
  Afferents from baroreceptors: arterial, atrial, ventric-
    ular, pulmonary

arterial blood or a decline in its P_{O_2}. After denervation of the carotid chemoreceptors, the response to a drop in P_{O_2} is abolished; the predominant effect of hypoxia after denervation of the carotid bodies is a direct depression of the respiratory center. The response to changes in arterial blood H^+ concentration in the pH 7.3–7.5 range is also abolished, although larger changes exert some effect. The response to changes in arterial P_{CO_2}, on the other hand, is affected only slightly; it is reduced no more than 30–35%.

Carotid & Aortic Bodies

There is a carotid body near the carotid bifurcation on each side, and there are usually 2 or more aortic bodies near the arch of the aorta (Fig 36–3). Each carotid and aortic body (**glomus**) contains islands of 2 types of cells, type I and type II cells, surrounded by fenestrated sinusoidal capillaries. The type I or **glomus cells** are closely associated with cuplike endings of the afferent nerves (Fig 36–4). The glomus cells resemble adrenal chromaffin cells and have dense-core granules containing catecholamines that are released upon exposure to hypoxia and cyanide (see

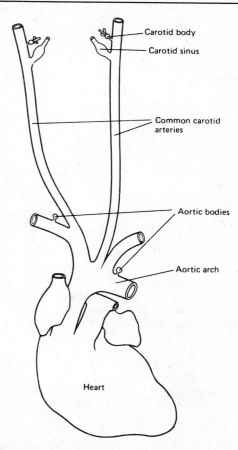

Figure 36–3. Location of carotid and aortic bodies.

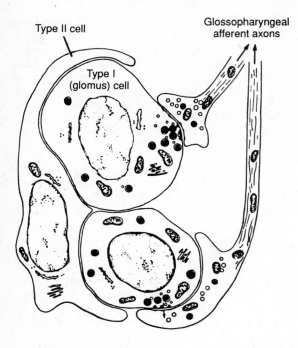

Figure 36–4. Organization of the carotid body. Type I (glomus) cells contain catecholamines. When exposed to hypoxia, they release catecholamines, and these stimulate the cuplike endings of the carotid sinus nerve fibers in the glossopharyngeal nerve. The glialike type II cells surround the type I cells and probably have a sustentacular function.

Figure 36–5. Change in the rate of discharge of a single afferent fiber from the carotid body when P_{aCO_2} is reduced. (Courtesy of S Sampson.)

below). The cells are excited by hypoxia, and the amines they release excite the nerve endings, although it is uncertain which catecholamine is involved. The type II cells are glialike, and each surrounds 4 to 6 type I cells. Their function is probably sustentacular.

Outside the capsule of each body, the nerve fibers acquire a myelin sheath; however, they are only 2–5 μm in diameter and conduct at the relatively low rate of 7–12 m/s. Afferents from the carotid bodies ascend to the medulla via the carotid sinus and glossopharyngeal nerves, and fibers from the aortic bodies ascend in the vagi. Studies in which one carotid body has been isolated and perfused while recordings are being taken from its afferent nerve fibers show that there is a graded increase in impulse traffic in these afferent fibers as the P_{O_2} of the perfusing blood is lowered (Fig 36–5) or the P_{CO_2} raised.

The exact mechanism by which hypoxia causes release of catecholamines from type I cells is unsettled. However, an attractive hypothesis is that exocytosis is triggered by release of Ca^{2+} from the mitochondria. The mitochondria contain **cytochrome aa$_3$** which maintains a negative membrane potential in these organelles relative to the cytoplasm by electron transport and expulsion of H^+. Hypoxia inhibits the cyto-

chrome, leading to a decrease in the membrane potential, and this triggers Ca^{2+} efflux.

The blood flow in each 2-mg carotid body is about 0.04 mL/min, or 2000 mL per 100 g of tissue per minute, compared with a blood flow per 100 g per minute of 54 mL in the brain and 420 mL in the kidneys (see Table 32–1). Because the blood flow per unit of tissue is so enormous, the O_2 needs of the cells can be met largely by dissolved O_2 alone. Therefore, the receptors are not stimulated in conditions such as anemia and carbon monoxide poisoning, in which the amount of dissolved O_2 in the blood reaching the receptors is generally normal even though the combined O_2 in the blood is markedly decreased. The receptors are stimulated when the arterial P_{O_2} is low or when, because of vascular stasis, the amount of O_2 delivered to the receptors per unit of time is decreased. Powerful stimulation is also produced by drugs such as cyanide, which prevent O_2 utilization at the tissue level. In sufficient doses, nicotine and lobeline activate the chemoreceptors. It has also been reported that infusion of K^+ increases the discharge rate in chemoreceptor afferents, and since the plasma K^+ level is increased during exercise, the increase may contribute to exercise-induced hyperpnea.

Because of their anatomic location, the aortic bodies have not been studied in as great detail as the carotid bodies. Their responses are probably similar but of lesser magnitude. In humans in whom both carotid bodies have been removed but the aortic bodies have been left intact, the responses are essentially the same as those following denervation of both carotid and aortic bodies in animals; there is little change in ventilation at rest, but the ventilatory response to hypoxia is lost and there is a 30% reduction in the ventilatory responses to CO_2. The humans who were studied had their carotid bodies removed in an effort to relieve severe asthma, but it is now clear that bilateral carotid body resection is of little benefit in this disease.

Chemoreceptors in the Brain Stem

The chemoreceptors that mediate the hyperventilation produced by increases in arterial P_{CO_2} after the carotid and aortic bodies are denervated are located in the medulla oblongata and consequently are called **medullary chemoreceptors.** They are near the respiratory center but separate from it. The response to CO_2, for example, is depressed during anesthesia and natural sleep, but the response to hypoxia is unchanged. This experimental observation indicates that CO_2 does not act directly on the inspiratory neurons.

The location of the medullary chemoreceptors on the ventral surface of the brain stem is shown in Fig 36–6. They monitor the H^+ concentration of the cerebrospinal fluid, including the brain interstitial fluid. CO_2 readily penetrates membranes, including the blood-brain barrier, whereas H^+ and HCO_3^- penetrate slowly. The CO_2 that enters the brain and CSF is promptly hydrated. The H_2CO_3 dissociates, so that the local H^+ concentration rises. The H^+ concentration in brain interstitial fluid parallels the arterial P_{CO_2}. Experimentally produced changes in the P_{CO_2} of CSF have minor, variable effects on respiration as long as the H^+ concentration is held constant, but any increase in spinal fluid H^+ concentration stimulates respiration. The magnitude of the stimulation is proportionate to the rise in H^+ concentration. Thus, the effects of CO_2 on respiration are mainly due to its movement into the CSF and brain interstitial fluid, where it increases the H^+ concentration and stimulates receptors sensitive to H^+.

Ventilatory Responses to Changes in Acid-Base Balance

In metabolic acidosis due, for example, to the accumulation of the acid ketone bodies in the circula-

tion in diabetes mellitus, there is pronounced respiratory stimulation (Kussmaul breathing; see Chapter 19). The hyperventilation decreases alveolar P_{CO_2} ("blows off CO_2") and thus produces a compensatory fall in blood H^+ concentration (see Chapter 39). Conversely, in metabolic alkalosis due, for example, to protracted vomiting with loss of HCl from the body, ventilation is depressed and the arterial P_{CO_2} rises, raising the H^+ concentration toward normal (see Chapter 39). If there is an increase in ventilation that is not secondary to a rise in arterial H^+ concentration, the drop in P_{CO_2} lowers the H^+ concentration below normal **(respiratory alkalosis);** conversely, hypoventilation that is not secondary to a fall in plasma H^+ concentration causes **respiratory acidosis.**

Ventilatory Responses to CO$_2$

The arterial P_{CO_2} is normally maintained at 40 mm Hg. When there is a rise in arterial P_{CO_2} as a result of increased tissue metabolism, ventilation is stimulated and the rate of pulmonary excretion of CO_2 is increased until the arterial P_{CO_2} falls to normal, shutting off the stimulus. The operation of this feedback mechanism keeps CO_2 excretion and production in balance.

When a gas mixture containing CO_2 is inhaled, the alveolar P_{CO_2} rises, elevating the arterial P_{CO_2} and stimulating ventilation as soon as the blood that contains more CO_2 reaches the medulla. CO_2 elimination is increased, and the alveolar P_{CO_2} drops toward normal. This is why relatively large increments in the P_{CO_2} of inspired air (eg, 15 mm Hg) produce relatively slight increments in alveolar P_{CO_2} (eg, 3 mm Hg). However, the P_{CO_2} does not drop to normal, and a new equilibrium is reached at which the alveolar P_{CO_2} is slightly elevated and the hyperventilation persists as long as CO_2 is inhaled. The essentially linear relationship between respiratory minute volume and the alveolar P_{CO_2} is shown in Fig 36–7.

There is, of course, an upper limit to this linearity. When the P_{CO_2} of the inspired gas is close to the alveolar P_{CO_2}, elimination of CO_2 becomes difficult. When the CO_2 content of the inspired gas is more than 7%, the alveolar and arterial P_{CO_2} begin to rise abruptly in spite of hyperventilation. The resultant accumulation of CO_2 in the body **(hypercapnia)** depresses the central nervous system, including the respiratory center, and produces headache, confusion, and eventually, coma **(CO$_2$ narcosis).**

Ventilatory Response to Oxygen Lack

When the O_2 content of the inspired air is decreased, there is an increase in respiratory minute volume. The stimulation is slight when the P_{O_2} of the inspired air is more than 60 mm Hg, and marked stimulation of respiration occurs only at lower P_{O_2} values (Fig 36–8). However, any decline in arterial P_{O_2} below 100 mm Hg produces increased discharge in the nerves from the carotid and aortic chemorecep-

Figure 36–6. Rostral (R) and caudal (C) chemosensitive areas on the ventral surface of the medulla.

Figure 36–7. Responses of normal subjects to inhaling O_2 and approximately 2, 4, and 6% CO_2. The increase in respiratory minute volume is due to an increase in both the depth and rate of respiration. (Reproduced, with permission, from Lambertsen CJ in: *Medical Physiology,* 13th ed. Mountcastle VB [editor]. Mosby, 1974.)

Figure 36–8. *Top:* Average respiratory minute volume during the first half hour of exposure to gases containing various amounts of O_2. The horizontal line in each case indicates the mean; the vertical bar indicates one standard deviation. *Bottom:* Alveolar P_{O_2} and P_{CO_2} values when breathing air at various barometric pressures. The 2 graphs are aligned so that the P_{O_2} of the inspired gas mixtures in the upper graph correspond to the P_{O_2} at the various barometric pressures in the lower graph. (Data compiled by RH Kellogg.)

tors. There are 2 reasons why in normal individuals this increase in impulse traffic does not increase ventilation to any extent until the P_{O_2} is less than 60 mm Hg. Because Hb is a weaker acid than HbO_2 (see Chapter 35), there is a slight decrease in the H^+ concentration of arterial blood when the arterial P_{O_2} falls and hemoglobin becomes less saturated with O_2. The fall in H^+ concentration tends to inhibit respiration. In addition, any increase in ventilation that does occur lowers the alveolar P_{CO_2}, and this also tends to inhibit respiration. Therefore, the stimulatory effects of hypoxia on ventilation are not clearly manifest until they become strong enough to override the counterbalancing inhibitory effects of a decline in arterial H^+ concentration and P_{CO_2}.

The effects on ventilation of decreasing the alveolar P_{O_2} while holding the alveolar P_{CO_2} constant are shown in Fig 36–9. When the alveolar P_{CO_2} is stabilized at a level 2–3 mm Hg above normal, there is an inverse relationship between ventilation and the alveolar P_{O_2} even in the 90–110 mm Hg range; but when the alveolar P_{CO_2} is fixed at lower than normal values, there is no stimulation of ventilation by hypoxia until the alveolar P_{O_2} falls below 60 mm Hg.

Figure 36–9. Ventilation at various alveolar P_{O_2} values when P_{CO_2} is held constant at 49, 44, or 37 mm Hg. (Modified from Loeschke HH, Gertz KH: Einfluss des O_2-Druckes in der Einatmungsluft auf die Atemtätigkeit des Menschen, geprüft unter Konstanthaltung des alveolaren CO_2-Druckes. *Arch Ges Physiol Pflügers* 1958;**267**:460.)

Effects of Hypoxia on the CO₂ Response Curve

When the converse experiment is performed, ie, when the alveolar P_{O_2} is held constant while the response to varying amounts of inspired CO_2 is tested, a linear response is obtained (Fig 36–10). When the CO_2 response is tested at different fixed P_{O_2} values, the slope of the response curve changes, with the slope increased when alveolar P_{O_2} is decreased. In other words, hypoxia makes the individual more sensitive to increases in arterial P_{CO_2}. However, the alveolar P_{CO_2} level at which the curves in Fig 36–10 intersect is unaffected. In the normal individual, this threshold value is just below the normal alveolar P_{CO_2}, indicating that normally there is a very slight but definite "CO_2 drive" of the respiratory center.

Effect of H⁺ on the CO₂ Response

The stimulatory effects of H⁺ and CO_2 on respiration appear to be additive and not, like those of CO_2 and O_2, complexly interrelated. In metabolic acidosis, the CO_2 response curves are similar to those in Fig 36–10, except that they are shifted to the left. In other words, the same amount of respiratory stimulation is produced by lower arterial P_{CO_2} levels. It has been calculated that the CO_2 response curve shifts 0.8 mm Hg to the left for each nanomole rise in arterial H⁺. About 40% of the ventilatory response to CO_2 is removed if the increase in arterial H⁺ produced by CO_2 is prevented. As noted above, the remaining 60% is probably due to the effect of CO_2 on spinal fluid or brain interstitial fluid H⁺ concentration.

Breath Holding

Respiration can be voluntarily inhibited for some time, but eventually the voluntary control is overridden. The point at which breathing can no longer be voluntarily inhibited is called the **breaking point.** Breaking is due to the rise in arterial P_{CO_2} and the fall in P_{O_2}. Individuals can hold their breath longer after removal of the carotid bodies. Breathing 100% oxygen before breath holding raises alveolar P_{O_2} initially, so that the breaking point is delayed. The same is true of hyperventilating room air, because CO_2 is blown off and arterial P_{CO_2} is lower at the start. Reflex or mechanical factors appear to influence the breaking point, since subjects who hold their breath as long as possible and then breathe a gas mixture low in O_2 and high in CO_2 can hold their breath for an additional 20 seconds or more. Psychologic factors also play a role, and subjects can hold their breath longer when they are told their performance is very good than when they are not.

NONCHEMICAL INFLUENCES ON RESPIRATION

Responses Mediated by Receptors in the Airways & Lungs

Receptors in the airways and lungs are innervated by myelinated and unmyelinated vagal fibers. The unmyelinated fibers are C fibers. The receptors innervated by myelinated fibers are commonly divided into **slowly adapting receptors** and **rapidly adapting receptors** on the basis of whether sustained stimulation leads to prolonged or transient discharge in their afferent nerve fibers (Table 36–2). The other group of receptors are presumably the endings of C fibers, and they are divided into pulmonary and bronchial subgroups on the basis of their location.

The shortening of inspiration produced by vagal afferent activity (Fig 36–2) is mediated by slowly adapting receptors. So are the **Hering-Breuer reflexes.** The Hering-Breuer inflation reflex is an increase in the duration of expiration produced by steady lung inflation, and the Hering-Breuer deflation reflex is a decrease in the duration of expiration produced by marked deflation of the lung.

Because the rapidly adapting receptors are stimulated by chemicals such as histamine, they have been called **irritant receptors.** Activation of rapidly adapting receptors in the trachea causes coughing, bronchoconstriction, and mucus secretion, and activation of rapidly adapting receptors in the lung may produce hyperpnea.

Because the C fiber endings are close to pulmonary vessels, they have been called J (juxtacapillary) receptors. They are stimulated by hyperinflation of the lung, but they respond as well to intravenous or intracardiac administration of chemicals such as capsaicin. The reflex response that is produced is apnea followed by rapid breathing, bradycardia, and hypotension (**pulmonary chemoreflex**). A similar re-

Figure 36–10. Fan of lines showing CO_2 response curves at various fixed values for alveolar P_{O_2}.

Table 36-2. Airway and lung receptors.[1]

Vagal Innervation	Type	Location in Interstitium	Stimulus	Response
Myelinated	Slowly adapting	Among airway smooth muscle cells(?)	Lung inflation	Inspiratory time-shortening Hering-Breuer inflation and deflation reflexes Bronchodilation Tachycardia
	Rapidly adapting	Among airway epithelial cells	Lung hyperinflation Exogenous and endogenous substances (eg, histamine, prostaglandins)	Hyperpnea Cough Bronchoconstriction Mucus secretion
Unmyelinated C fibers	Pulmonary C fibers Bronchial C fibers	Close to blood vessels	Lung hyperinflation Exogenous and endogenous substances (eg, capsaicin, bradykinin, serotonin)	Apnea followed by rapid breathing Bronchoconstriction Bradycardia Hypotension Mucus secretion

[1]Modified and reproduced, with permission, from Berger AJ, Hornbein TF: Control of respiration. In: *Textbook of Physiology*, 21st ed. Vol. 2, Patton HD et al (editors). Saunders, 1989.

sponse is produced by receptors in the heart (**Bezold-Jarisch reflex** or the **coronary chemoreflex;** see Chapter 31). The physiologic role of this reflex is uncertain, but it probably occurs in pathologic states such as pulmonary congestion or embolization, in which it is produced by endogenously released substances.

Coughing & Sneezing

Coughing begins with a deep inspiration followed by forced expiration against a closed glottis. This increases the intrapleural pressure to 100 mm Hg or more. The glottis is then suddenly opened, producing an explosive outflow of air at velocities up to 965 km (600 miles) per hour. Sneezing is a similar expiratory effort with a continuously open glottis (see Chapter 14). These reflexes help expel irritants and keep airways clear.

Responses in Patients
With Heart-Lung Transplants

Transplantation of the heart and lungs has progressed beyond the experimental stage and is now an established treatment for severe pulmonary disease and some other conditions. In individuals with transplants, the recipient's right atrium and ascending aorta are sutured to the donor heart, and the donor heart does not reinnervate, so the resting heart rate is elevated. The donor trachea is sutured to the recipient's just above to carina, and afferent fibers from the lungs do not regrow. Consequently, healthy patients with heart-lung transplants provide an opportunity to evaluate the role of lung innervation in normal physiology. Their cough responses to stimulation of the trachea are normal, because the trachea remains innervated, but their cough responses to stimu-

lation of the smaller airways are absent. Their bronchi tend to be dilated to a greater degree than normal. In addition, they have the normal number of yawns and sighs, indicating that these do not depend on innervation of the lungs. Finally, they lack Hering-Breuer reflexes, but their pattern of breathing at rest is normal, indicating that these reflexes do not play an important role in the regulation of resting respiration in humans.

Afferents From "Higher Centers"

Pain and emotional stimuli affect respiration, so there must also be afferents from the limbic system and hypothalamus to the respiratory neurons in the brain stem. In addition, even though breathing is not usually a conscious event, both inspiration and expiration are under voluntary control. The pathways for voluntary control pass from the neocortex to the motor neurons innervating the respiratory muscles, bypassing the medullary neurons.

Since voluntary and automatic control of respiration are separate, automatic control is sometimes disrupted without loss of voluntary control. The clinical condition that results has been called **Ondine's curse.** In German legend, Ondine was a water nymph who had an unfaithful mortal lover. The king of the water nymphs punished the lover by casting a curse upon him that took away all his automatic functions. In this state, he could stay alive only by staying awake and remembering to breathe. He eventually fell asleep from sheer exhaustion, and his respiration stopped. Patients with this intriguing condition generally have bulbar poliomyelitis or disease processes that compress the medulla. The condition has also been inadvertently produced in patients who have been subjected to bilateral anterolateral cervical cordotomy for pain (see Chapter 7). This cuts the pathways that bring about automatic respiration while leaving the voluntary efferent pathways in the corticospinal and rubrospinal tracts intact.

Afferents From Proprioceptors

Carefully controlled experiments have shown that active and passive movements of joints stimulate respiration, presumably because impulses in afferent pathways from proprioceptors in muscles, tendons, and joints stimulate the inspiratory neurons. This effect probably helps increase ventilation during exercise.

Respiratory Components of Visceral Reflexes

The respiratory adjustments during vomiting, swallowing, and gagging are discussed in Chapters 14 and 26. Inhibition of respiration and closure of the glottis during these activities not only prevent the aspiration of food or vomitus into the trachea but, in the case of vomiting, fix the chest so that contraction of the abdominal muscles increases the intra-abdominal pressure. Similar glottic closure and inhibition of respiration occur during voluntary and involuntary straining.

Hiccup is a spasmodic contraction of the diaphragm that produces an inspiration during which the glottis suddenly closes. The glottic closure is responsible for the characteristic sensation and sound. Yawning is a peculiar "infectious" respiratory act whose physiologic basis and significance are uncertain. However, underventilated alveoli have a tendency to collapse, and it has been suggested that the deep inspiration and stretching open them and prevent the development of atelectasis. Sighing may have a similar function. Yawning also increases venous return to the heart.

Respiratory Effects of Baroreceptor Stimulation

Afferent fibers from the baroreceptors in the carotid sinuses, aortic arch, atria, and ventricles relay to the respiratory neurons as well as the vasomotor and cardioinhibitory neurons in the medulla. Impulses in them inhibit respiration, but the inhibitory effect is slight and of little physiologic importance. The hyperventilation in shock is due to chemoreceptor stimulation caused by acidosis and hypoxia secondary to local stagnation of blood flow and is not baroreceptor-mediated. The activity of inspiratory neurons affects blood pressure and heart rate (see Chapters 28 and 31), and activity in the vasomotor and cardiac areas in the medulla may have minor effects on respiration.

Effects of Sleep

Respiration is less rigorously controlled during sleep than in the waking state, and brief periods of

Figure 36–11. Effect of graded increases in arterial P_{CO_2} on ventilation ($\dot{V}$) during sleep in a healthy young adult. The solid circles are the responses while the subject was awake, and the open squares are the responses during slow-wave sleep. (Reproduced, with permission, from Cherniak NS: Respiratory dysrhythmias during sleep. *N Engl J Med* 1981;**305**:325.)

apnea occur in normal sleeping adults. The causes of the changes include a decreased sensitivity to CO_2 during slow-wave sleep (Fig 36–11). There are variable changes in the ventilatory response to hypoxia. If the P_{CO_2} falls during the waking state, various stimuli from proprioceptors and the environment maintain respiration, but during sleep, these stimuli are decreased and a decrease in P_{CO_2} can cause apnea. During REM sleep, the response to CO_2 appears to be reduced to an even greater degree, but breathing is irregular and the CO_2 response is highly variable.

In a small percentage of individuals, apnea during sleep can constitute a clinical problem of appreciable magnitude. This **sleep-apnea syndrome** can occur at any age. The symptoms include morning headaches, fatigue, and, in more advanced cases, the clinical picture of respiratory failure with normal lungs. These patients are polycythemic, hypoxemic, and hypercapnic. The causes of sleep apnea are not yet clearly defined, but one cause appears to be failure during sleep of the genioglossus muscles to contract during inspiration. These muscles pull the tongue forward, and when they do not contract, the tongue falls back and obstructs the airway.

Sudden Infant Death Syndrome

The sudden infant death syndrome (SIDS) may be a form of sleep apnea. This disorder, in which apparently healthy infants are found dead, often in their cribs, has attracted a great deal of attention. Apneic spells are common in premature infants. However, periods of prolonged apnea do not correlate with the subsequent occurrence of death, and none of the known tests of chemoresponsiveness reliably predict which infants will subsequently have difficulty.

37

Respiratory Adjustments in Health & Disease

INTRODUCTION

This chapter is concerned with the respiratory adjustments to exercise, high altitude, hypoxia, hypercapnia, and other respiratory abnormalities. In the same way that the compensatory adjustments of the cardiovascular system to environmental changes and disease illustrate the integrated operation of the cardiovascular regulatory mechanisms (see Chapter 33), these respiratory adjustments highlight the operation of the respiratory regulatory mechanisms discussed in Chapters 34–36.

EFFECTS OF EXERCISE

Many cardiovascular and respiratory mechanisms must operate in an integrated fashion if the O_2 needs of the active tissue are to be met and the extra CO_2 and heat removed from the body during exercise. Circulatory changes increase muscle blood flow while maintaining adequate circulation in the rest of the body (see Chapter 33). There is in addition an increase in the extraction of O_2 from the blood in exercising muscles and an increase in ventilation that provides extra O_2, eliminates some of the heat, and excretes extra CO_2.

Changes in Ventilation

During exercise, the amount of O_2 entering the blood in the lungs is increased because the amount of O_2 added to each unit of blood and the pulmonary blood flow per minute are increased. The P_{O_2} of blood flowing into the pulmonary capillaries falls from 40 to 25 mm Hg or less, so that the alveolar-capillary P_{O_2} gradient is increased and more O_2 enters the blood. Blood flow per minute is increased from 5.5 L/min to as much as 20–35 L/min. The total amount of O_2 entering the blood therefore increases from 250 mL/min at rest to values as high as 4000 mL/min. The amount of CO_2 removed from each unit of blood is increased, and CO_2 excretion increases from 200 mL/min to as much as 8000 mL/min. The increase in O_2 uptake is proportionate to work load up to a maximum. Above this maximum, O_2 consumption levels off and the blood lactate level continues to rise (Fig 37–1). The

lactate comes from muscles in which aerobic resynthesis of energy stores cannot keep pace with their utilization and an **oxygen debt** is being incurred (see Chapter 3).

There is an abrupt increase in ventilation with the onset of exercise, followed after a brief pause by a further, more gradual increase (Fig 37–2). With moderate exercise, the increase is due mostly to an increase in the depth of respiration; this is accompanied by an increase in the respiratory rate when the exercise is more strenuous. There is an abrupt decrease in ventilation when exercise ceases, followed after a brief pause by a more gradual decline to preexercise values. The abrupt increase at the start of exercise is presumably due to psychic stimuli and afferent impulses from proprioceptors in muscles, tendons, and joints. The more gradual increase is presumably humoral even though arterial pH, P_{CO_2}, and P_{O_2} remain constant during moderate exercise. The increase in ventilation is proportionate to the increase in O_2 consumption, but the mechanisms responsible for the stimulation of respiration are still the subject of much debate. The increase in body temperature may play a role. As noted in Chapter 36, exercise increases the plasma K^+ level, and this increase may stimulate the peripheral chemoreceptors. In addition, it may be that the sensitivity of the respiratory center to CO_2 is increased or that the respiratory fluctuations in arterial P_{CO_2} increase so that, even though the mean arterial P_{CO_2} does not rise, it is CO_2 that is responsible for the increase in ventilation. O_2 also seems to play some role despite the lack of a decrease in arterial P_{O_2}, since, during the performance of a given amount of work, the increase in ventilation while breathing 100% O_2 is 10–20% less than the increase while breathing air (Fig 37–3). Thus, it currently appears that a number of different factors combine to produce the increase in ventilation seen during moderate exercise.

When exercise becomes more vigorous, buffering of the increased amounts of lactic acid that are produced liberates more CO_2, and this further increases ventilation. The response to graded exercise is shown in Fig 37–4. With increased production of acid, the increases in ventilation and CO_2 production remain proportionate, so alveolar and arterial CO_2 change

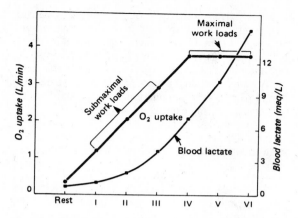

Figure 37–1. Relation between work load, blood lactate level, and O_2 uptake. I–VI, increasing work loads produced by increasing the speed and grade of a treadmill on which the subjects worked. (Reproduced, with permission, from Mitchell JH, Blomqvist G: Maximal oxygen uptake. *N Engl J Med* 1971;**284**:1018.)

Figure 37–3. Effect of 100% O_2 on ventilation during work at various levels on the bicycle ergometer. (Reproduced, with permission, from Astrand P-O: The respiratory activity in man exposed to prolonged hypoxia. *Acta Physiol Scand* 1954;**30**:343.)

relatively little **(isocapnic buffering).** Because of the hyperventilation, alveolar P_{O_2} increases. With further accumulation of lactic acid, the increase in ventilation outstrips CO_2 production and alveolar P_{CO_2} falls, as does arterial P_{CO_2}. The decline in arterial P_{CO_2} provides respiratory compensation (see Chapter 39) for the metabolic acidosis produced by the additional lactic acid. The additional increase in ventilation produced by the acidosis is dependent on the carotid bodies and does not occur if they are removed.

The respiratory rate after exercise does not reach basal levels until the O_2 debt is repaid. This may take as long as 90 minutes. The stimulus to ventilation after exercise is not the arterial P_{CO_2}, which is normal or low, or the arterial P_{O_2}, which is normal or high, but the elevated arterial H^+ concentration due to the lactic acidemia. The magnitude of the O_2 debt is the amount by which O_2 consumption exceeds basal con-

Figure 37–2. Changes in ventilation during exercise. (Modified from Wasserman K, Whipp BJ, Casaburi R: Respiratory control during exercise. In: *Handbook of Physiology.* Section 3, *The Respiratory System,* vol II, part 2. Fishman AP [editor]. American Physiological Society, 1986.)

Figure 37–4. Changes in alveolar P_{CO_2}, alveolar P_{O_2}, ventilation ($\dot{V}_E$), CO_2 production ($\dot{V}_{CO_2}$), O_2 consumption ($\dot{V}_{O_2}$), arterial HCO_3^-, and arterial pH with graded increases in work on a bicycle ergometer. The subject was a normal adult male. (Reproduced, with permission, from Wasserman K, Whipp BJ, Casaburi R: Respiratory control during exercise. In: *Handbook of Physiology.* Section 3, *The Respiratory System,* vol II, part 2. Fishman AP [editor]. American Physiological Society, 1986.)

sumption from the end of exertion until the O_2 consumption has returned to preexercise basal levels. During repayment of the O_2 debt, there is a small rise in the O_2 in muscle myoglobin. ATP and phosphorylcreatine are resynthesized, and lactic acid is removed. Eighty percent of the lactic acid is converted to glycogen and 20% is metabolized to CO_2 and H_2O.

Because of the extra CO_2 produced by the buffering of lactic acid during strenuous exercise, the R rises, reaching 1.5–2.0. After exertion, while the O_2 debt is being repaid, the R falls to 0.5 or less.

Changes in the Tissues

Maximum O_2 uptake during exercise is limited by the maximum rate at which O_2 is transported to the mitochondria in the exercising muscle. However, this limitation is not normally due to deficient O_2 uptake in the lungs, and hemoglobin in arterial blood is saturated even during the most severe exercise.

During exercise, the contracting muscles use more O_2, and the tissue P_{O_2} and the P_{O_2} in venous blood from exercising muscle fall nearly to zero. More O_2 diffuses from the blood, the blood P_{O_2} drops, and more O_2 is removed from hemoglobin. Because the capillary bed is dilated and many previously closed capillaries are open, the mean distance from the blood to the tissue cells is greatly decreased; this facilitates the movement of O_2 from blood to cells. The oxygen-hemoglobin dissociation curve is steep in the P_{O_2} range below 60 mm Hg, and a relatively large amount of O_2 is supplied for each drop of 1 mm Hg in P_{O_2} (see Fig 35–2). Additional O_2 is supplied because, as a result of the accumulation of CO_2 and the rise in temperature in active tissues—and perhaps because of a rise in red blood cell 2,3-DPG—the dissociation curve shifts to the right (see Fig 35–3). The net effect is a 3-fold increase in O_2 extraction from each unit of blood. Since this increase is accompanied by a 30-fold or greater increase in blood flow, it permits the metabolic rate of muscle to rise as much as 100-fold during exercise (see Chapter 3).

Fatigue

Fatigue is a poorly understood phenomenon that is a normal consequence of intense exercise or mental effort. In addition, it is a symptom of many different diseases. During exercise, acidosis and other factors contribute to its production. The subjective hardness or "heaviness" of exercise correlates with the rate of O_2 consumption, not with the actual work performed in kg-m/min. Barrages of impulses in afferents from proprioceptors in muscles are said to make one feel "tired." The effects of the acidosis on the brain may contribute to the sensation of fatigue. Prolonged exercise produces hypoglycemia in normal individuals, but prevention of the hypoglycemia does not affect endurance or delay the onset of exhaustion. On the other hand, there is a correlation in humans between exhaustion and the degree of depletion of muscle glycogen, as determined by muscle biopsy. Sustained muscle contractions are painful, because the muscle becomes ischemic and a substance that stimulates pain endings accumulates ("P factor"; see Chapter 7), but intermittent contractions are not painful, because the P factor is washed away. Muscle stiffness may be due in part to the accumulation of interstitial fluid in the muscles during exertion.

Heat dissipation during exercise is discussed in Chapter 33 and summarized in Fig 33–4. The changes in acid-base balance associated with respiration are reviewed in Chapter 39.

HYPOXIA

Hypoxia is O_2 deficiency at the tissue level. It is a more correct term than **anoxia,** there rarely being no O_2 at all left in the tissues.

Traditionally, hypoxia has been divided into 4 types. Numerous other classifications have been used, but the 4-type system still has considerable utility if the definitions of the terms are kept clearly in mind. The 4 categories are as follows: (1) **hypoxic hypoxia (anoxic anoxia),** in which the P_{O_2} of the arterial blood is reduced; (2) **anemic hypoxia,** in which the arterial P_{O_2} is normal but the amount of hemoglobin available to carry O_2 is reduced; (3) **stagnant** or **ischemic hypoxia,** in which the blood flow to a tissue is so low that adequate O_2 is not delivered to it despite a normal P_{O_2} and hemoglobin concentration; and (4) **histotoxic hypoxia,** in which the amount of O_2 delivered to a tissue is adequate but, because of the action of a toxic agent, the tissue cells cannot make use of the O_2 supplied to them.

Effects of Hypoxia

The effects of stagnant hypoxia depend upon the tissue affected. In hypoxic hypoxia and the other generalized forms of hypoxia, the brain is affected first. A sudden drop in the inspired P_{O_2} to less than 20 mm Hg, which occurs, for example, when cabin pressure is suddenly lost in a plane flying above 16,000 m, causes loss of consciousness in about 20 seconds (Fig 37–5) and death in 4–5 minutes. Less severe hypoxia causes a variety of mental aberrations not unlike those produced by alcohol: impaired judgment, drowsiness, dulled pain sensibility, excitement, disorientation, loss of time sense, and headache. Other symptoms include anorexia, nausea, vomiting, tachycardia, and, when the hypoxia is severe, hypertension. The rate of ventilation is increased in proportion to the severity of the hypoxia of the carotid chemoreceptor cells.

Respiratory Stimulation

Dyspnea is by definition breathing in which the subject is conscious of shortness of breath; **hyperpnea** is the general term for an increase in the rate or

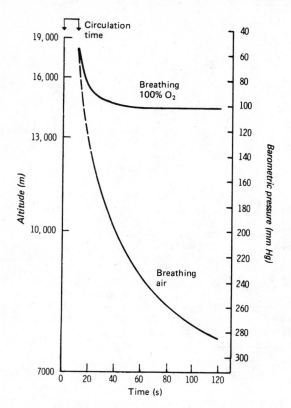

Figure 37–5. Duration of useful consciousness upon sudden exposure to the ambient pressure at various altitudes. Ten seconds is the approximate lung-to-brain circulation time.

depth of breathing regardless of the patient's subjective sensations. **Tachypnea** is rapid, shallow breathing. In general, a normal individual is not conscious of respiration until ventilation is doubled, and breathing is not uncomfortable (ie, the **dyspnea point** is not reached) until ventilation is tripled or quadrupled. Whether or not a given level of ventilation is uncomfortable also depends upon the respiratory reserve. This factor is taken into account in the **dyspneic index,** the percentage of the respiratory capacity not being used at a given respiratory minute volume.

When the dyspneic index is less than about 70%, dyspnea is usually present. An additional factor is the effort involved in moving the air in and out of the lungs (the work of breathing). In asthma, for example, the bronchi are constricted, especially during expiration, and breathing against this increased airway resistance is uncomfortable work. Another factor producing dyspnea may be increased discharge in vagal afferents from lung rapidly adapting receptors.

Cyanosis

Reduced hemoglobin has a dark color, and a dusky bluish discoloration of the tissues, called **cyanosis,** appears when the reduced hemoglobin concentration of the blood in the capillaries is more than 5 g/dL. Cyanosis is most easily seen in the nail beds and mucous membranes and in the earlobes, lips, and fingers, where the skin is thin. Its occurrence depends upon the total amount of hemoglobin in the blood, the degree of hemoglobin unsaturation, and the state of the capillary circulation.

One might think that cyanosis would be more marked when the cutaneous vessels were dilated. However, when there is cutaneous arteriolar and venous constriction, blood flow through the capillaries is very slow and more O_2 is removed from the hemoglobin. This is why moderate cold causes cyanosis in exposed areas even in normal individuals. In very cold weather cyanosis does not develop, because the drop in skin temperature inhibits the dissociation of oxyhemoglobin and the O_2 consumption of the cold tissues is decreased. Cyanosis does not occur in anemic hypoxia when the total hemoglobin content is low; in carbon monoxide poisoning, because the color of reduced hemoglobin is obscured by the cherry-red color of carbonmonoxyhemoglobin (see below); or in histotoxic hypoxia, because the blood gas content is normal. A discoloration of the skin and mucous membranes similar to cyanosis is produced by high circulating levels of methemoglobin (see Chapter 27).

HYPOXIC HYPOXIA

Hypoxic hypoxia is a problem in normal individuals at high altitudes and is a complication of pneumonia and a variety of other diseases of the respiratory system.

Effects of Decreased Barometric Pressure

The composition of air stays the same, but the total barometric pressure falls with increasing altitude (Fig 37–6). Therefore, the P_{O_2} also falls. At 3000 m (approximately 10,000 ft) above sea level, the alveolar P_{O_2} is about 60 mm Hg and there is enough hypoxic stimulation of the chemoreceptors to definitely increase ventilation. As one ascends higher, the alveolar P_{O_2} falls less rapidly and the alveolar P_{CO_2} declines somewhat because of the hyperventilation. The resulting fall in arterial P_{CO_2} produces respiratory alkalosis.

Hypoxic Symptoms Breathing Air

There are a number of compensatory mechanisms that operate over a period of time to increase altitude tolerance (**acclimatization),** but in unacclimatized subjects, mental symptoms such as irritability appear at about 3700 m. At 5500 m, the hypoxic symptoms are severe; and at altitudes above 6100 m (20,000 ft), consciousness is usually lost (Fig 37–7).

Figure 37–6. Composition of alveolar air in individuals breathing air (0–6100 m) and 100% O_2 (6100–13,700 m). The minimal alveolar P_{O_2} that an unacclimatized subject can tolerate without loss of consciousness is about 35–40 mm Hg. Note that with increasing altitude, the alveolar P_{CO_2} drops because of the hyperventilation due to hypoxic stimulation of the carotid and aortic chemoreceptors. The fall in barometric pressure with increasing altitude is not linear, because air is compressible.

Hypoxic Symptoms Breathing Oxygen

The total atmospheric pressure becomes the limiting factor in altitude tolerance when breathing 100% O_2. The partial pressure of water vapor in the alveolar air is constant at 47 mm Hg, and that of CO_2 is normally 40 mm Hg, so that the lowest barometric pressure at which a normal alveolar P_{O_2} of 100 mm Hg is possible is 187 mm Hg, the pressure at about 10,400 m (34,000 ft). At greater altitudes, the increased ventilation due to the decline in alveolar P_{O_2} lowers the alveolar P_{CO_2} somewhat, but the maximum alveolar P_{O_2} that can be attained breathing 100% O_2 at the ambient barometric pressure of 100 mm Hg at 13,700 m is about 40 mm Hg. At about 14,000 m, consciousness is lost in spite of the administration of 100% O_2 (Fig 37–5). However, an artificial atmosphere can be created around an individual; in a pressurized suit or cabin supplied with O_2 and a system to remove CO_2, it is possible to ascend to any altitude and to live in the vacuum of interplanetary space.

At 19,200 m, the barometric pressure is 47 mm Hg, and at or below this pressure the body fluids boil at body temperature. The point is largely academic, however, because any individual exposed to such a low pressure would be dead of hypoxia before the bubbles of steam could cause death.

Delayed Effects of High Altitude

Many individuals, when they first arrive at a high altitude, develop transient "mountain sickness." This syndrome develops 8–24 hours after arrival at altitude and lasts 4–8 days. It is characterized by headache, irritability, insomnia, breathlessness, and nausea and vomiting. Its cause is unsettled, but it appears to be associated with cerebral edema. The low P_{O_2} at high altitude causes arteriolar dilation, and if cerebral autoregulation does not compensate, there is an increase in capillary pressure that favors increased transudation of fluid into brain tissue. Individuals who do not develop mountain sickness have a diuresis at high altitude, and urine volume is decreased in individuals who develop the condition. However, treatment with diuretics does not prevent mountain sickness. The

Figure 37–7. Acute effects of hypoxia in individuals breathing air at various altitudes.

symptoms are reduced if alkalosis is reduced by treatment with acetazolamide or if the cerebral edema is reduced by administration of large doses of glucocorticoids.

High-altitude pulmonary edema and cerebral edema are serious forms of mountain sickness. Pulmonary edema is prone to occur in individuals who ascend quickly to altitudes above 2500 m and engage in heavy physical activity during the first 3 days after arrival. It is also seen in individuals acclimatized to high altitudes who spend 2 weeks or more at sea level and then reascend. It occurs in the absence of cardiovascular or pulmonary disease. It is associated with marked pulmonary hypertension, but left atrial pressures are normal. The protein content of the edema fluid is high, but the mechanism responsible for its production is unknown. However, the Ca^{2+} channel-blocking drug nifedipine, which lowers pulmonary arterial pressure, is of value in the treatment and prevention of the condition. Pulmonary edema also responds to rest and O_2 treatment and generally does not develop in individuals who ascend to high altitudes gradually and avoid physical exertion for the first few days of high-altitude exposure.

Acclimatization

Acclimatization to altitude is due to the operation of a variety of compensatory mechanisms. The respiratory alkalosis produced by the hyperventilation shifts the oxygen hemoglobin dissociation curve to the left, but there is a concomitant increase in red blood cell 2,3-DPG, which tends to decrease the O_2 affinity of hemoglobin. The net effect is an increase in P_{50} (see Chapter 35). The decrease in O_2 affinity makes more O_2 available to the tissues. However, it should be noted that the value of the increase in P_{50} is lost at very high altitudes, because when the arterial P_{O_2} is markedly reduced, the decreased O_2 affinity also interferes with O_2 uptake by hemoglobin in the lungs.

The initial ventilatory response to increased altitude is relatively small, because the alkalosis tends to counteract the stimulating effect of hypoxia. However, there is a steady increase in ventilation over the next 4 days (Fig 37–8) because the active transport of H^+ into CSF, or possibly a developing lactic acidosis in the brain, causes a fall in CSF pH that increases the response to hypoxia. After 4 days, the ventilatory response begins to decline slowly, but it takes years of residence at higher altitudes for it to decline to the initial level. Associated with this decline is a gradual desensitization to the stimulatory effects of hypoxia.

Erythropoietin secretion increases promptly on ascent to high altitude (see Chapter 24) and then falls somewhat over the following 4 days as the ventilatory response increases and the arterial P_{O_2} rises. The increase in circulating red blood cells triggered by the erythropoietin begins in 2–3 days and is sustained as long as the individual remains at high altitude.

Figure 37–8. Effect of acclimatization on the ventilatory response at various altitudes. $\dot{V}_E/\dot{V}_{O_2}$ is the ventilatory equivalent, the ratio of expired minute volume ($\dot{V}_E$) to the O_2 consumption ($\dot{V}_{O_2}$). (Reproduced, with permission, from Lenfant C, Sullivan K: Adaptation to high altitude. *N Engl J Med* 1971;**284:**1298.)

There are also compensatory changes in the tissues. The mitochondria, which are the site of oxidative reactions, increase in number, and there is an increase in myoglobin (see Chapter 35) that facilitates the movement of O_2 in the tissues. There is also an increase in the tissue content of cytochrome oxidase.

The effectiveness of the acclimatization process is indicated by the fact that in the Andes and Himalayas there are permanent human habitations at elevations above 5500 m (18,000 ft). The natives who live in these villages are barrel-chested and markedly polycythemic. They have low alveolar P_{O_2} values, but in most other ways they are remarkably normal.

Diseases Causing Hypoxic Hypoxia

Hypoxic hypoxia is the most common form of hypoxia seen clinically. The diseases that cause it can be roughly divided into those in which the gas exchange apparatus fails, those such as congenital heart disease in which large amounts of blood are shunted from the venous to the arterial side of the circulation, and those in which the respiratory pump fails (Table 37–1). Lung failure occurs when conditions such as pulmonary fibrosis produce alveolar-capillary block or there is ventilation-perfusion imbalance. Pump

Table 37–1. Disorders causing hypoxic hypoxia.

Lung failure (gas exchange failure)
  Pulmonary fibrosis
  Ventilation-perfusion imbalance
Shunt
Pump failure (ventilatory failure)
  Fatigue
  Mechanical defects
  Depression of respiratory controller in the brain

failure can be due to fatigue of the respiratory muscles in conditions in which the work of breathing is increased or to a variety of mechanical defects such as pneumothorax or bronchial obstruction that limit ventilation. It can also be caused by abnormalities of the neural mechanisms that control ventilation, such as depression of the respiratory neurons in the medulla by morphine and other drugs.

Ventilation-Perfusion Imbalance

Patchy ventilation-perfusion imbalance is by far the most common cause of hypoxic hypoxia in clinical situations. The physiologic effects of ventilation-perfusion imbalance and their role in producing the alterations in alveolar gas due to gravity are discussed in Chapter 34.

In disease processes that prevent ventilation of some of the alveoli, the ventilation-blood flow ratios in different parts of the lung determine the extent to which systemic arterial P_{O_2} declines. If nonventilated alveoli are perfused, the nonventilated but perfused portion of the lung is in effect a right-to-left shunt, dumping unoxygenated blood into the left side of the heart. Lesser degrees of ventilation-perfusion imbalance are more common. In the example illustrated in Fig 37–9, the underventilated alveoli (B) have a low alveolar P_{O_2}, whereas the overventilated alveoli (A) have a high alveolar P_{O_2}. However, the unsaturation of the hemoglobin of the blood coming from B is not completely compensated by the greater saturation of the blood coming from A, because hemoglobin is normally nearly saturated in the lungs and the higher alveolar P_{O_2} adds only a little more O_2 to the hemoglobin than it normally carries. Consequently, the arterial blood is unsaturated. On the other hand, the CO_2 content of the arterial blood is generally normal in such situations, since extra loss of CO_2 in overventilated regions can balance diminished loss in underventilated areas.

Venous-to-Arterial Shunts

When a cardiovascular abnormality such as an interatrial septal defect permits large amounts of unoxygenated venous blood to bypass the pulmonary capillaries and dilute the oxygenated blood in the systemic arteries ("right-to-left shunt"), chronic hypoxic

	A	B	A + B		A	B	A + B
Alveolar ventilation (L/min)	2.0	2.0	4.0	Alveolar ventilation (L/min)	3.2	0.8	4.0
Pulmonary blood flow (L/min)	2.5	2.5	5.0	Pulmonary blood flow (L/min)	2.5	2.5	5.0
Ventilation/blood flow ratio	0.8	0.8	0.8	Ventilation/blood flow ratio	1.3	0.3	0.8
Mixed venous O_2 saturation (%)	75.0	75.0	75.0	Mixed venous O_2 saturation (%)	75.0	75.0	75.0
Arterial O_2 saturation (%)	97.4	97.4	97.4	Arterial O_2 saturation (%)	98.2	91.7	95.0
Mixed venous O_2 tension (mm Hg)	40.0	40.0	40.0	Mixed venous O_2 tension (mm Hg)	40.0	40.0	40.0
Alveolar O_2 tension (mm Hg)	104.0	104.0	104.0	Alveolar O_2 tension (mm Hg)	116.0	66.0	106.0
Arterial O_2 tension (mm Hg)	104.0	104.0	104.0	Arterial O_2 tension (mm Hg)	116.0	66.0	84.0

Figure 37–9. *Left:* "Ideal" ventilation/blood flow relationship. *Right:* Nonuniform ventilation and uniform blood flow, uncompensated. $\dot{V}_A$, alveolar ventilation; MV, respiratory minute volume. (Reproduced, with permission, from Comroe JH Jr et al: *The Lung: Clinical Physiology and Pulmonary Function Tests,* 2nd ed. Year Book, 1962.)

hypoxia and cyanosis (**cyanotic congenital heart disease**) result. Administration of 100% O_2 raises the O_2 content of alveolar air and improves the hypoxia due to hypoventilation, impaired diffusion, or ventilation-perfusion imbalance (short of perfusion of totally unventilated segments) by increasing the amount of O_2 in the blood leaving the lungs. However, in patients with venous-to-arterial shunts and normal lungs, any beneficial effect of 100% O_2 is slight and is due solely to an increase in the amount of dissolved O_2 in the blood.

Collapse of the Lung

When a bronchus or bronchiole is obstructed, the gas in the alveoli beyond the obstruction is absorbed and the lung segment collapses. Collapse of alveoli is called **atelectasis.** The atelectatic area may range in size from a small patch to a whole lung. Some blood is diverted from the collapsed area to better ventilated portions of the lung, and this reduces the magnitude of the decline in arterial P_{O_2}.

When a large part of the lung is collapsed, there is an appreciable decrease in lung volume. The intrapleural pressure therefore becomes more negative and pulls the mediastinum, which in humans is a fairly flexible structure, to the affected side.

Another cause of atelectasis is absence or inactivation of surfactant, the surface-tension-depressing agent normally found in the thin fluid lining the alveoli (see Chapter 34). This abnormality is a major cause of failure of the lungs to expand normally at birth. Collapse of the lung may also be due to the presence of the pleural space of air (**pneumothorax**), tissue fluids (**hydrothorax, chylothorax**), or blood (**hemothorax**).

Pneumothorax

When air is admitted to the pleural space, through either a rupture in the lung or a hole in the chest wall, the lung on the affected side collapses because of its elastic recoil. Since the intrapleural pressure on the affected side is atmospheric, the mediastinum shifts toward the normal side. If the communication between the pleural space and the exterior remains open (**open** or **sucking pneumothorax**), more air moves in and out of the pleural space each time the patient breathes. If the hole is large, the resistance to air flow into the pleural cavity is less than the resistance to air flow into the intact lung, and little air enters the lung during inspiration. The mediastinum shifts farther to the intact side, kinking the great vessels until it flaps back during expiration. There is marked stimulation of respiration due to hypoxia, hypercapnia, and activation of pulmonary deflation receptors. Respiratory distress is severe.

If there is a flap of tissue over the hole in the lung or chest wall that acts as a flutter valve, permitting air to enter during inspiration but preventing its exit during expiration, the pressure in the pleural space rises

above atmospheric pressure (**tension pneumothorax**). The hypoxic stimulus to respiration causes deeper inspiratory efforts, which further increase the pressure in the pleural cavity, kinking the great veins and causing further hypoxia and shock. Intrapleural pressure in such cases may rise to 20–30 mm Hg. The peripheral veins become distended, there is intense cyanosis, and the condition is potentially fatal if the pneumothorax is not decompressed by removing the air.

On the other hand, if the hole through which air enters the pleural space seals off (**closed pneumothorax**), respiratory distress is not great because, with each inspiration, air flows into the lung on the unaffected side rather than into the pleural space. Because the vascular resistance is increased in the collapsed lung, blood is diverted to the other lung. Consequently, unless the pneumothorax is very large, it does not cause much hypoxia.

The air in a closed pneumothorax is absorbed. Since it is at atmospheric pressure, its total pressure, P_{O_2} and P_{N_2} are greater than the corresponding values in venous blood (compare values for air and venous blood in Fig 34–14). Gas diffuses down these gradients into the blood, and after 1–2 weeks all of the gas disappears.

Asthma

Asthma is characterized by episodic or chronic wheezing, cough, and a feeling of tightness in the chest as a result of bronchoconstriction. Hypoxia is observed in moderate and severe cases. Most if not all individuals with the disease are believed to have an underlying inflammatory process which causes epithelial desquamation, infiltration of the walls and lumen of the airways with eosinophils and lymphocytes, and an increased number of mast cells, and hyperresponsiveness of the airways to stimuli that make them constrict. Some have argued that there is an imbalance between the muscarinic receptors mediating bronchoconstriction and the β adrenergic receptors mediating bronchodilation, and the disease generally responds well to inhaled $β_2$ adrenergic agonists. However, leukotrienes and other bronchoconstricting inflammatory mediators appear to be involved, and there is some evidence for deficiency of VIP, the mediator in the noncholinergic nonadrenergic system that produces bronchodilation (see above). Asthma attacks are more severe in the late night and early morning hours because, as noted above, this is the period of maximal constriction in the circadian rhythm of bronchial tone. Cool air and exercise, both of which normally cause bronchoconstriction, also trigger asthma attacks, and the effects of both are inhibited by inhibitors of leukotriene synthesis or action.

Emphysema

In the degenerative and potentially fatal pulmonary disease called **emphysema,** the lungs lose their elasticity as a result of disruption of elastic tissue and the walls between the alveoli break down so that the alveoli are replaced by large air sacs. The physiologic dead space is greatly increased, and because of inadequate and uneven alveolar ventilation and perfusion of underventilated alveoli, severe hypoxia develops. Late in the disease, hypercapnia also develops. Inspiration and expiration are labored, and the work of breathing is greatly increased. The changes in the pressure-volume curve of the lungs are shown in Fig 34–7. The chest becomes enlarged and barrel-shaped because the chest wall expands as the opposing elastic recoil of the lungs declines. The hypoxia leads to polycythemia. Pulmonary hypertension develops, and the right side of the heart enlarges (**cor pulmonale**) and then fails.

The most common cause of emphysema is heavy cigarette smoking. The smoke causes an increase in the number of pulmonary alveolar macrophages, and these macrophages release a chemical substance that attracts leukocytes to the lungs. The leukocytes in turn release proteases including elastase, which attacks the elastic tissue in the lungs. At the same time, α_1-antitrypsin, a plasma protein that normally inactivates elastase and other proteases, is itself inhibited. The α_1-antitrypsin is inactivated by oxygen radicals, and these are released by the leukocytes. In addition, oxidants are released by the macrophages and there is some direct oxidative inactivation of α_1-antitrypsin by the smoke. The final result is a protease-antiprotease imbalance with increased destruction of lung tissue.

In about 2% of cases of emphysema, there is also a congenital deficiency of active α_1-antitrypsin. If individuals who are homozygous for this deficiency smoke, they develop crippling emphysema early in life and have a 20-year reduction in their life span. If they do not smoke, they may still develop emphysema, but they do much better and their life expectancy is much improved. Thus, α_1-antitrypsin deficiency provides an interesting example of the interaction between genetic factors and environmental factors in the production of disease.

Cystic Fibrosis

Cystic fibrosis is another condition that leads to repeated pulmonary infections and progressive, eventually fatal destruction of the lungs. In this congenital recessive disorder, a Cl^- channel in the apical membrane of airway epithelial cells is not activated in a normal fashion by cyclic AMP. An additional Cl^- channel in the membrane is normal, but loss of the cyclic AMP-activated channel results in a decrease in the transport of Cl^- and the water and Na^+ that moves with it into the airways, making the mucus that they contain thick and inspissated. Among Caucasians, cystic fibrosis is one of the most common genetic disorders: 5% of the population carries the defective

gene on chromosome 7, and the disease occurs in one of every 2000 births.

The gene that is abnormal in cystic fibrosis has been cloned. It encodes a Cl^- channel called the **cystic fibrosis transmembrane conductance regulator (CFTR),** which has 12 membrane-spanning domains, 2 ATP-binding sites, and a region containing phosphorylation sites for cyclic AMP-dependent protein kinase (protein kinase A) (see Chapter 1). This channel resembles members of a superfamily of transporters that mediate, among other things, export of the α-factor mating pheromone in yeast. The number of different mutations in the CFTR gene that cause cystic fibrosis has now passed 24, but this is not surprising in a gene encoding such a complex protein. In most cases, the mutations interfere with ATP binding or the conformational change that is presumably produced by the binding.

OTHER FORMS OF HYPOXIA

Anemic Hypoxia

Hypoxia due to anemia is not severe at rest unless the hemoglobin deficiency is marked, because red blood cell 2,3-DPG increases. However, anemic patients may have considerable difficulty during exercise because of limited ability to increase O_2 delivery to the active tissues.

Carbon Monoxide Poisoning

Carbon monoxide (CO), a gas formed by incomplete combustion of carbon, was used by the Greeks and Romans to execute criminals. Today it causes more deaths than any other gas. CO poisoning has become less common in the USA since natural gas, which does not contain CO, replaced artificial gases such as coal gas, which contains large amounts. However, the exhaust of gasoline engines is 6% or more CO.

CO is toxic because it reacts with hemoglobin to form **carbonmonoxyhemoglobin (carboxyhemoglobin, COHb),** and COHb cannot take up O_2. Carbon monoxide poisoning is often listed as a form of anemic hypoxia because there is a deficiency of hemoglobin that can carry O_2, but the total hemoglobin content of the blood is unaffected by CO. The affinity of hemoglobin for CO is 210 times its affinity for O_2, and COHb liberates CO very slowly. An additional difficulty is that when COHb is present the dissociation curve of the remaining HbO_2 shifts to the left, decreasing the amount of O_2 released. This is why an anemic individual who has 50% of the normal amount of HbO_2 may be able to perform moderate work, whereas an individual whose HbO_2 is reduced to the same level because of the formation of COHb is seriously incapacitated.

Because of the affinity of CO for hemoglobin, there is progressive COHb formation when the alveolar P_{CO} is greater than 0.4 mm Hg. However, the amount of COHb formed depends upon the duration of exposure to CO as well as the concentration of CO in the inspired air and the alveolar ventilation.

CO is also toxic to the cytochromes in the tissues, but the amount of CO required to poison the cytochromes is 1000 times the lethal dose; tissue toxicity thus plays no role in clinical CO poisoning.

The symptoms of CO poisoning are those of any type of hypoxia, especially headache and nausea, but there is little stimulation of respiration, since in the arterial blood, P_{O_2} remains normal and the carotid and aortic chemoreceptors are not stimulated (see Chapter 36). The cherry-red color of COHb is visible in the skin, nail beds, and mucous membranes. Death results when about 70–80% of the circulating hemoglobin is converted to COHb. The symptoms produced by chronic exposure to sublethal concentrations of CO are those of progressive brain damage, including mental changes and, sometimes, a parkinsonismlike state (see Chapter 32).

Treatment of CO poisoning consists of immediate termination of the exposure and adequate ventilation, by artificial respiration if necessary. Ventilation with O_2 is preferable to ventilation with fresh air, since O_2 hastens the dissociation of COHb. Hyperbaric oxygenation (see below) is useful in this condition.

Stagnant Hypoxia

Hypoxia due to slow circulation is a problem in organs such as the kidneys and heart during shock (see Chapter 33). The liver and possibly the brain are damaged by stagnant hypoxia in congestive heart failure. The blood flow to the lung is normally very large, and it takes prolonged hypotension to produce significant damage. However, **shock lung** can develop in prolonged circulatory collapse, particularly in those areas of the lung that are higher than the heart. In this condition, production of surfactant is decreased in the underperfused areas.

Histotoxic Hypoxia

Hypoxia due to inhibition of tissue oxidative processes is most commonly the result of cyanide poisoning. Cyanide inhibits cytochrome oxidase and possibly other enzymes. Methylene blue or nitrites are used to treat cyanide poisoning. They act by forming **methemoglobin**, which then reacts with cyanide to form **cyanmethemoglobin**, a nontoxic compound. The extent of treatment with these compounds is, of course, limited by the amount of methemoglobin that can be safely formed. Hyperbaric oxygenation may also be useful.

OXYGEN TREATMENT

Administration of oxygen-rich gas mixtures is of very limited value in stagnant, anemic, and histotoxic hypoxia because all that can be accomplished in this way is an increase in the amount of dissolved O_2 in the arterial blood. This is also true in hypoxic hypoxia when it is due to shunting of unoxygenated venous blood past the lungs. In other forms of hypoxic hypoxia, O_2 is of great benefit. However, it must be remembered that in hypercapnic patients in severe pulmonary failure, the CO_2 level may be so high that it depresses rather than stimulates respiration. Some of these patients keep breathing only because the carotid and aortic chemoreceptors drive the respiratory center. If the hypoxic drive is withdrawn by administering O_2, breathing may stop. During the resultant apnea, the arterial P_{O_2} drops, but breathing may not start again, because the increase in P_{CO_2} further depresses the respiratory center. Therefore, O_2 therapy in this situation may be fatal.

When 100% O_2 is first inhaled, there may be a slight decrease in respiration in normal individuals, suggesting that there is normally some hypoxic chemoreceptor drive. However, the effect is minor and can be demonstrated only by special techniques. In addition, it is offset by a slight accumulation of H^+ ions, since the concentration of deoxygenated hemoglobin in the blood is reduced and Hb is a better buffer than HbO_2 (see Chapter 35).

Oxygen Toxicity

It is interesting that while O_2 is necessary for life in aerobic organisms, it is also toxic. Indeed, 100% O_2 has been demonstrated to exert toxic effects not only in animals but also in bacteria, fungi, cultured animal cells, and plants. The toxicity seems to be due to the production of the superoxide anion (O_2^-), which is a free radical, and H_2O_2. When 80–100% O_2 is administered to humans for periods of 8 hours or more, the respiratory passages become irritated, causing substernal distress, nasal congestion, sore throat, and coughing. Exposure for 24–48 hours causes lung damage as well. The reason O_2 produces the irritation is not completely understood, although O_2 treatment appears to inhibit the ability of lung macrophages to kill bacteria, and surfactant production is reduced. In animals, more prolonged administration without irritation is possible if treatment is briefly interrupted from time to time, but it is not certain that periodic interruptions are of benefit to humans.

Some infants treated with O_2 for respiratory distress syndrome develop a chronic condition characterized by lung cysts and densities (**bronchopulmonary dysplasia**). There is evidence that this syndrome is a manifestation of O_2 toxicity. Another complication in these infants is **retrolental fibroplasia,** the formation of opaque vascular tissue in the eyes, which can lead to serious visual defects. The retinal receptors mature from the center to the periphery of the retina, and they use considerable O_2. This causes the retina to become vascularized in an orderly fashion. Oxygen treatment before maturation is complete provides the needed O_2 to the photoreceptors, and consequently the normal vascular pattern fails to develop. There is evidence that this condition can be prevented or ameliorated by treatment with vitamin E, which exerts an antioxidant effect.

Administration of 100% O_2 at increased pressure accelerates the onset of O_2 toxicity, with the production not only of tracheobronchial irritation but also of muscle twitching, ringing in the ears, dizziness, convulsions, and coma. The speed with which these symptoms develop is proportionate to the pressure at which the O_2 is administered; eg, at 4 atmospheres, symptoms develop in half the subjects in 30 minutes, whereas at 6 atmospheres, convulsions develop in a few minutes. Administration of other gases at increased pressure also causes central nervous system symptoms (see below). Administration of O_2 at elevated pressures to rats decreases their brain GABA content (see Chapter 4) and their brain, liver, and kidney ATP content.

Exposure to O_2 at increased pressures (**hyperbaric oxygenation**) can produce marked increases in the dissolved O_2 in blood. It is for this reason that surgery for certain forms of congenital heart disease is sometimes carried out in high-pressure tanks. Hyperbaric O_2 is also of value in the treatment of gas gangrene, carbon monoxide poisoning, and probably cyanide poisoning. However, O_2 toxicity limits exposures to less than 5 hours and pressures to 3 atmospheres or less.

HYPERCAPNIA & HYPOCAPNIA

Hypercapnia

Retention of CO_2 in the body (hypercapnia) initially stimulates respiration. Retention of larger amounts produces symptoms due to depression of the central nervous system: confusion, diminished sensory acuity, and, eventually, coma with respiratory depression and death. In patients with these symptoms, the P_{CO_2} is markedly elevated, there is severe respiratory acidosis, and the plasma HCO_3^- may exceed 40 meq/L. Large amounts of HCO_3^- are excreted, but more HCO_3^- is reabsorbed, raising the plasma HCO_3^- and partially compensating for the acidosis (see Chapter 39).

CO_2 is so much more soluble than O_2 that hypercapnia is rarely a problem in patients with pulmonary fibrosis. However, it does occur in ventilation–perfusion inequality and when for any reason alveolar ventilation is inadequate in the various forms of pump failure (Table 37–1). It is exacerbated when CO_2 production is increased. For example, in febrile patients there is a 13% increase in CO_2 production for each 1 °C rise in temperature, and a high carbohydrate intake increases CO_2 production because of the increase in RQ (see Chapter 17). Normally alveolar ventilation increases and the extra CO_2 is expired, but it accumulates when ventilation is compromised.

Hypocapnia

Hypocapnia is the result of hyperventilation. During voluntary hyperventilation, the arterial P_{CO_2} falls from 40 to as low as 15 mm Hg while the alveolar P_{O_2} rises to 120–140 mm Hg.

The more chronic effects of hypocapnia are seen in neurotic patients who chronically hyperventilate. Cerebral blood flow may be reduced 30% or more because of the direct constrictor effect of hypocapnia on the cerebral vessels (see Chapter 32). The cerebral ischemia causes light-headedness, dizziness, and paresthesias. Hypocapnia also increases cardiac output. It has a direct constrictor effect on many vessels, but it depresses the vasomotor center, so that the blood pressure is usually unchanged or only slightly elevated.

Other consequences of hypocapnia are due to the associated respiratory alkalosis, the blood pH being increased to 7.5 or 7.6. The plasma HCO_3^- level is low, but bicarbonate reabsorption is decreased because of the inhibition of renal acid secretion by the low P_{CO_2}. The plasma total calcium level does not change, but the plasma Ca^{2+} level falls and hypocapnic individuals develop carpopedal spasm, a positive Chvostek sign, and other signs of tetany (see Chapter 21).

OTHER RESPIRATORY ABNORMALITIES

Asphyxia

In asphyxia produced by occlusion of the airway, acute hypercapnia and hypoxia develop together. There is pronounced stimulation of respiration, with violent respiratory efforts. Blood pressure and heart rate rise sharply, catecholamine secretion is increased, and blood pH drops. Eventually the respiratory efforts cease, the blood pressure falls, and the heart slows. Asphyxiated animals can still be revived at this point by artificial respiration, although they are prone to ventricular fibrillation, probably because of the combination of hypoxic myocardial damage and high circulating catecholamine levels. If artificial respiration is not started, cardiac arrest occurs in 4–5 minutes.

Drowning

In about 10% of drownings, the first gasp of water after the losing struggle not to breathe triggers laryngospasm, and death results from asphyxia without any water in the lungs. In the remaining cases, the glottic muscles eventually relax and the lungs are flooded. Fresh water is rapidly absorbed, diluting the plasma and causing intravascular hemolysis. Ocean water is markedly hypertonic and draws fluid from the vascular system into the lungs, decreasing plasma volume. The immediate goal in the treatment of drowning is, of course, resuscitation, but long-term treatment must also take into account the circulatory effects of the water in the lungs.

Periodic Breathing

The acute effects of voluntary hyperventilation demonstrate the interaction of the chemical respiratory regulating mechanisms. When an individual hyperventilates for 2–3 minutes, then stops and permits respiration to continue without exerting any voluntary control over it, there is a period of apnea. This is followed by a few shallow breaths and then by another period of apnea, followed again by a few breaths (**periodic breathing**). The cycles may last for some time before normal breathing is resumed (Fig 37–10). The apnea apparently is due to CO_2 lack because it does not occur following hyperventilation with gas mixtures containing 5% CO_2. During the apnea, the alveolar P_{O_2} falls and the P_{CO_2} rises. Breathing resumes because of hypoxic stimulation of the carotid and aortic chemoreceptors before the CO_2 level has returned to normal. A few breaths eliminate the hypoxic stimulus, and breathing stops until the alveolar P_{O_2} falls again. Gradually, however, the P_{CO_2} returns to normal, and normal breathing resumes.

Cheyne-Stokes Respiration

Periodic breathing occurs in various disease states and is often called **Cheyne-Stokes respiration.** It is seen most commonly in patients with congestive heart failure and uremia, but it occurs also in patients with brain disease and during sleep in some normal individuals (Fig 37–11). Some of the patients with Cheyne-Stokes respiration have increased sensitivity to CO_2. The increased response is apparently due to disruption of neural pathways that normally inhibit respiration. In these individuals, CO_2 causes relative hyperventilation, lowering the arterial P_{CO_2}. During the resultant apnea, the arterial P_{CO_2} again rises to normal, but the respiratory mechanism again overresponds to CO_2. Breathing ceases, and the cycle repeats.

Another cause of periodic breathing in patients with cardiac disease is prolongation of the lung-to-brain circulation time, so that it takes longer for changes in arterial gas tensions to affect the respiratory center. When individuals with a slower circulation hyperventilate, they lower the P_{CO_2} of the blood in their lungs, but it takes longer than normal for the blood with a low P_{CO_2} to reach the brain. During this time, the P_{CO_2} in the pulmonary capillary blood continues to be lowered, and when this blood reaches the brain, the low P_{CO_2} inhibits the respiratory center, producing apnea. In other words, the respiratory control system oscillates because the negative feedback loop from lungs to brain is abnormally long.

EFFECTS OF INCREASED BAROMETRIC PRESSURE

The ambient pressure increases by 1 atmosphere for every 10 m of depth in seawater and every 10.4 m of depth in fresh water. Therefore, at a depth of 31 m (100 ft) in the ocean, a diver is exposed to a pressure of 4 atmospheres. Those who dig underwater tunnels are also exposed to the same hazards because the pressure in the chambers (caissons) in which they work is increased to keep out the water.

The hazards of exposure to increased barometric pressure used to be the concern largely of the specialists who cared for deep-sea divers and tunnel workers. However, the invention of SCUBA gear (self-contained underwater breathing apparatus, a tank and valve system carried by the diver) transformed diving from a business into a sport. The popularity of skin diving is so great that all physicians should be aware of its potential dangers. The interest in performing certain surgical procedures in high-pressure tanks

Figure 37–10. Changes in breathing and composition of alveolar air after forced hyperventilation for 2 minutes.

Figure 37–11. Cheyne-Stokes breathing during sleep. Two periods of apnea are separated by an increase and then a smooth decrease in tidal volume (V_T). (Reproduced, with permission, from Cherniack NS: Respiratory dysrhythmias during sleep. *N Engl J Med* 1981;**305**:325.)

(see above) also focuses attention on the effects of increased barometric pressure.

Nitrogen Narcosis & the High-Pressure Nervous Syndrome

A diver must breathe air or other gases at increased pressure to equalize the increased pressure on the chest wall and abdomen. CO_2 is routinely removed to prevent its accumulation. At increased pressure, 100% O_2 causes central nervous system symptoms of oxygen toxicity (Table 37–2). Since the harmful effects of breathing O_2 (see above) are proportionate to the P_{O_2}, they can be prevented by decreasing the concentration of O_2 in the gas mixture to 20% or less.

If a diver breathes compressed air, the increased P_{N_2} can cause **nitrogen narcosis,** a condition also known as "rapture of the deep" (Table 37–2). At pressures of 4–5 atmospheres (ie, at depths of 30–40 m in the ocean), 80% N_2 produces definite euphoria. At greater pressures, the symptoms resemble alcohol intoxication. Manual dexterity is maintained, but intellectual functions are impaired.

The problem of nitrogen narcosis can be avoided by breathing mixtures of O_2 and helium, and deeper dives can be made. However, the **high-pressure nervous syndrome** (HPNS) develops during deep dives with such mixtures. This condition is characterized by tremors, drowsiness, and a depression of the α activity in the EEG. Unlike nitrogen narcosis, intellectual functions are not severely affected but manual dexterity is impaired. The cause of HPNS is not settled, but it is worth noting that a variety of gases that are physiologically inert at atmospheric pressure are anesthetics at increased pressure. This is true of N_2 and also of xenon, krypton, argon, neon, and helium. Their anesthetic activity parallels their lipid solubility, and the anesthesia may be due to an action on nerve cell membranes.

Decompression Sickness

As a diver breathing 80% N_2 ascends from a dive, the elevated alveolar P_{N_2} falls. N_2 diffuses from the tissues into the lungs along the partial pressure gradient. If the return to atmospheric pressure (decompression) is gradual, no harmful effects are observed; but if the ascent is rapid, N_2 escapes from solution. Bubbles form in the tissues and blood, causing the symptoms of **decompression sickness (the bends, caisson disease).** Bubbles in the tissues cause severe pains, particularly around joints, and neurologic symptoms that include paresthesias and itching. Bubbles in the bloodstream, which occur in more severe cases, obstruct the arteries to the brain, causing major paralyses and respiratory failure. Bubbles in the pulmonary capillaries are apparently responsible for the dyspnea that divers call "the chokes," and bubbles in the coronary arteries may cause myocardial damage.

Treatment of this disease is prompt recompression in a pressure chamber, followed by slow decompression. Recompression is frequently lifesaving. Recovery is often complete, but there may be residual neurologic sequelae due to irreversible damage to the nervous system.

It should be noted that ascent in an airplane is equivalent to ascent from a dive. The decompression during a climb from sea level to 8550 m in the unpressurized cabin of an airplane (pressure drop from 1 down to 0.33 atmosphere) is the same as that during ascent to the surface after spending time in 20 m of seawater (pressure drop from 3 down to 1 atmosphere). A rapid ascent in either situation can cause decompression sickness.

Air Embolism

If a diver breathing from a tank at increased pressure during a dive holds his or her breath and suddenly heads for the surface—as may occur if the diver gets into trouble and panics—the gas in the lungs may expand rapidly enough to rupture the pulmonary veins. This drives air into the vessels, causing air embolism. Fatal air embolism has occurred during rapid ascent from as shallow a depth as 5 m. The consequences of air in the circulatory system are discussed in Chapter 30. This cannot happen, of course, to an individual who takes a breath on the surface, dives and returns to the surface still holding that breath, no matter how deep the dive.

Air embolism also occurs as a result of rapid expansion of the gas in the lungs when the external pressure is suddenly reduced from atmospheric to subatmospheric, as it is when the wall of the pressurized cabin of an airplane or rocket at high altitude is breached (**explosive decompression**).

ARTIFICIAL RESPIRATION

In acute asphyxia due to drowning, CO or other forms of gas poisoning, electrocution, anesthetic accidents, and other similar causes, artificial respiration

---

Table 37–2. Potential problems associated with exposure to increased barometric pressure.

Oxygen toxicity
  Lung damage
  Convulsions
Nitrogen narcosis
  Euphoria
  Impaired performance
High-pressure nervous syndrome
  Tremors
  Somnolence
Decompression sickness
  Pain
  Paralyses
Air embolism
  Sudden death

after breathing has ceased may be lifesaving. It should always be attempted, because respiration stops before the heart stops. There are numerous methods of emergency artificial respiration, but the method presently recommended to produce adequate ventilation in all cases is mouth-to-mouth breathing. The advantages of mouth-to-mouth resuscitation lie not only in its simplicity but also in the fact that it works by expanding the lungs.

Mouth-to-Mouth Breathing

In this form of resuscitation, the operator first places the victim in the supine position and opens the airway by placing a hand under the neck and lifting, while keeping pressure with the other hand on the victim's forehead. This extends the neck and lifts the tongue away from the back of the throat. The victim's mouth is covered by the operator's mouth while the fingers of the hand already on the forehead occlude the nostrils (Fig 37–12). About 12 times a minute, the operator blows into the victim's mouth a volume about twice the tidal volume, then permits the elastic recoil of the victim's lungs to produce passive expiration. The victim's neck is kept extended. Any gas blown into the stomach can be expelled by applying upward pressure on the abdomen from time to time. In apneic individuals in whom there is no detectable heartbeat, mouth-to-mouth breathing should be alternated with cardiac massage (see Chapter 28).

Mechanical Respirators

For treatment of chronic respiratory insufficiency due to inadequate ventilation, mechanical respirators are available. These are airtight metal or plastic containers that used to enclose the body except for the head but are now portable and cover only the chest. By means of a motor, negative pressure is applied to the chest at regular intervals, moving the chest wall in a way that resembles normal breathing. Intermittent-positive-pressure breathing machines are also used. These machines produce intermittent increases in intrapulmonary pressure by means of "pulses" of air delivered via a face mask.

Breathing against a positive end-expiratory pressure (PEEP) is of value in pulmonary edema, probably because it recruits previously unventilated alveoli. It also improves oxygenation in **adult respiratory distress syndrome.** This condition, which has a high mortality rate, is characterized by noncardiac pulmonary edema and inflammation produced by aspiration or other direct injuries to the lung, shock, or, in some instances, sepsis elsewhere in the body.

It is worth noting that it can be difficult to wean patients from mechanical ventilation. Often they are weak and malnourished. In addition, when the respirator takes over the work of breathing, there is disuse atrophy of the respiratory muscles, and they need to regain their strength.

Figure 37–12. Proper performance of mouth-to-mouth resuscitation. **A:** Open airway by positioning neck anteriorly in extension. Inserts show airway obstructed when the neck is in resting flexed position and open when neck is extended. **B:** Close victim's nose with fingers, seal mouth around victim's mouth, and deliver breath by vigorous expiration. **C:** Allow victim to exhale passively by unsealing mouth and nose. Rescuer should listen and feel for expiratory air flow. (Reproduced, with permission, from Schroeder SA, Krupp MA, Tierney LM Jr [editors]: *Current Medical Diagnosis & Treatment 1990.* Appleton & Lange, 1990.)

REFERENCES

Section VII: Respiration

Barnes PJ: Airway neuropeptides: Roles in fine tuning and in disease? *News Physiol Sci* 1989;**4:**116.

Bärtsch P et al: Prevention of high-altitude pulmonary edema by nifedipine. *N Engl J Med* 1991;**325:**1284.

Biscoe TJ, Duchem MR: Monitoring P_{O_2} by the carotid chemoreceptor. *News Physiol Sci* 1990;**5:**229.

Bove AA, Davis JC (editors): *Diving Medicine,* 2nd ed. Williams & Wilkins, 1990.

Crystal RG, West JB (editors): *The Lung: Scientific Foundations.* 2 vols. Raven, 1991.

Deneke SM, Fonburg BL: Normobaric oxygen toxicity of the lung. *N Engl J Med* 1980;**303:**76.

Fels AOS, Cohn ZA: The alveolar macrophage. *J Appl Physiol* 1986;**60:**353.

Fishman AP (editor): *Handbook of Physiology,* Section 3, *The Respiratory System.* Vol I–IV, American Physiological Society, 1985.

Fishman AP, Renkin EM (editors): *Pulmonary Edema.* American Physiological Society, 1979.

Golding J, Limerick S, Macfarlane A: *Sudden Infant Death: Patterns, Puzzles and Problems.* Univ of Washington Press, 1985.

Gross I: Regulation of fetal lung maturation. *Am J Physiol* 1990;**259:**L337.

Hawgood S: Pulmonary surfactant apoproteins: Review of protein and genomic structure. *Am J Physiol* 1989;**257:**L13.

Higenbottam T, Otulana BA, Wallwork J: The physiology of heart-lung transplantation in humans. *News Physiol Sci* 1990;**5:**71.

Insel PA, Wasserman SI: Asthma: A disorder of adrenergic receptors? *FASEB J* 1990;**4:**2732.

Johnson TS, Rock PB: Acute mountain sickness. *N Engl J Med* 1988;**319:**841.

Macklem PT: The act of breathing. *News Physiol Sci* 1990;**5:**233.

Melamed Y, Shupak A, Bitterman H: Medical problems associated with underwater diving. *N Engl J Med* 1992;**326:**30.

Millhorn DE, Eldridge FL: Role of ventrolateral medulla in regulation of respiratory and cardiovascular systems. *J Appl Physiol* 1986;**61:**1249.

Moffat K, Deatherage JF, Seybert DW: A structural model for the kinetic behavior of hemoglobin. *Science* 1979;**206:**1035.

Murray JF: *The Normal Lung: The Basis for Diagnosis and Treatment of Pulmonary Disease,* 2nd ed. Saunders, 1985.

Murray JF, Nadel J (editors): *Respiratory Medicine.* Saunders, 1986.

Newhouse M, Sanchis J, Bienenstock J: Lung defense mechanisms. *N Engl J Med* 1976;**295:**990.

Perlmutter DH, Pierce JA: The α_1-antitrypsin gene and emphysema. *Am J Physiol* 1989;**257:**L147.

Rannels DE et al: Use of radioisotopes in quantitative studies of lung metabolism. *Fed Proc* 1982;**41:**2833.

Reynolds HY: Immunologic system in the respiratory tract. *Physiol Rev* 1991;**71:**1117.

Rigatto H: Control of ventilation in the newborn. *Annu Rev Physiol* 1984;**46:**661.

Rounds S, Brody JS: Putting PEEP in perspective. *N Engl J Med* 1984;**311:**323.

Ryan JW: Peptidase enzymes of the pulmonary vascular surface. *Am J Physiol* 1989;**257:**L53.

Said SI (editor): *The Pulmonary Circulation and Acute Lung Injury.* Futura, 1985.

Sears TA, Berger AJ, Phillipson EA: Reciprocal tonic activation of inspiratory and expiratory motoneurones by chemical drives. *Nature* 1982;**299:**728.

Sugar O: In search of Ondine's curse. *JAMA* 1978;**240:**236.

Taylor AE et al: *Clinical Respiratory Physiology.* Saunders, 1989.

Wagner PD: Diffusion and chemical reaction in pulmonary gas exchange. *Physiol Rev* 1977;**57:**257.

Ward MP, Milledge JS, West JB: *High Altitude Medicine and Physiology.* Univ of Pennsylvania Press, 1989.

Weibel ER: *The Pathway for Oxygen: Structure and Function in the Mammalian Respiratory System.* Harvard Univ Press, 1984.

Weinberger SE, Schwartzstein RM, Weiss JW: Hypercapnia. *N Engl J Med* 1980;**321:**1223.

Welsh MJ: Electrolyte transport by airway epithelial. *Physiol Rev* 1987;**67:**1143.

West JB: *Physiology: The Essentials,* 4th ed. Williams & Wilkins, 1991.

Whittenberger JL: Myoglobin-facilitated oxygen diffusion: Role of myoglobin in oxygen entry into muscle. *Physiol Rev* 1970;**50:**559.

Wright JR, Dobbs LG: Regulation of pulmonary surfactant secretion and clearance. *Annu Rev Physiol* 1991;**53:**395.

Symposium: Mucus secretion and ion transport in airways. *Fed Proc* 1980;**39:**3061.

Section VIII.
Formation & Excretion of Urine

Renal Function & Micturition

38

INTRODUCTION

In the kidneys, a fluid that resembles plasma is filtered through the glomerular capillaries into the renal tubules (**glomerular filtration**). As this glomerular filtrate passes down the tubules, its volume is reduced and its composition altered by the processes of **tubular reabsorption** (removal of water and solutes from the tubular fluid) and **tubular secretion** (secretion of solutes into the tubular fluid) to form the urine that enters the renal pelvis. A comparison of the composition of the plasma and an average urine specimen illustrates the magnitude of some of these changes (Table 38–1) and emphasizes the manner in which wastes are eliminated while water and important electrolytes and metabolites are conserved. Furthermore, the composition of the urine can be varied, and many homeostatic regulatory mechanisms minimize or prevent changes in the composition of the ECF by changing the amount of water and various specific solutes in the urine. From the renal pelvis, the urine passes to the bladder and is expelled to the exterior by the process of urination, or **micturition.** The kidneys are also endocrine organs, making kinins (see Chapter 31), secreting renin (see Chapter 24) and erythropoietin (see Chapters 24 and 27), and forming 1,25-dihydroxycholecalciferol (see Chapter 21).

FUNCTIONAL ANATOMY

The Nephron

Each individual renal tubule and its glomerulus is a unit (**nephron**). The size of the kidneys in various species is determined largely by the number of nephrons they contain. There are approximately 1.3 million nephrons in each human kidney. The parts of the nephron are shown in diagrammatic fashion in Fig 38–1. On the right side of Fig 38–2, the scale drawing also shows the long, thin nature of the nephrons.

The glomerulus, which is about 200 μm in diameter, is formed by the invagination of a tuft of capillaries into the dilated, blind end of the nephron (**Bowman's capsule**). The capillaries are supplied by an **afferent arteriole** and drained by a slightly

Table 38–1. Urinary and plasma concentrations of some physiologically important substances.

Substance	Concentration in		U/P Ratio
	Urine (U)	Plasma (P)	
Glucose (mg/dL)	0	100	0
Na$^+$ (meq/L)	90	150	0.6
Urea (mg/dL)	900	15	60
Creatinine (mg/dL)	150	1	150

Figure 38–1. Diagram of a juxtamedullary nephron. The main histologic features of the cells that make up each portion of the tubule are also shown.

Figure 38–2. Renal circulation in the dog. The pattern in human kidneys is essentially identical. The vascular structures are simplified, and on the left the horizontal scale is expanded to show details. The same nephrons are shown undistorted on the right. AA, afferent arteriole; C, cortex; CD, collecting duct; EA, efferent arteriole; G, glomerulus; IM, inner medulla; OM, outer medulla; PCT, proximal convoluted tubule; VR, vasa recta. The distal convoluted tubules are shaded. (Reproduced, with permission, from Beeuwkes R III: The vascular organization of the kidney. *Annu Rev Physiol* 1980;**42:**531. Copyright © 1980 by Annual Reviews, Inc.)

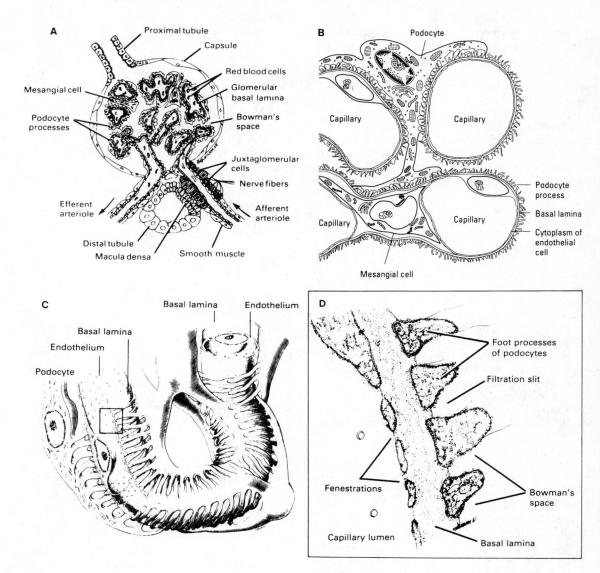

Figure 38–3. Structural details of glomerulus. **A:** Section through vascular pole, showing capillary loops. **B:** Relation of mesangial cells and podocytes to glomerular capillaries. **C:** Detail of the way podocytes form filtration slits on the basil lamina, and the relation of the lamina to the capillary endothelium. **D:** Enlargement of the rectangle in C to show the podocyte processes. The fuzzy material on their surfaces is glomerular polyanion. (Modified from Ham AW: Histology, 7th ed, Lippincott, 1974; Junqueira LC, Carneiro J, Kelly RO: Basic Histology, 7th ed, Appleton & Lange, 1991; and Wyngaarden JB, Smith LH Jr [Editors]: Cecil Textbook of Medicine, 18th ed, Saunders, 1988).

smaller **efferent arteriole** (Figs 38–1 and 38–2). There are 2 cellular layers separating the blood from the glomerular filtrate in Bowman's capsule: the capillary endothelium and the specialized epithelium that lies on top of the glomerular capillaries (Fig 38–3). These layers are separated by a basal lamina. Stellate cells called **mesangial cells** are located between the basal lamina and the endothelium. They are similar to cells called **pericytes,** which are found in the walls of capillaries elsewhere in the body. Mesangial cells are especially common between 2 neighboring capillaries, and in these locations the basal membrane forms a sheath shared by both capillaries (Fig 38–3). The mesangial cells are contractile and play a role in the regulation of glomerular filtration (see below). They also secrete various substances, take up immune complexes, and are involved in the production of glomerular disease.

The endothelium of the glomerular capillaries is fenestrated, with pores that are approximately 100 nm in diameter. The cells of the epithelium (**podocytes**) have numerous pseudopodia that interdigitate (Fig 38–3) to form **filtration slits** along the capillary wall. The slits are approximately 25 nm wide, and each is closed by a thin membrane. The basal lamina does not contain visible gaps or pores.

Functionally, the glomerular membrane permits the free passage of neutral substances up to 4 nm in diameter and almost totally excludes those with diameters greater than 8 nm. However, the charges on molecules as well as their diameters affect their passage into Bowman's capsule (see below). The average capillary area in each glomerulus is about 0.4 mm^2, and the total area of glomerular capillary endothelium across which filtration occurs in humans is about 0.8 m^2.

The general features of the cells that make up the walls of the tubules are shown in Fig 38–1. However, there are cell subtypes in all segments, and the anatomic differences between them correlate with differences in function (see below).

The human **proximal convoluted tubule** is about 15 mm long and 55 μm in diameter. Its wall is made up of a single layer of cells that interdigitate with one another and are united by apical tight junctions. Between the bases of the cells, there are extensions of the extracellular space called the **lateral intercellular spaces.** The luminal edges of the cells have a striate **brush border** due to the presence of innumerable 1×0.7 μm microvilli.

The convoluted portion of the proximal tubule (pars convoluta) drains into the straight portion (pars recta), which forms the first part of the **loop of Henle** (Figs 38–1 and 38–2). The proximal tubule terminates in the thin segment of the descending limb of the loop of Henle, which has an epithelium made up of attenuated, flat cells. The nephrons with glomeruli in the outer portions of the renal cortex have short loops of Henle (**cortical nephrons**), whereas those with glomeruli in the juxtamedullary region of the cortex (**juxtamedullary nephrons**) have long loops extending down into the medullary pyramids. In humans, only 15% of the nephrons have long loops. The total length of the thin segment of the loop varies from 2 to 14 mm in length. It ends in the thick segment of the ascending limb, which is about 12 mm in length. The cells of the thick ascending limb are cuboid. They have numerous mitochondria, and the basilar portions of their cell membranes are extensively invaginated.

The thick ascending limb of the loop of Henle reaches the glomerulus of the nephron from which the tubule arose and passes close to its afferent arteriole and efferent arteriole. The walls of the afferent arterioles contain the renin-secreting juxtaglomerular cells. At this point, the tubular epithelium is modified histologically to form the **macula densa.** The juxtaglomerular cells, the macula densa, and the lacis cells near them are known collectively as the **juxtaglomerular apparatus** (see Fig 24–3).

The **distal convoluted tubule** is about 5 mm long. Its epithelium is lower than that of the proximal tubule, and although there are a few microvilli, there is no distinct brush border. The distal tubules coalesce to form **collecting ducts** that are about 20 mm long and pass through the renal cortex and medulla to empty into the pelvis of the kidney at the apexes of the medullary pyramids. The epithelium of the collecting ducts is made up of **principal cells (P cells),** which among other things respond to vasopressin with increased water permeability (see below and Chapter 14), and **intercalated cells (I cells),** which secrete acid. In addition, there are I cells in the epithelium of the distal tubules. The total length of the nephrons, including the collecting ducts, ranges from 45 to 65 mm.

Cells in the kidneys that appear to have a secretory function include not only the juxtaglomerular cells but also some of the cells in the interstitial tissue of the medulla. These cells are called **type I medullary interstitial cells.** They contain lipid droplets and probably secrete prostaglandins, predominantly PGE$_2$ (see Chapter 17). PGE$_2$ is also secreted by the cells in the collecting ducts, and prostacyclin (PGI$_2$), as well as other prostaglandins is secreted by the arterioles and glomeruli.

Blood Vessels

The renal circulation is diagrammed in Fig 38–2.

The afferent arterioles are short, straight branches of the interlobular arteries. Each divides into multiple capillary branches to form the tuft of vessels in the glomerulus. The capillaries coalesce to form the efferent arteriole, which in turn breaks up into capillaries that supply the tubules (**peritubular capillaries**) before draining into the interlobular veins. The arterial segments between glomeruli and tubules are thus technically a portal system, and the glomerular capillaries are the only capillaries in the body that drain into arterioles. However, there is relatively little smooth muscle in the efferent arterioles.

The capillaries draining the tubules of the cortical nephrons form a peritubular network, whereas the efferent arterioles from the juxtamedullary glomeruli drain not only into a peritubular network but also into vessels that form hairpin loops (the **vasa recta**). These loops dip into the medullary pyramids alongside the loops of Henle (Fig 38–2). The efferent arteriole from each glomerulus breaks up into capillaries that supply a number of different nephrons. Thus, the tubule of each nephron does not necessarily receive blood solely from the efferent arteriole of that nephron. In humans, the total surface of the renal capillaries is approximately equal to the total surface area of the tubules, both being about 12 m². The volume of blood in the renal capillaries at any given time is 30–40 mL.

Lymphatics

The kidneys have an abundant lymphatic supply that drains via the thoracic duct into the venous circulation in the thorax.

Capsule

The renal capsule is thin but tough. If the kidney becomes edematous, the capsule limits the swelling, and the tissue pressure (**renal interstitial pressure**) rises. This decreases the glomerular filtration rate and is claimed to enhance and prolong the anuria in acute renal failure (see Chapter 33).

Innervation of the Renal Vessels

The renal nerves travel along the renal blood vessels as they enter the kidney. They contain many postganglionic sympathetic efferent fibers and a few afferent fibers of unknown function. There also appears to be a cholinergic innervation via the vagus nerve, but its function is uncertain. The sympathetic preganglionic innervation comes primarily from the lower thoracic and upper lumbar segments of the spinal cord, and the cell bodies of the postganglionic neurons are in the sympathetic ganglion chain, in the superior mesenteric ganglion, and along the renal artery. The sympathetic fibers are distributed primarily to the afferent and efferent arterioles, the proximal and distal tubules, and the juxtaglomerular cells (see Chapter 24). In addition, there is a dense noradrenergic innervation of the thick ascending limb of the loop of Henle.

RENAL CIRCULATION

Blood Flow

In a resting adult, the kidneys receive 1.2–1.3 L of blood per minute, or just under 25% of the cardiac output (see Table 32–1). Renal blood flow can be measured with electromagnetic or other types of flow meters, or it can be determined by applying the Fick principle (see Chapter 29) to the kidney—ie, by measuring the amount of a given substance taken up per unit of time and dividing this value by the arteriovenous difference for the substance across the kidney. Since the kidney filters plasma, the **renal plasma flow** equals the amount of a substance excreted per unit of time divided by the renal arteriovenous difference as long as the amount in the red cells is unaltered during passage through the kidney. Any excreted substance can be used if its concentration in arterial and renal venous plasma can be measured and if it is

not metabolized, stored, or produced by the kidney and does not itself affect blood flow.

Renal plasma flow can be measured by infusing p-aminohippuric acid (PAH) and determining its urine and plasma concentrations. PAH is filtered by the glomeruli and secreted by the tubular cells, so that its **extraction ratio** (arterial concentration minus renal venous concentration divided by arterial concentration) is high. For example, when PAH is infused at low doses, 90% of the PAH in arterial blood is removed in a single circulation through the kidney. It has therefore become commonplace to calculate the "renal plasma flow" by dividing the amount of PAH in the urine by the plasma PAH level, ignoring the level in renal venous blood. Peripheral venous plasma can be used because its PAH concentration is essentially identical to that in the arterial plasma reaching the kidney. The value obtained should be called the **effective renal plasma flow (ERPF)** to indicate that the level in renal venous plasma was not measured. In humans, ERPF averages about 625 mL/min.

Effective renal plasma flow (ERPF) =

$$\frac{U_{PAH} \dot{V}}{P_{PAH}} = \text{Clearance of PAH } (C_{PAH})$$

Example:

Concentration of PAH in urine (U_{PAH}): 14 mg/mL
Urine flow ($\dot{V}$): 0.9 mL/min
Concentration of PAH in plasma (P_{PAH}): 0.02 mg/mL

$$\text{ERPF} = \frac{14 \times 0.9}{0.02} = 630 \text{ mL/min}$$

It should be noted that the ERPF determined in this way is the **clearance** of PAH. The concept of clearance is discussed in detail below.

ERPF can be converted to actual renal plasma flow (RPF):

Average PAH extraction ratio: 0.9

$$\frac{\text{ERPF}}{\text{Extraction ratio}} = \frac{630}{0.9} = \text{Actual RPF} = 700 \text{ mL/min}$$

From the renal plasma flow, the renal blood flow can be calculated by dividing by 1 minus the hematocrit:

Hematocrit (Hct): 45%

$$\text{Renal blood flow} = \text{RPF} \times \frac{1}{1 - \text{Hct}}$$

$$= 700 \times \frac{1}{0.55}$$

$$= 1273 \text{ mL/min}$$

Regional Blood Flow

The blood flow in the renal cortex is much greater than that in the medulla. Values obtained in dogs are 4–5 mL/g of kidney tissue/min in the cortex, which accounts for 93% of the blood flow; 0.2 mL/g/min in the outer medulla, which accounts for 6% of the blood flow; and 0.03 mL/g/min in the inner medulla, which accounts for the rest of the blood flow. However, even the medullary flow is relatively large on a tissue weight basis. For comparison, the normal blood flow of the brain is 0.5 mL/g/min (see Chapter 32).

Pressure in Renal Vessels

The pressure in the glomerular capillaries has been measured directly in rats and has been found to be considerably lower than predicted on the basis of indirect measurements. When the mean systemic arterial pressure is 100 mm Hg, the glomerular capillary pressure is about 45 mm Hg. The pressure drop across the glomerulus is only 1–3 mm Hg, but there is a further drop in the efferent arteriole so that the pressure in the peritubular capillaries is about 8 mm Hg. The pressure in the renal vein is about 4 mm Hg. Pressure gradients are similar in the squirrel monkey and presumably in humans, with a glomerular capillary pressure that is about 40% of systemic arterial pressure.

Regulation of the Renal Blood Flow

Catecholamines constrict the renal vessels, with the greatest effect of injected norepinephrine being exerted on the interlobular arteries and the afferent arterioles. Angiotensin II exerts a selective constrictor effect on the efferent arterioles. Prostaglandins increase blood flow in the renal cortex and decrease blood flow in the renal medulla. In conscious dogs, moderate hemorrhage actually causes renal vasodilation, and this response is prevented by indomethacin, a drug that blocks prostaglandin synthesis. Acetylcholine also produces renal vasodilation. For unknown reasons, a high-protein diet increases renal blood flow.

Functions of the Renal Nerves

Stimulation of the sympathetic noradrenergic nerves to the kidney causes a marked decrease in renal blood flow. This effect is mediated by α_1 adrenergic receptors and to a lesser extent by postsynaptic α_2 adrenergic receptors. Similar renal vasoconstriction can be produced by stimulating the vasomotor area in the medulla oblongata, parts of the brain stem, and parts of the cerebral cortex. There is some tonic discharge in the renal nerves at rest in animals and humans. When systemic blood pressure falls, the vasoconstrictor response produced by decreased discharge in the baroreceptor nerves includes renal vasoconstriction. Renal blood flow is decreased during exercise and, to a lesser extent, on rising from the supine to the standing position.

Stimulation of the renal nerves also increases renin secretion by a direct action of released norepinephrine on β_1 adrenergic receptors on the juxtaglomerular cells (see Chapter 24) and increases Na^+ reabsorption, probably by a direct action of norepinephrine on renal tubular cells. The proximal and distal tubules and the thick ascending limb of the loop of Henle are richly innervated. When the renal nerves are stimulated at increasing strengths in experimental animals, the first response is an increase in the sensitivity of the juxtaglomerular cells (Table 38–2), followed by increased renin secretion, then increased Na^+ reabsorption, and finally, at the highest threshold, renal vasoconstriction with decreased glomerular filtration and renal blood flow. It is still unsettled whether the effect on Na^+ reabsorption is mediated via α or β adrenergic receptors, and it may be mediated by both. The physiologic role of the renal nerves in Na^+ metabolism is also unsettled, in part because most renal functions appear to be normal in patients with transplanted kidneys, and it takes some time for transplanted kidneys to acquire a functional innervation.

Autoregulation of Renal Blood Flow

When the kidney is perfused at moderate pressures (90–220 mm Hg in the dog), the renal vascular resis-

Table 38–2 Renal responses to graded renal nerve stimulation.[1]

Renal Nerve Stimulation Frequency (Hz)	RSR[2]	$U_{Na}V$[2]	GFR	RBF[2]
0.25	No effect on basal values; augments RSR mediated by nonneural stimuli.	0	0	0
0.50	Increased without changing $U_{Na}V$, GFR, or RBF.	0	0	0
1.0	Increased with decreased $U_{Na}V$ without changing GFR or RBF.	↓	0	0
2.50	Increased with decreased $U_{Na}V$, GFR, and RBF.	↓	↓	↓

[1] Reproduced from DiBona GF: Neural control of renal function: Cardiovascular implications. *Hypertension* 1989;**13**:539. By permission of the American Heart Association, Inc.
[2] RSR, renin secretion rate; $U_{Na}V$, urinary sodium excretion; RBF, renal blood flow.

tance varies with the pressure so that renal blood flow is relatively constant (Fig 38–4). Autoregulation of this type occurs in other organs, and several factors contribute to it (see Chapter 31). Renal autoregulation is present in denervated and in isolated, perfused kidneys but is prevented by the administration of drugs that paralyze vascular smooth muscle. It is probably produced in part by a direct contractile response of the smooth muscle of the afferent arteriole to stretch. At low perfusion pressures, angiotensin II also appears to play a role by constricting the efferent arterioles, thus maintaining the glomerular filtration rate. This is believed to be the explanation of the renal failure that sometimes develops in patients with poor renal perfusion who are treated with drugs which inhibit angiotensin-converting enzyme.

Renal Oxygen Consumption

The O_2 consumption of the human kidney is about 18 mL/min. The renal blood flow per gram of tissue is very large, and it is therefore not surprising that the arteriovenous O_2 difference is only 14 mL/L of blood, compared with 62 mL/L for the brain and 114 mL/L for the heart (see Table 32–1). The renal function with which the O_2 consumption correlates best is the rate of active transport of Na^+. The O_2 consumption of the cortex is about 9 mL/100 g/min, whereas that of the inner medulla is only 0.4 mL/100 g/min. The tubular fluid passes through the medulla on its way to the renal pelvis, and the P_{O_2} of the urine is also low.

GLOMERULAR FILTRATION

Measuring GFR

The **glomerular filtration rate (GFR)** can be measured in intact animals and humans by measuring the excretion and plasma level of a substance that is freely filtered through the glomeruli and neither se-

creted nor reabsorbed by the tubules. The amount of such a substance in the urine per unit of time must have been provided by filtering exactly the number of milliliters of plasma that contained this amount. Therefore, if the substance is designated by the letter X, the GFR is equal to the **concentration** of X in urine (U_X) times the **urine flow** per unit of time (V) divided by the **arterial plasma level** of X (P_X), or $U_X V / P_X$. This value is called the **clearance** of X (C_X). P_X is, of course, the same in all parts of the arterial circulation, and if X is not metabolized to any extent in the tissues, the level of X in peripheral venous plasma can be substituted for the arterial plasma level.

Substances Used to Measure GFR

In addition to the requirement that it be freely filtered and neither reabsorbed nor secreted in the tubules, a substance suitable for measuring the GFR should meet other criteria (Table 38–3). Inulin, a polymer of fructose with a molecular weight of 5200 that is found in dahlia tubers, meets these criteria in humans and most animals and is extensively used to measure GFR. In practice, a loading dose of inulin is administered intravenously, followed by a sustaining infusion to keep the arterial plasma level constant. After the inulin has equilibrated with body fluids, an accurately timed urine specimen is collected and a plasma sample obtained halfway through the collection. Plasma and urinary inulin concentrations are determined and the clearance calculated.

Example:

$$U_{In} = 35 \text{ mg/mL}$$
$$\dot{V} = 0.9 \text{ mL/min}$$
$$P_{In} = 0.25 \text{ mg/mL}$$
$$C_{In} = \frac{U_{In}\dot{V}}{P_{In}} = \frac{35 \times 0.9}{0.25}$$
$$C_{In} = 126 \text{ mL/min}$$

In dogs, cats, rabbits, and a number of other mammalian species, clearance of creatinine (C_{Cr}) can also be used to determine the GFR, but in primates, including humans, some creatinine is secreted by the tubules and some may be reabsorbed. In addition, plasma creatinine determinations are inaccurate at

Figure 38–4. Autoregulation in the kidneys.

Table 38-3. Characteristics of a substance suitable for measuring the GFR by determining its clearance.

Freely filtered (ie, not bound to protein in plasma or sieved in the process of ultrafiltration)
Not reabsorbed or secreted by tubules
Not metabolized
Not stored in kidneys
Not toxic
No effect on filtration rate
Preferably easy to measure in plasma and urine

low creatinine levels because the method for determining creatinine measures small amounts of other plasma constituents. In spite of this, the clearance of endogenous creatinine is frequently measured in patients. The values agree quite well with the GFR values measured with inulin because, although the value for $U_{Cr}V$ is high as a result of tubular secretion, the value for P_{Cr} is also high as a result of nonspecific chromogens, and the errors thus tend to cancel. Endogenous creatinine clearance is easy to measure and is a worthwhile index of renal function, but when precise measurements of GFR are needed it seems unwise to rely on a method that owes what accuracy it has to compensating errors.

Normal GFR

The GFR in an average-sized normal man is approximately 125 mL/min. Its magnitude correlates fairly well with surface area, but values in women are 10% lower than those in men even after correction for surface area. It should be noted that 125 mL/min is 7.5 L/h, or 180 L/d, whereas the normal urine volume is about 1 L/d. Thus, 99% or more of the filtrate is normally reabsorbed. At the rate of 125 mL/min, the kidneys filter in 1 day an amount of fluid equal to 4 times the total body water, 15 times the ECF volume, and 60 times the plasma volume.

Control of GFR

The factors governing filtration across the glomerular capillaries are the same as those governing filtration across all other capillaries (see Chapter 30), ie, the size of the capillary bed, the permeability of the capillaries, and the hydrostatic and osmotic pressure gradients across the capillary wall. For each nephron:

$$\textbf{GFR} = \textbf{K}_f[(\textbf{P}_{GC} - \textbf{P}_T) - (\textbf{II}_{GC} - \Pi_T)]$$

K_f, the glomerular ultrafiltration coefficient, is the product of the glomerular capillary wall hydraulic conductivity (ie, its permeability) and the effective filtration surface area. P_{GC} is the mean hydrostatic pressure in the glomerular capillaries, P_T the mean hydrostatic pressure in the tubule, Π_{GC} the osmotic pressure of the plasma in the glomerular capillaries, and Π_T the osmotic pressure of the filtrate in the tubule.

Permeability

The permeability of the glomerular capillaries is about 50 times that of the capillaries in skeletal muscle. Neutral substances with effective molecular diameters of less than 4 nm are freely filtered, and the filtration of neutral substances with diameters of more than 8 nm approaches zero (Fig 38–5). Between these values, filtration is inversely proportionate to diameter. However, sialoproteins in the glomerular capillary wall are negatively charged, and studies with anionically charged and cationically charged

Figure 38–5. Effect of electrical charge on the fractional clearance of dextran molecules of various sizes in rats. The negative charges in the glomerular membrane retard the passage of negatively charged molecules (anionic dextran) and facilitate the passage of positively charged molecules (cationic dextran). (Reproduced, with permission, from Brenner BM, Beeuwkes R: The renal circulations. *Hosp Pract* [July] 1978;**13**:35.)

dextrans indicate that the negative charges repel negatively charged substances in blood, with the result that filtration of anionic substances 4 nm in diameter is less than half that of neutral substances of the same size. This probably explains why albumin, with an effective molecular diameter of approximately 7 nm, normally has a glomerular concentration only 0.2% of its plasma concentration rather than the higher concentration that would be expected on the basis of diameter alone; circulating albumin is negatively charged. Filtration of cationic substances is slightly greater than that of neutral substances.

The presence of albumin in the urine is called **albuminuria.** In nephritis, the negative charges in the glomerular wall are dissipated, and albuminuria can occur for this reason without an increase in the size of the "pores" in the membrane.

Because of the nondiffusibility of plasma proteins, there is a Donnan effect (see Chapter 1) on the monovalent diffusible ion distribution, the concentration of anions being about 5% greater in the glomerular filtrate than in plasma and the concentration of monovalent cations 5% less. However, for most purposes this effect can be ignored, and in other respects the composition of the filtrate is similar to that of plasma.

Size of the Capillary Bed

It is now clear that K_f can be altered by the mesangial cells, contraction of these cells producing a decrease in K_f that is largely due to a reduction in the area available for filtration. Contraction of points where the capillary loops bifurcate probably shifts flow away from some of the loops, and elsewhere, contracted mesangial cells distort and encroach on the capillary lumen. Agents that have been shown to affect the mesangial cells are listed in Table 38–4. Angiotensin II is an important regulator of mesangial contraction, and there are angiotensin II receptors in the glomeruli. In addition, there is some evidence that mesangial cells make renin.

Hydrostatic & Osmotic Pressure

The pressure in the glomerular capillaries is higher than that in other capillary beds because the afferent arterioles are short, straight branches of the interlobular arteries. Furthermore, the vessels "downstream" from the glomeruli, the efferent arterioles, have a relatively high resistance. The capillary hydrostatic pressure is opposed by the hydrostatic pressure in Bowman's capsule. It is also opposed by the osmotic pressure gradient across the glomerular capillaries ($\Pi_{GC} - \Pi_T$). Π_T is normally negligible, and the gradient is equal to the oncotic pressure of the plasma proteins.

The actual pressures in one strain of rats are shown in Fig 38–6. The net filtration pressure (P_{UF}) is 15 mm Hg at the afferent end of the glomerular capillaries, but it falls to zero—ie, **filtration equilibrium** is reached—proximal to the efferent end of the glomerular capillaries. This is because fluid leaves the plasma and the oncotic pressure rises as blood passes through the glomerular capillaries. The calculated change in $\Delta\Pi$ along an idealized glomerular capillary is shown in Fig 38–6. It is apparent that portions of the glomerular capillaries do not normally contribute to the formation of the glomerular ultrafiltrate; ie, exchange across the glomerular capillaries is flow-limited rather than diffusion-limited (see Chapter 30). It is also apparent that a decrease in the rate of rise of the $\Delta\Pi$ curve produced by an increase in renal plasma flow would increase filtration without any change in ΔP because it would increase the distance along the capillary in which filtration was taking place.

There is considerable species variation in whether filtration equilibrium is reached, and there are some

	(mm Hg)	
	Afferent end	Efferent end
P_{GC}	45	45
P_T	10	10
Π_{GC}	20	35
$P_{UF} = P_{GC} - P_T - \Pi_{GC}$	15	0

Figure 38–6. Hydrostatic pressure (P_{GC}) and osmotic pressure (Π_{GC}) in a glomerular capillary in the rat. P_T, pressure in Bowman's capsule; P_{UF}, net filtration pressure. Π_T is normally negligible, so $\Delta\Pi = P_{GC}$. $\Delta P = P_{GC} - P_T$. (Reproduced, with permission, from Mercer PF, Maddox DA, Brenner BM: Current concepts of sodium chloride and water transport by the mammalian nephron. *West J Med* 1974;**120**:33.)

uncertainties in the measurement of K_f. It is uncertain whether filtration equilibrium is reached in humans.

Changes in GFR

Variations in the factors discussed in the preceding paragraphs and listed in Table 38–5 have predictable effects on the GFR. Changes in renal vascular resistance as a result of autoregulation tend to stabilize filtration pressure, but when the mean systemic arterial pressure drops below 90 mm Hg, there is a sharp drop in GFR. The GFR tends to be maintained when efferent arteriolar constriction is greater than afferent constriction, but either type of constriction decreases the tubular blood flow.

Filtration Fraction

The ratio of the GFR to the renal plasma flow (RPF), the **filtration fraction,** is normally 0.16–0.20.

Table 38–4. Agents causing contraction or relaxation of mesangial cells.[1]

Contraction	Relaxation
Angiotensin II	ANP
Vasopressin	Dopamine
Norepinephrine	PGE_2
Platelet-activating factor	Cyclic AMP
Platelet-derived growth factor	
Thromboxane A_2	
PGF_2	
Leukotrienes C and D	
Histamine	

[1]Modified from Schlondorff D: The glomerular mesangial cell: An expanding role for a specialized pericyte. *FASEB J* 1987;**1**:272.

Table 38–5. Factors affecting the GFR.

Changes in renal blood flow
Changes in glomerular capillary hydrostatic pressure
  Changes in systemic blood pressure
  Afferent or efferent arteriolar constriction
Changes in hydrostatic pressure in Bowman's capsule
  Ureteral obstruction
  Edema of kidney inside tight renal capsule
Changes in concentration of plasma proteins:
  dehydration, hypoproteinemia, etc (minor factors)
Changes in K_f
  Changes in glomerular capillary permeability
  Changes in effective filtration surface area

The GFR varies less than the RPF. When there is a fall in systemic blood pressure, the GFR falls less than the RPF because of efferent arteriolar constriction, and consequently the filtration fraction rises.

TUBULAR FUNCTION

General Considerations

The amount of any substance that is filtered is the product of the GFR and the plasma level of the substance ($C_{In}P_X$). The tubular cells may add more of the substance to the filtrate (tubular secretion), may remove some or all of the substance from the filtrate (tubular reabsorption), or may do both. The amount of the substance excreted (U_XV) equals the amount filtered plus the **net amount transferred** by the tubules. This latter quantity is conveniently indicated by the symbol T_X. When there is net tubular secretion, T_X is positive; when there is net tubular absorption, T_X is negative (Fig 38–7). Since the clearance of any substance is UV/P, the clearance equals the GFR if there is no net tubular secretion or reabsorption, exceeds the GFR if there is net tubular secretion, and is less than the GFR if there is net tubular reabsorption. It should be noted that in the latter 2 cases the clearance is a "virtual volume," an index of renal function, rather than an actual volume.

Much of our knowledge about glomerular filtration and tubular function has been obtained by using micropuncture techniques. Micropipettes can be inserted into the tubules of the living kidney and the composition of aspirated tubular fluid determined by the use of microchemical techniques. Techniques such as insertion of 2 pipettes in a single nephron have made it possible to study tubular function in great detail. It has also been possible to study isolated perfused segments of renal tubules.

Membrane potentials of tubular cells have been measured with microelectrodes. The potential difference of –70 mV between the tubular lumen and the interior of proximal tubular cells is the same, or nearly the same, as that between the cells and the ECF. The potential between the tubular lumen and the ECF is –2 mV (tubular lumen negative) in the first part of the proximal tubule and +2 mV (tubular lumen positive) in the remainder of the proximal tubule. The lumen of the loop of Henle is negative to the ECF except for the thick ascending limb, where the potential difference is +7 mV. In the distal tubule, the potential difference again becomes lumen negative, and it increases to a peak of –45 mV in the late portion. The potential difference in the collecting duct is –35 mV (lumen negative).

The effects of tubular reabsorption and secretion on substances of major physiologic interest are summarized in Table 38–6.

Mechanisms of Tubular Reabsorption & Secretion

Small proteins and some peptide hormones are reabsorbed in the proximal tubules by endocytosis. Other substances are secreted or reabsorbed in the tubules by passive or facilitated diffusion down chemical or electrical gradients or actively transported against such gradients (see Chapter 1). Movement is by way of ion channels, exchangers, cotransporters, and pumps. A recent list of the transport mechanisms in the proximal tubule is shown in Fig 38–8, and new mechanisms are being discovered all the time. Note that the pumps and other units in the luminal membrane are different from those in the basolateral mem-

$$GFR \times P_X + T_X = U_X\dot{V}$$

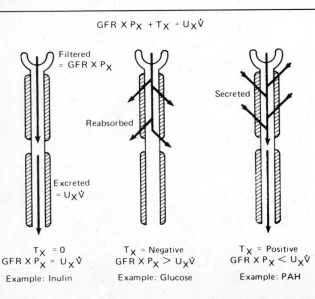

Figure 38–7. Tubular function. For explanation of symbols, see text.

Table 38–6. Renal handling of various plasma constituents in a normal adult human on an average diet.

Substance	Per 24 Hours				Percentage Reabsorbed	Location[1]
	Filtered	Reabsorbed	Secreted	Excreted		
Na⁺ (meq)	26,000	25,850		150	99.4	P, L, D, C
K⁺ (meq)	600	560[2]	50[2]	90	93.3	P, L, D, C
Cl⁻ (meq)	18,000	17,850		150	99.2	P, L, D, C
HCO₃⁻ (meq)	4,900	4,900		0	100	P, D
Urea (mmol)	870	460[3]		410	53	P, L, D, C
Creatinine (mmol)	12	1[4]	1[4]	12		
Uric acid (mmol)	50	49	4	5	98	P
Glucose (mmol)	800	800		0	100	P
Total solute (mosm)	54,000	53,400	100	700	87	P, L, D, C
Water (mL)	180,000	179,000		1000	99.4	P, L, D, C

[1]P, proximal tubules; L, loops of Henle; D, distal tubules; C, collecting ducts.
[2]K⁺ is both reabsorbed and secreted.
[3]Urea diffuses into as well as out of some portions of the nephron.
[4]Variable secretion and probable reabsorption of creatinine in humans.

brane. It is this different distribution that makes possible net movement of solutes across the epithelia.

Like transport systems elsewhere, renal active transport systems have a maximal rate, or **transport maximum (Tm),** at which they can transport a particular solute. Thus, the amount of a particular solute transported is proportionate to the amount present up to the Tm for the solute, but at higher concentrations, the transport mechanism is **saturated** and there is no appreciable increment in the amount transported. However, the Tm's for some systems are high, and it is difficult to saturate them.

Not only does the reabsorption of Na⁺ and Cl⁻ play a major role in body electrolyte and water metabolism, but Na⁺ transport is coupled to the movement of H⁺, other electrolytes, glucose, amino acids, organic acids, phosphate, and other substances across the tubule walls.

Na⁺ Reabsorption

In the proximal tubules, the thick portion of the ascending limb of the loop of Henle, the distal tubules, and the collecting ducts, Na⁺ moves by cotransport or exchange from the tubular lumen into the tubular epithelial cells down its concentration and electrical gradients and is actively pumped from these cells into the interstitial space. Thus, Na⁺ is actively transported out of all parts of the renal tubule except the thin portions of the loop of Henle. Na⁺ is pumped into the interstitium by Na⁺-K⁺ ATPase. The operation of this ubiquitous Na⁺ pump is considered in detail in Chapter 1. It extrudes 3 Na⁺ in exchange for 2 K⁺ that are pumped into the cell.

The tubular cells are connected by tight junctions at their luminal edges, but there is space between the

Figure 38–8. Transport mechanisms in the proximal tubule. (Courtesy of KJ Ullrich.)

cells along the rest of their lateral borders. Much of the Na$^+$ is actively transported into these extensions of the interstitial space, the **lateral intercellular spaces** (Fig 38–9). Proximal tubular reabsorbate is slightly hypertonic, and water moves passively along the osmotic gradient created by its absorption. The rate at which solutes and water move into the capillaries from the lateral intercellular spaces and the rest of the interstitium is determined by the Starling forces determining movement across the walls of all capillaries, ie, the hydrostatic and osmotic pressures in the interstitium and the capillaries (see Chapter 30). When the capillary hydrostatic pressure is increased or the plasma protein pressure is decreased, the movement of solute and water into the capillaries slows and the lateral intercellular spaces expand. Water can leak across the tight junctions even in the hydropenic state, and when the lateral intercellular spaces are expanded, the "leak" back into the tubules via this paracellular pathway is large. In this way, net Na$^+$ reabsorption is decreased.

Glucose Reabsorption

Glucose, amino acids, and bicarbonate are reabsorbed along with Na$^+$ in the early portion of the proximal tubule (Fig 38–10). Farther along the tubule, Na$^+$ is reabsorbed with Cl$^-$. Glucose is typical of substances removed from the urine by secondary active transport. It is filtered at a rate of approximately 100 mg/min (80 mg/dL of plasma × 125 mL/min). Essentially all of the glucose is reabsorbed, and no more than a few milligrams appear in the urine per 24 hours. The amount reabsorbed is proportionate to the amount filtered and hence to the plasma glucose level (P_G) times the GFR up to the transport maximum (Tm_G); but when the Tm_G is exceeded, the amount of glucose in the urine rises (Fig 38–11). The Tm_G is about 375 mg/min in men and 300 mg/min in women.

The **renal threshold** for glucose is the plasma level at which the glucose first appears in the urine in more than the normal minute amounts. One would predict that the renal threshold would be about 300 mg/dL—ie, 375 mg/min (Tm_G) divided by 125 mL/min (GFR). However, the actual renal threshold is about 200 mg/dL of arterial plasma, which corresponds to a venous level of about 180 mg/dL. Fig 38–11 shows why the actual renal threshold is less than the predicted threshold. The "ideal" curve shown in this diagram would be obtained if the Tm_G in all the tubules was identical and if all the glucose were removed from each tubule when the amount filtered was below the Tm_G. This is not the case, and in humans, for example, the actual curve is rounded and deviates considerably from the "ideal" curve. This deviation is called **splay.** The magnitude of the splay is inversely proportionate to the avidity with which the transport mechanism binds the substance it transports.

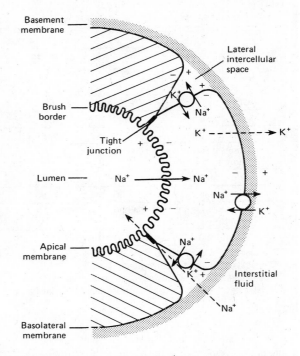

Figure 38–9. Mechanism for Na$^+$ reabsorption in the proximal tubule. Solid lines indicate active transport or, in the case of luminal Na$^+$, cotransport or exchange. Dashed lines indicate diffusion. Note that some Na$^+$ and H$_2$O leak back into the renal tubule through the tight junctions between the epithelial cells. (Reproduced, with permission, from Sullivan LP, Grantham JJ: *Physiology of the Kidney,* 2nd ed. Lea & Febiger, 1982.)

Glucose Transport Mechanism

Glucose reabsorption in the kidneys is similar to glucose reabsorption in the intestine (see Chapter 25). Glucose and Na$^+$ bind to the common carrier SGLT 1 in the luminal membrane (see Fig 25–3), and glucose is carried into the cell as Na$^+$ moves down its electrical and chemical gradient. The Na$^+$ is then pumped out of the cell into the lateral intercellular spaces, and the glucose is transported by GLUT 2 into the interstitial fluid. Thus, glucose transport in the kidneys as well as in the intestine is an example of secondary active transport; the energy for the active transport is provided by the Na$^+$-K$^+$ ATPase that pumps the Na$^+$ out of the cells.

The symport specifically binds the D isomer of glucose, and the rate of transport of D-glucose is many times greater than that of L-glucose. Glucose transport in the kidneys is inhibited, as it is in the intestine, by the plant glucoside **phlorhizin,** which competes with D-glucose for binding to the symport.

Figure 38–11. **Top:** Relation between the plasma level (P) and excretion (UV̇) of glucose and inulin. **Bottom:** Relation between the plasma glucose level (P_G) and amount of glucose reabsorbed (T_G).

Figure 38–10. Reabsorption of various solutes in the proximal tubule in relation to the potential difference (PD) along the tubule. TF/P, tubular fluid:plasma concentration ratio. (Courtesy of FC Rector Jr.)

Additional Examples of Secondary Active Transport

A variety of other substances are transported by secondary active transport via symports, with the energy provided by active transport of Na^+ out of the renal tubular cells. These substances include some amino acids, lactate, citrate, phosphate, H^+, and Cl^-.

Like glucose reabsorption, amino acid reabsorption is most marked in the early portion of the proximal convoluted tubule. Absorption in this location resembles absorption in the intestine (see Chapter 25). In general, the main carriers in the luminal membrane cotransport Na^+, whereas the carriers in the basolateral membranes are not Na^+-dependent. When cotransport occurs, Na^+ is pumped out of the cells by Na^+-K^+ ATPase and the amino acids leave by passive or facilitated diffusion to the interstitial fluid.

It appears that Cl^- is transported in 3 ways: by passive diffusion when the Cl^- concentration in the lumen is elevated or when the tubular lumen is negative relative to the interstitial fluid; by cotransport with Na^+ and K^+; and by a Cl^--OH^- antiport which operates in concert with a Na^+-H^+ antiport, putting H^+ and OH^- into the tubular lumen.

PAH Transport

The dynamics of PAH transport illustrate the operation of the active transport mechanisms that secrete substances into the tubular fluid. The filtered load of PAH is a linear function of the plasma level, but PAH secretion increases as P_{PAH} rises only until a maximal secretion rate (Tm_{PAH}) is reached (Fig 38–12). When P_{PAH} is low, C_{PAH} is high; but as P_{PAH} rises above Tm_{PAH}, C_{PAH} falls progressively. It eventually approaches the clearance of inulin (C_{In}) (Fig 38–13), because the amount of PAH secreted becomes a smaller and smaller fraction of the total amount excreted. Conversely, the clearance of glucose is essentially zero at P_G levels below the renal threshold; but above the threshold, C_G rises to approach C_{In} as P_G is raised.

The use of C_{PAH} to measure effective renal plasma flow (ERPF) is discussed above. It should be emphasized again that the excretion of any substance that is not stored or metabolized in the kidneys can be used to measure renal plasma flow if the level in arterial and renal venous plasma can be measured, and the only advantage of PAH is that its extraction ratio is so high that renal plasma flow can be approximated without measuring the PAH concentration in renal venous blood.

Other Substances Secreted by the Tubules

Derivatives of hippuric acid in addition to PAH, phenol red and other sulfonphthalein dyes, penicillin,

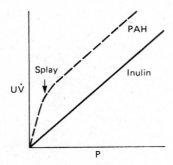

Figure 38–12. Relation between plasma levels (P) and excretion (UV̇) of PAH and inulin.

and a variety of iodinated dyes such as iodopyracet (Diodrast) are actively secreted into the tubular fluid. Substances that are normally produced in the body and secreted by the tubules include various ethereal sulfates, steroid and other glucuronides, and 5-hydroxyindoleacetic acid, the principal metabolite of serotonin (see Chapter 4).

Tubuloglomerular Feedback & Glomerulotubular Balance

Signals from the renal tubules feed back to affect glomerular filtration. As the rate of flow through the ascending limb of the loop of Henle and first part of the distal tubule increases, glomerular filtration in the same nephron decreases, and, conversely, a decrease in flow increases the GFR (Fig 38–14). This process, which is called **tubuloglomerular feedback,** tends to maintain the constancy of the load delivered to the distal tubule. The sensor for the response appears to

Figure 38–13. Clearance of inulin, glucose, and PAH at various plasma levels of each substance in humans.

Figure 38–14. Mechanisms of glomerulotubular balance and tubuloglomerular feedback.

be the macula densa, and GFR is adjusted by constriction or dilation of the afferent arteriole. Constriction may be mediated by thromboxane A_2.

Conversely, an increase in GFR causes an increase in the reabsorption of solutes, and consequently of water, primarily in the proximal tubule, so that in general the percentage of the solute reabsorbed is held constant. This process is called **glomerulotubular balance,** and it is particularly prominent for Na^+. The change in Na^+ reabsorption occurs within seconds after a change in filtration, so it seems unlikely that an extrarenal humoral factor is involved. One factor is the oncotic pressure in the peritubular capillaries. When the GFR is high, there is a relatively large increase in the oncotic pressure of the plasma by the time it reaches the efferent arterioles and their capillary branches. This increases the reabsorption of Na^+ from the tubule. However, other as yet unidentified intrarenal mechanisms are also involved.

WATER EXCRETION

Normally, 180 L of fluid are filtered through the glomeruli each day, while the average daily urine volume is about 1 L. The same load of solute can be excreted per 24 hours in a urine volume of 500 mL with a concentration of 1400 mosm/L or in a volume of 23.3 L with a concentration of 30 mosm/L (Table 38–7). These figures demonstrate 2 important facts: first, that at least 87% of the filtered water is reab-

Table 38–7. Alterations in water metabolism produced by vasopressin in humans. In each case, the osmotic load excreted is 700 mosm/d.

	GFR (mL/min)	Percentage of Filtered Water Reabsorbed	Urine Volume (L/d)	Urine Concentration (mosm/L)	Gain or Loss of Water in Excess of Solute (L/d)
Urine isotonic to plasma	125	98.7	2.4	290	...
Vasopressin (maximal antidiuresis)	125	99.7	0.5	1400	1.9 gain
No vasopressin ("complete" diabetes insipidus)	125	87.1	23.3	30	20.9 loss

sorbed, even when the urine volume is 23 L; and second, that the reabsorption of the remainder of the filtered water can be varied without affecting total solute excretion. Therefore, when the urine is concentrated, water is retained in excess of solute; and when it is dilute, water is lost from the body in excess of solute. Both facts have great importance in the body economy and the regulation of the osmolality of the body fluids.

Proximal Tubule

Many substances are actively transported out of the fluid in the proximal tubule, but fluid obtained by micropuncture remains essentially isoosmotic to the end of the proximal tubule (Fig 38–10). Therefore, in the proximal tubule, water moves passively out of the tubule along the osmotic gradients set up by active transport of solutes, and isotonicity is maintained. The osmotic gradient between renal interstitial fluid and proximal tubular fluid is very small, but it is effective in moving water out of the tubular lumen. Since the ratio of the concentration in tubular fluid to the concentration in plasma (TF/P) of the non-reabsorbable substance inulin is 2.5–3.3 at the end of the proximal tubule, it follows that 60–70% of the filtered solute and 60–70% of the filtered water have been removed by the time the filtrate reaches this point (Fig 38–15).

Loop of Henle

As noted above, the loops of Henle of the juxtamedullary nephrons dip deeply into the medullary pyramids before draining into the distal convoluted tubules in the cortex, and all of the collecting ducts descend back through the medullary pyramids to drain at the tips of the pyramids into the renal pelvis. There is a graded increase in the osmolality of the interstitium of the pyramids, the osmolality at the tips of the papillae normally being about 1200 mosm/L, approximately 4 times that of plasma. The descending limb of the loop of Henle is permeable to water, but the ascending limb is impermeable (Table 38–8). Na^+, K^+, and Cl^- are cotransported out of the thick segment of the ascending limb (see below). Therefore, the fluid in the descending limb of the loop of Henle becomes hypertonic as water moves into the

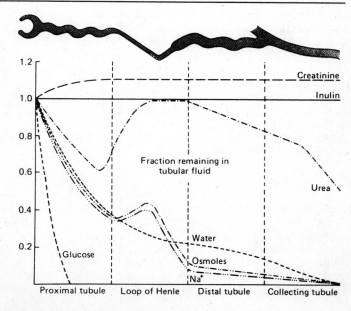

Figure 38–15. Changes in the fraction of the filtered amount of substances remaining in the tubular fluid along the length of the nephron. (Reproduced, with permission, from Sullivan LP, Grantham JJ: *Physiology of the Kidney,* 2nd ed. Lea & Febiger, 1982.)

Table 38-8. Permeability and transport in various segments of the nephron.[1]

	Permeability			Active Transport of Na⁺
	H₂O	Urea	NaCl	
Loop of Henle				
Thin descending limb	4+	+	±	0
Thin ascending limb	0	+	4+	0
Thick ascending limb	0	±	±	4+
Distal convoluted tubule	±	±	±	3+
Collecting tubule				
Cortical portion	3+*	0	±	2+
Outer medullary portion	3+*	0	±	1+
Inner medullary portion	3+*	3+	±	1+

[1]Data are based on studies of rabbit and human kidneys. Values indicated by asterisks are in the presence of vasopressin. These values are 1 + in the absence of vasopressin. (Modified and reproduced, with permission, from Kokko JP: Renal concentrating and diluting mechanisms. *Hosp Pract* [Feb] 1979;**110**:14.)

Figure 38–16. Summary of changes in the osmolality of tubular fluid in various parts of the nephron. The thickened wall of the ascending limb of the loop of Henle indicates relative impermeability of the tubular epithelium to water. In the presence of vasopressin, the fluid in the collecting ducts becomes hypertonic, whereas in the absence of this hormone, the fluid remains hypotonic throughout the collecting duct. Aldosterone promotes reabsorption of Na⁺ and secretion of H⁺ and K⁺ in the distal convoluted tubule. (Reproduced, with permission, from Cannon PJ: The kidney in heart failure. *N Engl J Med* 1977;**296**:26.)

hypertonic interstitium. In the ascending limb it becomes more dilute, and when it reaches the top it is hypotonic to plasma because of the movement of Na⁺ and Cl⁻ out of the tubular lumen (Fig 38–16). In passing through the loop of Henle, another 15% of the filtered water is removed, so approximately 20% of the filtered water enters the distal tubule, and the TF/P of inulin at this point is about 5.

The transport mechanism in the thick ascending limb depends on a carrier that transports one Na⁺, one K⁺, and 2 Cl⁻ from the tubular lumen into the cells (Fig 38–17). The Na⁺ is actively transported out of the cells into the interstitium by Na⁺-K⁺ ATPase, and this portion of the renal tubule has a higher Na⁺-K⁺ ATPase content than any other part. The K⁺ diffuses passively back into the tubular lumen, maintaining the transtubular potential difference of +7 mV (see above). One Cl⁻ diffuses passively into the interstitium, and one is cotransported with K⁺.

Distal Tubule

The distal tubule, particularly its first part, is in effect an extension of the thick segment of the ascending limb. It is relatively impermeable to water, and continued removal of the solute in excess of solvent further dilutes the tubular fluid. However, the reabsorption of Na⁺ in this segment is variable, being regulated by aldosterone. About 5% of the filtered water is removed in this segment.

Collecting Ducts

The collecting ducts have 2 portions: a cortical portion and a medullary portion through which the filtrate flows from the cortex to the renal pelvis. The changes in osmolality and volume in the collecting ducts depend on the amount of vasopressin acting on the ducts. This antidiuretic hormone from the posterior pituitary gland increases the permeability of the

Figure 38–17. Cotransport of Na⁺, K⁺, and Cl⁻ by secondary active transport in cells in the thick ascending limb of the loop of Henle. Solid lines across the membrane indicate active transport or secondary active transport; dashed lines indicate diffusion. Active transport by Na⁺-K⁺ ATPase is indicated by arrows in the circle.

collecting ducts to water by causing the rapid insertion of water channels into the luminal membrane of principal cells (see Chapter 14). In the presence of enough vasopressin to produce maximal antidiuresis, water moves out of the hypotonic fluid entering the cortical collecting ducts into the interstitium of the cortex, and the tubular fluid becomes isotonic. In this fashion, as much as 10% of the filtered water is removed. The isotonic fluid then enters the medullary collecting ducts with a TF/P inulin of about 20. An additional 4.7% or more of the filtrate is reabsorbed into the hypertonic interstitium of the medulla, producing a concentrated urine with a TF/P inulin of over 300. In humans, the osmolality of urine may reach 1400 mosm/L, almost 5 times the osmolality of plasma, with a total of 99.7% of the filtered water being reabsorbed and only 0.3% appearing in the urine (Table 38–7). In other species, the ability to concentrate urine is even greater. Maximal urine osmolality is about 2500 mosm/L in dogs, about 3200 mosm/L in laboratory rats, and as high as 5000 mosm/L in certain desert rodents.

When vasopressin is absent, the collecting duct epithelium is relatively impermeable to water. The fluid therefore remains hypotonic, and large amounts flow into the renal pelvis. In humans, the urine osmolality may be as low as 30 mosm/L. The impermeability of the distal portions of the nephron is not absolute; along with the salt that is pumped out of the collecting duct fluid, about 2% of the filtered water is reabsorbed in the absence of vasopressin. However, as much as 13% of the filtered water may be excreted, and urine flow may reach 15 mL/min or more. The effects on daily water metabolism of the absence and the presence of vasopressin are summarized in Table 38–7.

It is much easier to drop a hydrometer into a urine specimen than to determine its osmolality, and the **specific gravity** used to be measured as an index of urine concentration. The specific gravity of an ultrafiltrate of plasma is 1.010, while that of a maximally concentrated urine specimen is about 1.035. However, it should be remembered that the specific gravity of a solution depends upon the nature as well as the number of solute particles in it. Thus, for example, a subject excreting radiographic contrast medium may have a urine specific gravity of 1.040–1.050 with relatively little increase in osmolality. Consequently, it is more accurate to measure osmolality.

The Countercurrent Mechanism

The concentrating mechanism depends upon the maintenance of a gradient of increasing osmolality along the medullary pyramids. This gradient is produced by the operation of the loops of Henle as **countercurrent multipliers** and maintained by the operation of the vasa recta as **countercurrent exchangers.** A countercurrent system is a system in which the inflow runs parallel to, counter to, and in close proximity to the outflow for some distance. This occurs for both the loops of Henle and the vasa recta in the renal medulla (Fig 38–2).

The operation of each loop of Henle as a countercurrent multiplier depends of the active transport of Na^+ and Cl^- out of its thick ascending limb (see above), the high permeability of its thin descending limb to water (Table 38–8), and the inflow of tubular fluid from the proximal tubule, with outflow into the distal tubule. The process is best understood in terms of hypothetic steps leading to the normal equilibrium condition, although of course the steps do not occur in vivo, and equilibrium is maintained unless the osmotic gradient is washed out. These steps are summarized in Fig 38–18 for a cortical nephron with no thin ascending limb. Assume first a condition in which osmolality is 300 mosm/kg of H_2O throughout the descending and ascending limbs and the medullary interstitium (Fig 38–18A). Assume in addition that the pumps in the thick ascending limb can pump 100 mosm/kg of Na^+ and Cl^- from the tubular fluid to the interstitium, increasing interstitial osmolality to 400 mosm/kg of H_2O. Water then moves out of the thin descending limb, and its contents equilibrate with the interstitium (Fig 38–18B). However, fluid containing 300 mosm/kg of H_2O is continuously entering this limb from the proximal tubule (Fig 38–18C), so the gradient against which the Na^+ and Cl^- are pumped is reduced and more enters the interstitium (Fig 38–18D). Meanwhile, hypotonic fluid flows into the distal tubule, and isotonic and subsequently hypertonic fluid flows into the ascending thick limb. The process keeps repeating, and the final result, as shown in Fig 38–18, is a gradient of osmolality from the top to the bottom of the loop.

In juxtamedullary nephrons with longer loops and thin ascending limbs, the osmotic gradient is spread over a greater distance and the osmolality at the tip of the loop is greater. This is because the thin ascending limb is relatively impermeable to water but permeable to Na^+ and Cl^-. Therefore, Na^+ and Cl^- move down their concentration gradients into the interstitium, and there is additional passive countercurrent multiplication. The greater the length of the loop of Henle, the greater the osmolality that can be reached at the tip of the pyramid.

The osmotic gradient in the medullary pyramids would not last long if the Na^+ and urea in the interstitial spaces were removed by the circulation. These solutes remain in the pyramids primarily because the vasa recta operate as countercurrent exchangers (Fig 38–19). The solutes diffuse out of the vessels conducting blood toward the cortex and into the vessels descending into the pyramid. Conversely, water diffuses out of the descending vessels and into the ascending vessels. Therefore, the solutes tend to recirculate in the medulla and water tends to bypass it, so that hypertonicity is maintained. The water removed from the collecting ducts in the pyramids is also re-

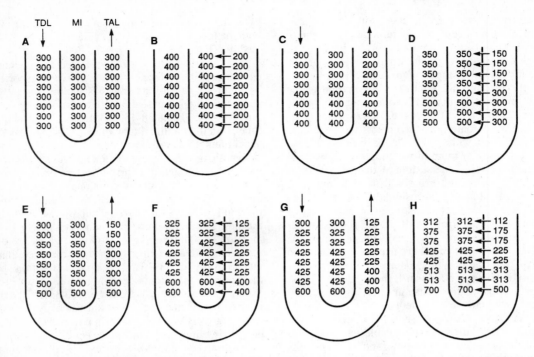

Figure 38–18. Operation of the loop of Henle as a countercurrent multiplier producing a gradient of hyperosmolarity in the medullary interstitium (MI). TDL, thin descending limb; TAL, thick ascending limb. The process of generation of the gradient is illustrated as occurring in hypothetic steps, starting at A, where osmolality in both limbs and the interstitium is 300 mosm/kg of water. The pumps in the thick ascending limb move Na^+ and Cl^- into the interstitium, increasing its osmolality to 400 mosm/kg, and this equilibrates with the fluid in the thin descending limb. However, isotonic fluid continues to flow into the thin descending limb and hypotonic fluid out of the thick ascending limb. Continued operation of the pumps makes the fluid leaving the thick ascending limb even more hypotonic, while hypertonicity accumulates at the apex of the loop. (Modified and reproduced, with permission, from Johnson LR [editor]: *Essential Medical Physiology.* Raven Press, 1992.)

moved by the vasa recta and enters the general circulation. It should be noted that countercurrent exchange is a passive process; it depends upon diffusion of water and solutes in both directions across the permeable walls of the vasa recta and could not maintain the osmotic gradient along the pyramids if the process of countercurrent multiplication in the loops of Henle were to cease.

It is worth noting that there is a very large osmotic gradient in the loop of Henle and, in the presence of vasopressin, in the collecting ducts. It is the countercurrent system that makes this gradient possible by spreading it along a system of tubules 1 cm or more in length rather than across a single layer of cells that is only a few micrometers thick. There are other examples of the operation of countercurrent exchangers in animals. One is the heat exchange between the arteries and venae comitantes of the limbs. To a minor degree in humans but to a major degree in mammals living in cold water, heat is transferred from the arterial blood flowing into the limbs to the adjacent veins draining blood back into the body, making the tips of the limbs cold while conserving body heat.

Role of Urea

Urea contributes to the establishment of the osmotic gradient in the medullary pyramids and to the ability to form a concentrated urine in the collecting ducts. Urea moves passively out of the proximal tubule, but except for the inner portion of the collecting duct, the rest of the tubular epithelium is virtually impermeable to this compound. Consequently, urea is increasingly concentrated in the fluid as water is removed in the loop and distal tubule. However, when the inner medullary portion of the collecting duct is reached, urea moves into the interstitium of the pyramids, adding to the hyperosmolarity. The movement of urea in this portion of the collecting duct is facilitated by vasopressin. Conversely, when the tubular fluid reaching this region is dilute in the absence of vasopressin, urea moves from the interstitium into the tubular lumen and the osmotic gradient in the pyramids is reduced. The amount of urea in the medullary interstitium and, consequently, in the urine varies with the amount of urea filtered, and this in turn varies with the dietary intake of protein. Therefore, a

Figure 38–19. Operation of the vasa recta as countercurrent exchangers in the kidney. NaCl and urea diffuse out of the ascending limb of the vessel and into the descending limb, whereas water diffuses out of the descending and into the ascending limb of the vascular loop.

high-protein diet increases the ability of the kidneys to concentrate the urine.

Water Diuresis

The feedback mechanism controlling vasopressin secretion and the way vasopressin secretion is stimulated by a rise and inhibited by a drop in the effective osmotic pressure of the plasma are discussed in Chapter 14. The **water diuresis** produced by drinking large amounts of hypotonic fluid begins about 15 minutes after ingestion of a water load and reaches its maximum in about 40 minutes. The act of drinking produces a small decrease in vasopressin secretion before the water is absorbed, but most of the inhibition is produced by the decrease in plasma osmolality after the water is absorbed.

Water Intoxication

During excretion of an average osmotic load, the maximal urine flow that can be produced during a water diuresis is about 16 mL/min. If water is ingested at a higher rate than this for any length of time, swelling of the cells because of the uptake of water from the hypotonic ECF becomes severe and, rarely, the symptoms of **water intoxication** may develop.

Swelling of the cells in the brain causes convulsions and coma and leads eventually to death. Water intoxication can also occur when water intake is not reduced after administration of exogenous vasopressin or secretion of endogenous vasopressin in response to nonosmotic stimuli such as surgical trauma.

Osmotic Diuresis

The presence of large quantities of unreabsorbed solutes in the renal tubules causes an increase in urine volume called **osmotic diuresis.** Solutes that are not reabsorbed in the proximal tubules exert an appreciable osmotic effect as the volume of tubular fluid decreases and their concentration rises. Therefore, they "hold water in the tubules." In addition, there is a limit to the concentration gradient against which Na^+ can be pumped out of the proximal tubules. Normally, the movement of water out of the proximal tubule prevents any appreciable gradient from developing, but Na^+ concentration in the fluid falls when water reabsorption is decreased because of the presence in the tubular fluid of increased amounts of unreabsorbable solutes. The limiting concentration gradient is reached, and further proximal reabsorption of Na^+ is prevented; more Na^+ remains in the tubule, and water stays with it. The result is that the loop of Henle is presented with a greatly increased volume of isotonic fluid. This fluid has a decreased Na^+ concentration, but the total amount of Na^+ reaching the loop per unit time is increased. In the loop, reabsorption of water and Na^+ is decreased because the medullary hypertonicity is decreased. The decrease is due primarily to decreased reabsorption of Na^+, K^+, and Cl^- in the ascending limb of the loop because the limiting concentration gradient for Na^+ reabsorption is reached. More fluid passes through the distal tubule, and because of the decrease in the osmotic gradient along the medullary pyramids, less water is reabsorbed in the collecting ducts. The result is a marked increase in urine volume and excretion of Na^+ and other electrolytes.

Osmotic diuresis is produced by the administration of compounds such as mannitol and related polysaccharides that are filtered but not reabsorbed. It is also produced by naturally occurring substances when they are present in amounts exceeding the capacity of the tubules to reabsorb them. In diabetes, for example, the glucose that remains in the tubules when the filtered load exceeds the Tm_G causes polyuria. Osmotic diuresis can also be produced by the infusion of large amounts of sodium chloride or urea.

It is important to recognize the difference between osmotic diuresis and water diuresis. In water diuresis, the amount of water reabsorbed in the proximal portions of the nephron is normal, and the maximal urine flow that can be produced is about 16 mL/min. In osmotic diuresis, increased urine flow is due to decreased water reabsorption in the proximal tubules and loops and very large urine flows can be produced.

As the load of excreted solute is increased, the concentration of the urine approaches that of plasma (Fig 38–20) in spite of maximal vasopressin secretion, because an increasingly large fraction of the excreted urine is isotonic proximal tubular fluid. If osmotic diuresis is produced in an animal with diabetes insipidus, the urine concentration rises for the same reason.

Relation of Urine Concentration to GFR

The magnitude of the osmotic gradient along the medullary pyramids is increased when the rate of flow of fluid through the loops of Henle is decreased. A reduction in GFR such as that caused by dehydration produces a decrease in the volume of fluid presented to the countercurrent mechanism, so that the rate of flow in the loops declines and the urine becomes more concentrated. When the GFR is low, the urine can become quite concentrated in the absence of vasopressin. If one renal artery is constricted in an animal with diabetes insipidus, the urine excreted on the side of the constriction becomes hypertonic because of the reduction in GFR, whereas that excreted on the opposite side remains hypotonic.

"Free Water Clearance"

In order to quantitate the gain or loss of water by excretion of a concentrated or dilute urine, the "free water clearance" (C_{H_2O}) is sometimes calculated. This is the difference between the urine volume and the clearance of osmoles (C_{osm}):

$$C_{H_2O} = \dot{V} - \frac{U_{osm}\dot{V}}{P_{osm}}$$

where $\dot{V}$ is the urine flow rate and U_{osm} and P_{osm} the urine and plasma osmolality, respectively. C_{osm} is the amount of water necessary to excrete the osmotic load in a urine that is isotonic with plasma. Therefore, C_{H_2O} is negative when the urine is hypertonic and positive when the urine is hypotonic. In Table 38–7, for example, the values for C_{H_2O} are –1.3 mL/min (–1.9 L/d) during maximal antidiuresis and 14.5 mL/min (20.9 L/d) in the absence of vasopressin.

ACIDIFICATION OF THE URINE & BICARBONATE EXCRETION

H+ Secretion

The cells of the proximal and distal tubules, like the cells of the gastric glands, secrete hydrogen ions (see Chapter 26). Acidification also occurs in the collecting ducts. The reaction that is primarily responsible for H+ secretion in the proximal tubules is Na+-H+ exchange (Fig 38–21). This is an example of secondary active transport; extrusion of Na+ from the cells into the interstitium by Na+-K+ ATPase lowers intracellular Na+, and this causes Na+ to enter the cell from the tubular lumen, with coupled extrusion of H+. The H+ comes from intracellular dissociation of H_2CO_3, and the HCO_3^- that is formed diffuses into the interstitial fluid. Thus, for each H+ ion secreted, one Na+ ion and one HCO_3^- ion enter the interstitial fluid.

Carbonic anhydrase catalyzes the formation of H_2CO_3, and drugs that inhibit carbonic anhydrase depress both secretion of acid by the proximal tubules and the reactions which depend on it.

There is some evidence that H+ is secreted in the proximal tubules by other types of pumps, but the evidence for these additional pumps is controversial, and in any case, their contribution is small relative to that of the Na+-H+ exchange mechanism. This is in contrast to what occurs in the distal tubules and collecting ducts, where H+ secretion is relatively independent of Na+ in the tubular lumen. In this part of the tubule, H+ is secreted by an ATP-driven proton pump. Aldosterone acts on this pump to increase dis-

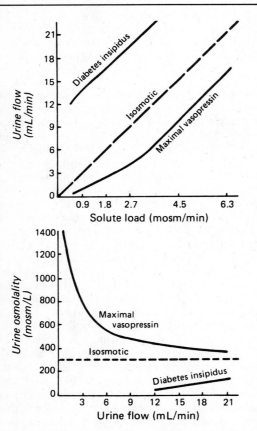

Figure 38–20. Approximate relationship between urine concentration and urine flow in osmotic diuresis in humans. The dashed line in the lower diagram indicates the concentration at which the urine is isosmotic with plasma. (Reproduced, with permission, from Berliner RW, Giebisch G in: *Best and Taylor's Physiological Basis of Medical Practice,* 9th ed. Brobeck JR [editor]. Williams & Wilkins, 1979.)

Figure 38–21. Secretion of acid by proximal tubular cells in the kidney. H^+ is transported into the tubular lumen by an antiport in exchange for Na^+. Active transport by Na^+-K^+ ATPase is indicated by arrows in the circle. Dashed arrows indicate diffusion. Compare with Fig 26–7.

Figure 38–22. Fate of H^+ secreted into a tubule in exchange for Na^+. **Top:** Reabsorption of filtered bicarbonate. **Middle:** Formation of monobasic phosphate. **Bottom:** Ammonium formation. Note that in each instance one Na^+ ion and one HCO_3^- ion enter the bloodstream for each H^+ ion secreted. A^-, anion.

tal H^+ secretion. The I cells in this part of the renal tubule secrete acid, and like the parietal cells in the stomach, the I cells contain abundant carbonic anhydrase and numerous tubulovesicle structures. There is evidence that the H^+-translocating ATPase which produces H^+ secretion is located on these vesicles as well as on the luminal cell membrane and that, in acidosis, the number of H^+ pumps is increased by insertion of these tubulovesicles into the luminal cell membrane. These cells also contain **Band 3,** an anion exchange protein, in their basolateral cell membranes, and this protein may function as a Cl^--HCO_3^- exchanger for the transport of HCO_3^- to the interstitial fluid.

Fate of H^+ in the Urine

The amount of acid secreted depends upon the subsequent events in the tubular urine. The maximal H^+ gradient against which the transport mechanisms can secrete in humans corresponds to a urine pH of about 4.5, ie, an H^+ concentration in the urine that is 1000 times the concentration in plasma. pH 4.5 is thus the **limiting pH.** If there were no buffers that "tied up" H^+ in the urine, this pH would be reached rapidly, and H^+ secretion would stop. However, 3 important reactions in the tubular fluid remove free H^+, permitting more acid to be secreted (Fig 38–22). These are the reactions with HCO_3^- to form CO_2 and H_2O, with HPO_4^{2-} to form $H_2PO_4^-$, and with NH_3 to form NH_4^+.

Reaction With Buffers

The dynamics of buffering are discussed in Chapter 35. The pK′ of the bicarbonate system is 6.1 (see

Chapter 35), that of the dibasic phosphate system is 6.8, and that of the ammonia system is 9.0. The concentration of HCO_3^- in the plasma, and consequently in the glomerular filtrate, is normally about 24 meq/L, whereas that of phosphate is only 1.5 meq/L. Therefore, in the proximal tubule, most of the secreted H^+ reacts with HCO_3^- to form H_2CO_3 (Fig 38–22). The H_2CO_3 breaks down to form CO_2 and H_2O. In the proximal (but not in the distal) tubule, there is carbonic anhydrase in the brush border of the cells; this facilitates the formation of CO_2 and H_2O in the tubular fluid. The CO_2, which diffuses readily across all biologic membranes, enters the tubular cells, where it adds to the pool of CO_2 available to form H_2CO_3. Since most of the H^+ is removed from the tubule, the pH of the fluid is changed very little. This is the mechanism by which HCO_3^- is reabsorbed; for each mole of HCO_3^- removed from the tubular fluid, 1 mole of HCO_3^- diffuses from the tubular cells into the blood, even though it is not the same mole that disappeared from the tubular fluid.

Secreted H^+ also reacts with dibasic phosphate (HPO_4^{2-}) to form monobasic phosphate ($H_2PO_4^-$).

This happens to the greatest extent in the distal tubules and collecting ducts, because it is here that the phosphate which escapes proximal reabsorption is greatly concentrated by the reabsorption of water. The reaction with NH_3 occurs in the proximal and distal tubules. H^+ also combines to a minor degree with other buffer anions.

Each H^+ ion that reacts with the buffers contributes to the urinary **titratable acidity,** which is measured by determining the amount of alkali that must be added to the urine to return its pH to 7.4, the pH of the glomerular filtrate. However, the titratable acidity obviously measures only a fraction of the acid secreted, since it does not account for the H_2CO_3 that has been converted to H_2O and CO_2. In addition, the pK′ of the ammonia system is 9.0, and the ammonia system is titrated only from the pH of the urine to pH 7.4, so it contributes very little to the titratable acidity.

Ammonia Secretion

Reactions in the renal tubular cells produce NH_4^+ and HCO_3^-. In proximal tubule cells, the NH_4^+ enters the tubular lumen, whereas the HCO_3^- enters the interstitium, helping to buffer acid loads. The NH_4^+ that enters the proximal tubular urine is reabsorbed in the thick ascending limb of Henle, and this maintains an NH_4^+ gradient by countercurrent multiplication with a high NH_4^+ concentration in the medullary pyramids. In the collecting ducts, NH_4^+ enters the tubular fluid and is excreted. In addition, NH_4^+ is in equilibrium with $NH_3 + H^+$ in the cells. Since the pK′ of this reaction is 9.0, the ratio of NH_3 to NH_4^+ at pH 7.0 is 1:100 (Fig 38–23). However, NH_3 is lipid-soluble and diffuses across the cell membranes down its concentration gradient into the interstitial fluid and tubular urine. In the urine it reacts with H^+ to form NH_4^+, and the NH_4^+ remains in the urine.

The principal reaction producing NH_4^+ in cells is conversion of glutamine to glutamate. This reaction is catalyzed by the enzyme **glutaminase,** which is abundant in renal tubular cells (Fig 38–23). **Glutamic dehydrogenase** catalyzes the conversion of glutamate to α-ketoglutarate, with the production of

more NH_4^+. Subsequent metabolism of α-ketoglutarate utilizes 2 H^+, freeing 2 HCO_3^-.

In chronic acidosis, the amount of NH_4^+ excreted at any given urine pH also increases, because more NH_3 enters the tubular urine. The effect of this **adaptation** of NH_3 secretion, whose the cause is unsettled, is a further removal of H^+ from the tubular fluid and consequently a further enhancement of H^+ secretion.

The process by which NH_3 is secreted into the urine and then changed to NH_4^+, maintaining the concentration gradient for diffusion of NH_3, is called **nonionic diffusion** (see Chapter 1). Salicylates and a number of other drugs that are weak bases or weak acids are also secreted by nonionic diffusion. They diffuse into the tubular fluid at a rate that is dependent upon the urine pH, and the amount of each drug excreted therefore varies with the pH of the urine.

pH Changes Along the Nephrons

Micropuncture studies indicate that an appreciable drop in pH occurs in the proximal tubule in mammalian kidneys, although there is further acidification in the distal tubule. Since most of the HCO_3^- is absorbed in the proximal tubule and the H^+ secretion responsible for its reabsorption has only a small effect on the pH of the fluid, it is apparent that there is considerably more acid secretion in the proximal than in the distal tubule. The acid-secreting mechanism in the distal tubules, and particularly in the collecting ducts, has a lower secretory capacity but is capable of generating a large pH difference between the tubular lumen and the cells.

Factors Affecting Acid Secretion

Renal acid secretion is altered by changes in the intracellular P_{CO_2}, K^+ concentration, carbonic anhydrase level, and adrenocortical hormone concentration. When the P_{CO_2} is high (respiratory acidosis), more intracellular H_2CO_3 is available to buffer the hydroxyl ions and acid secretion is enhanced, whereas the reverse is true when the P_{CO_2} falls. K^+ depletion enhances acid secretion, apparently because the loss of K^+ causes intracellular acidosis even though the plasma pH may be elevated. Conversely, K^+ excess in the cells inhibits acid secretion. When carbonic anhydrase is inhibited, acid secretion is inhibited, because the formation of H_2CO_3 is decreased. Aldosterone and the other adrenocortical steroids that enhance tubular reabsorption of Na^+ also increase the secretion of H^+ and K^+.

Bicarbonate Excretion

Although the process of HCO_3^- reabsorption does not actually involve transport of this ion into the tubular cells, HCO_3^- reabsorption is proportionate to the amount filtered over a relatively wide range. There is no Tm, but HCO_3^- reabsorption is decreased by an unknown mechanism when the ECF volume is expanded (Fig 38–24). When the plasma HCO_3^- con-

$$NH_4^+ \rightleftharpoons NH_3 + H^+$$

$$pH = pK' + \log \frac{[NH_3]}{[NH_4^+]}$$

$$\text{Glutamine} \xrightarrow{\text{Glutaminase}} \text{Glutamate} + NH_4^+$$

$$\text{Glutamate} \xrightarrow{\text{Glutamic dehydrogenase}} \alpha\text{-Ketoglutarate} + NH_4^+$$

Figure 38–23. Major reactions involved in ammonia production in the kidneys. See also Chapter 17.

Figure 38–24. Effect of ECF volume on HCO_3^- filtration, reabsorption, and excretion in rats. The pattern of HCO_3^- excretion is similar in humans. The plasma HCO_3^- concentration is normally about 24 meq/L. (Reproduced, with permission, from Valtin H: *Renal Function,* 2nd ed. Little, Brown, 1983. Copyright © Little, Brown and Co., 1983.)

centration is low, all the filtered HCO_3^- is reabsorbed; but when the plasma HCO_3^- concentration is high, above 26–28 meq/L (the renal threshold for HCO_3^-), HCO_3^- appears in the urine and the urine becomes alkaline. Conversely, when the plasma HCO_3^- falls below about 26 meq/L, the value at which all the secreted H^+ is being used to reabsorb HCO_3^-, more H^+ becomes available to combine with other buffer anions. Therefore, the lower the plasma HCO_3^- concentration drops, the more acidic the urine becomes and the greater its NH_4^+ content.

Implications of Urinary pH Changes

Depending upon the rates of the interrelated processes of acid secretion, NH_4^+ production, and HCO_3^- excretion, the pH of the urine in humans varies from 4.5 to 8.0. Excretion of a urine that is at a pH different from that of the body fluids has important implications for the body's electrolyte and acid-base economy which are discussed in detail in Chapter 39. Acids are buffered in the plasma and cells, the overall reaction being $HA + NaHCO_3 \rightarrow NaA + H_2CO_3$. The H_2CO_3 forms CO_2 and H_2O, and the CO_2 is expired, while the NaA appears in the glomerular filtrate. To the extent that the Na^+ is replaced by H^+ in the urine, Na^+ is conserved in the body. Furthermore, for each H^+ ion excreted with phosphate or as NH_4^+, there is a net gain of one HCO_3^- ion in the blood, replenishing the supply of this important buffer anion. Conversely, when base is added to the body fluids, the OH^- ions are buffered, raising the plasma HCO_3^-. When the

plasma level exceeds 28 meq/L, the urine becomes alkaline and the extra HCO_3^- is excreted in the urine. Because the rate of maximal H^+ secretion by the tubules varies directly with the arterial P_{CO_2}, HCO_3^- reabsorption also is affected by the P_{CO_2}. This relationship is discussed in detail in Chapter 39.

REGULATION OF Na^+ & Cl^- EXCRETION

Na^+ is filtered in large amounts, but it moves passively out of some portions of the tubules and is actively transported out of all portions except the thin loop of Henle. Normally, 96% to well over 99% of the filtered Na^+ is reabsorbed. Most of the Na^+ is reabsorbed with Cl^- (Table 38–9), but some is reabsorbed in the processes by which one Na^+ ion enters the bloodstream for each H^+ ion secreted by the tubules, and in the distal tubules a small amount is actively reabsorbed in association with the secretion of K^+.

Regulation of Na^+ Excretion

Because Na^+ is the most abundant cation in ECF and because sodium salts account for over 90% of the osmotically active solute in the plasma and interstitial fluid, the amount of Na^+ in the body is a prime determinant of the ECF volume. Therefore, it is not surprising that multiple regulatory mechanisms have evolved in terrestrial animals to control the excretion of this ion. Through the operation of these regulatory mechanisms, the amount of Na^+ excreted is adjusted to equal the amount ingested over a wide range of dietary intakes, and the individual stays in Na^+ balance. Thus, urinary Na^+ output ranges from less than 1 meq/d on a low-salt diet to 400 meq/d or more when the dietary Na^+ intake is high. In addition, there is a natriuresis when saline is infused intravenously and a decrease in Na^+ excretion when ECF volume is reduced. Variations in Na^+ excretion are effected by

Table 38–9. Quantitative aspects of Na^+ reabsorption in a normal man.

GFR = 125 mL/min	
Plasma HCO_3^- = 27 meq/L	
Plasma Na^+ = 145 meq/L	
Na^+ filtered per minute	18,125 μeq
Reabsorbed with Cl^-	14,585 μeq
Reabsorbed while reabsorbing 3375 μeq HCO_3^-	3,375 μeq
Reabsorbed in association with formation of titratable acidity and ammonia	50 μeq
Reabsorbed in association with secretion of K^+	50 μeq
Total Na^+ reabsorbed per minute	18,060 μeq

changes in the amount filtered (Table 38–10) and the amount reabsorbed in the tubules. The factors affecting the GFR, including tubuloglomerular feedback, are discussed above, as are glomerulotubular balance and the role of oncotic pressure in the peritubular capillaries in this process. Other factors affecting Na$^+$ reabsorption include the circulating level of aldosterone and other adrenocortical hormones, the circulating level of ANP and possibly other natriuretic hormones, the angiotensin II in the kidney, and the rate of tubular secretion of H$^+$ and K$^+$.

Effects of Adrenocortical Steroids

Adrenal mineralocorticoids such as aldosterone increase tubular reabsorption of Na$^+$ in association with secretion of K$^+$ and H$^+$ (Fig 38–16) and also Na$^+$ reabsorption with Cl$^-$ (see Chapter 20). When these hormones are injected into adrenalectomized animals, there is a latent period of 10–30 minutes before their effects on Na$^+$ reabsorption become manifest, because of the time required for the steroids to alter protein synthesis via their action on DNA. They act on the distal convoluted tubules and collecting ducts, and there is some evidence that they cause the reabsorption of a small amount of Na$^+$ from the bladder. The glucocorticoids such as cortisol also increase Na$^+$ reabsorption; but, unlike the mineralocorticoids, they increase the GFR. They may therefore increase rather than decrease Na$^+$ excretion, because the increase they produce in the filtered load of Na$^+$ may exceed the increase in reabsorption.

Reduction of the dietary intake of salt increases aldosterone secretion (see Fig 20–28). Changes in aldosterone secretion can explain changes in Na$^+$ excretion that occur over days or even hours, but, because of the latent period before its effects are manifest, rapid changes in Na$^+$ excretion cannot be attributed to this hormone.

Other Humoral Effects

As noted in Chapter 24, ANP probably increases Na$^+$ excretion by increasing GFR, although it may also exert an effect on the tubules. The Na$^+$-K$^+$ ATPase-inhibiting factor, which appears to be endogenously produced ouabain, also increases Na$^+$ excretion. Angiotensin II increases reabsorption of Na$^+$ and HCO$_3^-$ by an action on the proximal tubules. There is an appreciable amount of angiotensin-converting enzyme in the kidneys, and the kidneys convert 20% of the circulating angiotensin I reaching them to angiotensin II. In addition, angiotensin I is generated in the kidneys.

Relation to Acid & K$^+$ Secretion

Na$^+$ excretion is increased by drugs that decrease renal acid secretion by inhibiting carbonic anhydrase. After CO$_2$ or acid is buffered in the blood, Na$^+$ filtered with acid anions is lost in the urine if the amount in the filtrate exceeds the capacity of the tubules to exchange the Na$^+$ for H$^+$. Changes in Na$^+$ excretion due to changes in the rate of K$^+$ secretion are small (Table 38–9).

Chloride Excretion

Chloride reabsorption is increased when HCO$_3$ reabsorption is decreased, and vice versa, so that Cl$^-$ concentration in plasma varies inversely with the HCO$_3^-$ concentration, keeping the total anion concentration constant. Passive diffusion can explain the movements of Cl$^-$ in some situations. However, Cl$^-$ also leaves the tubular lumen by secondary active transport, as described above.

REGULATION OF K$^+$ EXCRETION

Much of the filtered K$^+$ is removed from the tubular fluid by active reabsorption in the proximal tubules (Table 38–6), and K$^+$ is then secreted into the fluid by the distal tubular cells. The rate of K$^+$ secretion is proportionate to the rate of flow of the tubular fluid through the distal portions of the nephron, because with rapid flow there is less opportunity for the tubular K$^+$ concentration to rise to a value that stops further secretion. In the absence of complicating factors, the amount secreted is approximately equal to the K$^+$ intake, and K$^+$ balance is maintained. In the distal tubules, Na$^+$ is generally reabsorbed and K$^+$ is secreted. There is no rigid one-for-one exchange, and much of the movement of K$^+$ is passive. However, there is electrical coupling in the sense that intracellular migration of Na$^+$ tends to lower the potential difference across the tubular cell, and this favors movement of K$^+$ into the tubular lumen. Since Na$^+$ is also reabsorbed in association with H$^+$ secretion, there is competition for the Na$^+$ in the tubular fluid. K$^+$ excretion is decreased when the amount of Na$^+$ reaching the distal tubule is small, and it is also decreased when H$^+$ secretion is increased. When total body K$^+$ is high, H$^+$ secretion is inhibited, apparently

Table 38–10. Changes in Na$^+$ excretion that would occur as a result of changes in GFR if there were no concomitant changes in Na$^+$ reabsorption.

GFR (mL/min)	Plasma Na$^+$ (μeq/mL)	Amount Filtered (μeq/min)	Amount Reabsorbed (μeq/min)	Amount Excreted (μeq/min)
125	145	18,125	18,000	125
127	145	18,415	18,000	415
124.1	145	18,000	18,000	0

because of intracellular alkalosis; and K⁺ secretion and excretion are therefore facilitated. Conversely, the cells are acid in K⁺ depletion, and K⁺ secretion declines. Apparently the K⁺ secretory mechanism is capable of "adaptation," because the amount of K⁺ excreted gradually increases when a constant large dose of a potassium salt is administered for a prolonged period.

DIURETICS

Although a detailed discussion of diuretic agents is outside the scope of this book, consideration of their mechanisms of action constitutes an informative review of the factors affecting urine volume and electrolyte excretion. These mechanisms are summarized in Table 38–11. Water, alcohol, osmotic diuretics, xanthines, and acidifying salts have limited clinical

Table 38–11. Mechanism of action of various diuretics.

Agent	Mechanism of Action
Water	Inhibits vasopressin secretion.
Ethanol	Inhibits vasopressin secretion.
Antagonists of V_2 vasopressin receptors	Inhibit action of vasopressin on collecting duct.
Large quantities of osmotically active substances such as mannitol and glucose	Produce osmotic diuresis.
Xanthines such as caffeine and theophylline	Decrease tubular reabsorption of Na⁺ and increase GFR.
Acidifying salts such as $CaCl_2$ and NH_4Cl	Supply acid load; H⁺ is buffered, an anion is excreted with Na⁺ when the ability of the kidneys to replace Na⁺ with H⁺ is exceeded.
Carbonic anhydrase inhibitors such as acetazolamide (Diamox)	Decrease H⁺ secretion, with resultant increase in Na⁺ and K⁺ excretion.
Metolazone (Zaroxolyn), thiazides such as chlorothiazide (Diuril)	Inhibit Na⁺ and K⁺ reabsorption in the early portion of the distal tubule.
Loop diuretics such as furosemide (Lasix), ethacrynic acid (Edecrin), and bumetanide	Inhibit Na⁺-K⁺-2Cl⁻ cotransport in the medullary thick ascending limb of the loop of Henle.
K⁺-retaining natriuretics such as spironolactone (Aldactone), triamterene (Dyrenium), and amiloride (Colectril)	Inhibit Na⁺-K⁺ "exchange" in the distal portion of the distal tubule and the collecting duct by inhibiting the action of aldosterone (spironolactone) or by inhibiting Na⁺ reabsorption (triamterene, amiloride).

usefulness, and the vasopressin antagonists are used primarily for research. However, many of the other agents on the list are used extensively in medical practice.

The carbonic anhydrase-inhibiting drugs are only moderately effective as diuretic agents, but because they inhibit acid secretion by decreasing the supply of carbonic acid, they have far-reaching effects. Not only is Na⁺ excretion increased because H⁺ secretion is decreased, but also HCO_3^- reabsorption is depressed; and because H⁺ and K⁺ compete with each other and with Na⁺, the decrease in H⁺ secretion facilitates the secretion and excretion of K⁺.

Another determinant of the rate of K⁺ secretion is the amount of Na⁺ delivered to the Na⁺-K⁺ "exchange" site in the distal tubule. Thiazides, furosemide, ethacrynic acid, and bumetanide act proximal to this site, and the resultant increase in Na⁺ delivery increases K⁺ secretion. The K⁺ loss is appreciable, and K⁺ depletion is one of the common complications of treatment with these agents. The thiazides inhibit Na⁺ and Cl⁻ transport in the proximal portion of the distal tubule (Fig 38–25). Furosemide, ethacrynic acid, and bumetanide inhibit Na⁺ and Cl⁻ reabsorption in the thick ascending limb of the loop of Henle by inhibiting Na⁺-K⁺-2Cl⁻ transport (Fig 38–17). Spironolactone, triamterene, and amiloride act on the Na⁺-K⁺ exchange mechanism itself, causing K⁺ retention and, in some cases, mild hyperkalemia.

Figure 38–25. Sites of action of various diuretics. ①: Furosemide and other loop diuretics act on the thick ascending limb of the loop of Henle. ②: Thiazides act on the early portion of the distal convoluted tubule. ③: Aldosterone antagonists, triamterene, and amiloride act primarily on the collecting ducts. ④: Antagonists to V_2 vasopressin receptors act on the collecting ducts.

EFFECTS OF DISORDERED RENAL FUNCTION

A number of abnormalities are common to many different types of renal disease. The secretion of renin by the kidneys and the relation of the kidneys to hypertension are discussed in Chapters 24 and 33. A frequent finding in various forms of renal disease is the presence in the urine of protein, leukocytes, red cells, and **casts,** which are bits of proteinaceous material precipitated in the tubules and washed into the bladder. Other important consequences of renal disease are loss of the ability to concentrate or dilute the urine, uremia, acidosis, and abnormal retention of Na^+.

Proteinuria

In many renal diseases and in one benign condition, the permeability of the glomerular capillaries is increased, and protein is found in the urine in more than the usual trace amounts (**proteinuria**). Most of this protein is albumin, and the defect is commonly called **albuminuria.** The relation of charges on the glomerular membrane to albuminuria is discussed above. The amount of protein in the urine may be very large, and, especially in nephrosis, the urinary protein loss may exceed the rate at which the liver can synthesize plasma proteins. The resulting hypoproteinemia reduces the oncotic pressure, and the plasma volume declines, sometimes to dangerously low levels, while edema fluid accumulates in the tissues.

A benign condition that causes proteinuria is a poorly understood change in renal hemodynamics which in some otherwise normal individuals causes protein to appear in urine formed when they are in the standing position (**orthostatic albuminuria**). Urine formed when these individuals are lying down is protein-free.

Loss of Concentrating & Diluting Ability

In renal disease, the urine becomes less concentrated and urine volume is often increased, producing the symptoms of **polyuria** and **nocturia** (waking up at night to void). The ability to form a dilute urine is often retained, but in advanced renal disease, the osmolality of the urine becomes fixed at about that of plasma, indicating that the diluting and concentrating functions of the kidney have both been lost. The loss is due in part to disruption of the countercurrent mechanism, but a more important cause is a loss of functioning nephrons. When one kidney is removed surgically, the number of functioning nephrons is halved. The number of osmoles excreted is not reduced to this extent, and so the remaining nephrons must each be filtering and excreting more osmotically active substances, producing what is in effect an osmotic diuresis. In osmotic diuresis, the osmolality of the urine approaches that of plasma (see above);

the same thing happens when the number of functioning nephrons is reduced by disease. Of course, when most of the nephrons are destroyed, urine volume falls and **oliguria** or even **anuria** is present.

Uremia

When the breakdown products of protein metabolism accumulate in the blood, the syndrome known as **uremia** develops. The symptoms of uremia include lethargy, anorexia, nausea and vomiting, mental deterioration and confusion, muscle twitching, convulsions, and coma. The BUN and creatinine levels are high, and the blood levels of these substances are used as an index of the severity of the uremia. It probably not the accumulation of urea and creatinine per se but rather the accumulation of other toxic substances—possibly organic acids or phenols—that produces the symptoms of uremia. Because urea crosses the blood-brain barrier slowly, it is administered intravenously to neurosurgical patients to make the ECF outside the brain hypertonic and shrink the brain during surgery. However, urea infusions cause changes in the electrical activity of the brain in experimental animals and may not be as innocuous as they appear to be. In dogs, prolonged infusions of urea cause anorexia, weakness, vomiting, and diarrhea.

The toxic substances that cause the symptoms of uremia can be removed by dialyzing the blood of uremic patients against a bath of suitable composition in an **artificial kidney.** By repeated dialysis, patients can be kept alive and in reasonable health for many months, even when they are completely anuric or have had both kidneys removed.

Other features of chronic renal failure include anemia, which is caused primarily by failure to produce erythropoietin (see Chapter 24), and secondary hyperparathyroidism due to 1,25-dihydroxycholecalciferol deficiency (see Chapter 21).

Acidosis

Acidosis is common in chronic renal disease because of failure to excrete the acid products of digestion and metabolism (see Chapter 39). In the rare syndrome of **renal tubular acidosis,** there is specific impairment of the ability to make the urine acidic, and other renal functions are usually normal. However, in most cases of chronic renal disease the urine is maximally acidified, and acidosis develops because the total amount of H^+ that can be secreted is reduced because of impaired renal tubular production of NH_4^+.

Abnormal Na⁺ Metabolism

Many patients with renal disease retain excessive amounts of Na^+ and become edematous. There are at least 3 causes of Na^+ retention in renal disease. In acute glomerulonephritis, a disease that affects primarily the glomeruli, there is a marked decrease in the amount of Na^+ filtered. In the nephrotic syndrome, an increase in aldosterone secretion contributes to the salt retention. The plasma protein level is low in this condition, and so fluid moves from the plasma into the interstitial spaces and the plasma volume falls. The decline in plasma volume triggers the increase in aldosterone secretion via the renin-angiotensin system. A third cause of Na^+ retention and edema in renal disease is **heart failure** (see Chapter 33). Renal disease predisposes to heart failure, partly because of the hypertension it frequently produces.

Variations in the Response to Aldosterone

As noted in Chapter 20, normal individuals "escape" when exposed to excess mineralocorticoids. They stop retaining Na^+ but lose large amounts of K^+. On the other hand, edematous patients with the nephrotic syndrome, cirrhosis, or heart failure continue to retain Na^+ and do not lose K^+. This is due in part to a reduction in the amount of Na^+ reaching the distal portions of the nephrons; Na^+ in the tubular fluid helps maintain the potential difference between the tubular lumen and the cells, and this favors K^+ secretion. The reduction in distal tubular Na^+ is due to avid Na^+ retention in the proximal tubules and a decrease in filtered Na^+.

FILLING OF THE BLADDER

The walls of the ureters contain smooth muscle arranged in spiral, longitudinal, and circular bundles, but distinct layers of muscle are not seen. Regular peristaltic contractions occurring 1–5 times per minute move the urine from the renal pelvis to the bladder, where it enters in spurts synchronous with each peristaltic wave. The ureters pass obliquely through the bladder wall and, although there are no ureteral sphincters as such, the oblique passage tends to keep the ureters closed except during peristaltic waves, preventing reflux of urine from the bladder.

EMPTYING OF THE BLADDER

Anatomic Considerations

The smooth muscle of the bladder, like that of the ureters, is arranged in spiral, longitudinal, and circular bundles. Contraction of this muscle, which is called the **detrusor muscle,** is mainly responsible for emptying the bladder during urination (micturition). Muscle bundles pass on either side of the urethra, and these fibers are sometimes called the **internal urethral sphincter,** although they do not encircle the urethra. Farther along the urethra is a sphincter of skeletal muscle, the sphincter of the membranous urethra (**external urethral sphincter**). The bladder epithelium is made up of a superficial layer of flat cells and a deep layer of cuboidal cells. The innervation of the bladder is summarized in Fig 38–26.

Figure 38–26. Innervation of the bladder. Dashed lines indicate sensory nerves. Parasympathetic innervation is shown at the left, sympathetic at the upper right, and somatic at the lower right.

Micturition

The physiology of micturition and its disorders are subjects about which there is much confusion. Micturition is fundamentally a spinal reflex facilitated and inhibited by higher brain centers and, like defecation, subject to voluntary facilitation and inhibition. Urine enters the bladder without producing much increase in intravesical pressure until the viscus is well filled. In addition, like other types of smooth muscle, the bladder muscle has the property of plasticity; when it is stretched, the tension initially produced is not maintained. The relation between intravesical pressure and volume can be studied by inserting a catheter and emptying the bladder, then recording the pressure while the bladder is filled with 50-mL increments of water or air (**cystometry**). A plot of intravesical pressure against the volume of fluid in the bladder is called a **cystometrogram** (Fig 38–27). The curve shows an initial slight rise in pressure when the first increments in volume are produced; a long, nearly flat segment as further increments are produced; and a sudden, sharp rise in pressure as the micturition reflex is triggered. These 3 components are sometimes called segments Ia, Ib, and II. The first urge to void is felt at a bladder volume of about 150 mL, and a marked sense of fullness at about 400 mL. The flatness of segment Ib is a manifestation of the law of Laplace (see Chapter 30). This law states that the pressure in a spherical viscus is equal to twice the wall tension divided by the radius. In the case of the bladder, the tension increases as the organ fills, but so

Figure 38–27. Cystometrogram in a normal human. The numerals identify the 3 components of the curve described in the text. The dotted line indicates the pressure-volume relations that would have been found had micturition not occurred and produced component II. (Modified and reproduced, with permission, from Tanagho EA, McAninch JW: *Smith's General Urology,* 13th ed. Appleton & Lange, 1992.)

does the radius. Therefore, the pressure increase is slight until the organ is relatively full.

During micturition, the perineal muscles and external urethral sphincter are relaxed; the detrusor muscle contracts; and urine passes out through the urethra. The bands of smooth muscle on either side of the urethra apparently play no role in micturition, and their main function is believed to be the prevention of reflux of semen into the bladder during ejaculation.

The mechanism by which voluntary urination is initiated remains unsettled. One of the initial events is relaxation of the muscles of the pelvic floor, and this may cause a sufficient downward tug on the detrusor muscle to initiate its contraction. The perineal muscles and external sphincter can be contracted voluntarily, preventing urine from passing down the urethra or interrupting the flow once urination has begun. It is through the learned ability to maintain the external sphincter in a contracted state that adults are able to delay urination until the opportunity to void presents itself. After urination, the female urethra empties by gravity. Urine remaining in the urethra of the male is expelled by several contractions of the bulbocavernosus muscle.

Reflex Control

The bladder smooth muscle has some inherent contractile activity; however, when its nerve supply is intact, stretch receptors in the bladder wall initiate a reflex contraction that has a lower threshold than the inherent contractile response of the muscle. Fibers in the pelvic nerves are the afferent limb of the voiding reflex, and the parasympathetic fibers to the

bladder that constitute the efferent limb also travel in these nerves. The reflex is integrated in the sacral portion of the spinal cord. In the adult, the volume of urine in the bladder that normally initiates a reflex contraction is about 300–400 mL. The sympathetic nerves to the bladder play no part in micturition, but they do mediate the contraction of the bladder muscle that prevents semen from entering the bladder during ejaculation (see Chapter 23).

There is no small motor nerve system to the stretch receptors in the bladder wall; but the threshold for the voiding reflex, like the stretch reflexes, is adjusted by the activity of facilitatory and inhibitory centers in the brain stem. There is a facilitatory area in the pontine region and an inhibitory area in the midbrain. After transection of the brain stem just above the pons, the threshold is lowered and less bladder filling is required to trigger it, whereas after transection at the top of the midbrain, the threshold for the reflex is essentially normal. There is another facilitatory area in the posterior hypothalamus. In humans with lesions in the superior frontal gyrus, the desire to urinate is reduced and there is also difficulty in stopping micturition once it has commenced. However, stimulation experiments in animals indicate that other cortical areas also affect the process. The bladder can be made to contract by voluntary facilitation of the spinal voiding reflex when it contains only a few milliliters of urine. Voluntary contraction of the abdominal muscles aids the expulsion of urine by increasing the intra-abdominal pressure, but voiding can be initiated without straining even when the bladder is nearly empty.

Abnormalities of Micturition

There are 3 major types of bladder dysfunction due to neural lesions: (1) the type due to interruption of the afferent nerves from the bladder; (2) the type due to interruption of both afferent and efferent nerves; and (3) the type due to interruption of facilitatory and inhibitory pathways descending from the brain. In all 3 types the bladder contracts, but the contractions are generally not sufficient to empty the viscus completely, and residual urine is left in the bladder.

Effects of Deafferentation

When the sacral dorsal roots are cut in experimental animals or interrupted by diseases of the dorsal roots such as **tabes dorsalis** in humans, all reflex contractions of the bladder are abolished. The bladder becomes distended, thin-walled, and hypotonic, but there are some contractions because of the intrinsic response of the smooth muscle to stretch.

Effects of Denervation

When the afferent and efferent nerves are both destroyed, as they may be by tumors of the cauda equina or filum terminale, the bladder is flaccid and distended for a while. Gradually, however, the muscle of the "decentralized bladder" becomes active, with many contraction waves that expel dribbles of urine out of the urethra. The bladder becomes shrunken and the bladder wall hypertrophied. The reason for the difference between the small, hypertrophic bladder seen in this condition and the distended, hypotonic bladder seen when only the afferent nerves are interrupted is not known. The hyperactive state in the former condition suggests the development of denervation hypersensitization even though the neurons interrupted are preganglionic rather than postganglionic.

Effects of Spinal Cord Transection

During spinal shock, the bladder is flaccid and un-responsive. It becomes overfilled, and urine dribbles through the sphincters (**overflow incontinence**). After spinal shock has passed, the voiding reflex returns, although there is, of course, no voluntary control and no inhibition or facilitation from higher centers when the spinal cord is transected. Some paraplegic patients train themselves to initiate voiding by pinching or stroking their thighs, provoking a mild mass reflex (see Chapter 12). In some instances, the voiding reflex becomes hyperactive. Bladder capacity is reduced, and the wall becomes hypertrophied. This type of bladder is sometimes called the **spastic neurogenic bladder.** The reflex hyperactivity is made worse by, and may possibly be caused by, infection in the bladder wall.

Regulation of Extracellular Fluid Composition & Volume

INTRODUCTION

This chapter is a review of the major homeostatic mechanisms that operate, primarily through the kidneys and the lungs, to maintain the **tonicity,** the **volume,** and the **specific ionic composition,** particularly the **H⁺** concentration, of the ECF. The interstitial portion of this fluid is the fluid environment of the cells, and life depends upon the constancy of this "internal sea" (see Chapter 1).

DEFENSE OF TONICITY

The defense of the tonicity of the ECF is primarily the function of the vasopressin-secreting and thirst mechanisms. The total body osmolality is directly proportionate to the total body sodium plus the total body potassium divided by the total body water, so that changes in the osmolality of the body fluids occur when there is a disproportion between the amount of these electrolytes and the amount of water ingested or lost from the body (see Chapter 1). When the effective osmotic pressure of the plasma rises, vasopressin secretion is increased and the thirst mechanism is stimulated. Water is retained in the body, diluting the hypertonic plasma, and water intake is increased (Fig 39–1). Conversely, when the plasma becomes hypotonic, vasopressin secretion is decreased and "solute-free water" (water in excess of solute) is excreted. In this way, the tonicity of the body fluids is maintained within a narrow normal range. In health, plasma osmolality ranges from 280 to 295 mosm/L, with vasopressin secretion maximally inhibited at 285 mosm/L and stimulated at higher values (see Fig 14–13). The details of the way the regulatory mechanisms operate and the disorders that result when their function is disrupted are considered in Chapters 14 and 38.

DEFENSE OF VOLUME

The volume of the ECF is determined primarily by the total amount of osmotically active solute in the ECF. The composition of the ECF is discussed in Chapter 1. Since Na⁺ and Cl⁻ are by far the most abundant osmotically active solutes in ECF, and since changes in Cl⁻ are to a great extent secondary to changes in Na⁺, the amount of Na⁺ in the ECF is the most important determinant of ECF volume, and the mechanisms that control Na⁺ balance are the major mechanisms defending ECF volume. There is, however, a volume control of water excretion as well; a rise in ECF volume inhibits vasopressin secretion, and a decline in ECF volume produces an increase in the secretion of this hormone. Volume stimuli override the osmotic regulation of vasopressin secretion. Angiotensin II stimulates aldosterone and vasopressin secretion. It also causes thirst and constricts blood vessels, which help to maintain blood pressure. Thus, angiotensin II plays a key role in the body's response to hypovolemia (Fig 39–2). In addition, expansion of the ECF volume increases the secretion of ANP by the heart, and this causes natriuresis and diuresis (see Chapter 24).

In disease states, loss of water from the body (**dehydration**) causes a moderate decrease in ECF volume, because water is lost from both the intracellular and extracellular fluid compartments; but loss of Na⁺ in the stools (diarrhea), urine (severe acidosis, adrenal insufficiency), or sweat (heat prostration) decreases ECF volume markedly and eventually leads to shock. The immediate compensations in shock operate principally to maintain intravascular volume (see Chapter 33), but they also affect Na⁺ balance. In adrenal insufficiency, the decline in ECF volume is due not only to loss of Na⁺ in the urine but also to its movement into cells (see Chapter 20).

Because of the key position of Na⁺ in volume homeostasis, it is not surprising that more than one mechanism has evolved to control the excretion of this ion. The filtration and reabsorption of Na⁺ in the kidneys and the effects of these processes on Na⁺ excretion are discussed in Chapter 38. When ECF volume is decreased, blood pressure falls. Glomerular capillary pressure declines, and the GFR therefore falls, reducing the amount of Na⁺ filtered. Tubular reabsorption of Na⁺ is increased, in part because the secretion of aldosterone is increased. Aldosterone secretion is controlled in part by a feedback system in which the change that initiates increased secretion is

Figure 39–1. Mechanisms for defending ECF tonicity. The dashed arrow indicates inhibition. (Courtesy of J Fitzsimmons.)

a decline in mean intravascular pressure (see Chapters 20 and 24). Other changes in Na$^+$ excretion occur too rapidly to be due solely to changes in aldosterone secretion. For example, rising from the supine to the standing position increases aldosterone secretion. However, Na$^+$ excretion is decreased within a few minutes, and this rapid change in Na$^+$ excretion occurs in adrenalectomized subjects. It is probably due

Figure 39–2. Defense of ECF volume by angiotensin II. ACE, angiotensin-converting enzyme. (Modified from Ramsay DJ, Ganong WF: CNS regulation of salt and water intake. In: *Neuroendocrinology.* Krieger DT, Hughes JC: (editors). Sinauer Associates, 1980.)

to hemodynamic changes and possibly to decreased ANP secretion.

DEFENSE OF SPECIFIC IONIC COMPOSITION

Special regulatory mechanisms maintain the levels of certain specific ions in the ECF as well as the levels of glucose and other nonionized substances important in metabolism (see Chapters 17 and 19). The feedback of Ca^{2+} on the parathyroids and the calcitonin-secreting cells to adjust their secretion maintains the ionized calcium level of the ECF (see Chapter 21). Magnesium concentration is subject to close regulation, but the mechanisms controlling Mg$^+$ metabolism are incompletely understood.

The mechanisms controlling Na$^+$ and K$^+$ content are part of those determining the volume and tonicity of ECF and are discussed above. The levels of these ions are also dependent upon the H$^+$ concentration, and pH is one of the major factors affecting the anion composition of ECF.

DEFENSE OF H$^+$ CONCENTRATION

The mystique that has grown up around the subject of acid-base balance makes it necessary to point out that the core of the problem is not "buffer base" or "fixed cation" or the like but simply the maintenance of the H$^+$ concentration of the ECF. The mechanisms regulating the composition of the ECF are particularly important as far as this specific ion is concerned, because the machinery of the cells is very sensitive to changes in H$^+$ concentration. Intracellular H$^+$ concentration, which can be measured by using microelectrodes, pH-sensitive fluorescent dyes, and phosphorus magnetic resonance, is different from extracellular pH and appears to regulate a variety of intracellular processes. However, it is sensitive to changes in ECF H$^+$ concentration.

The pH notation is a useful means of expressing H$^+$ concentrations in the body, because the H$^+$ concentrations happen to be low relative to those of other cations. Thus, the normal Na$^+$ concentration of arterial plasma that has been equilibrated with red blood cells is about 140 meq/L, whereas the H$^+$ concentration is 0.00004 meq/L (Table 39–1). The pH, the negative logarithm of 0.00004, is therefore 7.4. Of course, a decrease in pH of 1 unit, eg, from 7.0 to 6.0, represents a 10-fold increase in H$^+$ concentration. It is also important to remember that the pH of blood is the pH of **true plasma**—plasma that has been in equilibrium with red cells—because the red cells contain hemoglobin, which is quantitatively one of the most important blood buffers (see Chapter 35).

Table 39–1. H$^+$ concentration and pH of body fluids.

		H$^+$ Concentration		
		meq/L	mol/L	pH
Gastric HCl		150	0.15	0.8
Maximal urine acidity		0.03	3×10^{-5}	4.5
Plasma	Extreme acidosis	0.0001	1×10^{-7}	7.0
	Normal	0.00004	4×10^{-8}	7.4
	Extreme alkalosis	0.00002	2×10^{-8}	7.7
Pancreatic juice		0.00001	1×10^{-8}	8.0

H$^+$ Balance

The pH of the arterial plasma is normally 7.40 and that of venous plasma slightly lower. Technically, **acidosis** is present whenever the arterial pH is below 7.40, and **alkalosis** is present whenever it is above 7.40, although variations of up to 0.05 pH unit occur without untoward effects. The H$^+$ concentrations in the ECF that are compatible with life cover an approximately 5-fold range, from 0.00002 meq/L (pH 7.70) to 0.0001 meq/L (pH 7.00).

Amino acids are utilized in the liver for gluconeogenesis, leaving as products NH$_4^+$ and HCO$_3^-$ from their amino and carboxyl groups (Fig 39–3). The NH$_4^+$ is incorporated into urea and the protons that are formed are buffered intracellularly by HCO$_3^-$, so little NH$_4^+$ and HCO$_3^-$ escape into the circulation. However, metabolism of sulfur-containing amino acids produces H$_2$SO$_4$, and metabolism of phosphorylated amino acids such as phosphoserine produces H$_3$PO$_4$. These strong acids enter the circulation and present a major H$^+$ load to the buffers in the ECF. The

H$^+$ load from amino acid metabolism is normally about 50 meq/d. The CO$_2$ formed by metabolism in the tissues is in large part hydrated to H$_2$CO$_3$ (see Chapter 35), and the total H$^+$ load from this source is over 12,500 meq/d. However, most of the CO$_2$ is excreted in the lungs, and only small quantities of the H$^+$ remain to be excreted by the kidneys. Common sources of extra acid loads are strenuous exercise (lactic acid), diabetic ketosis (acetoacetic acid and β-hydroxybutyric acid), and ingestion of acidifying salts such as NH$_4$Cl and CaCl$_2$, which in effect add HCl to the body. Failure of diseased kidneys to excrete normal amounts of acid is also a cause of acidosis. Fruits are the main dietary source of alkali. They contain Na$^+$ and K$^+$ salts of weak organic acids, and the anions of these salts are metabolized to CO$_2$, leaving NaHCO$_3$ and KHCO$_3$ in the body. NaHCO$_3$ and other alkalinizing salts are sometimes ingested in large amounts, but a more common cause of alkalosis is loss of acid from the body as a result of vomiting of gastric juice rich in HCl. This is, of course, equivalent to adding alkali to the body.

Buffering

Buffering and the buffer systems in the body are discussed in Chapters 1 and 35. The principal buffers are hemoglobin (Hb), protein (Prot), phosphate (H$_2$PO$_4^-$), and H$_2$CO$_3$:

$$HHb \rightleftarrows H^+ + Hb^-$$
$$HProt \rightleftarrows H^+ + Prot^-$$
$$H_2PO_4^- \rightleftarrows H^+ + HPO_4^{2-}$$
$$H_2CO_3 \rightleftarrows H^+ + HCO_3^-$$

The position of H$_2$CO$_3$ is unique, because it is converted to H$_2$O and CO$_2$ and the CO$_2$ is then excreted in the lungs. As noted in Chapter 35, the clinically relevant form of the Henderson-Hasselbalch equation for the equilibrium in this system is

$$pH = 6.10 + \log \frac{[HCO_3^-]}{0.0301\ P_{CO_2}}$$

Respiratory Acidosis & Alkalosis

It is apparent from the Henderson-Hasselbalch equation for the bicarbonate system that the primary changes in arterial P_{CO_2} which occur with disorders of respiration change the [HCO$_3^-$]/[CO$_2$] ratio and, therefore, the pH. A rise in arterial P_{CO_2} due to decreased ventilation causes **respiratory acidosis.** The CO$_2$ that is retained is in equilibrium with H$_2$CO$_3$, which in turn is in equilibrium with HCO$_3^-$, so that the plasma HCO$_3^-$ rises and a new equilibrium is reached at a lower pH. This can be indicated graphically on a plot of plasma HCO$_3^-$ concentration versus pH (Fig 39–4). Conversely, a decline in P_{CO_2} causes **respiratory alkalosis.**

Figure 39–3. Role of the liver and kidneys in the handling of metabolically produced acid loads. Sites where regulation occurs are indicated by asterisks. (Modified and reproduced, with permission, from Knepper MA et al: Ammonium, urea, and systemic pH regulation. *Am J Physiol* 1987;**235:** F199.)

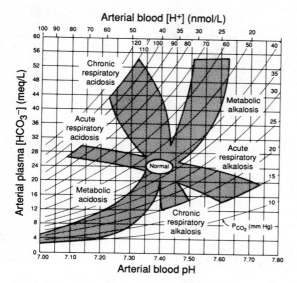

Arterial blood [H+] (nmol/L)

Arterial plasma [HCO$_3^-$] (meq/L)

P_{CO_2} (mm Hg)

Arterial blood pH

Figure 39–4. Acid-base nomogram showing changes in the P_{CO_2} (curved lines) plasma HCO$_3^-$, and pH of arterial blood in respiratory and metabolic acidosis. Note the shifts in HCO$_3^-$ and pH as acute respiratory acidosis and alkalosis are compensated, producing their chronic counterparts. (Reproduced, with permission, from Cogan MG, Rector FC Jr: Acid-base disorders. Pages 457–517. In: *The Kidney,* 4th ed. Brenner BM, Rector FC Jr [editors]. Saunders, 1991.)

The initial changes shown in Fig 39–4 are those that occur independently of any compensatory mechanism; ie, they are those of **uncompensated** respiratory acidosis or alkalosis. In either situation, changes are produced in the kidneys, which then tend to **compensate** for the acidosis or alkalosis, adjusting the pH toward normal.

Renal Compensation

HCO$_3^-$ reabsorption in the renal tubules depends not only on the filtered load of HCO$_3^-$, which is the product of the GFR and the plasma HCO$_3^-$ level, but also on the rate of H$^+$ secretion by the renal tubular cells, since HCO$_3^-$ is reabsorbed by exchange for H$^+$. The rate of H$^+$ secretion—and hence the rate of HCO$_3^-$ reabsorption—is proportionate to the arterial P_{CO_2} probably because the more CO$_2$ that is available to form H$_2$CO$_3$ in the cells, the more H$^+$ can be secreted (see Chapter 38). Furthermore, when the P_{CO_2} is high, the interior of most cells becomes more acidic (see Chapter 35). In respiratory acidosis, renal tubular H$^+$ secretion is therefore increased, removing H$^+$ from the body; and even though the plasma HCO$_3^-$ is elevated, HCO$_3^-$ reabsorption is increased, further raising the plasma HCO$_3^-$. This renal compensation for respiratory acidosis is shown graphically in the shift from acute to chronic respiratory acidosis in Fig 39–4. Cl$^-$ excretion is increased, and plasma Cl$^-$

falls as plasma HCO$_3^-$ is increased. Conversely, in respiratory alkalosis, the low P_{CO_2} hinders renal H$^+$ secretion, HCO$_3^-$ reabsorption is depressed, and HCO$_3^-$ is excreted, further reducing the already low plasma HCO$_3^-$ and lowering the pH toward normal (Fig 39–4).

Metabolic Acidosis

When acids stronger than HHb and the other buffer acids are added to blood, **metabolic acidosis** is produced; and when the free H$^+$ level falls as a result of addition of alkali or removal of acid, **metabolic alkalosis** results. If, for example, H$_2$SO$_4$ is added, the H$^+$ is buffered and the Hb$^-$, Prot$^-$, and HCO$_3^-$ levels in plasma drop. The H$_2$CO$_3$ formed is converted to H$_2$O and CO$_2$, and the CO$_2$ is rapidly excreted via the lungs. The importance of this fact is demonstrated in Fig 39–5. If enough acid were added to halve the HCO$_3^-$ level in plasma and CO$_2$ were not formed and excreted, the pH would fall to approximately 6.0 and death would result. If the H$_2$CO$_3$ level were so regulated that it remained constant, the pH would fall only to about 7.1. This is the situation in **uncompensated** metabolic acidosis (Fig 39–6). Actually, the rise in plasma H$^+$ stimulates respiration, so that the P_{CO_2}, instead of rising or remaining constant, is reduced. This **respiratory compensation** raises the pH even further. The **renal** compensatory mechanisms then bring about the excretion of the extra H$^+$ and return the buffer systems to normal.

Renal Compensation

The anions that replace HCO$_3^-$ in the plasma in

meq/L [HCO$_3^-$]

meq/L [H$_2$CO$_3$]

$\dfrac{[HCO_3^-]}{[H_2CO_3]}$ ratio	20	0.9	10	16
pH	7.4	6.0	7.1	7.3

Figure 39–5. Buffering by the H$_2$CO$_3$-HCO$_3^-$ system in blood. The bars are drawn as if buffering occurred in separate steps in order to show the effect of the initial reaction, the reduction of H$_2$CO$_3$ to its previous value, and its further reduction by the increase in ventilation. In this case, [H$_2$CO$_3$] is actually the concentration of dissolved CO$_2$, so that the meq/L values for it are arbitrary.

metabolic acidosis are filtered, each with a cation (principally Na$^+$), thus maintaining electrical neutrality. The renal tubular cells secrete H$^+$ into the tubular fluid in exchange for Na$^+$; and for each H$^+$ secreted, one Na$^+$ and one HCO$_3^-$ are added to the blood (see Chapter 38). The limiting urinary pH of 4.5 would be reached rapidly and the total amount of H$^+$ secreted would be small if there were no buffers in the urine that "tied up" H$^+$. However, secreted H$^+$ reacts with HCO$_3^-$ to form CO$_2$ and H$_2$O (bicarbonate reabsorption); with HPO$_4^{2-}$ to form H$_2$PO$_4^-$; and with NH$_3$ to form NH$_4^+$. In this way large amounts of H$^+$ can be secreted, permitting correspondingly large amounts of HCO$_3^-$ to be returned to (in the case of bicarbonate reabsorption) or added to the depleted body stores and large numbers of the cations to be reabsorbed. It is only when the acid load is very large that cations are lost with the anions, producing diuresis and depletion of body cation stores. In chronic acidosis, glutamine synthesis in the liver is increased, using some of the NH$_4^+$ that usually is converted to urea (Fig 39–3), and the glutamine provides the kidneys with an additional source of NH$_4^+$ (see Chapter 38). NH$_3$ secretion increases over a period of days (adaptation of NH$_3$ secretion; see Chapter 38), further improving the renal compensation for acidosis. In addition, the metabolism of glutamine in the kidneys produces α-ketoglutarate, and this in turn is decarboxylated, producing HCO$_3^-$, which enters the bloodstream and helps buffer the acid load (Fig 39–3).

The overall reaction in blood when a strong acid such as H$_2$SO$_4$ is added is

$$2NaHCO_3 + H_2SO_4 \rightarrow Na_2SO_4 + 2H_2CO_3$$

For each mole of H$^+$ added, 1 mol of NaHCO$_3$ is lost. The kidney in effect reverses the reaction:

$$Na_2SO_4 + 2H_2CO_3 \rightarrow 2NaHCO_3 + 2H^+ + SO_4^{2-}$$

and the H$^+$ and SO$_4^{2-}$ are excreted. Of course, H$_2$SO$_4$ is not excreted as such, the H$^+$ appearing in the urine as titratable acidity and NH$_4^+$.

In metabolic acidosis, the respiratory compensation tends to inhibit the renal response in the sense that the induced drop in P$_{CO_2}$ hinders acid secretion, but it also decreases the filtered load of HCO$_3^-$ and so its net inhibitory effect is not great.

Metabolic Alkalosis

In metabolic alkalosis, the plasma HCO$_3^-$ level and pH rise (Fig 39–6). The respiratory compensation is a decrease in ventilation produced by the decline in H$^+$ concentration, and this elevates the P$_{CO_2}$. This brings the pH back toward normal while elevating the plasma HCO$_3^-$ level still further. The magnitude of this compensation is limited by the carotid and aortic chemoreceptor mechanisms, which drive the respira-

Figure 39–6. Changes in true plasma pH, HCO$_3^-$, and P$_{CO_2}$ in metabolic acidosis and alkalosis. This is called a Davenport diagram and is based on: Davenport HW: *The ABC of Acid-Base Chemistry*, 6th ed. Univ of Chicago Press, 1974.

tory center if there is any appreciable fall in the arterial P$_{O_2}$. In metabolic alkalosis, more renal H$^+$ secretion is expended in reabsorbing the increased filtered load of HCO$_3^-$; and if the HCO$_3^-$ level in plasma exceeds 26–28 meq/L, HCO$_3^-$ appears in the urine. The rise in P$_{CO_2}$ inhibits the renal compensation by facilitating acid secretion, but its effect is relatively slight.

Clinical Evaluation of Acid-Base Status

In evaluating disturbances of acid-base balance, it is important to know the pH and HCO$_3^-$ content of arterial plasma. Reliable pH determinations can be made with a pH meter and a glass pH electrode. The HCO$_3^-$ content of plasma cannot be measured directly, but the P$_{CO_2}$ can be measured with a CO$_2$ electrode and the HCO$_3^-$ concentration calculated, as noted above. The P$_{CO_2}$ is 7–8 mm Hg higher and the pH 0.03–0.04 unit lower in venous than arterial plasma because venous blood contains the CO$_2$ being carried from the tissues to the lungs. Therefore, the calculated HCO$_3^-$ concentration is about 2 mmol/L higher. However, if this is kept in mind, free-flowing venous blood can be substituted for arterial blood in most clinical situations.

A measurement that is of some value in the differential diagnosis of metabolic acidosis is the **anion gap**. This gap, which is something of a misnomer, refers to the difference between the concentration of cations other than Na$^+$ and the concentration of anions other than Cl$^-$ and HCO$_3^-$ in the plasma. It consists for the most part of proteins in the anionic form, HPO$_4^{2-}$, SO$_4^{2-}$, and organic acids, and a normal value is about 12 meq/L. It is increased when the plasma concentration of K^+, Ca^{2+}, or Mg^{2+} is decreased; when the concentration of or the charge on plasma proteins is increased; or when organic anions such as lactate or foreign anions accumulate in blood. It is decreased when cations are increased or when plasma albumin is decreased. The anion gap is increased in metabolic acidosis due to ketoacidosis, lactic acidosis, and other forms of acidosis in which organic anions are increased. It is not increased in hyperchloremic acidosis due to ingestion of NH_4Cl or carbonic anhydrase inhibitors.

The Siggaard-Andersen Curve Nomogram

Use of the Siggaard-Andersen curve nomogram (Fig 39–7) to plot the acid-base characteristics of arterial blood is helpful in clinical situations. This nomogram has P_{CO_2} plotted on a log scale on the vertical axis and pH on the horizontal axis. Thus, any point to the left of a vertical line through pH 7.40 indicates

acidosis, and any point to the right indicates alkalosis. The position of the point above or below the horizontal line through a P_{CO_2} of 40 mm Hg defines the effective degree of hypoventilation or hyperventilation.

If a solution containing $NaHCO_3$ and no buffers were equilibrated with gas mixtures containing various amounts of CO_2, the pH and P_{CO_2} values at equilibrium would fall along the dashed line on the left in Fig 39–7 or a line parallel to it. If buffers were present, the slope of the line would be greater; and the greater the buffering capacity of the solution, the steeper the line. For normal blood containing 15 g of hemoglobin per deciliter, the **CO_2 titration line** passes through the 15 g/dL mark on the hemoglobin scale (on the underside of the upper curved scale) and the point where the P_{CO_2} = 40 mm Hg and pH = 7.40 lines intersect, as shown in Fig 39–7. When the hemoglobin content of the blood is low, there is signifi-

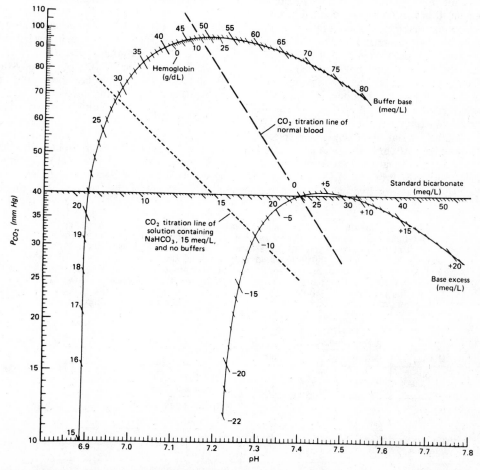

Figure 39–7. Siggaard-Andersen curve nomogram. (Courtesy of O Siggaard-Andersen and Radiometer, Copenhagen, Denmark.)

cant loss of buffering capacity, and the slope of the CO_2 titration line diminishes. However, blood of course contains buffers in addition to hemoglobin, so that even the line drawn from the zero point on the hemoglobin scale through the normal P_{CO_2}-pH intercept is steeper than the curve for a solution containing no buffers.

For clinical use, arterial blood or arterialized capillary blood is drawn anaerobically and its pH measured. The pHs of the same blood after equilibration with each of 2 gas mixtures containing different known amounts of CO_2 are also determined. The pH values at the known P_{CO_2} levels are plotted and connected to provide the CO_2 titration line for the blood sample. The pH of the blood sample before equilibration is plotted on this line, and the P_{CO_2} of the sample is read off the vertical scale. The **standard bicarbonate** content of the sample is indicated by the point at which the CO_2 titration line intersects the bicarbonate scale on the $P_{CO_2} = 40$ mm Hg line. The standard bicarbonate is not the actual bicarbonate concentration of the sample but, rather, what the bicarbonate concentration would be after elimination of any respiratory component. It is a measure of the alkali reserve of the blood, except that it is measured by determining the pH rather than the total CO_2 content of the sample after equilibration. Like the alkali reserve, it is an index of the degree of metabolic acidosis or alkalosis present.

Additional graduations on the upper curved scale of the nomogram (Fig 39–7) are provided for measuring **buffer base** content; the point where the CO_2 calibration line of the arterial blood sample intersects this scale shows the meq/L of buffer base in the sample. The buffer base is equal to the total number of buffer anions (principally Prot⁻, HCO_3^-, and Hb⁻; see Chapter 35) that can accept hydrogen ions in the

blood. The normal value in an individual with 15 g of hemoglobin per deciliter of blood is 48 meq/L.

The point at which the CO_2 calibration line intersects the lower curved scale on the nomogram indicates the **base excess**. This value, which is positive in alkalosis and negative in acidosis, is the amount of acid or base that would restore 1 L of blood to normal acid-base composition at a P_{CO_2} of 40 mm Hg. It should be noted that a base deficiency cannot be completely corrected simply by calculating the difference between the normal standard bicarbonate (24 meq/L) and the actual standard bicarbonate and administering this amount of $NaHCO_3$ per liter of blood; some of the added HCO_3^- is converted to CO_2 and H_2O, and the CO_2 is lost in the lungs. The actual amount that must be added is roughly 1.2 times the standard bicarbonate deficit, but the lower curved scale on the nomogram, which has been developed empirically by analyzing many blood samples, is more accurate.

Some examples of clinical acid-base disturbances are shown in Table 39–2.

In treating acid-base disturbances, one must, of course, consider not only the blood but also all the body fluid compartments. The other fluid compartments have markedly different concentrations of buffers. It has been determined empirically that administration of an amount of acid (in alkalosis) or base (in acidosis) equal to 50% of the body weight in kilograms times the blood base excess per liter will correct the acid-base disturbance in the whole body. At least when the abnormality is severe, however, it is unwise to attempt such a large correction in a single step; instead, about half the indicated amount should be given and the arterial blood acid-base values determined again. The amount required for final correction can then be calculated and administered. It is also worth noting that, at least in lactic acidosis,

Table 39–2. Plasma pH, HCO_3^-, and P_{CO_2} values in various typical disturbances of acid-base balance. In the diabetic acidosis and prolonged vomiting examples, respiratory compensation for primary metabolic acidosis and alkalosis has occurred, and the P_{CO_2} has shifted from 40 mm Hg. In the emphysema and high-altitude examples, renal compensation for primary respiratory acidosis and alkalosis has occurred and has made the deviations from normal of the plasma HCO_3^- larger than they would otherwise be. Data from various authors.

| Condition | | Arterial Plasma | | Cause |
	pH	HCO_3^- (meq/L)	P_{CO_2} (mm Hg)	
NORMAL	7.40	24.1	40	
Metabolic acidosis	7.28	18.1	40	NH₄Cl ingestion
	6.96	5.0	23	Diabetic acidosis
Metabolic alkalosis	7.50	30.1	40	$NaHCO_3$ ingestion
	7.56	49.8	58	Prolonged vomiting
Respiratory acidosis	7.34	25.0	48	Breathing 7% CO_2
	7.34	33.5	64	Emphysema
Respiratory alkalosis	7.53	22.0	27	Voluntary hyperventilation
	7.48	18.7	26	Three-week residence at 4000 m altitude

$NaHCO_3$ decreases cardiac output and lowers blood pressure, so it should be used with caution.

Relation Between Potassium Metabolism & Acid-Base Balance

The K^+ and H^+ concentrations of the ECF parallel each other, in part because of the effects of K^+ on renal H^+ secretion (see Chapter 38). K^+ depletion apparently produces an intracellular acidosis, promoting H^+ secretion into the urine. H^+ is thus removed from the body and HCO_3^- reabsorption increased, producing an extracellular alkalosis. Conversely, K^+ excess increases K^+ secretion by the renal tubular cells. Since both H^+ and K^+ are secreted in exchange for Na^+ and compete for the available Na^+ in the tubular fluid, H^+ secretion is inhibited, promoting extracellular acidosis.

REFERENCES

Section VIII: Formation & Excretion of Urine

Aronson PS: Mechanisms of active H^+ secretion in the proximal tubule. *Am J Physiol* 1983;**245**:F647.

Badr KF, Ichikawa T: Prerenal failure: A deleterious shift from renal compensation to decompensation. *N Engl J Med* 1988;**319**:623.

Bayliss C, Blantz RC: Glomerular hemodynamics. *News Physiol Sci* 1986;**1**:86.

Boyle J III, Robertson G: Acid-base balance: An educational computer game. *Bioscience* 1987;**37**:511.

Brenner BM, Rector FC Jr (editors): *The Kidney,* 4th ed. 2 vols. Saunders, 1991.

Cogan MG: *Fluid & Electrolytes: Physiology & Pathophysiology.* Appleton & Lange, 1991.

Davenport HW: *The ABC of Acid-Base Chemistry,* 6th ed. Univ of Chicago Press, 1974.

DiBona GF: Neural control of renal function: Cardiovascular implications. *Hypertension* 1989;**13**:539.

Dunn MJ (editor): *Renal Endocrinology.* Williams & Wilkins, 1983.

Forte JG, Reenstra WW: Plasma membrane proton pumps in animal cells. *Bioscience* 1985;**35**:38.

Frömter E: Viewing the kidney through microelectrodes. *Am J Physiol* 1984;**247**:F695.

Gebow PA: Disorders associated with an altered anion gap. *Kidney Int* 1985;**27**:472.

Handler JS: Antidiuretic hormone moves membranes. *Am J Physiol* 1988;**255**:F375.

Hayslett JP: Functional adaptation to reduction in renal mass. *Physiol Rev* 1979;**59**:137.

Hutch JA: *Anatomy and Physiology of the Bladder, Trigone and Urethra.* Appleton-Century-Crofts, 1972.

Jamison RL, Kriz W: *Urinary Concentrating Mechanism: Structure and Function.* Oxford Univ Press, 1982.

Knepper MA, Packer R, Good DW: Ammonium transport in the kidney. *Physiol Rev* 1989;**69**:179.

Loeb JN: The hyperosmolar state. *N Engl J Med* 1974;**290**:1184.

Madsen KM, Tisher CC: Structural-functional relationships along the distal nephron. *Am J Physiol* 1986;**250**:F1.

Mandel LJ: Metabolic substrates, cellular energy production, and the regulation of proximal tubular transport. *Annu Rev Physiol* 1985;**47**:85.

Mené P, Simonson MS, Dunn MJ: Physiology of the mesangial cell. *Physiol Rev* 1989;**19**:1347.

Morel F: Sites of hormone action in the mammalian nephron. *Am J Physiol* 1981;**240**:F159.

Orci L et al: Membrane ultrastructure in urinary tubules. *Int Rev Cytol* 1981;**73**:183.

Rector FC Jr: Sodium bicarbonate and chloride absorption by the proximal tubule. *Am J Physiol* 1983;**244**:F461.

Roos A, Boron WF: Intracellular pH. *Physiol Rev* 1981;**61**:296.

Schafer JA: Mechanisms coupling the absorption of solute and water in proximal nephron. *Kidney Int* 1984;**25**:708.

Schmidt-Nielson K: Countercurrent systems in animals. *Sci Am* (May) 1981;**244**:118.

Schrier RW: Pathogenesis of sodium and water retention in high-output and low-output cardiac failure, nephrotic syndrome, cirrhosis, and pregnancy. *N Engl J Med* 1988;**319**:1065.

Schuster VC, Stokes JB: Chloride transport by the cortical and outer medullary collecting duct. *Am J Physiol* 1987;**253**:F203.

Seldin DW, Giebisch G (editors): *The Kidney: Physiology and Pathophysiology,* 2nd ed. 3 vols. Raven, 1991.

Seldin DW, Giebisch G (editors): *The Regulation of Acid-Base Balance.* Raven, 1989.

Seldin DW, Giebisch G (editors): *The Regulation of Sodium and Chloride Balance.* Raven, 1989.

Stanton B, Giebisch G: Mechanism of urinary potassium excretion. *Miner Electrolyte Metab* 1981;**5**:100.

Stella A, Zanchetti A: Functional role of renal afferents. *Physiol Rev* 1991;**71**:659.

Valtin H: *Renal Function,* 2nd ed. Little, Brown, 1983.

Verkman A: Water channels in cell membranes. *Annu Rev Physiol* 1992;**54**:97.

Wright FS: Flow-dependent transport processes: Filtration, absorption, secretion. *Am J Physiol* 1982;**243**:F1.

Wright FS: Intrarenal regulation of glomerular filtration rate. *J Hypertension* 1984;**2(Suppl 1)**:105.

Appendix

GENERAL REFERENCES

Many large, comprehensive textbooks of physiology are available. The following are among the best of the recently published or revised volumes:

Berne RM, Levy MN (editors): *Physiology,* 2nd ed. Mosby, 1988.
Guyton AC: *Textbook of Medical Physiology,* 8th ed. Saunders, 1991.
Johnson LR (editor): *Essential Medical Physiology.* Raven, 1992.
Patton HD et al (editors): *Textbook of Physiology,* 21st ed. 2 vols. Saunders, 1989.
Schmidt RF, Thews G (editors): *Human Physiology,* 2nd ed. Springer-Verlag, 1989.
West JB (editor): *Best and Taylor's Physiological Basis of Medical Practice,* 12th ed. Williams & Wilkins, 1990.

Textbooks of pathophysiology include the following:

Frohlich ED (editor): *Pathophysiology,* 3rd ed. Lippincott, 1984.
Sodeman WA Jr, Sodeman WA: *Pathologic Physiology,* 6th ed. Saunders, 1979.

A standard anatomy reference is:

Williams PL et al (editors): *Gray's Anatomy,* 37th ed. Churchill Livingstone, 1989.

Excellent summaries of current research on selected aspects of physiology can be found in the *FASEB Journal* and in the *Annals of the New York Academy of Sciences.* Summary articles and various types of valuable reviews appear in the *New England Journal of Medicine.* These include articles that review current topics in physiology and biochemistry with the aim of providing up-to-date information for practicing physicians. The most pertinent serial review publications are *Physiological Reviews, Pharmacological Reviews, Recent Progress in Hormone Research* (the proceedings of each annual Laurentian Hormone Conference), *Annual Review of Physiology,* and other volumes of the Annual Reviews series. The *Handbook of Physiology,* which is now being published by Oxford University Press, New York, has separate volumes that cover all aspects of physiology.

Recently revised volumes cover the nervous system, the cardiovascular system, respiration, renal physiology, and the gastrointestinal system. Volumes on endocrinology, adaptation to the environment, and adipose tissue are also available. The articles in the handbook are valuable but extremely detailed reviews.

Powerful new techniques of imaging have added to the advance of physiology as well as clinical medicine. These include **magnetic resonance imaging (MRI),** a method that permits not only the production of pictures of the interior of the body but also noninvasive analysis of metabolic and other phenomena in living animals and humans. Various forms of **emission tomography** are also very valuable for localizing compounds such as dopamine and various sugars in living individuals. References that describe these important techniques include the following:

Haaga JR, Alfidi RJ (editors): *Computed Tomography of the Whole Body,* 2nd ed. Mosby, 1988.
Kean D, Smith M: *Magnetic Resonance Imaging.* Williams & Wilkins, 1986.
Moonen CTW et al: Functional magnetic resonance imaging in medicine and physiology. *Science* 1990;**250**:53.

NORMAL VALUES & THE STATISTICAL EVALUATION OF DATA

The approximate ranges of values in normal humans for some commonly measured plasma constituents are summarized in the table on the inside back cover. A worldwide attempt has been under way to convert to a single standard nomenclature by using SI (Système International) units. The system is based on the 7 dimensionally independent physical quantities summarized in Table 1. Units derived from the basic units are summarized in Table 2, and the prefixes used to refer to decimal fractions and multiples of these and other units are listed in Table 3. There are a number of complexities involved in the use of these units—for example, the problem of expressing enzyme units—and they are only slowly making their way into the medical literature. In this book, the values in the text are in traditional units, but they are followed in key instances by values in SI units. In addition, values in SI units are listed beside values in

Table 1. Basic units.

Quantity	Name	Symbol
Length	meter	m
Mass	kilogram	kg
Time	second	s
Electric current	ampere	A
Thermodynamic temperature	kelvin	K
Luminous intensity	candela	cd
Amount of substance	mole	mol

Table 3. Standard prefixes.

Prefix[1]	Abbreviation	Magnitude
exa-	E	10^{18}
peta-	P	10^{15}
tera-	T	10^{12}
giga-	G	10^9
mega-	M	10^6
kilo-	k	10^3
hecto-	h	10^2
deca-	da	10^1
deci-	d	10^{-1}
centi-	c	10^{-2}
milli-	m	10^{-3}
micro-	μ	10^{-6}
nano-	n, mμ	10^{-9}
pico-	p, $\mu\mu$	10^{-12}
femto-	f	10^{-15}
atto-	a	10^{-18}

[1]These prefixes are applied to SI and other units. For example, a micrometer (μm) is 10^{-6} meter (also called a micron); a picoliter (pL) is 10^{-12} liter, and a kilogram (kg) is 10^3 grams. Also applied to seconds, units, moles, hertz, volts, farads, ohms, curies, equivalents, osmoles, etc.

more traditional units in the table on the inside back cover of this book.

The accuracy of the methods used for laboratory measurements varies. It is important in evaluating any single measurement to know the possible errors in making the measurement. For chemical determinations on body fluids, these include errors in obtaining the sample and the inherent error of the chemical method. However, the values obtained by using even the most accurate methods vary from one normal individual to the next as a result of what is usually called **biologic variation.** This variation occurs because in any system as complex as a living organism or tissue there are many variables that affect the particular measurement. Variables such as age, sex, time of day, time since last meal, etc, can be taken into account. Numerous other variables cannot, and for this reason the values obtained differ from individual to individual.

The magnitude of the normal range for any given physiologic or clinical measurement can be calculated by standard statistical techniques if the measurement has been made on a suitable sample of the normal population (preferably more than 20 individuals). It is important to know not only the average value in this sample but also the extent of the deviation of the individual values from the average.

The average (**arithmetic mean, M**) of the series of values is readily calculated:

$$M = \frac{\Sigma X}{n}$$

where

Σ = **Sum of**
X = **Individual values**
n = **Number of individual values in the series**

The average deviation is the mean of the deviations of each of the values from the arithmetic mean. From a mathematical point of view, a better measure of the deviation is the **geometric mean** of the deviations from the mean. This is called the **standard deviation of the sample (s):**

$$s = \sqrt{\frac{\Sigma(M - X)^2}{n - 1}}$$

The term n−1, rather than n, is used for complex mathematical reasons. s should be distinguished from the standard deviation of the mean of the whole population, which is designated σ. However, if the sample is truly representative, s and σ will be comparable.

The following form of the equation for s can be derived algebraically and is convenient for those computing s.

Table 2. Some derived SI units.

Quantity	Unit Name	Unit Symbol
Area	square meter	m^2
Clearance	liter/second	L/s
Concentration		
Mass	kilogram/liter	kg/L
Substance	mole/liter	mol/L
Density	kilogram/liter	kg/L
Electric potential	volt	$V = kg\ m^2/s^3\ A$
Energy	joule	$J = kg\ m^2/s^2$
Force	newton	$N = kg\ m/s^2$
Frequency	hertz	Hz = 1 cycle/s
Pressure	pascal	$Pa = kg/m\ s^2$
Temperature	degree Celsius	$°C = °K - 273.15$
Volume	cubic meter	m^3
	liter	$L = dm^3$

$$s = \sqrt{\frac{\Sigma X^2 - \frac{(\Sigma X)^2}{n}}{n - 1}}$$

Here, ΣX^2 is the sum of the squares of the individual values, and $(\Sigma X)^2$ is the square of the sum of the items, ie, the items added together and then squared.

Another commonly used index of the variation is the **standard error of the mean (SEM)**:

$$SEM = \frac{\sigma}{\sqrt{n}}$$

Strictly speaking, the SEM indicates the reliability of the sample mean as representative of the true mean of the general population from which the sample was drawn.

A **frequency distribution** curve can be constructed from the individual values in a population by plotting the frequency with which any particular value occurs in the series against the values. If the group of individuals tested was homogeneous, the frequency distribution curve is usually symmetric (Fig 1), with the highest frequency corresponding to the mean and the width of the curve varying with σ (curve of **normal distribution**). Within an ideal curve of normal distribution, the percentage of observations that fall within various ranges is shown in Table 4. The mean and s of a representative sample are approximately the mean and σ of the whole population. It is therefore possible to predict from the mean and s of the sample the probability that any particular value in the general population is normal. For example, if the difference between such a value and the mean is equal to 1.96 s, the chances are 1 out of 20 (5 out of 100) that it is normal. Conversely, of course, the chances are 19 out of 20 that it is abnormal.

Statistical analysis is also useful in evaluating the significance of the difference between 2 means. In physiologic and clinical research, measurements are often made on a group of animals or patients given a particular treatment. These measurements are compared with similar measurements made on a control group that ideally has been exposed to exactly the same conditions except that the treatment has not been given. If a particular mean value in the treated

Table 4. Percentage of values in a population which will fall within various ranges within an ideal curve of normal distribution.

Mean ± 1 σ	68.27%
Mean ± 1.96 σ	95.00%
Mean ± 2 σ	95.45%
Mean ± 3 σ	99.73%

group is different from the corresponding mean for the control group, the question arises whether the difference is due to the treatment or to chance variation. The probability that the difference represents chance variation can be estimated in many instances by using **Student's t test.** The value t is the ratio of the difference in the means of 2 series (M_a and M_b) to the uncertainty in these means. The formula used to calculate t is

$$t = \frac{M_a - M_b}{\sqrt{\dfrac{(n_a + n_b)[(n_a - 1)s_a^{\,2} + (n_b - 1)s_b^{\,2}]}{n_a n_b (n_a + n_b - 2)}}}$$

where n_a and n_b are the number of individual values in series a and b, respectively. When $n_a = n_b$, the equation for t becomes simplified to

$$t = \frac{M_a - M_b}{\sqrt{(SEM_a)^2 + (SEM_b)^2}}$$

The higher the value of t, the less the probability that the difference represents chance variation. This probability also decreases as the number of individuals (n) in each group rises, because the greater the number of measurements, the smaller the error in the measurements. A mathematical expression of the probability (P) for any value of t at different values of n can be found in tables in most texts on statistics. P is a fraction which expresses the probability that the difference between 2 means was due to chance variation. Thus, for example, if the P value is 0.10, the probability that the difference was due to chance is 10% (1 chance in 10). A P value of <0.001 means that the chances that the difference was due to random variation are less than 1 in 1000. When the P value is <0.05, most investigators call the difference "statistically significant"; ie, it is concluded that the difference is due to the operation of some factor other than chance—presumably the treatment. The use of t tests is appropriate only for comparison of 2 groups. If they are used for comparison in experiments involving more than 2 groups, a systematic error is introduced that makes the probability of deciding that there is a significant difference too high. In this situation, **analysis of variance (ANOVA)** is the appropriate statistical test. This and other techniques are discussed in statistics texts.

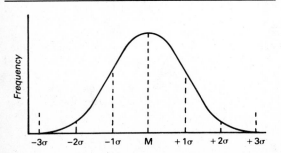

Figure 1. Curve of normal distribution (frequency distribution curve of values from a homogeneous population).

These elementary methods and many others available for statistical analysis in the research laboratory and the clinic provide a valuable objective means of evaluation. Statistical significance does not arbitrarily mean physiologic significance, and the reverse may sometimes be true; but replacement of evaluation by subjective impression with analysis by statistical methods is an important goal in the medical sciences.

REFERENCES

Appendix

Dawson-Saunders B, Trapp RG: *Basic and Clinical Biostatistics.* Appleton & Lange, 1990.

Rosner B: *Fundamentals of Biostatistics.* Duxbury, 1982.

The SI for the Health Professions: World Health Organization, 1977.

Zar JH: *Biostatistical Analysis.* Prentice-Hall, 1974.

ABBREVIATIONS & SYMBOLS COMMONLY USED IN PHYSIOLOGY

[ ]: Concentration of

Δ: Change in. (*Example:* ΔV = change in volume.) In steroid nomenclature, Δ followed by a number (eg, Δ^4-) indicates the position of a double bond

σ: Standard deviation

Ia, Ib, II, III, IV nerve fibers: Types of fibers in sensory nerves (see Chapter 2)

μ: Micro, 10^{-6}; see Table 3, above

a: Atto-, 10^{-18}; see Table 3, above

A(Å): Angstrom unit(s) (10^{-10} m, 0.1 nm); also alanine

A^-: General symbol for anion

A_1, A_2, A_{1B}, A_{2B}, B, O: Major blood groups

AASH: Adrenal androgen-stimulating hormone

A, B, and C nerve fibers: Types of fibers in peripheral nerves (see Chapter 2)

ABP: Androgen-binding protein

ACE: Angiotensin-converting enzyme

Acetyl-CoA: Acetyl-coenzyme A

ACH, Ach: Acetylcholine

ACTH: Adrenocorticotropic hormone

Acyl-CoA: General symbol for an organic compound-coenzyme A ester

ADH: Antidiuretic hormone (vasopressin)

ADP: Adenosine diphosphate

AHG: Antihemophilic globulin

Ala: Alanine

AMP: Adenosine 5'-monophosphate

ANOVA: Analysis of variance

ANP: Atrial natriuretic peptide

APUD cells: Amine precursor uptake and decarboxylation cells that secrete hormones

Arg: Arginine

Asn: Asparagine

Asp: Aspartic acid

atm: Atmosphere: 1 atm = 760 torr = mean atmospheric pressure at sea level

ATP: Adenosine triphosphate

A-V difference: Arteriovenous concentration difference of any given substance

AV node: Atrioventricular node

AVP: Arginine vasopressin

aVR, aVF, aVL: Augmented unipolar electrocardiographic leads

AV valves: Atrioventricular valves of heart

ATPD: See Table 34–1

ATPS: See Table 34–1

BDNF: Brain-derived neurotrophic factor

BER: Basic electrical rhythm

BGP: Bone Gla protein

BMR: Basal metabolic rate

BSP: Sulfobromophthalein

BTPS: See Table 34–1

BUN: Blood urea nitrogen

c: Centi-, 10^{-2}; see Table 3, above

C: Celsius; also cysteine

C followed by subscript: Clearance; eg, C_{In} clearance of inulin

C_{19} steroids: Steroids containing 19 carbon atoms

C_{21} steroids: Steroids containing 21 carbon atoms

cal: The calorie (gram calorie)

Cal: 1000 calories; kilocalorie

CAM: Cell adhesion molecule

cAMP: Cyclic adenosine 3',5'-monophosphate

CBF: Cerebral blood flow

CBG: Corticosteroid-binding globulin, transcortin

cc: Cubic centimeters

CCK, CCK-PZ: Cholecystokinin-pancreozymin

CFF: Critical fusion frequency

CFTR: Cystic fibrosis transmembrane conductance regulator

cGMP: Cyclic 3',5'-guanosine monophosphate

CGP: Chorionic growth hormone-prolactin (same as hCS)

CGRP: Calcitonin-gene-related peptide

C_{H_2O}: "Free water clearance"

Ci: Curie

CLIP: Corticotropinlike intermediate lobe polypeptide

$CMRO_2$: Cerebral metabolic rate for oxygen

CNS: Central nervous system

CoA: Coenzyme A

COHb: Carbonmonoxyhemoglobin

Compound A: 11-Dehydrocorticosterone

Compound B: Corticosterone

Compound E: Cortisone

Compound F: Cortisol

Compound S: 11-Deoxycortisol

COMT: Catechol-O-methyltransferase

cps: Cycles per second, hertz
CR: Conditioned reflex
Cr: Creatinine
CRH, CRF: Corticotropin-releasing hormone
CRO: Cathode-ray oscilloscope
Crp: Creatinine phosphate (phosphocreatine, PC)
CS: Conditioned stimulus
CSF: Cerebrospinal fluid; *also* colony-stimulating factor
CT: Computed tomography
C terminal: End of peptide or protein having a free –COOH group
CTP: Cytidine triphosphate
CV: Closing volume
CVR: Cerebral vascular resistance
cyclic AMP: Cyclic adenosine 3′,5′-monophosphate
CyS: Half-cystine
CZI: Crystalline zinc insulin
d: Day
D: Geometric isomer of L form of chemical compound
D: Aspartic acid
Da: Dalton, unit of mass equal to one-twelfth the mass of the carbon-12 atom, or about 1.66×10^{-24} g
DAG: Diacylglycerol
dB: Decibel
DBP: Vitamin D-binding protein
DDAVP: 1-Deamino-8-D-arginine vasopressin
DDD: Derivative of DDT that inhibits adrenocortical function
DEA, DHEA: Dehydroepiandrosterone
DEAS, DHEAS: Dehydroepiandrosterone sulfate
DFP: Diisopropyl fluorophosphate
DHT: Dihydrotestosterone
DIT: Diiodotyrosine
DNA: Deoxyribonucleic acid
D/N ratio: Ratio of dextrose (glucose) to nitrogen in the urine
D_2O: Deuterium oxide (heavy water)
DOCA: Desoxycorticosterone acetate
DOMA: 3,4-Dihydroxymandelic acid
Dopa: Dihydroxyphenylalanine, L-dopa
DOPAC: 3,4-Dihydroxyphenylacetic acid
DOPEG: 3,4-Dihydroxyphenylglycol
DOPET: 3,4-Dihydroxyphenylethanol
DPG, 2,3-DPG: 2,3-Diphosphoglycerate
DPL: Dipalmitoyl lecithin
DPN: Diphosphopyridine nucleotide
DPNH: Reduced diphosphopyridine nucleotide
DPPC: Dipalmitoylphosphatidylcholine
e: Base for natural logarithms = 2.7182818 . . .
E: Glutamic acid
E followed by subscript number: Esophageal electrocardiographic lead, followed by number of cm it is inserted in esophagus
E cells: Expiratory neurons
E_1: Estrone
E_2: Estradiol
EACA: Epsilon-aminocaproic acid
ECF: Extracellular fluid
eCG: Equine chorionic gonadotropin
ECG, EKG: Electrocardiogram
ECoG: Electrocorticogram
EDRF: Endothelium-derived relaxing factor
EDTA: Ethylenediaminetetraacetic acid
EEG: Electroencephalogram
EGF: Epidermal growth factor

EJP: Excitatory junction potential
EMG: Electromyogram
EP: Endogenous pyrogen
EPSP: Excitatory postsynaptic potential
eq: Equivalent(s)
ERG: Electroretinogram
ERPF: Effective renal plasma flow
ETP: Electron transport particle
f: Femto-, 10^{15}; see Table 3, above
F: Fahrenheit; also phenylalanine
FDP: Fibrinogen degradation products
FeV 1″: Forced expiratory volume in 1 second
FFA: Unesterified free fatty acid (also called NEFA, UFA)
FGF: Fibroblast growth factor
FRH, FSH-RH, FRF: FSH-releasing hormone
FSH: Follicle-stimulating hormone
ft: Foot or feet
g, gm: Gram(s)
g: Unit of force; 1 g equals the force of gravity on the earth's surface
G: Glucose; also giga-, 10^9; see Table 3, above; also glycine
GABA: Gamma-aminobutyrate
GAD: Glutamate decarboxylase
GBG: Gonadal steroid-binding globulin
G-CSF: Granulocyte colony-stimulating factor
GFR: Glomerular filtration rate
GH: Growth hormone
GIH, GIF: Growth hormone-inhibiting hormone
GIP: Gastric inhibitory peptide
Gla: Gamma-carboxyglutamic acid
GLI: Glucagonlike immuno-reactive factor from gastrointestinal mucosa
Gln: Glutamine
Glu: Glutamic acid
Gly: Glycine
GM-CSF: Granulocyte-macrophage colony-stimulating factor
GnRH: Gonadotropin-releasing hormone; same as LHRH
G6PD: Glucose 6-phosphate dehydrogenase
GRH, GRF: Growth hormone-releasing hormone
GRP: Gastrin-releasing polypeptide
GTP: Guanosine triphosphate
h: Hour(s)
H: Histidine
HA: General symbol for an acid
Hb: Deoxygenated hemoglobin
HBE: His bundle electrogram
HbO_2: Oxyhemoglobin
HCC, 25-HCC: 25-Hydroxycholecalciferol, an active metabolite of vitamin D_3
hCG: Human chorionic gonadotropin
hCS: Human chorionic somatomammotropin
Hct: Hematocrit
HDL: High-density lipoprotein
hGH: Human growth hormone
HIOMT: Hydroxyindole-O-methyltransferase
His: Histidine
HIV: Human immunodeficiency virus
HLA: Human leukocyte antigens
hPL: Human placental lactogen (same as hCS)
HS-CoA: Reduced coenzyme A
H substance: Histaminelike capillary vasodilator

5-HT: Serotonin

HVA: Homovanillic acid

Hyl: Hydroxylysine

Hyp: 4-Hydroxyproline

Hz: Hertz, unit of frequency. 1 cycle per second = 1 hertz

I: Isoleucine

I cells: Inspiratory neurons; also intercalated cells in the renal tubules

IDL: Intermediate-density lipoprotein

IFN: Interferon

IGF-I, IGF-II: Insulinlike growth factors I and II

^{123}I-IMP: ^{123}I-labeled iodoamphetamine

IJP: Inhibitory junction potential

IL: Interleukin

Ile, Ileu: Isoleucine

In: Inulin

IP_3:   Inositol 1,4,5-trisphosphate, inositol triphosphate

IPSP: Inhibitory postsynaptic potential

ITP: Inosine triphosphate

IU: International unit(s)

IUD: Intrauterine device

J: Joule (SI unit of energy)

k: Kilo-, 10^3; see Table 3, above

K: Lysine; also kilodalton

kcal (Cal):   Kilocalorie (1000 calories)

K_E: Exchangeable body potassium

Kf: Glomerular ultrafiltration coefficient

L: Geometric isomer of D form of chemical compound

L: Leucine

LATS: Long-acting thyroid stimulator

LCAT: Lecithin-cholesterol acyltransferase

LCR: Locus control region

LDL: Low-density lipoprotein

LES: Lower esophageal sphincter

Leu: Leucine

LH: Luteinizing hormone

ln: Natural logarithm

log: Logarithm to base 10

LPH: Lipotropin

LRH, LHRH, LRF: Luteinizing hormone-releasing hormone

LRP: LDL receptor-related protein

LSD: Lysergic acid diethylamide

LTH: Luteotropic hormone (prolactin)

LTP: Long-term potentiation

LVET: Left ventricular ejection time

Lys: Lysine

m: Meter(s); also milli-, 10^{-3}; see Table 3, above

M: Molarity (mol/L); also mega-, 10^6; see Table 3, above; also methionine

M cells: Microfold cells

MI: Motor cortex

MAO: Monoamine oxidase

MBC: Maximal breathing capacity (same as MVV)

M-CSF: Macrophage colony-stimulating factor

MDMA: 3,4-Methylene-dioxymethamphetamine

Met: Methionine

MGP: Matrix Gla protein

MHC: Major histocompatibility complex

mho: Unit of conductance; the reciprocal of the ohm

MHPG: 3-Methoxy-4-hydroxyphenylglycol

min: Minute(s)

MIS: Müllerian inhibiting substance; same as MRF, müllerian regression factor

MIT: Monoiodotyrosine

mol: Mole, gram-molecular weight

MPP^+:   1-Methyl-4-phenylpyridinium

MPR: Mannose-6-phosphate receptor

MPTP: 1-Methyl-4-phenyl-1,2,5,6-tetrahydropyridine

MRI: Magnetic resonance imaging

mRNA: Messenger RNA

MSH: Melanocyte-stimulating hormone

MT: 3-Methoxytyramine

multi-CSF: Multipotential colony-stimulating factor

MVV: Maximal voluntary ventilation

n: nano-, 10^{-9}; see Table 3, above

N: Normality (of a solution):   also newton (SI unit of force); also asparagine

NAD^+:   Nicotinamide adenine dinucleotide; same as DPN

NADH: Dihydronicotinamide adenine dinucleotide; same as DPNH

$NADP^+$:   Nicotinamide adenine dinucleotide phosphate; same as TPN

NADPH: Dihydronicotinamide adenine dinucleotide phosphate; same as TPNH

Na_E: Exchangeable body sodium

NEFA: Unesterified (nonesterified) free fatty acid (same as FFA)

NEP: Neutral endopeptidase

NGF: Nerve growth factor

NPH insulin: Neutral protamine Hagedorn insulin

NPN: Nonprotein nitrogen

NREM sleep: Nonrapid eye movement (spindle) sleep

NSILA: Nonsuppressible insulinlike activity

NSILP: Nonsuppressible insulinlike protein

N terminal: End of peptide or protein having a free $-NH_2$ group

NTS: Nucleus of the tractus solitarius

O: Indicates absence of a sex chromosome, eg, XO as opposed to XX or XY

OBP: Odorant-binding protein

OGF: Ovarian growth factor

osm: Osmole(s)

OVLT: Organum vasculosum of the lamina terminalis

p: Pico-, 10^{-12}; see Table 3, above

P: Proline

P followed by subscript: Plasma concentration, eg, P_{Cr} = plasma creatinine concentration; also permeability coefficient, eg, P_{Na+} = permeability coefficient for Na^+; also pressure (see respiratory symbols, below)

P_{50}:   Partial pressure of O_2 at which hemoglobin is half-saturated with O_2

Pa: Pascal (SI unit of pressure)

PAF: Platelet activating factor

PAH: para-Aminohippuric acid

PAM: Pulmonary alveolar macrophage

PBI: Protein-bound iodine

P cells: Principal cells in the renal tubules; also pacemaker cells of SA and AV nodes

PDGF: Platelet-derived growth factor

PEEP: Positive end-expiratory pressure breathing

PEP: Preejection period

PET: Positron emission tomography

P factor: Hypothetical pain producing substance produced in ischemic muscle

PGO spikes: Ponto-geniculo-occipital spikes in REM sleep

pH: Negative logarithm of the H^+ concentration of a solution

Phe: Phenylalanine

PHM-27: Peptide histidyl methionine, produced along with VIP from prepro-VIP in humans

Pi: Inorganic phosphate

PIH, PIF: Prolactin-inhibiting hormone

pK: Negative logarithm of the equilibrium constant for a chemical reaction

PLP: Parathyroid hormone-like protein

PMN: Polymorphonuclear neutrophilic leukocyte

PNMT: Phenylethanolamine-N-methyltransferase

POMC: Pro-opiomelanocortin

PRA: Plasma renin activity

PRC: Plasma renin concentration

PRH, PRF: Prolactin-releasing hormone

Pro: Proline

$Prot^-$: Protein anion

5-PRPP: 5-Phosphoribosyl pyrophosphate

PTA: Plasma thromboplastin antecedent (clotting factor XI)

PTC: Plasma thromboplastin component (clotting factor IX); also phenylthiocarbamide

PTH: Parathyroid hormone

(pyro)Glu: Pyroglutamic acid

PZI: Protamine zinc insulin

Q: Glutamine

QS_2: Total electrochemical systole

R: General symbol for remainder of a chemical formula, eg, an alcohol is R—OH; also gas constant; also respiratory exchange ratio; also arginine

RAS: Reticular activating system

rbc: Red blood cell(s)

RDS: Respiratory distress syndrome

Re: Reynolds' number

REF: Renal erythropoietic factor

REM sleep: Rapid eye movement (paradoxical) sleep

Reverse T_3: 3,3',5'-Triiodothyronine; isomer of triiodothyronine

RFLP: Restriction fragment length polymorphism

Rh factor: Rhesus group of red cell agglutinogens

RMV: Respiratory minute volume

RNA: Ribonucleic acid

RPF: Renal plasma flow

RQ: Respiratory quotient

R state: State of heme in hemoglobin that increases O_2 binding

R unit: Unit of resistance in cardiovascular system; mm Hg divided by mL/s

s: Second(s); also standard deviation of a sample

S: Serine

SA: Specific activity

SA node: Sinoatrial node

SCUBA: Self-contained underwater breathing apparatus

SDA: Specific dynamic action

SEM: Standard error of the mean

Ser: Serine

S_f units: Svedberg units of flotation

SGOT: Serum glutamic-oxaloacetic transaminase

SH: Sulfhydryl

SI units: Units of the Système International d'Unites

SIADH: Syndrome of inappropriate hypersecretion of antidiuretic hormone (vasopressin)

SIDS: Sudden infant death syndrome

SIF cells: Small, intensely fluorescent cells in sympathetic ganglia

smg: Small GTP-binding protein

SPCA: Proconvertin (clotting factor VII)

SPECT: Single photon emission computed tomography

sq cm: Square centimeter(s), cm^2

SRIF: Somatotropin release-inhibiting factor; same as GIH

sRNA: Soluble or transfer RNA

SRY: Sex-determining region of Y chromosome

SS 14: Somatostatin 14

SS 28: Somatostatin 28

SS 28 (1–12): Polypeptide related to somatostatin that is found in tissues

STH: Somatotropin, growth hormone

STPD: See Table 34–1

Substance P: Polypeptide found in brain, and other tissues

T: Absolute temperature; also threonine

T_3: 3,5,3'-Triiodothyronine

T_4: Thyroxine

TBG: Thyroxine-binding globulin

TBPA: Thyroxine-binding prealbumin

TBW: Total body water

^{99m}Tc-PYP: Technetium 99m stannous pyrophosphate

TDF: Testis-determining factor

TEA: Tetraethylammonium

TETRAC: Tetraiodothyroacetic acid

TF/P: Concentration of a substance in renal tubular fluid divided by its concentration in plasma

TGF-α: Transforming growth factor-α

Thr: Threonine

Tm: Renal tubular maximum

TNF: Tumor necrosis factor

torr: 1/760 atm = 1.00000014 mm Hg; unit for various pressures in the body

TPA: Tissue plasminogen activator

TPN: Triphosphopyridine nucleotide

TPNH: Reduced triphosphopyridine nucleotide

TRH, TRF: Thyrotropin-releasing hormone

tRNA: Transfer RNA; same as sRNA

Trp, Try, Tryp: Tryptophan

TSF: Thrombopoietic stimulating factor, thrombopoietin

TSH: Thyroid-stimulating hormone

TSI: Thyroid-stimulating immunoglobulins

T/S ratio: Thyroid/serum iodide ratio

T state: State of heme in hemoglobin that decreases O_2 binding

TTX: Tetrodotoxin

Tyr: Tyrosine

U: Unit(s)

U followed by subscript: Urine concentration, eg, U_{Cr} = urine creatinine concentration

UDPG: Uridine diphosphoglucose

UFA: Unesterified free fatty acid (same as FFA)

UL: Unstirred layer

URF: Uterine-relaxing factor; relaxin

US: Unconditioned stimulus

UTP: Uridine triphosphate

V: Urine volume/unit time; volt; also valine

V_1, V_2, etc: Unipolar chest electrocardiographic leads

Val: Valine

VF: Unipolar left leg electrocardiographic lead

VIP: Vasoactive intestinal polypeptide

VL: Left arm unipolar electrocardiographic lead

VLDL: Very low density lipoprotein
VMA: Vanillylmandelic acid (3-methoxy-4-hydroxyman-
delic acid)
$\dot{V}_{C_{2max}}$: Maximal oxygen consumption
VOR: Vestibulo-ocular reflex
VR: Unipolar right arm electrocardiographic lead
W: Tryptophan
wbc: White blood cell(s)

X chromosome: One of the sex chromosomes in humans
X zone: Inner zone of adrenal cortex in some young
mammals
Y: Tyrosine
Y chromosome: One of the sex chromosomes in humans

Some Standard Respiratory Symbols (See *Handbook of Physiology*, Section 3: *The Respiratory System*. American Physiological Society, 1986.)

General Variables

V Gas volume
V̇ Gas volume/unit of time. (Dot over a symbol indicates rate.)
P Gas pressure
P̄ Mean gas pressure
f Respiratory frequency (breaths/unit of time)
D Diffusing capacity
F Fractional concentration in dry gas phase
R Respiratory exchange ratio $= V_{CO_2}/V_{O_2}$
Q Volume of blood

Localization (Subscript letters)

I Inspired gas
E Expired gas

A Alveolar gas
T Tidal gas
D Dead space gas
B Barometric
a Arterial blood
c Capillary blood
v Venous blood

Molecular Species

Indicated by chemical formula printed as subscript

Examples

$P_{I_{O_2}}$ = Pressure of oxygen in inspired air
V_D = Dead space gas volume

Equivalents of Metric, United States, and English Measures
(Values rounded off to 2 decimal places.)

Length
1 kilometer = 0.62 mile
1 mile = 5280 feet = 1.61 kilometers
1 meter = 39.37 inches
1 inch = 1/12 foot = 2.54 centimeters

Volume
1 liter = 1.06 US liquid quart
1 US liquid quart = 32 fluid ounces = 1/4 US gallon = 0.95 liter
1 milliliter = 0.03 fluid ounce
1 fluid ounce = 29.57 milliliters
1 US gallon = 0.83 English (Imperial) gallon

Weight
1 kilogram = 2.20 pounds (avoirdupois) = 2.68 pounds (apothecaries')
1 pound (avoirdupois) = 16 ounces = 453.60 grams
1 grain = 65 milligrams

Energy
1 kilogram-meter = 7.25 foot-pounds
1 foot-pound = 0.14 kilogram-meters

Temperature
To convert Celsius degrees into Fahrenheit, multiply by 9/5 and add 32
To convert Fahrenheit degrees into Celsius, subtract 32 and multiply by 5/9

Greek Alphabet

Symbol		Name	Symbol		Name
A	α	alpha	N	ν	nu
B	β	beta	Ξ	ξ	xi
Γ	γ	gamma	O	o	omicron
Δ	δ	delta	Π	π	pi
E	ϵ	epsilon	P	ρ	rho
Z	ζ	zeta	Σ	σ, ς	sigma
H	η	eta	T	τ	tau
Θ	θ	theta	Υ	υ	upsilon
I	ι	iota	Φ	ϕ	phi
K	κ	kappa	X	χ	chi
Λ	λ	lambda	Ψ	ψ	psi
M	μ	mu	Ω	ω	omega

Self Study

Objectives, Essay Questions, & Multiple-Choice Questions

Chapter 1

This chapter is a review of the cellular, molecular, and general basis of medical physiology. The material in the chapter should help students to—

- Name the different fluid compartments in the body, the size of each, and ways in which their sizes can be measured.
- Define moles, equivalents, and osmoles.
- List and compare the various passive and active forces that produce movement of substances across cell membranes.
- Define osmosis, and give examples of its role in moving fluid from one location in the body to another.
- Describe and give examples of the coupling of active Na^+ transport to the movement of other electrolytes and nonelectrolytes across cell membranes.
- Define and explain the resting membrane potential.
- List the unique features of the capillary wall.
- Know the various parts of the cells and the functions of each.
- Know the chemical nature and physiologic significance of the compounds that make up the cell membrane.
- Understand in general terms the structure of DNA and RNA and the role these nucleotides and other substances in the cell play in the process of protein synthesis.
- Define the processes of exocytosis and endocytosis, and describe the contribution of each to normal cell function.
- Describe the principal ways that the chemical messengers in the extracellular fluid produce changes inside cells.
- List the principal types of receptors.

- Define homeostasis, and give examples of homeostatic mechanisms.

GENERAL QUESTIONS

1. Name the processes responsible for movement of substances across cell membranes. Which of them require the input of energy? How is this energy provided?

2. Why is Na^+-K^+ ATPase important in physiology?

3. Compare the compositions of plasma, interstitial fluid, and intracellular fluid. Explain the differences.

4. What happens to resting membrane potential when the extracellular K^+ concentration is increased from 5 to 10 meq/L? What happens when the extracellular Na^+ concentration is increased from 142 to 155 meq/L? Why?

5. Why do red blood cells swell and eventually burst when they are placed in a solution of 0.3% sodium chloride?

6. Genes dictate the formation of specific proteins. What are the steps between DNA and the formation of a protein that is secreted by the cell?

7. Discuss the functions of the proteins found in cell membranes.

8. What is the function of calmodulin? Why is this function important?

9. What is receptor-mediated endocytosis?

10. Discuss aging from the point of view of its cellular and molecular bases, its physiologic consequences in humans, and its implications for the practice of medicine.

MULTIPLE-CHOICE QUESTIONS

In questions 1–9, select the single best answer.

1. Cell membranes
   - (A) consist almost entirely of protein molecules
   - (B) are impermeable to fat-soluble substances
   - (C) in some tissues permit the transport of glucose at a greater rate in the presence of insulin
   - (D) are freely permeable to electrolytes but not to proteins
   - (E) have a stable composition throughout the life of the cell

2. The primary force moving water molecules from the blood plasma to the interstitial fluid is
   - (A) active transport
   - (B) cotransport with H^+
   - (C) facilitated diffusion
   - (D) cotransport with Na^+
   - (E) filtration

3. Second messengers
   - (A) are substances that interact with first messengers inside cells
   - (B) are substances that bind to first messengers in the cell membrane
   - (C) are hormones secreted by cells in response to stimulation by another hormone
   - (D) mediate the intracellular responses to many different hormones and neurotransmitters
   - (E) are not formed in the brain

4. The resting membrane potential of a cell
   - (A) is dependent on the permeability of the cell membrane to K^+ being greater than the permeability to Na^+
   - (B) falls to zero immediately

when Na^+-K^+ ATPase in the membrane is inhibited
- (C) is usually equal to the equilibrium potential for K^+
- (D) is usually equal to the equilibrium potential for Na^+
- (E) is markedly altered if the extracellular Na^+ concentration is increased

5. Proteins that are secreted by cells are generally
- (A) not synthesized on membrane-bound ribosomes
- (B) initially synthesized with a signal peptide or leader sequence at their C terminal
- (C) found in vesicles and secretory granules
- (D) moved across the cell membranes by endocytosis
- (E) secreted in a form that is larger than the form present in the endoplasmic reticulum

6. Osmosis is
- (A) movement of solvent across a semipermeable membrane from an area where the hydrostatic pressure is high to an area where the hydrostatic pressure is low
- (B) movement of solute across a semipermeable membrane from an area in which it is in low concentration to an area in which it is in high concentration
- (C) movement of solute across a semipermeable membrane from an area in which it is in high concentration to an area in which it is in low concentration
- (D) movement of solvent across a semipermeable membrane from an area in which it is in low concentration to an area in which it is in high concentration
- (E) movement of solvent across a semipermeable membrane from an area in which it is in high concentration to an area in which it is in low concentration

7. Deuterium oxide and inulin are injected into a normal 30-year-old man. The volume of distribution of deuterium oxide is found to be 42 L and that of inulin 14 L.
- (A) The man's intracellular fluid volume is about 14 L
- (B) The man's intracellular fluid volume is about 28 L
- (C) The man's plasma volume is about 7 L

- (D) The man's interstitial fluid volume is about 9 L
- (E) The man's total body water cannot be calculated from these data

8. Which of the following receptors does *not* span the cell membrane 7 times?
- (A) β-Adrenergic receptor
- (B) Rhodopsin
- (C) 5-HT_{IC} receptor
- (D) Mineralocorticoid receptor
- (E) LH receptor

9. Which of the following does *not* act intracellularly to produce physiologic effects?
- (A) Triiodothyronine
- (B) Inositol triphosphate
- (C) Aldosterone
- (D) Cyclic AMP
- (E) Dopamine

In questions 10–13, match the disease in each question with the lettered item that is most closely associated with it. Each lettered item may be selected once, more than once, or not at all.

- (A) G protein deficiency
- (B) Abnormal G protein
- (C) Antibodies against receptors
- (D) Deficiency of receptors for extracellular protein
- (E) Abnormal receptor for extracellular protein

10. Many cases of Graves' disease

11. Some cases of pseudohypoparathyroidism

12. Some cases of familial hypercholesterolemia

13. Some cases of acromegaly

In questions 14–20, indicate whether the first item is greater than (G), the same as (S), or less than (L), the second item.

14. Hydrogen ion concentration in lysosomes

G S L

Hydrogen ion concentration in cytoplasm of cells

15. ECF volume 30 minutes after intravenous infusion of 1000 mL of isotonic (5%) glucose

G S L

ECF volume 30 minutes after intravenous infusion of 1000 mL of isotonic (0.9%) sodium chloride

16. Total body water 30 minutes after intravenous infusion of 1000 mL isotonic (5%) glucose

G S L

Total body water 30 minutes after intravenous infusion of 1000 mL isotonic (0.9%) sodium chloride

17. Contribution of normal concentrations of plasma glucose to total plasma osmolality

G S L

Contribution of normal concentrations of plasma Na^+ to total plasma osmolality

18. Calculated plasma volume when some of the dye used to measure it is unknowingly injected subcutaneously instead of intravenously

G S L

Calculated plasma volume when all of the dye used to measure it is injected intravenously

19. Concentration of K^+ in intracellular fluid

G S L

Concentration of K^+ in interstitial fluid

20. Concentration of Ca^{2+} in intracellular fluid

G S L

Concentration of Ca^{2+} in interstitial fluid

Chapter 2

This is the first of 4 chapters concerned with nerves and muscles and the interactions between them. It deals with the morphology of neurons and the electrical and ionic events that underlie their excitation and their ability to conduct impulses. The material in the chapter should help students to—

- Name the various parts of a neuron and the functions of each.
- Distinguish between unmyelinated and myelinated neurons, describe the chemical nature of myelin, and summarize the differences in the ways in which unmyelinated and myelinated neurons conduct impulses.
- Define orthograde axoplasmic transport, retrograde axoplasmic transport, and wallerian degeneration.
- Define excitation and conduction, and describe the changes in membrane permeability and ionic movements that underlie electrotonic potentials and action potentials.

- Explain the following characteristics of a nerve impulse: latent period, firing level, spike potential, after-depolarization, and after-hyperpolarization.
- Describe and explain the compound action potential of mixed nerves.
- List the various nerve fiber types found in humans, and comment on their significance in terms of the normal and abnormal function of peripheral nerves.
- Describe the nerve growth factor molecule, and comment on the function of each of its subunits.
- Describe the various types of glial cells.

GENERAL QUESTIONS

1. Of what benefit to the animal is the evolution of myelinated neurons?

2. Compare the function of dendrites with the function of axons.

3. Describe and explain the effects of tetrodotoxin on the resting membrane potential and the action potential of nerve cells.

4. Which nerve fiber types would you expect to find in a sympathetic nerve that runs from the celiac ganglion to the intestine? Explain your answer.

5. How do the absolute and relative refractory periods correlate in time with the various phases of the action potential? Why is the nerve refractory during these periods?

6. Compare the electrical and ionic events in neurons with those in skeletal, cardiac, and smooth muscle.

7. What general types of ion channels are found in the cell membranes of neurons, and what is the function of each type?

MULTIPLE-CHOICE QUESTIONS

In questions 1–6, select the single best answer.

1. The action potential of a neuron
    (A) is initiated by efflux of Na^+
    (B) is terminated by efflux of K^+
    (C) declines in amplitude as it moves along the axon
    (D) results in a transient reversal of the concentration gradient of Na^+ across the cell membrane
    (E) is not associated with any net movement of Na^+ or K^+ across the cell membrane

Figure 2–A.

2. A squid axon is placed on stimulating electrodes, and an intracellular electrode is inserted and connected through a cathode-ray oscilloscope (CRO) to an indifferent electrode (Fig 2–A). When the axon is stimulated, the latent period is 1.5 ms. The intracellular electrode is 6 cm from the anode of the stimulator and 4.5 cm from the cathode of the stimulator. What is the conduction velocity of the axon?
    (A) 15 m/s
    (B) 30 m/s
    (C) 40 m/s
    (D) 67.5 m/s
    (E) This cannot be determined from the information given

3. Which of the following has the slowest conduction velocity?
    (A) A α fibers
    (B) A β fibers
    (C) A γ fibers
    (D) B fibers
    (E) C fibers

4. A man falls into a deep sleep with one arm under his head. This arm is paralyzed when he awakens, but it tingles, and pain sensation in it is still intact. The reason for the loss of motor function without loss of pain sensation is that in the nerves to his arm,
    (A) A fibers are more susceptible to hypoxia than B fibers
    (B) A fibers are more sensitive to pressure than C fibers
    (C) C fibers are more sensitive to pressure than A fibers
    (D) sensory nerves are nearer the bone than motor nerves and hence are less affected by cooling
    (E) sensory nerves are nearer the bone than motor nerves and hence are less affected by pressure

5. Which part of a neuron has the highest concentration of Na^+ channels per square millimeter of cell membrane?
    (A) Dendrites
    (B) Cell body near dendrites
    (C) Synaptic knobs
    (D) Axonal membrane under myelin
    (E) None of the above

6. Which of the following statements about nerve growth factor is *not* true?
    (A) It is made up of 3 polypeptide subunits
    (B) It is found in high concentration in the submaxillary salivary glands of female mice
    (C) It is necessary for the growth and development of the sympathetic nervous system
    (D) It is picked up by nerves from the organs they innervate
    (E) It is present in the brain

*In questions 7–10, indicate whether the first item is greater than (**G**), the same as (**S**), or less than (**L**), the second item.*

7. Increase in Na^+ permeability during the rising phase of the action potential

G S L

Increase in K^+ permeability during the rising phase of the action potential

8. Rate of conduction in small-diameter axons

G S L

Rate of conduction in large-diameter axons

9. Excitability of neuron when membrane potential is increased to –80 mV

G S L

Excitability of neuron when membrane potential is reduced to –60 mV

10. Total duration of compound action potential of a mixed nerve 5 mm from stimulating electrode

G S L

Total duration of compound action potential of a mixed nerve 50 mm from stimulating electrode

Chapter 3

This chapter compares and contrasts the morphologic and functional characteristics of the 3 types of muscle found in the body: skeletal, cardiac, and smooth muscle. The material in the chapter should help students to—

- Describe the gross and microscopic anatomy of skeletal muscle, including the cross-striations, the relation of actin to myosin, and the sarcotubular system.
- Compare the morphology of cardiac, visceral smooth muscle, and multi-unit smooth muscle with that of skeletal muscle.
- List the sequence of electrical and ionic events leading from an action potential in the motor nerve to contraction of a skeletal muscle, and discuss the significance of each.
- Compare isometric and isotonic contractions.
- Explain summation of contractions and the relation between initial muscle length and the strength of contraction.
- Describe the sources of energy for muscle contraction, and explain how energy is transferred to the contractile mechanism.
- Define oxygen debt, and describe its role in muscle function during exercise.
- Describe the differences between fast and slow skeletal muscles.
- Define the term motor unit, and discuss the effects of denervation on skeletal muscle.
- Describe the extracellularly and intracellularly recorded action potential of cardiac muscle and the relation of the electrical events to the contractile response of the muscle and the electrocardiogram.
- Describe the ionic events underlying the action potential in cardiac muscle.
- Describe the fluctuations in membrane potential seen in pacemaker tissue and the ionic events responsible for the prepotential.
- Compare the electrical and mechanical events in smooth muscle with those in skeletal muscle, and summarize the effects of acetylcholine and norepinephrine on the electrical and mechanical activity of visceral smooth muscle.

GENERAL QUESTIONS

1. Define total tension, passive tension, active tension, resting length, and equilibrium length. How does the relationship of actin to myosin explain the length-tension curve of skeletal muscle?

2. "Muscle is a machine for converting chemical into mechanical energy." Analyze and discuss this statement.

3. What is the minimum stimulation frequency at which tetanus occurs in a fast muscle with a twitch duration of 7.5 ms, and in a slow muscle with a twitch duration of 100 ms?

4. Discuss the factors that permit gradation of skeletal muscle responses in a living, intact animal.

5. Discuss the relation of the length-tension relationship in cardiac muscle to normal and abnormal function of the heart.

6. Compare the molecular events that initiate contraction in smooth muscle and skeletal muscle. What are the similarities and the differences?

7. Define plasticity as the term is applied to muscle. What are the practical physiologic consequences of this property of smooth muscle?

MULTIPLE-CHOICE QUESTIONS

In questions 1–5, select the single best answer.

1. The action potential of skeletal muscle
   (A) has a prolonged plateau phase
   (B) spreads inward to all parts of the muscle via the T tubules
   (C) causes the immediate uptake of Ca^{2+} into the lateral sacs of the sarcoplasmic reticulum
   (D) is longer than the action potential of cardiac muscle
   (E) is not essential for contraction

2. The functions of tropomyosin in skeletal muscle include
   (A) sliding on actin to produce shortening
   (B) releasing Ca^{2+} after initiation of contraction
   (C) binding to myosin during contraction
   (D) acting as a "relaxing protein" at rest by covering up the sites where myosin binds to actin
   (E) generating ATP, which it passes to the contractile mechanism

3. The cross-bridges of the sarcomere in skeletal muscle are made up of
   (A) actin
   (B) myosin
   (C) troponin
   (D) tropomyosin
   (E) myelin

4. The contractile response in skeletal muscle
   (A) starts after the action potential is over
   (B) does not last as long as the action potential
   (C) produces more tension when the muscle contracts isotonically than when the muscle contracts isometrically
   (D) produces more work when the muscle contracts isometrically than when the muscle contracts isotonically
   (E) can increase in magnitude with repeated stimulation

5. Gap junctions
   (A) are absent in cardiac muscle
   (B) are present but of little functional importance in cardiac muscle
   (C) are present and provide the pathway for rapid spread of excitation from one cardiac muscle fiber to another
   (D) are absent in smooth muscle
   (E) connect the sarcotubular system to individual skeletal muscle cells

In questions 6–10, select

A   if the item is associated with (a) below,
B   if the item is associated with (b) below,
C   if the item is associated with both (a) and (b), and
D   if the item is associated with neither (a) nor (b)

(a) Strength of contraction of a given muscle
(b) Duration of contraction of the muscle

6. Length of muscle at the start of contraction

7. Type of myosin heavy chain isoform in the muscle

8. Degree to which firing in the

motor nerve to the muscle is asynchronous

9. Troponin content of muscle

10. Dystrophin content of muscle

In questions 11–15, indicate whether the first item is greater than (**G**), *the same as* (**S**), *or less than* (**L**), *the second item.*

11. Duration of the increase in Ca^{2+} influx during the action potential of cardiac muscle

    G S L

    Duration of the increase in Na^+ influx during the action potential of cardiac muscle

12. In a normal man before breakfast, proportion of the caloric needs of the heart obtained from fat

    G S L

    In a normal man before breakfast, proportion of the caloric needs of the heart obtain from carbohydrate

13. Tension in wall of bladder 1 second after it is distended

    G S L

    Tension in wall of bladder 1 minute after it is distended to the same degree

14. Excitability of uterine smooth muscle after administration of progesterone

    G S L

    Excitability of uterine smooth muscle after administration of estrogen

15. ATPase activity of cardiac muscle after thyroidectomy

    G S L

    ATPase activity of cardiac muscle after thyroxine (T_4) treatment

Chapter 4

This chapter is concerned with the synaptic junctions between neurons, with the neuromuscular junction, and with the neurotransmitters and receptors involved in transmission at these junctions. The material in the chapter should help students to—

- Describe the main morphologic features of synapses.

- Distinguish between chemical and electrical transmission at synapses.
- Define convergence and divergence in neural networks, and discuss their implications.
- Describe fast and slow excitatory and inhibitory postsynaptic potentials, outline the ionic fluxes that underlie them, and explain how the potentials interact to generate action potentials.
- Define and give examples of direct inhibition, indirect inhibition, presynaptic inhibition, and postsynaptic inhibition.
- Define synaptic plasticity, and discuss its relation to learning.
- Know the types of chemicals that serve as synaptic transmitters.
- List 10 neurotransmitters and the principal sites in the nervous system at which they are released.
- Summarize the steps involved in the biosynthesis, release, action, and removal from the synaptic cleft of acetylcholine, dopamine, norepinephrine, epinephrine, and serotonin.
- Define the term opioid peptide, list the principal opioid peptides in the body, and name the precursor molecules from which they come.
- Define and explain habituation, sensitization, long-term potentiation (LTP), and long-term depression (LTD).
- Describe the neuromuscular junction, and explain how action potentials in the motor neuron at the junction lead to contraction of the skeletal muscle.
- Define and explain denervation hypersensitivity.

GENERAL QUESTIONS

1. Conduction can occur in either direction along an axon but in only one direction at a chemical synaptic junction. Why?

2. Discuss neuronal regeneration.

Figure 4–A. Patch clamp records obtained from 2 different neurons. Downward deflection indicates inward current pulses. Upward deflection indicates outward current pulses.

3. One of the records in Fig 4–A was obtained with acetylcholine acting as an excitatory transmitter, and the other was obtained with γ-aminobutyric acid (GABA) acting as an inhibitory transmitter. Which is which? Explain your answer.

4. What is the role of reuptake in the metabolism of amine neurotransmitters? How is it produced? How does its absence contribute to denervation supersensitivity?

5. "In the human forebrain, the ratio of synapses to neurons has been calculated to be 40,000 to 1." Explain this statement, and discuss its physiologic implications.

6. What is synaptic delay? Discuss its significance in terms of brain function.

7. Discuss the role of glutamate and the physiology of the glutamate receptors in the brain.

8. Although some neurons secrete neurotransmitters, others secrete hormones, ie, chemical messengers that enter the bloodstream. Name the hormones secreted in this fashion, and identify and locate the neurons involved.

9. The nerve gases that were developed for chemical warfare generally inhibit acetylcholinesterase. Why does inhibition of acetylcholinesterase produce death?

10. Compare the endings of somatic motor nerves in skeletal muscle with the endings of autonomic motor nerves in cardiac and smooth muscle.

MULTIPLE-CHOICE QUESTIONS

In questions 1–5, match the statement in each question with the lettered structure that is most closely associated with it. Each lettered structure may be selected once, more than once, or not at all.

(**A**) Microtubules
(**B**) Active zones
(**C**) Nuclei
(**D**) Clear vesicles
(**E**) Granulated vesicles

1. Contain acetylcholine

2. Contain catecholamines

3. Contain steroid receptors

4. Contain neuropeptide Y

5. Contain kinesin

In questions 6–14, select the single best answer.

6. Which of the following is *not* synthesized in postganglionic sympathetic neurons?
   (A) L-Dopa
   (B) Dopamine
   (C) Norepinephrine
   (D) Epinephrine
   (E) Acetylcholine

7. Fast inhibitory postsynaptic potentials (IPSPs)
   (A) are produced by the process of indirect inhibition
   (B) are a consequence of presynaptic inhibition
   (C) interact with other fast and slow potentials to move the membrane potential of the postsynaptic neuron toward or away from the firing level
   (D) are produced by a decrease in K^+ permeability
   (E) occur in skeletal muscle

8. Initiation of an action potential in skeletal muscle by stimulating its motor nerve
   (A) requires spatial facilitation
   (B) requires temporal facilitation
   (C) requires the release of norepinephrine
   (D) is inhibited by a high concentration of Ca^{2+} at the neuromuscular junction
   (E) requires the release of acetylcholine

9. Which of the following is a ligand-gated ion channel?
   (A) VIP receptor
   (B) Norepinephrine receptor
   (C) $GABA_A$ receptor
   (D) $GABA_B$ receptor
   (E) Metabotropic glutamine receptor

10. Which of the following is most likely *not* to be involved in production of LTP?
    (A) NO
    (B) Ca^{2+}
    (C) Spatial summation
    (D) NMDA receptors
    (E) Membrane depolarization

11. Which of the following synaptic transmitters is *not* a peptide, polypeptide, or protein?
    (A) Substance P
    (B) Met-enkephalin
    (C) β-Endorphin
    (D) Serotonin
    (E) Dynorphin

12. Activation of which of the following receptors would be expected to decrease anxiety?
    (A) Nicotinic cholinergic receptors
    (B) Glutamate receptors
    (C) $GABA_A$ receptors
    (D) Glucocorticoid receptors
    (E) α_1-Adrenergic receptors

13. Which of the following receptors is coupled to a G protein?
    (A) Glycine receptor
    (B) $GABA_A$ receptor
    (C) Nicotinic acetylcholine receptor
    (D) α_1-Adrenergic receptor
    (E) Glutamate receptor

14. Which of the following would *not* be expected to enhance noradrenergic transmission?
    (A) A drug that increases the entry of tyrosine into neurons
    (B) A drug that enhances tyrosine hydroxylase activity
    (C) A drug that enhances dopamine β-hydroxylase activity
    (D) A drug that inhibits monoamine oxidase
    (E) A drug that facilitates norepinephrine reuptake

In questions 15–20, indicate whether the first item is greater than (G), the same as (S), or less than (L), the second item.

15. Amount of norepinephrine released at a noradrenergic synapse when presynaptic α_2-adrenergic receptors are stimulated

G S L

Amount of norepinephrine released at a noradrenergic synapse when postsynaptic α_2-adrenergic receptors are stimulated

16. Number of acetylcholine receptors on the surface of denervated skeletal muscle before nerve regrows

G S L

Number of acetylcholine receptors on the surface of denervated skeletal muscle after nerve regrows

17. Amount of neurotransmitter released by a nerve impulse after habituation

G S L

Amount of neurotransmitter released by a nerve impulse after sensitization

18. Number of times a β_2-adrenergic receptor spans the cell membrane

G S L

Number of times an M_1 muscarinic acetylcholine receptor spans the cell membrane

19. Amount of acetylcholine in endings of preganglionic sympathetic neurons

G S L

Amount of norepinephrine in preganglionic sympathetic neurons

20. Amount of Nissl substance visible in the cell body of a neuron before transection of its axon

G S L

Amount of Nissl substance visible in the cell body of a neuron after transection of its axon

---

Chapter 5

This chapter describes generator potentials and the way they produce action potentials in sensory nerves, lists the senses, and summarizes the principles and laws that govern the coding of sensory information. The material in the chapter should help students to—

- Define the term sensory receptor.
- Define adequate stimulus, and give examples.
- Discuss the various classifications of the senses.
- Name the types of sensory receptors found in the skin, and discuss their relation to the 4 cutaneous senses: touch, cold, warmth, and pain.
- Describe the relationship between the size of a generator potential in a sensory receptor and the number of impulses generated in its sensory nerve.
- Define adaptation, the doctrine of specific nerve energies, and the law of projection.
- Explain how the intensity, location, and quality of stimuli are coded.

GENERAL QUESTIONS

1. Discuss sensory units and their recruitment as the intensity of a stimulus increases.

2. What is the difference between tonic and phasic receptors. Which physiologic functions are subserved by each type?

3. What part of a pacinian corpuscle produces its generator potential? Where are action potentials in the sensory nerve produced? What is the function of the multiple lamellas of connective tissue that cover the ending of the sensory nerve?

4. Why do people see lightning and hear thunder rather than hear lightning and see thunder?

5. Amputees may feel pain and other sensations in a limb that is no longer there, and this can be a distressing medical problem. Explain the occurrence of pain in the absent limb. How would you treat a patient with such a "phantom limb"?

MULTIPLE-CHOICE QUESTIONS

In questions 1–5, select the single best answer.

1. Pacinian corpuscles are
   (A) a type of temperature receptor
   (B) usually innervated by A δ nerve fibers
   (C) rapidly adapting touch receptors
   (D) slowly adapting touch receptors
   (E) pain receptors

2. Adaptation to a sensory stimulus produces
   (A) a diminished sensation when other types of sensory stimuli are applied near a given stimulus
   (B) a more intense sensation when a given stimulus is applied repeatedly
   (C) a sensation localized to the hand when the nerves of the brachial plexus are stimulated
   (D) a diminished sensation when a given stimulus is applied steadily over time
   (E) failure to perceive the stimulus when one's attention is directed to another matter

3. Which of the following sensations is *not* generated by impulses initiated in naked nerve endings?
   (A) Touch
   (B) Pain
   (C) Temperature
   (D) Tickle
   (E) Taste

4. A sensation felt in one hand *cannot* be produced by stimulation of

(A) the afferent neurons from the hand as they enter the spinal cord
(B) the spinothalamic tract neurons on which the afferent neurons from the hand terminate
(C) the hand area of the thalamus on the opposite side of the body
(D) the hand area of the sensory cortex on the opposite side of the body
(E) the hand area of the sensory cortex on the same side of the body

5. Which of the following receptors and sense organs are *incorrectly* paired?
   (A) Rods and cones : eye
   (B) Hair cells : olfactory epithelium
   (C) Hair cells : inner ear
   (D) Receptors sensitive to stretch : carotid sinus
   (E) Glomus cells : carotid body

In questions 6–10, indicate whether the first item is greater than (G), the same as (S), or less than (L), the second item.

6. Amplitude of generator potential with weak stimulus

   G S L

   Amplitude of generator potential with strong stimulus

7. Amplitude of action potentials in sensory nerve with weak stimulus

   G S L

   Amplitude of action potentials in sensory nerve with strong stimulus

8. Frequency of action potentials in sensory nerve with weak stimulus

   G S L

   Frequency of action potentials in sensory nerve with strong stimulus

9. Sensation of pain when warmth receptors are subjected to weak stimulation

   G S L

   Sensation of pain when warmth receptors are subjected to strong stimulation

10. Number of sensory units activated when a pin is pushed weakly on a spot in the skin

    G S L

    Number of sensory units activated

when a pin is pushed strongly on same spot in the skin

Chapter 6

This chapter describes the components that make up the reflex arc, the monosynaptic and polysynaptic spinal reflexes, and the properties and functions of muscle spindles. The material in the chapter should help students to—

- State the Bell-Magendie law, and summarize its physiologic implications.
- Distinguish between and compare monosynaptic and polysynaptic reflexes.
- Give examples of stretch reflexes, including those that are frequently tested clinically.
- Describe the muscle spindles and analyze their function, with particular attention to how they operate as part of a feedback system to maintain muscle length.
- Define reciprocal innervation, inverse stretch reflex, clonus, and lengthening reaction.
- Define and explain fractionation and occlusion.

GENERAL QUESTIONS

1. What is a "final common path"? Discuss its physiologic basis and importance.

2. Why do paraplegic patients sometimes urinate, defecate, and show marked withdrawal responses and fluctuations in blood pressure in response to relatively minor noxious stimuli distal to their level of cord transection?

3. Define local sign, and explain the mechanism responsible for it.

4. Why does a strong noxious stimulus produce a prolonged withdrawal response?

5. In a normal subject, the central delay for a given reflex response was found to be 0.9 ms. Is it likely that the response is mediated by a monosynaptic or a polysynaptic pathway? Explain your answer.

6. It is frequently stated that Golgi tendon organs are part of a servomechanism that controls muscle

force. Analyze and discuss this statement.

7. How are reflex responses graded and modified?

MULTIPLE-CHOICE QUESTIONS

In questions 1–5, select the single best answer.

1. The inverse stretch reflex
   - (A) has a lower threshold than the stretch reflex
   - (B) is probably a monosynaptic reflex
   - (C) is probably a disynaptic reflex with a single interneuron inserted between the afferent and efferent limbs
   - (D) is probably a polysynaptic reflex with many interneurons inserted between the afferent and efferent limbs
   - (E) requires the discharge of central neurons that release acetylcholine

2. When γ motor neuron discharge increases at the same time as α motor neuron discharge to muscle
   - (A) there is prompt inhibition of discharge in spindle Ia afferents
   - (B) the contraction of the muscle is prolonged
   - (C) the muscle will not contract
   - (D) the number of impulses in spindle Ia afferents is smaller than when α discharge alone is increased
   - (E) the number of impulses in spindle Ia afferents is greater than when α discharge alone is increased

3. Which of the following is *not* characteristic of a reflex action?
   - (A) Modification by impulses from various parts of the central nervous system
   - (B) May involve simultaneous contraction of some muscles and relaxation of others
   - (C) May involve either somatic or visceral responses but never both simultaneously
   - (D) Always involves transmission across at least one synapse
   - (E) Frequently occurs without conscious perception

Questions 4 and 5 refer to Fig 6–A.

In Fig 6–A, A, B, and C are records of action potentials in afferent nerve fibers from a muscle before and after the muscle was stretched, as shown by the change in muscle length.

Figure 6–A.

4. Which record is from a fiber that innervates a nuclear bag fiber?

5. Which record is from a fiber that innervates a nuclear chain fiber?

In questions 6–10, indicate whether the first item is greater than (G), the same as (S), or less than (L), the second item.

6. Response of a skeletal muscle to stretch after cutting the sensory nerves that supply it

   G S L

   Response of a skeletal muscle to stretch after cutting the motor nerves that supply it

7. During stretch reflex, rate of discharge in motor nerve to protagonist muscles

   G S L

   During stretch reflex, rate of discharge in motor nerve to antagonist muscles

8. Number of neurons involved in stretch reflex

   G S L

   Number of neurons involved in withdrawal reflex

9. Rhythmic contractile response to sustained stretch of a muscle when rate of γ efferent discharge is high

   G S L

   Rhythmic contractile response to sustained stretch of a muscle when rate of γ efferent discharge is low

10. Latency of a monosynaptic reflex response

    G S L

    Latency of a polysynaptic reflex response

Chapter 7

This chapter describes the sensory mechanisms responsible for touch, proprioception, warmth, cold, pain, itching, and sensations produced by combinations of these stimuli. Pain is considered in detail, including the characteristics of so-called fast pain and slow pain, deep pain, and pain from visceral structures; treatment of chronic pain is also discussed. The material in the chapter should help students to—

- Outline the neural connections and pathways from the skin, deep tissues, and viscera to the cerebral cortex that mediate sensations from them.
- Describe the structure and function of the substantia gelatinosa.
- Describe the areas in which the sensory projection neurons from the thalamus terminate in the sensory cortex.
- Name the types of nerve fibers that mediate warmth and cold in peripheral nerves.
- Name the receptors that mediate pain, and explain the differences between fast and slow pain.
- Compare superficial, deep, and visceral pain.
- Define hyperalgesia.
- Describe and explain referred pain.
- List the main procedures that have been used for relief of intractable pain. For each, give the rationale for its use and comment on its clinical effectiveness.
- Explain vibratory sensibility, 2-point discrimination, and stereognosis.

GENERAL QUESTIONS

1. What are the adequate stimuli for warmth, cold, and pain? Are pain receptors specific? Which chemical agents may initiate impulses in pain fibers?

2. Why do tumors that arise in the center of the upper spinal cord cause loss of pain and temperature sensations first in the upper segments of the body and only later in the lower segments? What would you expect to be the corresponding order of loss of proprioception?

3. Describe the location of the 2

touch pathways in the spinal cord. What type of touch information does each conduct?

4. Discuss P factor and the role it plays in muscle pain.

5. Why is visceral pain poorly localized? Why is it particularly unpleasant?

6. What are opioid peptides? How do they alter pain sensation?

MULTIPLE-CHOICE QUESTIONS

In questions 1–5, select the single best answer.

1. An anterolateral cordotomy is performed that produces relief of pain in the right leg. It is effective because it interrupts the
(A) left dorsal column
(B) left ventral spinothalamic tract
(C) right lateral spinothalamic tract
(D) left lateral spinothalamic tract
(E) right corticospinal tract

2. Stimulation of which of the following might be expected to produce itching?
(A) Dorsal root C fibers
(B) B fibers in peripheral nerves
(C) Dorsal columns of the spinal cord
(D) Touch receptors
(E) Motor fibers in the ventral roots of the spinal nerves

3. Visceral pain
(A) shows relatively rapid adaptation
(B) is mediated by B fibers in the dorsal roots of the spinal nerves
(C) can sometimes be relieved by applying an irritant to the skin
(D) most closely resembles "fast pain" produced by noxious stimulation of the skin
(E) can be produced by marked and prolonged stimulation of touch receptors

4. The distance by which 2 touch stimuli must be separated to be perceived as 2 separate stimuli is greatest on
(A) the lips
(B) the palm of the hand
(C) the back of the scapula
(D) the dorsum of the hand
(E) the tips of the fingers

5. Injecting a local anesthetic into an area to which visceral pain is referred sometimes abolishes the pain. When this occurs, it indicates that
(A) the sensory neurons involved send one branch to the viscus and another to the area to which the pain is referred
(B) the referral is psychosomatic
(C) the individual with the pain has previously had a painful injury in the area to which the pain is referred
(D) impulses in afferents from the viscus somehow facilitate transmission in the sensory pathway from the area to which the pain is referred
(E) the afferents from the viscus and those from the area to which the pain is referred end on the same second-order sensory neurons

In questions 6–10, select

A if the item is associated with (a) below,
B if the item is associated with (b) below,
C if the item is associated with both (a) and (b), and
D if the item is associated with neither (a) nor (b)

(a) Dorsal root C fibers
(b) A fibers

6. Cold

7. Warmth

8. Pain

9. Touch

10. Muscle strength

In questions 11–15, indicate whether the first item is greater than (G), *the same as* (S), *or less than* (L), *the second item.*

11. Vibration sensitivity in a patient with pernicious anemia

G S L

Vibration sensitivity in a patient with iron deficiency anemia

12. Rate of conduction in afferent neurons from warmth receptors

G S L

Rate of conduction in afferent neurons from pressure receptors

13. Loss of proprioceptive sensation in a patient with disease of the dorsal columns

G S L

Loss of pain sensation in the same patient

14. Deficit in processing of sensory information in sensory area I (SI) after lesion in SII

G S L

Deficit in processing of sensory information in sensory area II (SII) after lesion in SI

15. Effects of parietal lobe lesions on stereognosis

G S L

Effect of frontal lobe lesions on stereognosis

Chapter 8

This chapter reviews the functional anatomy of the visual system, then analyzes in order the processes involved in vision: formation of the visual image on the retina, conversion of light energy to electrical responses in the rods and cones, processing of impulses in the retina, transmission of impulses via the lateral geniculate bodies to the visual cortex, and the events that occur in the visual cortex. The material in the chapter should help students to—

• Describe the various parts of the eye, and list the functions of each.
• Trace the neural pathways that transmit visual information from the rods and cones to the visual cortex.
• Explain how light rays in the environment are brought to a focus on the retina and the role of accommodation in this process.
• Define hyperopia, myopia, astigmatism, presbyopia, and strabismus.
• Describe the electrical responses produced by rods and cones, and explain how these responses are produced.
• Describe the electrical responses seen in bipolar cells, horizontal cells, amacrine cells, and ganglion cells, and comment on the function of each type of cell.
• Describe the responses of cells in the visual cortex and the functional organization of the areas of the cerebral cortex that process visual information.
• Define and explain dark adaptation and visual acuity.
• Describe the functions of cones, the

neural pathways to the visual cortex, and the visual cortex in color vision.
- Name the 4 types of eye movements and the function of each.

GENERAL QUESTIONS

1. What type of visual field defect is produced by each of the following lesions and why?
   a. A lesion of one optic nerve
   b. A lesion of one optic tract
   c. A lesion of the optic chiasm
   d. Lesions in various parts of the geniculocalcarine tract
   e. Bilateral lesions destroying the visual cortex (area 17)

2. What is the near point of vision? Why does it recede throughout life?

3. How many different kinds of photosensitive pigments are found in the human retina? Discuss their chemistry, how they produce electrical responses in the retina, and their relation to color vision.

4. What are "on center" and "off center" cells? Define lateral inhibition, and comment on its general physiologic significance.

5. What are orientation columns and ocular dominance columns? How are they mapped experimentally? What are their functions?

6. Discuss parallel processing of visual information in the cerebral cortex.

7. What is binocular vision? Why is it limited to part of the visual field? Discuss its role and the role of monocular visual processes in depth perception.

8. What is amblyopia ex anopsia? Explain its pathogenesis.

MULTIPLE-CHOICE QUESTIONS

In the questions 1–6, select the single best answer.

1. If the principal focal distance of a lens is 0.75 m, its refractive power is
   (A) 0.25 diopter
   (B) 0.75 diopter
   (C) 1.0 diopter
   (D) 1.33 diopters
   (E) 10.3 diopters

2. Visual accommodation involves

   (A) increased tension on the lens ligaments
   (B) a decrease in the curvature of the lens
   (C) relaxation of the sphincter muscle of the iris
   (D) contraction of the ciliary muscle
   (E) increased intraocular pressure

3. When a visual stimulus falls on a given point in the retina for a long time
   (A) the image becomes more clearly focused
   (B) there is adaptation in the superior colliculus
   (C) the discharge rate in the bipolar cells increases
   (D) the pupils constrict
   (E) the image fades and disappears

4. The fovea of the eye
   (A) has the lowest light threshold
   (B) is the region of highest visual acuity
   (C) contains only red and green cones
   (D) contains only rods
   (E) is situated over the head of the optic nerve

5. Which of the following parts of the eye has the greatest concentration of rods?
   (A) Ciliary body
   (B) Iris
   (C) Optic disk
   (D) Fovea
   (E) Parafoveal region

6. The following events that occur in rods in response to light are listed in random sequence
   1. Activation of transducin
   2. Decreased release of synaptic transmitter
   3. Structural changes in rhodopsin
   4. Closure of Na^+ channels
   5. Decrease in intracellular cyclic GMP

   What is the sequence in which they normally occur?
   (A) 2, 1, 3, 5, 4
   (B) 1, 2, 3, 5, 4
   (C) 5, 3, 1, 4, 2
   (D) 3,1, 5, 4, 2
   (E) 3, 1, 4, 5, 2

In questions 7–10, one or more than one of the answers may be correct. Select

A if (1), (2), and (3) are correct;
B if (1) and (3) are correct;
C if (2) and (4) are correct;

D if only (4) is correct; and
E if all are correct

7. Vitamin A is a precursor for the synthesis of
   (1) somatostatin
   (2) cone pigments
   (3) the pigment of the iris
   (4) rhodopsin

8. The innervation of the eye and its muscles includes
   (1) afferent fibers mediating vision
   (2) somatic efferent fibers to the medial and lateral rectus muscles
   (3) visceral efferent fibers to the ciliary muscle
   (4) visceral efferent fibers to the iris muscle

9. The visual receptors connect *directly* to the ganglion cells in the retina via
   (1) horizontal cells
   (2) amacrine cells
   (3) cones
   (4) bipolar cells

10. Which of the following parts of the brain is/are concerned with processing visual information?
    (1) Medial geniculate body
    (2) Lateral portion of temporal lobe
    (3) Postcentral gyrus
    (4) Primary visual cortex (area 17)

In questions 11–15, indicate whether the first item is greater than (G), the same as (S), or less than (L), the second item.

11. Membrane potential of synaptic end of visual receptor in light

    G S L

    Membrane potential of synaptic end of visual receptor in darkness

12. Number of action potentials in right visual cortex after stimulation of medial half of right retina

    G S L

    Number of action potentials in left visual cortex after stimulation of medial half of right retina

13. Visual acuity in a person with 20/15 vision

    G S L

    Visual acuity in a person with 15/20 vision

14. Principal focal distance of lens used to correct defect in a patient with severe hyperopia

    G S L

Principal focal distance of lens used to correct defect in a patient with mild hyperopia

15. Response to colored stimuli of cells in visual cortex that stain for cytochrome oxidase

G S L

Response to colored stimuli of cells in visual cortex that do not stain for cytochrome oxidase.

Chapter 9

This chapter is concerned with the auditory system and the vestibular system. The material in the chapter should help students to—

- Describe the way that movements of molecules in the air are converted into impulses generated in hair cells in the cochlea.
- Describe how deformations in the basilar membrane are converted to impulses in auditory nerve fibers.
- Trace the path of auditory impulses in the neural pathways from the cochlear hair cells to the auditory cortex, and discuss the function of the auditory cortex.
- Explain how pitch, loudness, and timbre are coded in the auditory pathways.
- Describe the various forms of deafness.
- Explain how the receptors in the semicircular canals detect rotational acceleration and how the receptors in the saccule and utricle detect linear acceleration.
- List the major sensory inputs that provide the information which is synthesized in the brain into the sense of position in space.

GENERAL QUESTIONS

1. Otosclerosis is a disease in which the foot plate of the stapes becomes rigidly attached to the oval window. Why does this cause deafness?

2. What is the function of the tectorial membrane in the cochlea?

3. Which sound frequencies are audible to humans? At which sound frequency is the auditory threshold the lowest?

4. Define and explain masking.

5. What is the tympanic reflex, and what is its function?

6. Compare ossicular, air, and bone conduction. What is a fenestration operation, and why does it improve hearing in patients with conduction deafness?

7. Describe how a tuning fork can be used to distinguish between nerve deafness and conduction deafness.

8. If the vision of individuals lacking vestibular function is obscured while they are diving in water, they may swim away from rather than toward the surface and drown. Why does this occur?

MULTIPLE-CHOICE QUESTIONS

In questions 1–7, select the single best answer.

1. Sound intensity is measured in
   (A) diopters
   (B) daltons
   (C) torrs
   (D) decibels
   (E) pounds

2. In humans, the primary auditory cortex is located in the
   (A) limbic system
   (B) posterior part of the occipital lobe
   (C) posterior part of the parietal lobe
   (D) postcentral gyrus
   (E) superior part of the temporal lobe

3. Postrotatory nystagmus is caused by
   (A) continued movement of perilymph over hair cells that have their processes embedded in the tectorial membrane
   (B) continued movement of aqueous humor over the ciliary body in the eye
   (C) continued movement of cerebrospinal fluid over the parts of the brain stem that contain the vestibular nuclei
   (D) continued movement of endolymph in the semicircular canals, with consequent bending of the cupula and stimulation of hair cells
   (E) a disease process affecting the vestibular system

4. The basilar membrane of the cochlea

   (A) is unaffected by movement of fluid in the scala vestibuli
   (B) covers the oval window and the round window
   (C) vibrates in a pattern determined by the form of the traveling wave in the fluids in the cochlea
   (D) is under tension
   (E) vibrates when the body is subjected to linear acceleration

5. Which of the following are *incorrectly* paired?
   (A) Timbre (quality) of sound : harmonic vibrations
   (B) Pitch of sound : frequency of sound wave
   (C) Pitch of sound : point of greatest displacement of the basilar membrane
   (D) Direction of sound : frequency of impulses in an auditory nerve fiber
   (E) Loudness of sound : amplitude of sound wave

6. Which of the following are *incorrectly* paired?
   (A) Tympanic membrane : manubrium of malleus
   (B) Helicotrema : apex of cochlea
   (C) Foot plate of stapes : oval window
   (D) Otoliths : semicircular canals
   (E) Basement membrane : cochlea

7. The direction of nystagmus is vertical when a subject is rotated
   (A) after warm water is put in one ear
   (B) with the head tipped backward
   (C) after cold water is put in both ears
   (D) with the head tipped sideways
   (E) after section of one vestibular nerve

In questions 8–12, which refer to Fig 9–A, select

A if the item is associated with **(a)** below,
B if the item is associated with **(b)** below,
C if the item is associated with both **(a)** and **(b)**, and
D if the item is associated with neither **(a)** nor **(b)**

(a) Sound transmission
(b) Linear acceleration

8. Structure V

9. Structure W

10. Structure X

11. Structure Y

Figure 9–A. Human ear.

In questions 13–15, indicate whether the first item is greater than (G), the same as (S), or less than (L), the second item.

13. K$^+$ concentration in perilymph

G S L

K$^+$ concentration in endolymph

14. Maximum sound frequency audible to humans

G S L

Maximum sound frequency audible to dogs

15. Change in membrane potential when stereocilia of a saccular hair cell are pushed toward the kinocilium

G S L

Change in membrane potential when stereocilia of a saccular hair cell are pushed at a right angle to the kinocilium

Chapter 10

This chapter considers olfaction and taste from the point of view of signal transduction, pathways to the CNS, central representation, and the molecular and neural mechanisms that make it possible to discriminate among different odors and tastes. The material in the chapter should help students to—

• Describe the olfactory receptors and the way in which impulses are initiated in them.

• Outline the pathways by which impulses generated in the olfactory

mucous membrane reach the cerebral cortex.

• Describe and analyze olfactory sensitivity, discrimination, and adaptation.

• Describe the taste buds, and discuss their characteristics and distribution in relation to the 4 basic tastes.

• Outline the taste pathways.

• List the substances that produce sweet, sour, bitter, and salt tastes, and comment on how the signal for each is transduced.

GENERAL QUESTIONS

1. Compare and contrast smell and taste. How do they interact?

2. Humans can discriminate among many different odors and many different tastes. What are the mechanisms that make this possible?

3. Why does sniffing improve olfactory sensitivity and discrimination?

4. Discuss the relationship between olfaction and sexual behavior.

5. What part do pain fibers in the trigeminal nerves play in olfaction?

6. Name the chemoreceptors in the body.

7. A number of seemingly unrelated substances other than sugars taste sweet. List some of these compounds, and comment on their use in the treatment of various conditions and diseases.

8. Discuss taste as a motivation for aversive learning.

MULTIPLE-CHOICE QUESTIONS

In questions 1–3, select the single best answer.

1. Impulses generated by olfactory receptors in the nasal mucous membrane
   (A) are relayed to the thalamus
   (B) pass through the internal capsule
   (C) are relayed to the olfactory cortex via the hypothalamus
   (D) pass to the mitral cells and from there directly to the olfactory cortex
   (E) pass to the mitral cells and from there to the olfactory cortex via the taste area

2. Impulses generated in the taste buds of the tongue reach the cerebral cortex via the
   (A) thalamus
   (B) internal capsule
   (C) dorsal roots of the first cervical spinal nerves
   (D) trochlear nerve
   (E) hypoglossal nerve

3. The number of different odors that can be distinguished by humans is
   (A) less than 50
   (B) about 10
   (C) 200–500
   (D) 900–1000
   (E) 2000–4000

In questions 4–8, match the sensation in each question with the lettered signal transduction mechanism that is most closely associated with it. Each lettered mechanism may be selected once, more than once, or not at all.

(A) G protein-adenylate cyclase
(B) G protein-phospholipase C
(C) Activation of tyrosine kinase
(D) Na$^+$ influx via amiloride-sensitive channels
(E) H$^+$ block of K$^+$ channels

4. Salty

5. Bitter

6. Sour

7. Sweet

8. Odor

In questions 9 and 10, indicate whether the first item is greater than (G), the same as (S), or less than (L), the second item.

9. Olfactory sensitivity in a 20-year-old woman

G S L

Olfactory sensitivity in a 70-year-old woman

10. Taste threshold concentration of saccharin (in moles)

G S L

Taste threshold concentration of glucose (in moles)

Chapter 11

This chapter deals with alertness and sleep, the mechanisms that produce these states, and the correlation between them and the electrical activity of the brain. The material in the chapter should help students to—

- Describe the anatomy and functions of the reticular formation.
- Define and explain primary evoked potentials and diffuse secondary responses.
- Describe the primary types of rhythms that make up the EEG and the behavioral states that correlate with each.
- Define and explain synchronization, α block, and desynchronization.
- Summarize the behavioral and EEG characteristics of each of the stages of slow-wave sleep.
- Summarize the electroencephalographic and other characteristics of rapid eye movement (REM) sleep, and describe the mechanisms responsible for its production.
- Describe the pattern of normal nighttime sleep in adults and the variations in this pattern from birth to old age.
- List the main clinical uses of the EEG.

GENERAL QUESTIONS

1. What are the main electrical events in the cortex that produce the EEG?

2. Discuss general anesthesia. How is it produced? What happens to transmission in specific and nonspecific pathways during anesthesia?

3. Define somnambulism and narcolepsy, and discuss their relationship to normal sleep.

4. Compare REM and non-REM sleep.

5. What is the difference between the specific sensory relay nuclei and the nonspecific projection nuclei of the thalamus?

6. Discuss the relation of drugs and neurotransmitters to sleep.

7. What is recurrent collateral inhibition? What is its physiologic significance?

MULTIPLE-CHOICE QUESTIONS

In questions 1–6, select the single best answer.

1. In a healthy, alert adult sitting with the eyes closed, the dominant EEG rhythm observed with electrodes over the occipital lobes is
   (A) delta (0.5–4 Hz)
   (B) theta (4–7 Hz)
   (C) alpha (8–13 Hz)
   (D) beta (18–30 Hz)
   (E) fast, irregular low-voltage activity

Figure 11–A.

2. The EEG record in Fig 11–A is characteristic of
   (A) deep sleep
   (B) a learning response produced by a painful stimulus
   (C) REM sleep
   (D) a psychomotor seizure
   (E) an absence seizure

Electrooculogram
Electromyogram
Electroencephalogram

Figure 11–B. Reproduced, with permission, from Kales A et al: *Ann Intern Med* 1968;**68**:1078.

3. The electrical records in Fig 11–B are characteristic of
   (A) deep sleep
   (B) grand mal epilepsy
   (C) REM sleep
   (D) psychomotor seizures
   (E) an absence seizure

4. High-frequency stimulation of which of the following does *not* produce desynchronization?
   (A) Sciatic nerve
   (B) Lateral spinothalamic tract
   (C) Midbrain reticular formation
   (D) Medial lemniscus above the midbrain
   (E) Intralaminar nuclei of the thalamus

5. Stimulation of which of the following brain areas can produce sleep?
   (A) Posterior hypothalamus
   (B) Occipital cortex
   (C) Medullary reticular formation near the raphe nuclei
   (D) Gracile and cuneate nuclei
   (E) Deep nuclei of the cerebellum

6. Regular rhythmic fluctuations in electrical activity are observed in which part of the brain?
   (A) Mediobasal portion of the hypothalamus
   (B) Cerebellar cortex
   (C) Midbrain reticular formation
   (D) Amygdala
   (E) Pons

In questions 7–11, select—
   A if the item is associated with (**a**) below,
   B if the item is associated with (**b**) below,
   C if the item is associated with both (**a**) and (**b**), and
   D if the item is associated with neither (**a**) nor (**b**)

   (**a**) Slow-wave sleep
   (**b**) REM sleep

7. Dreaming

8. Sleep spindles

9. Sleepwalking

10. Bruxism

11. High threshold for arousal

*In questions 12–15, indicate whether the first item is greater than (**G**), the same as (**S**), or less than (**L**), the second item.*

12. Frequency of EEG waves in a subject in coma

   G S L

   Frequency of EEG waves in a subject in REM sleep

13. Frequency of episodes of REM sleep after first falling asleep

   G S L

   Frequency of episodes of REM sleep after 6 hours of sleep

14. Degree of synchrony during slow-wave sleep

   G S L

   Degree of synchrony during alert attention

15. Number of different types of sensory input producing responses in a thalamocortical neuron in the ventrobasal nuclei of the thalamus

   G S L

   Number of different types of sensory input producing responses in a thalamocortical neuron in the midline nuclei of the thalamus

Chapter 12

This chapter is a review of somatic motor function, with emphasis on the contributions of the spinal cord, medulla, midbrain, basal ganglia, cerebral cortex, and cerebellum. The material in the chapter should help students to—

- Describe in general terms how posture and movement are regulated, and outline the function of each of the main components of the regulatory systems. Define the concept of release.
- Discuss the function of the cerebral cortex, cerebellum, basal ganglia, and corticospinal and corticobulbar tracts in skilled voluntary movement.
- Define spinal shock, and explain the initial and long-term changes in spinal reflexes that follow transection of the spinal cord.
- List and explain the long-term complications that make it difficult to treat patients with spinal cord injuries.
- Define decerebrate and decorticate rigidity, and comment on the cause and physiologic significance of each.
- Describe the postural reflexes that are integrated in the medulla oblongata, the pons, the midbrain, and the cerebral cortex.
- Describe the basal ganglia, and list the pathways that interconnect them, along with the neurotransmitters in each pathway.
- Describe and explain the symptoms of Parkinson's disease.
- List the pathways to and from the cerebellum and the connections of each within the cerebellum.
- Discuss the functions of the cerebellum and the neurologic abnormalities produced by diseases of this part of the brain.

GENERAL QUESTIONS

1. What is meant by the terms upper motor neuron and lower motor neuron? Contrast the effects of lower motor neuron lesions with those of lesions affecting each of the types of upper motor neurons.

2. What is encephalization? Discuss its significance.

3. What is the Babinski sign? What is its physiologic and pathologic significance?

4. Describe the mass reflex, a distressing late complication of injuries that transect the spinal cord, and explain why it occurs.

5. What is the evidence that the hopping and placing reflexes are cortical in origin?

6. Define athetosis, ballism, and chorea, and describe the disease processes that produce each of them.

7. List 3 drugs used in the treatment of Parkinson's disease, and explain why each is of value.

8. List 5 types of neurons found in the cerebellar cortex, and describe the morphology and function of each.

9. What is an intention tremor? Why does it occur in cerebellar disease?

MULTIPLE-CHOICE QUESTIONS

Questions 1–4 refer to Fig 12–A. In each question, lettered items may be selected once or more than once.

Figure 12–A. Lateral view of the human cerebral cortex.

1. From which of the labeled structures do the corticospinal and corticobulbar tracts originate?

2. Which of the labeled structures is/are concerned with elaboration of somatic sensory perception?

3. Which of the labeled structures is/are concerned with perception of taste stimuli?

4. Which of the labeled structures is/are association areas?

In questions 5–10, select the single best answer.

5. A primary function of the basal ganglia is
(A) sensory integration
(B) short-term memory
(C) planning voluntary movement
(D) neuroendocrine control
(E) slow-wave sleep

6. Interruption of the motor pathways in the internal capsule on one side of the body causes
(A) spastic paralysis on the same side of the body
(B) spastic paralysis on the opposite side of the body
(C) loss of touch and pressure sensation on the opposite side of the body
(D) flaccid paralysis on the same side of the body
(E) flaccid paralysis on the opposite side of the body

7. Increased neural activity before a skilled voluntary movement is *first* seen in the
(A) spinal motor neurons
(B) precentral motor cortex
(C) basal ganglia
(D) cerebellum
(E) cortical association areas

8. After falling down a flight of stairs, a young woman is found to have partial loss of voluntary movement on the right side of her body and loss of pain and temperature sensation on the left side below the midthoracic region. It is probable that she has a lesion
(A) transecting the left half of the spinal cord in the lumbar region
(B) transecting the left half of the spinal cord in the upper thoracic region
(C) transecting sensory and motor pathways on the right side of the pons
(D) transecting the right half of the spinal cord in the upper thoracic region
(E) transecting the dorsal half of the spinal cord in the upper thoracic region

9. Patients with transected spinal cords frequently have a negative nitrogen balance because
(A) they develop hypercalcemia, and this causes dissolution of the protein in bone
(B) they are paralyzed below the level of the transection
(C) they lack the afferent input that normally maintains growth hormone secretion
(D) they have difficulty voiding, and this causes nitrogen to

accumulate in the urine in the bladder

(E) their ACTH response to stress is reduced

10. Which of the following is *not* involved in producing the continuous contraction of antigravity muscles that occurs during standing?

(A) Tonically increased discharge in the γ efferent neurons to the spindles in these muscles

(B) Increased discharge in the α efferent neurons to the extrafusal fibers in these muscles

(C) Increased discharge of neurons in the brain stem

(D) Otoliths

(E) Cerebral cortex

In questions 11–15, match the item in each question with the lettered structure that is most closely associated with it. Each lettered structure may be selected once, more than once, or not at all.

(A) Spinal cord
(B) Spinocerebellum
(C) Hypothalamus
(D) Limbic system
(E) Basal ganglia

11. Emotional responses

12. Maintenance of body temperature

13. Planning of voluntary movements

14. Intention tremor

15. Huntington's disease

Chapter 13

This chapter is concerned specifically with the motor component of the peripheral autonomic nervous system that innervates the viscera. Visceral afferents are discussed in Chapters 5 and 7, and the central regulation of visceral function in Chapter 14. The material in the chapter should help students to—

• Describe the location of preganglionic sympathetic and parasympathetic neurons in the CNS and identify the nerves by which their axons leave the CNS.

• Describe the location of postganglionic sympathetic and parasympathetic neurons and the pathways

they take to the visceral structures they innervate.

• Separate the autonomic nervous system into divisions on the basis of the neurotransmitters secreted by the neurons in it, and compare this classification to the anatomically defined sympathetic and parasympathetic divisions.

• Outline the functions of autonomic nervous system.

• List the ways that drugs act to increase or decrease sympathetic activity, parasympathetic activity, or both.

GENERAL QUESTIONS

1. Cannon called mass sympathetic discharge the "preparation for flight or fight." How does sympathetic discharge prepare the individual for flight or fight? Does the sympathetic nervous system have any other functions?

2. The cholinergic division of the autonomic nervous system has been called the anabolic nervous system. Discuss the actions of the cholinergic division that justify this label.

3. Where are the cell bodies of the postganglionic sympathetic neurons that supply the head? How do they reach the visceral effectors they innervate?

4. Which neurotransmitters or hormones are secreted by each of the following?
a. Preganglionic autonomic neurons
b. Postganglionic sympathetic neurons innervating intestinal smooth muscle
c. Postganglionic sympathetic neurons innervating sweat glands
d. Postganglionic parasympathetic neurons
e. Small, intensely fluorescent (SIF) cells
f. Adrenal medulla

MULTIPLE-CHOICE QUESTIONS

In questions 1 and 2, select the single best answer.

1. Complete denervation of the small intestine would be expected to
(A) increase the secretion of gastrointestinal hormones

(B) decrease the resting rate of intestinal peristalsis
(C) have little effect on the resting rate of intestinal peristalsis
(D) cause intestinal obstruction
(E) cause peristalsis to become chaotic and irregular

2. Which of the following drugs would *not* be expected to increase sympathetic discharge or mimic the effects of increased sympathetic discharge?
(A) Norepinephrine
(B) Neostigmine
(C) Amphetamine
(D) Isoproterenol
(E) Atropine

In questions 3–8, one or more than one of the answers may be correct. Select

A if (1), (2), and (3) are correct;
B if (1) and (3) are correct;
C if (2) and (4) are correct;
D if only (4) is correct; and
E if all are correct

3. Administration of physostigmine, a drug that inhibits acetylcholinesterase, would be expected to
(1) increase the secretion of gastric juice
(2) increase the rate of melatonin synthesis and secretion from the pineal gland
(3) increase the rate of epinephrine secretion from the adrenal medulla
(4) decrease the secretion of glucagon

4. In low doses, nicotine
(1) acts directly on intestinal smooth muscle, causing it to contract
(2) increases the rate of discharge of postganglionic parasympathetic neurons
(3) decreases the release of acetylcholine in sympathetic ganglia
(4) increases the rate of discharge of postganglionic sympathetic neurons

5. In the parasympathetic nervous system
(1) the axons of the preganglionic neurons are in the dorsal roots of the spinal nerves in the sacral region
(2) the axons of the preganglionic neurons are in the ventral roots of the spinal nerves in the lower thoracic and lumbar regions
(3) dopamine is the neurotrans-

mitter responsible for transmission from preganglionic neurons to postganglionic neurons

(4) the axons of postganglionic neurons are usually shorter than those of preganglionic neurons

6. Administration of a drug that blocks conversion of L-dopa to dopamine would be expected to
   (1) increase the diameter of the trachea and bronchi
   (2) disrupt the function of the SIF cells in the sympathetic ganglia
   (3) decrease peristaltic activity in the small intestine
   (4) decrease the amount of norepinephrine in the circulating blood

7. Administration of a drug that blocks β-adrenergic receptors would be expected to
   (1) decrease the heart rate
   (2) decrease the force of cardiac contraction
   (3) decrease the secretion of renin from the kidneys
   (4) decrease the secretion of insulin from the B cells in the pancreatic islets

8. Polypeptides found in the autonomic nervous system include
   (1) gonadotropin-releasing hormone (GnRH)
   (2) neuropeptide Y
   (3) gastrin-releasing peptide (GRP)
   (4) vasoactive intestinal polypeptide (VIP)

In questions 9 and 10, indicate whether the first item is greater than (G), the same as (S), or less than (L), the second item.

9. Number of C fibers in white ramus communicans

   G S L

   Number of C fibers in gray ramus communicans

10. Number of B fibers in white ramus communicans

   G S L

   Number of B fibers in gray ramus communicans

Chapter 14

This chapter reviews central integration of autonomic and related visceral functions at the level of the spinal cord, medulla oblongata, and hypothalamus. The material in the chapter should help students to—

- Describe the autonomic reflexes integrated at the level of the spinal cord.
- Describe vomiting, including its autonomic and somatic components, and discuss the initiation of the vomiting reflex by afferents from the gastrointestinal tract and the chemoreceptor trigger zone.
- Describe the anatomic connections between the hypothalamus and the pituitary gland and the functional significance of each connection.
- Discuss the relationship of the hypothalamus to sleep and to body rhythms.
- Describe the contribution of the hypothalamus and the role of other factors in the regulation of food intake.
- List the factors that control water intake, and outline the way they exert their effects.
- Discuss the synthesis, processing, storage, and secretion of the hormones of the posterior pituitary.
- List the effects of vasopressin, and discuss how vasopressin secretion is regulated.
- List the effects of oxytocin, and discuss how oxytocin secretion is regulated.
- Name the hypophyseotropic hormones, and outline the effects that each has on anterior pituitary function.
- List the mechanisms by which heat is produced in and lost from the body, and comment on the differences in temperature in the hypothalamus, rectum, oral cavity, and skin.
- List the temperature-regulating mechanisms, and describe the way in which they are integrated under hypothalamic control to maintain normal body temperature.
- Discuss the pathophysiology of fever.

GENERAL QUESTIONS

1. What is endogenous pyrogen? Describe its production, and discuss its effects.

2. What is behavioral thermoregulation? Give examples, and describe how they operate to maintain body temperature.

3. What functions do the hypophyseotropic hormones have in addition to regulation of anterior pituitary secretion?

4. What endocrine changes would you expect to see after section of the pituitary stalk, and why?

5. Draw graphs showing the relation of thirst to plasma osmolality, plasma vasopressin to plasma osmolality, and plasma vasopressin to extracellular fluid volume.

6. Why does food intake increase in untreated diabetes mellitus?

Figure 14–A.

7. Using the axes in Fig 14–A, draw the curve relating metabolic rate to body temperature from 95 to 106 °F (35–41 °C).

8. Some individuals drink large quantities of water for psychologic reasons (psychogenic polydipsia). How would you differentiate among psychogenic polydipsia, diabetes insipidus due to a hypothalamic lesion, and diabetes insipidus due to an inability of the kidneys to respond to vasopressin (nephrogenic diabetes insipidus)?

9. What is the effect of bilateral lesions of the area postrema on the vomiting reflex?

MULTIPLE-CHOICE QUESTIONS

In questions 1–3, select the single best answer.

1. Thirst is stimulated by
   (A) increases in plasma osmolality and volume

(B) an increase in plasma osmolality and a decrease in plasma volume

(C) a decrease in plasma osmolality and an increase in plasma volume

(D) decreases in plasma osmolality and volume

(E) injection of vasopressin into the hypothalamus

2. The appetite for food is *not* increased by
   (A) hypoglycemia
   (B) decreased glucose utilization in the ventromedial nuclei
   (C) insulin deficiency
   (D) distention of the stomach
   (E) a fall in body temperature

3. When an individual is standing naked in a room in which the air temperature is 21 °C (69.8 °F) and the humidity 80%, the greatest amount of heat will be lost from the body by
   (A) elevated metabolism
   (B) respiration
   (C) urination
   (D) vaporization of sweat
   (E) radiation and conduction

Questions 4–8 refer to Fig 14–B. Each lettered structure may be selected once, more than once, or not at all.

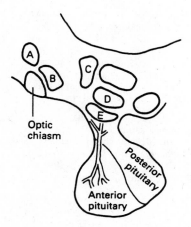

Figure 14–B. Hypothalamus and pituitary.

4. In which of the labeled structures is there the highest concentration of angiotensin II?

5. In which of the labeled structures is there a high concentration of corticotropin-releasing hormone (CRH)?

6. In which of the labeled structures

is there a high concentration of thyrotropin-releasing hormone (TRH)?

7. In which of the labeled structures is there a high concentration of dopamine?

8. In which of the labeled structures is there a high concentration of glucose-sensitive neurons?

In questions 9–15, match the functions in each question with the lettered part of the brain that is most closely associated with it. Each lettered part may be selected once, more than once, or not at all.

   (A) Spinal cord
   (B) Medulla oblongata
   (C) Midbrain
   (D) Hypothalamus
   (E) Amygdala

9. Fine control of blood pressure

10. Satiety

11. Primitive sexual reflexes

12. Vomiting

13. Fear

14. Water intake

15. Shivering

In questions 16–20, select

   A if the item is associated with **(a)** below,
   B if the item is associated with **(b)** below,
   C if the item is associated with both **(a)** and **(b)**, and
   D if the item is associated with neither **(a)** nor **(b)**

   (a) V_1 vasopressin receptors
   (b) V_2 vasopressin receptors

16. Activation of G_s

17. Vasoconstriction

18. Increase in intracellular inositol triphosphate

19. Antidiuresis

20. Glycosuria

Chapter 15

This chapter is concerned with the limbic system and with the neural circuits involved in motivation, emotional responses such as fear and rage, and sexual and other instinctive and

stereotype behaviors. The material in the chapter should help students to—

- Describe in general terms the structure and function of the limbic system.
- Discuss the brain regions and hormones involved in the regulation of sexual behavior in both sexes.
- Summarize the effects of sex hormones on the brain in fetal and early neonatal life.
- Describe the brain systems that mediate repeated self-stimulation and avoidance of stimulation.
- Outline the anatomy of the serotonergic pathways in the brain, and summarize their known and suspected functions.
- Outline the anatomy of the noradrenergic (norepinephrine-secreting) pathways in the brain, and summarize their known and suspected functions.
- Outline the anatomy of the adrenergic (epinephrine-secreting) pathways in the brain, and summarize their known and suspected functions.
- Outline the anatomy of the dopaminergic pathways in the brain, and summarize their known and suspected functions.
- Outline the anatomy of the histaminergic pathways in the brain, and summarize their known and suspected functions.
- Outline the opioid peptide-secreting pathways in the brain, and summarize their known and suspected functions.

GENERAL QUESTIONS

1. Discuss sham rage. How is it produced? What is the evidence indicating that the word sham is inappropriate?

2. What is a pheromone? What are the sources and function of pheromones in monkeys and humans?

3. Which brain areas and transmitters appear to mediate addiction to cocaine and other addicting drugs?

4. Discuss the role of testicular, ovarian, and adenocortical hormones in the regulation of sexual behavior in women.

5. Some scientists talk about the "female hypothalamus" and the "male hypothalamus." What are the differences between the hypo-

thalamus of the 2 sexes, and how are these differences produced?

6. What is the putative relationship between norepinephrine and mood? What pharmacologic agents have been used to treat depression, and why?

7. What are the relationships of brain dopamine and brain serotonin to mental abnormalities and other diseases?

MULTIPLE-CHOICE QUESTIONS

Questions 1–6 are made up of a statement and an explanation. Select

A if both the statement and the explanation are true and they are related,

B if both the statement and explanation are true but they are unrelated,

C if the statement is true but the explanation is false,

D if the statement is false but the explanation is true, and

E if the statement and the explanation are both false

1. L-dopa is of benefit in the treatment of Parkinson's disease because it increases the dopamine content of the tuberoinfundibular dopaminergic neurons

2. Drugs that decrease norepinephrine reuptake are probably of benefit in the treatment of some depressive illnesses because they increase extracellular norepinephrine in the brain

3. Procedures that increase the release of opioid peptides in the CNS decrease the effect of painful stimuli because the periaqueductal gray and related areas contain opioid receptors to which the peptides can bind

4. The antipsychotic activity of the major tranquilizers parallels their ability to block serotonin receptors because the serotonergic neural systems in the brain are concerned, among other things, with normal mental function

5. Women with a congenital adrenocortical enzyme defect that causes large amounts of androgens to be secreted have somewhat more masculine behavior than normal women because the androgens act on the brain during fetal development

6. "Sham rage" is sometimes seen in patients after encephalitis or brain surgery because these conditions can damage pathways from the neocortex to the limbic system

In questions 7–10, select the single best answer.

7. Which of the following psychotomimetic drugs does *not* appear to produce its effects via serotonin receptors?
   (A) Lysergic acid diethylamide (LSD)
   (B) Psilocin
   (C) N,N-dimethyltryptamine (DMT)
   (D) Mescaline
   (E) Amphetamine

8. Benzodiazepines such as diazepam (Valium) relieve anxiety because they
   (A) inhibit dopamine binding in the brain stem
   (B) inhibit glutamate binding in the limbic system
   (C) facilitate norepinephrine release in the hypothalamus
   (D) facilitate the effects of GABA throughout the nervous system
   (E) block muscarinic acetylcholine receptors in the limbic system

9. Which of the following would be expected to reduce maternal behavior?
   (A) Lesions of the occipital cortex
   (B) Lesions of the mediobasal hypothalamus
   (C) Hypophysectomy
   (D) Ovariectomy
   (E) Hysterectomy

10. The cell bodies of the principal histaminergic neurons in the brain are located in the
   (A) preoptic area of the hypothalamus
   (B) central nucleus of the amygdala
   (C) subfornical organ
   (D) locus ceruleus
   (E) mediobasal hypothalamus

In questions 11–15, indicate whether the first item is greater than (G), the same as (S), or less than (L), the second item.

11. Area of brain where animals will avoid stimulation through an implanted electrode

   G S L

   Area of brain where animals will

work to receive repeated stimulation through an implanted electrode

12. Amount of norepinephrine in the locus ceruleus

   G S L

   Amount of dopamine in the locus ceruleus

13. Amount of norepinephrine in the raphe nuclei

   G S L

   Amount of serotonin in the raphe nuclei

14. Amount of norepinephrine in the nigrostriatal pathway

   G S L

   Amount of dopamine in the nigrostriatal pathway

15. Rage behavior before removal of testes

   G S L

   Rage behavior after removal of testes

Chapter 16

This chapter is concerned with the "higher functions of the nervous system," and specifically with learning, memory, the specialized natures of the left and right cerebral hemispheres, and the language functions of the human brain. The material in the chapter should help students to—

- Give examples of habituation and sensitization occurring in humans, and discuss the molecular events that may underlie them.
- Define conditioned reflexes, describe their properties, and analyze their physiologic basis.
- List and explain the abnormalities of brain function produced by sectioning the corpus callosum.
- List the parts of the brain that appear to be involved in memory in mammals, and summarize the proposed role of each in memory processing and storage.
- Describe the abnormalities of brain structure and function found in Alzheimer's disease.
- Define the terms categorical hemisphere and representational hemisphere, and summarize the difference between the hemispheres and their relationship to handedness.

- Define and explain agnosia, unilateral neglect, dyslexia, and prosopagnosia.
- Summarize the differences between fluent and nonfluent aphasia, and explain each type on the basis of its pathophysiology.

GENERAL QUESTIONS

1. What methods are used to study learning and memory in experimental animals and humans?

2. Why does a mother often sleep through many different kinds of noise but wake when her baby cries?

3. What is biofeedback? Explain it in physiologic terms.

4. List and explain the differences between declarative and nondeclarative memory.

5. It is now established that the 2 cerebral hemispheres are not anatomically symmetrical. What are the anatomic differences between the 2 sides, and what is their relation to the physiologic differences between the hemispheres?

6. Discuss the functions of Wernicke's area, Broca's area, the angular gyrus, and the basal ganglia in the production of speech.

MULTIPLE-CHOICE QUESTIONS

In questions 1–8, select the single best answer.

1. Retrograde amnesia
   (A) is abolished by prefrontal lobotomy
   (B) responds to drugs that block dopamine receptors
   (C) is commonly precipitated by a blow on the head
   (D) is increased by administration of vasopressin
   (E) is due to damage to the brain stem

2. The representational hemisphere is
   (A) the right cerebral hemisphere in most right-handed individuals
   (B) the left cerebral hemisphere in most left-handed individuals
   (C) the part of the brain con-

cerned with language functions
   (D) the site of lesions in most patients with aphasia
   (E) morphologically identical to the opposite nonrepresentational hemisphere

3. The optic chiasm is sectioned in a dog, and with the right eye closed, the animal is trained to bark when it sees a red square. The right eye is then opened and the left eye closed. The animal will now
   (A) fail to respond to the red square because the square does not produce impulses that reach the right occipital cortex
   (B) fail to respond to the red square because the animal has bitemporal hemianopia
   (C) fail to respond to the red square if the posterior commissure is also sectioned
   (D) respond to the red square only after retraining
   (E) respond promptly to the red square in spite of the lack of input to the left occipital cortex

4. The effects of bilateral loss of hippocampal function include
   (A) disappearance of remote memories
   (B) inability to recall the events of the moment
   (C) loss of short-term memory, ie, inability to encode events of the recent past in long-term memory
   (D) loss of the ability to recall faces and forms but not the ability to recall printed or spoken words
   (E) production of inappropriate emotional responses when recalling events of the recent past

5. Which of the following are *incorrectly* paired?
   (A) Lesion of the inferior parietal lobule of the representational hemisphere : unilateral inattention and neglect
   (B) Loss of cholinergic neurons in the nucleus basalis of Meynert and related areas of the forebrain : loss of recent memory
   (C) Lesions of mamillary bodies : loss of recent memory
   (D) Lesion of the angular gyrus in the categorical hemisphere : nonfluent aphasia
   (E) Lesion of Broca's area in the

categorical hemisphere : slow speech

6. The representational hemisphere is better than the categorical hemisphere at
   (A) language functions
   (B) recognition of objects by their form
   (C) understanding printed words
   (D) understanding spoken words
   (E) mathematical calculations

7. Which of the following is not characteristic of conditioned reflexes?
   (A) Failure to form conditioned reflexes when the conditioned and unconditioned stimuli are separated by more than 2 minutes
   (B) Disappearance of the conditioned reflex if it is not reinforced from time to time
   (C) Rapid formation of conditioned reflexes when the conditioned stimulus is pleasant for the animal
   (D) Failure to form conditioned reflexes when the conditioned stimulus is unpleasant to the animal
   (E) Failure to form conditioned reflexes when after the conditioned stimulus, there is a distracting stimulus before the unconditioned stimulus

8. A lesion of Wernicke's area (the posterior end of the superior temporal gyrus) in the categorical hemisphere causes patients to
   (A) lose short-term memory
   (B) speak in a slow, halting voice
   (C) experience deja vu
   (D) talk rapidly but make little sense
   (E) lose the ability to recognize faces

In questions 9–15, match the structure in each question with the lettered item that is most clearly associated with it. Each lettered item may be selected once, more than once, or not at all.

   (A) Experimental neurosis
   (B) Short-term memory
   (C) Asteriognosis
   (D) Unilateral neglect
   (E) Nondeclarative memory

9. Hypothalamus

10. Frontal lobe

11. Hippocampus

12. Temporal lobe

13. Basal ganglia

14. Entorhinal cortex

15. Cerebellum

Chapter 17

This chapter provides a summary and overview of energy balance and metabolism as a background for consideration of the functions of the endocrine glands. The topics considered are energy balance; the metabolic rate; the genesis and functions of ATP; the metabolism of carbohydrates, proteins, and fats; important fat derivatives, including leukotrienes and prostaglandins; and nutrition. The material in the chapter should help students to—

- Define metabolic rate, calorie, and respiratory quotient.
- State the caloric value per unit weight of carbohydrate, protein, and fat.
- Describe how the basal metabolic rate is measured, and list some of the major diseases in which it is abnormal.
- Discuss the biosynthesis and functions of ATP.
- List the principal dietary hexoses, and summarize the main mechanisms that regulate glucose synthesis and breakdown.
- Describe the function of the citric acid cycle in the metabolism of glucose, amino acids, and fatty acids.
- Define the glucostatic function of the liver, and explain how the liver carries out this function.
- Define amino acid, polypeptide, and protein.
- List the main sources of uric acid in the body and the main metabolic products produced from uric acid.
- Describe the metabolic responses to starvation.
- List the major classes of lipids in the body, and describe the principal characteristics of each.
- Define ketone bodies, and describe their formation and metabolism.
- Describe the sources of the free fatty acids in plasma, their metabolic fate, and the principal factors regulating the fatty acid level in plasma.
- Describe the exogenous and endogenous pathways by which lipids are transported in the body, and summarize the processes involved in cholesterol metabolism.
- Outline the major pathways involved in the formation of leukotrienes, lipoxins, thromboxanes, prostacyclin, and prostaglandins, and list the functions of each of these eicosanoids.
- List the components in a normal diet that will maintain weight and health.
- Define the term vitamin, name the major vitamins, and summarize the effects of deficiency and excess of each vitamin.

GENERAL QUESTIONS

1. The oxygen consumption of a fasting animal at rest was found to be 30 mL/kg body weight/min. What is the basal metabolic rate (BMR) (in kcal/kg/24 h)? What is the approximate size of this animal?

2. What is specific dynamic action? How is it produced?

3. What is a "directional flow valve" in metabolism? Why are such valves important?

4. Discuss the changes in carbohydrate metabolism that occur during exercise.

5. Compare the metabolism of fructose with that of glucose. How do they interact?

6. Discuss gout.

7. What are essential amino acids?

8. What treatment would you prescribe to lower plasma cholesterol? Why is a reduced cholesterol intake important? Discuss the relation of cholesterol to low-density lipoproteins (LDL) and high-density lipoproteins (HDL).

9. Which metabolites of arachidonic acid are important in the body? Why has a major effort been devoted to the development of drugs that modify arachidonic acid metabolism?

10. Discuss the pathophysiology of obesity. How would you treat this condition?

MULTIPLE-CHOICE QUESTIONS

In questions 1–9, select the single best answer.

1. In the body, metabolism of 10 g of protein would produce approximately
   (A) 1 kcal
   (B) 41 kcal
   (C) 410 kcal
   (D) 4100 kcal
   (E) 41 cal

2. A man with a respiratory quotient (RQ) of 70
   (A) has been eating a high-fat diet
   (B) has been eating a high-protein diet
   (C) has been fasting for 24 hours
   (D) has eaten nothing but carbohydrates for 24 hours
   (E) is dehydrated

3. Which of the following daily intakes (in grams) of carbohydrates (C), protein (P), and fat (F) would be best for a healthy 80-kg male construction worker?
   (A) C: 420; P: 80; F: 125
   (B) C: 420; P: 130; F: 100
   (C) C: 325; P: 80; F: 65
   (D) C: 550; P: 80; F: 65
   (E) C: 340; P: 50; F: 65

4. The major lipoprotein source of the cholesterol used in cells is
   (A) chylomicrons
   (B) intermediate-density lipoproteins (IDL)
   (C) very-low-density lipoproteins (VLDL)
   (D) LDL
   (E) HDL

5. Which of the following produces the most high-energy phosphate compounds?
   (A) Aerobic metabolism of 1 mol of glucose
   (B) Anaerobic metabolism of 1 mol of glucose
   (C) Metabolism of 1 mol of galactose
   (D) Metabolism of 1 mol of amino acid
   (E) Metabolism of 1 mol of long-chain fatty acid

6. Which of the following produces the most glucose when metabolized in the body?
   (A) 1 mol of serotonin
   (B) 1 mol of alanine
   (C) 1 mol of oleic acid
   (D) 1 mol of thyroxine (T_4)
   (E) 1 mol of acetoacetic acid

7. Which of the following would *not* produce an increase in the plasma level of free fatty acids?
   (A) A drug that increases the level of intracellular cyclic AMP

(B) A drug that activates β_1-adrenergic receptors

(C) A drug that inhibits hormone-sensitive lipase

(D) A drug that decreases the metabolic clearance of glucagon

(E) A drug that inhibits phosphodiesterase

8. When LDL enters cells by receptor-mediated endocytosis, which of the following does *not* occur?

(A) Decrease in the formation of cholesterol from mevalonic acid

(B) Increase in the intracellular concentration of cholesteryl esters

(C) Increase in the transfer of cholesterol from the cell to HDL

(D) Decrease in the rate of synthesis of LDL receptors

(E) Decrease in the cholesterol in endosomes

9. Which of the following does *not* increase the output of glucose from the liver?

(A) Induction of phosphorylase

(B) Insulin

(C) Glucagon

(D) Increased intracellular protein kinase A

(E) Epinephrine

In questions 10–15, select

A if the item is associated with **(a)** below,

B if the item is associated with **(b)** below,

C if the item is associated with both **(a)** and **(b)**, and

D if the item is associated with neither **(a)** nor **(b)**

(a) Cyclooxygenase
(b) 5-Lipoxygenase

10. Arachidonic acid

11. Palmitic acid

12. Leukotrienes

13. Thromboxanes

14. Aspirin

15. Lipoxins

In questions 16–20, match each disease or condition in the questions with the lettered substance that is most closely associated with it. Each lettered substance may be selected once, more than once, or not at all.

(A) Vitamin A
(B) Vitamin C
(C) Vitamin B$_1$
(D) Vitamin K
(E) Uric acid

16. Primary gout

17. Secondary gout

18. Scurvy

19. Beriberi

20. Abnormal fetal development

Chapter 18

This chapter reviews the structure and function of the thyroid gland and its hormones, T$_4$ and T$_3$. The major diseases of the thyroid gland are also considered. The material in the chapter should help students to–

- Describe the gross and microscopic anatomy of the thyroid gland.
- Summarize the main features of iodine metabolism.
- Compare the structure of T$_4$, T$_3$, and RT$_3$.
- Outline the steps involved in the biosynthesis of thyroid hormones, their storage in the colloid, and the transfer of thyroid hormones from the colloid to the bloodstream.
- Name the proteins that bind thyroid hormones in the plasma, describe the relationship between bound and free thyroid hormones, and summarize the mechanisms that regulate thyroid hormone binding.
- Outline the principal pathways by which thyroid hormones are metabolized.
- List the main physiologic actions of thyroid hormones.
- Describe the mechanism of action of thyroid hormones.
- Outline the processes involved in the regulation of secretion of thyroid hormones.
- List the principal drugs that affect thyroid function, and describe the mechanism by which each exerts its effect.
- List the major diseases of the thyroid and their principal symptoms and signs.

GENERAL QUESTIONS

Figure 18–A.

1. The curves in Fig 18–A show the changes in basal metabolic rate that follow administration of a single dose of T$_3$ or T$_4$ to an individual with previously untreated hypothyroidism. Which curve is produced by T$_3$ and which by T$_4$? Explain the differences between the 2 curves.

2. What is the role of TRH in the day-to-day regulation of thyroid function?

3. Discuss the causes and the pathophysiology of goiter. Why do people who eat large amounts of cabbage develop goiters?

4. Why do children with cretinism have short stature and mental deficiency?

5. Patients with thyrotoxicosis often have dramatic remission of many of their signs and symptoms upon administration of β-adrenergic-blocking drugs such as propranolol. Which signs and symptoms are improved, and which are not? Why?

6. Which hormones produce an increase in metabolic rate? How are these effects produced?

7. What are the embryologic origins of the thyroid gland, the parathyroid glands, and the parafollicular cells of the thyroid gland? Why is it important for clinicians to know about the embryologic development of these tissues?

MULTIPLE-CHOICE QUESTIONS

In questions 1–10, select the single best answer.

1. In which of the following conditions is it *least* likely that the TSH

response to TRH will be greater than normal?
- (A) Hypothyroidism due to tissue resistance to thyroid hormone
- (B) Hypothyroidism due to disease destroying thyroid gland
- (C) Hyperthyroidism due to circulating antithyroid antibodies with TSH activity
- (D) Hyperthyroidism due to diffuse hyperplasia or thyrotropes of anterior pituitary
- (E) Iodine deficiency

2. Propylthiouracil causes
- (A) a decrease in thyroid size and an increase in plasma T_4
- (B) a decrease in both thyroid size and plasma T_4
- (C) an increase in both thyroid size and plasma T_4
- (D) an increase in thyroid size and a decrease in plasma T_4
- (E) no change in thyroid size and a decrease in plasma T_4

3. A young woman has puffy skin and a hoarse voice. Her plasma TSH concentration is low but increases markedly when she is given TRH. She probably has
- (A) hyperthyroidism due to a thyroid tumor
- (B) hypothyroidism due to a primary abnormality in the thyroid gland
- (C) hypothyroidism due to a primary abnormality in the pituitary gland
- (D) hypothyroidism due to a primary abnormality in the hypothalamus
- (E) hyperthyroidism due to a primary abnormality in the hypothalamus

4. The coupling of monoiodotyrosine and diiodotyrosine and the iodination of thyroglobulin are blocked by
- (A) divalent cations
- (B) monovalent anions such as perchlorate
- (C) TSH
- (D) TRH
- (E) thiocarbamides such as propylthiouracil

5. The thyroidal uptake of radioactive iodine is *not* affected by
- (A) calcitonin
- (B) T_3
- (C) TSH
- (D) propylthiouracil
- (E) the iodine content of the diet

6. The BMR is *least* affected by an increase in the plasma level

- (A) TSH
- (B) TRH
- (C) TBG
- (D) free T_4
- (E) free T_3

7. Which of the following is *not* essential for normal biosynthesis of thyroid hormones?
- (A) Iodine
- (B) Ferritin
- (C) Thyroglobulin
- (D) Protein synthesis
- (E) TSH

8. Which of the following would be *least* affected by injections of TSH?
- (A) Thyroidal uptake of iodine
- (B) Synthesis of thyroglobulin
- (C) Cyclic AMP in thyroid cells
- (D) Size of the thyroid
- (E) Tyrosine kinase in thyroid cells

9. Hypothyroidism due to disease of the thyroid gland is associated with increased plasma levels of
- (A) cholesterol
- (B) albumin
- (C) RT_3
- (D) iodide
- (E) TBG

10. Which of the following is most likely to bring about improvement in exophthalamos?
- (A) Administration of T_4
- (B) Administration of drugs that inhibit the production T lymphocytes
- (C) Administration of testosterone
- (D) Hypophysectomy
- (E) Thyroidectomy

In questions 11–15, indicate whether the first item is greater than (G), the same as (S), or less than (L), the second item.

11. Food intake in myxedema

G S L

Food intake in thyrotoxicosis

12. Number of β receptors in the heart in myxedema

G S L

Number of β receptors in the heart in thyrotoxicosis

13. Thyroidal uptake of radioactive iodine 4 hours after a single dose of propylthiouracil

G S L

Thyroidal uptake of radioactive iodine 4 hours after a single dose of perchlorate

14. Plasma TSH level in an infant upon exposure to cold

G S L

Plasma TSH level in an adult exposure to cold

15. Amount of endocytosis at the colloid-thyroid cell border following injection of TSH

G S L

Amount of endocytosis at the colloid-thyroid cell border following injection of T_4

Chapter 19

This chapter is concerned with insulin, the other hormones of the pancreatic islets, and the many additional hormones in the body that affect carbohydrate metabolism. Diabetes mellitus is discussed in detail in this context. The material in the chapter should help students to—

- List the hormones that affect the blood glucose concentration, and briefly describe the action of each.
- Describe the structure of the pancreatic islets, and name the hormones secreted by each of the cell types in the islets.
- Describe the structure of insulin, and outline the steps involved in its biosynthesis and release into the bloodstream.
- List the consequences of insulin deficiency, and explain how each of these abnormalities is produced.
- Describe insulin receptors, the way they mediate the effects of insulin, and the way they are regulated.
- Describe the effects of hypoglycemia, and summarize the homeostatic mechanisms that combat hypoglycemia.
- Describe the types of glucose transporters found in the body and the function of each.
- List the major factors that affect the secretion of insulin.
- Describe the structure of glucagon and its relation to glicentin.
- List the physiologically significant effects of glucagon.
- List the principal factors that affect the secretion of glucagon.
- Describe the probably physiologic effects of somatostatin in the pancreas.
- Outline the mechanisms by which thyroid hormones, adrenal gluco-

corticoids, catecholamines, and growth hormone affect carbohydrate metabolism.

- Define type I (insulin-dependent) and type II (non-insulin-dependent) diabetes mellitus, and describe the principal features and probable causes of each type.

GENERAL QUESTIONS

1. A patient has hypoglycemia. He may have an insulin-secreting tumor or functional hypoglycemia, but there is also reason to believe that he may be surreptitiously injecting himself with insulin. How would you differentiate among these possibilities? Which tests would you order, and what would each tell you?

Figure 19–A.

2. Which of the curves in Fig 19–A would you expect to see in a patient with diabetes mellitus? Explain your answer.

Figure 19–B.

3. Which of the curves in Fig 19–B would you expect to see (1) in a normal individual, (2) in an individual with type I (insulin-dependent) diabetes mellitus, and (3) in an individual with type II (non-insulin dependent) diabetes mellitus? Explain the differences in the curves.

4. Why is ketosis more common and more severe in type I diabetes than in type II diabetes?

5. It has been argued that the pancreatic islets are sites where hormones act in a paracrine fashion. Summarize and evaluate the evidence for this hypothesis.

6. Describe the changes in intermediary metabolism that occur in the first 3 hours after a meal containing large quantities of carbohydrate and protein. What roles of hormones play in these changes?

7. Discuss glucagon in terms of the biosynthesis of glucagon and glicentin from the preprohormone, the locations in which glucagon is found in the body, and the function of glucagon in each of these locations.

8. Insulin is sometimes described as "the hormone of plenty" and glucagon as "the hormone of starvation." Discuss the appropriateness of these terms in the context of the physiologic effects of each hormone and their interactions.

9. What is proinsulin? What is its relation to preproinsulin, and what is its fate?

MULTIPLE-CHOICE QUESTIONS

In questions 1–10, select the single best answer.

1. Which of the following are *incorrectly* paired?
   (A) B cells : insulin
   (B) D cells : somatostatin
   (C) A cells : glucagon
   (D) Pancreatic exocrine cells : chymotrypsinogen
   (E) F cells : gastrin

2. Which of the following are *incorrectly* paired?
   (A) Epinephrine : increased glycogenolysis in skeletal muscle
   (B) Insulin : increased protein synthesis

   (C) Glucagon : increased gluconeogenesis
   (D) Progesterone : increased plasma glucose level
   (E) Growth hormone : increased plasma glucose level

3. Which of the following would be *least* likely to be seen 14 days after a rat is injected with a drug that kills all of its pancreatic B cells?
   (A) A rise in the plasma H^+ concentration
   (B) A rise in the plasma glucagon concentration
   (C) A fall in the plasma HCO_3^- concentration
   (D) A fall in the plasma amino acid concentration
   (E) A rise in plasma osmolality

4. Glucagon is *not* normally found in the
   (A) brain
   (B) pancreas
   (C) gastrointestinal tract
   (D) plasma
   (E) adrenal glands

5. When the blood glucose concentration falls to low levels, a number of different hormones help combat the hypoglycemia. After intravenous administration of a large dose of insulin, the return of a low blood sugar level to normal is delayed in
   (A) adrenal medullary insufficiency
   (B) glucagon deficiency
   (C) combined adrenal medullary insufficiency and glucagon deficiency
   (D) thyrotoxicosis
   (E) diabetes mellitus

6. Which of the following is *not* produced by total pancreatectomy?
   (A) Steatorrhea
   (B) Increased plasma levels of free fatty acids
   (C) Decreased plasma P_{CO_2}
   (D) No change or a rise in the plasma insulin level
   (E) No change or a rise in the plasma glucagon level

7. Insulin increases the entry of glucose into
   (A) all tissues
   (B) renal tubular cells
   (C) the mucosa of the small intestine
   (D) most neurons in the cerebral cortex
   (E) skeletal muscle

8. The mechanisms by which glucagon produces an increase in the

blood glucose concentration involves
  (A) binding of glucagon to DNA in the nucleus of target cells
  (B) binding of glucagon to receptors in the cytoplasm of target cells
  (C) activation of G_s in target cells
  (D) increased binding of Ca^{2+} in the cytoplasm of target cells
  (E) inhibition of insulin secretion

9. Glucagon increases glycogenolysis in liver cells but ACTH does not because
  (A) cortisol increases the plasma glucose level
  (B) liver cells have an adenylate cyclase different from that in adrenocortical cells
  (C) ACTH cannot enter the nucleus of liver cells
  (D) the membranes of liver cells contain receptors different from that in adrenocortical cells
  (E) liver cells contain a protein that inhibits the action of ACTH

10. A meal rich in proteins containing the amino acids that stimulate insulin secretion but low in carbohydrate does not cause hypoglycemia because
  (A) the meal causes a compensatory increase in T_4 secretion
  (B) cortisol in the circulation prevents glucose from entering muscle
  (C) glucagon secretion is also stimulated by the meal
  (D) the amino acids in the meal are promptly converted to glucose
  (E) insulin does not bind to insulin receptors if the plasma concentration of amino acids is elevated

In questions 11–15, indicate whether the first item is greater than (G), the same as (S), or less than (L), the second item.

11. Plasma glucagon level after eating candy

  G S L

  Plasma glucagon level after eating meat

12. Insulin secretion after glucose is given intravenously

  G S L

  Insulin secretion after the same amount of glucose is given orally

13. Blood sugar level after surgical pancreatectomy

  G S L

  Blood sugar level after administration of a toxin such as alloxan that selectively kills B cells

14. Plasma concentration of C peptide in fasting subject

  G S L

  Plasma concentration of C peptide in fed subject

15. Increase in GLUT 4 glucose transporters in membrane of skeletal muscle cell following injection of insulin

  G S L

  Increase in GLUT 2 glucose transporters in membrane of skeletal muscle cell following injection of insulin

---

Chapter 20

This chapter is concerned with the multiple functions of the adrenal glands: the secretion of catecholamines by the adrenal medulla, glucocorticoids and sex hormones by the zona fasciculata and zona reticularis of the adrenal cortex, and aldosterone by the zona glomerulosa of the adrenal cortex. There is a brief review of the clinical abnormalities produced by excess and deficiency of each of these hormones. The material in the chapter should help students to—

- Name the 3 catecholamines secreted by the adrenal medulla, and summarize their biosynthesis, metabolism, and function.
- List the principal stimuli that increase adrenal medullary secretion, and describe the way they bring about the increase.
- Differentiate between C_{18}, C_{19}, and C_{21} steroids, and give examples of each.
- Outline the steps involved in steroid biosynthesis in the adrenal cortex.
- Name the plasma proteins that bind adrenocortical steroids, and discuss the role they play in adrenocortical physiology.
- Name the major site of adrenocortical hormone metabolism and the principal metabolites produced from glucocorticoids, adrenal androgens, and aldosterone.
- List and briefly describe the physio-

logic and pharmacologic effects of glucocorticoids.
- Contrast the physiologic and pathologic effects of adrenal androgens.
- Describe the mechanisms that regulate secretion of glucocorticoids and adrenal sex hormones.
- Outline the actions of aldosterone.
- Describe the mechanisms that regulate aldosterone secretion.
- Describe the main features of the diseases caused by excess or deficiency of each of the hormones of the adrenal gland.

GENERAL QUESTIONS

1. Which of the metabolites of norepinephrine and epinephrine would you measure to determine whether a patient with pheochromocytoma had a norepinephrine- or epinephrine-secreting tumor?

2. Which enzymes are unique to each of the 3 zones of the adrenal cortex, ie, found only in that particular zone? What are the physiologic consequences of this unique distribution?

3. What are the functions of the human fetal adrenal cortex?

4. What would you expect to be the consequences of 17α-hydroxylase deficiency? Explain your answer.

5. Why do some patients with congenital adrenal hyperplasia and virilization have high blood pressure but others have excessive loss of sodium and hypotension? Explain your answer.

6. Patient with nephrosis sometimes have low plasma 17-hydroxycorticoid levels but do not develop signs and symptoms of adrenal insufficiency. Why?

7. List the possible causes of Cushing's syndrome. How would you treat each of them?

8. What are the advantages and disadvantages of long-term, high-dose treatment with glucocorticoids in diseases such as rheumatoid arthritis and asthma?

9. After hypophysectomy, the responsiveness of the adrenal cortex to ACTH is reduced, and after ACTH treatment, the response to subsequent doses of ACTH is increased. Explain these alterations in adrenal responsiveness.

MULTIPLE-CHOICE QUESTIONS

In questions 1–6, select the single best answer.

1. Which of the following is produced only by large amounts of glucocorticoids?
   (A) Normal responsiveness of fat depots to norepinephrine
   (B) Maintenance of normal vascular reactivity
   (C) Increased excretion of a water load
   (D) Inhibition of the inflammatory response
   (E) Inhibition of ACTH secretion

2. Which of the following are *incorrectly* paired?
   (A) Gluconeogenesis : cortisol
   (B) Free fatty acid mobilization : dehydroepiandrosterone
   (C) Muscle glycogenolysis : epinephrine
   (D) Kaliuresis : aldosterone
   (E) Hepatic glycogenesis : insulin

3. Which of the following hormones has the shortest half-life (is most rapidly destroyed in the blood)?
   (A) Corticosterone
   (B) Renin
   (C) Dehydroepiandrosterone
   (D) Aldosterone
   (E) Norepinephrine

4. Mole for mole, which of the following has the greatest effect of Na^+ excretion?
   (A) Progesterone
   (B) Cortisol
   (C) Vasopressin
   (D) Aldosterone
   (E) Dehydroepiandrosterone

5. Mole for mole, which of the following has the greatest effect on plasma osmolality?
   (A) Progesterone
   (B) Cortisol
   (C) Vasopressin
   (D) Aldosterone
   (E) Dehydroepiandrosterone

6. The secretion of which of the following would be *least* affected by a decrease in ECF volume?
   (A) CRH
   (B) Arginine vasopressin
   (C) Dehydroepiandrosterone
   (D) Angiotensin-converting enzyme
   (E) Aldosterone

In questions 7–10, match the statement in each question with the lettered steroid conversion that is most closely associated with it. Each lettered conversion may be picked once, more than once, or not at all.

   (A) Progesterone → corticosterone
   (B) Corticosterone → aldosterone
   (C) 17-Hydroxypregnenolone → dehydroepiandrosterone
   (D) Pregnenolone → 17-hydroxypregnenolone
   (E) Cholesterol → pregnenolone

7. Primary site of action of ACTH

8. Blocked in congenital 17α-hydroxylase deficiency

9. Produces a 17-ketosteroid

10. Facilitated by angiotensin II but not by ACTH

In questions 11–15, which refer to Fig 20–A, select the letter designating the part of the steroid secreting cell in Fig 20–A that has the highest concentration of the material. Each lettered part may be selected once, more than once, or not at all.

Figure 20–A. Steroid-secreting cell. (Reproduced, with permission, from Fawcett DW, Long JA, Jones AL: The ultrastructure of endocrine glands. *Recent Prog Hormone Res* 1969;**25**:315.)

11. Pregnenolone

12. 11-Deoxycortisol

13. ACTH receptors

14. 3β-Hydroxysteroid dehydrogenase

15. P450c11

Chapter 21

This chapter reviews the metabolism of calcium and phosphorus, the physiology of bone, and the 3 hormones that are the primary regulators of calcium metabolism: 1,25-dihydroxycholecalciferol, parathyroid hormone, and calcitonin. In addition, there is a brief review of the effects of other hormones on calcium metabolism. The material in the chapter should help students to—

- Describe the distribution of calcium and phosphorus in the body and the forms in which they exist in plasma.
- Name the factors affecting plasma Ca^{2+} concentration, and discuss the mechanism by which each exerts its effects.
- Name the types of cells found in bone, and describe the function of each.
- Describe the formation of vitamin D in the skin, its subsequent hydroxylation in the liver and kidneys, and the actions of its biologically active metabolites.
- Name the factors affecting the 1α-hydroxylase that forms 1,25-dihydroxycholecalciferol in the kidneys, and describe the action of each.
- Describe the biosynthesis, actions, and metabolism of parathyroid hormone.
- Identify the source of calcitonin, its chemical nature, and its principal actions.
- Summarize the effects of glucocorticoids, growth hormone, and growth factors on Ca^{2+} metabolism.

GENERAL QUESTIONS

1. How would you identify the cells in which parathyroid hormone and calcitonin are produced?

2. Name the tissues in the body on which 1,25-dihydroxycholecalciferol, parathyroid hormone, and calcitonin act, and describe the effects of these hormones on each of these tissues.

3. How do bones grow? What is epiphyseal closure, and what produces it?

4. What are the signs and symptoms of hyperparathyroidism, and what causes each of them?

5. Define osteomalacia, osteosclerosis, and osteoporosis. Discuss the causes and treatment of osteoporosis.

6. There was a sharp increase in the incidence of bone disease with the onset of the Industrial Revolution in the 19th century, but the incidence has since decreased. Which bone disease was produced, and why? Why has the incidence declined in the 20th century?

Figure 21–A.

7. On the axes in Fig 21–A, plot the relationship of plasma calcitonin concentration to plasma Ca^{2+} concentration and that of plasma parathyroid hormone concentration to plasma Ca^{2+} concentration.

MULTIPLE-CHOICE QUESTIONS

In questions 1–8, select the single best answer.

1. A patient with parathyroid deficiency 10 days after inadvertent damage to the parathyroid glands during thyroid surgery would probably have
   (A) low plasma phosphate and Ca^{2+} levels and tetany
   (B) low plasma phosphate and Ca^{2+} levels and tetanus
   (C) a low plasma Ca^{2+} level, increased muscular excitability, and a characteristic spasm of the muscles of the upper extremity (Trousseau's sign)
   (D) high plasma phosphate and Ca^{2+} levels and bone demineralization
   (E) increased muscular excitability, a high plasma Ca^{2+} level, and bone demineralization

2. A high plasma Ca^{2+} level causes
   (A) bone demineralization
   (B) increased formation of 1,25-dihydroxycholecalciferol
   (C) decreased secretion of calcitonin
   (D) decreased blood coagulability
   (E) increased formation of 24,25-dihydroxycholecalciferol

3. Which of the following is *not* involved in regulating plasma Ca^{2+} levels?
   (A) Kidneys
   (B) Skin
   (C) Liver
   (D) Lungs
   (E) Intestine

4. Which of the following exerts the greatest effect on parathyroid hormone secretion?
   (A) Plasma phosphate concentration
   (B) Calcitonin
   (C) 1,25-Dihydroxycholecalciferol
   (D) Total plasma calcium concentration
   (E) Plasma Ca^{2+} concentration

5. 1,25-Dihydroxycholecalciferol affects intestinal Ca^{2+} absorption through a mechanism that
   (A) is comparable to the action of adrenocortical hormones
   (B) activates adenylate cyclase
   (C) increases cell turnover
   (D) changes gastric acid secretion
   (E) is comparable to the action of polypeptide hormones

6. Which of the following would you expect to find in a patient whose diet has been low in calcium for 2 months?
   (A) Increased formation of 24,25-dihydroxycholecalciferol
   (B) Decreased amounts of calcium-binding protein in intestinal epithelial cells
   (C) Increased parathyroid hormone secretion
   (D) A high plasma calcitonin concentration
   (E) Increased plasma phosphates

7. Which of the following would tend to *increase* the plasma Ca^{2+} level?
   (A) Decreased intestinal absorption of Ca^{2+}
   (B) Elevated plasma protein level
   (C) High plasma pH
   (D) Decreased phosphate reabsorption in the renal tubules

   (E) Severe renal disease

8. In which of the following diseases is plasma Ca^{2+} low?
   (A) Pseudohypoparathyroidism
   (B) Adrenal failure
   (C) Cancer
   (D) Precocious puberty
   (E) Renal hypertension

In questions 9 and 10, indicate whether the first item is greater than (G), the same as (S), or less than (L), the second item.

9. Percentage of ingested Ca^{2+} absorbed from the intestine in a subject fed a high-calcium diet

   G S L

   Percentage of ingested Ca^{2+} absorbed from the intestine in a subject fed a low-calcium diet

10. Incidence of osteoporosis in men

    G S L

    Incidence of osteoporosis in women

Chapter 22

This chapter is a review of the morphology of the pituitary gland, pro-opiomelanocortin and its processing, the intermediate lobe of the pituitary, growth hormone, the somatomedins IGF-I and IGF-II, and the physiology of growth. The material in the chapter should help students to—

- Name the hormones secreted by the pituitary gland in humans, and list the main functions of each.
- Describe the embryonic origin of the 3 lobes of the pituitary gland and their structure in adults
- Describe the mechanisms responsible for changes in skin coloration in species in which such changes take place, and name the hormones involved.
- Describe the pro-opiomelanocortin molecule and its products in the anterior and intermediate lobes of the pituitary gland.
- Describe the structure of growth hormone, and list its actions and the mechanism by which it acts.
- Describe the relation of IGF-I and IGF-II to the actions of growth hormone.
- List the 2 hypothalamic hypophyseotropic hormones that regulate growth hormone secretion and

the principal stimuli that bring about increases or decreases in growth hormone secretion.
- List the factors needed for normal growth, and describe the contribution of each during prenatal and postnatal development.
- Describe and explain the main features of hypopituitarism, acromegaly, and Nelson's syndrome.

GENERAL QUESTIONS

1. Which endocrine disease causes alterations in skin pigmentation in humans? Which hormones are involved?

2. Although the same pro-opiomelanocortin molecule is found in the anterior and intermediate lobes of the pituitary gland and in neurons in the brain, different products are formed from it in each location. How can different products be formed from the same molecule? What processes are involved?

3. How would you decide whether an increase in growth hormone secretion was due to decreased secretion of somatostatin or increased secretion of GRH?

4. Describe the growth hormone receptor. What is unique about it?

5. Why does destruction of the posterior lobe of the pituitary gland cause marked polyuria whereas destruction of the whole pituitary gland causes little if any polyuria?

6. Discuss the relationship of nutrition to growth.

7. List the endocrinologic causes of dwarfism, and explain how each leads to short stature.

8. How is human growth hormone (hGH) related to human chorionic somatomammotropin (hCS)?

MULTIPLE-CHOICE QUESTIONS

In questions 1–7, select the single best answer.

1. Which of the following hormones exerts the least effect on growth?
   (A) Growth hormone
   (B) Testosterone
   (C) T_4
   (D) Insulin
   (E) Vasopressin

2. Which of the following are *incorrectly* paired?
   (A) Intermediate lobe : corticotropinlike intermediate lobe peptide (CLIP)
   (B) Anterior lobe : growth hormone
   (C) Laron dwarfism : tissue resistance to growth hormone
   (D) African pygmies : low plasma growth hormone
   (E) Acromegaly : high plasma IGF-I concentration

3. Which of the following pituitary hormones is an opioid peptide?
   (A) α-Melanocyte-stimulating hormone (α-MSH)
   (B) β-MSH
   (C) ACTH
   (D) Growth hormone
   (E) β-Endorphin

4. Which of the following hormones is *not* made up of α- and β-subunits?
   (A) TSH
   (B) LH
   (C) FSH
   (D) hCG
   (E) Prolactin

5. Which of the following is *not* characteristic of hypopituitarism?
   (A) Cachexia
   (B) Infertility
   (C) Pallor
   (D) Low basal metabolic rate
   (E) Intolerance to stress

6. Which of the following is *not* synthesized in both endocrine glands and the brain?
   (A) Somatostatin
   (B) Cortisol
   (C) Dopamine
   (D) ACTH
   (E) Oxytocin

7. A scientist finds that infusion of growth hormone into the median eminence of the hypothalamus in experimental animals inhibits the secretion of growth hormone, and concludes that this proves that growth hormone feeds back to inhibit GRH secretion. Do you accept this conclusion?
   (A) No, because growth hormone does not cross the blood-brain barrier
   (B) No, because the infused growth hormone could be stimulating dopamine secretion
   (C) No, because substances placed in the median eminence are easily transported to the anterior pituitary

   (D) Yes, because systemically administered growth hormone inhibits growth hormone secretion
   (E) Yes, because growth hormone combines with GRH and inactivates it

*In questions 8–10, indicate whether the first item is greater than (**G**), the same as (**S**), or less than (**L**), the second item.*

8. Plasma growth hormone level after 30 minutes of sleep

   G   S   L

   Plasma growth hormone level after 8 hours of sleep

9. Age at the time of pubertal growth spurt in girls

   G   S   L

   Age at the time of pubertal growth spurt in boys

10. Insulin sensitivity in hypopituitarism

    G   S   L

    Insulin sensitivity in acromegaly

Chapter 23

This chapter is a review of the physiology of the reproductive system in adult males and females. It also considers lactation and pregnancy, sexual differentiation and development in the fetus, puberty, and the menopause. The material in the chapter should help students to—

- Name the important hormones secreted by the Leydig cells and Sertoli cells of the testes and by the graafian follicles and corpora of the ovaries.
- Outline the role of chromosomes, hormones, and related factors in sex determination and development.
- Summarize the hormonal and other changes that occur at puberty in males and females.
- Outline the hormonal and other changes that occur at menopause.
- List the principal stimuli that increase prolactin secretion and the drugs that decrease prolactin secretion.
- Outline the steps involved in spermatogenesis, from the primitive germ cells to mature, motile spermatozoa.

- Describe the mechanisms that produce erection and ejaculation.
- Know the general structure of testosterone, and describe its biosynthesis, transport, metabolism, and actions.
- Describe the processes involved in regulation of testosterone secretion.
- Describe the changes that occur in the ovaries, uterus, cervix, vagina, and breasts during the menstrual cycle.
- Know the general structures of 17β-estradiol and progesterone, and describe their biosynthesis, transport, metabolism, and actions.
- Describe the roles of the pituitary and the hypothalamus in the regulation of ovarian function, and the role of feedback loops in this process.
- Describe the hormonal changes that accompany pregnancy and parturition.
- Outline the processes involved in development of the breasts, production of milk, milk ejection, and termination of lactation.

GENERAL QUESTIONS

1. What are the effects on subsequent sexual development when a human female fetus is exposed to excess androgens in utero?

2. What are the effects on subsequent sexual development when normal testes fail to develop in a human male fetus?

3. What is the effect of hypophysectomy on the uterus? Which other operation would produce the same results?

4. What physiologic changes occur in females during sexual intercourse?

5. What is the function of the acrosome?

Day of menstrual cycle

Figure 23–A.

6. Using the axes in Fig 23–A, diagram the changes in plasma LH, FSH estrogen, progesterone, and inhibin occurring during the human menstrual cycle.

7. Describe the testicular feminizing syndrome, and discuss its possible causes.

8. What is the evidence that the onset of puberty is under neural control?

9. Using your knowledge of reproductive physiology, list the possible ways that contraception could be produced. How many of these are actually being used by the general public? Comment on the strengths and weaknesses of the 5 methods that you believe are most commonly used.

10. Discuss the relationship of hormones to cancer.

MULTIPLE-CHOICE QUESTIONS

In questions 1–14, select the single best answer.

1. If a young woman has high plasma levels of T_3, cortisol, and renin activity but her blood pressure is only slightly elevated and she has no symptoms or signs of thyrotoxicosis or Cushing's syndrome, the most likely explanation is that
   (A) she has been treated with TSH and ACTH
   (B) she has been treated with T_3 and cortisol
   (C) she is in the third trimester of pregnancy
   (D) she has an adrenocortical tumor
   (E) she has been subjected to chronic stress

2. Full development and function of the seminiferous tubules requires
   (A) somatostatin
   (B) LH
   (C) oxytocin
   (D) FSH
   (E) androgens and FSH

3. In humans, fertilization usually occurs in the
   (A) vagina
   (B) cervix
   (C) uterine cavity
   (D) uterine tubes
   (E) abdominal cavity

4. In human males, testosterone is produced mainly by the
   (A) Leydig cells
   (B) Sertoli cells
   (C) seminiferous tubules
   (D) epididymis
   (E) vas deferens

5. Home-use kits for determining a woman's fertile period depend on the detection of one hormone in the urine. This hormone is
   (A) FSH
   (B) progesterone
   (C) estradiol
   (D) hCG
   (E) LH

6. Puberty does not normally occur in humans under the age of 8 years, because before that age
   (A) the tissues are unresponsive to gonadal steroids
   (B) the ovaries and testes are unresponsive to gonadotropins
   (C) the pituitary cannot manufacture adequate amounts of gonadotropins
   (D) the brain secretes a substance that inhibits the responsiveness of the gonads to gonadotropins
   (E) the hypothalamus fails to secrete GnRH in a normal fashion

7. Castration cells are found in the
   (A) uterus
   (B) prostate
   (C) placenta
   (D) anterior pituitary
   (E) hypothalamus

8. Decidual cells are found in the
   (A) uterus
   (B) prostate
   (C) placenta
   (D) anterior pituitary
   (E) hypothalamus

Questions 9–12 refer to Fig 23–B.

Figure 23–B. Ovarian follicle.

9. Which of the labeled structures has a higher concentration of angiotensin II than plasma?

10. Which of the labeled structures produces the most androstenedione?

11. Which of the labeled structures produces estradiol that mainly enters the follicular fluid?

12. Which of the labeled structures contains the smallest amount of DNA per cell?

13. Which of the following is *not* a steroid?
(A) 17α-Hydroxyprogesterone
(B) Estrone
(C) Relaxin
(D) Pregnenolone
(E) Eticholanolone

14. Which of the following is *not* a male secondary sexual characteristic?
(A) A beard
(B) An increased incidence of acne
(C) A deep voice
(D) Increased fat in the buttocks
(E) An enlarged penis

In questions 15–20, indicate whether the first item is greater than (G), the same as (S), or less than (L), the second item.

15. Size of the clitoris in a woman with congenital 17α-hydroxylase deficiency

G S L

Size of the clitoris in a woman with congenital 21β-hydroxylase deficiency

16. Size of the clitoris in a woman with congenital 11β-hydroxylase deficiency

G S L

Size of the clitoris in a woman with congenital 21β-hydroxylase deficiency

17. Adult height of a boy castrated before puberty

G S L

Adult height of a boy with precocious puberty

18. Plasma LH concentration in a 59-year-old man

G S L

Plasma LH concentration in a 59-year-old woman

19. hCG production in the first trimester of pregnancy

G S L

hCG production in the third trimester of pregnancy

20. hCS production in the first trimester of pregnancy

G S L

hCS production in the third trimester of pregnancy

Chapter 24

This chapter is concerned with renin and erythropoietin, 2 of the 3 hormones produced by the kidneys; with ANP, BNP, and CNP, the natriuretic peptides produced by the heart, the brain, and other tissue; and with melatonin, the pineal hormone. The material in the chapter should help students to—

• Outline the cascade of reactions that lead to the formation of angiotensin II and its metabolites in the circulation, and describe the properties of the enzymes and substrates involved.
• Describe the juxtaglomerular apparatus, and list the factors that regulate its secretion.
• List the functions of angiotensin II and the receptors on which it acts to carry out these functions.
• Describe the site and mechanism of action of erythropoietin, and the feedback regulation of its secretion.
• Describe the structure and functions of ANP, BNP, and CNP and the receptors on which they act.
• Diagram the steps involved in the formation of melatonin from serotonin in the pineal gland, list the proposed functions of melatonin, and discuss the regulation of melatonin secretion.

GENERAL QUESTIONS

1. Discuss the renin angiotensin system from the point of view of the similarities and differences between it and the kallikrein system.

2. Why are patients with chronic renal failure anemic? How would you treat their anemia?

Figure 24–A.

3. Using the axes in Fig 24–A, plot the relationship between dietary sodium intake and (1) plasma renin activity, (2) the plasma concentration of ANP, and (3) the plasma concentration of cortisol.

4. Discuss the relationship of the renin-angiotensin system to clinical hypertension.

5. Define Goldblatt hypertension, Bartter's syndrome, and secondary hyperaldosteronism.

Figure 24–B. Aldosterone secretion by adrenocortical tissue in vitro.

6. One of the 2 curves in Fig 24–B was obtained without and one with 10^{-6} M losartan in the medium. Which curve is which? Explain the shape of the curves and the horizontal displacement of one from the other. What do the 2 curves tell you about the type of angiotensin receptor involved in the regulation of aldosterone secretion?

7. Discuss natriuretic hormone from the point of view of the types that may exist, their structure, their origins, and their effects.

8. Melatonin secretion is high at night and low during the day. How does the pineal gland "know" that it is day or night?

MULTIPLE-CHOICE QUESTIONS

In questions 1–8, select the single best answer.

1. Renin is secreted by
   (A) cells in the macula densa
   (B) cells in the proximal tubules
   (C) cells in the renal glomeruli
   (D) juxtaglomerular cells
   (E) cells in the cortical peritubular capillaries

2. Erythropoietin is secreted by
   (A) cells in the macula densa
   (B) cells in the proximal tubules
   (C) cells in the renal glomeruli
   (D) juxtaglomerular cells
   (E) cells in the cortical peritubular capillaries

3. Melatonin secretion would probably *not* be increased by
   (A) stimulation of the superior cervical ganglia
   (B) intravenous infusion of tryptophan
   (C) intravenous infusion of epinephrine
   (D) stimulation of the optic nerve

   (E) induction of pineal hydroxy-indole-O-methyltransferase

4. When a woman who has been on a low-sodium diet for 8 days is given an intravenous injection of captopril, a drug that inhibits angiotensin-converting enzyme, one would expect
   (A) her blood pressure to rise because her cardiac output would fall
   (B) her blood pressure to rise because her peripheral resistance would fall
   (C) her blood pressure to fall because her cardiac output would fall
   (D) her blood pressure to fall because her peripheral resistance would fall
   (E) her plasma renin activity to fall because her circulating angiotensin I level would rise

5. Which of the following would be expected to cause an increase in ANP secretion from the heart?
   (A) Prolactin
   (B) Growth hormone
   (C) Thiorphan
   (D) Constriction of the ascending aorta
   (E) Constriction of the inferior vena cava

6. Which of the following would *not* be expected to increase renin secretion?
   (A) Administration of a drug that blocks angiotensin-converting enzyme
   (B) Administration of a drug that blocks AT_1 receptors
   (C) Administration of a drug that blocks β-adrenergic receptors
   (D) Constriction of the aorta between the celiac artery and the renal arteries
   (E) Administration of a drug that reduces ECF volume

7. Activation of receptors for ANP increases target cell
   (A) cyclic AMP
   (B) IP_3
   (C) protein kinase A activity
   (D) guanylate cyclase activity
   (E) GTP

8. Erythropoietin
   (A) contains zinc
   (B) contains iron
   (C) is an important ligand for iron in the tissues
   (D) stimulates renin secretion
   (E) acts on some but not all stem cells in bone marrow

In questions 9 and 10, indicate whether the first item is greater than (G), the same as (S), or less than (L), the second item.

9. Plasma renin *activity* before injection of estrogen

   G S L

   Plasma renin *activity* after injection of estrogen

10. Plasma renin *concentration* before injection of estrogen

   G S L

   Plasma renin *concentration* after injection of estrogen

Chapter 25

This chapter is concerned with digestion and absorption, and Chapter 26 deals with the details of how the gastrointestinal tract and its associated glands function to aid these processes. The material in Chapter 25 should help students to—

- Name the principal digestive enzymes, their precursors, their substrates, and the products of the action of the enzymes.
- Define brush border, unstirred layer, and glycocalyx.
- Outline the processes involved in the conversion of dietary carbohydrates to glucose and other hexoses in the intestine.
- Summarize the processes involved in the absorption of hexoses from the intestine into the bloodstream.
- Describe the conversion of dietary protein into amino acids and small peptides in the intestine.
- Summarize the processes involved in the absorption of amino acids and small peptides from the intestine into the bloodstream.
- Outline the events occurring during digestion of fats.
- Describe the processes by which fatty acids and other lipids are absorbed from the intestine into the bloodstream.
- Summarize the processes that regulate the absorption of water, Na^+, K^+, HCO_3^-, Ca^{2+}, and iron from the gastrointestinal tract.

GENERAL QUESTIONS

1. Many adults who are not of western European origin have abdominal symptoms after ingestion of milk. What are the symptoms, what causes them, and how would you treat the condition?

2. Describe secondary active transport, the process producing movement of glucose from the intestinal lumen to the bloodstream. Why is it called secondary rather than primary active transport?

3. After a meal, the lumens of the stomach and the intestine contain powerful proteolytic enzymes. Why do these enzymes not digest the pancreas, which produces them, or the wall of the intestine?

4. Intestinal absorption of calcium and iron are both subject to feedback control but by very different mechanisms. Compare and contrast the mechanisms.

5. What are M cells? What is their function?

MULTIPLE-CHOICE QUESTIONS

In questions 1–5, select the single best answer.

1. The pathway from the intestinal lumen to the circulating blood for a short-chain fatty acid (<10 carbon atoms) is
   (A) intestinal mucosal cell → chylomicrons → lymphatic duct → systemic venous blood
   (B) intestinal mucosal cell → hepatic portal vein blood → systemic venous blood
   (C) space between mucosal cells → lymphatic duct → systemic venous blood
   (D) space between mucosal cells → chylomicrons → lymphatic duct → systemic venous blood
   (E) intestinal mucosal cell → LDL → hepatic portal vein blood → systemic venous blood

2. Water is absorbed in the jejunum, ileum, and colon and excreted in the feces. Arrange these in order of the amount of water absorbed or excreted from greatest to smallest.
   (A) Colon, jejunum, ileum, feces
   (B) Feces, colon, ileum, jejunum
   (C) Jejunum, ileum, colon, feces
   (D) Colon, ileum, jejunum, feces
   (E) Feces, jejunum, ileum, colon

3. Drugs and toxins that increase the cyclic AMP content of the intestinal mucosa cause diarrhea because they
   (A) increase Na^+-K^+ cotransport in the small intestine
   (B) increase K^+ secretion into the colon
   (C) inhibit K^+ absorption in the crypts of Lieberkühn
   (D) increase Na^+ absorption in the small intestine
   (E) increase Cl^- secretion into the intestinal lumen

4. Which of the following are *incorrectly* paired?
   (A) Pancreatic α-amylase : starch
   (B) Elastase : tissues rich in elastin
   (C) Enteropeptidase : pancreatic lipase
   (D) Renin : coagulated milk
   (E) Lingual lipase : digestion in the stomach

5. Calcium absorption is increased by
   (A) hypercalcemia
   (B) oxalates in the diet
   (C) iron overload
   (D) 1,25-dihydroxycholecalciferol
   (E) increased Na^+ absorption

In questions 6–10, select

   A if the item is affected by (a) below,
   B if the item is affected by (b) below,
   C if the item is affected by both (a) and (b), and
   D if the item is affected by neither (a) nor (b)

   (a) Absorption of glucose
   (b) Absorption of amino acids

6. Plasma Na^+

7. Na^+ in intestinal lumen

8. Plasma insulin

9. Phloridzin

10. Trypsin in intestinal lumen

Chapter 26

This chapter is a detailed consideration of the way in which the gastrointestinal tract and its associated glands, the salivary glands, liver, and pancreas, function in carrying out digestion and absorption of food. The material in the chapter should help students to—

- Describe the structure and function of the enteric nervous system.
- List the principal gastrointestinal hormones, the sites where each is secreted, and the main physiologic function of each of these hormones.
- Summarize the functions of the mouth, the salivary glands, and the esophagus.
- Outline the functional anatomy and histology of the stomach.
- Describe how acid is secreted by cells in the gastric mucosa.
- Describe the mechanisms that regulate the secretion and motility of the stomach.
- Describe the relationship between cyanocobalamin and intrinsic factor.
- List the main components of pancreatic juice, and outline the mechanisms that regulate its secretion.
- Describe the functional anatomy of the liver, and discuss the formation of bile.
- Discuss the function of the gallbladder and the processes that regulate the passage of bile to the intestinal lumen.
- List the types of movement seen in the small intestine and the function of each.
- Describe the events in the colon that lead to defecation, and outline the neural pathways that control this semireflex response.

GENERAL QUESTIONS

1. What factors regulate the secretion of the exocrine portion of the pancreas?

2. Discuss the abnormalities of gastrin secretion seen in disease states.

3. What abnormalities would you expect to be produced by resection of the terminal portion of the ileum? Why?

4. What is heartburn? Discuss its

Table 26–A

Pattern	Plasma Bilirubin Direct	Plasma Bilirubin Indirect	Plasma Alkaline Phosphatase	Hematocrit	Plasma Bile Acids
A	—	—	—	—	↓
B	↑	↑↑	—	—	—
C	↑↑	↑	↑	—	↓
D	↑	↑↑	—	↓	—
E	↑	↑↑	↑	—	↑

— = no change; ↑ = increase; ↑↑ = marked increase; ↓ = decrease.

pathophysiology and ways it might be treated.

5. What causes gallstones? How would you treat them?

6. Several surgical procedures have been recommended for the treatment of severe obesity that fails to respond to other forms of treatment. What are these procedures? How do they cause weight loss? What are their long-term complications?

7. The bacteria in the colon exist in a symbiotic relationship with the host. How does the host benefit from this relationship, and what harmful or potentially harmful effects may occur to the host?

MULTIPLE-CHOICE QUESTIONS

In questions 1–15, select the single best answer.

Questions 1–4 refer to Fig 26–A.

Figure 26–A. Cross section of small intestine. (Reproduced, with permission, from Bell GH, Emslie-Smith D, Paterson CR: *Textbook of Physiology and Biochemistry,* 9th ed. Churchill Livingstone, 1976.)

1. Which of the labeled structures is the site where maltase is found?

2. Which of the labeled structures is most concerned with integration of peristalsis?

3. Which of the labeled structures has the most rapid cell turnover?

4. Which of the labeled structures contains the cell bodies of chemoreceptors?

Questions 5–8 refer to Table 26–A.

5. Which set of data would be found in a patient with a gallstone obstructing the common bile duct?

6. Which set of data would be found in a patient with intravascular hemolysis?

7. Which set of data would be found in a patient with infectious hepatitis?

8. Which set of data would be found in a patient with a resection of the ileum?

9. Removal of the entire colon would be expected to cause
   (A) death
   (B) megaloblastic anemia
   (C) severe malnutrition
   (D) an increase in the blood level of ammonia in patients with cirrhosis of the liver
   (E) decreased urinary urobilinogen

10. In infants, defecation often follows a meal. The cause of colonic contractions in this situation is
    (A) the gastroileal reflex
    (B) increased circulating levels of CCK
    (C) the gastrocolic reflex
    (D) increased circulating levels of somatostatin
    (E) the enterogastric reflex

11. After a meal rich in carbohydrate is ingested, insulin secretion is probably stimulated by
    (A) GIP
    (B) CCK
    (C) serotonin
    (D) VIP
    (E) gastrin

12. The symptoms of the dumping syndrome (discomfort after meals in patients with intestinal short circuits such as anastamosis of the jejunum to the stomach) are caused in part by
    (A) increased blood pressure
    (B) increased secretion of glucagon
    (C) increased secretion of CCK
    (D) hypoglycemia
    (E) hyperglycemia

13. Which of the following has the highest pH?
    (A) Gastric juice
    (B) Bile in the gallbladder
    (C) Pancreatic juice
    (D) Saliva
    (E) Secretions of the intestinal glands

14. After complete hepatectomy, a rise would be expected in the blood level of
    (A) glucose
    (B) fibrinogen
    (C) 25-hydroxycholecalciferol
    (D) conjugated bilirubin
    (E) estrogens

15. Which of the following would *not* be produced by total pancreatectomy?
    (A) Vitamin E deficiency
    (B) Hyperglycemia
    (C) Metabolic acidosis
    (D) Weight gain
    (E) Decreased absorption of amino acids

Many different abnormalities cause diarrhea. In questions 16–20, match the diarrhea-producing disorder in each question with the lettered abnormality most closely associated with it. Each lettered abnormality may be selected once, more than once, or not at all.

   (A) Increased cyclic AMP in enterocytes

(B) Increased gastric acid secretion

(C) Increased bile acids in the colon

(D) Alteration in intestinal mucin composition

(E) Abnormal digestion of carbohydrates

16. Ulcerative colitis

17. Operations that bypass the terminal ileum

18. Cholera

19. Lactase deficiency

20. Zollinger-Ellison syndrome

Chapter 27

This chapter reviews the functions of blood and lymph plus the formed elements of the blood: white blood cells, red blood cells, and platelets. The material in the chapter should help students to—

- Name the types of bone marrow, and describe the function and location of each.
- List the various types of cells found in the blood and the precursor cells for each type.
- Describe the functions of neutrophils.
- Describe the functions of monocytes.
- Name the various types of lymphocytes, and describe the origin and function of each.
- Describe the processes responsible for the production of circulating antibodies and for the rejection of tissue transplanted from other individuals.
- Describe the structure and function of platelets and the way they discharge their granules.
- Summarize the functions of red blood cells.
- Name the common blood types, and describe how blood is typed and cross-matched.
- Outline the blood-clotting and anticlotting mechanisms.

GENERAL QUESTIONS

1. Compare and contrast the composition of blood plasma and lymph.

2. How is the production of white blood cells, platelets, and red blood cells adjusted to meet the varying needs of the individual?

3. What are the tissue macrophages, and what is their origin and function?

4. Mutant genes that cause the production of abnormal hemoglobins are common in humans. What determines whether a given mutation in hemoglobin is harmless or harmful?

5. Describe how blood and tissue typing can be used to determine paternity.

6. What is the clonal selection theory of antibody formation? What is the evidence for this theory?

7. Why is blood clotting abnormal in patients with vitamin K deficiency?

8. What are the functions of natural killer cells? How do these cells differ from other lymphocytes?

9. Compare and contrast humoral and cellular immunity.

MULTIPLE-CHOICE QUESTIONS

In questions 1–5, match the immunoglobulin in each question with the lettered characteristics that are most closely associated with it. Each lettered characteristic may be selected once, more than once, or not at all. Note that in some instances, the correct answer may be more than one letter.

(A) Complement fixation
(B) Secretory immunity
(C) Contains J chain
(D) Contains SC chain
(E) Monomer only

1. IgA

2. IgD

3. IgE

4. IgG

5. IgM

In questions 6–14, select the single best answer.

Questions 6–9 refer to Fig 27–A.

Figure 27–A. Red blood cells in plasma.

6. Which of the patterns would be seen if plasma from an individual with blood type B were mixed with red cells from an individual with blood type O?

7. Which of the patterns would be seen if plasma from an individual with blood type O were mixed with red cells from an individual with blood type B?

8. In which of the patterns would the hemoglobin in the plasma be highest?

9. Which of the following is *not* primarily a function of blood plasma?
   (A) Transport of hormones
   (B) Maintenance of red cell size
   (C) Transport of chylomicrons
   (D) Transport of antibodies
   (E) Transport of O_2

10. A hematocrit of 41% means that, in the sample of blood analyzed
    (A) 41% of the hemoglobin is in the plasma
    (B) 41% of the total blood volume is made up of blood plasma
    (C) 41% of the total blood volume is made up of red and white blood cells and platelets
    (D) 41% of the hemoglobin is in red blood cells
    (E) 41% of the formed elements in blood are red blood cells

11. In normal human blood
    (A) the eosinophil is the most common type of white blood cell
    (B) there are more lymphocytes than neutrophils
    (C) the iron is mostly in hemoglobin
    (D) there are more white cells than red cells
    (E) there are more platelets than red cells

12. Lymphocytes
    (A) all originate from the bone marrow

(B) are unaffected by hormones
(C) convert to monocytes in response to antigens
(D) interact with eosinophils to produce platelets
(E) are part of the body's defense against cancer

Figure 27–B.

13. Production of the type of cell shown in Fig 27–B is increased by
(A) interleukin-2 (IL-2)
(B) granulocyte colony-stimulating factor (G-CSF)
(C) erythropoietin
(D) interleukin-4 (IL-4)
(E) interleukin-5 (IL-5)

14. The cell shown in Fig 27–B does *not* contain
(A) actin
(B) free radicals
(C) myeloperoxidase
(D) cathepsins
(E) a Y chromosome

In questions 15–20, indicate whether the first item is greater than (G), the same as (S), or less than (L) the second item.

15. Blood hemoglobin concentration in anemia

G S L

Blood hemoglobin concentration in polycythemia

16. Number of light chains in a basic immunoglobulin molecule

G S L

Number of heavy chains in a basic immunoglobulin molecule

17. Tendency to hemorrhage when the platelet count is high

G S L

Tendency to hemorrhage when the platelet count is low

18. Percentage of total plasma osmolality due to circulating albumin

G S L

Percentage of total plasma osmolality due to circulating sodium chloride

19. Incidence of the sickle cell gene in North America

G S L

Incidence of the sickle cell gene in Africa

20. Total O_2 per unit blood in the aorta

G S L

Total O_2 per unit blood in the inferior vena cava

Chapter 28

This chapter is a review of the conduction system of the heart and the way the impulse that generates a normal heartbeat spreads from the SA node through the AV node, the bundle of His, and the Purkinje system to all parts of the ventricles. The genesis of the ECG, some of its abnormalities, and cardiac arrhythmias are also considered. The material in the chapter should help students to—

- Describe the structure and function of the conduction system of the heart, and compare the action potentials in each part of it with those in cardiac muscle.
- Describe the way the ECG is recorded, the waves of the ECG, and the relationship of the ECG to the electrical axis of the heart.
- Name the common cardiac arrhythmias, and describe the processes that produce them.
- Know how and when to carry out cardiac massage.
- List the principal early and late ECG manifestations of myocardial infarction, and explain the early changes in terms of the underlying ionic events that produce them.
- Describe the ECG changes and the changes in cardiac function produced by alterations in the ionic composition of the body fluids.

GENERAL QUESTIONS

1. What is sinus arrhythmia? How is it produced? What is its clinical significance?

2. What is an ectopic focus of excitation? What is its pathophysiologic significance?

3. List and explain the effects of slow Ca^{2+} channel blocking drugs on the heart. In which clinical conditions are they of value?

4. Compare and contrast the Wolff-Parkinson-White syndrome and the Lown-Ganong-Levine syndrome. What are the underlying defects that produce these 2 conditions?

5. Define calcium rigor, and explain its occurrence.

MULTIPLE-CHOICE QUESTIONS

In questions 1–7, select the single best answer.

Questions 1–4 refer to Fig 28–A.

Figure 28–A. ECG.

1. Which of the labels identifies the part of the ECG that corresponds to ventricular repolarization?

2. Which of the labels identifies the Q wave?

3. Which of the labels identifies the part of the ECG that corresponds to maximum opening of ventricular Na^+ channels?

4. Which of the labels identifies the part of the ECG that corresponds to maximum opening of ventricular Ca^{2+} channels?

5. Which of the following normally has the steepest prepotential?
(A) Sinoatrial node
(B) Atrioventricular node

(C) Bundle of His
(D) Terminals of Purkinje fibers
(E) Ventricular muscle cells

6. In second-degree heart block
   (A) the ventricular rate is lower than the atrial rate
   (B) the ventricular ECG complexes are distorted
   (C) there is a high incidence of ventricular tachycardia
   (D) stroke volume is decreased
   (E) cardiac output is increased

7. Carotid sinus massage sometimes stops supraventricular tachycardia because
   (A) it decreases sympathetic discharge to the SA node
   (B) it increases vagal discharge to the SA node
   (C) it increases vagal discharge to the conducting tissue between the atria and the ventricles
   (D) it decreases sympathetic discharge to the conducting tissue between the atria and the ventricles
   (E) it increases the refractory period of the ventricular myocardium

In questions 8–10, one or more than one of the answers may be correct. Select

   A if (1), (2), and (3) are correct;
   B if (1) and (3) are correct;
   C if (2) and (4) are correct;
   D if only (4) is correct; and
   E if all are correct

8. In complete heart block
   (1) fainting may occur because the atria are unable to pump blood into the ventricles
   (2) death due to ventricular fibrillation is common
   (3) the atrial rate is lower than the ventricular rate
   (4) fainting may occur because of prolonged periods during which the ventricles fail to contract

9. Reentry is a common cause of
   (1) paroxysmal atrial tachycardia
   (2) paroxysmal nodal tachycardia
   (3) atrial fibrillation
   (4) sinus arrhythmia

10. Currents caused by opening of which of the following channels contribute to the repolarization phase of the action potential of ventricular muscle fibers?
    (1) Na^+ channels
    (2) Cl^- channels
    (3) Ca^{2+} channels
    (4) K^+ channels

In questions 11-115, indicate whether the first item is greater than (G), the same as (S), or less than (L) the second item.

11. Height of T waves when the plasma K^+ level is 5.5 meq/L

    G S L

    Height of T waves when the plasma K^+ level is 8.5 meq/L.

12. In normal adults, height of the R wave in lead V_1 of the ECG

    G S L

    In normal adults, height of the R wave in lead V_5 of the ECG

13. Heart rate in atrial fibrillation

    G S L

    Heart rate in atrial tachycardia

14. Interval between an atrial premature beat and the next normal beat

    G S L

    Interval between a ventricular premature beat and the next normal beat

15. Slope of prepotential in SA node after stimulation of vagus

    G S L

    Slope of prepotential in SA node after stimulation of sympathetic nerves to heart

Chapter 29

This chapter considers the mechanical events in the cardiac cycle as the heart pumps the blood through the circulation, heart sounds and murmurs, and measurement and regulation of cardiac output. The material in the chapter should help students to—

- Describe in sequence the events that occur in the heart during the cardiac cycle.
- Outline the changes in the duration of systole and diastole that occur with changes in heart rate, and discuss their physiologic consequences.
- Describe the arterial pulse and jugular venous pulse.
- Describe and explain the first and second heart sounds and the occa-

sionally observed third and fourth heart sounds.
- State the timing of the murmurs produced by aortic stenosis, aortic insufficiency, mitral stenosis, and mitral insufficiency.
- List the factors affecting cardiac output and the effect of each.
- Summarize the factors governing oxygen consumption by the heart.

GENERAL QUESTIONS

1. List the methods that are commonly used to measure cardiac output. What are the advantages and disadvantages of each method?

2. It takes several months for nerves to grow into transplanted hearts. However, before the nerves regrow, patients with transplanted hearts are able to increase their cardiac output when they exercise. What is the mechanism involved, and how does it operate?

3. Explain postextrasystolic potentiation.

4. What is the ejection fraction? Which conditions cause it to increase, and which conditions cause it to decrease? Why?

Heart rate

Figure 29–A.

5. Using the axes shown in Fig 29–A, plot the changes in cardiac output as the heart rate increases. Explain the curve that you draw.

MULTIPLE-CHOICE QUESTIONS

In questions 1–6, select the single best answer.

1. The first heart sound is caused by
   (A) closure of the aortic and pulmonary valves

(B) vibrations in the ventricular wall during systole

(C) ventricular filling

(D) closure of the mitral and tricuspid valves

(E) retrograde flow in the vena cava

2. The fourth heart sound is caused by
   (A) closure of the aortic and pulmonary valves
   (B) vibrations in the ventricular wall during systole
   (C) ventricular filling
   (D) closure of the mitral and tricuspid valves
   (E) retrograde flow in the vena cava

3. The dicrotic notch on the aortic pressure curve is caused by
   (A) closure of the mitral valve
   (B) closure of the tricuspid valve
   (C) closure of the aortic valve
   (D) closure of the pulmonary valve
   (E) rapid filling of the left ventricle

4. During exercise, a man consumes 1.8 L of oxygen per minute. His arterial O_2 content is 190 mL/L, and the O_2 content of his mixed venous blood is 134 mL/L. His cardiac output is approximately
   (A) 3.2 L/min
   (B) 16 L/min
   (C) 32 L/min
   (D) 54 L/min
   (E) 160 mL/min

5. The work performed by the left ventricle is substantially greater than that performed by the right ventricle, because in the left ventricle
   (A) the contraction is slower
   (B) the wall is thicker
   (C) the stroke volume is greater
   (D) the preload is greater
   (E) the afterload is greater

6. Starling's law of the heart
   (A) does not operate in the failing heart
   (B) does not operate during exercise
   (C) explains the increase in heart rate produced by exercise
   (D) explains the increase in cardiac output that occurs when venous return is increased
   (E) explains the decrease in end-diastolic volume of the ventricles when venous return is increased

In questions 7–10, match the murmur in each question with the lettered abnormality that is most closely associated with it. Each lettered abnormality may be used once, more than once, or not at all.

   (A) Aortic stenosis
   (B) Aortic insufficiency
   (C) Mitral stenosis
   (D) Mitral insufficiency
   (E) Tricuspid insufficiency

7. Systolic murmur loudest over the apex of the heart

8. Diastolic murmur loudest over the base of the heart

9. Diastolic murmur loudest over the apex of the heart

10. Systolic murmur loudest over the base of the heart

Chapter 30

This chapter considers the blood vessels and lymphatics and the movement of fluids through them. It includes the functional anatomy of the vessels, the principles and forces that govern pressure and flow in them, and the factors affecting exchange across capillary walls. The material in the chapter should help students to—

- Describe in relative terms the diameter, wall thickness, and total cross-sectional area of the aorta, smaller arteries, arterioles, capillaries, venules, and veins.
- Describe the relationship between flow, pressure, and resistance in the vascular system.
- List and assess the methods commonly used to measure blood flow.
- Define laminar flow and critical closing pressure.
- State the Poiseuille-Hagen formula for flow in blood vessels, and explain on the basis of this formula why the radius of a vessel is such an important determinant of flow.
- Define the law of Laplace, and list 3 examples of its operation in the body.
- Describe in detail how blood pressure in humans is measured by the auscultatory method and the palpation method.
- Describe the Starling forces that determine the net movement of fluid across the capillary wall. Define flow-limited exchange and diffusion-limited exchange, and describe the variations in capillary perme-

ability and structure seen in different parts of the body.

GENERAL QUESTIONS

1. What is a Windkessel? Why are the aorta and large arteries called Windkessel vessels? What is the physiologic significance of the Windkessel effect?

2. What is Bernoulli's principle? Discuss its significance in cardiovascular physiology.

3. A young man has a blood pressure of 130/73 mm Hg. What is his pulse pressure? What is his mean arterial pressure? How did you calculate each of these pressures?

4. The pressure in a large artery is 60,000 dynes/cm^2, and the radius of the artery is 0.5 cm. What is the wall tension in the artery? What would the wall tension be if the diameter were 1 cm?

5. Define and explain Korotkow's sounds.

6. Where in the cardiovascular system is turbulent flow normally found? What factors make flow change from streamline to turbulent?

MULTIPLE-CHOICE QUESTIONS

In questions 1–10, select the single best answer.

Questions 1–4 refer to Table 30–A.

One hind limb of an anesthetized experimental animal is denervated and attached to instruments so that it is possible to measure the pressure at the arterial and venous ends of the capillaries, the weight of the limb, its blood flow, and the arteriovenous oxygen difference across it. Various substances are then administered. The results obtained are listed in Table 30–A.

1. Which pattern of changes would be observed if the substance produced arteriolar dilation?

2. Which pattern would be observed if the substance produced increased capillary permeability?

3. Which pattern would be observed if the substance produced constriction of the veins?

4. Which pattern would be observed

Table 30–A

Pattern	Pressure at		Weight of Limb	Blood Flow in Limb	Arteriovenous Oxygen Difference
	Arterial End of Capillary	Venous End of Capillary			
A	—	—	↑	—	—
B	—	↑	—	↓	—
C	↓	↓	—	↓	↑
D	↑	↑	↑	↑	↓
E	↑	↑	↑	↓	—

— = no change; ↑ = increase; ↓ = decrease.

if the substance produced a decrease in systemic blood pressure?

5. In which of the following lists of blood vessels is the sequence of vessels arranged from highest to lowest *total* cross-sectional area in the body?
   (A) Arteries, arterioles, capillaries, veins
   (B) Arterioles, capillaries, arteries, veins
   (C) Capillaries, arterioles, veins, arteries
   (D) Veins, capillaries, arterioles, arteries
   (E) Arteries, veins, arterioles, capillaries

6. Lymph flow from the foot is
   (A) increased when an individual rises from the supine to the standing position
   (B) increased by massaging the foot
   (C) increased when capillary permeability is decreased
   (D) decreased when the valves of the leg veins are incompetent
   (E) decreased by exercise

7. Movement of fluid from a capillary into tissue is increased by
   (A) a fall in arterial pressure
   (B) a rise in venous pressure
   (C) a rise in plasma oncotic pressure
   (D) constriction of the arterioles
   (E) closure of precapillary sphincters

8. The mean pressure in the artery supplying a given organ is 100 mm Hg, the mean pressure in the vein draining it is 10 mm Hg, and the blood flow through the organ is 600 mL/min. What is the vascular resistance across the organ?
   (A) 0.1 R unit
   (B) 0.9 R unit
   (C) 1 R unit
   (D) 9 R units
   (E) 11 R units

9. Which of the following does *not* occur as blood passes through the systemic capillaries?
   (A) Its hematocrit increases
   (B) Its hemoglobin dissociation curve shifts to the left
   (C) Its protein content increases
   (D) Its pH decreases
   (E) Its red blood cells increase in size

10. The velocity of blood flow
    (A) is higher in the capillaries, because the total cross-sectional area of the capillaries is large
    (B) is higher in the veins than in the venules
    (C) falls to zero in the descending aorta during diastole
    (D) is reduced in a constricted area of a blood vessel
    (E) is measured in cm^3 per second

In questions 11–15, indicate whether the first item is greater than (G), the same as (S), or less than (L) the second item.

11. Resistance of an arteriole with a diameter of 10μm

    G S L

    Resistance of an arteriole with a diameter of 7μm

12. Hydrostatic pressure in venules at the level of the heart

    G S L

    Hydrostatic pressure in veins at the level of the heart

13. Permeability of capillaries in the lungs

    G S L

    Permeability of capillaries in the liver

14. Pressure in capacitance vessels

    G S L

    Pressure in resistance vessels

15. Interstitial fluid volume in exercising muscle

    G S L

    Interstitial fluid volume in resting muscle

Chapter 31

This chapter reviews the local mechanisms, circulating humoral agents, and neural mechanisms that act together to maintain blood pressure and blood flow to the various organs at rest and during a wide variety of emergency situations. The material in the chapter should help students to—

- Define autoregulation, discuss its role in physiology, and summarize the theories that have been advanced to explain its occurrence.
- List the important vasodilator metabolites, and discuss their role in the regulation of tissue perfusion.
- Describe the role of EDRF and endothelins in normal and pathological states.
- List the principal hormones that affect blood pressure, and comment on the physiologic role of each.
- Outline the neural mechanisms that control blood pressure and heart rate, including the receptors, afferent and efferent pathways, central integrating pathways, and effector mechanisms involved.
- Describe the direct effects of CO_2 and hypoxia on the vasomotor areas in the medulla oblongata.

GENERAL QUESTIONS

1. Why does increased sympathetic activity cause an increase in right atrial pressure? What are the

mechanisms and pathways involved?

Figure 31–A.

2. Using the axes in Fig 31–A, plot the R-R interval against time before and after intravenous injection of the vasodilator nitroprusside at the arrow.

3. Compare the function of the carotid and aortic baroreceptors with the function of the baroreceptors in the atria and great veins.

4. Which hormones affect the heart rate and how do they produce their effects?

5. Which peptides are found in nerves innervating blood vessels? What is the function of these peptides?

6. What are the effects of baroreceptor denervation?

7. A pharmacologist injects a drug into a dog and finds that it produces a prompt increase in mean arterial pressure. List the possible mechanisms by which it could be acting to produce this increase, and outline a series of experiments that would permit you to decide the exact mechanism of action of the drug.

MULTIPLE-CHOICE QUESTIONS

In questions 1–5, select the single best answer.

1. Which of the following are *incorrectly* paired?
   (A) Angiotensin-converting enzyme : kinin metabolism
   (B) Kallikrein : plasma renin activity
   (C) Prostacyclin : vasodilation
   (D) Increased pressure in the carotid sinuses : decreased sympathetic discharge to arterioles
   (E) Increased H_2 in tissues : vasoconstriction in tissues

2. When a pheochromocytoma (tumor of the adrenal medulla) suddenly discharges a large amount of epinephrine into the circulation, the patient's heart rate would be expected to
   (A) increase because the increase in blood pressure stimulates the carotid and aortic baroreceptors
   (B) increase because epinephrine has a direct chronotropic effect on the heart
   (C) increase because of increased tonic parasympathetic discharge to the heart
   (D) decrease because the increase in blood pressure stimulates the carotid and aortic chemoreceptors
   (E) decrease because of increased tonic parasympathetic discharge to the heart

3. Vasopressin secretion is increased by
   (A) increased pressure in the right ventricle
   (B) decreased pressure in the right ventricle
   (C) increased pressure in the right atrium
   (D) decreased pressure in the right atrium
   (E) increased pressure in the aorta

4. Catecholamines acting on α-adrenergic receptors
   (A) increase the contractility of cardiac muscle
   (B) increase the rate of discharge of the sinoatrial node
   (C) increase cardiac output
   (D) constrict coronary arteries by a direct action on these blood vessels
   (E) dilate blood vessels in skeletal muscle

5. Which of the following would be expected to raise blood pressure?
   (A) Prostacyclin
   (B) A drug that inhibits angiotensin-converting enzyme
   (C) A drug that inhibits NO synthetase
   (D) A drug that inhibits V_1 vasopressin receptors
   (E) A drug that inhibits endothelin-converting enzyme

In questions 6–10, match the substance in each question with the lettered item that is most closely associated with it. Each lettered item may be used once, more than once, or not at all.

   (A) Vasoconstrictor metabolite produced in tissue
   (B) Vasodilator metabolite produced in tissue
   (C) Pressor hormone in circulation
   (D) Depressor hormone in circulation
   (E) Neurotransmitter in vasopressor pathway from brain

6. CO_2

7. K^+

8. ANP

9. Angiotensin II

10. Acetylcholine

Chapter 32

This chapter describes special aspects of the circulation of the brain including the blood-brain barrier, the heart, the splanchnic region (including the liver), the skin, and the placenta. The material in the chapter should help students to—

- Give approximate values for blood flow per unit weight and blood flow per organ at rest in the major organs of the body.
- List the unique gross and microscopic aspects of the circulation of the brain.
- Describe the formation, absorption, and functions of cerebrospinal fluid.
- Outline the characteristics of the blood-brain barrier, and comment on its importance in clinical medicine.
- Describe the circumventricular organs, and list their general functions.
- Summarize the main anatomic features of the coronary circulation.
- List the chemical and neural factors that regulate the coronary circulation, and describe the role of each.
- Outline the unique features of the circulation of the liver and the splanchnic bed, and comment on the reservoir function of the splanchnic circulation.
- Describe the triple response produced by firmly stroking the skin, and explain each of its components.

- Describe the operation of the placenta as the "fetal lung."
- Diagram the circulation of the fetus before birth, and list the changes that occur in it at birth.

GENERAL QUESTIONS

1. What is the axon reflex? What is the evidence for its existence? Which neurotransmitter is probably involved?

2. What is reactive hyperemia? Where does it occur?

3. What is a neurohemal organ? What are the principal neurohemal organs in the body?

4. How can positron emission tomography and MRI be used to study the metabolism and blood flow of the brain?

5. Discuss the use of radionuclides in the study of the coronary circulation.

6. What is the Monro-Kellie doctrine, and what are its physiologic consequences?

7. What is a hepatic acinus? Why is the acinar organization of the liver important, and what is its physiologic significance?

MULTIPLE-CHOICE QUESTIONS

In questions 1–11, select the single best answer.

Figure 32–A. Sagittal view of the human brain.

Questions 1–5 refer to Fig 32–A.

1. Main site at which hypothalamic hormones leave the brain

2. Main site at which vasopressin acts to decrease cardiac output

3. Chemoreceptor area that is a trigger zone for vomiting

4. Main site where changes in plasma osmolality act to alter vasopressin secretion

5. Main site at which angiotensin II acts to increase water intake

6. Blood in which of the following vessels normally has the lowest P_{O_2}?
   (A) Maternal artery
   (B) Maternal uterine vein
   (C) Maternal femoral vein
   (D) Umbilical artery
   (E) Umbilical vein

7. The pressure differential between the heart and the aorta is least in the
   (A) left ventricle during systole
   (B) left ventricle during diastole
   (C) right ventricle during systole
   (D) right ventricle during diastole
   (E) left atrium during systole

8. Injection of tissue plasminogen activator (TPA) would probably be most beneficial
   (A) after at least 1 year of uncomplicated recovery following occlusion of a coronary artery
   (B) after at least 2 months of rest and recuperation following occlusion of a coronary artery
   (C) during the second week after occlusion of a coronary artery
   (D) during the second day after occlusion of a coronary artery
   (E) during the second hour after occlusion of a coronary artery

9. Which of the following organs has the greatest blood flow per 100 g of tissue?
   (A) Brain
   (B) Heart muscle
   (C) Skin
   (D) Liver
   (E) Kidneys

10. Which of the following organs has the most permeable capillaries?
    (A) Brain
    (B) Posterior pituitary gland
    (C) Liver
    (D) Small intestine
    (E) Kidneys

11. Which of the following does *not* dilate arterioles in the skin?
    (A) Increased body temperature
    (B) Epinephrine
    (C) Bradykinin
    (D) Substance P
    (E) Vasopressin

*In questions 12–15, indicate whether the first item is greater than (**G**), the same as (**S**), or less than (**L**) the second item.*

12. Amount of water actively transported into the CSF

    G S L

    Amount of water actively transported out of the CSF

13. Blood flow in the motor cortex of the brain at rest

    G S L

    Blood flow in the motor cortex of the brain when the fists are clenched

14. Blood flow in the endocardial portion of the myocardium during systole

    G S L

    Blood flow in the epicardial portion of the myocardium during systole

15. Amount of blood brought to the liver by the hepatic artery

    G S L

    Amount of blood brought to the liver by the portal vein.

Chapter 33

This chapter summarizes the cardiovascular adjustments that occur on assuming the upright position, during space flight, and during exercise. It also describes the pathophysiology of shock, syncope (fainting), hypertension, and heart failure and the cardiovascular compensations that occur in these conditions. The material in the chapter should help students to—

- Outline the compensatory mechanisms that maintain blood pressure on rising from the supine to the standing position.
- Describe the effects of positive and negative gravitational acceleration (g) on the body.
- Describe and explain the circulatory changes that occur during exercise.
- Outline the main beneficial effects

of athletic training on the cardiovascular system and other systems in the body.

- Define shock, name its major causes, and summarize the main abnormalities that occur during each type of shock. Summarize the short- and long-term compensatory mechanisms that operate in each type.
- List the common causes of fainting.
- List the common causes of high blood pressure in humans, and divide them into those that are now curable and those that are not.
- List the main symptoms of heart failure, and describe how each is produced.

GENERAL QUESTIONS

1. Compare the composition of blood from exercising skeletal muscle with that of blood from resting muscle, and describe how the composition of blood in the systemic circulation changes during exercise.

2. What are the similarities and differences between hypovolemic shock and low-resistance shock?

3. What is refractory shock? How can it be prevented?

4. Discuss the physiologic advantages and disadvantages of treating shock by raising the foot of the bed.

5. Why do some people faint when they rise from the supine to the standing position?

6. "Hypertension is a syndrome, not a disease." Discuss this statement and its implications.

7. List 5 ways that chronic hypertension can be produced in experimental animals. Which mechanisms produce the elevation of blood pressure in each case? Discuss the similarities and differences between each form of experimental hypertension and human hypertension.

8. Why is vital capacity reduced in heart failure? What is the effect of posture on this reduction?

9. Why are angiotensin-converting enzyme inhibitors of value in the treatment of congestive heart failure?

10. Mr. Smith, a 55-year-old man who weighs 71 kg, feels tired after relatively mild exertion. On physical examination, he is found to have a moderately intense basal systolic murmur. During cardiac catheterization, the following data were obtained:

Heart rate (beats/min): 64

Ventilation (L/min): 4.98

O_2 consumption (mL/min): 246

Pulmonary arterial O_2 (mL of O_2/L of blood): 137

Pulmonary venous O_2 (mL of O_2/L of blood): 189

Right ventricular pressure (mm Hg): 27/2

Pulmonary artery pressure (mm Hg): 27/10

Left ventricular pressure (mm Hg): 169/0

Aortic pressure (mm Hg): 108/72

What is Mr. Smith's cardiac output? What is his stroke volume? What is the most likely diagnosis?

MULTIPLE-CHOICE QUESTIONS

Each of the patients in questions 1–5 has a blood pressure of 85/40 mm Hg. Match the condition in each question with the lettered emergency treatment that is most appropriate. Each lettered treatment may be selected once, more than once, or not at all.

(A) Injection of dopamine
(B) Infusion of concentrated human albumin
(C) Infusion of isotonic saline solution
(D) Injection of epinephrine
(E) Infusion of whole blood

1. Girl stung by a bee

2. Man bleeding from a stab wound

3. Woman burned over 35% of her body in a kitchen fire

4. Woman who had a myocardial infarction 24 hours previously

5. Man with severe diarrhea

In questions 6–12, select the single best answer.

6. Plasma renin activity is most likely to be lower than normal
   (A) in congestive heart failure
   (B) in hemorrhagic shock
   (C) in shock due to infection with gram-negative bacteria
   (D) in essential hypertension
   (E) during quiet standing

7. Which of the following takes longest to return to normal after 1 L of blood is removed from a normal individual?
   (A) Plasma aldosterone concentration
   (B) Blood pressure
   (C) Renin secretion
   (D) Plasma volume
   (E) Number of red blood cells in peripheral blood

8. Which of the following is *least* likely to cause sustained hypertension?
   (A) Chronically increased secretion of the adrenal medulla
   (B) Chronically increased secretion of the zona fasciculata and zona reticularis of the adrenal cortex
   (C) Chronically increased secretion of the zona glomerulosa of the adrenal cortex
   (D) Chronically increased secretion of the posterior pituitary gland
   (E) Chronic treatment with oral contraceptives

9. Which of the following is *not* increased during exercise?
   (A) Respiratory rate
   (B) Stroke volume
   (C) Heart rate
   (D) Total peripheral resistance
   (E) Systolic blood pressure

10. Which of the following are *incorrectly* paired?
   (A) Loss of blood : hypotension
   (B) Negative g : blackout
   (C) Increased cardiac output : exercise
   (D) Renal artery constriction : increased blood pressure
   (E) Decreased total peripheral resistance : fainting

11. Which of the following would you expect to decrease in a normal individual who stands quietly in the same position for 1 hour?
   (A) Hematocrit
   (B) Diameter of the thigh
   (C) Plasma renin activity
   (D) Plasma vasopressin concentration
   (E) Central venous pressure

12. Which of the following is *least* likely to cause fainting?
   (A) Pressure on the carotid body
   (B) Paroxysmal ventricular fibrillation

(C) Autonomic insufficiency
(D) Strong emotion
(E) Myocardial infarction

In questions 13–15, indicate whether the first item is greater than (G), the same as (S), or less than (L) the second item.

13. Interstitial fluid volume at rest

    G S L

    Interstitial fluid volume after hemorrhage of 15 mL/kg body weight

14. Blood flow to brain during exercise

    G S L

    Blood flow to brain at rest

15. Blood flow to kidneys during exercise

    G S L

    Blood flow to kidneys at rest

Chapter 34

This chapter is an analysis of the functions of the respiratory system, including the properties of gases, the way that the lungs and chest operate to produce inspiration and expiration, gas exchange in the lungs, the special features of the pulmonary circulation, lung defense mechanisms, and the metabolic functions of the lungs. The material in the chapter should help students to—

- Define partial pressure, and calculate the partial pressure of each of the important gases in the atmosphere at sea level.
- Draw a graph of the changes in intrapulmonary and intrapleural pressure and lung volume that occur during inspiration and expiration.
- Describe the passages through which air passes from the exterior to the alveoli and the cells that line them.
- List the major muscles involved in respiration, and state the role of each.
- Define tidal volume, inspiratory reserve volume, expiratory reserve volume, and residual volume, and give approximate values for each in a normal adult.
- Define compliance, and give examples of diseases in which it is abnormal.

- Describe the chemical composition and function of surfactant.
- List the factors that determine alveolar ventilation.
- Define diffusion capacity, and compare the diffusion of O_2 with that of CO_2 in the lungs.
- Compare the pulmonary and systemic circulations, listing the main differences between them.
- Describe the metabolic functions of the lungs.

GENERAL QUESTIONS

1. Discuss the law of Laplace as it relates to pulmonary function.

2. In the standing position, how does blood flow at the apex of each lung compare with blood flow at the base? How does ventilation compare? Explain the differences.

3. What is a hysteresis loop? Give an example in respiratory physiology, and explain its occurrence.

4. What are the functions of epithelial cells and pulmonary macrophages?

5. What is the difference between the anatomic and physiologic dead spaces? What role do they play in pulmonary disease?

6. What factors affect the diffusion capacity of the lungs for O_2?

7. Why is the P_{O_2} of blood in the aorta slightly less than the P_{O_2} of blood in the pulmonary veins?

8. What components contribute to the work of breathing? Discuss the normal and pathologic conditions that cause alterations in each component.

MULTIPLE-CHOICE QUESTIONS

In questions 1–8, select the single best answer.

1. On the summit of Mt. Everest, where the barometric pressure is about 250 mm Hg, the partial pressure of O_2 is about
   (A) 0.1 mm Hg
   (B) 0.5 mm Hg
   (C) 5 mm Hg
   (D) 50 mm Hg
   (E) 100 mm Hg

2. The approximate amount of gas left in the lungs after maximal

forced expiration in a normal woman is
   (A) zero
   (B) 0.1 L
   (C) 1.1 L
   (D) 3.1 L
   (E) 4.2 L

3. The tidal volume in a normal man at rest is about
   (A) 0.5 L
   (B) 1.2 L
   (C) 2.5 L
   (D) 4.9 L
   (E) 6.0 L

4. What is the approximate dead space of a normal 70-kg man breathing through a tube that has a radius of 5 mm and a length of 100 cm?
   (A) 150 mL
   (B) 180 mL
   (C) 230 mL
   (D) 280 mL
   (E) 350 mL

5. Which of the following is responsible for the movement of O_2 from the alveoli into the blood in the pulmonary capillaries?
   (A) Active transport
   (B) Filtration
   (C) Secondary active transport
   (D) Facilitated diffusion
   (E) Passive diffusion

Questions 6–8 refer to Fig 34–A.

Figure 34–A. Lung volumes at various transpulmonary pressures.

6. Which of the labeled curves would you expect to see in a normal individual?

7. Which of the labeled curves would you expect to see in a patient with severe pulmonary fibrosis?

8. Which of the labeled curves would you expect to see in a patient with advanced emphysema?

In questions 9–13, one or more than one of the answers may be correct. Select

A if **(1)**, **(2)**, and **(3)** are correct;
B if **(1)** and **(3)** are correct;
C if **(2)** and **(4)** are correct;
D if only **(4)** is correct; and
E if all are correct

9. Which of the following causes contraction of bronchial smooth muscle?
   (1) Leukotrienes
   (2) Vasoactive intestinal polypeptide (VIP)
   (3) Acetylcholine
   (4) Epinephrine

10. Surfactant lining the alveoli
    (1) helps prevent alveolar collapse
    (2) is decreased in hyaline membrane disease
    (3) is decreased in the lungs of heavy smokers
    (4) is a mixture of proteins and lipids

11. During inspiration, there is an increase in
    (1) intra-abdominal pressure
    (2) venous return to the heart
    (3) heart rate
    (4) intrapleural pressure

12. Which of the following would be expected to cause a reduction in pulmonary ventilation?
    (1) Transection of both phrenic nerves
    (2) Transection of the spinal cord at the first thoracic level
    (3) A large dose of morphine
    (4) Pulmonary fibrosis

13. Airway resistance
    (1) is increased if the lungs are removed and inflated with saline
    (2) does not affect the work of breathing
    (3) is increased in paraplegic patients
    (4) is increased in asthma

In questions 14 and 15, indicate whether the first item is greater than **(G)**, *the same as* **(S)**, *or less than* **(L)** *the second item.*

14. Alveolar ventilation when the respiratory rate is 25 breaths/min and the tidal volume is 200 mL.

    G S L

    Alveolar ventilation when the respiratory rate is 10 breaths/min and the tidal volume is 500 mL

15. Angiotensin II concentration in blood in pulmonary veins

    G S L

    Angiotensin II concentration in blood in renal veins

Chapter 35

This chapter describes the flow of O_2 from the lungs to the tissues and the flow of CO_2 from the tissues to the lungs, with emphasis on the physical and chemical mechanisms that greatly augment the ability of the blood to carry O_2 and CO_2. The material in the chapter should help students to—

- Describe the manner in which O_2 flows "downhill" from the lungs to the tissues and CO_2 flows "downhill" from the tissues to the lungs.
- Describe the reactions of O_2 with hemoglobin and the oxygen-hemoglobin dissociation curve.
- List the important factors affecting the affinity of hemoglobin for O_2, and comment on the physiologic significance of each.
- Describe myoglobin, and outline its physiologic role.
- Write the Henderson-Hasselbalch equation, and comment on its significance.
- List the principal buffers in blood, interstitial fluid, and intracellular fluid.
- List the reactions that increase the amount of CO_2 in the blood, and draw the CO_2 dissociation curve for arterial and venous blood.

GENERAL QUESTIONS

1. Why is the Cl^- concentration inside red blood cells in venous blood greater than that in red blood cells in arterial blood? Describe the mechanisms responsible for this difference.

2. Why can oxyhemoglobin bind less H^+ than can reduced hemoglobin?

3. What is the Bohr effect? How is it brought about, and what is its physiologic significance?

4. Discuss 2,3-diphosphoglycerate (2,3-DPG). Define P_{50}, and describe how 2,3-DPH affects the P_{50} in blood.

5. What is the role of carbonic anhydrase in red blood cells?

MULTIPLE-CHOICE QUESTIONS

In questions 1–6, select the single best answer.

1. Most of the CO_2 transported in the blood is
   (A) dissolved in plasma
   (B) in carbamino compounds formed from plasma proteins
   (C) in carbamino compounds formed from hemoglobin
   (D) bound to Cl^-
   (E) in HCO_3^-

2. Which of the following has the greatest effect on the ability of blood to transport oxygen?
   (A) Capacity of the blood to dissolve oxygen
   (B) Amount of hemoglobin in the blood
   (C) pH of plasma
   (D) CO_2 content of red blood cells
   (E) Temperature of the blood

3. Which of the following has the greatest buffering capacity in the interstitial fluid?
   (A) Hemoglobin system
   (B) Bicarbonate system
   (C) Phosphate system
   (D) Proteins in the interstitial fluid
   (E) Ammonium-ammonia system

4. Which of the following is *not* true of the system

   $$CO_2 + H_2O \overset{1}{\rightleftharpoons} H_2CO_3 \overset{2}{\rightleftharpoons} H^+ + HCO_3^- ?$$

   (A) reaction 1 is catalyzed by carbonic anhydrase
   (B) Because of reaction 2, the pH of blood declines during breath holding
   (C) Reaction 1 occurs in the kidneys
   (D) Reaction 1 occurs primarily in plasma
   (E) The reactions move to the left when there is excess H^+ in the tissues

In questions 5–8, select

A if the item is associated with **(a)** below,
B if the item is associated with **(b)** below,
C if the item is associated with both **(a)** and **(b)**, and

D if the item is associated with neither **(a)** nor **(b)**

(a) Involved in O_2 transport in blood

(b) Involved in CO_2 transport in blood

5. Carbamino compounds

6. Myoglobin

7. Hydrogen ions

8. Plasma proteins

In questions 9 and 10, indicate whether the first item is greater than **(G),** *the same as* **(S),** *or less than* **(L)** *the second item.*

9. Affinity of fetal hemoglobin (hemoglobin F) for O_2

   G S L

   Affinity of adult hemoglobin (hemoglobin A) for O_2

10. Number of osmotically active particles in red blood cells in arterial blood

    G S L

    Number of osmotically active particles in red blood cells in venous blood

Chapter 36

This chapter is a review of the mechanisms that regulate respiration. These include the fundamental rhythm generator in the medulla oblongata and the pons; chemical control via the carotid and aortic chemoreceptors and the chemoreceptors located on the ventral surface of the medulla oblongata; and nonchemical control via inputs from higher centers and receptors in the lungs, muscles, tendons, and joints. The material in the chapter should help students to—

- Locate and comment on the function of the dorsal and ventral groups of respiratory neurons, the pneumotaxic center, and the apneustic center in the brain stem.
- List the specific respiratory functions of the vagus nerves and the respiratory receptors in the carotid body, the aortic body, and the ventral surface of the medulla oblongata.
- Describe and explain the ventilatory responses to increased CO_2 concentrations in the inspired air.

- Describe and explain the ventilatory responses to decreased O_2 concentrations in the inspired air.
- Describe the effects of each of the main nonchemical factors that influence respiration.
- Describe the effect of sleep on respiration.

GENERAL QUESTIONS

1. Discuss the Hering-Breuer reflexes.

2. What role does the blood-brain barrier play in the regulation of respiration?

3. What is Ondine's curse? Discuss its relation to sudden infant death syndrome.

4. What is sleep apnea? Discuss the factors that affect it.

5. What is the breaking point? Discuss the factors that affect it.

6. How does hypoxia generate increased numbers of impulses in the afferent nerves from the carotid chemoreceptors? Discuss the cellular and molecular mechanisms that are involved.

MULTIPLE-CHOICE QUESTIONS

In questions 1–8, select the single best answer.

1. Which of the following discharge spontaneously during quiet breathing?
   (A) Inspiratory neurons
   (B) Motor neurons to respiratory muscles
   (C) Neurons in the apneustic center
   (D) Expiratory neurons
   (E) Stretch receptors in the lungs

2. Intravenous lactic acid increases ventilation. The receptors responsible for this effect are located in the
   (A) medulla oblongata
   (B) carotid bodies
   (C) lung parenchyma
   (D) aortic baroreceptors
   (E) trachea and large bronchi

3. Spontaneous respiration ceases after
   (A) transection of the brain stem above the pons
   (B) transection of the brain stem at the caudal end of the medulla

   (C) bilateral vagotomy
   (D) bilateral vagotomy combined with transection of the brain stem at the superior border of the pons
   (E) transection of the spinal cord at the level of the first thoracic segment

4. The following physiologic events that occur in the body are listed in random order:
   (1) Decreased CSF pH
   (2) Increased arterial P_{CO_2}
   (3) Increased CSF P_{CO_2}
   (4) Stimulation of medullary chemoreceptors
   (5) Increased alveolar P_{CO_2}

   What is the usual sequence in which they occur when they affect respiration
   (A) 1, 2, 3, 4, 5
   (B) 4, 1, 3, 2, 5
   (C) 3, 4, 5, 1, 2
   (D) 5, 2, 3, 1, 4
   (E) 5, 3, 2, 4, 1

5. Stimulation of the central (proximal) end of a cut vagus nerve would *not* be expected to
   (A) increase heart rate
   (B) inhibit inspiration
   (C) cause coughing
   (D) lower blood pressure
   (E) cause apnea

6. Injection of a drug that stimulates the carotid bodies would be expected to cause
   (A) a decrease in the pH of arterial blood
   (B) an increase in the P_{O_2} of arterial blood
   (C) an increase in the HCO_3^- concentration of arterial blood
   (D) an increase in urinary Na^+ excretion
   (E) an increase in plasma Cl^-

7. The main respiratory control neurons
   (A) send out regular bursts of impulses to expiratory muscles during quiet respiration
   (B) are unaffected by stimulation of pain receptors
   (C) are located in the pons
   (D) send out regular bursts of impulses to inspiratory muscles during quiet respiration
   (E) are unaffected by impulses from the cerebral cortex

8. Variations in which of the following components of blood or cerebrospinal fluid do *not* affect respiration?

(A) Arterial HCO_3^- concentration

(B) Arterial H^+ concentration

(C) Arterial Na^+ concentration

(D) Cerebrospinal fluid CO_2 concentration

(E) Cerebrospinal fluid H^+ concentration

In questions 9 and 10, indicate whether the first item is greater than (G), *the same as* (S), *or less than* (L) *the second item.*

9. Proportion of the alteration in ventilation produced by changes in arterial P_{CO_2} that is due to medullary chemoreceptors

   G S L

   Proportion of the alteration in ventilation produced by changes in arterial P_{CO_2} that is due to peripheral chemoreceptors

10. Blood flow per gram of tissue to carotid bodies

    G S L

    Blood flow per gram of tissue to brain

Chapter 37

This chapter analyzes the changes in respiration that occur with exercise, during exposure to altitude and other forms of hypoxia, during hypercapnia and hypocapnia, and with increased barometric pressure. Drowning and artificial respiration are also discussed. The material in the chapter should help students to—

- Describe the effects of exercise on ventilation and O_2 exchange in the tissues.
- Define hypoxia, and describe its 4 principal forms.
- Describe the acute effects of high altitude on respiration, and discuss acclimatization to altitude.
- Define and give examples of ventilation-perfusion imbalance.

- List and explain the effects of carbon monoxide on the body.
- Summarize the abnormalities that occur in emphysema and asthma.
- List and explain the adverse effects of excess O_2.
- Describe the effects of hypercapnia and hypocapnia, and give examples of conditions that can cause them.
- Define periodic breathing, and explain its occurrence in various disease states.
- Describe in detail the technique of mouth-to-mouth resuscitation, and explain how it works to maintain life.

GENERAL QUESTIONS

1. By means of a diagram, illustrate the relationship between the partial pressure of O_2 in blood and the percent saturation of hemoglobin at rest, and the changes produced in this relationship by exercise.

2. What is the oxygen debt mechanism? How is an oxygen debt measured? What is the value of the mechanism to the individual?

3. Discuss fatigue from the point of view of its causes, prevention, and physiologic significance.

4. What is Cheyne-Stokes respiration? Explain its occurrence.

5. In patients with severe respiratory failure who are hypercapnic and hypoxic, administration of O_2 may stop respiration and even cause death if artificial respiration is not instituted. Why?

6. Diving not only is performed by professionals but, with the development of scuba equipment, has become a popular sport. What medical problems can be produced by diving, and how would you treat them?

MULTIPLE-CHOICE QUESTIONS

In questions 1–12, select the single best answer.

1. In which of the following conditions is CO_2 retention most likely to occur?
   (A) Climbing a high mountain
   (B) Ventilatory failure
   (C) Carbon monoxide poisoning
   (D) Lung failure
   (E) Hysterical hyperventilation

Questions 2–6 refer to Table 37–A.

2. Which set of data would you expect to find in a normal person?

3. Which set of data would you expect to find in a person at rest breathing air at an altitude of 3000 m (10,000 feet; barometric pressure = 520 mm Hg)?

4. Which set of data would you expect to find in a patient with a large left-to-right shunt?

5. Which set of data would you expect to find in a patient with a collapsed lung?

6. Which set of data would you expect to find in a normal subject at rest breathing 100% O_2 in an unpressurized airplane cabin at 10,000 m (32,800 feet; barometric pressure = 180 mm Hg)?

Questions 7 and 8 refer to Table 37–B.

Table 37–B. Arterial blood values.

	P_{CO_2} (mm Hg)	pH	HCO_3^- (meq/L)
A	60	7.25	29
B	25	7.30	12
C	42	7.55	35
D	25	7.50	20
E	60	7.05	15

Table 37–A. P_{O_2} (mm Hg).

	Superior Vena Cava	Right Ventricle	Alveolar Gas	Left Ventricle
A	45	40	104	94
B	40	35	105	94
C	40	35	60	55
D	40	35	104	55
E	45	80	104	94

7. Which set of data would you expect to find in a man who is breathing through a tube that greatly increases his dead space?

8. Which set of data would you expect to find in a woman who has flown to an altitude of 4500 m (14,750 feet) in the open cockpit of an airplane?

9. Pulmonary fibrosis would be expected to produce
(A) histotoxic hypoxia
(B) stagnant hypoxia
(C) decreased vital capacity
(D) cyanosis
(E) emphysema

10. Which of the following deleterious effects would probably *not* be produced by chronic cigarette smoking?
(A) Patches of atelectasis
(B) Myocardial ischemia
(C) Loss of elastic tissue in the lung
(D) Decreased anatomic dead space
(E) Increased carbonmonoxyhemoglobin in blood

11. O_2 delivery to the tissues would be reduced to the greatest extent in
(A) a normal subject breathing 100% O_2 on top of Mt. Everest
(B) a normal subject running a marathon at sea level
(C) a patient with carbon monoxide poisoning
(D) a patient who has ingested cyanide
(E) a patient with moderately severe metabolic acidosis

12. Which of the following is *not* a manifestation of oxygen toxicity?
(A) Irritation of the respiratory tract
(B) Difficulty in seeing because of retrolental fibroplasia
(C) Convulsions
(D) Lung cysts in infants
(E) Rapture of the deep

In questions 13–15, select

A if the item is associated with (a) below,
B if the item is associated with (b) below,
C if the item is associated with both (a) and (b), and
D if the item is associated with neither (a) nor (b)

(a) Marked decrease in arterial P_{O_2}

(b) Air embolism

13. Sudden loss of cabin pressurization in a jet airplane at 12,500 m (41,000 feet)

14. Rapid ascent while holding breath after 35 minutes of scuba diving in 15 m (49 feet) of salt water

15. Hemothorax

Chapter 38

This chapter is concerned with renal function, including glomerular filtration and the factors affecting it, tubular reabsorption and secretion and their control, and the function of the ureters and bladder. The material in the chapter should help students to—

- Describe the morphology of a typical nephron and its blood supply.
- Define autoregulation, give normal values for renal blood flow, and list the major theories advanced to explain autoregulation in the kidneys.
- Outline the functions of the renal nerves.
- Define glomerular filtration rate, describe how it can be measured, and list the major factors affecting it.
- Define and discuss tubuloglomerular feedback and glomerulotubular balance.
- Discuss tubular reabsorption and secretion, using glucose and K^+ as examples.
- Outline tubular handling of Na^+.
- Summarize tubular handling of Cl^-, HCO_3^-, urea, and uric acid in terms of amounts filtered, secreted, reabsorbed, and excreted in urine.
- Describe how the countercurrent mechanism in the kidney operates to produce a hypertonic or hypotonic urine.
- Outline the processes involved in the secretion of H^+ into the tubules, and discuss the significance of these processes in the regulation of acid-base balance.
- List the major classes of diuretics and how each operates to increase urine flow.
- Describe the voiding reflex and draw a cystometrogram.

GENERAL QUESTIONS

1. Explain why, when the Tm of a substance that is secreted by the tubules is reached, clearance of the substance decreases as its plasma concentration increases.

2. Why does the clearance of glucose rise from zero and gradually approach the clearance of inulin as the plasma glucose level increases?

3. What is the physiologic role of the mesangial cells in the glomeruli, and how do they carry it out?

4. Compare the cellular mechanisms responsible for H^+ secretion in the proximal tubule with those in the distal tubule and those in the gastric mucosa.

5. Discuss the mechanisms responsible for the adaptation of NH_4^+ excretion that develops over a period of days in prolonged acidosis.

6. Why is acidosis a common complication of chronic renal disease? How would you treat it?

7. The following observations were made on a patient

Plasma $[HCO_3^-]$: 20 meq/L

GFR: 125 mL/min

24-hour urine volume: 1500 mL

Urinary $[HCO_3^-]$: 25 meq/L

Urinary $[NH_4^+]$: 75 meq/L

Urinary titratable acidity: 50 meq/L

    a. Approximately how much HCO_3^- is being reabsorbed per 24 hours?
    b. How much Na^+ is being reabsorbed with the HCO_3^-?
    c. How much H^+ is being secreted by the renal tubules per 24 hours?

MULTIPLE-CHOICE QUESTIONS

In questions 1–5, which refer to Table 38–A, match the disease or condition in each question with the lettered pattern of laboratory findings in Table 38–A that is most closely associated with it. Each lettered pattern may be selected once, more than once, or not at all.

Table 38–A

	24-Hour Urine Volume (L)	Ketones	Glucose	Protein
A	1.4	+	0	0
B	6.2	2+	4+	0
C	1.6	0	0	4+
D	6.4	0	0	0
E	0.4	0	0	0

1. Diabetes insipidus

2. Nephrosis

3. Fasting

4. Dehydration

5. Diabetes mellitus

In questions 6–15, select the single best answer.

6. In the presence of vasopressin, the greatest fraction of filtered water is absorbed in the
   (A) proximal tubule
   (B) loop of Henle
   (C) distal tubule
   (D) cortical collecting duct
   (E) medullary collecting duct

7. In the absence of vasopressin, the greatest fraction of filtered water is absorbed in the
   (A) proximal tubule
   (B) loop of Henle
   (C) distal tubule
   (D) cortical collecting duct
   (E) medullary collecting duct

8. If the clearance of a substance which is freely filtered is less than that of inulin
   (A) there is net reabsorption of the substance in the tubules
   (B) there is net secretion of the substance in the tubules
   (C) the substance is neither secreted nor reabsorbed in the tubules
   (D) the substance becomes

bound to protein in the tubules
   (E) the substance is secreted in the proximal tubule to a greater degree than in the distal tubule

9. Glucose reabsorption occurs in
   (A) proximal tubule
   (B) loop of Henle
   (C) distal tubule
   (D) cortical collecting duct
   (E) medullary collecting duct

10. On which of the following does aldosterone exert its greatest effect?
    (A) Glomerulus
    (B) Proximal tubule
    (C) Thin portion of the loop of Henle
    (D) Thick portion of the loop of Henle
    (E) Cortical collecting duct

11. What is the clearance of a substance when its concentration in the plasma is 10 mg/dL, its concentration in the urine is 100 mg/dL, and urine flow is 2 mL/min?
    (A) 2 mL/min
    (B) 10 mL/min
    (C) 20 mL/min
    (D) 200 mL/min
    (E) Clearance cannot be determined from the information given

12. As urine flow increases during osmotic diuresis

(A) the osmolality of urine falls below that of plasma
(B) the osmolality of urine increases because of the increased amounts of non-reabsorbable solute in the urine
(C) the osmolality of urine approaches that of plasma because plasma leaks into the tubules
(D) the osmolality of urine approaches that of plasma because an increasingly large fraction of the excreted urine is isotonic proximal tubular fluid
(E) the action of vasopressin on the renal tubules is inhibited

Questions 13–15 refer to the data in Table 38–B, which were obtained in a normal woman.

13. The glomerular filtration rate in the experimental period is
    (A) 4 times that in the control period
    (B) twice that in the control period
    (C) the same as in the control period
    (D) one-half that in the control period
    (E) one-quarter that in the control period

14. The clearance of urea in the experimental period is

Table 38–B

	Control Period	Experimental Period
Arterial plasma Inulin (mg/mL)	0.004	0.004
Glucose (mg/dL)	100 (1 mg/mL)	300 (3 mg/mL)
Urea (μmol/mL)	5	5
Urine Inulin (mg/mL)	0.4	0.2
Glucose (mg/mL)	0	5
Urea (μmol/mL)	300	160
Urine flow (mL/min)	1	2

(A) increased, probably because the urine flow is increased

(B) increased, probably because the urine glucose concentration is increased

(C) the same as in the control period

(D) decreased, probably because the urine flow is increased

(E) decreased, probably because the amount of urea filtered is decreased

15. The Tm for glucose in this woman
    (A) is 100 mg/min
    (B) is 130 mg/min
    (C) is 200 mg/min
    (D) is 290 mg/min
    (E) cannot be calculated from the data given

Questions 16–20 refer to Fig 38–A. Select the letter or letters identifying the appropriate part of the nephron. Each lettered item may be selected once, more than once, or not at all. Note that in some instances, the correct answer may be more than one letter.

Figure 38–A. Juxtamedullary nephron.

16. Site(s) at which furosemide acts

17. Site(s) as which thiazides act

18. Site(s) at which tubular fluid osmolality exceeds that of plasma in the absence of vasopressin

19. Site(s) at which Na$^+$ is actively reabsorbed

20. Site(s) at which K$^+$ is secreted

Chapter 39

This chapter is a review of the homeostatic mechanisms that operate to maintain the osmolality, volume, and ionic composition of the extracellular fluid within normal limits. This includes the concentration of H$^+$, with consideration of respiratory and metabolic acidosis and alkalosis. The material in the chapter should help students to—

• Describe how the tonicity (osmolality) of the extracellular fluid is maintained by alterations in water intake and vasopressin secretion.

• Describe how the volume of the extracellular fluid is maintained by alterations in renin and aldosterone secretion.

• Name the mechanisms that operate to maintain the constancy of plasma concentrations of glucose and Ca^{2+}.

• Define acidosis and alkalosis, and give (in meq/L and pH) the normal mean and the range of H$^+$ concentrations in blood that are compatible with health.

• List the principal buffers in blood, interstitial fluid, and intracellular fluid; and, using the Henderson-Hasselbalch equation, describe what is unique about the bicarbonate buffer system.

• Describe the changes in blood chemistry that occur during the development of metabolic acidosis and metabolic alkalosis, and the respiratory and renal compensations for these conditions.

• Describe the changes in blood chemistry that occur during the development of respiratory acidosis and respiratory alkalosis, and the renal compensation for these conditions.

GENERAL QUESTIONS

1. What are the main sources of the acid loads presented to the body in everyday living? What are some common diseases that cause increased acid loads in the body, and in each, how is the load produced?

2. Compare and contrast metabolic acidosis and respiratory acidosis.

3. Which conditions might be present in patients with the following blood chemistry values, and why?

   a. pH, 7.28; HCO$_3^-$, 18.1 meq/L; P$_{CO_2}$, 40 mm Hg.
   b. pH, 7.34; HCO$_3^-$, 25 meq/L; P$_{CO_2}$, 48 mm Hg.

4. Describe and explain the alterations in extracellular fluid volume and acid-base balance that occur in patients with chronically elevated plasma aldosterone concentrations due to primary hyperaldosteronism.

5. Describe by means of a diagram plotting plasma HCO$_3^-$ concentration against pH (a Davenport diagram) the immediate and more long-term changes that occur in the acid-base balance of a normal individual who hyperventilates for 5 minutes.

MULTIPLE-CHOICE QUESTIONS

Questions 1–4 refer to the numbered points on the nomogram in Fig 39–A. Match the numbered point in each question with the lettered condition that is most closely associated with it. Each lettered condition may be selected once, more than once, or not at all.

(A) Values seen in a mountain climber after several weeks at high altitude

(B) Values seen in long-standing pulmonary disease

(C) Values seen in diabetic coma

(D) Values seen after 5 minutes of hyperventilation

(E) Values seen after prolonged vomiting

1. Point 1

2. Point 2

3. Point 3

4. Point 4

In questions 5–8, select the single best answer.

5. Increasing alveolar ventilation increases the blood pH because
   (A) it activates neural mechanisms that remove acid from the blood
   (B) it makes hemoglobin a stronger acid
   (C) it increases the P$_{O_2}$ of the blood
   (D) it decreases the P$_{CO_2}$ in the alveoli
   (E) the increased muscle work of increased breathing generates more CO$_2$

Figure 39–A. Siggaard-Andersen curve nomogram. (Courtesy of O Siggaard-Andersen and Radiometer, Copenhagen, Denmark.)

6. In uncompensated respiratory alkalosis
  (A) the plasma pH, the plasma HCO_3^- concentration and the arterial P_{CO_2} are all low
  (B) the plasma pH is high and the plasma HCO_3^- concentration and the arterial P_{CO_2} are low
  (C) the plasma pH and the plasma HCO_3^- concentration are low and the arterial P_{CO_2} is normal
  (D) the plasma pH is high, the plasma HCO_3^- concentration is low, and the arterial P_{CO_2} is normal
  (E) the plasma pH is low, the plasma HCO_3^- concentration is high, and the arterial P_{CO_2} is normal

7. In a patient with a plasma pH of 7.10, the $[HCO_3^-]/[H_2CO_3]$ ratio in plasma is
  (A) 20
  (B) 10
  (C) 2
  (D) 1
  (E) 0.1

8. In a patient who has become dehydrated, body water should be replaced by intravenous infusion of
  (A) distilled water
  (B) 0.9% sodium chloride solution
  (C) 5% glucose solution
  (D) hyperoncotic albumin
  (E) 10% glucose solution

In questions 9 and 10, indicate whether the first item is greater than **(G)**, *the same as* **(S)**, *or less than* **(L)** *the second item.*

9. Na^+ concentration in normal plasma

  G S L
  Na^+ concentration in isotonic saline

10. Osmolality of extracellular fluid after drinking 1 L of isotonic glucose solution

  G S L
  Osmolality of extracellular fluid after drinking 1 L of isotonic saline solution

Answers to Quantitative
and Multiple-Choice Questions

Chapter 1

1. C
2. E
3. D
4. A
5. C
6. E
7. B
8. D
9. E
10. C
11. B
12. E
13. B
14. G
15. L
16. S
17. L
18. G
19. G
20. L

Chapter 2

1. B
2. B
3. E
4. B
5. E
6. B
7. G
8. L
9. L
10. L

Chapter 3

General question 3.
Fast muscle, >133; slow
muscle, >10.

1. B
2. D
3. B
4. E
5. C
6. C
7. C
8. C
9. D
10. C
11. G
12. G
13. G
14. L
15. L

Chapter 4

1. D
2. E
3. C
4. E
5. A
6. D
7. C
8. E
9. C
10. C
11. D
12. C

13. D
14. E
15. L
16. G
17. L
18. S
19. G
20. L

Chapter 5

1. C
2. D
3. E
4. E
5. B
6. L
7. S
8. L
9. S
10. L

Chapter 6

1. C
2. E
3. C
4. C
5. B
6. S
7. G
8. L
9. G
10. L

Chapter 7

1. D
2. A
3. C
4. C
5. D
6. C
7. A
8. C
9. C
10. B
11. L
12. L
13. G
14. L
15. G

Chapter 8

1. D
2. D
3. E
4. B
5. E
6. D
7. C
8. E
9. D
10. C
11. L
12. L
13. G
14. L
15. G

Chapter 9

1. D
2. E
3. D
4. C
5. D
6. D
7. D
8. A
9. D
10. B
11. B
12. A
13. L
14. L
15. G

Chapter 10

1. D
2. A
3. E
4. D
5. B
6. E
7. A
8. A
9. G
10. L

Chapter 11

1. C
2. E
3. C
4. D
5. A
6. B
7. B

8. A
9. A
10. B
11. B
12. L
13. L
14. G
15. L

Chapter 12

1. B, C, D, E
2. E
3. D
4. A
5. C
6. B
7. E
8. D
9. B
10. E
11. D
12. C
13. E
14. B
15. E

Chapter 13

1. C
2. E
3. A
4. C
5. D
6. C
7. E
8. E
9. L
10. G

Chapter 14

1. B
2. D
3. E
4. A
5. C
6. C
7. E
8. D
9. B
10. D
11. A
12. B
13. E
14. D
15. D
16. B
17. A
18. A
19. B
20. D

Chapter 15

1. B
2. A
3. A
4. D
5. A
6. B
7. E
8. D
9. C
10. E
11. L
12. G
13. L
14. L
15. G

Chapter 16

1. C
2. A
3. E
4. C
5. D
6. B
7. D
8. D
9. B
10. A
11. B
12. B
13. E
14. B
15. E

Chapter 17

General question 1.
BMR, 208 kcal/kg/24 h;
animal is small.

1. B
2. A
3. D
4. D
5. E
6. B
7. C
8. E
9. C
10. C
11. D
12. B
13. A
14. A
15. D
16. E
17. E
18. B
19. C
20. A

Chapter 18

1. C
2. D
3. D
4. E
5. A
6. C
7. B
8. E
9. A
10. B
11. L
12. L
13. G
14. G
15. G

Chapter 19

1. E
2. D
3. D
4. E
5. C
6. D
7. E
8. C
9. D
10. C
11. L
12. L
13. L
14. L
15. G

Chapter 20

1. D
2. B

3. E
4. D
5. C
6. D
7. E
8. D
9. C
10. B
11. B
12. A
13. D
14. A
15. B

Chapter 21

1. C
2. E
3. D
4. E
5. A
6. C
7. D
8. A
9. L
10. L

Chapter 22

1. E
2. D
3. E
4. E
5. A
6. B
7. C
8. G
9. L
10. G

Chapter 23

1. C
2. E
3. D
4. A
5. E
6. E
7. D
8. A
9. C
10. A
11. B
12. D
13. C
14. D
15. L
16. S
17. G
18. L
19. G
20. L

Chapter 24

1. D
2. E
3. D
4. D
5. D
6. C
7. D
8. E
9. L
10. G

Chapter 25

1. B
2. C

3. E
4. C
5. D
6. D
7. C
8. D
9. A
10. B

Chapter 26

1. D
2. E
3. D
4. C
5. C
6. D
7. B
8. A
9. E
10. C
11. A
12. D
13. C
14. E
15. D
16. D
17. C
18. A
19. E
20. B

Chapter 27

1. B, C, D
2. E
3. E
4. A, E
5. A, C
6. B
7. A

8. A
9. E
10. C
11. C
12. E
13. B
14. E
15. L
16. S
17. L
18. L
19. L
20. G

Chapter 28

1. E
2. B
3. C
4. D
5. A
6. A
7. C
8. D
9. A
10. D
11. L
12. L
13. L
14. L
15. L

Chapter 29

1. D
2. C
3. C
4. C
5. E
6. D
7. D

8. B
9. C
10. A

Chapter 30

General question 3.
Pulse pressure, 57 mm
Hg; mean arterial pressure, 92 mm Hg.
General question 4.
30,000 dynes/cm;
60,000 dynes/cm.

1. D
2. A
3. E
4. C
5. C
6. B
7. B
8. D
9. B
10. B
11. L
12. G
13. L
14. L
15. G

Chapter 31

1. E
2. B
3. D
4. D
5. C
6. B
7. B
8. D
9. C
10. E

Chapter 32

1. C
2. D
3. D
4. B
5. A
6. D
7. A
8. E
9. E
10. C
11. E
12. S
13. L
14. L
15. L

Chapter 33

General question 10.
Cardiac output, 4.73
L/min; stroke volume,
73.9 mL.

1. D
2. E
3. B
4. A
5. C
6. D
7. E
8. D
9. D
10. B
11. E
12. A
13. G
14. S
15. L

Chapter 34

1. D
2. C
3. A
4. C
5. E
6. B
7. C
8. A
9. B
10. E
11. A
12. A
13. D
14. L
15. G

Chapter 35

1. E
2. B
3. B
4. D
5. B
6. D
7. C
8. C
9. G
10. L

Chapter 36

1. A
2. B
3. B
4. D
5. A
6. B
7. D

8. C
9. G
10. G

Chapter 37

1. B
2. B
3. C
4. E
5. D
6. A
7. A
8. D
9. D

10. D
11. C
12. E
13. C
14. C
15. A

Chapter 38

General question 7.
(a) 3562.5 meq of
HCO_3^-; (b) 3562.5 meq
of Na^+; (c) 3562.5 +
105 + 75 = 3742.5 meq
of H^+.

1. D

2. C
3. A
4. E
5. B
6. A
7. A
8. A
9. A
10. E
11. C
12. D
13. C
14. A
15. D
16. C
17. D
18. B

19. A, C, D, E
20. D, E

Chapter 39

1. D
2. C
3. B
4. A
5. D
6. B
7. B
8. C
9. L
10. L

Index

A antigens, 487
AASH. *See* Adrenal androgen-stimulating hormone
A band, 56–57
Abbreviations, in physiology, 675–679
Abnormal red cell fragility, 482
Abortion, clonal, 478
ABO system, 485–486, 486*t*
  antigens of, 486*f*
ABP. *See* Androgen-binding protein
Absolute refractory period, 48, 69
Absorption, 427
  bicarbonate, 435
  calcium, 435–436
  carbohydrate, 429–430, *429*
  chloride, 435
  cholesterol, 433
  in colon, 463
  fat, 433, 435*f*
  iron, 436–437
    control of, 436*f*
  lipid, 433, 434*f*
  mineral, 435–437
  nucleic acid, 432
  potassium, 433–435
  protein, 431–432
  sodium, 433–435
  vitamin, 435
  water, 433–435
Acalculia, 250
Accelerated AV conduction, 505–506, 506*f*
Acceleration
  circulation and, 572–573
  linear, responses to, 164–165
  rotational, responses to, 164
Acclimatization
  altitude tolerance and, 623, 625
  ventilatory response and, 625*f*
Accommodation, 47, 137–138, 137*f*
  decline with age, 138*f*
ACE. *See* Angiotensin-converting enzyme
A cells, 302, 304*f*, 444
Acetate, cholesterol biosynthesis from, 279
Acetohexamide, insulin secretion and, 315

Acetylcholine, 85
  autonomic nervous system and, 201–202
  biosynthesis and catabolism of, 86*f*
  brain, 239–240
  excitatory postsynaptic potentials and, 78
  insulin secretion and, 315
  mechanism of action of, 86*t*
  neuromuscular transmission and, 100
  pancreatic juice secretion and, 454
  quantal release of, 100
  synthesis of, 87
  visceral smooth muscle and, 73
Acetylcholine receptor, 85–87
Acetylcholinesterase, 87, 239
Achalasia, 446
Acholic stools, 464
Acid, secretion of, factors affecting, 656
Acid-base balance
  changes in, ventilatory responses to, 615
  disturbances of, 670*t*
  potassium metabolism and, 671
Acid-base status, clinical evaluation of, 668–669
Acidification, urine, 654–57
Acidity, titratable, 656
Acidophils, 362
Acidosis, 312, 577, 666
  lactic, 312, 577, 670–671
  metabolic, 615, 667
    arterial blood values in, 667*f*
    plasma values in, 668*f*
    renal compensation and, 667–668
  myocardial contractility and, 518
  renal tubular, 660
  respiratory, 615, 666–667
    arterial blood values in, 667*f*
Acinar cells, 453*f*
Acini, 287
  as functional units of liver, 565*f*
Acquired immune deficiency syndrome (AIDS), 480
Acromegalic facies, 374
Acromegaly, 39, 366, 374
  findings in, 366*f*

Acrosin, 408
Acrosomal reaction, 408
Acrosome, 389
ACTH. *See* Adrenocorticotropic hormone
Actin, 8–9, 56
  isoforms of, 65
  muscle contraction and, 60
  sliding on myosin, 58*f*
Actin fibers, 6
Actinin, 56
Actinomycin D, protein synthesis and, 18*t*
Action potential, 45, 46, 47
  in auditory nerve fibers, 162
  biphasic, 49–50, 50*f*
  in cardiac muscle, 67–69, 69*f*
  changes in excitability during, 48, 49*f*
  compound, 52–53, 52*f*
  electrogenesis of, 48–49
  generation in postsynaptic neurons, 79–80
  ionic fluxes during, 50–51
  presynaptic, inhibition and facilitation and, 81*f*
  recording of, 47*f*
  sodium/potassium conductance during, 51*f*
Activated amino acid, 16
Activation heat, 64
Active tension, 62
Active transport, 22
  secondary, 24, 647, 650*f*
Active zone, 76
Acupuncture, 129
Acute pancreatitis, 453
Acute renal failure, 578
Adaptation, 106–107, 107*f*, 656
  dark, 148–149, 149*f*
  light, 148
  to noxious odors, 169
Addison's disease, 347
  pigmentation in, 346*f*
Addisonian crisis, 347
Adenoids, 602
Adenosine, 97–98
  mechanism of action of, 86*t*
  vasodilation and, 538

Note: An *f* following a page number refers to an illustration; a *t* refers to a table.

735

Adenosine derivatives, energy-rich, 257*f*
Adenosine triphosphatase (ATPase) muscle contraction and, 60
  sodium-potassium-activated, 22, 24, 24*f*
    cell volume/pressure and, 27
    resting membrane potential and, 28–29
Adenosine triphosphate (ATP), 7, 257
  hydrolysis of, energy released by, 64*f*
  muscle contraction and, 60
  resynthesis of, 63
    reactions in, 64*f*
Adenylate cyclase, 36
  activation of, 37
Adequate stimulus, 103, 118, 125
ADH. *See* Antidiuretic hormone
Adhesion receptors, 477
Adiado-chokinesia, 200
Adipsin, food intake and, 215
Adolescence, 382
Adrenal androgen-stimulating hormone (AASH), 383
Adrenal cortex, 323, 328–347
  fetal, 323–324
  hormones of
    principal, 330*t*
    structure/biosynthesis of, 328–332, 328*f*, 329*f*
    transport, metabolism, and excretion of, 332–34
  hypophysectomy and, 324*f*
  zones of, 323, 324*f*
Adrenalectomy, effects of, 342
Adrenal gland, morphology of, 323–324
Adrenal hyperplasia, congenital, 331
  hypertensive form of, 332
  salt-losing form of, 332
Adrenal insufficiency, 331–332
  glucocorticoids and, 334
  hyperpigmentation in, 364
Adrenal medulla, 323, 324*f*, 325–328
  hormones of
    secretion of, 327–328
    structure/function of, 325–327
  norepinephrine-secreting cells of, 324*f*
Adrenal sulfokinase, 331
Adrenarche, 383
Adrenergic neurons, 88
Adrenocortical disease, plasma electrolyte levels in, 342*f*
Adrenocortical hyperfunction, effects of, 346–347
Adrenocortical hypofunction, effects of, 346–347
Adrenocortical steroids, sodium excretion and, 658
Adrenocorticotropic hormone (ACTH), 222, 361
  adrenal responsiveness to, 338–339, 339*f*

adrenal venous output of steroids and, 344*f*
  aldosterone secretion and, 343–344
  chemistry and metabolism of, 338
  circadian rhythm and, 339
  cortisol output and, 339*f*
  glucocorticoid secretion and, 338
  hypersecretion of, 336
  mechanism of action of, 331, 331*f*
  plasma
    concentrations in clinical states, 340*f*
    fluctuations in, 339*f*
    pattern of values of, 341*f*
  secretion of, glucocorticoids and, 335
  stress and, 339–340
  structure of, 338*f*
Adrenogenital syndrome, 332, 332*f*, 346
Adult respiratory distress syndrome (ARDS), 579, 633
Adynamic ileus, 462
Aerobic glycolysis, 63, 262
Aerophagia, 446
Affect, 231
Afferent arteriole, 635
Afferent fibers
  central connections of, 113
  substance P-containing, 126, 126*f*
Afferent inhibition, 143
Afferent neurons, in dorsal horns, 121*f*
After-depolarization, 47
After-discharge, 117
After-hyperpolarization, 47
Afterload, 517
Afterloaded muscles, isotonic contraction of, model for, 517*f*
Agammaglobulinemia, Bruton's X-linked, 480
Aganglionic megacolon, 466
Ageusia, 172
Agglutination, red cell, 487*f*
Agglutinins, 486
Agglutinogens, 485, 487
Aggregated lymphatic nodules, 459
Aggressive behavior, gonadal hormones and, 235
Aging, 40
Agnosia, 247
AIDS. *See* Acquired immune deficiency syndrome
Air, hypoxic symptoms breathing, 623, 624*f*
Air conduction, 161
Air embolism, 536, 632
Airway, receptors in, 617–618, 618*t*
Airway closure, 597
Akinesia, 195
Albinism, 364
Albumin, 291, 488
  synthesis of, 489
Albuminuria, 642, 660
  orthostatic, 660

Alcohol
  gastric secretion and, 450
  vasopressin secretion and, 221
Alcoholism, recent-memory impairment and, 245
Aldosterone
  high blood pressure due to, 583
  mechanism of action of, 341–342
  metabolism and excretion of, 334
  responses to, variations in, 661
  secretion of, 329
    feedback and, 345*f*
    regulation of, 343–346
    second messengers and, 346*t*
    stimuli increasing, 343*t*
Aldosterone synthase, 331
Alerting response, 176
Alertness, catecholamines and, 326
ALG. *See* Antilymphocyte globulin
Alimentary glycosuria, 264
Alkalosis, 666
  metabolic, 615, 668
    plasma values in, 668*f*
  respiratory, 615, 666–667
Allen regimen, 312
Allergic response, leukotrienes and, 280
Allocortex, 231
Allopurinol, for gout, 271
All or none law, 47–48
Alloxan, diabetes mellitus due to, 308
Alpha block, 176
Alpha rhythm, 175, 176*f*
ALS. *See* Amyotrophic lateral sclerosis
Alternative pathway, 478
Altitude, delayed effects of, 624–625, 624*f*
Altitude tolerance, compensatory mechanisms for, 623
Aluminum hydroxide, 451
Aluminum poisoning, 284
Alveolar air
  composition of, 598
  sampling, 598
Alveolar ducts, 588, 589*f*
Alveolar sac, 589*f*
Alveolar ventilation, 597–598
  respiratory rate/depth and, 597*t*
Alveoli, 588, 589*f*
  surface tension of, 593
Alveolocapillary membrane, diffusion across, 598–599
Alzheimer's disease, 240, 246
Amacrine cells, 132
Amblyopia ex anopsia, 152
Amenorrhea, 407
  primary, 383, 407
  secondary, 407–408
Amination, 267–268
Amino acid(s), 266
  absorption of, 431–432
  activated, 16
  excitatory, 93–94
  inhibitory, 94–95

insulin secretion and, 314
metabolic functions of, 266–267
metabolism of, 267*f*
nutritionally essential, 266
oral administration of
   glucagon response to, 318
in proteins, 267*f*
structure of, 266*f*
Amino acid pool, 266, 267*f*
π-Aminohippuric acid (PAH), 639
clearance of, 648*f*
excretion of, plasma levels and,
   648*f*
transport of, 647
Aminolipids, 280
Aminopeptidase, formation and metabolism of, 415*f*
4-Aminopyridine, potassium channel
   and, 22
Ammonia
   production in kidneys, 656*f*
   secretion of, 656
Ammonia intoxication, 561
Amnesia, retrograde, 244
Amniocentesis, 382
AMPA receptor, 94
Amphetamine, psychosis due to, 239
Amphipathic, 456
Amplitude, 158
Ampulla, 155
Amygdala
   encoding declarative memory and,
      245
   olfactory stimuli and, 166
Amylase, 427
Amylopectin, 427
Amylose, 427
Amyotrophic lateral sclerosis (ALS),
   246
Anabolic nervous system, 206
Anabolism, 253
Anaerobic glycolysis, 63
Analgesia, stress, 128
Analysis of variance (ANOVA), 674
Anaphylactic shock, 580
Anastomoses, arteriovenous, 523
Anatomic dead space, 597
Androgen-binding protein (ABP), 389
Androgen(s), 375
   anabolic effects of, 392–393
   cancer and, 413
   2,3-diphosphoglycerate and, 606
   effects of, 334
   erythropoietin secretion and, 419
   growth spurt at puberty and, 370
Androgen resistance, 382
   receptor-positive, 382
Androgen-secreting tumors, 395
Androstenedione, secretion of, 329
Anelectrotonic potentials, 48
Anemia
   blood flow turbulence in, 526
   hemolytic, hereditary, 482
   iron deficiency, 485
   pernicious, 452

sickle cell, 485
Anemic hypoxia, 622, 628
Anergy, clonal, 478
Anesthesia, general, reticular activat-
   ing system and, 179–180
Aneuploidy, 13
Angina pectoris, 563–564
Angiogenesis, 523
   factors involved in, 525*t*
Angioplasty, coronary, 564
Angiotensin, 414–415
   action of, 415–416
   formation and metabolism of, 415*f*
Angiotensinase, 414
Angiotensin-converting enzyme
   (ACE), 414, 541
   extracellular fluid volume and, 665*f*
   formation and metabolism of, 415*f*
Angiotensin I, 414
   N-terminal end of, structure of,
      415*f*
Angiotensin II, 97, 414, 415–416
   action of, 331
   adrenal venous output of steroids
      and, 344*f*
   aldosterone secretion and, 344
   extracellular fluid volume and,
      665*f*
   hypovolemia and, 664
   N-terminal end of, structure of,
      415*f*
   thirst and, 216
   vasoconstrictor action of, 542
   vasopressin secretion and, 220
Angiotensin II receptor, 416
Angiotensin III, 414, 416
   N-terminal end of, structure of,
      415*f*
Angiotensinogen, 414
   N-terminal end of, structure of,
      415*f*
Angiotonin, 415
Angle-closure glaucoma, 132
Anion gap, 668
Ankyrin, 482
Annulospiral endings, 112
Anomic aphasia, 249
Anosmia, 169
ANOVA. *See* Analysis of variance
Anovulatory cycles, 398, 407
Anoxia, 622
   anoxic, 622
Anoxic anoxia, 622
ANP. *See* Atrial natriuretic peptide
ANS. *See* Autonomic nervous system
Antacids, 451
Anterior corticospinal tract, 184–185
Anterograde transport, 45
Anterolateral cordotomy, 129
Anterolateral system, 120, 121*f*
Antiarrhythmic drugs, 506
Antibiotics
   blood-brain barrier and, 556
   protein synthesis and, 17, 18*t*
Antibodies, monoclonal, 478

Anticholinergics, Parkinson's disease
   and, 196
Anticlotting mechanisms, 490–491
Anticoagulants, 491–492
Antidiuretic hormone (ADH), 218.
   See also Vasopressin
Antidromic conduction, 49
Antigen-presenting cells (APCs), 475
   interactions with T lymphocytes,
      476*f*
Antigens, 474
   blood group, 485
Antiketogenic, 274
Antilymphocyte globulin (ALG), 480
Antiports, 22
Antithrombin III, 491
Antithyroid drugs, 300–301
Antitrypsin deficiency, emphysema
   and, 628
Antrum formation, 396
Anuria, 660
Anxiety, gamma efferent discharge
   and, 114
Aorta
   coarctation of, 583
   pressures in, 562*f*
Aortic arch, 544, 545
   baroreceptor areas in, 545*f*
Aortic body, 613–614
   location of, 613*f*
Aortic insufficiency
   pulse in, 513
   sounds of Korotkow in, 532
AP–1 site, 15
APCs. *See* Antigen-presenting cells
Aphasia, 249, 249*t*
   anomic, 249
   conduction, 249
   fluent/nonfluent, 249
   global, 249
Apnea, sleep, 181
Apneusis, 612
Apoferritin, 436–437
Apomorphine, growth hormone secre-
   tion and, 370
Apoproteins, 275
Apoptosis, 77, 419
Appetite, regulation of, long-term,
   215
Approach system, 236
APUD cells, 441
Aqueous humor, 132
Arachidonic acid, 280
   biologically active substances pro-
      duced from, 280*f*
Arachnoid trabeculae, 553
Arcuate fasciculus, 248, 249*f*
ARDS. *See* Adult respiratory distress
   syndrome
Area postrema, 209, 416, 555
Arginine vasopressin (AVP), 216,
   217*f*. *See also* Vasopressin
Argyll Robertson pupil, 138
Arithmetic mean, 673
Aromatase, 389, 401

Arousal, cortical stimulation and, 178
Arousal response, 176
Arousal value, 241
Arrhythmias, 501–506
 hypothalamic stimulation and, 211
Arterial circulation, 529–532
Arterial plasma level, 641
Arterial pressure, 529–530
 brachial, 530*f*
 cardiac pressure and, 517*f*
 as function of age, 533*f*
 gravity and, 530, 531*f*
 normal, 532
Arterial pulse, 513
Arteries, 521
Arteriolar circulation, 529–532
Arterioles, 521
 afferent, 635
 caliber of, factors affecting, 543*t*
Arterioluminal vessels, 561
Arteriosinusoidal vessels, 561
Arteriovenous (A-V) anastomoses, 523
Artificial kidney, 660
Artificial respiration, 632–633, 633*f*
Ascending fibers, anterolateral sys-
  tem of, 120
Aspartate, 93–94
 mechanism of action of, 86*t*
Asphyxia, 630
Aspiration pneumonia, 592
Aspirin
 antipyretic effect of, 229
 thromboxane formation and, 481
 thromboxane-prostacyclin balance
  and, 539
Association areas, 247, 247*f*
Astereognosis, 247
Asthma, 627
Astigmatism, 139
Astrocytes, 54*f*, 55
 fibrous, 55
 protoplasmic, 55
Ataxia, 200
Atelectasis, 627
 patchy, 595
Atherosclerosis, 279
 endothelium-derived relaxing fac-
  tor and, 539
Athetosis, 195
ATP. *See* Adenosine triphosphate
ATPase. *See* Adenosine
   triphosphatase
ATP synthase, 258
Atretic follicles, 396
Atrial arrhythmias, 503–504
 consequences of, 504
Atrial extrasystole, 503*f*
Atrial fibrillation, 503*f*, 504
Atrial flutter, 503, 503*f*
Atrial natriuretic peptide (ANP), 419–
   422, 420*f*, 542
 actions of, 420–421
 secretion and metabolism of,
   421–422
 structure of, 419–420, 420*f*

Atrial natriuretic peptide receptor,
   421, 421*f*
Atrial pressure, jugular pulse and,
   513–514
Atrial stretch receptors, 546–547
Atrial systole, 494, 510
Atrial tachycardia, 503*f*
 paroxysmal with block, 504
Atrioventricular conduction, acceler-
   ated, 505–506, 506*f*
Atrioventricular nodal block, 502
Atrioventricular nodal delay, 496
Atrioventricular (AV) node, 494, 495*f*
Atropine
 autonomic nervous system and, 207
 insulin secretion and, 315
Attack reaction, 235
Attention, hemispheric specialization
   and, 248
Audibility curve, 160*f*
Audiometer, 163
Audiometry, 163
Auditory canal, external, 153
Auditory cortex, 154, 162–163
Auditory meatus, external, 153
Auditory nerve fibers
 action potentials in, 162
 spike rate in, frequency/intensity of
   sound and, 162*f*
Auditory ossicles, 153, 154*f*
Auditory pathways, central, 154–155,
   157*f*
Auditory tube, 153
Auerbach's plexus, 438
Augmented limb leads, 498
Auscultatory gap, 532
Auscultatory method, 531–532, 531*f*
Autocrine communication, 31
Autogenic inhibition, 115
Autoimmune disease, 480
Autologous transfusion, 487
Automaticity, increased of heart, 503
Autonomic failure, primary, forms
   of, 572*t*
Autonomic insufficiency, primary,
   postural hypotension due
   to, 572
Autonomic junctions, chemical transmission at, 201–202
Autonomic nerve impulses, effector
   organ responses to, 203–
   204, 204–205*t*
Autonomic nervous system (ANS),
   201–207, 202*f*
 anatomic organization of, 201
 catecholamine metabolism in, 206*f*
 chemical divisions of, 202
 chemical transmission in, 201–202
 cholinegic discharge in, 204–207
 effector organs of, 203–204, 204–
   205*t*
 insulin secretion and, 315
 noradrenergic discharge in, 206
 parasympathetic division of, 201,
   203*f*

pharmacology of, 207, 207*t*
 sympathetic division of, 201, 203*f*
Autonomic outflow, anatomic organi-
   zation of, 201
Autonomic reflexes, 190
 medullary, 208
Autonomic responses, limbic system
   and, 231–232
Autophagic vacuole, 8
Autoreceptors, 85
Autoregulation, 538, 559, 559*f*
 in kidneys, 641*f*
 metabolic theory of, 538
 myogenic theory of, 538
 of renal blood flow, 640–641
 tissue pressure hypothesis of, 538
Autosomes, 375
AV conduction, accelerated, 505–
   506, 506*f*
A-V difference, 515
Average velocity, 526
Aversion conditioning, food, 243
AV nodal block, 502
AV nodal delay, 496
Avoidance reaction, 235
Avoidance reflexes, conditioned, 243
Avoidance system, 236
AVP. *See* Arginine vasopressin
a wave, 143, 513
Axial muscle, control of, 184
Axon(s), 43
 electrical events of, 46
 fusiform
   dynamic, 113
   static, 113
 local current flow in, 49*f*
 myelinated, 44*f*
   saltatory conduction in, 49
 relation to Schwann cells in periph-
   eral nerves, 45*f*
 of spinothalamic tracts, 120
Axon hillock, 43
Axon reflex, 543, 566, 566*f*
Axon telodendria, 43
Axoplasmic flow, 45
5-Azacytidine, DNA methylation
   and, 15
Azotemia, 578

Babinski sign, 187
Bacteria, intestinal, 463–464
Bacterial infection, glucocorticoids
   and, 337–338
Bainbridge reflex, 547
Ballism, 195
Band 3, 655
B antigens, 487
Barbiturates
 myocardial contractility and, 518
 rapid eye movement sleep and, 181
 vasopressin secretion and, 221
Barometric pressure
 hypoxic hypoxia and, 623, 624*f*
 increased, effects of, 631–632, 632*t*

Baroreceptor(s), 544–545, 545f
  extracellular fluid volume and, 547
  respiratory effects of, 619
Baroreceptor reflex mechanism, 546
Baroreceptor resetting, 546
Barr body, 377, 377f
Bartter's syndrome, 418
Basal forebrain sleep zone, 179
Basal ganglia, 183, 193–196, 193f,
    194t
  anatomy of, 193–194
  connections of, 194f
  diseases of, 195–196
  function of, 194–195
  metabolism of, 194
Basal granule, 10
Basal lamina, 6
Basal metabolic rate (BMR), 255–256
Base excess, 670
Basement membrane, 6
Basic electric rhythm (BER), 449
Basic units, 673t
Basilar arteries, 397
Basilar membrane, 153
Basket cells, 198, 199f
Basophils, 362, 470
B cell(s), 302, 303f, 304f
B cell exhaustion, 315–316
Becker's muscular dystrophy, 67
Bed-wetting, 181
Bell-Magendie law, 111
Bels, 158
Bends, 632
Benedict apparatus, 255f
Benzodiazepines
  Alzheimer's disease and, 246
  gamma-aminobutyric acid receptor
      and, 94–95
BER. See Basic electric rhythm
Bernouilli's principle, 530–531, 531f
Beta-adrenergic receptor
  desensitization in, 85
  structure of, 37f
Beta rhythm, 175
Bezold-Jarisch reflex, 547, 618
Bicarbonate
  absorption of, 435
  excretion of, 656–657, 657f
  secretion of, secretin and, 443
Biguanides, insulin secretion and, 315
Bi-iodate, iodide transport and, 300
Bile, 454
  cholesterol solubility in, 459f
  composition of, 455–456, 456t
  secretion of, regulation of, 458
  substances excreted in, 458
Bile acids, 455–456, 456t
Bile canaliculi, 454
Bile pigments, 455
Bile salts, 455–456
  enterohepatic circulation of, 457f
Biliary system, 454–459
Bilirubin, 455, 485
  metabolism and excretion of, 457,
      457f

structure of, 486f
Biliverdin, 455, 485
Binocular vision, 150, 150f
Biofeedback, 243
Biologic half-life, 219
Biologic membrane, 6f
Biologic oxidation, 257–258
Biologic variation, 673
Biphasic action potential, 49–50, 50f
Bipolar, 175
Bipolar cells, 132
Bipolar leads, 497–498
Bipolar limb leads, 499–500
Bipolar recording, 497
Bitterness, 172
Bladder. See Urinary bladder
Blastocyst, 408
Blind loop syndrome, 464
Blindness
  color, 147
    inheritance of, 147–148
  night, 149
Blindsight, 151
Blind spot, 133
Blood, 469–492
  bone marrow, 469
  buffers in, 607–609
  carbon dioxide in, fate of, 609, 610t
  cells, glucocorticoids and, 336
  cellular elements in, normal values
      for, 472t
  formed elements of, development
      of, 471f
  gas content of, 605t
  hemostasis, 489–492
  immune mechanisms, 473–481
  insulin and insulinlike activity in,
      303–304
  insulin-glucagon molar ratios in,
      318, 318t
  ionic composition of, effects of
      changes in, 509
  plasma, 1, 488–489
    calcium distribution in, 349t
    composition of, 553t
    insulinlike activity in, 305t
    osmolal concentration of, 26
    true, 665
    viscosity and, 527–528
  platelets, 481
  protein molecules in, dimensions
      and weights of, 489f
  red blood cells, 481–485, 482f
    characteristics of, 482t
    formation and destruction of,
        482f
  substitutes, 607
  types, 485–488
  typing, 486
  venous, epinephrine/norepineph-
      rine levels in, 326f
  viscosity of, 527–528
    changes in hematocrit and, 527f
  white cells, 470–473
Blood-brain barrier, 553–556

angiotensin II and, 416
  development of, 555
  function of, 556
Blood-clotting factors, system for
    naming, 488t
Blood-clotting mechanism, 490, 490f
Blood-CSF barrier, 554
Blood flow
  cerebral, 556–558, 557t
    autoregulation of, 559, 559f
    factors affecting, 559f
    hemispheric, 558f
  coronary
    pressure gradients and, 561–562
    variations in, 562–563
  laminar, 526
  measurement of, 524–525
  muscle, exercise and, 573–574,
      573f
  physical principles applied, 525
  pulmonary, regulation of, 601–602
  regional, sedentary, 575t
  renal, 639
    autoregulation of, 640–641
    regulation of, 640
  resistance and, 523–524
  resting, of various organs, 552f
  uterine, changes during pregnancy,
      567f
Blood glucose homeostasis, 264f
  disordered in insulin deficiency,
      309f
Blood glucose level
  factors determining, 264–265
  insulin secretion and, 313–314
Blood group antigens, 485
Blood pressure
  angiotensin II and, 416
  control of, 208
  endothelin–1 and, 540
  gravitational effects on, compensa-
      tions for, 571–572
  measurement of, 530–532
  medullary control of, pathways in-
      volved in, 544f
  in pulmonary and systemic circula-
      tions, 600f
  renin and, 414
  systemic, intracranial pressure
      changes and, 559
  vasopressin and, 219, 220f
Blood-testis barrier, 388–389
Blood vessels
  cerebral, 551
    innervation of, 551–552
  characteristics of, 522t
  flow in, physical principles ap-
      plied, 525
  innervation of, 542–543
  pulmonary, 599
  renal, 638–639
    innervation of, 639
    pressure in, 640
  velocities in, effect of constriction
      on, 526f

Blood volume, total, 1, 2
B lymphocytes, 474
BMR. *See* Basal metabolic rate
BNP. *See* Brain natriuretic peptide
Body fluid(s)
  buffering capacity of, 4
  buffers in, 609*t*
  circulating, 469–493
  electrolyte composition of, 21*f*
  hydrogen ion concentration of, 666*t*
  pH of, 666*t*
Body fluid compartments, 1–3
  fluid movement in, 2*f*
  size of, 1
Body fluid volumes, measurement of,
    1–2
Body heat loss, 226–227, 226*t*
Body heat production, 226, 226*t*
Body mechanics, 66–67
Body of Luys, 193
Body on body righting reflexes, 192
Body on head righting reflexes, 192
Body temperature
  chart of, 226*f*
  normal, 225–226
  rectal and oral, ranges of, 226*f*
  regulation of
    exercise and, 575–576
    hypothalamus and, 225–230
    mechanisms of, 227–228, 228*t*
Body water, total, 1, 3
  related to age and sex, 3*t*
Body weight
  correlation with metabolic rate,
    256*f*
  hypothalamic lesions and, 213*f*
Bohr effect, 605
Bomb calorimeter, 254
Bone
  changes in mass with age, 353*f*
  compact, 350, 350*f*
  formation and resorption of,
    351–352
  growth of, 351
  growth factors in, 351*t*
  long, 350, 350*f*
  mineral uptake by, 352
  spongy, 350
  structure of, 349–351, 350*f*
  trabecular, 350, 352*f*
Bone conduction, 161
Bone disease, 352–353
Bone marrow, 469
Bony labyrinth, 153
Borborygmi, 446
Bowman's capsule, 635–637
Brachial artery, turbulent flow in,
    sounds of Korotkow and,
    532
Brachial artery pressure, 530*f*
Bradbury-Eggleston syndrome, 572*t*
Bradycardia, 501
Bradykinesia, 195
Bradykinin, 97, 541
  capillary permeability and, 534

Brain
  ammonia removal from, 561
  blood flow in, 557, 557*t*
  energy sources of, 560
  facial recognition areas of, 250*f*
  glial cells in, 54*f*
  hypoglycemia and, 560–561
  investing membranes of, 554*f*
  metabolism, 560–561
    cerebral blood vessels and, 559
  monoaminergic pathways in, 237–
    238, 237*f*
  oxygen requirements of, 560
  penetration of substances into, 555,
    555*f*
  sexual differentiation of, 380–381
  somatic sensation areas in, 122*f*
  uptake and release of substances
    by, 560, 560*f*
Brain natriuretic peptide (BNP), 420
  structure of, 420*f*
Brain stem
  chemoreceptors in, 615, 615*f*
  motor control and, 184
  respiratory neurons in, 611–612,
    612*f*
  vestibular pathways of, 158*f*
Breaking point, 617
Breasts
  cyclic changes in, 398–399
  development during pregnancy,
    411, 411*f*
  estrogen-dependent carcinomas of,
    413
  estrogens and, 402–403
Breath holding, 617
Breathing
  chemical control of, 613–617
  neural control of, 611–612
  periodic, 631
  work of, 595–596
    components of, 595*t*
Brittle bones disease, 352
Broca's aphasia, 249
Broca's area, 248, 249*f*
Brodmann's areas, 121
Bromocriptine
  Parkinson's disease and, 196
  prolactin secretion and, 387
Bronchi, innervation of, 588
Bronchial tone, 592–593
Bronchioles, 588, 589*f*
  respiratory, 588, 589*f*
Bronchoconstriction, 593
Bronchopulmonary dysplasia, 629
Brown fat, 215, 226, 275
  mitochondrial membrane in, proton
    transport across, 275*f*
Bruits, 514
Brunner's glands, 459
  secretions of, 461
Brush border, 427, 459, 638
Bruton's X-linked agammaglobuline-
    mia, 480
Buffer(s), 4, 607–609, 655–656

  in body fluids, 609*t*
Buffer base, 670
Buffering, 666, 667*f*
Buffering capacity, 4
Buffer nerves, 545
  activity of, 545–546, 545*f*
  section, effect of, 546*f*
Bulk, 464
Bulk flow, 25
Bundle branch block, 502
Bundle of His, 494
Bundle of Kent, 505
Burn shock, 580
Bursal equivalents, 474
Burst activity, 179
Butorphanol, vasopressin secretion
    and, 221
b wave, 143

Cachectin, 580
Cadherins, 7
Caerulein, 444
Caffeine
  gastric secretion and, 450
  myocardial contractility and, 518
Caisson disease, 632
Calbindin, 34
Calbindin-D proteins, 354
Calcarine fissure, projection of retina
    on, 135*f*
Calcitonin (CT), 348, 357–359
  actions of, 358–359
  appetite and, 215
  concentration as function of serum
    calcium, 359*f*
  secretion and metabolism of, 358
  structure of, 358, 358*f*
Calcitonin gene-related peptide
    (CGRP), 97
Calcium
  absorption of, 435–436
  distribution in plasma, 349*t*
  intracellular, 34
  metabolism, 348–349, 349*f*
    in cells, 34*f*
    humoral agents and, 359–360
    muscle contraction and, 60*f*
Calcium-binding proteins, 34
Calcium carbonate, 451
Calcium rigor, 509
Calmodulin, 34
  structure of, 35*f*
Calmodulin-dependent myosin light
    chain kinase, 72
Caloric intake, 282–284
Caloric stimulation, 165
Calorie (cal), 253
Calorigenesis, 294
Calorigenic action, 293–294, 293*f*
Calorimeter, bomb, 254
Calorimetry, 253–54
  direct, 253–54
  indirect, 254
Calpain, 397

cAMP. *See* Cyclic AMP
CAMs. *See* Cell adhesion molecules
Canal of Schlemm, 132
Cancer
 hypercalcemia and, 360
 mitosis and, 13
 oncogenes and, 15
Capacitance vessels, 529
Capacitation, 389
Capillary(ies), 521–522
 active and inactive, 534
 cerebral, transport across, 554f
 circulation, 532–534
 cross-sections of, 523f
 fenestrations in, 524f
 glomerular
  hydrostatic/osmotic pressure
   and, 643, 643f
  permeability of, 642
 hydraulic conductivity of, 524t
 peritubular, 638
 pressure, 533
  pulmonary, 600
 pulmonary, 588, 589f
 thoroughfare vessel, 521
 wall, 29
  flow-limited/diffusion-limited
   exchange across, 534f
  pressure gradients across, 534f
Cap site, 14, 16
Captopril, renin secretion and, 417
Carbamino compounds, 609
Carbohydrate(s)
 absorption of, 429–430, *429*
 digestion of, 427–429, 429f
 end products of, 429f
 as energy source, 63
 homeostasis, in exercise, 265
 metabolism, 259–265, 260f
  directional-flow valves in, 262,
   262f
  endocrine regulation of, 319–320
  growth hormone and, 367
  thyroid hormones and, 295
Carbon dioxide
 dissociation curve, 610f
 fate in blood, 609, 610t
 response, hydrogen and, 617
 response curve, effects of hypoxia
  on, 617, 617f
 transport, 609–610
 vasodilation and, 538
 ventilatory responses to, 615
Carbonic acid-bicarbonate system,
 608
Carbonic anhydrase, 608–609, 654
Carbon monoxide poisoning, 628–629
Carbon monoxyhemoglobin, 484, 628
Carboxyhemoglobin, 484, 628
Carcinoma
 androgen-dependent, 413
 estrogen-dependent, 413
Cardiac arrhythmias, 501–506
 hypothalamic stimulation and, 211
Cardiac conduction system, 494

Cardiac cycle
 blood flow changes during, 530f
 blood flow in heart during, 510f
 mechanical events of, 510–515,
  512f
Cardiac disease, electrocardiographic
  findings in, 506–509, 507f
Cardiac excitation
 ectopic foci of, 503
 origin and spread of, 494–497
Cardiac index, 516
Cardiac innervation, 543
Cardiac massage, 505, 505f
Cardiac muscle, 67–71, 68f. *See also*
  Heart
 contractile response of, 69–70, 69f
 depolarization of, 503f
 electrical properties of, 67–69
 ion channels in, gating of, 69t
 length-tension relationship for, 70f,
  517
 mechanical properties of, 69–70
 metabolism, 70
 morphology of, 67
 pacemaker tissue in, 70–71
 properties of, 494–495
 ventricular, factors increasing/de-
  creasing length of, 518t
Cardiac myosin heavy chain genes,
  295f
Cardiac output, 515–515–520
 components regulating, arterial
  pressure and, 517f
 congestive heart failure and, 584f
 determination by indicator dilution,
  516f
 effect of various conditions on,
  516, 516t
 factors controlling, 516–517
 integrated control of, 519–520
 measurement of, 515–516
 sedentary, 575t
Cardiac pacemaker, 494
 abnormal, 501–503
 potential, 495–496, 495f
Cardiac rate
 length of action potential and, 513t
 normal, 501
Cardiac tamponade, 581
Cardiac tissue, conduction speeds in,
  496t
Cardiac vector, 499–500, 500f
Cardiac ventricles, pressure-volume
  loop during exercise, 511f
Cardiogenic shock, 577, 581
Cardiovascular system
 exercise and, 573–576
 regulation of
  endothelium-secreted substances
   and, 539–541
  hormones and, 541–542
  local, 538–539
  nervous system and, 542–550
  rising from supine to upright position and, 572f

zero gravity and, 573
Carnitine, 272
Carotenemia, 294
Carotid body, 613–614
 location of, 613f
 organization of, 614f
Carotid clamping, effect of, 546f
Carotid sinus, 544, 545
 afferent nerve fiber from, discharges in, 545f
 baroreceptor areas in, 545f
 perfusion pressure in, blood pressure and, 546f
Carotid sinus nerve, 545
Carotid sinus syncope, 580
Carriers, 5, 21
Carrying angle, 403
Castration, 394
Catabolism, 253
 glucose, 259–261
 hemoglobin, 485
Catalase, 10
Catecholamines
 biosynthesis and release of, 88–89,
  89f
 carbohydrate metabolism and, 319
 cardiac muscle contraction and, 70
 catabolism of, 90
 effects of, 326–327
 erythropoietin secretion and, 419
 metabolism in sympathetic nervous
  system, 206f
 myocardial contractility and, 518
 positively inotropic effect of, 70
 secretion of, regulation of, 327–328
 structure and function of, 327
 thyroid hormones and, 295
Catechol-O-methyltransferase
  (COMT), 90
Categorical hemisphere, 247
Catelectrotonic potentials, 48
Cathartics, saline, 434
Cathode-ray oscilloscope (CRO), 46,
  46f
Caudate nucleus, 193, 193f
Caveolae, 603
CBG. *See* Corticosteroid-binding
  globulin
CCK. *See* Cholecystokinin
CCK-PZ. *See* Cholecystokinin-
  pancreozymin
CD3 polypeptides, 477
cdc2 kinase, 13
Celiac disease, 462
Cell
 calcium metabolism in, 34f
 functional morphology of, 4–8
 intercellular communication of,
  30–39
 intercellular connections of, 6
Cell adhesion molecules (CAMs), 7,
  477
 wound healing and, 576, 576f
Cell-attached patch clamp, 21
Cell cycle, 13, 13f

Cell membrane, 4–6
  permeability of, 20–22
  phosphatidylinositol metabolism
    in, 38f
  transport across, 20–25, 22f
    forces causing, 25–29
  transport proteins of, 20–22, 22f
Cellular immunity, 474, 476–477
Cellular lipids, 274–275
Central delay, 113
Central excitatory state, 118–119
Central inhibitory state, 118–119
Central nervous system (CNS), synap-
    tic transmission in, 237–240
Central venous pressure (CVP),
    536–537
Centrioles, 9
Centrosomes, 9
Cerebellar cortex, 198
Cerebellar hemispheres, 196
Cerebellar nuclei, deep, 198
Cerebellar peduncles, 196, 196f
Cerebellum, 196–200
  afferent pathways in, 198t
  anatomic divisions of, 196, 196f
  connections in, 198
  effects on movement, 200
  flocculonodular lobe of, 197, 199
  functional divisions of, 196–198,
    197f
  functions of, 200
  learning and, 200
  lobules of, 197f
  motion sickness and, 199–200
  neural circuits in, 198–199, 199f
  stretch reflexes and, 200
  voluntary movement and, 183
Cerebral blood flow, 556–558, 557t
  autoregulation of, 559, 559f
  factors affecting, 559f
  hemispheric, 558f
Cerebral blood vessels, 551
  brain metabolism and, 559
  innervation of, 551–552
Cerebral capillaries, transport across,
    554f
Cerebral circulation, 551–561
  regulation of, 558–560
Cerebral cortex, 173–174
  direct pathways to, 120
  motor areas of, 185–186
  posture-regulating systems and,
    192–193
  sensory, motor, and association
    areas of, 247f
  somatic sensory areas of, 121
Cerebral dominance, 247
Cerebral edema, high-altitude, 625
Cerebral hemispheres, complemen-
    tary specialization of,
    247–248
Cerebral metabolic rate, 560
Cerebral salt wasting, 221
Cerebral vascular resistance (CVR),
    557–558

Cerebrospinal fluid (CSF), 552–553
  composition of, 553t
  formation and absorption of, 554f
Cerulo-plasmin, 194
Cervical mucus, 399f
CF. See Cystic fibrosis
CFF. See Critical fusion frequency
CFTR. See Cystic fibrosis transmem-
    brane conductance regulator
CGP. See Chorionic growth hormone-
    prolactin
CGRP. See Calcitonin gene-related
    peptide
Chaperones, 17
Chelating agents, 492
Chemical gradients, 21–22, 25
Chemical messengers, 31
  intercellular communication by, 32f
  mechanisms of action, 32–33, 32t
  receptors for, 31–32
Chemoreceptor(s), 104, 612–613
  medullary, 615, 615f
Chemoreceptor trigger zone, 208f,
    209
Chemoreflex
  coronary, 547, 618
  pulmonary, 547, 617
Chemotaxis, 470
Chest wall, compliance of, 593
Chewing, 444
Cheyne-Stokes respiration, 631, 631f
CHF. See Congestive heart failure
Chiari-Frommel syndrome, 412
Chief cells, 355, 355f, 446, 447f
Chloramphenicol, protein synthesis
    and, 18t
Chlorate, iodide transport and, 300
Chlordiazepoxide, gamma-
    aminobutyric acid receptor
    and, 94
Chloride
  absorption of, 435
  secondary active transport of, 650f
Chloride channels, 22–24, 23f
Chloride excretion, 658
Chloride shift, 609–610
π-Chlorophenylalanine, serotonin
    and, 238
Chloroquine, protein synthesis and,
    18t
Chlorpromazine, prolactin secretion
    and, 387
Chlorpropamide, vasopressin secre-
    tion and, 221
Chlortetracycline, blood-brain barrier
    and, 556
Cholagogues, 458
Cholecalciferol, formation and hy-
    droxylation of, 353f
Cholecystectomy, effects of, 458
Cholecystography, 458
Cholecystokinin (CCK)
  appetite and, 215
  glucagon secretion and, 318
  insulin secretion and, 315

Cholecystokinin-A receptor, 215
Cholecystokinin-B receptor, 215
Cholecystokinin-pancreozymin
    (CCK-PZ), 442
Cholelithiasis, 458
Choleretics, 458
Cholesterol
  absorption of, 433
  atherosclerosis and, 279
  biosynthesis of, 279f
  cellular uptake of, 278f
  metabolism, 278–279, 278f
    diabetes mellitus and, 312
    thyroid hormones and, 295
  receptor-mediated endocytosis and,
    19
  solubility in bile, 459f
Cholesterol desmolase, 330
Cholesterol desmolase deficiency,
    331
Cholesterol ester hydrolase, 330
Cholic acid, 456
Choline acetyltransferase, 87, 239
Cholinergic division, 202
Cholinergic neurons, 85
Cholinesterases, 87
  nonspecific, 87
  specific, 87
  true, 87
Chondrodysplasias, 352
Chondrogenesis, growth hormone
    and, 366
Chorea, 195
Chorionic growth hormone-prolactin
    (CGP), 409
Chorionic villus sampling, 382
Choroid, 132
Choroid plexus, cells of, transport
    across, 554f
Chromatin, 11
  sex, 377
Chromium deficiency, 284
Chromogranin A, 88–89, 327
Chromophilic cells, 361–362
Chromosomal sex, 375–377
Chromosome(s), 11
  abnormalities of, 381–382
  sex, 375–376
    nondisjunction of, 381f
  somatic, 375
Chromosome 21, Down's syndrome
    and, 246
Chronotropic action, 517
Chvostek's sign, 357
Chylomicron(s), 275–276
Chylomicron remnants, 276
Chylothorax, 627
Chyme, 461
Chymosin, 431
Chymotrypsins, 431
Cigarette smoking, emphysema due
    to, 628
Cilia, 9–10, 588, 602
Ciliary body, 132
Cimetidine, peptic ulcer and, 451

Cingulin, 6
Circadian rhythm, adrenocortico-
    tropic hormone and, 339
Circhoral secretion, 385
Circulation
  acceleration effects on, 572–573
  arterial, 529–532
  arteriolar, 529–532
  capillary, 532–534
  cerebral, 551–561
    regulation of, 558–560
  coronary, 561–564, 561f
  diagram of, 521f
  fetal, 568, 568f, 569f
    changes at birth, 569–570
  gravitational effects on, 571–572
  hepatic, 565
  intestinal, 564–565
  lymphatic, 534
  placental, 567
  pulmonary, 588–590, 599–602
    blood pressures in, 600f
  renal, 636f, 639–641
  of skin, 566–567
  splanchnic, 564–566, 564f
    reservoir function of, 565–566
  systemic
    blood pressures in, 600f
    exercise and, 574–575, 574f
  uterine, 567, 567f
  venous, 535–537
Circulation time, 526
  arm-to-tongue, 527f
Circulatory system, 469
Circumventricular organs, 555–556,
    555f
Circus movement, 503
Citric acid cycle, 7, 261, 261f
  in transamination and gluconeogen-
    esis, 268f
Clasp-knife effect, 116
Classical conditioning, 243, 244
Classic pathway, 478
Clathrin, 19
Clearance, 639, 641
Climacteric, 385
Climbing fibers, 198, 199f
Clofibrate, vasopressin secretion and,
    221
Clonal abortion, 478
Clonal anergy, 478
Clonal selection, 475
Clone, 475, 478
Clonus, 116–117
Closed pneumothorax, 627
Closing volume (CV), 597
Clot formation, thromboxane-pro-
    stacyclin balance and, 539
Clotting-factor deficiencies, disease
    due to, 492t
Clotting mechanism, 490, 490f
CNS. See Central nervous system
CoA. See Coenzyme A
Coagulation, cell-adhesion molecules
    in, 576f

Coarctation of aorta, 583
Coated pits, 19
Cobalt deficiency, 284
Cobalt salts, erythropoietin secretion
    and, 419
Cochlea, 153, 156f
Cochlear nuclei
  dorsal, 154
  ventral, 154
Coenzyme A (CoA), 257
  reduced, formula of, 257f
Cognition, 231
Cog-wheel rigidity, 196
Colchicine
  for gout, 270
  microtubule assembly and, 8
  protein synthesis and, 18t
Cold shock, 577
Colectomy, effects of, 465
Colipase, 432
Collagen, 350
Collapsing pulse, 513
Collateral ganglia, 201
Collecting ducts, 638
  water excretion and, 650–651
Colloid, 287
Colon, 462–466, 462f
  absorption in, 463
  bacteria of, 463–464
  layers of, 438f
  motility and secretion of, 462–463
  secretory immunity in, 464
  transit time in, 463
Colony-stimulating factors (CSFs),
    472–473
  production of, 473t
Color
  characteristics of, 146
  complementary, 146
  primary, 146
Color blindness, 147
  inheritance of, 147–148
Color vision, 146–148
  neural mechanisms of, 147
  Young-Helmholtz theory of, 146
Coma
  hepatic, 561
  hyperosmolar, 312
Common metabolic pool, 257
Communicating hydrocephalus, 553
Compact bone, 350, 350f
Compensation, 200
Compensatory pause, 504
Compensatory reactions
  activated by hemorrhage, 577t
  long-term, 578
  rapid, 577–578
Complementary color, 146
Complement system, 478
Complete heart block, 502, 502f
Complete tetanus, 61
Complex cells, 145
Compliance, 593
Compound action potential, 52–53,
    52f

COMT. See Catechol-O-
    methyltransferase
Conation, 231
Concentration, 641
  osmolal, 3–4
    of plasma, 26
  units for measuring, 3–4
Concentration gradient, 25, 27
Concordance rate, 322
Conditioned avoidance reflexes, 243
Conditioned reflexes, 242–243,
    243f
Conditioned stimulus (CS), 242–243
Conditioning
  aversion, food, 243
  classical, 243, 244
  operant, 243
Conduction, 226–227
Conduction aphasia, 249
Conduction deafness, 163
Cones, 132, 134–136, 135f
  density curve for, 136f
  phototransduction in, sequence of
    events in, 142f
  pigments of, 142
    absorption spectra of, 146f
Congenital adrenal hyperplasia, 331
  hypertensive form of, 332
  salt-losing form of, 332
Congenital heart disease, cyanotic,
    627
Congenital hemolytic icterus, 482
Congenital hypothyroidism, 298f
Congestive heart failure (CHF),
    583–584
  cardiac output in, 584f
  events leading to, 585f
  pathogenesis of, 584t
Connection peptide, 303
Connexons, 7, 7f
Conn's syndrome, 342–343, 346
Consciousness, hypoxia and, 623f
Consensual light reflex, 138
Constipation, 466
Constitutive endocytosis, 19
Constitutive pathway, 19
Contraception, 407
  effectiveness of methods, 407t
Contraction. See Muscle contraction
Contracture, 60
Contrecoup injury, 553
Convection, 227
Convergence, 76
  referred pain and, 128, 128f
Convergence movements, 151
Copper deficiency, 284
Copper metabolism, disorder of, 194
Copulation, 232
Cordotomy, anterolateral, for pain,
    129
Cornea, 132
Coronary angioplasty, 564
Coronary artery(ies), 561f
  blood flow in, 562f
Coronary artery disease, 563–564

Coronary blood flow
  pressure gradients and, 561–562
  variations in, 562–563
Coronary chemoreflex, 547, 618
Coronary circulation, 561–564, 561*f*
Cor pulmonale, 628
Corpus albicans, 396
Corpus hemorrhagicum, 396
Corpus luteum, 396
  of pregnancy, 409
Corresponding points, 150
Corrigan pulse, 513
Cortical association areas, voluntary
    movement and, 183
Cortical lesions, somatic sensation
    and, 123
Cortical motor areas, 185–186
Cortical nephron, 638
Cortical neurons
  electrical responses of, 178*f*
  reticular activating system and, 178
Corticobulbar system, 183
  anatomy of, 184–186
  function of, 186–187
Corticofugal fibers, 178
Corticogonadotropes, 362
Corticospinal system, 183
  anatomy of, 184–186
  function of, 186–187
Corticospinal tract, 185*f*
  anterior, 184–185
  lateral, 184
  ventral, 184–185
Corticosteroid(s), potencies of, 330*t*
Corticosteroid-binding globulin
    (CBG), 332
Corticosterone, secretion of, 329
Corticostriate projection, 193
Corticothyrotropes, 362
Corticotropin, 222
Corticotropin-releasing hormone
    (CRH), 222–223
Cortisol
  distribution in plasma, 392*t*
  effects on blood cell counts, 336*t*
  free and bound, interrelationships
    of, 333*f*
  growth hormone secretion and, 370
  hepatic metabolism of, 333*f*
  hypoglycemia and, 313
  output of, adrenocorticotropic hor-
    mone and, 339*f*
  plasma, pattern of values of, 341*f*
  secretion of, 329
    feedback control of, 340
Cotransmitters, 85, 98
Cotransporter, 430
Coughing, 618
Cough syncope, 580
Countercurrent exchange, 228
Countercurrent exchanger, 651, 653*f*
Countercurrent mechanism, 651–652
Countercurrent multiplier, 651, 652*f*
Coupling ratio, 24
Coupling reaction, 289

C peptide, 303
Cranial nerves, nerve fibers in, 54*t*
Cranial outflow, 201
Creatine, 269
  structure of, 269*f*
Creatine phosphate, 257
Creatinine, 269
  structure of, 269*f*
Creatinuria, 269
Cretinism, 298
CRH. *See* Corticotropin-releasing
    hormone
Crista(e), 7
Crista ampullaris, 155
Critical closing pressure, 528, 528*f*
Critical fusion frequency (CFF),
    149–150
Critical micelle concentration, 456
Critical velocity, 526, 532
Crossed extensor response, 117
Crush syndrome, 579
Cryptorchidism, 395
Crypts of Lieberkuhn, 459
Crystalline lens, 132
CS. *See* Conditioned stimulus
CSF. *See* Cerebrospinal fluid
CSFs. *See* Colony-stimulating factors
CT. *See* Calcitonin
CTP. *See* Cytidine triphosphate
Cupula, 155, 155*f*
Cushing reflex, 548, 559
Cushing's syndrome, 336–337, 346, 374
  findings in, 337*f*
  hypertension in, 583
Cutaneous sense organs, 104–105
CV. *See* Closing volume
CVP. *See* Central venous pressure
CVR. *See* Cerebral vascular resistance
c wave, 143, 513
Cyanmethemoglobin, 629
Cyanocobalamin, 451–452
  structure of, 452*f*
Cyanosis, 623
Cyanotic congenital heart disease, 627
Cyclic AMP, 36–37
  formation and metabolism of, 39*f*
  insulin secretion and, 314–315
  protein kinase A activation and, 263
Cyclic AMP system, 39*f*
Cyclic GMP, 37
  intracellular, nitric oxide and, 539*f*
  resynthesis of, 142
Cyclic phenomena, hypothalamus
    and, 212–213
Cyclins, 13
Cycloheximide, protein synthesis
    and, 18*t*
Cyclopentanoperhydrophenanthrene
    nucleus, 328, 328*f*, 455–456
Cyclosporin A, 480
Cystic fibrosis (CF), 628
  chloride channel and, 24
Cystic fibrosis transmembrane con-
    ductance regulator (CFTR),
    628

Cystinuria, 431
Cystometrogram, 661, 662*f*
Cystometry, 661
Cytidine triphosphate (CTP), 257
Cytochrome aa3, 614
Cytochrome c, 484
Cytokines, 474, 479, 479*t*
Cytopempsis, 29
Cytotoxic T cells, 474
Cytotrophoblast, 408

Dactinomycin, protein synthesis and,
    18*t*
Dalmane. *See* Flurazepam
Damping, 159
Dark adaptation, 148–149, 149*f*
D cells, 302, 443
Dead space
  anatomic, 597
  physiologic, 597
  respiratory, 591, 597–598
Deafferentation, micturition and, 662
Deafness, 163
  conduction, 163
  nerve, 163
  tuning fork tests for, 163*t*
Deamination, 267–268
  oxidative, 267
Decerebrate rigidity, 190
  mechanism of, 190
  significance of, 191
Decerebration, 190
Decibel scale, 158, 160*f*
Declarative memory, 243–244
  encoding, 245–246, 246*f*
Decomposition of movement, 200
Decompression, explosive, 632
Decompression sickness, 632
Decorticate rigidity, 191, 191*f*,
    192–193
Decortication, effects of, 192
Decubitus ulcers, 189
Deep cerebellar nuclei, 198
Deep pain, 125
Defecation, 465
Deglutition, 445–446
Deglutition syncope, 580
Degranulation, 470
Dehydration, 664
Dehydroepiandrosterone (DHEA), se-
    cretion of, 329
Dehydroepiandrosterone sulfate
    (DHEAS), serum, changes
    with age, 383*f*
Deiodination, fluctuations in,
    292–293
Deja vu phenomenon, 245
Delta waves, 175
Demeclocycline, inappropriate secre-
    tion of vasopressin and, 221
Dementia, senile, 246
  cerebral blood flow in, 557
Dendrites, 43, 74, 75*f*
  function of, 80

Denervation
micturition and, 662–663
skeletal muscle and, 65
Denervation hypersensitivity,
101–102
Dentate nucleus, 198
Deoxycholate, 456
Deoxycholic acid, 456
Deoxycorticosterone
high blood pressure due to, 583
secretion of, 329
2-Deoxyglucose, growth hormone se-
cretion and, 370
Deoxyribonucleic acid. *See* DNA
Deprenyl, Parkinson's disease and,
196
Depression
extracellular norepinephrine and,
238
long-term, 99
Dermatomal rule, 127–128
DES. *See* Diethylstilbestrol
Desensitization, 85
heterologous, 85
homologous, 85
Desmosomes, 6, 6f
Desoxycorticosterone acetate
(DOCA), 342
sodium-retaining effect of, 343f
Desynchronization, 176, 176f
mechanisms of, 178
Detrusor muscle, 661
Deuteranomaly, 147
Deuteranopia, 147
Development. *See* Growth and development
Dexamethasone, potency of, 329–330
Dextran molecules, fractional clearance of, electrical charge
and, 642f
DFP. *See* Diisopropyl fluorophosphate
DHEA. *See* Dehydroepiandrosterone
DHEAS. *See*
Dehydroepiandrosterone
sulfate
Diabetes insipidus, 221, 309
nephrogenic, 39, 221
Diabetes mellitus, 308–309, 321–322
cholesterol metabolism in, 312
fat metabolism in, 310–312
insulin-dependent, 322
insulin requirements in, 322
intracellular glucose and, 274
juvenile, 322
maturity-onset, 322
non-insulin-dependent, 322
ophthalmoscopy and, 133
type I, 322
type II, 321–322
types of, 322t
uncontrolled, metabolic abnormalities in, 311f
Diabetic mephropathy, 321
Diabetic neuropathy, 321

Diabetic retinopathy, 321
Diacylglycerol
release of, 38f
as second messenger, 35–36
Diamine oxidase, 93
Diapedesis, 470
Diaphragm, 591
Diarrhea, 466
Diastole, 494
early, 511
late, events in, 510
length of, 511–513, 513t
Diastolic hypertension, causes of,
582t
Diastolic pressure, 510, 529–530
Dichromats, 147
Dicrotic notch, 513
Dicumarol, 492
Diencephalic sleep zone, 179
Diet, high-fat, low-carbohydrate, in-
tracellular glucose and,
274
Dietary allowances, recommended
daily, 283t
Dietary components, essential, 282
Dietary fiber, 464–465
Dietary hexose, structure of, 259f
Diethylstilbestrol (DES), structure of,
404f
Diffuse secondary response, 175
Diffusion, 25
across alveolocapillary membrane,
598–599
facilitated, 22, 306
Fick's law of, 25
nonionic, 26–27, 656
Diffusion capacity, 599
Diffusion-limited exchange, 534,
534f, 598
DiGeorge's syndrome, 480
Digestion, 427
carbohydrate, 427–429, 429f
end products of, 429f
fat, 432–433
protein, 431
Digestive enzymes, 427, 428t
Digitalis, myocardial contractility
and, 518
Dihydronicotinamide adenine dinucleotide phosphate
(NADPH), 258
Dihydropyridine receptor, in T tubule
membrane, 60
1,25-Dihydroxycholecalciferol, 348,
354
formation of, feedback control of,
355f
mechanism of action of, 354
Diiodotyrosine (DIT), 289–290
Diisopropyl fluorophosphate (DFP),
autonomic nervous system
and, 207
2,5-Dimethoxy–4-methylampheta-
mine (DOM), serotonin
and, 238

N,N-Dimethyltryptamine (DMT), se-
rotonin and, 238
Diopters, 137
2,3-Diphosphoglycerate (2,3-DPG),
605
formation and catabolism of, 606f
Diphtheria toxin
protein synthesis and, 18t
receptor-mediated endocytosis and,
19
Diploid number, 11
Diplopia, 150
Direct calorimetry, 253–54
Direct Fick method, 515
Direct inhibition, 79, 80
Direct oxidative pathway, 260
Disaccharidases, 429
Disaccharides, 427
Discharge zone, 82
Distal convoluted tubule, 638
Distal muscle, control of, 184
Distal tubule, water excretion and,
650
Distributive shock, 577, 580
DIT. *See* Diiodotyrosine
Diuresis
osmotic, 653–654
urine concentration-urine flow
in, 654f
water, 653
Diuretics, 659
mechanism of action of, 659t
sites of action of, 659f
thiazide, insulin secretion and, 315
Diurnal rhythm, 339
Divergence, 76
DMT. *See* N,N-Dimethyltryptamine
DNA, 11
double helix of, 11, 12f
methylation of, 15
mitochondrial, 8
nuclear, 7–8
structure and function of, 11–13,
12f
DNA fingerprint, 13
DNA polymerase, 13
D/N ratio, 310
DOCA. *See* Desoxycorticosterone acetate
Doctrine of specific nerve energies,
107
DOM. *See* 2,5-Dimethoxy–4-
methylamphetamine
Dominant follicle, 396
Dominant hemisphere, 247
Donnan effect, 27
Dopa decarboxylase, 88
Dopamine, 90, 323
agonists, Parkinson's disease and,
196
brain, 239
catabolism of, 92f
effects of, 327
mechanism of action of, 86t
structure and function of, 327

Dopamine *(cont.)*
transporter, 89
Dopamine beta-hydroxylase, 88
Dopamine beta-hydroxylase deficiency, 572*t*
Dopaminergic neurons, 88, 239*f*
Doppler flow meters, 524
Dorsal cochlear nuclei, 154
Dorsal column, 120
Dorsal group, 611
Dorsal horn, 120
afferent neurons in, 121*f*
gating in, pain transmission and, 129
Dorsal thalamus, 173
Double helix, 11, 12*f*
Double opponent cells, 147
Double vision, 150
Down regulation, 32
Down's syndrome, 382
Alzheimer-type dementia and, 246
2,3-DPG. *See* 2,3-Diphosphoglycerate
Drowning, 630–631
Drugs, autonomic nervous system and, 207, 207*t*
Duchenne's muscular dystrophy, 67
Ductosus arteriosus, 568, 569*f*
Ductosus venosus, 568, 569*f*
Dumping syndrome, 452
Duodenal glands, 459
Duodenal lumen, pancreatic proteases in, activation of, 453*f*
Duodenal ulcers, gastrin and, 442
Duplicity theory, 136
Dwarfism, 371–372
Laron, 372
Dynamic fusimotor axons, 113
Dynamic response, 114
Dynamin, 9
Dynein, 9
Dysgeusia, 172
Dyslexia, 248
Dysmenorrhea, 408
Dysmetria, 200
Dysosmia, 169
Dyspnea, 622–623
congestive heart failure and, 584
Dyspnea point, 623
Dyspneic index, 623
Dystrophin, muscular dystrophy and, 67

Ear
external, 153
inner, 153
middle, 153
sound transmission in, 159
transmission of vibrations in, 161*f*
Eardrum, 153
Ear dust, 155
Early endosome, 19
EC cells, 443
ECF. *See* Extracellular fluid
ECG. *See* Electrocardiogram

Echocardiography, 515
ECoG. *See* Electrocorticogram
Ectasy, 238
Ectopic ACTH syndrome, 336
Ectopic focus, 503
Edema, 535
causes of, 535*t*
cerebral, high-altitude, 625
pulmonary, 584
high-altitude, 625
EDRF. *See* Endothelium-derived relaxing factor
EEG. *See* Electroencephalogram
Effective perfusion pressure, 524
Effective renal plasma flow (ERPF), 639
Effector T cells, 474
Effort syncope, 580
EGF. *See* Epidermal growth factor
Ehlers-Danlos syndrome, 352
Eicosanoids, 280–282
Einthoven's triangle, 497
Ejaculation, 390
Ejaculatory ducts, 387
Ejection fraction, 511
EJPs. *See* Excitatory junction potentials
Elastase, 431
Electrical equivalence, 3
Electrical gradient, 22, 25, 27
Electrocardiogram (ECG), 497–501
bipolar leads, 497–498
cardiac vector and, 499–500, 500*f*
findings in cardiac disease, 506–509, 507*f*
intervals, 498*t*
monitoring, 501
normal, 498–499, 499*f*
plasma potassium level and, 508*f*
unipolar leads, 498, 498*f*
waves of, 497*f*
Electrocorticogram (ECoG), 175
Electroencephalogram (EEG), 175–176
clinical uses of, 181–182
desynchronization of, 176*f*
source of, 177–178
variations in, 176
Electrogenic pump, 24
Electrogram, His bundle, 500
normal, 501*f*
Electrolyte(s)
absorption of, 433–435
aldosterone secretion and, 344–345
metabolism, growth hormone and, 367
Electromagnetic flow meters, 524
Electromyogram (EMG), 66
Electromyography, 66, 66*f*
Electroretinogram (ERG), 143
Electrotonic potentials, 45, 48, 48*f*
changes in excitability during, 48, 49*f*
Elephantiasis, 535
Embden-Meyerhof pathway, 260, 262

Emboli, 492
Emboliform nucleus, 198
Embolism
air, 536, 632
pulmonary, 492
Embolization, pulmonary, 602
EMG. *See* Electromyogram
Emission, 390
Emission tomography, 672
Emmetropic, 137
Emphysema, 602, 627–628
expiratory pressure-volume curves in, 594*f*
Enalapril, renin secretion and, 417
Enalkiren, renin secretion and, 417
Encephalization, 183
Encephalopathy, hepatic, 464
Enchondral bone formation, 351
End-diastolic ventricular volume, 511
factors affecting, 518, 518*t*
Endocrine communication, 30
Endocytosis, 19–20
constitutive, 19
receptor-mediated, 19
Endogenous pyrogen (EP), 229
Endolymph, 153
composition of, 160*f*
Endometrium
changes during menstrual cycle, 397*f*
spiral arteries of, 397, 398*f*
Endopeptidases, 431
Endoplasmic reticulum, 5*f*, 11
rough (granular), 5*f*, 11
smooth (agranular), 11
translation of protein into, 17*f*
Endorphins
distribution of, 240*f*
pain and, 129
Endosomes
early, 19
late, 19
Endothelin(s), 97, 540–541
biologic actions of, 540*t*
structure of, 540*f*
Endothelin–1, 540
Endothelin–2, 540
Endothelin–3, 540
Endothelin-converting enzyme, 540
Endothelium, substances secreted by, 539–541
Endothelium-derived relaxing factor (EDRF), 539–540
Endotoxin, 580
End-plate potential, 100
miniature, 100
End pressure, 530
End-systolic ventricular volume, 511
Energy balance, 256
Energy metabolism, 253–256
Energy production, 261–262
Energy transfer, 257
Enkephalins, 96, 325–326, 444
pain and, 129
Enteric nervous system, 201, 438–439

Enterochromaffin cells, 443, 459
Enterocytes, ion movement across, 435f
Enterogastric reflex, 450
Enterogastrone, 443
Enterohepatic circulation, 455, 457f
Enterokinase, 453
  secretion of, 442
Enteropeptidase, 453
Enuresis, nocturnal, 181
Environmental temperature, metabolic rate and, 255
Enzyme(s), 5–6
  digestive, 427, 428t
  lysosomal, 8t
Enzyme deficiency, steriod biosynthesis and, 331–332
Eosinophils, 470
EP. *See* Endogenous pyrogen
Epidermal growth factor (EGF), 444
  receptor-mediated endocytosis and, 19
Epididymis, 387
Epilepsy
  cerebral blood flow in, 557
  petit mal, 181f
Epinephrine, 87–88, 323
  brain, 238
  catabolism of, 90, 91f
  effects of, 326–327
  hypoglycemia and, 313
  intravenous infusion of, circulatory changes due to, 327f
  levels in venous blood, 326f
  myocardial contractility and, 518
  phosphorylase activation due to, 263f
  reuptake of, 89
  secretion of, regulation of, 327–328
  structure and function of, 327
  visceral smooth muscle and, 72
Epineurium, 52
Epiphyseal closure, 351
Epiphyseal plate, 351
Epiphyses, 351
Epithalamus, 173
Epithelial cells, alveolar, 588
EPSPs. *See* Excitatory postsynaptic potentials
Equilibrium length, 62
Equilibrium potential, 27–28
Equivalents, 3
Erection, 390
ERG. *See* Electroretinogram
ERPF. *See* Effective renal plasma flow
Erythroblastosis fetalis, 488
Erythrocytes, 481–485, 482f
  agglutination of, 487f
  characteristics of, 482t
  formation and destruction of, 482f
  fragility of, 482–483
Erythromycin, blood-brain barrier and, 556
Erythropoiesis, 418, 481

cyanocobalamin and, 451
Erythropoietin, 418–419, 473, 481
  plasma levels of, 419f
  secretion of, 419
    high altitude and, 625
  sources of, 419
  structure and function of, 418–419
Escape phenomenon, 342, 343f
Esophageal sphincter, lower, 446
Esophagus, 444–446
  motor disorders of, 446
Essential dietary components, 282
Essential fatty acids, 279–280
Essential hypertension, 583
Essential trace elements, 284t
Essential vitamins, 285–286t
Estradiol, 400
  biosynthesis and metabolism of, 401f
  secretion of, 402
  synthesis and secretion of, 401f
Estriol, 400
  biosynthesis and metabolism of, 401f
Estrogen(s), 375
  behavioral effects of, 402
  biosynthesis and metabolism of, 401f
  calcium metabolism and, 360
  cancer and, 413
  effects of, 334
    effects on breasts, 402–403
    on endocrine organs, 402
    on female genitalia, 402
  growth hormone secretion and, 370
  growth spurt at puberty and, 370
  hypertension due to, 583
  mechanism of action of, 403
  osteoporosis and, 352
  receptor, structure of, 33f
  sexual activity and, 232
  synthetic, 403, 404f
  testicular production of, 393
  visceral smooth muscle and, 73
Estrone, 400
  biosynthesis and metabolism of, 401f
Estrous cycle, 234, 400
Estrus, 234
Ethereal sulfates, 267
Ethinyl estradiol, structure of, 404f
Etidronate, mineral content of bone and, 352–353
Etiocholanolone, 334
  interleukin–1 production and, 229
Eukaryotes, 5
Eukaryotic cells
  molecular motors from, 10t
  nucleus of, 10–11
Eumelanins, 363
Eunuchoidism, 383, 395
Euploidy, 13
Eustachian tube, 153
Euthyroid, 292
Evans blue (T–1824), 2

Evoked cortical potentials, 174–175
Evoked potentials
  cortical, 174–175
  primary, 175
Excitation-contraction coupling, 60
Excitatory amino acids, 93–94
Excitatory junction potentials (EJPs), 101
Excitatory postsynaptic potentials (EPSPs), 77–78
  fast, 202
  ionic basis of, 78
  slow, 202
Exercise
  carbohydrate homeostasis in, 265
  cardiac function and, 519t
  cardiac responses to, 520f
  cardiovascular system and, 573–576
  2,3-diphosphoglycerate and, 606
  energy exchange in, 575f
  glucagon secretion and, 317
  insulin secretion and, 315
  metabolic rate and, 255
  muscle blood flow and, 573–574, 573f
  respiratory adjustments to, 620–622, 621f
  systemic circulation and, 574–575, 574f
  thermoregulation and, 575–576
Exertional rhabdomyolysis, 574
Exocytosis, 18–19
  sequence of events in, 19f
Exons, 14
Exopeptidases, 431
Exophthalmic goiter, 298–299, 299f
Exophthalmos, 298–299, 299f
Experience, referred pain and, 128
Experimental hypertension, 581–582
  procedures producing, 582t
Experimental neurosis, 250
Expiration, 590
  changes in pressure during, 590f
  full, 592f
  quiet, pressure-volume changes during, 596f
Expiratory muscles, 592
Expiratory pressure-volume curves, 594f
Expiratory reserve volume, 591
Explosive decompression, 632
External auditory canal, 153
External auditory meatus, 153
External ear, 153
External hydrocephalus, 553
External inhibition, 243
External intercostal muscles, 592
External respiration, 587
External urethral sphincter, 661
Exteroceptors, 104
Extinction, 243
Extracellular fluid (ECF), 1
  hydrogen ion concentration of, defense of, 665–671

Extracellular fluid (ECF) *(cont.)*
specific ionic composition of, 664
defense of, 665
tonicity of, defense of, 664, 665*f*
Extracellular fluid volume, 2–3
baroreceptors and, 547
bicarbonate filtration, reabsorption,
and excretion and, 657*f*
defense of, 664–665, 665*f*
thirst and, 215–216, 216*f*
vasopressin secretion and, 220
Extraction ratio, 639
Extrafusal fibers, 111
Extramedullary hematopoiesis, 469
Extrapyramidal system, 184
Extrasystole, 503
atrial, 503*f*
Extrinsic system, 490
Eye
image-forming mechanism of,
136–139
defects of, 139
movements of, 151–152
types of, 151*f*
muscles of, 136, 136*f*
optical system of, defects of, 139*f*
protection of, 136
structures of, 132*f*
vitamin deficiency and, 149

Facial recognition areas, 250*f*
Facilitated diffusion, 22, 306
Facilitation, 77
presynaptic, 80–81
presynaptic action potential and,
81*f*
sequence of events in, 82*f*
referred pain and, 128, 128*f*
Factor I, 478
Factor VIII, 481, 490
Factor IX, 490
Factor X, 490
Factor XI, 490
Factor XII, 490
FAD. *See* Flavin adenine nucleotide
Fainting, 580–581
Familial hypercholesterolemia, 279
Farsightedness, 139
Fascicular block, 502–503
Fasciculation, 65
Fast EPSP, 202
Fastigial nucleus, 198
Fast pain, 124
Fat
absorption of, 433, 435*f*
brown, 215, 226, 275
proton transport in, 275*f*
digestion of, 432–433
effect on gastric emptying, 451*f*
insulin secretion and, 314
metabolism, 271–282
diabetes mellitus and, 310–312
growth hormone and, 367
neutral, 275

Fatigue, 622
Fatty acids
essential, 279–280
free, 275
diabetes mellitus and, 311
as energy source, 63
glucagon secretion and, 318
growth hormone secretion and,
370
metabolism of, 278
oxidation and synthesis of, 272,
273*f*
F body, 377
F cells, 302
Fear, limbic system and, 235–236
Feature detectors, 145
Feces, 463
composition of, 463*t*
Feedback inhibition, of spinal motor
neuron, 82*f*
Feeding
center, 213
gastrointestinal hormones and,
214–215
hypothalamic lesions and, 214*f*
hypothalamus and, 213
limbic system and, 231–232
Female pseudohermaphroditism, 332,
382
Female reproductive system, 395–
408, 395*f*
Fenestration(s), 522, 524*f*
Fenestration procedures, 163–164
Ferritin, 436–437
Fertilization, 408–409
in vitro, 409
Fetal adrenal cortex, 323–324
Fetal circulation, 568, 568*f*, 569*f*
changes at birth, 569–570
Fetal hemoglobin, 484
Fetal respiration, 569
changes at birth, 569–570
Fetoplacental unit, 410
Fever, 228–229
pathogenesis of, 229*f*
FFAs. *See* Free fatty acids
Fibrillation, 65
Fibrin, 490
Fibrin monomer, 490
Fibrinogen, 488, 490
Fibrinolysin, 491
Fibrinolytic system, 491, 491*f*
Fibrous astrocytes, 55
Fick's law of diffusion, 25
Fick method, indirect, 515
Fick principle, 515, 525, 556
Fighting reaction, 235
Filiform papillae, 170
Filtration, 29
Filtration equilibrium, 643
Filtration fraction, 643–644
Filtration slits, 638
Final common path, 118
Firing level, 47, 48, 48*f*
First-degree heart block, 502, 502*f*

First messengers, 33
First polar body, 397
First sound, 514
FK–506, 480
Flaccidity, 116
Flare, 125, 566
Flatus, 446
Flavin adenine nucleotide (FAD),
258
Flavoprotein-cytochrome system,
258
Flavor, 172
Fleeing reaction, 235
Flight or fight response, 206
Flocculonodular lobe, 197, 199
Flower-spray endings, 112
Flow-limited exchange, 533, 534*f*
Fluent aphasia, 249
Fluids. *See also* Body fluids
Fluoride, osteoporosis and, 353
Fluorine deficiency, 284
Fluorocortisol, potency of, 329–330
Flurazepam, gamma-aminobutyric
acid receptor and, 94
Focal distance, principal, 137
Folic acid deficiency, tropical sprue
and, 462
Follicles, 287
Follicle-stimulating hormone (FSH),
222, 361
episodic secretion of, 405*f*
two-unit structure of, 363
Follicular phase, 396
Folliculostellate cells, 362–363
Food aversion conditioning, 243
Food intake, factors regulating, 215
Foot plate, 153
Formed elements, development of,
471*f*
Fourth sound, 514
Fovea centralis, 133, 134*f*
Fractionation, 118
Frank-Starling curve, myocardial con-
tractility and, 519*f*
Frank-Starling law, 517
Free fatty acids (FFAs), 275
diabetes mellitus and, 311
as energy source, 63
glucagon secretion and, 318
growth hormone secretion and, 370
metabolism of, 278
Free radical, 470
Free water clearance, 654
Frequency, 158
Frequency distribution, 674
Frontal association area, 247, 247*f*
Fructose
metabolism, 265
transport, 430
FSH. *See* Follicle-stimulating hor-
mone
Functional hypoglycemia, 321
Fundoplication, 446
Fungiform papillae, 170
Fusion, 150

GABA. *See* Gamma-aminobutyric
acid
GABA-T. *See* Gamma-aminobutyric
acid transaminase
GAD. *See* Glutamate decarboxylase
Galactorrhea, 412
Galactose, metabolism of, 265
Galactosemia, 265
Galactose transport, 430
Galanin, insulin secretion and, 315
Gallbladder
ducts, connections of, 453*f*
functions of, 458
Gallbladder bile, hepatic duct bile
compared, 458*t*
Gallstones, 458–459
Gametogenesis, 375, 388–391
nondisjunction in, 381
Gamma-aminobutyric acid (GABA),
94–95
formation and metabolism of, 95*f*
glucagon secretion and, 317
mechanism of action of, 86*t*
receptor, 94–95
structure of, 95*f*
transporter, 89
Gamma-aminobutyric acid transaminase (GABA-T), 94
Gamma efferent discharge, 114
Gamma efferent fibers, 112
Ganglia
basal, 183, 193–196, 193*f*, 194*t*
anatomy of, 193–194
connections of, 194*f*
diseases of, 195–196
function of, 194–195
metabolism of, 194
collateral, 201
middle, 201
stellate, 201
superior, 201
sympathetic, transmission in, 202
Ganglion cells, 132
Gap junctions, 6, 6*f*, 7, 7*f*
Gases
partial pressures of, 587, 599*f*
properties of, 587–588
Gas exchange, in lungs, 598–599
Gastric acid, secretion of, 450*f*
gastrin and, 441
regulation of, 449–450, 449*f*
Gastric emptying, 448–449
cholecystokin and, 442
regulation of, 450–451
Gastric inhibitory peptide (GIP), 443
insulin secretion and, 315
Gastric juice, 447
contents of, 447*t*
Gastric lipase, 432
Gastric motility, 448–449
regulation of, 450–451
Gastric pits, 446, 447*f*
Gastric secretion, 447
Gastric slow wave, 449
Gastrin, 441–442

duodenal ulcers and, 442
glucagon secretion and, 318
insulin secretion and, 315
secretion of, stimuli affecting, 442*t*
Gastrinomas, 441
Gastrocnemius muscle, 57*f*
Gastrocolic reflex, 465
Gastroileal reflex, 463
Gastrointestinal hormones, 443–444
feeding and, 214–215
integrated action of, 443*f*
Gastrointestinal peptides, distribution
of, 441*f*
Gastrointestinal tract
organization of, 438
segments of, lengths of, 459*t*
water turnover in, 435*t*
GBG. *See* Gonadal steroid-binding
globulin
G cells, 441, 443
G-CSF. *See* Granulocyte colony-stimulating factor
Gelatinase, 431
General anesthesia, reticular activating system and, 179–180
General-onset seizures, 182
Generator potentials, 45, 105–106,
105*f*
muscle length-size relation and,
106*f*
source of, 106
Genes, 11, 13–15
activation by second messengers,
16*f*
expression of, regulation of, 15
immediate-early, 15
organization of, 14*f*
Genetic female, 376
Genetic male, 376
Geniculocalcarine tract, 134
Genitalia
embryology of, 377–380
embryonic differentiation of, 379*f*
external, differentiation of, 380*f*
female, estrogens and, 402
Genotype, 487
Geometric mean, 673
Gestagens, 405
GFR. *See* Glomerular filtration rate
GH. *See* Growth hormone
Gibbs-Donnan equation, 27
Gigantism, 374
GIH. *See* Growth hormone-inhibiting
hormone
GIP. *See* Gastric inhibitory peptide
Glaucoma, 132
angle-closure, 132
open-angle, 132
GLI. *See* Glucagonlike immunoreac-
tivity
Glia, 55
Gliadin, celiac disease and, 462
Glial cells, in brain, 54*f*
Glicentin, 316, 444
Glipizide, insulin secretion and, 315

Global aphasia, 249
Globin, 483
Globose nucleus, 198
Globulin, 488
Globus pallidus, 193, 193*f*
Glomerular capillaries
hydrostatic/osmotic pressure and,
643, 643*f*
permeability of, 642
Glomerular filtration, 635, 641–644
Glomerular filtration rate (GFR)
changes in, 643
control of, 642
factors affecting, 643*t*
measurement of, 641–642
normal, 642
renal plasma flow and, 643–644
sodium excretion and, 658*t*
urine concentration and, 654
Glomerulotubular balance, 648
mechanisms of, 648*f*
Glomerulus, 198, 637*f*
Glomus, 613
Glomus cells, 613
Glottis, 592
Glucagon, 302, 316–318
action of, 316–317
chemistry of, 316
gluconeogenesis and, 310
hepatic glucose output and, 317*f*
human substrate metabolism and,
316*f*
hypersecretion of, diabetes mellitus
and, 309
hypoglycemia and, 313
insulin secretion and, 315
metabolism of, 317
myocardial contractility and, 518
secretion of, 444
factors affecting, 317*t*
regulation of, 317–318
Glucagonlike immunoreactivity
(GLI), 444
Glucocorticoid(s), 323, 328
anti-inflammatory and antiallergic
effects of, 337–338
binding of, 332–333
carbohydrate metabolism and, 320,
320*t*
gluconeogenesis and, 310, 335*f*
metabolism and excretion of,
333–334
pharmacologic and pathologic effects of, 336–338
physiologic effects of, 334–336
plasma
concentrations in clinical states,
340*f*
fluctuations in, 339*f*
receptor, 342
structure of, 33*f*
secretion of
feedback control of, 340–341,
340*f*
regulation of, 338–341

Glucocorticoid(s) *(cont.)*
surfactant and, 595
Glucogenesis, hepatic, 309
Glucogenic, 268
Glucokinase, 259
Gluconeogenesis, 261
citric acid cycle in, 268*f*
diabetes mellitus and, 310
glucagon and, 316
glucocorticoids and, 335*f*
Gluconeogenic, 268
Glucose
blood
catecholamines and, 326
insulin secretion and, 313–314
blood levels, factors determining, 264–265
catabolism of, 259–261
clearance of, 648*f*
deficiency, intracellular, effects of, 310
excretion of, plasma level and, 647*f*
homeostasis, 264*f*
disordered in insulin deficiency, 309*f*
protein-sparing effect of, 271
reabsorption of, 645–646
renal threshold for, 264, 646
splanchnic output of, 265*f*
tolerance, 309–310
tolerance test, oral, 309, 309*f*
transport, 430
mechanism of, 430*f*, 646
transporter, 306, 307*t*
sodium-dependent, 430
transport protein, orientation in plasma membrane, 306*f*
uptake of
in brain, 560
effect on insulin on, 306*t*
Glucose-dependent insulinotropic polypeptide, 443
Glucose–6-phosphatase, 264
Glucose–6-phosphate, 259
Glucose–6-phosphate dehydrogenase deficiency, 483
congenital, 473
Glucostat, 213
hepatic, 264
Glucostatic hypothesis, 214
Glucuronyl transferase, 457
substances conjugated by, 457–458
Glutamate, 93–94
mechanism of action of, 86*t*
uptake by brain, 561
Glutamate decarboxylase (GAD), 94
Glutamic dehydrogenase, 656
Glutaminase, 656
Glyburide, insulin secretion and, 315
Glycine, 95
Glycocalyx, 427, 460
Glycocholic acid, 456
Glycogen
as energy source, 63
formation and breakdown of, 263*f*

Glycogenesis, 259
Glycogenolysis, 259
glucagon and, 316
vasopressin and, 219
Glycogenosis, myophosphorylase deficiency, 264
Glycogen synthase, 263
Glycolysis, 259
aerobic, 63, 262
anaerobic, 63
Glycosuria, 264
alimentary, 264
hyperglycemia and, 310
renal, 264
Glycosyl-phosphatidylinositol anchors, 5
GM-CSF. *See* Granulocyte-macrophage colony-stimulating factor
GnRH. *See* Gonadotropin-releasing hormone
Goiter, 296
exophthalmic, 298–299, 299*f*
iodine deficiency, 299
Goitrin, 301
structure of, 301*f*
Goitrogens, naturally occurring, 301
structure of, 301*f*
Goldblatt hypertension, 417, 582
Goldman constant-field equation, 28
Golgi apparatus, 5*f*, 17–18
protein processing in, 19*f*
Golgi cells, 198, 199*f*
Golgi tendon organ, 115, 115*f*
Gonadal dysgenesis, 372, 381
Gonadal hormones, aggressive behavior and, 235
Gonadal steroid-binding globulin (GBG), 392
Gonadal steroids, distribution in plasma, 392*t*
Gonadotropin-releasing hormone (GnRH), 222–223
secretion by interneurons, 201
Gonadotropins, 385–387
Gonads, development of, 377
Gout, 270–271
G protein(s), 34–35
heterotrimeric, 35
families of, 36*f*
metabotropic receptors and, 94
signal transduction by, 35*f*
small, 35
G protein-coupled receptor, ligands for, 36*t*
Graafian follicle, 396
Grade I proteins, 284
Grade II proteins, 284
Gram equivalent, 3
Grand mal seizures, 182
Granular pneumocytes, 588
Granule cells, 198, 199*f*
Granulocyte(s), 470–472
Granulocyte colony-stimulating factor (G-CSF), 472

Granulocyte-macrophage colony-stimulating factor (GM-CSF), 472
Granulosa cells, interactions of, 401*f*
Grasp reflex, 192
Graves' disease, 39, 298–299, 299*f*, 480
thyroid and pituitary function in, 299*f*
Gravity
arterial pressure and, 530, 531*f*
effects on blood pressure, compensations for, 571–572
pulmonary circulation and, 600–601, 600*f*
zero, effects of, 573
Gray rami communicantes, 201
Greek alphabet, 680
GRH. *See* Growth hormone-releasing hormone
Growth and development
of bone, 351
growth hormone and, 366–367
normal and abnormal, 372*f*
physiology of, 370–372
rate from birth to age 20, 370*f*
sexual, 375–385
thyroid hormones and, 295–296
Growth factors, 37–38, 367
in bone, 351*t*
Growth hormone (GH), 222, 361, 364–370
actions mediated by, 369*f*
biosynthesis and chemistry of, 364–365
calcium metabolism and, 359–360
carbohydrate metabolism and, 320
2,3-diphosphoglycerate and, 606
effects on growth, 366–367
hypersecretion of, 374
hypoglycemia and, 313
mechanism of action of, 368
plasma levels, binding, and metabolism of, 365
receptors, 366
secretion of
factors affecting, 386*t*
feedback control of, 369*f*
hypothalamic control of, 368
stimuli affecting, 368–370, 369*t*
species specificity of, 365
structure of, 365*f*
Growth hormone-inhibiting hormone (GIH), 222
Growth hormone insensitivity syndrome, 372
Growth hormone-releasing hormone (GRH), 222–224, 368
GRP, 443
GTP. *See* Guanosine triphosphate
GTP-binding proteins, 18
Guanosine triphosphate (GTP), 257
Guanylate cyclase, 37
structure of, 40*f*

Gustatory receptors, 171
Gynecomastia, 412

Habituation, 99, 241
of reflex responses, 119
Hair cells, 156–157
electrical responses of, 156
functions of, 161
inner, 153–154
outer, 153–154
structure of, 156, 159f
Hairline, 393f
Half-life, biologic, 219
Hallucinogens, synaptic transmission
in brain and, 237
Handedness, hemispheric specializa-
tion and, 248
Haploid number, 11
Hartnup disease, 431
Hashimoto's thyroiditis, 298
Haustra, 462
Haversian canals, 350
Haversian systems, 350
H band, 56
HBE. *See* His bundle electrogram
hCG. *See* Human chorionic gonadotropin
hCS. *See* Human chorionic
somatomammotropin
HDL. *See* High-density lipoprotein
Head, venous pressure in, 536
Head injury, 553
Hearing, 157–164
Heart. *See also* Cardiac muscle
conducting system of, 495f
electrical activity in, normal spread
of, 497f
endocrine function of, 419–422
excitation of, origin and spread of,
494–497
exercise and, 519t, 520f
increased automaticity of, 503
innervation of, 543
oxygen consumption of, 520
rate
baroreflex-mediated lowering of,
547f
control of, 208, 550
factors affecting, 550t
length of action potential and,
513t
normal, 501
sounds, 514
Starling's law of, 70, 517
thyroid hormones and, 294
ventricles of, pressure-volume loop
during exercise, 511f
Heartbeat, venous circulation and,
536
Heart block, 502f
complete, 502
first-degree, 502
incomplete, 502
second-degree, 502

third-degree, 502
Heart disease, congenital, cyanotic,
627
Heart failure, 583–585
congestive, 583–584
cardiac output in, 584f
events leading to, 585f
pathogenesis of, 584t
sodium retention and, 661
Heart-lung preparation, 517, 518f
Heart-lung transplantation, 618
Heat, 234
production in muscle, 64
Heat loss, 226–227, 226t
Heat production, 226, 226t
Heat shock protein, 33
Heavy chains, 475
Helicotrema, 153
Helper T cells, 474
Hematocrit, 527
changes in, blood viscosity and, 527f
Hematopoiesis
extramedullary, 469
factors regulating, 473f
Heme, 483, 604
reaction with oxygen, 484f
Hemianopia
heteronymous, 150
homonymous, 150
Hemiballism, 195
Hemiblock, 502
Hemidesmosomes, 6, 6f
Hemochromatosis, 437
iron overload and, 284
Hemoconcentration, 580
Hemoglobin, 483
affinity for oxygen, factors affecting, 605–607
amino acid composition of, 485t
catabolism of, 485
development of, 484f
abnormalities of, 484–485
dissociation curve of, 607f
fetal, 484
histidine residues of, buffering capacity of, 608
in maternal/fetal blood, dissociation curves for, 569f
reactions of, 483–484
with oxygen, 604–605
synthesis of, 485
titration curves of, 608f
Hemoglobin A, 310, 483
structure of, 483f
Hemoglobin F, 484
Hemoglobinopathy, 484–485
Hemolysis, 482
Hemolytic anemia, hereditary, 482
Hemolytic disease of newborn,
487–488
Hemolytic transfusion reaction, 486
Hemophilia A, 492
Hemorrhage
compensatory reactions activated
by, 577t

plasma-red cell volume changes
during, 579f
vasopressin secretion and, 219,
220
Hemorrhagic shock, 577
Hemosiderin, 437
Hemosiderosis, 437
Hemostasis, 489–492
abnormalities of, 492
reactions involved in, 489f
Hemostatic plug, 489
Hemothorax, 627
Henderson-Hasselbalch equation, 607
Heparin, 491
Hepatic circulation, 565
Hepatic coma, 561
Hepatic duct bile
composition of, 456t
gallbladder bile compared, 458t
Hepatic encephalopathy, 464
Hepatic glucogenesis, 309
Hepatic glucostat, 264
Hepatomegaly, congestive heart fail-
ure and, 584
Hereditary hemolytic anemia, 482
Hereditary methemoglobinemia, 483
Hereditary spherocytosis, 482
viscosity in, 528
Hering-Breuer reflex, 617
Hermaphroditism, true, 381
Herring bodies, 217
Heterologous desensitization, 85
Heterometric regulation, 517
Heteronymous hemianopia, 150
Heterotrimeric G proteins, 35
families of, 36f
Heterozygous, 487
Hexokinase, 259
Hexose
absorption of, 429
dietary, structure of, 259f
metabolism of, 265
Hexose monophosphate shunt, 260,
262
High-density lipoprotein (HDL), 276
High-energy phosphate compounds,
257
High-pressure nervous syndrome
(HPNS), 632
High-pressure system, 529
Hippocampal system, 233f
Hippocampus
electrical activity of, 242f
encoding declarative memory and,
245
Hirschsprung's disease, 466
His bundle electrogram (HBE), 500
normal, 501f
Histaminase, 93
Histamine, 92–93, 491–492
brain, 238–239
capillary permeability and, 534,
538
synthesis and catabolism of, 93f
Histotoxic hypoxia, 622, 629

HIV. *See* Human immunodeficiency virus
HLA-A2, structure of, 475*f*
HMG-CoA reductase, 279
Holter monitor, 501
Homeostasis, 40
  blood glucose, 264*f*
    disordered in insulin deficiency, 309*f*
  carbohydrate, in exercise, 265
Homeothermic, 225
Homing molecules, 7
Homologous desensitization, 85
Homometric regulation, 517
Homonymous hemianopia, 150
Homozygous, 487
Hopping reaction, 193
Horizontal cells, 132
Hormone(s), 253, 442
  adrenal medullary
    secretion of, 327–328
    structure and function of, 325–327
  adrenocortical
    principal, 330*t*
    structure and biosynthesis of, 328–332, 328*f*, 329*f*
    transport, metabolism, and excretion of, 332–34
  cardiovascular system regulation by, 541–542
  gastrointestinal, 443–444
    feeding and, 214–215
    integrated action of, 443*f*
  gonadal, aggressive behavior and, 235
  growth and development and, 370–371, 371*f*
  hypophyseotropic, 222–224
    effects on anterior pituitary hormone secretion, 223*f*
    neurons secreting, 224*f*
    structure of, 224*f*
  intestinal, insulin secretion and, 315
  islet cell, 318–319
    effects on hormone secretion, 319*f*
  melanocyte-stimulating, 363
  natriuretic, 419–420, 542
  neural, 217
  ovarian, 400–405
  pituitary, 362*t*
    anterior, 222, 223*f*
    intermediate lobe, 363–364
    placental, 409–410
  receptors for, 31–32
  sex, 323, 375
    effects of, 232–233
    in infancy on adult behavior, 234–235
  steroid
    adrenal venous output of, 344*f*
    biosynthesis of, 330–331
    secreted, 329
    sodium excretion and, 658

synthetic, 329–330
thyroid
  binding to plasma proteins, 291, 291*t*
  calcium metabolism and, 360
  carbohydrate metabolism and, 319–320
  2,3-diphosphoglycerate and, 606
  effects of, 293–296
  formation and secretion of, 288–290
  nonthyroidal diseases and, 301
  secretion and interconversion of, 291*f*
  structure of, 289*f*
  synthesis of, 289–290, 290*f*
  transport and metabolism of, 290–293
  tropic, 361
  vasoconstrictor, 541
Hormone-sensitive lipase, 278
Horripilation, 227
hPL. *See* Human placental lactogen
HPNS. *See* High-pressure nervous syndrome
Hue, 146
Human chorionic gonadotropin (hCG), 409
Human chorionic somatomammotropin (hCS), 409
Human histocompatiblity antigen, structure of, 475*f*
Human immunodeficiency virus (HIV), 480
Human placental lactogen (hPL), 409
Humoral agents, calcium metabolism and, 359–360
Humoral immunity, 474, 475
Hunger, hypothalamus and, 213–215
Hunger contractions, 215, 449
Huntington's disease, 195
  cerebral blood flow in, 557
Hyaline membrane disease, 595
H-Y antigen, 377
Hybridomas, 478
Hydraulic conductivity, 524*t*
Hydrocephalus
  communicating (external), 553
  noncommunicating (internal), 553
Hydrochloric acid, 447
  secretion of, 447–448, 448*f*
Hydrogen
  carbon dioxide response and, 617
  fate in urine, 655
  secretion of, 654–655
Hydrogen ion concentration, 665–671, 666*t*
  defense of, 665–671
Hydrophilic, 4
Hydrophobic, 4
Hydrops fetalis, 488
Hydrostatic pressure, glomerular capillaries and, 643, 643*f*
Hydrostatic pressure gradient, 533

Hydrothorax, 627
Hydroxyapatites, 350
25-Hydroxycholecalciferol, 354
Hydroxylase, 330–331
Hydroxylase deficiency, 331–332
Hydroxysteroid dehydrogenase, 342
5-Hydroxytryptamine (5-HT). *See* Serotonin
Hyperactive stretch reflexes, 184
Hyperaldosteronism
  primary, 342–343
    hypertension in, 583
  secondary, 346, 418
Hyperalgesia, 125
  primary, 125
  secondary, 125
Hyperammonemia, 464
Hyperbaric oxygenation, 630
Hyperbilirubinemia, 457
Hypercalcemia, 360, 509
Hypercalciuria, 360
Hypercapnia, 615, 630
  myocardial contractility and, 518
Hypercholesterolemia, familial, 279
Hyperemia, reactive, 563, 566–567
Hyperglucagonemia, diabetes mellitus and, 310
Hyperglycemia
  effects of, 310
  plasma vasopressin and, 221*f*
Hyperkalemia, 509
Hyperkinesis, 195
Hyperopia, 139
Hyperosmolality, 26
Hyperosmolar coma, 312
Hyperparathyroidism, secondary, 357
Hyperpigmentation, 364
Hyperpnea, 622–623
Hyperprolactinemia, 387
Hypersensitivity, denervation, 101–102
Hypersexuality, lesions causing, 233*f*
Hypertensin, 415
Hypertension, 532, 581–583
  causes of, 583
  diastolic, causes of, 582*t*
  endothelium-derived relaxing factor and, 539
  essential, 583
  experimental, 581–582
    procedures producing, 582*t*
  Goldblatt, 417, 582
  low-renin, 583
  malignant, 582–583
  ophthalmoscopy and, 133
  pulmonary, 581
  renal, 417, 582
  renin and, 417–418
Hyperthyroidism, 298–299
  causes of, 298*t*
  sounds of Korotkow in, 532
Hypertonicity, 26, 116
Hypervariable regions, 12, 477–478
Hyperventilation, changes in breathing after, 631*f*

Hypervitaminosis A, 286
Hypervitaminosis D, 286
Hypervitaminosis K, 286
Hypoaldosteronism, hyporeninemic, 347
Hypocalcemia, 509
Hypocalcemic tetany, 348
  position of hand in, 357f
Hypocapnia, 630
Hypogeusia, 172
Hypoglycemia, 320–321
  brain and, 560–561
  catecholamine secretion and, 328
  functional, 321
  glucagon secretion and, 317
  hypoketonemic, 272
  insulin excess and, 313
Hypoglycemic agents, oral, insulin secretion and, 315
Hypogonadism, male, 395
Hypokalemia, 509
Hypokalemic nephropathy, 343, 583
Hypoketonemic hypoglycemia, 272
Hypokinesis, 195
Hypomenorrhea, 408
Hypophysectomy
  diabetes mellitus and, 373
  effect on adrenal cortex, 324f
Hypophyseotropic hormones, 222–224
  effects on anterior pituitary hormone secretion, 223f
  neurons secreting, 224f
  structure of, 224f
Hypopituitarism, 372–374, 373f
Hypoproteinemia, 489
Hyporeninemic hypoaldosteronism, 347
Hyposmia, 169
Hypospadias, 331
Hypotension
  orthostatic, 572
  postural, 572
  vasopressin secretion and, 220
Hypothalamic obesity, 213, 214f
Hypothalamohypophyseal tract, 210
Hypothalamus, 209–230, 209f
  afferent/efferent connections of, 210, 211t
  anatomy of, 209–210
  anterior pituitary gland secretion and, 222–225
  autonomic function and, 210–212
  cyclic phenomena and, 212–213
  disease of, signs and symptoms of, 225t
  functions of, 210, 212t
  growth hormone secretion and, 368
  hunger and, 213–215
  lesions of, behavioral heat and, 234f
  ovarian function and, 405–406
  posterior pituitary secretion and, 216–222
  relation to pituitary gland, 210
  sexual activity in males and, 233

sleep and, 212
temperature regulation and, 225–230
thirst and, 215–216
Hypothermia, 229–230
Hypothyroidism, 297–298
  congenital, 298f
Hypotonicity, 26, 116
Hypovolemia
  angiotensin II and, 664
  plasma vasopressin and, 221f
  vasopressin secretion and, 220
Hypovolemic shock, 577, 579–580
Hypoxia, 622–629
  anemic, 622, 628
  carbon dioxide response curve and, 617, 617f
  effects of, 622
  erythropoietin secretion and, 419
  high altitude and, 624–625, 624f
  histotoxic, 622, 629
  hypoxic, 622, 623–628
    diseases causing, 625, 626t
  ischemic, 622
  myocardial contractility and, 518
  oxygen treatment for, 629–630
  stagnant, 629
Hypoxic hypoxia, 622, 623–628
  diseases causing, 625, 626t
Hysteresis loop, 595

Ia fibers, 112
I band, 56
I cells, 442, 638
IDDM. See Diabetes mellitus, insulin-dependent
Ideal solution, 26
Idioventricular rhythm, 502
IDL. See Intermediate-density lipoprotein
IGF. See Insulinlike growth factor
IGF-I. See Insulinlike growth factor-I
IGF-II. See Insulinlike growth factor-II
IgG superfamily, 7
IJPs. See Inhibitory junction potentials
Ileostomy, 465
Ileus, paralytic, 462
Image-forming mechanism, 136–139
  defects of, 139
Immediate-early genes, 15
Immediate recall, 244
Immune mechanisms, 473–481
Immune system
  activation of, 476f
  development of, 474, 474f
  diversity in, genetic basis of, 477–478
Immunity
  cellular, 474, 476–477
  humoral, 474, 475
  secretory, 432, 464, 476
Immunocytochemistry, 83

Immunoglobulin(s), 475–476, 477t
  adhesion receptors, 477
  secretory, 476
  thyroid-stimulating, 298
Immunoglobulin A, 445
Immunoglobulin G, structure of, 477f
Immunophilins, 480
Immunoreactivity, glucagonlike, 444
Immunosympathectomy, 55
Implantation, 408–409
Impotence, endothelium-derived relaxing factor secretion and, 540
Inactivated state, 51
Inborn errors of metabolism, 90, 265, 331
Incertohypothalamic system, 239
Incomplete heart block, 502
Incomplete tetanus, 61–62
Incontinence, overflow, 663
Incus, 153
Indicator dilution method, 515, 524
  determination of cardiac output by, 516f
Indirect calorimetry, 254
Indirect inhibition, 80
Indomethacin, renin secretion and, 417
Inducer T cells, 474
Infection, bacterial, glucocorticoids and, 337–338
Inferior colliculi, 154
Inferior peduncle, 196, 196f
Infertility, 409
Inflammation, 576
  glucocorticoids and, 337
  leukotrienes and, 280
Inflammatory response, 470
Infranodal block, 502
Inhibin, 375, 389, 394
  precursor proteins, 394f
Inhibition
  afferent, 143
  autogenic, 115
  collateral, recurrent, 179
  direct, 79, 80
  external, 243
  feedback, of spinal motor neuron, 82f
  indirect, 80
  internal, 243
  lateral, 143
  postsynaptic, 79, 80
    neurons producing, 79, 81f
  presynaptic, 80–81
    neurons producing, 81f
    presynaptic action potential and, 81f
  referred pain and, 128–129
Inhibitory amino acids, 94–95
Inhibitory junction potentials (IJPs), 101
Inhibitory postsynaptic potentials (IPSPs), 78–79, 79f
  ionic basis of, 79

Inhibitory postsynaptic potentials
(IPSPs) *(cont.)*
slow, 202
Initial heat, 64
Initial segment, 43, 79
Inner ear, 153
transmission of vibrations to, 161*f*
Inner hair cells, 153–154
functions of, 161
Innervation
pain, visceral, 127*f*
reciprocal, 80, 114–115, 611
Inosine triphosphate (ITP), 257
Inositol triphosphate
release of, 38*f*
as second messenger, 35–36
Inotropic action, 517
Insensible water loss, 227
Inside-out patch, 21
Insomnia, 181
Inspiration, 590
changes in pressure during, 590*f*
full, 592*f*
quiet, pressure-volume changes
during, 596*f*
Inspiratory muscles, 591–592
Inspiratory reserve volume, 590
Insufficiency, 514
Insulin, 302
biosynthesis of, 305*f*
calcium metabolism and, 360
crystalline zinc, blood-glucose-lowering effects of, 307*t*
deficiency
blood glucose homeostasis disorder in, 309*f*
consequences of, 308–313
effects of, 313*f*
effects of, 305–307, 305*t*, 306*t*,
307*t*
excess, 313
human substrate metabolism and,
316*f*
hypoglycemic action of, 305–306
hypophysectomy and, 373
insulinlike growth factors compared, 368*t*
intestinal transport of sugars and,
430
mechanism of action of, 307–308
metabolism of, 304–305
potassium and, 306
preparations, 306
receptor, 304, 307–308
structure of, 308*f*
receptor-mediated endocytosis and,
19
requirements for diabetes mellitus,
322
secreted, fate of, 303–305
secretion of, 303
factors affecting, 314*t*
regulation of, 313–316
sustained glucose infusion and,
314*f*

structure of, 303, 304*f*, 367*f*
Insulin-glucagon molar ratios, 318,
318*t*
Insulinlike growth factor(s), insulin
compared, 368*t*
Insulinlike growth factor-I (IGF-I),
304, 366
actions mediated by, 369*f*
structure of, 367*f*
Insulinlike growth factor-II (IGF-II),
304
structure of, 367*f*, 368*f*
Insulinoma, 321
Insulin protease, 304
Integral proteins, 5
Integrin(s), 7
Integrin family, 477
Intensity, 146
Intention tremor, 200
Interatrial septum, 494
Intercalated cells, 638
Intercalated disks, 67, 68*f*
Intercellular communication, 30–39
Intercostal muscles, external, 592
Interleukin(s), 472, 474, 479, 479*t*
Interleukin–1
fever and, 228–229
recombinant, systemic effects of,
480*t*
Intermediary metabolism, 256–258
glucocorticoids and, 335
Intermediate-density lipoprotein
(IDL), 276
Intermediate filaments, 8
Intermediate olfactory stria, 166
Internal hydrocephalus, 553
Internal inhibition, 243
Internalization, 32
Internal respiration, 587
Internal urethral sphincter, 661
Interneurons, 76
Interoceptors, 104
Interpositus nucleus, 198
Interstitial cells of Leydig, 387
Interstitial fluid, 1
capillary circulation and, 533–534
Interstitial fluid volume, 3, 535
increased, causes of, 535*t*
Intestinal adaptation, 461
Intestinal circulation, 564–565
Intestinal colic, 462
Intestinal gas, 446
Intestinal glands, 459
Intestinal hormones, insulin secretion
and, 315
Intestinal motility, 460–461
Intestinal mucus, 460
Intestine. *See also* Colon; Small intes-
tine
secretory immunity in, 464
substance transport by, 430*t*
Intoxication
ammonia, 561
vitamin A, 286
water, 653

Intracellular fluid volume, 3
Intracellular glucose deficiency, ef-
fects of, 310
Intracranial pressure
cerebral circulation and, 558–559
systemic blood pressure and, 559
Intracristal space, 7
Intrafusal fibers, 111–112
Intramembranous bone formation,
351
Intrapleural pressure, 590*f*
ventilation and, 596*f*
Intrapulmonary pressure, 590*f*
Intrauterine devices (IUDs), 407
Intrinsic factor, 451–452
deficiency, 451
Intrinsic system, 490
Introns, 14
Inulin
clearance of, 648*f*
excretion of, plasma level and, 647*f*
Inverse stretch reflex, 115
pathways responsible for, 115*f*
In vitro fertilization, 409
Iodide trapping, 289
Iodine
deficiency, 284, 299
deficiency goiter, 299
metabolism, 288–289, 289*f*
protein-bound, 290
radioactive, 299–300
distribution of, 300*f*
tracer doses of, 289
Ion(s)
concentration in spinal motor neurons, 20*t*
forces acting on, 27–28
hydrated, size of, 21*t*
Ion channels, 5, 22–24, 23*f*
cardiac, gating of, 69*t*
distribution in myelinated neurons,
51–52
ligand-gated, 21, 94
in neurons, 51*t*
voltage-gated, 21
IPSPs. *See* Inhibitory postsynaptic potentials
Iridophores, 363
Iris, 132
Iron
absorption of, 436–437
control of, 436*f*
deficiency, 284
deficiency anemia, 485
Irradiation of stimulus, 117
Irreversible shock, 579
Irritant receptors, 617
Ischemic hypoxia, 622
Islet cell hormones, 318–319
effects on hormone secretion, 319*f*
Islets of Langerhans, 302, 303*f*
Isoforms, 65
Isometric contraction, 61, 62*f*
Isotonic contraction, 61, 62*f*
model for, 517*f*

Isotonicity, 26
Isovolumetric ventricular contraction, 510
Isovolumetric ventricular relaxation, 511
Itch, 130
ITP. *See* Inosine triphosphate
IUDs. *See* Intrauterine devices

Jaundice, 457
Jendrassik's maneuver, 114
JG cells, 416
Jugular pulse, atrial pressure and, 513–514
Jugular venous pressure (JVP), 514*f*
Junctional folds, 100
Juvenile diabetes, 322
Juxtacrine communication, 31
Juxtaglomerular apparatus, 416, 417*f*, 638
Juxtaglomerular cells, 416
Juxtallocortex, 231
Juxtamedullary nephron, 635*f*, 638
JVP. *See* Jugular venous pressure

Kainate receptor, 94
Kalidin, 541
Kallikrein, 490, 541
  plasma, 541
  tissue, 303, 541
Kanamycin, protein synthesis and, 18*t*
Kartagener's syndrome, 602–603
Karyotype, 376, 376*f*
Kernicterus, 488
Ketoacidosis, 312
Ketogenesis, glucagon and, 316
Ketogenic, 268
Ketone(s), glucagon secretion and, 318
Ketone bodies, 272–274
  formation and metabolism of, 274*f*
Ketosis, 274, 312
17-Ketosteroids, 328
  metabolism and excretion of, 334
Kety method, 556
Kidney(s)
  ammonia production in, 656*f*
  artificial, 660
  autoregulation in, 641*f*
  blood vessels of, 638–639
  circulation in, 636*f*
  disordered function of, effects of, 660–661
  functional anatomy of, 635–639
  lymphatics of, 639
  metabolically produced acid loads and, 666*f*
Killer cells, 474, 478–479
Kilocalorie (kcal), 253
Kinesin, 9
Kinin(s), 541
  formation from kininogens, 541*f*
  structure of, 541*f*

Kininase I, 541
Kininase II, 541
Kininogen, 490
  high-molecular-weight, 541
  low-molecular-weight, 541
Kinocilium, 156, 159*f*
Klinefelter's syndrome, 381
Knee jerk reflex, 111
Krause's end-bulbs, 104, 104*f*
Krebs cycle, 261, 261*f*
Kupffer cells, 454, 455*f*
Kussmaul breathing, 312
Kyphosis, 352

Labyrinth, 153
  bony, 153
  membranous, 153, 155*f*
Labyrinthectomy, effects of, 165
Labyrinthine righting reflexes, 192
Lacis cells, 416
Lacrimal duct, 136
Lacrimal gland, 136
Lactase, 429
Lactate
  blood, work load and, 621*f*
  vasodilation and, 538
Lactation, 411–413
  hormonal control of, 411*f*
  initiation of, 412
  menstrual cycle and, 412
Lacteal, 459
Lactic acidosis, 312, 577, 670–671
Lactoferrin, 445
Lactose, 427
Lambert-Eaton syndrome, 100–101
Lamellar bodies, 594
Laminar flow, 526, 526*f*
Language disorders, physiology and pathology of, 248–249
Laron dwarfism, 372
Latch bridge mechanism, 72
Late endosome, 19
Latent period, 46–47
Lateral corticospinal tract, 184
Lateral geniculate body, 134
Lateral geniculate nucleus, 144*f*
Lateral inhibition, 143
Lateral intercellular spaces, 638, 645, 646*f*
Lateral olfactory stria, 166
Lateral tegmental system, 238
Law of Laplace, 528–529, 661
Law of projection, 107
LCAT. *See* Lecithin-cholesterol acyltransferase
L channels, 496
LDL. *See* Low-density lipoprotein
Lead, uptake by bone, 352
Leader sequence, 17
Lead-pipe rigidity, 196
Learning, 241–246
  biologic basis of, 244
  cerebellum and, 200
  forms of, 241–242

intercortical, 243
  mating behavior and, 232
  nonassociative, 242
  synaptic plasticity and, 98–99
LEC-CAMs, 7
Lecithin-cholesterol acyltransferase (LCAT), 276
Left axis deviation, 500
Left bundle branch block, 502, 502*f*
Left ventricle
  pressures in, 562*f*
  receptors, 547
Left ventricular ejection time (LVET), 513
Lemniscal system, 120
Lengthening reaction, 116
Lens
  accommodation by, 137–138, 137*f*
  light ray refraction by, 137*f*
Lens ligament, 132
Lenticular nucleus, 193
LES. *See* Lower esophageal sphincter
Leucine zipper, 15
Leu-enkephalin, 96
Leukotrienes, 280
  biosynthesis of, 281*f*
  bronchoconstriction due to, 593
Leydig cells, 377, 387
LH. *See* Luteinizing hormone
LHRH. *See* Luteinizing hormone-releasing hormone
Librium. *See* Chlordiazepoxide
Ligand(s)
  for G protein-coupled receptors, 36*t*
  receptors for, 31–32
Ligand-gated ion channels, 94
Light adaptation, 148
Light chains, 475
Light reflex
  consensual, 138
  pupillary, 138
Limb(s), touch, pain, and temperature pathways from, 121*f*
Limbic cortex, 231
  relation to neocortex, 232*f*
Limbic lobe, 231
Limbic system, 231
  afferent/efferent connections of, 231, 233*f*
  fear and, 235–236
  functions of, 231–232
  motivation and, 236–237
  rage and, 235–236
  sexual behavior and, 232–235
Limb leads
  augmented, 498
  bipolar, 499–500
  standard, 497, 498
Limiting pH, 655
Linear acceleration, responses to, 164–165
Lingual lipase, 432, 444
Lipemia, 275
Lipids, 271–272, 272*t*
  absorption of, 433, 434*f*

Lipids *(cont.)*
  cellular, 274–275
  digestion of, 432–433, 432*f*
  as energy source, 63
  plasma, 275–278
  structural, 274–275
  transport of, 275–278
    systems for, 277*f*
Lipolysis, glucagon and, 316
Lipoprotein(s), 275, 276*t*
  high-density, 276
  intermediate-density, 276
  low-density, 276–278
    structure of, 276*f*
  receptor-mediated endocytosis and,
    19
  very low-density, 276
Lipoprotein lipase, 275–276
Lipostatic hypothesis, 215
Lipoxins, 280
  biosynthesis of, 281*f*
Lithocholate, 456
Lithocholic acid, 456
Liver, 454–459
  acinus as functional unit of, 565*f*
  bilirubin metabolism in, 457, 457*f*
  cells, 454, 455*f*
  cortisol metabolism in, 333*f*
  ducts of, connections of, 453*f*
  functions of, 454–455, 455*t*
  glucostatic function of, hyperglyce-
    mia in diabetes and, 309
  metabolic abnormalities in, diabe-
    tes mellitus and, 311*f*
  metabolically produced acid loads
    and, 666*f*
  organization of, 454, 455*f*
Lobectomy, temporal, 250–251
Lobotomy, prefrontal, 250
Local response, 48, 48*f*
Local seizures, 182
Local sign, 117–118, 118*f*
Locomotor generator, 189–190
Locus ceruleus system, 238
Long-term depression (LTD), 99
Long-term memory, encoding, brain
    areas concerned with, 245*f*
Long-term potentiation (LTP), 94, 99
Loop of Henle, 638
  operation as countercurrent multi-
    plier, 652*f*
  water excretion and, 649–650
Loudness, 158
Lou Gehrig's disease, 246
Low-density lipoprotein (LDL),
    276–278
  structure of, 276*f*
Lower esophageal sphincter (LES), 446
Lower motor neuron(s), 184
  lesions of, 66
Lown-Ganong-Levine syndrome, 506
Low-pressure system, 529
Low-renin hypertension, 583
Low-resistance shock, 577
LSD. *See* Lysergic acid diethylamide

LTD. *See* Long-term depression
LTH. *See* Luteotropic hormone
LTP. *See* Long-term potentiation
Luminance, range of, 148*f*
Lumirubin, 485
Lung(s)
  anatomy of, 588–590
  collapse of, 627
  compliance of, 593
  defense mechanisms of, 602–603
  gas exchange in, 598–599
  metabolic and endocrine functions
    of, 603
  pressure-volume relations in, 594*f*
  receptors in, 617–618, 618*t*
  substances metabolized by, 603*t*
  ventilation in different parts of,
    596–597
  volumes of, 590–591, 591*f*
Luteal cells, 396
Luteal phase, 396, 397
Luteinizing hormone (LH), 222, 361
  episodic secretion of, 405*f*
  two-unit structure of, 363
Luteinizing hormone-releasing hor-
    mone (LHRH), 222
Luteolysis, 406
Luteotropic hormone (LTH), 222
LVET. *See* Left ventricular ejection
    time
17,20-Lyase, 330
Lymph, 493
  protein content of, 493*t*
Lymphagogues, 534
Lymphatic circulation, 469, 534
Lymphatic nodules, aggregated, 459
Lymphatic organs, glucocorticoids
    and, 336
Lymphatics, 522–523
  renal, 639
Lymphatic system, functions of,
    534–535
Lymph nodules, solitary, 459
Lymphocytes, 470, 473
  B, 474
  T, 474
    interactions with antigen-present-
      ing cells, 476*f*
    receptors, 477
Lymphokines, 479
Lysergic acid diethylamide (LSD), se-
    rotonin and, 238
Lysine vasopressin, 216
Lysolecithin, 453
Lysosomal storage disease, 8
Lysosomes, 8
  enzymes of, 8*t*
Lysozyme, 445
Lysylbradykinin, 541

Macroheterogeneity, 441
Macrophage(s)
  pulmonary alveolar, 602
  tissue, 472, 472*t*

Macrophage colony-stimulating fac-
    tor (M-CSF), 472
Macrosomia, 321
Macula, 155
Macula densa, 638
Macula lutea, 133, 134*f*
Macular sparing, 150–151
Magnesium hydroxide, 451
Magnetic resonance imaging (MRI),
    672
Magnet reaction, 189
Magnocellular neurons
  electrical activity of, 217–218
  response to stimulation, 218*f*
Major circulation, 469
Major histocompatibility complex
    (MHC), 474–475
Malabsorption, disease processes as-
    sociated with, 461*t*
Malabsorption syndrome, 461–462
Male hypogonadism, 395
Male pseudohermaphroditism, 382,
    393
Male reproductive system, 387–395,
    387*f*
Malignant hypertension, 582–583
Malleus, 153
Maltose, 427
Maltotriose, 428
Mammosomatotropes, 362
Manic depression, cerebral blood
    flow in, 557
Mannose–6-phosphate receptor
    (MPR), 17–18
Manubrium, 153
MAO. *See* Monoamine oxidase
Marijuana, 98
Masking, 159
Mass action contraction, 463
Mass reflex, 119, 190
Mast cells, 93, 491–492
Mastectomy, radical, edema due to,
    535
Mastication, 444
Maternal behavior, limbic lesions
    and, 235
Mathematical ability, impairment of,
    250
Mating, 232
Matrix, striatal, 194
Matrix space, 7
Maximal breathing capacity, 591
Maximal stimulus, 52
Maximal voluntary ventilation
    (MVV), 591
Mayer waves, 548
McArdle's disease, 264
McCune-Albright syndrome, 39
M cells, 432, 443
M-CSF. *See* Macrophage colony-
    stimulating factor
MDMA. *See* 3,4-Methylenedioxy-
    methamphetamine
Mean
  arithmetic, 673

geometric, 673
standard error of, 674
Mean pressure, 530
Mean QRS vector, 499–500, 500*f*
Measures, equivalents of, 679
Mechanical respirators, 633
Mechlorethamine, protein synthesis and, 18*t*
Medial geniculate body, 154
Medial longitudinal fasciculi, 156
Median eminence, 210
Medroxyprogesterone, growth hormone secretion and, 370
Medulla, posture-regulating systems and, 190–192
Medulla oblongata, 208–209
neurons in, auditory responses of, 162
Medullary autonomic reflexes, 208
Medullary chemoreceptors, 615, 615*f*
Medullary hormones
secretion of, 327–328
structure and function of, 325–327
Medullary interstitial cells, type I, 638
Medullary synchronizing zone, 179
Megacolon, 466
aganglionic, 466
Megakaryocytes, 481
Megaloblasts, 451
Meiosis, 11, 13
Meissner's corpuscles, 104–105, 104*f*, 123
Meissner's plexus, 438
Melanins, 363
Melanocytes, 364
Melanocyte-stimulating hormones (MSHs), 363
Melanophores, 363
Melanosomes, 363
Melanotropins, 363
Melatonin, 422–423
diurnal rhythms of, 423*f*
formation and metabolism of, 423*f*
Membrane potential, 20
genesis of, 28–29
resting, 20, 46
in cardiac muscle, 67–69
ionic basis of, 50
in skeletal muscle, 59
variations in, 29
Membranous labyrinth, 153, 155*f*
Memory, 243–244
biologic basis of, 244
declarative, 243–244
encoding, 245–246, 246*f*
forms of, 244*t*
long-term, brain areas concerned with, 245*f*
nondeclarative, 243–244
encoding, 244
recent, 244
alcoholism and, 245
reflexive, 243
remote, 244
smell and, 168

Memory B cells, 474, 475
Memory T cells, 474
Menarche, 382
Menopause, 384–385
Menorrhagia, 408
Menstrual cycle, 395–400
abnormalities of, 407–408
basal body temperature during, 400*f*
endometrial changes during, 397*f*
lactation and, 412
sex steroid production rates during, 402*t*
Menstruation, 396
normal, 398
Merkel's disks, 104–105, 104*f*, 123
generator potential in, 106
Mesangial cells, 637, 637*f*
contraction/relaxation of, agents causing, 643*t*
Mescaline, serotonin and, 238
Mesocortical system, 239
Messenger RNA. *See* RNA
Metabolic acidosis, 615, 667
arterial blood values in, 667*f*
plasma values in, 668*f*
renal compensation and, 667–668
Metabolic alkalosis, 615, 668
plasma values in, 668*f*
Metabolic insufficiency, thyroid therapy and, 301
Metabolic pool, common, 257
Metabolic rate, 253
basal, 255–256
catecholamines and, 326–327
correlation with body weight, 256*f*
factors affecting, 255–256, 255*t*
measurement of, 254–255
Metabolic theory of autoregulation, 538
Metabolism, 253
adrenocorticotropic hormone, 338
amino acid, 267*f*
bilirubin, 457, 457*f*
brain, 560–561
cerebral blood vessels and, 559
calcitonin, 358
calcium, 348–349, 349*f*
humoral agents and, 359–360
carbohydrate, 259–265, 260*f*
directional-flow valves in, 262, 262*f*
endocrine regulation of, 319–320
growth hormone and, 367
thyroid hormones and, 295
cholesterol, 278–279, 278*f*
diabetes mellitus and, 312
thyroid hormones and, 295
cortisol, 333*f*
electrolyte, growth hormone and, 367
energy, 253–256
estrogen, 401*f*
fat, 271–282
diabetes mellitus and, 310–312

growth hormone and, 367
free fatty acid, 278
glucagon, 317
glucocorticoid, 333–334
growth hormone, 365
hexose, 265
inborn errors of, 90, 265, 331
insulin, 304–305
intermediary, 256–258
glucocorticoids and, 335
iodine, 288–289, 289*f*
melatonin, 423*f*
phosphorus, 349
potassium, acid-base balance and, 671
progesterone, 403–404, 404*f*
protein, 266–271
diabetes mellitus and, 310
growth hormone and, 367
sodium, abnormal, 660
surfactant, 595*f*
testosterone, 391, 391*f*
thyroid hormone, 292
uric acid, 270*f*
water
adrenal insufficiency and, 335–336
pituitary insufficiency and, 373
vasopressin and, 649*t*
Metabotropic receptor, 94
Metarhodopsin II, 141
Metarterioles, 521
Met-enkephalin, 96
Methemoglobin, 483, 629
Methemoglobinemia, hereditary, 483
Methimazole, structure of, 300, 301*f*
3,4-Methylenedioxymethamphetamine (MDMA), serotonin and, 238
1-Methyl-4-phenyl-1,2,5,6-tetrahydropyridine
conversion to MPP, 195*f*
parkinsonism due to, 195
Metrorrhagia, 408
MHC. *See* Major histocompatibility complex
Micelles, 432, 456
cross section of, 456*f*
Microadenoma, 336
Microcirculation, 522*f*
Microfilaments, 8–9, 9*f*
Microfold cells, 432
Microglia, 54*f*, 55
Microheterogeneity, 441
Microsomes, 10
fatty acid synthesis in, 273*f*
Microtubules, 8, 9*f*
assembly and disassembly of, 10*f*
Microvilli, 287, 459–460, 460*f*
Micturition, 635, 661–663
abnormalities of, 662
deafferentation and, 662
denervation and, 662–663
spinal cord transection and, 663
Micturition syncope, 580

Midbrain, posture-regulating systems and, 192
Midbrain animal, 192
Middle ear, 153
Middle ganglia, 201
Middle peduncle, 196, 196f
Mifepristone (RU–486), 405
Milk ejection, 222, 411–412
Milk ejection reflex, 222, 412
Millisecond (ms), 45
Millivolt (mV), 46
Mineralocorticoid(s), 323, 328
    effects of, 341–343
    excess, effects of, 342
    receptor, 342
        structure of, 33f
    salt balance and, 346
Minerals
    absorption of, 435–437
    requirements for, 284
Miniature end-plate potential, 100
Minor circulation, 469
Miraculin, 172
MIS. See Mullerian inhibiting substance
MIT. See Monoiodotyrosine
Mithramycin, protein synthesis and, 18t
Mitochondria, 7–8
    oxidative phosphorylation in, 259f
Mitochondrial DNA, 8
Mitochondrial membrane, proton transport across, 259f
Mitomycin C, protein synthesis and, 18t
Mitosis, 11, 13
Mitral cells, 166
Mixed nerves, properties of, 52–53
M line, 58
Modiolus, 154
Mole(s), 3
Molecular mimicry, 481
Molecular motors, 9, 10t
Monoamine oxidase (MAO), 90
Monochromats, 147
Monoclonal antibodies, 478
Monocular vision, 150, 150f
Monocytes, 470, 472
Monoiodotyrosine (MIT), 288, 289–290
Monophasic, 47
Monosaccharides, 427
Monosynaptic reflexes, 111–117
Monro-Kellie doctrine, 558
Mood, extracellular norepinephrine and, 238
Morphine
    pain and, 129
    vasopressin secretion and, 220
Morphogens, 34
Mossy fibers, 198, 199f
Motilin, 443
Motion sickness, 199–200
Motivation, limbic system and, 236–237

Motor control. See Voluntary movement
Motor control system, peripheral, 116f
Motor cortex, 185–186, 185f
Motor end-plate, 99
Motor integration, levels of, 187, 188t
Motor neuron(s)
    discharge frequency of, 116f
    lower, 184
        lesions of, 66
    membrane, ion fluxes across, 28f
    myelinated axon of, 44f
    negative feedback inhibition of, 82f
    spinal, somatic motor activity and, 183
    synapses on, 75f
    synaptic activity in, recording, 77f
    upper, 184
Motor unit, 65–66
    recruitment of, 66, 117
Mottled enamel, 352
Mountain sickness, 624–625
Mouth, 444–446
Mouth-to-mouth resuscitation, 633, 633f
Movement. See Voluntary movement
MPRs. See Mannose–6-phosphate receptors
MPTP. See 1-Methyl–4-phenyl–1,2,5,6-tetrahydropyridine
MRI. See Magnetic resonance imaging
MSHs. See Melanocyte-stimulating hormones
Mucins, 445
Mucous membrane, olfactory, 166, 167f
Mucus, 447
    intestinal, 460
Muller cells, 133
Mullerian inhibiting substance (MIS), 377–380, 389
Multiple sclerosis, 480
Multi-unit smooth muscle, 73
Mural thrombus, 492
Murmurs, 514–515, 514t
Muscarinic actions, 85
Muscarinic receptor, 85
Muscle. See also Cardiac muscle; Skeletal muscle; Smooth muscle
    axial, control of, 184
    blood flow, exercise and, 573–574, 573f
    distal, control of, 184
    equilibrium length of, 62
    expiratory, 592
    fibers, types of, 64–65, 65t
    heat production in, 64
    inspiratory, 591–592
    pain, 125
    properties of, 65–67
    protein isoforms in, 65
    pump, 535, 536

red, 64–65
respiratory, 591–592
resting length of, 62
spasm, visceral pain and, 126–127
spindles
    discharge of, 113f
    function of, 113–114
    muscle stretch and, 114f
    sensory endings in, 112
    structure of, 111–113, 112f
tone, 115–116
twitch, 59
white, 65
Muscle contraction
    initiation by calcium, 60f
    isometric, 61, 62f
    isotonic, 61, 62f
    length-tension relationship and, 62–63, 63f
    molecular basis of, 59–61
    summation of, 61–62
    tetanic, 61
    types of, 61
Muscular dystrophy, 67
    Becker's, 67
    Duchenne's, 67
Muscular exercise. See Exercise
MVV. See Maximal voluntary ventilation
Myasthenia gravis, 39, 100–101, 480
Myelin, 43
    tubular, 594
Myelination, 43
Myelin sheath, 43
Myeloperoxidase, 470
    deficiency, congenital, 473
Myenteric plexus, 438
Myocardial contractility, 518
    Frank-Starling curve and, 519f
Myocardial infarction (MI), 563–564
    electrocardiographic findings in, 506–509, 507f
    membrane polarization in, abnormalities of, 506f
Myocardial perfusion, normal pattern of, 563f
Myoepithelial cells, oxytocin and, 222
Myofibrils, 56, 57f
Myogenic theory of autoregulation, 538
Myoglobin, 484, 607
    dissociation curve of, 607f
Myoneural junction, 100
Myopathy, 269
    thyrotoxic, 295
Myophosphorylase deficiency glycogenosis, 264
Myopia, 139
Myosin, 8–9, 56
    heavy chain(s), 65
    heavy chain genes, cardiac, 295f
    isoforms, 65
    light chain kinase, 34
        calmodulin-dependent, 72
    muscle contraction and, 60

Myosin-I, 9, 10*f*
Myosin-II, 9, 10*f*, 57, 58*f*
Myxedema, 297–298

NAD+. *See* Nicotinamide adenine dinucleotide
NADP+. *See* Nicotinamide adenine dinucleotide phosphate
NADPH. *See* Dihydronicotinamide adenine dinucleotide phosphate
NADPH oxidase, 470
Narcolepsy, 181
Natriuresis, 420
Natriuretic hormones, 419–420, 542
Natural killer cells, 478–479
Nausea, vasopressin secretion and, 220–221
Near point of vision, 138
Near response, 138
Nearsightedness, 139
Neck righting reflexes, 192
Negative reinforcement, 243
Negative supporting reaction, 189
Nelson's syndrome, 374
Neocerebellum, 183, 197
Neocortex, 231
    functions of, 247–251
    long-term memory storage in, 245
    neuronal connections in, 174*f*
    organization of, 174
    relation to limbic cortex, 232*f*
Neocortical pyramidal cell, 174*f*
Neomycin, protein synthesis and, 18*t*
Nephrogenic diabetes insipidus, 39, 221
Nephron, 635–638
    cortical, 638
    juxtamedullary, 635*f*, 638
    permeability and transport in, 650*t*
    pH changes along, 656
    tubular fluid osmolality in, changes in, 650*f*
Nephropathy
    diabetic, 321
    hypokalemic, 343, 583
Nernst equation, 27
Nerve(s). *See also* Neurons
    deafness, 163
    energy sources and metabolism of, 52
    impulses
        autonomic, effector organ responses to, 203–204, 204–205*t*
        conduction of, 45
    mixed, properties of, 52–53
    net, 82*f*
Nerve fibers, 53, 53*t*
    auditory
        action potentials in, 162
        frequency and intensity of sound and, 162*f*
    noradrenergic, 542, 542*f*

in peripheral and cranial nerves, 54*t*
    susceptibility to conduction block, 54*t*
Nerve growth factor (NGF), 54–55
    receptor-mediated endocytosis and, 19
Nervi erigentes, 390
Nervous system. *See also* Autonomic nervous system
    adrenal insufficiency and, 335
    anabolic, 206
    cardiovascular system regulation and, 542–550
    central, synaptic transmission in, 237–240
    enteric, 201, 438–439
    nonadrenergic noncholinergic, 202, 593
    organization of, 242*f*
    thyroid hormones and, 294
Net amount transferred, 644
Net flux, 25
Neural communication, 30
Neural control, mechanisms of, 217*f*
Neural hormones, 217
Neurofibrillary tangles, 246
Neurogenic bladder, spastic, 663
Neurogenic shock, 580
Neurohemal organs, 556
Neurokinin B gene, 96
Neurokinin receptor, 96
Neuromuscular junction, 74, 99–101, 100*f*
Neuromuscular transmission, 99–102
    sequence of events in, 100
Neurons. *See also* Motor neurons
    adrenergic, 88
    afferent, in dorsal horns, 121*f*
    beta-endorphin, distribution of, 240*f*
    cholinergic, 85
    cortical
        electrical responses of, 178*f*
        reticular activating system and, 178
    dopaminergic, 88, 239*f*
    electrical events in, measurement of, 45–50
    excitation of, 45
    functional organization of, 45*f*
    hypophyseotropic hormone-secreting, location of, 224*f*
    ion channels in, 51*t*
    magnocellular
        electrical activity of, 217–218
        response to stimulation, 218*f*
    morphology of, 43–44
    myelinated, ion channels in, 51–52
    noradrenergic, 88
        reuptake mechanism of, 89
    postganglionic, 101, 101*f*, 201, 202*f*
        fast and slow responses of, 204*t*
    postsynaptic, action potential generation in, 79–80
    preganglionic, 201, 202*f*

protein synthesis and, 44–45
    respiratory, 611–612, 612*f*
    sensory, numerical classification for, 53*t*
    short noradrenergic, 201
    types in nervous system, 44*f*
Neuropathy, diabetic, 321
Neuropeptide Y, 97, 202, 206, 318–319
Neurophysin, 217
Neurosecretion, 217
Neurosis, experimental, 250
Neurosteroids, 98
Neurotensin, 443
Neurotransmitters, 74
    chemistry of, 83–85
    localization of, 83–84*t*
    mechanism of action of, 86*t*
    purinergic, 97–98
    receptors, 31–32, 85
Neutral fat, 275
Neutral sulfur, 267
Neutrophils, 470
    hypomotility of, 473
Newborn, hemolytic disease of, 487–488
NGF. *See* Nerve growth factor
Nicotinamide adenine dinucleotide (NAD+), 258
    formula of, 258*f*
Nicotinamide adenine dinucleotide phosphate (NADP+), 258
Nicotine
    Alzheimer's disease and, 246
    vasopressin secretion and, 220
Nicotinic actions, 85
Nicotinic receptor, 85
    fetal, 87*f*
NIDDM. *See* Diabetes mellitus, non-insulin-dependent
Nifedipine, high-altitude pulmonary edema and, 625
Night blindness, 149
Nigrostriatal system, 239
Nitrates
    angina pectoris and, 564
    iodide transport and, 300
Nitric oxide (NO), 98, 539–540
    synthesis from arginine, 539*f*
Nitrogen balance, 271
    negative, diabetes mellitus and, 310
Nitrogen mustards, protein synthesis and, 18*t*
Nitrogen narcosis, 632
Nitroglycerin
    angina pectoris and, 564
    guanylate cyclase and, 530–540
Nitroprusside, guanylate cyclase and, 540
Nizatidine, peptic ulcer and, 451
NK cells, 478–479
NMDA receptor, 94
NO. *See* Nitric oxide
Nociceptive stimulus, 117
Nociceptors, 104

Nocturia, 660
Nocturnal enuresis, 181
Nodal point, 138, 139*f*
Nodes of Ranvier, 43
Nonadrenergic noncholinergic nervous system, 202, 593
Noncommunicating hydrocephalus, 553
Nonconstitutive pathway, 18–19
Nondeclarative memory, 243–244
encoding, 244
Nondisjunction, 381, 381*f*
Nonfluent aphasia, 249
Nonionic diffusion, 26–27, 656
Non-REM sleep, 176
Nonspecific cholinesterase, 87
Nonspecific projection nuclei, 173
Nonsuppressible insulinlike activity (NSILA), 304
Noradrenergic division, 202
Noradrenergic nerve fibers, 542, 542*f*
Noradrenergic neurons, 88
reuptake mechanism of, 89
Norepinephrine, 87–88, 323
autonomic nervous system and, 201–202
brain, 238
catabolism of, 90, 91*f*
effects of, 326–327
hypoglycemia and, 313
intravenous infusion of, circulatory changes due to, 327*f*
levels in venous blood, 326*f*
mechanism of action of, 86*t*
myocardial contractility and, 518
reuptake of, 89
secretion of, regulation of, 327–328
structure and function of, 327
transporter, 89
vasoconstrictor action of, 542
visceral smooth muscle and, 72–73
Norethindrone, 407
structure of, 407*f*
Normal distribution, 674, 674*f*
Normal sinus rhythm (NSR), 501
Nose, pain fibers in, 169
Novobiocin, protein synthesis and, 18*t*
NSILA. *See* Nonsuppressible insulinlike activity
NSR. *See* Normal sinus rhythm
Nuclear bag fiber, 112, 112*f*
Nuclear chain fiber, 112, 112*f*
Nuclear membrane, 11
Nucleic acids, 269
absorption of, 432
Nucleolus, 5*f*, 11
Nucleosides, 269
Nucleus, 10–11
Nucleus basalis of Meynert, 245
Nucleus of tractus solitarius, 170
Nutrition, 282–286
growth and development and, 370
Nutritionally essential amino acids, 266

Nyctalopia, 149
Nystagmus, 164, 192
physiologic, 149
postrotatory, 164

Obesity, 282
brown fat and, 215
hypothalamic, 213, 214*f*
Objective scotoma, 150
OBPs. *See* Odorant-binding proteins
Obstructive shock, 577, 581
Occlusion, 82–83, 118
Ocular dominance columns, 145–146, 146*f*
Ocular muscles, external, actions of, 151*t*
Oculocardiac reflex, 504
Odorant-binding proteins (OBPs), 168
Olfaction. *See also* Smell
physiology of, 167–169
Olfactory bulbs, 166–167
dopaminergic neurons in, 239
neural elements in, 167*f*
Olfactory cortex, 166
Olfactory glomeruli, 166
Olfactory mucosa, signal transduction in, 169*f*
Olfactory mucous membrane, 166, 167*f*
Olfactory stria
intermediate, 166
lateral, 166
Olfactory system, signal transduction in, 168, 169*f*
Olfactory thresholds, 167–168, 168*t*
Oligodendrogliocytes, 43, 55
Oligomenorrhea, 408
Oligophrenia, phenylpyruvic, 89–90
Oligosaccharidases, 428
Oliguria, 660
Olivocochlear bundle, 155
Omeprazole, peptic ulcer and, 451
Oncogenes, 15
Oncotic pressure, 29, 488
Ondine's curse, 618
Oocytes, secondary, 397
Open-angle glaucoma, 132
Open pneumothorax, 627
Operant conditioning, 243
Ophthalmoscopy, 133
Opiate receptor, 96
physiologic effects produced by, 97*t*
Opioid peptides, 96–97, 97*t*
brain, 240
Opsin, 141*f*
Opsonization, 470
Optical righting reflexes, 192
Optic chiasm, 134
Optic disk, 133, 134*f*
Optic nerve, 133, 134
Optic pathways, lesions in, effects of, 150–151
Optics, principles of, 137

Optic tract, 134
Oral contraceptives, hypertension due to, 583
Oral glucose tolerance test, 309, 309*f*
Oral hypoglycemic agents, insulin secretion and, 315
Orbit, muscles of, 136*f*
Organelles, 4, 5*f*
Organ of Corti, 153–154, 156*f*
Organum vasculosum of lamina terminalis (OVLT), 216, 416, 555
Orientation columns, 145
Orthodromic conduction, 49
Orthopnea, congestive heart failure and, 584
Orthostatic albuminuria, 660
Orthostatic hypotension, 572
Osmolal concentration, 3–4
of plasma, 26
Osmolality, 4
plasma
plasma vasopressin and, 220*f*
thirst and, 215, 216*f*
total body, 30
Osmolarity, 4
Osmoles, 3–4
Osmoreceptors, 215
Osmosis, 25–26, 25*f*
Osmotic diuresis, 653–654
urine concentration-urine flow in, 654*f*
Osmotic fragility, 482
Osmotic pressure, 25
glomerular capillaries and, 643, 643*f*
Osmotic pressure gradient, 533
Ossicles, functions of, 159–160
Ossicular conduction, 161
Osteitis cystica, 357
Osteoblasts, 350–351
Osteoclacin, 352
Osteoclasts, 350–351
Osteocytes, 350, 350*f*
Osteogenesis imperfecta, 352
Osteomalacia, 352
Osteonectin, 352
Osteons, 350
Osteopetrosis, 352
Osteopontin, 352
Osteoporosis, 337, 352, 352*f*, 360
Otoconia, 155
Otolith(s), 155
Otolithic organ, 155
Otosclerosis, 163
Ouabain, 422
Outer ear, transmission of vibrations from, 161*f*
Outer hair cells, 153–154
functions of, 161
Oval window, 153
Ovarian agenesis, 381
Ovarian cycle, 396–397
control of, 406
Ovary, 396*f*

development of, 378f
function of
abnormalities of, 407–408
control of, 405–407
feedback regulation of, 406, 406f
primordial follicles in, number at
various ages, 385f
tumors of, 408
Overflow incontinence, 663
Overshoot, 47
Overtones, 158
OVLT. *See* Organum vasculosum of
lamina terminalis
Ovulation, 396
indicators of, 399–400
reflex, 406–407
Oxidation, 257–258
fatty acid, 272, 273f
at substrate level, 258
Oxidative deamination, 267
Oxidative pathway, direct, 260
Oxidative phosphorylation, 7, 258
in mitochondria, 259f
Oxilorphan, vasopressin secretion
and, 221
Oxygen
consumption
renal, 641
of various organs, 552f
debt, 64, 620
debt mechanism, 63–64
deficiency, 622–629
hemoglobin affinity for, factors affecting, 605–607
hypoxic symptoms breathing, 624
lack, ventilatory responses to,
615–616
reaction with hemoglobin, 604–605
toxicity, 629–630
transport, 604–607
treatment, 629–630
uptake, work load and, 621f
Oxygenation, 604
hyperbaric, 630
Oxygen-hemoglobin dissociation
curve, 604–605, 605f
effect of temperature and pH on,
606f
Oxyhemoglobin, 483
Oxyntic cells, 446
Oxyphil cells, 355, 355f
Oxytocin, 97, 216, 217f, 361
biosynthesis, transport, and secretion of, 216–217
effects of, 221–222
localization of, 218
milk ejection reflex and, 412
role in parturition, 411f

P450c11, 331
P450c17, 330
P450c21, 331
P450scc, 330
Pacemaker, 494

abnormal, 501–503
potential, 495–496, 495f
tissue, 70–71
Pacinian corpuscles, 104–105, 104f,
123
generator potential in, 105f
PAF. *See* Platelet-activating factor
PAH. *See* π-Aminohippuric acid
Pain, 124–125
chronic, causes and treatment of,
129
deep, 125
fast, 124
fibers
in nose, 169
visceral, stimulation of, 126
inhibition and facilitation of,
128–129
lobotomy for, 250
muscle, 125
pathways, 121f
referred, 127–128
slow, 124
surgical procedures for, 130f
vasopressin secretion and, 220
visceral, 126–127
Palpation method, 532
PAMs. *See* Pulmonary alveolar macrophages
Pancreas, exocrine portion of,
452–454
Pancreatectomy, diabetes mellitus
due to, 308
Pancreatic amylase, 453
Pancreatic ducts
acinar cells surrounding, 453f
connections of, 453f
Pancreatic islets
capillaries in, fenestrations in, 524f
organization of, 319
Pancreatic juice
composition of, 452–453, 453t
secretion of
cholecystokin and, 442
regulation of, 453–454
secretin and, 443, 453–454
Pancreatic lipase, 432
Pancreatic polypeptide, 302, 318–
319, 444
Pancreatic proteases, activation of,
453f
Pancreatitis, acute, 453
Panting, 227
Paracrine communication, 30–31
Paradoxical sleep, 176
Parafollicular cells, 357, 358f
Paraganglia, 323
Parallel fibers, 198, 199f
Paralysis, spastic, 184
Paralysis agitans, 195–196
Paralytic ileus, 462
Paraplegia, complications of, 189
Parasympathetic division, 201, 203f
Parathion, autonomic nervous system
and, 207

Parathyroidectomy, effects of, 357
Parathyroid glands, 355–357, 355f
Parathyroid hormone (PTH), 348
concentration as function of serum
calcium, 359f
excess, 357
mechanism of action of, 356
secretion of, 356–357
structure of, 356f
synthesis and metabolism of,
355–356
Parathyroid hormone-like protein
(PLP), 360
Paravertebral sympathetic ganglion
chain, 201
Paresis, 184
Parietal cells, 446, 447f, 448f
gastric acid secretion by, 449f
hydrochloric acid secretion by, 448f
Parietal cortex, posterior, 186
Parietal-temporal-occipital associa-
tion area, 247, 247f
Parkinson's disease, 195–196
Paroxysmal atrial tachycardia with
block, 504
Paroxysmal tachycardia, 503
Paroxysmal ventricular tachycardia,
504–505
Pars compacta, 193
Pars reticularis, 193
Partial pressures, 587, 599f
Partial seizures, 182
Parturition, 410–411
oxytocin in, 411f
Passive tension, 62
Past-pointing, 200
Patch clamping, 21
Patches, striatal, 194
Patchy atelectasis, 595
Pattern generator, 189–190
respiratory control, 611
PBI. *See* Protein-bound iodine
P cells, 494, 638
PCR. *See* Polymerase chain reaction
PDGF. *See* Platelet-derived growth
factor
PEEP. *See* Positive end-expiratory
pressure
Penicillin, blood-brain barrier and,
556
Penis, erection of, 390
Pentoses, absorption of, 429
Pepsin, 431
secretion of, gastrin and, 441
Pepsinogen, 431
secretion of, 447
Pepsinogen I, peptic ulcer and, 451
Pepstatin, renin secretion and, 417
Peptic cells, 446, 447f
Peptic ulcer, 451
Peptide bonds, 266, 266f
Peptides, 266
opioid, 96–97, 97t
brain, 240
Perception, subcortical, 124–125

Perchlorate, iodide transport and, 300
Perforins, 477
Perfusion-limited exchange, 598
Perfusion pressure, effective, 524
Pericardium, 511
Pericytes, 637
Perilymph, 153
    composition of, 160*f*
Perimeter, 150
Perimetry, 150
Perinuclear cisterns, 11
Periodate, iodide transport and, 300
Periodic breathing, 631
Peripheral motor control system, 116*f*
Peripheral nerves
    nerve fibers in, 54*t*
    relation of axons to Schwann cells
        in, 45*f*
Peripheral proteins, 5
Peripheral resistance, 517, 524
Peripheral venous pressure, 536–537
Peristaltic rushes, 461
Peristaltic waves, 461
Peritubular capillaries, 638
Periventricular system, 239
Permeability, glomerular capillary, 642
Pernicious anemia, 452
Peroxisome proliferators, 10
Peroxisomes, 10
Personality, lobotomy and, 250
Pertechnetate, iodide transport and,
        300
PET. *See* Positron emission tomogra-
        phy
Petit mal epilepsy, 181*f*
Petit mal seizures, 182
Peyer's patches, 459
P factor, 125
pH, 4, 4*f*
    of body fluids, 666*t*
    limiting, 655
    oxygen-hemoglobin dissociation
        curve and, 606*f*
    urinary, implications of, 657
Phagocytic vacuole, 8
Phagocytosis, 19, 470
    disorders of, 473
Phantom limb, 107
Phasic bursting, 218
Phasic receptors, 106
Phasic reflexes, 187
Phenformin, insulin secretion and,
        315
Phenothiazines, parkinsonism due to,
        195
Phenotype, 487
Phenylalanine hydroxylase, 88
Phenylbutazone, for gout, 271
Phenylethanolamine-N-methyltrans-
        ferase (PNMT), 88
    adrenal medullary, 327
Phenylketonuria (PKU), 90
Phenylpyruvic oligophrenia, 89–90
Phenylthiocarbamide (PTC), ability
        to taste, 172

Pheochromocytoma, 327
    hypertension in, 583
Pheomelanins, 363
Pheromones, 234
Phlorhizin, 646
PHM–27, 443
Phosphate compounds, high-energy,
        257
Phosphatidylinositol, metabolism in
        cell membranes, 38*f*
Phosphaturic action, 356
Phosphoglyceraldehyde, conversion
        to phosphoglycerate, 262
Phospholipase, 453
Phospholipids, 4–5, 271–272
Phosphorus metabolism, 349
Phosphorylase, 262–264
    activation of, 263*f*
Phosphorylase kinase, 34, 36
Phosphorylation, oxidative, 258
    in mitochondria, 259*f*
Phosphorylcreatine, 63, 257, 269
    structure of, 269*f*
Photopic vision, 136
Photoreceptor mechanism, 139
Photoreceptor potentials, ionic basis
        of, 140
Photosensitive compounds, 141
Phototherapy, 485
Phototransduction
    initial steps in, 142*f*
    sequence of events in, 142*f*
Physiologic dead space, 597
Physiologic nystagmus, 149
Physiologic shunt, 600
Physiologic tremor, 114
Physiology, abbreviations and sym-
        bols used in, 675–679
Physostigmine, Alzheimer's disease
        and, 246
Picrotoxin, gamma-aminobutyric
        acid receptor and, 94
Piebaldism, 364
Pigment abnormalities, 364
PIH. *See* Prolactin-inhibiting hormone
Pineal, 422–424, 422*f*
    anatomy of, 422
    function of, 423–424
Pineal sand, 422
Pinocytosis, 19
Pirenzepine, peptic ulcer and, 451
Piriform cortex, olfactory discrimina-
        tion and, 166
Pitch, 158
Pituicytes, 361
Pituitary gland, 361–374
    anterior
        cell types in, 361–363
        hormones of, 222, 223*f*
        secretions of, 222–225
        ultrastructure of, 363*f*
    formation of, 362*f*
    gonadotropins, 385–387
    hormones of, 362*t*
    intermediate lobe, 363–364

    hyperfunction, 374
    insufficiency, 372–374, 373*f*
        causes of, 373–374
        partial, 374
    morphology of, 361–363
    posterior, 555
        secretions of, 216–222
    relation to hypothalamus, 210
PKU. *See* Phenylketonuria
Placenta
    circulation in, 567
    gaseous interchange across, 569*t*
    hormones of, 409–410
    interactions with fetal adrenal cor-
        tex, 410*f*
    section through, 568*f*
Placidity, limbic system and, 235
Placing reaction, 193
Planum temporale, 248, 248*f*
Plasma, 1, 488–489
    calcium distribution in, 349*t*
    cells, 474, 475
    composition of, 553*t*
    constituents, renal handling of, 645*t*
    insulinlike activity in, 305*t*
    kallikrein, 541
    lipids, 275–278
    membrane, 4–6
        glucose transport protein in, 306*f*
    osmolal concentration of, 26
    osmolality
        plasma vasopressin and, 220*f*
        thirst and, 215, 216*f*
    proteins, 488–489, 607
        origin of, 489
        thyroid hormones binding to,
            291, 291*t*
    skimming, 528
    true, 665
    viscosity and, 527–528
    volume, 2
Plasma renin activity (PRA), 415
Plasma renin concentration (PRC),
        415
Plasmin, 491
Plasminogen, structure of, 491*f*
Plasticity, 73
    synaptic, learning and, 98–99
Plate endings, 113
Platelet(s), 481
    activation, 481
    adhesion, 481
    aggregation, 481
        prostacyclin and, 539
        thromboxane and, 280, 539
Platelet-activating factor (PAF), 481
Platelet-derived growth factor
        (PDGF), 481
Plethysmograph, 525
Plethysmography, 525, 525*f*
PLP. *See* Parathyroid hormone-like
        protein
Plutonium, uptake by bone, 352
PMNs. *See* Polymorphonuclear leuko-
        cytes

Pneumocytes, granular, 588
Pneumonia, aspiration, 592
Pneumotaxic center, 612
Pneumothorax, 627
　closed, 627
　open, 627
　tension, 627
PNMT. *See* Phenylethanolamine-N-
　　methyltransferase
Podocytes, 637f, 638
Poikilothermic, 225
Poiseuille-Hagen formula, 526–527
Poisoning
　aluminum, 284
　carbon monoxide, 628–629
　radiation, 464
Polar bodies
　first, 397
　second, 397
Polarization, 6
Poly(A) tail, 16
Polycystic ovary syndrome, 408
Polydipsia, 221
Polymerase chain reaction (PCR), 13
Polymorphonuclear leukocytes
　　(PMNs), 470
Polypeptides, 266
　in brain, 97
Polypeptide YY (PYY), 318
Polyribosomes, 16
Polysaccharides, 427
Polysomes, 16
Polysynaptic reflexes, 111, 117–118
Polyuria, 221, 660
POMC. *See* Pro-opiomelanocortin
Ponto-geniculo-occipital (PGO)
　　spikes, 176
Portal hypophyseal vessels, 210
Positive end-expiratory pressure
　　(PEEP), 633
Positive reinforcement, 243
Positive supporting reaction, 189
Positron emission tomography
　　(PET), 557
Postcentral gyrus, somatic sensory
　　area I in, 121
Posterior parietal cortex, 186
Postextrasystolic potentiation, 518
Postganglionic neurons, 101, 101f,
　　201, 202f
　fast and slow responses of, 204t
Postprandial alkaline tide, 448
Postrotatory nystagmus, 164
Postsynaptic cell, 74
Postsynaptic inhibition, 79, 80
　neurons producing, 79, 81f
Postsynaptic neurons, action poten-
　　tial generation in, 79–80
Postsynaptic potentials
　excitatory, 77–78
　　ionic basis of, 78
　inhibitory, 78–79
　　ionic basis of, 79
　slow, 79
Postsynaptic receptor, 85f

Posttetanic potentiation, 99
Posttranscriptional modification, 13,
　　15f, 16–17
Postural control, 187
Postural hypotension, 572
Postural reflexes, 188t
Postural syncope, 580
Posture-regulating systems, 183,
　　187–193
　cortical components of, 192–193
　medullary components of, 190–192
　midbrain components of, 192
　spinal integration in, 189–190
Potassium
　absorption of, 433–435
　conductance, during action poten-
　　tial, 51f
　depletion, insulin secretion and,
　　315
　excretion, regulation of, 658–659
　insulin and, 306
　ion(s), active transport of, 24–25
　ion channels, 22–24, 23f
　metabolism, acid-base balance and,
　　671
　secondary active transport of, 650f
　total body, 30, 30t
　vasodilation and, 538
Potentiation
　long-term, 94, 99
　postextrasystolic, 518
　posttetanic, 99
PRA. *See* Plasma renin activity
PRC. *See* Plasma renin concentration
Precapillary sphincters, 521
Precocious pseudopuberty, 334, 383
　true, 383
Prednisolone, potency of, 329–330
Preejection period, 513
Prefrontal lobotomy, 250
Preganglionic neurons, 201, 202f
Pregnancy, 408–411
　diabetes mellitus and, 321
　hormone levels during, 409f
　oxytocin and, 222
　thyroxine-binding globulin levels
　　in, 292
　toxemia of, 582
　uterine blood flow in, 567f
Prekallikrein, 541
Preload, 517
Premature beat, 503
　ventricular, 504f, 505f
Premenstrual syndrome, 408
Premotor cortex, 185, 185f, 186
Prepotency, 117
Prepotential, 495
Preproenkephalin, 325
Preprogastrin, 441
Preproglucagon, 316
Prepro-opiomelanocortin molecule,
　　364f
Prepro-oxyphysin, 217

　structure of, 218f
Prepropressophysin, 217
　structure of, 218f
Preproproteins, maturation of, 20f
Prepro-PTH, 355
Preprorenin, 414
Presbyopia, 138, 139
Pressure, blood flow and, 523–524
Pressure gradients, coronary blood
　　flow and, 561–562
Presynaptic cell, 74
Presynaptic facilitation, 80–81
　presynaptic action potential and,
　　81f
　sequence of events in, 82f
Presynaptic inhibition, 80–81
　neurons producing, 81f
　presynaptic action potential and,
　　81f
Presynaptic receptor, 85, 85f
PRH. *See* Prolactin-releasing hor-
　　mone
Primary amenorrhea, 383, 407
Primary autonomic failure, forms of,
　　572t
Primary autonomic insufficiency, pos-
　　tural hypotension due to,
　　572
Primary colors, 146
Primary endings, 112
Primary evoked potential, 175
Primary gout, 270–271
Primary hyperaldosteronism, 342–343
　hypertension in, 583
Primary hyperalgesia, 125
Primary plexus, 210
Primary spermatocytes, 388f, 389
Primary structure, 266
Primordial follicles, 396, 396f
Principal axis, 137
Principal cells, 638
Principal focal distance, 137
Principal focus, 137
Proarrhythmic, 506
Probenecid, for gout, 271
Procainamide, myocardial contractil-
　　ity and, 518
Prodynorphin, 96
Pro-enkephalin, 96
Proenzymes, 431
Progestational agents, 405
Progesterone, 375
　actions of, 404–405
　biosynthesis and metabolism of,
　　403–404, 404f
　secretion of, 404
　sexual activity and, 232
Progestins, 405
　hypertension due to, 583
Proglumide, peptic ulcer and, 451
Prognathism, 374
Progoitrin, 301
Prohormones, 15f, 17
Proinsulin, 303
Projection, law of, 107

Prokaryotes, 5
Prolactin, 222, 361, 385–387
  hypersecretion of, 374
  secretion of
    factors affecting, 386t
    regulation of, 386–387
    serotonin and, 238
  structure of, 386f
Prolactin-inhibiting hormone (PIH), 222
Prolactin-releasing hormone (PRH), 222–223, 387
Proliferative phase, 397
Promoter region, 14
Pro-opiomelanocortin (POMC), 96, 363
Properdin, 478
Properdin pathway, 478
Propranolol, renin secretion and, 417
Proprioception, 123–124
Proprioceptors, 104
Pro-PTH, 356
Propylthiouracil, 300
  structure of, 301f
Prorenin, 414
Prosopagnosia, 250
Prostacyclin, 280, 481, 539
  antiaggregating effect of, 491
Prostaglandins, 98, 280
  biosynthesis of, 281f
  fever and, 229
  in semen, 390
Prostate, 387
  androgen-dependent carcinoma of, 413
Protanomaly, 147
Protanopia, 147
Protein(s), 266
  absorption of, 431–432
  amino acids found in, 267f
  cell membrane, 5
  degradation, 270
  digestion of, 431
  effect on gastric emptying, 451f
  folding, 17
  grade I, 284
  grade II, 284
  insulin secretion and, 314
  metabolism, 266–271
    diabetes mellitus and, 310
    growth hormone and, 367
  plasma, 488–489, 607
    origin of, 489
    thyroid hormones binding to, 291, 291t
  primary structure of, 266
  processing in Golgi apparatus, 19f
  quaternary structure of, 266
  secondary structure of, 266
  synthesis, 15–16
    antibiotics and, 17, 18t
    diagram of, 14f
    neurons and, 44–45
  tertiary structure of, 266

translation into endoplasmic reticu-
  lum, 17f
Protein-bound iodine (PBI), 290
Protein C activator, 491
Protein kinase(s), 33, 33t
  activation of, 36
Protein kinase A, 36
  activation of, cyclic AMP and, 263
Protein-sparing effect, 271
Proteinuria, 660
Protodiastole, 511
Proto-oncogenes, 15
Protoplasmic astrocytes, 55
Proximal convoluted tubule, 638
Proximal tubular cells, secretion of
  acid by, 655f
Proximal tubule
  sodium reabsorption in, mecha-
    nism for, 646f
  solute reabsorption in, 647f
  transport mechanisms in, 646f
  water excretion and, 649
Pseudocholinesterase, 87, 239
Pseudohermaphroditism
  female, 332, 382
  male, 382, 393
Pseudohypoaldosteronism, 347
Pseudohypoparathyroidism, 39, 356
Pseudo-H zone, 56, 58
Psilocin, serotonin and, 238
Psoriasis, 1,25-
    dihydroxycholecalciferol
    and, 354
Psychic energizers, synaptic transmis-
    sion in brain and, 237
Psychosis, amphetamine and, 239
PTC. See Phenylthiocarbamide
PTH. See Parathyroid hormone
Ptyalin, 427, 444
Pubarche, 382
Puberty, 382–383
  delayed or absent, 383
  pineal and, 424
  plasma hormone concentrations in, 383f
  precocious, 383–384, 384f
    causes of, 384t
Pulmonary alveolar macrophages
    (PAMs), 602
Pulmonary blood flow, regulation of, 601–602
Pulmonary blood vessels, 599
Pulmonary capillary(ies), 588, 589f
Pulmonary capillary pressure, 600
Pulmonary chemoreflex, 547, 617
Pulmonary circulation, 469, 588–590, 599–602
  blood pressures in, 600f
Pulmonary edema, 584
  high-altitude, 625
Pulmonary embolism, 492
Pulmonary embolization, 602
Pulmonary fibrosis, expiratory pres-
    sure-volume curves in, 594f
Pulmonary hypertension, 581

Pulmonary receptors, 547
Pulmonary reservoir, 601
Pulmonary salt wasting, 221
Pulmonary ventilation, 591
Pulse
  arterial, 513
  jugular, atrial pressure and, 513–514
Pulse pressure, 530
Pumps, 5
Punctate representation, 123
Punishment system, 236
Pupil, 132
  Argyll Robertson, 138
Pupillary light reflex, 138
Pupillary reflexes, 138
Purine(s), 269–270
  physiologically important, 269f
Purine-containing compounds, 270t
Purinergic transmitters, 97–98
Purkinje cells, 198, 199f
Purkinje system, 494
Puromycin, protein synthesis and, 18t
Putamen, 193, 193f
P wave, 497, 498
Pyloric sphincter, 449
Pyramidal cells, neocortical, 174f
Pyramidal lobe, 287, 287f
Pyramidal system, 184
Pyrimidine(s), 269–270
  physiologically important, 269f
Pyrimidine-containing compounds, 270t
PYY. See Polypeptide YY

QRS complex, 67, 497, 498
QRS vector, mean, 499–500, 500f
Quadriplegia, complications of, 189
Quaternary structure, 266
Quetelet index, 282
Quinidine, myocardial contractility
  and, 518
Q wave, 498

Radiation, 227
Radiation poisoning, 464
Radical mastectomy, edema due to, 535
Radioactive iodine
  distribution of, 300f
  tracer doses of, 289
  uptake of, 299–300
Radioimmunoassay, 31
Radionuclide, 562
Radionuclide scan, 458
Radium, uptake by bone, 352
Rage
  limbic system and, 235–236
  sham, 235
Ranitidine, peptic ulcer and, 451
Rapamycin, 480
Rapid eye movement (REM) sleep, 176

concomitants of, 180
genesis of, 181
growth hormone secretion and, 370
Rapidly adapting receptors, 617
RAS. *See* Reticular activating system
Rathke's pouch, 361
Rayleigh match, 147
RDS. *See* Respiratory distress syndrome
Reaction time, 113
Reactive hyperemia, 563, 566–567
Rebound phenomenon, 200
Recall. *See also* Memory
immediate, 244
Recent memory, 244
alcoholism and, 245
Receptive field, 108, 143*f*
Receptive relaxation, 449
Receptor(s), 5
acetylcholine, 85–87
adhesion, 477
AMPA, 94
angiotensin II, 416
atrial natriuretic peptide, 421, 421*f*
atrial stretch, 546–547
beta-adrenergic
desensitization in, 85
structure of, 37*f*
chemical messenger, 31–32
cholecystokinin-A, 215
cholecystokinin-B, 215
dihydropyridine, in T tubule membrane, 60
diseases of, 39
estrogen, structure of, 33*f*
gamma-aminobutyric acid, 94–95
glucagon, 317
glucocorticoid, 342
structure of, 33*f*
G protein-coupled, ligands for, 36*t*
growth hormone, 366
gustatory, 171
insulin, 304, 307–308
structure of, 308*f*
interleukin, 479
irritant, 617
kainate, 94
left ventricular, 547
mannose–6-phosphate, 17–18
metabotropic, 94
mineralocorticoid, 342
structure of, 33*f*
muscarinic, 85
neurokinin, 96
neurotransmitter, 85
nicotinic, 85
fetal, 87*f*
NMDA, 94
opiate, 96
physiologic effects produced by, 97*t*
pain, 124–125
phasic, 106
postsynaptic, 85*f*
potential, 105

presynaptic, 85, 85*f*
pulmonary, 547
rapidly adapting, 617
ryanodine, 60
scavenger, 277
sensory, 103
electrical and ionic events in, 105–106
excitation of, 106–107
in skin, 104*f*
serotonin, 92
serpentine, 35, 358
slowly adapting, 617
smell, 166–167
structure of, 33–34, 33*f*
taste, 170
T cell, 477
thyroid-stimulating hormone, 296–297
tonic, 106
touch, 123
vasopressin, 219
visual, 134–136
effect of light on, 140*f*
Receptor-mediated endocytosis, 19
Reciprocal innervation, 80, 114–115, 611
Recoverin, 142
Recruiting response, 179, 179*f*
Recruitment of motor units, 66, 117
Rectum, distention of, responses to, 465*f*
Recurrent collateral inhibition, 179
Red blood cells, 481–485, 482*f*
agglutination of, 487*f*
characteristics of, 482*t*
formation and destruction of, 482*f*
fragility of, 482–483
abnormal, 482
Red cell volume, 2
Red marrow, 469
changes in cellularity, 470*f*
Red muscles, 64–65
Red reaction, 566
Reduction, 258
Reentry, 503
Referred pain, 127–128
Reflex arc, 111, 112*f*
monosynaptic, 111
polysynaptic, 111
Reflex(es)
autonomic, 190
medullary, 208
avoidance, conditioned, 243
axon, 543, 566, 566*f*
Bainbridge, 547
Bezold-Jarisch, 547, 618
conditioned, 242–243, 243*f*
Cushing, 548, 559
enterogastric, 450
gastrocolic, 465
gastroileal, 463
grasp, 192
habituation and sensitization of, 119

Hering-Breuer, 617
knee jerk, 111
mass, 119, 190
milk ejection, 222, 412
monosynaptic, 111–117
oculocardiac, 504
phasic, 187
polysynaptic, 117–118
postural, 188*t*
properties of, 118–119
pupillary, 138
righting, 192
sexual, 190
somatosympathetic, 549, 549*f*
static, 187
stretch, 111
cerebellum and, 200
corticospinal/corticobulbar systems and, 187
hyperactive, 184
inverse, 115, 115*f*
supraspinal regulation of, 190–191, 190*f*, 191*f*
tonic labyrinthine, 191
tonic neck, 191–192
transcortical load-compensating, 187
tympanic, 160–161
vestibulo-ocular, 164, 192
visceral, respiratory components of, 619
vomiting, afferent pathways for, 208*f*
withdrawal, 117
local sign and, 118*f*
Reflexive memory, 243
Reflex ovulation, 406–407
Refractory period, 48
absolute, 48, 69
relative, 48
Refractory shock, 579
Regulated pathway, 19
Regurgitation, 514
Reinforcement
negative, 243
positive, 243
Reissner's membrane, 153
Relative refractory period, 48
Relaxation heat, 64
Relaxation pressure curve, 593, 593*f*, 596*f*
Relaxation volume, 593
Relaxin, 375, 405
REM behavior disorder, 181
Remote memory, 244
REM sleep, 176
concomitants of, 180
genesis of, 181
Renal blood flow, 639
autoregulation of, 640–641
regulation of, 640
Renal capsule, 639
Renal circulation, 636*f*, 639–641
Renal compensation
metabolic acidosis and, 667–668

Renal compensation *(cont.)*
  respiratory alkalosis/acidosis and, 667
Renal cortex, blood flow in, 640
Renal failure, acute, 578
Renal function, disordered, effects of, 660–661
Renal glycosuria, 264
Renal hypertension, 417, 582
Renal interstitial pressure, 639
Renal nerves
  functions of, 640
  stimulation of, renal responses to, 640*t*
Renal plasma flow (RPF), 639
  effective, 639
  glomerular filtration rate and, 643–644
Renal threshold, 264, 646
Renal tubular acidosis, 660
Renal tubules
  function of, 644–648, 644*f*
  substances secreted by, 647–648
Renal vessels
  innervation of, 639
  pressure in, 640
Renin, 414–415
  aldosterone secretion and, 344
  clinical hypertension and, 417–418
  secretion of
    factors affecting, 418*f*
    regulation of, 416–417
    stimuli increasing, 417*t*
  substrate, 414
Renin-angiotensin system, 414–418
  pharmacologic manipulation of, 417
Rennin, 431
Reperfusion-induced injury, 580
Representational hemisphere, 247
Reproductive system
  embryology of, 377–381
  female, 395–408, 395*f*
  male, 387–395, 387*f*
Reserpine, slow-wave sleep and, 181
Residual volume, 591
Resistance, 524, 593
  peripheral, 524
Resistance vessels, 529
Resonator, 159
Respiration
  artificial, 632–633, 633*f*
  chemical control of, 613–617
  Cheyne-Stokes, 631, 631*f*
  control of, 208
  external, 587
  fetal, 569
    changes at birth, 569–570
  internal, 587
  mechanics of, 590–598
  nonchemical influences on, 617–619
  regulation of, 611–619
Respirators, mechanical, 633
Respiratory acidosis, 615, 666–667

arterial blood values in, 667*f*
Respiratory activity, regulation of, 612–613
Respiratory alkalosis, 615, 666–667
Respiratory bronchioles, 588, 589*f*
Respiratory burst, 470
Respiratory center, 611
  stimuli affecting, 613*t*
Respiratory compensation, metabolic acidosis and, 667
Respiratory control pattern generator, 611
Respiratory dead space, 591, 597–598
Respiratory distress syndrome (RDS), 595
  adult, 579, 633
Respiratory exchange ratio, 254
Respiratory minute volume, 591, 616*f*
Respiratory muscles, 591–592
Respiratory neurons, 611–612, 612*f*
Respiratory phenomena, quantitating, 588
Respiratory quotient (RQ), 254
Respiratory rate, alveolar ventilation and, 597*t*
Respiratory symbols, 679
Respiratory system
  functions of, 602–603
  relaxation pressure curve of, 596*f*
Resting heat, 64
Resting length, 62
Resting membrane potential, 20, 46
  in cardiac muscle, 67–69
  ionic basis of, 50
  in skeletal muscle, 59
Restriction fragment length polymorphism (RFLP), 13
Resuscitation, mouth-to-mouth, 633, 633*f*
Reticular activating system (RAS), 173
  ascending, 174*f*
  cortical neurons and, 178
  general anesthesia and, 179–180
Reticular formation, 173
Reticular lamina, 153
Reticuloendothelial system, 472
Retina, 132–134
  cells of, electrical responses of, 140–143, 140*f*
  color vision and, 146–147
  dopaminergic neurons in, 239
  image formation in, 143
  neural components of, 133*f*
  projection on calcarine fissure, 135*f*
  synaptic mediators in, 143
Retinal image, 138–139
Retinals, 141
Retinene, 141
  structure of, 141*f*
Retinitis pigmentosa, 142
Retinoic acid, limb development and, 34
Retinols, 141
Retinopathy, diabetic, 321

Retrograde amnesia, 244
Retrograde transport, 45
Retrolental fibroplasia, 629
Reuptake mechanism, 89
Reverberating circuits, 117
Reverse triiodothyronine, 288
Reward system, 236
RFLP. *See* Restriction fragment length polymorphism
Rhabdomyolysis, 579
  exertional, 574
Rhodopsin, 141
  gene for, 147
  sodium channels and, 141–142
  structure of, 37*f*, 141*f*
Rh system, 487
Ribonucleic acid. *See* RNA
Ribosomal RNA. *See* RNA
Ribosomes, 11
Rickets, 352, 354
Right axis deviation, 500
Right bundle branch block, 502
Righting reflexes, 192
Right ventricle, pressures in, 562*f*
Rigidity
  cog-wheel, 196
  decerebrate, 190
    mechanism of, 190
    significance of, 191
  decorticate, 191, 191*f*, 192–193
  due to Parkinson's disease, 195–196
  lead-pipe, 196
Rigor, 64
Rigor mortis, 64
Riley-Day syndrome, 572*t*
Rinne test, 163*t*
Riva-Rocci cuff, 531
RNA, 11
  messenger, 13
  ribosomal, 8, 13
  structure and function of, 11–13
  transfer, 8, 13
RNA polymerase, 13
Rods, 132, 134–136, 135*f*
  density curve for, 136*f*
  phototransduction in
    initial steps in, 142*f*
    sequence of events in, 142*f*
Rods of Corti, 153
Rotation, ampullary responses to, 164*f*
Rotational acceleration, responses to, 164
Round window, 153
RPF. *See* Renal plasma flow
RQ. *See* Respiratory quotient
R state, 604
RU–486, 405
Ruffini endings, 104–105, 104*f*, 123
R units, 524
Ryanodine receptor, 60

Saccades, 151
Saccule, 155

Sacral outflow, 201
Saline cathartics, 434
Saliva, 444–445
  ionic composition of, 445
  secretion of, control of, 445
Salivary amylase, 444–445
Salivary glands, 444–445
  characteristics of, 445*t*
Saltatory conduction, 49
Salt balance, mineralocorticoids and, 346
Saltiness, 172
Salt wasting, 221
Saralasin, renin secretion and, 417
Sarcolemma, 56, 57*f*
Sarcomere, 56
Sarcoplasmic reticulum, 58
Sarcotubular system, 58
Satiety, hypothalamus and, 213
Satiety center, 213
Saturation, 146, 645
Saxitoxin (STX), sodium channel and, 22
Scala(e), 153, 156*f*
Scala media, 153
Scala tympani, 153
Scala vestibuli, 153
Scanning speech, 200
Scavenger receptor, 277
S cells, 442
Schizophrenia
  cerebral blood flow in, 557
  genetic defect in, 239
Schwabach test, 163
Schwann cells, 43
  relation to axons in peripheral nerves, 45*f*
Sciatic nerve, ether anesthesia and, 180*f*
Sclera, 132
Scopolamine, autonomic nervous system and, 207
Scotoma
  objective, 150
  suppression, 152
Scotopic vision, 136
Scotopsin, 141
SDA. *See* Specific dynamic action
Secondary active transport, 24
Secondary amenorrhea, 407–408
Secondary endings, 112
Secondary gout, 270–271
Secondary hyperaldosteronism, 346, 418
Secondary hyperalgesia, 125
Secondary hyperparathyroidism, 357
Secondary oocyte, 397
Secondary sex characteristics
  female, 403
  male, 392, 392*t*
Secondary spermatocytes, 389
Secondary structure, 266
Secondary tympanic membrane, 153
Second-degree heart block, 502, 502*f*
Second messengers, 33

  activation of genes by, 16*f*
  aldosterone secretion and, 346*t*
  diacylglycerol as, 35–36
  inositol triphosphate as, 35–36
  metabotropic receptors and, 94
Second polar body, 397
Second sound, 514
Secosteroids, 354
Secretin, 442–443
  glucagon secretion and, 318
  insulin secretion and, 315
  pancreatic juice secretion and, 443, 453–454, 454*f*
Secretory immunity, 432, 464, 476
Secretory immunoglobulins, 476
Secretory phase, 397
Segmentation contractions, 461, 461*f*
Seizures
  cerebral blood flow during, 557
  general-onset, 182
  grand mal, 182
  partial, 182
  petit mal, 182
Selectin family, 477
Self, recognition of, 478
SEM. *See* Standard error of mean
Semen, 390
  composition of, 390*t*
Semicircular canals, 155, 155*f*
Seminiferous epithelium, 388*f*
Seminiferous tubule dysgenesis, 381
Seminiferous tubules, 387
Senile dementia, 246
  cerebral blood flow in, 557
Senile plaques, 246
Sensation, intensity of, 107–108
Sense organs, 103
  classification of, 103–104
  cutaneous, 104–105
  pain, 124–125
  proprioception, 123–124
  temperature, 124
  touch, 123
Senses, 103–105
Sensitization, 99, 241
  of reflex responses, 119
Sensory homunculus, 122*f*
Sensory information, coding of, 107–108
Sensory modalities, 103, 104*t*
Sensory neurons, numerical classification for, 53*t*
Sensory physiology, principles of, 123
Sensory receptors, 103
  electrical and ionic events in, 105–106
  excitation of, 106–107
  in skin, 104*f*
Sensory units, 108
Septic shock, 580
Serotonin, 90–92
  biosynthesis and metabolism of, 92*f*
  brain, functions of, 238

  mechanism of action of, 86*t*
  receptor, 92
  secretion of, 444
  transporter, 89
Serpentine receptors, 35, 358
Sertoli cells, 377, 388*f*, 389
Serum, 488
Serum glutamic-oxaloacetic transaminase (SGOT), 267
Sex chromatin, 377
Sex chromosomes, 375–376
  nondisjunction of, 381*f*
Sex determination, basis of, 376*f*
Sex differentiation, 375–385
  aberrant, 381–382
  disorders of, 381*t*
  normal, 378*f*
Sex hormones, 323, 375
  effects of, 232–233
    in infancy on adult behavior, 234–235
Sex steroid(s), production rates during menstrual cycle, 402*t*
Sex steroid-binding globulin, 392
Sexual behavior
  female, 234
  limbic system and, 232–235
  male, 233–234
Sexual function, smell and, 168
Sexual intercourse, female reproductive system and, 399
Sexual precocity, 383–384, 384*f*
  causes of, 384*t*
Sexual reflexes, 190
SFO. *See* Subfornical organ
SGLT. *See* Sodium-dependent glucose transporter
SGOT. *See* Serum glutamic-oxaloacetic transaminase
Sham rage, 235
Shivering, 228
Shock, 576–581
  anaphylactic, 580
  burn, 580
  cardiogenic, 577, 581
  distributive, 577, 580
  hemorrhagic, 577
  hypovolemic, 577, 579–580
  irreversible, 579
  low-resistance, 577
  neurogenic, 580
  obstructive, 577, 581
  refractory, 579
  septic, 580
  spinal, 189
  surgical, 580
  traumatic, 579
  treatment of, 581
  types of, 577*t*
  vasogenic, 577
Shock lung, 629
Shortening heat, 64
Short noradrenergic neurons, 201
Shunts, 523
Shy-Drager syndrome, 572*t*

SIADH. *See* Syndrome of inappropriate hypersecretion of antidiuretic hormone
Sickle cell anemia, 485
Sickle cell trait, 485
Sick sinus syndrome, 501
Side-chain cleavage enzyme, 330
Side pressure, 530
SIDS. *See* Sudden infant death syndrome
SIF cells, 202
Siggaard-Andersen curve nomogram, 669–670, 669*f*
Signal hypothesis, 17*f*
Signal peptide, 17
Signal recognition particle (SRP), 17
Signal transduction
  by G proteins, 35*f*
  in olfactory system, 168, 169*f*
Simple cells, 145
Single-breath nitrogen curve, 597*f*
Single photon emission computed tomography (SPECT), 557
Sinoatrial (SA) node, 494, 495*f*
  membrane potential of, effect of stimulation on, 496*f*
Sinus arrhythmia, 501, 501*f*
Sinus rhythm, normal, 501
Sinus venosus, 501
SI units, 673*t*
Size principle, 65
Skeletal muscle, 56–67, 57*f*. *See also* Muscle
  contractile responses of, 59–63
  contraction and relaxation of, sequence of events in, 61*f*
  electrical characteristics of, 59, 59*f*
  energy sources and metabolism of, 62–65
  filament arrangement in, 58*f*
  ion distribution in, 59*t*
  morphology of, 56–58
  organization of, 56
  properties of, 65–67
  sarcotubular system of, 58
  strength of, 66
  striations in, 56–58
  thyroid hormones and, 294–295
  visceral pain and, 126–127
Skin
  circulation of, 566–567
  coloration
    abnormalities in, 364
    control of, 363–364
  sensory receptors in, 104*f*
Sleep
  disorders, 181
  EEG and muscle activity during, 177*f*
  hypothalamus and, 212
  non-REM, 176
  paradoxical, 176
  peptide, 181
  rapid eye movement, 176
    concomitants of, 180

genesis of, 181
growth hormone secretion and, 370
respiration and, 619
slow-wave, genesis of, 180–181
stages of, distribution of, 180, 180*f*
zone
  basal forebrain, 179
  diencephalic, 179
Sleep apnea, 181
Sleep-apnea syndrome, 619
Sleepwalking, 181
Slow EPSP, 202
Slow IPSP, 202
Slowly adapting receptors, 617
Slow pain, 124
Slow postsynaptic potentials, 79
Slow-wave sleep, 176
  genesis of, 180–181
Small bowel slow wave, 460–461
Small G proteins, 35
Small intestine, 459–462
  enterocytes from, 460*f*
  layers of, 438*f*
  mechanical obstruction of, 462
  motility of, 460–461
  mucus of, 460
  secretions of, regulation of, 461
  structure of, 460*f*
  transit time in, 463
Small motor nerve system, 112
Smell, 166–169. *See also* Olfaction
  memory and, 168
  receptors and pathways of, 166–167
  sexual function and, 168
Smooth muscle, 71–73
  electrical activity in, 71*f*
  morphology of, 71
  multi-unit, 73
  nerve endings in, 101, 101*f*
  nerve supply to, function of, 73
  plasticity of, 73
  visceral, 71–73
    contraction and relaxation of, sequence of events in, 72*t*
    electrical and mechanical activity of, 71
    membrane potential of, effects of agents on, 72*f*
    molecular basis for contraction of, 71–72
    stimulation of, 72–73
Smooth pursuit movements, 151
Sneezing, 618
Sniffing, 168–169
Sodium
  absorption of, 433–435
  conductance, during action potential, 51*f*
  excretion
    glomerular filtration rate and, 658*t*
    regulation of, 657–658
  ion(s), active transport of, 24–25
  ion channels, 22–24, 23*f*

activation of, 50
  rhodopsin and, 141–142
  metabolism, abnormal, 660–661
  reabsorption of, 645
    mechanism for, 646*f*
    quantitative aspects of, 657*t*
  secondary active transport of, 650*f*
  total body, 29–30, 30*t*
Sodium-dependent glucose transporter (SGLT), 430
Sodium-potassium-activated adenosine triphosphatase, 22, 24, 24*f*
  cell volume and pressure and, 27
  resting membrane potential and, 28–29
Sodium-potassium pump, 22
Solitary lymph nodules, 459
Solubility product, 351
Solute, 25
  concentration of, units for measuring, 3–4
Solution, ideal, 26
Solvent, 25
Solvent drag, 25
Somatic chromosomes, 375
Somatic motor activity, spinal motor neurons and, 183
Somatic sensation
  brain areas concerned with, 122*f*
  visceral sensation compared, 125–126
Somatic sensory areas, 121
Somatomedins, 367–368
Somatostatin, 222, 302, 318, 443
Somatosympathetic reflex, 549, 549*f*
Somnambulism, 181
Sound
  audible, 158
  localization of, 163
  transmission in ear, 159
Sounds of Korotkow, 531–532
Sound waves, 157–159
  characteristics of, 160*f*
  conduction in ear, 161
  traveling, 161, 161*f*
Sourness, 171–172
Space, orientation in, 165
Spasticity, 116
Spastic neurogenic bladder, 663
Spastic paralysis, 184
Spatial summation, 77, 78*f*
Specific activity, 29–30
Specific cholinesterase, 87
Specific dynamic action (SDA), 255
Specific gravity, 651
Specific ionic composition, 664
  defense of, 665
Specific sensory relay nuclei, 173
SPECT. *See* Single photon emission computed tomography
Spectrin, 482
Speech, scanning, 200
Spermatids, 389
Spermatocytes

primary, 388f, 389
  secondary, 389
Spermatogenesis, 387, 389
  temperature and, 389–390
Spermatogonia, 389
Spermatozoa, 389, 389f
Spherocytosis, hereditary, 482
  viscosity in, 528
Sphygmomanometer, 531
Spike potential, 47
Spinal animal, 117, 189
Spinal cord
  afferent and efferent neurons in,
    connections between, 117f
  afferent neurons in, 121f
  motor control and, 184
  pathways in, 122f
  pattern generator in, 189
  postsynaptic inhibition in, 80
  transection
    complications of, 189
    micturition and, 663
Spinal integration, 189–190
Spinal motor neurons, somatic motor
    activity and, 183
Spinal shock, 189
Spinnbarkeit, 398
Spinocerebellum, 183, 197
Spinothalamic tracts, axons of, 120
Spiral arteries, 397, 398f
Spiral ganglion, 154
Spirometer, 254, 593
Splanchnic circulation, 564–566, 564f
  reservoir function of, 565–566
Splay, 646
Spleen, function of, 483
Spliceosomes, 14
Spongy bone, 350
Sprue, 451
  tropical, 462
Squint, 151–152
SRP. See Signal recognition particle
Stagnant hypoxia, 622, 629
Staircase phenomenon, 62
Standard deviation, 673
Standard error of mean (SEM), 674
Standard limb leads, 497, 498
Standard prefixes, 673t
Stapedius muscle, 153
Stapes, 153
Starling forces, 533
Starling's law of heart, 70, 517
Starvation
  acromegaly and, 308
  insulin receptors and, 308
  intracellular glucose and, 274
  obesity and, 308
  plasma levels of thyroid hormones
    and, 293f
  responses to, 271
Static fusiform axons, 113
Static reflexes, 187
Static response, 114
Steady potential, 20
Steatorrhea, 433, 461

Stein-Leventhal syndrome, 408
Stellate cells, 198, 199f, 637
Stellate ganglia, 201
Stenosis, 514
Stereocilia, 156, 159f
Stereognosis, 131
Steroid hormones
  adrenal venous output of, 344f
  biosynthesis of, 330–331
  brain function and, 98
  mechanism of action of, 33f
  rapid action of, 34
  secreted, 329
  sodium excretion and, 658
  synthetic, 329–330
Steroid-secreting cells, cytologic fea-
    tures of, 325f
Sterols, 272
  absorption of, 433
Stimulus
  adequate, 103, 118, 125
  arousal value of, 241
  conditioned, 242–243
  irradiation of, 117
  maximal, 52
  nociceptive, 117
  supramaximal, 52
  unconditioned, 242–243
Stimulus artifact, 46
Stokes-Adams syndrome, 502
Stomach, 446–452
  anatomy of, 446–447, 446f
  functions of, 451–452
  hunger contractions of, 215
  layers of, 438f
  motility and emptying of, 448–449
    regulation of, 450–451
  secretions of, 447–448
Strabismus, 139, 151–152
Stratum basale, 397
Stratum functionale, 397
Strength-duration curve, 47
Streptokinase, 564
Streptomycin, protein synthesis and,
    18t
Streptozocin, diabetes mellitus due
    to, 308
Stress
  adrenocorticotropic hormone and,
    339–340
  glucocorticoids and, 336
Stress analgesia, 128
Stretch reflexes, 111
  cerebellum and, 200
  corticospinal/corticobulbar systems
    and, 187
  hyperactive, 184
  inverse, 115
    pathways responsible for, 115f
  supraspinal regulation of, 190–191,
    190f, 191f
Striae, 337
Striatum, 193
Striosomes, 194
Stroke volume, 516

Structural lipids, 274–275
Structural proteins, 5
Strychnine, glycine and, 95
ST segment, 497
Student's t test, 674
STX. See Saxitoxin
Subcortical perception, 124–125
Subfornical organ (SFO), 216, 416,
    555
Subliminal firing, 82
Submandibular gland, structure of,
    444f
Submaxillary gland, structure of, 444f
Submucous plexus, 438
Substance P, 95–96, 443
  capillary permeability and, 534
Substance P-containing afferent fi-
    bers, 126, 126f
Substance P/neurokinin A gene, 96
Substantia gelatinosa, 120
Substantia nigra, 193, 193f
Subthalamic nucleus, 193, 193f
  function of, 196
Sucking pneumothorax, 627
Sucralfate, peptic ulcer and, 451
Sucrose, 427
Sucrose space, 2
Sudden infant death syndrome
    (SIDS), 619
Sulfadiazine, blood-brain barrier and,
    556
Sulfates
  ethereal, 267
  urinary, 267
Sulfonamides, carbonic anhydrase
    and, 609
Sulfonylurea, insulin secretion and,
    315
Sulfur, neutral, 267
Sulfur dioxide, bronchoconstriction
    due to, 593
Summation, 82–83
  of contractions, 61–62
Superior ganglia, 201
Superior peduncle, 196, 196f
Supersensitivity, 102
Supplementary motor area, 185, 186
Suppression scotoma, 152
Suppressor T cells, 474
Supramaximal stimulus, 52
Surface tension, alveolar, 593
Surfactant, 594–595
  composition of, 594t
  formation and metabolism of, 595f
Surgical shock, 580
Surgical stress, vasopressin secretion
    and, 220
Sustaining collateral, 102
Swallowing, 445–446
Sweetness, 172
Sylvian fissure, somatic sensory area
    II in, 121
Symbols
  in physiology, 675–679
  respiratory, 679

Sympathectomy, 542
Sympathetic division, 201, 203*f*
Sympathetic ganglia, transmission in, 202
Sympathetic vasodilator system, 542, 549–550, 549*f*
Symports, 22, 430
Synapse en passant, 101
Synapses, 74, 75*f*
  activity of
    chemical transmission of, 83–98
      recording, 77*f*
  axo-axonal, 76*f*, 81
  axodendritic, 76*f*
  axosomatic, 76*f*
  development of, 77
  electrical events at, 77–80
  events occurring at, 78*f*
  inhibition and facilitation at, 80–83
  one-way conduction in, 76
  serotonergic, biochemical events at, 93*f*
Synapsins, 76
Synaptic cleft, 75
Synaptic delay, 78
Synaptic knobs, 43, 75–76, 75*f*
Synaptic plasticity, learning and, 98–99
Synaptic potentials, 45
Synaptic transmission, 74–99
Synchronization, 176
  mechanisms of, 178–179
Syncope, 580–581
  carotid sinus, 580
  cough, 580
  deglutition, 580
  effort, 580
  micturition, 580
  postural, 580
  vasovagal, 580
Syncytiotrophoblast, 408
Syndrome of inappropriate hypersecretion of antidiuretic hormone (SIADH), 221
Synthetic senses, 130
Systeme International units, 672, 673*t*
Systemic circulation, 469
  blood flow through, pressure-velocity changes and, 529*f*
  blood pressures in, 600*f*
  exercise and, 574–575, 574*f*
Systole
  atrial, 494, 510
  length of, 511–513, 513*t*
  total electromechanical, 513
  ventricular, 494, 510–511
Systolic pressure, 510, 529–530

Tabes dorsalis, 662
Tachycardia, 501
  atrial, 503*f*
  paroxysmal, 503
  ventricular, 504*f*

paroxysmal, 504–505
Tachykinins, 95–96, 96*t*
Tachypnea, 602, 623
Tangent screen, 150
Taste, 170–172
  modalities, 171
  modifier proteins, 172
  physiology of, 171–172
  receptor organs and pathways of, 170, 171*f*
  sensations, primary, substances evoking, 171–172
  thresholds, 171, 171*t*
Taste buds, 170, 170*f*
TATA box, 15
Taurocholic acid, 456
Tay-Sachs disease, 8
TBG. *See* Thyroxine-binding globulin
TBPA. *See* Thyroxine-binding prealbumin
TBW. *See* Total body water
T channels, 496
Tectorial membrane, 154
Teeth, enamel of, mottled, 352
Teleceptors, 104
Temperature
  oxygen-hemoglobin dissociation curve and, 606*f*
  pathways, 121*f*
  regulation
    exercise and, 575–576
    hypothalamus and, 225–230
    mechanisms of, 227–228, 228*t*
  sense organs, 124
  spermatogenesis and, 389–390
Temporal lobectomy, 250–251
Temporal summation, 77, 78*f*
Teniae coli, 462
Tension
  active, 62
  muscle contraction and, 62–63
  passive, 62
  total, 62
Tension pneumothorax, 627
Tensor tympani muscle, 153
Terminal buttons, 43, 74
Terminal cisterns, 58
Tertiary structure, 266
Testes
  duct system of, 387*f*
  endocrine function of, 391–393
  function of
    abnormalities of, 395
    control of, 393–394
  interrelationships of, 394*f*
  section of, 388*f*
Testicular feminizing syndrome, 382
Testosterone, 375
  biosynthesis and metabolism of, 391, 391*f*
  male contraception and, 394
  mechanism of action of, 393
  plasma, at various ages, 392*f*
  secretion of, 392
  sexual activity and, 232

transport of, 392
Tetanic contraction, 61
Tetanus, 61, 62*f*
  complete, 61
  incomplete, 61–62
Tetanus toxin, glycine and, 95
Tetany, hypocalcemic, 348
  position of hand in, 357*f*
Tetracycline
  protein synthesis and, 18*t*
  tropical sprue and, 462
Tetraethylammonium (TEA), potassium channel and, 22
Tetrahydrobiopterin, 88
Tetraploidy, 13
Tetrodotoxin (TTX), sodium channel and, 22
TG cell, 441
TGF. *See* Transforming growth factor
Thalamic fasciculus, 194
Thalamic nuclei, 173–174
Thalamic syndrome, 125
Thalamus, 173–174
  dorsal, 173
  ventral, 173
Thalassemia, 484
Thebesian veins, 561
Theca cells, interactions of, 401*f*
Theca interna, 396
Thelarche, 382
Theophylline, myocardial contractility and, 518
Thermal gradient, 227
Thermodilution, 516
Thermogenesis, thyroid hormone, 297
Thermoregulation
  exercise and, 575–576
  hypothalamus and, 225–230
  mechanisms of, 227–228, 228*t*
Theta rhythm, 175
Thiazide diuretics, insulin secretion and, 315
Thick filaments, 56, 58
  arrangement in skeletal muscle, 58*f*
Thin filaments, 56–58
  arrangement in skeletal muscle, 58*f*
  muscle contraction and, 59–60
Thiocarbamides, antithyroid, 300–301
  structure of, 301*f*
Thiocyanate, iodide transport and, 300
Third-degree heart block, 502, 502*f*
Third sound, 514
Thirst, hypothalamus and, 215–216
Thoracic pump, 536
Thoroughfare vessel, 521
Threshold, 47
Threshold intensity, 47
Thrombasthenic purpura, 481
Thrombi, mural, 492
Thrombin, 490
Thrombocytopenic purpura, 481
Thrombocytosis, 481
Thrombomodulin, 491
Thromboplastin, 490

Thrombosis, 492
Thromboxane, 280, 481, 539
    biosynthesis of, 281f
    platelet-aggregating effect of, 491
Thymus, congenital absence of, 480
Thyroglobulin, 288
Thyroglossal duct, 287
Thyroid cell, 288f
Thyroid gland, 287–301, 287f
    anatomy of, 287
    histology, 288f
    isthmus, 287
    parafollicular cells in, 357, 358f
    secretion of
        feedback control of, 297f
        regulation of, 296–297
Thyroid hormone(s)
    binding to plasma proteins, 291,
        291t
    calcium metabolism and, 360
    carbohydrate metabolism and,
        319–320
    2,3-diphosphoglycerate and, 606
    effects of, 293–296
    formation and secretion of,
        288–290
    mechanism of action of, 33f
    nonthyroidal diseases and, 301
    secretion and interconversion of,
        291f
    structure of, 289f
    synthesis of, 289–290, 290f
    thermogenesis, 297
    transport and metabolism of,
        290–293
Thyroiditis, Hashimoto's, 298
Thyroid peroxidase, 289
Thyroid-stimulating hormone (TSH),
        222, 361
    chemistry and metabolism of, 296
    effects on thyroid, 296
    receptors, 296–297
    two-unit structure of, 363
Thyroid-stimulating immunoglobulin
        (TSI), 298
Thyrotoxic myopathy, 295
Thyrotoxicosis, 298–299
Thyrotropin, 222, 361
Thyrotropin-releasing hormone
        (TRH), 222–224
Thyroxine, 288
    conversion to triiodothyronine, 292f
    distribution in body, 291f
    index, free, 291
    mechanism of action of, 293
    metabolism of, 292
    plasma levels of, 290
Thyroxine-binding globulin (TBG),
        291
Thyroxine-binding prealbumin
        (TBPA), 291
TIAs. See Transient ischemic attacks
Tickle, 130
Tidal volume, 590
Tight junctions, 6, 6f

Timbre, 158
Time constant, 77
Timed vital capacity, 591
Tissue conductance, 227
Tissue kallikrein, 303, 541
Tissue macrophage(s), 472, 472t
Tissue macrophage system, 472
Tissue oxygenation, 604
Tissue plasminogen activator (TPA),
        491, 564
Tissue pressure hypothesis of autoreg-
        ulation, 538
Tissue transplantation, 479–480
Titratable acidity, 656
T lymphocytes, 474
    interactions with antigen-present-
        ing cells, 476f
    receptors, 477
Tm. See Transport maximum
TNF. See Tumor necrosis factor
Tolazamide, insulin secretion and,
        315
Tolbutamide, insulin secretion and,
        315
Tone, 71, 115–116
Tongue, sensory innervation of, 170f
Tonicity, 26
    defense of, 664, 665f
Tonic labyrinthine reflexes, 191
Tonic neck reflexes, 191–192
Tonic receptors, 106
Tonsils, 602
Tonus, 71, 115–116
Total blood volume, 1, 2
Total body osmolality, 30
Total body potassium, 30, 30t
Total body sodium, 29–30, 30t
Total body water (TBW), 1, 3
    related to age and sex, 3t
Total dead space, 597
Total electromechanical systole, 513
Total tension, 62
Touch pathways, 121f
Touch receptors, 123
Touch stimulus, magnitude of, action
        potential frequency and,
        108f
Toxemia of pregnancy, 582
Toxins, autonomic nervous system
        and, 207, 207t
TPA. See Tissue plasminogen activa-
        tor
Trabecular bone, 350, 352f
Trace elements, essential, 284t
Tracer doses, 289
Trachea, 588
Tractus solitarius, nucleus of, 170
Trail endings, 113
Tranquilizers
    parkinsonism due to, 195
    synaptic transmission in brain and,
        237
Transaminases, 267
Transamination, 267–268
    citric acid cycle in, 268f

Transcellular fluids, 2
Transcortical load-compensating re-
        flex, 187
Transcortin, 332
Transcription, 13, 15f
    stimulation of, 33
Transcription factors, 15
Transcription unit, 14
Transcytosis, 29
Transducin, 141
Transferrin, 436–437
Transfer RNA. See RNA
Transforming growth factor (TGF),
        31
Transfusion, autologous, 487
Transfusion reactions, 486–487
    hemolytic, 486
Transient ischemic attacks (TIAs),
        481
Translation, 13, 15f
Transmural pressure, 528–529
Transplantation
    heart-lung, 618
    tissue, 479–480
Transport maximum (Tm), 645
Transport proteins, 20–22, 22f
Transverse tubules, 58
Traube-Hering waves, 548
Traumatic shock, 579
Tremor
    due to Parkinson's disease, 195
    intention, 200
    physiologic, 114
Treppe, 62
TRH. See Thyrotropin-releasing hor-
        mone
Triads, 58
Tricarboxylic acid cycle, 261, 261f
Triceps muscle, length-tension rela-
        tionship for, 63f
Trichromats, 147
Triglycerides, 271
Triiodothyronine, 288
    index, free, 291
    mechanism of action of, 293
    metabolism of, 292
    plasma levels of, 290
    reverse, 288
Triple response, 566
Trisomy 21, 246, 382
Tritanomaly, 147
Tritanopia, 147
Trophic action, 441
Tropical sprue, 462
Tropic hormones, 361
Tropomyosin, 56, 58, 58f
    isoforms, 65
Troponin, 34, 56
Troponin C, 56, 58, 60
Troponin I, 56, 58, 60
Troponin T, 56, 58
    isoforms, 65
Troponin-tropomyosin complex, 60
Trousseau's sign, 357, 357f
True cholinesterase, 87

True hermaphroditism, 381
True plasma, 665
True precocious pseudopuberty, 383
Trunk, touch, pain, and temperature pathways from, 121*f*
Trypsin, 431, 453
TSH. *See* Thyroid-stimulating hormone
TSI. *See* Thyroid-stimulating immunoglobulin
T state, 604
T system, 58
T tubule membrane, dihydropyridine receptor in, 60
TTX. *See* Tetrodotoxin
Tuberohypophyseal system, 239
Tuberoinfundibular system, 239
Tubular function, 644–648, 644*f*
Tubular myelin, 594
Tubular reabsorption, 635
   mechanisms of, 644–645
Tubular secretion, 635
   mechanisms of, 644–645
Tubulin, 8
Tubuloglomerular feedback, 648
   mechanisms of, 648*f*
Tumor necrosis factor (TNF), 580
Tumors
   androgen-secreting, 395
   gastrin-secreting, 441
   hormone-secreting, 374
   insulin-secreting, 321
   ovarian, 408
   vasoactive intestinal peptide-secreting, 443
Tuning fork tests, 163*t*
Turbulence, blood flow, 526
Turner's syndrome, 381
T wave, 67, 497, 498
Two-point discrimination, 130–131
Two-point threshold, 130
Tympanic membrane, 153
   functions of, 159–160
   secondary, 153
Tympanic reflex, 160–161
Type I cells, 588
Type I medullary interstitial cells, 638
Type II cells, 588
Tyrosine hydroxylase, 88
Tyrosine kinases, structure of, 40*f*
Tyrosine phosphatases, structure of, 40*f*

Ubiquitin, 270
UDPGA. *See* Uridine diphosphoglucuronic acid
Ulcer
   decubitus, 189
   duodenal, gastrin and, 442
   peptic, 451
Ultimobranchial bodies, 357
Ultrasonography, 458
Unconditioned stimulus (US), 242–243

Unilateral inattention and neglect, 247
Unipolar, 175
Unipolar leads, 498, 498*f*
Unipolar recording, 497
Uniports, 22
Unstirred layer, 25, 427
Upper motor neurons, 184
Up regulation, 32
Urea
   formation of, 268–269
   penetration into muscle, brain, and spinal cord, 555
   water excretion and, 652–653
Urea cycle, 269*f*
Uremia, 578, 660
Urethral sphincter
   external, 661
   internal, 661
Uric acid, 270
Uric acid metabolism, 270*f*
Uridine diphosphoglucuronic acid (UDPGA), 457
Uridine triphosphate (UTP), 257
Urinary bladder
   contraction of, hypothalamic stimulation and, 211
   emptying of, 661–663
   filling of, 661
   innervation of, 661*f*
   reflex control of, 662
   spastic neurogenic, 663
Urinary sulfates, 267
Urination. *See* Micturition
Urine
   acidification of, 654–657
   concentration
      glomerular filtration rate and, 654
      urine flow in osmotic diuresis and, 654*f*
   fate of hydrogen in, 655
   flow, 641
   pH of, implications of, 657
Urogastrone, 444
Uropepsinogen, 447
US. *See* Unconditioned stimulus
Uterine blood flow, changes during pregnancy, 567*f*
Uterine cervix, cyclic changes in, 398
Uterine circulation, 567, 567*f*
Uterine cycle, 397–398
Uterus, smooth muscle of, oxytocin and, 222
UTP. *See* Uridine triphosphate
Utricle, 155
U wave, 497

Vagal tone, 543
Vaginal cycle, 398
Vaginal smear, 399*f*
Vallate papillae, 170
Valsalva maneuver, 548
   response to, 548*f*
Valvulae conniventes, 459–460
Valvular insufficiency, 514

Valvular regurgitation, 514
Varicose veins, 536
Varicosities, 101
Vasa recta, 638
   as countercurrent exchanger, 653*f*
Vascular hindrance, 527
Vascular system
   flow in, 524
   pressure-flow relation in, 528*f*
Vas deferens, 387
Vasectomy, 390–391
Vasoactive intestinal contractor (VIC), 540
Vasoactive intestinal peptide (VIP), 202, 206, 443
Vasoconstriction, 538
   local, 490
   localized, 539
   thromboxane and, 280, 539
   vasopressin and, 219
Vasoconstrictors, circulating, 542
Vasodilation, 538
   prostacyclin and, 539
Vasodilator metabolites, 538
Vasogenic shock, 577
Vasomotor area, 543
   activity of, factors affecting, 545*t*
   afferents to, 543–544
   chemoreceptor stimulation and, 548
   direct effects on, 548–549
Vasomotor center, 543
Vasomotor nerves, cerebral circulation and, 559
Vasopressin, 97, 361
   agonists and antagonists of, 219
   arterial blood pressure and, 220*f*
   biosynthesis, transport, and secretion of, 216–217
   effects of, 218–219
   localization of, 218
   mechanism of action of, 219
   metabolism of, 219
   plasma osmolality and, 220*f*
   receptors, 219
   secretion of, control of, 219–221, 219*t*
   vasoconstrictor action of, 542
   water excretion and, 651
   water metabolism and, 649*t*
Vasovagal syncope, 580
Vectorcardiography, 500
Veins, 523
   valves of, function of, 525*f*
   varicose, 536
Venae comitantes, 228
Venoconstriction, 538
Venodilation, 538
Venous circulation, 535–537
Venous pressure, 535–536
   central, 536–537
   gravity and, 531*f*
   measurement of, 536
   peripheral, 536–537
Venous-to-arterial shunts, 626–627

Ventilation
alveolar, 597–598
respiratory rate/depth and, 597t
in different parts of lung, 596–597
exercise and, 620–622, 621f
intrapleural pressures and, 596f
maximal voluntary, 591
pulmonary, 591
Ventilation/blood flow relationship, 626f
Ventilation/perfusion imbalance, 625–626
Ventilation/perfusion ratio, 601, 601f
Ventilatory response, acclimatization to altitude and, 625f
Ventral cochlear nuclei, 154
Ventral corticospinal tract, 184–185
Ventral group, 611
Ventral thalamus, 173
Ventricular arrhythmias, 504–505
Ventricular contraction, isovolumetric, 510
Ventricular ejection, 510–511
Ventricular fibrillation, 505, 505f
Ventricular premature beat (VPB), 504f, 505f
Ventricular relaxation, isovolumetric, 511
Ventricular systole, 494, 510–511
Ventricular tachycardia, 504f
paroxysmal, 504–505
Ventricular volume
end-diastolic, 511
factors affecting, 518, 518t
end-systolic, 511
Venules, 523
Vermis, 196
Vertigo, 165
Very low-density lipoprotein (VLDL), 276
Vesicular transport, 29, 522
Vestibular function, 164–165
Vestibular movement, 151
Vestibular pathway, 158f
Vestibular placing reaction, 192
Vestibulocerebellum, 197
Vestibulo-ocular reflex (VOR), 164, 192
Vestibulospinal pathway, stretch reflexes and, 191
Vibratory sensibility, 130
VIC. See Vasoactive intestinal contractor
Villi, 459–460, 460f
Vinblastine, microtubule assembly and, 8
VIP. See Vasoactive intestinal peptide
VIPomas, 443
Virilization, 332
Viruses, receptor-mediated endocytosis and, 19
Viscera, pain innervation of, 127f
Visceral pain, 126–127
Visceral reflexes, respiratory components of, 619

Visceral sensation, somatic sensation compared, 125–126
Visceral smooth muscle, 71–73
contraction and relaxation of, sequence of events in, 72t
electrical and mechanical activity of, 71
membrane potential of, effects of agents on, 72f
molecular basis for contraction of, 71–72
stimulation of, 72–73
Viscosity, 527–528
Viscus, hollow, distending pressure-wall tension relation in, 528f
Vision
binocular, 150
color, 146–148
neural mechanisms of, 147
Young-Helmholtz theory of, 146
double, 150
monocular, 150
near point of, 138
photopic, 136
scotopic, 136
Visual acuity, 149
Visual angle, 138
Visual cortex, 134
orientation preferences in, 145f
pathways to, 144
Visual fields, 150
monocular/binocular, 150f
Visual information, parallel processing of, 148
Visual pathways, 134, 135f
organization of, 144f
receptive fields of cells in, 143f
responses in, 144
Visual purple, 141
Visual receptors, current flow in, effect of light on, 140f
Visual threshold, 149
Visual vortex, responses in, 144–146
Vital capacity, 591
timed, 591
Vital centers, 208
Vitamin(s), 284–286
absorption of, 435
deficiency, effects on eye, 149
essential, 285–286t
Vitamin A, rod function and, 149
Vitamin A intoxication, 286
Vitamin D, 353–355
formation and hydroxylation of, 353f
Vitiligo, 364
Vitreous humor, 132
VLDL. See Very low-density lipoprotein
Volley effect, 162
Volume conductor, 50, 497
Volume of distribution, 1
Voluntary movement
cerebellum and, 200

commands for, 183
control of, 184f
corticospinal/corticobulbar systems and, 186–187
Vomiting, 209
Vomiting reflex, afferent pathways for, 208f
von Willebrand's disease, 492
von Willebrand factor, 481, 492
VOR. See Vestibulo-ocular reflex
VPBs. See Ventricular premature beats
Vulnerable period, 505
v wave, 513

Walking, pattern generator for, 189–190
Wallerian degeneration, 45, 102
Warfarin, 492
Water
absorption of, 433–435
clearance, free, 654
diuresis, 653
excretion, 648–654
intake, factors regulating, 216
intoxication, 653
loss, insensible, 227
metabolism
adrenal insufficiency and, 335–336
pituitary insufficiency and, 373
vasopressin and, 649t
retention, vasopressin and, 218–219
turnover in gastrointestinal tract, 435t
Waterfall effect, 601
Water-hammer pulse, 513
Weber-Fechner law, 108
Weber test, 163
Wenckebach phenomenon, 502, 502f
Wernicke's area, 248, 249f
Wheal, 566
White blood cells, 470–473
White muscles, 65
White rami communicantes, 201
White reaction, 566
Widow's hump, 352
Wilson's disease, 194
copper excess and, 284
Windkessel effect, 529
Withdrawal bleeding, 402
Withdrawal reflex, 117
local sign and, 118f
Wolff-Chaikoff effect, 301
Wolff-Parkinson-White syndrome, 505
Work load, oxygen uptake and, 621f
Wound healing, 576
cell-adhesion molecules and, 576f

Xanthines, myocardial contractility and, 518
X chromosome, 375–376

X chromosome *(cont.)*
    color blindness and, 147–148
Xerostomia, 445

Y chromosome, 375–376
Yellow marrow, 469
Young-Helmholtz theory of color vision, 146

Zero gravity, effects of, 573
Zinc deficiency, 284
Zinc finger, 15, 16*f*
Z lines, 56
    muscle contraction and, 59
ZO–1, 6
Zollinger-Ellison syndrome, 358, 451
Zona fasciculata, 323–324, 324*f*
    hormone biosynthesis in, 329*f*
Zona glomerulosa, 323–324, 324*f*
    hormone biosynthesis in, 330*f*

Zona pellucida, 408
Zona reticularis, 323–324, 324*f*
    hormone biosynthesis in, 329*f*
Zonulae adherens, 6, 6*f*
Zonulae occludens, 6, 6*f*
Zonule, 132
Zygote, 11
Zymogen, 444
Zymogen cells, 446, 447*f*
Zymogen granules, 452

LANGE
medical books

**LANGE medical books
are available at your local
health science bookstore
or by calling
Appleton & Lange
toll-free
1-800-423-1359**
(in CT 838-4400).

See reverse side for a complete listing of LANGE titles.

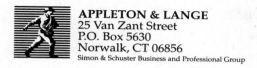
a smart investment in your medical career

Basic Science Textbooks

Color Atlas of Basic Histology
Berman
1993, ISBN 0-8385-0445-0, A0445-5
Jawetz, Melnick & Adelberg's
Medical Microbiology, 19/e
Brooks, Butel & Ornston
1991, ISBN 0-8385-6241-8, A6241-2
Concise Pathology
Chandrasoma & Taylor
1991, ISBN 0-8385-1320-4, A1320-9
Correlative Neuroanatomy, 21/e
deGroot & Chusid
1991, ISBN 0-8385-1332-8, A1332-4
Review of Medical Physiology, 16/e
Ganong
1993, ISBN 0-8385-8426-8, A8426-7
Basic Histology, 7/e
Junqueira, Carniero & Kelly
1992, ISBN 0-8385-0576-7, A0576-7
Basic & Clinical Pharmacology, 5/e
Katzung
1992, ISBN 0-8385-0562-7, A0562-7
Pharmacology
Examination & Review, 3/e
Katzung & Trevor
1993, ISBN 0-8385-7807-1, A7807-9
Medical Microbiology & Immunology
Examination & Board Review, 2/e
Levinson & Jawetz
1992, ISBN 0-8385-6262-0, A6262-8
Harper's Biochemistry, 23/e
Murray, et al.
1993, ISBN 0-8385-3562-3, A3562-4
Basic Histology
Examination & Board Review, 2/e
Paulsen
1993, ISBN 0-8385-0569-4, A0569-2
Basic & Clinical Immunology, 7/e
Stites & Terr
1991, ISBN 0-8385-0544-9, A0544-5
Basic Human Immunology
Stites & Terr
1991, ISBN 0-8385-0543-0, A0543-7

Clinical Science Textbooks

Clinical Cardiology, 6/e
Cheitlin & Sokolow
1993, ISBN 0-8385-1093-0, A1093-2
Fluid & Electrolytes
Physiology & Pathophysiology
Cogan
1991, ISBN 0-8385-2546-6, A2546-8
Basic and Clinical Biostatistics
Dawson-Saunders & Trapp
1990, ISBN 0-8385-6200-0, A6200-8
Basic Gynecology and Obstetrics
Gant & Cunningham
1993, ISBN 0-8385-9633-9, A9633-7
Review of General Psychiatry, 3/e
Goldman
1992, ISBN 0-8385-8428-4, A8428-3

Principles of Clinical Electrocardiography, 13/e
Goldschlager & Goldman
1989, ISBN 0-8385-7951-5, A7951-5
Medical Epidemiology
Greenberg
1993, ISBN 0-8385-6204-3, A6204-0
Clinical Neurology, 2/e
Greenberg, Aminoff & Simon
1993, ISBN 0-8385-1311-5, A1311-8
Basic and Clinical Endocrinology, 3/e
Greenspan
1991, ISBN 0-8385-0545-7, A0545-2
Occupational Medicine
LaDou
1990, ISBN 0-8385-7207-3, A7207-2
Clinical Anatomy
Lindner
1989, ISBN 0-8385-1259-3, A1259-9
Clinical Anesthesiology
Morgan & Mikhail
1992, ISBN 0-8385-1324-7, A1324-1
Dermatology
Orkin, Maibach & Dahl
1991, ISBN 0-8385-1288-7, A1288-8
Clinical Thinking in Surgery
Sterns
1988, ISBN 0-8385-5686-8, A5686-9
Smith's General Urology, 13/e
Tanagho & McAninch
1992, ISBN 0-8385-8608-2, A8608-0
General Ophthalmology, 13/e
Vaughan, Asbury & Riordan-Eva
1992, ISBN 0-8385-3115-6, A3115-1

CURRENT Clinical References

CURRENT Pediatric Diagnosis & Treatment, 11/e
Hathaway, et al.
1993, ISBN 0-8385-1440-5, A1440-5
CURRENT Obstetric & Gynecologic Diagnosis & Treatment, 7/e
Pernoll
1991, ISBN 0-8385-1424-3, A1424-9
CURRENT Emergency Diagnosis & Treatment, 4/e
Saunders & Ho
1992, ISBN 0-8385-1347-6, A1347-2
CURRENT Medical Diagnosis & Treatment 1993
Tierney, et al.
1993, ISBN 0-8385-1350-6, A1350-6
CURRENT Surgical Diagnosis & Treatment, 9/e
Way
1991, ISBN 0-8385-1426-X, A1426-4

LANGE Clinical Manuals

Dermatology
Diagnosis and Therapy
Bondi, Jegasothy & Lazarus
1991, ISBN 0-8385-1274-7, A1274-8
Office & Bedside Procedures
Chesnutt, Dewar & Locksley
1992, ISBN 0-8385-1095-7, A1095-7
Psychiatry
Diagnosis & Treatment, 2/e
Flaherty, Davis & Janicak
1993, ISBN 0-8385-1267-4, A1267-2
Neonatology
Management, Procedures, On-Call Problems, Diseases, Drugs, 2/e
Gomella
1992, ISBN 0-8385-1284-4, A1284-7
Drug Therapy, 2/e
Katzung
1991, ISBN 0-8385-1312-3, A1312-6
Ambulatory Medicine
The Primary Care of Families
Mengel & Schwiebert
1993, ISBN 0-8385-1294-1, A1294-6
Poisoning & Drug Overdose
Olson
1990, ISBN 0-8385-1297-6, A1297-9
Internal Medicine
Diagnosis and Therapy, 3/e
Stein
1993, ISBN 0-8385-1112-0, A1112-0
Surgery
Diagnosis & Therapy
Stillman
1989, ISBN 0-8385-1283-6, A1283-9
Medical Perioperative Management
Wolfsthal
1989, ISBN 0-8385-1298-4, A1298-7

LANGE Handbooks

Handbook of Gynecology & Obstetrics
Brown & Crombleholme
1993, ISBN 0-8385-3608-5, A3608-5
Pocket Guide to Diagnostic Tests
Detmer, et al.
1992, ISBN 0-8385-8020-3, A8020-8
Handbook of Poisoning
Prevention, Diagnosis, and Treatment
Dreisbach & Robertson
1987, ISBN 0-8385-3643-3, A3643-2
Handbook of Clinical Endocrinology, 2/e
Fitzgerald
1992, ISBN 0-8385-3615-8, A3615-0
Pocket Guide to Commonly Prescribed Drugs
Levine
1993, ISBN 0-8385-8023-8, A8023-2
Silver, Kempe, Bruyn & Fulginiti's
Handbook of Pediatrics, 16/e
Merenstein, Kaplan & Rosenberg
1991, ISBN 0-8385-3639-5, A3639-0